NETWORK FLOWS

Theory, Algorithms, and Applications

RAVINDRA K. AHUJA
Department of Industrial & Management Engineering
Indian Institute of Technology, Kanpur

THOMAS L. MAGNANTI
Sloan School of Management
Massachusetts Institute of Technology, Cambridge

JAMES B. ORLIN
Sloan School of Management
Massachusetts Institute of Technology, Cambridge

PRENTICE HALL, Englewood Cliffs, New Jersey 07632

Library of Congress Cataloging-in-Publication Data

Ahuja, Ravindra K. (date)
Network flows : theory, algorithms, and applications / Ravindra K. Ahuja, Thomas L. Magnanti, James B. Orlin.
p. cm.
Includes bibliographical references and index.
ISBN 0-13-617549-X
1. Network analysis (Planning) 2. Mathematical optimization.
I. Magnanti, Thomas L. II. Orlin, James B., (date). III. Title.
T57.85.A37 1993
658.4'032—dc20 92-26702
CIP

Acquisitions editor: Pete Janzow
Production editor: Merrill Peterson
Cover designer: Design Source
Prepress buyer: Linda Behrens
Manufacturing buyer: David Dickey
Editorial assistant: Phyllis Morgan

Printed in the United States of America

10 9 8 7 6 5 4 3 2

ISBN 0-13-617549-X

Prentice-Hall International (UK) Limited, *London*
Prentice-Hall of Australia Pty. Limited, *Sydney*
Prentice-Hall Canada Inc., *Toronto*
Prentice-Hall Hispanoamericana, S.A., *Mexico*
Prentice-Hall of India Private Limited, *New Delhi*
Prentice-Hall of Japan, Inc., *Tokyo*
Simon & Schuster Asia Pte. Ltd., *Singapore*
Editora Prentice-Hall do Brasil, Ltda., *Rio de Janeiro*

Ravi dedicates this book to his spiritual master,
Revered Sri Ranaji Saheb.

Tom dedicates this book to his favorite network,
Beverly and Randy.

Jim dedicates this book to Donna,
who inspired him in so many ways.

Collectively, we offer this book as a tribute
to Lester Ford and Ray Fulkerson, whose pioneering research
and seminal text in network flows have been an enduring
inspiration to us and to a generation
of researchers and practitioners.

CONTENTS

PREFACE

If you would not be forgotten,
As soon as you are dead and rotten,
Either write things worthy reading,
Or do things worth the writing.
—Benjamin Franklin

Network flows is an exciting field that brings together what many students, practitioners, and researchers like best about the mathematical and computational sciences. It couples deep intellectual content with a remarkable range of applicability, covering literally thousands of applications in such wide-ranging fields as chemistry and physics, computer networking, most branches of engineering, manufacturing, public policy and social systems, scheduling and routing, telecommunications, and transportation. It is classical, dating from the work of Gustav Kirchhoff and other eminent physical scientists of the last century, and yet vibrant and current, bursting with new results and new approaches. Its heritage is rooted in the traditional fields of mechanics, engineering, and applied mathematics as well as the contemporary fields of computer science and operations research.

In writing this book we have attempted to capture these varied perspectives and in doing so to fill a need that we perceived for a comprehensive text on network flows that would bring together the old and the new, and provide an integrative view of theory, algorithms, and applications. We have attempted to design a book that could be used either as an introductory or advanced text for upper-level undergraduate or graduate students or as a reference for researchers and practitioners. We have also strived to make the coverage of this material as readable, accessible, and insightful as possible, particularly for readers with a limited background in computer science and optimization.

The book has the following features:

- In-depth and self-contained treatment of shortest path, maximum flow, and minimum cost flow problems, including descriptions of new and novel polynomial-time algorithms for these core models.
- Emphasis on powerful algorithmic strategies and analysis tools, such as data scaling, geometric improvement arguments, and potential function arguments.
- An easy-to-understand description of several important data structures, including *d*-heaps, Fibonacci heaps, and dynamic trees.
- Treatment of other important topics in network optimization and of practical solution techniques such as Lagrangian relaxation.

- Each new topic introduced by a set of applications and an entire chapter devoted to applications.
- A special chapter devoted to conducting empirical testing of algorithms.
- Over 150 applications of network flows to a variety of engineering, management, and scientific domains.
- Over 800 exercises that vary in difficulty, including many that develop extensions of material covered in the text.
- Approximately 400 figures that illustrate the material presented in the text.
- Extensive reference notes that provide readers with historical contexts and with guides to the literature.

As indicated by this list, we have not attempted to cover all topics at the same level of depth, but instead, have treated certain core topics more extensively than others. Moreover, although we have focused on the design and analysis of efficient algorithms, we have also placed considerable emphasis on applications.

In attempting to streamline the material in this book and present it in an integrated fashion, we have devoted considerable time, and in several cases conducted research, to improve our presentation. As a result, our coverage of some topics differs from the discussion in the current literature. We hope that this approach will not lead to any confusion, and that it will, in fact, promote a better understanding of the material and uncover new connections between topics that might not appear to be so related in the literature.

TO INSTRUCTORS AND STUDENTS

We have attempted to write this book so that it is accessible to students with many backgrounds. Although students require some mathematical maturity—for example, a basic understanding of proof techniques—and some familiarity with computer programming, they need not be specialists in mathematics, computer science, or optimization. Some basic knowledge of these topics would undoubtedly prove to be useful in using this book. In Chapter 3 and Appendices A, B, and C, we have provided some of this general background.

The book contains far more material than anyone could possibly cover in a one-semester course. Chapters 1 to 3, 19, selected material from Chapters 5 to 12, 16, and 17, and portions of Chapters 13 to 15 and 18 would serve as a broad-based course on network flows and network optimization. Because the chapters are generally modular, instructors might use the introductory material in the first few sections of each chapter as well as a selection of additional material for chapters that they would not want to cover in their entirety. An advanced course on algorithms might focus on Chapter 4 to 12, covering the material in these chapters in their entirety.

In teaching the algorithms in this book, we feel that it is important to understand the underlying methods of algorithm design and analysis as well as specific results. Therefore, we encourage both instructors and students to refer frequently back to Chapter 3 and its discussion of algorithm design and analysis.

Many of the topics that we have examined in this book are specially structured linear programs. Therefore, we could have adopted a linear programming approach while presenting much of the material. Instead, with the exception of Chapter 17 and parts of Chapters 15 and 16, we have argued almost exclusively from first principles and adopted a network or graphical viewpoint. We believe that this approach, while occasionally imposing a less streamlined development than might be possible using linear programming, offers several advantages. First, the material is readily accessible to a wider audience. Second, this approach permits students who are not optimization specialists to learn many of the ideas of linear programming in a concrete setting with easy geometric and algebraic interpretations; it also permits students with prior knowledge of linear programming to refine their understanding by seeing material in a different light. In fact, when the audience for a course has a background in linear programming, we would encourage instructors and students to make explicit connections between our coverage and more general results in linear programming.

Although we have included some numerical exercises that test basic understanding of material presented in the text, many of the exercises address applications or theory. Instructors might like to use this material, with suitable amplification, as lecture material. Instructors wishing more numerical examples might modify the ones we have provided.

TO OUR GENERAL READERS

Professionals in applied mathematics, computer science, engineering, and management science/operations research as well as practitioners in a variety of application domains might wish to extract specific information from the text without covering the book in detail. Since the book is organized primarily by model type (e.g., shortest path or spanning tree problems), readers have ready access to the material along these dimensions. For a guide to applications, readers might consult Section 19.10, which contains a set of tables summarizing the various network flow applications that we have considered in the text. The end-of-chapter reference notes also contain references to a number of applications that we have either not discussed or considered only in the exercises. For the most part, with the exception of applications, we have been selective in our citations to the literature. Many of the general references that we have mentioned at the end of Chapter 1 contain more detailed references to the literature.

We have described many algorithms in this book using a pseudocode that should be understandable to readers with any passing familiarity with computer programming. This approach provides us with a certain degree of universality in describing algorithms, but also requires that those wishing to use the algorithms must translate them into specific programming languages and add material such as input/output and error handling procedures that will be implementation dependent.

FEEDBACK

Any book of this size and complexity will undoubtedly contain errors; moreover, in writing this book, we might have inadvertently not given proper credit to everyone deserving recognition for specific results. We would be pleased to learn about any comments that you might have about this book, including errors that you might find. Please direct any feedback as follows:

Professor James B. Orlin
Sloan School of Management, MIT
Cambridge, MA 02139, USA
e-mail: jorlin@eagle.mit.edu
fax: 617-258-7579

ACKNOWLEDGMENTS

Many individuals have contributed to this text by enhancing our understanding of network flows, by advising us in general about the book's content, or by providing constructive feedback on early drafts.

We owe an intellectual debt to several individuals. From direct collaborations or through the writings of E. Dinic, Jack Edmonds, Hal Gabow, Fred Glover, Matsao Iri, Bruce Golden, Richard Karp, A. Karzanov, Darwin Klingman, Eugene Lawler, Robert Tarjan, Eva Tardos, Richard Wong, and so many others, we have learned much about the general topic of networks and network optimization; the pioneering efforts of George Dantzig, Lester Ford, and Delbert Fulkerson in the 1950s defined the field of network flows as we know it today. We hope that our treatment of the subject is true to the spirit of these individuals. Many of our colleagues, too numerous to mention by name, responded to a questionnaire that we distributed soliciting advice on the coverage for this book. Their comments were very helpful in refining our ideas about the book's overall design. Anant Balakrishnan and Janny Leung read and commented on portions of the manuscript; S. K. Gupta, Leslie Hall, Prakash Mirchandani, and Steffano Pallottino each commented in detail about large segments of the manuscript. We are especially indebted to Steffano Pallottino for several careful reviews of the manuscript and for identifying numerous corrections. Bill Cunningham offered detailed suggestions on an earlier book chapter that served as the starting point for this book. Each of these individual's advice has greatly improved our final product. Over the years, many of our doctoral students have helped us to test and refine our ideas. Several recent students—including Murali Kodialam, Yusin Lee, Tim Magee, S. Raghavan, Rina Schneur, and Jim Walton—have helped us in developing exercises. These students and many others in our classes have discovered errors and offered constructive criticisms on the manuscript. In this regard, we are particularly grateful to the Spring 1991 class at MIT in Network Optimization. Thanks also to Prentice Hall reviewer Leslie Hall, Princeton University. Charu Aggarwal and Ajay Mishra also provided us valuable assistance in debugging the book, proofreading it, and preparing the index; our special thanks go to them.

Ghanshyam Hoshing (I.I.T., Kanpur) and Karen Martel and Laura Terrell (both at M.I.T., Cambridge) each did a superb job in typing portions of the manuscript. Ghanshyam Hoshing deserves much credit for typing/drawing and editing most of the text and figures.

We are indebted to the Industrial and Management Engineering Department at I.I.T., Kanpur and to the Sloan School of Management and Operations Research Center at M.I.T. for providing us with an environment conducive to conducting research in network flows and to writing this book. We are also grateful to the National Science Foundation, the Office of Naval Research, the Department of Transportation, and GTE Laboratories for supporting our research that underlies much of this book.

We'd like to acknowledge our parents, Kailash and Ganesh Das Ahuja, Florence and Lee Magnanti, and Roslyn and Albert Orlin, for their affection and encouragement and for instilling in us a love for learning. Finally, we offer our heartfelt thanks to our wives—Smita Ahuja, Beverly Magnanti, Donna Orlin—and our children—Saumya and Shaman Ahuja; Randy Magnanti; and Jenna, Ben, and Caroline Orlin—for their love and understanding as we wrote this book. This book started as a far less ambitious project and so none of our families could possibly have realized how much of our time and energies we would be wresting from them as we wrote it. This work is as much theirs as it is ours!

Kanpur and Cambridge

R. K. Ahuja
T. L. Magnanti
J. B. Orlin

1

INTRODUCTION

Begin at the beginning . . . and go on till you come to the end: then stop.
—Lewis Carroll

Chapter Outline

1.1 INTRODUCTION

Everywhere we look in our daily lives, networks are apparent. Electrical and power networks bring lighting and entertainment into our homes. Telephone networks permit us to communicate with each other almost effortlessly within our local communities and across regional and international borders. National highway systems, rail networks, and airline service networks provide us with the means to cross great geographical distances to accomplish our work, to see our loved ones, and to visit new places and enjoy new experiences. Manufacturing and distribution networks give us access to life's essential foodstock and to consumer products. And computer networks, such as airline reservation systems, have changed the way we share information and conduct our business and personal lives.

In all of these problem domains, and in many more, we wish to move some entity (electricity, a consumer product, a person or a vehicle, a message) from one point to another in an underlying network, and to do so as efficiently as possible, both to provide good service to the users of the network and to use the underlying (and typically expensive) transmission facilities effectively. In the most general sense, this objective is what this book is all about. We want to learn how to model application settings as mathematical objects known as network flow problems and to study various ways (algorithms) to solve the resulting models.

Network flows is a problem domain that lies at the cusp between several fields of inquiry, including applied mathematics, computer science, engineering, management, and operations research. The field has a rich and long tradition, tracing its roots back to the work of Gustav Kirchhof and other early pioneers of electrical engineering and mechanics who first systematically analyzed electrical circuits. This early work set the foundations of many of the key ideas of network flow theory and established networks (graphs) as useful mathematical objects for representing many physical systems. Much of this early work was descriptive in nature, answering such questions as: If we apply a set of voltages to a given network, what will be the resulting current flow? The set of questions that we address in this book are a bit different: If we have alternative ways to use a network (i.e., send flow), which alternative will be most cost-effective? Our intellectual heritage for answering such questions is much more recent and can be traced to the late 1940s and early 1950s when the research and practitioner communities simultaneously developed optimization as an independent field of inquiry and launched the computer revolution, leading to the powerful instruments we know today for performing scientific and managerial computations.

For the most part, in this book we wish to address the following basic questions:

1. *Shortest path problem.* What is the best way to traverse a network to get from one point to another as cheaply as possible?
2. *Maximum flow problem.* If a network has capacities on arc flows, how can we send as much flow as possible between two points in the network while honoring the arc flow capacities?
3. *Minimum cost flow problem.* If we incur a cost per unit flow on a network with arc capacities and we need to send units of a good that reside at one or more points in the network to one or more other points, how can we send the material at minimum possible cost?

In the sense of traditional applied and pure mathematics, each of these problems is trivial to solve. It is not very difficult (but not at all obvious for the later two problems) to see that we need only consider a finite number of alternatives for each problem. So a traditional mathematician might say that the problems are well solved: Simply enumerate the set of possible solutions and choose the one that is best. Unfortunately, this approach is far from pragmatic, since the number of possible alternatives can be very large—more than the number of atoms in the universe for many practical problems! So instead, we would like to devise algorithms that are in a sense "good," that is, whose computation time is small, or at least reasonable, for problems met in practice. One way to ensure this objective is to devise algorithms whose running time is guaranteed not to grow very fast as the underlying network becomes larger (the computer science, operations research, and applied mathematics communities refer to the development of algorithms with such performance guarantees as *worst-case analysis*). Developing algorithms that are good in this sense is another major theme throughout this book, and our development builds heavily on the theory of computational complexity that began to develop within computer science, applied mathematics, and operations research circles in the 1970s, and has flourished ever since.

The field of computational complexity theory combines both craftsmanship and theory; it builds on a confluence of mathematical insight, creative algorithm design, and the careful, and often very clever use of data structures to devise solution methods that are provably good in the sense that we have just mentioned. In the field of network flows, researchers devised the first, seminal contributions of this nature in the 1950s before the field of computational complexity theory even existed as a separate discipline as we know it today. And throughout the last three decades, researchers have made a steady stream of innovations that have resulted in new solution methods and in improvements to known methods. In the past few years, however, researchers have made contributions to the design and analysis of network flow algorithms with improved worst-case performance guarantees at an explosive, almost dizzying pace; moreover, these contributions were very surprising: Throughout the 1950s, 1960s, and 1970s, network flows had evolved into a rather mature field, so much so that most of the research and practitioner communities believed that the core models that we study in this book were so very well understood that further innovations would be hard to come by and would be few and far between. As it turns out, nothing could have been further from the truth.

Our presentation is intended to reflect these new developments; accordingly, we place a heavy emphasis on designing and analyzing good algorithms for solving the core optimization models that arise in the context of network flows. Our intention is to bring together and synthesize the many new contributions concerning efficient network flow algorithms with traditional material that has evolved over the past four decades. We have attempted to distill and highlight some of the essential core ideas (e.g., scaling and potential function arguments) that underlie many of the recent innovations and in doing so to give a unified account of the many algorithms that are now available. We hope that this treatment will provide our readers not only with an accessible entrée to these exciting new developments, but also with an understanding of the most recent and advanced contributions from the literature. Although we are bringing together ideas and methodologies from applied mathematics, computer science, and operations research, our approach has a decidedly computer science orientation as applied to certain types of models that have traditionally arisen in the context of managing a variety of operational systems (the foodstuff of operations research).

We feel that a full understanding of network flow algorithms and a full appreciation for their use requires more than an in-depth knowledge of good algorithms for core models. Consequently, even though this topic is our central thrust, we also devote considerable attention to describing applications of network flow problems. Indeed, we feel that our discussion of applications throughout the text, in the exercises, and in a concluding chapter is one of the major distinguishing features of our coverage.

We have not adopted a linear programming perspective throughout the book, however, because we feel there is much to be gained from a more direct approach, and because we would like the material we cover to be readily accessible to readers who are not optimization specialists. Moreover, we feel that an understanding of network flow problems from first principles provides a useful concrete setting from which to draw considerable insight about more general linear programs.

Similarly, since several important variations of the basic network flow problems are important in practice, or in placing network flows in the broader context of the field of combinatorial optimization, we have also included several chapters on additional topics: assignments and matchings, minimum spanning trees, models with convex (instead of linear) costs, networks with losses and gains, and multicommodity flows. In each of these chapters we have not attempted to be comprehensive, but rather, have tried to provide an introduction to the essential ideas of the topics.

The Lagrangian relaxation chapter permits us to show how the core network models arise in broader problem contexts and how the algorithms that we have developed for the core models can be used in conjunction with other methods to solve more complex problems that arise frequently in practice. In particular, this discussion permits us to introduce and describe the basic ideas of decomposition methods for several important network optimization models—constrained shortest paths, the traveling salesman problem, vehicle routing problem, multicommodity flows, and network design.

Since the proof of the pudding is in the eating, we have also included a chapter on some aspects of computational testing of algorithms. We devote much of our discussion to devising the best possible algorithms for solving network flow problems, in the theoretical sense of computational complexity theory. Although the theoretical model of computation that we are using has proven to be a valuable guide for modeling and predicting the performance of algorithms in practice, it is not a perfect model, and therefore algorithms that are not theoretically superior often perform best in practice. Although empirical testing of algorithms has traditionally been a valuable means for investigating algorithmic ideas, the applied mathematics, computer science, and operations research communities have not yet reached a consensus on how to measure algorithmic performance empirically. So in this chapter we not only report on computational experience with an algorithm we have presented, but also offer some thoughts on how to measure computational performance and compare algorithms.

1.2 NETWORK FLOW PROBLEMS

In this section we introduce the network flow models we study in this book, and in the next section we present several applications that illustrate the practical importance of these models. In both the text and exercises throughout the remaining chapters, we introduce many other applications. In particular, Chapter 19 contains a more comprehensive summary of applications with illustrations drawn from several specialties in applied mathematics, engineering, logistics, manufacturing, and the physical sciences.

Minimum Cost Flow Problem

The minimum cost flow model is the most fundamental of all network flow problems. Indeed, we devote most of this book to the minimum cost flow problem, special cases of it, and several of its generalizations. The problem is easy to state: We wish to determine a least cost shipment of a commodity through a network in order to satisfy demands at certain nodes from available supplies at other nodes. This model has a number of familiar applications: the distribution of a product from manufacturing plants to warehouses, or from warehouses to retailers; the flow of raw material and intermediate goods through the various machining stations in a production line; the routing of automobiles through an urban street network; and the routing of calls through the telephone system. As we will see later in this chapter and in Chapters 9 and 19, the minimum cost flow model also has many less transparent applications.

In this section we present a mathematical programming formulation of the minimum cost flow problem and then describe several of its specializations and variants as well as other basic models that we consider in later chapters. We assume our readers are familiar with the basic notation and definitions of graph theory; those readers without this background might consult Section 2.2 for a brief account of this material.

Let $G = (N, A)$ be a directed network defined by a set N of n *nodes* and a set A of m *directed arcs*. Each arc $(i, j) \in A$ has an associated *cost* c_{ij} that denotes the cost per unit flow on that arc. We assume that the flow cost varies linearly with the amount of flow. We also associate with each arc $(i, j) \in A$ a *capacity* u_{ij} that denotes the maximum amount that can flow on the arc and a *lower bound* l_{ij} that denotes the minimum amount that must flow on the arc. We associate with each node $i \in N$ an integer number $b(i)$ representing its supply/demand. If $b(i) > 0$, node i is a *supply node*; if $b(i) < 0$, node i is a *demand node* with a demand of $-b(i)$; and if $b(i) = 0$, node i is a *transshipment node*. The decision variables in the minimum cost flow problem are arc flows and we represent the flow on an arc $(i, j) \in A$ by x_{ij}. The minimum cost flow problem is an optimization model formulated as follows:

$$\text{Minimize} \sum_{(i,j)\in A} c_{ij}x_{ij} \tag{1.1a}$$

subject to

$$\sum_{\{j:(i,j)\in A\}} x_{ij} - \sum_{\{j:(j,i)\in A\}} x_{ji} = b(i) \qquad \text{for all } i \in N, \tag{1.1b}$$

$$l_{ij} \le x_{ij} \le u_{ij} \qquad \text{for all } (i,j) \in A, \tag{1.1c}$$

where $\sum_{i=1}^{n} b(i) = 0$. In matrix form, we represent the minimum cost flow problem as follows:

$$\text{Minimize } cx \tag{1.2a}$$

subject to

$$\mathcal{N}x = b, \tag{1.2b}$$

$$l \le x \le u. \tag{1.2c}$$

In this formulation, $\mathcal{N}$ is an $n \times m$ matrix, called the *node–arc incidence matrix* of the minimum cost flow problem. Each column $\mathcal{N}_{ij}$ in the matrix corresponds to the variable x_{ij}. The column $\mathcal{N}_{ij}$ has a $+1$ in the ith row, a -1 in the jth row; the rest of its entries are zero.

We refer to the constraints in (1.1b) as *mass balance constraints*. The first term in this constraint for a node represents the total *outflow* of the node (i.e., the flow emanating from the node) and the second term represents the total *inflow* of the node (i.e., the flow entering the node). The mass balance constraint states that the outflow minus inflow must equal the supply/demand of the node. If the node is a supply node, its outflow exceeds its inflow; if the node is a demand node, its inflow exceeds its outflow; and if the node is a transshipment node, its outflow equals its inflow. The flow must also satisfy the lower bound and capacity constraints (1.1c), which we refer to as *flow bound constraints*. The flow bounds typically model physical capacities or restrictions imposed on the flows' operating ranges. In most applications, the lower bounds on arc flows are zero; therefore, if we do not state lower bounds for any problem, we assume that they have value zero.

In most parts of the book we assume that the data are integral (i.e., all arc capacities, arc costs, and supplies/demands of nodes are integral). We refer to this assumption as the *integrality assumption*. The integrality assumption is not restrictive for most applications because we can always transform rational data to integer data by multiplying them by a suitably large number. Moreover, we necessarily need to convert irrational numbers to rational numbers to represent them on a computer.

The following special versions of the minimum cost flow problem play a central role in the theory and applications of network flows.

Shortest path problem. The shortest path problem is perhaps the simplest of all network flow problems. For this problem we wish to find a path of minimum cost (or length) from a specified *source node* s to another specified *sink node* t, assuming that each arc $(i, j) \in A$ has an associated cost (or length) c_{ij}. Some of the simplest applications of the shortest path problem are to determine a path between two specified nodes of a network that has minimum length, or a path that takes least time to traverse, or a path that has the maximum reliability. As we will see in our later discussions, this basic model has applications in many different problem domains, such as equipment replacement, project scheduling, cash flow management, message routing in communication systems, and traffic flow through congested cities. If we set $b(s) = 1$, $b(t) = -1$, and $b(i) = 0$ for all other nodes in the minimum cost flow problem, the solution to the problem will send 1 unit of flow from node s to node t along the shortest path. The shortest path problem also models situations in which we wish to send flow from a single-source node to a single-sink node in an uncapacitated network. That is, if we wish to send v units of flow from node s to node t and the capacity of each arc of the network is at least v, we would send the flow along a shortest path from node s to node t. If we want to determine shortest paths from the source node s to every other node in the network, then in the minimum cost flow problem we set $b(s) = (n - 1)$ and $b(i) = -1$ for all other nodes. [We can set each arc capacity u_{ij} to any number larger than $(n - 1)$.] The minimum cost flow solution would then send unit flow from node s to every other node i along a shortest path.

Maximum flow problem. The maximum flow problem is in a sense a complementary model to the shortest path problem. The shortest path problem models situations in which flow incurs a cost but is not restricted by any capacities; in contrast, in the maximum flow problem flow incurs no costs but is restricted by flow bounds. The maximum flow problem seeks a feasible solution that sends the maximum amount of flow from a specified source node s to another specified sink node t. If we interpret u_{ij} as the maximum flow rate of arc (i, j), the maximum flow problem identifies the maximum steady-state flow that the network can send from node s to node t per unit time. Examples of the maximum flow problem include determining the maximum steady-state flow of (1) petroleum products in a pipeline network, (2) cars in a road network, (3) messages in a telecommunication network, and (4) electricity in an electrical network. We can formulate this problem as a minimum cost flow problem in the following manner. We set $b(i) = 0$ for all $i \in N$, $c_{ij} = 0$ for all $(i, j) \in A$, and introduce an additional arc (t, s) with cost $c_{ts} = -1$ and flow bound $u_{ts} = \infty$. Then the minimum cost flow solution maximizes the flow on arc (t, s); but since any flow on arc (t, s) must travel from node s to node t through the arcs in A [since each $b(i) = 0$], the solution to the minimum cost flow problem will maximize the flow from node s to node t in the original network.

Assignment problem. The data of the assignment problem consist of two equally sized sets N_1 and N_2 (i.e., $| N_1 | = | N_2 |$), a collection of pairs $A \subseteq N_1 \times N_2$ representing possible assignments, and a cost c_{ij} associated with each element $(i, j) \in A$. In the assignment problem we wish to pair, at minimum possible cost, each object in N_1 with exactly one object in N_2. Examples of the assignment problem include assigning people to projects, jobs to machines, tenants to apartments, swimmers to events in a swimming meet, and medical school graduates to available internships. The assignment problem is a minimum cost flow problem in a network $G = (N_1 \cup N_2, A)$ with $b(i) = 1$ for all $i \in N_1$, $b(i) = -1$ for all $i \in N_2$, and $u_{ij} = 1$ for all $(i, j) \in A$.

Transportation problem. The transportation problem is a special case of the minimum cost flow problem with the property that the node set N is partitioned into two subsets N_1 and N_2 (of possibly unequal cardinality) so that (1) each node in N_1 is a supply node, (2) each node N_2 is a demand node, and (3) for each arc (i, j) in A, $i \in N_1$ and $j \in N_2$. The classical example of this problem is the distribution of goods from warehouses to customers. In this context the nodes in N_1 represent the warehouses, the nodes in N_2 represent customers (or, more typically, customer zones), and an arc (i, j) in A represents a distribution channel from warehouse i to customer j.

Circulation problem. The circulation problem is a minimum cost flow problem with only transshipment nodes; that is, $b(i) = 0$ for all $i \in N$. In this instance we wish to find a feasible flow that honors the lower and upper bounds l_{ij} and u_{ij} imposed on the arc flows x_{ij}. Since we never introduce any exogenous flow into the network or extract any flow from it, all the flow circulates around the network. We wish to find the circulation that has the minimum cost. The design of a routing schedule of a commercial airline provides one example of a circulation problem. In this setting, any airplane circulates among the airports of various cities; the lower bound l_{ij} imposed on an arc (i, j) is 1 if the airline needs to provide service between cities i and j, and so must dispatch an airplane on this arc (actually, the nodes will represent a combination of both a physical location and a time of day so that an arc connects, for example, New York City at 8 A.M. with Boston at 9 A.M.).

In this book, we also study the following generalizations of the minimum cost flow problem.

Convex cost flow problems. In the minimum cost flow problem, we assume that the cost of the flow on any arc varies linearly with the amount of flow. Convex cost flow problems have a more general cost structure: The cost is a convex function of the amount of flow. Flow costs vary in a convex manner in numerous problem settings, including (1) power losses in an electrical network due to resistance, (2) congestion costs in a city transportation network, and (3) expansion costs of a communication network.

Generalized flow problems. In the minimum cost flow problem, arcs conserve flows (i.e., the flow entering an arc equals the flow leaving the arc). In generalized flow problems, arcs might "consume" or "generate" flow. If x_{ij} units of flow enter an arc (i, j), then $\mu_{ij}x_{ij}$ units arrive at node j; μ_{ij} is a positive *multiplier* associated with the arc. If $0 < \mu_{ij} < 1$, the arc is *lossy*, and if $1 < \mu_{ij} < \infty$, the arc is *gainy*. Generalized network flow problems arise in several application contexts: for example, (1) power transmission through electric lines, with power lost with distance traveled, (2) flow of water through pipelines or canals that lose water due to seepage or evaporation, (3) transportation of a perishable commodity, and (4) cash management scenarios in which arcs represent investment opportunities and multipliers represent appreciation or depreciation of an investment's value.

Multicommodity flow problems. The minimum cost flow problem models the flow of a single commodity over a network. Multicommodity flow problems arise when several commodities use the same underlying network. The commodities may either be differentiated by their physical characteristics or simply by their origin–destination pairs. Different commodities have different origins and destinations, and commodities have separate mass balance constraints at each node. However, the sharing of the common arc capacities binds the different commodities together. In fact, the essential issue addressed by the multicommodity flow problem is the allocation of the capacity of each arc to the individual commodities in a way that minimizes overall flow costs. Multicommodity flow problems arise in many practical situations, including (1) the transportation of passengers from different origins to different destinations within a city; (2) the routing of nonhomogeneous tankers (nonhomogeneous in terms of speed, carrying capability, and operating costs); (3) the worldwide shipment of different varieties of grains (such as corn, wheat, rice, and soybeans) from countries that produce grains to those that consume it; and (4) the transmission of messages in a communication network between different origin–destination pairs.

Other Models

In this book we also study two other important network models: the *minimum spanning tree problem* and the *matching problem*. Although these two models are not flow problems per se, because of their practical and mathematical significance and because of their close connection with several flow problems, we have included them as part of our treatment of network flows.

Minimum spanning tree problem. A spanning tree is a tree (i.e., a connected acyclic graph) that spans (touches) all the nodes of an undirected network. The cost of a spanning tree is the sum of the costs (or lengths) of its arcs. In the minimum spanning tree problem, we wish to identify a spanning tree of minimum cost (or length). The applications of the minimum spanning tree problem are varied and include (1) constructing highways or railroads spanning several cities; (2) laying pipelines connecting offshore drilling sites, refineries, and consumer markets; (3) designing local access networks; and (4) making electric wire connections on a control panel.

Matching problems. A *matching* in a graph $G = (N, A)$ is a set of arcs with the property that every node is incident to at most one arc in the set; thus a matching induces a pairing of (some of) the nodes in the graph using the arcs in A. In a matching, each node is matched with at most one other node, and some nodes might not be matched with any other node. The *matching problem* seeks a matching that optimizes some criteria. Matching problems on a bipartite graphs (i.e., those with two sets of nodes and with arcs that join only nodes between the two sets, as in the assignment and transportation problems) are called *bipartite matching problems*, and those on nonbipartite graphs are called *nonbipartite matching problems*. There are two additional ways of categorizing matching problems: *cardinality matching problems*, which maximize the number of pairs of nodes matched, and *weighted matching problems*, which maximize or minimize the weight of the matching. The weighted matching problem on a bipartite graph is also known as the *assignment problem*. Applications of matching problems arise in matching roommates to hostels, matching pilots to compatible airplanes, scheduling airline crews for available flight legs, and assigning duties to bus drivers.

1.3 APPLICATIONS

Networks are pervasive. They arise in numerous application settings and in many forms. Physical networks are perhaps the most common and the most readily identifiable classes of networks; and among physical networks, transportation networks are perhaps the most visible in our everyday lives. Often, these networks model homogeneous facilities such as railbeds or highways. But on other occasions, they correspond to composite entities that model, for example, complex distribution and logistics decisions. The traditional operations research "transportation problem" is illustrative. In the transportation problem, a shipper with inventory of goods at its warehouses must ship these goods to geographically dispersed retail centers, each with a given customer demand, and the shipper would like to meet these demands incurring the minimum possible transportation costs. In this setting, a transportation link in the underlying network might correspond to a complex distribution channel with, for example, a trucking shipment from the warehouse to a railhead, a rail shipment, and another trucking leg from the destination rail yard to the customer's site.

Physical networks are not limited to transportation settings; they also arise in several other disciplines of applied science and engineering, such as mathematics, chemistry, and electrical, communications, mechanical, and civil engineering. When physical networks occur in these different disciplines, their nodes, arcs, and flows model many different types of physical entities. For example, in a typical communication network, nodes will represent telephone exchanges and transmission facilities, arcs will denote copper cables or fiber optic links, and flow would signify the transmission of voice messages or of data. Figure 1.1 shows some typical associations for the nodes, arcs, and flows in a variety of physical networks.

Network flow problems also arise in surprising ways for problems that on the surface might not appear to involve networks at all. Sometimes these applications are linked to a physical entity, and at other times they are not. Sometimes the nodes and arcs have a temporal dimension that models activities that take place over time.

Applications	Physical analog of nodes	Physical analog of arcs	Flow
Communication systems	Telephone exchanges, computers, transmission facilities, satellites	Cables, fiber optic links, microwave relay links	Voice messages, data, video transmissions
Hydraulic systems	Pumping stations, reservoirs, lakes	Pipelines	Water, gas, oil, hydraulic fluids
Integrated computer circuits	Gates, registers, processors	Wires	Electrical current
Mechanical systems	Joints	Rods, beams, springs	Heat, energy
Transportation systems	Intersections, airports, rail yards	Highways, railbeds, airline routes	Passengers, freight, vehicles, operators

Figure 1.1 Ingredients of some common physical networks.

Many scheduling applications have this flavor. In any event, networks model a variety of problems in project, machine, and crew scheduling; location and layout theory; warehousing and distribution; production planning and control; and social, medical, and defense contexts. Indeed, these various applications of network flow problems seem to be more widespread than are the applications of physical networks. We present many such applications throughout the text and in the exercises; Chapter 19, in particular, brings together and summarizes many applications. In the following discussion, to set a backdrop for the next few chapters, we describe several sample applications that are intended to illustrate a range of problem contexts and to be suggestive of how network flow problems arise in practice. This set of applications provides at least one example of each of the network models that we introduced in the preceding section.

Application 1.1 Reallocation of Housing

A housing authority has a number of houses at its disposal that it lets to tenants. Each house has its own particular attributes. For example, a house might or might not have a garage, it has a certain number of bedrooms, and its rent falls within a particular range. These variable attributes permit us to group the house into several categories, which we index by $i = 1, 2, \ldots, n$.

Over a period of time a number of tenants will surrender their tenancies as they move or choose to live in alternative accommodations. Furthermore, the requirements of the tenants will change with time (because new families arrive, children leave home, incomes and jobs change, and other considerations). As these changes occur, the housing authority would like to relocate each tenant to a house of his or her choice category. While the authority can often accomplish this objective by simple exchanges, it will sometimes encounter situations requiring multiple moves: moving one tenant would replace another tenant from a house in a different category, who, in turn, would replace a tenant from a house in another category, and so on, thus creating a cycle of changes. We call such a change a *cyclic change*. The decision problem is to identify a cyclic change, if it exists, or to show that no such change exists.

To solve this problem as a network problem, we first create a *relocation graph* G whose nodes represent various categories of houses. We include arc (i, j) in the graph whenever a person living in a house of category i wishes to move to a house of category j. A directed cycle in G specifies a cycle of changes that will satisfy the requirements of one person in each of the categories contained in the cycle. Applying this method iteratively, we can satisfy the requirements of an increasing number of persons.

This application requires a method for identifying directed cycles in a network, if they exist. A well-known method, known as *topological sorting*, will identify such cycles. We discuss topological sorting in Section 3.4. In general, many tenant reassignments might be possible, so the relocation graph G might contain several cycles. In that case the authority's management would typically want to find a cycle containing as few arcs as possible, since fewer moves are easier to handle administratively. We can solve this problem using a shortest path algorithm (see Exercise 5.38).

Application 1.2 Assortment of Structural Steel Beams

In its various construction projects, a construction company needs structural steel beams of a uniform cross section but of varying lengths. For each $i = 1, \ldots, n$, let $D_i > 0$ denote the demand of the steel beam of length L_i, and assume that $L_1 < L_2 < \cdots < L_n$. The company could meet its needs by maintaining and drawing upon an inventory of stock containing exactly D_i units of the steel beam of length L_i. It might not be economical to carry all the demanded lengths in inventory, however, because of the high cost of setting up the inventory facility to store and handle each length. In that case, if the company needs a beam of length L_i not carried in inventory, it can cut a beam of longer length down to the desired length. The cutting operation will typically produce unusable steel as scrap. Let K_i denote the cost for setting up the inventory facility to handle beams of length L_i, and let C_i denote the cost of a beam of length L_i. The company wants to determine the lengths of beams to be carried in inventory so that it will minimize the total cost of (1) setting up the inventory facility, and (2) discarding usable steel lost as scrap.

We formulate this problem as a shortest path problem as follows. We construct a directed network G on $(n + 1)$ nodes numbered $0, 1, 2, \ldots, n$; the nodes in this network correspond to various beam lengths. Node 0 corresponds to a beam of length zero and node n corresponds to the longest beam. For each node i, the network contains a directed arc to every node $j = i + 1, i + 2, \ldots, n$. We interpret the arc (i, j) as representing a storage strategy in which we hold beams of length L_j in inventory and use them to satisfy the demand of all the beams of lengths L_{i+1}, $L_{i+2}, \ldots, L_j$. The cost c_{ij} of the arc (i, j) is

$$c_{ij} = K_j + C_j \sum_{k=i+1}^{j} D_k.$$

The cost of arc (i, j) has two components: (1) the fixed cost K_j of setting up the inventory facility to handle beams of length L_j, and (2) the cost of using beams of length L_j to meet the demands of beams of lengths $L_{i+1}, \ldots, L_j$. A directed path from node 0 to node n specifies an assortment of beams to carry in inventory and the cost of the path equals the cost associated with this inventory scheme. For example, the path 0–4–6–9 corresponds to the situation in which we set up the inventory facility for handling beams of lengths L_4, L_6, and L_9. Consequently, the shortest path from node 0 to node n would prescribe the least cost assortment of structural steel beams.

Application 1.3 Tournament Problem

Consider a round-robin tournament between n teams, assuming that each team plays against every other team c times. Assume that no game ends in a draw. A person claims that α_i for $1 \le i \le n$ denotes the number of victories accrued by the ith team at the end of the tournament. How can we determine whether the given set of nonnegative integers $\alpha_1, \alpha_2, \ldots, \alpha_n$ represents a possible winning record for the n teams?

Define a directed network $G = (N, A)$ with node set $N = \{1, 2, \ldots, n\}$ and arc set $A = \{(i, j) \in N \times N : i < j\}$. Therefore, each node i is connected to the nodes $i + 1, i + 2, \ldots, n$. Let x_{ij} for $i < j$ represent the number of times team i defeats team j. Observe that the total number of times team i beats teams $i + 1, i + 2, \ldots, n$ is $\sum_{\{j:(i,j)\in A\}} x_{ij}$. Also observe that the number of times that team i beats a team $j < i$ is $c - x_{ji}$. Consequently, the total number of times that team i beats teams $1, 2, \ldots, i - 1$ is $(i - 1)c - \sum_{\{j:(j,i)\in A\}} x_{ji}$. The total number of wins of team i must equal the total number of times it beats the teams $1, 2, \ldots, n$. The preceding observations show that

$$\sum_{\{j:(i,j)\in A\}} x_{ij} - \sum_{\{j:(j,i)\in A\}} x_{ji} = \alpha_i - (i - 1)c \qquad \text{for all } i \in N. \tag{1.3}$$

In addition, a possible winning record must also satisfy the following lower and upper bound conditions:

$$0 \le x_{ij} \le c \qquad \text{for all } (i, j) \in A. \tag{1.4}$$

This discussion shows that the record α_i is a possible winning record if the constraints defined by (1.3) and (1.4) have a feasible solution x. Let $b(i) = \alpha_i - (i - 1)c$. Observe that the expressions $\sum_{i\in N}\alpha_i$ and $\sum_{i\in N}(i - 1)c$ are both equal to $cn(n - 1)/2$, which is the total number of games played. Consequently, $\sum_{i\in N}b(i) = 0$. The problem of finding a feasible solution of a network flow system like (1.3) and (1.4) is called a *feasible flow problem* and can be solved by solving a maximum flow problem (see Section 6.2).

Application 1.4 Leveling Mountainous Terrain

This application was inspired by a common problem facing civil engineers when they are building road networks through hilly or mountainous terrain. The problem concerns the distribution of earth from high points to low points of the terrain to produce a leveled roadbed. The engineer must determine a plan for leveling the route by specifying the number of truckloads of earth to move between various locations along the proposed road network.

We first construct a *terrain graph*: it is an undirected graph whose nodes represent locations with a demand for earth (low points) or locations with a supply of earth (high points). An arc of this graph represents an available route for distributing the earth, and the cost of this arc represents the cost per truckload of moving earth between the two points. (A *truckload* is the basic unit for redistributing the earth.) Figure 1.2 shows a portion of the terrain graph.

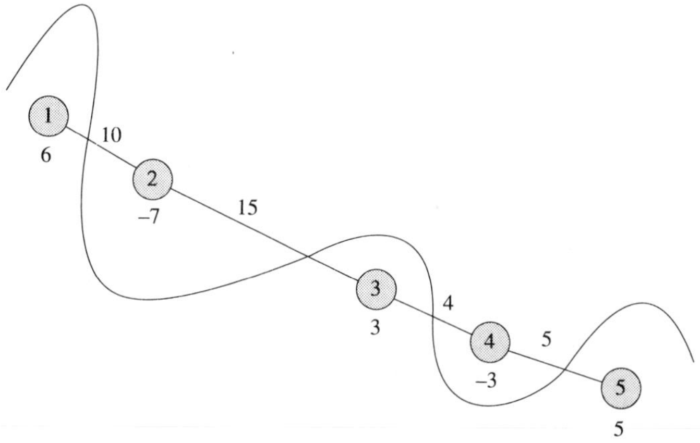

Figure 1.2 Portion of the terrain graph.

A leveling plan for a terrain graph is a flow (set of truckloads) that meets the demands at nodes (levels the low points) by the available supplies (by earth obtained from high points) at the minimum cost (for the truck movements). This model is clearly a minimum cost flow problem in the terrain graph.

Application 1.5 Rewiring of Typewriters

For several years, a company had been using special electric typewriters to prepare punched paper tapes to enter data into a digital computer. Because the typewriter is used to punch a six-hole paper tape, it can prepare $2^6 = 64$ binary hole/no-hole patterns. The typewriters have 46 characters, and each punches one of the 64 patterns. The company acquired a new digital computer that uses a different coding hole/no-hole patterns to represent characters. For example, using 1 to represent a hole and 0 to represent a no-hole, the letter A is 111100 in the code for the old computer and 011010 in the code for the new computer. The typewriter presently punches the former and must be modified to punch the latter.

Each key in the typewriter is connected to a steel code bar, so changing the code of that key requires mechanical changes in the steel bar system. The extent of the changes depends on how close the new and old characters are to each other. For the letter A, the second, third, and sixth bits are identical in the old and new codes and no changes need be made for these bits; however, the first, fourth, and fifth bits are different, so we would need to make three changes in the steel code bar connected to the A-key. Each change involves removing metal at one place and adding metal at another place. When a key is pressed, its steel code bar activates six cross-bars (which are used by all the keys) that are connected electrically to six hole punches. If we interchange the fourth and fifth wires of the cross-bars to the hole punches (which is essentially equivalent to interchanging the fourth and fifth bits of all characters in the old code), we would reduce the number of mechanical changes needed for the A-key from three to one. However, this change of wires might increase the number of changes for some of the other 45 keys. The problem, then, is how to optimally connect the wires from the six cross-bars to the six punches so that we can minimize the number of mechanical changes on the steel code bars.

We formulate this problem as an assignment problem as follows. Define a network $G = (N_1 \cup N_2, A)$ with node sets $N_1 = \{1, 2, \ldots, 6\}$ and $N_2 = \{1', 2', \ldots, 6'\}$, and an arc set $A = N_1 \times N_2$; the cost of the arc $(i, j') \in A$ is the number of keys (out of 46) for which the ith bit in the old code differs from the jth bit in the new code. Thus if we assign cross-bar i to the punch j, the number of mechanical changes needed to print the ith bit of each symbol correctly is c_{ij}. Consequently, the minimum cost assignment will minimize the number of mechanical changes.

Application 1.6 Pairing Stereo Speakers

As a part of its manufacturing process, a manufacturer of stereo speakers must pair individual speakers before it can sell them as a set. The performance of the two speakers depends on their frequency response. To measure the quality of the pairs, the company generates matching coefficients for each possible pair. It calculates these coefficients by summing the absolute differences between the responses of the two speakers at 20 discrete frequencies, thus giving a matching coefficient value between 0 and 30,000. Bad matches yield a large coefficient, and a good pairing produces a low coefficient.

The manufacturer typically uses two different objectives in pairing the speakers: (1) finding as many pairs as possible whose matching coefficients do not exceed a specification limit, or (2) pairing speakers within specification limits to minimize the total sum of the matching coefficients. The first objective minimizes the number of pairs outside specification, and so the number of speakers that the firm must sell at a reduced price. This model is an application of the nonbipartite cardinality matching problem on an undirected graph: the nodes of this graph represent speakers and arcs join two nodes if the matching coefficients of the corresponding speakers are within the specification limit. The second model is an application of the nonbipartite weighted matching problem.

Application 1.7 Measuring Homogeneity of Bimetallic Objects

This application shows how a minimum spanning tree problem can be used to determine the degree to which a bimetallic object is homogeneous in composition. To use this approach, we measure the composition of the bimetallic object at a set of sample points. We then construct a network with nodes corresponding to the sample points and with an arc connecting physically adjacent sample points. We assign a cost with each arc (i, j) equal to the product of the physical (Euclidean) distance between the sample points i and j and a homogeneity factor between 0 and 1. This homogeneity factor is 0 if the composition of the corresponding samples is exactly alike, and is 1 if the composition is very different; otherwise, it is a number between 0 and 1. Note that this measure gives greater weight to two points if they are different and are far apart. The cost of the minimum spanning tree is a measure of the homogeneity of the bimetallic object. The cost of the tree is 0 if all the sample points are exactly alike, and high cost values imply that the material is quite nonhomogeneous.

Application 1.8 Electrical Networks

The electrical network shown in Figure 1.3 has eight resistors, two current sources (at nodes 1 and 6), and one current sink (at node 7). In this network we wish to determine the equilibrium current flows through the resistors. A popular method for solving this problem is to introduce a variable x_{ij} representing the current flow on the arc (i, j) of the electrical network and write a set of equilibrium relationships for these flows; that is, the voltage–current relationship equations (using Ohm's law) and the current balance equations (using Kirchhof's law). The solution of these equations gives the arc currents x_{ij}. An alternative, and possibly more efficient approach is to formulate this problem as a convex cost flow problem. This formulation uses the well-known result that the equilibrium currents on resistors are those flows for which the resistors dissipate the least amount of the total power supplied by the voltage sources (i.e., the electric current follows the path of least resistance). Ohm's law shows that a resistor of resistance r_{ij} dissipates $r_{ij}x_{ij}^2$ ohms of power. Therefore, we can obtain the optimal currents by solving the following convex cost flow problem:

$$\text{Minimize} \sum_{(i,j)\in A} r_{ij}x_{ij}^2$$

subject to

$$\sum_{\{j:(i,j)\in A\}} x_{ij} - \sum_{\{j:(j,i)\in A\}} x_{ji} = b(i) \qquad \text{for each node } i \in N,$$

$$x_{ij} \geq 0 \qquad \text{for each arc } (i,j) \in A.$$

In this model $b(i)$ represents the supply/demand of a current source or sink.

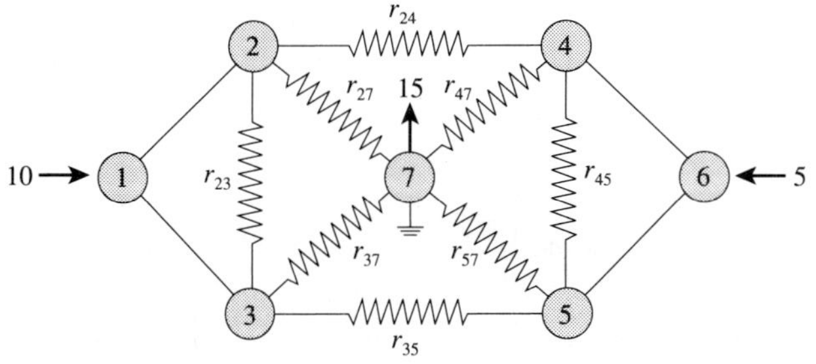

Figure 1.3 Electrical network.

The formulation of a set of equilibrium conditions as an equivalent optimization model is a powerful idea in the physical sciences, dating from the last century, which has become known as so-called *variational principles*. The term ''variational'' arises because the equilibrium conditions are the ''optimality conditions'' for the equivalent optimization model that tell us that we cannot improve the optimal solution by varying (hence the term ''variational'') the optimal solution to this optimization model.

Application 1.9 Determining an Optimal Energy Policy

As part of their national planning effort, most countries need to decide on an energy policy (i.e., how to utilize the available raw materials to satisfy their energy needs). Assume, for simplicity, that a particular country has four basic raw materials: crude oil, coal, uranium, and hydropower; and it has four basic energy needs: electricity, domestic oil, petroleum, and gas. The country has the technological base and infrastructure to convert each raw material into one or more energy forms. For example, it can convert crude oil into domestic oil or petrol, coal into electricity, and so on. The available technology base specifies the efficiency and the cost of each conversion. The objective is to satisfy, at the least possible cost of energy conversion, a certain annual consumption level of various energy needs from a given annual production of raw materials.

Figure 1.4 shows the formulation of this problem as a generalized network flow problem. The network has three types of arcs: (1) *source arcs* (s, i) emanating from the source node s, (2) *sink arcs* (j, t) entering the sink node t, and (3) *conversion arcs* (i, j). The source arc (s, i) has a capacity equal to the availability $\alpha(i)$ of the raw material i and a flow multiplier of value 1. The sink arc (j, t) has capacity equal to the demand $\beta(j)$ of type j energy need and flow multiplier of value 1. Each conversion arc (i, j) represents the conversion of raw material i into the energy form j; the multiplier of this arc is the efficiency of the conversion (i.e., units of energy j obtained from 1 unit of raw material i); and the cost of the arc (i, j) is the cost of this conversion. In this model, since $\alpha(i)$ is an upper bound on the use of raw material

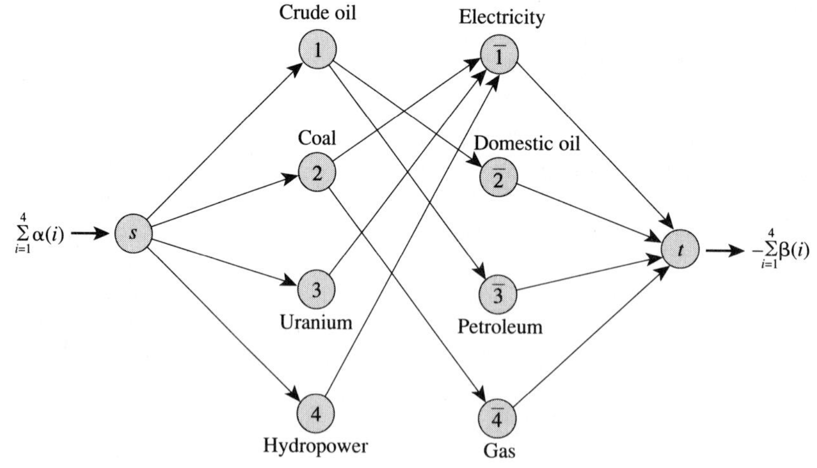

Figure 1.4 Energy problem as a generalized network flow problem.

i, $\sum_{i=1}^{4} \alpha(i)$ is an upper bound on the flow out of node s. Similarly, $\sum_{i=1}^{4} \beta(i)$ is a lower bound on the flow into node t. In Exercise 15.29, we show how to convert this problem into a standard form without bounds on supplies and demands.

Application 1.10 Racial Balancing of Schools

In 1968, the U.S. Supreme Court ruled that all school systems in the country should begin admitting students to schools on a nondiscriminatory basis and should employ faster techniques to promote desegregated schools across the nation. This decision made it necessary for many school systems to develop radically different procedures for assigning students to schools. Since the Supreme Court did not specify what constitutes an acceptable racial balance, the individual school boards used their own best judgments to arrive at acceptable criteria on which to base their desegregation plans. This application describes a multicommodity flow model for determining an optimal assignment of students to schools that minimizes the total distance traveled by the students, given a specification of lower and upper limits on the required racial balance in each school.

Suppose that a school district has S schools and school j has capacity u_j. For the purpose of this formulation, we divide the school district into L population centers. These locations might, for example, be census tracts, bus stops, or city blocks. The only restriction on the population centers is that they be finite in number and that a single distance measure reasonably approximates the distance any student at center i must travel if he or she is assigned to school j. Let S_{ik} denote the available number of students of the kth ethnic group at the ith population center. The objective is to assign students to schools in a way that achieves the desired ethnic composition for each school and minimizes the total distance traveled by the students. Each school j has the ethnic requirement that it must enroll at least l_{jk} and no more than u_{jk} students from the kth ethnic group.

We can model this problem as a multicommodity flow problem on an appropriately defined network. Figure 1.5 shows this network representation for a problem with three population centers and three schools. This network has one node for each population center and for each school as well as a "source" and a "sink" node for each ethnic group. The flow commodities represent the students of different ethnic groups. The students of the kth ethnic group flow from source a_k to sink e_k via population center and school nodes. We set the upper bound on arc (a_k, b_i) connecting the kth ethnic group source node and the ith population center equal to S_{ik} and the cost of the arc (b_i, b_j) connecting the ith population center and jth school equal to f_{ij}, the distance between that population center and that school. By setting the capacity of the arc (c_j, d_j) equal to u_j, we ensure that the total number of students (of all ethnic groups) allocated to the jth school does not exceed the maximum student population for this school. The students of all ethnic groups must share the capacity of each school. Finally, we incorporate constraints on the ethnic compositions of the schools by setting the lower and upper bounds on the arc (d_j, e_k) equal to l_{jk} and u_{jk}. It is fairly easy to verify that the multicommodity flow problem models the racial balancing problem, so a minimum multicommodity flow will specify an optimal assignment of students to the schools.

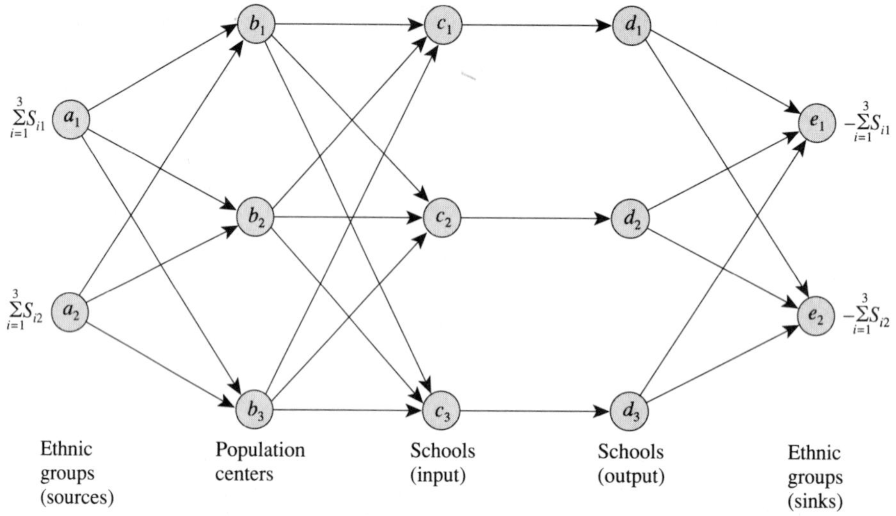

Figure 1.5 Formulating the racial balancing problem as a multicommodity flow problem.

1.4 SUMMARY

In this chapter we introduced the network flow problems that we study in this book and described a few scenarios in which these problems arise. We began by giving a linear programming formulation of the minimum cost flow problem and identifying several special cases: the shortest path problem, the maximum flow problem, the assignment problem, the transportation problem, and the circulation problem. We next described several generalizations of the minimum cost flow problem: the convex cost flow problem, the generalized network flow problem, and the multicommodity flow problem. Finally, we described two other important network models: the minimum spanning tree problem and the matching problem. Although these two problems are not network flow problems per se, we have included them in this book because they are closely related to several network flow problems and because they arise often in the context of network optimization.

Networks are pervasive and arise in numerous application settings. Physical networks, which are the most readily identifiable classes of networks, arise in many applications in many different types of systems: communications, hydraulic, mechanical, electronic, and transportation. Network flow problems also arise in surprising ways in optimization problems that on the surface might not appear to involve networks at all. We described several of these ''indirect'' applications of network flow problems, in such problem settings as urban housing, production planning, electrical networks, racial balancing, leveling mountainous terrain, evaluating tournaments, matching stereo speakers, wiring typewriters, assessing the homogeneity of physical materials, and energy planning. The applications we have considered offer only a brief glimpse of the wide-ranging practical importance of network flows; although our discussion of applications in this chapter is limited, it does provide at least one example of each of the network models that we have introduced in this chapter.

REFERENCE NOTES

The study of network flow models predates the development of linear programming. The first studies in this problem domain, conducted by Kantorovich [1939], Hitchcock [1941], and Koopmans [1947], considered the transportation problem, a special case of the minimum cost flow problem. These studies provided insight into the problem structure and yielded algorithmic approaches. Interest in the network flow problems grew with the advent of the simplex method by Dantzig in 1947, who also specialized this algorithm for the transportation problem (see Dantzig [1951]).

During the 1950s, researchers began to exhibit increasing interest in the minimum cost flow problem and its specializations—the shortest path problem, the maximum flow problem, and the assignment problem—mainly because of the importance of these models in real-world applications. Soon researchers developed special algorithms for solving these problems. Dantzig, Ford, and Fulkerson pioneered these efforts. Whereas Dantzig focused on the simplex-based methods, Ford and Fulkerson developed primal–dual combinatorial algorithms. The landmark books by Dantzig [1962] and Ford and Fulkerson [1962] present thorough discussions of these early contributions.

In the years following this groundbreaking work, network flow problems and their generalizations emerged as major research topics in thousands of papers and numerous text and reference books. The following books summarize developments in the field and serve as a guide to the literature:

1. *Flows in Networks* (Ford and Fulkerson [1962])
2. *Programming, Games and Transportation Networks* (Berge and Ghouila-Houri [1962])
3. *Finite Graphs and Networks* (Busacker and Saaty [1965])
4. *Network Flow, Transportation and Scheduling* (Iri [1969])
5. *Integer Programming and Network Flows* (Hu [1969])
6. *Communication, Transmission, and Transportation Networks* (Frank and Frisch [1971])
7. *Flows in Transportation Networks* (Potts and Oliver [1972])
8. *Graph Theory: An Algorithmic Approach* (Christophides [1975])
9. *Flow Algorithms* (Adel'son-Vel'ski, Dinics, and Karzanov [1975])
10. *Graph Theory with Applications* (Bondy and Murty [1976])
11. *Combinatorial Optimization: Networks and Matroids* (Lawler [1976])
12. *Optimization Algorithms for Networks and Graphs* (Minieka [1978])
13. *Graph Algorithms* (Even [1979])
14. *Algorithms for Network Programming* (Kennington and Helgason [1980])
15. *Network Flow Programming* (Jensen and Barnes [1980])
16. *Fundamentals of Network Analysis* (Phillips and Garcia-Diaz [1981])
17. *Combinatorial Optimization: Algorithms and Complexity* (Papadimitriou and Steiglitz [1982])
18. *Discrete Optimization Algorithms* (Syslo, Deo, and Kowalik [1983])
19. *Data Structures and Network Algorithms* (Tarjan [1983])

20. *Graphs and Algorithms* (Gondran and Minoux [1984])
21. *Network Flows and Monotropic Optimization* (Rockafellar [1984])
22. *Linear Programming and Network Models* (Gupta [1985])
23. *Programming in Networks and Graphs* (Derigs [1988])
24. *Linear Programming and Network Flows*, 2nd ed. (Bazaraa, Jarvis, and Sherali [1990])

As an additional source of references, the reader might consult the bibliographies on network optimization prepared by Golden and Magnanti [1977], Ahuja, Magnanti, and Orlin [1989, 1991], Bazaraa, Jarvis, and Sherali [1990], and the extensive set of references on integer programming compiled by researchers at the University of Bonn (Kastning [1976], Hausman [1978], and Von Randow [1982, 1985]).

Since the applications of network flow models are so pervasive, no single source provides a comprehensive account of network flow models and their impact on practice. Several researchers have prepared general surveys of selected application areas. Notable among these are the papers by Bennington [1974], Glover and Klingman [1976], Bodin, Golden, Assad, and Ball [1983], Aronson [1989], and Glover, Klingman, and Phillips [1990]. The book by Gondran and Minoux [1984] also describes a variety of applications of network flow problems. In this book we describe or cite over 150 selected applications of network flow problems. We provide the references for these problems in the reference notes given at the end of Chapters 4, 6, 9, 12, 13, 14, 15, 16, 17, and 19. We have adapted many of these applications from the paper of Ahuja, Magnanti, Orlin, and Reddy [1992].

The applications we present in Section 1.3 are adapted from the following references:

1. Reallocation of housing (Wright [1975])
2. Assortment of structural steel beams (Frank [1965])
3. Tournament problem (Ford and Johnson [1959])
4. Leveling mountainous terrain (Farley [1980])
5. Rewiring of typewriters (Machol [1961])
6. Pairing stereo speakers (Mason and Philpott [1988])
7. Measuring homogeneity of bimetallic objects (Shier [1982])
8. Electrical networks (Hu [1966])
9. Determining an optimal energy policy (Gondran and Minoux [1984])
10. Racial balancing of schools (Clarke and Surkis [1968])

EXERCISES

1.1. Formulate the following problems as circulation problems: (1) the shortest path problem; (2) the assignment problem; and (3) the transportation problem.

1.2. Consider a variant of the transportation problem for which (1) the sum of demands exceeds the sum of supplies, and (2) we incur a penalty p_j for every unit of unfulfilled demand at demand node j. Formulate this problem as a standard transportation problem with total supply equal to total demand.

1.3. In this exercise we examine a generalization of Application 1.2, concerning assortment of structural steel beams. In the discussion of that application, we assumed that if we must cut a beam of length 5 units to a length of 2 units, we obtain a single beam of length 2 units; the remaining 3 units have no value. However, in practice, from a beam of length 5 we can cut two beams of length 2; the remaining length of 1 unit will have some scrap value. Explain how we might incorporate the possibility of cutting multiple beam lengths (of the same length) from a single piece and assigning some salvage value to the scrap. Assume that the scrap has a value of β per unit length.

1.4. Large-scale personnel assignment. A recurring problem in the U.S. armed forces is efficient distribution and utilization of skilled personnel. Each month thousands of individuals in the U.S. military vacate jobs, and thousands of personnel become available for assignment. Each job has particular characteristics and skill requirements, while each person from the pool of available personnel has specific skills and preferences. Suppose that we use this information to compute the utility (or desirability) d_{ij} of each possible assignment of a person to a job. The decision problem is to assign personnel to the vacancies in a way that maximizes the total utility of all the assignments. Explain how to formulate this problem as a network flow problem.

1.5. Dating problem. A dating service receives data from p men and p women. These data determine what pairs of men and women are mutually compatible. Since the dating service's commission is proportional to the number of dates it arranges, it would like to determine the maximum number of compatible couples that can be formed. Formulate this problem as a matching problem.

1.6. Pruned chessboard problem. A chessboard consists of 64 squares arranged in eight rows and eight columns. A *domino* is a wooden or plastic piece consisting of two squares joined on a side. Show that it is possible to fully cover the chessboard using 32 dominos (i.e., each domino covers two squares of the board, no two dominos overlap, and some domino covers each square). A *pruned board* is a chessboard with some squares removed.

(a) Suppose that we want to know whether it is possible to fully cover a pruned board, and if not, to find the maximum number of dominos we can place on the pruned board so that each domino covers two squares and no two dominos overlap. Formulate this problem as a bipartite cardinality matching problem.

(b) Suppose that we prune only two diagonally opposite corners of the chessboard. Show that we cannot cover the resulting board with 31 dominos.

1.7. Paragraph problem. The well-known document processing program T_eX uses an optimization procedure to decompose a paragraph into several lines so that when lines are left- and right-adjusted, the appearance of the paragraph will be the most attractive. Suppose that a paragraph consists of n words and that each word is assigned a sequence number. Let c_{ij} denote the attractiveness of a line if it begins with the word i and ends with the word $j - 1$. The program T_eX uses formulas to compute the value of each c_{ij}. Given the c_{ij}'s, show how to formulate the problem of decomposing the paragraph into several lines of text in order to maximize the total attractiveness (of all lines) as a shortest path problem.

1.8. Seat-sharing problem. Several families are planning a shared car trip on scenic drives in the White Mountains, New Hampshire. To minimize the possibility of any quarrels, they want to assign individuals to cars so that no two members of a family are in the same car. Formulate this problem as a network flow problem.

1.9. Police patrol problem (Khan [1979]). A police department in a small city consists of three precincts denoted p_1, p_2, and p_3. Each precinct is assigned a number of patrol cars equipped with two-way radios and first-aid equipment. The department operates with three shifts. Figure 1.6(a) and (b) shows the minimum and maximum number of patrol cars needed in each shift. Administrative constraints require that (1) shifts 1, 2, and 3 have, respectively, at least cars 10, 20, and 18 cars available; and (2) precincts p_1, p_2, and p_3 are, respectively, allocated at least 10, 14, and 13 cars. The police department wants to determine an allocation of patrol units that will meet all the require-

	Shift 1	Shift 2	Shift 3
p_1	2	4	3
p_2	3	6	5
p_3	5	7	6

(a)

	Shift 1	Shift 2	Shift 3
p_1	3	7	5
p_2	5	7	10
p_3	8	12	10

(b)

Figure 1.6 Patrol car requirements: (a) minimum required per shift; (b) maximum required per shift.

ments with the fewest possible units committed to the field. Formulate this problem as a circulation problem.

1.10. Forest scheduling problem. Paper and wood products companies need to define cutting schedules that will maximize the total wood yield of their forests over some planning period. Suppose that a company with control of p forest units wants to identify the best cutting schedule over a planning horizon of k years. Forest unit i has a total acreage of a_i units, and studies that the company has undertaken predict that this unit will have w_{ij} tons of woods available for harvesting in the jth year. Based on its prediction of economic conditions, the company believes that it should harvest at least l_j tons of wood in year j. Due to the availability of equipment and personnel, the company can harvest at most u_j tons of wood in year j. Formulate the problem of determining a schedule with maximum wood yield as a network flow problem.

2

PATHS, TREES, AND CYCLES

I hate definitions.
—Benjamin Disraeli

Chapter Outline

2.1 INTRODUCTION

Because graphs and networks arise everywhere and in a variety of alternative forms, several professional disciplines have contributed important ideas to the evolution of network flows. This diversity has yielded numerous benefits, including the infusion of many rich and varied perspectives. It has also, however, imposed costs: For example, the literature on networks and graph theory lacks unity and authors have adopted a wide variety of conventions, customs, and notation. If we so desired, we could formulate network flow problems in several different standard forms and could use many alternative sets of definitions and terminology. We have chosen to adopt a set of common, but not uniformly accepted, definitions: for example, arcs and nodes instead of edges and vertices (or points). We have also chosen to use models with capacitated arcs and with exogenous supplies and demands at the nodes. The circulation problem we introduced in Chapter 1, without exogenous supplies and demands, is an alternative model and so is the capacitated transportation problem. Another special case is the uncapacitated network flow problem. In Chapter 1 we viewed each of these models as special cases of the minimum cost network flow problem. Perhaps somewhat surprisingly, we could have started with any of these models and shown that all the others were special cases. In this sense, each of these models offers another way to capture the mathematical essence of network flows.

In this chapter we have three objectives. First, we bring together many basic definitions of network flows and graph theory, and in doing so, we set the notation that we will be using throughout this book. Second, we introduce several different data structures used to represent networks within a computer and discuss the relative advantages and disadvantages of each of these structures. In a very real sense, data structures are the life blood of most network flow algorithms, and choosing among alternative data structures can greatly influence the efficiency of an algorithm, both in practice and in theory. Consequently, it is important to have a good understanding of the various available data structures and an idea of how and when to use them. Third, we discuss a number of different ways to transform a network flow problem and obtain an equivalent model. For example, we show how to eliminate flow bounds and formulate any model as an uncapacitated problem. As another example, we show how to formulate the minimum cost flow problem as a transportation problem (i.e., how to define it over a bipartite graph). This discussion is of theoretical interest, because it establishes the equivalence between several alternative models and therefore shows that by developing algorithms and theory for any particular model, we will have at hand algorithms and theory for several other models. That is, our results enjoy a certain universality. This development is also of practical value since on various occasions throughout our discussion in this book we will find it more convenient to work with one modeling assumption rather than another—our discussion of network transformations shows that there is no loss in generality in doing so. Moreover, since algorithms developed for one set of modeling assumptions also apply to models formulated in other ways, this discussion provides us with one very reassuring fact: We need not develop separate computer implementations for every alternative formulation, since by using the transformations, we can use an algorithm developed for any one model to solve any problem formulated as one of the alternative models.

We might note that many of the definitions we introduce in this chapter are quite intuitive, and much of our subsequent discussion does not require a complete understanding of all the material in this chapter. Therefore, the reader might simply wish to skim this chapter on first reading to develop a general overview of its content and then return to the chapter on an "as needed" basis later as we draw on the concepts introduced at this point.

2.2 NOTATION AND DEFINITIONS

In this section we give several basic definitions from graph theory and present some basic notation. We also state some elementary properties of graphs. We begin by defining directed and undirected graphs.

Directed Graphs and Networks: A *directed graph* $G = (N, A)$ consists of a set N of nodes and a set A of arcs whose elements are ordered pairs of distinct nodes. Figure 2.1 gives an example of a directed graph. For this graph, $N = \{1, 2, 3, 4, 5, 6, 7\}$ and $A = \{(1, 2), (1, 3), (2, 3), (2, 4), (3, 6), (4, 5), (4, 7), (5, 2), (5, 3), (5, 7), (6, 7)\}$. A *directed network* is a directed graph whose nodes and/or arcs have associated numerical values (typically,

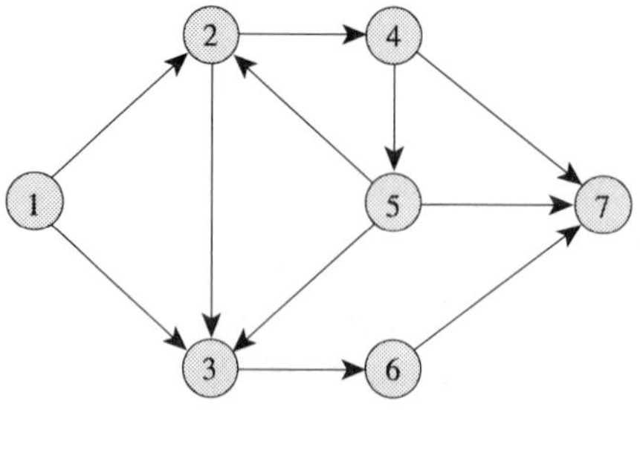

Figure 2.1 Directed graph.

costs, capacities, and/or supplies and demands). In this book we often make no distinction between graphs and networks, so we use the terms "graph" and "network" synonymously. As before, we let n denote the number of nodes and m denote the number of arcs in G.

Undirected Graphs and Networks: We define an undirected graph in the same manner as we define a directed graph except that arcs are unordered pairs of distinct nodes. Figure 2.2 gives an example of an undirected graph. In an undirected graph, we can refer to an arc joining the node pair i and j as either (i, j) or (j, i). An undirected arc (i, j) can be regarded as a two-way street with flow permitted in both directions: either from node i to node j or from node j to node i. On the other hand, a directed arc (i, j) behaves like a one-way street and permits flow only from node i to node j.

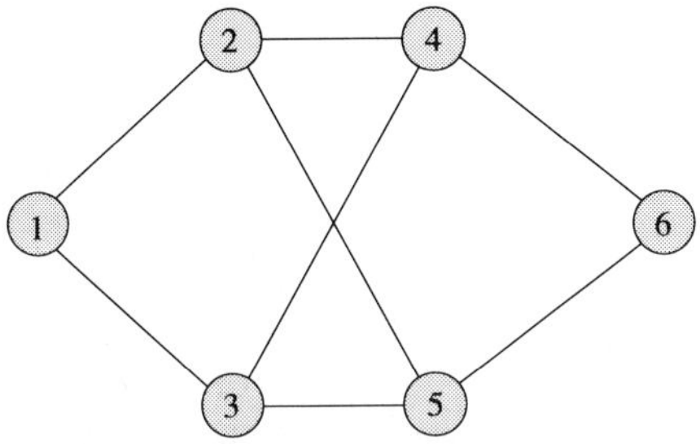

Figure 2.2 Undirected graph.

In most of the material in this book, we assume that the underlying network is directed. Therefore, we present our subsequent notation and definitions for directed networks. The corresponding definitions for undirected networks should be transparent to the reader; nevertheless, we comment briefly on some definitions for undirected networks at the end of this section.

Tails and Heads: A directed arc (i, j) has two *endpoints* i and j. We refer to node i as the *tail* of arc (i, j) and node j as its *head*. We say that the arc (i, j) *emanates* from node i and *terminates* at node j. An arc (i, j) is *incident to* nodes i and j. The arc (i, j) is an *outgoing arc* of node i and an *incoming arc* of node j. Whenever an arc $(i, j) \in A$, we say that node j is *adjacent* to node i.

Degrees: The *indegree* of a node is the number of incoming arcs of that node and its *outdegree* is the number of its outgoing arcs. The *degree* of a node is the sum of its indegree and outdegree. For example, in Figure 2.1, node 3 has an indegree of 3, an outdegree of 1, and a degree of 4. It is easy to see that the sum of indegrees of all nodes equals the sum of outdegrees of all nodes and both are equal to the number of arcs m in the network.

Adjacency List: The *arc adjacency list* $A(i)$ of a node i is the set of arcs emanating from that node, that is, $A(i) = \{(i, j) \in A: j \in N\}$. The *node adjacency list* $A(i)$ is the set of nodes adjacent to that node; in this case, $A(i) = \{j \in N: (i, j) \in A\}$. Often, we shall omit the terms "arc" and "node" and simply refer to the adjacency list; in all cases it will be clear from context whether we mean arc adjacency list or node adjacency list. We assume that arcs in the adjacency list $A(i)$ are arranged so that the head nodes of arcs are in increasing order. Notice that $|A(i)|$ equals the outdegree of node i. Since the sum of all node outdegrees equals m, we immediately obtain the following property:

Property 2.1. $\sum_{i \in N} |A(i)| = m.$

Multiarcs and Loops: *Multiarcs* are two or more arcs with the same tail and head nodes. A *loop* is an arc whose tail node is the same as its head node. In most of the chapters in this book, we assume that graphs contain no multiarcs or loops.

Subgraph: A graph $G' = (N', A')$ is a *subgraph* of $G = (N, A)$ if $N' \subseteq N$ and $A' \subseteq A$. We say that $G' = (N', A')$ is the subgraph of G *induced* by N' if A' contains each arc of A with both endpoints in N'. A graph $G' = (N', A')$ is a *spanning subgraph* of $G = (N, A)$ if $N' = N$ and $A' \subseteq A$.

Walk: A *walk* in a directed graph $G = (N, A)$ is a subgraph of G consisting of a sequence of nodes and arcs $i_1 - a_1 - i_2 - a_2 - \cdots - i_{r-1} - a_{r-1} - i_r$ satisfying the property that for all $1 \le k \le r - 1$, either $a_k = (i_k, i_{k+1}) \in A$ or $a_k = (i_{k+1}, i_k) \in A$. Alternatively, we shall sometimes refer to a walk as a set of (sequence of) arcs (or of nodes) without any explicit mention of the nodes (without explicit mention of arcs). We illustrate this definition using the graph shown in Figure 2.1. Figure 2.3(a) and (b) illustrates two walks in this graph: 1–2–5–7 and 1–2–4–5–2–3.

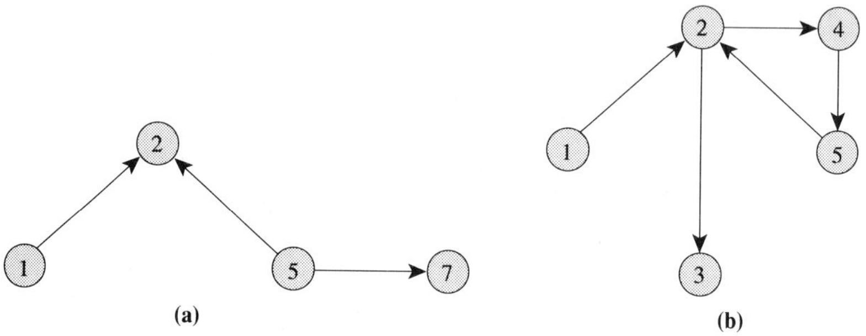

Figure 2.3 Examples of walks.

Directed Walk: A *directed walk* is an "oriented" version of a walk in the sense that for any two consecutive nodes i_k and i_{k+1} on the walk, $(i_k, i_{k+1}) \in A$. The walk shown in Figure 2.3(a) is not directed; the walk shown in Figure 2.3(b) is directed.

Path: A *path* is a walk without any repetition of nodes. The walk shown in Figure 2.3(a) is also a path, but the walk shown in Figure 2.3(b) is not because it repeats node 2 twice. We can partition the arcs of a path into two groups: forward arcs and backward arcs. An arc (i, j) in the path is a *forward arc* if the path visits node i prior to visiting node j, and is a *backward arc* otherwise. For example, in the path shown in Figure 2.3(a), the arcs (1, 2) and (5, 7) are forward arcs and the arc (5, 2) is a backward arc.

Directed Path: A *directed path* is a directed walk without any repetition of nodes. In other words, a directed path has no backward arcs. We can store a path (or a directed path) easily within a computer by defining a *predecessor* index *pred*(j) for every node j in the path. If i and j are two consecutive nodes on the path (along its orientation), pred(j) = i. For the path 1–2–5–7 shown in Figure 2.3(a), pred(7) = 5, pred(5) = 2, pred(2) = 1, and pred(1) = 0. (Frequently, we shall use the convention of setting the predecessor index of the initial node of a path equal to zero to indicate the beginning of the path.) Notice that we cannot use predecessor indices to store a walk since a walk may visit a node more than once, and a single predecessor index of a node cannot store the multiple predecessors of any node that a walk visits more than once.

Cycle: A *cycle* is a path $i_1 - i_2 - \cdots - i_r$ together with the arc (i_r, i_1) or (i_1, i_r). We shall often refer to a cycle using the notation $i_1 - i_2 - \cdots - i_r - i_1$. Just as we did for paths, we can define forward and backward arcs in a cycle. In Figure 2.4(a) the arcs (5, 3) and (3, 2) are forward arcs and the arc (5, 2) is a backward arc of the cycle 2–5–3.

Directed Cycle: A *directed cycle* is a directed path $i_1 - i_2 - \cdots - i_r$ together with the arc (i_r, i_1). The graph shown in Figure 2.4(a) is a cycle, but not a directed cycle; the graph shown in Figure 2.4(b) is a directed cycle.

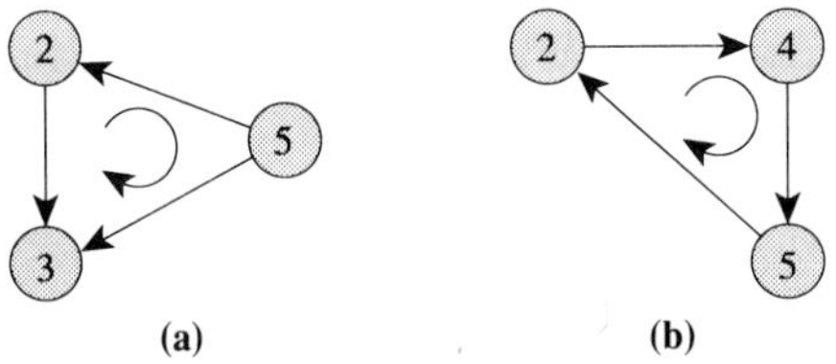

Figure 2.4 Examples of cycles.

Acyclic Network: A graph is *acyclic* if it contains no directed cycle.

Connectivity: We will say that two nodes i and j are *connected* if the graph contains at least one path from node i to node j. A graph is *connected* if every pair of its nodes is connected; otherwise, the graph is *disconnected.* We refer to the maximal connected subgraphs of a disconnected network as its *components*. For instance, the graph shown in Figure 2.5(a) is connected, and the graph shown in Figure 2.5(b) is disconnected. The latter graph has two components consisting of the node sets {1, 2, 3, 4} and {5, 6}. In Section 3.4 we describe a method for determining whether a graph is connected or not, and in Exercise 3.41 we discuss a method for identifying all components of a graph.

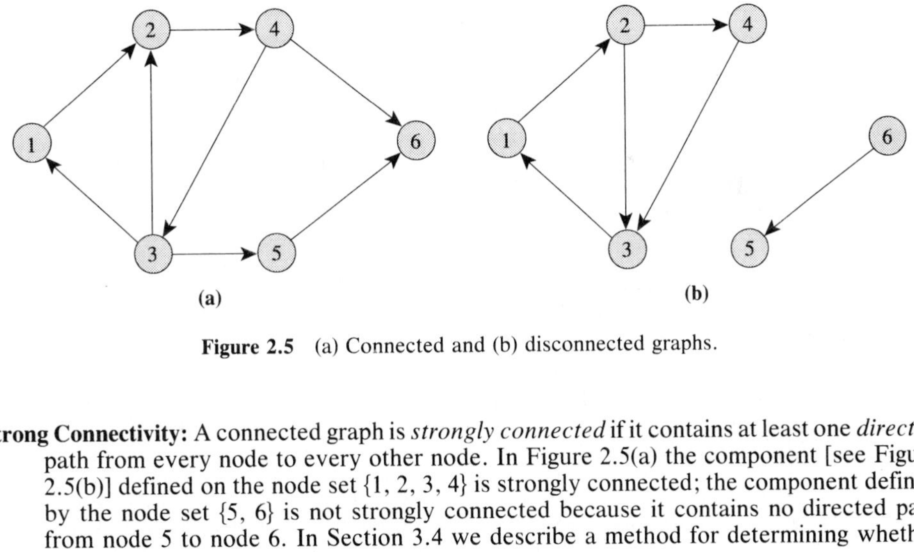

Figure 2.5 (a) Connected and (b) disconnected graphs.

Strong Connectivity: A connected graph is *strongly connected* if it contains at least one *directed* path from every node to every other node. In Figure 2.5(a) the component [see Figure 2.5(b)] defined on the node set {1, 2, 3, 4} is strongly connected; the component defined by the node set {5, 6} is not strongly connected because it contains no directed path from node 5 to node 6. In Section 3.4 we describe a method for determining whether or not a graph is strongly connected.

Cut: A *cut* is a partition of the node set N into two parts, S and $\bar{S} = N - S$. Each cut defines a set of arcs consisting of those arcs that have one endpoint in S and another endpoint in $\bar{S}$. Therefore, we refer to this set of arcs as a cut and represent it by the notation $[S, \bar{S}]$. Figure 2.6 illustrates a cut with $S = \{1, 2, 3\}$ and $\bar{S} = \{4, 5, 6, 7\}$. The set of arcs in this cut are {(2, 4), (5, 2), (5, 3), (3, 6)}.

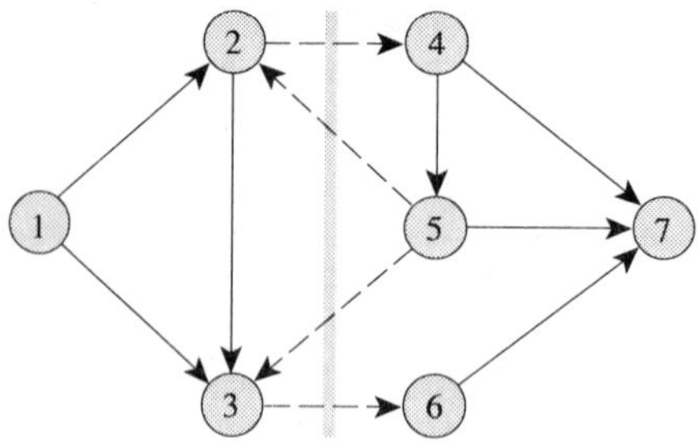

Figure 2.6 Cut.

***s–t* Cut:** An *s–t cut* is defined with respect to two distinguished nodes s and t, and is a cut $[S, \bar{S}]$ satisfying the property that $s \in S$ and $t \in \bar{S}$. For instance, if $s = 1$ and $t = 6$, the cut depicted in Figure 2.6 is an $s–t$ cut; but if $s = 1$ and $t = 3$, this cut is not an $s–t$ cut.

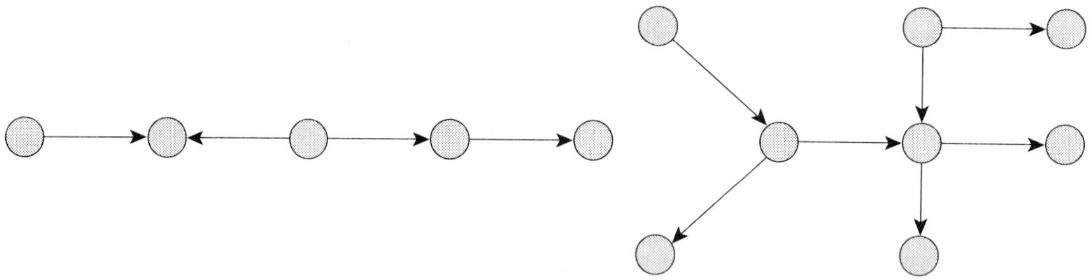

Figure 2.7 Example of two trees.

Tree. A *tree* is a connected graph that contains no cycle. Figure 2.7 shows two examples of trees.

A tree is a very important graph theoretic concept that arises in a variety of network flow algorithms studied in this book. In our subsequent discussion in later chapters, we use some of the following elementary properties of trees.

Property 2.2

(a) *A tree on n nodes contains exactly* $n - 1$ *arcs.*
(b) *A tree has at least two leaf nodes* (*i.e., nodes with degree* 1).
(c) *Every two nodes of a tree are connected by a unique path.*

Proof. See Exercise 2.13.

Forest: A graph that contains no cycle is a *forest*. Alternatively, a forest is a collection of trees. Figure 2.8 gives an example of a forest.

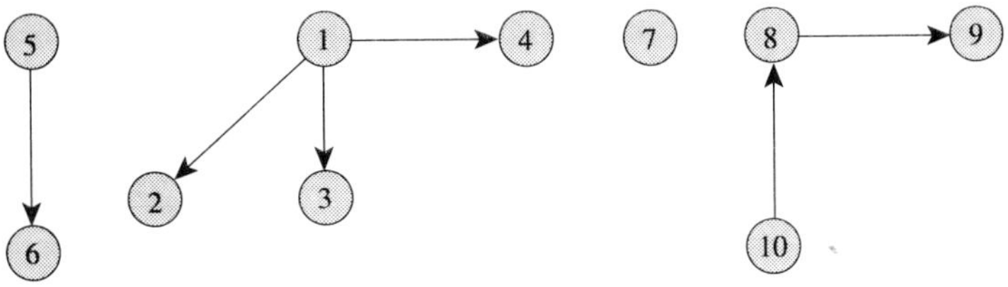

Figure 2.8 Forest.

Subtree: A connected subgraph of a tree is a *subtree*.

Rooted Tree: A rooted tree is a tree with a specially designated node, called its *root*; we regard a rooted tree as though it were hanging from its root. Figure 2.9 gives an instance of a rooted tree; in this instance, node 1 is the root node.

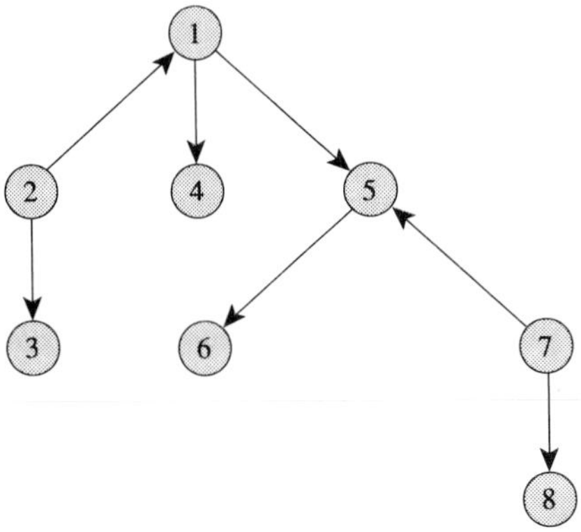

Figure 2.9 Rooted tree.

We often view the arcs in a rooted tree as defining predecessor–successor (or parent–child) relationships. For example, in Figure 2.9, node 5 is the predecessor of nodes 6 and 7, and node 1 is the predecessor of nodes 2, 4, and 5. Each node i (except the root node) has a unique predecessor, which is the next node on the unique path in the tree from that node to the root; we store the predecessor of node i using a predecessor index *pred(i)*. If j = pred(i), we say that node j is the predecessor of node i and node i is a successor of node j. These predecessor indices uniquely define a rooted tree and also allow us to trace out the unique path from any node back to the root. The *descendants* of a node i consist of the node itself, its successors, successors of its successors, and so on. For example, in Figure 2.9 the node set {5, 6, 7, 8} is the set of descendants of node 5. We say that a node is an *ancestor* of all of its descendants. For example, in the same figure, node 2 is an ancestor of itself and node 3.

In this book we occasionally use two special type of rooted trees, called a *directed in-tree* and a *directed out-tree*.

Directed-out-Tree: A tree is a *directed out-tree rooted at node s* if the unique path in the tree from node s to every other node is a directed path. Figure 2.10(a) shows an instance of a directed out-tree rooted at node 1. Observe that every node in the directed out-tree (except node 1) has indegree 1.

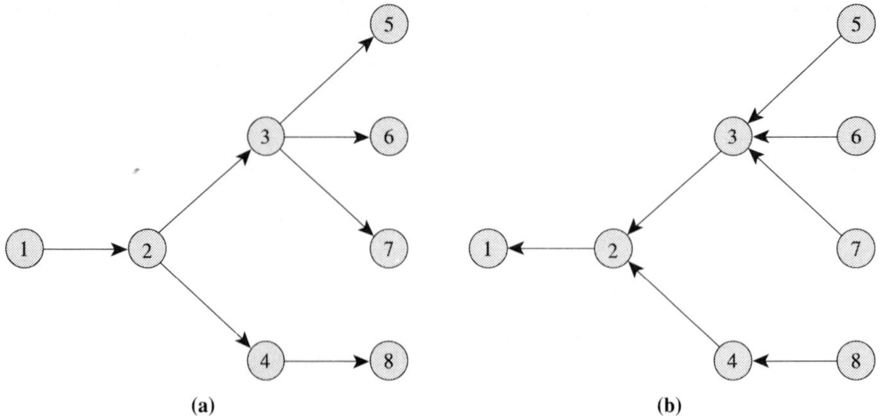

Figure 2.10 Instances of directed out-tree and directed in-tree.

Directed-in-Tree: A tree is a *directed in-tree rooted at node s* if the unique path in the tree from any node to node s is a directed path. Figure 2.10(b) shows an instance of a directed in-tree rooted at node 1. Observe that every node in the directed in-tree (except node 1) has outdegree 1.

Spanning Tree: A tree T is a spanning tree of G if T is a spanning subgraph of G. Figure 2.11 shows two spanning trees of the graph shown in Figure 2.1. Every spanning tree of a connected n-node graph G has $(n - 1)$ arcs. We refer to the arcs belonging to a spanning tree T as *tree arcs* and arcs not belonging to T as *nontree arcs*.

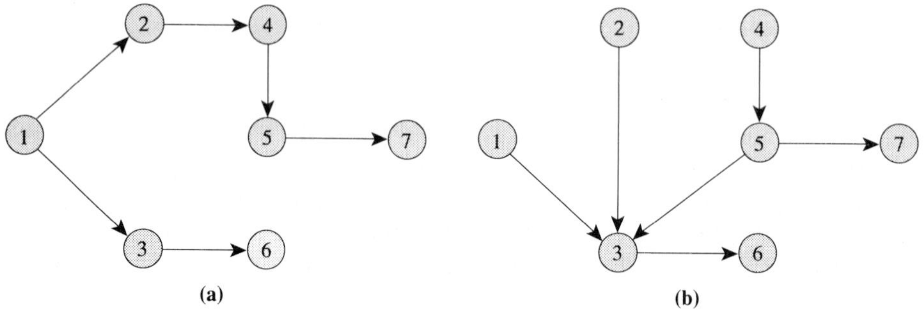

Figure 2.11 Two spanning trees of the network in Figure 2.1.

Fundamental Cycles: Let T be a spanning tree of the graph G. The addition of any nontree arc to the spanning tree T creates exactly one cycle. We refer to any such cycle as a *fundamental cycle* of G with respect to the tree T. Since the network contains $m - n + 1$ nontree arcs, it has $m - n + 1$ fundamental cycles. Observe that if we delete any arc in a fundamental cycle, we again obtain a spanning tree.

Fundamental Cuts: Let T be a spanning tree of the graph G. The deletion of any tree arc of the spanning tree T produces a disconnected graph containing two subtrees T_1 and T_2. Arcs whose endpoints belong to the different subtrees constitute a cut. We refer to any such cut as a *fundamental cut* of G with respect to the tree T. Since a spanning tree contains $n - 1$ arcs, the network has $n - 1$ fundamental cuts with respect to any tree. Observe that when we add any arc in the fundamental cut to the two subtrees T_1 and T_2, we again obtain a spanning tree.

Bipartite Graph: A graph $G = (N, A)$ is a *bipartite graph* if we can partition its node set into two subsets N_1 and N_2 so that for each arc (i, j) in A either (i) $i \in N_1$ and $j \in N_2$, or (ii) $i \in N_2$ and $j \in N_1$. Figure 2.12 gives two examples of bipartite graphs. Although it might not be immediately evident whether or not the graph in Figure 2.12(b) is bipartite, if we define $N_1 = \{1, 2, 3, 4\}$ and $N_2 = \{5, 6, 7, 8\}$, we see that it is.

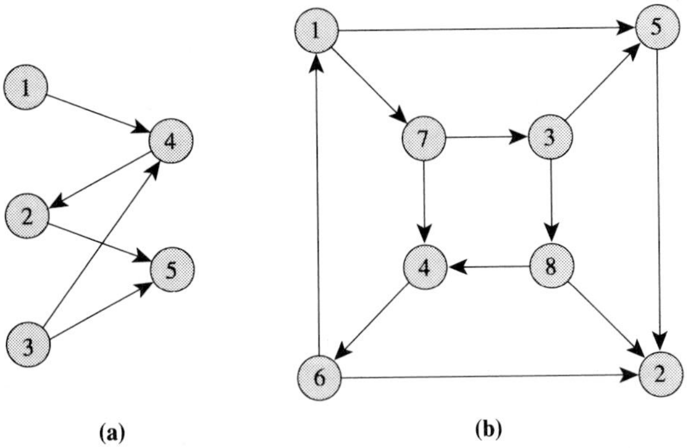

Figure 2.12 Examples of bipartite graphs.

Frequently, we wish to discover whether or not a given graph is bipartite. Fortunately, there is a very simple method for resolving this issue. We discuss this method in Exercise 3.42, which is based on the following well-known characterization of bipartite graphs.

Property 2.3. *A graph G is a bipartite graph if and only if every cycle in G contains an even number of arcs.*

Proof. See Exercise 2.21.

Definitions for undirected networks. The definitions for directed networks easily translate into those for undirected networks. An undirected arc (i, j) has two endpoints, i and j, but its tail and head nodes are undefined. If the network contains the arc (i, j), node i is adjacent to node j, and node j is adjacent to node i. The arc adjacency list (as well as the node adjacency list) is defined similarly except that arc (i, j) appears in $A(i)$ as well as $A(j)$. Consequently, $\sum_{i \in N} |A(i)| = 2m$. The degree of a node is the number of nodes adjacent to node i. Each of the graph theoretic concepts we have defined so far—walks, paths, cycles, cuts and trees—has essentially the same definition for undirected networks except that we do not distinguish between a path and a directed path, a cycle and a directed cycle, and so on.

2.3 NETWORK REPRESENTATIONS

The performance of a network algorithm depends not only on the algorithm, but also on the manner used to represent the network within a computer and the storage scheme used for maintaining and updating the intermediate results. By representing a network more cleverly and by using improved data structures, we can often improve the running time of an algorithm. In this section we discuss some popular ways of representing a network. In representing a network, we typically need to store two types of information: (1) the network topology, that is, the network's node and arc structure; and (2) data such as costs, capacities, and supplies/demands associated with the network's nodes and arcs. As we will see, usually the scheme we use to store the network's topology will suggest a natural way for storing the associated node and arc information. In this section we describe in detail representations for directed graphs. The corresponding representations for undirected networks should be apparent to the reader. At the end of the section, however, we briefly discuss representations for undirected networks.

Node–Arc Incidence Matrix

The *node–arc incidence matrix* representation, or simply the *incidence matrix* representation, represents a network as the constraint matrix of the minimum cost flow problem that we discussed in Section 1.2. This representation stores the network as an $n \times m$ matrix $\mathcal{N}$ which contains one row for each node of the network and one column for each arc. The column corresponding to arc (i, j) has only two nonzero elements: It has a $+1$ in the row corresponding to node i and a -1 in the row corresponding to node j. Figure 2.14 gives this representation for the network shown in Figure 2.13.

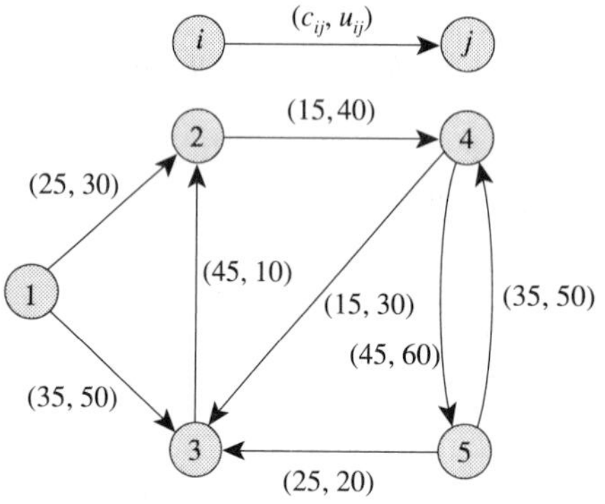

Figure 2.13 Network example.

	(1, 2)	(1, 3)	(2, 4)	(3, 2)	(4, 3)	(4, 5)	(5, 3)	(5, 4)
1	1	1	0	0	0	0	0	0
2	−1	0	1	−1	0	0	0	0
3	0	−1	0	1	−1	0	−1	0
4	0	0	−1	0	1	1	0	−1
5	0	0	0	0	0	−1	1	1

Figure 2.14 Node–arc incidence matrix of the network example.

The node–arc incidence matrix has a very special structure: Only $2m$ out of its nm entries are nonzero, all of its nonzero entries are $+1$ or -1, and each column has exactly one $+1$ and one -1. Furthermore, the number of $+1$'s in a row equals the outdegree of the corresponding node and the number of -1's in the row equals the indegree of the node.

Because the node–arc incidence matrix $\mathcal{N}$ contains so few nonzero coefficients, the incidence matrix representation of a network is not space efficient. More efficient schemes, such as those that we consider later in this section would merely keep track of the nonzero entries in the matrix. Because of its inefficiency in storing the underlying network topology, use of the node–arc incidence matrix rarely produces efficient algorithms. This representation is important, however, because it represents the constraint matrix of the minimum cost flow problem and because the node–arc incidence matrix possesses several interesting theoretical properties. We study some of these properties in Sections 11.11 and 11.12.

Node–Node Adjacency Matrix

The node–node adjacency matrix representation, or simply the adjacency matrix representation, stores the network as an $n \times n$ matrix $\mathcal{H} = \{h_{ij}\}$. The matrix has a row and a column corresponding to every node, and its ijth entry h_{ij} equals 1 if $(i, j) \in A$ and equals 0 otherwise. Figure 2.15 specifies this representation for the network shown in Figure 2.13. If we wish to store arc costs and capacities as well as the network topology, we can store this information in two additional $n \times n$ matrices, $\mathcal{C}$ and $\mathcal{U}$.

The adjacency matrix has n^2 elements, only m of which are nonzero. Consequently, this representation is space efficient only if the network is sufficiently dense; for sparse networks this representation wastes considerable space. Nevertheless, the simplicity of the adjacency representation permits us to use it to implement most network algorithms rather easily. We can determine the cost or capacity of any arc (i, j) simply by looking up the ijth element in the matrix $\mathcal{C}$ or $\mathcal{U}$. We can obtain the arcs emanating from node i by scanning row i: If the jth element in this row has a nonzero entry, (i, j) is an arc of the network. Similarly, we can obtain the arcs entering node j by scanning column j: If the ith element of this column has a nonzero entry, (i, j) is an arc of the network. These steps permit us to identify all the outgoing or incoming arcs of a node in time proportional to n. For dense networks we can usually afford to spend this time to identify the incoming or outgoing arcs, but for

	1	2	3	4	5
1	0	1	1	0	0
2	0	0	0	1	0
3	0	1	0	0	0
4	0	0	1	0	1
5	0	0	1	1	0

Figure 2.15 Node–node adjacency matrix of the network example.

sparse networks these steps might be the bottleneck operations for an algorithm. The two representations we discuss next permit us to identify the set of outgoing arcs $A(i)$ of any node in time proportional to $|A(i)|$.

Adjacency Lists

Earlier we defined the *arc adjacency list* $A(i)$ of a node i as the set of arcs emanating from that node, that is, the set of arcs $(i, j) \in A$ obtained as j ranges over the nodes of the network. Similarly, we defined the *node adjacency list* of a node i as the set of nodes j for which $(i, j) \in A$. The *adjacency list representation* stores the node adjacency list of each node as a singly linked list (we refer the reader to Appendix A for a description of singly linked lists). A linked list is a collection of cells each containing one or more fields. The node adjacency list for node i will be a linked list having $|A(i)|$ cells and each cell will correspond to an arc $(i, j) \in A$. The cell corresponding to the arc (i, j) will have as many fields as the amount of information we wish to store. One data field will store node j. We might use two other data fields to store the arc cost c_{ij} and the arc capacity u_{ij}. Each cell will contain one additional field, called the *link*, which stores a pointer to the next cell in the adjacency list. If a cell happens to be the last cell in the adjacency list, by convention we set its link to value zero.

Since we need to be able to store and access n linked lists, one for each node, we also need an array of pointers that point to the first cell in each linked list. We accomplish this objective by defining an n-dimensional array, *first*, whose element *first(i)* stores a pointer to the first cell in the adjacency list of node i. If the adjacency list of node i is empty, we set *first(i)* $= 0$. Figure 2.16 specifies the adjacency list representation of the network shown in Figure 2.13.

In this book we sometimes assume that whenever arc (i, j) belongs to a network, so does the reverse arc (j, i). In these situations, while updating some information about arc (i, j), we typically will also need to update information about arc (j, i). Since we will store arc (i, j) in the adjacency list of node i and arc (j, i) in the adjacency list of node j, we can carry out any operation on both arcs efficiently if we know where to find the reversal (j, i) of each arc (i, j). We can access both arcs

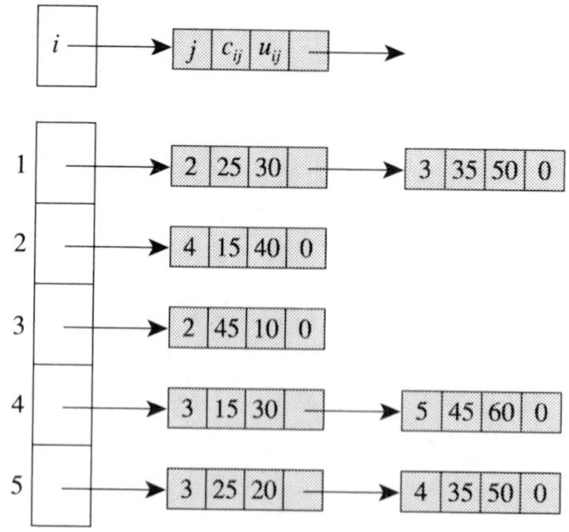

Figure 2.16 Adjacency list representation of the network example.

easily if we define an additional field, *mate*, that contains a pointer to the cell containing data for the reversal of each arc. The mate of arc (i, j) points to the cell of arc (j, i) and the mate of arc (j, i) points to the cell of arc (i, j).

Forward and Reverse Star Representations

The *forward star representation* of a network is similar to the adjacency list representation in the sense that it also stores the node adjacency list of each node. But instead of maintaining these lists as linked lists, it stores them in a single array. To develop this representation, we first associate a unique sequence number with each arc, thus defining an ordering of the arc list. We number the arcs in a specific order: first those emanating from node 1, then those emanating from node 2, and so on. We number the arcs emanating from the same node in an arbitrary fashion. We then sequentially store information about each arc in the arc list. We store the tails, heads, costs, and capacities of the arcs in four arrays: *tail*, *head*, *cost*, and *capacity*. So if arc (i, j) is arc number 20, we store the tail, head, cost, and capacity data for this arc in the array positions tail(20), head(20), cost(20), and capacity(20). We also maintain a pointer with each node i, denoted by *point(i)*, that indicates the smallest-numbered arc in the arc list that emanates from node i. [If node i has no outgoing arcs, we set point(i) equal to point($i + 1$).] Therefore, the forward star representation will store the outgoing arcs of node i at positions point(i) to (point($i + 1$) $-$ 1) in the arc list. If point(i) $>$ point($i + 1$) $-$ 1, node i has no outgoing arc. For consistency, we set point(1) $=$ 1 and point($n + 1$) $= m + 1$. Figure 2.17(a) specifies the forward star representation of the network given in Figure 2.13.

The forward star representation provides us with an efficient means for determining the set of outgoing arcs of any node. To determine, simultaneously, the set of incoming arcs of any node efficiently, we need an additional data structure known as the *reverse star representation*. Starting from a forward star representation, we can create a reverse star representation as follows. We examine the nodes $i = 1$ to n in order and sequentially store the heads, tails, costs, and capacities of the incoming arcs at node i. We also maintain a reverse pointer with each node i, denoted by *rpoint(i)*, which denotes the first position in these arrays that contains information about an incoming arc at node i. [If node i has no incoming arc, we set rpoint(i) equal to rpoint($i + 1$).] For sake of consistency, we set rpoint(1) $=$ 1 and rpoint($n + 1$) $= m + 1$. As before, we store the incoming arcs at node i at positions rpoint(i) to (rpoint($i + 1$) $-$ 1). This data structure gives us the representation shown in Figure 2.17(b).

Observe that by storing both the forward and reverse star representations, we will maintain a significant amount of duplicate information. We can avoid this duplication by storing arc numbers in the reverse star instead of the tails, heads, costs, and capacities of the arcs. As an illustration, for our example, arc (3, 2) has arc number 4 in the forward star representation and arc (1, 2) has an arc number 1. So instead of storing the tails, costs, and capacities of the arcs, we simply store arc numbers; and once we know the arc numbers, we can always retrieve the associated information from the forward star representation. We store arc numbers in an array *trace* of size m. Figure 2.18 gives the complete *trace* array of our example.

In our discussion of the adjacency list representation, we noted that sometimes

	point
1	1
2	3
3	4
4	5
5	7
6	9

	tail	head	cost	capacity
1	1	2	25	30
2	1	3	35	50
3	2	4	15	40
4	3	2	45	10
5	4	3	15	30
6	4	5	45	60
7	5	3	25	20
8	5	4	35	50

(a)

cost	capacity	tail	head	
45	10	3	2	1
25	30	1	2	2
35	50	1	3	3
15	30	4	3	4
25	20	5	3	5
35	50	5	4	6
15	40	2	4	7
45	60	4	5	8

rpoint	
1	1
1	2
3	3
6	4
8	5
9	6

(b)

Figure 2.17 (a) Forward star and (b) reverse star representations of the network example.

while updating data for an arc (i, j), we also need to update data for its reversal (j, i). Just as we did in the adjacency list representation, we can accomplish this task by defining an array *mate* of size m, which stores the arc number of the reversal of an arc. For example, the forward star representation shown in Figure 2.17(a) assigns the arc number 6 to arc (4, 5) and assigns the arc number 8 to arc (5, 4).

	point
1	1
2	3
3	4
4	5
5	7
6	9

	tail	head	cost	capacity
1	1	2	25	30
2	1	3	35	50
3	2	4	15	40
4	3	2	45	10
5	4	3	15	30
6	4	5	45	60
7	5	3	25	20
8	5	4	35	50

trace	
4	1
1	2
2	3
5	4
7	5
8	6
3	7
6	8

rpoint	
1	1
1	2
3	3
6	4
8	5
9	6

Figure 2.18 Compact forward and reverse star representation of the network example.

Therefore, if we were using the *mate* array, we would set *mate(6)* = 8 and *mate(8)* = 6.

Comparison of Forward Star and Adjacency List Representations

The major advantage of the forward star representation is its space efficiency. It requires less storage than does the adjacency list representation. In addition, it is much easier to implement in languages such as FORTRAN that have no natural provisions for using linked lists. The major advantage of adjacency list representation is its ease of implementation in languages such as Pascal or C that are able to manipulate linked lists efficiently. Further, using an adjacency list representation, we can add or delete arcs (as well as nodes) in constant time. On the other hand, in the forward star representation these steps require time proportional to m, which can be too time consuming.

Storing Parallel Arcs

In this book we assume that the network does not contain parallel arcs; that is, no two arcs have the same tail and head nodes. By allowing parallel arcs, we encounter some notational difficulties, since (i, j) will not specify the arc uniquely. For networks with parallel arcs, we need more complex notation to specify arcs, arc costs, and capacities. This difficulty is merely notational, however, and poses no problems computationally: both the adjacency list representation and the forward star representation data structures are capable of handling parallel arcs. If a node i has two outgoing arcs with the same head node but (possibly) different costs and capacities, the linked list of node i will contain two cells corresponding to these two arcs. Similarly, the forward star representation allows several entries with the same tail and head nodes but different costs and capacities.

Representing Undirected Networks

We can represent undirected networks using the same representations we have just described for directed networks. However, we must remember one fact: Whenever arc (i, j) belongs to an undirected network, we need to include both of the pairs (i, j) and (j, i) in the representations we have discussed. Consequently, we will store each arc (i, j) of an undirected network twice in the adjacency lists, once in the list for node i and once in the list for node j. Some other obvious modifications are needed. For example, in the node–arc incidence matrix representation, the column corresponding to arc (i, j) will have $+1$ in both rows i and j. The node–node adjacency matrix will have $+1$ in position h_{ij} and h_{ji} for every arc $(i, j) \in A$. Since this matrix will be symmetric, we might as well store half of the matrix. In the adjacency list representation, the arc (i, j) will be present in the linked lists of both nodes i and j. Consequently, whenever we update information for one arc, we must update it for the other arc as well. We can accomplish this task by storing for each arc the address of its other occurrence in an additional mate array. The forward star representation requires this additional storage as well. Finally, observe that undirected networks do not require the reverse star representation.

2.4 NETWORK TRANSFORMATIONS

Frequently, we require network transformations to simplify a network, to show equivalences between different network problems, or to state a network problem in a standard form required by a computer code. In this section, we describe some of these important transformations. In describing these transformations, we assume that the network problem is a minimum cost flow problem as formulated in Section 1.2. Needless to say, these transformations also apply to special cases of the minimum cost flow problem, such as the shortest path, maximum flow, and assignment problems, wherever the transformations are appropriate. We first recall the formulation of the minimum cost flow problem for convenience in discussing the network transformations.

$$\text{Minimize} \sum_{(i,j)\in A} c_{ij}x_{ij} \tag{2.1a}$$

subject to

$$\sum_{\{j:(i,j)\in A\}} x_{ij} - \sum_{\{j:(j,i)\in A\}} x_{ji} = b(i) \qquad \text{for all } i \in N, \tag{2.1b}$$

$$l_{ij} \le x_{ij} \le u_{ij} \qquad \text{for all } (i, j) \in A. \tag{2.1c}$$

Undirected Arcs to Directed Arcs

Sometimes minimum cost flow problems contain undirected arcs. An undirected arc (i, j) with cost $c_{ij} \geq 0$ and capacity u_{ij} permits flow from node i to node j and also from node j to node i; a unit of flow in either direction costs c_{ij}, and the total flow (i.e., from node i to node j plus from node j to node i) has an upper bound u_{ij}. That is, the undirected model has the constraint $x_{ij} + x_{ji} \leq u_{ij}$ and the term $c_{ij}x_{ij} + c_{ij}x_{ji}$ in the objective function. Since the cost $c_{ij} \geq 0$, in some optimal solution one of x_{ij} and x_{ji} will be zero. We refer to any such solution as non-overlapping.

For notational convenience, in this discussion we refer to the undirected arc (i, j) as $\{i, j\}$. We assume (with some loss of generality) that the arc flow in either direction on arc $\{i, j\}$ has a lower bound of value 0; our transformation is not valid if the arc flow has a nonzero lower bound or the arc cost c_{ij} is negative (why?). To transform the undirected case to the directed case, we replace each undirected arc $\{i, j\}$ by two directed arcs, (i, j) and (j, i), both with cost c_{ij} and capacity u_{ij}. To establish the correctness of this transformation, we show that every non-overlapping flow in the original network has an associated flow in the transformed network with the same cost, and vice versa. If the undirected arc $\{i, j\}$ carries α units of flow from node i to node j, in the transformed network $x_{ij} = \alpha$ and $x_{ji} = 0$. If the undirected arc $\{i, j\}$ carries α units of flow from node j to node i, in the transformed network $x_{ij} = 0$ and $x_{ji} = \alpha$. Conversely, if x_{ij} and x_{ji} are the flows on arcs (i, j) and (j, i) in the directed network, $x_{ij} - x_{ji}$ or $x_{ji} - x_{ij}$ is the associated flow on arc $\{i, j\}$ in the undirected network, whichever is positive. If $x_{ij} - x_{ji}$ is positive, the flow from node i to node j on arc $\{i, j\}$ equals this amount. If $x_{ji} - x_{ij}$ is positive, the flow from node j to node i on arc $\{i, j\}$ equals $x_{ji} - x_{ij}$. In either case, the flow in the opposite direction is zero. If $x_{ji} - x_{ij}$ is zero, the flow on arc $\{i, j\}$ is 0.

Removing Nonzero Lower Bounds

If an arc (i, j) has a nonzero lower bound l_{ij} on the arc flow x_{ij}, we replace x_{ij} by $x'_{ij} + l_{ij}$ in the problem formulation. The flow bound constraint then becomes $l_{ij} \leq x'_{ij} + l_{ij} \leq u_{ij}$, or $0 \leq x'_{ij} \leq (u_{ij} - l_{ij})$. Making this substitution in the mass balance constraints decreases $b(i)$ by l_{ij} units and increases $b(j)$ by l_{ij} units [recall from Section 1.2 that the flow variable x_{ij} appears in the mass balance constraint (2.1b) of only nodes i and j]. This substitution changes the objective function value by a constant that we can record separately and then ignore when solving the problem. Figure 2.19 illustrates this transformation graphically. We can view this transformation as a two-step flow process: We begin by sending l_{ij} units of flow on arc (i, j), which decreases $b(i)$ by l_{ij} units and increases $b(j)$ by l_{ij} units, and then we measure (by the variable x'_{ij}) the incremental flow on the arc beyond the flow value l_{ij}.

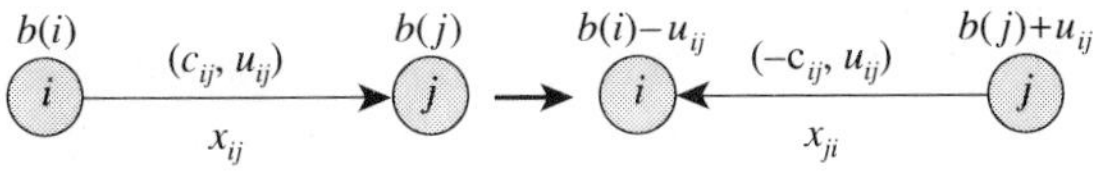

Figure 2.19 Removing nonzero lower bounds.

Arc Reversal

The arc reversal transformation is typically used to remove arcs with negative costs. Let u_{ij} denote the capacity of the arc (i, j) or an upper bound on the arc's flow if the arc is uncapacitated. In this transformation we replace the variable x_{ij} by $u_{ij} - x_{ji}$. Doing so replaces the arc (i, j), which has an associated cost c_{ij}, by the arc (j, i) with an associated cost $-c_{ij}$. As shown in Figure 2.20, the transformation has the following network interpretation. We first send u_{ij} units of flow on the arc (which decreases $b(i)$ by u_{ij} units and increases $b(j)$ by u_{ij} units) and then we replace arc (i, j) by arc (j, i) with cost $-c_{ij}$. The new flow x_{ji} measures the amount of flow we "remove" from the "full capacity" flow of u_{ij}.

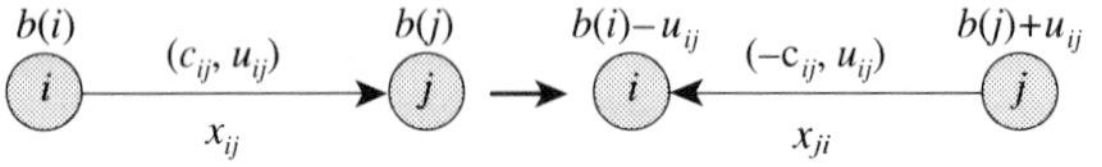

Figure 2.20 Arc reversal transformation.

Removing Arc Capacities

If an arc (i, j) has a positive capacity u_{ij}, we can remove the capacity, making the arc *uncapacitated*, by using the following idea: We introduce an additional node so that the capacity constraint on arc (i, j) becomes the mass balance constraint of the new mode. Suppose that we introduce a slack variable $s_{ij} \geq 0$, and write the capacity constraint $x_{ij} \leq u_{ij}$ in an equality form as $x_{ij} + s_{ij} = u_{ij}$. Multiplying both sides of the equality by -1, we obtain

$$-x_{ij} - s_{ij} = -u_{ij} \tag{2.2}$$

We now treat constraint (2.2) as the mass balance constraint of an additional node k. Observe that the flow variable x_{ij} now appears in three mass balance constraints and s_{ij} in only one. By subtracting (2.2) from the mass balance constraint of node j (which contains the flow variable x_{ij} with a negative sign), we assure that each of x_{ij} and s_{ij} appears in exactly two constraints—in one with a positive sign and in the other with a negative sign. These algebraic manipulations correspond to the network transformation shown in Figure 2.21.

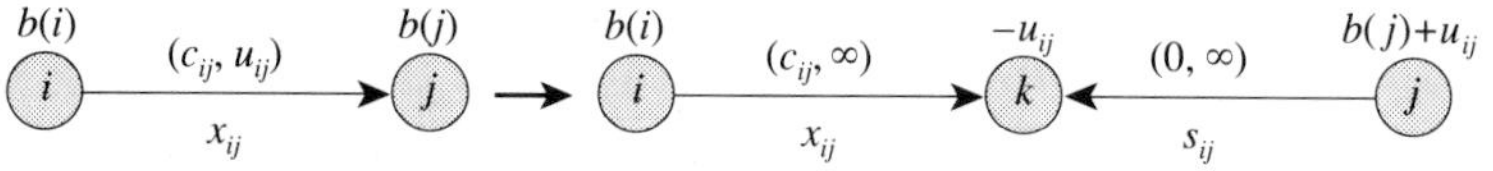

Figure 2.21 Transformation for removing an arc capacity.

To see the relationships between the flows in the original and transformed networks, we make the following observations. If x_{ij} is the flow on arc (i, j) in the original network, the corresponding flow in the transformed network is $x'_{ik} = x_{ij}$ and $x'_{jk} = u_{ij} - x_{ij}$. Notice that both the flows x and x' have the same cost. Similarly, a flow x'_{ik}, x'_{jk} in the transformed network yields a flow $x_{ij} = x'_{ik}$ of the same cost in the original network. Furthermore, since $x'_{ik} + x'_{jk} = u_{ij}$ and x'_{ik} and x'_{jk} are both nonnegative, $x_{ij} = x'_{ik} \leq u_{ij}$. Therefore, the flow x_{ij} satisfies the arc capacity, and the transformation does correctly model arc capacities.

Suppose that every arc in a given network $G = (N, A)$ is capacitated. If we apply the preceding transformation to every arc, we obtain a bipartite uncapacitated network G' (see Figure 2.22 for an illustration). In this network (1) each node i on the left corresponds to a node $i \in N$ of the original network and has a supply equal to $b(i) + \sum_{\{k:(k,i)\in A\}} u_{ki}$, and (2) each node i–j on the right corresponds to an arc $(i, j) \in A$ in the original network and has a demand equal to u_{ij}; this node has exactly two incoming arcs, originating at nodes i and j from the left. Consequently, the transformed network has $(n + m)$ nodes and $2m$ arcs.

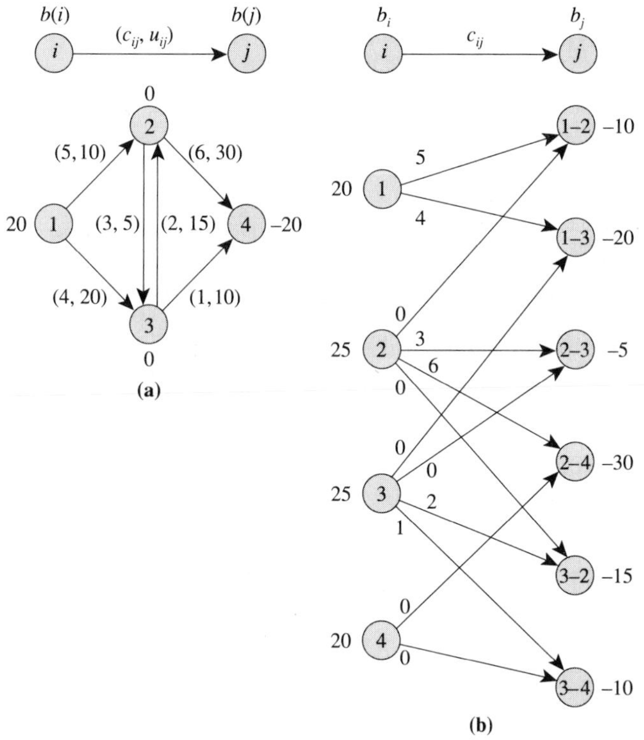

Figure 2.22 Transformation for removing arc capacities: (a) original network; (b) transformed network with uncapacitated arcs.

At first glance we might be tempted to believe that this technique for removing arc capacities would be unattractive computationally since the transformation substantially increases the number of nodes in the network. However, on most occasions the original and transformed networks have algorithms with the same complexity, because the transformed network possesses a special structure that permits us to design more efficient algorithms.

Node Splitting

The node splitting transformation splits each node i into two nodes i' and i'' corresponding to the node's *output* and *input* functions. This transformation replaces each original arc (i, j) by an arc (i', j'') of the same cost and capacity. It also adds an arc (i'', i') of zero cost and with infinite capacity for each i. The input side of node i (i.e., node i'') receives all the node's inflow, the output side (i.e., node i') sends all the node's outflow, and the additional arc (i'', i') carries flow from the input side to the output side. Figure 2.23 illustrates the resulting network when we carry out the node splitting transformation for all the nodes of a network. We define the supplies/demands of nodes in the transformed network in accordance with the following three cases:

1. If $b(i) > 0$, then $b(i'') = b(i)$ and $b(i') = 0$.
2. If $b(i) < 0$, then $b(i'') = 0$ and $b(i') = b(i)$.
3. If $b(i) = 0$, then $b(i') = b(i'') = 0$.

It is easy to show a one-to-one correspondence between a flow in the original network and the corresponding flow in the transformed network; moreover, the flows in both networks have the same cost.

The node splitting transformation permits us to model numerous applications in a variety of practical problem domains, yet maintain the form of the network flow model that we introduced in Section 1.2. For example, we can use the transformation to handle situations in which nodes as well as arcs have associated capacities and costs. In these situations, each flow unit passing through a node i incurs a cost c_i and the maximum flow that can pass through the node is u_i. We can reduce this problem to the standard "arc flow" form of the network flow problem by performing the node splitting transformation and letting c_i and u_i be the cost and capacity of arc

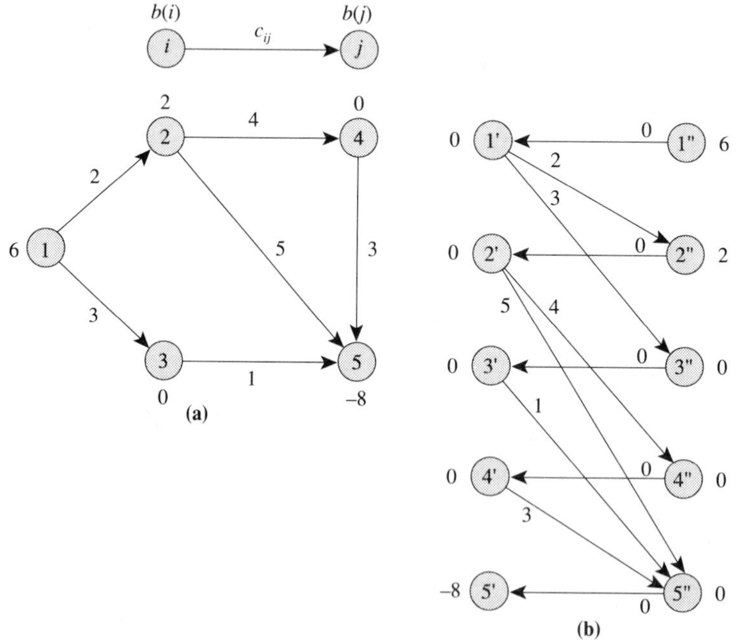

Figure 2.23 Node splitting transformation: (a) original network; (b) transformed network.

(i'', i'). We shall study more applications of the node splitting transformation in Sections 6.6 and 12.7 and in several exercises.

Working with Reduced Costs

In many of the network flow algorithms discussed in this book, we measure the cost of an arc relative to "imputed" costs associated with its incident nodes. These imputed costs typically are intermediate data that we compute within the context of an algorithm. Suppose that we associate with each node $i \in N$ a number $\pi(i)$, which we refer to as the *potential* of that node. With respect to the node potentials $\pi = (\pi(1), \pi(2), \ldots, \pi(n))$, we define the *reduced cost* c_{ij}^{π} of an arc (i, j) as

$$c_{ij}^{\pi} = c_{ij} - \pi(i) + \pi(j). \tag{2.3}$$

In many algorithms discussed later, we often work with reduced costs c_{ij}^{π} instead of the actual costs c_{ij}. Consequently, it is important to understand the relationship between the objective functions $z(\pi) = \sum_{(i,j)\in A} c_{ij}^{\pi} x_{ij}$ and $z(0) = \sum_{(i,j)\in A} c_{ij}x_{ij}$. Suppose, initially, that $\pi = 0$ and we then increase the node potential of node k to $\pi(k)$. The definition (2.3) of reduced costs implies that this change reduces the reduced cost of each unit of flow leaving node k by $\pi(k)$ and increases the reduced cost of each flow unit entering node k by $\pi(k)$. Thus the total decrease in the objective function equals $\pi(k)$ times the outflow of node k minus the inflow of node k. By definition (see Section 1.2), the outflow minus inflow equals the supply/demand of the node. Consequently, increasing the potential of node k by $\pi(k)$ decreases the objective function value by $\pi(k)b(k)$ units. Repeating this argument iteratively for each node establishes that

$$z(0) - z(\pi) = \sum_{i\in N} \pi(i)b(i) = \pi b.$$

For a given node potential π, πb is a constant. Therefore, a flow that minimizes $z(\pi)$ also minimizes $z(0)$. We formalize this result for easy future reference.

Property 2.4. *Minimum cost flow problems with arc costs c_{ij} or c_{ij}^{π} have the same optimal solutions. Moreover, $z(\pi) = z(0) - \pi b$.*

We next study the effect of working with reduced costs on the cost of cycles and paths. Let W be a directed cycle in G. Then

$$\begin{aligned}\sum_{(i,j)\in W} c_{ij}^{\pi} &= \sum_{(i,j)\in W} (c_{ij} - \pi(i) + \pi(j)), \\ &= \sum_{(i,j)\in W} c_{ij} + \sum_{(i,j)\in W} (\pi(j) - \pi(i)), \\ &= \sum_{(i,j)\in W} c_{ij}.\end{aligned}$$

The last equality follows from the fact that for any directed cycle W, the expression $\sum_{(i,j)\in W} (\pi(j) - \pi(i))$ sums to zero because for each node i in the cycle W, $\pi(i)$ occurs once with a positive sign and once with a negative sign. Similarly, if P

is a directed path from node k to node l, then

$$\sum_{(i,j)\in P} c_{ij}^{\pi} = \sum_{(i,j)\in P} (c_{ij} - \pi(i) + \pi(j)),$$

$$= \sum_{(i,j)\in P} c_{ij} - \sum_{(i,j)\in P} (\pi(i) - \pi(j)),$$

$$= \sum_{(i,j)\in P} c_{ij} - \pi(k) + \pi(l),$$

because all $\pi(\cdot)$ corresponding to the nodes in the path, other than the terminal nodes k and l, cancel each other in the expression $\sum_{(i,j)\in P} (\pi(i) - \pi(j))$. We record these results for future reference.

Property 2.5

(a) *For any directed cycle W and for any node potentials π, $\sum_{(i,j)\in W} c_{ij}^{\pi} = \sum_{(i,j)\in W} c_{ij}$.*

(b) *For any directed path P from node k to node l and for any node potentials π, $\sum_{(i,j)\in P} c_{ij}^{\pi} = \sum_{(i,j)\in P} c_{ij} - \pi(k) + \pi(l)$.*

Working with Residual Networks

In designing, developing, and implementing network flow algorithms, it is often convenient to measure flow not in absolute terms, but rather in terms of incremental flow about some given feasible solution—typically, the solution at some intermediate point in an algorithm. Doing so leads us to define a new, ancillary network, known as the *residual network*, that functions as a "remaining flow network" for carrying the incremental flow. We show that formulations of the problem in the original network and in the residual network are equivalent in the sense that they give a one-to-one correspondence between feasible solutions to the two problems that preserves the value of the cost of solutions.

The concept of residual network is based on the following intuitive idea. Suppose that arc (i, j) carries x_{ij}° units of flow. Then we can send an additional $u_{ij} - x_{ij}^{\circ}$ units of flow from node i to node j along arc (i, j). Also notice that we can send up to x_{ij}° units of flow from node j to node i over the arc (i, j), which amounts to canceling the existing flow on the arc. Whereas sending a unit flow from node i to node j on arc (i, j) increases the flow cost by c_{ij} units, sending flow from node j to node i on the same arc decreases the flow cost by c_{ij} units (since we are saving the cost that we used to incur in sending the flow from node i to node j).

Using these ideas, we define the residual network with respect to a given flow x° as follows. We replace each arc (i, j) in the original network by two arcs, (i, j) and (j, i): the arc (i, j) has cost c_{ij} and *residual capacity* $r_{ij} = u_{ij} - x_{ij}^{\circ}$, and the arc (j, i) has cost $-c_{ij}$ and *residual capacity* $r_{ji} = x_{ij}^{\circ}$ (see Figure 2.24). The residual network consists of only the arcs with a positive residual capacity. We use the notation $G(x^{\circ})$ to represent the residual network corresponding to the flow x°.

In general, the concept of residual network poses some notational difficulties. If for some pair i and j of nodes, the network G contains both the arcs (i, j) and

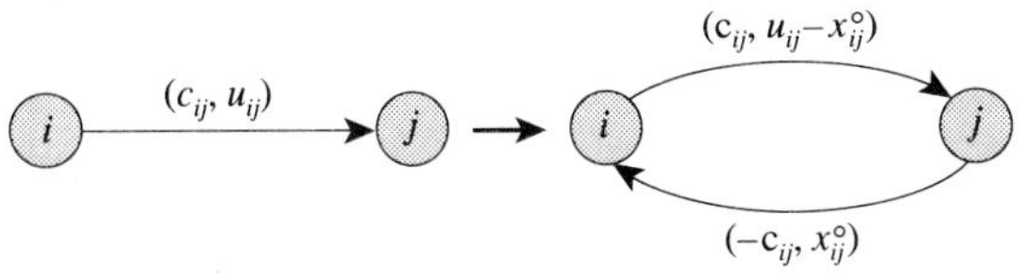

Figure 2.24 Constructing the residual network $G(x^0)$.

(j, i), the residual network may contain two (parallel) arcs from node i to node j with different costs and residual capacities, and/or two (parallel) arcs from node j to node i with different costs and residual capacities. In these instances, any reference to arc (i, j) will be ambiguous and will not define a unique arc cost and residual capacity. We can overcome this difficulty by assuming that for any pair of nodes i and j, the graph G does not contain both arc (i, j) and arc (j, i); then the residual network will contain no parallel arcs. We might note that this assumption is merely a notational convenience; it does not impose any loss of generality, because by suitable transformations we can always define a network that is equivalent to any given network and that will satisfy this assumption (see Exercise 2.47). However, we need not actually make this transformation in practice, since the network representations described in Section 2.3 are capable of handling parallel arcs.

We note further that although the construction and use of the residual network poses some notational difficulties for the general minimum cost flow problem, the difficulties might not arise for some special cases. In particular, for the maximum flow problem, the parallel arcs have the same cost (of zero), so we can merge both of the parallel arcs into a single arc and set its residual capacity equal to the sum of the residual capacities of the two arcs. For this reason, in our discussion of the maximum flow problem, we will permit the underlying network to contain arcs joining any two nodes in both directions.

We now show that every flow x in the network G corresponds to a flow x' in the residual network $G(x^{\circ})$. We define the flow $x' \geq 0$ as follows:

$$x'_{ij} - x'_{ji} = x_{ij} - x^{\circ}_{ij}, \tag{2.4}$$

and

$$x'_{ij}x'_{ji} = 0. \tag{2.5}$$

The condition (2.5) implies that x'_{ij} and x'_{ji} cannot both be positive at the same time. If $x_{ij} \geq x^{\circ}_{ij}$, we set $x'_{ij} = (x_{ij} - x^{\circ}_{ij})$ and $x'_{ji} = 0$. Notice that if $x_{ij} \leq u_{ij}$, then $x'_{ij} \leq u_{ij} - x^{\circ}_{ij} = r_{ij}$. Therefore, the flow x'_{ij} satisfies the flow bound constraints. Similarly, if $x_{ij} < x^{\circ}_{ij}$, we set $x'_{ij} = 0$ and $x'_{ji} = x^{\circ}_{ij} - x_{ij}$. Observe that $0 \leq x'_{ji} \leq x^{\circ}_{ij} = r_{ji}$, so the flow x'_{ji} also satisfies the flow bound constraints. These observations show that if x is a feasible flow in G, its corresponding flow x' is a feasible flow in $G(x^{\circ})$.

We next establish a relationship between the cost of a flow x in G and the cost of the corresponding flow x' in $G(x^{\circ})$. Let c' denote the arc costs in the residual network. Then for every arc $(i, j) \in A$, $c'_{ij} = c_{ij}$ and $c'_{ji} = -c_{ij}$. For a flow x_{ij} on arc (i, j) in the original network G, the cost of flow on the pair of arcs (i, j) and (j, i) in the residual network $G(x^{\circ})$ is $c'_{ij}x'_{ij} + c'_{ji}x'_{ji} = c'_{ij}(x'_{ij} - x'_{ji}) = c_{ij}x_{ij} - c_{ij}x^{\circ}_{ij}$; the last equality follows from (2.4). We have thus shown that

$$c'x' = cx - cx^{\circ}.$$

Similarly, we can show the converse result that if x' is a feasible flow in the residual network $G(x°)$, the solution given by $x_{ij} = (x'_{ij} - x'_{ji}) + x^{\circ}_{ij}$ is a feasible flow in G. Moreover, the costs of these two flows is related by the equality $cx = c'x' + cx°$. We ask the reader to prove these results in Exercise 2.48. We summarize the preceding discussion as the following property.

Property 2.6. *A flow x is a feasible flow in the network G if and only if its corresponding flow x', defined by $x'_{ij} - x'_{ji} = x_{ij} - x^{\circ}_{ij}$ and $x'_{ij}x'_{ji} = 0$, is feasible in the residual network $G(x°)$. Furthermore, $cx = c'x' + cx°$.*

One important consequence of Property 2.6 is the flexibility it provides us. Instead of working with the original network G, we can work with the residual network $G(x°)$ for some $x°$: Once we have determined an optimal solution in the residual network, we can immediately convert it into an optimal solution in the original network. Many of the maximum flow and minimum cost flow algorithms discussed in the subsequent chapters use this result.

2.5 SUMMARY

In this chapter we brought together many basic definitions of network flows and graph theory and presented basic notation that we will use throughout this book. We defined several common graph theoretic terms, including adjacency lists, walks, paths, cycles, cuts, and trees. We also defined acyclic and bipartite networks.

Although networks are often geometric entities, optimization algorithms require computer representations of them. The following four representations are the most common: (1) the node–arc incidence matrix, (2) the node–node adjacency matrix, (3) adjacency lists, and (4) forward and reverse star representations. Figure 2.25 summarizes the basic features of these representations.

Network representations	Storage space	Features
Node–arc incidence matrix	nm	1. Space inefficient 2. Too expensive to manipulate 3. Important because it represents the constraint matrix of the minimum cost flow problem
Node–node adjacency matrix	kn^2 for some constant k	1. Suited for dense networks 2. Easy to implement
Adjacency list	$k_1n + k_2m$ for some constants k_1 and k_2	1. Space efficient 2. Efficient to manipulate 3. Suited for dense as well as sparse networks
Forward and reverse star	$k_3n + k_4m$ for some constants k_3 and k_4	1. Space efficient 2. Efficient to manipulate 3. Suited for dense as well as sparse networks

Figure 2.25 Comparison of various network representations.

The field of network flows is replete with transformations that allow us to transform one problem to another, often transforming a problem that appears to include new complexities into a simplified "standard" format. In this chapter we described some of the most common transformations: (1) transforming undirected networks to directed networks, (2) removing nonzero lower flow bounds (which permits us to assume, without any loss of generality, that flow problems have zero lower bounds on arc flows), (3) performing arc reversals (which often permits us to assume, without any loss of generality, that arcs have nonnegative arc costs), (4) removing arc capacities (which allows us to transform capacitated networks to uncapacitated networks), (5) splitting nodes (which permits us to transform networks with constraints and/or cost associated with "node flows" into our formulation with all data and constraints imposed upon arc flows), and (6) replacing costs with reduced costs (which permits us to alter the cost coefficients, yet retain the same optimal solutions).

The last transformation we studied in this chapter permits us to work with residual networks, which is a concept of critical importance in the development of maximum flow and minimum cost flow algorithms. With respect to an existing flow x, the residual network $G(x)$ represents the capacity and cost information in the network for carrying incremental flows on the arcs. As our discussion has shown, working with residual networks is equivalent to working with the original network.

REFERENCE NOTES

The applied mathematics, computer science, engineering, and operations research communities have developed no standard notation of graph concepts; different researchers and authors use different names to denote the same object (e.g., some authors refer to nodes as vertices or points). The notation and definitions we have discussed in Section 2.2 and adopted throughout this book are among the most popular in the literature. The network representations and transformation that we described in Sections 2.3 and 2.4 are part of the folklore; it is difficult to pinpoint their origins. The books by Aho, Hopcroft, and Ullman [1974], Gondran and Minoux [1984], and Cormen, Leiserson, and Rivest [1990] contain additional information on network representations. The classic book by Ford and Fulkerson [1962] discusses many transformations of network flow problems.

EXERCISES

Note: If any of the following exercises does not state whether a graph is undirected or directed, assume either option, whichever is more convenient.

2.1 Consider the two graphs shown in Figure 2.26.
- **(a)** List the indegree and outdegree of every node.
- **(b)** Give the node adjacency list of each node. (Arrange each list in the increasing order of node numbers.)
- **(c)** Specify a directed walk containing six arcs. Also, specify a walk containing eight arcs.
- **(d)** Specify a cycle containing nine arcs and a directed cycle containing seven arcs.

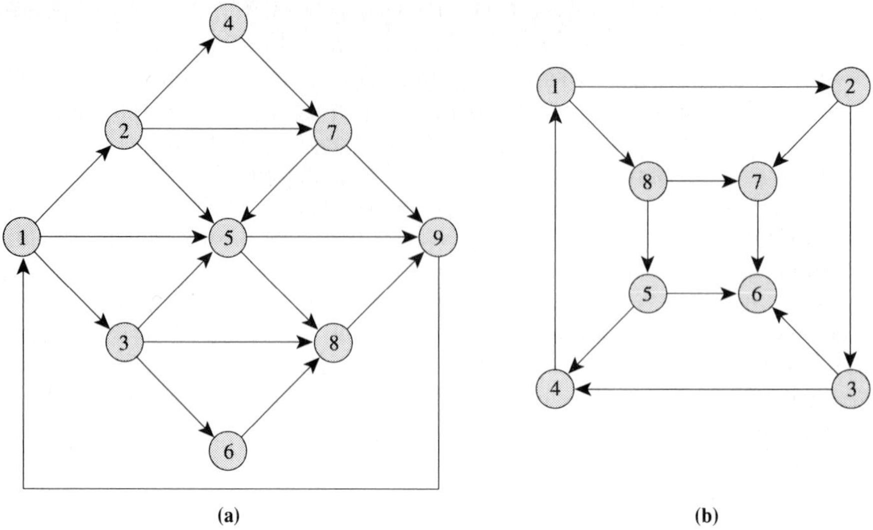

Figure 2.26 Example networks for Exercises 2.1 to 2.4.

2.2. Specify a spanning tree of the graph in Figure 2.26(a) with six leaves. Specify a cut of the graph in Figure 2.26(a) containing six arcs.

2.3. For the graphs shown in Figure 2.26, answer the following questions.
  (a) Are the graphs acyclic?
  (b) Are the graphs bipartite?
  (c) Are the graphs strongly connected?

2.4. Consider the graphs shown in Figure 2.26.
  (a) Do the graphs contain a directed in-tree for some root node?
  (b) Do the graphs contain a directed out-tree for some root node?
  (c) In Figure 2.26(a), list all fundamental cycles with respect to the following spanning tree $T = \{(1, 5), (1, 3), (2, 5), (4, 7), (7, 5), (7, 9), (5, 8), (6, 8)\}$.
  (d) For the spanning tree given in part (c), list all fundamental cuts. Which of these are the s–t cuts when $s = 1$ and $t = 9$?

2.5. **(a)** Construct a directed strongly connected graph with five nodes and five arcs.
  (b) Construct a directed bipartite graph with six nodes and nine arcs.
  (c) Construct an acyclic directed graph with five nodes and ten arcs.

2.6. **Bridges of Königsberg.** The first paper on graph theory was written by Leonhard Euler in 1736. In this paper, he started with the following mathematical puzzle: The city of Königsburg has seven bridges, arranged as shown in Figure 2.27. Is it possible to start at some place in the city, cross every bridge exactly once, and return to the starting place? Either specify such a tour or prove that it is impossible to do so.

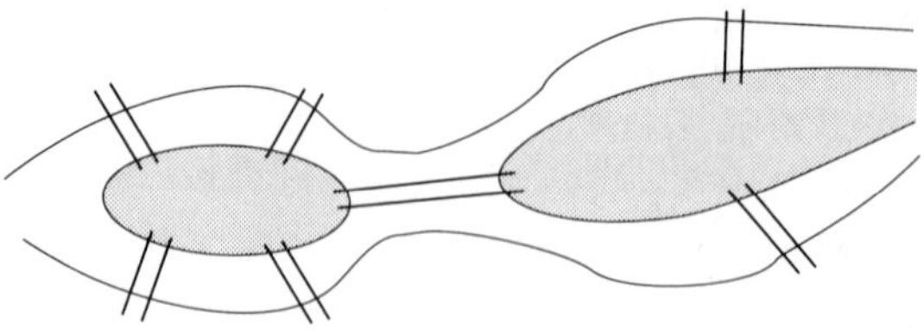

Figure 2.27 Bridges of Königsberg.

2.7. At the beginning of a dinner party, several participants shake hands with each other. Show that the participants that shook hands an odd number of times must be even in number.

2.8. Show that in a directed strongly connected graph containing more than one node, no node can have a zero indegree or a zero outdegree.

2.9. Suppose that every node in a directed graph has a positive indegree. Show that the graph must contain a directed cycle.

2.10. Show that a graph G remains connected even after deleting an arc (i, j) if and only if arc (i, j) belongs to some cycle in G.

2.11. Show that an undirected graph $G = (N, A)$ is connected if and only if for every partition of N into subsets N_1 and N_2, some arc has one endpoint in N_1 and the other endpoint in N_2.

2.12. Let d_{min} denote the minimum degree of a node in an undirected graph. Show that the graph contains a path containing at least d_{min} arcs.

2.13. Prove the following properties of trees.

(a) A tree on n nodes contains exactly $(n - 1)$ arcs.

(b) A tree has at least two leaf nodes (i.e., nodes with degree 1).

(c) Every two nodes of a tree are connected by a unique path.

2.14. Show that every tree is a bipartite graph.

2.15. Show that a forest consisting of k components has $m = n - k$ arcs.

2.16. Let d_{max} denote the maximum degree of a node in a tree. Show that the tree contains at least d_{max} nodes of degree 1. (*Hint*: Use the fact that the sum of the degrees of all nodes in a tree is $2m = 2n - 2$.)

2.17. Let Q be any cut of a connected graph and T be any spanning tree. Show that $Q \cap T$ is nonempty.

2.18. Show that a closed directed walk containing an odd number of arcs contains a directed cycle having an odd number of arcs. Is it true that a closed directed walk containing an even number of arcs also contains a directed cycle having an even number of arcs?

2.19. Show that any cycle of a graph G contains an even number of arcs (possibly zero) in common with any cut of G.

2.20. Let d_{min} denote the minimum degree of a node in an undirected graph G. Show that if $d_{min} \geq 2$, then G must contain a cycle.

2.21. **(a)** Show that in a bipartite graph every cycle contains an even number of arcs.

(b) Show that a (connected) graph, in which every cycle contains an even number of arcs, must be bipartite. Conclude that a graph is bipartite if and only if every cycle has an even number of arcs.

2.22. The *k-color problem* on an undirected graph $G = (N, A)$ is defined as follows: Color all the nodes in N using at most k colors so that for every arc $(i, j) \in A$, nodes i and j have a different color.

(a) Given a world map, we want to color countries using at most k colors so that the countries having common boundaries have a different color. Show how to formulate this problem as a k-color problem.

(b) Show that a graph is bipartite if and only if it is *2-colorable* (i.e., can be colored using at most two colors).

2.23. Two undirected graphs $G = (N, A)$ and $G' = (N', A')$ are said to be *isomorphic* if we can number the nodes of the graph G so that G becomes identical to G'. Equivalently, G is isomorphic to G' if some one-to-one function f maps N onto N' so that (i, j) is an arc in A if and only if $(f(i), f(j))$ is an arc in A'. Give several necessary conditions for two undirected graphs to be isomorphic. (*Hint*: For example, they must have the same number of nodes and arcs.)

2.24. **(a)** List all nonisomorphic trees having four nodes.

(b) List all nonisomorphic trees having five nodes. (*Hint:* There are three such trees.)

2.25. For any undirected graph $G = (N, A)$, we define its *complement* $G^c = (N, A^c)$ as follows: If $(i, j) \in A$, then $(i, j) \notin A^c$, and if $(i, j) \notin A$, then $(i, j) \in A^c$. Show that if the graph G is disconnected, its complement G^c is connected.

2.26. Let $G = (N, A)$ be an undirected graph. We refer to a subset $N_1 \subseteq N$ as *independent* if no two nodes in N_1 are adjacent. Let $\beta(G)$ denote the maximum cardinality of any independent set of G. We refer to a subset $N_2 \subseteq N$ as a *node cover* if each arc in A has at least one of its endpoints in N_2. Let $\eta(G)$ denote the minimum cardinality of any node cover G. Show that $\beta(G) + \eta(G) = n$. (*Hint:* Show that the complement of an independent set is a node cover.)

2.27. Problem of queens. Consider the problem of determining the maximum number of queens that can be placed on a chessboard so that none of the queens can be taken by another. Show how to transform this problem into an independent set problem defined in Exercise 2.26.

2.28. Consider a directed graph $G = (N, A)$. For any subset $S \subseteq N$, let *neighbor(S)* denote the set of neighbors of S [i.e., neighbor$(S) = \{j \in N$: for some $i \in S$, $(i, j) \in A$ and $j \notin S\}$]. Show that G is strongly connected if and only if for every proper nonempty subset $S \subset N$, neighbor$(S) \neq \emptyset$.

2.29. A subset $N_1 \subseteq N$ of nodes in an undirected graph $G = (N, A)$ is said to be a *clique* if every pair of nodes in N_1 is connected by an arc. Show that the set N_1 is a clique in G if and only if N_1 is independent in its complement G^c.

2.30. Specify the node–arc incidence matrix and the node–node adjacency matrix for the graph shown in Figure 2.28.

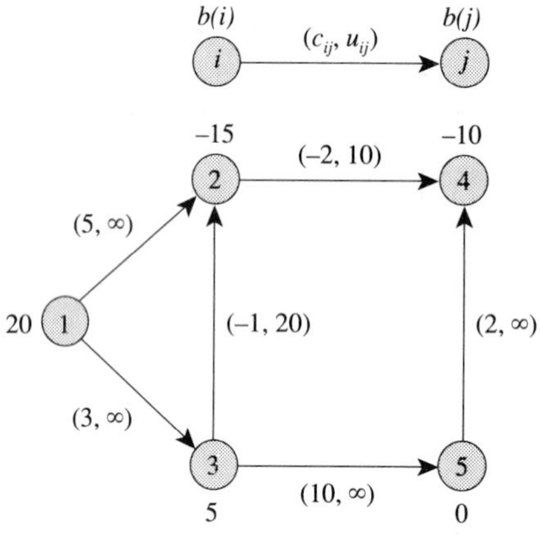

Figure 2.28 Network example.

2.31. **(a)** Specify the forward star representation of the graph shown in Figure 2.28.
(b) Specify the forward and reverse star representations of the graph shown in Figure 2.28.

2.32. Let $\mathcal{N}$ denote the node–arc incidence matrix of an undirected graph and let $\mathcal{N}^T$ denote its transpose. Let "$\cdot$" denote the operation of taking a product of two matrices. Show how to interpret the diagonal elements of $\mathcal{N} \cdot \mathcal{N}^T$?

2.33. Let $\mathcal{H}$ denote the node–node adjacency matrix of a directed network, and let $\mathcal{N}$ denote the node–arc incidence matrix of this network. Can $\mathcal{H} = \mathcal{N} \cdot \mathcal{N}^T$?

2.34. Let $\mathcal{H}$ be the node–node adjacency matrix of a directed graph $G = (N, A)$. Let $\mathcal{H}^T$ be the transpose of $\mathcal{H}$, and let G^T be the graph corresponding to $\mathcal{H}^T$. How is the graph G^T related to G?

2.35. Let G be a bipartite graph. Show that we can always renumber the nodes of G so that the node–node adjacency matrix $\mathcal{H}$ of G has the following form:

0	F
E	0

2.36. Show that a directed graph G is acyclic if and only if we can renumber its nodes so that its node–node adjacency matrix is a lower triangular matrix.

2.37. Let $\mathcal{H}$ denote the node–node adjacency matrix of a network G. Define $\mathcal{H}^k = \mathcal{H} \cdot \mathcal{H}^{k-1}$ for each $k = 2, 3, \ldots, n$. Show that the ijth entry of the matrix $\mathcal{H}^2$ is the number of directed paths consisting of two arcs from node i to node j. Then using induction, show that the ijth entry of matrix $\mathcal{H}^k$ is the number of distinct walks from node i to node j containing exactly k arcs. In making this assessment, assume that two walks are *distinct* if their sequences of arcs are different (even if the unordered set of arcs are the same).

2.38. Let $\mathcal{H}$ denote the node–node adjacency matrix of a network G. Show that G is strongly connected if and only if the matrix $\mathcal{R}$ defined by $\mathcal{R} = \mathcal{H} + \mathcal{H}^2 + \mathcal{H}^3 + \cdots + \mathcal{H}^n$ has no zero entry.

2.39. Write a pseudocode that takes as an input the node–node adjacency matrix representation of a network and produces as an output the forward and reverse star representations of the network. Your pseudocode should run in $O(n^2)$ time.

2.40. Write a pseudocode that accepts as an input the forward star representation of a network and produces as an output the network's node–node adjacency matrix representation.

2.41. Write a pseudocode that takes as an input the forward star representation of a network and produces the reverse star representation. Your pseudocode should run in $O(m)$ time.

2.42. Consider the minimum cost flow problem shown in Figure 2.28. Suppose that arcs (1, 2) and (3, 5) have lower bounds equal to $l_{12} = l_{35} = 5$. Transform this problem to one where all arcs have zero lower bounds.

2.43. In the network shown in Figure 2.28, some arcs have finite capacities. Transform this problem to one where all arcs are uncapacitated.

2.44. Consider the minimum cost flow problem shown in Figure 2.28 (note that some arcs have negative arc costs). Modify the problem so that all arcs have nonnegative arc costs.

2.45. Construct the residual network for the minimum cost flow problem shown in Figure 2.28 with respect to the following flow: $x_{12} = x_{13} = x_{32} = 10$ and $x_{24} = x_{35} = x_{54} = 5$.

2.46. For the minimum cost flow problem shown in Figure 2.28, specify a vector π of node potentials so that $c_{ij}^{\pi} \geq 0$ for every arc $(i, j) \in A$. Compute cx, $c^{\pi}x$, and πb for the flow given in Exercise 2.45 and verify that $cx = c^{\pi}x + \pi b$.

2.47. Suppose that a minimum cost flow problem contains both arcs (i, j) and (j, i) for some pair of nodes. Transform this problem to one in which the network contains either arc (i, j) or arc (j, i), but not both.

2.48. Show that if x' is a feasible flow in the residual network $G(x^{\circ})$, the solution given by $x_{ij} = (x'_{ij} - x'_{ji}) + x_{ij}^{\circ}$ is a feasible flow in G and satisfies $cx = c'x' + cx^{\circ}$.

2.49. Suppose that you are given a minimum cost flow code that requires that its input data be specified so that $l_{ij} = u_{ij}$ for no arc (i, j). How would you eliminate such arcs?

2.50. Show how to transform a minimum cost flow problem stated in (2.1) into a circulation problem. Establish a one-to-one correspondence between the feasible solutions of these two problems. (*Hint:* Introduce two new nodes and some arcs.)

2.51. Show that by adding an extra node and appropriate arcs, we can formulate any minimum cost flow problem with one or more inequalities for supplies and demands (i.e., the mass balance constraints are stated as "$\leq b(i)$" for a supply node i, and/or "$\geq b(j)$" for a demand node j) into an equivalent problem with all equality constriants (i.e., "$= b(k)$" for all nodes k).

3

ALGORITHM DESIGN AND ANALYSIS

Numerical precision is the very soul of science.
—Sir D'Arcy Wentworth Thompson

Chapter Outline

3.1 INTRODUCTION

Scientific computation is a unifying theme that cuts across many disciplines, including computer science, operations research, and many fields within applied mathematics and engineering. Within the realm of computational problem solving, we almost always combine three essential building blocks: (1) a recipe, or algorithm, for solving a particular class of problems; (2) a means for encoding this procedure in a computational device (e.g., a calculator, a computer, or even our own minds); and (3) the application of the method to the data of a specific problem. For example, to divide one number by another, we might use the iterative algorithm of long division, which is a systematic procedure for dividing any two numbers. To solve a specific problem, we could use a calculator that has this algorithm already built into its circuitry. As a first step, we would enter the data into storage locations on the calculator; then we would instruct the calculator to apply the algorithm to our data.

Although dividing two numbers is an easy task, the essential steps required to solve this very simple problem—designing, encoding, and applying an algorithm—are similar to those that we need to address when solving complex network flow problems. We need to develop an algorithm, or a mathematical prescription, for solving a class of network flow problems that contains our problem—for example, to solve a particular shortest path problem, we might use an algorithm that is known to solve any shortest path problem with nonnegative arc lengths. Since solving a network flow problem typically requires the solution of an optimization model with hundreds or thousands of variables, equations, and inequalities, we will invariably solve the problem on a computer. Doing so requires that we not only express the mathematical steps of the algorithm as a computer program, but that we also develop data structures for manipulating the large amounts of information required to represent the problem. We also need a method for entering the data into the computer and for performing the necessary operations on it during the course of the solution procedure.

In Chapter 2 we considered the lower-level steps of the computational problem-solving hierarchy; that is, we saw how to represent network data and therefore how to encode and manipulate the data within a computer. In this chapter we consider the highest level of the solution hierarchy: How do we design algorithms, and how do we measure their effectiveness? Although the idea of an algorithm is an old one—Chinese mathematicians in the third century B.C. had already devised algorithms for solving small systems of simultaneous equations—researchers did not begin to explore the notion of algorithmic efficiency as discussed in this book in any systematic and theoretical sense until the early 1970s. This particular subject matter, known as computational complexity theory, provides a framework and a set of analysis tools for gauging the work performed by an algorithm as measured by the elementary operations (e.g., addition, multiplication) it performs. One major stream of research in computational complexity theory has focused on developing performance guarantees or worst-case analyses that address the following basic question: When we apply an algorithm to a class of problems, can we specify an upper bound on the amount of computations that the algorithm will require? Typically, the performance guarantee is measured with respect to the size of the underlying problem: for example, for network flow problems, the number n of nodes and the number m of arcs in the underlying graph. For example, we might state that the complexity of an algorithm for solving shortest path problems with nonnegative arc lengths is $2n^2$, meaning that the number of computations grow no faster than twice the square of the number of nodes. In this case we say that the algorithm is "good" because its computations are bounded by a polynomial in the problem size (as measured by the number of nodes). In contrast, the computational time for a "bad" algorithm would grow exponentially when applied to a certain class of problems. With the theoretical worst-case bound in hand, we can now assess the amount of work required to solve (nonnegative length) shortest path problems as a function of their size. We also have a tool for comparing any two algorithms: the one with the smaller complexity bound is preferred from the viewpoint of a worst-case analysis.

Network optimization problems have been the core and influential subject matter in the evolution of computational complexity theory. Researchers and analysts have developed many creative ideas for designing efficient network flow algorithms based on the concepts and results emerging in the study of complexity theory; at the same time, many ideas originating in the study of network flow problems have proven to be useful in developing and analyzing a wide variety of algorithms in many other problem domains. Although network optimization has been a constant subject of study throughout the years, researchers have developed many new results concerning complexity bounds for network flow algorithms at a remarkable pace in recent years. Many of these recent innovations draw on a small set of common ideas, which are simultaneously simple and powerful.

Our intention in this chapter is to bring together some of the most important of these ideas. We begin by reviewing the essential ingredients of computational complexity theory, including the definition and computational implications of good algorithms. We then describe several key ideas that appear to be mainstays in the development and analysis of good network flow algorithms. One idea is an approximation strategy, known as *scaling*, that solves a sequence of "simple" approximate versions of a given problem (determined by scaling the problem data) in such a way that the problems gradually become better approximations of the original problem. A second idea is a *geometric improvement argument* that is quite useful in analyzing algorithms; it shows that whenever we make sufficient (i.e., fixed percentage) improvements in the objective function at every iteration, an algorithm is good.

We also describe some important tools that can be used in analyzing or streamlining algorithms: (1) a *potential function method* that provides us with a scalar integer-valued function that summarizes the progress of an algorithm in such a way that we can use it to bound the number of steps that the algorithm takes, and (2) a *parameter balancing technique* that permits us to devise an algorithm based on some underlying parameter and then to set the parameter so that we minimize the number of steps required by the algorithm. Next, we introduce the idea of *dynamic programming*, which is a useful algorithmic strategy for developing good algorithms. The dynamic programming technique decomposes the problem into stages and uses a recursive relationship to go from one stage to another. Finally, we introduce the *binary search* technique, another well-known technique for obtaining efficient algorithms. Binary search performs a search over the feasible values of the objective function and solves an easier problem at each search point.

In this chapter we also describe important and efficient (i.e., good) algorithms that we use often within the context of network optimization: *search algorithms* that permit us to find all the nodes in a network that satisfy a particular property. Often in the middle of a network flow algorithm, we need to discover all nodes that share a particular attribute; for example, in solving a maximum flow problem, we might want to find all nodes that are reachable from the designated source node along a directed path in the residual network. Search algorithms provide us with a mechanism to perform these important computations efficiently. As such, they are essential, core algorithms used to design other more complex algorithms.

Finally, we study *network decomposition algorithms* that permit us to decompose a solution to a network flow problem, formulated in terms of arc flows, into a set of flows on paths and cycles. In our treatment of network flow problems, we have chosen to use a model with flows defined on arcs. An alternative modeling approach is to view all flows as being carried along paths and cycles in the network. In this model, the variables are the amount of flow that we send on any path or cycle. Although the arc flow formulation suffices for most of the topics that we consider in this book, on a few occasions such as our discussion of multicommodity flows in Chapter 17, we will find it more convenient to work with a path and cycle flow model. Moreover, even if we do not use the path and cycle flow formulation per se, understanding this model provides additional insight about the nature of network flow problems. The network decomposition algorithms show that the arc flow model and the path and cycle flow model are equivalent, so we could use any of these models for formulating network flow problems; in addition, these algorithms provide us with an efficient computational procedure for finding a set of path and cycle flows that is equivalent to any given set of arc flows.

3.2 COMPLEXITY ANALYSIS

An algorithm is a step-by-step procedure for solving a problem. By a *problem* we mean a generic model such as the shortest path problem or the minimum cost flow problem. Problems can be subsets of one another: For example, not only does the set of all shortest path problems define a problem, but so does the class of all shortest path problems with nonnegative arc costs. An *instance* is a special case of a problem with data specified for all the problem parameters. For example, to define an instance of the shortest path problem we would need to specify the network topology $G = (N, A)$, the source and destination nodes, and the values of the arc costs. An algorithm is said to *solve* a problem P if when applied to any instance of P, the algorithm is guaranteed to produce a solution. Generally, we are interested in finding the most "efficient" algorithm for solving a problem. In the broadest sense, the notion of efficiency involves all the various computing resources needed for executing an algorithm. However, in this book since time is often a dominant computing resource, we use the time taken by an algorithm as our metric for measuring the "most efficient" algorithm.

Different Complexity Measures

As already stated, an algorithm is a step-by-step procedure for solving a problem. The different steps an algorithm typically performs are (1) assignment steps (such as assigning some value to a variable), (2) arithmetic steps (such as addition, subtraction, multiplication, and division), and (3) logical steps (such as comparison of two numbers). The number of steps performed (or taken) by the algorithm is said to be the sum total of all steps it performs. The number of steps taken by an algorithm, which to a large extent determines the time it requires, will differ from one instance of the problem to another. Although an algorithm might solve some "good" instances of the problem quickly, it might take a long time to solve some "bad" instances. This range of possible outcomes raises the question of how we should measure the performance of an algorithm so that we can select the "best" algorithm from among several competing algorithms for solving a problem. The literature has widely adopted three basic approaches for measuring the performance of an algorithm:

1. *Empirical analysis.* The objective of empirical analysis is to estimate how algorithms behave in practice. In this analysis we write a computer program for the algorithm and test the performance of the program on some classes of problem instances.
2. *Average-case analysis.* The objective of average-case analysis is to estimate the expected number of steps an algorithm takes. In this analysis we choose a probability distribution for the problem instances and using statistical analysis derive asymptotic expected running times for the algorithm.
3. *Worst-case analysis.* Worst-case analysis provides upper bounds on the number of steps that a given algorithm can take on *any* problem instance. In this analysis we count the largest possible number of steps; consequently, this analysis provides a "guarantee" on the number of steps an algorithm will take to solve any problem instance.

Each of these three performance measures has its relative merits and drawbacks. Empirical analysis has several major drawbacks: (1) an algorithm's performance depends on the programming language, compiler, and computer used for the computational experiments, as well as the skills of the programmer who wrote the program; (2) often this analysis is too time consuming and expensive to perform; and (3) the comparison of algorithms is often inconclusive in the sense that different algorithms perform better on different classes of problem instances and different empirical studies report contradictory results.

Average-case analysis has major drawbacks as well: (1) the analysis depends crucially on the probability distribution chosen to represent the problem instances, and different choices might lead to different assessments as to the relative merits of the algorithms under consideration; (2) it is often difficult to determine appropriate probability distributions for problems met in practice; and (3) the analysis often requires quite intricate mathematics even for assessing the simplest type of algorithm—the analysis typically is extremely difficult to carry out for more complex algorithms. Furthermore, the prediction of an algorithm's performance, based on its average-case analysis, is tailored for situations in which the analyst needs to solve a large number of problem instances; it does not provide information about the distribution of outcomes. In particular, although the average-case performance of an algorithm might be good, we might encounter exceptions with little statistical significance on which the algorithm performs very badly.

Worst-case analysis avoids many of these drawbacks. The analysis is independent of the computing environment, is relatively easier to perform, provides a guarantee on the steps (and time) taken by an algorithm, and is definitive in the sense that it provides conclusive proof that an algorithm is superior to another for the worst possible problem instances that an analyst might encounter. Worst-case analysis is not perfect, though: One major drawback of worst-case analysis is that it permits ''pathological'' instances to determine the performance of an algorithm, even though they might be exceedingly rare in practice. However, the advantages of the worst-case analysis have traditionally outweighed its shortcomings, and this analysis has become the most popular method for measuring algorithmic performance in the scientific literature. The emergence of the worst-case analysis as a tool for assessing algorithms has also had a great impact on the field of network flows, stimulating considerable research and fostering many algorithmic innovations. In this book, too, we focus primarily on worst-case analysis. We also try to provide insight about the empirical performance, particularly in Chapter 18, since we believe that the empirical behavior of algorithms provides important information for guiding the use of algorithms in practice.

Problem Size

To express the time requirement of an algorithm, we would like to define some measure of the ''complexity'' of the problem instances we encounter. Having a single performance measure for all problem instances rarely makes sense since as the problem instances become larger, they typically become more difficult to solve (i.e., take more time); often the effort required to solve problem instances varies roughly with their size. Hence to measure the complexity of problem instances, we must consider the "size" of the problem instance. But what is the size of a problem?

Before we address this question, let us discuss what is the size of a data item whose value is x. We can make one of the two plausible assumptions: (1) assume that the size of the data item is x, or (2) assume that the size of the data item is $\log x$. Of these, for several reasons the second assumption is more common. The primary reason is that $\log x$ reflects the way that computers work. Most modern computers represent numbers in binary form (i.e., in bits) and store them in memory locations of fixed bit size. The binary representation of item x requires $\log x$ bits, and hence the space required to store x is proportional to $\log x$.

The size of a network problem is a function of how the problem is stated. For a network problem, the input might be in the form of one of the representations discussed in Section 2.3. Suppose that we specify the network in the adjacency list representation, which is the most space-efficient representation we could use. Then the size of the problem is the number of bits needed to store its adjacency list representation. Since the adjacency list representation stores one pointer for each node and arc, and one data element for each arc cost coefficient and each arc capacity, it requires approximately $n \log n + m \log m + m \log C + m \log U$ bits to store all of the problem data for a minimum cost network flow problem (recall that C represents the largest arc cost and U represents the largest arc capacity). Since $m \leq n^2$, $\log m \leq \log n^2 = 2 \log n$. For this reason, when citing the size of problems using a "big O" complexity notation that ignores constants (see the subsection entitled "big O" to follow), we can (and usually do) replace each occurrence of $\log m$ by the term $\log n$.

In principle, we could express the running time of an algorithm as a function of the problem size; however, that would be unnecessarily awkward. Typically, we will express the running time more simply and more directly as a function of the network parameters n, m, $\log C$, and $\log U$.

Worst-Case Complexity

The time taken by an algorithm, which is also called the *running time* of the algorithm, depends on both the nature and size of the input. Larger problems require more solution time, and different problems of the same size typically require different solution times due to differences in the data. A *time complexity function* for an algorithm is a function of the problem size and specifies the largest amount of time needed by the algorithm to solve any problem instance of a given size. In other words, the time complexity function measures the rate of growth in solution time as the problem size increases. For example, if the time complexity function of a network algorithm is cnm for some constant $c \geq 0$, the running time needed to solve *any* network problem with n nodes and m arcs is at most cnm. Notice that the time complexity function accounts for the dependence of the running time on the problem size by measuring the *largest* time needed to solve any problem instance of a given size; at this level of detail in measuring algorithmic performance, the complexity function provides a performance guarantee that depends on the appropriate measure of the problem's input data. Accordingly, we also refer to the time complexity function as the *worst-case complexity* (or, simply, the *complexity*) of the algorithm. We also refer to the worst-case complexity of an algorithm as its *worst-case bound*, for it states an upper bound on the time taken by the algorithm.

Big O Notation

To define the complexity of an algorithm completely, we need to specify the values for one or more constants. In most cases the determination of these constants is a nontrivial task; moreover, the determination might depend heavily on the computer, and other factors. Consider, for example, the following segment of an algorithm, which adds two $p \times q$ arrays:

```
for i: = 1 to p do
  for j: = 1 to q do
    c_ij: = a_ij + b_ij;
```

At first glance, this program segment seems to perform exactly pq additions and the same number of assignments of values to the computer locations storing the values of the variables c_{ij}. This accounting, however, ignores many computations that the computer would actually perform. A computer generally stores a two-dimensional array of size $p \times q$ as a single array of length pq and so would typically store the element a_{ij} at the location $(i - 1)q + j$ of the array a. Thus each time we retrieve the value of a_{ij} and b_{ij} we would need to perform one subtraction, one multiplication, and one addition. Further, whenever, the computer would increment the index i (or j), it would perform a comparison to determine whether $i > p$ (or $j > q$). Needless to say, such a detailed analysis of an algorithm is very time consuming and not particularly illuminating.

The dependence of the complexity function on the constants poses yet another problem: How do we compare an algorithm that performs $5n$ additions and $3n$ comparisons with an algorithm that performs n multiplications and $2n$ subtractions? Different computers perform mathematical and logical operations at different speeds, so neither of these algorithms might be universally better.

We can overcome these difficulties by ignoring the constants in the complexity analysis. We do so by using "big O" notation, which has become commonplace in computational mathematics, and replace the lengthy and somewhat awkward expression "the algorithm required cnm time for some constant c" by the equivalent expression "the algorithm requires $O(nm)$ time." We formalize this definition as follows:

An algorithm is said to run in $O(f(n))$ time if for some numbers c and n_0, the time taken by the algorithm is at most $cf(n)$ for all $n \geq n_0$.

Although we have stated this definition in terms of a single measure n of a problem-size parameter, we can easily incorporate other size parameters m, C, and U in the definition.

The big O notation has several implications. The complexity of an algorithm is an upper bound on the running time of the algorithm for sufficiently large values of n. Therefore, this complexity measure states the asymptotic growth rate of the running time. We can justify this feature of the complexity measure from practical considerations since we are more interested about the behavior of the algorithm on very large inputs, as these inputs determine the limits of applicability of the algorithm. Furthermore, the big O notation indicates only the most dominant term in the running time, because for sufficiently large n, terms with a smaller growth rate become insignificant as compared to terms with a higher growth rate. For example, if the running time of an algorithm is $100n + n^2 + 0.0001n^3$, then for all $n \geq 100$, the second term dominates the first term, and for all $n \geq 10{,}000$, the third term dominates the second term. Therefore, the complexity of the algorithm is $O(n^3)$.

Another important implication of ignoring constants in the complexity analysis is that we can assume that each elementary mathematical operation, such as addition, subtraction, multiplication, division, assignment, and logical operations, requires an equal amount of time. A computer typically performs these operations at different speeds, but the variation in speeds can typically be bounded by a constant (provided the numbers are not too large), which is insignificant in big O notation. For example, a computer typically multiplies two numbers by repeated additions and the number of such additions are equal to number of bits in the smaller number. Assuming that the largest number can have 32 bits, the multiplication can be at most 32 times more expensive than addition. These observations imply that we can summarize the running time of an algorithm by recording the number of elementary mathematical operations it performs, viewing every operation as requiring an equivalent amount of time.

Similarity Assumption

The assumption that each arithmetic operation takes one step might lead us to underestimate the asymptotic running time of arithmetic operations involving very large numbers on real computers since, in practice, a computer must store such numbers in several words of its memory. Therefore, to perform each operation on very large numbers, a computer must access a number of words of data and thus take more than a constant number of steps. Thus the reader should be forewarned that the running times are misleading if the numbers are exponentially large. To avoid this systematic underestimation of the running time, in comparing two running times, we will sometimes assume that both C (i.e., the largest arc cost) and U (i.e., the largest arc capacity) are polynomially bounded in n [i.e., $C = O(n^k)$ and $U = O(n^k)$, for some constant k]. We refer to this assumption as the *similarity assumption*.

Polynomial- and Exponential-Time Algorithms

We now consider the question of whether or not an algorithm is "good." Ideally, we would like to say that an algorithm is good if it is sufficiently efficient to be usable in practice, but this definition is imprecise and has no theoretical grounding. An idea that has gained wide acceptance in recent years is to consider a network algorithm "good" if its worst-case complexity is bounded by a polynomial function of the problem's parameters (i.e., it is a polynomial function of n, m, $\log C$, and $\log U$). Any such algorithm is said to be a *polynomial-time algorithm*. Some examples of polynomial-time bounds are $O(n^2)$, $O(nm)$, $O(m + n \log C)$, $O(nm \log(n^2/m))$, and $O(nm + n^2 \log U)$. (Note that $\log n$ is polynomially bounded because its growth rate is slower than n.) A polynomial-time algorithm is said to be a *strongly polynomial-time algorithm* if its running time is bounded by a polynomial function in only n and m, and does not involve $\log C$ or $\log U$, and is a *weakly polynomial-time algorithm* otherwise. Some strongly polynomial time bounds are $O(n^2m)$ and $O(n \log n)$. In principle, strongly polynomial-time algorithms are preferred to weakly polynomial-time algorithms because they can solve problems with arbitrary large values for the cost and capacity data.

Note that in this discussion we have said that an algorithm is polynomial time if its running time is bounded by a polynomial in the network parameters n, m, $\log C$, and $\log U$. Typically, in computational complexity we say that an algorithm is polynomial time if its running time is bounded by a polynomial in the problem size, in this case $n \log n + m \log m + n \log C + m \log U$; however, it is easy to see that the running time of a network problem is bounded by a polynomial in its problem size if and only if it is also bounded by a polynomial in the problem parameters. For example, if the running time is bounded by n^{100}, it is strictly less than the problem size to the 100th power. Similarly, if the running time is bounded by the problem size to the 100th power, it is less than $(n \log n + m \log m + n \log C + m \log U)^{100}$, which in turn is bounded by $(n^2 + m^2 + n \log C + m \log U)^{100}$, which is a polynomial in n, m, $\log C$, and $\log U$.

An algorithm is said to be an *exponential-time algorithm* if its worst-case running time grows as a function that cannot be polynomially bounded by the input length. Some examples of exponential time bounds are $O(nC)$, $O(2^n)$, $O(n!)$, and $O(n^{\log n})$. (Observe that nC cannot be bounded by a polynomial function of n and $\log C$.) We say that an algorithm is a *pseudopolynomial-time algorithm* if its running time is polynomially bounded in n, m, C, and U. The class of pseudopolynomial-time algorithms is an important subclass of exponential-time algorithms. Some examples of pseudopolynomial-time bounds are $O(m + nC)$ and $O(mC)$. For problems that satisfy the similarity assumption, pseudopolynomial-time algorithms become polynomial-time algorithms, but the algorithms will not be attractive if C and U are high-degree polynomials in n.

There are several reasons for preferring polynomial-time algorithms to exponential-time algorithms. Any polynomial-time algorithm is asymptotically superior to any exponential-time algorithm, even in extreme cases. For example, n^{4000} is smaller than $n^{0.1 \log n}$ if n is sufficiently large (i.e., $n \geq 2^{100,000}$). Figure 3.1 illustrates the growth rates of several typical complexity functions. The exponential complexity functions have an explosive growth rate and, in general, they are able to solve only small problems. Further, much practical experience has shown that the polynomials encountered in practice typically have a small degree, and generally, polynomial-time algorithms perform better than exponential-time algorithms.

n	$\log n$	$n^{0.5}$	n^2	n^3	2^n	$n!$
10	3.32	3.16	10^2	10^3	10^3	3.6×10^6
100	6.64	10.00	10^4	10^6	1.27×10^{30}	9.33×10^{157}
1000	9.97	31.62	10^6	10^9	1.07×10^{301}	$4.02 \times 10^{2,567}$
10,000	13.29	100.00	10^8	10^{12}	$0.99 \times 10^{3,010}$	$2.85 \times 10^{35,659}$

Figure 3.1 Growth rates of some polynomial and exponential functions.

A brief examination of the effects of improved computer technology on algorithms is even more revealing in understanding the impact of various complexity functions. Consider a polynomial-time algorithm whose complexity is $O(n^2)$. Suppose that the algorithm is able to solve a problem of size n_1 in 1 hour on a computer with speed of s_1 instructions per second. If we increase the speed of the computer to s_2, then $(n_2/n_1)^2 = s_2/s_1$ specifies the size n_2 of the problem that the algorithm can solve in the same time. Consequently, a 100-fold increase in computer speed would allow us to solve problems that are 10 *times* larger. Now consider an exponential-time algorithm with a complexity of $O(2^n)$. As before, let n_1 and n_2 denote the problem sizes solved on a computer with speeds s_1 and s_2 in 1 hour of computation time. Then $s_2/s_1 = 2^{n_2}/2^{n_1}$. Alternatively, $n_2 = n_1 + \log(s_2/s_1)$. In this case, a 100-fold increase in computer speed would allow us to solve problems that are only about 7 *units* larger. This discussion shows that a substantial increase in computer speed allows us to solve problems by polynomial-time algorithms that are larger by a multiplicative factor; for exponential-time algorithms we obtain only additive improvements. Consequently, improved hardware capabilities of computers can have only a marginal impact on the problem-solving ability of exponential-time algorithms.

Let us pause to summarize our discussion of polynomial and exponential-time algorithms. In the realm of complexity theory, our objective is to obtain polynomial-time algorithms, and within this domain our objective is to obtain an algorithm with the smallest possible growth rate, because an algorithm with smaller growth rate is likely to permit us to solve larger problems in the same amount of computer time (depending on the associated constants). For example, we prefer $O(\log n)$ to $O(n^k)$ for any $k > 0$, and we prefer $O(n^2)$ to $O(n^3)$. However, running times involving more than one parameter, such as $O(n\, m \log n)$ and $O(n^3)$, might not be comparable. If $m < n^2/\log n$, then $O(n\, m \log n)$ is superior; otherwise, $O(n^3)$ is superior.

Can we say that a polynomial-time algorithm with a smaller growth rate would run faster in practice, or even that a polynomial-time algorithm would empirically outperform an exponential-time algorithm? Although this statement is generally true, there are many exceptions to the rule. A classical exception is provided by the simplex method and Khachian's "ellipsoid" algorithm for solving linear programming problems. The simplex algorithm is known to be an exponential-time algorithm, but in practice it runs much faster than Khachian's polynomial-time algorithm. Many of these exceptions can be explained by the fact that the worst-case complexity is greatly inferior to the average complexity of some algorithms, while for other algorithms the worst-case complexity and the average complexity might be comparable. As a consequence, considering worst-case complexity as synonymous with average complexity can lead to incorrect conclusions.

Sometimes, we might not succeed in developing a polynomial-time algorithm for a problem. Indeed, despite their best efforts spanning several decades, researchers have been unable to develop polynomial time algorithm for a huge collection of important combinatorial problems; all known algorithms for these problems are exponential-time algorithms. However, the research community has been able to show that most of these problems belong to a class of problems, called *$\mathcal{NP}$-complete problems*, that are equivalent in the sense that if there exists a polynomial-time algorithm for one problem, there exists a polynomial-time algorithm for every other $\mathcal{NP}$-complete problem. Needless to say, developing a polynomial-time algorithm for some $\mathcal{NP}$-complete problem is one of the most challenging and intriguing issues facing the research community; the available evidence suggests that no such algorithm exists. We discuss the theory of $\mathcal{NP}$-completeness in greater detail in Appendix B.

Big Ω and Big Θ Notation

The big O notation that we introduced earlier in this section is but one of several convenient notational devices that researchers use in the analysis of algorithms. In this subsection we introduce two related notational constructs: the big Ω (big omega) notation and the big Θ (big theta) notation.

Just as the big O notation specifies an upper bound on an algorithm's performance, the big Ω notation specifies a lower bound on the running time.

An algorithm is said to be $\Omega(f(n))$ if for some numbers c' and n_0 and all $n \geq n_0$, the algorithm takes at least $c'f(n)$ time on some problem instance.

The reader should carefully note that the big O notation and the big Ω notation are defined in somewhat different ways. If an algorithm runs in $O(f(n))$ time, *every* instance of the problem of size n takes *at most* $cf(n)$ time for a constant c. On the other hand, if an algorithm runs in $\Omega(f(n))$ time, *some* instance of size n takes *at least* $c'f(n)$ time for a constant c'.

The big Θ (big theta) notation provides both a lower and an upper bound on an algorithm's performance.

An algorithm is said to be $\Theta(f(n))$ if the algorithm is both $O(f(n))$ and $\Omega(f(n))$.

We generally prove an algorithm to be an $O(f(n))$ algorithm and then try to see whether it is also an $\Omega(f(n))$ algorithm. Notice that the proof that the algorithm requires $O(f(n))$ time does not imply that it would actually take $cf(n)$ time to solve all classes of problems of the type we are studying. The upper bound of $cf(n)$ could be "too loose" and might never be achieved. There is always a distinct possibility that by conducting a more clever analysis of the algorithm we might be able to improve the upper bound of $cf(n)$, replacing it by a "tighter" bound. However, if we prove that the algorithm is also $\Omega(f(n))$, we know that the upper bound of $cf(n)$ is "tight" and cannot be improved by more than a constant factor. This result would imply that the algorithm can actually achieve its upper bound and no tighter bound on the algorithm's running time is possible.

Potential Functions and Amortized Complexity

An algorithm typically performs some basic operations repetitively with each operation performing a sequence of steps. To bound the running time of the algorithm we must bound the running time of each of its basic operations. We typically bound the total number of steps associated with an operation using the following approach: We obtain a bound on the number of steps per operation, obtain a bound on the number of operations, and then take a product of the two bounds. In some of the algorithms that we study in this book, the time required for a certain operation might vary depending on the problem data and/or the stage the algorithm is in while solving a problem. Although the operation might be easy to perform most of the time, occasionally it might be quite expensive. When this happens and we consider the time for the operation corresponding to the worst-case situation, we could greatly overestimate the running time of the algorithm. In this situation, a more global analysis is required to obtain a "tighter" bound on the running time of the operation. Rather than bounding the number of steps per operation and the number of operations executed in the algorithm, we should try to bound the total number of steps over all executions of these operations. We often carry out this type of worst-case analysis using a *potential function* technique.

We illustrate this concept on a problem of inserting and removing data from a data structure known as a *stack* (see Appendix A for a discussion of this data structure). On a stack S, we perform two operations:

push(x, S). Add element x to the *top* of the stack S.

popall(S). Pop (i.e., take out) every element of S.

The operation *push(x, S)* requires $O(1)$ time and the operation popall(S) requires $O(|S|)$ time. Now assume that starting with an empty state, we perform a sequence of n operations in which push and popall operations occur in a random order. What is the worst-case complexity of performing this sequence of n operations?

A naive worst-case analysis of this problem might proceed as follows. Since we require at most n *push* operations, and each push takes $O(1)$ time, the push operations require a total of $O(n)$ time. A popall requires $O(|S|)$ time and since $|S| \leq n$, the complexity of this operation is $O(n)$. Since our algorithm can invoke at most n popall operations, these operations take a total of $O(n^2)$ time. Consequently, a random sequence of n push and popall operations has a worst-case complexity of $O(n^2)$.

However, if we look closely at the arguments we will find that the bound of $O(n^2)$ is a substantial overestimate of the algorithm's computational requirements. A popall operation pops $|S|$ items from the stack, one by one until the stack becomes empty. Now notice that any element that is popped from the stack must have been pushed into the stack at some point, and since the number of push operations is at most n, the total number of elements popped out of the stack must be at most n. Consequently, the total time taken by all popall operations is $O(n)$. We can therefore conclude that a random sequence of n push and popall operations has a worst-case complexity of $O(n)$.

Let us provide a formal framework, using *potential functions*, for conducting the preceding arguments. Potential function techniques are general-purpose techniques for establishing the complexity of an algorithm by analyzing the effects of different operations on an appropriately defined function. The use of potential functions enables us to define an "accounting" relationship between the occurrences of various operations of an algorithm so that we can obtain a bound on the operations that might be difficult to obtain using other arguments.

Let $\phi(k) = |S|$ denote the number of items in the stack at the end of the kth step; for the purpose of this argument we define a step as either a push or a popall operation. We assume that we perform the popall step on a nonempty stack; for otherwise, it requires $O(1)$ time. Initially, $\phi(0) = 0$. Each push operation increases $\phi(k)$ by 1 unit and takes 1 unit of time. Each popall step decreases $\phi(k)$ by at least 1 unit and requires time proportional to $|\phi(k)|$. Since the total increase in ϕ is at most n (because we invoke at most n push steps), the total decrease in ϕ is also at most n. Consequently, the total time taken by all push and popall steps is $O(n)$.

This argument is fairly representative of the potential function arguments. Our objective was to bound the time for the popalls. We did so by defining a potential function that decreases whenever we perform a popall. The potential increases only when we perform a push. Thus we can bound the total decrease by the total increase in ϕ. In general, we bound the number of steps of one type by using known bounds on the number of steps of other types.

The analysis we have just discussed is related to the concept known as *amortized complexity*. An operation is said to be of amortized complexity $O(f(n))$ if the time to perform a sequence of k operations is $O(kf(n))$ for sufficiently large k. In our preceding example, the worst-case complexity of performing k popalls for $k \geq n$ is $O(k)$; hence the amortized complexity of the popall operation is $O(1)$. Roughly speaking, the amortized complexity of an operation is the "average" worst-case complexity of the operation so that the total obtained using this average will indeed be an upper bound on the number of steps performed by the algorithm.

Parameter Balancing

We frequently use the parameter balancing technique in situations when the running time of an algorithm is a function of a parameter k and we wish to determine the value of k that gives the smallest running time. To be more specific, suppose that the running time of an algorithm is $O(f(n, m, k) + g(n, m, k))$ and we wish to determine an optimal value of k. We shall assume that $f(n, m, k) \geq 0$ and $g(n, m, k) \geq 0$ for all feasible values of k. The optimization problem is easy to solve if the functions $f(n, m, k)$ and $g(n, m, k)$ are both either monotonically increasing or monotonically decreasing in k. In the former case, we set k to the smallest possible value, and in the latter case, we set k to the largest possible value. Finding the optimal value of k is more complex if one function is monotonically decreasing and the other function is monotonically increasing. So let us assume that $f(n, m, k)$ is monotonically decreasing in k and $g(n, m, k)$ is monotonically increasing in k.

One method for selecting the optimal value of k is to use differential calculus. That is, we differentiate $f(n, m, k) + g(n, m, k)$ with respect to k, set the resulting expression equal to zero, and solve for k. A major drawback of this approach is that finding a value of k that will set the expression to value zero, and so determine the optimal value of k, is often a difficult task. Consider, for example, a shortest path algorithm (which we discuss in Section 4.7) that runs in time $O(m \log_k n + nk \log_k n)$. In this case, choosing the optimal value of k is not trivial. We can restate the algorithm's time bound as $O((m \log n + nk \log n)/\log k)$. The derivative of this expression with respect to k is

$$(nk \log n \log k - m \log n - nk \log n)/k(\log k)^2.$$

Setting this expression to zero, we obtain

$$m + nk - nk \log k = 0.$$

Unfortunately, we cannot solve this equation in closed form.

The parameter balancing technique is an alternative method for determining the "optimal value" of k and is based on the idea that it is not necessary to select a value of k that minimizes $f(n, m, k) + g(n, m, k)$. Since we are evaluating the performance of algorithms in terms of their worst-case complexity, it is sufficient to select a value of k for which $f(n, m, k) + g(n, m, k)$ is within a constant factor of the optimal value. The parameter balancing technique determines a value of k so that $f(n, m, k) + g(n, m, k)$ is at most twice the minimum value.

In the parameter balancing technique, we select k^* so that $f(n, m, k^*) = g(n, m, k^*)$. Before giving a justification of this approach, we illustrate it on two examples. We first consider the $O(m \log_k n + nk \log_k n)$ time shortest path algorithm that we mentioned earlier. We first note that $m \log_k n$ is a decreasing function of k and $nk \log_k n$ is an increasing function of k. Therefore, the parameter balancing technique is appropriate. We set $m \log_{k^*} n = nk^* \log_{k^*} n$, which gives $k^* = m/n$. Consequently, we achieve the best running time of the algorithm, $O(m \log_{m/n} n)$, by setting $k = m/n$.

Our second example concerns a maximum flow algorithm whose running time is $O((n^3/k)(\log k) + nm(\log k))$. We set

$$\frac{n^3}{k^*} \log k^* = nm \log k^*,$$

which gives $k^* = n^2/m$. Therefore, the best running time of this maximum flow algorithm is $O(nm \log(n^2/m))$. In Exercise 3.13 we discuss more examples of the parameter balancing technique.

We now justify the parameter balancing technique. Suppose we select k^* so that $f(n, m, k^*) = g(n, m, k^*)$. Let $\lambda^* = f(n, m, k^*) + g(n, m, k^*)$. Then for any $k < k^*$,

$$f(n, m, k) + g(n, m, k) \geq f(n, m, k) \geq f(n, m, k^*) = \lambda^*/2. \tag{3.1}$$

The second inequality follows from the fact that the function $f(n, m, k)$ is monotonically decreasing in k. Similarly, for any $k > k^*$,

$$f(n, m, k) + g(n, m, k) \geq g(n, m, k) \geq g(n, m, k^*) = \lambda^*/2. \tag{3.2}$$

The expressions (3.1) and (3.2) imply that for any k,

$$f(n, m, k) + g(n, m, k) \geq \lambda^*/2.$$

This result establishes the fact that $\lambda^* = f(n, m, k^*) + g(n, m, k^*)$ is within a factor of 2 of the minimum value of $f(n, m, k) + g(n, m, k)$.

3.3 DEVELOPING POLYNOMIAL-TIME ALGORITHMS

Researchers frequently employ four important approaches for obtaining polynomial-time algorithms for network flow problems: (1) *a geometric improvement* approach, (2) *a scaling* approach, (3) *a dynamic programming* approach, and (4) *a binary search* approach. In this section we briefly outline the basic ideas underlying these four approaches.

Geometric Improvement Approach

The geometric improvement approach permits us to show that an algorithm runs in polynomial time if at every iteration it makes an improvement in the objective function value proportional to the difference between the objective values of the current and optimal solutions. Let H be the difference between the maximum and minimum objective function values of an optimization problem. For most network problems, H is a function of n, m, C, and U. For example, in the maximum flow problem $H = mU$, and in the minimum cost flow problem $H = mCU$. We also assume that the optimal objective function value is integer.

Theorem 3.1. *Suppose that z^k is the objective function value of some solution of a minimization problem at the kth iteration of an algorithm and z^* is the minimum objective function value. Furthermore, suppose that the algorithm guarantees that for every iteration k,*

$$(z^k - z^{k+1}) \geq \alpha(z^k - z^*) \tag{3.3}$$

(i.e., the improvement at iteration $k + 1$ is at least α times the total possible improvement) for some constant α with $0 < \alpha < 1$ (which is independent of the problem data). Then the algorithm terminates in $O((\log H)/\alpha)$ iterations.

Proof. The quantity $(z^k - z^*)$ represents the total possible improvement in the objective function value after the kth iteration. Consider a consecutive sequence of $2/\alpha$ iterations starting from iteration k. If each iteration of the algorithm improves the objective function value by at least $\alpha(z^k - z^*)/2$ units, the algorithm would determine an optimal solution within these $2/\alpha$ iterations. Suppose, instead, that at some iteration $q + 1$, the algorithm improves the objective function value by less than $\alpha(z^k - z^*)/2$ units. In other words,

$$z^q - z^{q+1} \leq \alpha(z^k - z^*)/2. \tag{3.4}$$

The inequality (3.3) implies that

$$\alpha(z^q - z^*) \leq z^q - z^{q+1}. \tag{3.5}$$

The inequalities (3.4) and (3.5) imply that

$$(z^q - z^*) \leq (z^k - z^*)/2,$$

so the algorithm has reduced the total possible improvement $(z^k - z^*)$ by a factor at least 2. We have thus shown that within $2/\alpha$ consecutive iterations, the algorithm either obtains an optimal solution or reduces the total possible improvement by a factor of at least 2. Since H is the maximum possible improvement and every objective function value is an integer, the algorithm must terminate within $O((\log H)/\alpha)$ iterations. ◆

We have stated this result for the minimization version of optimization problems. A similar result applies to the maximization problems.

The geometric improvement approach might be summarized by the statement "network algorithms that have a geometric convergence rate are polynomial-time algorithms." To develop polynomial-time algorithms using this approach, we look for local improvement techniques that lead to large (i.e., fixed percentage) improvements in the objective function at every iteration. The maximum augmenting path algorithm for the maximum flow problem discussed in Section 7.3 and the maximum improvement algorithm for the minimum cost flow problem discussed in Section 9.6 provide two examples of this approach.

Scaling Approach

Researchers have used scaling methods extensively to derive polynomial-time algorithms for a wide variety of network and combinatorial optimization problems. Indeed, for problems that satisfy the similarity assumption, the scaling-based algorithms achieve the best worst-case running time for most of the network optimization problems we consider in this book.

We shall describe the simplest form of scaling, which we call *bit-scaling*. In the bit-scaling technique, we represent the data as binary numbers and solve a problem P parametrically as a sequence of problems $P_1, P_2, P_3, \ldots, P_K$: The problem P_1 approximates data to the first most significant bit, the problem P_2 approximates data to the first two most significant bits, and each successive problem is a better approximation, until $P_K = P$. Moreover, for each $k = 2, \ldots, K$, the optimal solution of problem P_{k-1} serves as the starting solution for problem P_k. The scaling technique is useful whenever reoptimization from a good starting solution is more efficient than solving the problem from scratch.

For example, consider a network flow problem whose largest arc capacity has value U. Let $K = \lceil \log U \rceil$ and suppose that we represent each arc capacity as a K-bit binary number, adding leading zeros if necessary to make each capacity K bits long. Then the problem P_k would consider the capacity of each arc as the k leading bits in its binary representation. Figure 3.2 illustrates an example of this type of scaling.

The manner of defining arc capacities easily implies the following property.

Property 3.2. *The capacity of an arc in P_k is twice that in P_{k-1} plus* 0 *or* 1.

The algorithm shown in Figure 3.3 encodes a generic version of the bit-scaling technique.

This approach is very robust, and variants of it have led to improved algorithms for both the maximum flow and minimum cost flow problems. This approach works well for these applications, in part, for the following reasons:

1. The problem P_1 is generally easy to solve.
2. The optimal solution of problem P_{k-1} is an excellent starting solution for problem P_k since P_{k-1} and P_k are quite similar. Therefore, we can easily reoptimize the problem starting from the optimal solution of P_{k-1} to obtain an optimal solution of P_k.

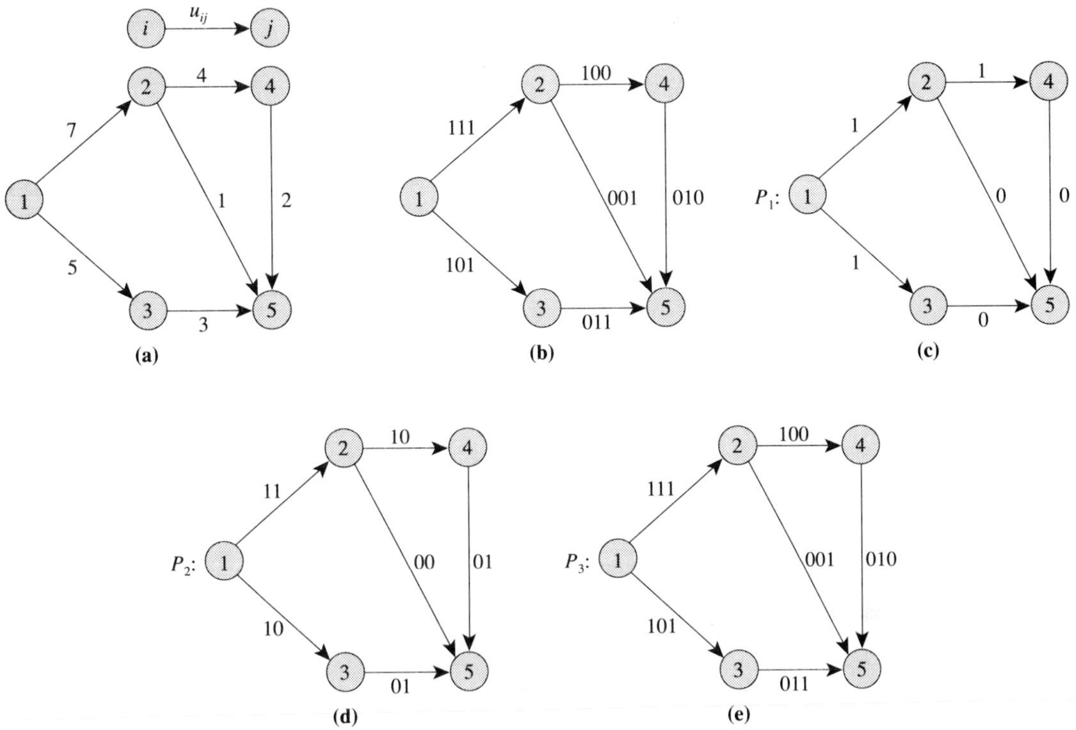

Figure 3.2 Examples of a bit-scaling technique: (a) network with arc capacities; (b) network with binary expansions of arc capacities; (c)–(e) problems P_1, P_2, and P_3.

3. The number of reoptimization problems we solve is $O(\log C)$ or $O(\log U)$. Thus for this approach to work, reoptimization needs to be only a little more efficient (i.e., by a factor of $\log C$ or $\log U$) than optimization.

Consider, for example, the maximum flow problem. Let v_k denote the maximum flow value for problem P_k and let x_k denote an arc flow corresponding to v_k. In the problem P_k, the capacity of an arc is twice its capacity in P_{k-1} plus 0 or 1. If we multiply the optimal flow x_{k-1} of P_{k-1} by 2, we obtain a feasible flow for P_k. Moreover, $v_k - 2v_{k-1} \le m$ because multiplying the flow x_{k-1} by 2 takes care of the doubling of the capacities and the additional 1's can increase the maximum flow value by at most m units (if we add 1 to the capacity of any arc, we increase the

```
algorithm bit-scaling;
begin
    obtain an optimal solution of P_1;
    for k: = 2 to K do
    begin
        reoptimize using the optimal solution of P_{k-1} to obtain an optimal solution of P_k;
    end;
end;
```

Figure 3.3 Typical bit-scaling algorithm.

maximum flow from the source to the sink by at most 1). In general, it is easier to reoptimize such a maximum flow problem than to solve a general problem from scratch. For example, the classical labeling algorithm as discussed in Section 6.5 would perform the reoptimization in at most m augmentations, requiring $O(m^2)$ time. Therefore, the scaling version of the labeling algorithm runs in $O(m^2 \log U)$ time, improving on the running time $O(nmU)$ of the nonscaling version. The former time bound is polynomial and the latter bound is only pseudopolynomial. Thus this simple bit-scaling algorithm improves the running time dramatically.

An alternative approach to scaling considers a sequence of problems $P(1)$, $P(2)$, . . . , $P(K)$, each involving the original data, but in this case we do not solve the problem $P(k)$ optimally, but solve it approximately, with an error of Δ_k. Initially, Δ_1 is quite large, and it subsequently converges geometrically to 0. Usually, we can interpret an error of Δ_k as follows. From the current nearly optimal solution x^k, there is a way of modifying some or all of the data by at most Δ_k units so that the resulting solution is optimal. Our discussion of the capacity scaling algorithm for the maximum flow problem in Section 7.3 illustrates this type of scaling.

Dynamic Programming

Researchers originally conceived of dynamic programming as a stagewise optimization technique. However, for our purposes in this book, we prefer to view it as a "table-filling" approach in which we complete the entries of a two-dimensional tableau using a recursive relationship. Perhaps the best way to explain this approach is through several illustrations.

Computing Binomial Coefficients

In many application of combinatorics, for example in elementary probability, we frequently wish to determine the number ${}^pC_q$ of different combinations of p objects taken q at a time for some given values of p and q ($p \geq q$). As is well known, ${}^pC_q = p!/((p - q)!q!)$. Suppose that we wish to make this computation using only the mathematical operation of addition and using the fact that the combination function ${}^pC_q$ satisfies the following recursive relationship:

$$ {}^iC_j = {}^{i-1}C_j + {}^{i-1}C_{j-1}. \tag{3.6} $$

To solve this problem, we define a lower triangular table $D = \{d(i, j)\}$ with p rows and q columns: Its entries, which we would like to compute, will be $d(i, j) = {}^iC_j$ for $i \geq j$. We will fill in the entries in the table by scanning the rows in the order 1 through p; when scanning each row i, we scan its columns in the order 1 through i. Note that we can start the computations by setting the ith entry $d(i, 1) = {}^iC_1$ in the first column to value i since there are exactly i ways to select one object from a collection of i objects. Observe that whenever we scan the element (i, j) in the table, we have already computed the entries ${}^{i-1}C_j$ and ${}^{i-1}C_{j-1}$, and their sum yields $d(i, j)$. So we always have the available information to compute the entries in the table as we reach them. When we have filled the entire table, the entry $d(p, q)$ gives us the desired answer to our problem.

Knapsack Problem

We can also illustrate the dynamic programming approach on another problem, known as the *knapsack problem*, which is a classical model in the operations research literature. A hiker must decide which goods to include in her knapsack on a forthcoming trip. She must choose from among p objects: Object i has weight w_i (in pounds) and a utility u_i to the hiker. The objective is to maximize the utility of the hiker's trip subject to the weight limitation that she can carry no more than W pounds. This knapsack problem has the following formulation as an integer program:

$$\text{Maximize} \sum_{i=1}^{p} u_i x_i \tag{3.7a}$$

subject to

$$\sum_{i=1}^{p} w_i x_i \leq W, \tag{3.7b}$$

$$x_i = \{0, 1\} \qquad \text{for all } i. \tag{3.7c}$$

To solve the knapsack problem, we construct a $p \times W$ table D whose elements $d(i, j)$ are defined as follows:

$d(i, j)$: The maximum utility of the selected items if we restrict our selection to the items 1 through i and impose a weight restriction of j.

Clearly, our objective is to determine $d(p, W)$. We determine this value by computing $d(i, j)$ for increasing values of i and, for a fixed value of i, for increasing values of j. We now develop the recursive relationship that would allow us to compute $d(i, j)$ from those elements of the tableau that we have already computed. Note that any solution restricted to the items 1 through i, either (1) does not use item i, or (2) uses this item. In case (1), $d(i, j) = d(i - 1, j)$. In case (2), $d(i, j) = u_i + d(i - 1, j - w_i)$ for the following reason. The first term in this expression represents the value of including item i in the knapsack and the second term denotes the optimal value obtained by allocating the remaining capacity of $j - w_i$ among the items 1 through $i - 1$. We have thus shown that

$$d(i, j) = \max\{d(i - 1, j), u_i + d(i - 1, j - w_i)\}.$$

When carrying out these computations, we also record the decision corresponding to each $d(i, j)$ (i.e., whether $x_i = 0$ or $x_i = 1$). These decisions allow us to construct the solution for any $d(i, j)$, including the desired solution for $d(p, W)$.

In both these illustrations of dynamic programming, we scanned rows of the table in ascending order and for each fixed row, we scanned columns in ascending order. In general, we could scan the rows and columns of the table in either ascending or descending order as long as the recursive relationship permits us to determine the entries needed in the recursion from those we have already computed.

To conclude this brief discussion, we might note that much of the traditional literature in dynamic programming views the problem as being composed of "stages" and "states" (or possible outcomes within each state). Frequently, the stages correspond to points in time (this is the reason that this topic has become known as dynamic programming). To reconceptualize our tabular approach in this stage and state framework, we would view each row as a stage and each column within each row as a possible state at that stage. For both the binomial coefficient and knapsack applications that we have considered, each stage corresponds to a restricted set of objects (items): In each case stage i corresponds to a restricted problem containing only the first i objects. In the binomial coefficient problem, the states are the number of elements in a subset of the i objects; in the knapsack problem, the states are the possible weights that we could hold in a knapsack containing only the first i items.

Binary Search

Binary search is another popular technique for obtaining polynomial-time algorithms for a variety of network problems. Analysts use this search technique to find, from among a set of feasible solutions, a solution satisfying "desired properties." At every iteration, binary search eliminates a fixed percentage (as the name binary implies, typically, 50 percent) of the solution set, until the solution set becomes so small that each of its feasible solutions is guaranteed to be a solution with the desired properties.

Perhaps the best way to describe the binary search technique is through examples. We describe two examples. In the first example, we wish to find the telephone number of a person, say James Morris, in a phone book. Suppose that the phone book contains p pages and we wish to find the page containing James Morris's phone number. The following "divide and conquer" search strategy is a natural approach. We open the phone book to the middle page, which we suppose is page x. By viewing the first and last names on this page, we reach one of the following three conclusions: (1) page x contains James Morris's telephone number, (2) the desired page is one of pages 1 through $x - 1$, or (3) the desired page is one of pages $x + 1$ to p. In the second case, we would next turn to the middle of the pages 1 through $x - 1$, and in the third case, we would next turn to the middle of the pages $x + 1$ through p. In general, at every iteration, we maintain an interval $[a, b]$ of pages that are guaranteed to contain the desired phone number. Our next trial page is the middle page of this interval, and based on the information contained on this page, we eliminate half of the pages from further consideration. Clearly, after $O(\log p)$ iterations, we will be left with just one page and our search would terminate. If we are fortunate, the search would terminate even earlier.

As another example, suppose that we are given a continuous function $f(x)$ satisfying the properties that $f(0) < 0$ and $f(1) > 0$. We want to determine an interval of size at most $\epsilon > 0$ that contains a *zero* of the function, that is, a value of x for which $f(x) = 0$ (to within the accuracy of the computer we are using). In the first iteration, the interval $[0, 1]$ contains a zero of the function $f(x)$, and we evaluate the function at the midpoint of this interval, that is, at the point 0.5. Three outcomes are possible: (1) $f(0.5) = 0$, (2) $f(0.5) < 0$, and (3) $f(0.5) > 0$. In the first case, we have found a zero x and we terminate the search. In the second case, the continuity property of the function $f(x)$ implies that the interval $[0.5, 1]$ contains a zero of the function, and in the third case the interval $[0, 0.5]$ contains a zero. In the second and third cases, our next trial point is the midpoint of the resulting interval. We repeat this process, and eventually, when the interval size is less than ϵ, we discontinue the search. As the reader can verify, this method will terminate within $O(\log(1/\epsilon))$ iterations.

In general, we use the binary search technique to identify a desired value of a parameter among an interval of possible values. The interval $[l, u]$ is defined by a lower limit l and an upper limit u. In the phone book example, we wanted to identify a page that contains a specific name, and in the zero value problem we wanted to identify a value of x in the range [0, 1] for which $f(x)$ is zero. At every iteration we perform a test at the midpoint $(l + u)/2$ of the interval, and determine whether the desired parameter lies in the range $[l, (l + u)/2]$ or in the range $[(l + u)/2, u]$. In the former case, we reset the upper limit to $(l + u)/2$, and in the latter case, we reset the lower limit to $(l + u)/2$. We might note that eliminating one-half of the interval requires that the problem satisfy certain properties. For instance, in the phone book example, we used the fact that the names in the book are arranged alphabetically, and in the zero-value problem we used the fact that the function $f(x)$ is continuous. We repeat this process with the reduced interval and keep reapplying the procedure until the interval becomes so small that it contains only points that are desired solutions. If w_{max} denotes the maximum (i.e., starting) width of the interval (i.e., $u - l$) and w_{min} denotes the minimum width of the interval, the binary search technique required $O(\log(w_{max}/w_{min}))$ iterations.

In most applications of the binary search technique, we perform a single test and eliminate half of the feasible interval. The worst-case complexity of the technique remains the same, however, even if we perform several, but a constant number, of tests at each step and eliminate a constant portion (not necessarily 50 percent) of the feasible interval (in Exercise 3.23 we discuss one such application). Although we typically use the binary search technique to perform a search over a single parameter, a generalized version of the method would permit us to search over multiple parameters.

3.4 SEARCH ALGORITHMS

Search algorithms are fundamental graph techniques that attempt to find all the nodes in a network satisfying a particular property. Different variants of search algorithms lie at the heart of many maximum flow and minimum cost flow algorithms. The applications of search algorithms include (1) finding all nodes in a network that are reachable by directed paths from a specific node, (2) finding all the nodes in a network that can reach a specific node t along directed paths, (3) identifying all connected components of a network, and (4) determining whether a given network is bipartite. To illustrate some of the basic ideas of search algorithms, in this section we discuss only the first two of these applications; Exercises 3.41 and 3.42 consider the other two applications.

Another important application of search algorithms is to identify a directed cycle in a network, and if the network is acyclic, to reorder the nodes $1, 2, \ldots, n$ so that for each arc $(i, j) \in A$, $i < j$. We refer to any such order as a *topological ordering*. Topological orderings prove to be essential constructs in several applications, such as project scheduling (see Chapter 19). They are also useful in the design of certain algorithms (see Section 10.5). We discuss topological ordering later in this section.

To illustrate the basic ideas of search algorithms, suppose that we wish to find all the nodes in a network $G = (N, A)$ that are reachable along directed paths from a distinguished node s, called the *source*. A search algorithm *fans out* from the source and identifies an increasing number of nodes that are reachable from the source. At every intermediate point in its execution, the search algorithm designates all the nodes in the network as being in one of the two states: *marked* or *unmarked*. The marked nodes are known to be reachable from the source, and the status of unmarked nodes has yet to be determined. Note that if node i is marked, node j is unmarked, and the network contains the arc (i, j), we can mark node j; it is reachable from source via a directed path to node i plus arc (i, j). Let us refer to arc (i, j) as *admissible* if node i is marked and node j is unmarked, and refer to it as *inadmissible* otherwise. Initially, we mark only the source node. Subsequently, by examining admissible arcs, the search algorithm will mark additional nodes. Whenever the procedure marks a new node j by examining an admissible arc (i, j), we say that node i is a *predecessor* of node j [i.e., $pred(j) = i$]. The algorithm terminates when the network contains no admissible arcs.

The search algorithm traverses the marked nodes in a certain order. We record this traversal order in an array *order*: the entry *order(i)* is the ith node in the traversal. Figure 3.4 gives a formal description of the search algorithm. In the algorithmic description, *LIST* represents the set of marked nodes that the algorithm has yet to examine in the sense that some admissible arcs might emanate from them. When the algorithm terminates, it has marked all the nodes in G that are reachable from s via a directed path. The predecessor indices define a tree consisting of marked nodes. We call this tree a *search tree*. Figure 3.5(b) and (c), respectively, depict two search trees for the network shown in Figure 3.5(a).

To identify admissible arcs, we need to be able to access the arcs of the network and determine whether or not they connect a marked and unmarked node. To do so we must design a data structure for storing the arcs and assessing the status of

```
algorithm search;
begin
    unmark all nodes in N;
    mark node s;
    pred(s) : = 0;
    next : = 1;
    order(next) : = s;
    LIST : = {s}
    while LIST ≠ Ø do
    begin
        select a node i in LIST;
        if node i is incident to an admissible arc (i, j) then
        begin
            mark node j;
            pred(j) : = i;
            next : = next + 1;
            order(j) : = next
            add node j to LIST;
        end
        else delete node i from LIST;
    end;
end;
```

Figure 3.4 Search algorithm.

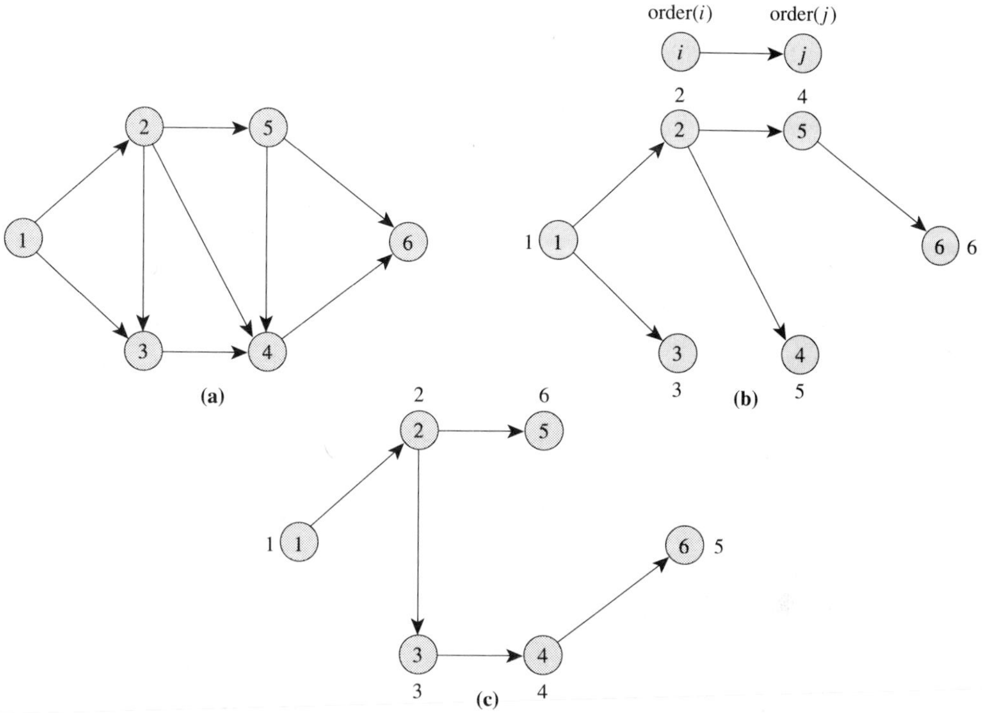

Figure 3.5 Two search trees of a network.

their incident nodes. In later chapters, too, we need the same data structure to implement maximum flow and minimum cost flow algorithms. We use the *current-arc* data structure, defined as follows, for this purpose. We maintain with each node i the adjacency list $A(i)$ of arcs emanating from it (see Section 2.2 for the definition of adjacency list). For each node i, we define a *current arc* (i, j), which is the next candidate arc that we wish to examine. Initially, the current arc of node i is the first arc in $A(i)$. The search algorithm examines the list $A(i)$ sequentially: At any stage, if the current arc is inadmissible, the algorithm designates the next arc in the arc list as the current arc. When the algorithm reaches the end of the arc list, it declares that the node has no admissible arc. Note that the order in which the algorithm examines the nodes depends on how we have arranged the arcs in the arc adjacency lists $A(i)$. We assume here, as well as elsewhere in this book, that we have ordered the arcs in $A(i)$ in the increasing order of their head nodes [i.e., if (i, j) and (i, k) are two consecutive arcs in $A(i)$, then $j < k$].

It is easy to show that the search algorithm runs in $O(m + n) = O(m)$ time. Each iteration of the while loop either finds an admissible arc or does not. In the former case, the algorithm marks a new node and adds it to LIST, and in the latter case it deletes a marked node from LIST. Since the algorithm marks any node at most once, it executes the while loop at most $2n$ times. Now consider the effort spent in identifying the admissible arcs. For each node i, we scan the arcs in $A(i)$ at most once. Therefore, the search algorithm examines a total of $\sum_{i \in N} |A(i)| = m$ arcs, and thus terminates in $O(m)$ time.

The algorithm, as described, does not specify the manner for examining the nodes or for adding the nodes to LIST. Different rules give rise to different search techniques. Two data structures have proven to be the most popular for maintaining LIST—a *queue* and a *stack* (see Appendix A for a discussion of these data structures)—and they give rise to two fundamental search strategies: *breadth-first search* and *depth-first search*.

Breadth-First Search

If we maintain the set LIST as a queue, we always select nodes from the front of LIST and add them to the rear. In this case the search algorithm selects the marked nodes in a first-in, first-out order. If we define the distance of a node i as the minimum number of arcs in a directed path from node s to node i, this kind of search first marks nodes with distance 1, then those with distance 2, and so on. Therefore, this version of search is called a *breadth-first search* and the resulting search tree is a *breadth-first search tree*. Figure 3.5(b) specifies the breadth-first search tree for the network shown in Figure 3.5(a). In subsequent chapters we use the following property of the breadth-first search tree whose proof is left as an exercise (see Exercise 3.30).

Property 3.3. *In the breadth-first search tree, the tree path from the source node s to any node i is a shortest path (i.e., contains the fewest number of arcs among all paths joining these two nodes).*

Depth-First Search

If we maintain the set LIST as a stack, we always select the nodes from the front of LIST and also add them to the front. In this case the search algorithm selects the marked node in a last-in, first-out order. This algorithm performs a deep probe, creating a path as long as possible, and backs up one node to initiate a new probe when it can mark no new node from the tip of the path. Consequently, we call this version of search a *depth-first search* and the resulting tree a *depth-first search tree*. The depth-first traversal of a network is also called its *preorder traversal*. Figure 3.5(c) gives the depth-first search tree for the network shown in Figure 3.5(a).

In subsequent chapters we use the following property of the depth-first search tree, which can be easily proved using induction arguments (see Exercise 3.32).

Property 3.4

(a) *If node j is a descendant of node i and $j \neq i$, then $order(j) > order(i)$.*
(b) *All the descendants of any node are ordered consecutively in sequence.*

Reverse Search Algorithm

The search algorithm described in Figure 3.4 allows us to identify all the nodes in a network that are reachable from a given node s by directed paths. Suppose that we wish to identify all the nodes in a network from which we can reach a given node t along directed paths. We can solve this problem by using the algorithm we have just described with three slight changes: (1) we initialize LIST as LIST $= \{t\}$; (2) while examining a node, we scan the incoming arcs of the node instead of its outgoing arcs; and (3) we designate an arc (i, j) as admissible if i is unmarked and j is marked. We subsequently refer to this algorithm as a *reverse search algorithm*. Whereas the (forward) search algorithm gives us a directed out-tree rooted at node s, the reverse search algorithm gives us a directed in-tree rooted at node t.

Determining Strong Connectivity

Recall from Section 2.2 that a network is strongly connected if for every pair of nodes i and j, the network contains a directed path from node i to node j. This definition implies that a network is strongly connected if and only if for any arbitrary node s, every node in G is reachable from s along a directed path and, conversely, node s is reachable from every other node in G along a directed path. Clearly, we can determine the strong connectivity of a network by two applications of the search algorithm, once applying the (forward) search algorithm and then the reverse search algorithm.

We next consider the problem of finding a topological ordering of the nodes of an acyclic network. We will show how to solve this problem by using a minor modification of the search algorithm.

Topological Ordering

Let us label the nodes of a network $G = (N, A)$ by distinct numbers from 1 through n and represent the labeling by an array *order* [i.e., order(i) gives the label of node i]. We say that this labeling is a *topological ordering* of nodes if every arc joins a lower-labeled node to a higher-labeled node. That is, for every arc $(i, j) \in A$, order(i) $<$ order(j). For example, for the network shown in Figure 3.6(a), the labeling shown in Figure 3.6(b) is not a topological ordering because (5, 4) is an arc and order(5) $>$ order(4). However, the labelings shown in Figure 3.6(c) and (d) are topological orderings. As shown in this example, a network might have several topological orderings.

Some networks cannot be topologically ordered. For example, the network shown in Figure 3.7 has no such ordering. This network is cyclic because it contains a directed cycle and for any directed cycle W we can never satisfy the condition order(i) $<$ order(j) for each $(i, j) \in W$. Indeed, acyclic networks and topological ordering are closely related. A network that contains a directed cycle has no topological ordering, and conversely, a network that possesses a topological order cannot contain a cycle. This observation shows that a network is acyclic if and only if it possesses a topological ordering of its nodes.

By using a search algorithm, we can either detect the presence of a directed cycle or produce a topological ordering of the nodes. The algorithm is fairly easy to describe. In the network G, select any node of zero indegree. Give it a label of 1, and then delete it and all the arcs emanating from it. In the remaining subnetwork select *any* node of zero indegree, give it a label of 2, and then delete it and all arcs emanating from it. Repeat this process until no node has a zero indegree. At this point, if the remaining subnetwork contains some nodes and arcs, the network G

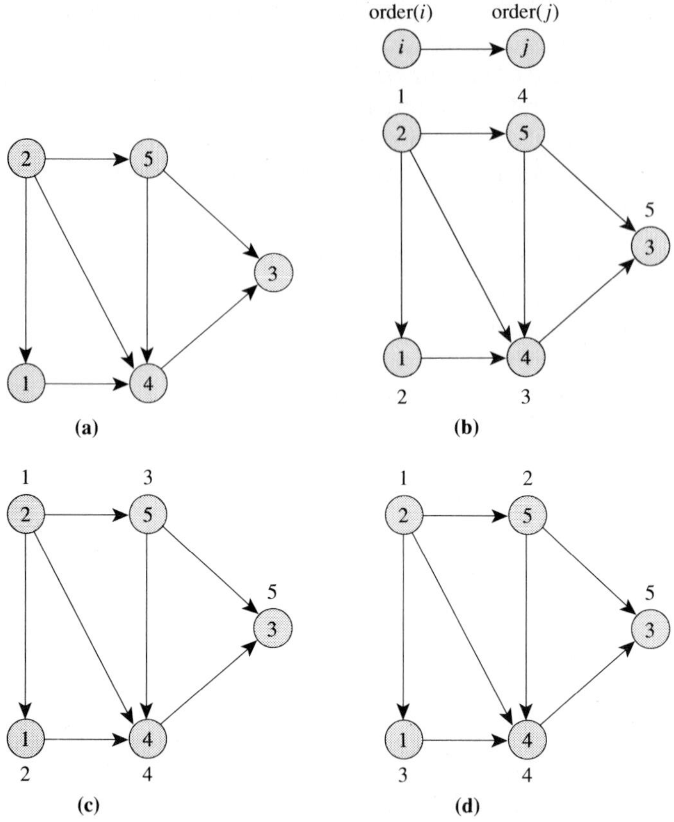

Figure 3.6 Topological ordering of nodes.

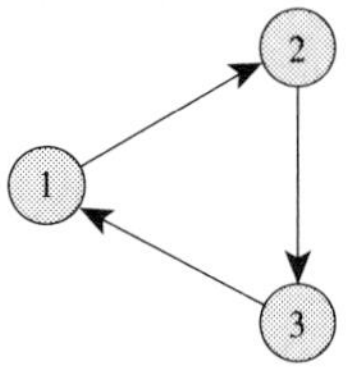

Figure 3.7 Network without a topological ordering of the nodes.

contains a directed cycle (see Exercise 3.38). Otherwise, the network is acyclic and we have assigned labels to all the nodes. Now notice that whenever we assign a label to a node at an iteration, the node has only outgoing arcs and they all must necessarily point to nodes that will be assigned higher labels in subsequent iterations. Consequently, this labeling gives a topological ordering of nodes.

We now describe an efficient implementation of this algorithm that runs in $O(m)$ time. Figure 3.8 specifies this implementation. This algorithm first computes the indegrees of all nodes and forms a set LIST consisting of all nodes with zero indegrees. At every iteration we select a node i from LIST, for every arc $(i, j) \in A(i)$ we reduce the indegree of node j by 1 unit, and if indegree of node j becomes zero, we add node j to the set LIST. [Observe that deleting the arc (i, j) from the

```
algorithm topological ordering;
begin
    for all i ∈ N do indegree(i) : = 0;
    for all (i, j) ∈ A do indegree( j) : = indegree( j) + 1;
    LIST : = Ø;
    next : = 0;
    for all i ∈ N do
        if indegree(i) = 0 then LIST : = LIST ∪ {i};
    while LIST ≠ Ø do
    begin
        select a node i from LIST and delete it;
        next : = next + 1;
        order(i) : = next;
        for all (i, j) ∈ A(i) do
        begin
            indegree( j) : = indegree( j) − 1;
            if indegree( j) = 0 then LIST : = LIST ∪ { j};
        end;
    end;
    if next < n then the network contains a directed cycle
    else the network is acyclic and the array order gives a topological order of nodes;
end;
```

Figure 3.8 Topological ordering algorithm.

network is equivalent to decreasing the indegree of node j by 1 unit.] Since the algorithm examines each node and each arc of the network $O(1)$ times, it runs in $O(m)$ time.

3.5 FLOW DECOMPOSITION ALGORITHMS

In formulating network flow problems, we can adopt either of two equivalent modeling approaches: We can define flows on arcs (as discussed in Section 1.2) or define flows on paths and cycles. For example, the arc flow shown in Figure 3.9(a) sends 7 units of flow from node 1 to node 6. Figure 3.9(b) shows a path and cycle flow

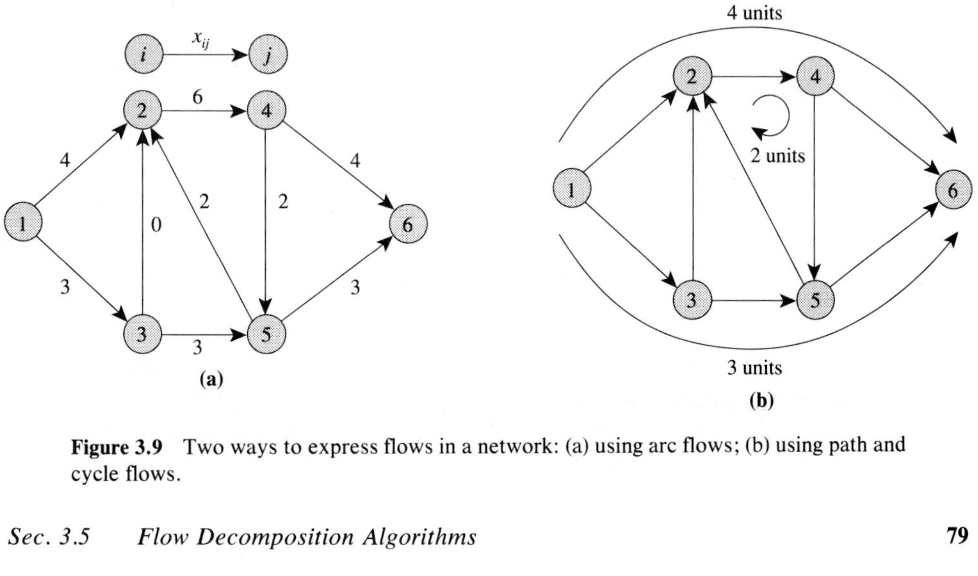

Figure 3.9 Two ways to express flows in a network: (a) using arc flows; (b) using path and cycle flows.

corresponding to this arc flow: In the path and cycle flow, we send 4 units along the path 1–2–4–6, 3 units along the path 1–3–5–6, and 2 units along the cycle 2–4–5–2. Throughout most of this book, we use the arc flow formulation; on a few occasions, however, we need to use the path and cycle flow formulation or results that stem from this modeling perspective. In this section we develop several connections between these two alternative formulations.

In this discussion, by an "arc flow" we mean a vector $x = \{x_{ij}\}$ that satisfies the following constraints:

$$\sum_{\{j:(i,j)\in A\}} x_{ij} - \sum_{\{j:(j,i)\in A\}} x_{ji} = -e(i) \qquad \text{for all } i \in N, \tag{3.8a}$$

$$0 \le x_{ij} \le u_{ij} \qquad \text{for all } (i, j) \in A. \tag{3.8b}$$

where $\sum_{i=1}^{n} e(i) = 0$. Notice that in this model we have replaced the supply/demand $b(i)$ of node i by another term, $-e(i)$; we refer to $e(i)$ as the node's *imbalance*. We have chosen this alternative modeling format purposely because some of the maximum flow and minimum cost flow algorithms described in this book maintain a solution that satisfies the flow bound constraints, but not necessarily the supply/demand constraints. The term $e(i)$ represents the inflow minus outflow of node i. If the inflow is more than outflow, $e(i) > 0$ and we say that node i is an *excess node*. If inflow is less than the outflow, $e(i) < 0$ and we say that node i is a *deficit node*. If the inflow equals outflow, we say that node i is a *balanced* node. Observe that if $e = -b$, the flow x is feasible for the minimum cost flow problem.

In the arc flow formulation discussed in Section 1.2, the basic decision variables are flows x_{ij} on the arcs $(i, j) \in A$. The path and cycle flow formulation starts with an enumeration of all directed paths P between any pair of nodes and all directed cycles W of the network. We let $\mathcal{P}$ denote the collection of all paths and $\mathcal{W}$ the collection of all cycles. The decision variables in the path and cycle flow formulation are $f(P)$, the flow on path P, and $f(W)$, the flow on cycle W; we define these variables for every directed path P in $\mathcal{P}$ and every directed cycle W in $\mathcal{W}$.

Notice that every set of path and cycle flow uniquely determines arc flows in a natural way: The flow x_{ij} on arc (i, j) equals the sum of the flows $f(P)$ and $f(W)$ for all paths P and cycles W that contain this arc. We formalize this observation by defining some new notation: $\delta_{ij}(P)$ equals 1 if arc (i, j) is contained in the path P, and is 0 otherwise. Similarly, $\delta_{ij}(W)$ equals 1 if arc (i, j) is contained in the cycle W, and is 0 otherwise. Then

$$x_{ij} = \sum_{P\in\mathcal{P}} \delta_{ij}(P)f(P) + \sum_{W\in\mathcal{W}} \delta_{ij}(W)f(W).$$

Thus each path and cycle flow determines arc flows uniquely. Can we reverse this process? That is, can we decompose any arc flow into (i.e., represent it as) path and cycle flow? The following theorem provides an affirmative answer to this question.

Theorem 3.5 (Flow Decomposition Theorem). *Every path and cycle flow has a unique representation as nonnegative arc flows. Conversely, every nonnegative arc flow x can be represented as a path and cycle flow (though not necessarily uniquely) with the following two properties:*

(a) *Every directed path with positive flow connects a deficit node to an excess node.*
(b) *At most $n + m$ paths and cycles have nonzero flow; out of these, at most m cycles have nonzero flow.*

Proof. In the light of our previous observations, we need to establish only the converse assertions. We give an algorithmic proof to show how to decompose any arc flow x into a path and cycle flow. Suppose that i_0 is a deficit node. Then some arc (i_0, i_1) carries a positive flow. If i_1 is an excess node, we stop; otherwise, the mass balance constraint (3.8a) of node i_1 implies that some other arc (i_1, i_2) carries positive flow. We repeat this argument until we encounter an excess node or we revisit a previously examined node. Note that one of these two cases will occur within n steps. In the former case we obtain a directed path P from the deficit node i_0 to some excess node i_k, and in the latter case we obtain a directed cycle W. In either case the path or the cycle consists solely of arcs with positive flow. If we obtain a directed path, we let $f(P) = \min\{-e(i_0), e(i_k), \min\{x_{ij} : (i, j) \in P\}\}$ and redefine $e(i_0) = e(i_0) + f(P)$, $e(i_k) = e(i_k) - f(P)$, and $x_{ij} = x_{ij} - f(P)$ for each arc (i, j) in P. If we obtain a directed cycle W, we let $f(W) = \min\{x_{ij} : (i, j) \in W\}$ and redefine $x_{ij} = x_{ij} - f(W)$ for each (i, j) in W.

We repeat this process with the redefined problem until all node imbalances are zero. Then we select any node with at least one outgoing arc with a positive flow as the starting node, and repeat the procedure, which in this case must find a directed cycle. We terminate when $x = 0$ for the redefined problem. Clearly, the original flow is the sum of flows on the paths and cycles identified by this method. Now observe that each time we identify a directed path, we reduce the excess/deficit of some node to zero or the flow on some arc to zero; and each time we identify a directed cycle, we reduce the flow on some arc to zero. Consequently, the path and cycle representation of the given flow x contains at most $n + m$ directed paths and cycles, and at most m of these are directed cycles. ◆

Let us consider a flow x for which $e(i) = 0$ for all $i \in N$. Recall from Section 1.2 that we call any such flow a circulation. When we apply the flow decomposition algorithm to a circulation, each iteration discovers a directed cycle consisting solely of arcs with positive flow, and subsequently reduces the flow on at least one arc to zero. Consequently, a circulation decomposes into flows along at most m directed cycles.

Property 3.6. *A circulation x can be represented as cycle flow along at most m directed cycles.*

We illustrate the flow decomposition algorithm on the example shown in Figure 3.10(a). Initially, nodes 1 and 5 are deficit nodes. Suppose that the algorithm selects node 5. We would then obtain the directed path 5–3–2–4–6 and the flow on this path is 3 units. Removing this path flow gives the flow given in Figure 3.10(b). The algorithm selects node 1 as the starting node and obtains the path flow of 2 units along the directed path 1–2–4–5–6. In the third iteration, the algorithm identifies a cycle flow of 4 units along the directed cycle 5–3–4–5. Now the flow becomes zero and the algorithm terminates.

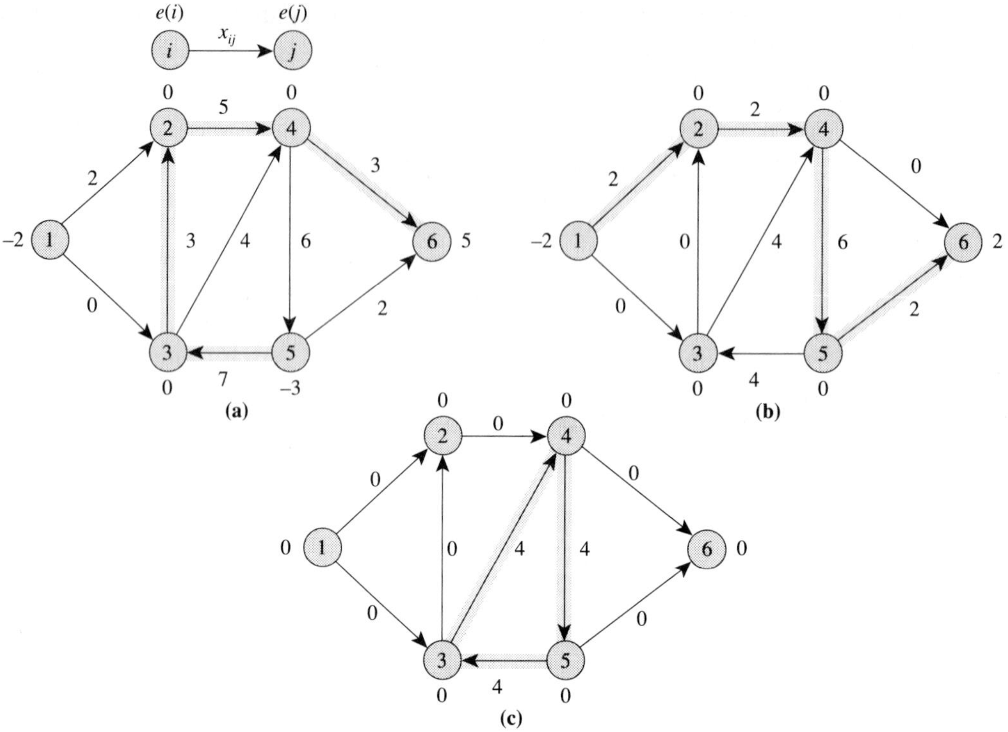

Figure 3.10 Illustrating the flow decomposition theorem.

What is the time required for the flow decomposition algorithm described in the proof of Theorem 3.5? In the algorithm, we first construct a set LIST of deficit nodes. We maintain LIST as a doubly linked list (see Appendix A for a description of this data structure) so that selection of an element as well as addition and deletion of an element require $O(1)$ time. As the algorithm proceeds, it removes nodes from LIST. When LIST eventually becomes empty, we initialize it as the set of arcs with positive flow. Consider now another basic operation in the flow decomposition algorithm: identifying an arc with positive flow emanating from a node. We refer to such arcs as *admissible* arcs. We use the *current-arc* data structure (described in Section 3.4) to identify an admissible arc emanating from a node. Notice that in any iteration, the flow decomposition algorithm requires $O(n)$ time plus the time spent in scanning arcs to identify admissible arcs. Also notice that since arc flows are nonincreasing, an arc found to be inadmissible in one iteration remains inadmissible in subsequent iterations. Consequently, we preserve the current arcs of the nodes in the current-arc data structure when we proceed from one iteration to the next. Since the current-arc data structure requires a total of $O(m)$ time in arc scanning to identify admissible arcs and the algorithm performs at most $(n + m)$ iterations, the flow decomposition algorithm runs in $O(m + (n + m)n) = O(nm)$ time.

The flow decomposition theorem has a number of important consequences. As one example, it enables us to compare any two solutions of a network flow problem in a particularly convenient way and to show how we can build one solution from another by a sequence of simple operations. The augmenting cycle theorem, to be discussed next, highlights these ideas.

We begin by introducing the concept of augmenting cycles with respect to a flow x. A cycle W (not necessarily directed) in G is called an *augmenting cycle* with respect to the flow x if by augmenting a positive amount of flow $f(W)$ around the cycle, the flow remains feasible. The augmentation increases the flow on forward arcs in the cycle W and decreases the flow on backward arcs in the cycle. Therefore, a cycle W is an augmenting cycle in G if $x_{ij} < u_{ij}$ for every forward arc (i, j) and $x_{ij} > 0$ for every backward arc (i, j). We next extend the notation of $\delta_{ij}(W)$ for cycles that are not necessarily directed. We define $\delta_{ij}(W)$ equal to 1 if arc (i, j) is a forward arc in the cycle W, $\delta_{ij}(W)$ equal to -1 if arc (i, j) is a backward arc in the cycle W, and equal to 0 otherwise.

Notice that in terms of residual networks (defined in Section 2.4), each augmenting cycle W with respect to a flow x corresponds to a directed cycle W in $G(x)$, and vice versa. We define the cost of an augmenting cycle W as $c(W) = \sum_{(i,j)\in W} c_{ij}\delta_{ij}(W)$. The cost of an augmenting cycle represents the change in the cost of a feasible solution if we augment 1 unit of flow along the cycle. The change in flow cost for augmenting $f(W)$ units along the cycle W is $c(W)f(W)$.

We next use the flow decomposition theorem to prove an augmenting cycle theorem formulated in terms of residual networks. Suppose that x and x° are any two feasible solutions of the minimum cost flow problem. We have seen earlier that some feasible circulation x^1 in $G(x^\circ)$ satisfies the property that $cx = cx^\circ + cx^1$. Property 3.6 implies that we can represent the circulation x^1 as cycle flows $f(W_1)$, $f(W_2), \ldots, f(W_r)$, with $r \le m$. Notice that each of the cycles $W_1, W_2, \ldots, W_r$ is an augmenting cycle in $G(x^\circ)$. Furthermore, we see that

$$\sum_{(i,j)\in A} c_{ij}x_{ij} = \sum_{(i,j)\in A} c_{ij}x^\circ_{ij} + \sum_{(i,j)\in G(x^\circ)} c_{ij}x^1_{ij}$$

$$= \sum_{(i,j)\in A} c_{ij}x^\circ_{ij} + \sum_{(i,j)\in G(x^\circ)} c_{ij}\left[\sum_{k=1}^{r} \delta_{ij}(W_k)f(W_k)\right]$$

$$= \sum_{(i,j)\in A} c_{ij}x^\circ_{ij} + \sum_{k=1}^{r} c(W_k)f(W_k).$$

We have thus established the following result:

Theorem 3.7 (Augmenting Cycle Theorem). *Let x and x° be any two feasible solutions of a network flow problem. Then x equals x° plus the flow on at most m directed cycles in $G(x^\circ)$. Furthermore, the cost of x equals the cost of x° plus the cost of flow on these augmenting cycles.* ◆

In Section 9.3 we see that the augmenting cycle theorem permits us to obtain the following novel characterization of the optimal solutions of the minimum cost flow problem.

Theorem 3.8 (Negative Cycle Optimality Theorem). *A feasible solution x^* of the minimum cost flow problem is an optimal solution if and only if the residual network $G(x^*)$ contains no negative cost directed cycle.*

3.6 SUMMARY

The design and analysis of algorithms is an expansive topic that has grown in importance over the past 30 years as computers have become more central to scientific and administrative computing. In this chapter we described several fundamental techniques that are widely used for this purpose. Having some way to measure the performance of algorithms is critical for comparing algorithms and for determining how well they perform. The research community has adopted three basic approaches for measuring the performance of an algorithm: empirical analysis, average-case analysis, and worst-case analysis. Each of these three performance measures has its own merits and drawbacks. Worst-case analysis has become a widely used approach, due in large part to the simplicity and theoretical appeal of this type of analysis. A worst-case analysis typically assumes that each arithmetic and logical operation requires unit time, and it provides an upper bound on the time taken by an algorithm (correct to within a constant factor) for solving *any* instance of a problem. We refer to this bound, which we state in big O notation as a function of the problem's size parameters n, m, $\log C$, and $\log U$, as the worst-case complexity of the algorithm. This bound gives the growth rate (in the worst case) that the algorithm requires for solving successively larger problems. If the worst-case complexity of an algorithm is a polynomial function of n, m, $\log C$, and $\log U$, we say that the algorithm is a polynomial-time algorithm; otherwise, we say that it is an exponential-time algorithm. Polynomial-time algorithms are preferred to exponential-time algorithms because polynomial-time algorithms are asymptotically (i.e., for sufficiently large networks) faster than exponential-time algorithms. Among several polynomial-time algorithms for the same problem, we prefer an algorithm with the least order polynomial running time because this algorithm will be asymptotically fastest.

A commonly used approach for obtaining the worst-case complexity of an iterative algorithm is to obtain a bound on the number of iterations, a bound on the number of steps per iteration, and take the product of these two bounds. Sometimes this method overestimates the actual number of steps, especially when an iteration might be easy most of the time, but expensive occasionally. In these situations, arguments based on potential functions (see Section 3.3) often allow us to obtain a tighter bound on an algorithm's required computations.

In this chapter we described four important approaches that researchers frequently use to obtain polynomial-time algorithms for network flow problems: (1) geometric improvement, (2) scaling, (3) dynamic programming, and (4) binary search. Researchers have recently found the scaling approach to be particularly useful for solving network flow problems efficiently, and currently many of the fastest network flow algorithms use scaling as an algorithmic strategy.

Search algorithms lie at the core of many network flow algorithms. We described search algorithms for performing the following tasks: (1) identifying all nodes that are reachable from a specified source node via directed paths, (2) identifying all nodes that can reach a specified sink node via directed paths, and (3) identifying whether a network is strongly connected. Another important application of search algorithms is to determine whether a given directed network is acyclic and, if so, to number the nodes in a topological order [i.e., so that $i < j$ for every arc $(i, j) \in A$]. This algorithm is a core subroutine in methods for project planning (so called

CPM/PERT models) that practitioners used extensively in many industrial settings. All of these search algorithms run in $O(m)$ time. Other $O(m)$ search algorithms are able (1) to identify whether a network is disconnected and if so to identify all of its components, and (2) to identify whether a network is bipartite. We discuss these algorithms in the exercises for this chapter.

We concluded this chapter by studying flow decomposition theory. This theory shows that we can formulate flows in a network in two alternative ways: (1) flows on arcs, or (2) flows along directed paths and directed cycles. Although we use the arc flow formulation throughout most of this book, sometimes we need to rely on the path and cycle flow formulation. Given a path and cycle flow, we can obtain the corresponding arc flow in a straightforward manner (to obtain the flow on any arc, add the flow on this arc in each path and cycle); finding path and cycle flows that corresponds to a set of given arc flows is more difficult. We described an $O(nm)$ algorithm that permits us to find these path and cycle flows. One important consequence of flow decomposition theory is the fact that we can transform any feasible flow of the minimum cost flow problem into any other feasible flow by sending flows along at most m augmenting cycles. We used this result to derive a negative cycle optimality condition for characterizing optimal solutions for the minimum cost flow problem. These conditions state that a flow x is optimal if and only if the residual network $G(x)$ contains no negative cost augmenting cycle.

REFERENCE NOTES

Over the past two decades, worst-case complexity (see Section 3.2) has become a very popular approach for analyzing algorithms. A number of books provide excellent treatments of this topic. The book by Garey and Johnson [1979] is an especially good source of information concerning the topics we have considered. Books by Aho, Hopcroft, and Ullman [1974], Papadimitriou and Steiglitz [1982], Tarjan [1983], and Cormen, Leiserson, and Rivest [1990] provide other valuable treatments of this subject matter.

The techniques used to develop polynomial-time algorithms (see Section 3.3) fall within the broad domain of algorithm design. Books on algorithms and data structures offer extensive coverage of this topic. Edmonds and Karp [1972] and Dinic [1973] independently discovered the scaling technique and its use for obtaining polynomial-time algorithms for the minimum cost flow problem. Gabow [1985] popularized the scaling technique by developing scaling-based algorithms for the shortest path, maximum flow, assignment, and matching problems. This book is the first that emphasizes scaling as a generic algorithmic tool. The geometric improvement technique is a combinatorial analog of linear convergence in the domain of nonlinear programming. For a study of linear convergence, we refer the reader to any book in nonlinear programming. Dynamic programming, which was first developed by Richard Bellman, has proven to be a very successful algorithmic tool. Some important sources of information on dynamic programming are books by Bellman [1957], Bertsekas [1976], and Denardo [1982]. Binary search is a standard technique in searching and sorting; Knuth [1973b] and many other books on data structures and algorithms develop this subject.

Search algorithms are important subroutines for network optimization algorithms. The books by Aho, Hopcroft, and Ullman [1974], Even [1979], Tarjan [1983], and Cormen, Leiserson, and Rivest [1990] present insightful treatments of search algorithms. Ford and Fulkerson [1962] developed flow decomposition theory; their book contains additional material on this topic.

EXERCISES

3.1. Write a pseudocode that, for any integer n, computes n^n by performing at most 2 log n multiplications. Assume that multiplying two numbers, no matter how large, requires one operation.

3.2. Compare the following functions for various values of n and determine the approximate values of n when the second function becomes larger than the first.
(a) $1000n^2$ and $2^n/100$.
(b) $(\log n)^3$ and $n^{0.001}$.
(c) $10{,}000n$ and $0.1n^2$.

3.3. Rank the following functions in increasing order of their growth rates.
(a) $2^{\log \log n}$, $n!$, n^2, 2^n, $(1.5)^{(\log n)^2}$.
(b) $1000(\log n)^2$, $0.005n^{0.0001}$, $\log \log n$, $(\log n)(\log \log n)$.

3.4. Rank the following functions in increasing order of their growth rates for two cases: (1) when a network containing n nodes and m arcs is connected and very sparse [i.e., $m = O(n)$]; and (2) when the network is very dense [i.e., $m = \Omega(n^2)$].
(a) $n^2m^{1/2}$, $nm + n^2 \log n$, $nm \log n$, $nm \log(n^2/m)$.
(b) n^2, $m \log n$, $m + n \log n$, $m \log \log n$.
(c) $n^3 \log n$, $(m \log n)(m + n \log n)$, $nm(\log \log n)\log n$.

3.5. We say that a function $f(n)$ is $O(g(n))$ if for some numbers c and n_0, $f(n) \le cg(n)$ for all $n \ge n_0$. Similarly, we say that a function is $\Omega(g(n))$ if for some numbers c' and n_0, $f(n) \ge c'g(n)$ for infinitely many $n \ge n_0$. Finally, we say that a function $f(n)$ is $\Theta(g(n))$ if $f(n) = O(g(n))$ and $f(n) = \Omega(g(n))$. For each of the functions $f(n)$ and $g(n)$ specified below, indicate whether $f(n)$ is $O(g(n))$, $\Omega(g(n))$, $\Theta(g(n))$, or none of these.

(a) $f(n) = \begin{cases} n & \text{if } n \text{ is odd} \\ n^2 & \text{if } n \text{ is even} \end{cases}; \quad g(n) = \begin{cases} n & \text{if } n \text{ is even} \\ n^2 & \text{if } n \text{ is odd} \end{cases}$

(b) $f(n) = \begin{cases} n & \text{if } n \text{ is odd} \\ n^2 & \text{if } n \text{ is even} \end{cases}; \quad g(n) = \begin{cases} n & \text{if } n \text{ is prime} \\ n^2 & \text{if } n \text{ is not prime} \end{cases}$

(c) $f(n) = 3 + 1/(\log n)$; $g(n) = (n + 4)/(n + 3)$

3.6. Are the following statements true or false?
(a) $(\log n)^{100} = O(n^\epsilon)$ for any $\epsilon > 0$.
(b) $2^{n+1} = O(2^n)$.
(c) $f(n) + g(n) = O(\max(f(n), g(n)))$.
(d) If $f(n) = O(g(n))$, then $g(n) = \Omega(f(n))$.

3.7. Let $g(n, m) = m \log_d n$, where $d = \lceil m/n + 2 \rceil$. Show that for any $\epsilon > 0$, $g(n, m) = O(m^{1+\epsilon})$.

3.8. Show that if $f(n) = O(g(n))$ and $g(n) = O(h(n))$, then $f(n) = O(h(n))$. Is it true that if $f(n) = \Omega(g(n))$ and $g(n) = \Omega(h(n))$, then $f(n) = \Omega(h(n))$? Prove or disprove this statement.

3.9. **Bubble sort.** The *bubble sort algorithm* is a popular method for sorting n numbers in nondecreasing order of their magnitudes. The algorithm maintains an ordered set of the numbers $\{a_1, a_2, \ldots, a_n\}$ that it rearranges through a sequence of several passes over the set. In each pass, the algorithm examines every pairs of elements (a_k, a_{k+1}) for each $k = 1, \ldots, n$, and if the pair is out of order (i.e., $a_k > a_{k+1}$), it swaps the positions of these elements. The algorithm terminates when it makes no swap during one entire pass. Show that the algorithm performs at most n passes and runs in $O(n^2)$ time. For every n, construct a sorting problem (i.e., the initial ordered set of numbers $\{a_1,$

$a_2, \ldots, a_n\}$ so that the algorithm performs $\Omega(n^2)$ operations. Conclude that the bubble sort is a $\Theta(n^2)$ algorithm.

3.10. **Bin packing problem.** The bin packing problem requires that we pack n items of lengths $a_1, a_2, \ldots, a_n$ (assume that each $a_i \le 1$) into bins of unit length using the minimum possible number of bins. Several approximate methods, called *heuristics*, are available for solving the bin packing problem. The *first-fit heuristic* is one of the more popular of these heuristics. It works as follows. Arrange items in an arbitrary order and examine them one by one in this order. For an item being examined, scan the bins one by one and put the item in the bin where it fits first. If an items fits in none of the bins that currently contain an item, we introduce a new bin and place the item in it. Write a pseudocode for the first-fit heuristic and show that it runs in $O(n^2)$ time. For every n, construct an instance of the bin packing problem for which your first-fit heuristic runs in $\Omega(n^2)$ time. Conclude that the first-fit heuristic runs in $\Theta(n^2)$ time.

3.11. Consider a queue of elements on which we perform two operations: (1) *insert(i)*, which adds an element i to the rear of the queue; and (2) *delete(k)*, which deletes the k frontmost elements from the queue. Show that an arbitrary sequence of n insert and delete operations, starting with an empty queue, requires a total of $O(n)$ time.

3.12. An algorithm performs three different operations. The first and second operations are executed $O(nm)$ and $O(n^2)$ times respectively and the number of executions of the third operation is yet to be determined. These operations have the following impact on an appropriately defined potential function ϕ: Each execution of operation 1 increases ϕ by at most n units, each execution of operation 2 increases ϕ by 1 unit, and each execution of operation 3 decreases ϕ by at least 1 unit. Suppose we know that $1 \le \phi \le n^2$. Obtain a bound on the number of executions of the third operation.

3.13. **Parameter balancing.** For each of the time bounds stated below as a function of the parameter k, use the parameter balancing technique to determine the value of k that yields the minimum time bound. Also try to determine the optimal value of k using differential calculus.

(a) $O\left(\frac{n^3}{k} + knm\right)$

(b) $O\left(nk + \frac{m}{k}\right)$

(c) $O\left(\frac{m \log n}{\log k} + \frac{n\, k \log n}{\log k}\right)$

3.14. **Generalized parameter balancing.** In Section 3.3 we discussed the parameter balancing technique for situations when the time bound contains two expressions. In this exercise we generalize the technique to bounds containing three expressions. Suppose that the running time of an algorithm is $O(f(n, k) + g(n, k) + h(n, k))$ and we wish to determine the optimal value of k—that is, the value of k producing the smallest possible overall time. Assume that for all k, $f(n, k)$, $g(n, k)$, and $h(n, k)$ are all nonnegative, $f(n, k)$ is monotonically increasing, and both $g(n, k)$ and $h(n, k)$ are monotonically decreasing. Show how to obtain the optimal value of k and prove that your method is valid. Illustrate your technique on the following time bounds: (1) $kn^2 + n^3/k + n^4/k^2$; (2) $nm/k + kn^2 + n^2 \log_k U$.

3.15. In each of the algorithms described below, use Theorem 3.1 to obtain an upper bound on the total number of iterations the algorithm performs.

(a) Let v^* denote the maximum flow value and v the flow value of the current solution in a maximum flow algorithm. This algorithm increases the flow value by an amount $(v^* - v)/m$ at each iteration. How many iterations will this algorithm perform?

(b) Let z^* and z represent the optimal objective function value and objective function value of the current solution in an application of the some algorithm for solving the shortest path problem. Suppose that this algorithm ensures that each iteration decreases the objective function value by at least $(z - z^*)/2n^2$. How many iterations will the algorithm perform?

3.16. Consider a function $f(n, m)$, defined inductively as follows:

$$f(n, 0) = n, \quad f(0, m) = 2m, \quad \text{and}$$

$$f(n, m) = f(n - 1, m) + f(n, m - 1) - f(n - 1, m - 1).$$

Derive the values of $f(n, m)$ for all values of $n, m \le 4$. Simplify the definition of $f(n, m)$ and prove your result using inductive arguments.

3.17. In Section 3.3 we described a dynamic programming algorithm for the 0–1 knapsack problem. Generalize this approach so that it can be used to solve a knapsack problem in which we can place more than one item of the same type in the knapsack.

3.18. **Shortest paths in layered networks.** We say that a directed network $G = (N, A)$ with a specified source node s and a specified sink node t is *layered* if we can partition its node set N into k layers $N_1, N_2, \ldots, N_k$ so that $N_1 = \{s\}$, $N_k = \{t\}$, and for every arc $(i, j) \in A$, nodes i and j belong to adjacent layers (i.e., $i \in N_l$ and $j \in N_{l+1}$ for some $1 \le l \le k - 1$). Suggest a dynamic programming algorithm for solving the shortest path problem in a layered network. What is the running time of your algorithm? (*Hint*: Examine nodes in the layers $N_1, N_2, \ldots, N_k$, in order and compute shortest path distances.)

3.19. Let $G = (N, A)$ be a directed network. We want to determine whether G contains an odd-length directed cycle passing through node i. Show how to solve this problem using dynamic programming. [*Hint:* Define $d^k(j)$ as equal to 1 if the network contains a walk from node i to node j with exactly k arcs, and as 0 otherwise. Use recursion on k.]

3.20. Now consider the problem of determining whether a network contains an even-length directed cycle passing through node i. Explain why the approach described in Exercise 3.19 does not work in this case.

3.21. Consider a network with a length c_{ij} associated with each arc (i, j). Give a dynamic programming algorithm for finding a shortest walk (i.e., of minimum total length) containing exactly k arcs from a specified node s to every other node j in a network. Does this algorithm work in the presence of negative cycles? [*Hint*: Define $d^k(j)$ as the length of the shortest walk from node s to node j containing exactly k arcs and write a recursive relationship for $d^k(j)$ in terms of $d^{k-1}(j)$ and c_{ij}'s.]

3.22. Professor May B. Wright suggests the following sorting method utilizing a binary search technique. Consider a list of n numbers and suppose that we have already sorted the first k numbers in the list (i.e., arranged them in the nondecreasing order). At the $(k + 1)$th iteration, select the $(k + 1)$th number in the list, perform binary search over the first k numbers to identify the position of this number, and then insert it to produce the sorted list of the first $k + 1$ elements. Professor Wright claims that this method runs in $O(n \log n)$ time because it performs n iterations and each binary search requires $O(\log n)$ time. Unfortunately, Professor Wright's claim is false and it is not possible to implement the algorithm in $O(n \log n)$ time. Explain why. (*Hint*: Work out the details of this implementation including the required data structures.)

3.23. Given a convex function $f(x)$ of the form shown in Figure 3.11, suppose that we want to find a value of x that minimizes $f(x)$. Since locating the exact minima is a difficult task, we allow some approximation and wish to determine a value x so that the interval $(x - \epsilon, x + \epsilon)$ contains a value that minimizes $f(x)$. Suppose that we know that $f(x)$

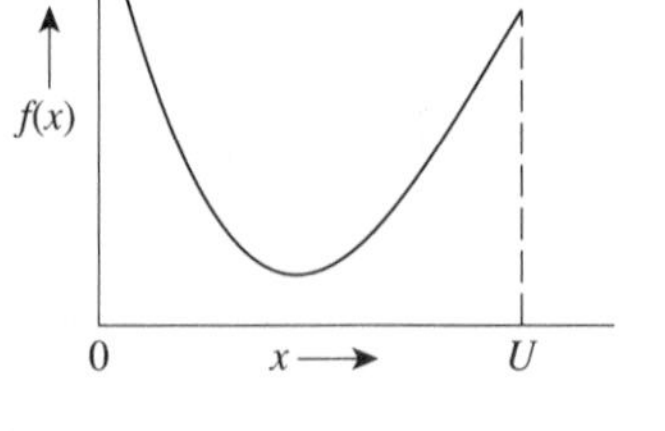

Figure 3.11 Convex function.

attains its minimum value in the interval [0, U]. Develop a binary search algorithm for solving this problem that runs in $O(\log(U/\epsilon))$ time. (*Hint*: At any iteration when $[a, b]$ is the feasible interval, evaluate $f(x)$ at the points $(a + b)/4$ and $3(a + b)/4$, and exclude the region $[a, (a + b)/4]$ or $[3(a + b)/4, b]$.)

3.24. **(a)** Determine the breadth-first and depth-first search trees with $s = 1$ as the source node for the graph shown in Figure 3.12.

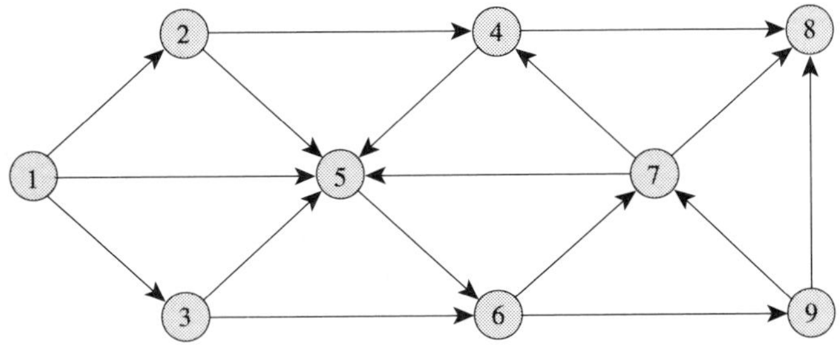

Figure 3.12 Example for Exercise 3.24.

(b) Is the graph shown in Figure 3.12 acyclic? If not, what is the minimum number of arcs whose deletion will produce an acyclic graph? Determine a topological ordering of the nodes in the resulting graph. Is the topological ordering unique?

3.25. **Knight's tour problem.** Consider the chessboard shown in Figure 3.13. Note that some squares are shaded. We wish to determine a knight's tour, if one exists, that starts at the square designated by s and, after visiting the minimum number of squares, ends at the square designated by t. The tour must not visit any shaded square. Formulate this problem as a reachability problem on an appropriately defined graph.

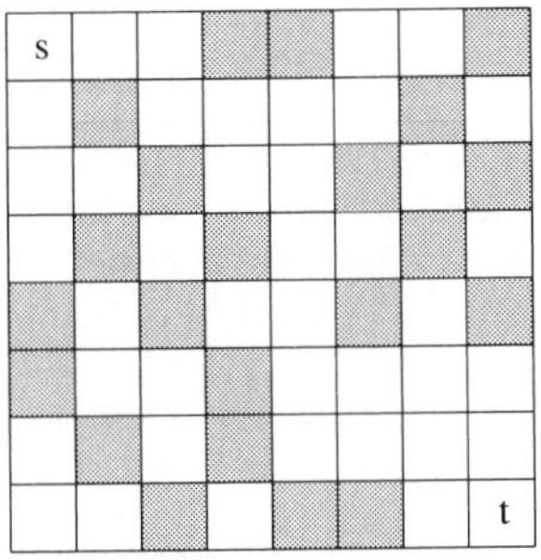

Figure 3.13 Chessboard.

3.26. **Maze problem.** Show how to formulate a maze problem as a reachability problem in a directed network. Illustrate your method on the maze problem shown in Figure 3.14. (*Hint*: Define rectangular segments in the maze as *cords* and represent cords by nodes.)

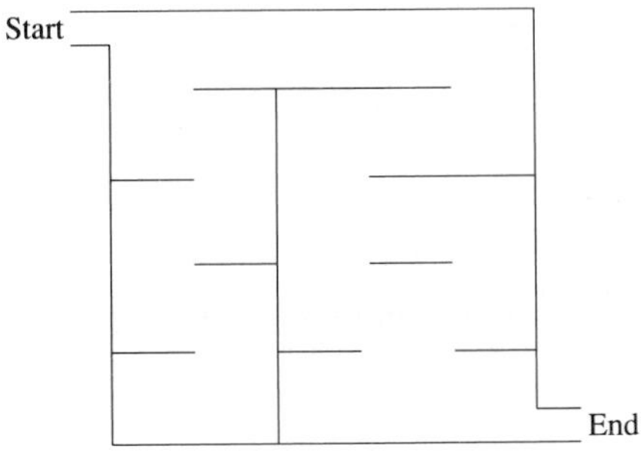

Figure 3.14 Maze.

3.27. Wine division problem. Two men have an 8-gallon jug full of wine and two empty jugs with a capacity of 5 and 3 gallons. They want to divide the wine into two equal parts. Suppose that when shifting the wine from one jug to another, in order to know how much they have transferred, the men must always empty out the first jug or fill the second, or both. Formulate this problem as a reachability problem in an appropriately defined graph. (*Hint*: Let a, b, and c, respectively, denote a partitioning of the 8 gallons of wine into the jugs of 8, 5, and 3 gallons capacity. Refer to any such partitioning as a feasible state of the jugs. Since at least one of the jugs is always empty or full, we can define 16 possible feasible states. Suppose that we represent these states by nodes and connect two nodes by an arc when we can permissibly move wine from one jug to another to move from one state to the other.)

3.28. Give a five-node network for which a breadth-first traversal examines the nodes in the same order as a depth-first traversal.

3.29. Let T be a depth-first search tree of an undirected graph G. Show that for every nontree arc (k, l) in G, either node k is an ancestor of node l in T or node l is an ancestor of node k in T. Show by a counterexample that a breadth-first search tree need not satisfy this property.

3.30. Show that in a breadth-first search tree, the tree path from the source node to any node i is a shortest path (i.e., contains the fewest number of arcs among all paths joining these two nodes). (*Hint*: Use induction on the number of labeled nodes.)

3.31. In an undirected graph $G = (N, A)$, a set of nodes $S \subseteq N$ defines a *clique* if for every pair of nodes i, j in S, $(i, j) \in A$. Show that in the depth-first tree of G, all nodes in any clique S appear on one path. Do all the nodes in a clique S appear consecutively on the path?

3.32. Show that a depth-first order of a network satisfies the following properties.
(a) If node j is a descendant of node i, order(j) > order(i).
(b) All the descendants of any node are ordered consecutively in the order sequence.

3.33. Show that a directed network G is either strongly connected or contains a cut $[S, \bar{S}]$ having no arc (i, j) with $i \in S$ and $j \in \bar{S}$.

3.34. We define the diameter of a graph as a longest path (i.e., one containing the largest number of arcs) in the graph: The path can start and end at any node. Construct a graph whose diameter equals the longest path in a depth-first search tree (you can select any node as the source node). Construct another graph whose diameter is strictly less than the longest path in any depth-first search tree, no matter which node is selected as the source node.

3.35. Transitive closure. A *transitive closure* of a graph $G = (N, A)$ is a matrix $\tau = \{\tau_{ij}\}$ defined as follows:

$$\tau_{ij} = \begin{cases} 1 & \text{if the graph G contains a directed path from node } i \text{ to node } j \\ 0 & \text{otherwise.} \end{cases}$$

Give an $O(nm)$ algorithm for constructing the transitive closure of a (possibly cyclic) graph G.

3.36. Let $\mathcal{H} = \{h_{ij}\}$ denote the node–node adjacency matrix of a graph G. Consider the following set of statements:

```
for l: = 1 to n - 1 do
  for k: = 1 to n do
    for j: = 1 to n do
      for i: = 1 to n do
        h_ij: = max{h_ij, h_ik, h_kj};
```

Show that at the end of these computations, the matrix $\mathcal{H}$ represents the transitive closure of G.

3.37. Given the transitive closure of a graph G, describe an $O(n^2)$ algorithm for determining all strongly connected components of the graph.

3.38. Show that in a directed network, if each node has indegree at least one, the network contains a directed cycle.

3.39. Show through an example that a network might have several topological orderings of its nodes. Show that the topological ordering of a network is unique if and only if the network contains a simple directed path passing through all of its nodes.

3.40. Given two n-vectors $(\alpha(1), \alpha(2), \ldots, \alpha(n))$ and $(\beta(1), \beta(2), \ldots, \beta(n))$, we say that α is *lexicographically smaller* than β (i.e., $\alpha \leq \beta$) if for the first index k for which $\alpha(k) \neq \beta(k)$, $\alpha(k)$ is less than $\beta(k)$. [For example, (2, 4, 8) is lexicographically smaller than (2, 5, 1).] Modify the algorithm given in Figure 3.8 so that it gives the lexico-minimum topological ordering of its nodes (i.e., a topological ordering that is lexicographically smaller than every other topological ordering).

3.41. Suggest an $O(m)$ algorithm for identifying all components of a (possibly) disconnected graph. Design the algorithm so that it will assign a label 1 to all nodes in the first component, a label 2 to all nodes in the second component, and so on. (*Hint*: Maintain a doubly linked list of all unlabeled node.)

3.42. Consider an (arbitrary) spanning tree T of a graph G. Show how to label each node in T as 0 or 1 so that whenever arc (i, j) is contained in the tree, nodes i and j have different labels. Using this result, prove that G is bipartite if and only if for every nontree arc (k, l), nodes k and l have different labels. Using this characterization, describe an $O(m)$ algorithm for determining whether a graph is bipartite or not.

3.43. In an acyclic network $G = (N, A)$ with a specified source node s, let $\alpha(i)$ denote the number of distinct paths from node s to node i. Give an $O(m)$ algorithm that determines $\alpha(i)$ for all $i \in N$. (*Hint*: Examine nodes in a topological order.)

3.44. For an acyclic network G with a specified source node s, outline an algorithm that enumerates *all* distinct directed paths from the source node to every other node in the network. The running time of your algorithm should be proportional to the total length of all the paths enumerated (i.e., linear in terms of the output length.) (*Hint*: Extend your method developed in Exercise 3.43.)

3.45. In an undirected connected graph $G = (N, A)$, an *Euler tour* is a walk that starts at some node, visits each arc exactly once, and returns to the starting node. A graph is *Eulerian* if it contains an Euler tour. Show that in an Eulerian graph, the degree of every node is even. Next, show that if every node in a connected graph has an even degree, the graph is Eulerian. Establish the second result by describing an $O(m)$ algorithm for determining whether a graph is Eulerian and, if so, will construct an Euler tour. (*Hint*: Describe an algorithm that decomposes any graph with only even-degree nodes into a collection of arc-disjoint cycles, and then converts the cycles into an Euler tour.)

3.46. Let T be a depth-first search tree of a graph. Let $D(i)$ denote an ordered set of descendants of the node $i \in T$, arranged in the same order in which the depth-first search method labeled them. Define $last(i)$ as the last element in the set $D(i)$. Modify the depth-first search algorithm so that while computing the depth-first traversal of the network G, it also computes the last index of every node. Your algorithm should run in $O(m)$ time.

3.47. **Longest path in a tree** (Handler, 1973). A longest path in an undirected tree T is a path containing the maximum number of arcs. The longest path can start and end anywhere. Show that we can determine a longest path in T as follows: Select any node i and use a search algorithm to find a node k farthest from node i. Then use a search algorithm to find a node l farthest from node k. Show that the tree path from node k to node l is a longest path in T. (*Hint*: Consider the midmost node or arc on any longest path in the tree depending on whether the path contains an even or odd number or arcs. Need the longest path starting from any node j pass through this node or arc?)

3.48. Consider the flow given in Figure 3.15(a). Compute the imbalance $e(i)$ for each node $i \in N$ and decompose the flow into a path and cycle flow. Is this decomposition unique?

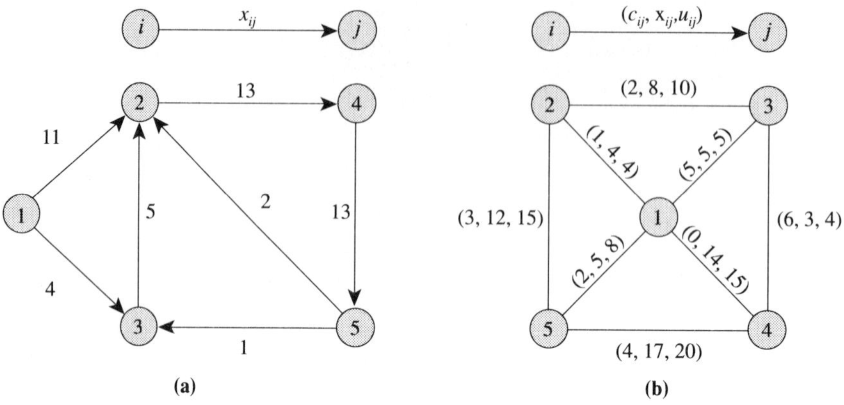

Figure 3.15 Examples for Exercises 3.48 and 3.49.

3.49. Consider the circulation given in Figure 3.15(b). Decompose this circulation into flows along directed cycles. Draw the residual network and use Theorem 3.8 to check whether the flow is an optimal solution of the minimum cost flow problem.

3.50. Consider the circulation shown in Figure 3.16. Show that there are $k!$ distinct flow decompositions of this circulation.

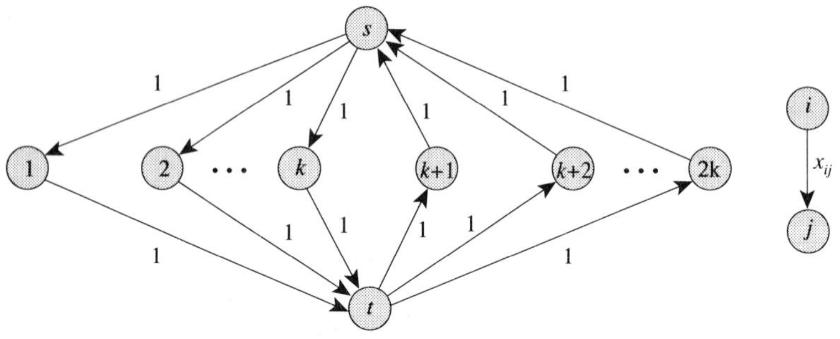

Figure 3.16 Example for Exercise 3.50.

3.51. Show that a unit flow along directed walk from node i to node j $(i \neq j)$ can be decomposed into a directed path from node i to node j plus some arc-disjoint directed cycles. Next, show that a unit flow along a closed directed walk can be decomposed into unit flows along arc-disjoint directed cycles.

3.52. Show that if an undirected connected graph $G = (N, A)$ contains exactly $2k$ odd-degree nodes, the graph contains k arc-disjoint walks $P_1, P_2, \ldots, P_k$ satisfying the property that $A = P_1 \cup P_2 \cup \cdots \cup P_k$.

3.53. Let $G = (N, A)$ be a connected network in which every arc $(i, j) \in A$ has positive lower bound $l_{ij} > 0$ and an infinite upper bound $u_{ij} = \infty$. Show that G contains a feasible circulation (i.e., a flow in which the inflow equals the outflow for every node) if and only if G is strongly connected.

3.54. Show that a solution x satisfying the flow bound constraints is a circulation if and only if the net flow across any cut is zero.

4

SHORTEST PATHS: LABEL-SETTING ALGORITHMS

A journey of a thousand miles starts with a single step and if that step is the right step, it becomes the last step.
—Lao Tzu

Chapter Outline

4.1 INTRODUCTION

Shortest path problems lie at the heart of network flows. They are alluring to both researchers and to practitioners for several reasons: (1) they arise frequently in practice since in a wide variety of application settings we wish to send some material (e.g., a computer data packet, a telephone call, a vehicle) between two specified points in a network as quickly, as cheaply, or as reliably as possible; (2) they are easy to solve efficiently; (3) as the simplest network models, they capture many of the most salient core ingredients of network flows and so they provide both a benchmark and a point of departure for studying more complex network models; and (4) they arise frequently as subproblems when solving many combinatorial and network optimization problems. Even though shortest path problems are relatively easy to solve, the design and analysis of most efficient algorithms for solving them requires considerable ingenuity. Consequently, the study of shortest path problems is a natural starting point for introducing many key ideas from network flows, including the use of clever data structures and ideas such as data scaling to improve the worst-case algorithmic performance. Therefore, in this and the next chapter, we begin our discussion of network flow algorithms by studying shortest path problems.

We first set our notation and describe several assumptions that we will invoke throughout our discussion.

Notation and Assumptions

We consider a directed network $G = (N, A)$ with an *arc length* (or *arc cost*) c_{ij} associated with each arc $(i, j) \in A$. The network has a distinguished node s, called the *source*. Let $A(i)$ represent the arc adjacency list of node i and let $C = \max\{c_{ij} : (i, j) \in A\}$. We define the *length of a directed path* as the sum of the lengths of arcs in the path. The shortest path problem is to determine for every nonsource node $i \in N$ a shortest length directed path from node s to node i. Alternatively, we might view the problem as sending 1 unit of flow as cheaply as possible (with arc flow costs as c_{ij}) from node s to each of the nodes in $N - \{s\}$ in an uncapacitated network. This viewpoint gives rise to the following linear programming formulation of the shortest path problem.

$$\text{Minimize} \sum_{(i,j)\in A} c_{ij}x_{ij} \tag{4.1a}$$

subject to

$$\sum_{\{j:(i,j)\in A\}} x_{ij} - \sum_{\{j:(j,i)\in A\}} x_{ji} = \begin{cases} n-1 & \text{for } i = s \\ -1 & \text{for all } i \in N - \{s\} \end{cases} \tag{4.1b}$$

$$x_{ij} \geq 0 \quad \text{for all } (i, j) \in A. \tag{4.1c}$$

In our study of the shortest path problem, we will impose several assumptions.

Assumption 4.1. *All arc lengths are integers.*

The integrality assumption imposed on arc lengths is necessary for some algorithms and unnecessary for others. That is, for some algorithms we can relax it and still perform the same analysis. Algorithms whose complexity bound depends on C assume integrality of the data. Note that we can always transform rational arc capacities to integer arc capacities by multiplying them by a suitably large number. Moreover, we necessarily need to convert irrational numbers to rational numbers to represent them on a computer. Therefore, the integrality assumption is really not a restrictive assumption in practice.

Assumption 4.2. *The network contains a directed path from node s to every other node in the network.*

We can always satisfy this assumption by adding a "fictitious" arc (s, i) of suitably large cost for each node i that is not connected to node s by a directed path.

Assumption 4.3. *The network does not contain a negative cycle (i.e., a directed cycle of negative length).*

Observe that for any network containing a negative cycle W, the linear programming formulation (4.1) has an unbounded solution because we can send an infinite amount of flow along W. The shortest path problem with a negative cycle is substantially harder to solve than is the shortest path problem without a negative cycle. Indeed, because the shortest path problem with a negative cycle is an $\mathcal{NP}$-complete problem, no polynomial-time algorithm for this problem is likely to exist (see Appendix B for the definition of $\mathcal{NP}$-complete problems). Negative cycles complicate matters, in part, for the following reason. All algorithms that are capable of solving shortest path problems with negative length arcs essentially determine shortest length directed walks from the source to other nodes. If the network contains no negative cycle, then some shortest length directed walk is a path (i.e., does not repeat nodes), since we can eliminate directed cycles from this walk without increasing its length. The situation for networks with negative cycles is quite different; in these situations, the shortest length directed walk might traverse a negative cycle an infinite number of times since each such repetition reduces the length of the walk. In these cases we need to prohibit walks that revisit nodes; the addition of this apparently mild stipulation has significant computational implications: With it, the shortest path problem becomes substantially more difficult to solve.

Assumption 4.4. *The network is directed.*

If the network were undirected and all arc lengths were nonnegative, we could transform this shortest path problem to one on a directed network. We described this transformation in Section 2.4. If we wish to solve the shortest path problem on an undirected network and some arc lengths are negative, the transformation described in Section 2.4 does not work because each arc with negative length would produce a negative cycle. We need a more complex transformation to handle this situation, which we describe in Section 12.7.

Various Types of Shortest Path Problems

Researchers have studied several different types of (directed) shortest path problems:

1. Finding shortest paths from one node to all other nodes when arc lengths are nonnegative
2. Finding shortest paths from one node to all other nodes for networks with arbitrary arc lengths
3. Finding shortest paths from every node to every other node
4. Various generalizations of the shortest path problem

In this and the following chapter we discuss the first three of these problem types. We refer to problem types (1) and (2) as the *single-source shortest path problem* (or, simply, the *shortest path problem*), and the problem type (3) as the *all-pairs shortest path problem*. In the exercises of this chapter we consider the following variations of the shortest path problem: (1) the maximum capacity path problem, (2) the maximum reliability path problem, (3) shortest paths with turn penalties, (4) shortest paths with an additional constraint, and (5) the resource-constrained shortest path problem.

Analog Solution of the Shortest Path Problem

The shortest path problem has a particularly simple structure that has allowed researchers to develop several intuitively appealing algorithms for solving it. The following analog mode for the shortest path problem (with nonnegative arc lengths) provides valuable insight that helps in understanding some of the essential features of the shortest path problem. Consider a shortest path problem between a specified pair of nodes s and t (this discussion extends easily for the general shortest path model with multiple destination nodes and with nonnegative arc lengths). We construct a string model with nodes represented by knots, and for any arc (i, j) in A, a string with length equal to c_{ij} joining the two knots i and j. We assume that none of the strings can be stretched. After constructing the model, we hold the knot representing node s in one hand, the knot representing node t in the other hand, and pull our hands apart. One or more paths will be held tight; these are the shortest paths from node s to node t.

We can extract several insights about the shortest path problem from this simple string model:

1. For any arc on a shortest path, the string will be taut. Therefore, the shortest path distance between any two successive nodes i and j on this path will equal the length c_{ij} of the arc (i, j) between these nodes.
2. For any two nodes i and j on the shortest path (which need not be successive nodes on the path) that are connected by an arc (i, j) in A, the shortest path distance from the source to node i plus c_{ij} (a composite distance) is always as large as the shortest path distance from the source to node j. The composite distance might be larger because the string between nodes i and j might not be taut.
3. To solve the shortest path problem, we have solved an associated *maximization* problem (by pulling the string apart). As we will see in our later discussions, in general, all network flow problems modeled as minimization problems have an associated "dual" maximization problem; by solving one problem, we generally solve the other as well.

Label-Setting and Label-Correcting Algorithms

The network flow literature typically classifies algorithmic approaches for solving shortest path problems into two groups: *label setting* and *label correcting*. Both approaches are iterative. They assign tentative distance labels to nodes at each step; the distance labels are estimates of (i.e., upper bounds on) the shortest path distances. The approaches vary in how they update the distance labels from step to step and how they "converge" toward the shortest path distances. Label-setting algorithms designate one label as permanent (optimal) at each iteration. In contrast, label-correcting algorithms consider all labels as temporary until the final step, when they all become permanent. One distinguishing feature of these approaches is the class of problems that they solve. Label-setting algorithms are applicable only to (1) shortest path problems defined on acyclic networks with arbitrary arc lengths, and to (2) shortest path problems with nonnegative arc lengths. The label-correcting algorithms are more general and apply to all classes of problems, including those with negative arc lengths. The label-setting algorithms are much more efficient, that is, have much better worst-case complexity bounds; on the other hand, the label-correcting algorithms not only apply to more general classes of problems, but as we will see, they also offer more algorithmic flexibility. In fact, we can view the label-setting algorithms as special cases of the label-correcting algorithms.

In this chapter we study label-setting algorithms; in Chapter 5 we study label-correcting algorithms. We have divided our discussion in two parts for several reasons. First, we wish to emphasize the difference between these two solution approaches and the different algorithmic strategies that they employ. The two problem approaches also differ in the types of data structures that they employ. Moreover, the analysis of the two types of algorithms is quite different. The convergence proofs for label-setting algorithms are much simpler and rely on elementary combinatorial arguments. The proofs for the label-correcting algorithms tend to be much more subtle and require more careful analysis.

Chapter Overview

The basic label-setting algorithm has become known as *Dijkstra's algorithm* because Dijkstra was one of several people to discover it independently. In this chapter we study several variants of Dijkstra's algorithm. We first describe a simple implementation that achieves a time bound of $O(n^2)$. Other implementations improve on this implementation either empirically or theoretically. We describe an implementation due to Dial that achieves an excellent running time in practice. We also consider several versions of Dijkstra's algorithm that improve upon its worst-case complexity. Each of these implementations uses a *heap* (or *priority queue*) data structure. We consider several such implementations, using data structures known as binary heaps, d-heaps, Fibonacci heaps, and the recently developed radix heap. Before examining these various algorithmic approaches, we first describe some applications of the shortest path problem.

4.2 APPLICATIONS

Shortest path problems arise in a wide variety of practical problem settings, both as stand-alone models and as subproblems in more complex problem settings. For example, they arise in the telecommunications and transportation industries whenever we want to send a message or a vehicle between two geographical locations as quickly or as cheaply as possible. Urban traffic planning provides another important example: The models that urban planners use for computing traffic flow patterns are complex nonlinear optimization problems or complex equilibrium models; they build, however, on the behavioral assumption that users of the transportation system travel, with respect to prevailing traffic congestion, along shortest paths from their origins to their destinations. Consequently, most algorithmic approaches for finding urban traffic patterns solve a large number of shortest path problems as subproblems (one for each origin–destination pair in the network).

In this book we consider many other applications like this with embedded shortest path models. These many and varied applications attest to the importance of shortest path problems in practice. In Chapters 1 and 19 we discuss a number of stand-alone shortest path models in such problem contexts as urban housing, project management, inventory planning, and DNA sequencing. In this section and in the exercises in this chapter, we consider several other applications of shortest paths that are indicative of the range of applications of this core network flow model. These applications include generic mathematical applications—approximating functions, solving certain types of difference equations, and solving the so-called knapsack problem—as well as direct applications in the domains of production planning, telephone operator scheduling, and vehicle fleet planning.

Application 4.1 Approximating Piecewise Linear Functions

Numerous applications encountered within many different scientific fields use piecewise linear functions. On several occasions, these functions contain a large number of breakpoints; hence they are expensive to store and to manipulate (e.g., even to evaluate). In these situations it might be advantageous to replace the piecewise linear function by another approximating function that uses fewer breakpoints. By approximating the function we will generally be able to save on storage space and on the cost of using the function; we will, however, incur a cost because of the inaccuracy of the approximating function. In making the approximation, we would like to make the best possible trade-off between these conflicting costs and benefits.

Let $f_1(x)$ be a piecewise linear function of a scalar x. We represent the function in the two-dimensional plane: It passes through n points $a_1 = (x_1, y_1)$, $a_2 = (x_2, y_2), \ldots, a_n = (x_n, y_n)$. Suppose that we have ordered the points so that $x_1 \le x_2 \le \cdots \le x_n$. We assume that the function varies linearly between every two consecutive points x_i and x_{i+1}. We consider situations in which n is very large and for practical reasons we wish to approximate the function $f_1(x)$ by another function $f_2(x)$ that passes through only a subset of the points $a_1, a_2, \ldots, a_n$ (including a_1 and a_n). As an example, consider Figure 4.1(a): In this figure we have approximated a function $f_1(x)$ passing through 10 points by a function $f_2(x)$ drawn with dashed lines) passing through only five of the points.

This approximation results in a savings in storage space and in the use of the function. For purposes of illustration, assume that we can measure these costs by a per unit cost α associated with any single interval used in the approximation (which

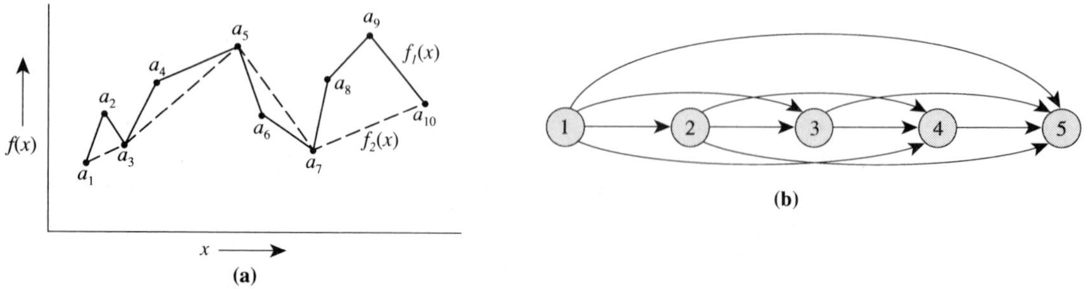

Figure 4.1 Illustrating Application 4.1: (a) approximating the function $f_1(x)$ passing through 10 points by the function $f_2(x)$; (b) corresponding shortest problem.

is defined by two points, a_i and a_j). As we have noted, the approximation also introduces errors that have an associated penalty. We assume that the error of an approximation is proportional to the sum of the squared errors between the actual data points and the estimated points (i.e., the penalty is $\beta \sum_{i=1}^{n} [f_1(x_i) - f_2(x_i)]^2$ for some constant β). Our decision problem is to identify the subset of points to be used to define the approximation function $f_2(x)$ so that we incur the minimum total cost as measured by the sum of the cost of storing and using the approximating function and the cost of the errors imposed by the approximation.

We will formulate this problem as a shortest path problem on a network G with n nodes, numbered 1 through n, as follows. The network contains an arc (i, j) for each pair of nodes i and j such that $i < j$. Figure 4.1(b) gives an example of the network with $n = 5$ nodes. The arc (i, j) in this network signifies that we approximate the linear segments of the function $f_1(x)$ between the points $a_i, a_{i+1}, \ldots, a_j$ by one linear segment joining the points a_i and a_j. The cost c_{ij} of the arc (i, j) has two components: the storage cost α and the penalty associated with approximating all the points between a_i and a_j by the corresponding points lying on the line joining a_i and a_j. In the interval $[x_i, x_j]$, the approximating function is $f_2(x) = f_1(x_i) + (x - x_i)[f_1(x_j) - f_1(x_i)]/(x_j - x_i)$, so the total cost in this interval is

$$c_{ij} = \alpha + \beta\left[\sum_{k=i}^{j} (f_1(x_k) - f_2(x_k))^2\right].$$

Each directed path from node 1 to node n in G corresponds to a function $f_2(x)$, and the cost of this path equals the total cost for storing this function and for using it to approximate the original function. For example, the path 1–3–5 corresponds to the function $f_2(x)$ passing through the points a_1, a_3, and a_5. As a consequence of these observations, we see that the shortest path from node 1 to node n specifies the optimal set of points needed to define the approximating function $f_2(x)$.

Application 4.2 Allocating Inspection Effort on a Production Line

A production line consists of an ordered sequence of n production stages, and each stage has a manufacturing operation followed by a potential inspection. The product enters stage 1 of the production line in batches of size $B \geq 1$. As the items within a batch move through the manufacturing stages, the operations might introduce defects. The probability of producing a defect at stage i is α_i. We assume that all of the defects are nonrepairable, so we must scrap any defective item. After each stage, we can either inspect all of the items or none of them (we do not sample the items); we assume that the inspection identifies every defective item. The production line must end with an inspection station so that we do not ship any defective units. Our decision problem is to find an optimal inspection plan that specifies at which stages we should inspect the items so that we minimize the total cost of production and inspection. Using fewer inspection stations might decrease the inspection costs, but will increase the production costs because we might perform unnecessary manufacturing operations on some units that are already defective. The optimal number of inspection stations will achieve an appropriate trade-off between these two conflicting cost considerations.

Suppose that the following cost data are available: (1) p_i, the manufacturing cost per unit in stage i; (2) f_{ij}, the fixed cost of inspecting a batch after stage j, given that we last inspected the batch after stage i; and (3) g_{ij}, the variable per unit cost for inspecting an item after stage j, given that we last inspected the batch after stage i. The inspection costs at station j depend on when the batch was inspected last, say at station i, because the inspector needs to look for defects incurred at any of the intermediate stages $i + 1, i + 2, \ldots, j$.

We can formulate this inspection problem as a shortest path problem on a network with $(n + 1)$ nodes, numbered $0, 1, \ldots, n$. The network contains an arc (i, j) for each node pair i and j for which $i < j$. Figure 4.2 shows the network for an

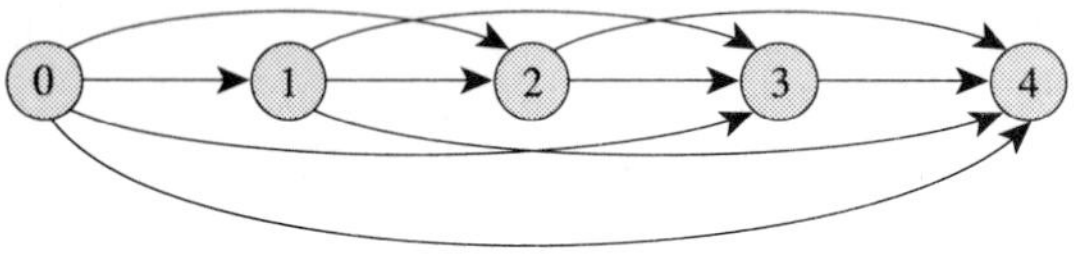

Figure 4.2 Shortest path network associated with the inspection problem.

inspection problem with four stations. Each path in the network from node 0 to node 4 defines an inspection plan. For example, the path 0–2–4 implies that we inspect the batches after the second and fourth stages. Letting $B(i) = B \prod_{k=1}^{i} (1 - \alpha_k)$ denote the expected number of nondefective units at the end of stage i, we associate the following cost c_{ij} with any arc (i, j) in the network:

$$c_{ij} = f_{ij} + B(i)g_{ij} + B(i) \sum_{k=i+1}^{j} p_k. \tag{4.2}$$

It is easy to see that c_{ij} denotes the total cost incurred in the stages $i + 1, i + 2, \ldots, j$; the first two terms on the right-hand side of (4.2) are the fixed and variable inspection costs, and the third term is the production cost incurred in these stages. This shortest path formulation permits us to solve the inspection application as a network flow problem.

Application 4.3 Knapsack Problem

In Section 3.3 we introduced the knapsack problem and formulated this classical operations research model as an integer program. For convenience, let us recall the underlying motivation for this problem. A hiker must decide which goods to include in her knapsack on a forthcoming trip. She must choose from among p objects: Object i has weight w_i (in pounds) and a utility u_i to the hiker. The objective is to maximize the utility of the hiker's trip subject to the weight limitation that she can carry no more than W pounds. In Section 3.3 we described a dynamic programming algorithm for solving this problem. Here we formulate the knapsack problem as a longest path problem on an acyclic network and then show how to transform the longest path problem into a shortest path problem. This application illustrates an intimate connection between dynamic programming and shortest path problems on acyclic networks. By making the appropriate identification between the stages and "states" of any dynamic program and the nodes of a network, we can formulate essentially all deterministic dynamic programming problems as equivalent shortest path problems. For these reasons, the range of applications of shortest path problems includes most applications of dynamic programming, which is a large and extensive field in its own right.

We illustrate our formulation using a knapsack problem with four items that have the weights and utilities indicated in the accompanying table:

j	1	2	3	4
u_j	40	15	20	10
w_j	4	2	3	1

Figure 4.3 shows the longest path formulation for this sample knapsack problem, assuming that the knapsack has a capacity of $W = 6$. The network in the formulation has several layers of nodes: It has one layer corresponding to each item and one layer corresponding to a source node s and another corresponding to a sink node t. The layer corresponding to an item i has $W + 1$ nodes, $i^0, i^1, \ldots, i^W$. Node

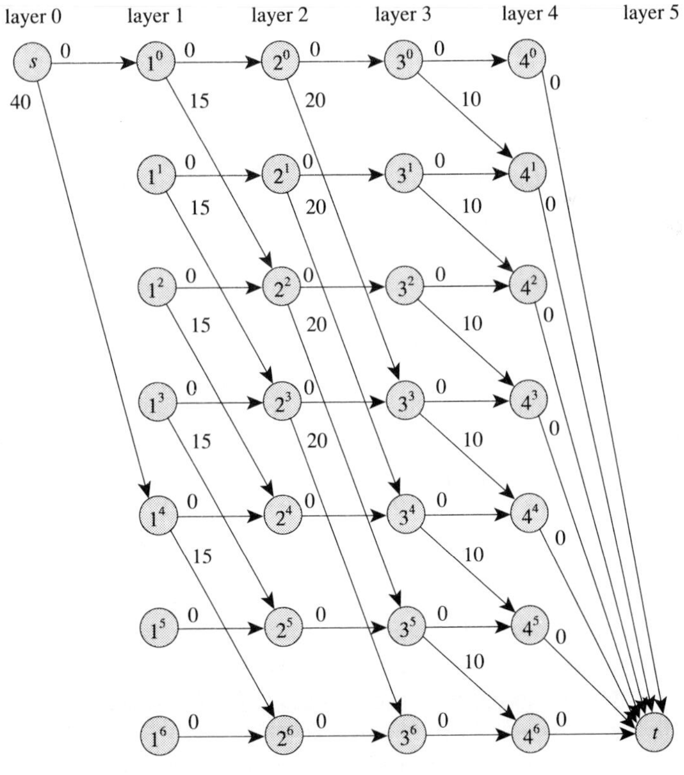

Figure 4.3 Longest path formulation of the knapsack problem.

i^k in the network signifies that the items 1, 2, . . . , i have consumed k units of the knapsack's capacity. The node i^k has at most two outgoing arcs, corresponding to two decisions: (1) do not include item $(i + 1)$ in the knapsack, or (2) include item $i + 1$ in the knapsack. [Notice that we can choose the second of these alternatives only when the knapsack has sufficient spare capacity to accommodate item $(i + 1)$, i.e., $k + w_{i+1} \leq W$.] The arc corresponding to the first decision is $(i^k, (i + 1)^k)$ with zero utility and the arc corresponding to the second decision (provided that $k + w_{i+1} \leq W$) is $(i^k, (i + 1)^{k+w_{i+1}})$ with utility u_{i+1}. The source node has two incident arcs, $(s, 1^0)$ and $(s, 1^{w_1})$, corresponding to the choices of whether or not to include item 1 in an empty knapsack. Finally, we connect all the nodes in the layer corresponding to the last item to the sink node t with arcs of zero utility.

Every feasible solution of the knapsack problem defines a directed path from node s to node t; both the feasible solution and the path have the same utility. Conversely, every path from node s to node t defines a feasible solution to the knapsack problem with the same utility. For example, the path s–1^0–2^2–3^5–4^5–t implies the solution in which we include items 2 and 3 in the knapsack and exclude items 1 and 4. This correspondence shows that we can find the maximum utility selection of items by finding a maximum utility path, that is, a longest path in the network.

The longest path problem and the shortest path problem are closely related. We can transform the longest path problem to a shortest path problem by defining arc costs equal to the negative of the arc utilities. If the longest path problem contains any positive length directed cycle, the resulting shortest path problem contains a negative cycle and we cannot solve it using any of the techniques discussed in the book. However, if all directed cycles in the longest path problem have nonpositive lengths, then in the corresponding shortest path problem all directed cycles have nonnegative lengths and this problem can be solved efficiently. Notice that in the longest path formulation of the knapsack problem, the network is acyclic; so the resulting shortest path problem is efficiently solvable.

To conclude our discussion of this application, we offer a couple of concluding remarks concerning the relationship between shortest paths and dynamic programming. In Section 3.3 we solved the knapsack problem by using a recursive relationship for computing a quantity $d(i, j)$ that we defined as the maximum utility of selecting items if we restrict our selection to items 1 through i and impose a weight restriction of j. Note that $d(i, j)$ can be interpreted as the longest path length from node s to node i^j. Moreover, as we will see, the recursion that we used to solve the dynamic programming formulation of the knapsack problem is just a special implementation of one of the standard algorithms for solving shortest path problems on acyclic networks (we describe this algorithm in Section 4.4). This observation provides us with a concrete illustration of the meta statement that "(deterministic) dynamic programming is a special case of the shortest path problem."

Second, as we show in Section 4.4, shortest path problems on acyclic networks are *very* easy to solve—by methods that are *linear* in the number n of nodes and number m of arcs. Since the nodes of the network representation correspond to the "stages" and "states" of the dynamic programming formulation, the dynamic programming model will be easy to solve if the number of states and stages is not very large (i.e., do not grow exponentially fast in some underlying problem parameter).

Application 4.4 Tramp Steamer Problem

A tramp steamer travels from port to port carrying cargo and passengers. A voyage of the steamer from port i to port j earns p_{ij} units of profit and requires τ_{ij} units of time. The captain of the steamer would like to know which tour W of the steamer (i.e., a directed cycle) achieves the largest possible mean daily profit when we define the daily profit for any tour W by the expression

$$\mu(W) = \frac{\sum_{(i,j)\in W} p_{ij}}{\sum_{(i,j)\in W} \tau_{ij}}.$$

We assume that $\tau_{ij} \geq 0$ for every arc $(i, j) \in A$, and that $\sum_{(i,j)\in W} \tau_{ij} > 0$ for every directed cycle W in the network.

In Section 5.7 we study the tramp steamer problem. In this application we examine a more restricted version of the tramp steamer problem: The captain of the steamer wants to know whether some tour W will be able to achieve a mean daily profit greater than a specified threshold μ_0. We will show how to formulate this problem as a negative cycle detection problem. In this restricted version of the tramp steamer problem, we wish to determine whether the underlying network G contains a directed cycle W satisfying the following condition:

$$\frac{\sum_{(i,j)\in W} p_{ij}}{\sum_{(i,j)\in W} \tau_{ij}} > \mu_0.$$

By writing this inequality as $\sum_{(i,j)\in W} (\mu_0\tau_{ij} - p_{ij}) < 0$, we see that G contains a directed cycle W in G whose mean profit exceeds μ_0 if and only if the network contains a negative cycle when the cost of arc (i, j) is $(\mu_0\tau_{ij} - p_{ij})$. In Section 5.5 we show that label-correcting algorithms for solving the shortest path problem are able to detect negative cycles, which implies that we can solve this restricted version of the tramp steamer problem by applying a shortest path algorithm.

Application 4.5 System of Difference Constraints

In some linear programming applications, with constraints of the form $\mathcal{A}x \leq b$, the $n \times m$ constraint matrix $\mathcal{A}$ contains one $+1$ and one -1 in each row; all the other entries are zero. Suppose that the kth row has a $+1$ entry in column j_k and a -1 entry in column i_k; the entries in the vector b have arbitrary signs. Then this linear program defines the following set of m *difference constraints* in the n variables $x = (x(1), x(2), \ldots, x(n))$:

$$x(j_k) - x(i_k) \leq b(k) \qquad \text{for each } k = 1, \ldots, m. \tag{4.3}$$

We wish to determine whether the system of difference constraints given by (4.3) has a feasible solution, and if so, we want to identify a feasible solution. This model arises in a variety of applications; in Application 4.6 we describe the use of this model in the telephone operator scheduling, and in Application 19.6 we describe the use of this model in the scaling of data.

Each system of difference constraints has an associated graph G, which we call a *constraint graph*. The constraint graph has n nodes corresponding to the n variables and m arcs corresponding to the m difference constraints. We associate an arc (i_k, j_k) of length $b(k)$ in G with the constraint $x(j_k) - x(i_k) \leq b(k)$. As an example, consider the following system of constraints whose corresponding graph is shown in Figure 4.4(a):

$$x(3) - x(4) \leq 5, \tag{4.4a}$$

$$x(4) - x(1) \leq -10, \tag{4.4b}$$

$$x(1) - x(3) \leq 8, \tag{4.4c}$$

$$x(2) - x(1) \leq -11, \tag{4.4d}$$

$$x(3) - x(2) \leq 2. \tag{4.4e}$$

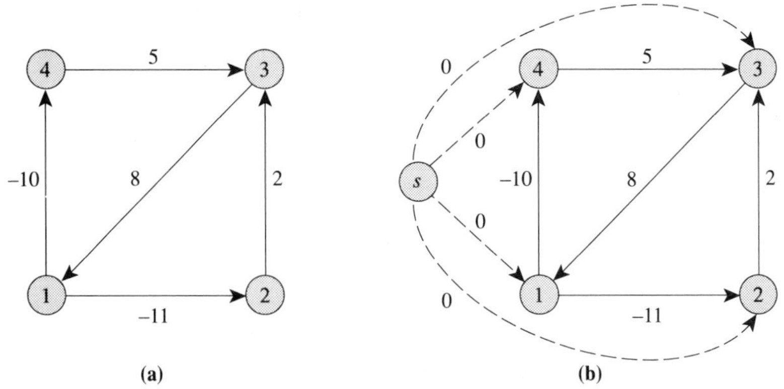

Figure 4.4 Graph corresponding to a system of difference constraints.

In Section 5.2 we show that the constraints (4.4) are identical with the optimality conditions for the shortest path problem in Figure 4.4(a) and that we can satisfy these conditions if and only if the network contains no negative (cost) cycle. The network shown in Figure 4.4(a) contains a negative cycle 1–2–3 of length -1, and the corresponding constraints [i.e., $x(2) - x(1) \leq -11$, $x(3) - x(2) \leq 2$, and $x(1) - x(3) \leq 8$] are inconsistent because summing these constraints yields the invalid inequality $0 \leq -1$.

As noted previously, we can detect the presence of a negative cycle in a network by using the label-correcting algorithms described in Chapter 5. The label-correcting algorithms do require that all the nodes are reachable by a directed path from some node, which we use as the source node for the shortest path problem. To satisfy this requirement, we introduce a new node s and join it to all the nodes in the network with arcs of zero cost. For our example, Figure 4.4(b) shows the modified network. Since all the arcs incident to node s are directed out of this node, node s is not contained in any directed cycle, so the modification does not create any new directed cycles and so does not introduce any cycles with negative costs. The label-correcting algorithms either indicate the presence of a negative cycle or provide the shortest path distances. In the former case the system of difference constraints has no solution, and in the latter case the shortest path distances constitute a solution of (4.4).

Application 4.6 Telephone Operator Scheduling

As an application of the system of difference constraints, consider the following telephone operator scheduling problem. A telephone company needs to schedule operators around the clock. Let $b(i)$ for $i = 0, 1, 2, \ldots, 23$, denote the minimum number of operators needed for the ith hour of the day [here $b(0)$ denotes number of operators required between midnight and 1 A.M.]. Each telephone operator works in a shift of 8 consecutive hours and a shift can begin at any hour of the day. The telephone company wants to determine a "cyclic schedule" that repeats daily (i.e., the number of operators assigned to the shift starting at 6 A.M. and ending at 2 P.M. is the same for each day). The optimization problem requires that we identify the fewest operators needed to satisfy the minimum operator requirement for each hour of the day. Letting y_i denote the number of workers whose shift begins at the ith hour, we can state the telephone operator scheduling problem as the following optimization model:

$$\text{Minimize } \sum_{i=0}^{23} y_i \tag{4.5a}$$

subject to

$$y_{i-7} + y_{i-6} + \cdots + y_i \geq b(i) \qquad \text{for all } i = 8 \text{ to } 23, \tag{4.5b}$$

$$y_{17+i} + \cdots + y_{23} + y_0 + \cdots + y_i \geq b(i) \qquad \text{for all } i = 0 \text{ to } 7, \tag{4.5c}$$

$$y_i \geq 0 \qquad \text{for all } i = 0 \text{ to } 23. \tag{4.5d}$$

Notice that this linear program has a very special structure because the associated constraint matrix contains only 0 and 1 elements and the 1's in each row appear consecutively. In this application we study a restricted version of the telephone operator scheduling problem: We wish to determine whether some feasible schedule uses p or fewer operators. We convert this restricted problem into a system of difference constraints by redefining the variables. Let $x(0) = y_0$, $x(1) = y_0 + y_1$, $x(2) = y_0 + y_1 + y_2, \ldots$, and $x(23) = y_0 + y_2 + \cdots + y_{23} = p$. Now notice that we can rewrite each constraint in (4.5b) as

$$x(i) - x(i - 8) \geq b(i) \qquad \text{for all } i = 8 \text{ to } 23, \tag{4.6a}$$

and each constraints in (4.5c) as

$$\begin{aligned} &x(23) - x(16 + i) + x(i) \\ &\qquad = p - x(16 + i) + x(i) \geq b(i) \qquad \text{for all } i = 0 \text{ to } 7. \end{aligned} \tag{4.6b}$$

Finally, the nonnegativity constraints (4.5d) become

$$x(i) - x(i - 1) \geq 0. \tag{4.6c}$$

By virtue of this transformation, we have reduced the restricted version of the telephone operator scheduling problem into a problem of finding a feasible solution of the system of difference constraints. We discuss a solution method for the general problem in Exercise 4.12. Exercise 9.9 considers a further generalization that incorporates costs associated with various shifts.

In the telephone operator scheduling problem, the rows of the underlying optimization model (in the variables y) satisfy a "wraparound consecutive 1's property"; that is, the variables in each row have only 0 and 1 coefficients and all of the variables with 1 coefficients are consecutive (if we consider the first and last variables to be consecutive). In the telephone operator scheduling problem, each row has exactly eight variables with coefficients of value 1. In general, as long as any optimization model satisfies the wraparound consecutive 1's property, even if the rows have different numbers of variables with coefficients of value 1, the transformation we have described would permit us to model the problem as a network flow model.

4.3 TREE OF SHORTEST PATHS

In the shortest path problem, we wish to determine a shortest path from the source node to all other $(n - 1)$ nodes. How much storage would we need to store these paths? One naive answer would be an upper bound of $(n - 1)^2$ since each path could contain at most $(n - 1)$ arcs. Fortunately, we need not use this much storage: $(n - 1)$ storage locations are sufficient to represent all these paths. This result follows from the fact that we can always find a directed out-tree rooted from the source with the property that the unique path from the source to any node is a shortest path to that node. For obvious reasons we refer to such a tree as a *shortest path tree*. Each shortest path algorithm discussed in this book is capable of determining this tree as it computes the shortest path distances. The existence of the shortest path tree relies on the following property.

Property 4.1. *If the path $s = i_1 - i_2 - \cdots - i_h = k$ is a shortest path from node s to node k, then for every $q = 2, 3, \ldots, h - 1$, the subpath $s = i_1 - i_2 - \cdots - i_q$ is a shortest path from the source node to node i_q.*

This property is fairly easy to establish. In Figure 4.5 we assume that the shortest path P_1–P_3 from node s to node k passes through some node p, but the subpath P_1 up to node p is not a shortest path to node p; suppose instead that path P_2 is a shorter path to node p. Notice that P_2–P_3 is a directed walk whose length is less than that of path P_1–P_3. Also, notice that any directed walk from node s to node k decomposes into a directed path plus some directed cycles (see Exercise 3.51), and these cycles, by our assumption, must have nonnegative length. As a result, some directed path from node s to node k is shorter than the path P_1–P_3, contradicting its optimality.

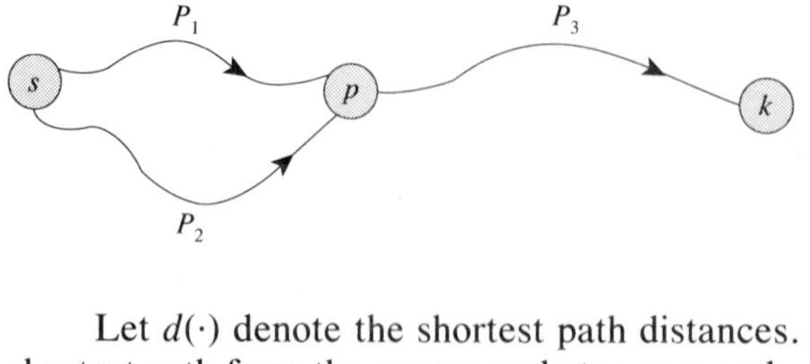

Figure 4.5 Proving Property 4.1.

Let $d(\cdot)$ denote the shortest path distances. Property 4.1 implies that if P is a shortest path from the source node to some node k, then $d(j) = d(i) + c_{ij}$ for every arc $(i, j) \in P$. The converse of this result is also true; that is, if $d(j) = d(i) + c_{ij}$

for every arc in a directed path P from the source to node k, then P must be a shortest path. To establish this result, let $s = i_1 - i_2 - \cdots - i_h = k$ be the node sequence in P. Then

$$d(k) = d(i_h) = (d(i_h) - d(i_{h-1})) + (d(i_{h-1}) - d(i_{h-2})) + \cdots + (d(i_2) - d(i_1)),$$

where we use the fact that $d(i_1) = 0$. By assumption, $d(j) - d(i) = c_{ij}$ for every arc $(i, j) \in P$. Using this equality we see that

$$d(k) = c_{i_{h-1}i_h} + c_{i_{h-2}i_{h-1}} + \cdots + c_{i_1i_2} = \sum_{(i,j)\in P} c_{ij}.$$

Consequently, P is a directed path from the source node to node k of length $d(k)$. Since, by assumption, $d(k)$ is the shortest path distance to node k, P must be a shortest path to node k. We have thus established the following result.

Property 4.2. *Let the vector d represent the shortest path distances. Then a directed path P from the source node to node k is a shortest path if and only if* $d(j) = d(i) + c_{ij}$ *for every arc* $(i, j) \in P$.

We are now in a position to prove the existence of a shortest path tree. Since only a finite number of paths connect the source to every node, the network contains a shortest path to every node. Property 4.2 implies that we can always find a shortest path from the source to every other node satisfying the property that for every arc (i, j) on the path, $d(j) = d(i) + c_{ij}$. Therefore, if we perform a breadth-first search of the network using the arcs satisfying the equality $d(j) = d(i) + c_{ij}$, we must be able to reach every node. The breadth-first search tree contains a unique path from the source to every other node, which by Property 4.2 must be a shortest path to that node.

4.4 SHORTEST PATH PROBLEMS IN ACYCLIC NETWORKS

Recall that a network is said to be *acyclic* if it contains no directed cycle. In this section we show how to solve the shortest path problem on an acyclic network in $O(m)$ time even though the arc lengths might be negative. Note that no other algorithm for solving the shortest path problem on acyclic networks could be any faster (in terms of the worst-case complexity) become any algorithm for solving the problem must examine every arc, which itself would take $O(m)$ time.

Recall from Section 3.4 that we can always number (or order) nodes in an acyclic network $G = (N, A)$ in $O(m)$ time so that $i < j$ for every arc $(i, j) \in A$. This ordering of nodes is called a *topological ordering*. Conceptually, once we have determined the topological ordering, the shortest path problem is quite easy to solve by a simple dynamic programming algorithm. Suppose that we have determined the shortest path distances $d(i)$ from the source node to nodes $i = 1, 2, \ldots, k - 1$. Consider node k. The topological ordering implies that all the arcs directed into this node emanate from one of the nodes 1 through $k - 1$. By Property 4.1, the shortest path to node k is composed of a shortest path to one of the nodes $i = 1, 2, \ldots, k - 1$ together with the arc (i, k). Therefore, to compute the shortest path distance to node k, we need only select the minimum of $d(i) + c_{ik}$ for all incoming arcs (i, k). This algorithm is a *pulling* algorithm in that to find the shortest path distance to any node, it "pulls" shortest path distances forward from lower-numbered nodes. Notice that to implement this algorithm, we need to access conveniently all the arcs directed into each node. Since we frequently store the adjacency list $A(i)$ of each node i, which gives the arcs emanating out of a node, we might also like to implement a *reaching* algorithm that propagates information from each node to higher-indexed nodes, and so uses the usual adjacency list. We next describe one such algorithm.

We first set $d(s) = 0$ and the remaining distance labels to a very large number. Then we examine nodes in the topological order and for each node i being examined, we scan arcs in $A(i)$. If for any arc $(i, j) \in A(i)$, we find that $d(j) > d(i) + c_{ij}$, then we set $d(j) = d(i) + c_{ij}$. When the algorithm has examined all the nodes once in this order, the distance labels are optimal.

We use induction to show that whenever the algorithm examines a node, its distance label is optimal. Suppose that the algorithm has examined nodes $1, 2, \ldots, k$ and their distance labels are optimal. Consider the point at which the algorithm examines node $k + 1$. Let the shortest path from the source to node $k + 1$ be $s = i_1 - i_2 - \cdots - i_h - (k + 1)$. Observe that the path $i_1 - i_2 - \cdots - i_h$ must be a shortest path from the source to node i_h (by Property 4.1). The facts that the nodes are topologically ordered and that the arc $(i_h, k + 1) \in A$ imply that $i_h \in \{1, 2, \ldots, k\}$ and, by the inductive hypothesis, the distance label of node i_h is equal to the length of the path $i_1 - i_2 - \cdots - i_h$. Consequently, while examining node i_h, the algorithm must have scanned the arc $(i_h, k + 1)$ and set the distance label of node $(k + 1)$ equal to the length of the path $i_1 - i_2 - \cdots - i_h - (k + 1)$. Therefore, when the algorithm examines the node $k + 1$, its distance label is optimal. The following result is now immediate.

Theorem 4.3. *The reaching algorithm solves the shortest path problem on acyclic networks in $O(m)$ time.*

In this section we have seen how we can solve the shortest path problem on acyclic networks very efficiently using the simplest possible algorithm. Unfortunately, we cannot apply this one-pass algorithm, and examine each node and each arc exactly once, for networks containing cycles; nevertheless, we can utilize the same basic reaching strategy used in this algorithm and solve any shortest path problem with nonnegative arc lengths using a modest additional amount of work. As we will see, we incur additional work because we no longer have a set order for examining the nodes, so at each step we will need to investigate several nodes in order to determine which node to reach out from next.

4.5 DIJKSTRA'S ALGORITHM

As noted previously, Dijkstra's algorithm finds shortest paths from the source node s to all other nodes in a network with nonnegative arc lengths. Dijkstra's algorithm maintains a distance label $d(i)$ with each node i, which is an upper bound on the shortest path length to node i. At any intermediate step, the algorithm divides the nodes into two groups: those which it designates as *permanently labeled* (or permanent) and those it designates as *temporarily labeled* (or temporary). The distance label to any permanent node represents the shortest distance from the source to that node. For any temporary node, the distance label is an upper bound on the shortest path distance to that node. The basic idea of the algorithm is to fan out from node s and permanently label nodes in the order of their distances from node s. Initially, we give node s a permanent label of zero, and each other node j a temporary label equal to ∞. At each iteration, the label of a node i is its shortest distance from the source node along a path whose internal nodes (i.e., nodes other than s or the node i itself) are all permanently labeled. The algorithm selects a node i with the minimum temporary label (breaking ties arbitrarily), makes it permanent, and reaches out from that node—that is, scans arcs in $A(i)$ to update the distance labels of adjacent nodes. The algorithm terminates when it has designated all nodes as permanent. The correctness of the algorithm relies on the key observation (which we prove later) that we can always designate the node with the minimum temporary label as permanent.

Dijkstra's algorithm maintains a directed out-tree T rooted at the source that spans the nodes with finite distance labels. The algorithm maintains this tree using predecessor indices [i.e., if $(i, j) \in T$, then $pred(j) = i$]. The algorithm maintains the invariant property that every tree arc (i, j) satisfies the condition $d(j) = d(i) + c_{ij}$ with respect to the current distance labels. At termination, when distance labels represent shortest path distances, T is a shortest path tree (from Property 4.2).

Figure 4.6 gives a formal algorithmic description of Dijkstra's algorithm.

In Dijkstra's algorithm, we refer to the operation of selecting a minimum temporary distance label as a *node selection* operation. We also refer to the operation of checking whether the current labels for nodes i and j satisfy the condition $d(j) > d(i) + c_{ij}$ and, if so, then setting $d(j) = d(i) + c_{ij}$ as a *distance update* operation.

We illustrate Dijkstra's algorithm using the numerical example given in Figure 4.7(a). The algorithm permanently labels the nodes 3, 4, 2, and 5 in the given sequence: Figure 4.7(b) to (e) illustrate the operations for these iterations. Figure 4.7(f) shows the shortest path tree for this example.

```
algorithm Dijkstra;
begin
    S : = Ø; S̄: = N;
    d(i) : = ∞ for each node i ∈ N;
    d(s) : = 0 and pred(s) : = 0;
    while |S| < n do
    begin
        let i ∈ S̄ be a node for which d(i) = min{d(j) : j ∈ S̄};
        S : = S ∪ {i};
        S̄ : = S̄ − {i};
        for each (i, j) ∈ A(i) do
            if d(j) > d(i) + c_ij then d(j) : = d(i) + c_ij and pred(j) : = i;
    end;
end;
```

Figure 4.6 Dijkstra's algorithm.

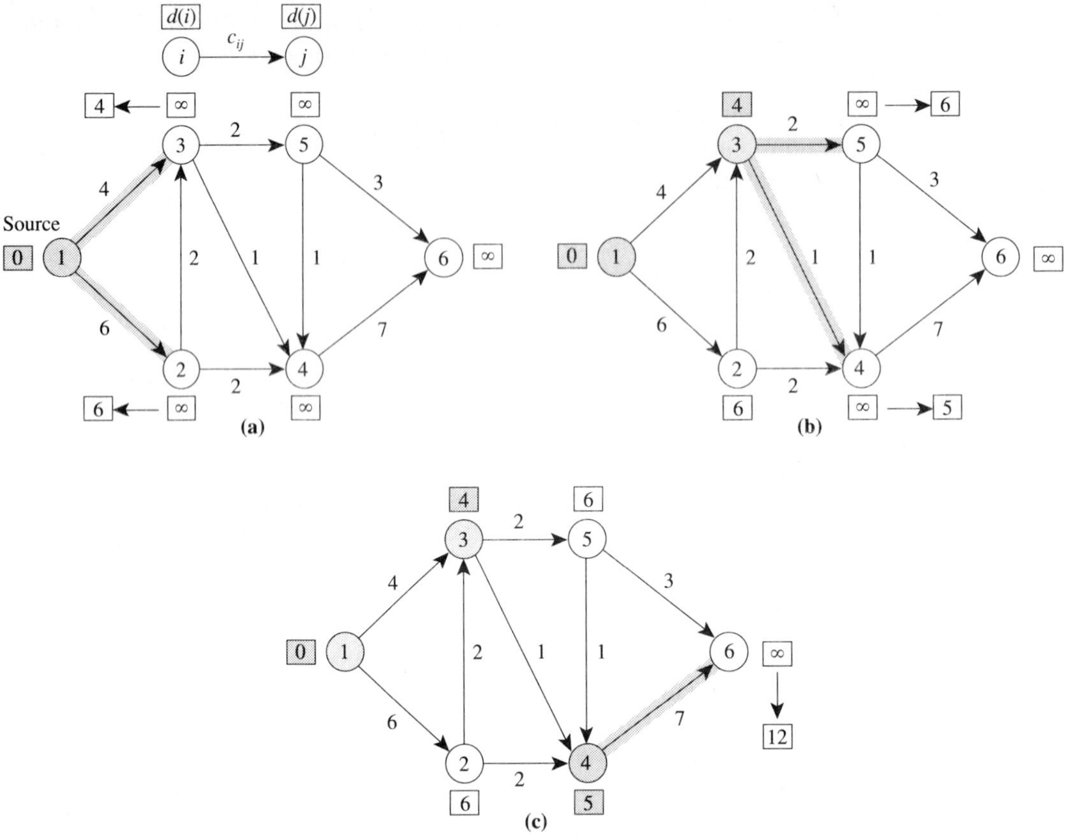

Figure 4.7 Illustrating Dijkstra's algorithm.

Correctness of Dijkstra's Algorithm

We use inductive arguments to establish the validity of Dijkstra's algorithm. At any iteration, the algorithm has partitioned the nodes into two sets, S and $\bar{S}$. Our induction hypothesis are (1) that the distance label of each node in S is optimal, and (2) that the distance label of each node in $\bar{S}$ is the shortest path length from the source provided that each internal node in the path lies in S. We perform induction on the cardinality of the set S.

To prove the first inductive hypothesis, recall that at each iteration the algorithm transfers a node i in the set $\bar{S}$ with smallest distance label to the set S. We need to show that the distance label $d(i)$ of node i is optimal. Notice that by our induction hypothesis, $d(i)$ is the length of a shortest path to node i among all paths that do not contain any node in $\bar{S}$ as an internal node. We now show that the length of any path from s to i that contains some nodes in $\bar{S}$ as an internal node will be at least $d(i)$. Consider any path P from the source to node i that contains at least one node in $\bar{S}$ as an internal node. The path P can be decomposed into two segments P_1 and P_2: the path segment P_1 does not contain any node in $\bar{S}$ as an internal node, but terminates at a node k in $\bar{S}$ (see Figure 4.8). By the induction hypothesis, the length of the path P_1 is at least $d(k)$ and since node i is the smallest distance label

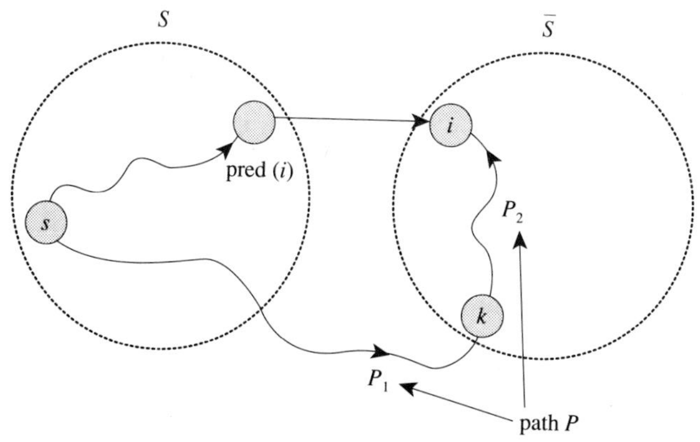

Figure 4.8 Proving Dijkstra's algorithm.

in $\bar{S}$, $d(k) \geq d(i)$. Therefore, the path segment P_1 has length at least $d(i)$. Furthermore, since all arc lengths are nonnegative, the length of the path segment P_2 is nonnegative. Consequently, length of the path P is at least $d(i)$. This result establishes the fact that $d(i)$ is the shortest path length of node i from the source node.

We next show that the algorithm preserves the second induction hypothesis. After the algorithm has labeled a new node i permanently, the distance labels of some nodes in $\bar{S} - \{i\}$ might decrease, because node i could become an internal node in the tentative shortest paths to these nodes. But recall that after permanently labeling node i, the algorithm examines each arc $(i, j) \in A(i)$ and if $d(j) > d(i) + c_{ij}$, then it sets $d(j) = d(i) + c_{ij}$ and pred$(j) = i$. Therefore, after the distance update operation, by the induction hypothesis the path from node j to the source node defined by the predecessor indices satisfies Property 4.2 and so the distance label of each node in $\bar{S} - \{i\}$ is the length of a shortest path subject to the restriction that each internal node in the path must belong to $S \cup \{i\}$.

Running Time of Dijkstra's Algorithm

We now study the worst-case complexity of Dijkstra's algorithm. We might view the computational time for Dijkstra's algorithm as allocated to the following two basic operations:

1. *Node selections*. The algorithm performs this operation n times and each such operation requires that it scans each temporarily labeled node. Therefore, the total node selection time is $n + (n - 1) + (n - 2) + \cdots + 1 = O(n^2)$.
2. *Distance updates*. The algorithm performs this operation $|A(i)|$ times for node i. Overall, the algorithm performs this operation $\sum_{i \in N} |A(i)| = m$ times. Since each distance update operation requires $O(1)$ time, the algorithm requires $O(m)$ total time for updating all distance labels.

We have established the following result.

Theorem 4.4. *Dijkstra's algorithm solves the shortest path problem in $O(n^2)$ time.*

The $O(n^2)$ time bound for Dijkstra's algorithm is the best possible for completely dense networks [i.e., $m = \Omega(n^2)$], but can be improved for sparse networks. Notice that the times required by the node selections and distance updates are not balanced. The node selections require a total of $O(n^2)$ time, and the distance updates require only $O(m)$ time. Researchers have attempted to reduce the node selection time without substantially increasing the time for updating the distances. Consequently, they have, using clever data structures, suggested several implementations of the algorithm. These implementations have either dramatically reduced the running time of the algorithm in practice or improved its worst-case complexity. In Section 4.6 we describe Dial's algorithm, which is an excellent implementation of Dijkstra's algorithm in practice. Sections 4.7 and 4.8 describe several implementations of Dijkstra's algorithm with improved worst-case complexity.

Reverse Dijkstra's Algorithm

In the (forward) Dijkstra's algorithm, we determine a shortest path from node s to every other node in $N - \{s\}$. Suppose that we wish to determine a shortest path from every node in $N - \{t\}$ to a sink node t. To solve this problem, we use a slight modification of Dijkstra's algorithm, which we refer to as the *reverse Dijkstra's algorithm*. The reverse Dijkstra's algorithm maintains a distance $d'(j)$ with each node j, which is an upper bound on the shortest path length from node j to node t. As before, the algorithm designates a set of nodes, say S', as permanently labeled and the remaining set of nodes, say $\bar{S}'$, as temporarily labeled. At each iteration, the algorithm designates a node with the minimum temporary distance label, say $d'(j)$, as permanent. It then examines each incoming arc (i, j) and modifies the distance label of node i to $\min\{d'(i), c_{ij} + d'(j)\}$. The algorithm terminates when all the nodes have become permanently labeled.

Bidirectional Dijkstra's Algorithm

In some applications of the shortest path problem, we need not determine a shortest path from node s to every other node in the network. Suppose, instead, that we want to determine a shortest path from node s to a specified node t. To solve this problem and eliminate some computations, we could terminate Dijkstra's algorithm as soon as it has selected t from $\bar{S}$ (even though some nodes are still temporarily labeled). The bidirectional Dijkstra's algorithm, which we describe next, allows us to solve this problem even faster in practice (though not in the worst case).

In the bidirectional Dijkstra's algorithm, we simultaneously apply the forward Dijkstra's algorithm from node s and reverse Dijkstra's algorithm from node t. The algorithm alternatively designates a node in $\bar{S}$ and a node in $\bar{S}'$ as permanent until both the forward and reverse algorithms have permanently labeled the same node, say node k (i.e., $S \cap S' = \{k\}$). At this point, let $P(i)$ denote the shortest path from node s to node $i \in S$ found by the forward Dijkstra's algorithm, and let $P'(j)$ denote the shortest path from node $j \in S'$ to node t found by the reverse Dijkstra's algorithm. A straightforward argument (see Exercise 4.52) shows that the shortest path from node s to node t is either the path $P(k) \cup P'(k)$ or a path $P(i) \cup \{(i, j)\} \cup P'(j)$ for some arc (i, j), $i \in S$ and $j \in S'$. This algorithm is very efficient because it tends to permanently label few nodes and hence never examines the arcs incident to a large number of nodes.

4.6 DIAL'S IMPLEMENTATION

The bottleneck operation in Dijkstra's algorithm is node selection. To improve the algorithm's performance, we need to address the following question. Instead of scanning all temporarily labeled nodes at each iteration to find the one with the minimum distance label, can we reduce the computation time by maintaining distances in some sorted fashion? Dial's algorithm tries to accomplish this objective, and reduces the algorithm's computation time in practice, using the following fact:

Property 4.5. *The distance labels that Dijkstra's algorithm designates as permanent are nondecreasing.*

This property follows from the fact that the algorithm permanently labels a node i with a smallest temporary label $d(i)$, and while scanning arcs in $A(i)$ during the distance update operations, never decreases the distance label of any temporarily labeled node below $d(i)$ because arc lengths are nonnegative.

Dial's algorithm stores nodes with finite temporary labels in a sorted fashion. It maintains $nC + 1$ sets, called *buckets*, numbered $0, 1, 2, \ldots, nC$: Bucket k stores all nodes with temporary distance label equal to k. Recall that C represents the largest arc length in the network, and therefore nC is an upper bound on the distance label of any finitely labeled node. We need not store nodes with infinite temporary distance labels in any of the buckets—we can add them to a bucket when they first receive a finite distance label. We represent the content of bucket k by the set *content(k)*.

In the node selection operation, we scan buckets numbered $0, 1, 2, \ldots,$ until we identify the first nonempty bucket. Suppose that bucket k is the first nonempty bucket. Then each node in *content(k)* has the minimum distance label. One by one, we delete these nodes from the bucket, designate them as permanently labeled, and scan their arc lists to update the distance labels of adjacent nodes. Whenever we update the distance label of a node i from d_1 to d_2, we move node i from $content(d_1)$ to $content(d_2)$. In the next node selection operation, we resume the scanning of buckets numbered $k + 1, k + 2, \ldots$ to select the next nonempty bucket. Property 4.5 implies that the buckets numbered $0, 1, 2, \ldots, k$ will always be empty in the subsequent iterations and the algorithm need not examine them again.

As a data structure for storing the content of the buckets, we store each set *content(k)* as a doubly linked list (see Appendix A). This data structure permits us to perform each of the following operations in $O(1)$ time: (1) checking whether a bucket is empty or nonempty, (2) deleting an element from a bucket, and (3) adding an element to a bucket. With this data structure, the algorithm requires $O(1)$ time for each distance update, and thus a total of $O(m)$ time for all distance updates. The bottleneck operation in this implementation is scanning $nC + 1$ buckets during node selections. Consequently, the running time of Dial's algorithm is $O(m + nC)$.

Since Dial's algorithm uses $nC + 1$ buckets, its memory requirements can be prohibitively large. The following fact allows us to reduce the number of buckets to $C + 1$.

Property 4.6. *If $d(i)$ is the distance label that the algorithm designates as permanent at the beginning of an iteration, then at the end of that iteration, $d(j) \leq d(i) + C$ for each finitely labeled node j in $\bar{S}$.*

This fact follows by noting that (1) $d(l) \leq d(i)$ for each node $l \in S$ (by Property 4.5), and (2) for each finitely labeled node j in $\bar{S}$, $d(j) = d(l) + c_{lj}$ for some node $l \in S$ (by the property of distance updates). Therefore, $d(j) = d(l) + c_{lj} \leq d(i) + C$. In other words, all finite temporary labels are bracketed from below by $d(i)$ and from above by $d(i) + C$. Consequently, $C + 1$ buckets suffice to store nodes with finite temporary distance labels.

Dial's algorithm uses $C + 1$ buckets numbered $0, 1, 2, \ldots, C$, which we might view as arranged in a circular fashion as in Figure 4.9. We store a temporarily labeled node j with distance label $d(j)$ in the bucket $d(j) \bmod (C + 1)$. Consequently, during the entire execution of the algorithm, bucket k stores nodes with temporary distance labels k, $k + (C + 1)$, $k + 2(C + 1)$, and so on; however, because of Property 4.6, at any point in time, this bucket will hold only nodes with the same distance label. This storage scheme also implies that if bucket k contains a node with the minimum distance label, then buckets $k + 1, k + 2, \ldots, C, 0, 1, 2, \ldots, k - 1$ store nodes in increasing values of the distance labels.

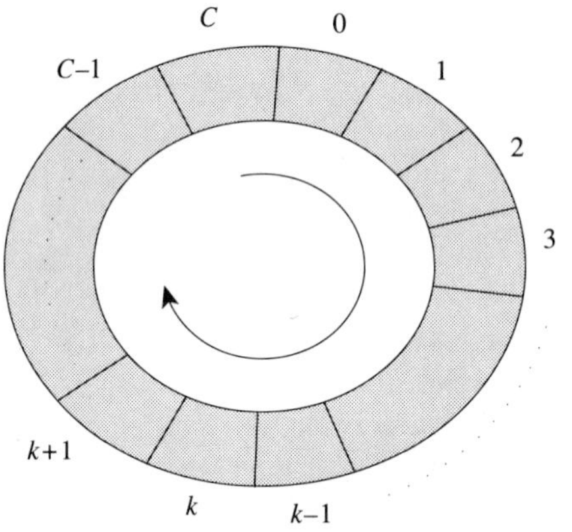

Figure 4.9 Bucket arrangement in Dial's algorithm.

Dial's algorithm examines the buckets sequentially, in a wraparound fashion, to identify the first nonempty bucket. In the next iteration, it reexamines the buckets starting at the place where it left off previously. A potential disadvantage of Dial's algorithm compared to the original $O(n^2)$ implementation of Dijkstra's algorithm is that it requires a large amount of storage when C is very large. In addition, because the algorithm might wrap around as many as $n - 1$ times, the computational time could be large. The algorithm runs in $O(m + nC)$ time, which is not even polynomial, but rather, is pseudopolynomial. For example, if $C = n^4$, the algorithm runs in $O(n^5)$ time, and if $C = 2^n$, the algorithm requires exponential time in the worst case. However, the algorithm typically does not achieve the bound of $O(m + nC)$ time. For most applications, C is modest in size, and the number of passes through all of the buckets is much less than $n - 1$. Consequently, the running time of Dial's algorithm is much better than that indicated by its worst-case complexity.

4.7 HEAP IMPLEMENTATIONS

This section requires that the reader is familiar with heap data structures. We refer an unfamiliar reader to Appendix A, where we describe several such data structures.

A *heap* (or *priority queue*) is a data structure that allows us to perform the following operations on a collection H of *objects*, each with an associated real number called its *key*. More properly, a priority queue is an abstract data type, and is usually implemented using one of several heap data structures. However, in this treatment we are using the words "heap" and "priority queue" interchangeably.

create-heap(H). Create an empty heap.

find-min(i, H). Find and return an object i of minimum key.

insert(i, H). Insert a new object i with a predefined key.

decrease-key(value, i, H). Reduce the key of an object i from its current value to *value*, which must be smaller than the key it is replacing.

delete-min(i, H). Delete an object i of minimum key.

If we implement Dijkstra's algorithm using a heap, H would be the collection of nodes with finite temporary distance labels and the key of a node would be its distance label. Using a heap, we could implement Dijkstra's algorithm as described in Figure 4.10.

As is clear from this description, the heap implementation of Dijkstra's algorithm performs the operations find-min, delete-min, and insert at most n times and the operation decrease-key at most m times. We now analyze the running times of Dijkstra's algorithm implemented using different types of heaps: binary heaps, d-heaps, Fibonacci heaps, and another data structure suggested by Johnson. We describe the first three of these four data structures in Appendix A and provide a reference for the fourth data structure in the reference notes.

```
algorithm heap-Dijkstra;
begin
    create-heap(H);
    d(j) : = ∞ for all j ∈ N;
    d(s) : = 0 and pred(s) : = 0;
    insert(s, H);
    while H ≠ Ø do
    begin
        find-min(i, H);
        delete-min(i, H);
        for each (i, j) ∈ A(i) do
        begin
            value : = d(i) + c_ij;
            if d(j) > value then
                if d(j) = ∞ then d(j) : = value, pred(j) : = i, and insert (j, H)
                else set d(j) : = value, pred(j) : = i, and decrease-key(value, i, H);
        end;
    end;
end;
```

Figure 4.10 Dijkstra's algorithm using a heap.

Binary heap implementation. As discussed in Appendix A, a binary heap data structure requires $O(\log n)$ time to perform insert, decrease-key, and delete-min, and it requires $O(1)$ time for the other heap operations. Consequently, the binary heap version of Dijkstra's algorithm runs in $O(m \log n)$ time. Notice that the binary heap implementation is slower than the original implementation of Dijkstra's algorithm for completely dense networks [i.e., $m = \Omega(n^2)$], but is faster when $m = O(n^2/\log n)$.

***d*-Heap implementation.** For a given parameter $d \geq 2$, the d-heap data structure requires $O(\log_d n)$ time to perform the insert and decrease-key operations; it requires $O(d \log_d n)$ time for delete-min, and it requires $O(1)$ steps for the other heap operations. Consequently, the running time of this version of Dijkstra's algorithm is $O(m \log_d n + nd \log_d n)$. To obtain an optimal choice of d, we equate the two terms (see Section 3.2), giving $d = \max\{2, \lceil m/n \rceil\}$. The resulting running time is $O(m \log_d n)$. Observe that for very sparse networks [i.e., $m = O(n)$], the running time of the d-heap implementation is $O(n \log n)$. For nonsparse networks [i.e., $m = \Omega(n^{1+\epsilon})$ for some $\epsilon > 0$], the running time of d-heap implementation is $O(m \log_d n) = O((m \log n)/(\log d)) = O((m \log n)/(\log n^{\epsilon})) = O((m \log n)/(\epsilon \log n)) = O(m/\epsilon) = O(m)$. The last equality is true since ϵ is a constant. Thus the running time is $O(m)$, which is optimal.

Fibonacci heap implementation. The Fibonacci heap data structure performs every heap operation in $O(1)$ amortized time except delete-min, which requires $O(\log n)$ time. Consequently the running time of this version of Dijkstra's algorithm is $O(m + n \log n)$. This time bound is consistently better than that of binary heap and d-heap implementations for all network densities. This implementation is also currently the best strongly polynomial-time algorithm for solving the shortest path problem.

Johnson's implementation. Johnson's data structure (see the reference notes) is applicable only when all arc lengths are integer. This data structure requires $O(\log \log C)$ time to perform each heap operation. Consequently, this implementation of Dijkstra's algorithm runs in $O(m \log \log C)$ time.

We next discuss one more heap implementation of Dijkstra's algorithm, known as the *radix heap implementation*. The radix heap implementation is one of the more recent implementations; its running time is $O(m + n \log(nC))$.

4.8 RADIX HEAP IMPLEMENTATION

The radix heap implementation of Dijkstra's algorithm is a hybrid of the original $O(n^2)$ implementation and Dial's implementation (the one that uses $nC + 1$ buckets). These two implementations represent two extremes. The original implementation considers all the temporarily labeled nodes together (in one large bucket, so to speak) and searches for a node with the smallest label. Dial's algorithm uses a large number of buckets and separates nodes by storing any two nodes with different labels in different buckets. The radix heap implementation improves on these methods by adopting an intermediate approach: It stores many, but not all, labels in a bucket. For example, instead of storing only nodes with a temporary label k in the kth bucket, as in Dial's implementation, we might store temporary labels from $100k$ to $100k + 99$ in bucket k. The different temporary labels that can be stored in a bucket make up the *range* of the bucket; the cardinality of the range is called its *width*. For the preceding example, the range of bucket k is $[100k, 100k + 99]$ and its width is 100. Using widths of size k permits us to reduce the number of buckets needed by a factor of k. But to find the smallest distance label, we need to search all of the elements in the smallest indexed nonempty bucket. Indeed, if k is arbitrarily large, we need only one bucket, and the resulting algorithm reduces to Dijkstra's original implementation.

Using a width of 100, say, for each bucket reduces the number of buckets, but still requires us to search through the lowest-numbered nonempty bucket to find the node with minimum temporary label. If we could devise a variable width scheme, with a width of 1 for the lowest-numbered bucket, we could conceivably retain the advantages of both the wide bucket and narrow bucket approaches. The radix heap algorithm we consider next uses variable widths and changes the ranges dynamically. In the version of the radix heap that we present:

1. The widths of the buckets are 1, 1, 2, 4, 8, 16, . . . , so that the number of buckets needed is only $O(\log(nC))$.
2. We dynamically modify the ranges of the buckets and we reallocate nodes with temporary distance labels in a way that stores the minimum distance label in a bucket whose width is 1.

Property 1 allows us to maintain only $O(\log(nC))$ buckets and thereby overcomes the drawback of Dial's implementation of using too many buckets. Property 2 permits us, as in Dial's algorithm, to avoid the need to search the entire bucket to find a node with the minimum distance label. When implemented in this way, this version of the radix heap algorithm has a running time of $O(m + n \log(nC))$.

To describe the radix heap in more detail, we first set some notation. For a given shortest path problem, the radix heap consists of $1 + \lceil \log(nC) \rceil$ buckets. The buckets are numbered $0, 1, 2, \ldots, K = \lceil \log(nC) \rceil$. We represent the range of bucket k by *range(k)* which is a (possibly empty) closed interval of integers. We store a temporary node i in bucket k if $d(i) \in \text{range}(k)$. We do not store permanent nodes. The set *content(k)* denotes the nodes in bucket k. The algorithm will change the ranges of the buckets dynamically, and each time it changes the ranges, it redistributes the nodes in the buckets. Initially, the buckets have the following ranges:

$$
\begin{aligned}
\text{range}(0) &= [0];\\
\text{range}(1) &= [1];\\
\text{range}(2) &= [2, 3];\\
\text{range}(3) &= [4, 7];\\
\text{range}(4) &= [8, 15];\\
&\vdots\\
\text{range}(K) &= [2^{K-1}, 2^K - 1].
\end{aligned}
$$

These ranges change as the algorithm proceeds; however, the widths of the buckets never increase beyond their initial widths.

As we have noted the fundamental difficulty associated with using bucket widths larger than 1, as in the radix heap algorithm, is that we have to examine every node in the bucket containing a node with the minimum distance label and this time might be "too large" from a worst-case perspective. The radix heap algorithm overcomes this difficulty in the following manner. Suppose that at some stage the minimum indexed nonempty bucket is bucket 4, whose range is [8, 15]. The algorithm would examine every node in content(4) to identify a node with the smallest distance label. Suppose that the smallest distance label of a node in content(4) is 9. Property 4.5 implies that no temporary distance label will ever again be less than 9 and, consequently, we will never again need the buckets 0 to 3. Rather than leaving these buckets idle, the algorithm redistributes the range [9, 15] to the previous buckets, resulting in the ranges range(0) = [9], range(1) = [10], range(2) = [11, 12], range(3) = [13,15] and range(4) = $\varnothing$. Since the range of bucket 4 is now empty, the algorithm shifts (or redistributes) the nodes in content(4) into the appropriate buckets (0, 1, 2, and 3). Thus each of the nodes in bucket 4 moves to a lower-indexed bucket and all nodes with the smallest distance label move to bucket 0, which has width 1.

To summarize, whenever the algorithm finds that nodes with the minimum distance label are in a bucket with width larger than 1, it examines all nodes in the bucket to identify a node with minimum distance label. Then the algorithm redistributes the bucket ranges and shifts each node in the bucket to the lower-indexed bucket. Since the radix heap contains K buckets, a node can shift at most K times, and consequently, the algorithm will examine any node at most K times. Hence the total number of node examinations is $O(nK)$, which is not "too large."

We now illustrate the radix heap data structure on the shortest path example given in Figure 4.11 with $s = 1$. In the figure, the number beside each arc indicates its length. For this problem $C = 20$ and $K = \lceil \log(120) \rceil = 7$. Figure 4.12 specifies the distance labels determined by Dijkstra's algorithm after it has examined node 1; it also shows the corresponding radix heap.

To select the node with the smallest distance label, we scan the buckets 0, 1, 2, . . . , K to find the first nonempty bucket. In our example, bucket 0 is nonempty. Since bucket 0 has width 1, every node in this bucket has the same (minimum) distance label. So the algorithm designates node 3 as permanent, deletes node 3 from the radix heap, and scans the arc (3, 5) to change the distance label of node 5 from

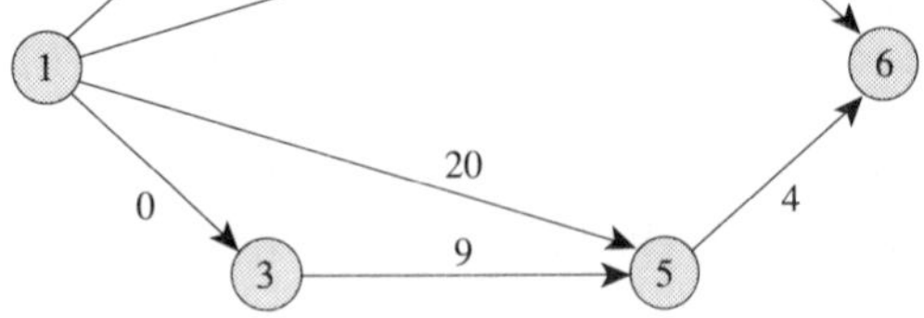

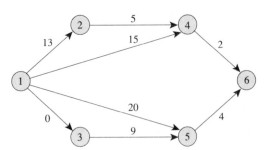

Figure 4.11 Shortest path example.

node i	1	2	3	4	5	6
label $d(i)$	0	13	0	15	20	∞

bucket k	0	1	2	3	4	5	6	7
range(k)	[0]	[1]	[2, 3]	[4, 7]	[8, 15]	[16, 31]	[32, 63]	[64, 127]
content(k)	{3}	∅	∅	∅	{2, 4}	{5}	∅	

Figure 4.12 Initial radix heap.

20 to 9. We check whether the new distance label of node 5 is contained in the range of its present bucket, which is bucket 5. It is not. Since its distance label has decreased, node 5 should move to a lower-indexed bucket. So we sequentially scan the buckets from right to left, starting at bucket 5, to identify the first bucket whose range contains the number 9, which is bucket 4. Node 5 moves from bucket 5 to bucket 4. Figure 4.13 shows the new radix heap.

node i	2	4	5	6
label $d(i)$	13	15	9	∞

bucket k	0	1	2	3	4	5	6	7
range(k)	[0]	[1]	[2, 3]	[4, 7]	[8, 15]	[16, 31]	[32, 63]	[64, 127]
content(k)	∅	∅	∅	∅	{2, 4, 5}	∅	∅	∅

Figure 4.13 Radix heap at the end of iteration 1.

We again look for the node with the smallest distance label. Scanning the buckets sequentially, we find that bucket $k = 4$ is the first nonempty bucket. Since the range of this bucket contains more than one integer, the first node in the bucket need not have the minimum distance label. Since the algorithm will never use the ranges range(0), . . . , range($k - 1$) for storing temporary distance labels, we can redistribute the range of bucket k into the buckets 0, 1, . . . , $k - 1$, and reinsert its nodes into the lower-indexed buckets. In our example, the range of bucket 4 is [8, 15], but the smallest distance label in this bucket is 9. We therefore redistribute the range [9, 15] over the lower-indexed buckets in the following manner:

$$\begin{aligned}
\text{range}(0) &= [9], \\
\text{range}(1) &= [10], \\
\text{range}(2) &= [11, 12], \\
\text{range}(3) &= [13, 15], \\
\text{range}(4) &= \varnothing.
\end{aligned}$$

Other ranges do not change. The range of bucket 4 is now empty, and we must reassign the contents of bucket 4 to buckets 0 through 3. We do so by successively selecting nodes in bucket 4, sequentially scanning the buckets 3, 2, 1, 0 and inserting the node in the appropriate bucket. The resulting buckets have the following contents:

$$\begin{aligned}
\text{content}(0) &= \{5\}, \\
\text{content}(1) &= \varnothing, \\
\text{content}(2) &= \varnothing, \\
\text{content}(3) &= \{2, 4\}, \\
\text{content}(4) &= \varnothing.
\end{aligned}$$

This redistribution necessarily empties bucket 4 and moves the node with the smallest distance label to bucket 0.

We are now in a position to outline the general algorithm and analyze its complexity. We first consider moving nodes between the buckets. Suppose that $j \in$ content(k) and that we are re-assigning node j to a lower-numbered bucket (because either $d(j)$ decreases or we are redistributing the useful range of bucket k and removing the nodes from this bucket). If $d(j) \notin$ range(k), we sequentially scan lower-numbered buckets from right to left and add the node to the appropriate bucket. Overall, this operation requires $O(m + nK)$ time. The term m reflects the number of distance updates, and the term nK arises because every time a node moves, it moves to a lower-indexed bucket: Since there are $K + 1$ buckets, a node can move at most K times. Therefore, $O(nK)$ is a bound on the total number of node movements.

Next we consider the node selection operation. Node selection begins by scanning the buckets from left to right to identify the first nonempty bucket, say bucket k. This operation requires $O(K)$ time per iteration and $O(nK)$ time in total. If $k = 0$ or $k = 1$, any node in the selected bucket has the minimum distance label. If $k \geq 2$, we redistribute the "useful" range of bucket k into the buckets $0, 1, \ldots, k - 1$ and reinsert its contents in those buckets. If the range of bucket k is $[l, u]$ and the smallest distance label of a node in the bucket is $d_{\min}$, the useful range of the bucket is $[d_{\min}, u]$.

The algorithm redistributes the useful range in the following manner: We assign the first integer to bucket 0, the next integer to bucket 1, the next two integers to bucket 2, the next four integers to bucket 3, and so on. Since bucket k has width less than 2^{k-1}, and since the widths of the first k buckets can be as large as $1, 1, 2, \ldots, 2^{k-2}$ for a total potential width of 2^{k-1}, we can redistribute the useful range of bucket k over the buckets $0, 1, \ldots, k - 1$ in the manner described. This redistribution of ranges and the subsequent reinsertions of nodes empties bucket k and moves the nodes with the smallest distance labels to bucket 0. The redistribution of ranges requires $O(K)$ time per iteration and $O(nK)$ time over all the iterations. As we have already shown, the algorithm requires $O(nK)$ time in total to move nodes and reinsert them in lower-indexed buckets. Consequently, the running time of the algorithm is $O(m + nK)$. Since $K = \lceil \log(nC) \rceil$, the algorithm runs in $O(m + n \log(nC))$ time. We summarize our discussion as follows.

Theorem 4.7. *The radix heap implementation of Dijkstra's algorithm solves the shortest path problem in $O(m + n \log(nC))$ time.*

This algorithm requires $1 + \lceil \log(nC) \rceil$ buckets. As in Dial's algorithm, Property 4.6 permits us to reduce the number of buckets to $1 + \lceil \log C \rceil$. This refined implementation of the algorithm runs in $O(m + n \log C)$ time. Using a Fibonacci heap data structure within the radix heap implementation, it is possible to reduce this bound further to $O(m + n \sqrt{\log C})$, which gives one of the fastest polynomial-time algorithm to solve the shortest path problem with nonnegative arc lengths.

4.9 SUMMARY

The shortest path problem is a core model that lies at the heart of network optimization. After describing several applications, we developed several algorithms for solving shortest path problems with nonnegative arc lengths. These algorithms, known as *label-setting algorithms*, assign tentative distance labels to the nodes and then iteratively identify a true shortest path distance (a permanent label) to one or more nodes at each step. The shortest path problem with arbitrary arc lengths requires different solution approaches; we address this problem class in Chapter 5.

The basic shortest path problem that we studied requires that we determine a shortest (directed) path from a source node s to each node $i \in N - \{s\}$. We showed how to store these $(n - 1)$ shortest paths compactly in the form of a directed out-tree rooted at node s, called the tree of shortest paths. This result uses the fact that if P is a shortest path from node s to some node j, then any subpath of P from node s to any of its internal nodes is also a shortest path to this node.

We began our discussion of shortest path algorithms by describing an $O(m)$ algorithm for solving the shortest path problem in acyclic networks. This algorithm computes shortest path distances to the nodes as it examines them in a topological order. This discussion illustrates a fact that we will revisit many times throughout this book: It is often possible to develop very efficient algorithms when we restrict the underlying network by imposing special structure on the data or on the network's topological structure (as in this case).

We next studied Dijkstra's algorithm, which is a natural and simple algorithm for solving shortest path problems with nonnegative arc lengths. After describing the original implementation of Dijkstra's algorithm, we examined several other implementations that either improve on its running time in practice or improve on its worst-case complexity. We considered the following implementations: Dial's implementation, a d-heap implementation, a Fibonacci heap implementation, and a radix heap implementation. Figure 4.14 summarizes the basic features of these implementations.

Algorithm	Running time	Features
Original implementation	$O(n^2)$	1. Selects a node with the minimum temporary distance label, designating it as permanent, and examines arcs incident to it to modify other distance labels. 2. Very easy to implement. 3. Achieves the best available running time for dense networks.
Dial's implementation	$O(m + nC)$	1. Stores the temporary labeled nodes in a sorted order in unit length buckets and identifies the minimum temporary distance label by sequentially examining the buckets. 2. Easy to implement and has excellent empirical behavior. 3. The algorithm's running time is pseudopolynomial and hence is theoretically unattractive.
d-Heap implementation	$O(m \log_d n)$, where $d = m/n$	1. Uses the d-heap data structure to maintain temporary labeled nodes. 2. Linear running time whenever $m = \Omega(n^{1+\epsilon})$ for any positive $\epsilon > 0$.
Fibonacci heap implementation	$O(m + n \log n)$	1. Uses the Fibonacci heap data structure to maintain temporary labeled nodes. 2. Achieves the best available strongly polynomial running time for solving shortest paths problems. 3. Intricate and difficult to implement.
Radix heap implementation	$O(m + n \log(nC))$	1. Uses a radix heap to implement Dijkstra's algorithm. 2. Improves Dial's algorithm by storing temporarily labeled nodes in buckets with varied widths. 3. Achieves an excellent running time for problems that satisfy the similarity assumption.

Figure 4.14 Summary of different implementations of Dijkstra's algorithm.

REFERENCE NOTES

The shortest path problem and its generalizations have a voluminous research literature. As a guide to these results before 1984, we refer the reader to the extensive bibliography compiled by Deo and Pang [1984]. In this discussion we present some selected references; additional references can be found in the survey papers of Ahuja, Magnanti, and Orlin [1989, 1991].

The first label-setting algorithm was suggested by Dijkstra [1959] and, independently, by Dantzig [1960], and Whiting and Hillier [1960]. The original implementation of Dijkstra's algorithm runs in $O(n^2)$ time, which is the optimal running time for fully dense networks [those with $m = \Omega(n^2)$] because any algorithm must examine every arc. However, the use of heaps permits us to obtain improved running times for sparse networks. The d-heap implementation of Dijkstra's algorithm with

$d = \max\{2, \lceil m/n \rceil\}$ runs in $O(m \log_d n)$ time and is due to Johnson [1977a]. The Fibonacci heap implementation, due to Fredman and Tarjan [1984], runs in $O(m + n \log n)$ time. Johnson [1982] suggested the $O(m \log \log C)$ implementation of Dijkstra's algorithm, based on earlier work by Boas, Kaas, and Zijlstra [1977]. Gabow's [1985] scaling algorithm, discussed in Exercise 5.51, is another efficient shortest path algorithm.

Dial [1969] (and also, independently, Wagner [1976]) suggested the $O(m + nC)$ implementation of Dijkstra's algorithm that we discussed in Section 4.6. Dial, Glover, Karney, and Klingman [1979] proposed an improved version of Dial's implementation that runs better in practice. Although Dial's implementation is only pseudopolynomial time, it has led to algorithms with better worst-case behavior. Denardo and Fox [1979] suggested several such improvements. The radix heap implementation that we described in Section 4.8 is due to Ahuja, Mehlhorn, Orlin, and Tarjan [1990]; we can view it as an improved version of Denardo and Fox's implementations. Our description of the radix heap implementation runs in $O(m + n \log(nC))$ time. Ahuja et al. [1990] also suggested several improved versions of the radix heap implementation that run in $O(m + n \log C)$, $O(m + (n \log C)/(\log \log C))$, $O(m + n \sqrt{\log C})$ time.

Currently, the best time bound for solving the shortest path problem with nonnegative arc lengths is $O(\min\{m + n \log n, m \log \log C, m + n \sqrt{\log C}\})$; this expression contains three terms because different time bounds are better for different values of n, m, and C. We refer to the overall time bound as $S(n, m, C)$; Fredman and Tarjan [1984], Johnson [1982], and Ahuja et al. [1990] have obtained the three bounds it contains. The best strongly polynomial-time bound for solving the shortest path problem with nonnegative arc lengths is $O(m + n \log n)$, which we subsequently refer to as $S(n, m)$.

Researchers have extensively tested label-setting algorithms empirically. Some of the more recent computational results can be found in Gallo and Pallottino [1988], Hung and Divoky [1988], and Divoky and Hung [1990]. These results suggest that Dial's implementation is the fastest label-setting algorithm for most classes of networks tested. Dial's implementation is, however, slower than some of the label-correcting algorithms that we discuss in Chapter 5.

The applications of the shortest path problem that we described in Section 4.2 are adapted from the following papers:

1. Approximating piecewise linear functions (Imai and Iri [1986])
2. Allocating inspection effort on a production line (White [1969])
3. Knapsack problem (Fulkerson [1966])
4. Tramp steamer problem (Lawler [1966])
5. System of difference constraints (Bellman [1958])
6. Telephone operator scheduling (Bartholdi, Orlin, and Ratliff [1980])

Elsewhere in this book we have described other applications of the shortest path problem. These applications include (1) reallocation of housing (Application 1.1, Wright [1975]), (2) assortment of steel beams (Application 1.2, Frank [1965]), (3) the paragraph problem (Exercise 1.7), (4) compact book storage in libraries (Exercise 4.3, Ravindran [1971]), (5) the money-changing problem (Exercise 4.5), (6) cluster analysis (Exercise 4.6), (7) concentrator location on a line (Exercises 4.7 and 4.8, Balakrishnan, Magnanti, and Wong [1989b]), (8) the personnel planning problem (Exercise 4.9, Clark and Hastings [1977]), (9) single-duty crew scheduling (Exercise 4.13, Veinott and Wagner [1962]), (10) equipment replacement (Application 9.6, Veinott and Wagner [1962]), (11) asymmetric data scaling with lower and upper bounds (Application 19.5, Orlin and Rothblum [1985]), (12) DNA sequence alignment (Application 19.7, Waterman [1988]), (13) determining minimum project duration (Application 19.9), (14) just-in-time scheduling (Application 19.10, Elmaghraby [1978], Levner and Nemirovsky [1991]), (15) dynamic lot sizing (Applications 19.19, Application 19.20, Application 19.21, Veinott and Wagner [1962], Zangwill [1969]), and (16) dynamic facility location (Exercise 19.22).

The literature considers many other applications of shortest paths that we do not cover in this book. These applications include (1) assembly line balancing (Gutjahr and Nemhauser [1964]), (2) optimal improvement of transportation networks (Goldman and Nemhauser [1967]), (3) machining process optimization (Szadkowski [1970]), (4) capacity expansion (Luss [1979]), (5) routing in computer communication networks (Schwartz and Stern [1980]), (6) scaling of matrices (Golitschek and Schneider [1984]), (7) city traffic congestion (Zawack and Thompson [1987]), (8) molecular confirmation (Dress and Havel [1988]), (9) order picking in an isle (Goetschalckx and Ratliff [1988]), and (10) robot design (Haymond, Thornton, and Warner [1988]).

Shortest path problems often arise as important subroutines within algorithms for solving many different types of network optimization problems. These applications are too numerous to mention. We do describe several such applications in subsequent chapters, however, when we show that shortest path problems are key subroutines in algorithms for the minimum cost flow problem (see Chapter 9), the assignment problem (see Section 12.4), the constrained shortest path problem (see Section 16.4), and the network design problem (see Application 16.4).

EXERCISES

4.1. Mr. Dow Jones, 50 years old, wishes to place his IRA (Individual Retirement Account) funds in various investment opportunities so that at the age of 65 years, when he withdraws the funds, he has accrued maximum possible amount of money. Assume that Mr. Jones knows the investment alternatives for the next 15 years: their maturity (in years) and the appreciation they offer. How would you formulate this investment problem as a shortest path problem, assuming that at any point in time, Mr. Jones invests all his funds in a single investment alternative.

4.2. Beverly owns a vacation home in Cape Cod that she wishes to rent for the period May 1 to August 31. She has solicited a number of bids, each having the following form: the day the rental starts (a rental day starts at 3 P.M.), the day the rental ends (checkout time is noon), and the total amount of the bid (in dollars). Beverly wants to identify a selection of bids that would maximize her total revenue. Can you help her find the best bids to accept?

4.3. **Compact book storage in libraries** (Ravindran [1971]). A library can store books according to their subject or author classification, or by their size, or by any other method that permits an orderly retrieval of the books. This exercise concerns an optimal storage of books by their size to minimize the storage cost for a given collection of books.

Suppose that we know the heights and thicknesses of all the books in a collection (assuming that all widths fit on the same shelving, we consider only a two-dimensional problem and ignore book widths). Suppose that we have arranged the book heights in ascending order of their n known heights $H_1, H_2, \ldots, H_n$; that is, $H_1 < H_2 < \cdots < H_n$. Since we know the thicknesses of the books, we can compute the required length of shelving for each height class. Let L_i denote the length of shelving for books of height H_i. If we order shelves of height H_i for length x_i, we incur cost equal to $F_i + C_i x_i$; F_i is a fixed ordering cost (and is independent of the length ordered) and C_i is the cost of the shelf per unit length. Notice that in order to save the fixed cost of ordering, we might not order shelves of every possible height because we can use a shelf of height H_i to store books of smaller heights. We want to determine the length of shelving for each height class that would minimize the total cost of the shelving. Formulate this problem as a shortest path problem.

4.4. Consider the compact book storage problem discussed in Exercise 4.3. Show that the storage problem is trivial if the fixed cost of ordering shelves is zero. Next, solve the compact book storage problem with the following data.

i	1	2	3	4	5	6
H_i	5 in.	6 in.	7 in.	9 in.	12 in.	14 in.
L_i	100	300	200	300	500	100
E_i	1000	1200	1100	1600	1800	2000
C_i	5	6	7	9	12	14

4.5. Money-changing problem. The money-changing problem requires that we determine whether we can change a given number p into coins of known denominations $a_1, a_2, \ldots, a_k$. For example, if $k = 3$, $a_1 = 3$, $a_2 = 5$, $a_3 = 7$, we can change all the numbers in the set {8, 12, 54}; on the other hand, we cannot change the number 4. In general, the money-changing problem asks whether $p = \sum_{i=1}^{k} a_i x_i$ for some nonnegative integers $x_1, x_2, \ldots, x_k$.

(a) Describe a method for identifying all numbers in a given range of numbers $[l, u]$ that we can change.

(b) Describe a method that identifies whether we can change a given number p, and if so, then identifies a denomination with the least number of coins.

4.6. Cluster analysis. Consider a set of n scalar numbers $a_1, a_2, \ldots, a_n$ arranged in nondecreasing order of their values. We wish to partition these numbers into clusters (or groups) so that (1) each cluster contains at least p numbers; (2) each cluster contains consecutive numbers from the list $a_1, a_2, \ldots, a_n$; and (3) the sum of the squared deviation of the numbers from their cluster means is as small as possible. Let $\bar{a}(S) = (\sum_{i \in S} a_i)/|S|$ denote the mean of a set S of numbers defining a cluster. If the number a_k belongs to cluster S, the squared deviation of the number a_k from the cluster mean is $(a_k - \bar{a}(S))^2$. Show how to formulate this problem as a shortest path problem. Illustrate your formulation using the following data: $p = 2$, $n = 6$, $a_1 = 0.5$, $a_2 = 0.8$, $a_3 = 1.1$, $a_4 = 1.5$, $a_5 = 1.6$, and $a_6 = 2.0$.

4.7. Concentrator location on a line (Balakrishnan, Magnanti, and Wong [1989]). In the telecommunication industry, telephone companies typically connect each customer directly to a switching center, which is a device that routes calls between the users in the system. Alternatively, to use fewer cables for routing the telephone calls, a company can combine the calls of several customers in a message compression device known as a *concentrator* and then use a single cable to route all of the calls transmitted by those users to the switching center. Constructing a concentrator at any node in the telephone network incurs a node-specific cost and assigning each customer to any concentrator incurs a "homing cost" that depends on the customer and the concentrator location. Suppose that all of the customers lie on a path and that we wish to identify the optimal location of concentrators to service these customers (assume that we must assign each customer to one of the concentrators). Suppose further that the set of customers allocated to any concentrator must be contiguous on the path (many telephone companies use this customer grouping policy). How would you find the optimal location of a single concentrator that serves any contiguous set of customers? Show how to use the solution of these single-location subproblems (one for each interval of customers) to solve the concentrator location problem on the path as a shortest path problem.

4.8. Modified concentrator location problem. Show how to formulate each of the following variants of the concentrator location problem that we consider in Exercise 4.7 as a shortest path problem. Assume in each case that all the customer lie on a path.

(a) The cost of connecting each customer to a concentrator is negligible, but each concentrator can handle at most five customers.

(b) Several types of concentrators are available at each node; each type of concentrator has its own cost and its own capacity (which is the maximum number of customers it can accommodate).

(c) In the situations considered in Exercise 4.7 and in parts (a) and (b) of this exercise, no customer can be assigned to a concentrator more that 1200 meters from the concentrator (because of line degradation of transmitted signals).

4.9. Personnel planning problem (Clark and Hastings [1977]). A construction company's work schedule on a certain site requires the following number of skilled personnel, called *steel erectors*, in the months of March through August:

Month	Mar.	Apr.	May	June	July	Aug.
Personnel	4	6	7	4	6	2

Personnel work at the site on the monthly basis. Suppose that three steel erectors are on the site in February and three steel erectors must be on site in September. The problem is to determine how many workers to have on site in each month in order to minimize costs, subject to the following conditions:

Transfer costs. Adding a worker to this site costs \$100 per worker and redeploying a worker to another site costs \$160.

Transfer rules. The company can transfer no more than three workers at the start of any month, and under a union agreement, it can redeploy no more than one-third of the current workers in any trade from a site at the end of any month.

Shortage time and overtime. The company incurs a cost of \$200 per worker per month for having a surplus of steel erectors on site and a cost of \$200 per worker per month for having a shortage of workers at the site (which must be made up in overtime). Overtime cannot exceed 25 percent of the regular work time.

Formulate this problem as a shortest path problem and solve it. (*Hint*: Give a dynamic programming-based formulation and use as many nodes for each month as the maximum possible number of steel erectors.)

4.10. Multiple-knapsack problem. In the shortest path formulation of the knapsack problem discussed in Application 4.3, an item is either placed in the knapsack or not. Consequently, each $x_j \in \{0, 1\}$. Consider a situation in which the hiker can place multiple copies of an item in her knapsack (i.e., $x_j \in \{0, 1, 2, 3, \ldots\}$). How would you formulate this problem as a shortest path problem? Illustrate your formulation on the example given in Application 4.3.

4.11. Modified system of difference constraints. In discussing system of difference constraints in Application 4.5, we assumed that each constraint is of the form $x(j_k) - x(i_k) \le b(k)$. Suppose, instead, that some constraints are of the form $x(j_k) \le b(k)$ or $x(i_k) \ge b(k)$. Describe how you would solve this modified system of constraints using a shortest path algorithm.

4.12. Telephone operator scheduling. In our discussion of the telephone operator scheduling problem in Application 4.6, we described a method for solving a restricted problem of determining whether some feasible schedule uses at most p operators. Describe a polynomial-time algorithm for determining a schedule with the fewest operators that uses the restricted problem as a subproblem.

4.13. Single-duty crew scheduling. The following table illustrates a number of possible duties for the drivers of a bus company. We wish to ensure, at the lowest possible cost, that at least one driver is on duty for each hour of the planning period (9 A.M. to 5 P.M.). Formulate and solve this scheduling problem as a shortest path problem.

Duty hours	9–1	9–11	12–3	12–5	2–5	1–4	4–5
Cost	30	18	21	38	20	22	9

4.14. Solve the shortest path problems shown in Figure 4.15 using the original implementation of Dijkstra's algorithm. Count the number of distance updates.

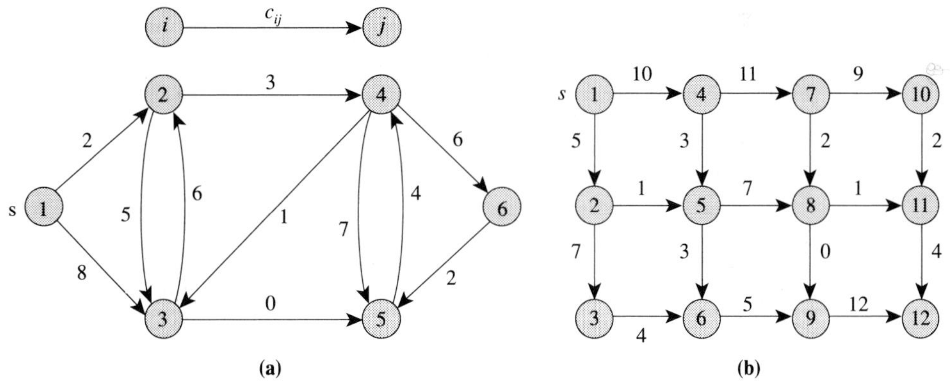

Figure 4.15 Some shortest path networks.

4.15. Solve the shortest path problem shown in Figure 4.15(a) using Dial's implementation of Dijkstra's algorithm. Show all of the buckets along with their content after the algorithm has examined the most recent permanently labeled node at each step.

4.16. Solve the shortest path problem shown in Figure 4.15(a) using the radix heap algorithm.

4.17. Consider the network shown in Figure 4.16. Assign integer lengths to the arcs in the network so that for every $k \in [0, 2^K - 1]$, the network contains a directed path of length k from the source node to sink node.

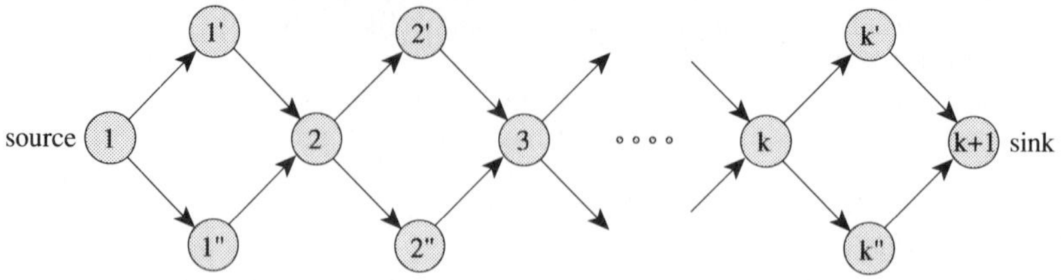

Figure 4.16 Network for Exercise 4.17.

4.18. Suppose that all the arcs in a network G have length 1. Show that Dijkstra's algorithm examines nodes for this network in the same order as the breadth-first search algorithm described in Section 3.4. Consequently, show that it is possible to solve the shortest path problem in this unit length network in $O(m)$ time.

4.19. Construct an example of the shortest path problem with some negative arc lengths, but no negative cycle, that Dijkstra's algorithm will solve correctly. Construct another example that Dijkstra's algorithm will solve incorrectly.

4.20. (Malik, Mittal, and Gupta [1989]) Consider a network without any negative cost cycle. For every node $j \in N$, let $d^s(j)$ denote the length of a shortest path from node s to node j and let $d^t(j)$ denote the length of a shortest path from node j to node t.

(a) Show that an arc (i, j) is on a shortest path from node s to node t if and only if $d^s(t) = d^s(i) + c_{ij} + d^t(j)$.

(b) Show that $d^s(t) = \min\{d^s(i) + c_{ij} + d^t(j) : (i, j) \in A\}$.

4.21. Which of the following claims are true and which are false? Justify your answer by giving a proof or by constructing a counterexample.

(a) If all arcs in a network have different costs, the network has a unique shortest path tree.

(b) In a directed network with positive arc lengths, if we eliminate the direction on every arc (i.e., make it undirected), the shortest path distances will not change.

(c) In a shortest path problem, if each arc length increases by k units, shortest path distances increase by a multiple of k.

(d) In a shortest path problem, if each arc length decreases by k units, shortest path distances decrease by a multiple of k.

(e) Among all shortest paths in a network, Dijkstra's algorithm always finds a shortest path with the least number of arcs.

4.22. Suppose that you are given a shortest path problem in which all arc lengths are the same. How will you solve this problem in the least possible time?

4.23. In our discussion of shortest path algorithms, we often assumed that the underlying network has no parallel arcs (i.e., at most one arc has the same tail and head nodes). How would you solve a problem with parallel arcs? (*Hint*: If the network contains k parallel arcs directed from node i to node j, show that we can eliminate all but one of these arcs.)

4.24. Suppose that you want to determine a path of shortest length that can start at either of the nodes s_1 or s_2 and can terminate at either of the nodes t_1 and t_2. How would you solve this problem?

4.25. Show that in the shortest path problem if the length of some arc decreases by k units, the shortest path distance between any pair of nodes decreases by at most k units.

4.26. **Most vital arc problem.** A *vital arc* of a network is an arc whose removal from the network causes the shortest distance between two specified nodes, say node s and node t, to increase. A most vital arc is a vital arc whose removal yields the greatest increase in the shortest distance from node s to node t. Assume that the network is directed, arc lengths are positive, and some arc is vital. Prove that the following statements are true or show through counterexamples that they are false.

(a) A most vital arc is an arc with the maximum value of c_{ij}.

(b) A most vital arc is an arc with the maximum value of c_{ij} on some shortest path from node s to node t.

(c) An arc that does not belong to any shortest path from node s to node t cannot be a most vital arc.

(d) A network might contain several most vital arcs.

4.27. Describe an algorithm for determining a most vital arc in a directed network. What is the running time of your algorithm?

4.28. A *longest path* is a directed path from node s to node t with the maximum length. Suggest an $O(m)$ algorithm for determining a longest path in an acyclic network with nonnegative arc lengths. Will your algorithm work if the network contains directed cycles?

4.29. Dijkstra's algorithm, as stated in Figure 4.6, identifies a shortest directed path from node s to every node $j \in N - \{s\}$. Modify this algorithm so that it identifies a shortest directed path from each node $j \in N - \{t\}$ to node t.

4.30. Show that if we add a constant α to the length of every arc emanating from the source node, the shortest path tree remains the same. What is the relationship between the shortest path distances of the modified problem and those of the original problem?

4.31. Can adding a constant α to the length of every arc emanating from a nonsource node produce a change in the shortest path tree? Justify your answer.

4.32. Show that Dijkstra's algorithm runs correctly even when a network contains negative cost arcs, provided that all such arcs emanate from the source node. (*Hint*: Use the result of Exercise 4.30.)

4.33. **Improved Dial's implementation** (Denardo and Fox [1979]). This problem discusses a practical speed-up of Dial's implementation. Let $c_{\min} = \min\{c_{ij} : (i, j) \in A\}$ and $w = \max\{1, c_{\min}\}$. Consider a version of Dial's implementation in which we use buckets of width w. Show that the algorithm will never decrease the distance label of any node in the least index nonempty bucket; consequently, we can permanently label any node in this bucket. What is the running time of this version of Dial's implementation?

4.34. Suppose that we arrange all directed paths from node s to node t in nondecreasing order of their lengths, breaking ties arbitrarily. The kth shortest path problem is to identify a path that can be at the kth place in this order. Describe an algorithm to find the kth shortest path for $k = 2$. (*Hint*: The second shortest path must differ from the first shortest path by at least one arc.)

4.35. Suppose that every directed cycle in a graph G has a positive length. Show that a shortest directed walk from node s to node t is always a path. Construct an example for which the first shortest directed walk is a path, but the second shortest directed walk is not a path.

4.36. Describe a method for identifying the first K shortest paths from node s to node t in an acyclic directed network. The running time of your algorithm should be polynomial in terms of n, m, and K. (*Hint*: For each node j, keep track of the first K shortest paths from node s to node j. Also, use the results in Exercise 4.34.)

4.37. **Maximum capacity path problem.** Let $c_{ij} \geq 0$ denote the capacity of an arc in a given network. Define the *capacity* of a directed path P as the minimum arc capacity in P. The *maximum capacity path problem* is to determine a maximum capacity path from a specified source node s to every other node in the network. Modify Dijkstra's algorithm so that it solves the maximum capacity path problem. Justify your algorithm.

4.38. Let $(i_1, j_1), (i_2, j_2), \ldots, (i_m, j_m)$ denote the arcs of a network in nondecreasing order of their arc capacities. Show that the maximum capacity path from node s to any node j remains unchanged if we modify some or all of the arc capacities but maintain the same (capacity) order for the arcs. Use this result to show that if we already have a sorted list of the arcs, we can solve the maximum capacity path problem in $O(m)$ time. (*Hint*: Modify arc capacities so that they are all between 1 and m. Then use a variation of Dial's implementation.)

4.39. Maximum reliability path problems. In the network G we associate a reliability $0 < \mu_{ij} \leq 1$ with every arc $(i, j) \in A$; the reliability measures the probability that the arc will be operational. We define the reliability of a directed path P as the product of the reliability of arcs in the path [i.e., $\mu(P) = \prod_{(i,j) \in P} \mu_{ij}$]. The maximum reliability path problem is to identify a directed path of maximum reliability from the source node s to every other node in the network.

(a) Show that if we are allowed to take logarithms, we can reduce the maximum reliability path problem to a shortest path problem.

(b) Suppose that you are not allowed to take logarithms because they yield irrational data. Specify an $O(n^2)$ algorithm for solving the maximum reliability path problem and prove the correctness of this algorithm. (*Hint*: Modify Dijkstra's algorithm.)

(c) Will your algorithms in parts (a) and (b) work if some of the coefficients μ_{ij} are strictly greater than 1?

4.40. Shortest paths with turn penalties. Figure 4.15(b) gives a road network in which all road segments are parallel to either the x-axis or the y-axis. The figure also gives the traversal costs of arcs. Suppose that we incur an additional cost (or penalty) of α units every time we make a left turn. Describe an algorithm for solving the shortest path problem with these turn penalties and apply it to the shortest path example in Figure 4.15(b). Assume that $\alpha = 5$. [*Hint*: Create a new graph G^* with a node $i - j$ corresponding to each arc $(i, j) \in A$ and with each pair of nodes $i - j$ and $j - k$ in N joined by an arc. Assign appropriate arc lengths to the new graph.]

4.41. Max-min result. We develop a max-min type of result for the maximum capacity path problem that we defined in Exercise 4.37. As in that exercise, suppose that we wish to find the maximum capacity path from node s to node t. We say that a cut $[S, \bar{S}]$ is an *s–t cut* if $s \in S$ and $t \in \bar{S}$. Define the *bottleneck value* of an s–t cut as the largest arc capacity in the cut. Show that the capacity of the maximum capacity path from node s to node t equals the minimum bottleneck value of a cut.

4.42. A farmer wishes to transport a truckload of eggs from one city to another city through a given road network. The truck will incur a certain amount of breakage on each road segment; let w_{ij} denote the fraction of the eggs broken if the truck traverses the road segment (i, j). How should the truck be routed to minimize the total breakage? How would you formulate this problem as a shortest path problem.

4.43. A^* algorithm. Suppose that we want to identify a shortest path from node s to node t, and not necessarily from s to any other node, in a network with nonnegative arc lengths. In this case we can terminate Dijkstra's algorithm whenever we permanently label node t. This exercise studies a modification of Dijkstra's algorithm that would speed up the algorithm in practice by designating node t as a permanent labeled node more quickly. Let $h(i)$ be a lower bound on the length of the shortest path from node i to node t and suppose that the lower bounds satisfy the conditions $h(i) \leq h(j) + c_{ij}$ for all $(i, j) \in A$. For instance, if nodes are points in a two-dimensional plane with coordinates (x_i, y_i) and arc lengths equal Euclidean distances between points, then $h(i) = [(x_i - x_t)^2 + (y_i - y_t)^2]^{1/2}$ (i.e., the Euclidean distance from i to t) is a valid lower bound on the length of the shortest path from node i to node t.

(a) Let $c^h_{ij} = c_{ij} + h(j) - h(i)$ for all $(i, j) \in A$. Show that replacing the arc lengths c_{ij} by c^h_{ij} does not affect the shortest paths between any pair of nodes.

(b) If we apply Dijkstra's algorithm with c^h_{ij} as arc lengths, why should this modification improve the empirical behavior of the algorithm? [*Hint*: What is its impact if each $h(i)$ represents actual shortest path distances from node i to node t?]

4.44. Arc tolerances. Let T be a shortest path tree of a network. Define the *tolerances* of an arc (i, j) as the maximum increase, α_{ij}, and the maximum decrease, β_{ij}, that the arc can tolerate without changing the tree of shortest paths.

(a) Show that if the arc $(i, j) \notin T$, then $\alpha_{ij} = +\infty$ and β_{ij} will be a finite number. Describe an $O(1)$ method for computing β_{ij}.

(b) Show that if the arc $(i, j) \in T$, then $\beta_{ij} = +\infty$ and α_{ij} will be a finite number. Describe an $O(m)$ method for computing α_{ij}.

4.45. **(a)** Describe an algorithm that will determine a shortest walk from a source node s to a sink node t subject to the additional condition that the walk must visit a specified node p. Will this walk always be a path?

(b) Describe an algorithm for determining a shortest walk from node s to node t that must visit a specified arc (p, q).

4.46. **Constrained shortest path problem.** Suppose that we associate two integer numbers with each arc in a network G: the arc's length c_{ij} and its traversal time $\tau_{ij} > 0$ (we assume that the traversal times are integers). The *constrained shortest path problem* is to determine a shortest length path from a source node s to every other node with the additional constraint that the traversal time of the path does not exceed τ_0. In this exercise we describe a dynamic programming algorithm for solving the constrained shortest path problem. Let $d_j(\tau)$ denote the length of a shortest path from node s to node j subject to the condition that the traversal time of the path does not exceed τ. Suppose that we set $d_j(\tau) = \infty$ for $\tau < 0$. Justify the following equations:

$$d_s(0) = 0,$$
$$d_j(\tau) = \min[d_j(\tau - 1), \min_k\{d_k(\tau - \tau_{kj}) + c_{kj}\}].$$

Use these equations to design an algorithm for the constrained shortest path problem and analyze its running time.

4.47. **Generalized knapsack problem.** In the knapsack problem discussed in Application 4.3, suppose that each item j has three associated numbers: *value* v_j, *weight* w_j, and *volume* r_j. We want to maximize the value of the items put in the knapsack subject to the condition that the total weight of the items is at most W and the total volume is at most R. Formulate this problem as a shortest path problem with an additional constraint.

4.48. Consider the generalized knapsack problem studied in Exercise 4.47. Extend the formulation in Application 4.3 in order to transform this problem into a longest path problem in an acyclic network.

4.49. Suppose that we associate two numbers with each arc (i, j) in a directed network $G = (N, A)$: the arc's length c_{ij} and its reliability r_{ij}. We define the reliability of a directed path P as the product of the reliabilities of arcs in the path. Describe a method for identifying a shortest length path from node s to node t whose reliability is at least r.

4.50. **Resource-constrained shortest path problem.** Suppose that the traversal time τ_{ij} of an arc (i, j) in a network is a function $f_{ij}(d)$ of the discrete amount of a resource d that we consume while traversing the arc. Suppose that we want to identify the shortest directed path from node s to node t subject to a budget D on the amount of the resource we can consume. (For example, we might be able to reduce the traversal time of an arc by using more fuel, and we want to travel from node s to node t before we run out of fuel.) Show how to formulate this problem as a shortest path problem. Assume that $d = 3$. (*Hint*: Give a dynamic programming-based formulation.)

4.51. **Modified function approximation problem.** In the function approximation problem that we studied in Application 4.1, we approximated a given piecewise linear function $f_1(x)$ by another piecewise linear function $f_2(x)$ in order to minimize a weighted function of the two costs: (1) the cost required to store the data needed to represent the function $f_2(x)$, and (2) the errors introduced by the approximating $f_1(x)$ by $f_2(x)$. Suppose that, instead, we wish to identify a subset of at most p points so that the function $f_2(x)$ defined by these points minimizes the errors of the approximation (i.e., $\sum_{k=1}^{n} [f_1(x_k) - f_2(x_k)]^2$). That is, instead of imposing a cost on the use of any breakpoint in the approximation, we impose a limit on the number of breakpoints we can use. How would you solve this problem?

4.52. Bidirectional Dijkstra's algorithm (Helgason, Kennington, and Stewart [1988]). Show that the bidirectional shortest path algorithm described in Section 4.5 correctly determines a shortest path from node s to node t. [*Hint*: At the termination of the algorithm, let S and S' be the sets of nodes that the forward and reverse versions of Dijkstra's algorithm have designated as permanently labeled. Let $k \in S \cap S'$. Let P^* be some shortest path from node s to node t; suppose that the first q nodes of P^* are in S and that the $(q + 1)$st node of P^* is not in S. Show first that some shortest path from node s to node t has the same first q nodes as P^* and has its $(q + 1)$st node in S'. Next show that some shortest path has the same first q nodes as P^* and each subsequent node in S'.]

4.53. Shortest paths in bipartite networks (Orlin [1988]). In this exercise we discuss an improved algorithm for solving shortest path problem in "unbalanced" bipartite networks $G = (N_1 \cup N_2, A)$, that is, those satisfying the condition that $n_1 = |N_1| \ll |N_2| = n_2$. Assume that the degree of any node in N_2 is at most K for some constant K, and that all arc costs are nonnegative. Shortest path problems with this structure arise in the context of solving the minimum cost flow problem (see Section 10.6). Let us define a graph $G' = (N_1, A')$ whose arc set A' is defined as the following set of arcs: For every pair of arcs (i, j) and (j, k) in A, A' has an arc (i, k) of cost equal to $c_{ij} + c_{jk}$.

(a) Show how to solve the shortest path problem in G by solving a shortest path problem in G'. What is the resulting running time of solving the shortest path problem in G in terms of the parameters n, m and K?

(b) A network G is *semi-bipartite* if we can partition its node set N into the subsets N_1 and N_2 so that no arc has both of its endpoints in N_2. Assume again that $|N_1| \ll |N_2|$ and the degree of any node in N_2 is at most K. Suggest an improved algorithm for solving shortest path problems in semi-bipartite networks.

5

SHORTEST PATHS: LABEL-CORRECTING ALGORITHMS

To get to heaven, turn right and keep straight ahead.
—Anonymous

Chapter Outline

5.1 INTRODUCTION

In Chapter 4 we saw how to solve shortest path problems very efficiently when they have special structure: either a special network topology (acyclic networks) or a special cost structure (nonnegative arc lengths). When networks have arbitrary costs and arbitrary topology, the situation becomes more complicated. As we noted in Chapter 4, for the most general situations—that is, general networks with negative cycles—finding shortest paths appears to be very difficult. In the parlance of computational complexity theory, these problems are NP-complete, so they are equivalent to solving many of the most noted and elusive problems encountered in the realm of combinatorial optimization and integer programming. Consequently, we have little hope of devising polynomial-time algorithms for the most general problem setting. Instead, we consider a tractable compromise somewhere between the special cases we examined in Chapter 4 and the most general situations: namely, algorithms that either identify a negative cycle, when one exists, or if the underlying network contains no negative cycle, solves the shortest path problem.

Essentially, all shortest path algorithms rely on the same important concept: distance labels. At any point during the execution of an algorithm, we associate a numerical value, or distance label, with each node. If the label of any node is infinite, we have yet to find a path joining the source node and that node. If the label is finite, it is the distance from the source node to that node along some path. The most basic algorithm that we consider in this chapter, the generic label-correcting algorithm, reduces the distance label of one node at each iteration by considering only local information, namely the length of the single arc and the current distance labels of its incident nodes. Since we can bound the sum of the distance labels from above and below in terms of the problem data, then under the assumption of integral costs, the distance labels will be integral and so the generic algorithm will always be finite. As is our penchant in this book, however, we wish to discover algorithms that are not only finite but that require a number of computations that grow as a (small) polynomial in the problem's size.

We begin the chapter by describing optimality conditions that permit us to assess when a set of distance labels are optimal—that is, are the shortest path distances from the source node. These conditions provide us with a termination criterion, or optimality certificate, for telling when a feasible solution to our problem is optimal and so we need perform no further computations. The concept of optimality conditions is a central theme in the field of optimization and will be a recurring theme throughout our treatment of network flows in this book. Typically, optimality conditions provide us with much more than a termination condition; they often provide considerable problem insight and also frequently suggest algorithms for solving optimization problems. When a tentative solution does not satisfy the optimality conditions, the conditions often suggest how we might modify the current solution so that it becomes "closer" to an optimal solution, as measured by some underlying metric. Our use of the shortest path optimality conditions in this chapter for developing label-correcting algorithms demonstrates the power of optimality conditions in guiding the design of solution algorithms.

Although the general label-correcting algorithm is finite, it requires $O(n^2C)$ computations to solve shortest path problems on networks with n nodes and with a bound of C on the maximum absolute value of any arc length. This bound is not very satisfactory because it depends linearly on the values of the arc costs. One of the advantages of the generic label-correcting algorithm is its flexibility: It offers considerable freedom in the tactics used for choosing arcs that will lead to improvements in the shortest path distances. To develop algorithms that are better in theory and in practice, we consider specific strategies for examining the arcs. One "balancing" strategy that considers arcs in a sequential wraparound fashion requires only $O(nm)$ computations. Another implementation that gives priority to arcs emanating from nodes whose labels were changed most recently, the so-called dequeue implementation, has performed very well in practice even though it has poor worst-case performance. In Section 5.4 we study both of these modified versions of the generic label-correcting algorithm.

We next consider networks with negative cycles and show how to make several types of modifications to the various label-correcting algorithms so that they can detect the presence of negative cycles, if the underlying network contains any. One nice feature of these methods is that they do not add to the worst-case computational complexity of any of the label-correcting algorithms.

We conclude this chapter by considering algorithms for finding shortest paths between all pairs of nodes in a network. We consider two approaches to this problem. One approach repeatedly applies the label-setting algorithm that we considered in Chapter 4, with each node serving as the source node. As the first step in this procedure, we apply the label-correcting algorithm to find the shortest paths from one arbitrary node, and use the results of this shortest path computation to redefine the costs so that they are all nonnegative and so that the subsequent n single-source problems are all in a form so that we can apply more efficient label-setting algorithms. The computational requirements for this algorithm is essentially the same as that required to solve n shortest path problems with nonnegative arc lengths and depends on which label-setting algorithm we adopt from those that we described in Chapter 4. The second approach is a label-correcting algorithm that simultaneously finds the shortest path distances between all pairs of nodes. This algorithm is very easy to implement; it uses a clever dynamic programming recursion and is able to solve the all-pairs shortest path problem in $O(n^3)$ computations.

5.2 OPTIMALITY CONDITIONS

As noted previously, label-correcting algorithms maintain a distance label $d(j)$ for every node $j \in N$. At intermediate stages of computation, the distance label $d(j)$ is an estimate of (an upper bound on) the shortest path distance from the source node s to node j, and at termination it is the shortest path distance. In this section we develop necessary and sufficient conditions for a set of distance labels to represent shortest path distances. Let $d(j)$ for $j \neq s$ denote the length of a shortest path from the source node to the node j [we set $d(s) = 0$]. If the distance labels are shortest path distances, they must satisfy the following necessary optimality conditions:

$$d(j) \leq d(i) + c_{ij}, \qquad \text{for all } (i, j) \in A. \tag{5.1}$$

These inequalities state that for every arc (i, j) in the network, the length of the shortest path to node j is no greater than the length of the shortest path to node i plus the length of the arc (i, j). For, if not, some arc $(i, j) \in A$ must satisfy the condition $d(j) > d(i) + c_{ij}$; in this case, we could improve the length of the shortest path to node j by passing through node i, thereby contradicting the optimality of distance labels $d(j)$.

These conditions also are sufficient for optimality, in the sense that if each $d(j)$ represents the length of some directed path from the source node to node j and this solution satisfies the conditions (5.1), then it must be optimal. To establish this result, consider any solution $d(j)$ satisfying (5.1). Let $s = i_1 - i_2 - \ldots - i_k = j$ be any directed path P from the source to node j. The conditions (5.1) imply that

$$\begin{aligned}
d(j) = d(i_k) &\leq d(i_{k-1}) + c_{i_{k-1}i_k}, \\
d(i_{k-1}) &\leq d(i_{k-2}) + c_{i_{k-2}i_{k-1}}, \\
&\vdots \\
d(i_2) &\leq d(i_1) + c_{i_1i_2} = c_{i_1i_2}.
\end{aligned}$$

The last equality follows from the fact that $d(i_1) = d(s) = 0$. Adding these inequalities, we find that

$$d(j) = d(i_k) \leq c_{i_{k-1}i_k} + c_{i_{k-2}i_{k-1}} + c_{i_{k-3}i_{k-2}} + \ldots + c_{i_1i_2} = \sum_{(i,j)\in P} c_{ij}.$$

Thus $d(j)$ is a lower bound on the length of any directed path from the source to node j. Since $d(j)$ is the length of some directed path from the source to node j, it also is an upper bound on the shortest path length. Therefore, $d(j)$ is the shortest path length, and we have established the following result.

Theorem 5.1 (Shortest Path Optimality Conditions). *For every node $j \in N$, let $d(j)$ denote the length of some directed path from the source node to node j. Then the numbers $d(j)$ represent shortest path distances if and only if they satisfy the following shortest path optimality conditions:*

$$d(j) \le d(i) + c_{ij} \qquad \text{for all } (i, j) \in A. \qquad (5.2)$$ ◆

Let us define the reduced arc length c_{ij}^d of an arc (i, j) with respect to the distance labels $d(\cdot)$ as $c_{ij}^d = c_{ij} + d(i) - d(j)$. The following properties about the reduced arc lengths will prove to be useful in our later development.

Property 5.2

(a) *For any directed cycle W, $\sum_{(i,j)\in W} c_{ij}^d = \sum_{(i,j)\in W} c_{ij}$.*
(b) *For any directed path P from node k to node l, $\sum_{(i,j)\in P} c_{ij}^d = \sum_{(i,j)\in P} c_{ij} + d(k) - d(l)$.*
(c) *If $d(\cdot)$ represent shortest path distances, $c_{ij}^d \ge 0$ for every arc $(i, j) \in A$.*

The proof of the first two results is similar to the proof of Property 2.5 in Section 2.4. The third result follows directly from Theorem 5.1.

We next note that if the network contains a negative cycle, then no set of distance labels $d(\cdot)$ satisfies (5.2). For suppose that W is a directed cycle in G. Property 5.2(c) implies that $\sum_{(i,j)\in W} c_{ij}^d \ge 0$. Property 5.2(a) implies that $\sum_{(i,j)\in W} c_{ij}^d = \sum_{(i,j)\in W} c_{ij} \ge 0$, and therefore W cannot be a negative cycle. Thus if the network were to contain a negative cycle, no distance labels could satisfy (5.2). We show in the next section that if the network does not contain a negative cycle, some shortest path distances do satisfy (5.2).

For those familiar with linear programming, we point out that the shortest path optimality conditions can also be viewed as the linear programming optimality conditions. In the linear programming formulation of the shortest path problem, the negative of the shortest path distances [i.e., $-d(j)$] define the optimal dual variables, and the conditions (5.2) are equivalent to the fact that in the optimal solution, reduced costs of all primal variables are nonnegative. The presence of a negative cycle implies the unboundedness of the primal problem and hence the infeasibility of the dual problem.

5.3 GENERIC LABEL-CORRECTING ALGORITHMS

In this section we study the generic label-correcting algorithm. We shall study several special implementations of the generic algorithm in the next section. Our discussion in this and the next section assumes that the network does not contain any negative cycle; we consider the case of negative cycles in Section 5.5.

The generic label-correcting algorithm maintains a set of distance labels $d(\cdot)$ at every stage. The label $d(j)$ is either ∞, indicating that we have yet to discover a directed path from the source to node j, or it is the length of some directed path from the source to node j. For each node j we also maintain a predecessor index, $pred(j)$, which records the node prior to node j in the current directed path of length $d(j)$. At termination, the predecessor indices allow us to trace the shortest path from the source node back to node j. The generic label-correcting algorithm is a general procedure for successively updating the distance labels until they satisfy the shortest path optimality conditions (5.2). Figure 5.1 gives a formal description of the generic label-correcting algorithm.

```
algorithm label-correcting;
begin
    d(s) : = 0 and pred(s) : = 0;
    d(j) : = ∞ for each j ∈ N − {s};
    while some arc (i, j) satisfies d(j) > d(i) + c_ij do
    begin
        d(j) : = d(i) + c_ij;
        pred(j) : = i;
    end;
end;
```

Figure 5.1 Generic label-correcting algorithm.

By definition of reduced costs, the distance labels $d(\cdot)$ satisfy the optimality conditions if $c_{ij}^d \geq 0$ for all $(i, j) \in A$. The generic label-correcting algorithm selects an arc (i, j) violating its optimality condition (i.e., $c_{ij}^d < 0$) and uses it to update the distance label of node j. This operation decreases the distance label of node j and makes the reduced arc length of arc (i, j) equal to zero.

We illustrate the generic label correcting algorithm on the network shown in Figure 5.2(a). If the algorithm selects the arcs (1, 3), (1, 2), (2, 4), (4, 5), (2, 5), and (3, 5) in this sequence, we obtain the distance labels shown in Figure 5.2(b) through (g). At this point, no arc violates its optimality condition and the algorithm terminates.

The algorithm maintains a predecessor index for every finitely labeled node. We refer to the collection of arcs $(\text{pred}(j), j)$ for every finitely labeled node j (except the source node) as the *predecessor graph*. The predecessor graph is a directed out-tree T rooted at the source that spans all nodes with finite distance labels. Each distance update using the arc (i, j) produces a new predecessor graph by deleting the arc $(\text{pred}(j), j)$ and adding the arc (i, j). Consider, for example, the graph shown in Figure 5.3(a): the arc (6, 5) enters, the arc (3, 5) leaves, and we obtain the graph shown in Figure 5.3(b).

The label-correcting algorithm satisfies the invariant property that for every arc (i, j) in the predecessor graph, $c_{ij}^d \leq 0$. We establish this result by performing induction on the number of iterations. Notice that the algorithm adds an arc (i, j) to the predecessor graph during a distance update, which implies that after this update $d(j) = d(i) + c_{ij}$, or $c_{ij} + d(i) - d(j) = c_{ij}^d = 0$. In subsequent iterations, $d(i)$ might decrease and so c_{ij}^d might become negative. Next observe that if $d(j)$ decreases during the algorithm, then for some arc (i, j) in the predecessor graph c_{ij}^d may become positive, thereby contradicting the invariant property. But observe that in this case, we immediately delete arc (i, j) from the graph and so maintain the invariant property. For an illustration, see Figure 5.3: in this example, adding arc (6, 5) to the graph decreases $d(5)$, thereby making $c_{58}^d < 0$. This step increases c_{35}^d, but arc (3, 5) immediately leaves the tree.

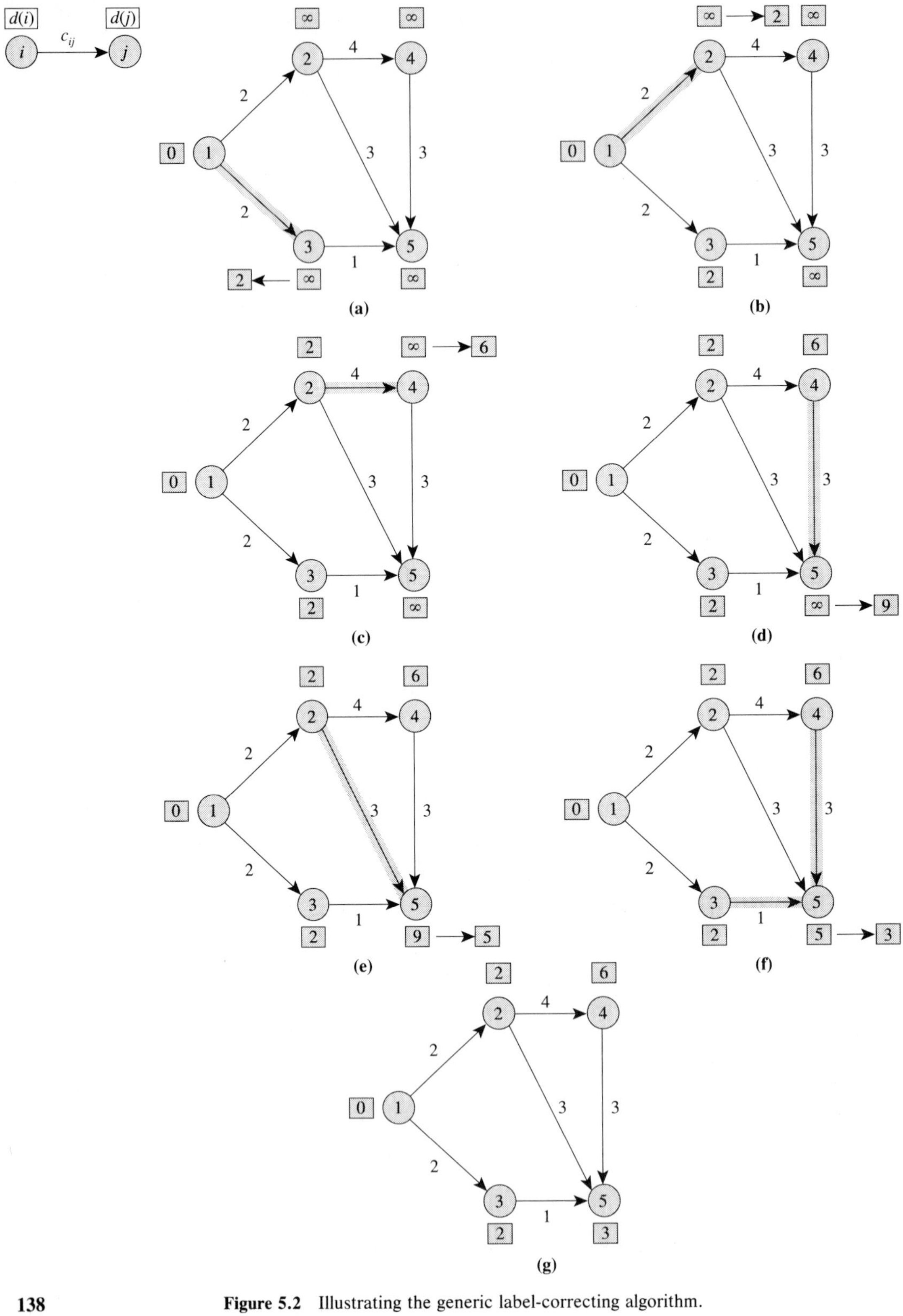

Figure 5.2 Illustrating the generic label-correcting algorithm.

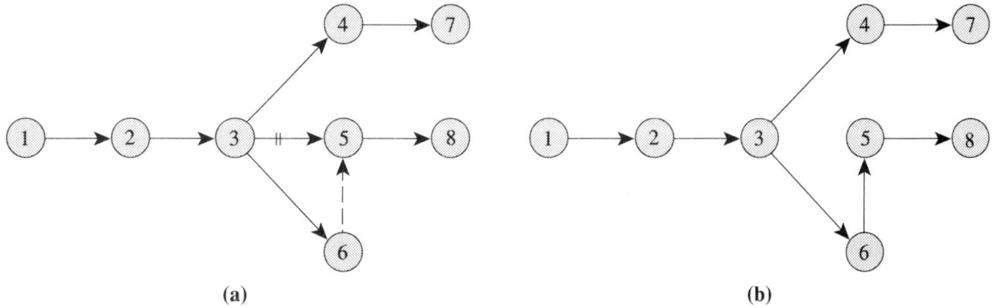

Figure 5.3 Showing that the predecessor graph is a directed out-tree.

We note that the predecessor indices might not necessarily define a tree. To illustrate this possibility, we use the situation shown in Figure 5.4(a). Suppose that arc (6, 2) satisfies $d(2) > d(6) + c_{62}$ (or $c_{62}^d < 0$) and we update the distance label of node 2. This operation modifies the predecessor index of node 2 from 1 to 6 and the graph defined by the predecessor indices is no longer a tree. Why has this happened? The predecessor indices do not define a tree because the network contained a negative cycle. To see that this is the case, notice from Property 5.1 that for the cycle 2–3–6–2, $c_{23} + c_{36} + c_{62} = c_{23}^d + c_{36}^d + c_{62}^d < 0$, because $c_{23}^d \le 0$, $c_{36}^d \le 0$, and $c_{62}^d < 0$. Therefore, the cycle 2–3–6–2 is a negative cycle. This discussion shows that in the absence of negative cycles, we will never encounter a situation shown in Figure 5.4(b) and the predecessor graph will always be a tree.

The predecessor graph contains a unique directed path from the source node to every node k and the length of this path is at most $d(k)$. To verify this result, let P be the path from the source to node k. Since every arc in the predecessor graph has a nonpositive reduced arc length, $\sum_{(i,j)\in P} c_{ij}^d \le 0$. Property 5.2(b) implies that $0 \ge \sum_{(i,j)\in P} c_{ij}^d = \sum_{(i,j)\in P} c_{ij} + d(s) - d(k) = \sum_{(i,j)\in P} c_{ij} - d(k)$. Alternatively, $\sum_{(i,j)\in P} c_{ij} \le d(k)$. When the label-correcting algorithm terminates, each arc in the predecessor graph has a zero reduced arc length (why?), which implies that the length of the path from the source to every node k equals $d(k)$. Consequently, when the algorithm terminates, the predecessor graph is a shortest path tree. Recall from Section 4.3 that a shortest path tree is a directed out-tree rooted at the source with the property that the unique path from the source to any node is a shortest path to that node.

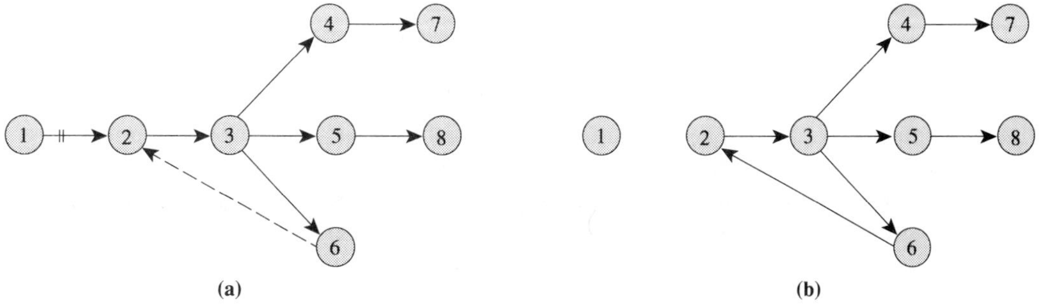

Figure 5.4 Formation of a cycle in a predecessor graph.

It is easy to show that the algorithm terminates in a finite number of iterations. We prove this result when the data are integral; Exercise 5.8 discusses situations when the data are nonintegral. Observe that each $d(j)$ is bounded from above by nC (because a path contains at most $n - 1$ arcs, each of length at most C) and is bounded from below by $-nC$. Therefore, the algorithm updates any label $d(j)$ at most $2nC$ times because each update of $d(j)$ decreases it by at least 1 unit. Consequently, the total number of distance label updates is at most $2n^2C$. Each iteration updates a distance label, so the algorithm performs $O(n^2C)$ iterations. The algorithm also terminates in $O(2^n)$ steps. (See Exercise 5.8.)

Modified Label-Correcting Algorithm

The generic label-correcting algorithm does not specify any method for selecting an arc violating the optimality condition. One obvious approach is to scan the arc list sequentially and identify any arc violating this condition. This procedure is very time consuming because it requires $O(m)$ time per iteration. We shall now describe an improved approach that reduces the workload to an average of $O(m/n)$ time per iteration.

Suppose that we maintain a list, LIST, of all arcs that *might* violate their optimality conditions. If LIST is empty, clearly we have an optimal solution. Otherwise, we examine this list to select an arc, say (i, j), violating its optimality condition. We remove arc (i, j) from LIST, and if this arc violates its optimality condition we use it to update the distance label of node j. Notice that any decrease in the distance label of node j decreases the reduced lengths of all arcs emanating from node j and some of these arcs might violate the optimality condition. Also notice that decreasing $d(j)$ maintains the optimality condition for all incoming arcs at node j. Therefore, if $d(j)$ decreases, we must add arcs in $A(j)$ to the set LIST. Next, observe that whenever we add arcs to LIST, we add *all* arcs emanating from a single node (whose distance label decreases). This suggests that instead of maintaining a list of all arcs that might violate their optimality conditions, we may maintain a list of *nodes* with the property that if an arc (i, j) violates the optimality condition, LIST must contain node i. Maintaining a node list rather than the arc list requires less work and leads to faster algorithms in practice. This is the essential idea behind the modified label-correcting algorithm whose formal description is given in Figure 5.5.

We call this algorithm the *modified label-correcting algorithm*. The correctness of the algorithm follows from the property that the set LIST contains every node i that is incident to an arc (i, j) violating the optimality condition. By performing induction on the number of iterations, it is easy to establish the fact that this property remains valid throughout the algorithm. To analyze the complexity of the algorithm, we make several observations. Notice that whenever the algorithm updates $d(j)$, it adds node j to LIST. The algorithm selects this node in a later iteration and scans its arc list $A(j)$. Since the algorithm can update the distance label $d(j)$ at most $2nC$ times, we obtain a bound of $\sum_{i \in N} (2nC) \mid A(i) \mid = O(nmC)$ on the total number of arc scannings. Therefore, this version of the generic label-correcting algorithm runs in $O(nmC)$ time. When C is exponentially large, the running time is $O(2^n)$. (See Exercise 5.8.)

```
algorithm modified label-correcting;
begin
    d(s) : = 0 and pred(s) : = 0;
    d(j) : = ∞ for each node j ∈ N − {s};
    LIST : = {s};
    while LIST ≠ Ø do
    begin
        remove an element i from LIST;
        for each arc (i, j) ∈ A(i) do
        if d(j) > d(i) + c_ij then
        begin
            d(j) : = d(i) + c_ij;
            pred(j) : = i;
            if j ∉ LIST then add node j to LIST;
        end;
    end;
end;
```

Figure 5.5 Modified label-correcting algorithm.

5.4 SPECIAL IMPLEMENTATIONS OF THE MODIFIED LABEL-CORRECTING ALGORITHM

One nice feature of the generic (or the modified) label-correcting algorithm is its flexibility: We can select arcs that do not satisfy the optimality condition in any order and still assure finite convergence of the algorithm. One drawback of this general algorithmic strategy, however, is that without a further restriction on the choice of arcs in the generic label-correcting algorithm (or nodes in the modified label-correcting algorithm), the algorithm does not necessarily run in polynomial time. Indeed, if we apply the algorithm to a pathological set of data and make a poor choice at every iteration, the number of steps can grow exponentially with n. (Since the algorithm is a pseudopolynomial-time algorithm, these instances must have exponentially large values of C. See Exercises 5.27 and 5.28 for a family of such instances.) These examples show that to obtain polynomially bounded label-correcting algorithms, we must organize the computations carefully. If we apply the modified label-correcting algorithm to a problem with nonnegative arc lengths and we always examine a node from LIST with the minimum distance label, the resulting algorithm is the same as Dijkstra's algorithm discussed in Section 4.5. In this case our selection rule guarantees that the algorithm examines at most n nodes, and the algorithm can be implemented to run in $O(n^2)$ time. Similarly, when applying the modified label-correcting algorithm to acyclic networks, if we examine nodes in LIST in the topological order, shortest path algorithm becomes the one that we discussed in Section 4.4, so it is a polynomial-time algorithm.

In this section we study two new implementations of the modified label-correcting algorithm. The first implementation runs in $O(nm)$ time and is currently the best strongly polynomial-time implementation for solving the shortest path problem with negative arc lengths. The second implementation is not a polynomial-time method, but is very efficient in practice.

O(nm) *Implementation*

We first describe this implementation for the generic label-correcting algorithm. In this implementation, we arrange arcs in A in some specified (possibly arbitrary) order. We then make passes through A. In each pass we scan arcs in A, one by one, and check the condition $d(j) > d(i) + c_{ij}$. If the arc satisfies this condition, we update $d(j) = d(i) + c_{ij}$. We stop when no distance label changes during an entire pass.

Let us show that this algorithm performs at most $n - 1$ passes through the arc list. Since each pass requires $O(1)$ computations for each arc, this conclusion implies the $O(nm)$ time bound for the algorithm. We claim that at the end of the kth pass, the algorithm will compute shortest path distances for all nodes that are connected to the source node by a shortest path consisting of k or fewer arcs. We prove this claim by performing induction on the number of passes. Our claim is surely true for $k = 1$. Now suppose that the claim is true for the kth pass. Thus $d(j)$ is the shortest path length to node j provided that some shortest path to node j contains k or fewer arcs, and is an upper bound on the shortest path length otherwise.

Consider a node j that is connected to the source node by a shortest path $s = i_0 - i_1 - i_2 - \cdots - i_k - i_{k+1} = j$ consisting of $k + 1$ arcs, but has no shortest path containing fewer than $k + 1$ arcs. Notice that the path $i_0 - i_1 - \cdots - i_k$ must be a shortest path from the source to node i_k, and by the induction hypothesis, the distance label of node i_k at the end of the kth pass must be equal to the length of this path. Consequently, when we examine arc (i_k, i_{k+1}) in the $(k + 1)$th pass, we set the distance label of node i_{k+1} equal to the length of the path $i_0 - i_1 - \cdots - i_k - i_{k+1}$. This observation establishes that our induction hypothesis will be true for the $(k + 1)$th pass as well.

We have shown that the label correcting algorithm requires $O(nm)$ time as long as at each pass we examine all the arcs. It is not necessary to examine the arcs in any particular order.

The version of the label-correcting algorithm we have discussed considers every arc in A during every pass. It need not do so. Suppose that we order the arcs in the arc list by their tail nodes so that all arcs with the same tail node appear consecutively on the list. Thus, while scanning arcs, we consider one node at a time, say node i, scan arcs in $A(i)$, and test the optimality condition. Now suppose that during one pass through the arc list, the algorithm does not change the distance label of node i. Then during the next pass, $d(j) \le d(i) + c_{ij}$ for every $(i, j) \in A(i)$ and the algorithm need not test these conditions. Consequently, we can store all nodes whose distance labels change during a pass, and consider (or examine) only those nodes in the next pass. One plausible way to implement this approach is to store the nodes in a list whose distance labels change in a pass and examine this list in the first-in, first-out (FIFO) order in the next pass. If we follow this strategy in every pass, the resulting implementation is exactly the same as the modified label-correcting algorithm stated in Figure 5.5 provided that we maintain LIST as a queue (i.e., select nodes from the front of LIST and add nodes to the rear of LIST). We call this algorithm the *FIFO label-correcting algorithm* and summarize the preceding discussion as the following theorem.

Theorem 5.3. *The FIFO label-correcting algorithm solves the shortest path problem in O(nm) time.*

Dequeue Implementation

The modification of the modified label-correcting algorithm we discuss next has a pseudopolynomial worst-case behavior but is very efficient in practice. Indeed, this version of the modified label-correcting algorithm has proven in practice to be one of the fastest algorithms for solving the shortest path problems in sparse networks. We refer to this implementation of the modified label-correcting algorithm as the *dequeue implementation.*

This implementation maintains LIST as a *dequeue.* A dequeue is a data structure that permits us to store a list so that we can add or delete elements from the front as well as the rear of the list. A dequeue can easily be implemented using an array or a linked list (see Appendix A). The dequeue implementation always selects nodes from the front of the dequeue, but adds nodes either at the front or at the rear. If the node has been in the LIST earlier, the algorithm adds it to the front; otherwise, it adds the node to the rear. This heuristic rule has the following intuitive justification. If a node i has appeared previously in LIST, some nodes, say i_1, i_2, . . . , i_k, might have node i as its predecessor. Suppose further that LIST contains the nodes i_1, i_2, . . . , i_k when the algorithm updates $d(i)$ again. It is then advantageous to update the distance labels of nodes i_1, i_2, . . . , i_k from node i as soon as possible rather than first examining the nodes i_1, i_2, . . . , i_k and then reexamine them when their distance labels eventually decrease due to decrease in $d(i)$. Adding node i to the front of LIST tends to correct the distance labels of nodes i_1, i_2, . . . , i_k quickly and reduces the need to reexamine nodes. Empirical studies have observed similar behavior and found that the dequeue implementation examines fewer nodes than do most other label-correcting algorithms.

5.5 DETECTING NEGATIVE CYCLES

So far we have assumed that the network contains no negative cycle and described algorithms that solve the shortest path problem. We now describe modifications required in these algorithms that would permit us to detect the presence of a negative cycle, if one exists.

We first study the modifications required in the generic label-correcting algorithm. We have observed in Section 5.2 that if the network contains a negative cycle, no set of distance labels will satisfy the optimality condition. Therefore, the label-correcting algorithm will keep decreasing distance labels indefinitely and will never terminate. But notice that $-nC$ is a lower bound on any distance label whenever the network contains no negative cycle. Consequently, if we find that the distance label of some node k has fallen below $-nC$, we can terminate any further computation. We can obtain the negative cycle by tracing the predecessor indices starting at node k.

Let us describe yet another negative cycle detection algorithm. This algorithm checks at repeated intervals to see whether the predecessor graph contains a directed cycle. Recall from the illustration shown in Figure 5.4 how the predecessor graph might contain a directed cycle. This algorithm works as follows. We first designate the source node as marked and all other nodes as unmarked. Then, one by one, we examine each unmarked node k and perform the following operation: We mark node k, trace the predecessor indices starting at node k, and mark all the nodes encountered until we reach the first already marked node, say node l. If $k = l$, the predecessor graph contains a cycle, which must be a negative cycle (why?). The reader can verify that this algorithm requires $O(n)$ time to check the presence of a directed cycle in the predecessor graph. Consequently, if we apply this algorithm after every αn distance updates for some constant α, the computations it performs will not add to the worst-case complexity of any label-correcting algorithm.

In general, at the time that the algorithm relabels node j, $d(j) = d(i) + c_{ij}$ for some node i which is the predecessor of j. We refer to the arc (i, j) as a *predecessor arc*. Subsequently, $d(i)$ might decrease, and the labels will satisfy the condition $d(j) \geq d(i) + c_{ij}$ as long as pred$(j) = i$. Suppose that P is a path of predecessor arcs from node 1 to node j. The inequalities $d(k) \geq d(l) + c_{kl}$ for all arcs (k, l) on this path imply that $d(j)$ is at least the length of this path. Consequently, no node j with $d(j) \leq -nC$ is connected to node 1 on a path consisting only of predecessor arcs. We conclude that tracing back predecessor arcs from node j must lead to a cycle, and by Exercise 5.56, any such cycle must be negative.

The FIFO label-correcting algorithm is also capable of easily detecting the presence of a negative cycle. Recall that we can partition the node examinations in the FIFO algorithm into several passes and that the algorithm examines any node at most once within each pass. To implement this algorithm, we record the number of times that the algorithm examines each node. If the network contains no negative cycle, it examines any node at most $(n - 1)$ times [because it makes at most $(n - 1)$ passes]. Therefore, if it examines a node more than $(n - 1)$ times, the network must contain a negative cycle. We can also use the technique described in the preceding paragraph to identify negative cycles.

The FIFO label-correcting algorithm detects the presence of negative cycles or obtains shortest path distances in a network in $O(nm)$ time, which is the fastest available strongly polynomial-time algorithm for networks with nonnegative arc lengths. However, for problems that satisfy the similarity assumption, other weakly polynomial-time algorithms run faster than the FIFO algorithm. These approaches formulate the shortest path problem as an assignment problem (as described in Section 12.7) and then use an $O(n^{1/2}m \log(nC))$ time assignment algorithm to solve the problem (i.e., either finds a shortest path or detects a negative cycle).

5.6 ALL-PAIRS SHORTEST PATH PROBLEM

The all-pairs shortest path problem requires that we determine shortest path distances between every pair of nodes in a network. In this section we suggest two approaches for solving this problem. The first approach, called the *repeated shortest path algorithm*, is well suited for sparse networks. The second approach is a generalization of the label-correcting algorithm discussed in previous sections; we refer to this procedure as the *all-pairs label-correcting algorithm*. It is especially well suited for dense networks. In this section we describe the generic all-pairs label-correcting algorithm and then develop a special implementation of this generic algorithm, known as the *Floyd–Warshall algorithm*, that runs in $O(n^3)$ time.

In this section we assume that the underlying network is strongly connected (i.e., it contains a directed path from any node to every other node). We can easily satisfy this assumption by selecting an arbitrary node, say node s, and adding arcs (s, i) and (i, s) of sufficiently large cost for all $i \in N - \{s\}$, if these arcs do not already exist. For reasons explained earlier, we also assume that the network does not contain a negative cycle. All the algorithms we discuss, however, are capable of detecting the presence of a negative cycle. We discuss situations with negative cycles at the end of this section.

Repeated Shortest Path Algorithm

If the network has nonnegative arc lengths, we can solve the all-pairs shortest path problem by applying any single-source shortest path algorithm n times, considering each node as the source node once. If $S(n, m, C)$ denotes the time needed to solve a shortest path problem with nonnegative arc lengths, this approach solves the all-pairs shortest path problem in $O(n\, S(n, m, C))$ time.

If the network contains some negative arcs, we first transform the network to one with nonnegative arc lengths. We select a node s and use the FIFO label-correcting algorithm, described in Section 5.4, to compute the shortest distances from node s to all other nodes. The algorithm either detects the presence of a negative cycle or terminates with the shortest path distances $d(j)$. In the first case, the all-pairs shortest path problem has no solution, and in the second case, we consider the shortest path problem with arc lengths equal to their reduced arc lengths with respect to the distance labels $d(j)$. Recall from Section 5.2 that the reduced arc length of an arc (i, j) with respect to the distance labels $d(j)$ is $c_{ij}^d = c_{ij} + d(i) - d(j)$, and if the distance labels are shortest path distances, then $c_{ij}^d \geq 0$ for all arcs (i, j) in A [see Property 5.2(c)]. Since this transformation produces nonnegative reduced arc lengths, we can then apply the single-source shortest path algorithm for problems with nonnegative arc lengths n times (by considering each node as a source once) to determine shortest path distances between all pairs of nodes in the transformed network. We obtain the shortest path distance between nodes k and l in the original network by adding $d(l) - d(k)$ to the corresponding shortest path distance in the transformed network [see Property 5.2(b)]. This approach requires $O(nm)$ time to solve the first shortest path problem, and if the network contains no negative cycles, it requires an extra $O(n\, S(n, m, C))$ time to compute the remaining shortest path distances. Therefore, this approach determines all pairs shortest path distances in $O(nm + n\, S(n, m, C)) = O(n\, S(n, m, C))$ time. We have established the following result.

Theorem 5.4. *The repeated shortest path algorithm solves the all-pairs shortest path problem in $O(n\, S(n, m, C))$ time.*

In the remainder of this section we study the generic all-pairs label-correcting algorithm. Just as the generic label-correcting algorithm relies on shortest path optimality conditions, the all-pairs label-correcting algorithm relies on all-pairs shortest path optimality conditions, which we study next.

All-Pairs Shortest Path Optimality Conditions

Let $[i, j]$ denote a pair of nodes i and j in the network. The all-pairs label-correcting algorithm maintains a distance label $d[i, j]$ for every pair of nodes; this distance label represents the length of some directed *walk* from node i to node j and hence will be an upper bound on the shortest path length from node i to node j. The algorithm updates the matrix of distance labels until they represent shortest path distances. It uses the following generalization of Theorem 5.1:

Theorem 5.5 (All-Pairs Shortest Path Optimality Conditions). *For every pair of nodes $[i, j] \in N \times N$, let $d[i, j]$ represent the length of some directed path from node i to node j. These distances represent all-pairs shortest path distances if and only if they satisfy the following all-pairs shortest path optimality conditions:*

$$d[i, j] \leq d[i, k] + d[k, j] \qquad \textit{for all nodes } i, j, \textit{ and } k. \tag{5.3}$$

Proof. We use a contradiction argument to establish that the shortest path distances $d[i, j]$ must satisfy the conditions (5.3). Suppose that $d[i, k] + d[k, j] < d[i, j]$ for nodes i, j, and k. The union of the shortest paths from node i to node k and node k to node j is a directed walk of length $d[i, k] + d[k, j]$ from node i to node j. This directed walk decomposes into a directed path, say P, from node i to node j and some directed cycles (see Exercise 3.51). Since each directed cycle in the network has nonnegative length, the length of the path P is at most $d[i, k] + d[k, j] < d[i, j]$, contradicting the optimality of $d[i, j]$.

We now show that if the distance labels $d[i, j]$ satisfy the conditions in (5.3), they represent shortest path distances. We use an argument similar to the one we used in proving Theorem 5.1. Let P be a directed path of length $d[i, j]$ consisting of the sequence of nodes $i = i_1 - i_2 - i_3 - \cdots - i_k = j$. The condition (5.3) implies that

$$d[i, j] = d[i_1, i_k] \leq d[i_1, i_2] + d[i_2, i_k] \leq c_{i_1 i_2} + d[i_2, i_k],$$

$$d[i_2, i_k] \leq c_{i_2 i_3} + d[i_3, i_k],$$

$$\vdots$$

$$d[i_{k-1}, i_k] \leq c_{i_{k-1} i_k}.$$

These inequalities, in turn, imply that

$$d[i, j] \leq c_{i_1 i_2} + c_{i_2 i_3} + \cdots + c_{i_{k-1} i_k} = \sum_{(i,j) \in P} c_{ij}.$$

Therefore, $d[i, j]$ is a lower bound on the length of any directed path from node i to node j. By assumption, $d[i, j]$ is also an upper bound on the shortest path length from node i to node j. Consequently, $d[i, j]$ must be the shortest path length between these nodes which is the derived conclusion of the theorem. ◆

All-Pairs Generic Label Correcting Algorithm

The all-pairs shortest path optimality conditions (throughout the remainder of this section we refer to these conditions simply as the optimality conditions) immediately yield the following generic all-pairs label-correcting algorithm: Start with some distance labels $d[i, j]$ and successively update these until they satisfy the optimality conditions. Figure 5.6 gives a formal statement of the algorithm. In the algorithm we refer to the operation of checking whether $d[i, j] > d[i, k] + d[k, j]$, and if so, then setting $d[i, j] = d[i, k] + d[k, j]$ as a *triple operation*.

```
algorithm all-pairs label-correcting;
begin
    set d[i, j] : = ∞ for all [i, j] ∈ N × N;
    set d[i, i] : = 0 for all i ∈ N;
    for each (i, j) ∈ A do d[i, j] : = c_ij;
    while the network contains three nodes i, j, and k
        satisfying d[i, j] > d[i, k] + d[k, j] do d[i, j] : = d[i, k] + d[k, j];
end;
```

Figure 5.6 Generic all-pairs label-correcting algorithm.

To establish the finiteness and correctness of the generic all-pairs label-correcting algorithm, we assume that the data are integral and that the network contains no negative cycle. We first consider the correctness of the algorithm. At every step the algorithm maintains the invariant property that whenever $d[i, j] < \infty$, the network contains a directed walk of length $d[i, j]$ from node i to node j. We can use induction on the number of iterations to show that this property holds at every step. Now consider the directed walk of length $d[i, j]$ from node i to node j at the point when the algorithm terminates. This directed walk decomposes into a directed path, say P, from node i to node j, and possibly some directed cycles. None of these cycles could have a positive length, for otherwise we would contradict the optimality of $d[i, j]$.

Therefore, all of these cycles must have length zero. Consequently, the path P must have length $d[i, j]$. The distance labels $d[i, j]$ also satisfy the optimality conditions (5.3), for these conditions are the termination criteria of the algorithm. This conclusion establishes the fact that when the algorithm terminates, the distance labels represent shortest path distances.

Now consider the finiteness of the algorithm. Since all arc lengths are integer and C is the largest magnitude of any arc length, the maximum (finite) distance label is bounded from above by nC and the minimum distance label is bounded from below by $-nC$. Each iteration of the generic all-pairs label-correcting algorithm decreases some $d[i, j]$. Consequently, the algorithm terminates within $O(n^3C)$ iterations. This bound on the algorithm's running time is pseudopolynomial and is not attractive from the viewpoint of worst-case complexity. We next describe a specific implementation of the generic algorithm, known as the *Floyd–Warshall algorithm*, that solves the all-pairs shortest path problem in $O(n^3)$ time.

Floyd–Warshall Algorithm

Notice that given a matrix of distances $d[i, j]$, we need to perform $\Omega(n^3)$ triple operations in order to test the optimality of this solution. It is therefore surprising that the Floyd–Warshall algorithm obtains a matrix of shortest path distances within $O(n^3)$ computations. The algorithm achieves this bound by applying the triple operations cleverly. The algorithm is based on inductive arguments developed by an application of a dynamic programming technique.

Let $d^k[i, j]$ represent the length of a shortest path from node i to node j subject to the condition that this path uses only the nodes $1, 2, \ldots, k-1$ as internal nodes. Clearly, $d^{n+1}[i, j]$ represents the actual shortest path distance from node i to node j. The Floyd–Warshall algorithm first computes $d^1[i, j]$ for all node pairs i and j. Using $d^1[i, j]$, it then computes $d^2[i, j]$ for all node pairs i and j. It repeats this process until it obtains $d^{n+1}[i, j]$ for all node pairs i and j, when it terminates. Given $d^k[i, j]$, the algorithm computes $d^{k+1}[i, j]$ using the following property.

Property 5.6. $d^{k+1}[i, j] = min\{d^k[i, j], d^k[i, k] + d^k[k, i]\}$.

This property is valid for the following reason. A shortest path that uses only the nodes $1, 2, \ldots, k$ as internal nodes either (1) does not pass through node k, in which case $d^{k+1}[i, j] = d^k[i, j]$, or (2) does pass through node k, in which case $d^{k+1}[i, j] = d^k[i, k] + d^k[k, j]$. Therefore, $d^{k+1}[i, j] = \min\{d^k[i, j], d^k[i, k] + d^k[k, j]\}$.

Figure 5.7 gives a formal description of the Floyd–Warshall algorithm.

```
algorithm Floyd-Warshall;
begin
    for all node pairs [i, j] ∈ N × N do
            d[i, j] : = ∞ and pred[i, j] : = 0;
    for all nodes i ∈ N do d[i, i] : = 0;
    for each arc (i, j) ∈ A do d[i, j] : = c_ij and pred[i, j] : = i;
    for each k : = 1 to n do
            for each [i, j] ∈ N × N do
                if d[i, j] > d[i, k] + d[k, j] then
                begin
                    d[i, j] : = d[i, k] + d[k, j];
                    pred[i, j] : = pred[k, j];
                end;
end;
```

Figure 5.7 Floyd–Warshall algorithm.

The Floyd–Warshall algorithm uses predecessor indices, *pred*$[i, j]$, for each node pair $[i, j]$. The index pred$[i, j]$ denotes the last node prior to node j in the tentative shortest path from node i to node j. The algorithm maintains the invariant property that when $d[i, j]$ is finite, the network contains a path from node i to node j of length $d[i, j]$. Using the predecessor indices, we can obtain this path, say P, from node k to node l as follows. We backtrack along the path P starting at node l. Let $g =$ pred$[k, l]$. Then g is the node prior to node l in P. Similarly, $h =$ pred$[k, g]$ is the node prior to node g in P, and so on. We repeat this process until we reach node k.

The Floyd–Warshall algorithm clearly performs n major iterations, one for each k, and within each major iteration, it performs $O(1)$ computations for each node pair. Consequently, it runs in $O(n^3)$ time. We thus have established the following result.

Theorem 5.7. *The Floyd–Warshall algorithm computes shortest path distances between all pairs of nodes in $O(n^3)$ time.* ◆

Detection of Negative Cycles

We now address the issue of detecting a negative cycle in the network if one exists. In the generic all-pairs label-correcting algorithm, we incorporate the following two tests whenever the algorithm updates a distance label $d[i, j]$ during a triple iteration:

1. If $i = j$, check whether $d[i, i] < 0$.
2. If $i \neq j$, check whether $d[i, j] < -nC$.

If either of these two tests is true, the network contains a negative cycle. To verify this claim, consider the first time during a triple iteration when $d[i, i] < 0$ for some node i. At this time $d[i, i] = d[i, k] + d[k, i]$ for some node $k \neq i$. This condition implies that the network contains a directed walk from node i to node k, and a directed walk from node k to node i, and that the sum of the lengths of these two walks is $d[i, i]$, which is negative. The union of these two walks is a closed walk, which can be decomposed into a set of directed cycles (see Exercise 3.51). Since $d[i, i] < 0$, at least one of these directed cycles must be negative.

We next consider the situation in which $d[i, j] < -nC$ for some node pair i and j. Consider the first time during a triple iteration when $d[i, j] < -nC$. At this time the network contains a directed walk from node i to node j of length $-nC$. As we observed previously, we can decompose this walk into a directed path P from node i to node j and some directed cycles. Since the path P must have a length of at least $-(n - 1)C$, at least one of these cycles must be a negative cycle.

Finally, we observe that if the network contains a negative cycle, then eventually $d[i, i] < 0$ for some node i or $d[i, j] < -nC$ for some node pair $[i, j]$, because the distance labels continue to decrease by an integer amount at every iteration. Therefore, the generic label-correcting algorithm will always determine a negative cycle if one exists.

In the Floyd–Warshall algorithm, we detect the presence of a negative cycle simply by checking the condition $d[i, i] < 0$ whenever we update $d[i, i]$ for some node i. It is easy to see that whenever $d[i, i] < 0$, we have detected the presence of a negative cycle. In Exercise 5.37 we show that whenever the network contains a negative cycle, then during the computations we will eventually satisfy the condition $d[i, i] < 0$ for some i.

We can also use an extension of the method described in Section 5.5, using the predecessor graph, to identify a negative cycle in the Floyd–Warshall algorithm. The Floyd–Warshall algorithm maintains a predecessor graph for each node k in the network, which in the absence of a negative cycle is a directed out-tree rooted at node k (see Section 5.3). If the network contains a negative cycle, eventually the predecessor graph contains a cycle. For any node k, the predecessor graph consists of the arcs $\{(\text{pred}[k, i], i) : i \in N - \{k\}\}$. Using the method described in Section 5.5, we can determine whether or not any predecessor graph contains a cycle. Checking this condition for every node requires $O(n^2)$ time. Consequently, if we use this method after every αn^2 triple operations for some constant α, the computations will not add to the worst-case complexity of the Floyd–Warshall algorithm.

Comparison of the Two Methods

The generic all-pairs label-correcting algorithm and its specific implementation as the Floyd–Warshall algorithm are matrix manipulation algorithms. They maintain a matrix of tentative shortest path distances between all pairs of nodes and perform repeated updates of this matrix. The major advantages of this approach, compared to the repeated shortest path algorithm discussed at the beginning of this section, are its simplicity, intuitive appeal, and ease of implementation. The major drawbacks of this approach are its significant storage requirements and its poorer worst-case complexity for all network densities except completely dense networks. The matrix manipulation algorithms require $\Omega(n^2)$ intermediate storage space, which could prohibit its application in some situations. Despite these disadvantages, the matrix manipulation algorithms have proven to be fairly popular computational methods for solving all-pairs shortest path problems.

5.7 MINIMUM COST-TO-TIME RATIO CYCLE PROBLEM

The *minimum cost-to-time ratio cycle problem* is defined on a directed graph G with both a cost and a travel time associated with each arc: we wish to find a directed cycle in the graph with the smallest ratio of its cost to its travel time. The minimum cost-to-time ratio cycle problem arises in an application known as the *tramp steamer problem*, which we defined in Application 4.4. A tramp steamer travels from port to port, carrying cargo and passengers. A voyage of the steamer from port i to port j earns p_{ij} units of profit and requires time τ_{ij}. The captain of the steamer wants to know what ports the steamer should visit, and in which order, in order to maximize its mean daily profit. We can solve this problem by identifying a directed cycle with the largest possible ratio of total profit to total travel time. The tramp steamer then continues to sail indefinitely around this cycle.

In the tramp steamer problem, we wish to identify a directed cycle W of G with the maximum ratio $(\sum_{(i,j)\in W} p_{ij})/(\sum_{(i,j)\in W} \tau_{ij})$. We can convert this problem into a minimization problem by defining the cost c_{ij} of each arc (i, j) as $c_{ij} = -p_{ij}$. We then seek a directed cycle W with the minimum value for the ratio

$$\mu(W) = \frac{\sum_{(i,j)\in W} c_{ij}}{\sum_{(i,j)\in W} \tau_{ij}}.$$

We assume in this section that all data are integral, that $\tau_{ij} \geq 0$ for every arc $(i, j) \in A$, and that $\sum_{(i,j)\in W} \tau_{ij} > 0$ for every directed cycle W in G.

We can solve the minimum cost-to-time ratio cycle problem (or, simply, the minimum ratio problem) by repeated applications of the negative cycle detection algorithm. Let μ^* denote the optimal objective function value of the minimum cost-to-time ratio cycle problem. For any arbitrary value of μ, let us define the length of each arc as $l_{ij} = c_{ij} - \mu\tau_{ij}$. With respect to these arc lengths, we could encounter three situations:

are faster than those available for the general minimum cost-to-time ratio cycle problem. In this section we describe an $O(nm)$-time dynamic programming algorithm for solving the minimum mean cycle problem.

In the subsequent discussion, we assume that the network is strongly connected (i.e., contains a directed path between every pair of nodes). We can always satisfy this assumption by adding arcs of sufficiently large cost; the minimum mean cycle will contain no such arcs unless the network is acyclic.

Let $d^k(j)$ denote the length, with respect to the arc lengths c_{ij}, of a shortest directed *walk* containing *exactly* k arcs from a specially designated node s to node j. We can choose any node s as the specially designated node. We emphasize that $d^k(j)$ is the length of a directed walk to node j; it might contain directed cycles. We can compute $d^k(j)$ for every node j and for every $k = 1, \ldots, n$, by using the following recursive relationship:

$$d^k(j) = \min_{\{i:(i,j)\in A\}} \{d^{k-1}(i) + c_{ij}\}. \tag{5.9}$$

We initialize the recursion by setting $d^0(j) = \infty$ for each node j. Given $d^{k-1}(j)$ for all j, using (5.9) we compute $d^k(j)$ for all j, which requires a total of $O(m)$ time. By repeating this process for all $k = 1, 2, \ldots, n$, within $O(nm)$ computations we determine $d^k(j)$ for every node j and for every k. As the next result shows, we are able to obtain a bound on the cost μ^* of the minimum mean cycle in terms of the walk lengths $d^k(j)$.

Theorem 5.8

$$\mu^* = \min_{j\in N} \max_{0\le k\le n-1} \left[\frac{d^n(j) - d^k(j)}{n - k}\right]. \tag{5.10}$$

Proof. We prove this theorem for two cases: when $\mu^* = 0$ and $\mu^* \neq 0$.

Case 1. $\mu^* = 0$. In this case the network does not contain a negative cycle (for otherwise, $\mu^* < 0$), but does contain a zero cost cycle W. For each node $j \in N$, let $d(j)$ denote the shortest path distance from node s to node j. We next replace each arc cost c_{ij} by its reduced cost $c_{ij}^d = c_{ij} + d(i) - d(j)$. Property 5.2 implies that as a result of this transformation, the network satisfies the following properties:

1. All arc costs are nonnegative.
2. All arc costs in W are zero.
3. For each node j, every arc in the shortest path from node s to node j has zero cost.
4. For each node j, the shortest path distances $d^k(j)$, for any $1 \le k \le n$, differ by a constant amount from their values before the transformation.

Let $\bar{d}^k(j)$ denote the length of the shortest walk from node s to node j with respect to the reduced costs c_{ij}^d. Condition 4 implies that the expression (5.10) remains valid even if we replace $d^n(j)$ by $\bar{d}^n(j)$ and $d^k(j)$ by $\bar{d}^k(j)$. Next, notice that for each node $j \in N$,

$$\max_{1\le k\le n-1} [\bar{d}^n(j) - \bar{d}^k(j)] \ge 0, \tag{5.11}$$

because for some k, $\bar{d}^k(j)$ will equal the shortest path length $\bar{d}(j)$, and $\bar{d}^n(j)$ will be at least as large. We now show that for some node p, the left-hand side of (5.11) will be zero, which will establish the theorem. We choose some node j in the cycle W and construct a directed walk containing n arcs in the following manner. First, we traverse the shortest path from node s to node j and then we traverse the arcs in W from node j until the walk contains n arcs. Let node p be the node where this walk ends. Conditions 2 and 3 imply that this walk from node s to node p has a zero length. This walk must contain one or more directed cycle because it contains n arcs. Removing the directed cycles from this walk gives a path, say of length $k \leq n - 1$, from node s to node p of zero length. We have thus shown that $\bar{d}^n(p) = \bar{d}^k(p) = 0$. For node p the left-hand side of (5.11) is zero, so this node satisfies the condition

$$\mu^* = \max_{0 \leq k \leq n-1} \left[\frac{d^n(p) - d^k(p)}{n - k}\right] = 0,$$

as required by the theorem.

Case 2. $\mu^* \neq 0$. Suppose that Δ is a real number. We study the effect of decreasing each arc cost c_{ij} by an amount Δ. Clearly, this change in the arc costs reduces μ^* by Δ, each $d^k(j)$ by $k\Delta$, and therefore the ratio $(d^n(v) - d^k(v))/(n - k)$, and so the right-hand side of (5.10), by an amount Δ. Consequently, translating the costs by a constant affects both sides of (5.10) equally. Choosing the translation to make $\mu^* = 0$ and then using the result of Case 1 provides a proof of the theorem. ◆

We ask the reader to show in Exercise 5.55 that how to use the $d^k(j)$'s to obtain a minimum mean cycle.

5.8 SUMMARY

In this chapter we developed several algorithms, known as the *label-correcting algorithms*, for solving shortest path problems with arbitrary arc lengths. The shortest path optimality conditions, which provide necessary and sufficient conditions for a set of distance labels to define shortest path lengths, play a central role in the development of label-correcting algorithms. The label-correcting algorithms maintain a distance label with each node and iteratively update these labels until the distance labels satisfy the optimality conditions. The generic label-correcting algorithm selects any arc violating its optimality condition and uses it to update the distance labels. Typically, identifying an arc violating its optimality condition will be a time-consuming component of the generic label-correcting algorithm. To improve upon this feature of the algorithm, we modified the algorithm so that we could quickly select an arc violating its optimality condition. We presented two specific implementations of this *modified label-correcting algorithm*: A FIFO implementation improves on its running time in theory and a dequeue implementation improves on its running time in practice. Figure 5.8 summarizes the important features of all the label-correcting algorithms that we have discussed.

The label-correcting algorithms determine shortest path distances only if the network contains no negative cycle. These algorithms are, however, capable of de-

Algorithm	Running Time	Features
Generic label-correcting algorithm	$O(\min\{n^2mC, m2^n\})$	1. Selects arcs violating their optimality conditions and updates distance labels. 2. Requires $O(m)$ time to identify an arc violating its optimality condition. 3. Very general: most shortest path algorithms can be viewed as special cases of this algorithm. 4. The running time is pseudopolynomial and so is unattractive.
Modified label-correcting algorithm	$O(\min\{nmC, m2^n\})$	1. An improved implementation of the generic label-correcting algorithm. 2. The algorithm maintains a set, LIST, of nodes: whenever a distance label $d(j)$ changes, we add node j to LIST. The algorithm removes a node i from LIST and examines arcs in $A(i)$ to update distance labels. 3. Very flexible since we can maintain LIST in a variety of ways. 4. The running time is still unattractive.
FIFO implementation	$O(nm)$	1. A specific implementation of the modified label-correcting algorithm. 2. Maintains the set LIST as a queue and hence examines nodes in LIST in first-in, first-out order. 3. Achieves the best strongly polynomial running time for solving the shortest path problem with arbitrary arc lengths. 4. Quite efficient in practice. 5. In $O(nm)$ time, can also identify the presence of negative cycles.
Dequeue implementation	$O(\min\{nmC, m2^n\})$	1. Another specific implementation of the modified label-correcting algorithm. 2. Maintains the set LIST as a dequeue. Adds a node to the front of dequeue if the algorithm has previously updated its distance label, and to the rear otherwise. 3. Very efficient in practice (possibly, linear time). 4. The worst-case running time is unattractive.

Figure 5.8 Summary of label-correcting algorithms.

tecting the presence of a negative cycle. We described two methods for identifying such a situation: the more efficient method checks at repeated intervals whether the predecessor graphs (i.e., the graph defined by the predecessor indices) contains a directed cycle. This computation requires $O(n)$ time.

To conclude this chapter we studied algorithms for the all-pairs shortest path problem. We considered two basic approaches: a repeated shortest path algorithm and an all-pairs label-correcting algorithm. We described two versions of the latter approach: the generic version and a special implementation known as the Floyd–Warshall algorithm. Figure 5.9 summarizes the basic features of the all-pairs shortest path algorithms that we studied.

Algorithm	Running Time	Features
Repeated shortest path algorithm	$O(nS(n,m,C))$	1. Preprocesses the network so that all (reduced) arc lengths are nonnegative. Then applies Dijkstra's algorithm n times with each node $i \in N$ as the source node. 2. Flexible in the sense that we can use an implementation of Dijkstra's algorithm. 3. Achieves the best available running time for all network densities. 4. Low intermediate storage.
Floyd–Warshall algorithm	$O(n^3)$	1. Corrects distance labels in a systematic way until they represent the shortest path distances. 2. Very easy to implement. 3. Achieves the best available running time for dense networks. 4. Requires $\Omega(n^2)$ intermediate storage.

Figure 5.9 Summary of all pairs shortest path algorithms. [$S(n, m, C)$ is the time required to solve a shortest path problem with nonnegative arc lengths.]

REFERENCE NOTES

Researchers, especially those within the operations research community, have actively studied label-correcting algorithms for many years; much of this development has focused on designing computationally efficient algorithms. Ford [1956] outlined the first label-correcting algorithm for the shortest path problem. Subsequently, several researchers, including Moore [1957] and Ford and Fulkerson [1962], studied properties of the generic label-correcting algorithms. Bellman's [1958] dynamic programming algorithm for the shortest path problem can also be viewed as a label-correcting algorithm. The FIFO implementation of the generic label-correcting algorithm is also due to Bellman [1958]. Although Bellman developed this algorithm more than three decades ago, it is still the best strongly polynomial-time algorithm for solving shortest path problems with arbitrary arc lengths.

In Section 12.7 we show how to transform the shortest path problem into an assignment problem and then solve it using any assignment algorithm. As we note in the reference notes of Chapter 12, we can solve the assignment problem in $O(n^{1/2}m \log(nC))$ time using either the algorithms reported by Gabow and Tarjan [1989a] or the algorithm developed by Orlin and Ahuja [1992]. These developments show that we can solve shortest path problems with arbitrary arc lengths in $O(n^{1/2}m \log(nC))$ time. Thus the best available time bound for solving the shortest path problem with arbitrary arc lengths is $O(\min\{nm, n^{1/2}m \log(nC)\})$: The first bound is due to Bellman [1958], and the second bound is due to Gabow and Tarjan [1989a] and Orlin and Ahuja [1992].

Researchers have exploited the inherent flexibility of the generic label-correcting algorithm to design algorithms that are very efficient in practice. Pape's implementation, described in Section 5.4, is based on an idea due to D'Esopo that was later refined and tested by Pape [1974]. Pape [1980] gave a FORTRAN listing of this algorithm. Pape's algorithm runs in pseudopolynomial time. Gallo and Pallottino [1986] describe a two-queue implementation that retains the computational efficiency of Pape's algorithm and still runs in polynomial time. The papers by Glover, Klingman, and Phillips [1985] and Glover, Klingman, Phillips, and Schneider [1985] have described a variety of specific implementations of the generic label-correcting algorithm and studied their theoretical and computational behavior. These two papers, along with those by Hung and Divoky [1988], Divoky and Hung [1990], and Gallo and Pallottino [1984, 1988], have presented extensive computational results of label-setting and label-correcting algorithms. These studies conclude that for a majority of shortest path problems with nonnegative or arbitrary arc lengths, the label-correcting algorithms, known as *Thresh X1* and *Thresh X2*, suggested by Glover, Klingman, and Phillips [1985], are the fastest shortest path algorithms. The reference notes of Chapter 11 provide references for simplex-based approaches for the shortest path problem.

The generic all-pairs label-correcting algorithm, discussed in Section 5.3, is a generalization of the single source shortest path problem. The Floyd–Warshall algorithm, which was published in Floyd [1962], was based on Warshall's [1962] algorithm for finding transitive closure of graphs.

Lawler [1966] and Dantzig, Blattner, and Rao [1966] are early and important references on the minimum cost-to-time ratio cycle problem. The binary search algorithm described by us in Section 5.7 is due to Lawler [1966]. Dantzig, Blattner, and Rao [1966] presented a primal simplex approach that uses the linear programming formulation of the minimum ratio problems; we discuss this approach in Exercise 5.47. Meggido [1979] describes a general approach for solving minimum ratio problems, which as a special case yields a strongly polynomial-time algorithm for the minimum cost-to-time ratio cycle problem.

The $O(nm)$-time minimum mean cycle algorithm, described in Section 5.7, is due to Karp [1978]. Several other algorithms are available for solving the minimum mean cycle problem: (1) an $O(nm \log n)$ parametric network simplex algorithm proposed by Karp and Orlin [1981], (2) an $O(n^{1/2}m \log(nC))$ algorithm developed by Orlin and Ahuja [1992], and (3) an $O(nm + n^2 \log n)$ algorithm designed by Young, Tarjan, and Orlin [1990]. The best available time bound for solving the minimum mean cycle problem is $O(\min\{nm, n^{1/2}m \log(nC))$: The two bounds contained in this expression are due to Karp [1978] and Orlin and Ahuja [1992]. However, we believe that the parametric network simplex algorithm by Karp and Orlin [1981] would prove to be the most efficient algorithm empirically. We describe an application of the minimum mean cycle problem in Application 19.6. The minimum mean cycle problem also arises in solving minimum cost flow problems (see Goldberg and Tarjan [1987, 1988]).

EXERCISES

5.1. Select a directed cycle in Figure 5.10(a) and verify that it satisfies Property 5.2(a). Similarly, select a directed path from node 1 to node 6 and verify that it satisfies Property 5.2(b). Does the network contain a zero-length cycle?

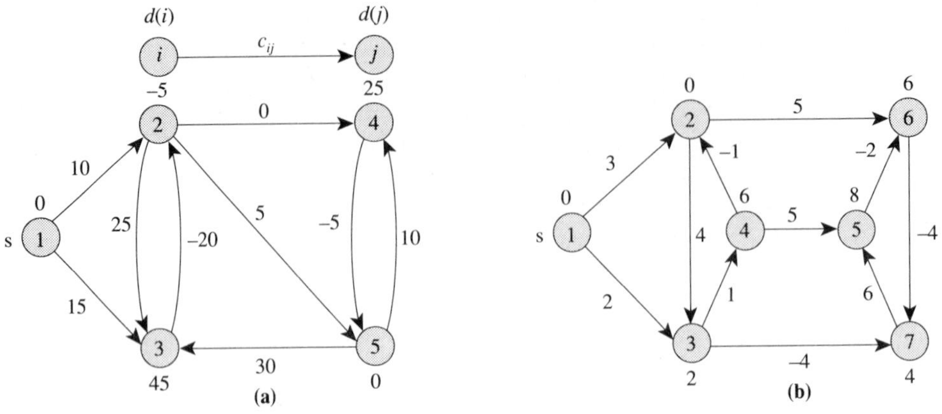

Figure 5.10 Examples for Exercises 5.1 to 5.5.

5.2. Consider the shortest path problems shown in Figure 5.10. Check whether or not the distance label $d(j)$ given next to each node j represents the length of some path. If your answer is yes for every node, list all the arcs that do not satisfy the shortest path optimality conditions.

5.3. Apply the modified label-correcting algorithm to the shortest path problem shown in Figure 5.10(a). Assume that the adjacency list of each node is arranged in increasing order of the head node numbers. Always examine a node with the minimum number in LIST. Specify the predecessor graph after examining each node and count the number of distance updates.

5.4. Apply the FIFO label-correcting algorithm to the example shown in Figure 5.10(b). Perform two passes of the arc list and specify the distance labels and the predecessor graph at the end of the second pass.

5.5. Consider the shortest path problem given in Figure 5.10(a) with the modification that the length of arc (4, 5) is -15 instead of -5. Verify that the network contains a negative cycle. Apply the dequeue implementation of the label-correcting algorithm; after every three distance updates, check whether the predecessor graph contains a directed cycle. How many distance updates did you perform before detecting a negative cycle?

5.6. Construct a shortest path problem whose shortest path tree contains a largest cost arc in the network but does not contain the smallest cost arc.

5.7. Bellman's equations

(a) Show that the shortest path distances $d(\cdot)$ must satisfy the following equations, known as *Bellman's equations*:

$$d(j) = \min\{d(i) + c_{ij} : (i, j) \in A(i)\} \qquad \text{for all } j \in N.$$

(b) Show that if a set of distance labels $d(i)$'s satisfy Bellman's equations and the network contains no zero-length cycle, these distance labels are shortest path distances.

(c) Verify that for the shortest path problem shown in Figure 5.11, the distance labels

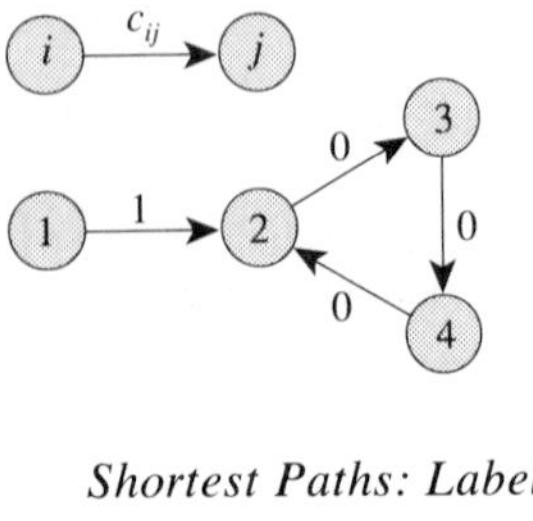

Figure 5.11 Example for Exercise 5.7.

$d = (0, 0, 0, 0)$ satisfy Bellman's equations but do not represent shortest path distances. This example shows that in the presence of a zero-length cycle, Bellman's equations are not sufficient for characterizing the optimality of distance labels.

5.8. Our termination argument of the generic label-correcting algorithm relies on the fact that the data are integral. Suppose that in the shortest path problem, some arc lengths are irrational numbers.
(a) Prove that for this case too, the generic label-correcting algorithm will terminate finitely. (*Hint*: Use arguments based on the predecessor graph.)
(b) (Gallo and Pallottino [1986]). Assuming that the network has no negative cost cycles, show the total number of relabels is $O(2^n)$. (*Hint*: Show first that if the algorithms uses the path 1-2-3-4 to label node 4, then it never uses the path 1-3-2-4 to label node 4. Then generalize this observation.)
(c) Show that the generic label correcting algorithm requires $O(2^n)$ iterations.

5.9. In Dijkstra's algorithm for the shortest path problem, let S denote the set of permanently labeled nodes at some stage. Show that for all node pairs $[i, j]$ for which $i \in S, j \in N$ and $(i, j) \in A, d(j) \leq d(i) + c_{ij}$. Use this result to give an alternative proof of correctness for Dijkstra's algorithm.

5.10. We define an *in-tree of shortest paths* as a directed in-tree rooted at a sink node t for which the tree path from any node i to node t is a shortest path. State a modification of the generic label-correcting algorithm that produces an in-tree of shortest paths.

5.11. Let $G = (N_1 \cup N_2, A)$ be a bipartite network. Suppose that $n_1 = |N_1|, n_2 = |N_2|$ and $n_1 \leq n_2$. Show that the FIFO label-correcting algorithm solves the shortest path problem in this network in $O(n_1 m)$ time.

5.12. Let $d^k(j)$ denote the shortest path length from a source node s to node j subject to the condition that the path contains at most k arcs. Consider the $O(nm)$ implementation of the label-correcting algorithm discussed in Section 5.4; let $D^k(j)$ denote the distance label of node j at the end of the kth pass. Show that $D^k(j) \leq d^k(j)$ for every node $j \in N$.

5.13. In the shortest path problem with nonnegative arc lengths, suppose that we know that the shortest path distance of nodes $i_1, i_2, \ldots, i_n$ are in nondecreasing order. Can we use this information to help us determine shortest path distances more efficiently than the algorithms discussed in Chapter 4? If we allow arc lengths to be negative, can you solve the shortest path problem faster than $O(nm)$ time?

5.14. Show that in the FIFO label-correcting algorithm, if the kth pass of the arc list decreases the distances of at least $n - k + 1$ nodes, the network must contain a negative cycle. (*Hint*: Use the arguments required in the complexity proof of the FIFO algorithm.)

5.15. **Modified FIFO algorithm** (Goldfarb and Hao [1988]). This exercise describes a modification of the FIFO label-correcting algorithm that is very efficient in practice. The generic label-correcting algorithm described in Figure 5.1 maintains a predecessor graph. Let $f(j)$ denote the number of arcs in the predecessor graph from the source node to node j. We can easily maintain these values by using the update formula $f(j) = f(i) + 1$ whenever we make the distance label update $d(j) = d(i) + c_{ij}$. Suppose that in the algorithm we always examine a node i in LIST with the minimum value of $f(i)$. Show that the algorithm examines the nodes with nondecreasing values of $f(\cdot)$ and that it examines no node more than $n - 1$ times. Use this result to specify an $O(nm)$ implementation of this algorithm.

5.16. Suppose after solving a shortest path problem, you realize that you underestimated each arc length by k units. Suggest an $O(m)$ algorithm for solving the original problem with the correct arc lengths. The running time of your algorithm should be independent of the value of k (*Hint*: Use Dial's implementation described in Section 4.6 on a modified problem.)

5.17. Suppose that after solving a shortest path problem, you realize that you underestimated some arc lengths. The actual arc lengths were $c'_{ij} \geq c_{ij}$ for all $(i, j) \in A$. Let $L = \sum_{(i,j)\in A}(c'_{ij} - c_{ij})$. Suggest an $O(m + L)$ algorithm for reoptimizing the solution obtained for the shortest path problem with arc lengths c_{ij}. (*Hint*: See the hint for Exercise 5.16.)

5.18. Suppose that after solving a shortest path problem, you realize that you underestimated some arc lengths and overestimated some other arc lengths. The actual arc lengths are c'_{ij} instead of c_{ij} for all $(i, j) \in A$. Let $L = \sum_{(i,j) \in A} | c_{ij} - c'_{ij} |$. Suggest an $O(mL)$ algorithm for reoptimizing the shortest path solution obtained with the arc lengths c_{ij}. (*Hint*: Apply the label-correcting algorithm on a modified problem.)

5.19. Identifying zero-length cycles. In a directed network G with arc lengths c_{ij}, let $d(j)$ denote the shortest path distance from the source node s to node j. Define reduced arc lengths as $c^d_{ij} = c_{ij} + d(i) - d(j)$ and define the *zero-residual network* G^0 as the subnetwork of G consisting only of arcs with zero reduced arc lengths. Show that there is a one-to-one correspondence between zero-length cycles in G and directed cycles in G^0. Explain how you can identify a directed cycle in G^0 in $O(m)$ time.

5.20. Enumerating all shortest paths. Define the zero-residual network G^0 as in Exercise 5.19, and assume that G^0 is acyclic. Show that a directed path from node s to node t in G is a shortest path if and only if it is a directed path from node s to node t in G^0. Using this result, describe an algorithm for enumerating all shortest paths in G from node s to node t. (*Hint*: Use the algorithm in Exercise 3.44.)

5.21. Professor May B. Wright suggests the following method for solving the shortest path problem with arbitrary arc lengths. Let $c_{\min} = \min\{c_{ij} : (i, j) \in A\}$. If $c_{\min} < 0$, add $| c_{\min} |$ to the length each arc in the network so that they all become nonnegative. Then use Dijkstra's algorithm to solve the shortest path problem. Professor Wright claims that the optimal solution of the transformed problem is also an optimal solution of the original problem. Prove or disprove her claim.

5.22. Describe algorithms for updating the shortest path distances from node s to every other node if we add a new node $(n + 1)$ and some arcs incident to this node. Consider the following three cases: (1) all arc lengths are nonnegative and node $(n + 1)$ has only incoming arcs; (2) all arc lengths are nonnegative and node $(n + 1)$ has incoming as well as outgoing arcs; and (3) arc lengths are arbitrary, but node $(n + 1)$ has only incoming arcs. Specify the time required for the reoptimization.

5.23. Maximum multiplier path problem. The *maximum multiplier path problem* is an extension of the maximum reliability path problem that we discussed in Exercise 4.39, obtained by permitting the constants μ_{ij} to be arbitrary positive numbers. Suppose that we are not allowed to use logarithms. State optimality conditions for the maximum multiplier path problem and show that if the network contains a positive multiplier directed cycle, no path can satisfy the optimality conditions. Specify an $O(nm)$ algorithm for solving the maximum multiplier path problem for networks that contain no positive multiplier directed cycles.

5.24. Sharp distance labels. The generic label-correcting algorithm maintains a predecessor graph at every step. We say that a distance label $d(i)$ is *sharp* if it equals the length of the unique path from node s to node i in the predecessor graph. We refer to an algorithm as *sharp* if every node examined by the algorithm has a sharp distance label. (A sharp algorithm might have nodes with nonsharp distances, but the algorithm never examines them.)

(a) Show by an example that the FIFO implementation of the generic label-correcting algorithm is not a sharp algorithm.

(b) Show that the dequeue implementation of the generic label correcting is a sharp algorithm. (*Hint*: Perform induction on the number of nodes the algorithm examines. Use the fact that the distance label of a node becomes nonsharp only when the distance label of one of its ancestors in the predecessor graph decreases.)

5.25. Partitioning algorithm (Glover, Klingman, and Phillips [1985]). The *partitioning algorithm* is a special case of the generic label-correcting algorithm which divides the set LIST of nodes into two subsets: NOW and NEXT. Initially, NOW $= \{s\}$ and NEXT $= \varnothing$. When examining nodes, the algorithm selects *any* node i in NOW and adds to NEXT any node whose distance label decreases, provided that the node is not already in NOW or NEXT. When NOW becomes empty, the algorithm transfers all the nodes from NEXT to NOW. The algorithm terminates when both NOW and NEXT become empty.

(a) Show that the FIFO label-correcting algorithm is a special case of the partitioning algorithm. (*Hint*: Specify rules for selecting the nodes in NOW, adding nodes to NEXT, and transferring nodes from NEXT to NOW.)

(b) Show that the partitioning algorithm runs in $O(nm)$ time. (*Hint*: Call the steps between two consecutive replenishments of NOW a *phase*. Extend the proof of the FIFO label-correcting algorithm to show that at the end of the kth phase, the algorithm determines optimal distances for all nodes whose shortest paths have no more than k arcs.)

5.26. Threshold algorithm (Glover, Klingman, and Phillips [1985]). The threshold algorithm is a variation of the partitioning algorithm discussed in Exercise 5.25. When NOW becomes empty, the threshold algorithm does not transfer all the nodes from NEXT to NOW; instead, it transfers only those nodes i for which $d(i) \leq t$ for some threshold value t. At each iteration, the algorithm choses the threshold value t to be at least as large as the minimum distance label in NEXT (before the transfer), so it transfers all those nodes with the minimum distance label, and possibly other nodes as well, from NEXT to NOW. (Note that we have considerable flexibility in choosing t at each step.)

(a) Show that if all arc lengths are nonnegative, the threshold algorithm runs in $O(nm)$ time. (*Hint*: Use the proof of Dijkstra's algorithm.)

(b) Show that if all arc lengths are nonnegative and the threshold algorithm transfers at most five nodes from NEXT to NOW at each step, including a node with the minimum distance label, then it runs in $O(n^2)$ time.

5.27. Pathological instances of the label-correcting algorithm (Pallottino [1991]). We noted in Section 5.4 that the dequeue implementation of the generic label-correcting algorithm has excellent empirical behavior. However, for some problem instances, the algorithm performs an exponential number of iterations. In this exercise we describe a method for constructing one such pathological instance for every n. Let $G = (N, A)$ be an acyclic graph with n nodes and an arc (i, j) for every node pair i and j satisfying $i > j$. Let node n be the source node. We define the cost of each arc (i, j) as $c_{ij} = 2^{i-2} - 2^{j-1} \geq 0$. Assume that the adjacency list of each node $i \in N - \{n\}$ is arranged in decreasing order of the head nodes and the adjacency list of the source node n is arranged in the increasing order of the head nodes.

(a) Verify that for $n = 6$, the method generates the instance shown in Figure 5.12.

(b) Consider the instance shown in Figure 5.12. Show that every time the dequeue implementation examines any node (other than node 1), it updates the distance label of node 1. Show that the label of node 1 assumes all values between 15 and 0.

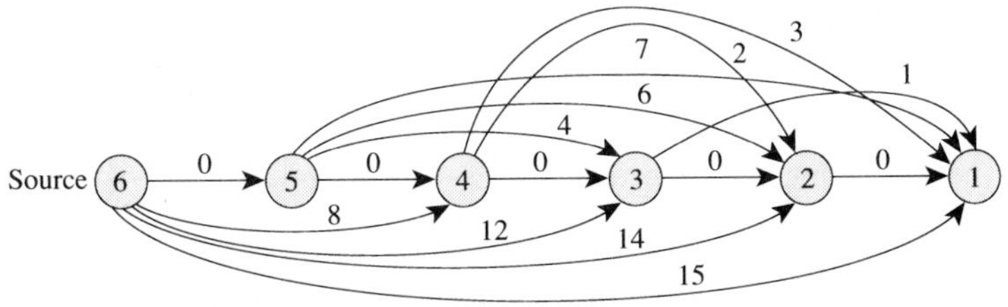

Figure 5.12 Pathological example of the label-correcting algorithm.

5.28. Using induction arguments, show that for an instance with n nodes constructed using the method described in Exercise 5.27, the dequeue implementation of the label-correcting algorithm assigns to node 1 all labels between $2^{n-2} - 1$ to 0 and therefore runs in exponential time.

5.29. Apply the first three iterations (i.e., $k = 1, 2, 3$) of the Floyd–Warshall algorithm to the all-pairs shortest path problems shown in Figure 5.13(a). List four triplets (i, j, k) that violate the all-pairs shortest path optimality conditions at the conclusion of these iterations.

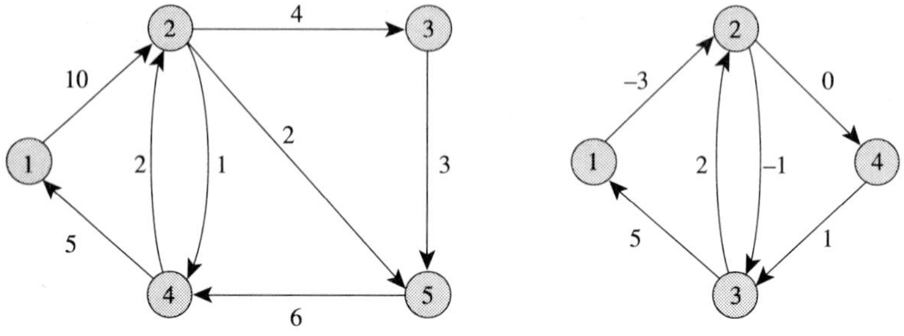

Figure 5.13 Example for Exercises 5.29 to 5.31.

5.30. Solve the all-pairs shortest path problem shown in Figure 5.13(b).

5.31. Consider the shortest path problem shown in Figure 5.13(b), except with c_{31} equal to 3. What is the least number of triple operations required in the Floyd–Warshall algorithm before the node pair distances $d^k[i,j]$ satisfy one of the negative cycle detection conditions?

5.32. Show that if a network contains a negative cycle, the generic all-pairs label-correcting algorithm will never terminate.

5.33. Suppose that the Floyd–Warshall algorithm terminates after detecting the presence of a negative cycle. At this time, how would you detect a negative cycle using the predecessor indices?

5.34. In an all-pairs shortest path problem, suppose that several shortest paths connect node i and node j. If we use the Floyd–Warshall algorithm to solve this problem, which path will the algorithm choose? Will this path be the one with the least number of arcs?

5.35. Consider the maximum capacity path problem defined in Exercise 4.37. Modify the Floyd–Warshall algorithm so that it finds maximum capacity paths between all pairs of nodes.

5.36. Modify the Floyd–Warshall all-pairs shortest path algorithm so that it determines maximum multiplier paths between all pairs of nodes.

5.37. Show that if we use the Floyd–Warshall algorithm to solve the all-pairs shortest path problem in a network containing a negative cycle, then at some stage $d^k[i, i] < 0$ for some node i. [*Hint*: Let i be the least indexed node satisfying the property that the network contains a negative cycle using only nodes 1 through i (not necessarily all of these nodes).]

5.38. Suppose that a network G contains no negative cycle. Let $d^{n+1}(i, j)$ denote the node pair distances at the end of the Floyd–Warshall algorithm. Show that $\min\{d^{n+1}[i, i] : 1 \le i \le n\}$ is the minimum length of a directed cycle in G.

5.39. In this exercise we discuss another dynamic programming algorithm for solving the all-pairs shortest path problem. Let d_{ij}^k denote the length of a shortest path from node i to node j subject to the condition that the path contains no more than k arcs. Express d_{ij}^k in terms of d_{ij}^{k-1} and the $c_{ij}'s$ and suggest an all-pairs shortest path algorithm that uses this relationship. Analyze the running time of your algorithm.

5.40. Sensitivity analysis. Let d_{ij} denote the shortest path distances between the pair $[i, j]$ of nodes in a directed network $G = (N, A)$ with arc lengths c_{ij}. Suppose that the length of one arc (p, q) changes to value $c'_{pq} < c_{pq}$. Show that the following set of statements finds the modified all-pairs shortest path distances:

```
if d_qp + c'_pq < 0, then the network has a negative cycle
else
  for each pair [i, j] of nodes do
    d_ij : = min {d_ij, d_ip + c'_pq + d_qj};
```

5.41. In Exercise 5.40 we described an $O(n^2)$ method for updating shortest path distances between all-pairs of nodes when we decrease the length of one arc (p, q). Suppose that we increase the length of the arc (p, q). Can you modify the method so that it reoptimizes the shortest path distances in $O(n^2)$ time? If your answer is yes, specify an algorithm for performing the reoptimization and provide a justification for it; and if your answer is no, outline the difficulties encountered.

5.42. Arc addition. After solving an all-pairs shortest path problem, you realize that you omitted five arcs from the network G. Can you reoptimize the shortest path distances with the addition of these arcs in $O(n^2)$ time? (*Hint*: Reduce this problem to the one in Exercise 5.40.)

5.43. Consider the reallocation of housing problem that we discussed in Application 1.1.

(a) The housing authority prefers to use short cyclic changes since they are easier to handle administratively. Suggest a method for identifying a cyclic change involving the least number of changes. (*Hint*: Use the result of one of the preceding exercises.)

(b) Suppose that the person presently residing in a house of category i desperately wants to move to his choice category and that the chair of the housing authority wants to help him. Can the chair identify a cyclic change that allocates the person to his choice category or prove that no such change is possible? (*Hint*: Use the result of one of the preceding exercises.)

5.44. Let $G = (N, A)$ denote the road network of the greater Boston area. Four people living in the suburbs form a car pool. They drive in separate cars to a common meeting point and drive from there in a van to a common point in downtown Boston. Suggest a method for identifying the common meeting point that minimizes the total driving time of all the participants. Also, suggest a method for identifying the common meeting point that minimizes the maximum travel time of any one person.

5.45. Location problems. In a directed $G = (N, A)$ with arc lengths c_{ij}, we define the distance between a pair of nodes i and j as the length of the shortest path from node i to node j.

(a) Define the *radial distance* from node i as the length of the distance from node i to the node farthest from it. We say that a node p is a *center* of the graph G if node p has as small a radial distance as any node in the network. Suggest a straightforward polynomial-time algorithm for identifying a center of G.

(b) Define the *star distance* of node i as the total distance from node i to all the nodes in the network. We refer to a node q as a *median* of G if node q has as small a star distance as any node in the network. Suggest a straightforward polynomial-time algorithm for identifying a median of G.

5.46. Suppose that a network $G = (N, A)$ contains no negative cycle. In this network, let f_{ij} denote the maximum amount we can decrease the length of arc (i, j) without creating any negative cycle, assuming that all other arc lengths remain intact. Design an efficient algorithm for determining f_{ij} for each arc $(i, j) \in A$. (*Hint*: Use the all-pairs shortest path distances.)

5.47. Consider the following linear programming formulation of the minimum cost-to-time ratio cycle problem:

$$\text{Minimize } z = \sum_{(i,j)\in A} c_{ij}x_{ij} \tag{5.12a}$$

subject to

$$\sum_{\{j:(i,j)\in A\}} x_{ij} - \sum_{\{j:(j,i)\in A\}} x_{ji} = 0 \quad \text{for all } i \in N, \tag{5.12b}$$

$$\sum_{(i,j)\in A} \tau_{ij}x_{ij} = 1, \tag{5.12c}$$

$$x_{ij} \geq 0 \quad \text{for all } (i, j) \in A. \tag{5.12d}$$

Show that each directed cycle in G defines a feasible solution of (5.12) and that each feasible solution of (5.12) defines a set of one or more directed cycles with the same ratio. Use this result to show that we can obtain an optimal solution of the minimum cost-to-time ratio problem from an optimal solution of the linear program (5.12).

5.48. Obtain a worst-case bound on the number of iterations performed by the sequential search algorithm discussed in Section 5.7 to solve the minimum cost-to-time ratio cycle problem.

5.49. In Section 5.7 we saw how to solve the minimum cost-to-time ratio cycle problem efficiently. This development might lead us to believe that we could also determine efficiently a minimum ratio directed path between two designated nodes s and t (i.e., a path P for which $(\sum_{(i,j)\in P} c_{ij})/(\sum_{(i,j)\in P} \tau_{ij})$ is minimum). This assertion is not valid. Outline the difficulties you would encounter in adapting the algorithm so that it would solve the minimum ratio path problem.

5.50. Use the minimum mean cycle algorithm to identify the minimum mean cycle in Figure 5.13(b).

5.51. **Bit-scaling algorithm** (Gabow [1985]). The bit-scaling algorithm for solving the shortest path problem works as follows. Let $K = \lceil \log C \rceil$. We represent each arc length as a K-bit binary number, adding leading zeros if necessary to make each arc length K bits long. The problem P_k considers the length of each arc as the k leading bits (see Section 3.3). Let d_k^* denote the shortest path distances in problem P_k. The bit-scaling algorithm solves a sequence of problems $P_1, P_2, \ldots, P_k$, using the solution of problem P_{k-1} as the starting solution of problem P_k.

(a) Consider problem P_k and define reduced arc lengths with respect to the distances $2d_{k-1}^*$. Show that the network contains a path from the source node to every other node whose reduced length is at most n. (*Hint*: Consider the shortest path tree of problem P_{k-1}.)

(b) Show how to solve each problem P_k in $O(m)$ time. Use this result to show that the bit-scaling algorithm runs in $O(m \log C)$ time.

5.52. **Modified bit-scaling algorithm.** Consider Exercise 5.51 but using a base β representation of arc cost c_{ij} in place of the binary representation. In problem P_k we use the k leading base β digits of the arc lengths as the lengths of the arcs. Let d_{k-1}^* denote the shortest path distances in Problem P_{k-1}.

(a) Show that if we define reduced arc lengths in problem P_k with respect to the distances βd_{k-1}^*, the network contains a path from the source to every other node whose reduced length is at most $\beta\, n$.

(b) Show how to solve each problem P_k in $O(m + \beta n)$ time and, consequently, show that the modified bit-scaling algorithm runs in $O((m + \beta n) \log_\beta C)$ time. What value of β achieves the least running time?

5.53. **Parametric shortest path problem.** In the parametric shortest path problem, the cost c_{ij} of each arc (i, j) is a linear function of a parameter λ (i.e., $c_{ij} = c_{ij}^\circ + \lambda c_{ij}^*$) and we want to obtain a tree of shortest paths for all values of λ from 0 to $+\infty$. Let T^λ denote a tree of shortest paths for a specific value of λ.

(a) Consider T^λ for some λ. Show that if $d^\circ(j)$ and $d^*(j)$ are the distances in T^λ with respect to the arc lengths c_{ij}° and c_{ij}^*, respectively, then $d^\circ(j) + \lambda d^*(j)$ are the distances with respect to the arc lengths $c_{ij}^{\circ} + \lambda c_{ij}^{*}$ in T^{λ}. Use this result to describe a method for determining the largest value of λ, say $\bar{\lambda}$, for which T^{λ} is a shortest path tree for all λ, $1 \le \lambda \le \bar{\lambda}$. Show that at $\lambda = \bar{\lambda}$, the network contains an alternative shortest path tree. (*Hint*: Use the shortest path optimality conditions.)

(b) Describe an algorithm for determining T^{λ} for all $0 \le \lambda \le \infty$. Show that T^{∞} is shortest path tree with the arc lengths as c_{ij}^{*}.

5.54. Consider a special case of the parametric shortest path problem in which each $c_{ij}^{*} = 0$ or 1. Show that as we vary λ from 0 to $+\infty$, we obtain at most n^2 trees of shortest paths. How many trees of shortest paths do you think we can obtain for the general case? Is it polynomial or exponential? [*Hint*: Let $f(j)$ denote the number of arcs with $c_{ij}^{*} = 1$ in the tree of shortest paths from node s to node j. Consider the effect on the potential function $\Phi = \sum_{j \in N} f(j)$ of the changes in the tree of shortest paths.]

5.55. Let $d^k(j)$ denote the length of the shortest path from node s to node j using at most k arcs in a network G. Suppose that $d^k(j)$ are available for all nodes $j \in N$ and all $k = 1, \ldots, n$. Show how to determine a minimum mean cycle in G. (*Hint*: Use some result contained in Theorem 5.8.)

5.56. Show that if the predecessor graph at any point in the execution of the label-correcting algorithm contains a directed cycle, then the network contains a negative cycle.

6

MAXIMUM FLOWS: BASIC IDEAS

*You get the maxx for the minimum at T. J. Maxx.**
—Advertisement for a clothing store.

Chapter Outline

6.1 INTRODUCTION

The maximum flow problem and the shortest path problem are complementary. They are similar because they are both pervasive in practice and because they both arise as subproblems in algorithms for the minimum cost flow problem. The two problems differ, however, because they capture different aspects of the minimum cost flow problem: Shortest path problems model arc costs but not arc capacities; maximum flow problems model capacities but not costs. Taken together, the shortest path problem and the maximum flow problem combine all the basic ingredients of network flows. As such, they have become the nuclei of network optimization. Our study of the shortest path problem in the preceding two chapters has introduced us to some of the basic building blocks of network optimization, such as distance labels, optimality conditions, and some core strategies for designing iterative solution methods and for improving the performance of these methods. Our discussion of maximum flows, which we begin in this chapter, builds on these ideas and introduces several other key ideas that reoccur often in the study of network flows.

The *maximum flow problem* is very easy to state: In a capacitated network, we wish to send as much flow as possible between two special nodes, a source node s and a sink node t, without exceeding the capacity of any arc. In this and the following two chapters, we discuss a number of algorithms for solving the maximum flow problem. These algorithms are of two types:

1. Augmenting path algorithms that maintain mass balance constraints at every node of the network other than the source and sink nodes. These algorithms incrementally augment flow along paths from the source node to the sink node.
2. Preflow-push algorithms that flood the network so that some nodes have excesses (or buildup of flow). These algorithms incrementally relieve flow from nodes with excesses by sending flow from the node forward toward the sink node or backward toward the source node.

We discuss the simplest version of the first type of algorithm in this chapter and more elaborate algorithms of both types in Chapter 7. To help us to understand the importance of the maximum flow problem, we begin by describing several applications. This discussion shows how maximum flow problems arise in settings as diverse as manufacturing, communication systems, distribution planning, matrix rounding, and scheduling.

We begin our algorithmic discussion by considering a *generic augmenting path algorithm* for solving the maximum flow problem and describing an important special implementation of the generic approach, known as the *labeling algorithm*. The labeling algorithm is a pseudopolynomial-time algorithm. In Chapter 7 we develop improved versions of this generic approach with better theoretical behavior. The correctness of these algorithms rests on the renowned *max-flow min-cut theorem* of network flows (recall from Section 2.2 that a cut is a set of arcs whose deletion disconnects the network into two parts). This central theorem in the study of network flows (indeed, perhaps the most significant theorem in this problem domain) not only provides us with an instrument for analyzing algorithms, but also permits us to model a variety of applications in machine and vehicle scheduling, communication systems planning, and several other settings, as maximum flow problems, even though on the surface these problems do not appear to have a network flow structure. In Section 6.6 we describe several such applications.

The max-flow min-cut theorem establishes an important correspondence between flows and cuts in networks. Indeed, as we will see, by solving a maximum flow problem, we also solve a complementary *minimum cut problem*: From among all cuts in the network that separate the source and sink nodes, find the cut with the minimum capacity. The relationship between maximum flows and minimum cuts is important for several reasons. First, it embodies a fundamental duality result that arises in many problem settings in discrete mathematics and that underlies linear programming as well as mathematical optimization in general. In fact, the max-flow min-cut theorem, which shows the equivalence between the maximum flow and minimum cut problems, is a special case of the well-known strong duality theorem of linear programming. The fact that maximum flow problems and minimum cut problems are equivalent has practical implications as well. It means that the theory and algorithms that we develop for the maximum flow problem are also applicable to many practical problems that are naturally cast as minimum cut problems. Our discussion of combinatorial applications in the text and exercises of this chapter and our discussion of applications in Chapter 19 features several applications of this nature.

Notation and Assumptions

We consider a capacitated network $G = (N, A)$ with a *nonnegative* capacity u_{ij} associated with each arc $(i, j) \in A$. Let $U = \max\{u_{ij}:(i, j) \in A\}$. As before, the arc adjacency list $A(i) = \{(i, k):(i, k) \in A\}$ contains all the arcs emanating from node i. To define the maximum flow problem, we distinguish two special nodes in the network G: a *source node* s and a *sink node* t. We wish to find the maximum flow from the source node s to the sink node t that satisfies the arc capacities and mass balance constraints at all nodes. We can state the problem formally as follows.

$$\text{Maximize } v \tag{6.1a}$$

subject to

$$\sum_{\{j:(i,j)\in A\}} x_{ij} - \sum_{\{j:(j,i)\in A\}} x_{ji} = \begin{cases} v & \text{for } i = s, \\ 0 & \text{for all } i \in N - \{s \text{ and } t\} \\ -v & \text{for } i = t \end{cases} \tag{6.1b}$$

$$0 \le x_{ij} \le u_{ij} \qquad \text{for each } (i, j) \in A. \tag{6.1c}$$

We refer to a vector $x = \{x_{ij}\}$ satisfying (6.1b) and (6.1c) as a *flow* and the corresponding value of the scalar variable v as the *value* of the flow. We consider the maximum flow problem subject to the following assumptions.

Assumption 6.1. *The network is directed.*

As explained in Section 2.4, we can always fulfill this assumption by transforming any undirected network into a directed network.

Assumption 6.2. *All capacities are nonnegative integers.*

Although it is possible to relax the integrality assumption on arc capacities for some algorithms, this assumption is necessary for others. Algorithms whose complexity bounds involve U assume integrality of the data. In reality, the integrality assumption is not a restrictive assumption because all modern computers store capacities as rational numbers and we can always transform rational numbers to integer numbers by multiplying them by a suitably large number.

Assumption 6.3. *The network does not contain a directed path from node s to node t composed only of infinite capacity arcs.*

Whenever every arc on a directed path P from s to t has infinite capacity, we can send an infinite amount of flow along this path, and therefore the maximum flow value is unbounded. Notice that we can detect the presence of an infinite capacity path using the search algorithm described in Section 3.4.

Assumption 6.4. *Whenever an arc (i, j) belongs to A, arc (j, i) also belongs to A.*

This assumption is nonrestrictive because we allow arcs with zero capacity.

Assumption 6.5. The network does not contain parallel arcs (i.e., two or more arcs with the same tail and head nodes).

This assumption is essentially a notational convenience. In Exercise 6.24 we ask the reader to show that this assumption imposes no loss of generality.

Before considering the theory underlying the maximum flow problem and algorithms for solving it, and to provide some background and motivation for studying the problem, we first describe some applications.

6.2 APPLICATIONS

The maximum flow problem, and the minimum cut problem, arise in a wide variety of situations and in several forms. For example, sometimes the maximum flow problem occurs as a subproblem in the solution of more difficult network problems, such as the minimum cost flow problem or the generalized flow problem. As we will see in Section 6.6, the maximum flow problem also arises in a number of combinatorial applications that on the surface might not appear to be maximum flow problems at all. The problem also arises directly in problems as far reaching as machine scheduling, the assignment of computer modules to computer processors, the rounding of census data to retain the confidentiality of individual households, and tanker scheduling. In this section we describe a few such applications; in Chapter 19 we discuss several other applications.

Application 6.1 Feasible Flow Problem

The feasible flow problem requires that we identify a flow x in a network $G = (N, A)$ satisfying the following constraints:

$$\sum_{\{j:(i,j)\in A\}} x_{ij} - \sum_{\{j:(j,i)\in A\}} x_{ji} = b(i) \qquad \text{for } i \in N, \tag{6.2a}$$

$$0 \le x_{ij} \le u_{ij} \qquad \text{for all } (i, j) \in A. \tag{6.2b}$$

As before, we assume that $\sum_{i\in N} b(i) = 0$. The following distribution scenario illustrates how the feasible flow problem arises in practice. Suppose that merchandise is available at some seaports and is desired by other ports. We know the stock of merchandise available at the ports, the amount required at the other ports, and the maximum quantity of merchandise that can be shipped on a particular sea route. We wish to know whether we can satisfy all of the demands by using the available supplies.

We can solve the feasible flow problem by solving a maximum flow problem defined on an augmented network as follows. We introduce two new nodes, a source node s and a sink node t. For each node i with $b(i) > 0$, we add an arc (s, i) with capacity $b(i)$, and for each node i with $b(i) < 0$, we add an arc (i, t) with capacity $-b(i)$. We refer to the new network as the *transformed network*. Then we solve a maximum flow problem from node s to node t in the transformed network. If the maximum flow saturates all the source and sink arcs, problem (6.2) has a feasible solution; otherwise, it is infeasible. (In Section 6.7 we give necessary and sufficient conditions for a feasible flow problem to have a feasible solution.)

It is easy to verify why this algorithm works. If x is a flow satisfying (6.2a) and (6.2b), the same flow with $x_{si} = b(i)$ for each source arc (s, i) and $x_{it} = -b(i)$ for each sink arc (i, t) is a maximum flow in the transformed network (since it saturates all the source and the sink arcs). Similarly, if x is a maximum flow in the transformed network that saturates all the source and the sink arcs, this flow in the original network satisfies (6.2a) and (6.2b). Therefore, the original network contains a feasible flow if and only if the transformed network contains a flow that saturates all the source and sink arcs. This observation shows how the maximum flow problem arises whenever we need to find a feasible solution in a network.

Application 6.2 Problem of Representatives

A town has r residents $R_1, R_2, \ldots, R_r$; q clubs $C_1, C_2, \ldots, C_q$; and p political parties $P_1, P_2, \ldots, P_p$. Each resident is a member of at least one club and can belong to exactly one political party. Each club must nominate one of its members to represent it on the town's governing council so that the number of council members belonging to the political party P_k is at most u_k. Is it possible to find a council that satisfies this "balancing" property?

We illustrate this formulation with an example. We consider a problem with $r = 7$, $q = 4$, $p = 3$, and formulate it as a maximum flow problem in Figure 6.1. The nodes $R_1, R_2, \ldots, R_7$ represent the residents, the nodes $C_1, C_2, \ldots, C_4$ represent the clubs, and the nodes $P_1, P_2, \ldots, P_3$ represent the political parties.

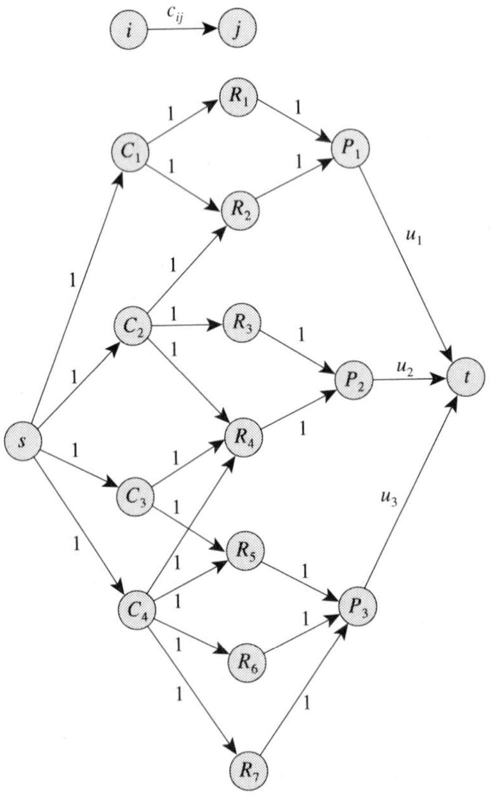

Figure 6.1 System of distinct representatives.

The network also contains a source node s and a sink node t. It contains an arc (s, C_i) for each node C_i denoting a club, an arc (C_i, R_j) whenever the resident R_j is a member of the club C_i, and an arc (R_j, P_k) if the resident R_j belongs to the political party P_k. Finally, we add an arc (P_k, t) for each $k = 1, \ldots, 3$ of capacity u_k; all other arcs have unit capacity.

We next find a maximum flow in this network. If the maximum flow value equals q, the town has a balanced council; otherwise, it does not. The proof of this assertion is easy to establish by showing that (1) any flow of value q in the network corresponds to a balanced council, and that (2) any balanced council implies a flow of value q in the network.

This type of model has applications in several resource assignment settings. For example, suppose that the residents are skilled craftsmen, the club C_i is the set of craftsmen with a particular skill, and the political party P_k corresponds to a particular seniority class. In this instance, a balanced town council corresponds to an assignment of craftsmen to a union governing board so that every skill class has representation on the board and no seniority class has a dominant representation.

Application 6.3 Matrix Rounding Problem

This application is concerned with consistent rounding of the elements, row sums, and column sums of a matrix. We are given a $p \times q$ matrix of *real* numbers $D = \{d_{ij}\}$, with row sums α_i and column sums β_j. We can round any real number a to the next smaller integer $\lfloor a \rfloor$ or to the next larger integer $\lceil a \rceil$, and the decision to round up or down is entirely up to us. The matrix rounding problem requires that we round the matrix elements, and the row and column sums of the matrix so that the sum of the rounded elements in each row equals the rounded row sum and the sum of the rounded elements in each column equals the rounded column sum. We refer to such a rounding as a *consistent rounding*.

We shall show how we can discover such a rounding scheme, if it exists, by solving a feasible flow problem for a network with nonnegative lower bounds on arc flows. (As shown in Section 6.7, we can solve this problem by solving two maximum flow problems with zero lower bounds on arc flows.) We illustrate our method using the matrix rounding problem shown in Figure 6.2. Figure 6.3 shows the maximum flow network for this problem. This network contains a node i corresponding to each row i and a node j' corresponding to each column j. Observe that this network

				Row sum
	3.1	6.8	7.3	17.2
	9.6	2.4	0.7	12.7
	3.6	1.2	6.5	11.3
Column sum	16.3	10.4	14.5	

Figure 6.2 Matrix rounding problem.

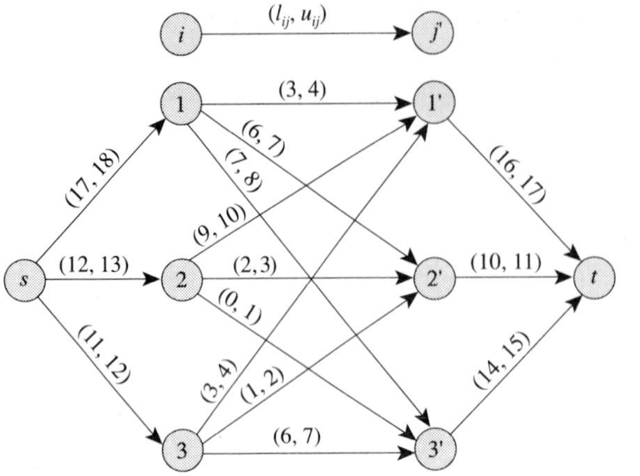

Figure 6.3 Network for the matrix rounding problem.

contains an arc (i, j') for each matrix element d_{ij}, an arc (s, i) for each row sum, and an arc (j', t) for each column sum. The lower and the upper bounds of each arc (i, j') are $\lfloor d_{ij} \rfloor$ and $\lceil d_{ij} \rceil$, respectively. It is easy to establish a one-to-one correspondence between the consistent roundings of the matrix and feasible flows in the corresponding network. Consequently, we can find a consistent rounding by solving a maximum flow problem on the corresponding network.

This matrix rounding problem arises in several application contexts. For example, the U.S. Census Bureau uses census information to construct millions of tables for a wide variety of purposes. By law, the bureau has an obligation to protect the source of its information and not disclose statistics that could be attributed to any particular person. We might disguise the information in a table as follows. We round off each entry in the table, including the row and column sums, either up or down to a multiple of a constant k (for some suitable value of k), so that the entries in the table continue to add to the (rounded) row and column sums, and the overall sum of the entries in the new table adds to a rounded version of the overall sums in the original table. This Census Bureau problem is the same as the matrix rounding problem discussed earlier except that we need to round each element to a multiple of $k \geq 1$ instead of rounding it to a multiple of 1. We solve this problem by defining the associated network as before, but now defining the lower and upper bounds for any arc with an associated real number α as the greatest multiple of k less than or equal to α and the smallest multiple of k greater than or equal to α.

Application 6.4 Scheduling on Uniform Parallel Machines

In this application we consider the problem of scheduling of a set J of jobs on M uniform parallel machines. Each job $j \in J$ has a processing requirement p_j (denoting the number of machine days required to complete the job), a release date r_j (representing the beginning of the day when job j becomes available for processing), and a due date $d_j \geq r_j + p_j$ (representing the beginning of the day by which the job must be completed). We assume that a machine can work on only one job at a time and that each job can be processed by at most one machine at a time. However, we allow *preemptions* (i.e., we can interrupt a job and process it on different machines on different days). The scheduling problem is to determine a feasible schedule that completes all jobs before their due dates or to show that no such schedule exists.

Scheduling problems like this arise in batch processing systems involving batches with a large number of units. The feasible scheduling problem, described in the preceding paragraph, is a fundamental problem in this situation and can be used as a subroutine for more general scheduling problems, such as the maximum lateness problem, the (weighted) minimum completion time problem, and the (weighted) maximum utilization problem.

Let us formulate the feasible scheduling problem as a maximum flow problem. We illustrate the formulation using the scheduling problem described in Figure 6.4 with M = 3 machines. First, we rank all the release and due dates, r_j and d_j for all j, in ascending order and determine $P \le 2 \mid J \mid - 1$ mutually disjoint intervals of dates between consecutive milestones. Let $T_{k,l}$ denote the interval that starts at the beginning of date k and ends at the beginning of date $l + 1$. For our example, this order of release and due dates is 1, 3, 4, 5, 7, 9. We have five intervals, represented by $T_{1,2}$, $T_{3,3}$, $T_{4,4}$, $T_{5,6}$, and $T_{7,8}$. Notice that within each interval $T_{k,l}$, the set of available jobs (i.e., those released but not yet due) does not change: we can process all jobs j with $r_j \le k$ and $d_j \ge l + 1$ in the interval.

Job (j)	1	2	3	4
Processing time (p_j)	1.5	1.25	2.1	3.6
Release time (r_j)	3	1	3	5
Due date (d_j)	5	4	7	9

Figure 6.4 Scheduling problem.

We formulate the scheduling problem as a maximum flow problem on a bipartite network G as follows. We introduce a source node s, a sink node t, a node corresponding to each job j, and a node corresponding to each interval $T_{k,l}$, as shown in Figure 6.5. We connect the source node to every job node j with an arc with capacity p_j, indicating that we need to assign p_j days of machine time to job j. We connect each interval node $T_{k,l}$ to the sink node t by an arc with capacity $(l - k + 1)M$, representing the total number of machine days available on the days from k to l. Finally, we connect a job node j to every interval node $T_{k,l}$ if $r_j \le k$ and $d_j \ge l + 1$ by an arc with capacity $(l - k + 1)$ which represents the maximum number of machines days we can allot to job j on the days from k to l. We next solve a maximum flow problem on this network: The scheduling problem has a feasible schedule if and only if the maximum flow value equals $\sum_{j \in J} p_j$ [alternatively, the flow on every arc (s, j) is p_j]. The validity of this formulation is easy to establish by showing a one-to-one correspondence between feasible schedules and flows of value $\sum_{j \in J} p_j$ from the source to the sink.

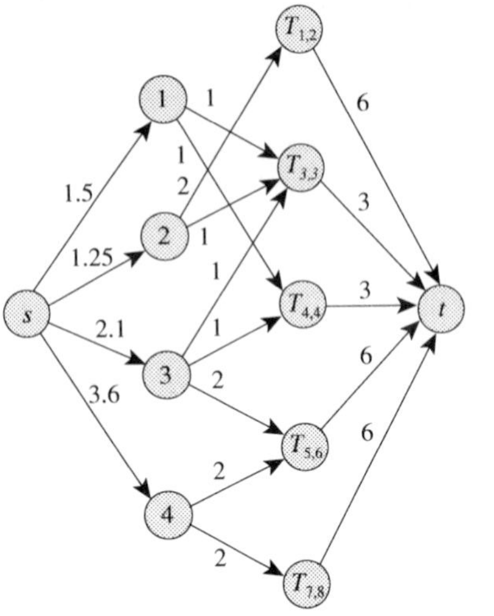

Figure 6.5 Network for scheduling uniform parallel machines.

Application 6.5 Distributed Computing on a Two-Processor Computer

This application concerns assigning different modules (subroutines) of a program to two processors in a way that minimizes the collective costs of interprocessor communication and computation. We consider a computer system with two processors; they need not be identical. We wish to execute a large program on this computer system. Each program contains several modules that interact with each other during the program's execution. The cost of executing each module on the two processes is known in advance and might vary from one processor to the other because of differences in the processors' memory, control, speed, and arithmetic capabilities. Let α_i and β_i denote the cost of computation of module i on processors 1 and 2, respectively. Assigning different modules to different processors incurs relatively high overhead costs due to interprocessor communication. Let c_{ij} denote the interprocessor communication cost if modules i and j are assigned to different processors; we do not incur this cost if we assign modules i and j to the same processor. The cost structure might suggest that we allocate two jobs to different processors—we need to balance this cost against the communication costs that we incur by allocating the jobs to different processors. Therefore, we wish to allocate modules of the program on the two processors so that we minimize the total cost of processing and interprocessor communication.

We formulate this problem as a minimum cut problem on an undirected network as follows. We define a source node s representing processor 1, a sink node t representing processor 2, and a node for every module of the program. For every node i, other than the source and sink nodes, we include an arc (s, i) of capacity β_i and an arc (i, t) of capacity α_i. Finally, if module i interacts with module j during program execution, we include the arc (i, j) with a capacity equal to c_{ij}. Figures 6.6 and 6.7 give an example of this construction. Figure 6.6 gives the data for this problem, and Figure 6.7 gives the corresponding network.

We now observe a one-to-one correspondence between s–t cuts in the network

i	1	2	3	4
α_i	6	5	10	4
β_i	4	10	3	8

(a)

	1	2	3	4
1	0	5	0	0
$\{c_{ij}\}$ = 2	5	0	6	2
3	0	6	0	1
4	0	2	1	0

(b)

Figure 6.6 Data for the distributed computing model.

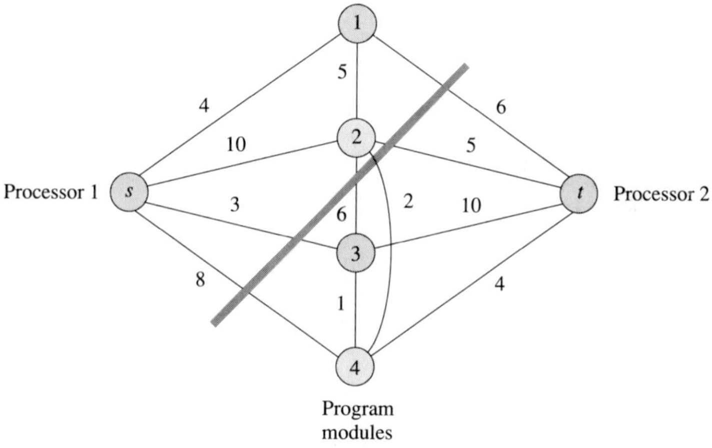

Figure 6.7 Network for the distributed computing model.

and assignments of modules to the two processors; moreover, the capacity of a cut equals the cost of the corresponding assignment. To establish this result, let A_1 and A_2 be an assignment of modules to processors 1 and 2, respectively. The cost of this assignment is $\sum_{i \in A_1} \alpha_i + \sum_{i \in A_2} \beta_i + \sum_{(i,j) \in A_1 x A_2} c_{ij}$. The s–t cut corresponding to this assignment is $(\{s\} \cup A_1, \{t\} \cup A_2)$. The approach we used to construct the network implies that this cut contains an arc (i, t) for every $i \in A_1$ of capacity α_i, an arc (s, i) for every $i \in A_2$ of capacity β_i, and all arcs (i, j) with $i \in A_1$ and $j \in A_2$ with capacity c_{ij}. The cost of the assignment A_1 and A_2 equals the capacity of the cut $(\{s\} \cup A_1, \{t\} \cup A_2)$. (We suggest that readers verify this conclusion using the example given in Figure 6.7 with $A_1 = \{1, 2\}$ and $A_2 = \{3, 4\}$.) Consequently, the minimum s–t cut in the network gives the minimum cost assignment of the modules to the two processors.

Application 6.6 Tanker Scheduling Problem

A steamship company has contracted to deliver perishable goods between several different origin–destination pairs. Since the cargo is perishable, the customers have specified precise dates (i.e., delivery dates) when the shipments must reach their destinations. (The cargoes may not arrive early or late.) The steamship company wants to determine the minimum number of ships needed to meet the delivery dates of the shiploads.

To illustrate a modeling approach for this problem, we consider an example with four shipments; each shipment is a full shipload with the characteristics shown in Figure 6.8(a). For example, as specified by the first row in this figure, the company must deliver one shipload available at port A and destined for port C on day 3. Figure 6.8(b) and (c) show the transit times for the shipments (including allowances for loading and unloading the ships) and the return times (without a cargo) between the ports.

Shipment	Origin	Destination	Delivery date
1	Port A	Port C	3
2	Port A	Port C	8
3	Port B	Port D	3
4	Port B	Port C	6

(a)

	C	D
A	3	2
B	2	3

(b)

	A	B
C	2	1
D	1	2

(c)

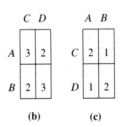

Figure 6.8 Data for the tanker scheduling problem: (a) shipment characteristics; (b) shipment transit times; (c) return times.

We solve this problem by constructing a network shown in Figure 6.9(a). This network contains a node for each shipment and an arc from node i to node j if it is possible to deliver shipment j after completing shipment i; that is, the start time of shipment j is no earlier than the delivery time of shipment i plus the travel time from the destination of shipment i to the origin of shipment j. A directed path in this network corresponds to a feasible sequence of shipment pickups and deliveries. The tanker scheduling problem requires that we identify the minimum number of directed paths that will contain each node in the network on exactly one path.

We can transform this problem to the framework of the maximum flow problem as follows. We split each node i into two nodes i' and i'' and add the arc (i', i''). We set the lower bound on each arc (i', i''), called the *shipment arc*, equal to 1 so that at least one unit of flow passes through this arc. We also add a source node s and connect it to the origin of each shipment (to represent putting a ship into service),

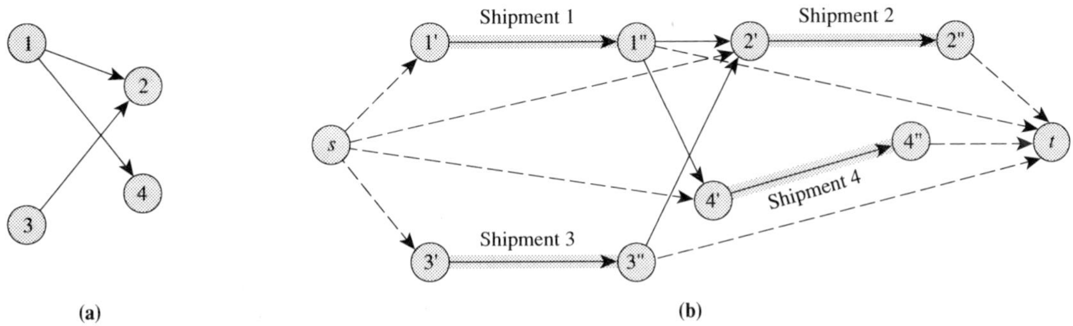

Figure 6.9 Network formulation of the tanker scheduling problem: (a) network of feasible sequences of two consecutive shipments; (b) maximum flow model.

and we add a sink node t and connect each destination node to it (to represent taking a ship out of service). We set the capacity of each arc in the network to value 1. Figure 6.9(b) shows the resulting network for our example. In this network, each directed path from the source s to the sink t corresponds to a feasible schedule for a single ship. As a result, a feasible flow of value v in this network decomposes into schedules of v ships and our problem reduces to identifying a feasible flow of minimum value. We note that the zero flow is not feasible because shipment arcs have unit lower bounds. We can solve this problem, which is known as the *minimum value problem*, using any maximum flow algorithm (see Exercise 6.18).

6.3 FLOWS AND CUTS

In this section we discuss some elementary properties of flows and cuts. We use these properties to prove the max-flow min-cut theorem to establish the correctness of the generic augmenting path algorithm. We first review some of our previous notation and introduce a few new ideas.

Residual network. The concept of *residual network* plays a central role in the development of all the maximum flow algorithms we consider. Earlier in Section 2.4 we defined residual networks and discussed several of its properties. Given a flow x, the residual capacity r_{ij} of any arc $(i, j) \in A$ is the maximum additional flow that can be sent from node i to node j using the arcs (i, j) and (j, i). [Recall our assumption from Section 6.1 that whenever the network contains arc (i, j), it also contains arc (j, i).] The residual capacity r_{ij} has two components: (1) $u_{ij} - x_{ij}$, the unused capacity of arc (i, j), and (2) the current flow x_{ji} on arc (j, i), which we can cancel to increase the flow from node i to node j. Consequently, $r_{ij} = u_{ij} - x_{ij} + x_{ji}$. We refer to the network $G(x)$ consisting of the arcs with positive residual capacities as the *residual network* (with respect to the flow x). Figure 6.10 gives an example of a residual network.

***s–t* cut.** We now review notation about cuts. Recall from Section 2.2 that a cut is a partition of the node set N into two subsets S and $\bar{S} = N - S$; we represent this cut using the notation $[S, \bar{S}]$. Alternatively, we can define a cut as the set of

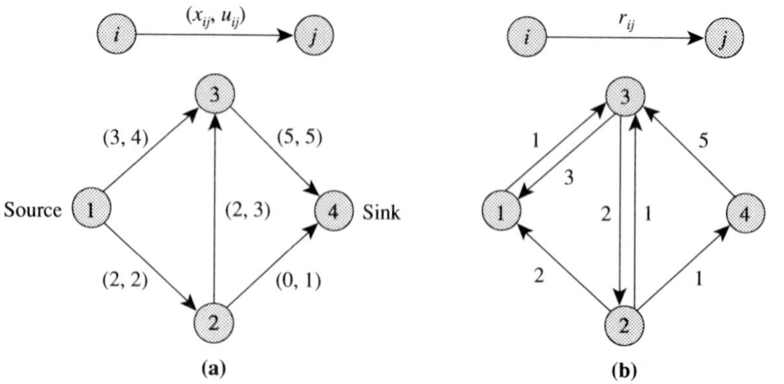

Figure 6.10 Illustrating a residual network: (a) original network G with a flow x; (b) residual network $G(x)$.

arcs whose endpoints belong to the different subsets S and $\bar{S}$. We refer to a cut as an *s–t cut* if $s \in S$ and $t \in \bar{S}$. We also refer to an arc (i, j) with $i \in S$ and $j \in \bar{S}$ as a *forward arc* of the cut, and an arc (i, j) with $i \in \bar{S}$ and $j \in S$ as a *backward arc* of the cut $[S, \bar{S}]$. Let $(S, \bar{S})$ denote the set of forward arcs in the cut, and let $(\bar{S}, S)$ denote the set of backward arcs. For example, in Figure 6.11, the dashed arcs constitute an s–t cut. For this cut, $(S, \bar{S}) = \{(1, 2), (3, 4), (5, 6)\}$, and $(\bar{S}, S) = \{(2, 3), (4, 5)\}$.

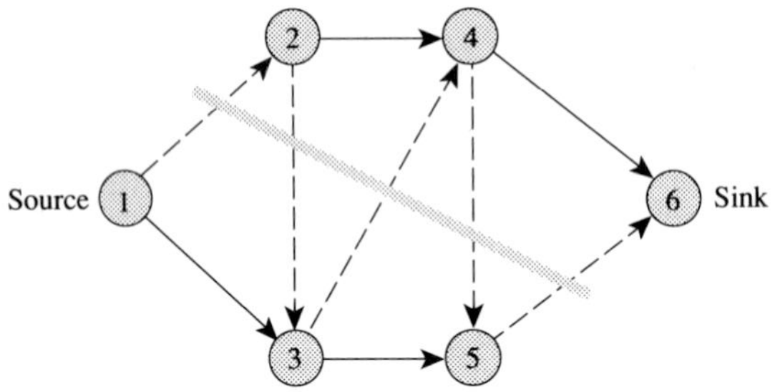

Figure 6.11 Example of an s–t cut.

Capacity of an s–t cut. We define the capacity $u[S, \bar{S}]$ of an s–t cut $[S, \bar{S}]$ as the sum of the capacities of the forward arcs in the cut. That is,

$$u[S, \bar{S}] = \sum_{(i,j) \in (S, \bar{S})} u_{ij}.$$

Clearly, the capacity of a cut is an upper bound on the maximum amount of flow we can send from the nodes in S to the nodes in $\bar{S}$ while honoring arc flow bounds.

Minimum cut. We refer to an s–t cut whose capacity is minimum among all s–t cuts as a *minimum cut*.

Residual capacity of an s–t cut. We define the residual capacity $r[S, \bar{S}]$ of an s–t cut $[S, \bar{S}]$ as the sum of the residual capacities of forward arcs in the cut. That is,

$$r[S, \bar{S}] = \sum_{(i,j)\in(S,\bar{S})} r_{ij}.$$

Flow across an *s–t* cut. Let x be a flow in the network. Adding the mass balance constraint (6.1b) for the nodes in S, we see that

$$v = \sum_{i\in S} \left[\sum_{\{j:(i,j)\in A\}} x_{ij} - \sum_{\{j:(j,i)\in A\}} x_{ji} \right].$$

We can simplify this expression by noting that whenever both the nodes p and q belong to S and $(p, q) \in A$, the variable x_{pq} in the first term within the brackets (for node $i = p$) cancels the variable $-x_{pq}$ in the second term within the brackets (for node $j = q$). Moreover, if both the nodes p and q belong to $\bar{S}$, then x_{pq} does not appear in the expression. This observation implies that

$$v = \sum_{(i,j)\in(S,\bar{S})} x_{ij} - \sum_{(i,j)\in(\bar{S},S)} x_{ij}. \tag{6.3}$$

The first expression on the right-hand side of (6.3) denotes the amount of flow from the nodes in S to nodes in $\bar{S}$, and the second expression denotes the amount of flow returning from the nodes in $\bar{S}$ to the nodes in S. Therefore, the right-hand side denotes the total (net) flow across the cut, and (6.3) implies that the flow across *any* s–t cut $[S, \bar{S}]$ equals v. Substituting $x_{ij} \le u_{ij}$ in the first expression of (6.3) and $x_{ij} \ge 0$ in the second expression shows that

$$v \le \sum_{(i,j)\in(S,\bar{S})} u_{ij} = u[S, \bar{S}]. \tag{6.4}$$

This expression indicates that the value of *any* flow is less than or equal to the capacity of *any* s–t cut in the network. This result is also quite intuitive. Any flow from node s to node t must pass through every s–t cut in the network (because any cut divides the network into two disjoint components), and therefore the value of the flow can never exceed the capacity of the cut. Let us formally record this result.

Property 6.1. *The value of any flow is less than or equal to the capacity of any cut in the network.*

This property implies that if we discover a flow x whose value equals the capacity of some cut $[S, \bar{S}]$, then x is a maximum flow and the cut $[S, \bar{S}]$ is a minimum cut. The max-flow min-cut theorem, proved in the next section, states that some flow always has a flow value equal to the capacity of some cut.

We next restate Property 6.1 in terms of the residual capacities. Suppose that x is a flow of value v. Moreover, suppose that that x' is a flow of value $v + \Delta v$ for some $\Delta v \ge 0$. The inequality (6.4) implies that

$$v + \Delta v \le \sum_{(i,j)\in(S,\bar{S})} u_{ij}. \tag{6.5}$$

Subtracting (6.3) from (6.5) shows that

$$\Delta v \le \sum_{(i,j)\in(S,\bar{S})} (u_{ij} - x_{ij}) + \sum_{(i,j)\in(\bar{S},S)} x_{ij}. \tag{6.6}$$

We now use Assumption 6.4 to note that we can rewrite $\sum_{(i,j)\in(\bar{S},S)} x_{ij}$ as $\sum_{(i,j)\in(S,\bar{S})} x_{ji}$. Consequently,

$$\Delta v \leq \sum_{(i,j)\in(S,\bar{S})} (u_{ij} - x_{ij} + x_{ji}) = \sum_{(S,\bar{S})} r_{ij}.$$

The following property is now immediate.

Property 6.2. *For any flow x of value v in a network, the additional flow that can be sent from the source node s to the sink node t is less than or equal to the residual capacity of any s–t cut.*

6.4 GENERIC AUGMENTING PATH ALGORITHM

In this section, we describe one of the simplest and most intuitive algorithms for solving the maximum flow problem. This algorithm is known as the *augmenting path algorithm*.

We refer to a directed path from the source to the sink in the residual network as an *augmenting path*. We define the residual capacity of an augmenting path as the minimum residual capacity of any arc in the path. For example, the residual network in Figure 6.10(b), contains exactly one augmenting path 1–3–2–4, and the residual capacity of this path is $\delta = \min\{r_{13}, r_{32}, r_{24}\} = \min\{1, 2, 1\} = 1$. Observe that, by definition, the capacity δ of an augmenting path is always positive. Consequently, whenever the network contains an augmenting path, we can send additional flow from the source to the sink. The generic augmenting path algorithm is essentially based on this simple observation. The algorithm proceeds by identifying augmenting paths and augmenting flows on these paths until the network contains no such path. Figure 6.12 describes the generic augmenting path algorithm.

```
algorithm augmenting path;
begin
    x : = 0;
    while G(x) contains a directed path from node s to node t do
    begin
        identify an augmenting path P from node s to node t;
        δ : = min{r_ij : (i, j) ∈ P};
        augment δ units of flow along P and update G(x);
    end;
end;
```

Figure 6.12 Generic augmenting path algorithm.

We use the maximum flow problem given in Figure 6.13(a) to illustrate the algorithm. Suppose that the algorithm selects the path 1–3–4 for augmentation. The residual capacity of this path is $\delta = \min\{r_{13}, r_{34}\} = \min\{4, 5\} = 4$. This augmentation reduces the residual capacity of arc (1, 3) to zero (thus we delete it from the residual network) and increases the residual capacity of arc (3, 1) to 4 (so we add this arc to the residual network). The augmentation also decreases the residual capacity of arc (3, 4) from 5 to 1 and increases the residual capacity of arc (4, 3) from 0 to 4. Figure 6.13(b) shows the residual network at this stage. In the second iteration, suppose that the algorithm selects the path 1–2–3–4. The residual capacity of this

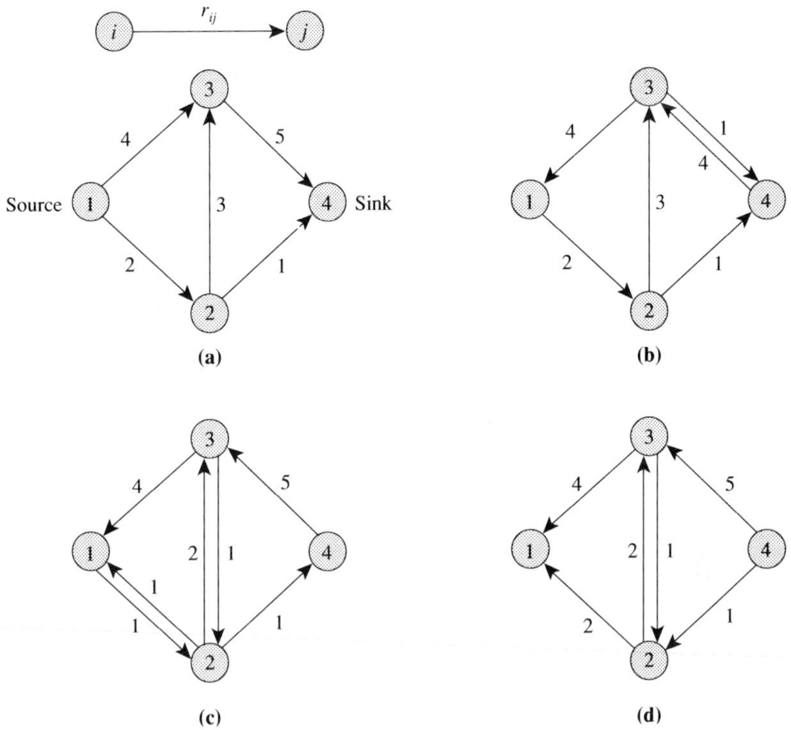

Figure 6.13 Illustrating the generic augmenting path algorithm: (a) residual network for the zero flow; (b) network after augmenting four units along the path 1–3–4; (c) network after augmenting one unit along the path 1–2–3–4; (d) network after augmenting one unit along the path 1–2–4.

path is $\delta = \min\{2, 3, 1\} = 1$. Augmenting 1 unit of flow along this path yields the residual network shown in Figure 6.13(c). In the third iteration, the algorithm augments 1 unit of flow along the path 1–2–4. Figure 6.13(d) shows the corresponding residual network. Now the residual network contains no augmenting path, so the algorithm terminates.

Relationship between the Original and Residual Networks

In implementing any version of the generic augmenting path algorithm, we have the option of working directly on the original network with the flows x_{ij}, or maintaining the residual network $G(x)$ and keeping track of the residual capacities r_{ij} and, when the algorithm terminates, recovering the actual flow variables x_{ij}. To see how we can use either alternative, it is helpful to understand the relationship between arc flows in the original network and residual capacities in the residual network.

First, let us consider the concept of an augmenting path in the original network. An augmenting path in the original network G is a path P (not necessarily directed) from the source to the sink with $x_{ij} < u_{ij}$ on every forward arc (i, j) and $x_{ij} > 0$ on every backward arc (i, j). It is easy to show that the original network G contains an augmenting path with respect to a flow x if and only if the residual network $G(x)$ contains a directed path from the source to the sink.

Now suppose that we update the residual capacities at some point in the algorithm. What is the effect on the arc flows x_{ij}? The definition of the residual capacity (i.e., $r_{ij} = u_{ij} - x_{ij} + x_{ji}$) implies that an additional flow of δ units on arc (i, j) in the residual network corresponds to (1) an increase in x_{ij} by δ units in the original network, or (2) a decrease in x_{ji} by δ units in the original network, or (3) a convex combination of (1) and (2). We use the example given in Figure 6.14(a) and the corresponding residual network in Figure 6.14(b) to illustrate these possibilities. Augmenting 1 unit of flow on the path 1–2–4–3–5–6 in the network produces the residual network in Figure 6.14(c) with the corresponding arc flows shown in Figure 6.14(d). Comparing the solution in Figure 6.14(d) with that in Figure 6.14(a), we find that the flow augmentation increases the flow on arcs (1, 2), (2, 4), (3, 5), (5, 6) and decreases the flow on arc (3, 4).

Finally, suppose that we are given values for the residual capacities. How should we determine the flows x_{ij}? Observe that since $r_{ij} = u_{ij} - x_{ij} + x_{ji}$, many combinations of x_{ij} and x_{ji} correspond to the same value of r_{ij}. We can determine

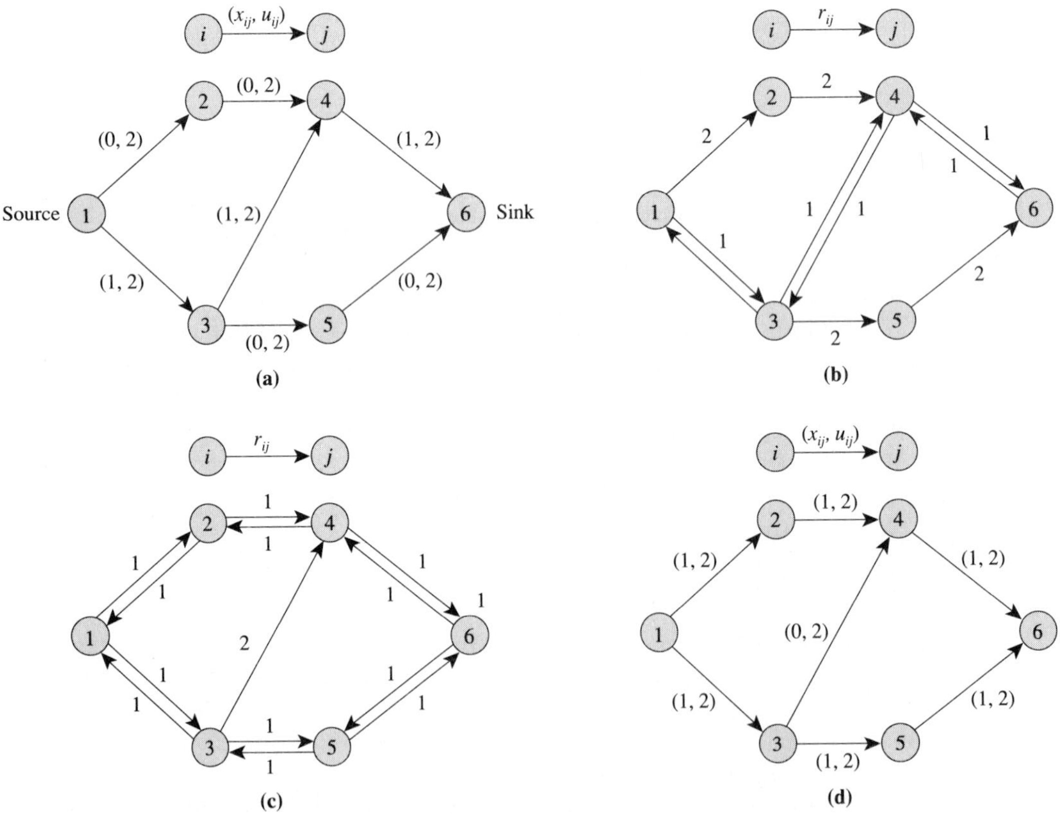

Figure 6.14 The effect of augmentation on flow decomposition: (a) original network with a flow x; (b) residual network for flow x; (c) residual network after augmenting one unit along the path 1–2–4–3–5–6; (d) flow in the original network after the augmentation.

one such choice as follows. To highlight this choice, let us rewrite $r_{ij} = u_{ij} - x_{ij} + x_{ji}$ as $x_{ij} - x_{ji} = u_{ij} - r_{ij}$. Now, if $u_{ij} \geq r_{ij}$, we set $x_{ij} = u_{ij} - r_{ij}$ and $x_{ji} = 0$; otherwise, we set $x_{ij} = 0$ and $x_{ji} = r_{ij} - u_{ij}$.

Effect of Augmentation on Flow Decomposition

To obtain better insight concerning the augmenting path algorithm, let us illustrate the effect of an augmentation on the flow decomposition on the preceding example. Figure 6.15(a) gives the decomposition of the initial flow and Figure 6.15(b) gives the decomposition of the flow after we have augmented 1 unit of flow on the path 1–2–4–3–5–6. Although we augmented 1 unit of flow along the path 1–2–4–3–5–6, the flow decomposition contains no such path. Why?

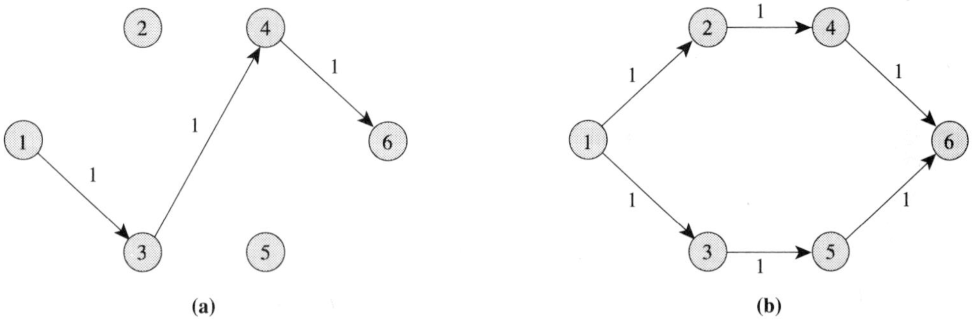

Figure 6.15 Flow decomposition of the solution in (a) Figure 6.14(a) and (b) Figure 6.14(d).

The path 1–3–4–6 defining the flow in Figure 6.14(a) contains three segments: the path up to node 3, arc (3, 4) as a forward arc, and the path up to node 6. We can view this path as an augmentation on the zero flow. Similarly, the path 1–2–4–3–5–6 contains three segments: the path up to node 4, arc (3, 4) as a backward arc, and the path up to node 6. We can view the augmentation on the path 1–2–4–3–5–6 as linking the initial segment of the path 1–3–4–6 with the last segment of the augmentation, linking the last segment of the path 1–3–4–6 with the initial segment of the augmentation, and canceling the flow on arc (3, 4), which then drops from both the path 1–3–4–6 and the augmentation (see Figure 6.16). In general, we can

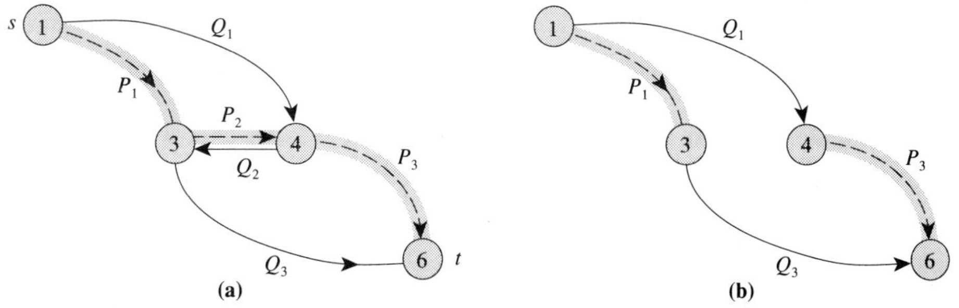

Figure 6.16 The effect of augmentation on flow decomposition: (a) the two augmentations P_1–P_2–P_3 and Q_1–Q_2–Q_3; (b) net effect of these augmentations.

view each augmentation as "pasting together" segments of the current flow decomposition to obtain a new flow decomposition.

6.5 LABELING ALGORITHM AND THE MAX-FLOW MIN-CUT THEOREM

In this section we discuss the augmenting path algorithm in more detail. In our discussion of this algorithm in the preceding section, we did not discuss some important details, such as (1) how to identify an augmenting path or show that the network contains no such path, and (2) whether the algorithm terminates in finite number of iterations, and when it terminates, whether it has obtained a maximum flow. In this section we consider these issues for a specific implementation of the generic augmenting path algorithm known as the *labeling algorithm.* The labeling algorithm is not a polynomial-time algorithm. In Chapter 7, building on the ideas established in this chapter, we describe two polynomial-time implementations of this algorithm.

The labeling algorithm uses a search technique (as described in Section 3.4) to identify a directed path in $G(x)$ from the source to the sink. The algorithm *fans out* from the source node to find all nodes that are reachable from the source along a directed path in the residual network. At any step the algorithm has partitioned the nodes in the network into two groups: *labeled* and *unlabeled.* Labeled nodes are those nodes that the algorithm has reached in the fanning out process and so the algorithm has determined a directed path from the source to these nodes in the residual network; the unlabeled nodes are those nodes that the algorithm has not reached as yet by the fanning-out process. The algorithm iteratively selects a labeled node and scans its arc adjacency list (in the residual network) to reach and label additional nodes. Eventually, the sink becomes labeled and the algorithm sends the maximum possible flow on the path from node s to node t. It then erases the labels and repeats this process. The algorithm terminates when it has scanned all the labeled nodes and the sink remains unlabeled, implying that the source node is not connected to the sink node in the residual network. Figure 6.17 gives an algorithmic description of the labeling algorithm.

Correctness of the Labeling Algorithm and Related Results

To study the correctness of the labeling algorithm, note that in each iteration (i.e., an execution of the whole loop), the algorithm either performs an augmentation or terminates because it cannot label the sink. In the latter case we must show that the current flow x is a maximum flow. Suppose at this stage that S is the set of labeled nodes and $\bar{S} = N - S$ is the set of unlabeled nodes. Clearly, $s \in S$ and $t \in \bar{S}$. Since the algorithm cannot label any node in $\bar{S}$ from any node in S, $r_{ij} = 0$ for each $(i, j) \in (S, \bar{S})$. Furthermore, since $r_{ij} = (u_{ij} - x_{ij}) + x_{ji}$, $x_{ij} \le u_{ij}$ and $x_{ji} \ge 0$, the condition $r_{ij} = 0$ implies that $x_{ij} = u_{ij}$ for every arc $(i, j) \in (S, \bar{S})$ and $x_{ij} = 0$ for every arc $(i, j) \in (\bar{S}, S)$. [Recall our assumption that for each arc $(i, j) \in A$,

```
algorithm labeling;
begin
    label node t;
    while t is labeled do
    begin
        unlabel all nodes;
        set pred(j) : = 0 for each j ∈ N;
        label node s and set LIST : = {s};
        while LIST ≠ Ø or t is unlabeled do
        begin
            remove a node i from LIST;
            for each arc (i, j) in the residual network emanating from node i do
                if r_ij > 0 and node j is unlabeled then set pred(j) : = i, label node j, and
                    add j to LIST;
        end;
        if t is labeled then augment
    end;
end;

procedure augment;
begin
    use the predecessor labels to trace back from the sink to the source to
        obtain an augmenting path P from node s to node t;
    δ : = min{r_ij : (i, j) ∈ P};
    augment δ units of flow along P and update the residual capacities;
end;
```

Figure 6.17 Labeling algorithm.

$(j, i) \in A$.] Substituting these flow values in (6.3), we find that

$$v = \sum_{(i,j)\in(S,\bar{S})} x_{ij} - \sum_{(i,j)\in(\bar{S},S)} x_{ij} = \sum_{(i,j)\in(S,\bar{S})} u_{ij} = u[S, \bar{S}].$$

This discussion shows that the value of the current flow x equals the capacity of the cut $[S, \bar{S}]$. But then Property 6.1 implies that x is a maximum flow and $[S, \bar{S}]$ is a minimum cut. This conclusion establishes the correctness of the labeling algorithm and, as a by-product, proves the following max-flow min-cut theorem.

Theorem 6.3 (Max-Flow Min-Cut Theorem). *The maximum value of the flow from a source node s to a sink node t in a capacitated network equals the minimum capacity among all s–t cuts.* ◆

The proof of the max-flow min-cut theorem shows that when the labeling algorithm terminates, it has also discovered a minimum cut. The labeling algorithm also proves the following augmenting path theorem.

Theorem 6.4 (Augmenting Path Theorem). *A flow x^* is a maximum flow if and only if the residual network $G(x^*)$ contains no augmenting path.*

Proof. If the residual network $G(x^*)$ contains an augmenting path, clearly the flow x^* is not a maximum flow. Conversely, if the residual network $G(x^*)$ contains no augmenting path, the set of nodes S labeled by the labeling algorithm defines an

s–t cut $[S, \bar{S}]$ whose capacity equals the flow value, thereby implying that the flow must be maximum. ◆

The labeling algorithm establishes one more important result.

Theorem 6.5 (Integrality Theorem). *If all arc capacities are integer, the maximum flow problem has an integer maximum flow.*

Proof. This result follows from an induction argument applied to the number of augmentations. Since the labeling algorithm starts with a zero flow and all arc capacities are integer, the initial residual capacities are all integer. The flow augmented in any iteration equals the minimum residual capacity of some path, which by the induction hypothesis is integer. Consequently, the residual capacities in the next iteration will again be integer. Since the residual capacities r_{ij} and the arc capacities u_{ij} are all integer, when we convert the residual capacities into flows by the method described previously, the arc flows x_{ij} will be integer valued as well. Since the capacities are integer, each augmentation adds at least one unit to the flow value. Since the maximum flow cannot exceed the capacity of any cut, the algorithm will terminate in a finite number of iterations. ◆

The integrality theorem does not imply that every optimal solution of the maximum flow problem is integer. The maximum flow problem may have noninteger solutions and, most often, has such solutions. The integrality theorem shows that the problem always has at least one integer optimal solution.

Complexity of the Labeling Algorithm

To study the worst-case complexity of the labeling algorithm, recall that in each iteration, except the last, when the sink cannot be labeled, the algorithm performs an augmentation. It is easy to see that each augmentation requires $O(m)$ time because the search method examines any arc or any node at most once. Therefore, the complexity of the labeling algorithm is $O(m)$ times the number of augmentations. How many augmentations can the algorithm perform? If all arc capacities are integral and bounded by a finite number U, the capacity of the cut $(s, N - \{s\})$ is at most nU. Therefore, the maximum flow value is bounded by nU. The labeling algorithm increases the value of the flow by at least 1 unit in any augmentation. Consequently, it will terminate within nU augmentations, so $O(nmU)$ is a bound on the running time of the labeling algorithm. Let us formally record this observation.

Theorem 6.6. *The labeling algorithm solves the maximum flow problem in $O(nmU)$ time.* ◆

Throughout this section, we have assumed that each arc capacity is finite. In some applications, it will be convenient to model problems with infinite capacities on some arcs. If we assume that some s–t cut has a finite capacity and let U denote the maximum capacity across this cut, Theorem 6.6 and, indeed, all the other results in this section remain valid. Another approach for addressing situations with infinite capacity arcs would be to impose a capacity on these arcs, chosen sufficiently large as to not affect the maximum flow value (see Exercise 6.23). In defining the residual capacities and developing algorithms to handle situations with infinite arc capacities, we adopt this approach rather than modifying the definitions of residual capacities.

Drawbacks of the Labeling Algorithm

The labeling algorithm is possibly the simplest algorithm for solving the maximum flow problem. Empirically, the algorithm performs reasonably well. However, the worst-case bound on the number of iterations is not entirely satisfactory for large values of U. For example, if $U = 2^n$, the bound is exponential in the number of nodes. Moreover, the algorithm can indeed perform this many iterations, as the example given in Figure 6.18 illustrates. For this example, the algorithm can select the augmenting paths s–a–b–t and s–b–a–t alternatively 10^6 times, each time augmenting unit flow along the path. This example illustrates one shortcoming of the algorithm.

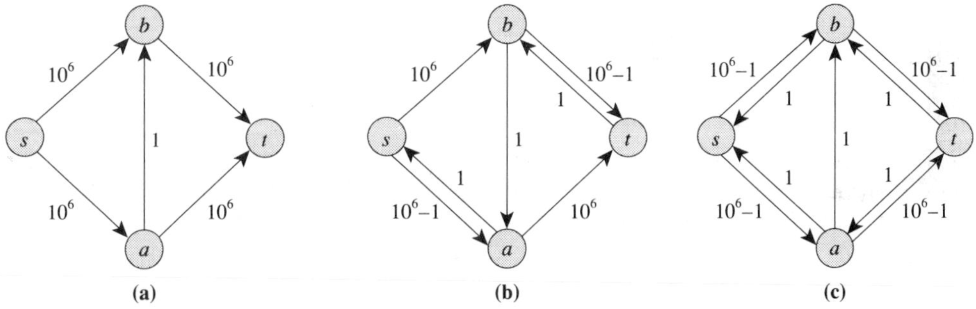

Figure 6.18 Pathological example of the labeling algorithm: (a) residual network for the zero flow; (b) network after augmenting unit flow along the path s–a–b–t; (c) network after augmenting unit flow along the path s–b–a–t.

A second drawback of the labeling algorithm is that if the capacities are irrational, the algorithm might not terminate. For some pathological instances of the maximum flow problem (see Exercise 6.48), the labeling algorithm does not terminate, and although the successive flow values converge, they converge to a value strictly less than the maximum flow value. (Note, however, that the max-flow min-cut theorem holds even if arc capacities are irrational.) Therefore, if the labeling algorithm is guaranteed to be effective, it must select augmenting paths carefully.

A third drawback of the labeling algorithm is its *"forgetfulness."* In each iteration, the algorithm generates node labels that contain information about augmenting paths from the source to other nodes. The implementation we have described erases the labels as it moves from one iteration to the next, even though much of this information might be valid in the next iteration. Erasing the labels therefore destroys potentially useful information. Ideally, we should retain a label when we can use it profitably in later computations.

In Chapter 7 we describe several improvements of the labeling algorithm that overcomes some or all of these drawbacks. Before discussing these improvements, we discuss some interesting implications of the max-flow min-cut theorem.

6.6 COMBINATORIAL IMPLICATIONS OF THE MAX-FLOW MIN-CUT THEOREM

As we noted in Section 6.2 when we discussed several applications of the maximum flow problem, in some applications we wish to find a minimum cut in a network, which we now know is equivalent to finding a maximum flow in the network. In fact, the relationship between maximum flows and minimum cuts permits us to view many problems from either of two dual perspectives: a flow perspective or a cut perspective. At times this dual perspective provides novel insight about an underlying problem. In particular, when applied in various ways, the max-flow min-cut theorem reduces to a number of min-max duality relationships in combinatorial theory. In this section we illustrate this use of network flow theory by developing several results in combinatorics. We might note that these results are fairly deep and demonstrate the power of the max-flow min-cut theorem. To appreciate the power of the max-flow min-cut theorem, we would encourage the reader to try to prove the following results without using network flow theory.

Network Connectivity

We first study some connectivity issues about networks that arise, for example, in the design of communication networks. We first define some notation. We refer to two directed paths from node s to node t as *arc disjoint* if they do not have any arc in common. Similarly, we refer to two directed paths from node s to node t as *node disjoint* if they do not have any node in common, except the source and the sink nodes. Given a directed network $G = (N, A)$ and two specified nodes s and t, we are interested in the following two questions: (1) What is the maximum number of arc-disjoint (directed) paths from node s to node t; and (2) what is the minimum number of arcs that we should remove from the network so that it contains no directed paths from node s to node t? The following theorem shows that these two questions are really alternative ways to address the same issue.

Theorem 6.7. *The maximum number of arc-disjoint paths from node s to node t equals the minimum number of arcs whose removal from the network disconnects all paths from node s to node t.*

Proof. Define the capacity of each arc in the network as equal to 1. Consider any feasible flow x of value v in the resulting unit capacity network. The flow decomposition theorem (Theorem 3.5) implies that we can decompose the flow x into flows along paths and cycles. Since flows around cycles do not affect the flow value, the flows on the paths sum to v. Furthermore, since each arc capacity is 1, these paths are arc disjoint and each carries 1 unit of flow. Consequently, the network contains v arc-disjoint paths from s to t.

Now consider any s–t cut $[S, \bar{S}]$ in the network. Since each arc capacity is 1, the capacity of this cut is $|(S, \bar{S})|$ (i.e., it equals the number of forward arcs in the cut). Since each path from node s to node t contains at least one arc in $(S, \bar{S})$, the removal of the arcs in $(S, \bar{S})$ disconnects all the paths from node s to node t. Consequently, the network contains a disconnecting set of arcs of cardinality equal to the capacity of any s–t cut $[S, \bar{S}]$. The max-flow min-cut theorem immediately implies that the maximum number of arc-disjoint paths from s to t equals the minimum number of arcs whose removal will disconnect all paths from node s to node t. ◆

We next discuss the node-disjoint version of the preceding theorem.

Theorem 6.8. *The maximum number of node-disjoint paths from node s to node t equals the minimum number of nodes whose removal from the network disconnects all paths from nodes s to node t.*

Proof. Split each node i in G, other than s and t, into two nodes i' and i'' and add a "node-splitting" arc (i', i'') of unit capacity. All the arcs in G entering node i now enter node i' and all the arcs emanating from node i now emanate from node i''. Let G' denote this transformed network. Assign a capacity of ∞ to each arc in the network except the node-splitting arcs, which have unit capacity. It is easy to see that there is one-to-one correspondence between the arc-disjoint paths in G' and the node-disjoint paths in G. Therefore, the maximum number of arc-disjoint paths in G' equals the maximum number of node-disjoint paths in G.

As in the proof of Theorem 6.7, flow decomposition implies that a flow of v units from node s to node t in G' decomposes into v arc-disjoint paths each carrying unit flow; and these v arc-disjoint paths in G' correspond to v node-disjoint paths in G. Moreover, note that any s–t cut with finite capacity contains only node-splitting arcs since all other arcs have infinite capacity. Therefore, any s–t cut in G' with capacity k corresponds to a set of k nodes whose removal from G destroys all paths from node s to node t. Applying the max-flow min-cut theorem to G' and using the preceding observations establishes that the maximum number of node-disjoint paths in G from node s to node t equals the minimum number of nodes whose removal from G disconnects nodes s and t. ◆

Matchings and Covers

We next state some results about matchings and node covers in a bipartite network. For a directed bipartite network $G = (N_1 \cup N_2, A)$ we refer to a subset $A' \subseteq A$ as a *matching* if no two arcs in A' are incident to the same node (i.e., they do not have any common endpoint). We refer to a subset $N' \subseteq N = N_1 \cup N_2$ as a *node cover* if every arc in A is incident to one of the nodes in N'. For illustrations of these definitions, consider the bipartite network shown in Figure 6.19. In this network the set of arcs $\{(1, 1'), (3, 3'), (4, 5'), (5, 2')\}$ is a matching but the set of arcs $\{(1, 2'), (3, 1'), (3, 4')\}$ is not because the arcs $(3, 1')$ and $(3, 4')$ are incident to the same node 3. In the same network the set of nodes $\{1, 2', 3, 5'\}$ is a node cover, but the set of nodes $\{2', 3', 4, 5\}$ is not because the arcs $(1, 1')$, $(3, 1')$, and $(3, 4')$ are not incident to any node in the set.

Theorem 6.9. *In a bipartite network $G = (N_1 \cup N_2, A)$, the maximum cardinality of any matching equals the minimum cardinality of any node cover of G.*

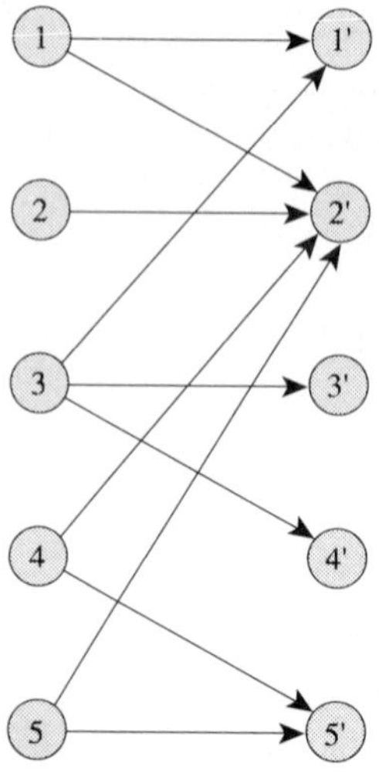

Figure 6.19 Bipartite network.

Proof. Augment the network by adding a source node s and an arc (s, i) of unit capacity for each $i \in N_1$. Similarly, add a sink node t and an arc (j, t) of unit capacity for each $j \in N_2$. Denote the resulting network by G'. We refer to the arcs in A as *original arcs* and the additional arcs as *artificial arcs*. We set the capacity of each artificial arc equal to 1 and the capacity of each original arc equal to ∞.

Now consider any flow x of value v from node s to node t in the network G'. We can decompose the flow x into v paths of the form s–i–j–t each carrying 1 unit of flow. Thus v arcs of the original network have a positive flow. Furthermore, these arcs constitute a matching, for otherwise the flow on some artificial arc would exceed 1 unit. Consequently, a flow of value v corresponds to a matching of cardinality v. Similarly, a matching of cardinality v defines a flow of value v.

We next show that any node cover H of $G = (N_1 \cup N_2, A)$ defines an s–t cut of capacity $|H|$ in G'. Given the node cover H, construct a set of arcs Q as follows: For each $i \in H$, if $i \in N_1$, add arc (s, i) to Q, and if $i \in N_2$, add arc (i, t) to Q. Since H is a node cover, each directed path from node s to node t in G' contains one arc in Q; therefore, Q is a valid s–t cut of capacity $|H|$.

We now show the converse result; that is, for a given s–t cut Q of capacity k in G', the network G contains a node cover of cardinality k. We first note that the cut Q consists solely of artificial arcs because the original arcs have infinite capacity. From Q we construct a set H of nodes as follows: if $(s, i) \in Q$ and $(i, t) \in Q$, we add i to H. Now observe that each original arc (i, j) defines a directed path s–i–j–t in G'. Since Q is an s–t cut, either $(s, i) \in Q$ or $(j, t) \in Q$ or both. By the preceding construction, either $i \in H$ or $j \in H$ or both. Consequently, H must be a node cover. We have thus established a one-to-one correspondence between node covers in G and s–t cuts in G'.

The max-flow min-cut theorem implies that the maximum flow value equals the capacity of a minimum cut. In view of the max-flow min-cut theorem, the preceding observations imply that the maximum number of independent arcs in G equals the minimum number of nodes in a node cover of G. The theorem thus follows. ◆

Figure 6.20 gives a further illustration of Theorem 6.9. In this figure, we have transformed the matching problem of Figure 6.19 into a maximum flow problem, and we have identified the minimum cut. The minimum cut consists of the arcs $(s, 1)$, $(s, 3)$, $(2', t)$ and $(5', t)$. Correspondingly, the set $\{1, 3, 2', 5'\}$ is a minimum cardinality node cover, and a maximum cardinality matching is $(1, 1')$, $(2, 2')$, $(3, 3')$ and $(5, 5')$.

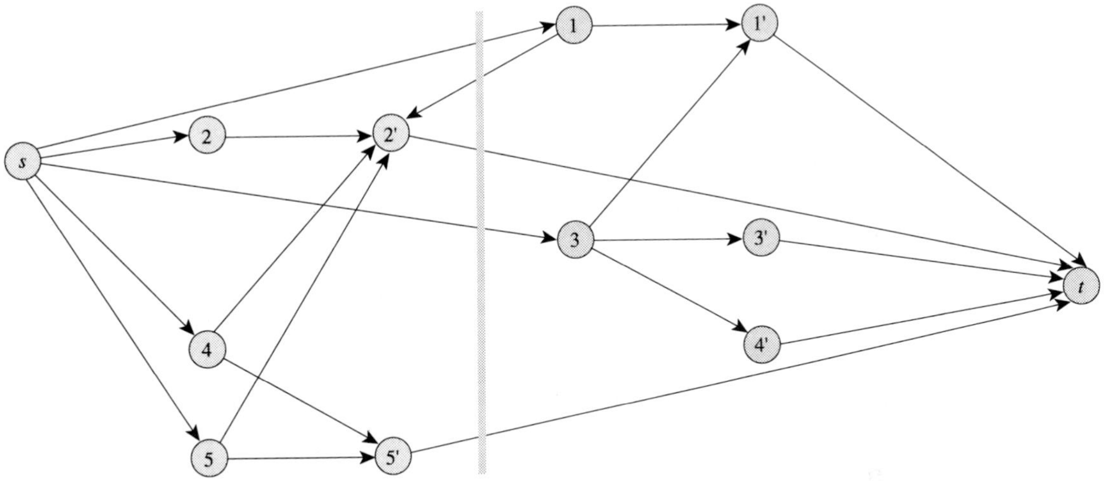

Figure 6.20 Minimum cut for the maximum flow problem defined in Figure 6.19.

As we have seen in the discussion throughout this section, the max-flow min-cut theorem is a powerful tool for establishing a number of results in the field of combinatorics. Indeed, the range of applications of the max-flow min-cut theorem and the ability of this theorem to encapsulate so many subtle duality (i.e., max-min) results as special cases is quite surprising, given the simplicity of the labeling algorithm and of the proof of the max-flow min-cut theorem. The wide range of applications reflects the fact that flows and cuts, and the relationship between them, embody central combinatorial results in many problem domains within applied mathematics.

6.7 FLOWS WITH LOWER BOUNDS

In this section we consider maximum flow problems with nonnegative lower bounds imposed on the arc flows; that is, the flow on any arc $(i, j) \in A$ must be at least $l_{ij} \geq 0$. The following formulation models this problem:

Maximize v

subject to

$$\sum_{\{j:(i,j)\in A\}} x_{ij} - \sum_{\{j:(j,i)\in A\}} x_{ji} = \begin{cases} v & \text{for } i = s, \\ 0 & \text{for all } i \in N - \{s, t\}, \\ -v & \text{for } i = t, \end{cases}$$

$$l_{ij} \leq x_{ij} \leq u_{ij} \qquad \text{for each } (i, j) \in A.$$

In previous sections we studied a special case of this problem with only zero lower bounds. Whereas the maximum flow problem with zero lower bounds always has a feasible solution (since the zero flow is feasible), the problem with nonnegative lower bounds could be infeasible. For example, consider the maximum flow problem given in Figure 6.21. This problem does not have a feasible solution because arc (1, 2) must carry at least 5 units of flow into node 2 and arc (2, 3) can remove at most 4 units of flow; therefore, we can never satisfy the mass balance constraint of node 2.

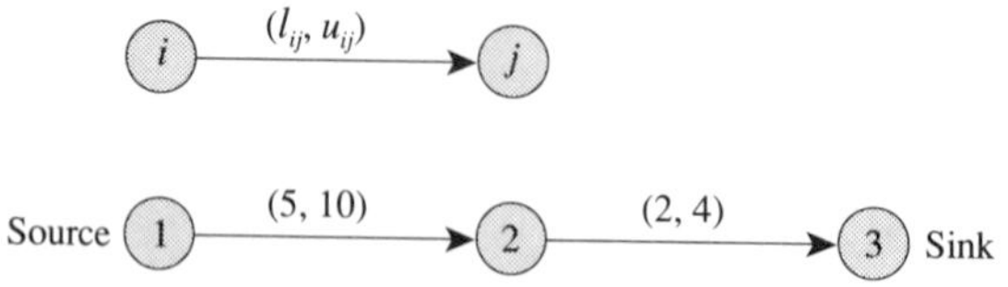

Figure 6.21 Maximum flow problem with no feasible solution.

As illustrated by this example, any maximum flow algorithm for problems with nonnegative lower bounds has two objectives: (1) to determine whether the problem is feasible, and (2) if so, to establish a maximum flow. It therefore comes as no surprise that most algorithms use a two-phase approach. The first phase determines a feasible flow if one exists, and the second phase converts a feasible flow into a maximum flow. We shall soon see that the problem in each phase essentially reduces to solving a maximum flow problem with zero lower bounds. Consequently, it is possible to solve the maximum flow problem with nonnegative lower bounds by solving two maximum flow problems, each with zero lower bounds. For convenience, we consider the second phase prior to the first phase.

Determining a Maximum Flow

Suppose that we have a feasible flow x in the network. We can then modify any maximum flow algorithm designed for the zero lower bound case to obtain a maximum flow. In these algorithms, we make only one modification: We define the residual capacity of an arc (i, j) as $r_{ij} = (u_{ij} - x_{ij}) + (x_{ji} - l_{ji})$; the first term in this expression denotes the maximum increase in flow from node i to node j using the remaining capacity of arc (i, j), and the second term denotes the maximum increase in flow from node i to node j by canceling the existing flow on arc (j, i). Notice that since each arc flow is within its lower and upper bounds, each residual capacity is nonnegative. Recall that the maximum flow algorithm described in this chapter (and the ones described in Chapter 7) works with only residual capacities and does not need arc flows, capacities, or lower bounds. Therefore we can use any of these algorithms to establish a maximum flow in the network. These algorithms terminate with optimal residual capacities. From these residual capacities we can construct maximum flow in a large number of ways. For example, through a change of variables we can reduce the computations to a situation we have considered before. For all arcs (i, j), let $u'_{ij} = u_{ij} - l_{ij}$, $r'_{ij} = r_{ij}$, and $x'_{ij} = x_{ij} - l_{ij}$. The residual capacity for arc (i, j) is $r_{ij} = (u_{ij} - x_{ij}) + (x_{ji} - l_{ji})$. Equivalently, $r'_{ij} = u'_{ij} - x'_{ij} + x'_{ji}$. Similarly, $r'_{ji} = u'_{ji} - x'_{ji} + x'_{ij}$. If we compute the x' values in terms of r' and u', we obtain the same expression as before, i.e., $x'_{ij} = \max(u'_{ij} - r'_{ij}, 0)$ and $x'_{ji} = \max(u'_{ji} -$

$r'_{ij}, 0)$. Converting back into the original variables, we obtain the following formulae:

$$x_{ij} = l_{ij} + \max(u_{ij} - r_{ij} - l_{ij}, 0),$$

$$x_{ji} = l_{ji} + \max(u_{ji} - r_{ji} - l_{ji}, 0).$$

We now show that the solution determined by this modified procedure solves the maximum flow problem with nonnegative lower bounds. Let x denote a feasible flow in G with value equal to v. Moreover, let $[S, \bar{S}]$ denote an s–t cut. We define the *capacity* of an s–t cut $[S, \bar{S}]$ as

$$u[S, \bar{S}] = \sum_{(i,j)\in(S,\bar{S})} u_{ij} - \sum_{(i,j)\in(\bar{S},S)} l_{ij}. \tag{6.7}$$

The capacity of the cut denotes the maximum amount of "net" flow that can be sent out of the node set S. We next use equality (6.3), which we restate for convenience.

$$v = \sum_{(i,j)\in(S,\bar{S})} x_{ij} - \sum_{(i,j)\in(\bar{S},S)} x_{ij}. \tag{6.8}$$

Substituting $x_{ij} \le u_{ij}$ in the first summation and $l_{ij} \le x_{ij}$ in the second summation of this inequality shows that

$$v \le \sum_{(i,j)\in(S,\bar{S})} u_{ij} - \sum_{(i,j)\in(\bar{S},S)} l_{ij} = u[S, \bar{S}]. \tag{6.9}$$

Inequality (6.9) indicates that the maximum flow value is less than or equal to the capacity of any s–t cut. At termination, the maximum flow algorithm obtains an s–t cut $[S, \bar{S}]$ with $r_{ij} = 0$ for every arc $(i, j) \in (S, \bar{S})$. Let x denote the corresponding flow with value equal to v. Since $r_{ij} = (u_{ij} - x_{ij}) + (x_{ji} - l_{ji})$, the conditions $x_{ij} \le u_{ij}$ and $l_{ji} \le x_{ji}$ imply that $x_{ij} = u_{ij}$ and $x_{ji} = l_{ji}$. Consequently, $x_{ij} = u_{ij}$ for every arc $(i, j) \in (S, \bar{S})$ and $x_{ij} = l_{ij}$ for every arc $(i, j) \in (\bar{S}, S)$. Substituting these values in (6.8), we find that

$$v = u[S, \bar{S}] = \sum_{(i,j)\in(S,\bar{S})} u_{ij} - \sum_{(i,j)\in(\bar{S},S)} l_{ij}. \tag{6.10}$$

In view of inequality (6.9), equation (6.10) implies that $[S, \bar{S}]$ is a minimum s–t cut and x is a maximum flow. As a by-product of this result, we have proved a generalization of the max-flow min-cut theorem for problems with nonnegative lower bounds.

Theorem 6.10 (Generalized Max-Flow Min-Cut Theorem). *If the capacity of an s–t cut $[S, \bar{S}]$ in a network with both lower and upper bounds on arc flows is defined by* (6.7), *the maximum value of flow from node s to node t equals the minimum capacity among all s–t cuts.* ◆

Establishing a Feasible Flow

We now address the issue of determining a feasible flow in the network. We first transform the maximum flow problem into a circulation problem by adding an arc (t, s) of infinite capacity. This arc carries the flow sent from node s to node t back to node s. Consequently, in the circulation formulation of the problem, the outflow of each node, including nodes s and t, equals its flow. Clearly, the maximum flow problem admits a feasible flow if and only if the circulation problem admits a feasible flow. Given the possibility of making this transformation, we now focus our intention on finding a feasible circulation, and characterizing conditions when an arbitrary circulation problem, with lower and upper bounds of flows, possesses a feasible solution.

The feasible circulation problem is to identify a flow x satisfying the following constraints:

$$\sum_{\{j:(i,j)\in A\}} x_{ij} - \sum_{\{j:(j,i)\in A\}} x_{ji} = 0 \qquad \text{for all } i \in N, \tag{6.11a}$$

$$l_{ij} \le x_{ij} \le u_{ij} \qquad \text{for all } (i, j) \in A. \tag{6.11b}$$

By replacing $x_{ij} = x'_{ij} + l_{ij}$ in constraints (6.11a) and (6.11b), we obtain the following transformed problem:

$$\sum_{\{j:(i,j)\in A\}} x'_{ij} - \sum_{\{j:(j,i)\in A\}} x'_{ji} = b(i) \qquad \text{for all } i \in N, \tag{6.12a}$$

$$0 \le x'_{ij} \le u_{ij} - l_{ij} \qquad \text{for all } (i, j) \in A, \tag{6.12b}$$

with supplies/demands $b(\cdot)$ at the nodes defined by

$$b(i) = \sum_{\{j:(j,i)\in A\}} l_{ji} - \sum_{\{j:(i,j)\in A\}} l_{ij}.$$

Observe that $\sum_{i\in N} b(i) = 0$ since each l_{ij} occurs twice in this expression, once with a positive sign and once with a negative sign. The feasible circulation problem is then equivalent to determining whether the transformed problem has a solution x' satisfying (6.12).

Notice that this problem is essentially the same as the feasible flow problem discussed in Application 6.1. In discussing this application we showed that by solving a maximum flow problem we either determine a solution satisfying (6.12) or show that no solution satisfies (6.12). If x'_{ij} is a feasible solution of (6.12), $x_{ij} = x'_{ij} + l_{ij}$ is a feasible solution of (6.11).

Characterizing a Feasible Flow

We next characterize feasible circulation problems (i.e., derive the necessary and sufficiency conditions for a circulation problem to possess a feasible solution). Let S be any set of nodes in the network. By summing the mass balance constraints of the nodes in S, we obtain the expression

$$\sum_{(i,j)\in(S,\bar{S})} x_{ij} - \sum_{(i,j)\in(\bar{S},S)} x_{ij} = 0. \tag{6.13}$$

Using the inequalities $x_{ij} \le u_{ij}$ in the first term of (6.13) and the inequalities $x_{ij} \ge l_{ij}$ in the second term, we find that

$$\sum_{(i,j)\in(\bar{S},S)} l_{ij} \le \sum_{(i,j)\in(S,\bar{S})} u_{ij}. \tag{6.14}$$

The expression in (6.14), which is a necessary condition for feasibility, states that the maximum amount of flow that we can send out from a set S of nodes must be at least as large as the minimum amount of flow that the nodes in S must receive. Clearly, if a set of nodes must receive more than what the other nodes can send them, the network has no feasible circulation. As we will see, these conditions are also sufficient for ensuring feasibility [i.e., if the network data satisfies the conditions (6.14) for every set S of nodes, the network has a feasible circulation that satisfies the flow bounds on all its arcs].

We give an algorithmic proof for the sufficiency of condition (6.14). The algorithm starts with a circulation x that satisfies the mass balance and capacity constraints, but might violate some of the lower bound constraints. The algorithm gradually converts this circulation into a feasible flow or identifies a node set S that violates condition (6.14).

With respect to a flow x, we refer to an arc (i, j) as *infeasible* if $x_{ij} < l_{ij}$ and *feasible* if $l_{ij} \leq x_{ij}$. The algorithm selects an infeasible arc (p, q) and attempts to make it feasible by increasing the flow on this arc. The mass balance constraints imply that in order to increase the flow on the arc, we must augment flow along one or more cycles in the residual network that contain arc (p, q) as a forward arc. We define the residual network $G(x)$ with respect to a flow x the same way we defined it previously except that we set the residual capacity of any infeasible arc (i, j) to the value $u_{ij} - x_{ij}$. Any augmenting cycle containing arc (p, q) as a forward arc must consist of a directed path in the residual network $G(x)$ from node q to node p plus the arc (p, q). We can use a labeling algorithm to identify a directed path from node q to node p.

We apply this procedure to one infeasible arc at a time, at each step decreasing the infeasibility of the arcs until we either identify a feasible flow or the labeling algorithm is unable to identify a directed path from node q to node p for some infeasible arc (p, q). We show that in the latter case, the maximum flow problem must be infeasible. Let S be the set of nodes labeled by the last application of the labeling algorithm. Clearly, $q \in S$ and $p \in \bar{S} \equiv N - S$. Since the labeling algorithm cannot label any node not in S, every arc (i, j) from S to $\bar{S}$ has a residual capacity of value zero. Therefore, $x_{ij} = u_{ij}$ for every arc $(i, j) \in (S, \bar{S})$ and $x_{ij} \leq l_{ij}$ for every arc $(i, j) \in (\bar{S}, S)$. Also observe that $(p, q) \in (\bar{S}, S)$ and $x_{pq} < l_{pq}$. Substituting these values in (6.13), we find that

$$\sum_{(i,j)\in(\bar{S},S)} l_{ij} > \sum_{(i,j)\in(S,\bar{S})} u_{ij},$$

contradicting condition (6.14), which we have already shown is necessary for feasibility. We have thus established the following fundamental theorem.

Theorem 6.11 (Circulation Feasibility Conditions). *A circulation problem with nonnegative lower bounds on arc flows is feasible if and only if for every set S of nodes*

$$\sum_{(i,j)\in(\bar{S},S)} l_{ij} \leq \sum_{(i,j)\in(S,\bar{S})} u_{ij}.$$ ◆

Note that the proof of this theorem specifies a one pass algorithm, starting with the zero flow, for finding a feasible solution to any circulation problem whose arc upper bounds u_{ij} are all nonnegative. In Exercise 6.7 we ask the reader to specify a one pass algorithm for any situation (i.e., even when some upper bounds are negative).

A by-product of Theorem 6.11 is the following result, which states necessary and sufficiency conditions for the existence of a feasible solution for the feasible flow problem stated in (6.2). (Recall that a feasible flow problem is the feasibility version of the minimum cost flow problem.) We discuss the proof of this result in Exercise 6.43.

Theorem 6.12. *The feasible flow problem stated in* (6.2) *has a feasible solution if and only if for every subset* $S \subseteq N$, $b(S) - u[S, \bar{S}] \leq 0$, *where* $b(S) = \sum_{i \in S} b(i)$. ◆

6.8 SUMMARY

In this chapter we studied two closely related problems: the maximum flow problem and the minimum cut problem. After illustrating a variety of applications of these problems, we showed that the maximum flow and the minimum cut problems are closely related of each other (in fact, they are dual problems) and solving the maximum flow problem also solves the minimum cut problem. We began by showing that the value of any flow is less than or equal to the capacity of any cut in the network (i.e., this is a "weak duality" result). The fact that the value of some flow equals the capacity of some cut in the network (i.e., the "strong duality" result) is a deeper result. This result is known as the *max-flow min-cut theorem*. We establish it by specifying a labeling algorithm that maintains a feasible flow x in the network and sends additional flow along directed paths from the source node to the sink node in the residual network $G(x)$. Eventually, $G(x)$ contains no directed path from the source to the sink. At this point, the value of the flow x equals the capacity of some cut $[S, \bar{S}]$ in the network. The weak duality result implies that x is a maximum flow and $[S, \bar{S}]$ is a minimum cut. Since the labeling algorithm maintains an integer flow at every step (assuming integral capacity data), the optimal flow that it finds is integral. This result is a special case of a more general network flow integrality result that we establish in Chapter 9. The labeling algorithm runs in $O(nmU)$ time. This time bound is not attractive from the worst-case perspective. In Chapter 7 we develop two polynomial-time implementations of the labeling algorithm.

The max-flow min-cut theorem has far-reaching implications. It allows us to prove several important results in combinatorics that appear difficult to prove using other means. We proved the following results: (1) the maximum number of arc-disjoint (or node-disjoint) paths connecting two nodes s and t in a network equals the minimum number of arcs (or nodes) whose removal from the network leaves no directed path from node s to node t; and (2) in a bipartite network, the maximum cardinality of any matching equals the minimum cardinality of any node cover. In the exercises we ask the reader to prove other implications of the max-flow min-cut theorem.

To conclude this chapter we studied the maximum flow problem with nonnegative lower bounds on arc flows. We can solve this problem using a two-phase approach. The first phase determines a feasible flow if one exists, and the second phase converts this flow into a maximum flow; in both phases we solve a maximum flow problem with zero lower bounds. We also described a theoretical result for characterizing when a maximum flow problem with nonnegative lower bounds has a feasible solution. Roughly speaking, this characterization states that the maximum flow problem has a feasible solution if and only if the maximum possible outflow of every cut is at least as large as the minimum required inflow for that cut.

REFERENCE NOTES

The seminal paper of Ford and Fulkerson [1956a] on the maximum flow problem established the celebrated max-flow min-cut theorem. Fulkerson and Dantzig [1955], and Elias, Feinstein, and Shannon [1956] independently established this result. Ford and Fulkerson [1956a] and Elias et al. [1956] solved the maximum flow problem by augmenting path algorithms, whereas Fulkerson and Dantzig [1955] solved it by specializing the simplex method for linear programming. The labeling algorithm that we described in Section 6.5 is due to Ford and Fulkerson [1956a]; their classical book, Ford and Fulkerson [1962], offers an extensive treatment of this algorithm. Unfortunately, the labeling algorithm runs in pseudopolynomial time; moreover, as shown by Ford and Fulkerson [1956a], for networks with arbitrary irrational arc capacities, the algorithm can perform an infinite sequence of augmentations and might converge to a value different from the maximum flow value. Several improved versions of the labeling algorithm overcome this limitation. We provide citations to these algorithms and to their improvements in the reference notes of Chapter 7. In Chapter 7 we also discuss computational properties of maximum flow algorithms.

In Section 6.6 we studied the combinatorial implications of the max-flow min-cut theorem. Theorems 6.7 and 6.8 are known as Menger's theorem. Theorem 6.9 is known as the König-Egerváry theorem. Ford and Fulkerson [1962] discuss these and several additional combinatorial results that are provable using the max-flow min-cut theorem.

In Section 6.7 we studied the feasibility of a network flow problem with nonnegative lower bounds imposed on the arc flows. Theorem 6.11 is due to Hoffman [1960], and Theorem 6.12 is due to Gale [1957]. The book by Ford and Fulkerson [1962] discusses these and some additional feasibility results extensively. The algorithm we have presented for identifying a feasible flow in a network with nonnegative lower bounds is adapted from this book.

The applications of the maximum flow problem that we described in Section 6.2 are adapted from the following papers:

1. Feasible flow problem (Berge and Ghouila-Houri [1962])
2. Problem of representatives (Hall [1956])
3. Matrix rounding problem (Bacharach [1966])
4. Scheduling on uniform parallel machines (Federgruen and Groenevelt [1986])
5. Distributed computing on a two-processor model (Stone [1977])
6. Tanker scheduling problem (Dantzig and Fulkerson [1954])

Elsewhere in this book we describe other applications of the maximum flow problem. These applications include: (1) the tournament problem (Application 1.3, Ford and Johnson [1959]), (2) the police patrol problem (Exercise 1.9, Khan [1979]),

(3) nurse staff scheduling (Exercise 6.2, Khan and Lewis [1987]), (4) solving a system of equations (Exercise 6.4, Lin [1986]), (5) statistical security of data (Exercises 6.5, Application 8.3, Gusfield [1988], Kelly, Golden, and Assad [1990]), (6) the minimax transportation problem (Exercise 6.6, Ahuja [1986]), (7) the baseball elimination problem (Application 8.1, Schwartz [1966]), (8) network reliability testing (Application 8.2, Van Slyke and Frank [1972]), (9) open pit mining (Application 19.1, Johnson [1968]), (10) selecting freight handling terminals (Application 19.2, Rhys [1970]), (11) optimal destruction of military targets (Application 19.3, Orlin [1987]), (12) the flyaway kit problem (Application 19.4, Mamer and Smith [1982]), (13) maximum dynamic flows (Application 19.12, Ford and Fulkerson [1958a]), and (14) models for building evacuation (Application 19.13, Chalmet, Francis, and Saunders [1982]).

Two other interesting applications of the maximum flow problem are preemptive scheduling on machines with different speeds (Martel [1982]), and the multifacility rectilinear distance location problem (Picard and Ratliff [1978]). The following papers describe additional applications or provide additional references: McGinnis and Nuttle [1978], Picard and Queyranne [1982], Abdallaoui [1987], Gusfield, Martel, and Fernandez-Baca [1987], Gusfield and Martel [1989], and Gallo, Grigoriadis, and Tarjan [1989].

EXERCISES

6.1. Dining problem. Several families go out to dinner together. To increase their social interaction, they would like to sit at tables so that no two members of the same family are at the same table. Show how to formulate finding a seating arrangement that meets this objective as a maximum flow problem. Assume that the dinner contingent has p families and that the ith family has $a(i)$ members. Also assume that q tables are available and that the jth table has a seating capacity of $b(j)$.

6.2. Nurse staff scheduling (Khan and Lewis [1987]). To provide adequate medical service to its constituents at a reasonable cost, hospital administrators must constantly seek ways to hold staff levels as low as possible while maintaining sufficient staffing to provide satisfactory levels of health care. An urban hospital has three departments: the emergency room (department 1), the neonatal intensive care nursery (department 2), and the orthopedics (department 3). The hospital has three work shifts, each with different levels of necessary staffing for nurses. The hospital would like to identify the minimum number of nurses required to meet the following three constraints: (1) the hospital must allocate at least 13, 32, and 22 nurses to the three departments (over all shifts); (2) the hospital must assign at least 26, 24, and 19 nurses to the three shifts (over all departments); and (3) the minimum and maximum number of nurses allocated to each department in a specific shift must satisfy the following limits:

		Department		
		1	2	3
Shift	1	(6, 8)	(11, 12)	(7, 12)
	2	(4, 6)	(11, 12)	(7, 12)
	3	(2, 4)	(10, 12)	(5, 7)

Suggest a method using maximum flows to identify the minimum number of nurses required to satisfy all the constraints.

6.3. A commander is located at one node p in a communication network G and his subordinates are located at nodes denoted by the set S. Let u_{ij} be the effort required to eliminate arc (i, j) from the network. The problem is to determine the minimal effort required to block all communications between the commander and his subordinates. How can you solve this problem in polynomial time?

6.4. Solving a system of equations (Lin [1986]). Let $F = \{f_{ij}\}$ be a given $p \times q$ matrix and consider the following system of $p + q$ equations in the (possibly fractional) variables y:

$$\sum_{j=1}^{q} f_{ij}y_{ij} = u_i, \qquad 1 \le i \le p, \tag{6.15a}$$

$$\sum_{i=1}^{p} f_{ij}y_{ij} = v_j, \qquad 1 \le j \le q. \tag{6.15b}$$

In this system $u_i \ge 0$ and $v_j \ge 0$ are given constants satisfying the condition $\sum_{i=1}^{p} u_i = \sum_{j=1}^{q} v_j$.

(a) Define a matrix $D = \{d_{ij}\}$ as follows: $d_{ij} = 0$ if $f_{ij} = 0$, and $d_{ij} = 1$ if $f_{ij} \neq 0$. Show that (6.15) has a feasible solution if and only if the following system of $p + q$ equations has a feasible solution x:

$$\sum_{j=1}^{q} d_{ij}x_{ij} = u_i, \qquad 1 \le i \le p, \tag{6.16a}$$

$$\sum_{i=1}^{p} d_{ij}x_{ij} = v_j, \qquad 1 \le j \le q. \tag{6.16b}$$

(b) Show how to formulate the problem of identifying a feasible solution of the system (6.16) as a feasible circulation problem (i.e., identifying a circulation in some network with lower and upper bounds imposed on the arc flows). [*Hint*: The network has a node i for the ith row in (6.16a), a node $\bar{j}$ for the jth row in (6.16b), and one extra node s.]

6.5. Statistical security of data (Kelly, Golden, and Assad [1990], and Gusfield [1988]). The U.S. Census Bureau produces a variety of tables from its census data. Suppose that it wishes to produce a $p \times q$ table $D = \{d_{ij}\}$ of nonnegative integers. Let $r(i)$ denote the sum of the matrix elements in the ith row and let $c(j)$ denote the sum of the matrix elements in the jth column. Assume that each sum $r(i)$ and $c(j)$ is strictly positive. The Census Bureau often wishes to disclose all the row and column sums along with some matrix elements (denoted by a set Y) and yet suppress the remaining elements to ensure the confidentiality of privileged information. Unless it exercises care, by disclosing the elements in Y, the Bureau might permit someone to deduce the exact value of one or more of the suppressed elements. It is possible to deduce a suppressed element d_{ij} if only one value of d_{ij} is consistent with the row and column sums and the disclosed elements in Y. We say that any such suppressed element is *unprotected*. Describe a polynomial-time algorithm for identifying all the unprotected elements of the matrix and their values.

6.6. Minimax transportation problem (Ahuja [1986]). Suppose that $G = (N, A)$ is an uncapacitated transportation problem (as defined in Section 1.2) and that we want to find an integer flow x that minimizes the objective function $\max\{c_{ij}x_{ij} : (i, j) \in A\}$ among all feasible integer flows.

(a) Consider a relaxed version of the minimax transportation problem: Given a parameter λ, we want to know whether some feasible flow satisfies the condition $\max\{c_{ij}x_{ij} : (i, j) \in A\} \le \lambda$. Show how to solve this problem as a maximum flow problem. [*Hint*: Use the condition $\max\{c_{ij}x_{ij}:(i,j) \in A\} \leq \lambda$ to formulate the problem as a feasible flow problem.]

(b) Use the result in part (a) to develop a polynomial-time algorithm for solving the minimax transportation problem. What is the running time of your algorithm?

6.7. Consider a generalization of the feasible flow problem discussed in Application 6.1. Suppose that the flow bounds constraints are $l_{ij} \leq x_{ij} \leq u_{ij}$ instead of $0 \leq x_{ij} \leq u_{ij}$ for some nonnegative l_{ij}. How would you solve this generalization of the feasible flow problem as a single maximum flow problem?

6.8 Consider a generalization of the problem that we discussed in Application 6.2. Suppose that each club must nominate one of its members as a town representative so that the number of representatives belonging to the political party P_k is between l_k and u_k. Formulate this problem as a maximum flow problem with nonnegative lower bounds on arc flows.

6.9. In the example concerning the scheduling of uniform parallel machines (Application 6.4), we assumed that the same number of machines are available each day. How would you model a situation when the number of available machines varies from day to day? Illustrate your method on the example given in Application 6.4. Assume that three machines are available on days 1, 2, 4, and 5; two machines on days 3 and 6; and four machines on the rest of the days.

6.10. Can you solve the police patrol problem described in Exercise 1.9 using a maximum flow algorithm. If so, how?

6.11. Suppose that we wish to partition an undirected graph into two components with the minimum number of arcs between the components. How would you solve this problem?

6.12. Consider the network shown in Figure 6.22(a) together with the feasible flow x given in the figure.

(a) Specify four s–t cuts in the network, each containing four forward arcs. List the capacity, residual capacity, and the flow across each cut.

(b) Draw the residual network for the network given in Figure 6.22(a) and list four augmenting paths from node s to node t.

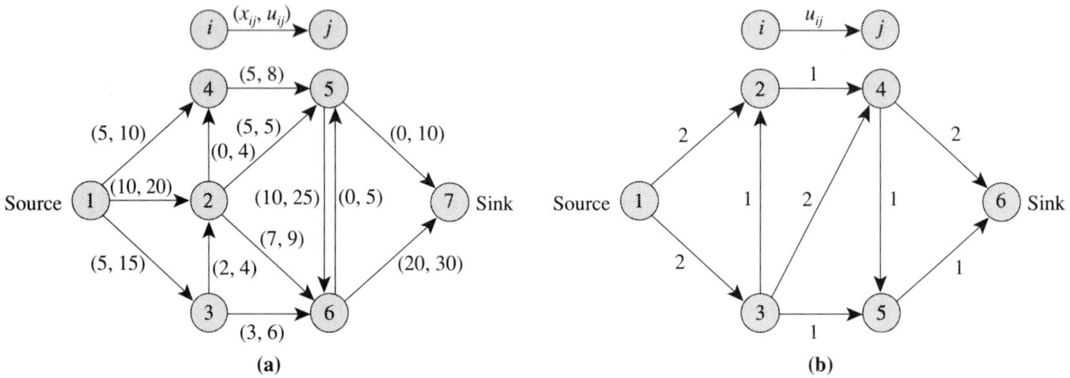

Figure 6.22 Examples for Exercises 6.12 and 6.13.

6.13. Solve the maximum flow problem shown in Figure 6.22(b) by the labeling algorithm, augmenting flow along the longest path in the residual network (i.e., the path containing maximum number of arcs). Specify the residual network before each augmentation. After every augmentation, decompose the flow into flows along directed paths from node s to node t. Finally, specify the minimum cut in the network obtained by the labeling algorithm.

6.14. Use the labeling algorithm to establish a maximum flow in the undirected network shown in Figure 6.23. Show the residual network at the end of each augmentation and specify the minimum cut that the algorithm obtains when it terminates.

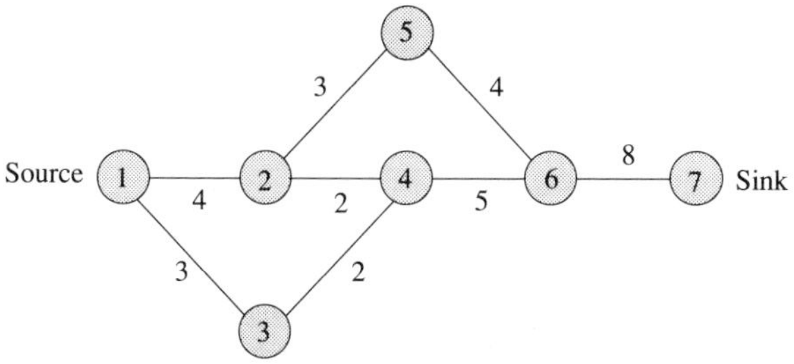

Figure 6.23 Example for Exercise 6.14.

6.15. Consider the network given in Figure 6.24; assume that each arc has capacity 1.

(a) Compute the maximum number of arc-disjoint paths from the source node to the sink node. (You might do so by inspection.)

(b) Enumerate all s–t cuts in the network. For each s–t cut $[S, \bar{S}]$, list the node partition and the sets of forward and backward arcs.

(c) Verify that the maximum number of arc-disjoint paths from node s to node t equals the minimum number of forward arcs in an s–t cut.

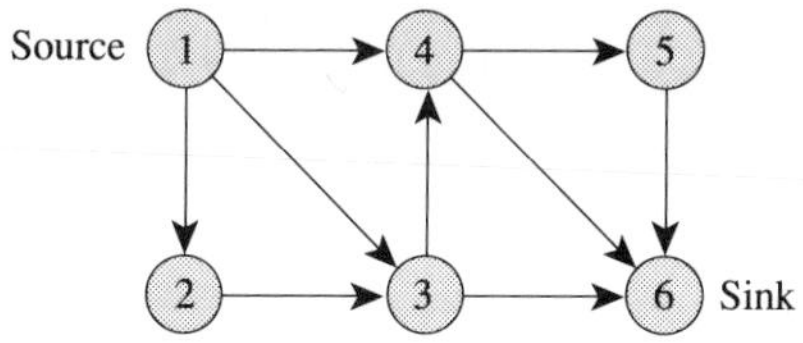

Figure 6.24 Example for Exercise 6.15.

6.16. Consider the matrix rounding problem given below (see Application 6.3). We want to round each element in the matrix, and also the row and column sums, to the nearest multiple of 2 so that the sum of the rounded elements in each row equals the rounded row sum and the sum of the rounded elements in each column equals the rounded column sum. Formulate this problem as a maximum flow problem and solve it.

				Row sum
	7.5	6.3	15.4	29.2
	3.9	9.1	3.6	16.6
	15.0	5.5	21.5	42.0
Column sum	26.4	20.9	40.5	

6.17. Formulate the following example of the scheduling problem on uniform parallel machines that we discussed in Application 6.4 as a maximum flow problem. Solve the problem by the labeling algorithm, assuming that two machines are available each day.

Job (j)	1	2	3	4
Processing time (p_j) (in days)	2.5	3.1	5.0	1.8
Release time (r_j)	1	5	0	2
Due date (d_j)	3	7	6	5

6.18. Minimum flow problem. The *minimum flow problem* is a close relative of the maximum flow problem with nonnegative lower bounds on arc flows. In the minimum flow problem, we wish to send the minimum amount of flow from the source to the sink, while satisfying given lower and upper bounds on arc flows.

(a) Show how to solve the minimum flow problem by using two applications of any maximum flow algorithm that applies to problems with zero lower bounds on arc flows. (*Hint*: First construct a feasible flow and then convert it into a minimum flow.)

(b) Prove the following *min-flow max-cut theorem*. Let the *floor* (or lower bound on the cut capacity) of an s–t cut $[S, \bar{S}]$ be defined as $\sum_{(i,j)\in(S,\bar{S})} l_{ij} - \sum_{(i,j)\in(\bar{S},S)} u_{ij}$. Show that the minimum value of all the flows from node s to node t equals the maximum floor of all s–t cuts.

6.19. Machine setup problem. A job shop needs to perform eight tasks on a particular day. Figure 6.25(a) shows the start and end times of each task. The workers must perform

Task	Start time	End time
1	1:00 P.M.	1:30 P.M.
2	6:00 P.M.	8:00 P.M.
3	10:00 P.M.	11:00 P.M.
4	4:00 P.M.	5:00 P.M.
5	4:00 P.M.	7:00 P.M.
6	12:00 noon	1:00 P.M.
7	2:00 P.M.	5:00 P.M.
8	11:00 P.M.	12:00 midnight

(a)

	1	2	3	4	5	6	7	8
1	—	60	10	25	30	20	15	40
2	10	—	40	55	40	5	30	35
3	65	30	—	0	45	30	20	5
4	0	50	35	—	20	15	10	20
5	20	24	40	50	—	15	5	23
6	10	8	9	35	12	—	30	30
7	15	30	6	18	15	30	—	10
8	20	35	15	12	75	13	25	—

(b)

Figure 6.25 Machine setup data: (a) task start and end times; (b) setup times in transforming between tasks.

these tasks according to this schedule so that exactly one worker performs each task. A worker cannot work on two jobs at the same time. Figure 6.25(b) shows the setup time (in minutes) required for a worker to go from one task to another. We wish to find the minimum number of workers to perform the tasks. Formulate this problem as a minimum flow problem (see Exercise 6.18).

6.20. Show how to transform a maximum flow problem having several source nodes and several sink nodes to one with only one source node and one sink node.

6.21. Show that if we add any number of incoming arcs, with any capacities, to the source node, the maximum flow value remains unchanged. Similarly, show that if we add any number of outgoing arcs, with any capacities, at the sink node, the maximum flow value remains unchanged.

6.22. Show that the maximum flow problem with integral data has a finite optimal solution if and only if the network contains no infinite capacity directed path from the source node to the sink node.

6.23. Suppose that a network has some infinite capacity arcs but no infinite capacity paths from the source to the sink. Let A^0 denote the set of arcs with finite capacities. Show that we can replace the capacity of each infinite capacity arc by a finite number $M \geq \sum_{(i,j)\in A^0} u_{ij}$ without affecting the maximum flow value.

6.24. Suppose that you want to solve a maximum flow problem containing parallel arcs, but the maximum flow code you own cannot handle parallel arcs. How would you use the code to solve your maximum flow problem?

6.25. **Networks with node capacities.** In some networks, in addition to arc capacities, each node i, other than the source and the sink, might have an upper bound, say $w(i)$, on the flow that can pass through it. For example, the nodes might be airports with limited runway capacity for takeoff and landings, or might be switches in a communication network with a limited number of ports. In these networks we are interested in determining the maximum flow satisfying both the arc and node capacities. Transform this problem to the standard maximum flow problem. From the perspective of worst-case complexity, is the maximum flow problem with upper bounds on nodes more difficult to solve than the standard maximum flow problem?

6.26. Suppose that a maximum flow network contains a node, other than the source node, with no incoming arc. Can we delete this node without affecting the maximum flow value? Similarly, can we delete a node, other than the sink node, with no outgoing arc?

6.27. Suppose that you are asked to solve a maximum flow problem in a directed network subject to the absolute value flow bound constraints $-u_{ij} \leq x_{ij} \leq u_{ij}$ imposed on some arcs (i, j). How would you solve this problem?

6.28. Suppose that a maximum flow is available. Show how you would find a minimum cut in $O(m)$ additional time. Suppose, instead, that a minimum cut is available. Could you use this cut to obtain a maximum flow faster than applying a maximum flow algorithm?

6.29. **Painted network theorem.** Let G be a directed network with a distinguished arc (s, t). Suppose that we paint each arc in the network as green, yellow, or red, with arc (s, t) painted yellow. Show that the painted network satisfies exactly one of the following two cases: (1) arc (s, t) is contained in a cycle of yellow and green arcs in which all yellow arcs have the same direction but green arcs can have arbitrary directions; (2) arc (s, t) is contained in a cut of yellow and red arcs in which all yellow arcs have the same direction but red arcs can have arbitrary directions.

6.30. Show that if $x_{ij} = u_{ij}$ for some arc (i, j) in every maximum flow, this arc must be a forward arc in some minimum cut.

6.31. An engineering department consisting of p faculty members, $F_1, F_2, \ldots, F_p$, will offer p courses, $C_1, C_2, \ldots, C_p$, in the coming semester and each faculty member will teach exactly one course. Each faculty member ranks two courses he (or she) would like to teach, ranking them according to his (or her) preference.

(a) We say that a course assignment is a *feasible* assignment if every faculty member teaches a course within his (or her) preference list. How would you determine whether the department can find a feasible assignment? (For a related problem see Exercise 12.46.)

(b) A feasible assignment is said to be *k-feasible* if it assigns at most k faculty members to their second most preferred courses. For a given k, suggest an algorithm for determining a k-feasible assignment.

(c) We say that a feasible assignment is an *optimal assignment* if it maximizes the number of faculty members assigned to their most preferred course. Suggest an algorithm for determining an optimal assignment and analyze its complexity. [*Hint*: Use the algorithm in part (b) as a subroutine.]

6.32. **Airline scheduling problem.** An airline has p flight legs that it wishes to service by the fewest possible planes. To do so, it must determine the most efficient way to combine these legs into flight schedules. The starting time for flight i is a_i and the finishing time is b_i. The plane requires r_{ij} hours to return from the point of destination of flight i to the point of origin of flight j. Suggest a method for solving this problem.

6.33. A flow x is *even* if for every arc $(i, j) \in A$, x_{ij} is an even number; it is *odd* if for every $(i, j) \in A$, x_{ij} is an odd number. Either prove that each of the following claims are true or give a counterexample for them.

(a) If all arc capacities are even, the network has an even maximum flow.

(b) If all arc capacities are odd, the network has an odd maximum flow.

6.34. Which of the following claims are true and which are false. Justify your answer either by giving a proof or by constructing a counterexample.

(a) If x_{ij} is a maximum flow, either $x_{ij} = 0$ or $x_{ji} = 0$ for every arc $(i, j) \in A$.

(b) Any network always has a maximum flow x for which, for every arc $(i, j) \in A$, either $x_{ij} = 0$ or $x_{ji} = 0$.

(c) If all arcs in a network have different capacities, the network has a unique minimum cut.

(d) In a directed network, if we replace each directed arc by an undirected arc, the maximum flow value remains unchanged.

(e) If we multiply each arc capacity by a positive number λ, the minimum cut remains unchanged.

(f) If we add a positive number λ to each arc capacity, the minimum cut remains unchanged.

6.35. (a) Suppose that after solving a maximum flow problem you realize that you have underestimated the capacity of an arc (p, q) by k units. Show that the labeling algorithm can reoptimize the problem in $O(km)$ time.

(b) Suppose that instead of underestimating the capacity of the arc (p, q), you had overestimated its capacity by k units. Can you reoptimize the problem in $O(km)$ time?

6.36. (a) Construct a family of networks with the number of s–t cuts growing exponentially with n.

(b) Construct a family of networks with the number of minimum cuts growing exponentially with n.

6.37. (a) Given a maximum flow in a network, describe an algorithm for determining the minimum cut $[S, \bar{S}]$ with the property that for every other minimum cut $[R, \bar{R}]$, $R \subseteq S$.

(b) Describe an algorithm for determining the minimum cut $[S, \bar{S}]$ with the property that for every other minimum cut $[R, \bar{R}]$, $S \subseteq R$.

(c) Describe an algorithm for determining whether the maximum flow problem has a unique minimum cut.

6.38. Let $[S, \bar{S}]$ and $[T, \bar{T}]$ be two s–t cuts in the directed network G. Show that the cut capacity function $u[.\,,\,.]$ is *submodular*, that is, $u[S, \bar{S}] + u[T, \bar{T}] \geq u[S \cup T, \overline{S \cup T}] + u[S \cap T, \overline{S \cap T}]$. (*Hint*: Prove this result by case analysis.)

6.39. Show that if $[S, \bar{S}]$ and $[T, \bar{T}]$ are both minimum cuts, so are $[S \cup T, \overline{S \cup T}]$ and $[S \cap T, \overline{S \cap T}]$.

6.40. Suppose that we know a noninteger maximum flow in a directed network with integer arc capacities. Suggest an algorithm for converting this flow into an integer maximum flow. What is the running time of your algorithm? (*Hint*: Send flows along cycles.)

6.41. Optimal coverage of sporting events. A group of reporters want to cover a set of sporting events in an olympiad. The sports events are held in several stadiums throughout a city. We known the starting time of each event, its duration, and the stadium where it is held. We are also given the travel times between different stadiums. We want to determine the least number of reporters required to cover the sporting events. How would you solve this problem?

6.42. In Section 6.7 we showed how to solve the maximum flow problem in directed networks with nonnegative lower bounds by solving two maximum flow problems with zero lower flow bounds. Try to generalize this approach for undirected networks in which the flow on any arc (i, j) is permitted in either direction, but whichever direction is chosen the amount of flow is at least l_{ij}. If you succeed in developing an algorithm, state the algorithm along with a proof that it correctly solves the problem; if you do not succeed in developing an algorithm state reasons why the generalization does not work.

6.43. Feasibility of the feasible flow problem (Gale [1957]). Show that the feasible flow problem, discussed in Application 6.1, has a feasible solution if and only if for every subset $S \subseteq N$, $b(S) - u[S, \bar{S}] \leq 0$. (*Hint*: Transform the feasible flow problem into a circulation problem with nonzero lower bounds and use the result of Theorem 6.11.)

6.44. Prove Theorems 6.7 and 6.8 for undirected networks.

6.45. Let N^+ and N^- be two nonempty disjoint node sets in G. Describe a method for determining the maximum number of arc-disjoint paths from N^+ to N^- (i.e., each path can start at any node in N^+ and can end at any node in N^-). What is the implication of the max-flow min-cut theorem in this case? (*Hint*: Generalize the statement of Theorem 6.7.)

6.46. Consider a 0–1 matrix **H** with n_1 rows and n_2 columns. We refer to a row or a column of the matrix **H** as a line. We say that a set of 1's in the matrix **H** is *independent* if no two of them appear in the same line. We also say that a set of lines in the matrix is a *cover* of **H** if they include (i.e., "cover") all the 1's in the matrix. Show that the maximum number of independent 1's equals the minimum number of lines in a cover. (*Hint*: Use the max-flow min-cut theorem on an appropriately defined network.)

6.47. In a directed acyclic network G, certain arcs are colored blue, while others are colored red. Consider the problem of covering the blue arcs by directed paths, which can start and end at any node (these paths can contain arcs of any color). Show that the minimum number of directed paths needed to cover the blue arcs is equal to the maximum number of blue arcs that satisfy the property that no two of these arcs belong to the same path. Will this result be valid if G contains directed cycles? (*Hint*: Use the min-flow max-cut theorem stated in Exercise 6.18.)

6.48. Pathological example for the labeling algorithm. In the residual network $G(x)$ corresponding to a flow x, we define an *augmenting walk* as a directed walk from node s to node t that visits any arc at most once (it might visit nodes multiple times—in particular, an augmenting walk might visit nodes s and t multiple times.)

(a) Consider the network shown in Figure 6.26(a) with the arcs labeled a, b, c and d; note that one arc capacity is irrational. Show that this network contains an infinite sequence of augmenting walks whose residual capacities sum to the maximum flow value. (*Hint*: Each augmenting walk of the sequence contains exactly two arcs from node s to node t with finite residual capacities.)

(b) Now consider the network shown in Figure 6.26(b). Show that this network contains an infinite sequence of augmenting walks whose residual capacities sum to a value different than the maximum flow value.

(c) Next consider the network shown in Figure 6.26(c); in addition to the arcs shown, the network contain an infinite capacity arc connecting each node pair in the set

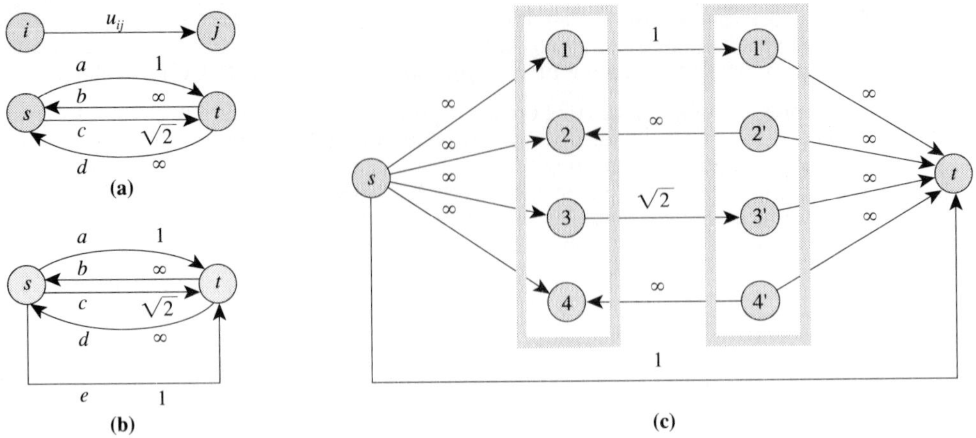

Figure 6.26 A subgraph of a pathological instance for labeling algorithm. The fall graph contains an infinite capacity are connecting each pair of nodes i and j as well as each pair of nodes i′ and j′.

{1, 2, 3, 4} and each node pair in the set {1′, 2′, 3′, 4′}. Show that each augmenting walk in the solution of part (b) corresponds to an augmenting path in Figure 6.26(c). Conclude that the labeling algorithm, when applied to a maximum flow problem with irrational capacities, might perform an infinite sequence of augmentations and the terminal flow value might be different than the maximum flow value.

7

MAXIMUM FLOWS: POLYNOMIAL ALGORITHMS

Every day, in every way, I am getting better and better.
—Émile Coué

Chapter Outline

7.1 INTRODUCTION

The generic augmenting path algorithm that we discussed in Chapter 6 is a powerful tool for solving maximum flow problems. Not only is it guaranteed to solve any maximum flow problem with integral capacity data, it also provides us with a constructive tool for establishing the fundamental max-flow min-cut theorem and therefore for developing many useful combinatorial applications of network flow theory.

As we noted in Chapter 6, however, the generic augmenting path algorithm has two significant computational limitations: (1) its worst-case computational complexity of $O(nmU)$ is quite unattractive for problems with large capacities; and (2) from a theoretical perspective, for problems with irrational capacity data, the algorithm might converge to a nonoptimal solution. These limitations suggest that the algorithm is not entirely satisfactory, in theory. Unfortunately, the algorithm is not very satisfactory in practice as well: On very large problems, it can require too much solution time.

Motivated by a desire to develop methods with improved worst-case complexity and empirical behavior, in this chapter we study several refinements of the generic augmenting path algorithm. We also introduce and study another class of algorithms, known as *preflow-push algorithms*, that have recently emerged as the most powerful techniques, both theoretically and computationally, for solving maximum flow problems.

Before describing these algorithms and analyzing them in detail, let us pause to reflect briefly on the theoretical limits of maximum flow algorithms and to introduce the general solution strategies employed by the refined augmenting path algorithms that we consider in this chapter. Flow decomposition theory shows that, in principle, we might be able to design augmenting path algorithms that are capable of finding a maximum flow in no more than m augmentations. For suppose that x is an optimal flow and x° is any initial flow (possibly the zero flow). By the flow decomposition property (see Section 3.5), we can obtain x from x° by a sequence of (1) at most m augmentations on augmenting paths from node s to node t, plus (2) flows around augmenting cycles. If we define x' as the flow vector obtained from x° by sending flows along only the augmenting paths, x' is also a maximum flow (because flows around augmenting cycles do not change the flow value into the sink node). This observation demonstrates a theoretical possibility of finding a maximum flow using at most m augmentations. Unfortunately, to apply this flow decomposition argument, we need to know a maximum flow. As a consequence, no algorithm developed in the literature achieves this theoretical bound of m augmentations. Nevertheless, it is possible to improve considerably on the $O(nU)$ bound on the number of augmentations required by the generic augmenting path algorithm.

How might we attempt to reduce the number of augmentations or even eliminate them altogether? In this chapter we consider three basic approaches:

1. Augmenting in "large" increments of flow
2. Using a combinatorial strategy that limits the type of augmenting paths we can use at each step
3. Relaxing the mass balance constraint at intermediate steps of the algorithm, and thus not requiring that each flow change must be an augmentation that starts at the source node and terminates at the sink node

Let us now consider each of these approaches. As we have seen in Chapter 6, the generic augmenting path algorithm could be slow because it might perform a large number of augmentations, each carrying a small amount of flow. This observation suggests one natural strategy for improving the augmenting path algorithm: Augment flow along a path with a *large* residual capacity so that the number of augmentations remains relatively *small*. The *maximum capacity augmenting path algorithm* uses this idea: It always augments flow along a path with the maximum residual capacity. In Section 7.3 we show that this algorithm performs $O(m \log U)$ augmentations. A variation of this algorithm that augments flows along a path with a *sufficiently large*, but not necessarily maximum residual capacity also performs $O(m \log U)$ augmentations and is easier to implement. We call this algorithm the *capacity scaling algorithm* and describe it in Section 7.3.

Another possible strategy for implementing and improving the augmenting path algorithm would be to develop an approach whose implementation is entirely independent of the arc capacity data and relies on a combinatorial argument for its convergence. One such approach would be somehow to restrict the choice of augmenting paths in some way. In one such approach we might always augment flow along a "shortest path" from the source to the sink, defining a shortest path as a directed path in the residual network consisting of the fewest number of arcs. If we augment flow along a shortest path, the length of any shortest path either stays the same or increases. Moreover, within m augmentations, the length of the shortest path is guaranteed to increase. (We prove these assertions in Section 7.4.) Since no path contains more than $n - 1$ arcs, this result guarantees that the number of augmentations is at most $(n - 1)m$. We call this algorithm the *shortest augmenting path algorithm* and discuss it in Section 7.4.

The preflow-push algorithms use the third strategy we have identified: They seek out "shortest paths" as in the shortest augmenting path algorithm, but do not send flow along paths from the source to the sink. Instead, they send flows on individual arcs. This "localized" strategy, together with clever rules for implementing this strategy, permits these algorithms to obtain a speed-up not obtained by any augmenting path algorithm. We study these preflow-push algorithms in Sections 7.6 through 7.9.

The concept of distance labels is an important construct used to implement the shortest augmenting path algorithm and the preflow-push algorithms that we consider in this chapter. So before describing the improved algorithms, we begin by discussing this topic.

7.2 DISTANCE LABELS

A *distance function* $d: N \rightarrow Z^+ \cup \{0\}$ with respect to the residual capacities r_{ij} is a function from the set of nodes to the set of nonnegative integers. We say that a distance function is *valid* with respect to a flow x if it satisfies the following two conditions:

$$d(t) = 0; \tag{7.1}$$

$$d(i) \leq d(j) + 1 \quad \text{for every arc } (i, j) \text{ in the residual network } G(x). \tag{7.2}$$

We refer to $d(i)$ as the *distance label* of node i and conditions (7.1) and (7.2) as the *validity conditions*. The following properties show why the distance labels might be of use in designing network flow algorithms.

Property 7.1. *If the distance labels are valid, the distance label $d(i)$ is a lower bound on the length of the shortest (directed) path from node i to node t in the residual network.*

To establish the validity of this observation, let $i = i_1 - i_2 - \cdots - i_k - i_{k+1} = t$ be any path of length k from node i to node t in the residual network. The validity conditions imply that

$$\begin{aligned}
d(i_k) &\leq d(i_{k+1}) + 1 = d(t) + 1 = 1, \\
d(i_{k-1}) &\leq d(i_k) + 1 \leq 2, \\
d(i_{k-2}) &\leq d(i_{k-1}) + 1 \leq 3, \\
&\vdots \\
d(i) = d(i_1) &\leq d(i_2) + 1 \leq k.
\end{aligned}$$

Property 7.2. *If $d(s) \geq n$, the residual network contains no directed path from the source node to the sink node.*

The correctness of this observation follows from the facts that $d(s)$ is a lower bound on the length of the shortest path from s to t in the residual network, and therefore no directed path can contain more than $(n - 1)$ arcs. Therefore, if $d(s) \geq n$, the residual network contains no directed path from node s to node t.

We now introduce some additional notation. We say that the distance labels are *exact* if for each node i, $d(i)$ equals the length of the shortest path from node i to node t in the residual network. For example, in Figure 7.1, if node 1 is the source node and node 4 is the sink node, then $d = (0, 0, 0, 0)$ is a valid vector of distance label, and $d = (3, 1, 2, 0)$ is a vector of exact distance labels. We can determine exact distance labels for all nodes in $O(m)$ time by performing a backward breadth-first search of the network starting at the sink node (see Section 3.4).

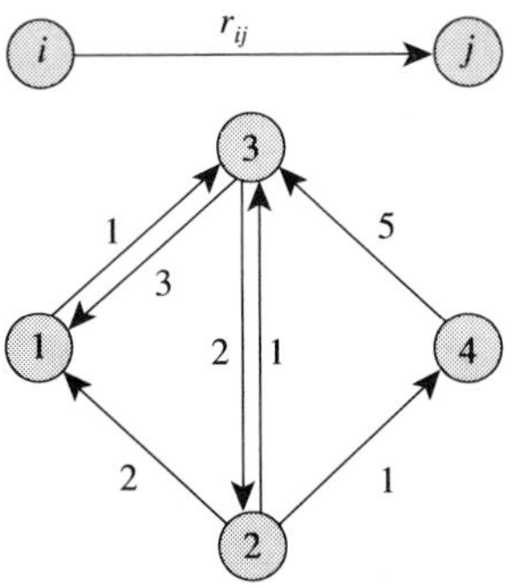

Figure 7.1 Residual network.

Admissible Arcs and Admissible Paths

We say that an arc (i, j) in the residual network is *admissible* if it satisfies the condition that $d(i) = d(j) + 1$; we refer to all other arcs as *inadmissible*. We also refer to a path from node s to node t consisting entirely of admissible arcs as an *admissible path*. Later, we use the following property of admissible paths.

Property 7.3. *An admissible path is a shortest augmenting path from the source to the sink.*

Since every arc (i, j) in an admissible path P is admissible, the residual capacity of this arc and the distance labels of its end nodes satisfy the conditions (1) $r_{ij} > 0$, and (2) $d(i) = d(j) + 1$. Condition (1) implies that P is an augmenting path and condition (2) implies that if P contains k arcs, then $d(s) = k$. Since $d(s)$ is a lower bound on the length of any path from the source to the sink in the residual network (from Property 7.1), the path P must be a shortest augmenting path.

7.3 CAPACITY SCALING ALGORITHM

We begin by describing the maximum capacity augmenting path algorithm and noting its computational complexity. This algorithm always augments flow along a path with the maximum residual capacity. Let x be any flow and let v be its flow value.

As before, let v^* be the maximum flow value. The flow decomposition property (i.e., Theorem 3.5), as applied to the residual network $G(x)$, implies that we can find m or fewer directed paths from the source to the sink whose residual capacities sum to $(v^* - v)$. Thus the maximum capacity augmenting path has residual capacity at least $(v^* - v)/m$. Now consider a sequence of $2m$ consecutive maximum capacity augmentations starting with the flow x. If each of these augmentations augments at least $(v^* - v)/2m$ units of flow, then within $2m$ or fewer iterations we will establish a maximum flow. Note, however, that if one of these $2m$ consecutive augmentations carries less than $(v^* - v)/2m$ units of flow, then from the initial flow vector x, we have reduced the residual capacity of the maximum capacity augmenting path by a factor of at least 2. This argument shows that within $2m$ consecutive iterations, the algorithm either establishes a maximum flow or reduces the residual capacity of the maximum capacity augmenting path by a factor of at least 2. Since the residual capacity of any augmenting path is at most $2U$ and is at least 1, after $O(m \log U)$ iterations, the flow must be maximum. (Note that we are essentially repeating the argument used to establish the geometric improvement approach discussed in Section 3.3.)

As we have seen, the maximum capacity augmentation algorithm reduces the number of augmentations in the generic labeling algorithm from $O(nU)$ to $O(m \log U)$. However, the algorithm performs more computations per iteration, since it needs to identify an augmenting path with the maximum residual capacity, not just any augmenting path. We now suggest a variation of the maximum capacity augmentation algorithm that does not perform more computations per iteration and yet establishes a maximum flow within $O(m \log U)$. Since this algorithm scales the arc capacities implicitly, we refer to it as the *capacity scaling algorithm*.

The essential idea underlying the capacity scaling algorithm is conceptually quite simple: We augment flow along a path with a *sufficiently large* residual capacity, instead of a path with the maximum augmenting capacity because we can obtain a path with a sufficiently large residual capacity fairly easily—in $O(m)$ time. To define the capacity scaling algorithm, let us introduce a parameter Δ and, with respect to a given flow x, define the *Δ-residual network* as a network containing arcs whose residual capacity is at least Δ. Let $G(x, \Delta)$ denote the Δ-residual network. Note that $G(x, 1) = G(x)$ and $G(x, \Delta)$ is a subgraph of $G(x)$. Figure 7.2 illustrates this definition. Figure 7.2(a) gives the residual network $G(x)$ and Figure 7.2(b) gives the Δ-residual network $G(x, \Delta)$ for $\Delta = 8$. Figure 7.3 specifies the capacity scaling algorithm.

Let us refer to a phase of the algorithm during which Δ remains constant as a *scaling phase* and a scaling phase with a specific value of Δ as a *Δ-scaling phase*. Observe that in a Δ-scaling phase, each augmentation carries at least Δ units of flow. The algorithm starts with $\Delta = 2^{\lfloor \log U \rfloor}$ and halves its value in every scaling phase until $\Delta = 1$. Consequently, the algorithm performs $1 + \lfloor \log U \rfloor = O(\log U)$ scaling phases. In the last scaling phase, $\Delta = 1$, so $G(x, \Delta) = G(x)$. This result shows that the algorithm terminates with a maximum flow.

The efficiency of the algorithm depends on the fact that it performs at most $2m$ augmentations per scaling phase. To establish this result, consider the flow at the end of the Δ-scaling phase. Let x' be this flow and let v' denote its flow value. Furthermore, let S be the set of nodes reachable from node s in $G(x', \Delta)$. Since

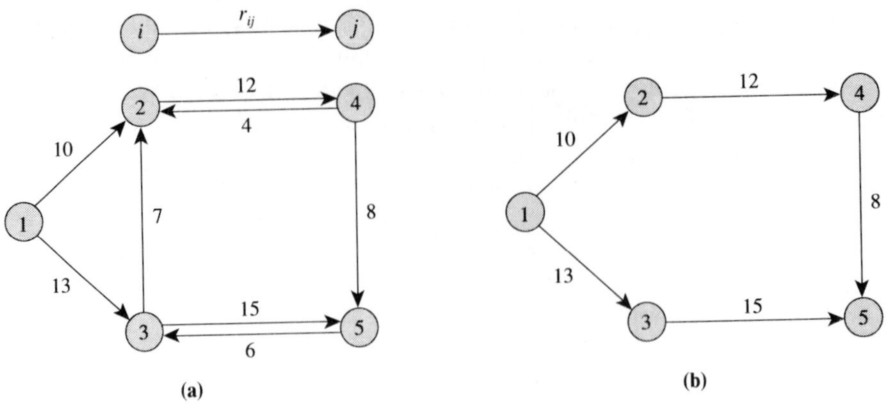

Figure 7.2 Illustrating the Δ-residual network: (a) residual network $G(x)$; (b) Δ-residual network $G(x, \Delta)$ for $\Delta = 8$.

```
algorithm capacity scaling;
begin
    x : = 0;
    Δ : = 2 ⌊log U⌋;
    while Δ ≥ 1 do
    begin
        while G(x, Δ) contains a path from node s to node t do
        begin
            identify a path P in G(x, Δ);
            δ : = min{r_ij : (i, j) ∈ P};
            augment δ units of flow along P and update G(x, Δ);
        end;
        Δ : = Δ/2;
    end;
end;
```

Figure 7.3 Capacity scaling algorithm.

$G(x', \Delta)$ contains no augmenting path from the source to the sink, $t \notin S$. Therefore, $[S, \bar{S}]$ forms an s–t cut. The definition of S implies that the residual capacity of every arc in $[S, \bar{S}]$ is strictly less than Δ, so the residual capacity of the cut $[S, \bar{S}]$ is at most $m\Delta$. Consequently, $v^* - v' \le m\Delta$ (from Property 6.2). In the next scaling phase, each augmentation carries at least $\Delta/2$ units of flow, so this scaling phase can perform at most $2m$ such augmentations. The labeling algorithm described in Section 6.5 requires $O(m)$ time to identify an augmenting path, and updating the Δ-residual network also requires $O(m)$ time. These arguments establish the following result.

Theorem 7.4. *The capacity scaling algorithm solves the maximum flow problem within $O(m \log U)$ augmentations and runs in $O(m^2 \log U)$ time.* ◆

It is possible to reduce the complexity of the capacity scaling algorithm even further—to $O(nm \log U)$—using ideas of the shortest augmenting path algorithm, described in the next section.

7.4 SHORTEST AUGMENTING PATH ALGORITHM

The shortest augmenting path algorithm always augments flow along a shortest path from the source to the sink in the residual network. A natural approach for implementing this approach would be to look for shortest paths by performing a breadth first search in the residual network. If the labeling algorithm maintains the set L of labeled nodes as a queue, then by examining the labeled nodes in a first-in, first-out order, it would obtain a shortest path in the residual network (see Exercise 3.30). Each of these iterations would require $O(m)$ steps in the worst case, and (by our subsequent observations) the resulting computation time would be $O(nm^2)$. Unfortunately, this computation time is excessive. We can improve it by exploiting the fact that the minimum distance from any node i to the sink node t is monotonically nondecreasing over all augmentations. By fully exploiting this property, we can reduce the average time per augmentation to $O(n)$.

The shortest augmenting path algorithm proceeds by augmenting flows along admissible paths. It constructs an admissible path incrementally by adding one arc at a time. The algorithm maintains a *partial admissible path* (i.e., a path from s to some node i consisting solely of admissible arcs) and iteratively performs *advance* or *retreat* operations from the last node (i.e., the tip) of the partial admissible path, which we refer to as the *current node*. If the current node i has (i.e., is incident to) an admissible arc (i, j), we perform an advance operation and add arc (i, j) to the partial admissible path; otherwise, we perform a retreat operation and backtrack one arc. We repeat these operations until the partial admissible path reaches the sink node at which time we perform an augmentation. We repeat this process until the flow is maximum. Before presenting a formal description of the algorithm, we illustrate it on the numerical example given in Figure 7.4(a).

We first compute the initial distance labels by performing the backward breadth-first search of the residual network starting at the sink node. The numbers next to the nodes in Figure 7.4(a) specify these values of the distance labels. In this example we adopt the convention of selecting the arc (i, j) with the smallest value of j whenever node i has several admissible arcs. We start at the source node with a null partial admissible path. The source node has several admissible arcs, so we perform an advance operation. This operation adds the arc (1, 2) to the partial admissible path. We store this path using predecessor indices, so we set pred(2) = 1. Now node 2 is the current node and the algorithm performs an advance operation at node 2. In doing so, it adds arc (2, 7) to the partial admissible path, which now becomes 1–2–7. We also set pred(7) = 2. In the next iteration, the algorithm adds arc (7, 12) to the partial admissible path obtaining 1–2–7–12, which is an admissible path to the sink node. We perform an augmentation of value $\min\{r_{12}, r_{27}, r_{7,12}\} = \min\{2, 1, 2\} = 1$, and thus saturate the arc (2, 7). Figure 7.4(b) specifies the residual network at this stage.

We again start at the source node with a null partial admissible path. The algorithm adds the arc (1, 2) and node 2 becomes the new current node. Now we find that node 2 has no admissible arc. To create new admissible arcs, we must increase the distance label of node 2. We thus increase $d(2)$ to the value $\min\{d(j) + 1 : (i, j) \in A(i) \text{ and } r_{ij} > 0\} = \min\{d(1) + 1\} = 4$. We refer to this operation as a *relabel* operation. We will later show that a relabel operation preserves the validity

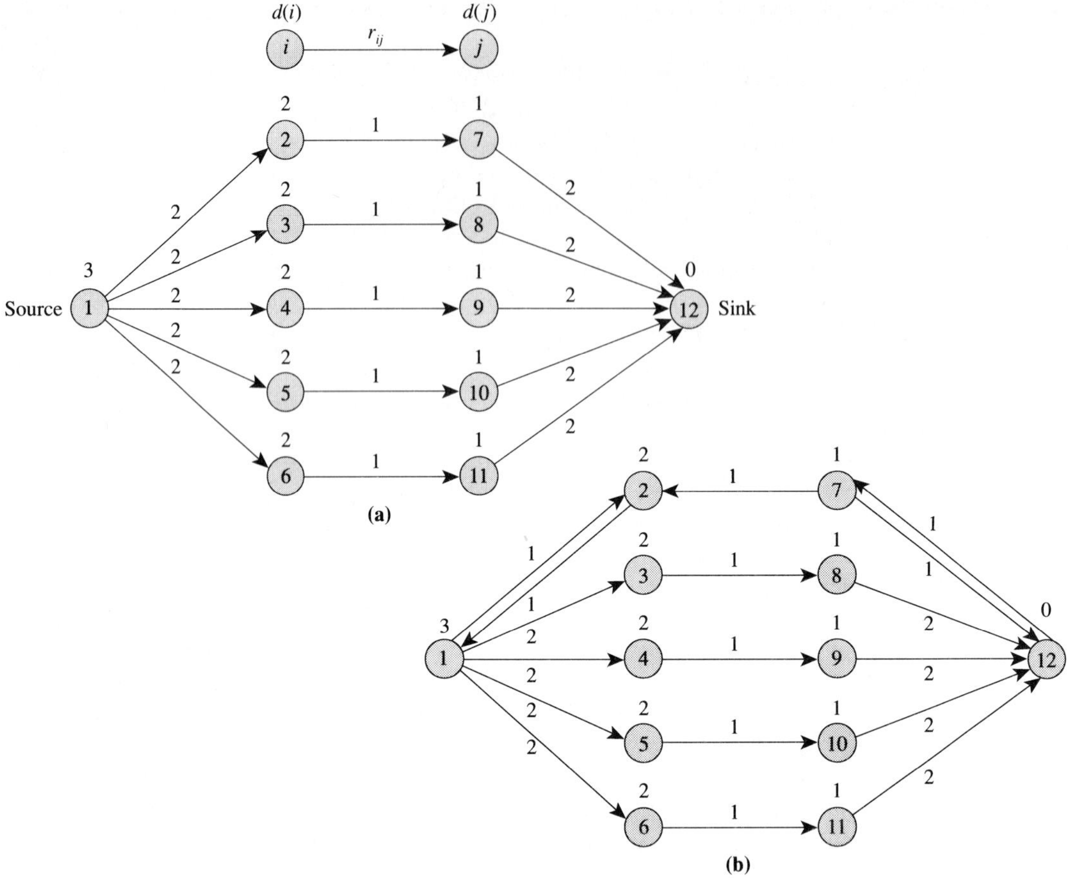

Figure 7.4 Illustrating the shortest augmenting path algorithm.

conditions imposed upon the distance labels. Observe that the increase in $d(2)$ causes arc $(1, 2)$ to become inadmissible. Thus we delete arc $(1, 2)$ from the partial admissible path which again becomes a null path. In the subsequent operations, the algorithm identifies the admissible paths 1–3–8–12, 1–4–9–12, 1–5–10–12, and 1–6–11–12 and augments unit flows on these paths. We encourage the reader to carry out the details of these operations. Figures 7.5 and 7.6 specify the details of the algorithm.

Correctness of the Algorithm

In our analysis of the shortest augmenting path algorithm we first show that it correctly solves the maximum flow problem.

Lemma 7.5. *The shortest augmenting path algorithm maintains valid distance labels at each step. Moreover, each relabel (or, retreat) operation strictly increases the distance label of a node.*

```
algorithm shortest augmenting path;
begin
    x : = 0;
    obtain the exact distance labels d(i);
    i : = s;
    while d(s) < n do
    begin
        if i has an admissible arc then
            begin
                advance(i);
                if i = t then augment and set i = s
            end
        else retreat(i)
    end;
end;
```

Figure 7.5 Shortest augmenting path algorithm.

```
procedure advance(i);
begin
    let (i, j) be an admissible arc in A(i);
    pred(j) : = i and i : = j;
end;
```

(a)

```
procedure retreat(i);
begin
    d(i) : = min{d(j) + 1 : (i, j) ∈ A(i) and r_ij > 0};
    if i ≠ s then i : = pred(i);
end;
```

(b)

```
procedure augment;
begin
    using the predecessor indices identify an augmenting
    path P from the source to the sink;
    δ : = min{r_ij : (i, j) ∈ P};
    augment δ units of flow along path P;
end;
```

(c)

Figure 7.6 Procedures of the shortest augmenting path algorithm.

Proof. We show that the algorithm maintains valid distance labels at every step by performing induction on the number of augment and relabel operations. (The advance operation does not affect the admissibility of any arc because it does not change any residual capacity or distance label.) Initially, the algorithm constructs valid distance labels. Assume, inductively, that the distance labels are valid prior to an operation (i.e., they satisfy the validity conditions). We need to check whether these conditions remain valid (a) after an augment operation, and (b) after a relabel operation.

(a) Although a flow augmentation on arc (i, j) might remove this arc from the residual network, this modification to the residual network does not affect the validity of the distance labels for this arc. An augmentation on arc (i, j) might, however, create an additional arc (j, i) with $r_{ji} > 0$ and therefore also create an additional inequality $d(j) \le d(i) + 1$ that the distance labels must satisfy. The distance labels satisfy this validity condition, though, since $d(i) = d(j) + 1$ by the admissibility property of the augmenting path.

(b) The relabel operation modifies $d(i)$; therefore, we must show that each incoming and outgoing arc at node i satisfies the validity conditions with respect to the new distance labels, say $d'(i)$. The algorithm performs a relabel operation at node i when it has no admissible arc; that is, no arc $(i, j) \in A(i)$ satisfies the conditions $d(i) = d(j) + 1$ and $r_{ij} > 0$. This observation, in light of the validity condition $d(i) \le d(j) + 1$, implies that $d(i) < d(j) + 1$ for all arcs $(i, j) \in A$ with a positive residual capacity. Therefore, $d(i) < \min\{d(j) + 1 : (i, j) \in A(i) \text{ and } r_{ij} > 0\} = d'(i)$, which is the new distance label after the relabel operation. We have thus shown that relabeling preserves the validity condition for all arcs emanating from node i, and that each relabel operation strictly increases the value of $d(i)$. Finally, note that every incoming arc (k, i) satisfies the inequality $d(k) \le d(i) + 1$ (by the induction hypothesis). Since $d(i) < d'(i)$, the relabel operation again preserves validity condition for arc (k, i). ◆

The shortest augmenting path algorithm terminates when $d(s) \ge n$, indicating that the network contains no augmenting path from the source to the sink (from Property 7.2). Consequently, the flow obtained at the end of the algorithm is a maximum flow. We have thus proved the following theorem.

Theorem 7.6. *The shortest augmenting path algorithm correctly computes a maximum flow.* ◆

Complexity of the Algorithm

We now show that the shortest augmenting path algorithm runs in $O(n^2m)$ time. We first describe a data structure used to select an admissible arc emanating from a given node. We call this data structure the *current-arc data structure*. Recall that we used this data structure in Section 3.4 in our discussion of search algorithms. We also use this data structure in almost all the maximum flow algorithms that we describe in subsequent sections. Therefore, we review this data structure before proceeding.

Recall that we maintain the arc list $A(i)$ which contains all the arcs emanating from node i. We can arrange the arcs in these lists arbitrarily, but the order, once decided, remains unchanged throughout the algorithm. Each node i has a *current arc*, which is an arc in $A(i)$ and is the next candidate for admissibility testing. Initially, the current arc of node i is the first arc in $A(i)$. Whenever the algorithm attempts to find an admissible arc emanating from node i, it tests whether the node's current arc is admissible. If not, it designates the next arc in the arc list as the current arc. The algorithm repeats this process until either it finds an admissible arc or reaches the end of the arc list.

Consider, for example, the arc list of node 1 in Figure 7.7. In this instance, $A(1) = \{(1, 2), (1, 3), (1, 4), (1, 5), (1, 6)\}$. Initially, the current arc of node 1 is arc (1, 2). Suppose that the algorithm attempts to find an admissible arc emanating from node 1. It checks whether the node's current arc, arc (1, 2), is admissible. Since it is not, the algorithm designates arc (1, 3) as the current arc of node 1. The arc (1, 3) is also inadmissible, so the current arc becomes arc (1, 4), which is admissible. From this point on, arc (1, 4) remains the current arc of node 1 until it becomes inadmissible because the algorithm has increased the value of $d(4)$ or decreased the value of the residual capacity of arc (1, 4) to zero.

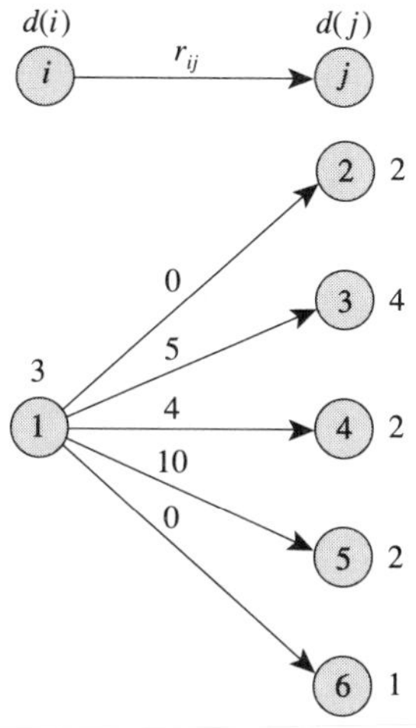

Figure 7.7 Selecting admissible arcs emanating from a node.

Let us consider the situation when the algorithm reaches the end of the arc list without finding any admissible arc. Can we say that $A(i)$ has no admissible arc? We can, because it is possible to show that if an arc (i, j) is inadmissible in previous iterations, it remains inadmissible until $d(i)$ increases (see Exercise 7.13). So if we reach the end of the arc list, we perform a relabel operation and again set the current arc of node i to be the first arc in $A(i)$. The relabel operation also examines each arc in $A(i)$ once to compute the new distance label, which is same as the time it spends in identifying admissible arcs at node i in one scan of the arc list. We have thus established the following result.

***Property** 7.7. If the algorithm relabels any node at most k times, the total time spent in finding admissible arcs and relabeling the nodes is* $O(k \sum_{i \in N} |A(i)|) = O(km)$.

We shall be using this result several times in this and the following chapters. We also use the following result in several places.

***Lemma** 7.8. If the algorithm relabels any node at most k times, the algorithm saturates arcs (i.e., reduces their residual capacity to zero) at most km/2 times.*

Proof. We show that between two consecutive saturations of an arc (i, j), both $d(i)$ and $d(j)$ must increase by at least 2 units. Since, by our hypothesis, the algorithm increases each distance label at most k times, this result would imply that the algorithm could saturate any arc at most $k/2$ times. Therefore, the total number of arc saturations would be $km/2$, which is the assertion of the lemma.

Suppose that an augmentation saturates an arc (i, j). Since the arc (i, j) is admissible,

$$d(i) = d(j) + 1. \tag{7.3}$$

Before the algorithm saturates this arc again, it must send flow back from node j to node i. At this time, the distance labels $d'(i)$ and $d'(j)$ satisfy the equality

$$d'(j) = d'(i) + 1. \tag{7.4}$$

In the next saturation of arc (i, j), we must have

$$d''(i) = d''(j) + 1. \tag{7.5}$$

Using (7.3) and (7.4) in (7.5), we see that

$$d''(i) = d''(j) + 1 \geq d'(j) + 1 = d'(i) + 2 \geq d(i) + 2.$$

The inequalities in this expression follow from Lemma 7.5. Similarly, it is possible to show that $d''(j) \geq d(j) + 2$. As a result, between two consecutive saturations of the arc (i, j), both $d(i)$ and $d(j)$ increase by at least 2 units, which is the conclusion of the lemma. ◆

Lemma 7.9.

(a) *In the shortest augmenting path algorithm each distance label increases at most n times. Consequently, the total number of relabel operations is at most n^2.*

(b) *The number of augment operations is at most $nm/2$.*

Proof. Each relabel operation at node i increases the value of $d(i)$ by at least 1 unit. After the algorithm has relabeled node i at most n times, $d(i) \geq n$. From this point on, the algorithm never again selects node i during an advance operation since for every node k in the partial admissible path, $d(k) < d(s) < n$. Thus the algorithm relabels a node at most n times and the total number of relabel operations is bounded by n^2. In view of Lemma 7.8, the preceding result implies that the algorithm saturates at most $nm/2$ arcs. Since each augmentation saturates at least one arc, we immediately obtain a bound of $nm/2$ on the number of augmentations. ◆

Theorem 7.10. *The shortest augmenting path algorithm runs in $O(n^2m)$ time.*

Proof. Using Lemmas 7.9 and 7.7 we find that the total effort spent in finding admissible arcs and in relabeling the nodes is $O(nm)$. Lemma 7.9 implies that the total number of augmentations is $O(nm)$. Since each augmentation requires $O(n)$ time, the total effort for the augmentation operations is $O(n^2m)$. Each retreat operation relabels a node, so the total number of retreat operations is $O(n^2)$. Each advance operation adds one arc to the partial admissible path, and each retreat operation deletes one arc from it. Since each partial admissible path has length at most n, the algorithm requires at most $O(n^2 + n^2m)$ advance operations. The first term comes from the number of retreat (relabel) operations, and the second term from the number of augmentations. The combination of these bounds establishes the theorem. ◆

A Practical Improvement

The shortest augmenting path algorithm terminates when $d(s) \geq n$. This termination criteria is satisfactory for the worst-case analysis but might not be efficient in practice. Empirical investigations have revealed that the algorithm spends too much time relabeling nodes and that a major portion of this effort is performed after the algorithm has established a maximum flow. This happens because the algorithm does not know that it has found a maximum flow. We next suggest a technique that is capable of detecting the presence of a minimum cut and so the existence of a maximum flow much before the label of node s satisfies the condition $d(s) \geq n$. Incorporating this technique in the shortest augmenting path algorithm improves its performance substantially in practice.

We illustrate this technique by applying it to the numerical example we used earlier to illustrate the shortest augmenting path algorithm. Figure 7.8 gives the residual network immediately after the last augmentation. Although the flow is now a maximum flow, since the source is not connected to the sink in the residual network, the termination criteria of $d(1) \geq 12$ is far from being satisfied. The reader can verify that after the last augmentation, the algorithm would increase the distance labels of nodes 6, 1, 2, 3, 4, 5, in the given order, each time by 2 units. Eventually, $d(1) \geq 12$ and the algorithm terminates. Observe that the node set S of the minimum cut $[S, \bar{S}]$ equals {6, 1, 2, 3, 4, 5}, and the algorithm increases the distance labels of all the nodes in S without performing any augmentation. The technique we describe essentially detects a situation like this one.

To implement this approach, we maintain an n-dimensional additional array,

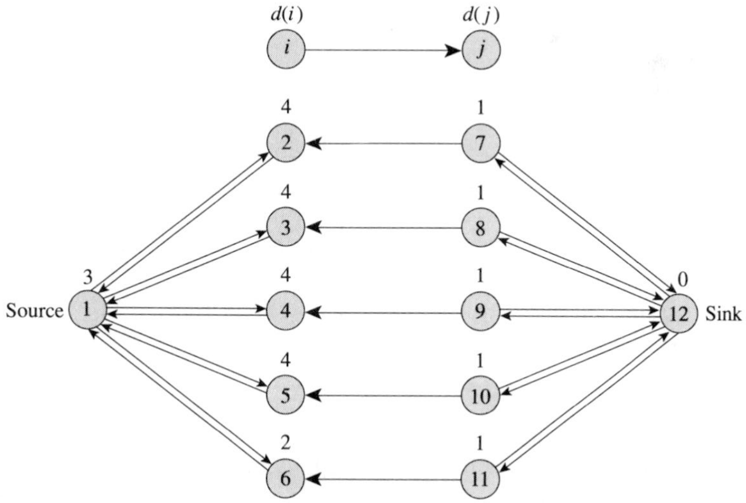

Figure 7.8 Bad example for the shortest augmenting path algorithm.

numb, whose indices vary from 0 to $(n - 1)$. The value numb(k) is the number of nodes whose distance label equals k. The algorithm initializes this array while computing the initial distance labels using a breadth first search. At this point, the positive entries in the array numb are consecutive [i.e., the entries numb(0), numb(1), . . . , numb(l) will be positive up to some index l and the remaining entries will all be zero]. For example, the numb array for the distance labels shown in Figure 7.8 is numb(0) = 1, numb(1) = 5, numb(2) = 1, numb(3) = 1, numb(4) = 4 and the remaining entries are zero. Subsequently, whenever the algorithm increases the distance label of a node from k_1 to k_2, it subtracts 1 from numb(k_1), adds 1 to numb(k_2) and checks whether numb$(k_1) = 0$. If numb(k_1) does equal zero, the algorithm terminates. As seen earlier, the shortest augmenting path algorithm augments unit flow along the paths 1–2–7–12, 1–3–8–12, 1–4–9–12, 1–5–10–12, and 1–6–11–12. At the end of these augmentations, we obtain the residual network shown in Figure 7.8. When we continue the shortest augmenting path algorithm from this point, it constructs the partial admissible path 1–6. Next it relabels node 6 and its distance label increases from 2 to 4. The algorithm finds that numb(2) = 0 and it terminates.

To see why this termination criterion works, let $S = \{i \in N: d(i) > k_1\}$ and $\bar{S} = \{i \in N: d(i) < k_1\}$. It is easy to verify that $s \in S$ and $t \in \bar{S}$. Now consider the s–t cut $[S, \bar{S}]$. The definitions of the sets S and $\bar{S}$ imply that $d(i) > d(j) + 1$ for all $(i, j) \in [S, \bar{S}]$. The validity condition (7.2) implies that $r_{ij} = 0$ for each arc $(i, j) \in [S, \bar{S}]$. Therefore, $[S, \bar{S}]$ is a minimum cut and the current flow is a maximum flow.

Application to Capacity Scaling Algorithm

In the preceding section we described an $O(m^2 \log U)$ time capacity scaling algorithm for the maximum flow problem. We can improve the running time of this algorithm to $O(nm \log U)$ by using the shortest augmenting path as a subroutine in the capacity scaling algorithm. Recall that the capacity scaling algorithm performs a number of Δ-scaling phases and in the Δ-scaling phase sends the maximum possible flow in the Δ-residual network $G(x, \Delta)$, using the labeling algorithm as a subroutine. In the improved implementation, we use the shortest augmenting path algorithm to send the maximum possible flow from node s to node t. We accomplish this by defining the distance labels with respect to the network $G(x, \Delta)$ and augmenting flow along the shortest augmenting path in $G(x, \Delta)$. Recall from the preceding section that a scaling phase contains $O(m)$ augmentations. The complexity analysis of the shortest augmenting path algorithm implies that if the algorithm is guaranteed to perform $O(m)$ augmentations, it would run in $O(nm)$ time because the time for augmentations reduces from $O(n^2m)$ to $O(nm)$ and all other operations, as before, require $O(nm)$ time. These observations immediately yield a bound of $O(nm \log U)$ on the running time of the capacity scaling algorithm.

Further Worst-Case Improvements

The idea of augmenting flows along shortest paths is intuitively appealing and easy to implement in practice. The resulting algorithms identify at most $O(nm)$ augmenting paths and this bound is tight [i.e., on particular examples these algorithms perform $\Omega(nm)$ augmentations]. The only way to improve the running time of the shortest augmenting path algorithm is to perform fewer computations per augmentation. The use of a sophisticated data structure, called *dynamic trees*, reduces the average time for each augmentation from $O(n)$ to $O(\log n)$. This implementation of the shortest augmenting path algorithm runs in $O(nm \log n)$ time, and obtaining further improvements appears quite difficult except in very dense networks. We describe the dynamic tree implementation of the shortest augmenting path algorithm in Section 8.5.

7.5 DISTANCE LABELS AND LAYERED NETWORKS

Like the shortest augmenting path algorithm, several other maximum flow algorithms send flow along shortest paths from the source to the sink. Dinic's algorithm is a popular algorithm in this class. This algorithm constructs shortest path networks, called *layered networks*, and establishes *blocking flows* (to be defined later) in these networks. In this section we point out the relationship between layered networks and distance labels. By developing a modification of the shortest augmenting path algorithm that reduces to Dinic's algorithm, we show how to use distance labels to simulate layered networks.

With respect to a given flow x, we define the *layered network* V as follows. We determine the exact distance labels d in $G(x)$. The layered network consists of those arcs (i, j) in $G(x)$ satisfying the condition $d(i) = d(j) + 1$. For example, consider the residual network $G(x)$ given in Figure 7.9(a). The number beside each node represents its exact distance label. Figure 7.9(b) shows the layered network of $G(x)$. Observe that by definition every path from the source to the sink in the layered network V is a shortest path in $G(x)$. Observe further that some arc in V might not be contained in any path from the source to the sink. For example, in Figure 7.9(b), arcs (5, 7) and (6, 7) do not lie on any path in V from the source to the sink. Since these arcs do not participate in any flow augmentation, we typically delete them from the layered network; doing so gives us Figure 7.9(c). In the resulting layered network, the nodes are partitioned into layers of nodes $V_0, V_1, V_2, \ldots, V_l$; layer k contains the nodes whose distance labels equal k. Furthermore, for every

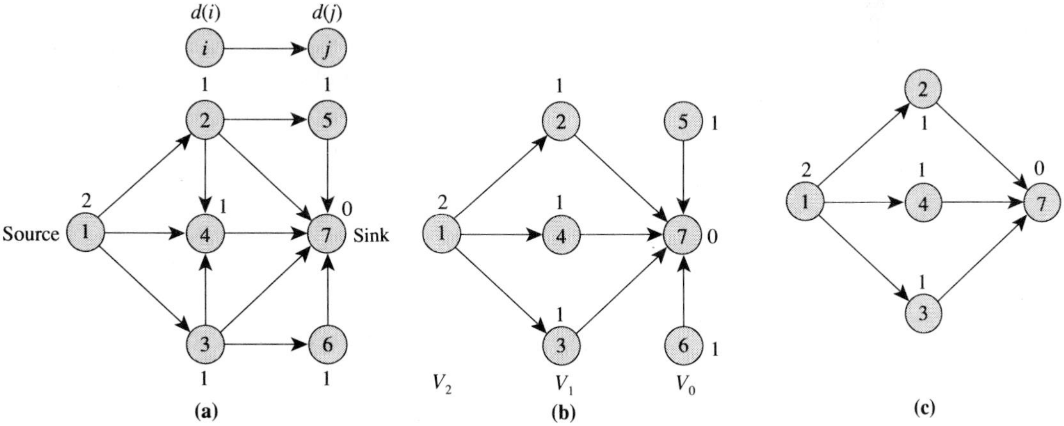

Figure 7.9 Forming layered networks: (a) residual network; (b) corresponding layered network; (c) layered network after deleting redundant arcs.

arc (i, j) in the layered network, $i \in V_k$ and $j \in V_{k-1}$ for some k. Let the source node have the distance label l.

Dinic's algorithm proceeds by augmenting flows along directed paths from the source to the sink in the layered network. The augmentation of flow along an arc (i, j) reduces the residual capacity of arc (i, j) and increases the residual capacity of the reversal arc (j, i); however, each arc of the layered network is admissible, and therefore Dinic's algorithm does not add reversal arcs to the layered network. Consequently, the length of every augmenting path is $d(s)$ and in an augmenting path every arc (i, j) has $i \in V_k$ and $j \in V_{k-1}$ for some k. The latter fact allows us to determine an augmenting path in the layered network, on average, in $O(n)$ time. (The argument used to establish the $O(n)$ time bound is the same as that used in our analysis of the shortest augmenting path algorithm.) Each augmentation saturates at least one arc in the layered network, and after at most m augmentations the layered network contains no augmenting path. We call the flow at this stage a *blocking flow*. We have shown that we can establish a blocking flow in a layered network in $O(nm)$ time.

When a blocking flow x has been established in a network, Dinic's algorithm recomputes the exact distance labels, forms a new layered network, and repeats these computations. The algorithm terminates when as it is forming the new layered networks, it finds that the source is not connected to the sink. It is possible to show that every time Dinic's algorithm forms a new layered network, the distance label of the source node strictly increases. Consequently, Dinic's algorithm forms at most n layered networks and runs in $O(n^2m)$ time.

We now show how to view Dinic's algorithm as a somewhat modified version of the shortest augmenting path algorithm. We make the following three modifications to the shortest augmenting path algorithm.

> *Modification 1*. In operation retreat(i), we do not change the distance label of node i, but subsequently term node i as *blocked*. A blocked node has no admissible path to the sink node.
>
> *Modification 2*. We define an arc (i, j) to be admissible if $d(i) = d(j) + 1$, $r_{ij} > 0$, and node j is not blocked.
>
> *Modification 3*. When the source node is blocked, by performing a backward breadth-first search we recompute the distance labels of all nodes exactly.

We term the computations within two successive recomputations of distance labels as occurring within a single *phase*. We note the following facts about the modified shortest augmenting path algorithm.

1. At the beginning of a phase, when the algorithm recomputes the distance labels $d(\cdot)$, the set of admissible arcs defines a layered network.
2. Each arc (i, j) in the admissible path satisfies $d(i) = d(j) + 1$; therefore, arc (i, j) joins two successive layers of the layered network. As a result, every admissible path is an augmenting path in the layered network.
3. Since we do not update distance labels within a phase, every admissible path has length equal to $d(s)$.

4. The algorithm performs at most m augmentations within a phase because each augmentation causes at least one arc to become inadmissible by reducing its residual capacity to zero, and the algorithm does not create new admissible arcs.
5. A phase ends when the network contains no admissible path from node s to node t. Hence, when the algorithm recomputes distance labels at the beginning of the next phase, $d(s)$ must increase (why?).

The preceding facts show that the modified shortest augmenting path algorithm essentially reduces to Dinic's algorithm. They also show that the distance labels are sufficiently powerful to simulate layered networks. Further, they are simpler to understand than layered networks, easier to manipulate, and lead to more efficient algorithms in practice. Distance labels are also attractive because they are generic solution approaches that find applications in several different algorithms; for example, the generic preflow-push algorithm described next uses distance labels, as does many of its variants described later.

7.6 GENERIC PREFLOW-PUSH ALGORITHM

We now study a class of algorithms, known as *preflow-push* algorithms, for solving the maximum flow problem. These algorithms are more general, more powerful, and more flexible than augmenting path algorithms. The best preflow-push algorithms currently outperform the best augmenting path algorithms in theory as well as in practice. In this section we study the generic preflow-push algorithm. In the following sections we describe special implementations of the generic approach with improved worst-case complexity.

The inherent drawback of the augmenting path algorithms is the computationally expensive operation of sending flow along a path, which requires $O(n)$ time in the worst case. Preflow-push algorithms do not suffer from this drawback and obtain dramatic improvements in the worst-case complexity. To understand this point better, consider the (artificially extreme) example shown in Figure 7.10. When applied to this problem, any augmenting path algorithm would discover 10 augmenting paths, each of length 10, and would augment 1 unit of flow along each of these paths. Observe, however, that although all of these paths share the same first eight arcs, each augmentation traverses all of these arcs. If we could have sent 10 units of flow from node 1 to node 9, and then sent 1 unit of flow along 10 different paths of length 2, we would have saved the repetitive computations in traversing the common set of arcs. This is the essential idea underlying the preflow-push algorithms.

Augmenting path algorithms send flow by augmenting along a path. This basic operation further decomposes into the more elementary operation of sending flow along individual arcs. Thus sending a flow of δ units along a path of k arcs decomposes into k basic operations of sending a flow of δ units along each of the arcs of the path. We shall refer to each of these basic operations as a *push*. The preflow-push algorithms push flows on individual arcs instead of augmenting paths.

Because the preflow-push algorithms push flows along the individual arcs, these algorithms do not satisfy the mass balance constraints (6.1b) at intermediate stages. In fact, these algorithms permit the flow entering a node to exceed the flow leaving

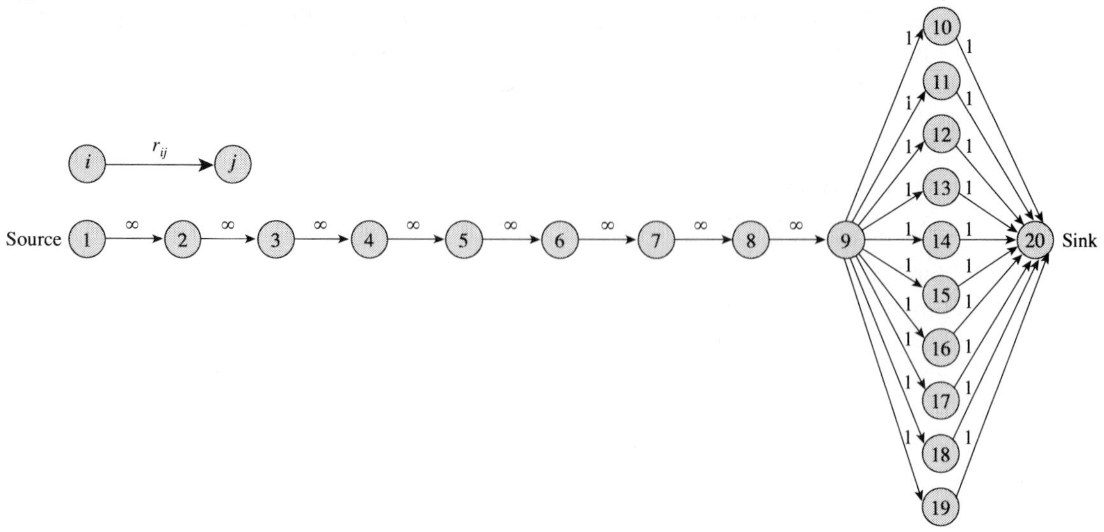

Figure 7.10 Drawback of the augmenting path algorithm.

the node. We refer to any such solution as a *preflow*. More formally, a *preflow* is a function $x: A \rightarrow \mathbf{R}$ that satisfies the flow bound constraint (6.1c) and the following relaxation of (6.1b).

$$\sum_{\{j:(j,i)\in A\}} x_{ji} - \sum_{\{j:(i,j)\in A\}} x_{ij} \geq 0 \qquad \text{for all } i \in N - \{s, t\}.$$

The preflow-push algorithms maintain a preflow at each intermediate stage. For a given preflow x, we define the *excess* of each node $i \in N$ as

$$e(i) = \sum_{\{j:(j,i)\in A\}} x_{ji} - \sum_{\{j:(i,j)\in A\}} x_{ij}.$$

In a preflow, $e(i) \geq 0$ for each $i \in N - \{s, t\}$. Moreover, because no arc emanates from node t in the preflow push algorithms, $e(t) \geq 0$ as well. Therefore node s is the only node with negative excess.

We refer to a node with a (strictly) positive excess as an *active* node and adopt the convention that the source and sink nodes are never active. The augmenting path algorithms always maintain feasibility of the solution and strive toward optimality. In contrast, preflow-push algorithms strive to achieve feasibility. In a preflow-push algorithm, the presence of active nodes indicates that the solution is infeasible. Consequently, the basic operation in this algorithm is to select an active node and try to remove its excess by pushing flow to its neighbors. But to which nodes should the flow be sent? Since ultimately we want to send flow to the sink, we push flow to the nodes that are *closer* to sink. As in the shortest augmenting path algorithm, we measure closeness with respect to the current distance labels, so sending flow closer to the sink is equivalent to pushing flow on admissible arcs. Thus we send flow only on admissible arcs. If the active node we are currently considering has no admissible arc, we increase its distance label so that we create at least one admissible arc. The algorithm terminates when the network contains no active node. The preflow-push algorithm uses the subroutines shown in Figure 7.11.

```
procedure preprocess;
begin
    x : = 0;
    compute the exact distance labels d(i);
    x_sj : = u_sj for each arc (s, j) ∈ A(s);
    d(s) : = n;
end;
```

(a)

```
procedure push/relabel(i);
begin
    if the network contains an admissible arc (i, j) then
        push δ : = min{e(i), r_ij} units of flow from node i to node j
    else replace d(i) by min{d(j) + 1 : (i, j) ∈ A(i) and r_ij > 0};
end;
```

(b)

Figure 7.11 Subroutines of the preflow-push algorithm.

A push of δ units from node i to node j decreases both $e(i)$ and r_{ij} by δ units and increases both $e(j)$ and r_{ji} by δ units. We say that a push of δ units of flow on an arc (i, j) is *saturating* if $\delta = r_{ij}$ and is *nonsaturating* otherwise. A nonsaturating push at node i reduces $e(i)$ to zero. We refer to the process of increasing the distance label of a node as a *relabel* operation. The purpose of the relabel operation is to create at least one admissible arc on which the algorithm can perform further pushes.

The generic version of the preflow-push algorithm (Figure 7.12) combines the subroutines just described.

```
algorithm preflow-push;
begin
    preprocess;
    while the network contains an active node do
    begin
        select an active node i;
        push/relabel(i);
    end;
end;
```

Figure 7.12 Generic preflow-push algorithm.

It might be instructive to visualize the generic preflow-push algorithm in terms of a physical network: Arcs represent flexible water pipes, nodes represent joints, and the distance function measures how far nodes are above the ground. In this network we wish to send water from the source to the sink. In addition, we visualize flow in an admissible arc as water flowing downhill. Initially, we move the source node upward, and water flows to its neighbors. In general, water flows downhill towards the sink; however, occasionally flow becomes trapped locally at a node that has no downhill neighbors. At this point we move the node upward, and again water flows downhill toward the sink. Eventually, no more flow can reach the sink. As we continue to move nodes upward, the remaining excess flow eventually flows back toward the source. The algorithm terminates when all the water flows either into the sink or flows back to the source.

The preprocessing operation accomplishes several important tasks. First, it gives each node adjacent to node s a positive excess, so that the algorithm can begin by selecting some node with a positive excess. Second, since the preprocessing operation saturates all the arcs incident to node s, none of these arcs is admissible and setting $d(s) = n$ will satisfy the validity condition (7.2). Third, since $d(s) = n$, Property 7.2 implies that the residual network contains no directed path from node s to node t. Since distance labels are nondecreasing, we also guarantee that in subsequent iterations the residual network will never contain a directed path from node s to node t, and so we will never need to push flow from node s again.

To illustrate the generic preflow-push algorithm, consider the example given in Figure 7.13(a). Figure 7.13(b) specifies the preflow determined by the preprocess operation.

Iteration 1. Suppose that the algorithm selects node 2 for the push/relabel operation. Arc (2, 4) is the only admissible arc and the algorithm performs a push of value $\delta = \min\{e(2), r_{24}\} = \min\{2, 1\} = 1$. This push is saturating. Figure 7.13(c) gives the residual network at this stage.

Iteration 2. Suppose that the algorithm again selects node 2. Since no admissible arc emanates from node 2, the algorithm performs a relabel operation and gives node 2 a new distance label $d(2) = \min\{d(3) + 1, d(1) + 1\} = \min\{2, 5\} = 2$. The new residual network is the same as the one shown in Figure 7.13(c) except that $d(2) = 2$ instead of 1.

Iteration 3. Suppose that this time the algorithm selects node 3. Arc (3, 4) is the only admissible arc emanating from node 3, the algorithm performs a push of value $\delta = \min\{e(3), r_{34}\} = \min\{4, 5\} = 4$. This push is nonsaturating. Figure 7.13(d) gives the residual network at the end of this iteration.

Iteration 4. The algorithm selects node 2 and performs a nonsaturating push of value $\delta = \min\{1, 3\} = 1$, obtaining the residual network given in Figure 7.13(e).

Iteration 5. The algorithm selects node 3 and performs a saturating push of value $\delta = \min\{1, 1\} = 1$ on arc (3, 4), obtaining the residual network given in Figure 7.13(f).

Now the network contains no active node and the algorithm terminates. The maximum flow value in the network is $e(4) = 6$.

Assuming that the generic preflow-push algorithm terminates, we can easily show that it finds a maximum flow. The algorithm terminates when the excess resides at the source or at the sink, implying that the current preflow is a flow. Since $d(s) = n$, the residual network contains no path from the source to the sink. This condition is the termination criterion of the augmenting path algorithm, and the excess residing at the sink is the maximum flow value.

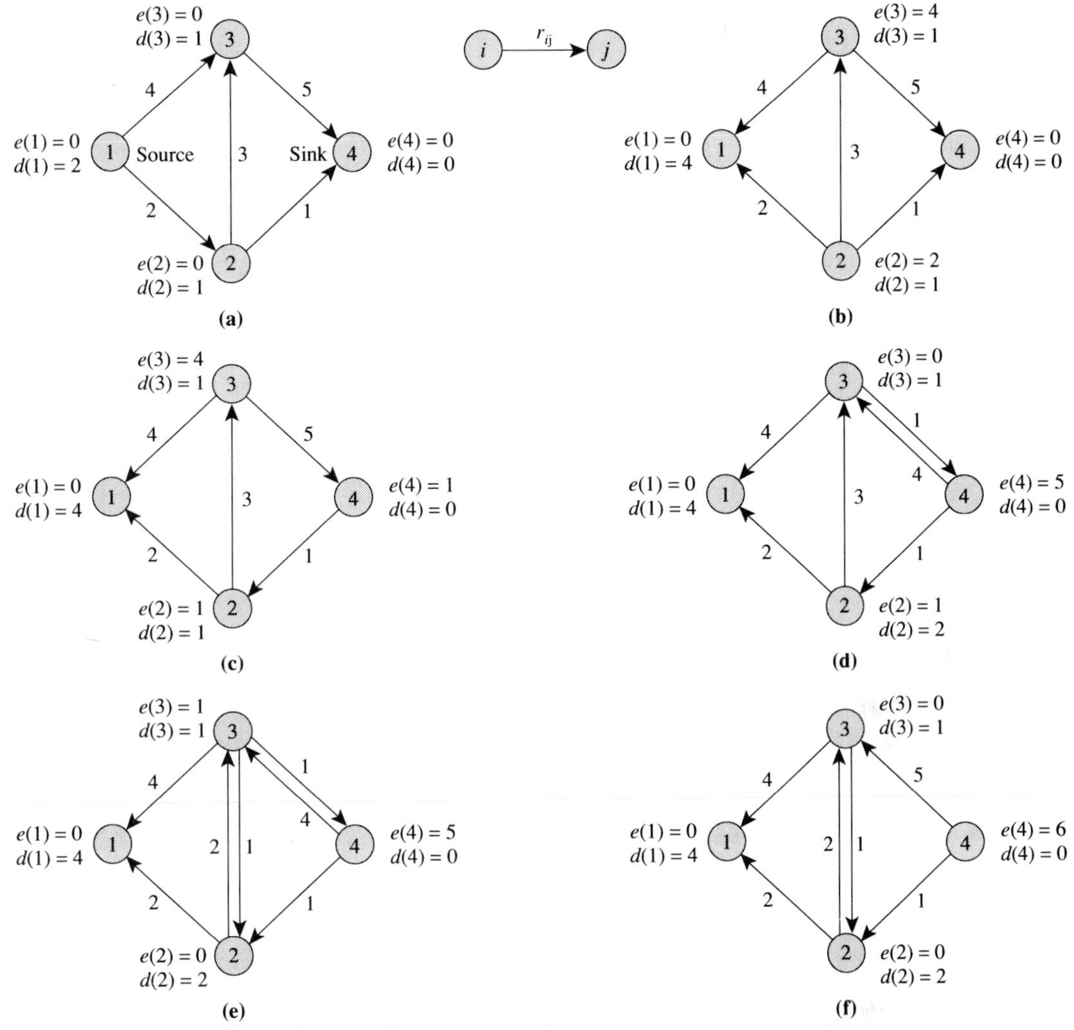

Figure 7.13 Illustrating the generic preflow-push algorithm.

Complexity of the Algorithm

To analyze the complexity of the algorithm, we begin by establishing one important result: distance labels are always valid and do not increase "too many" times. The first of these conclusions follows from Lemma 7.5, because, as in the shortest augmenting path algorithm, the preflow-push algorithm pushes flow only on admissible arcs and relabels a node only when no admissible arc emanates from it. The second conclusion follows from the following lemma.

Lemma 7.11. *At any stage of the preflow-push algorithm, each node i with positive excess is connected to node s by a directed path from node i to node s in the residual network.*

Proof. Notice that for a preflow x, $e(s) \leq 0$ and $e(i) \geq 0$ for all $i \in N - \{s\}$. By the flow decomposition theorem (see Theorem 3.5), we can decompose any preflow x with respect to the original network G into nonnegative flows along (1) paths from node s to node t, (2) paths from node s to active nodes, and (3) flows around directed cycles. Let i be an active node relative to the preflow x in G. The flow decomposition of x must contain a path P from node s to node i, since the paths from node s to node t and the flows around cycles do not contribute to the excess at node i. The residual network contains the reversal of P (P with the orientation of each arc reversed), so a directed path from node i to node s. ◆

This lemma implies that during a relabel operation, the algorithm does not minimize over an empty set.

Lemma 7.12. *For each node $i \in N$, $d(i) < 2n$.*

Proof. The last time the algorithm relabeled node i, the node had a positive excess, so the residual network contained a path P of length at most $n - 2$ from node i to node s. The fact that $d(s) = n$ and that $d(k) \leq d(l) + 1$ for every arc (k, l) in the path P implies that $d(i) \leq d(s) + |P| < 2n$. ◆

Since each time the algorithm relabels node i, $d(i)$ increases by at least 1 unit, we have established the following result.

Lemma 7.13. *Each distance label increases at most $2n$ times. Consequently, the total number of relabel operations is at most $2n^2$.* ◆

Lemma 7.14. *The algorithm performs at most nm saturating pushes.*

Proof. This result follows directly from Lemmas 7.12 and 7.8. ◆

In view of Lemma 7.7, Lemma 7.13 implies that the total time needed to identify admissible arcs and to perform relabel operations is $O(nm)$. We next count the number of nonsaturating pushes performed by the algorithm.

Lemma 7.15. *The generic preflow-push algorithm performs $O(n^2m)$ nonsaturating pushes.*

Proof. We prove the lemma using an argument based on potential functions (see Section 3.2). Let I denote the set of active nodes. Consider the potential function $\Phi = \sum_{i \in I} d(i)$. Since $|I| < n$, and $d(i) < 2n$ for all $i \in I$, the initial value of Φ (after the preprocess operation) is at most $2n^2$. At the termination of the algorithm, Φ is zero. During the push/relabel(i) operation, one of the following two cases must apply:

Case 1. The algorithm is unable to find an admissible arc along which it can push flow. In this case the distance label of node i increases by $\epsilon \geq 1$ units. This operation increases Φ by at most ϵ units. Since the total increase in $d(i)$ for each node i throughout the execution of the algorithm is bounded by $2n$, the total increase in Φ due to increases in distance labels is bounded by $2n^2$.

Case 2. The algorithm is able to identify an arc on which it can push flow, so it performs a saturating push or a nonsaturating push. A saturating push on arc (i, j) might create a new excess at node j, thereby increasing the number of active nodes by 1, and increasing Φ by $d(j)$, which could be as much as $2n$ per saturating push, and so $2n^2m$ over all saturating pushes. Next note that a nonsaturating push on arc (i, j) does not increase $|I|$. The nonsaturating push will decrease Φ by $d(i)$ since i becomes inactive, but it simultaneously increases Φ by $d(j) = d(i) - 1$ if the push causes node j to become active, the total decrease in Φ being of value 1. If node j was active before the push, Φ decreases by an amount $d(i)$. Consequently, net decrease in Φ is at least 1 unit per nonsaturating push.

We summarize these facts. The initial value of Φ is at most $2n^2$ and the maximum possible increase in Φ is $2n^2 + 2n^2m$. Each nonsaturating push decreases Φ by at least 1 unit and Φ always remains nonnegative. Consequently, the algorithm can perform at most $2n^2 + 2n^2 + 2n^2m = O(n^2m)$ nonsaturating pushes, proving the lemma. ◆

Finally, we indicate how the algorithm keeps track of active nodes for the push/relabel operations. The algorithm maintains a set LIST of active nodes. It adds to LIST those nodes that become active following a push and are not already in LIST, and deletes from LIST nodes that become inactive following a nonsaturating push. Several data structures (e.g., doubly linked lists) are available for storing LIST so that the algorithm can add, delete, or select elements from it in $O(1)$ time. Consequently, it is easy to implement the preflow-push algorithm in $O(n^2m)$ time. We have thus established the following theorem.

Theorem 7.16. *The generic preflow-push algorithm runs in $O(n^2m)$ time.* ◆

Several modifications to the generic preflow-push algorithm might improve its empirical performance. We define a maximum preflow as a preflow with the maximum possible flow into the sink. As stated, the generic preflow-push algorithm performs push/relabel operations at active nodes until all the excess reaches the sink node or returns to the source node. Typically, the algorithm establishes a maximum preflow long before it establishes a maximum flow; the subsequent push/relabel operations increase the distance labels of the active nodes until they are sufficiently higher than n so they can push their excesses back to the source node (whose distance label is n). One possible modification in the preflow-push algorithm is to maintain a set N' of nodes that satisfy the property that the residual network contains no path from a node in N' to the sink node t. Initially, $N' = \{s\}$ and, subsequently, whenever the distance label of a node is greater than or equal to n, we add it to N'. Further, we do not perform push/relabel operations for nodes in N' and terminate the algorithm when all nodes in $N - N'$ are inactive. At termination, the current preflow x is also an optimal preflow. At this point we convert the maximum preflow x into a maximum flow using any of the methods described in Exercise 7.11. Empirical tests have found that this two-phase approach often substantially reduces the running times of preflow push algorithms.

One sufficient condition for adding a node j to N' is $d(j) \geq n$. Unfortunately, this simple approach is not very effective and does not substantially reduce the running time of the algorithm. Another approach is to occasionally perform a reverse breadth-first search of the residual network to obtain exact distance labels and add all those nodes to N' that do not have any directed path to the sink. Performing this search occasionally, that is, after αn relabel operations for some constant α, does not effect the worst-case complexity of the preflow-push algorithm (why?) but improves the empirical behavior of the algorithm substantially.

A third approach is to let numb(k) denote the number of nodes whose distance label is k. As discussed in Section 7.4, we can update the array numb($\cdot$) in $O(1)$ steps per relabel operation. Moreover, whenever numb(k') $= 0$ for some k', any node j with $d(j) > k'$ is disconnected from the set of nodes i with $d(i) < k'$ in the residual network. At this point, we can increase the distance labels of each of these nodes to n and the distance labels will still be valid (why?). Equivalently, we can add any node j with $d(j) > k'$ to the set N'. The array numb($\cdot$) is easy to implement, and its use is quite effective in practice.

Specific Implementations of Generic Preflow-Push Algorithm

The running time of the generic preflow-push algorithm is comparable to the bound of the shortest augmenting path algorithm. However, the preflow-push algorithm has several nice features: in particular, its flexibility and its potential for further improvements. By specifying different rules for selecting active nodes for the push/relabel operations, we can derive many different algorithms, each with different worst-case complexity than the generic version of the algorithm. The bottleneck operation in the generic preflow-push algorithm is the number of nonsaturating pushes and many specific rules for examining active nodes can produce substantial reductions in the number of nonsaturating pushes. We consider the following three implementations.

1. *FIFO preflow-push algorithm.* This algorithm examines the active nodes in the first-in, first-out (FIFO) order. We shall show that this algorithm runs in $O(n^3)$ time.
2. *Highest-label preflow-push algorithm.* This algorithm always pushes from an active node with the highest value of the distance label. We shall show that this algorithm runs in $O(n^2 m^{1/2})$ time. Observe that this time is better than $O(n^3)$ for all problem densities.
3. *Excess scaling algorithm.* This algorithm pushes flow from a node with sufficiently large excess to a node with sufficiently small excess. We shall show that the excess scaling algorithm runs in $O(nm + n^2 \log U)$ time. For problems that satisfy the similarity assumption (see Section 3.2), this time bound is better than that of the two preceding algorithms.

We might note that the time bounds for all these preflow-push algorithms are tight (except the excess scaling algorithm); that is, for some classes of networks the generic preflow-push algorithm, the FIFO algorithm, and the highest-label preflow-push algorithms do perform as many computations as indicated by their worst-case time bounds. These examples show that we cannot improve the time bounds of these algorithms by a more clever analysis.

7.7 *FIFO PREFLOW-PUSH ALGORITHM*

Before we describe the FIFO implementation of the preflow-push algorithm, we define the concept of a *node examination*. In an iteration, the generic preflow-push algorithm selects a node, say node i, and performs a saturating push or a nonsaturating push, or relabels the node. If the algorithm performs a saturating push, then node i might still be active, but it is not mandatory for the algorithm to select this node again in the next iteration. The algorithm might select another node for the next push/relabel operation. However, it is easy to incorporate the rule that whenever the algorithm selects an active node, it keeps pushing flow from that node until either the node's excess becomes zero or the algorithm relabels the node. Consequently, the algorithm might perform several saturating pushes followed either by a nonsaturating push or a relabel operation. We refer to this sequence of operations as a node examination. We shall henceforth assume that every preflow-push algorithm adopts this rule for selecting nodes for the push/relabel operation.

The FIFO preflow-push algorithm examines active nodes in the FIFO order. The algorithm maintains the set LIST as a queue. It selects a node i from the front of LIST, performs pushes from this node, and adds newly active nodes to the rear of LIST. The algorithm examines node i until either it becomes inactive or it is relabeled. In the latter case, we add node i to the rear of the queue. The algorithm terminates when the queue of active nodes is empty.

We illustrate the FIFO preflow-push algorithm using the example shown in Figure 7.14(a). The preprocess operation creates an excess of 10 units at each of

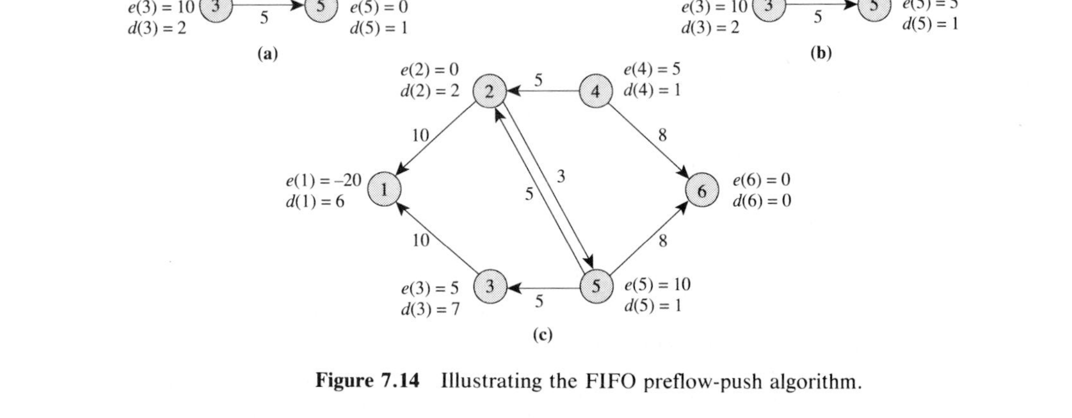

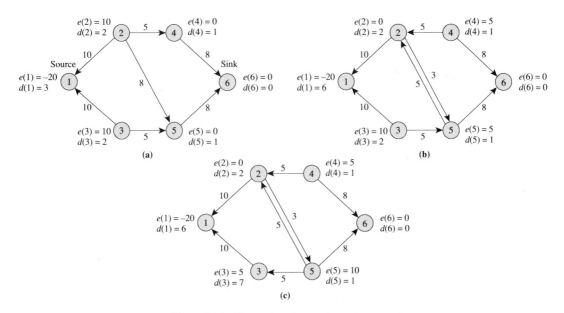

Figure 7.14 Illustrating the FIFO preflow-push algorithm.

the nodes 2 and 3. Suppose that the queue of active nodes at this stage is LIST = {2, 3}. The algorithm removes node 2 from the queue and examines it. Suppose that it performs a saturating push of 5 units on arc (2, 4) and a nonsaturating push of 5 units on arc (2, 5) [see Figure 7.14(b)]. As a result of these pushes, nodes 4 and 5 become active and we add these nodes to the queue in this order, obtaining LIST = {3, 4, 5}. The algorithm next removes node 3 from the queue. While examining node 3, the algorithm performs a saturating push of 5 units on arc (3, 5), followed by a relabel operation of node 3 [see Figure 7.14(c)]. The algorithm adds node 3 to the queue, obtaining LIST = {4, 5, 3}. We encourage the reader to complete the solution of this example.

To analyze the worst-case complexity of the FIFO preflow-push algorithm, we partition the total number of node examinations into different phases. The first phase consists of node examinations for those nodes that become active during the preprocess operation. The second phase consists of the node examinations of all the nodes that are in the queue after the algorithm has examined the nodes in the first phase. Similarly, the third phase consists of the node examinations of all the nodes that are in the queue after the algorithm has examined the nodes in the second phase, and so on. For example, in the preceding illustration, the first phase consists of the node examinations of the set {2, 3}, and the second phase consists of the node examinations of the set {4, 5, 3}. Observe that the algorithm examines any node at most once during a phase.

We will now show that the algorithm performs at most $2n^2 + n$ phases. Each phase examines any node at most once and each node examination performs at most one nonsaturating push. Therefore, a bound of $2n^2 + n$ on the total number of phases would imply a bound of $O(n^3)$ on the number of nonsaturating pushes. This result would also imply that the FIFO preflow-push algorithm runs in $O(n^3)$ time because the bottleneck operation in the generic preflow-push algorithm is the number of nonsaturating pushes.

To bound the number of phases in the algorithm, we consider the total change in the potential function $\Phi = \max\{d(i): i \text{ is active}\}$ over an entire phase. By the "total change" we mean the difference between the initial and final values of the potential function during a phase. We consider two cases.

Case 1. The algorithm performs at least one relabel operation during a phase. Then Φ might increase by as much as the maximum increase in any distance label. Lemma 7.13 implies that the total increase in Φ over all the phases is at most $2n^2$.

Case 2. The algorithm performs no relabel operation during a phase. In this case the excess of every node that was active at the beginning of the phase moves to nodes with smaller distance labels. Consequently, Φ decreases by at least 1 unit.

Combining Cases 1 and 2, we find that the total number of phases is at most $2n^2 + n$; the second term corresponds to the initial value of Φ, which could be at most n. We have thus proved the following theorem.

Theorem 7.17. *The FIFO preflow-push algorithm runs in $O(n^3)$ time.* ◆

7.8 HIGHEST-LABEL PREFLOW-PUSH ALGORITHM

The highest-label preflow-push algorithm always pushes flow from an active node with the highest distance label. It is easy to develop an $O(n^3)$ bound on the number of nonsaturating pushes for this algorithm. Let $h^* = \max\{d(i): i \text{ is active}\}$. The algorithm first examines nodes with distance labels equal to h^* and pushes flow to nodes with distance labels equal to $h^* - 1$, and these nodes, in turn, push flow to nodes with distance labels equal to $h^* - 2$, and so on, until either the algorithm relabels a node or it has exhausted all the active nodes. When it has relabeled a node, the algorithm repeats the same process. Note that if the algorithm does not relabel any node during n consecutive node examinations, all the excess reaches the sink (or the source) and the algorithm terminates. Since the algorithm performs at most $2n^2$ relabel operations (by Lemma 7.13), we immediately obtain a bound of $O(n^3)$ on the number of node examinations. Since each node examination entails at most one nonsaturating push, the highest-label preflow-push algorithm performs $O(n^3)$ nonsaturating pushes. (In Exercise 7.20 we consider a potential function argument that gives the same bound on the number of nonsaturating pushes.)

The preceding discussion is missing one important detail: How do we select a node with the highest distance label without expending too much effort? We use the following data structure. For each $k = 1, 2, \ldots, 2n - 1$, we maintain the list

$$\text{LIST}(k) = \{i : i \text{ is active and } d(i) = k\},$$

in the form of either linked stacks or linked queues (see Appendix A). We define a variable *level* that is an upper bound on the highest value of k for which LIST(k) is nonempty. To determine a node with the highest distance label, we examine the lists LIST(level), LIST(level-1), . . . , until we find a nonempty list, say LIST(p). We set level equal to p and select any node in LIST(p). Moreover, if the distance label of a node increases while the algorithm is examining it, we set level equal to the new distance label of the node. Observe that the total increase in level is at most $2n^2$ (from Lemma 7.13), so the total decrease is at most $2n^2 + n$. Consequently, scanning the lists LIST(level), LIST(level-1), . . . , in order to find the first nonempty list is not a bottleneck operation.

The highest-label preflow-push algorithm is currently the most efficient method for solving the maximum flow problem in practice because it performs the least number of nonsaturating pushes. To illustrate intuitively why the algorithm performs so well in practice, we consider the maximum flow problem given in Figure 7.15(a). The preprocess operation creates an excess of 1 unit at each node $2, 3, \ldots, n - 1$ [see Figure 7.15(b)]. The highest-label preflow-push algorithm examines nodes $2, 3, \ldots, n - 1$, in this order and pushes all the excess to the sink node. In contrast, the FIFO preflow-push algorithm might perform many more pushes. Suppose that at the end of the preprocess operation, the queue of active nodes is LIST $= \{n - 1, n - 2, \ldots, 3, 2\}$. Then the algorithm would examine each of these nodes in the first phase and would obtain the solution depicted in Figure 7.15(c). At this point, LIST $= \{n - 1, n - 2, \ldots, 4, 3\}$. It is easy to show that overall the algorithm would perform $n - 2$ phases and use $(n - 2) + (n - 3) + \ldots + 1 = \Omega(n^2)$ nonsaturating pushes.

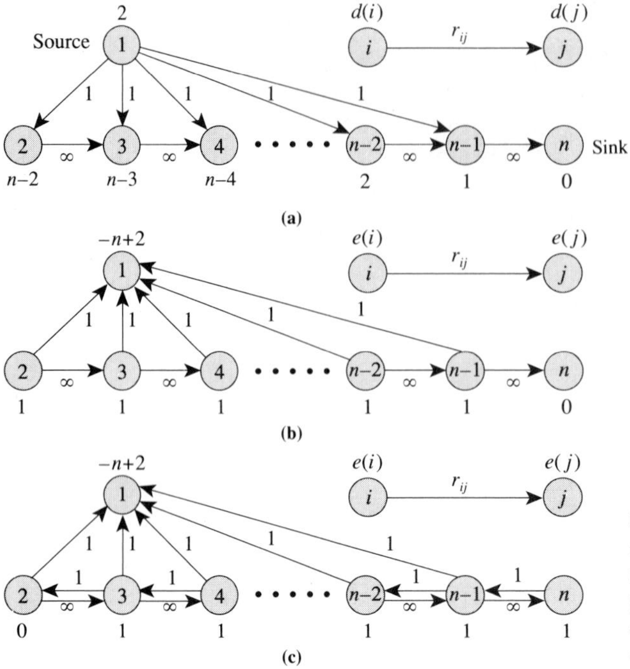

Figure 7.15 Bad example for the FIFO preflow-push algorithm: (a) initial residual network; (b) network after the preprocess operation; (c) network after one phase of the FIFO preflow-push algorithm.

Although the preceding example is rather extreme, it does illustrate the advantage in pushing flows from active nodes with the highest distance label. In our example the FIFO algorithm selects an excess and pushes it all the way to the sink. Then it selects another excess and pushes it to the sink, and repeats this process until no node contains any more excess. On the other hand, the highest-label preflow-push algorithm starts at the highest level and pushes all the excess at this level to the next lower level and repeats this process. As the algorithm examines nodes with lower and lower distance labels, it accumulates the excesses and pushes this accumulated excess toward the sink. Consequently, the highest-label preflow-push algorithm avoids repetitive pushes on arcs carrying a small amount of flow.

This nice feature of the highest-label preflow-push algorithm also translates into a tighter bound on the number of nonsaturating pushes. The bound of $O(n^3)$ on the number of nonsaturating pushes performed by the algorithm is rather loose and can be improved by a more clever analysis. We now show that the algorithm in fact performs $O(n^2m^{1/2})$ nonsaturating pushes. The proof of this theorem is somewhat complex and the reader can skip it without any loss of continuity.

At every state of the preflow-push algorithm, each node other than the sink has at most one current arc which, by definition, must be admissible. We denote this collection of current arcs by the set F. The set F has at most $n - 1$ arcs, has at most one outgoing arc per node, and does not contain any cycle (why?). These results imply that F defines a *forest*, which we subsequently refer to as the *current forest*. Figure 7.16 gives an example of a current forest. Notice that each tree in the forest is a rooted tree, the root being a node with no outgoing arc.

Before continuing, let us introduce some additional notation. For any node $i \in N$, we let $D(i)$ denote the set of descendants of that node in F (we refer the

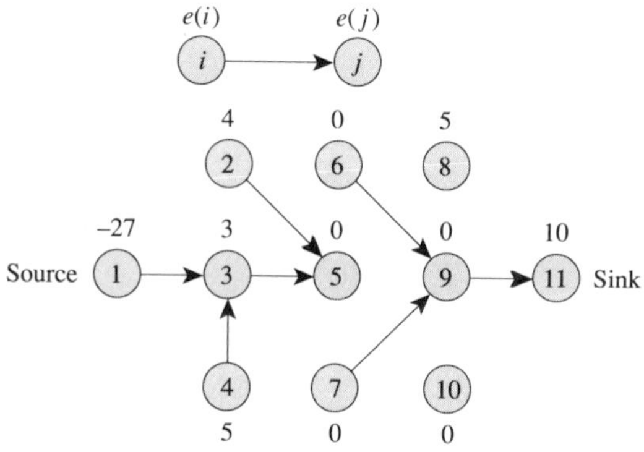

Figure 7.16 Example of a current-forest.

reader to Section 2.2 for the definition of descendants in a rooted tree). For example, in Figure 7.16, $D(1) = \{1\}$, $D(2) = \{2\}$, $D(3) = \{1, 3, 4\}$, $D(4) = \{4\}$, and $D(5) = \{1, 2, 3, 4, 5\}$. Notice that distance label of the descendants of any node i will be higher than $d(i)$. We refer to an active node with no active descendants (other than itself) as a *maximal active node*. In Figure 7.16 the nodes 2, 4, and 8 are the only maximal active nodes. Let H denote the set of maximal active nodes. Notice that two maximal active nodes have distinct descendants. Also notice that the highest-label preflow-push algorithm always pushes flow from a maximal active node.

We obtain the time bound of $O(n^2m^{1/2})$ for the highest-label preflow-push algorithm using a potential function argument. The argument relies on a parameter K, whose optimal value we select later. Our potential function is $\Phi = \sum_{i \in H} \Phi(i)$, with $\Phi(i)$ defined as $\Phi(i) = \max\{0, K + 1 - |D(i)|\}$. Observe that for any node i, $\Phi(i)$ is at most K [because $|D(i)| \geq 1$]. Also observe that Φ changes whenever the set H of maximal active nodes changes or $|D(i)|$ changes for a maximal active node i.

We now study the effect of various operations performed by the preflow-push algorithm on the potential function Φ. As the algorithm proceeds, it changes the set of current arcs, performs saturating and nonsaturating pushes, and relabels nodes. All these operations have an effect on the value of Φ. By observing the consequence of all these operations on Φ, we will obtain a bound on the number of nonsaturating pushes.

First, consider a nonsaturating push on an arc (i, j) emanating from a maximal active node i. Notice that a nonsaturating push takes place on a current arc and does not change the current forest; it simply moves the excess from node i to node j [see Figure 7.17(a) for a nonsaturating push on the arc (3, 4)]. As a result of the push, node i becomes inactive and node j might become a new maximal active node. Since $|D(j)| > |D(i)|$, this push decreases $\Phi(i) + \Phi(j)$ by at least 1 unit if $|D(i)| \leq K$ and does not change $\Phi(i) + \Phi(j)$ otherwise.

Now consider a saturating push on the arc (i, j) emanating from a maximal active node i. As a result of the push, arc (i, j) becomes inadmissible and drops out of the current forest [see Figure 7.17(b) for a saturating push on the arc (1, 3)]. Node i remains a maximal active node and node j might also become a maximal active node. Consequently, this operation might increase Φ by upto K units.

Next consider the relabeling of a maximal active node i. We relabel a node when it has no admissible arc; therefore, no current arc emanates from this node.

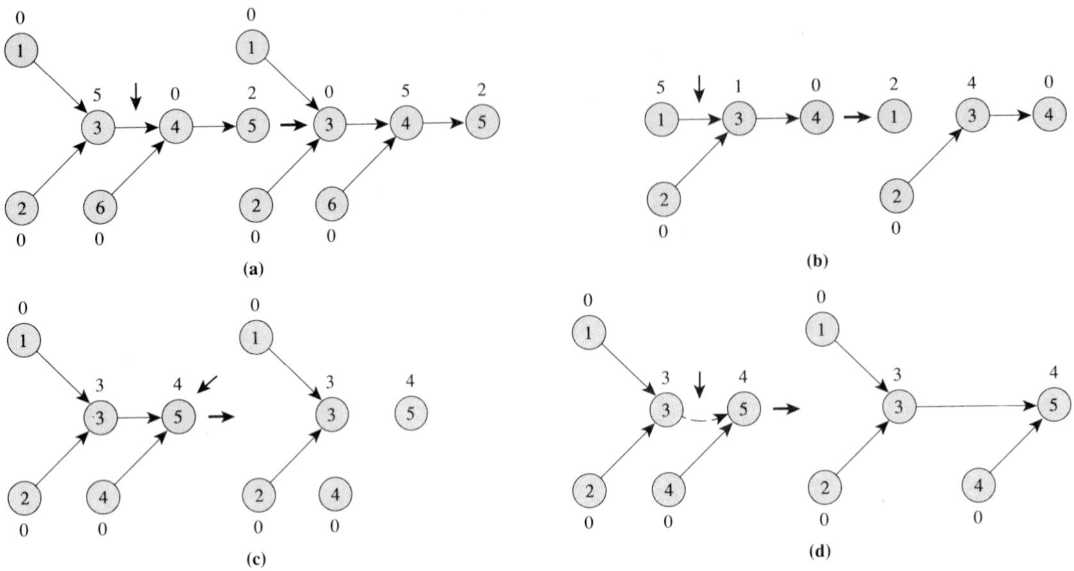

Figure 7.17 (a) Nonsaturating push on arc (d, 4); (b) saturating push on arc (1, 3); (c) relabel of node 5; (d) addition of the arc (3, 5) to the forest.

As a consequence, node i must be a root node in the current forest. Moreover, since node i is a maximal active node, none of its proper descendants can be active. After the algorithm has relabeled node i, all incoming arcs at node i become inadmissible; therefore, all the current arcs entering node i will no longer belong to the current forest [see Figure 7.17(c)]. Clearly, this change cannot create any new maximal active nodes. The relabel operation, however, decreases the number of descendants of node i to one. Consequently, $\Phi(i)$ can increase by at most K.

Finally, consider the introduction of new current arcs in the current forest. The addition of new arcs to the forest does not create any new maximal active nodes. It might, in fact, remove some maximal active nodes and increase the number of descendants of some nodes [see Figure 7.17(d)]. In both cases the potential Φ does not increase. We summarize the preceding discussion in the form of the following property.

Property 7.18

(a) *A nonsaturating push from a maximal active node i does not increase* Φ; *it decreases* Φ *by at least 1 unit if* $|D(i)| \leq K$.

(b) *A saturating push from a maximal active node i can increase* Φ *by at most K units.*

(c) *The relabeling of a maximal active node i can increase* Φ *by at most K units.*

(d) *Introducing current arcs does not increase* Φ.

For the purpose of worst-case analysis, we define the concept of phases. A *phase* consists of the sequence of pushes between two consecutive relabel operations. Lemma 7.13 implies that the algorithm contains $O(n^2)$ phases. We call a phase *cheap* if it performs at most $2n/K$ nonsaturating pushes, and *expensive* otherwise.

Clearly, the number of nonsaturating pushes in cheap phases is at most $O(n^2 \cdot 2n/K) = O(n^3/K)$. To obtain a bound on the nonsaturating pushes in expensive phases, we use an argument based on the potential function Φ.

By definition, an expensive phase performs at least $2n/K$ nonsaturating pushes. Since the network can contain at most n/K nodes with K descendants or more, at least n/K nonsaturating pushes must be from nodes with fewer than K descendants. The highest-label preflow-push algorithm always performs push/relabel operation on a maximal active node; consequently, Property 7.18 applies. Property 7.18(a) implies that each of these nonsaturating pushes produces a decrease in Φ of at least 1. So Properties 7.18(b) and (c) imply that the total increase in Φ due to saturating pushes and relabels is at most $O(nmK)$. Therefore, the algorithm can perform $O(nmK)$ nonsaturating pushes in expensive phases.

To summarize this discussion, we note that cheap phases perform $O(n^3/K)$ nonsaturating pushes and expensive phases perform $O(nmK)$ nonsaturating pushes. We obtain the optimal value of K by balancing both terms (see Section 3.2), that is, when both the terms are equal: $n^3/K = nmK$ or $K = n/m^{1/2}$. For this value of K, the number of nonsaturating pushes is $O(n^2m^{1/2})$. We have thus established the following result.

Theorem 7.19. *The highest-label preflow-push algorithm performs $O(n^2m^{1/2})$ nonsaturating pushes and runs in the same time.* ◆

7.9 EXCESS SCALING ALGORITHM

The generic preflow-push algorithm allows flow at each intermediate step to violate the mass balance equations. By pushing flows from active nodes, the algorithm attempts to satisfy the mass balance equations. The function $e_{\max} = \max\{e(i): i$ is an active node$\}$ provides one measure of the infeasibility of a preflow. Note that during the execution of the generic algorithm, we would observe no particular pattern in the values of $e_{\max}$, except that $e_{\max}$ eventually decreases to value 0. In this section we develop an *excess scaling technique* that systematically reduces the value of $e_{\max}$ to 0.

The excess scaling algorithm is similar to the capacity scaling algorithm we discussed in Section 7.3. Recall that the generic augmenting path algorithm performs $O(nU)$ augmentations and the capacity scaling algorithm reduces this number to $O(m \log U)$ by assuring that each augmentation carries a "sufficiently large" amount of flow. Similarly, in the generic preflow-push algorithm, nonsaturating pushes carrying small amount of flow bottleneck the algorithm in theory. The excess scaling algorithm assures that each nonsaturating push carries a "sufficiently large" amount of flow and so the number of nonsaturating pushes is "sufficiently small."

Let Δ denote an upper bound on $e_{\max}$; we refer to this bound as the *excess dominator*. We refer to a node with $e(i) \geq \Delta/2 \geq e_{\max}/2$ as a node with *large excess*, and as a node with *small excess* otherwise. The excess scaling algorithm always pushes flow from a node with a large excess. This choice assures that during nonsaturating pushes, the algorithm sends relatively large excess closer to the sink.

The excess scaling algorithm also does not allow the maximum excess to increase beyond Δ. This algorithmic strategy might prove to be useful for the following reason. Suppose that several nodes send flow to a single node j, creating a very large excess. It is likely that node j cannot send the accumulated flow closer to the sink, and thus the algorithm will need to increase its distance label and return much of its excess back to the nodes it came from. Thus pushing too much flow to any node is also likely to be a wasted effort.

The two conditions we have discussed—that each nonsaturating push must carry at least $\Delta/2$ units of flow and that no excess should exceed Δ—imply that we need to select the active nodes for push/relabel operations carefully. The following selection rule is one that assures that we achieve these objectives.

Node Selection Rule. *Among all nodes with a large excess, select a node with the smallest distance label (breaking ties arbitrarily).*

We are now in a position to give, in Figure 7.18, a formal description of the excess scaling algorithm.

The excess scaling algorithm uses the same push/relabel(i) operation as the generic preflow-push algorithm, but with one slight difference. Instead of pushing $\delta = \min\{e(i), r_{ij}\}$ units of flow, it pushes $\delta = \min\{e(i), r_{ij}, \Delta - e(j)\}$ units. This change ensures that the algorithm permits no excess to exceed Δ.

The algorithm performs a number of scaling phases with the value of the excess dominator Δ decreasing from phase to phase. We refer to a specific scaling phase with a particular value of Δ as a *Δ-scaling phase*. Initially, $\Delta = 2^{\lceil \log U \rceil}$. Since the logarithm is of base 2, $U \le \Delta \le 2U$. During the Δ-scaling phase, $\Delta/2 < e_{\max} \le \Delta$; the value of $e_{\max}$ might increase or decrease during the phase. When $e_{\max} \le \Delta/2$, we begin a new scaling phase. After the algorithm has performed $\lceil \log U \rceil + 1$ scaling phases, $e_{\max}$ decreases to value 0 and we obtain the maximum flow.

Lemma 7.20. *The algorithm satisfies the following two conditions:*
(a) *Each nonsaturating push sends at least $\Delta/2$ units of flow.*
(b) *No excess ever exceeds Δ.*

Proof. Consider a nonsaturating push on arc (i, j). Since arc (i, j) is admissible, $d(j) < d(i)$. Moreover, since node i is a node with the smallest distance label among all nodes with a large excess, $e(i) \ge \Delta/2$ and $e(j) < \Delta/2$. Since this push is non-

```
algorithm excess scaling;
begin
    preprocess;
    Δ : = 2^⌈log U⌉;
    while Δ ≥ 1 do
    begin (Δ-scaling phase)
        while the network contains a node i with a large excess do
            begin
                among all nodes with a large excess, select a node i with
                the smallest distance label;
                perform push/relabel(i) while ensuring that no node excess exceeds Δ;
            end;
        Δ : = Δ/2;
    end;
end;
```

Figure 7.18 Excess scaling algorithm.

saturating, it sends $\min\{e(i), \Delta - e(j)\} \geq \Delta/2$ units of flow, proving the first part of the lemma. This push operation increases the excess of only node j. The new excess of node j is $e(j) + \min\{e(i), \Delta - e(j)\} \leq e(j) + \{\Delta - e(j)\} \leq \Delta$. So all the node excesses remain less than or equal to Δ. This proves the second part of the lemma. ◆

Lemma 7.21. *The excess scaling algorithm performs $O(n^2)$ nonsaturating pushes per scaling phase and $O(n^2 \log U)$ pushes in total.*

Proof. Consider the potential function $\Phi = \sum_{i \in N} e(i)d(i)/\Delta$. Using this potential function, we will establish the first assertion of the lemma. Since the algorithm performs $O(\log U)$ scaling phases, the second assertion is a consequence of the first. The initial value of Φ at the beginning of the Δ-scaling phase is bounded by $2n^2$ because $e(i)$ is bounded by Δ and $d(i)$ is bounded by $2n$. During the push/relabel(i) operation, one of the following two cases must apply:

Case 1. The algorithm is unable to find an admissible arc along which it can push flow. In this case the distance label of node i increases by $\epsilon \geq 1$ units. This relabeling operation increases Φ by at most ϵ units because $e(i) \leq \Delta$. Since for each i the total increase in $d(i)$ throughout the running of the algorithm is bounded by $2n$ (by Lemma 7.13), the total increase in Φ due to the relabeling of nodes is bounded by $2n^2$ in the Δ-scaling phase (actually, the increase in Φ due to node relabelings is at most $2n^2$ over *all* scaling phases).

Case 2. The algorithm is able to identify an arc on which it can push flow, so it performs either a saturating or a nonsaturating push. In either case, Φ decreases. A nonsaturating push on arc (i, j) sends at least $\Delta/2$ units of flow from node i to node j and since $d(j) = d(i) - 1$, after this operation decreases Φ by at least $\frac{1}{2}$ unit. Since the initial value of Φ at the beginning of a Δ-scaling phase is at most $2n^2$ and the increases in Φ during this scaling phase sum to at most $2n^2$ (from Case 1), the number of nonsaturating pushes is bounded by $8n^2$. ◆

This lemma implies a bound of $O(nm + n^2 \log U)$ on the excess scaling algorithm since we have already seen that all the other operations—such as saturating pushes, relabel operations, and finding admissible arcs—require $O(nm)$ time. Up to this point we have ignored the method needed to identify a node with the minimum distance label among nodes with excess more than $\Delta/2$. Making this identification is easy if we use a scheme similar to the one used in the highest-label preflow-push algorithm in Section 7.8 to find a node with the highest distance label. We maintain the lists $LIST(k) = \{i \in N: e(i) > \Delta/2 \text{ and } d(i) = k\}$, and a variable *level* that is a lower bound on the smallest index k for which LIST(k) is nonempty. We identify the lowest-indexed nonempty list by starting at LIST(level) and sequentially scanning the higher-indexed lists. We leave as an exercise to show that the overall effort needed to scan the lists is bounded by the number of pushes performed by the algorithm plus $O(n \log U)$, so these computations are not a bottleneck operation. With this observation we can summarize our discussion as follows.

Theorem 7.22. *The excess scaling algorithm runs in $O(nm + n^2 \log U)$ time.* ◆

Algorithm	Running time	Features
Labeling algorithm	$O(nmU)$	1. Maintains a feasible flow and augments flows along directed paths in the residual network from node s to node t. 2. Easy to implement and very flexible. 3. Running time is pseudopolynomial: the algorithm is not very efficient in practice.
Capacity scaling algorithm	$O(nm \log U)$	1. A special implementation of the labeling algorithm. 2. Augments flows along paths from node s to node t with sufficiently large residual capacity. 3. Unlikely to be efficient in practice.
Successive shortest path algorithm	$O(n^2m)$	1. Another special implementation of the labeling algorithm. 2. Augments flows along shortest directed paths from node s to node t in the residual network. 3. Uses distance labels to identify shortest paths from node s to node t. 4. Relatively easy to implement and very efficient in practice.
Generic preflow-push algorithm	$O(n^2m)$	1. Maintains a pseudoflow; performs push/relabel operations at active nodes. 2. Very flexible; can examine active nodes in any order. 3. Relatively difficult to implement because an efficient implementation requires the use of several heuristics.
FIFO preflow-push algorithm	$O(n^3)$	1. A special implementation of the generic preflow-push algorithm. 2. Examines active nodes in the FIFO order. 3. Very efficient in practice.
Highest-label preflow-push algorithm	$O(n^2\sqrt{m})$	1. Another special implementation of the generic preflow-push algorithm. 2. Examines active nodes with the highest distance label. 3. Possibly the most efficient maximum flow algorithm in practice.
Excess scaling algorithm	$O(nm + n^2 \log U)$	1. A special implementation of the generic preflow-push algorithm. 2. Performs push/relabel operations at nodes with sufficiently large excesses and, among these nodes, selects a node with the smallest distance label. 3. Achieves an excellent running time without using sophisticated data structures.

Figure 7.19 Summary of maximum flow algorithms.

7.10 SUMMARY

Building on the labeling algorithm described in Chapter 6, in this chapter we described several polynomial-time algorithms for the maximum flow problem. The labeling algorithm can perform as many as nU augmentations because each augmentation might carry a small amount of flow. We studied two natural strategies for reducing the number of augmentations and thus for improving the algorithm's running time; these strategies lead to the capacity scaling algorithm and the shortest augmenting path algorithm. One inherent drawback of these augmenting path algorithms is the computationally expensive operation of sending flows along paths. These algorithms might repeatedly augment flows along common path segments. The preflow-push algorithms that we described next overcome this drawback; we can conceive of them as sending flows along several paths simultaneously. In our development we considered both a generic implementation and several specific implementations of the preflow-push algorithm. The FIFO and highest-label preflow-push algorithms choose the nodes for pushing/relabeling in a specific order. The excess scaling algorithm ensures that the push operations, and subsequent augmentations, do not carry small amounts of flow (with "small" defined dynamically throughout the algorithm). Figure 7.19 summarizes the running times and basic features of these algorithms.

REFERENCE NOTES

The maximum flow problem is distinguished by the long succession of research contributions that have improved on the worst-case complexity of the best known algorithms. Indeed, no other network flow problem has witnessed as many incremental improvements. The following discussion provides a brief survey of selective improvements; Ahuja, Magnanti, and Orlin [1989, 1991] give a more complete survey of the developments in this field.

The labeling algorithm of Ford and Fulkerson [1956a] runs in pseudopolynomial time. Edmonds and Karp [1972] suggested two polynomial-time implementations of this algorithm. The first implementation, which augments flow along paths with the maximum residual capacity, performs $O(m \log U)$ iterations. The second implementation, which augments flow along shortest paths, performs $O(nm)$ iterations and runs in $O(nm^2)$ time. Independently, Dinic [1970] introduced a concept of shortest path networks (in number of arcs), called *layered networks*, and obtained an $O(n^2m)$-time algorithm. Until this point all maximum flow algorithms were augmenting path algorithms. Karzanov [1974] introduced the first preflow-push algorithm on layered networks; he obtained an $O(n^3)$ algorithm. Shiloach and Vishkin [1982] described another $O(n^3)$ preflow-push algorithm for the maximum flow problem, which is a precursor of the FIFO preflow-push algorithm that we described in Section 7.7.

The capacity scaling described in Section 7.3 is due to Ahuja and Orlin [1991]; this algorithm is similar to the bit-scaling algorithm due to Gabow [1985] that we describe in Exercise 7.19. The shortest augmenting path algorithm described in Section 7.4 is also due to Ahuja and Orlin [1991]; this algorithm can be regarded as a variant of Dinic's [1970] algorithm and uses distance labels instead of layered networks.

Researchers obtained further improvements in the running times of the maximum flow algorithms by using distance labels instead of layered networks. Goldberg [1985] first introduced distance labels; by incorporating them in the algorithm of Shiloach and Vishkin [1982], he obtained the $O(n^3)$-time FIFO implementation that we described in Section 7.7. The generic preflow-push algorithm and its highest-label preflow-push implementation that we described in Sections 7.8 are due to Goldberg and Tarjan [1986]. Using a dynamic tree data structure developed by Sleator and Tarjan [1983], Goldberg and Tarjan [1986] improved the running time of the FIFO implementation to $O(nm \log(n^2/m))$. Using a clever analysis, Cheriyan and Maheshwari [1989] show that the highest-label preflow-push algorithm in fact runs in $O(n^2 \sqrt{m})$ time. Our discussion in Section 7.7 presents a simplified proof of Cheriyan and Maheshwari approach. Ahuja and Orlin [1989] developed the excess scaling algorithm described in Section 7.9; this algorithm runs in $O(nm + n^2 \log U)$ time and obtains dramatic improvements over the FIFO and highest-label preflow-push algorithms without using sophisticated data structures. Ahuja, Orlin, and Tarjan [1989] further improved the excess scaling algorithm and obtained several algorithms: the best time bound of these algorithms is $O(nm \log(n \sqrt{\log U}/m + 2))$.

Cheriyan and Hagerup [1989] proposed a randomized algorithm for the maximum flow problem that has an expected running time of $O(nm)$ for all $m \geq n \log^2 n$. Alon [1990] developed a nonrandomized version of this algorithm and obtained a (deterministic) maximum flow algorithm that runs in (1) $O(nm)$ time for all $m = \Omega(n^{5/3} \log n)$, and (2) $O(nm \log n)$ time for all other values of n and m. Cheriyan, Hagerup, and Mehlhorn [1990] obtained an $O(n^3/\log n)$ algorithm for the maximum flow problem. Currently, the best available time bounds for solving the maximum flow problem are due to Alon [1990], Ahuja, Orlin, and Tarjan [1989], and Cheriyan, Hagerup, and Mehlhorn [1990].

Researchers have also investigated whether the worst-case bounds of the maximum flow algorithms are "tight" (i.e., whether algorithms achieve their worst-case bounds for some families of networks). Galil [1981] constructed a family of networks and showed that the algorithms of Edmonds and Karp [1972], Dinic [1970], Karzanov [1974], and a few other maximum flow algorithms achieve their worst-case bounds. Using this family of networks, it is possible to show that the shortest augmenting path algorithm also runs in $\Omega(n^2 m)$ time. Cheriyan and Maheshwari [1989] have shown that the generic preflow-push algorithm and its FIFO and the highest-label preflow-push implementations run in $\Omega(n^2 m)$, $\Omega(n^3)$, and $\Omega(n^2 \sqrt{m})$ times, respectively. Thus the worst-case time bounds of these algorithms are tight.

Several computational studies have assessed the empirical behavior of maximum flow algorithms. Among these, the studies by Imai [1983], Glover, Klingman, Mote, and Whitman [1984], Derigs and Meier [1989], and Ahuja, Kodialam, Mishra, and Orlin [1992] are noteworthy. These studies find that preflow-push algorithms are faster than augmenting path algorithms. Among the augmenting path algorithms, the shortest augmenting path algorithm is the fastest, and among the preflow-push algorithms, the performance of the highest-label preflow-push algorithm is the most attractive.

EXERCISES

7.1. Consider the network shown in Figure 7.20. The network depicts only those arcs with a positive capacity. Specify the residual network with respect to the current flow and compute exact distance labels in the residual network. Next change the arc flows (without changing the flow into the sink) so that the exact distance label of the source node (1) decreases by 1 unit; (2) increases by 1 unit.

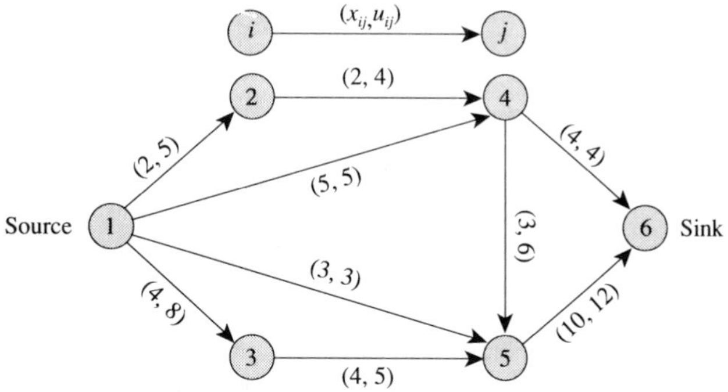

Figure 7.20 Example for Exercise 7.1.

7.2. Using the capacity scaling algorithm described in Section 7.3, find a maximum flow in the network given in Figure 7.21(b).

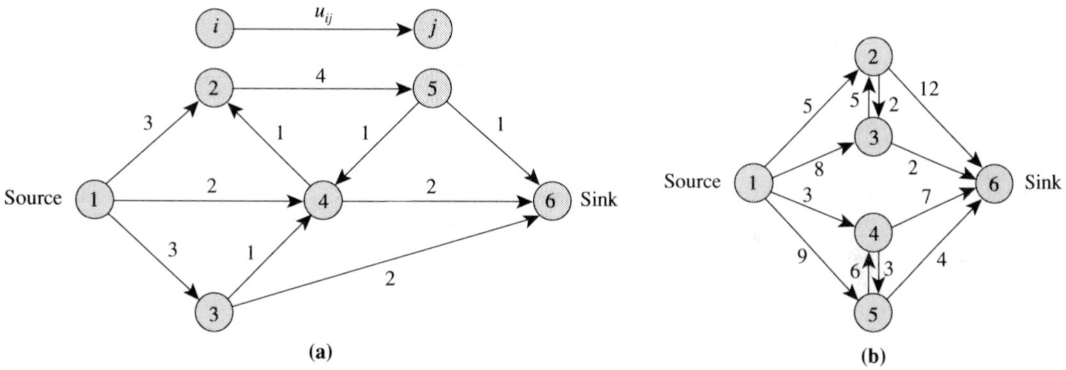

Figure 7.21 Examples for Exercises 7.2, 7.3, 7.5, and 7.6.

7.3. Using the shortest augmenting path algorithm, solve the maximum flow problem shown in Figure 7.21(a).

7.4. Solve the maximum flow problem shown in Figure 7.22 using the generic preflow-push algorithm. Incorporate the following rules to maintain uniformity of your computations: (1) Select an active node with the smallest index. [For example, if nodes 2 and 3 are active, select node 2.] (2) Examine the adjacency list of any node in the increasing order of the head node indices. [For example, if $A(1) = \{(1, 5), (1, 2), (1, 7)\}$, then examine arc (1, 2) first.] Show your computations on the residual networks encountered during the intermediate iterations of the algorithm.

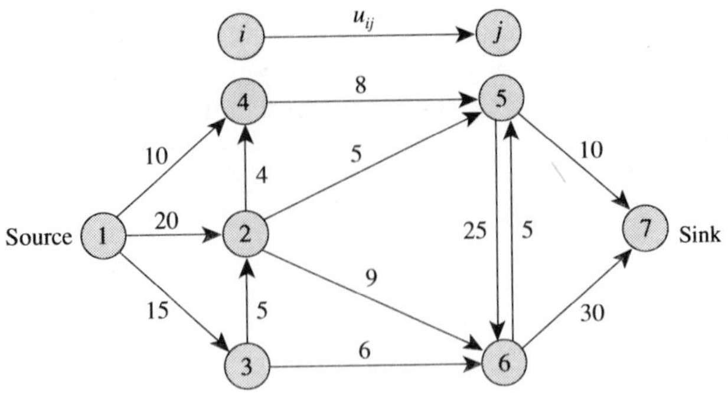

Figure 7.22 Example for Exercise 7.4.

7.5. Solve the maximum flow problem shown in Figure 7.21(a) using the FIFO preflow-push algorithm. Count the number of saturating and nonsaturating pushes and the number of relabel operations. Next, solve the same problem using the highest-label preflow-push algorithm. Compare the number of saturating pushes, nonsaturating pushes, and relabel operations with those of the FIFO preflow-push algorithm.

7.6. Using the excess scaling algorithm, determine a maximum flow in the network given in Figure 7.21(b).

7.7. **Most vital arcs.** We define a *most vital arc* of a network as an arc whose deletion causes the largest decrease in the maximum flow value. Either prove the following claims or show through counterexample that they are false.

(a) A most vital arc is an arc with the maximum value of u_{ij}.

(b) A most vital arc is an arc with the maximum value of x_{ij}.

(c) A most vital arc is an arc with the maximum value of x_{ij} among arcs belonging to some minimum cut.

(d) An arc that does not belong to some minimum cut cannot be a most vital arc.

(e) A network might contain several most vital arcs.

7.8. **Least vital arcs.** A *least vital arc* in a network is an arc whose deletion causes the least decrease in the maximum flow value. Either prove the following claims or show that they are false.

(a) Any arc with $x_{ij} = 0$ in any maximum flow is a least vital arc.

(b) A least vital arc is an arc with the minimum value of x_{ij} in a maximum flow.

(c) Any arc in a minimum cut cannot be a least vital arc.

7.9. Indicate which of the following claims are true or false. Justify your answer by giving a proof or by constructing a counterexample.

(a) If the capacity of every arc in a network is a multiple of α, then in every maximum flow, each arc flow will be a multiple of α.

(b) In a network G, if the capacity of every arc increases by α units, the maximum flow value will increase by a multiple of α.

(c) Let v^* denote the maximum flow value of a given maximum flow problem. Let v' denote the flow into the sink node t at some stage of the preflow-push algorithm. Then $v^* - v' \leq \sum_{i \text{ is active}} e(i)$.

(d) By the flow decomposition theory, some sequence of at most $m + n$ augmentations would always convert any preflow into a maximum flow.

(e) In the excess scaling algorithm, $e_{\max} = \max\{e(i): i \text{ is active}\}$ is a nonincreasing function of the number of push/relabel steps.

(f) The capacity of the augmenting paths generated by the maximum capacity augmenting path algorithm is nonincreasing.

(g) If each distance label $d(i)$ is a lower bound on the length of a shortest path from node i to node t in the residual network, the distance labels are valid.

7.10. Suppose that the capacity of every arc in a network is a multiple of α and is in the range $[0, \alpha K]$ for some integer K. Does this information improve the worst-case complexity of the labeling algorithm, FIFO preflow-push algorithm, and the excess scaling algorithm?

7.11. Converting a maximum preflow to a maximum flow. We define a *maximum preflow* x° as a preflow with the maximum possible flow into the sink.

(a) Show that for a given maximum preflow x^0, some maximum flow x^* with the same flow value as x^0, satisfies the condition that $x^*_{ij} \le x^0_{ij}$ for all arcs $(i, j) \in A$. (*Hint*: Use flow decomposition.)

(b) Suggest a labeling algorithm that converts a maximum preflow into a maximum flow in at most $n + m$ augmentations.

(c) Suggest a variant of the shortest augmenting path algorithm that would convert a maximum preflow into a maximum flow in $O(nm)$ time. (*Hint*: Define distance labels from the source node and show that the algorithm will create at most m arc saturations.)

(d) Suggest a variant of the highest-label preflow-push algorithm that would convert a maximum preflow into a maximum flow. Show that the running time of this algorithm is $O(nm)$. (*Hint*: Use the fact that we can delete an arc with zero flow from the network.)

7.12. **(a)** An arc is *upward critical* if increasing the capacity of this arc increases the maximum flow value. Does every network have an upward critical arc? Describe an algorithm for identifying all upward critical arcs in a network. The worst-case complexity of your algorithm should be substantially better than that of solving m maximum flow problems.

(b) An arc is *downward critical* if decreasing the capacity of this arc decreases the maximum flow value. Is the set of upward critical arcs the same as the set of downward critical arcs? If not, describe an algorithm for identifying all downward critical arcs; analyze your algorithm's worst-case complexity.

7.13. Show that in the shortest augmenting path algorithm or in the preflow-push algorithm, if an arc (i, j) is inadmissible at some stage, it remains inadmissible until the algorithm relabels node i.

7.14. Apply the generic preflow-push algorithm to the maximum flow problem shown in Figure 7.23. Always examine a node with the smallest distance label and break ties in favor of a node with the smallest node number. How many saturating and nonsaturating pushes does the algorithm perform?

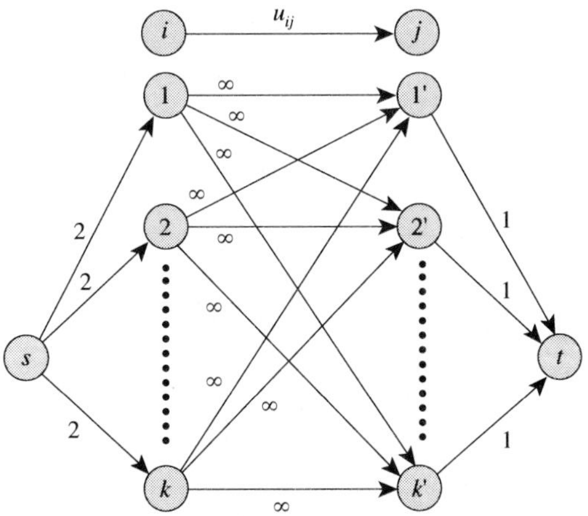

Figure 7.23 Example for Exercise 7.14.

7.15. Apply the FIFO preflow-push algorithm to the network shown in Figure 7.24. Determine the number of pushes as a function of the parameters W and L (correct within a constant factor). For a given value of n, what values of W and L produce the largest number of pushes?

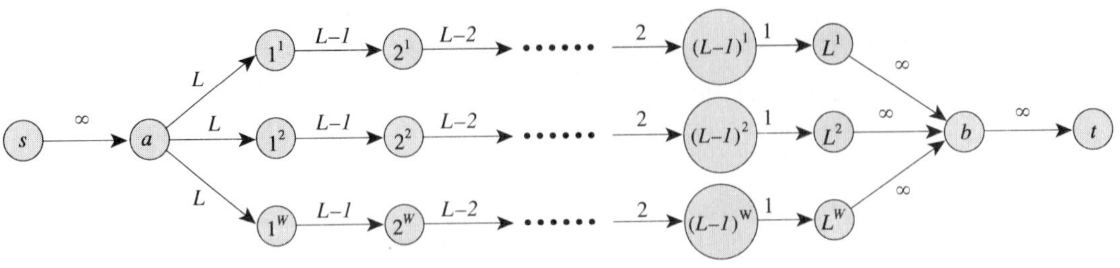

Figure 7.24 Example for Exercise 7.15.

7.16. Apply the highest-label preflow-push algorithm on the network shown in Figure 7.24. Determine the number of pushes as a function of the parameters W and L (correct within a constant factor). For a given n, what values of W and L produce the largest number of pushes?

7.17. Describe a more general version of the capacity scaling algorithm discussed in Section 7.3, one that scales Δ at each scaling phase by a factor of some integer number $\beta \geq 2$. Initially, $\Delta = \beta^{\lceil \log_\beta U \rceil}$ and each scaling phase reduces Δ by a factor of β. Analyze the worst-case complexity of this algorithm and determine the optimal value of β.

7.18. **Partially capacitated networks** (Ahuja and Orlin [1991]). Suppose that we wish to speed up the capacity scaling algorithm discussed in Exercise 7.17 for networks with some, but not all, arcs capacitated. Suppose that the network G has p arcs with finite capacities and $\beta = \max\{2, \lceil m/p \rceil\}$. Consider a version of the capacity scaling algorithm that scales Δ by a factor of β in each scaling phase. Show that the algorithm would perform at most $2m$ augmentations per scaling phase. [*Hint*: At the end of the Δ-scaling phase, the s–t cut in $G(\Delta)$ contains only arcs with finite capacities.] Conclude that the capacity scaling algorithm would solve the maximum flow problem in $O(m^2 \log_\beta U)$ time. Finally, show that this algorithm would run in $O(m^2)$ time if $U = O(n^k)$ for some k and $m = O(n^{1+\epsilon})$ for some $\epsilon > 0$.

7.19. **Bit-scaling algorithm** (Gabow [1985]). Let $K = \lceil \log U \rceil$. In the bit-scaling algorithm for the maximum flow problem works, we represent each arc capacity as a K-bit binary number, adding leading zeros if necessary to make each capacity K bits long. The problem P_k considers the capacity of each arc as the k leading bits. Let x_k^* denote a maximum flow and let v_k^* denote the maximum flow value in the problem P_k. The algorithm solves a sequence of problems $P_1, P_2, P_3, \ldots, P_K$, using $2x_{k-1}^*$ as a starting solution for the problem P_k.

(a) Show that $2x_{k-1}^*$ is feasible for P_k and that $v_k^* - 2v_{k-1}^* \leq m$.

(b) Show that the shortest augmenting path algorithm for solving problem P_k, starting with $2x_{k-1}^*$ as the initial solution, requires $O(nm)$ time. Conclude that the bit-scaling algorithm solves the maximum flow problem in $O(nm \log U)$ time.

7.20. Using the potential function $\Phi = \max\{d(i) : i \text{ is active}\}$, show that the highest-label preflow-push algorithm performs $O(n^3)$ nonsaturating pushes.

7.21. The *wave algorithm*, which is a hybrid version of the highest-label and FIFO preflow-push algorithms, performs passes over active nodes. In each pass it examines *all* the active nodes in nonincreasing order of the distance labels. While examining a node, it pushes flow from a node until either its excess becomes zero or the node is relabeled. If during a pass, the algorithm relabels no node, it terminates; otherwise, in the next pass, it again examines active nodes in nonincreasing order of their new distance labels.

Discuss the similarities and differences between the wave algorithm with the highest label and FIFO preflow-push algorithms. Show that the wave algorithm runs in $O(n^3)$ time.

7.22. Several rules listed below are possible options for selecting an active node to perform the push/relabel operation in the preflow-push algorithm. Describe the data structure and the implementation details for each rule. Obtain the tightest possible bound on the numbers of pushes performed by the algorithm and the resulting running time of the algorithm.

(a) Select an active node with the minimum distance label.
(b) Select an active node with the largest amount of excess.
(c) Select an active node whose excess is at least 50 percent of the maximum excess at any node.
(d) Select the active node that the algorithm had selected most recently.
(e) Select the active node that the algorithm had selected least recently.
(f) Select an active node randomly. (Assume that for any integer k, you can in $O(1)$ steps generate a random integer uniformly in the range $[1, k]$.)

7.23. In the excess scaling algorithm, suppose we require that each nonsaturating push pushes exactly $\Delta/2$ units of flow. Show how to modify the push/relabel step to meet this requirement. Does this modification change the worst-case running time of the algorithm?

7.24. In our development in this chapter we showed that the excess scaling algorithm performs $O(n^2 \log U^*)$ nonsaturating pushes if U^* is set equal to the largest arc capacity among the arcs emanating from the source node. However, we can often select smaller values of U^* and still show that the number of nonsaturating pushes is $O(n^2 \log U^*)$. Prove that we can also use the following values of U^* without affecting the worst-case complexity of the algorithm: (1) $U^* = \sum_{(s,j)\in A(s)} u_{sj}/ \mid A(s) \mid$; (2) $U^* = v^{\text{ub}}/ \mid A(s) \mid$ for any upper bound v^{ub} on the maximum flow value. (*Hint*: In the first scaling phase, set $\Delta = 2^{\lceil \log U^* \rceil}$ and forbid nodes except those adjacent to the source from having excess more than Δ.)

7.25. The excess scaling algorithm described in Section 7.9 scales excesses by a factor of 2. It starts with the value of the excess dominator Δ equal to the smallest power of 2 that is greater than or equal to U; in every scaling phase, it reduces Δ by a factor of 2. An alternative is to scale the excesses by a factor of some integer number $\beta \geq 2$. This algorithm would run as follows. It would start with $\Delta = \beta^{\lceil \log_\beta U \rceil}$; it would then reduce Δ by a factor of β in every scaling phase. In the Δ-scaling phase, we refer to a node with an excess of at least Δ/β as a node with a *large* excess. The algorithm pushes flow from a node with a large excess and among these nodes it chooses the node with the smallest distance label. The algorithm also ensures that no excess exceeds Δ. Determine the number of nonsaturating pushes performed by the algorithm as a function of n, β, and U. For what value of β would the algorithm perform the least number of nonsaturating pushes?

7.26. For any pair $[i, j] \in N \times N$, we define $\alpha[i, j]$ in the following manner: (1) if $(i, j) \in A$, then $\alpha[i, j]$ is the increase in the maximum flow value obtained by setting $u_{ij} = \infty$; and (2) if $(i, j) \notin A$, then $\alpha[i, j]$ is the increase in the maximum flow value obtained by introducing an infinite capacity arc (i, j) in the network.

(a) Show that $\alpha[i, j] \leq \alpha[s, j]$ and $\alpha[i, j] \leq \alpha[i, t]$.
(b) Show that $\alpha[i, j] = \min\{\alpha[s, j], \alpha[i, t]\}$.
(c) Show that we can compute $\alpha[i, j]$ for all node pairs by solving $O(n)$ maximum flow problems.

7.27. **Minimum cut with the fewest number of arcs.** Suppose that we wish to identify from among all minimum cuts, a minimum cut containing the least number of arcs. Show that if we replace u_{ij} by $u'_{ij} = mu_{ij} + 1$, the minimum cut with respect to the capacities u'_{ij} is a minimum cut with respect to the capacities u_{ij} containing the fewest number of arcs.

7.28. Parametric network feasibility problem. In a capacitated network G with arc capacities u_{ij}, suppose that the supply/demands of nodes are linear functions of time τ. Let each $b(i) = b^0(i) + \tau b^*(i)$ and suppose that $\sum_{i \in N} b^0(i) = 0$ and $\sum_{i \in N} b^*(i) = 0$. The network is currently (i.e., at time $\tau = 0$) able to fulfill the demands by the existing supplies but might not be able to do so at some point in future. You want to determine the largest integral value of τ up to which the network will admit a feasible flow. How would you solve this problem?

7.29. Source parametric maximum flow problem (Gallo, Grigoriadis, and Tarjan [1989]). In the *source parametric maximum flow problem*, the capacity of every source arc (s, j) is a nondecreasing linear function of a parameter λ (i.e., $u_{sj} = u^{\circ}_{sj} + \lambda u^*_{sj}$ for some constant $u^*_{sj} \geq 0$); the capacity of every other arc is fixed, and we wish to determine a maximum flow for p values $0 = \lambda_1, \lambda_2, \ldots, \lambda_p$ of the parameter λ. Assume that $\lambda_1 \leq \lambda_2 \leq \cdots \leq \lambda_p$ and $p \leq n$. As an application of the source-parametric maximum flow problem, consider the following variation of Application 6.5. Suppose that processor 1 is a shared multiprogrammed system and processor 2 is a graphic processor dedicated to a single user. Suppose further that we can accurately determine the times required for processing modules on processor 2, but the times for processing modules on processor 1 are affected by a general work load on the processor. As the work load on processor 1 changes, the optimal distribution of modules between processor 1 and 2 changes. The source–parametric maximum flow problem determines these distributions for different work loads on processor 1.

Let MF(λ) denote the maximum flow problem for a specific value of λ. Let $v(\lambda)$ denote the maximum flow value of MF(λ) and let $[S(\lambda), \bar{S}(\lambda)]$ denote an associated minimum cut. Clearly, the zero flow is optimal for MF(λ_1). Given an optimal flow $x(\lambda_k)$ of MF(λ_k), we solve MF(λ_{k+1}) as follows: With $x(\lambda_k)$ as the starting flow and the corresponding distance labels as the initial distance labels, we perform a preprocess step by sending additional flow along the source arcs so that they all become saturated. Then we apply the FIFO preflow-push algorithm until the network contains no more active nodes. We repeat this process until we have solved MF(λ_p).

(a) Show that the ending distance labels of MF(λ_k) are valid distances for MF(λ_{k+1}) in the residual network $G(x(\lambda_k))$ after the preprocess step.

(b) Use the result in part (a) to show that overall [i.e., in solving all the problems MF(λ_1), MF(λ_2), . . . , MF(λ_p)], the algorithm performs $O(n^2)$ relabels, $O(nm)$ saturating pushes, and $O(n^3)$ nonsaturating pushes. Conclude that the FIFO preflow-push algorithm solves the source parametric maximum flow problem in $O(n^3)$ time, which is the same time required to solve a single maximum flow problem.

(c) Show that $v(\lambda_1) \leq v(\lambda_2) \leq \cdots \leq v(\lambda_p)$ and some associated minimum cuts satisfy the nesting condition $S_1 \subseteq S_2 \subseteq \cdots \subseteq S_p$.

7.30. Source–sink parametric maximum flow problem. In the source–sink parametric maximum flow problem, the capacity of every source arc is a *nondecreasing* linear function of a parameter λ and capacity of every sink arc is a *nonincreasing* linear function of λ, and we want to determine a maximum flow for several values of parameter λ_1, $\lambda_2, \ldots, \lambda_p$, for $p \leq n$, that satisfy the condition $0 = \lambda_1 < \lambda_2 < \cdots < \lambda_p$. Show how to solve this problem in a total of $O(n^3)$ time. (*Hint*: The algorithm is same as the one considered in Exercise 7.29 except that in the preprocess step if some sink arc has flow greater than its new capacity, we decrease the flow.)

7.31. Ryser's theorem. Let Q be a $p \times p$ matrix consisting of 0–1 elements. Let α denote the vector of row sums of Q and β denote the vector of column sums. Suppose that the rows and columns are ordered so that $\alpha_1 \geq \alpha_2 \geq \cdots \geq \alpha_p$, and $\beta_1 \geq \beta_2 \geq \cdots \geq \beta_p$.

(a) Show that the vectors α and β must satisfy the following conditions: (1) $\sum_{i=1}^{p} \alpha_i = \sum_{i=1}^{p} \beta_i$ and (2) $\sum_{i=1}^{k} \min(\alpha_i, k) \leq \sum_{i=1}^{k} \beta_i$, for all $k = 1, \ldots, p$. [*Hint*: $\min(\alpha_i, k)$ is an upper bound on the sum of the first k components of row i.]

(b) Given the nonnegative integer vector α and β, show how to formulate the problem of determining whether some 0–1 matrix Q has a row sum vector α and a column sum vector β as a maximum flow problem. Use the max-flow min-cut theorem to show that the conditions stated in part (a) are sufficient for the existence of such a matrix Q.

8

MAXIMUM FLOWS: ADDITIONAL TOPICS

This was the most unkindest cut of all.
—Shakespeare in Julius Caeser Act III

Chapter Outline

8.1 INTRODUCTION

In all scientific disciplines, researchers are always making trade-offs between the generality and the specificity of their results. Network flows embodies these considerations. In studying minimum cost flow problems, we could consider optimization models with varying degrees of generality: for example, in increasing order of specialization, (1) general constrained optimization problems, (2) linear programs, (3) network flows, (4) particular network flow models (e.g., shortest path and maximum flow problems), and (5) the same models defined on problems with specific topologies and/or cost structures. The trade-offs in choosing where to study across the hierarchy of possible models is apparent. As models become broader, so does the range of their applications. As the models become more narrow, available results often become refined and more powerful. For example, as shown by our discussion in previous chapters, algorithms for shortest path and maximum flow problems have very attractive worst-case and empirical behavior. In particular, the computational complexity of these algorithms grows rather slowly in the number of underlying constraints (i.e., nodes) and decision variables (arcs). For more general linear programs, or even for more general minimum cost flow problems, the best algorithms are not nearly as good.

In considering what class of problems to study, we typically prefer models that are generic enough to be rich, both in applications and in theory. As evidenced by the coverage in this book, network flows is a topic that meets this criterion. Yet, through further specialization, we can develop a number of more refined results. Our study of shortest paths and maximum flow problems in the last four chapters has illustrated this fact. Even within these more particular problem classes, we have seen the effect of further specialization, which has led to us to discover more efficient shortest path algorithms for models with nonnegative costs and for models defined on acyclic graphs. In this chapter we carry out a similar program for maximum flow problems. We consider maximum flow problems with both (1) specialized data, that is, networks with unit capacity arcs, and (2) specialized topologies, namely, bipartite and planar networks.

For general maximum flow problems, the labeling algorithm requires $O(nmU)$ computations and the shortest augmenting path algorithm requires $O(n^2m)$ computations. When applied to unit capacity networks, these algorithms are guaranteed to perform even better. Both require $O(nm)$ computations. We obtain this improvement simply because of the special nature of unit capacity networks. By designing specialized algorithms, however, we can improve even further on these results. Combining features of both the labeling algorithm and the shortest augmenting path algorithm, the unit capacity maximum flow algorithm that we consider in this chapter requires only $O(\min\{n^{2/3}m, m^{3/2}\})$ computations.

Network connectivity is an important application context for the unit capacity maximum flow problem. The arc connectivity between any two nodes of a network is the maximum number of arc-disjoint paths that connect these nodes; the arc connectivity of the network as a whole is the minimum arc connectivity between any pair of nodes. To determine this important reliability measure of a network, we could solve a unit capacity maximum flow problem between every pair of nodes, thus requiring $O(\min\{n^{2/3}m, m^{3/2}\})$ computations. As we will see in this chapter, by exploiting the special structure of the arc connectivity problem, we can reduce this complexity bound considerably—to $O(nm)$.

For networks with specialized bipartite and planar topologies, we can also obtain more efficient algorithms. Recall that bipartite networks are composed of two node sets, N_1 with n_1 nodes and N_2 with n_2 nodes. Assume that $n_1 \leq n_2$. For these problems we develop a specialization of the generic preflow-push algorithm that requires $O(n_1^2 m)$ instead of $O((n_1 + n_2)^2 m)$ time. Whenever the bipartite network is unbalanced in the sense that $n_1 \ll (n_1 + n_2) = n$, the new implementation has a much better complexity than the general preflow-push algorithm. Planar networks are those that we can draw on the plane so that no two arcs intersect each other. For this class of networks, we develop a specialized maximum flow algorithm that requires only $O(n \log n)$ computations.

In this chapter we also consider two other additional topics: a dynamic tree implementation and the all-pairs minimum value cut problem. Dynamic trees is a special type of data structure that permits us to implicitly send flow on paths of length n in $O(\log n)$ steps on average. By doing so we are able to reduce the computational requirement of the shortest augmenting path algorithm for maximum flows from $O(n^2m)$ to $O(nm \log n)$.

In some application contexts, we need to find the maximum flow between every pair of nodes in a network. The max-flow min-cut theorem shows that this problem is equivalent to finding the minimum cut separating all pairs of nodes. The most naive way to solve this problem would be to solve the maximum flow problem $n(n - 1)$ times, once between every pair of nodes. Can we do better? In Section 8.7 we show that how to exploit the relationship of the cut problems between various node pairs to reduce the computational complexity of the all-pairs minimum value cut problem considerably in undirected networks. This algorithm requires solving only $(n - 1)$ minimum cut problems in undirected networks. Moreover, the techniques used in this development extend to a broader class of problems: they permit us to solve the all-pairs minimum cut problem for situations when the value of a cut might be different than the sum of the arc capacities across the cut.

The algorithms we examine in this chapter demonstrate the advantage of exploiting special structures to improve on the design of algorithms. This theme not only resurfaces on several other occasions in this book, but also is an important thread throughout the entire field of large-scale optimization. Indeed, we might view the field of large-scale optimization, and the field of network flows for that matter, as the study of theory and algorithms for exploiting special problem structure. In this sense this chapter is, in its orientation and overall approach, a microcosm of this entire book and of much of the field of optimization itself.

8.2 FLOWS IN UNIT CAPACITY NETWORKS

Certain combinatorial problems are naturally formulated as zero–one optimization models. When viewed as flow problems, these models yield networks whose arc capacities are all 1. We will refer to these networks as *unit capacity networks*. Frequently, it is possible to solve flow problems on these networks more efficiently than those defined on general networks. In this section we describe an efficient algorithm for solving the maximum flow problem on unit capacity networks. We subsequently refer to this algorithm as the *unit capacity maximum flow algorithm*.

In a unit capacity network, the maximum flow value is at most n, since the capacity of the s–t cut $[\{s\}, S - \{s\}]$ is at most n. The labeling algorithm therefore determines a maximum flow within n augmentations and requires $O(nm)$ effort. The shortest augmenting path algorithm also solves this problem in $O(nm)$ time since its bottleneck operation, which is the augmentation step, requires $O(nm)$ time instead of $O(n^2m)$ time. The unit capacity maximum flow algorithm that we describe is a hybrid version of these two algorithms. This unit capacity maximum flow algorithm is noteworthy because by combining features of both algorithms, it requires only $O(\min\{n^{2/3}m, m^{3/2}\})$ time, which is consistently better than the $O(nm)$ bound of either algorithm by itself.

The unit capacity maximum flow algorithm is a two-phase algorithm. In the first phase it applies the shortest augmenting path algorithm, although not until completion: rather, this phase terminates whenever the distance label of the source node satisfies the condition $d(s) \geq d^* = \min\{\lceil 2n^{2/3} \rceil, \lceil m^{1/2} \rceil\}$. Although the algorithm might terminate with a nonoptimal solution, the solution is probably nearly-optimal (its value is within d^* of the optimal flow value). In its second phase, the algorithm applies the labeling algorithm to convert this near-optimal flow into a maximum flow. As we will see, this two-phase approach works well for unit capacity networks because the shortest augmenting path algorithm obtains a near-optimal flow quickly (when augmenting paths are "short") but then takes a long time to convert this solution into a maximum flow (when augmenting paths become "long"). It so happens that the labeling algorithm converts this near-optimal flow into a maximum flow far more quickly than the shortest augmenting path algorithm.

Let us examine the behavior of the shortest augmenting path algorithm for $d^* = \min\{\lceil 2n^{2/3} \rceil, \lceil m^{1/2} \rceil\}$. Suppose the algorithm terminates with a flow vector x' with a flow value equal to v'. What can we say about $v^* - v'$? (Recall that v^* denotes the maximum flow value.) We shall answer this question in two parts: (1) when $d^* = \lceil 2n^{2/3} \rceil$, and (2) when $d^* = \lceil m^{1/2} \rceil$.

Suppose that $d^* = \lceil 2n^{2/3} \rceil$. For each $k = 0, 1, 2, \ldots, d^*$, let V_k denote the set of nodes with a distance label equal to k [i.e., $V_k = \{i \in N : d(i) = k\}$]. We refer to V_k as the set of nodes in the kth *layer* of the residual network. Consider the situation when each of the sets $V_1, V_2, \ldots, V_{d^*}$ is nonempty. It is possible to show that each arc (i, j) in the residual network $G(x')$ connects a node in the kth layer to a node in the $(k + 1)$th layer for some k, for otherwise $d(i) > d(j) + 1$, which contradicts the distance label validity conditions (7.2). Therefore, for each $k = 1, 2, \ldots, d^*$, the set of arcs joining the node sets V_k to V_{k-1} form an s–t cut in the residual network. In case one of the sets, say V_k, is empty, our discussion in Section 7.4 implies that the cut $[S, \bar{S}]$ defined by $S = V_{k+1} \cup V_{k+2} \cup \cdots \cup V_{d^*}$ is a minimum cut.

Note that $|V_1| + |V_2| + \cdots + |V_{d^*}| \le n - 1$, because the sink node does not belong to any of these sets. We claim that the residual network contains at least two consecutive layers V_k and V_{k-1}, each with at most $n^{1/3}$ nodes. For if not, every alternate layer (say, $V_1, V_3, V_5, \ldots$) must contain more than $n^{1/3}$ nodes and the total number of nodes in these layers would be strictly greater than $n^{1/3} d^*/2 \ge n$, leading to a contradiction. Consequently, $|V_k| \le n^{1/3}$ and $|V_{k-1}| \le n^{1/3}$ for some of the two layers V_k and V_{k-1}. The residual capacity of the s–t cut defined by the arcs connecting V_k to V_{k-1} is at most $|V_k||V_{k-1}| \le n^{2/3}$ (since at most one arc of unit residual capacity joins any pair of nodes). Therefore, by Property 6.2, $v^* - v' \le n^{2/3} \le d^*$.

Next consider the situation when $d^* = \lceil m^{1/2} \rceil$. The layers of nodes $V_1, V_2, \ldots, V_{d^*}$ define d^* s–t cuts in the residual network and these cuts are arc disjoint in $G(x)$. The sum of the residual capacities of these cuts is at most m since each arc contributes at most one to the residual capacity of any such cut. Thus some s–t cut must have residual capacity at most $\lceil m^{1/2} \rceil$. This conclusion proves that $v^* - v' \le \lceil m^{1/2} \rceil = d^*$.

In both cases, whenever $d^* = \lceil 2n^{2/3} \rceil$ or $d^* = \lceil m^{1/2} \rceil$, we find that the first phase obtains a flow whose value differs from the maximum flow value by at most d^* units. The second phase converts this flow into a maximum flow in $O(d^*m)$ time since each augmentation requires $O(m)$ time and carries a unit flow. We now show that the first phase also requires $O(d^*m)$ time.

In the first phase, whenever the distance label of a node k exceeds d^*, this node never occurs as an intermediate node in any subsequent augmenting path since $d(k) < d(s) < d^*$. So the algorithm relabels any node at most d^* times. This observation gives a bound of $O(d^*n)$ on the number of retreat operations and a bound of $O(d^*m)$ on the time to perform the retreat operations. Consider next the augmentation time. Since each arc capacity is 1, flow augmentation over an arc immediately saturates that arc. During two consecutive saturations of any arc (i, j), the distance labels of both the nodes i and j must increase by at least 2 units. Thus the algorithm can saturate any arc at most $\lfloor d^*/2 \rfloor$ times, giving an $O(d^*m)$ bound on the total time needed for flow augmentations. The total number of advance operations is bounded by the augmentation time plus the number of retreat operations and is again $O(d^*m)$. We have established the following result.

Theorem 8.1. *The unit capacity maximum flow algorithm solves a maximum flow problem on unit capacity networks in $O(min\{n^{2/3}m, m^{3/2}\})$ time.* ◆

The justification of this two-phase procedure should now be clear. If $d^* = \lceil 2n^{2/3} \rceil \le m^{1/2}$, the preceding discussion shows that the shortest augmenting path algorithm requires $O(n^{2/3}m)$ computations to obtain a flow within $n^{2/3}$ of the optimal. If we allow this algorithm to run until it achieves optimality [i.e., until $d(s) \ge n$], the algorithm could require an additional $O((n - n^{2/3})m)$ time to convert this flow into an optimal flow. For $n = 1000$, these observations imply that if the algorithm achieves these bounds, it requires 10 percent of the time to send 90 percent of the maximum flow and the remaining 90 percent of the time to send 10 percent of the maximum flow. (Empirical investigations have observed a similar behavior in practice as well.) On the other hand, the use of labeling algorithm in the second phase establishes a maximum flow in $O(n^{2/3}m)$ time and substantially speeds up the overall performance of the algorithm.

Another special case of unit capacity networks, called *unit capacity simple networks*, also arises in practice and is of interest to researchers. For this class of unit capacity networks, every node in the network, except the source and sink nodes, has at most one incoming arc or at most one outgoing arc. The unit capacity maximum flow algorithm runs even faster for this class of networks. We achieve this improvement by setting $d^* = \lceil n^{1/2} \rceil$ in the algorithm.

Theorem 8.2. *The unit capacity maximum flow algorithm establishes a maximum flow in unit capacity simple networks in $O(n^{1/2}m)$ time.*

Proof. Consider the layers of nodes $V_1, V_2, \ldots, V_{d^*}$ at the end of the first phase. Note first that $d(s) > d^*$ since otherwise we could find yet another augmentation in Phase 1. Suppose that layer V_h contains the smallest number of nodes. Then $|V_h| \le n^{1/2}$, since otherwise the number of nodes in all layers would be strictly greater than n. Let N' be the nodes in N with at most one outgoing arc. We define a cut $[S, N\text{-}S]$ as follows: $S = \{j{:}d(j) \ge h\} \cup \{j{:}d(j) = h \text{ and } j \in N'\}$. Since $d(s) > d^* \ge h$ and $d(t) = 0$, $[S, N\text{-}S]$ is an s–t cut. Each arc with residual capacity in the cut $[S, N\text{-}S]$ is either directed into a node in $V_h \cap (N\text{-}N')$ or else it is directed from a node in $V_h \cap N'$. Therefore, the residual capacity of the cut Q is at most $|V_h| \le n^{1/2}$. Consequently, at the termination of the first phase, the flow value differs from the maximum flow value by at most $n^{1/2}$ units. Using arguments similar to those we have just used, we can now easily show that the algorithm would run in $O(n^{1/2}m)$ time. ◆

The proof of the Theorem 8.2 relies on the fact that only 1 unit of flow can pass through each node in the network (except the source and sink nodes). If we satisfy this condition but allow some arc capacities to be larger than 1, the unit capacity maximum flow algorithm would still require only $O(n^{1/2}m)$ time. Networks with this structure do arise on occasion; for example, we encountered this type of network when we computed the maximum number of node-disjoint paths from the source node to the sink node in the proof of Theorem 6.8. Theorem 8.2 has another by-product: It permits us to solve the maximum bipartite matching problem in $O(n^{1/2}m)$ time since we can formulate this problem as a maximum flow problem on a unit capacity simple network. We study this transformation in Section 12.3.

8.3 FLOWS IN BIPARTITE NETWORKS

A *bipartite network* is a network $G = (N, A)$ with a node set N partitioned into two subsets N_1 and N_2 so that for every arc $(i, j) \in A$, either (1) $i \in N_1$ and $j \in N_2$, or (2) $i \in N_2$ and $j \in N_1$. We often represent a bipartite network using the notation $G = (N_1 \cup N_2, A)$. Let $n_1 = |N_1|$ and $n_2 = |N_2|$. Figure 8.1 gives an example of a bipartite network; in this case, we can let $N_1 = \{1, 2, 3, 9\}$ and $N_2 = \{4, 5, 6, 7, 8\}$.

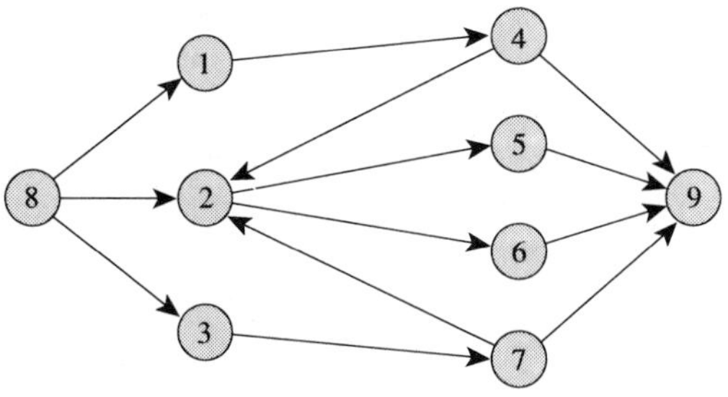

Figure 8.1 Bipartite network.

In this section we describe a specialization of the preflow-push algorithms that we considered in Chapter 7, but now adapt it to solve maximum flow problems on bipartite networks. The worst-case behavior of these special-purpose algorithms is similar to those of the original algorithms if the node sets N_1 and N_2 are of comparable size; the new algorithms are considerably faster than the original algorithms, however, whenever one of the sets N_1 or N_2 is substantially larger than the other. Without any loss of generality, we assume that $n_1 \le n_2$. We also assume that the source node belongs to N_2. [If the source node s belonged to N_1, then we could create a new source node $s' \in N_2$, and we could add an arc (s', s) with capacity M for sufficiently large M.] As one example of the type of results we will obtain, we show that the specialization of the generic preflow-push algorithm solves a maximum flow problem on bipartite networks in $O(n_1^2 m)$ time. If $n_1 \ll (n_1 + n_2) = n$, the new implementation is considerably faster than the original algorithm.

In this section we examine only the generic preflow-push algorithm for bipartite networks; we refer to this algorithm as the *bipartite preflow-push algorithm*. The ideas we consider also apply in a straightforward manner to the FIFO, highest-label preflow-push and excess-scaling algorithms and yield algorithms with improved worst-case complexity. We consider these improvements in the exercises.

We first show that a slightly modified version of the generic preflow-push algorithm requires less than $O(n^2 m)$ time to solve problems defined on bipartite networks. To establish this result, we change the preprocess operation by setting $d(s) = 2n_1 + 1$ instead of $d(s) = n$. The modification stems from the observation that any path in the residual network can have at most $2n_1$ arcs since every alternate node in the path must be in N_1 (because the residual network is also bipartite) and no path can repeat a node in N_1. Therefore, if we set $d(s) = 2n_1 + 1$, the residual network will never contain a directed path from node s to node t, and the algorithm will terminate with a maximum flow.

Lemma 8.3. *For each node $i \in N$, $d(i) < 4n_1 + 1$.*

Proof. The proof is similar to that of Lemma 7.12. ◆

The following result is a direct consequence of this lemma.

Lemma 8.4.

(a) *Each distance label increases at most $O(n_1)$ times. Consequently, the total number of relabel operations is $O(n_1(n_1 + n_2)) = O(n_1 n_2)$.*
(b) *The number of saturating pushes is $O(n_1 m)$.*

Proof. The proofs are similar to those of Lemmas 7.13 and 7.14. ◆

It is possible to show that the results of Lemma 8.4 yield a bound of $O((n_1 m)n)$ on the number of nonsaturating pushes, as well as on the complexity of the generic preflow-push algorithm. Instead of considering the details of this approach, we next develop a modification of the generic algorithm that runs in $O(n_1^2 m)$ time.

This modification builds on the following idea. To bound the nonsaturating pushes of any preflow-push algorithm, we typically use a potential function defined in terms of all the active nodes in the network. If every node in the network can be active, the algorithm will perform a certain number of nonsaturating pushes. However, if we permit only the nodes in N_1 to be active, because $n_1 \leq n$ we can obtain a tighter bound on the number of nonsaturating pushes. Fortunately, the special structure of bipartite networks permits us to devise an algorithm that always manages to keep the nodes in N_2 inactive. We accomplish this objective by starting with a solution whose only active nodes are in N_1, and by performing pushes of length 2; that is, we push flow over two consecutive admissible arcs so that any excess always returns to a node in N_1 and no node in N_2 ever becomes active.

Consider the residual network of a bipartite network given in Figure 8.2(a), with node excesses and distance labels displayed next to the nodes, and residual capacities displayed next to the arcs. The bipartite preflow-push algorithm first pushes flow from node 1 because it is the only active node in the network. The algorithm then identifies an admissible arc emanating from node 1. Suppose that it selects the arc (1, 3). Since we want to find a path of length 2, we now look for an admissible arc emanating from node 3. The arc (3, 2) is one such arc. We perform a push on this path, pushing $\delta = \min\{e(1), r_{13}, r_{32}\} = \min\{6, 5, 4\} = 4$ units. This push saturates arc (3, 2), and completes one iteration of the algorithm. Figure 8.2(b) gives the solution at this point.

In the second iteration, suppose that the algorithm again selects node 1 as an active node and arc (1, 3) as an admissible arc emanating from this node. We would also like to find an admissible arc emanating from node 3, but the network has none. So we relabel node 3. As a result of this relabel operation, arc (1, 3) becomes in-

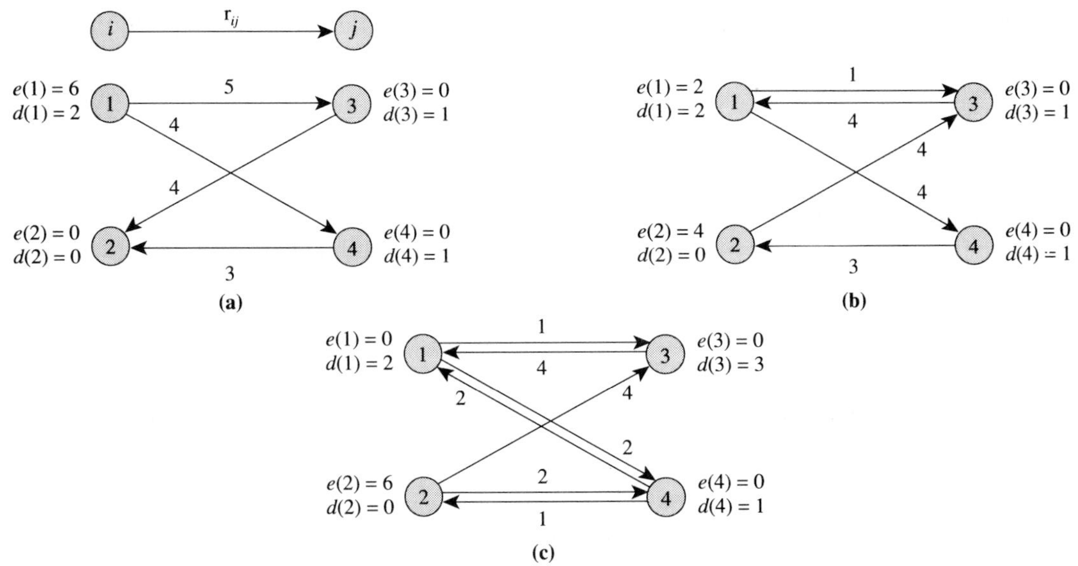

Figure 8.2 Illustrating the bipartite preflow-flow algorithm.

admissible. This operation completes the second iteration and Figure 8.2(b) gives the solution at this stage except that $d(3)$ is 3 instead of 1.

Suppose that the algorithm again selects node 1 as an active node in the third iteration. Then it selects the two consecutive admissible arcs (1, 4) and (4, 2), and pushes $\delta = \min\{e(1), r_{14}, r_{42}\} = \min\{2, 4, 3\} = 2$ units of flow over these arcs. This push is nonsaturating and eliminates the excess at node 1. Figure 8.2(c) depicts the solution at this stage.

As we have illustrated in this numerical example, the bipartite preflow-push algorithm is a simple generalization of the generic preflow-push algorithm. The bipartite algorithm is the same as the generic algorithm given in Figure 7.12 except that we replace the procedure push/relabel(i) by the procedure given in Figure 8.3.

```
procedure bipartite push/relabel(i);
begin
    if the residual network contains an admissible arc (i, j) then
        if the residual network contains an admissible arc (j, k) then
            push δ = min{e(i), r_ij, r_jk} units of flow over the path i–j–k
        else replace d(j) by min{d(k) + 1 : (j, k) ∈ A(j) and r_jk > 0}
    else replace d(i) by min{d(j) + 1 : (i, j) ∈ A(i) and r_ij > 0};
end;
```

Figure 8.3 Push/relabel operation for bipartite networks.

Lemma 8.5. *The bipartite preflow-push algorithm performs $O(n_1^2 m)$ nonsaturating pushes and runs in $O(n_1^2 m)$ time.*

Proof. The proof is same as that of Lemma 7.15. We consider the potential function $\Phi = \sum_{i \in I} d(i)$ whose index set I is the set of active nodes. Since we allow only the nodes in N_1 to be active, and $d(i) \leq 4n_1$ for all $i \in N_1$, the initial value of

Φ is at most $4n_1^2$. Let us observe the effect of executing the procedure bipartite push/relabel(i) on the potential function Φ. The procedure produces one of the following four outcomes: (1) it increases the distance label of node i; (2) it increases the distance label of a node $j \in N_2$; (3) it pushes flow over the arcs (i, j) and (j, k), saturating one of these two arcs; or (4) it performs a nonsaturating push. In case 1, the potential function Φ increases, but the total increase over all such iterations is only $O(n_1^2)$. In case 2, Φ remains unchanged. In case 3, Φ can increase by as much as $4n_1 + 1$ units since a new node might become active; Lemma 8.4 shows that the total increase over all iterations is $O(n_1^2 m)$. Finally, a nonsaturating push decreases the potential function by at least 2 units since it makes node i inactive, can make node k newly active and $d(k) = d(i) - 2$. This fact, in view of the preceding arguments, implies that the algorithm performs $O(n_1^2 m)$ nonsaturating pushes. Since all the other operations, such as the relabel operations and finding admissible arcs, require only $O(n_1 m)$ time, we have established the theorem. ◆

We complete this section by giving two applications of the maximum flow problem on bipartite networks with $n_1 \ll n_2$.

Application 8.1 Baseball Elimination Problem

At a particular point in the baseball season, each of $n + 1$ teams in the American League, which we number as $0, 1, \ldots, n$, has played several games. Suppose that team i has won w_i of the games that it has already played and that g_{ij} is the number of games that teams i and j have yet to play with each other. No game ends in a tie. An avid and optimistic fan of one of the teams, the Boston Red Sox, wishes to know if his team still has a chance to win the league title. We say that we can *eliminate* a specific team 0, the Red Sox, if for every possible outcome of the unplayed games, at least one team will have more wins than the Red Sox. Let $w_{\max}$ denote w_0 plus the total number of games team 0 has yet to play, which, in the best of all possible worlds, is the number of victories the Red Sox can achieve. Then we cannot eliminate team 0 if in some outcome of the remaining games to be played throughout the league, $w_{\max}$ is at least as large as the possible victories of every other team. We want to determine whether we can or cannot eliminate team 0.

We can transform this baseball elimination problem into a feasible flow problem on a bipartite network with two sets with n_1 and $n_2 = \Omega(n_1^2)$. As discussed in Section 6.2, we can represent the feasible flow problem as a maximum flow problem, as shown in Figure 8.4. The maximum flow network associated with this problem contains n *team nodes* 1 through n, $n(n - 1)/2$ *game nodes* of the type i–j for each $1 \le i \le j \le n$, and *source node* s. Each game node i–j has two incoming arcs $(i, i - j)$ and $(j, i - j)$, and the flows on these arcs represent the number of victories for team i and team j, respectively, among the additional g_{ij} games that these two teams have yet to play against each other (which is the required flow into the game node i–j). The flow x_{si} on the source arc (s, i) represents the total number of additional games that team i wins. We cannot eliminate team 0 if this network contains a feasible flow x satisfying the conditions

$$w_{\max} \ge w_i + x_{si} \qquad \text{for all } i = 1, \ldots, n,$$

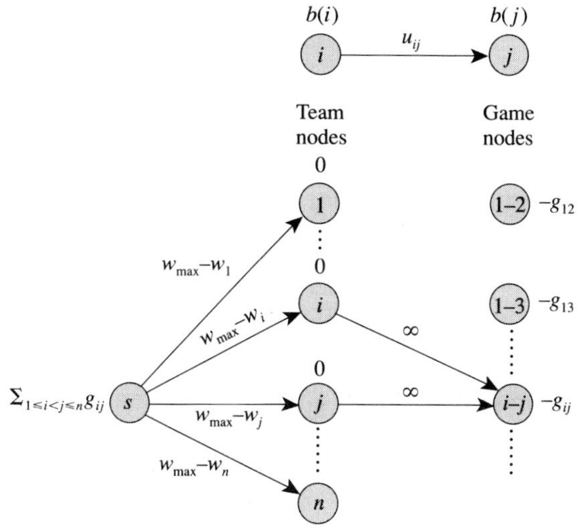

Figure 8.4 Network formulation of the baseball elimination problem.

which we can rewrite as

$$x_{si} \le w_{\max} - w_i \qquad \text{for all } i = 1, \ldots, n.$$

This observation explains the capacities of arcs shown in the figure. We have thus shown that if the feasible flow problem shown in Figure 8.4 admits a feasible flow, we cannot eliminate team 0; otherwise, we can eliminate this team and our avid fan can turn his attention to other matters.

Application 8.2 Network Reliability Testing

In many application contexts, we need to test or monitor the arcs of a network (e.g., the tracks in a rail network) to ensure that the arcs are in good working condition. As a practical illustration, suppose that we wish to test each arc $(i, j) \in A$ in an undirected communication network $G = (N, A)$ α_{ij} times; due to resource limitations, however, each day we can test at most β_j arcs incident to any communication node $j \in N$. The problem is to find a schedule that completes the testing of all the arcs in the fewest number of days.

We solve this problem on a bipartite network $G' = (\{s\} \cup \{t\} \cup N_1 \cup N_2, A')$ defined as follows: The network contains a node $i \in N_1$ for every node $i \in N$ in the communication network and a node i–$j \in N_2$ for every arc $(i, j) \in A$ in the communication network. Each i–j node has two incoming arcs from the nodes in N_1, one from node i and the other from node j; all these arcs have infinite capacity. The source node s is connected to every node $i \in N_1$ with an arc of capacity $\lambda\beta_j$, and every node i–$j \in N_2$ is connected to the sink node t with an arc of capacity α_{ij}. The reliability testing problem is to determine the smallest integral value of the days λ so that the maximum flow in the network saturates all the sink arcs. We can solve this problem by performing binary search on λ and solving a maximum flow problem at each search point. In these maximum flow problem, $|N_1| = n$ and $|N_2| = m$, and m can be as large as $n(n - 1)/2$.

8.4 FLOWS IN PLANAR UNDIRECTED NETWORKS

A network is said to be *planar* if we can draw it in a two-dimensional (Euclidean) plane so that no two arcs cross (or intersect each other); that is, we allow the arcs to touch one another only at the nodes. Planar networks are an important special class of networks that arise in several application contexts. Because of the special structure of planar networks, network flow algorithms often run faster on these networks than they do on more general networks. Indeed, several network optimization problems are NP-complete on general networks (e.g., the maximum cut problem) but can be solved in polynomial time on planar networks. In this section we study some properties of planar networks and describe an algorithm that solves a maximum flow problem in planar networks in $O(n \log n)$ time. In this section we restrict our attention to undirected networks. We remind the reader that the undirected networks we consider contain at most one arc between any pair i and j of nodes. The capacity u_{ij} of arc (i, j) denotes the maximum amount that can flow from node i to node j or from node j to node i.

Figure 8.5 gives some examples of planar networks. The network shown in Figure 8.5(a) does not appear to be planar because arcs (1, 3) and (2, 4) cross one another at the point D, which is not a node. But, in fact, the network is planar because, as shown in Figure 8.5(b), we can redraw it, maintaining the network structure (i.e., node, arc structure), so that the arcs do not cross. For some networks, however, no matter how we draw them, some arcs will always cross. We refer to such networks as *nonplanar*. Figure 8.6 gives two instances of nonplanar networks. In both instances we could draw all but one arc without any arcs intersecting; if we add the last arc, though, at least one intersection is essential. Needless to say, determining whether a network is planar or not a straightforward task. However,

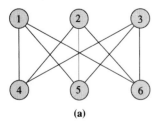

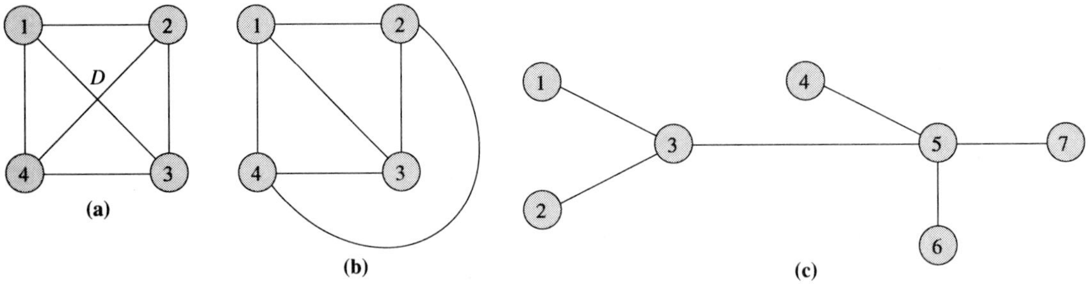

Figure 8.5 Instances of planar networks.

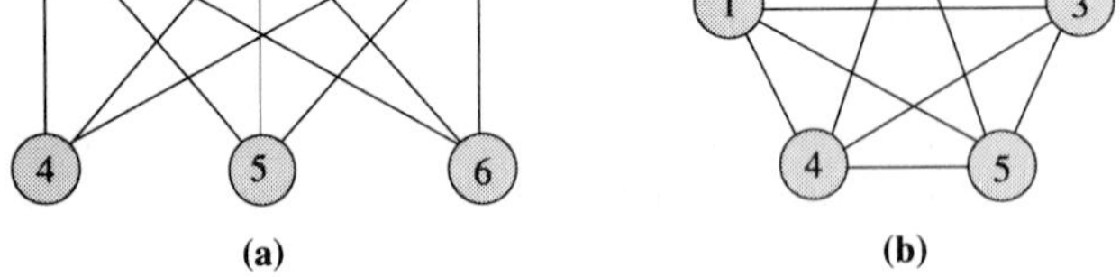

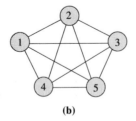

Figure 8.6 Instances of two nonplanar graphs.

researchers have developed very efficient algorithms (in fact, linear time algorithms) for testing the planarity of a network. Several theoretical characterizations of planar networks are also available.

Let $G = (N, A)$ be a planar network. A *face* z of G is a region of the (two-dimensional) plane bounded by arcs that satisfies the condition that any two points in the region can be connected by a continuous curve that meets no nodes and arcs. It is possible to draw a planar graph in several ways and each such representation might have a different set of faces. The *boundary* of a face z is the set of all arcs that enclose it. It is convenient to represent the boundary of a face by a cycle. Observe that each arc in the network belongs to the boundary of at most two faces. Faces z and z' are said to be *adjacent* if their boundaries contain a common arc. If two faces touch each other only at a node, we do not consider them to be adjacent. The network shown in Figure 8.5(b) illustrates these definitions. This network has four faces. The boundaries 1–2–3–1, 1–3–4–1, and 2–4–3–2 define the first three faces of the network. The fourth face is unbounded and consists of the remaining region; its boundary is 1–2–4–1. In Figure 8.5(b) each face is adjacent to every other face. The network shown in Figure 8.5(c) is a very special type of planar network. It has one unbounded face and its boundary includes all the arcs.

Next we discuss two well-known properties of planar networks.

Property 8.6 (Euler's Formula). *If a connected planar network has n nodes, m arcs, and f faces, then* $f = m - n + 2$.

Proof. We prove this property by performing induction on the value of f. For $f = 1$, $m = n - 1$, because a connected graph with just one face (which is the unbounded face) must be a spanning tree. Now assume, inductively, that Euler's formula is valid for every graph with k or fewer faces; we prove that the formula is valid for every graph with $k + 1$ faces. Consider a graph G with $k + 1$ faces and n nodes. We select any arc (i, j) that belongs to two faces, say z_1 and z_2 (show that the network always contains such an arc!). If we delete this arc from G, the two faces z_1 and z_2 merge into a single face. The resulting graph G' has m arcs, k faces, and n nodes, and by the induction hypothesis, $k = m - n + 2$. Therefore, if we reintroduce the arc (i, j) into G', we see that $k + 1 = (m + 1) - n + 2$, so Euler's formula remains valid. We have thus completed the inductive step and established Euler's formula in general. ◆

Property 8.7. *In a planar network,* $m < 3n$.

Proof. We prove this property by contradiction. Suppose that $m \geq 3n$. Alternatively,

$$n \geq m/3. \tag{8.1}$$

We next obtain a relationship between f and m. Since the network contains no parallel arcs, the boundary of each face contains at least three arcs. Therefore, if we traverse the boundaries of all the faces one by one, we traverse at least $3f$ arcs. Now notice that we would have traversed each arc in the network at most twice because it belongs to the boundaries of at most two faces. These observations show that $3f \leq 2m$. Alternatively,

$$f \leq 2m/3. \tag{8.2}$$

Using (8.1) and (8.2) in the formula $f = m - n + 2$, we obtain

$$2 = n - m + f \leq m/3 - m + 2m/3 = 0, \tag{8.3}$$

which is a contradiction. ◆

Property 8.7 shows that every planar graph is very sparse [i.e., $m = O(n)$]. This result, by itself, improves the running times for most network flow algorithms. For instance, as shown in Section 8.5, the shortest augmenting path algorithm for the maximum flow problem, implemented using the dynamic tree data structure, runs in $O(nm \log n)$ time. For planar networks, this time bound becomes $O(n^2 \log n)$. We can, in fact, develop even better algorithms by using the special properties of planar networks. To illustrate this point, we prove some results that apply to planar networks, but not to nonplanar networks. We show that we can obtain a minimum cut and a maximum flow for any planar network in $O(n \log n)$ time by solving a shortest path problem.

Finding Minimum Cuts Using Shortest Paths

Planar networks have many special properties. In particular, every connected planar network $G = (N, A)$ has an associated "twin" planar network $G^* = (N^*, A^*)$, which we refer to as the *dual* of G. We construct the dual G^* for a given graph G as follows. We first place a node f^* inside each face f of G. Each arc in G has a corresponding arc in G^*. Every arc (i, j) in G belongs to the boundaries of either (1) two faces, say f_1 and f_2; or (2) one face, say f_1. In case 1, G^* contains the arc (f_1^*, f_2^*); in case 2, G^* contains the loop (f_1^*, f_1^*). Figure 8.7, which illustrates this construction, depicts the dual network by dashed lines.

For notational convenience, we refer to the original network G as the *primal network*. The number of nodes in the dual network equals the number of faces in the primal network, and conversely, the number of faces in the dual network equals the number of nodes in the primal network. Both the primal and dual networks have

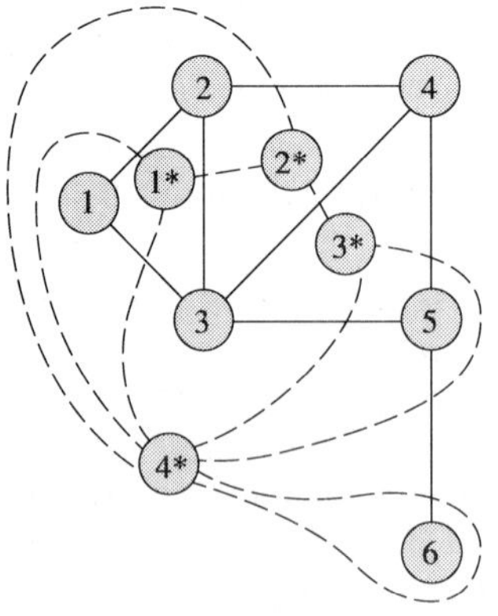

Figure 8.7 Constructing the dual of a planar network.

the same number of arcs. Furthermore, the dual of the dual network is the primal network. It is easy to show that a cycle in the dual network defines a cut in the primal network, and vice versa. For example, the cycle 4*–1*–2*–4* in the dual network shown in Figure 8.7 [with (4*, 1*) denoting the arc from 4* to 1* that also passes through arc (1, 2)], defines the cut {(2, 1), (2, 3), (2, 4)} in the primal network.

Our subsequent discussion in this section applies to a special class of planar networks known as *s–t planar* networks. A planar network with a source node s and a sink node t is called *s–t planar* if nodes s and t both lie on the boundary of the unbounded face. For example, the network shown in Figure 8.8(a) is s–t planar if $s = 1$ and $t = 8$; however, it is not s–t planar if (1) $s = 1$ and $t = 6$, or (2) $s = 3$ and $t = 8$.

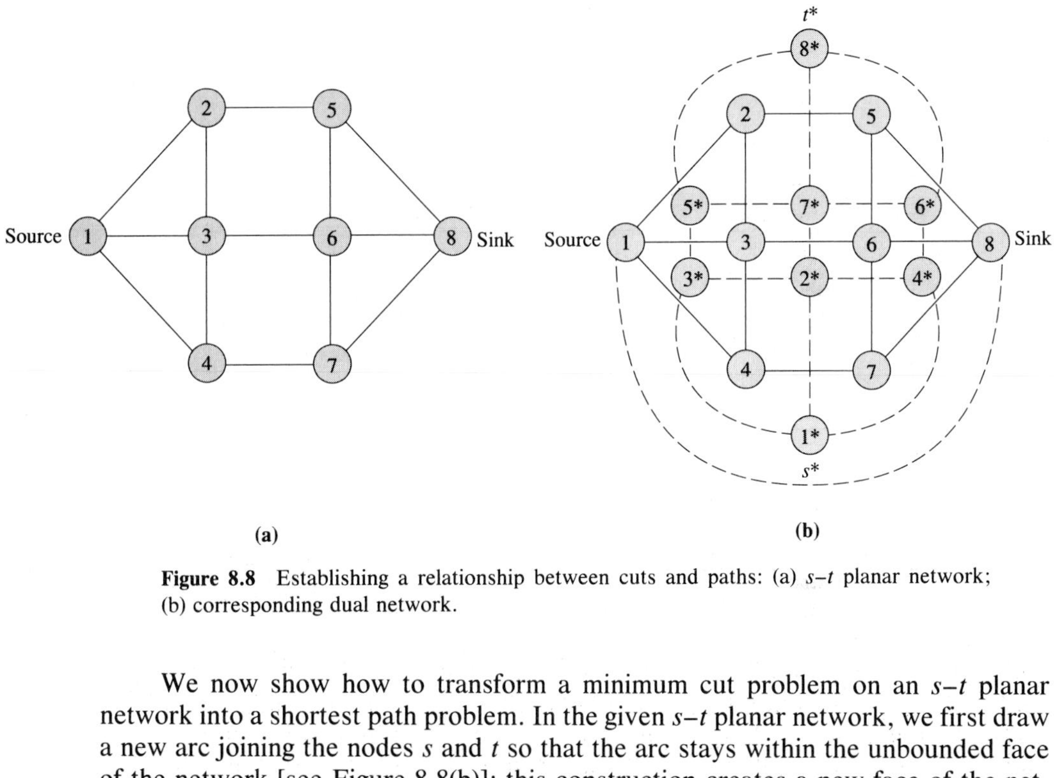

Figure 8.8 Establishing a relationship between cuts and paths: (a) s–t planar network; (b) corresponding dual network.

We now show how to transform a minimum cut problem on an s–t planar network into a shortest path problem. In the given s–t planar network, we first draw a new arc joining the nodes s and t so that the arc stays within the unbounded face of the network [see Figure 8.8(b)]; this construction creates a new face of the network, which we call the *additional face*, but maintains the network's planarity. We then construct the dual of this network; we designate the node corresponding to the additional face as the dual source s^* and the node corresponding to the unbounded face as the dual sink t^*. We set the cost of an arc in the dual network equal to the capacity of the corresponding arc in the primal network. The dual network contains the arc (s^*, t^*) which we delete from the network. Figure 8.8(b) shows this construction: the dashed lines are the arcs in the dual network. It is easy to establish a one-to-one correspondence between s–t cuts in the primal network and paths from node s^* to node t^* in the dual network; moreover, the capacity of the cut equals the cost of the corresponding path. Consequently, we can obtain a minimum s–t cut in the primal network by determining a shortest path from node s^* to node t^* in the dual network.

In the preceding discussion we showed that by solving a shortest path problem in the dual network, we can identify a minimum cut in a primal s–t planar network. Since we can solve the shortest path problem in the dual network in $O(m \log n) = O(n \log n)$ using the binary heap implementation of Dijkstra's algorithm (see Section 4.7), this development provides us with an $O(n \log n)$ algorithm for identifying a minimum cut in a planar network. Notice that this bound is substantially better than the one we would obtain for a general network. We now give a generalization of this result, obtaining a rather surprising result that the shortest path distances in the dual network provide a maximum flow in the primal network.

Let $d(j^*)$ denote the shortest path distance from node s^* to node j^* in the dual network. Recall from Section 5.2 that the shortest path distances satisfy the following conditions:

$$d(j^*) \le d(i^*) + c_{i^*j^*} \qquad \text{for each } (i^*, j^*) \in A^*. \tag{8.4}$$

Each arc (i, j) in the primal network corresponds to an arc (i^*, j^*) in the dual network. Let us define a function x_{ij} for each $(i, j) \in A$ in the following manner:

$$x_{ij} = d(j^*) - d(i^*). \tag{8.5}$$

Note that $x_{ij} = -x_{ji}$. Now notice that the network G is undirected so that the arc set A contains both the arc (i, j) and the arc (j, i). Hence we can regard a negative flow on arc (j, i) as a positive flow on arc (i, j). Consequently, the flow vector x will always nonnegative.

The expressions (8.5) and (8.4) imply that

$$x_{ij} = d(j^*) - d(i^*) \le c_{i^*j^*}. \tag{8.6}$$

Therefore, the flow x satisfies the arc capacity constraints. We next show that x also satisfies the mass balance constraints. Each node k in G, except node s and node t, defines a cut $Q = [\{k\}, N - \{k\}]$ consisting of all of the arcs incident to that node. The arcs in G^* corresponding to arcs in Q define a cycle, say W^*. For example, in Figure 8.7, the cycle corresponding to the cut for $k = 3$ is 1^*–2^*–3^*–4^*–1^*. Clearly,

$$\sum_{(i^*, j^*) \in W^*} (d(j^*) - d(i^*)) = 0, \tag{8.7}$$

because the terms cancel each other. Using (8.5) in (8.7) shows that

$$\sum_{(i, j) \in Q} x_{ij} = 0,$$

which implies that inflow equals outflow at node k. Finally, we show that the flow x is a maximum flow. Let P^* be a shortest path from node s^* to node t^* in G^*. The definition of P^* implies that

$$d(j^*) - d(i^*) = c_{i^*j^*} \qquad \text{for each } (i^*, j^*) \in P^*. \tag{8.8}$$

The arcs corresponding to P^* define an s–t cut Q in the primal network. Using (8.5) in expression (8.8) and using the fact that $c_{i^*j^*} = u_{ij}$, we get

$$x_{ij} = u_{ij} \qquad \text{for each } (i, j) \in Q. \tag{8.9}$$

Consequently, the flow saturates all the arcs in an s–t cut and must be a maximum flow. The following theorem summarizes our discussion.

Theorem 8.8. *It is possible to determine a maximum flow in an s–t planar network in O(n log n) time.* ◆

8.5 DYNAMIC TREE IMPLEMENTATIONS

A dynamic tree is an important data structure that researchers have used extensively to improve the worst-case complexity of several network algorithms. In this section we describe the use of this data structure for the shortest augmenting path algorithm. We do not describe how to actually implement the dynamic tree data structure; rather, we show how to use this data structure as a "black box" to improve the computational complexity of certain algorithms. Our objective is to familiarize readers with this important data structure and enable them to use it as a black box module in the design of network algorithms.

The following observation serves as a motivation for the dynamic tree structure. The shortest augmenting path algorithm repeatedly identifies a path consisting solely of admissible arcs and augments flows on these paths. Each augmentation saturates some arcs on this path, and by deleting all the saturated arcs from this path we obtain a set of *path fragments*: sets of partial paths of admissible arcs. The path fragments contain valuable information. If we reach a node in any of these path fragments using any augmenting path, we know that we can immediately extend the augmenting path along the path fragment. The standard implementation of the shortest augmenting path algorithm discards this information and possibly regenerates it again at future steps. The dynamic tree data structure cleverly stores these path fragments and uses them later to identify augmenting paths quickly.

The dynamic tree data structure maintains a collection of node-disjoint rooted trees, each arc with an associated value, called *val*. See Figure 8.9(a) for an example of the node-disjoint rooted trees. Each rooted tree is a directed in-tree with a unique root. We refer to the nodes of the tree by using the terminology of a predecessor–successor (or parent–child) relationship. For example, node 5 is the predecessor (parent) of nodes 2 and 3, and nodes 9, 10, and 11 are successors (children) of node 12. Similarly, we define the ancestors and descendants of a node (see Section 2.2 for these definitions). For example, in Figure 8.9(a) node 6 has nodes 1, 2, 3, 4, 5, 6 as its descendants, and nodes 2, 5, and 6 are the ancestors of node 2. Notice that, according to our definitions, each node is its own ancestor and descendant.

This data structure supports the following six operations:

find-root(i). Find and return the root of the tree containing node i.

find-value(i). Find and return the value of the tree arc leaving node i. If i is a root node, return the value ∞.

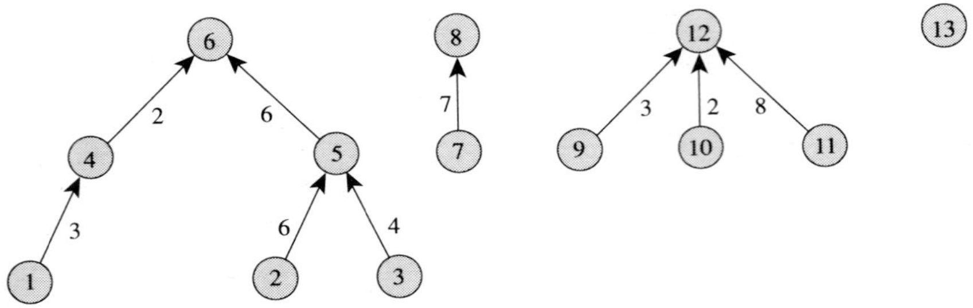

(a)

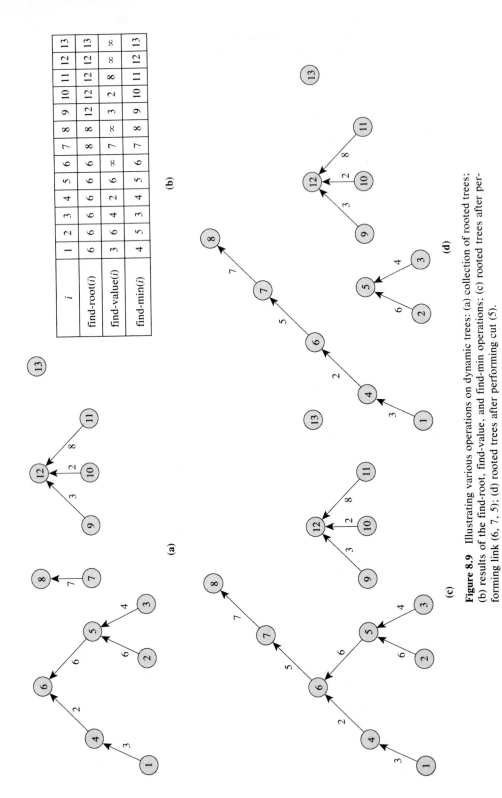

i	1	2	3	4	5	6	7	8	9	10	11	12	13
find-root(i)	6	6	6	6	6	6	8	8	12	12	12	12	13
find-value(i)	3	6	4	2	6	∞	7	∞	3	2	8	∞	∞
find-min(i)	4	5	3	4	5	6	7	8	9	10	11	12	13

(b)

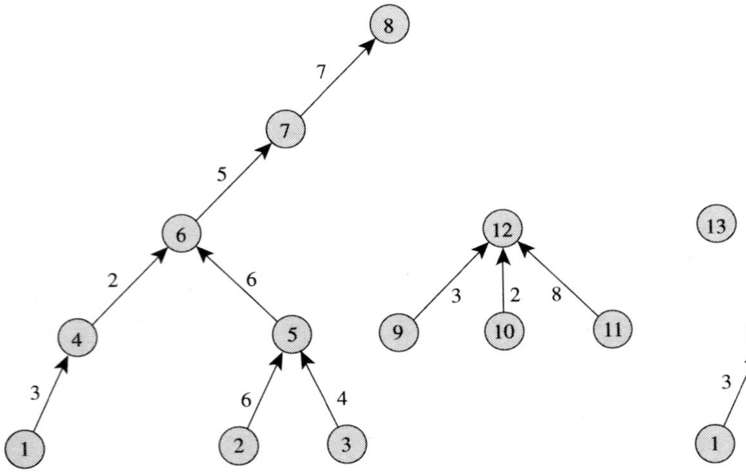

(c)

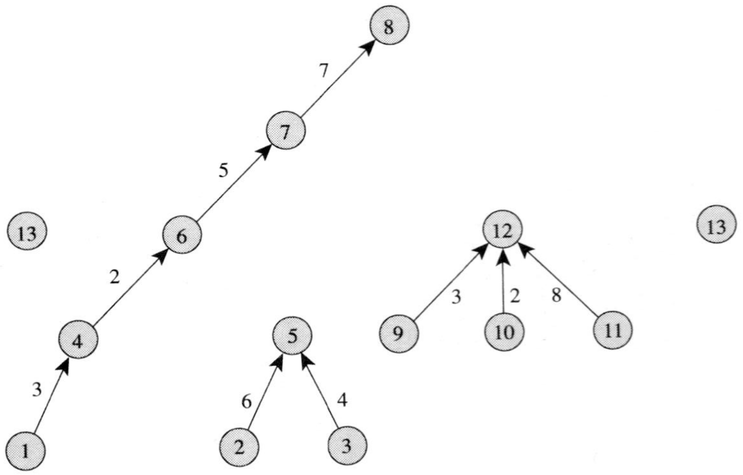

(d)

Figure 8.9 Illustrating various operations on dynamic trees: (a) collection of rooted trees; (b) results of the find-root, find-value, and find-min operations; (c) rooted trees after performing link (6, 7, 5); (d) rooted trees after performing cut (5).

find-min(i). Find and return the ancestor w of i with the minimum value of find-value(w). In case of a tie, chose the node w closest to the tree root.

Figure 8.9(b) shows the results of the operations find-root(i), find-value(i), and find-min(i) performed for different nodes i.

change-value(i, val). Add a real number *val* to the value of every arc along the path from node i to find-root(i). For example, if we execute change-value(1, 3) for the dynamic tree shown in Figure 8.9(a), we add 3 to the values of arcs (1, 4) and (4, 6) and these values become 6 and 5, respectively.

link(i, j, val). This operation assumes that i is a tree root and that i and j belong to different trees. The operation combines the trees containing nodes i and j by making node j the parent of node i and giving arc (i, j) the value *val*. As an illustration, if we perform the operation link (6, 7, 5) on our example, we obtain the trees shown in Figure 8.9(c).

cut(i). Break the tree containing node i into two trees by deleting the arc joining node i to its parent and returning the value of the deleted arc. We perform this operation when i is not a tree root. For example, if we execute cut(5) on trees in Figure 8.9(c), we delete arc (5, 6) and return its value 6. Figure 8.9(d) gives the new collection of trees.

The following important result, which we state without proof, lies at the heart of the efficiency of the dynamic tree data structure.

Lemma 8.9. *If z is the maximum tree size (i.e., maximum number of nodes in any tree), a sequence of l tree operations, starting with an initial collection of singleton trees, requires a total of $O(l \log(z + l))$ time.* ◆

The dynamic tree implementation stores the values of tree arcs only implicitly. If we were to store these values explicitly, the operation change-value on a tree of size z might require $O(z)$ time (if this tree happens to be a path), which is computationally excessive for most applications. Storing the values implicitly allows us to update the values in only $O(\log z)$ time. How the values are actually stored and manipulated is beyond the scope of this book.

How might we use the dynamic tree data structure to improve the computational performance of network flow algorithms. Let us use the shortest augmenting path algorithm as an illustration. The following basic idea underlies the algorithmic speed-up. In the dynamic tree implementation, each arc in the rooted tree is an admissible arc [recall that an arc (i, j) is admissible if $r_{ij} > 0$ and $d(i) = d(j) + 1$]. The value of an arc is its residual capacity. For example, consider the residual network given in Figure 8.10(a), which shows the distance labels next to the nodes and residual capacities next to the arcs. Observe that in this network, every arc, except the arc (12, 13), is admissible; moreover, the residual capacity of every arc is 2, except for the arc (12, 14) whose residual capacity is 1. Figure 8.10(b) shows one collection of rooted trees for this example. Notice that although every tree arc is admissible, every admissible arc need not be in some tree. Consequently, for a given set of admissible arcs, many collections of rooted trees are possible.

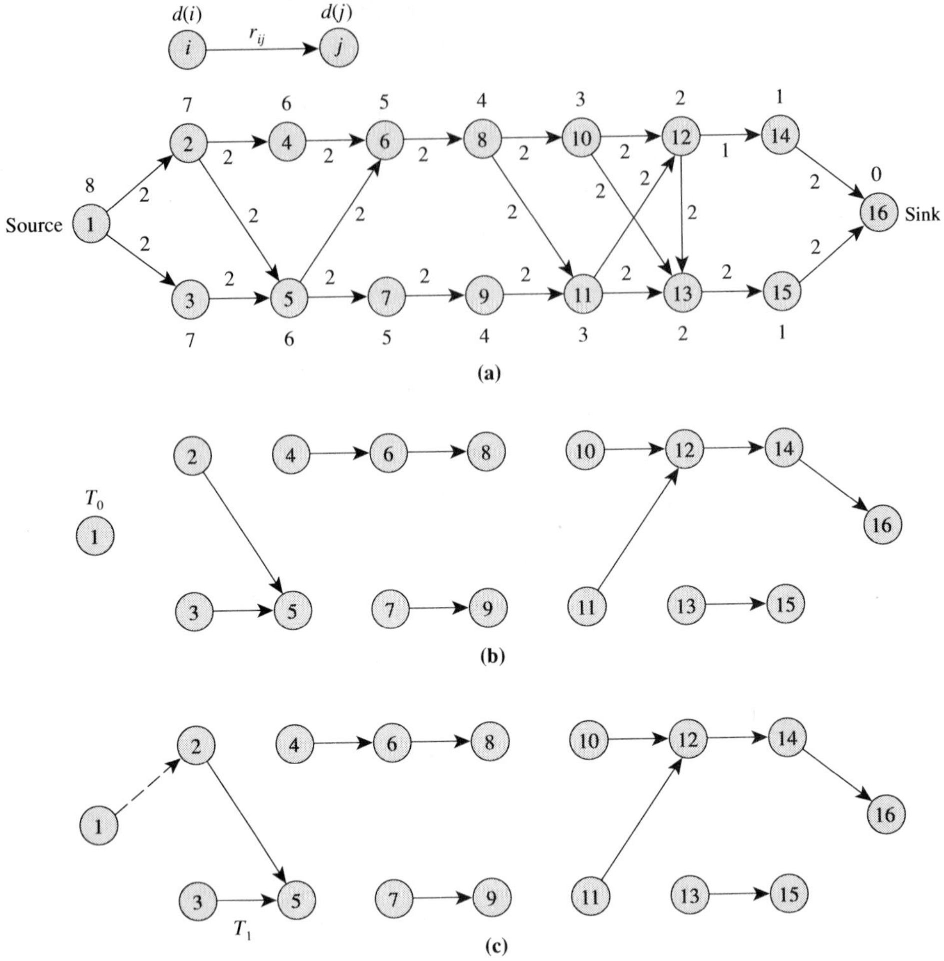

Figure 8.10 Finding an augmenting path in the dynamic tree implementations.

Before describing this implementation formally, we first show how the algorithm works on our example. It maintains a rooted tree containing the source node and progressively expands this tree until it contains the sink node, at which point the algorithm performs an augmentation. To grow the tree containing the source, the algorithm repeatedly performs link operations. In our example, the algorithm starts with the singleton tree T_0 containing only the source node 1 [see Figure 8.10(b)]. It identifies an admissible arc emanating from node 1. Suppose that we select arc (1, 2). The algorithm performs the operation link(1, 2, 2), which joins two rooted trees, giving us a larger tree T_1 containing node 1 [see Figure 8.10(c)]. The algorithm then identifies the root of T_1, by performing the operation find-root(1), which identifies node 5. The algorithm tries to find an admissible arc emanating from node 5. Suppose that the algorithm selects the arc (5, 6). The algorithm performs the operation link(5, 6, 2) and obtains a larger tree T_2 containing node 1 [see Figure 8.10(d)]. In the next iteration, the algorithm identifies node 8 as the root of T_2. Suppose that

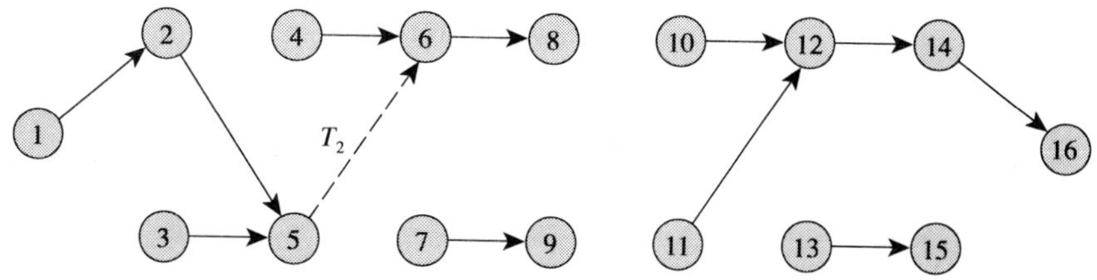

(d)

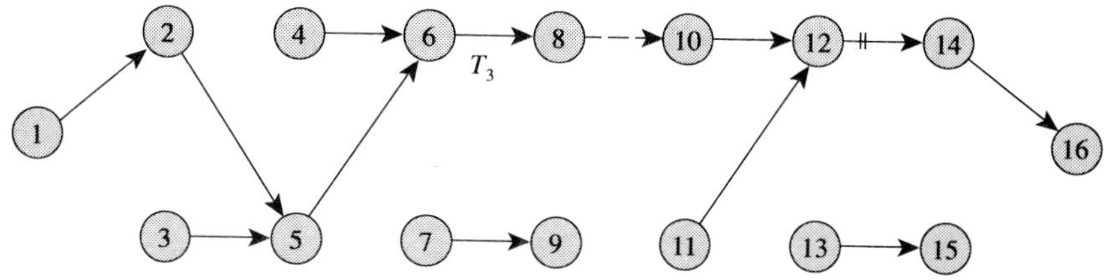

(e)

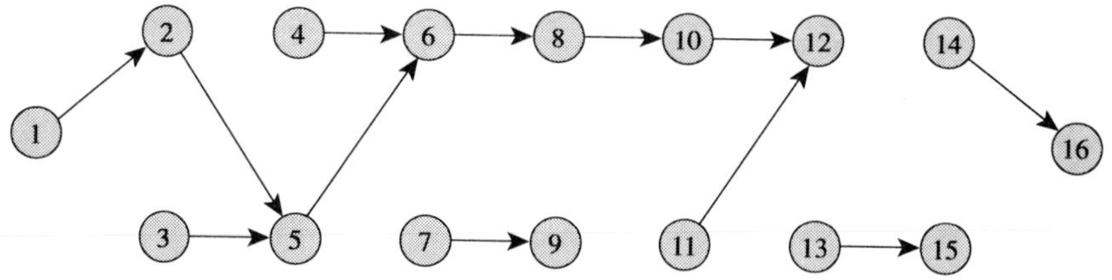

(f)

Figure 8.10 (*Continued*)

the algorithm selects arc (8, 10) as the admissible arc emanating from node 8. The algorithm performs the operation link(8, 10, 2) and obtains a rooted tree T_3 that contains both the source and sink nodes [see Figure 8.10(e)].

Observe that the unique path from the source to the sink in T_3 is an admissible path since by construction every arc in a rooted tree is admissible. The residual capacity of this path is the minimum value of the arcs in this path. How can we determine this value? Recall that the operation find-min(1) would determine an ancestor of node 1 with the minimum value of find-value, which is node 12 in our example. Performing find-value(12) will give us the residual capacity of this path, which is 1 in this case. We have thus discovered the possibility of augmenting 1 unit of flow along the admissible path and that arc (12, 14) is the blocking arc. We perform the augmentation by executing change-value(1, -1). This augmentation reduces the residual capacity of arc (12, 14) to zero. The arc (12, 14) now becomes inadmissible and we must drop it from the collection of rooted trees. We do so by performing cut(12). This operation gives us the collection of rooted trees shown in Figure 8.10(f).

To better understand other situations that might occur, let us execute the algorithm for one more iteration. We apply the dynamic tree algorithm starting with the collection of rooted trees given in Figure 8.10(f). Node 12 is the root of the tree containing node 1. But node 12 has no outgoing admissible arc; so we relabel node 12. This relabeling increases the distance label of node 12 to 3. Consequently, arcs (10, 12) and (11, 12) become inadmissible and we must drop them from the collection of rooted trees. We do so by performing cut(10) and cut(11), giving us the rooted trees shown in Figure 8.11(a). The algorithm again executes find-root(1) and finds node 10 as the root of the tree containing node 1. In the next two operations, the algorithm adds arcs (10, 13) and (15, 16); Figure 8.11(b) and 8.11(c) shows the corresponding trees.

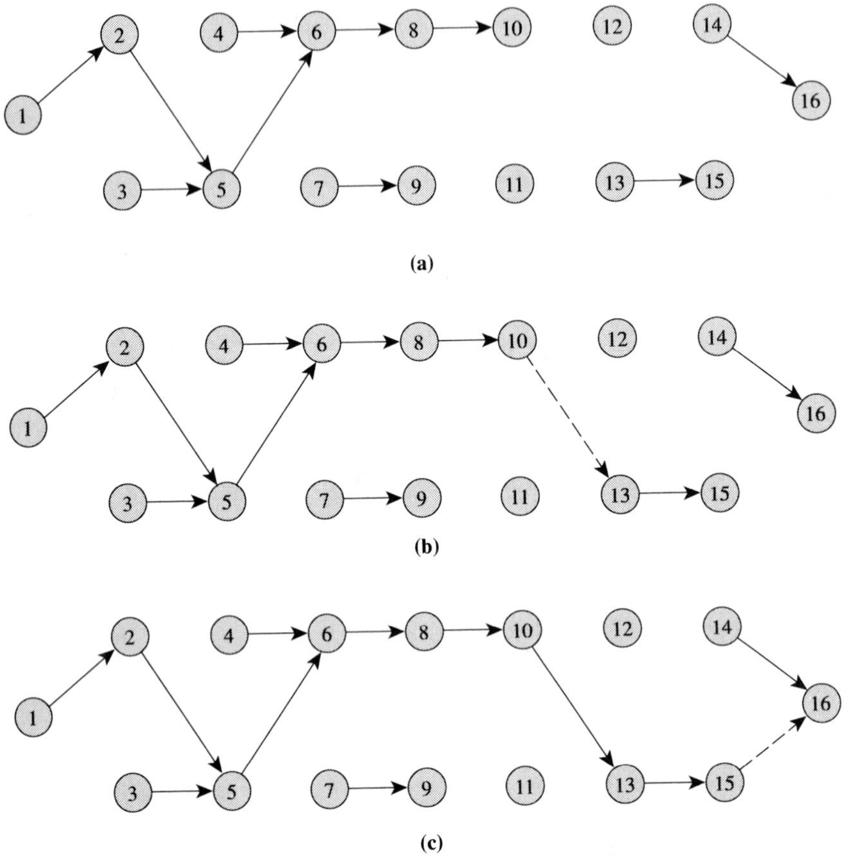

Figure 8.11 Another augmentation using dynamic trees.

Figures 8.12 and 8.13 give a formal statement of the algorithm. After offering explanatory comments, we consider a worst-case analysis of the algorithm. The algorithm is same as the one we presented in Section 7.4 except that it performs the procedures advance, retreat, and augment differently using trees. The first two procedures, tree-advance and tree-retreat, are straightforward, but the tree-augment procedure requires some explanation. If node p is an ancestor of node s with the minimum value of find-value(p) and, among such nodes in the path, it is closest to the sink, then find-value(p) gives the residual capacity of the augmenting path. The operation change(s, $-\delta$) implicitly updates the residual capacities of all the arcs in

```
algorithm tree-augmenting-path;
begin
    x : = 0;
    perform a reverse breadth-first search of the residual network
        from node t to obtain the distance labels d(i);
    let T be the collection of all singleton nodes;
    i : = s;
    while d(s) < n do
    begin
        if i has an admissible arc then tree-advance(i)
        else tree-retreat(i);
        if i = t then tree-augment;
    end;
end;
```

Figure 8.12 Dynamic tree implementation of the shortest augmenting path algorithm.

the augmenting path. This augmentation might cause the capacity of more than one arc in the path to become zero. The **while** loop identifies all such arcs, one by one, and deletes them from the collection of rooted trees.

We now consider the worst-case complexity of the algorithm. Why is the dynamic tree implementation more efficient than the original implementation of the shortest augmenting path algorithm? The bottleneck operations in the original shortest augmenting path algorithm are the advance and augment operations, which require $O(n^2m)$ time. Each advance operation in the original algorithm adds one arc;

```
procedure tree-advance(i);
begin
    let (i, j) be an admissible arc in A(i);
    link(i, j, r_ij);
    i : = find-root(j);
end;
```

(a)

```
procedure tree-retreat(i);
begin
    d(i) : = min{d(j) + 1 : (i, j) ∈ A(i) and r_ij > 0};
    for each tree arc (k, i) do cut(k);
    i : = find-root(s);
end;
```

(b)

```
procedure tree-augment;
begin
    p : = find-min(s);
    δ : = find-value(p);
    change-value(s, -δ);
    while find-value(p) = 0 do cut(p) and set p : = find-min(s);
    i : = find-root(s);
end;
```

(c)

Figure 8.13 Procedures of the tree-augmenting-path algorithm.

in contrast, the tree implementation adds a collection of arcs using the link operation. Thus the dynamic tree implementation substantially reduces the number of executions of the link operation. Similarly, while augmenting flow, the tree implementation augments flow over a collection of arcs by performing the operation change-value, thus again substantially reducing the number of required updates.

We now obtain a bound on the number of times the algorithm performs various tree operations. We will show that the algorithm performs each of the tree operations $O(nm)$ times. In deriving these bounds, we make use of the results of Lemma 7.9, proved in Section 7.4.

cut(j). The algorithm performs this operation during the tree-retreat and tree-augment operations. During the tree-retreat(i) operation, the algorithm might perform this operation as many times as the number of incoming arcs at node i. Since this operation relabels node i, and we can relabel a node at most n times, these operations sum to $O(n^2)$ over all nodes. Furthermore, during the tree-augment operation, we perform the cut operation for each arc saturated during an augmentation. Since the total number of arc saturations is $O(nm)$, the number of these operations sums to $O(nm)$.

link(i, j, val). Each link operation adds an arc to the collection of rooted trees. Observe that if an arc enters a rooted tree, it remains there until a cut operation deletes it from the tree. Therefore, the number of link operations is at most $(n - 1)$ plus the number of cut operations. The term $(n - 1)$ arises because initially the collection might contain no arc, and finally, it might contain as many as $(n - 1)$ arcs. Consequently, the total number of link operations is also $O(nm)$. Since each tree-advance operation performs a link operation, the previous result also implies an $O(nm)$ bound on the total number of tree-advance operations.

change-value(i, val). The algorithm performs this operation once per augmentation. Since the number of augmentations is at most $nm/2$, we immediately obtain a bound of $O(nm)$ on the number of change-value operations.

find-value(i) and find-min(i). The algorithm performs each of these two operations once per augmentation and once for each arc saturated during the augmentation. These observations imply a bound of $O(nm)$ on the number of executions of these two operations.

find-root(i). The algorithm performs this operation once during each execution of the tree-advance, tree-augment, and tree-retreat operations. Since the algorithm executes the first two operations $O(nm)$ times and the third operation $O(n^2)$ times, it executes the find-root operation $O(nm)$ times.

Using simple arguments, we have now shown that the algorithm performs each of the six tree operations $O(nm)$ times. It performs each tree operation on a tree of maximum size n. The use of Lemma 8.9 establishes the following important result.

Theorem 8.10. *The dynamic tree implementation of the shortest augmenting path algorithm solves the maximum flow problem in $O(nm \log n)$ time.* ◆

Although this result establishes the theoretical utility of the dynamic tree data structure for improving the worst-case complexity of the shortest augmenting path algorithm, the practical value of this data structure is doubtful. The dynamic tree implementation reduces the time for performing advance and augment operations from $O(n^2m)$ to $O(nm \log n)$, but simultaneously increases the time of performing retreat operations from $O(nm)$ to $O(nm \log n)$. Empirical experience shows that the retreat operation is one of the bottleneck operations in practice. Since the dynamic tree data structure increases the running time of a bottleneck operation, the use of this data structure actually slows down the algorithm in practice. Furthermore, this data structure introduces substantial overhead (i.e., a large-constant factor of work is associated with each tree operation), thus making it of limited practical utility.

8.6 NETWORK CONNECTIVITY

The connectivity of a network is an important measure of the network's reliability or stability. The *arc connectivity* of a connected network is the minimum number of arcs whose removal from the network disconnects it into two or more components. In this section we suggest algorithms for solving the arc connectivity problem on undirected networks. The algorithms for solving connectivity problems on directed networks are different from those we will be discussing; we consider these algorithms in the exercises for this chapter. To conform with this choice of coverage, in this section by the term "network" we will invariably mean an undirected (connected) network.

The *node connectivity* of a network equals the minimum number of nodes whose deletion from the network disconnects it into two or more components. We discuss issues related to node connectivity in Exercise 8.35. We begin by defining several terms. A *disconnecting set* is a set of arcs whose deletion from the network disconnects it into two or more components. Therefore, the arc connectivity of a network equals the minimum cardinality of any disconnecting set; we refer to this set of arcs as a *minimum disconnecting set.*

The arc connectivity of a pair of nodes i and j is the minimum number of arcs whose removal from the network disconnects these two nodes. We represent the pair of nodes i and j by $[i, j]$ and the arc connectivity of this pair by $\alpha[i, j]$. We also let $\alpha(G)$ denote the arc connectivity of the network G. Consequently,

$$\alpha(G) = \min\{\alpha[i, j] : [i, j] \in N \times N\}. \tag{8.10}$$

We first bring together some elementary facts concerning the arc connectivity of a network; we ask the reader to prove these properties in Exercise 8.29.

Property 8.11.

(a) $\alpha[i, j] = \alpha[j, i]$ *for every pair of nodes* $[i, j]$.

(b) *The arc connectivity of a network cannot exceed the minimum degree of nodes in the network. Therefore,* $\alpha(G) \le \lfloor m/n \rfloor$.

(c) *Any minimum disconnecting set partitions the network into exactly two components.*
(d) *The arc connectivity of a spanning tree equals* 1.
(e) *The arc connectivity of a cycle equals* 2.

Let δ denote the minimum degree of a node in the network and let node p be a node with degree equal to δ. Property 8.11(b) implies that $\alpha(G) \leq \delta \leq \lfloor m/n \rfloor$. Since a minimum disconnecting set partitions the node set into exactly two components $S^* \subset N$ and $\bar{S}^* = N - S^*$, we can represent this cut by the notation $[S^*, \bar{S}^*]$. We assume, without any loss of generality, that node $p \in S^*$.

Our development in Chapter 6 provides us with a means for determining the arc connectivity of a network. Theorem 6.7 states that the minimum number of arcs in a network whose removal disconnects a specified pair of source and sink nodes equals the maximum number of arc-disjoint paths from the source to the sink. Furthermore, the proof of this theorem shows that we can obtain the maximum number of arc-disjoint paths from the source to the sink by solving a maximum flow problem in a network G whose arcs all have capacity equal to 1. Thus, to determine the arc connectivity of a network, we need to solve a unit capacity maximum flow problem between every pair of nodes (by varying the source and sink nodes); the minimum value among such flows is $\alpha(G)$. Since solving a unit capacity maximum flow problem requires $O(\min\{n^{2/3}m, m^{3/2}\})$ time (see Section 8.2), this approach produces an algorithm running in time $O(n^2 \min\{n^{2/3}m, m^{3/2}\})$.

We can easily improve on this approach by a factor of n using the following idea. Consider a node $k \in \bar{S}^*$ and recall that node $p \in S^*$. Since the cut $[S^*, \bar{S}^*]$ disconnects nodes p and k, the minimum cardinality of a set of arcs that will disconnect these two nodes is at most $|[S^*, \bar{S}^*]|$. That is,

$$\alpha[p, k] \leq |[S^*, \bar{S}^*]|. \tag{8.11}$$

Next observe that $[S^*, \bar{S}^*]$ is a minimum disconnecting set of the network. The definition (8.10) of $\alpha(G)$ implies that

$$\alpha[p, k] \geq |[S^*, \bar{S}^*]|. \tag{8.12}$$

Using (8.11) and (8.12), we see that

$$\alpha[p, k] = |[S^*, \bar{S}^*]|. \tag{8.13}$$

The preceding observations imply that if we compute $\alpha[p, j]$ for all j, the minimum among these numbers equals $\alpha(G)$. To summarize the discussion, we can write

$$\alpha(G) = \min\{\alpha[p, j] : j \in N - \{p\}\}.$$

The preceding approach permits us to determine the arc connectivity of a network by solving $(n - 1)$ unit capacity maximum flow problems, requiring $O(\min\{n^{5/3}m, nm^{3/2}\})$ time. We can improve this time bound for sparse networks by solving these maximum flow problems using the labeling algorithm described in Section 6.5 instead of the specialized unit capacity algorithms described in Section 8.2. The labeling algorithm will perform at most $\lfloor m/n \rfloor$ augmentations to solve each maximum flow problem (because the degree of node p is $\delta \leq \lfloor m/n \rfloor$) and would require $O(m^2/n)$ time. This approach requires $O(m^2)$ time to solve all the maximum flow problems. Since $nm^{3/2} \geq m^2$, we can determine the arc connectivity of a network in $O(\min\{n^{5/3}m, m^2\})$ time. This algorithm is by no means the best algorithm for determining the arc connectivity of a network. We next describe an algorithm that computes arc connectivity in only $O(nm)$ time.

Just as the preceding algorithm determines the minimum cardinality of a set of arcs between pairs of nodes, the improved algorithm determines the minimum cardinality of a set of arcs that disconnects every node in a set S from some node $k \in \bar{S}$; we denote this number by $\alpha[S, k]$. We can compute $\alpha[S, k]$ using the labeling algorithm as follows: We allow the augmentation to start at any node in S but end only at node k. When the labeling algorithm terminates, the network contains no directed path from any node in S to node k. At this point the set of labeled nodes defines a cut in the network and the number of forward arcs in the cut is $\alpha[S, k]$.

Our preceding algorithm determines the arc connectivity of a network by computing $\alpha[p, j]$ for each node $j \in N - \{p\}$ and taking the minimum of these numbers. The correctness of this approach uses the fact that $\alpha(G)$ equals $\alpha[p, j]$ for some choice of the nodes p and j. Our improved algorithm determines the arc connectivity of a network by computing $\alpha[S, k]$ for at most $(n - 1)$ combinations of S and k and taking the minimum of these numbers. The algorithm selects the combinations S and k quite cleverly so that (1) for at least one combination of S and k, $\alpha[S, k] = \alpha(G)$; and (2) the labeling algorithm can compute $\alpha[S, k]$ for every combination in an average of $O(m)$ time because most augmentations involve only two arcs. Therefore this algorithm determines the arc connectivity of a network in $O(nm)$ time.

Before describing the algorithm, we first introduce some notation. For any set S of nodes, we let *neighbor*(S) denote the set of nodes in $\bar{S}$ that are adjacent to some node in S, and *nonneighbor*(S) as the set of nodes in $\bar{S}$ that are not adjacent to any node in S. Consequently, $N = S \cup \text{neighbor}(S) \cup \text{nonneighbor}(S)$. Our improved arc connectivity algorithm depends on the following crucial result.

Lemma 8.12. *Let δ be the minimum node degree of a network G and let $[S^*, \bar{S}^*]$ denote a minimum disconnecting set of the network. Suppose that $\alpha(G) \leq \delta - 1$. Then for any set $S \subseteq S^*$, nonneighbor(S) is nonempty.*

Proof. We first notice that the maximum number of arcs emanating from nodes in $\bar{S}^*$ is $|\bar{S}^*|(|\bar{S}^*| - 1) + \alpha(G)$ because any such arc either has both its endpoints in $\bar{S}^*$ or belongs to the minimum disconnecting set. Next notice that the minimum number of arcs emanating from the nodes in $\bar{S}^*$ is $\delta|\bar{S}^*|$ because δ is the minimum node degree. Therefore,

$$|\bar{S}^*|(|\bar{S}^*| - 1) + \alpha(G) \geq |\bar{S}^*|\delta.$$

Adding δ to both the sides of this inequality and simplifying the expression gives

$$(|\bar{S}^*| - 1)(|\bar{S}^*| - \delta) \geq \delta - \alpha(G) \geq 1.$$

The last inequality in this expression follows from the fact $\alpha(G) \leq \delta - 1$. Notice that the inequality $(|\bar{S}^*| - 1)(|\bar{S}^*| - \delta) \geq 1$ implies that both the terms to the left are at least one. Thus $|\bar{S}^*| \geq \delta + 1$; that is, the set $\bar{S}^*$ contains at least $\delta + 1$ nodes. Since the cut $[S^*, \bar{S}^*]$ contains fewer than δ arcs, at least one of the nodes in $\bar{S}^*$ is not adjacent to any node in S^*. Consequently, the set nonneighbor(S) must be nonempty, which establishes the lemma. ◆

The improved arc connectivity algorithm works as follows. It starts with $S = \{p\}$, selects a node $k \in$ nonneighbor(S), and computes $\alpha[S, k]$. It then adds node k to S, updates the sets neighbor(S) and nonneighbor(S), selects another node $k \in$ nonneighbor(S) and computes $\alpha[S, k]$. It repeats this operation until the set nonneighbor(S) is empty. The minimum value of $\alpha[S, k]$, obtained over all the iterations, is $\alpha(G)$. Figure 8.14 gives a formal description of this algorithm.

```
algorithm arc connectivity;
begin
    let p be a minimum degree node in the network and δ be its degree;
    set S* : = {p} and α* : = δ;
    set S : = {p};
    initialize neighbor(S) and nonneighbor(S);
    while nonneighbor(S) is nonempty do
    begin
        select a node k ∈ nonneighbor(S);
        compute α[S, k] using the labeling algorithm for the maximum flow
            problem and let [R, R̄] be the corresponding disconnecting cut;
        if α* > α[S, k] then set α* : = α[S, k] and [S*, S̄*] : = [R, R̄];
        add node k to S and update neighbor(S), nonneighbor(S);
    end;
end;
```

Figure 8.14 Arc connectivity algorithm.

To establish the correctness of the arc connectivity algorithm, let $[S^*, \bar{S}^*]$ denote the minimum disconnecting set. We consider two cases: when $\alpha(G) = \delta$ and when $\alpha(G) \le \delta - 1$. If $\alpha(G) = \delta$, the algorithmic description in Figure 8.14 implies that the algorithm would terminate with $[p, N - \{p\}]$ as the minimum disconnecting set. Now suppose $\alpha(G) \le \delta - 1$. During its execution, the arc connectivity algorithm determines $\alpha[S, k]$ for different combinations of S and k; we need to show that at some iteration, $\alpha(S, k)$ would equal $\alpha(G)$. We establish this result by proving that at some iteration, $S \subseteq S^*$ and $k \in \bar{S}^*$, in which case $\alpha[S, k] = \alpha(G)$ because the cut $[S^*, \bar{S}^*]$ disconnects every node in S from node k. Notice that initially S^* contains S (because both start with p as their only element), and finally it does not because, from Lemma 8.9, as long as S^* contains S, nonneighbor(S) is nonempty and the algorithm can add nodes to S. Now consider the last iteration for which $S \subseteq S^*$. At this iteration, the algorithm selects a node k that must be in $\bar{S}^*$ because $S \cup \{k\} \not\subseteq S^*$. But then $\alpha[S, k] = \alpha(G)$ because the cut $[S^*, \bar{S}^*]$ disconnects S from node k. This conclusion shows that the arc connectivity algorithm correctly solves the connectivity problem.

We next analyze the complexity of the arc connectivity algorithm. The algorithm uses the labeling algorithm described in Section 6.5 to compute $\alpha[S, k]$; suppose that the labeling algorithm examines labeled nodes in the first-in, first-out order so that it augments flow along shortest paths in the residual network. Each augmenting path starts at a node in S, terminates at the node k, and is one of two types: Its last internal node is in neighbor(S) or it is in nonneighbor(S). All augmenting paths of the first type are of length 2; within an iteration, we can find any such path in a total of $O(n)$ time (why?), so augmentations of the first type require a total of $O(n^2)$ time in all iterations. Detecting an augmenting path of the second type requires

$O(m)$ time; it is possible to show, however, that in all the applications of the labeling algorithm in various iterations, we never encounter more than n such augmenting paths. To see this, consider an augmenting path of the second type which contains node $l \in$ nonneighbor(S) as the last internal node in the path. At the end of this iteration, the algorithm will add node k to S; as a result, it adds node l to neighbor(S); the node will stay there until the algorithm terminates. So each time the algorithm performs an augmentation of the second type, it moves a node from the set nonneighbor(S) to neighbor(S). Consequently, the algorithm performs at most n augmentations of the second type and the total time for these augmentations will be $O(nm)$. The following theorem summarizes this discussion.

Theorem 8.13. *In $O(nm)$ time the arc connectivity algorithm correctly determines the arc connectivity of a network.* ◆

8.7 ALL-PAIRS MINIMUM VALUE CUT PROBLEM

In this section we study the all-pairs minimum value cut problem in undirected network, which is defined in the following manner. For a specific pair of nodes i and j, we define an $[i, j]$ *cut* as a set of arcs whose deletion from the network disconnects the network into two components S_{ij} and $\bar{S}_{ij}$ so that nodes i and j belong to different components (i.e., if $i \in S_{ij}$, then $j \in \bar{S}_{ij}$; and if $i \in \bar{S}_{ij}$, then $j \in S_{ij}$). We refer to this $[i, j]$ cut as $[S_{ij}, \bar{S}_{ij}]$ and say that this cut *separates* nodes i and j. We associate with a cut $[S_{ij}, \bar{S}_{ij}]$, a *value* that is a function of arcs in the cut. A *minimum* $[i, j]$ *cut* is a cut whose value is minimum among all $[i, j]$ cuts. We let $[S^*_{ij}, \bar{S}^*_{ij}]$ denote a minimum value $[i, j]$ cut and let $v[i, j]$ denote its value. The all-pairs minimum value cut problem requires us to determine for all pairs of nodes i and j, a minimum value $[i, j]$ cut $[S^*_{ij}, \bar{S}^*_{ij}]$ and its value $v[i, j]$.

The definition of a cut implies that if $[S_{ij}, \bar{S}_{ij}]$ is an $[i, j]$ cut, it is also a $[j, i]$ cut. Therefore, $v[i, j] = v[j, i]$ for all pairs i and j of nodes. This observation implies that we can solve the all-pairs minimum value cut problem by invoking $n(n - 1)/2$ applications of any algorithm for the single pair minimum value cut problem. We can, however, do better. In this section we show that we can solve the all-pairs minimum value cut problem by invoking only $(n - 1)$ applications of the single-pair minimum value cut problem.

We first mention some specializations of the all-pairs minimum value cut problem on undirected networks. If we define the value of a cut as its capacity (i.e., the sum of capacities of arcs in the cut), the all-pairs minimum value cut problem would identify minimum cuts (as defined in Chapter 6) between all pairs of nodes. Since the minimum cut capacity equals the maximum flow value, we also obtain the maximum flow values between all pairs of nodes. Several other functions defined on a cut $[S_{ij}, \bar{S}_{ij}]$ are plausible, including (1) the number of arcs in the cut, (2) the capacity of the cut divided by the number of arcs in the cut, and (3) the capacity of the cut divided by $|S_{ij}||\bar{S}_{ij}|$.

We first state and prove an elementary lemma concerning minimum value cuts.

Lemma 8.14. *Let $i_1, i_2, \ldots, i_k$ be an (arbitrary) sequence of nodes. Then $v[i_1, i_k] \geq \min\{v[i_1, i_2], v[i_2, i_3], \ldots, v[i_{k-1}, i_k]\}$.*

Proof. Let $i = i_1, j = i_k$, and $[S_{ij}^*, \bar{S}_{ij}^*]$ be the minimum value $[i, j]$ cut. Consider the sequence of nodes $i_1, i_2, \ldots, i_k$ in order and identify the smallest index r satisfying the property that i_r and i_{r+1} are in different components of the cut $[S_{ij}^*, \bar{S}_{ij}^*]$. Such an index must exist because, by definition, $i_1 = i \in S_{ij}^*$ and $i_k = j \notin S_{ij}^*$. Therefore, $[S_{ij}^*, \bar{S}_{ij}^*]$ is also an $[i_r, i_{r+1}]$ cut, which implies that the value of the minimum value $[i_r, i_{r+1}]$ cut will be no more than the value of the cut $[S_{ij}^*, \bar{S}_{ij}^*]$. In other words,

$$v[i_1, i_k] \geq v[i_r, i_{r+1}] \geq \min\{v[i_1, i_2], v[i_2, i_3], \ldots, v[i_{k-1}, i_k]\}, \tag{8.14}$$

which is the desired conclusion of the lemma. ◆

Lemma 8.14 has several interesting implications. Select any three nodes i, j, and k of the network and consider the minimum cut values $v[i, j]$, $v[j, k]$, and $v[k, i]$ between them. The inequality (8.14) implies that at least two of the values must be equal. For if these three values are distinct, then placing the smallest value on the left-hand side of (8.14) would contradict this inequality. Furthermore, it is possible to show that one of these values that is not equal to the other two must be the largest. Since for every three nodes, two of the three minimum cut values must be equal, it is conceivable that many of the $n(n - 1)/2$ cut values will be equal. Indeed, it is possible to show that the number of distinct minimum cut values is at most $(n - 1)$. This result is the subject of our next lemma. This lemma requires some background concerning the *maximum spanning tree problem* that we discuss in Chapter 13. In an undirected network G, with an associated value (or, profit) for each arc, the maximum spanning tree problem seeks a spanning tree T^*, from among all spanning trees, with the largest sum of the values of its arcs. In Theorem 13.4 we state the following optimality condition for the maximum spanning tree problem: A spanning tree T^* is a maximum spanning tree if and only if for every nontree arc (k, l), the value of the arc (k, l) is less than or equal to the value of every arc in the unique tree path from node k to node l.

Lemma 8.15. *In the $n(n - 1)/2$ minimum cut values between all pairs of nodes, at most $(n - 1)$ values are distinct.*

Proof. We construct a *complete* undirected graph $G' = (N, A')$ with n nodes. We set the value of each arc $(i, j) \in A'$ equal to $v[i, j]$ and associate the cut $[S_{ij}^*, \bar{S}_{ij}^*]$ with this arc. Let T^* be a maximum spanning tree of G'. Clearly, $|T^*| = n - 1$. We shall prove that the value of every nontree arc is equal to the value of some tree arc in T^* and this result would imply the conclusion of the lemma.

Consider a nontree arc (k, l) of value $v[k, l]$. Let P denote the unique path in T^* between nodes k and l. The fact that T^* is a maximum spanning tree implies that the value of arc (k, l) is less than or equal to the value of every arc $(i, j) \in P$. Therefore,

$$v[k, l] \leq \min[v[i, j] : (i, j) \in P]. \tag{8.15}$$

Now consider the sequence of nodes in P that starts at node k and ends at node l. Lemma 8.14 implies that

$$v[k, l] \geq \min[v[i, j] : (i, j) \in P]. \tag{8.16}$$

The inequalities (8.15) and (8.16) together imply that

$$v[k, l] = \min[v[i, j]:(i, j) \in P].$$

Consequently, the value of arc (k, l) equals the minimum value of an arc in P, which completes the proof of the lemma. ◆

The preceding lemma implies that we can store the $n(n - 1)/2$ minimum cut values in the form of a spanning tree T^* with a cut value associated with every arc in the tree. To determine the minimum cut values $v[k, l]$ between a pair k and l of nodes, we simply traverse the tree path from node k to node l; the cut value $v[k, l]$ equals the minimum value of any arc encountered in this path. Note that the preceding lemma only establishes the fact that there are at most $(n - 1)$ distinct minimum cut values and shows how to store them compactly in the form of a spanning tree. It does not, however, tell us whether we can determine these distinct cut values by solving $(n - 1)$ minimum cut problems, because the proof of the lemma requires the availability of minimum cut values between all node pairs which we do not have.

Now, we ask a related question. Just as we can concisely store the minimum cut values between all pairs of nodes by storing only $(n - 1)$ values, can we also store the minimum value *cuts* between all pairs of nodes concisely by storing only $(n - 1)$ cuts? Because the network has at most $(n - 1)$ distinct minimum cut values between $n(n - 1)/2$ pairs of nodes, does it have at most $(n - 1)$ distinct cuts that define the minimum cuts between all node pairs? In the following discussion we provide an affirmative answer to this question. Consider a pair k and l of nodes. Suppose that arc (i, j) is a minimum value arc in the path from node k to node l in T^*. Our preceding observations imply that $v[k, l] = v[i, j]$. Also notice that we have associated a cut $[S^*_{ij}, \bar{S}^*_{ij}]$ of value $v[i, j] = v[k, l]$ with the arc (i, j); this cut separates nodes i and j. Is $[S^*_{ij}, \bar{S}^*_{ij}]$ a minimum $[k, l]$ cut? It is if $[S^*_{ij}, \bar{S}^*_{ij}]$ separates nodes k and l, and it is not otherwise. If, indeed, $[S^*_{ij}, \bar{S}^*_{ij}]$ separates nodes k and l, and if the same result is true for every pair of nodes in the network, the cuts associated with arcs in T^* concisely store minimum value cuts between all pairs of nodes. We refer to such a tree T^* as a *separator tree*. In this section we show that every network G has a separator tree and that we can construct the separator tree by evaluating $(n - 1)$ single-pair minimum cut values. Before we describe this method, we restate the definition of the separator tree for easy future reference.

Separator tree. *An undirected spanning tree T^*, with a minimum $[i, j]$ cut $[S^*_{ij}, \bar{S}^*_{ij}]$ of value $v[i, j]$ associated with each arc (i, j), is a separator tree if it satisfies the following property for every nontree arc (k, l): If arc (i, j) is the minimum value arc from node k to node l in T^* (breaking ties in a manner to be described later), $[S^*_{ij}, \bar{S}^*_{ij}]$ separates nodes k and l.*

Our preceding discussion shows that we have reduced the all-pairs minimum value cut problem to a problem of obtaining a separator tree. Given the separator tree T^*, we determine the minimum $[k, l]$ cut as follows: We traverse the tree path from node k to node l; the cut corresponding to the minimum value in the path (breaking ties appropriately) is a minimum $[k, l]$ cut.

We call a subtree of a separator tree a *separator subtree*. Our algorithm for constructing a separator tree proceeds by constructing a separator subtree that spans an expanding set for nodes. It starts with a singleton node, adds one additional node to the separator subtree at every iteration, and terminates when the separator tree spans all the nodes. We add nodes to the separator subtree in the order 1, 2, 3, . . . , n. Let T^{p-1} denote the separator subtree for the node set $\{1, 2, \ldots, p-1\}$ and T^p denote the separator subtree for the node set $\{1, 2, \ldots, p\}$. We obtain T^p from T^{p-1} by adding an arc, say (p, k). The essential problem is to locate the node k incident to node p in T^p. Once we have located the node k, we identify a minimum value cut $[S^*_{pk}, \bar{S}^*_{pk}]$ between nodes p and k, and associate it with the arc (p, k). We set the value of the arc (p, k) equal to $v[p, k]$.

As already mentioned, our algorithm for constructing the separator tree adds arcs to the separator subtree one by one. We associate an index, called an *order* index, with every arc in the separator subtree. The first arc added to the separator subtree has order index 1, the second arc added has order index 2, and so on. We use the order index to resolve ties while finding a minimum value arc between a pair of nodes. As a rule, whether we specify so or not in the subsequent discussion, we always resolve any tie in favor of the arc with the least order index (i.e., the arc that we added first to the separator subtree).

Figure 8.15 describes the procedure we use to locate the node $k \in T^{p-1}$ on which the arc (p, k) will be incident in T^p.

Note that in every iteration of the locate procedure, $T_\alpha \subseteq N_\alpha$ and $T_\beta \subseteq N_\beta$. This fact follows from the observation that for every $k \in T_\alpha$ and $l \in T_\beta$, the arc (α, β) is the minimum value arc in the path from node k to node l in the separator subtree T and, by its definition, the cut must separate node k and node l.

We illustrate the procedure locate using a numerical example. Consider a nine-node network with nodes numbered 1, 2, 3, . . . , 9. Suppose that after five iterations, the separator subtree $T^{p-1} = T^6$ is as shown in Figure 8.16(a). The figure also shows the cut associated with each arc in the separator subtree (here we specify only $S^*_{\alpha\beta}$, because we can compute $\bar{S}^*_{\alpha\beta}$ by using $\bar{S}^*_{\alpha\beta} = N - S^*_{\alpha\beta}$). We next consider adding node 7 to the subtree. At this point, $T = T^6$ and the minimum value arc in T is (4, 5). Examining S^*_{45} reveals that node 7 is on the same side of the cut as node

procedure *locate*(T^{p-1}, p, k);
begin
    $T := T^{p-1}$;
    while T is not a singleton node do
    begin
        let (α, β) be the minimum value arc in T (we break ties in favor of the arc with the smallest order index);
        let $[S^*_{\alpha\beta}, \bar{S}^*_{\alpha\beta}]$ be the cut associated with the arc (α, β);
        let the arc (α, β) partition the tree T into the subtrees T_α and T_β so that $\alpha \in T_\alpha$ and $\beta \in T_\beta$;
        let the cut $[S^*_{\alpha\beta}, \bar{S}^*_{\alpha\beta}]$ partition the node set N into the subsets N_α and N_β so that $\alpha \in N_\alpha$ and $\beta \in N_\beta$;
        if $p \in N_\alpha$ **then** set $T := T_\alpha$ **else** set $T := T_\beta$;
    end;
    set k equal to the singleton node in T;
end;

Figure 8.15 Locate procedure.

4; therefore, we update T to be the subtree containing node 4. Figure 8.16(b) shows the tree T. Now arc (2, 3) is the minimum value arc in T. Examining S^*_{23} reveals that node 7 is on the same side of the cut as node 2; so we update T so that it is the subtree containing node 2. At this point, the tree T is a singleton, node 2. We set $k = 2$ and terminate the procedure. We next add arc (2, 7) to the separator subtree, obtain a minimum value cut between the nodes 7 and 2, and associate this cut with the arc (2, 7). Let $v[7, 2] = 5$ and $S^*_{72} = \{7\}$. Figure 8.16(c) shows the separator subtree spanning the nodes 1 through 7.

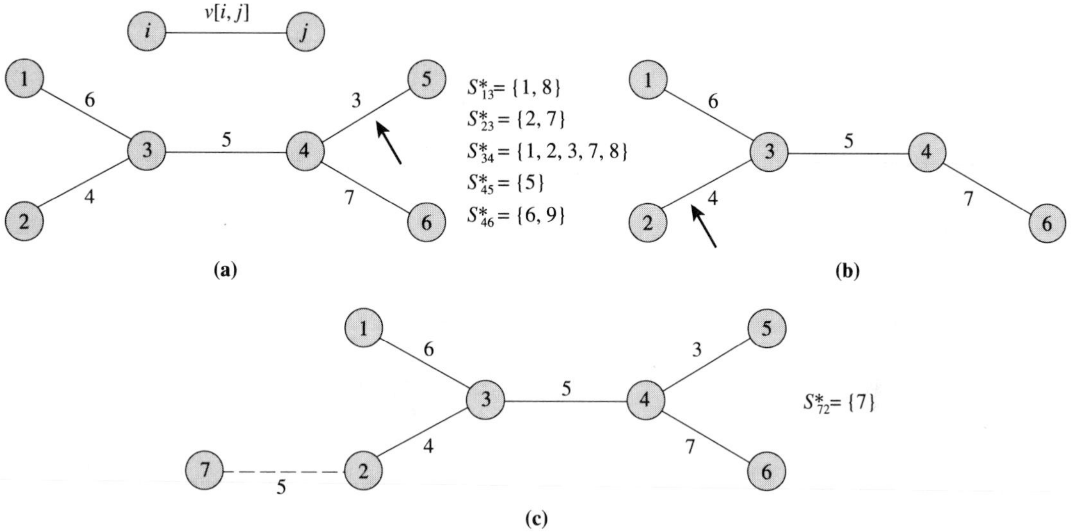

Figure 8.16 Illustrating the procedure locate.

We are now in a position to prove that the subtree T^p is a separator subtree spanning the nodes $\{1, 2, \ldots, p\}$. Since, by our inductive hypothesis, T^{p-1} is a separator subtree on the nodes $\{1, 2, \ldots, p-1\}$, our proof amounts to establishing the following result for every node $l \in \{1, 2, \ldots, p-1\}$: If (i, j) is the minimum value arc in T^p in the path from node p to node l (with ties broken appropriately), the cut $[S^*_{ij}, \bar{S}^*_{ij}]$ separates the nodes p and l. We prove this result in the following lemma.

Lemma 8.16. *For any node $l \in T^{p-1}$, if (i, j) is the minimum value arc in T^p in the path from node p to node l (when we break ties in favor of the arc with the least order index), $[S^*_{ij}, \bar{S}^*_{ij}]$ separates nodes p and l.*

Proof. We consider two possibilities for the arc (i, j).

Case 1: $(i, j) = (p, k)$. Let P denote the tree path in T^p from node k to node l. The situation $(i, j) = (p, k)$ can occur only when arc (p, k) is the unique minimum value arc in T^p in P, for otherwise, the tie will be broken in favor of an arc other than the arc (p, k) (why?). Thus

$$v[p, k] < v[g, h] \qquad \text{for every arc } (g, h) \in P. \tag{8.17}$$

Next consider any arc $(g, h) \in P$. We claim that both the nodes g and h must belong to the same component of the cut $[S^*_{pk}, \bar{S}^*_{pk}]$; for otherwise, $[S^*_{pk}, \bar{S}^*_{pk}]$ will also separate nodes g and h, so $v[p, k] \geq v[g, h]$, contradicting (8.17). Using this argument inductively for all arcs in P, we see that all the nodes in the path P (that starts at node k and ends at node l) must belong to the same component of the cut $[S^*_{pk}, \bar{S}^*_{pk}]$. Since the cut $[S^*_{pk}, \bar{S}^*_{pk}]$ separates nodes p and k, it also separates nodes p and l.

Case 2: $(i, j) \neq (p, k)$. We examine this case using the locate procedure. At the beginning of the locate procedure, $T = T^{p-1}$, and at every iteration the size of the tree becomes smaller, until finally, $T = \{k\}$. Consider the iteration when T contains both the nodes k and l, but in the next iteration the tree does not contain node l. Let P denote the path in T from node k to node l in this iteration. It is easy to see that the arc (α, β) selected by the locate procedure in this iteration must belong to the path P. By definition, (α, β) is the minimum value arc in T, with ties broken appropriately. Since T contains the path P, (α, β) is also a minimum value arc in P. Now notice from the statement of the lemma that arc (i, j) is defined as the minimum value arc in P, and since we break the tie in precisely the same manner, $(i, j) = (\alpha, \beta)$.

We next show that the cut $[S^*_{\alpha\beta}, \bar{S}^*_{\alpha\beta}]$ separates node p and node l. We recommend that the reader refers to Figure 8.17 while reading the remainder of the proof. Consider the same iteration of the locate procedure considered in the preceding paragraph, and let T_α, N_α, T_β, N_β be defined as in Figure 8.15. We have observed previously that $T_\alpha \subseteq N_\alpha$ and $T_\beta \subseteq N_\beta$. We assume that $p \in N_\alpha$; a similar argument applies when $p \in N_\beta$. The procedure implies that when $p \in N_\alpha$, we set $T = T_\alpha$, implying that $k \in T_\alpha \subseteq N_\alpha$. Since the cut $[S^*_{\alpha\beta}, \bar{S}^*_{\alpha\beta}]$ separates node k from node l it also separates node p from node l. The proof of the lemma is complete. ◆

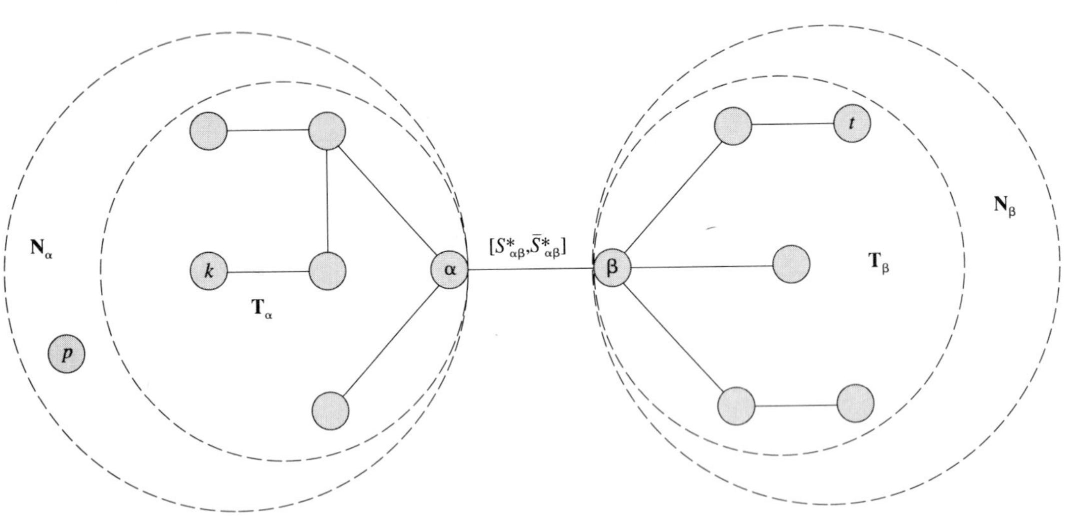

Figure 8.17 Proving Lemma 8.16.

Having proved the correctness of the all-pairs minimum cut algorithm, we next analyze its running time. The algorithm performs $(n - 1)$ iterations. In each iteration it executes the locate procedure and identifies the arc (p, k) to be added to the separator subtree. The reader can easily verify that an execution of the locate procedure requires $O(n^2)$ time. The algorithm then solves a minimum value cut problem and associates this cut and its value with the arc (p, k). The following theorem is immediate.

Theorem 8.17. *Solving the all-pairs minimum value cut problem requires $O(n^3)$ time plus the time required to solve $(n - 1)$ instances of the single-pair minimum value cut problem.* ◆

To summarize, we have shown that the all-pairs minimum cut algorithm finds the minimum capacity cut separating node i and node j (i.e., with node i on one side of the cut and node j on the other side) for every node pair $[i, j]$. The max-flow min-cut theorem shows that this algorithm also determines the maximum flow values between every pair of nodes.

Suppose, instead, that we are given a directed graph. Let $f[i, j]$ denote the value of the minimum cut from node i to node j. We cannot use the all-pairs minimum cut algorithm to determine $f[i, j]$ for all node pairs $[i, j]$ for the following simple reason: The algorithm would determine the minimum value of a cut separating node i from node j, and this value is $\min\{f[i, j], f[j, i]\}$ because the minimum cut from node i to node j separates nodes i and j and so does the minimum cut from node j to node i. (We did not face this problem for undirected networks because the minimum cut from node i to node j is also a minimum cut from node j to node i.) If we let $v[i, j] = \min\{f[i, j], f[j, i]\}$, we can use the all-pairs minimum cut algorithm to determine $v[i, j]$ for each node pair $[i, j]$. Moreover, this algorithm relies on evaluating $v[i, j]$ for only $(n - 1)$ pairs of nodes. Since we can determine $v[i, j]$ by finding a maximum flow from node i to node j and from node j to node i, we can compute $v[i, j]$ for all node pairs by solving $(2n - 2)$ maximum flow problems.

We complete this section by describing an application of the all-pairs minimum value cut problem.

Application 8.3 Maximum and Minimum Arc Flows in a Feasible Flow

Consider the feasible flow problem that we discussed in Application 6.1. Assume that the network is uncapacitated and that it admits a feasible flow. For each arc $(i, j) \in A$, let α_{ij} denote the minimum arc flow that (i, j) can have in some feasible flow, and let β_{ij} denote the maximum arc flow that (i, j) can have in some feasible flow. We will show that we can determine α_{ij} for all node pairs $[i, j]$ by solving at most n maximum flow problems, and we can determine β_{ij} for all node pairs $[i, j]$ by an application of the all-pairs minimum value cut problem.

The problem of determining the maximum and minimum values of the arc flows in a feasible flow arises in the context of determining the statistical security of data (e.g., census data). Given a two-dimensional table $\mathbf{A}$ of size $p \times q$, suppose that we want to disclose the row sums r_i, the column sums c_j, and a subset of the matrix elements. For security reasons (or to ensure confidentiality of the data), we would like to ensure that we have "disguised" the remaining matrix elements (or "hidden" entries). We wish to address the following question: Once we have disclosed the row and column sums and some matrix elements, how secure are the hidden entries? We address a related question: For each hidden element a_{ij}, what are the minimum and maximum values that a_{ij} can assume consistent with the data we have disclosed? If these two bounds are quite close, the element a_{ij} is not secure.

We assume that each revealed matrix element has value 0. We incur no loss of generality in making this assumption since we can replace a nonzero element a_{ij} by 0, replace r_i by $r_i - a_{ij}$, and replace c_j by $c_j - a_{ij}$. To conduct our analysis, we begin by constructing the bipartite network shown in Figure 8.18; in this network, each unrevealed matrix element a_{ij} corresponds to an arc from node i to node $\bar{j}$. It is easy to see that every feasible flow in this network gives values of the matrix elements that are consistent with the row and column sums.

How might we compute the α_{ij} values? Let x^* be any feasible flow in the network (which we can determine by solving a maximum flow problem). In Section 11.2 we show how to convert each feasible flow into a feasible spanning tree solution; in this solution at most $(n - 1)$ arcs have positive flow. As a consequence, if x^* is a spanning tree solution, at least $(m - n + 1)$ arcs have zero flow; and therefore, $\alpha_{ij} = 0$ for each of these arcs. So we need to find the α_{ij} values for only the remaining $(n - 1)$ arcs. We claim that we can accomplish this task by solving at most $(n - 1)$ maximum flow problems. Let us consider a specific arc (i, j). To determine the minimum flow on arc (i, j), we find the maximum flow from node i to node j in $G(x^*)$ when we have deleted arc (i, j). If the maximum flow from node i to node j has value k, we can reroute up to k units of flow from node i to node j and reduce the flow an arc (i, j) by the same amount. As a result, $\alpha_{ij} = \max\{0, x^*_{ij} - k\}$.

To determine the maximum possible flow on arc (i, j), we determine the maximum flow from node j to node i in $G(x^*)$. If we can send k units from node j to node i in $G(x^*)$, then $\beta_{ij} = k$. To establish this result, suppose that the k units consist of x^*_{ij} units on arc (j, i) [which is the reversal of arc (i, j)], and $k - x^*_{ij}$ units that do not use arc (i, j). Then, to determine the maximum flow on the arc (i, j), we can

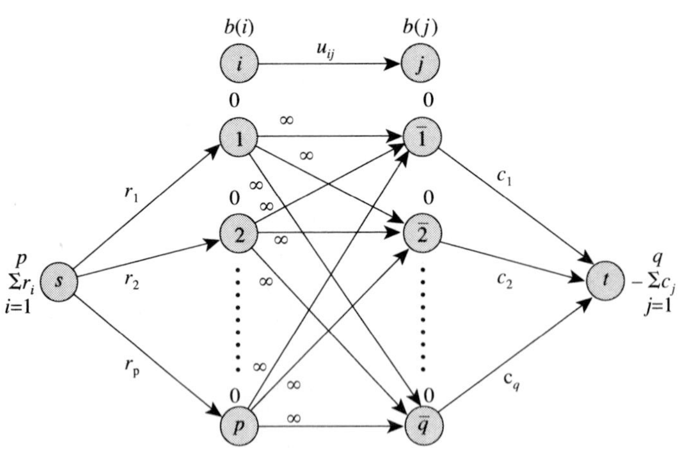

Figure 8.18 Feasible flow network for ensuring the statistical security of data.

send $k - x_{ij}^*$ units of flow from node j to node i and increase the flow in arc (i, j) by $k - x_{ij}^*$, leading to a total of k units.

Thus, to find β_{ij}, we need to compute the maximum flow from node j to node i in $G(x^*)$ for each arc $(i, j) \in A$. Equivalently, we want to find the minimum capacity cut $f[j, i]$ from node j to node i in $G(x^*)$. As stated previously, we cannot find $f[i, j]$ for all node pairs $[i, j]$ in a directed network; but we can determine $\min\{f[i, j], f[j, i]\}$. We now use the fact that we need to compute $f[j, i]$ when $(i, j) \in A$, in which case $f[i, j] = \infty$ (because the network is uncapacitated). Therefore, for each arc $(i, j) \in A$, $f[i, j] = \min\{f[i, j], f[j, i]\}$, and we can use the all-pairs minimum value cut algorithm to determine all of the β_{ij} values by solving $(2n - 2)$ maximum flow problems.

8.8 SUMMARY

As we have noted in our study of shortest path problems in our earlier chapters, we can sometimes develop more efficient algorithms by restricting the class of networks that we wish to consider (e.g., networks with nonnegative costs, or acyclic networks). In this chapter we have developed efficient special purpose algorithms for several maximum flow problems with specialized structure: (1) unit capacity networks, in which all arcs have unit capacities; (2) unit capacity simple networks, in which each node has a single incoming or outgoing arc; (3) bipartite networks; and (4) planar networks. We also considered one specialization of the maximum flow problem, the problem of finding the maximum number of arc-disjoint paths between two nodes in a network, and one generalization of the minimum cut problem, finding minimum value cuts between all pairs of nodes. For this last problem we permitted ourselves to measure the value of any cut by a function that is more general than the sum of the capacities of the arcs in the cut. Finally, we considered one other advanced topic: the use of the dynamic trees data structure to efficiently implement the shortest augmenting path algorithm for the maximum flow problem.

Figure 8.19, which summarizes the basic features of these various algorithms,

Algorithm	Running time	Features
Maximum flow algorithm for unit capacity networks	$O(\min\{n^{2/3}m, m^{3/2}\})$	1. Fastest available algorithm for solving the maximum flow problem in unit capacity networks. 2. Uses two phases: first applies the shortest augmenting path algorithm until the distance label of node s satisfies the condition $d(s) \geq d^* = \min\{\lceil 2n^{2/3} \rceil, \lceil m^{1/2} \rceil\}$. At this point, uses the labeling algorithm until it establishes a maximum flow. 3. Easy to implement and is likely to be efficient in practice.
Maximum flow algorithm for unit capacity simple networks	$O(n^{1/2}m)$	1. Fastest available algorithm for solving the maximum flow problem in unit capacity simple networks. 2. Same two phase approach as the preceding algorithm, except $d^* = \lceil n^{1/2} \rceil$.

Figure 8.19 Summary of algorithms discussed in this chapter.

Algorithm	Running time	Features
Bipartite preflow-push algorithm	$O(n_1^2 m)$	1. Faster approach for solving maximum flow problems in bipartite networks satisfying the condition $n_1 < n_2$. 2. Improved implementation of the generic preflow-push algorithm discussed in Section 7.6. 3. Uses "two-arc" push rule in which we always push flow from an active node over two consecutive admissible arcs. 4. As discussed in the exercises, significant further improvements are possible if we examine active nodes in some specific order.
Planar maximum flow algorithm	$O(n \log n)$	1. Highly efficient algorithm for solving the maximum flow problem in s–t planar networks. 2. Constructs the dual network and solves a shortest path problem over it. The shortest path in the dual network yields a minimum cut in the original network and the shortest path distances yield a maximum flow. 3. Applicable only to undirected s–t planar networks.
Dynamic tree algorithm	$O(nm \log n)$	1. Uses the dynamic tree data structure to implement the shortest augmenting path algorithm for the maximum flow problem. 2. Improves the running time of the shortest augmenting path algorithm from $O(n^2m)$ to $O(nm \log n)$. 3. Similar, though not as dramatic, improvements can be obtained by using this data structure in preflow-push algorithms. 4. The dynamic tree data structure is quite sophisticated, has substantial overhead and its practical usefulness has not yet been established.
Arc connectivity algorithm	$O(nm)$	1. Fastest available algorithm for obtaining the arc connectivity of a network. 2. Uses the labeling algorithm for the maximum flow problem as a subroutine. 3. Likely to be very efficient in practice. 4. Applicable only to undirected networks.
All-pairs minimum cut algorithm	$O(nM(n, m, U) + n^3)$	1. Fastest available algorithm for solving the all-pairs minimum cut problem. ($M(n, m, U)$ is the time needed for solving the maximum flow problem on a network with n nodes, m arcs, and U as the largest arc capacity.) 2. Determines minimum cuts between all pairs of nodes in the network by solving $(n - 1)$ maximum flow problems. 3. Can be used to determine the minimum value cuts between all pairs of nodes in the case in which we define the value of a cut differently than the capacity of the cut. 4. The $O(n^3)$ term in the worst-case bound can be reduced to $O(n^2)$ using different data structures. 5. Applicable to undirected networks only.

Figure 8.19 *(Continued)*

shows that by exploiting specialized structures or advanced data structures, we can improve on the running time of maximum flow computations, sometimes dramatically.

REFERENCE NOTES

We present the reference notes in this chapter separately for each of the several topics related to maximum flows that we have studied in this chapter.

Flows in unit capacity networks. Even and Tarjan [1975] showed that Dinic's algorithm solves the maximum flow problem in unit capacity and unit capacity simple networks in $O(\min\{n^{3/2}m, m^{3/2}\})$ and $O(n^{1/2}m)$ time, respectively. The algorithms we presented in Section 8.2 are due to Ahuja and Orlin [1991]; they use similar ideas and have the same running times. Fernandez-Baca and Martel [1989] presented and analyzed algorithms for solving more general maximum flow problems with "small" integer capacities.

Flows in bipartite networks. By improving on the running times of Dinic's [1970] and Karzanov's [1974] algorithms, Gusfield, Martel, and Fernandez-Baca [1987] developed the first specializations of maximum flow algorithms for bipartite networks. Ahuja, Orlin, Stein, and Tarjan [1990] provided further improvements and showed that it is possible to substitute n_1 for n in the time bounds of almost all preflow-push algorithms to obtain new time bounds for bipartite networks (recall that n_1 is the number of nodes on the smaller side of the bipartite network). This result implies that the generic preflow-push algorithm, the FIFO implementation, the highest-label implementation, and the excess scaling algorithm can solve the maximum flow problem in bipartite networks in $O(n_1^2m)$, $O(n_1m + n_1^3)$, $O(n_1m + n_1^2\sqrt{m})$, and $O(n_1m + n_1^2 \log U)$ time. Our discussion of the bipartite preflow-push algorithm in Section 8.3 is adapted from this paper. We have adapted the baseball elimination application from Schwartz [1966], and the network reliability application from Van Slyke and Frank [1972]. The paper by Gusfield, Martel, and Fernandez-Baca [1987] describes additional applications of bipartite maximum flow problems.

Flows in planar networks. In Section 8.4 we discussed the relationship between minimum s–t cuts in a network and shortest paths in its dual. Given a planar network G, the algorithm of Hopcroft and Tarjan [1974] constructs a planar representation in $O(n)$ time; from this representation, we can construct the dual network in $O(n)$ time. Berge [1957] showed that augmenting flow along certain paths, called *superior paths*, provides an algorithm that finds a maximum flow within n augmentations. Itai and Shiloach [1979] described an $O(n \log n)$ implementation of this algorithm. Hassin [1981] showed how to compute a maximum flow from the shortest path distances in the dual network. We have presented this method in our discussion in Section 8.4. For faster maximum flow algorithms in planar (but not necessarily s–t planar) undirected and directed networks, see Johnson and Venkatesan [1982] and Hassin and Johnson [1985].

Dynamic tree implementation. Sleator and Tarjan [1983] developed the dynamic tree data structure and used it to improve the worst-case complexity of Dinic's algorithm from $O(n^2m)$ to $O(nm \log n)$. Since then, researchers have used this data structure on many occasions to improve the performance of a range of network flow algorithms. Using the dynamic tree data structure, Goldberg and Tarjan [1986] improved the complexity of the FIFO preflow-push algorithm (described in Section 7.7) from $O(n^3)$ to $O(nm \log (n^2/m))$, and Ahuja, Orlin, and Tarjan [1989] improved the complexity of the excess scaling algorithm (described in Section 7.9) and several of its variants.

Network connectivity. Even and Tarjan [1975] offered an early discussion of arc connectivity of networks. Some of our discussion in Section 8.6 uses their results. The book by Even [1979] also contains a good discussion on node connectivity of a network. The $O(nm)$ time arc connectivity algorithm (for undirected networks) that we presented in Section 8.6 is due to Matula [1987] and is currently the fastest available algorithm. Mansour and Schieber [1988] presented an $O(nm)$ algorithm for determining the arc connectivity of a directed network.

All-pairs minimum value cut problem. Gomory and Hu [1961] developed the first algorithm for solving the all-pairs minimum cut problem on undirected networks that solves a sequence of $(n - 1)$ maximum flow problems. Gusfield [1990] presented an alternate all-pairs minimum cut algorithm that is very easy to implement using a code for the maximum flow problem. Talluri [1991] described yet a third approach. The algorithm we described in Section 8.7, which is due to Cheng and Hu [1990], is more general since it can handle cases when the value of a cut is defined differently than its capacity. Unfortunately, no one yet knows how to solve the all-pairs minimum value cut problem in directed networks as efficiently. No available algorithm is more efficient than solving $\Omega(n^2)$ maximum flow problems. The application of the all-pairs minimum value cut problem that we described at the end of Section 8.7 is due to Gusfield [1988]. Hu [1974] describes an additional application of the all-pairs minimum value cut problem that arises in network design.

EXERCISES

8.1 **(a)** Show that it is always possible to decompose a circulation in a unit capacity network into unit flows along arc-disjoint directed cycles.
(b) Show that it is always possible to decompose a circulation in a simple network into unit flows along node-disjoint directed cycles.

8.2. Let $G = (N, A)$ be a directed network. Show that it is possible to decompose the arc set A into an arc-disjoint union of directed cycles if and only if G has a circulation x with $x_{ij} = 1$ for every arc $(i, j) \in A$. Moreover, show that we can find such a solution if and only if the indegree of each node equals its outdegree.

8.3. An undirected network is *biconnected* if it contains two node disjoint paths between every pair of nodes (except, of course, at the starting and terminal points). Show that a biconnected network must satisfy the following three properties: (1) for every two nodes p and q, and any arc (k, l), some path from p to q contains arc (k, l); (2) for every three nodes p, q, and r, some path from p to r contains node q; (3) for every three nodes p, q, and r, some path from p to r does not contain q.

8.4. Suppose that you are given a maximum flow problem in which all arc capacities are the same. What is the most efficient method (from the worst-case complexity point of view) to solve this problem?

8.5. Using the unit capacity maximum flow algorithm, establish a maximum flow in the network shown in Figure 8.20.

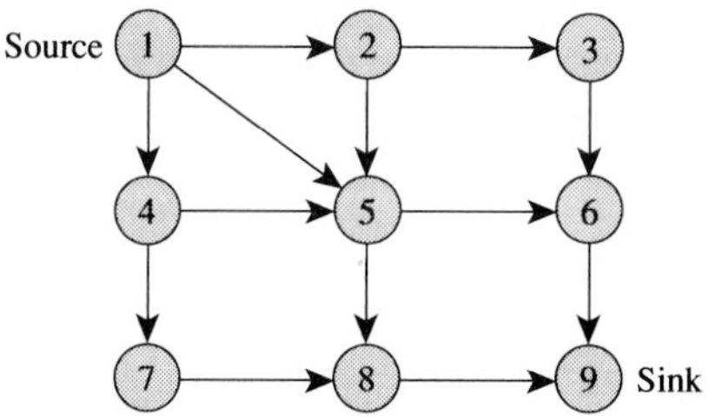

Figure 8.20 Example for Exercise 8.5.

8.6. Adapt the unit capacity maximum flow algorithm for bipartite networks. In doing so, try to obtain the best possible running time. Describe your algorithm and analyze its worst-case complexity.

8.7. Consider a generalization of unit capacity networks in which arcs incident to the source and the sink nodes can have arbitrary capacities, but the remainder of the arcs have unit capacities. Will the unit capacity maximum flow algorithm still solve the problem in $O(\min\{n^{2/3}m, m^{3/2}\})$ time, or might the algorithm require more time? Consider a further generalization of the problem in which arcs incident to the source, the sink, and one other node have arbitrary capacities. What will be the complexity of the unit capacity maximum flow algorithm when applied to this problem?

8.8. We define a class of networks to be *small-capacity networks* if each arc capacity is between 1 and 4. Describe a generalization of the unit capacity maximum flow algorithm that would solve the maximum flow problems on small-capacity networks in $O(\min\{n^{2/3}\, m, m^{3/2}\})$ time.

8.9. What is the best possible bound you can obtain on the running time of the generic preflow-push algorithm applied to unit capacity networks?

8.10. Suppose we apply the preflow-push algorithm on a unit capacity simple network with the modification that we do not perform push/relabel operations on any node whose distance label exceeds $n^{1/2}$.
(a) Show that the modified preflow-push algorithm terminates within $O(n^{1/2}m)$ time.
(b) Show that at the termination of the algorithm, the nodes have a total excess of at most $n^{1/2}$.
(c) Can you convert this preflow into a maximum flow in $O(n^{1/2}m)$ time? If yes, then how?

8.11. Let x be a flow in a unit capacity directed network G; assume that x is not a maximum flow. Let P denote a shortest augmenting path in $G(x)$ and let $\overline{P}$ denote its reversal. In addition, let x' denote the flow obtained by sending a unit flow along P and let P' denote the shortest augmenting path in $G(x')$. Show that $|P'| \geq |P| + 2|\overline{P} \cap P'|$. (*Hint*: Use the distance validity conditions for the arcs in P'.)

8.12. This exercise concerns the baseball elimination problem discussed in Application 8.1. Show that we can eliminate team 0 if and only if some nonempty set S of nodes satisfies the condition that

$$w_{\max} < \frac{\sum_{i \in S} w_i + \sum_{1 \leq i \leq j \leq n} g_{ij}}{|S|}.$$

(*Hint*: Use the max-flow min-cut theorem.)

8.13. Given two n-element arrays α and β, we want to know whether we can construct an n-node directed graph so that node i has outdegree equal to $\alpha(i)$ and an indegree equal to $\beta(i)$. Show how to solve this problem by solving a maximum flow problem. [*Hint*: Transform this problem to a feasible flow problem, as described in Application 6.1, on a complete bipartite graph $G = (N_1 \cup N_2, A)$ with $N_1 = \{1, 2, \ldots, n\}$, $N_2 = \{1', 2', \ldots, n'\}$, and $A = N_1 \times N_2$.]

8.14. Given an n-node graph G and two n-element arrays α and β, we wish to determine whether some subgraph G' of G satisfies the property that for each node i, $\alpha(i)$ and $\beta(i)$ are the outdegree and indegree of node i. Formulate this problem as a maximum flow problem. (*Hint*: The transformation is similar to the one used in Exercise 8.13.)

8.15. Apply the bipartite preflow-push algorithm to the maximum flow problem given in Figure 8.21. Among all the active nodes, push flow from a node with the smallest distance label.

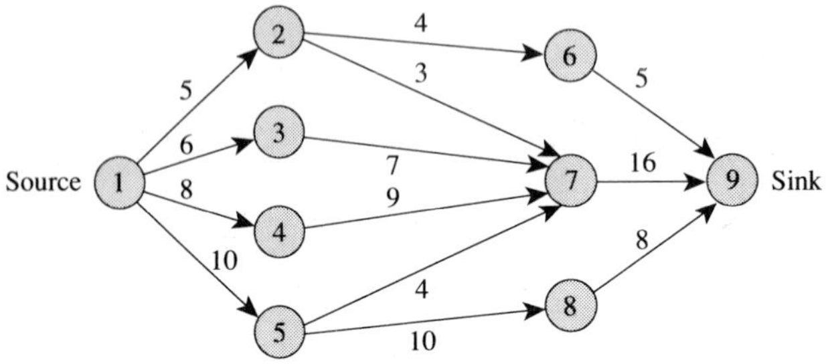

Figure 8.21 Example for Exercise 8.15.

8.16. Consider a bipartite network $G = (N_1 \cup N_2, A)$ with $n_1 = |N_1| \le |N_2| = n_2$. Show that when applied to this network, the shortest augmenting path algorithm performs $O(n_1(n_1 + n_2)) = O(n_1 n_2)$ relabel steps and $O(n_1 m)$ augmentations, and runs in $O(n_1^2 m)$ time.

8.17. Suppose that we wish to find the maximum flow between two nodes in a bipartite network $G = (N_1 \cup N_2, A)$ with $n_1 = |N_1|$ and $n_2 = |N_2|$. This exercise considers the development of faster special implementations of the generic bipartite preflow-push algorithms.

(a) Show that if the algorithm always pushes flow from an active node with the highest distance label, it runs in $O(n_1^3 + n_1 m)$ time.

(b) Show that if the algorithm examines active nodes in a FIFO order, it runs in $O(n_1^3 + n_1 m)$ time.

(c) Develop an excess scaling version of the generic bipartite flow algorithm and show that it runs in $O(n_1 m + n_1^2 \log U)$ time.

8.18. A *semi-bipartite network* is defined as a network $G = (N, A)$ whose node set N can be partitioned into two subsets N_1 and N_2 so that no arc has both of its endpoints in N_2 (i.e., we allow arcs with both of their endpoints in N_1). Let $n_1 = |N_1|$, $n_2 = |N_2|$, and $n_1 \le n_2$. Show how to modify the generic bipartite preflow-push algorithm so that it solves the maximum flow problem on semi-bipartite networks in $O(n_1^2 m)$ time.

8.19. (a) Prove that the graph shown in Figure 8.6(a) cannot be planar. (*Hint*: Use Euler's formula.)

(b) Prove that the graph shown in Figure 8.6(b) cannot be planar. (*Hint*: Use Euler's formula.)

8.20. Show that an undirected planar network always contains a node with degree at most 5.

8.21. Apply the planar maximum flow algorithm to identify a maximum flow in the network shown in Figure 8.22.

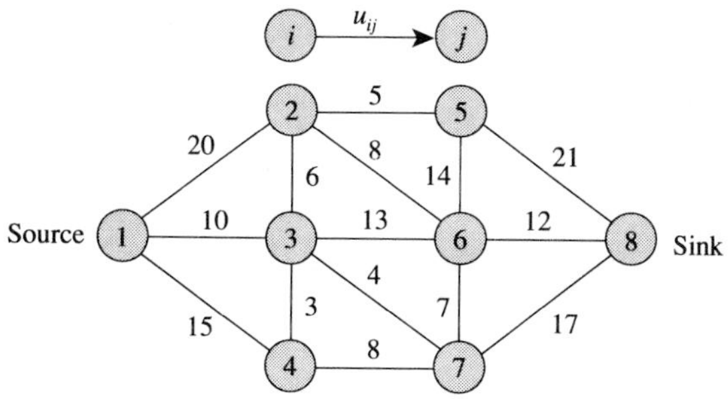

Figure 8.22 Example for Exercise 8.21.

8.22. Duals of directed s–t planar networks. We define the dual graph of a directed s–t planar network $G = (N, A)$ as follows. We first draw an arc (t, s) of zero capacity, which divides the unbounded face into two faces: a new unbounded face and a new bounded face. We then place a node f^* inside each face f of the primal network G. Let s^* and t^*, respectively, denote the nodes in the dual network corresponding to the new bounded face and the new unbounded face. Each arc $(i, j) \in A$ lies on the boundary of the two faces f_1 and f_2; corresponding to this arc, the dual graph contains two oppositely directed arcs (f_1, f_2) and (f_2, f_1). If arc (i, j) is a clockwise arc in the face f_1, we define the cost of arc (f_1, f_2) as u_{ij} and the cost of (f_2, f_1) as zero. We define arc costs in the opposite manner if arc (i, j) is a counterclockwise arc in the face f_1. Construct the dual of the s–t planar network shown in Figure 8.20. Next show that there is a one-to-one correspondence between s–t cuts in the primal network and directed paths from node s^* to node t^* in the dual network; moreover, show that the capacity of the cut equals the cost of the corresponding path.

8.23. Show that if G is an s–t planar directed network, the minimum number of arcs in a directed path from s to t is equal to the maximum number of arc-disjoint s–t cuts. (*Hint*: Apply Theorem 6.7 to the dual of G.)

8.24. Node coloring algorithm. In *the node coloring problem*, we wish to color the nodes of a network so that the endpoints of each arc have a different color. In this exercise we discuss an algorithm for coloring a planar undirected graph using at most six colors. The algorithm first orders the nodes of the network using the following iterative loop: It selects a node with degree at most 5 (from Exercise 8.20, we can always find such a node), deletes this node and its incident arcs from the network, and updates the degrees of all the nodes affected. The algorithm then examines nodes in the reverse order and assigns colors to them.

(a) Explain how to assign colors to the nodes to create a valid 6-coloring (i.e., the endpoints of every arc have a different color). Justify your method.

(b) Show how to implement the node coloring algorithm so that it runs in $O(m)$ time.

8.25. Consider the numerical example used in Section 8.5 to illustrate the dynamic tree implementation of the shortest augmenting path algorithm. Perform further iterations of the algorithm starting from Figure 8.11(c) until you find a maximum flow that has a value of 4.

8.26. Solve the maximum flow problem given in Figure 8.23 by the dynamic tree implementation of the shortest augmenting path algorithm.

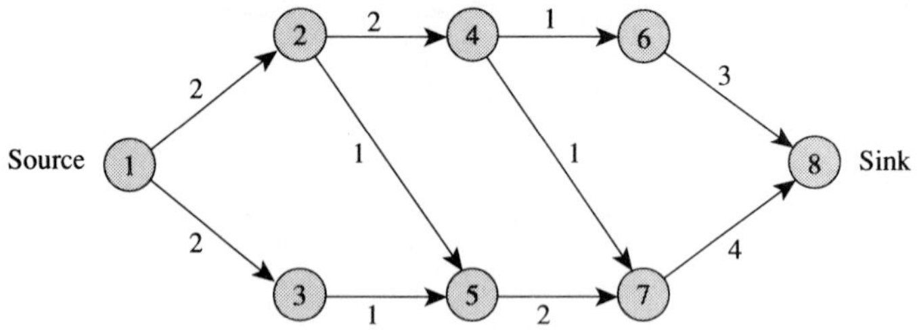

Figure 8.23 Example for Exercise 8.26.

8.27. Show how to use the dynamic tree data structure to implement in $O(m \log n)$ time the algorithm described in Exercise 7.11 for converting a maximum preflow into a maximum flow.

8.28. In Section 3.5 we showed how we can determine the flow decomposition of any flow in $O(nm)$ time. Show how to use the dynamic tree data structure to determine the flow decomposition in $O(m \log n)$ time.

8.29. Prove Property 8.11.

8.30. Compute the arc connectivity for the networks shown in Figure 8.24. Feel free to determine the maximum number of arc-disjoint paths between any pairs of nodes by inspection.

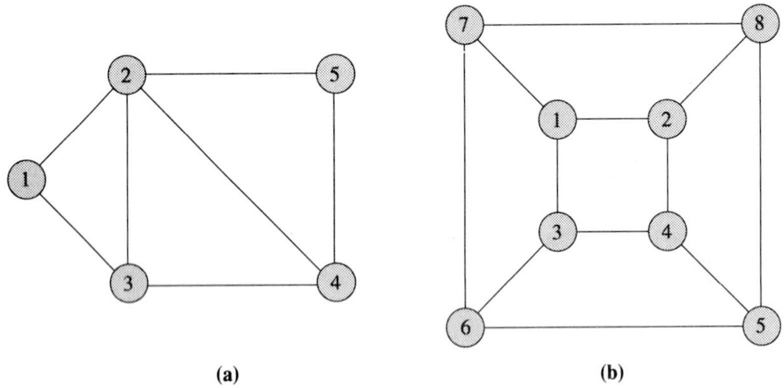

Figure 8.24 Example for Exercises 8.30 and 8.36.

8.31. Construct an undirected network whose nodes all have degree at least 3, but whose arc connectivity is 2.

8.32. An undirected network is said to be *k-connected* if every pair of nodes are connected by at least k arc-disjoint paths. Describe an $O(m)$ algorithm for determining whether a network is 1-connected. Use this algorithm to describe a simple $O(knm)$ algorithm for determining whether a network is k-connected. This algorithm should be different than those described in Section 8.6.

8.33. In a directed graph G, we define the *arc connectivity*, $\alpha[i, j]$, of an ordered pair $[i, j]$ of nodes as the minimum number of arcs whose deletion from the network eliminates all the directed path from node i to node j. We define the arc connectivity of a network G as $\alpha(G) = \min\{\alpha[i, j] : [i, j] \in N \times N\}$. Describe an algorithm for determining $\alpha(G)$ in $O(\min\{n^{5/3}m, m^2\})$ time and prove that this algorithm correctly determines the arc connectivity of any directed network. (*Hint*: Let p be a node in G of minimum degree. Determine $\alpha[p, j]$ and $\alpha[j, p]$ for all j, and take the minimum of these numbers.)

8.34. Arc connectivity of directed networks (Schnorr [1979]). In Exercise 8.33 we showed how to determine the arc connectivity $\alpha(G)$ of a directed network G by solving at most $2n$ maximum flow problems. Prove the following result, which would enable us to determine the arc connectivity by solving n maximum flow problems. Let $1, 2, \ldots, n$ be any ordering of the nodes in the network, and let node $(n + 1) = 1$. Show that $\alpha(G) = \min\{\alpha[i, i + 1]:i = 1, \ldots, n\}$.

8.35. Node connectivity of undirected networks. We define the *node connectivity*, $\beta[i, j]$, of a pair $[i, j]$ of nodes in an undirected graph $G = (N, A)$ as the minimum number of nodes whose deletion from the network eliminates all directed paths from node i to node j.

(a) Show that if $(i, j) \in A$, then $\beta[i, j]$ is not defined.

(b) Let $H = \{[i, j] \in N \times N:(i, j) \notin A\}$ and let $\beta(G) = \min\{\beta[i, j]:[i, j] \in H\}$ denote the node connectivity of a network G. Show that $\beta(G) \leq 2\lfloor m/n \rfloor$.

(c) Show that the node connectivity of a network is no more than its arc connectivity.

(d) A natural strategy for determining the node connectivity of a network would be to generalize an arc connectivity algorithm described in Section 8.6. We fix a node p (of minimum degree) and determine $\beta[p, j]$ for each j for which $(p, j) \notin A$. Using Figure 8.25 show that the minimum of these values will not be equal to $\beta(G)$. Explain why this approach fails for finding node connectivity even though it works for finding arc connectivity.

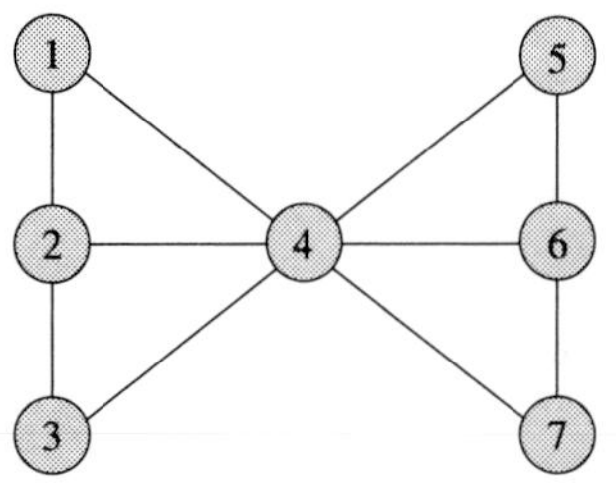

Figure 8.25 Example for Exercise 8.35.

8.36. Solve the all-pairs minimum cut problems given in Figure 8.24. Obtain the separator tree for each problem.

9

MINIMUM COST FLOWS: BASIC ALGORITHMS

. . . men must walk, at least, before they dance.
—Alexander Pope

Chapter Outline

9.1 INTRODUCTION

The minimum cost flow problem is the central object of study in this book. In the last five chapters, we have considered two special cases of this problem: the shortest path problem and the maximum flow problem. Our discussion has been multifaceted: (1) We have seen how these problems arise in application settings as diverse as equipment replacement, project planning, production scheduling, census rounding, and analyzing round-robin tournaments; (2) we have developed a number of algorithmic approaches for solving these problems and studied their computational complexity; and (3) we have shown connections between these problems and more general problems in combinatorial optimization such as the minimum cut problem and a variety of min-max duality results. As we have seen, it is easy to understand the basic nature of shortest path and maximum flow problems and to develop core algorithms for solving them; nevertheless, designing and analyzing efficient algorithms is a very challenging task, requiring considerable ingenuity and considerable insight concerning both basic algorithmic strategies and their implementations.

As we begin to study more general minimum cost flow problems, we might ask ourselves a number of questions.

1. How much more difficult is it to solve the minimum cost flow problem than its shortest path and maximum flow specializations?
2. Can we use some of the same basic algorithmic strategies, such as label-setting and label-correcting methods, and the many variants of the augmenting path methods (e.g., shortest augmenting paths, scaling methods) for solving minimum cost flow problems?
3. The shortest path problem and the maximum flow problem address different components of the overall minimum cost flow problem: Shortest path problems consider arc flow costs but no flow capacities; maximum flow problems consider capacities but only the simplest cost structure. Since the minimum cost flow problem combines these problem ingredients, can we somehow combine the material that we have examined for shortest path and maximum flow problems to develop optimality conditions, algorithms, and underlying theory for the minimum cost flow problem?

In this and the next two chapters, we provide (partial) answers to these questions. We develop a number of algorithms for solving the minimum cost flow problem. Although these algorithms are not as efficient as those for the shortest path and maximum flow problems, they still are quite efficient, and indeed, are among the most efficient algorithms known in applied mathematics, computer science, and operations research for solving large-scale optimization problems.

We also show that we can develop considerable insight and useful tools and methods of analysis by drawing on the material that we have developed already. For example, in order to give us a firm foundation for developing algorithms for solving minimum cost flow problems, in Section 9.3 we establish optimality conditions for minimum cost flow problems based on the notion of node potentials associated with the nodes in the underlying network. These node potentials are generalizations of the concept of distance labels that we used in our study of shortest path problems. Recall that we were able to use distance labels to characterize optimal shortest paths; in addition, we used the distance label optimality conditions as a starting point for developing the basic iterative label-setting and label-correcting algorithms for solving shortest path problems. We use the node potential in a similar fashion for minimum cost flow problems. The connection with shortest paths is much deeper, however, than this simple analogy between node potentials and distance labels. For example, we show how to interpret and find the optimal node potentials for a minimum cost flow problem by solving an appropriate shortest path problem: The optimal node potentials are equal to the negative of the optimal distance labels from this shortest path problem.

In addition, many algorithms for solving the minimum cost flow problem combine ingredients of both shortest path and maximum flow algorithms. Many of these algorithms solve a sequence of shortest path problems with respect to maximum flow-like residual networks and augmenting paths. (Actually, to define the residual network, we consider both cost and capacity considerations.) We consider four such algorithms in this chapter. The *cycle-canceling algorithm* uses shortest path computations to find augmenting cycles with negative flow costs; it then augments flows along these cycles and iteratively repeats these computations for detecting negative cost cycles and augmenting flows. The *successive shortest path algorithm* incrementally loads flow on the network from some source node to some sink node, each time selecting an appropriately defined shortest path. The *primal-dual* and *out-of-kilter algorithms* use a similar algorithmic strategy: at every iteration, they solve a shortest path problem and augment flow along one or more shortest paths. They vary, however, in their tactics. The primal–dual algorithm uses a maximum flow computation to augment flow simultaneously along several shortest paths. Unlike all the other algorithms, the out-of-kilter algorithm permits arc flows to violate their flow bounds. It uses shortest path computations to find flows that satisfy both the flow bounds and the cost and capacity based optimality conditions.

The fact that we can implement iterative shortest path algorithms in so many ways demonstrates the versatility that we have in solving minimum cost flow problems. Indeed, as we shall see in the next two chapters, we have even more versatility. Each of the algorithms that we discuss in this chapter is pseudopolynomial for problems with integer data. As we shall see in Chapter 10, by using ideas such as scaling of the problem data, we can also develop polynomial-time algorithms.

Since minimum cost flow problems are linear programs, it is not surprising to discover that we can also use linear programming methodologies to solve minimum cost flow problems. Indeed, many of the various optimality conditions that we have introduced in previous chapters and that we consider in this chapter are special cases of the more general optimality conditions of linear programming. Moreover, we can interpret many of these results in the context of a general theory of duality for linear programs. In this chapter we develop these duality results for minimum cost flow problems. In Chapter 11 we study the application of the key algorithmic approach from linear programming, the simplex method, for the minimum cost flow problem. In this chapter we consider one other algorithm, known as the *relaxation algorithm*, for solving the minimum cost flow problem.

To begin our discussion of the minimum cost flow problem, we first consider some additional applications, which help to show the importance of this problem in practice. Before doing so, however, let us set our notation and some underlying definitions that we use throughout our discussion.

Notation and Assumptions

Let $G = (N, A)$ be a directed network with a *cost* c_{ij} and a *capacity* u_{ij} associated with every arc $(i, j) \in A$. We associate with each node $i \in N$ a number $b(i)$ which indicates its supply or demand depending on whether $b(i) > 0$ or $b(i) < 0$. The minimum cost flow problem can be stated as follows:

$$\text{Minimize} \quad z(x) = \sum_{(i,j) \in A} c_{ij} x_{ij} \tag{9.1a}$$

subject to

$$\sum_{\{j:(i,j) \in A\}} x_{ij} - \sum_{\{j:(j,i) \in A\}} x_{ji} = b(i) \qquad \text{for all } i \in N, \tag{9.1b}$$

$$0 \le x_{ij} \le u_{ij} \qquad \text{for all } (i, j) \in A. \tag{9.1c}$$

Let C denote the largest magnitude of any arc cost. Further, let U denote the largest magnitude of any supply/demand or finite arc capacity. We assume that the lower bounds l_{ij} on arc flows are all zero. We further make the following assumptions:

Assumption 9.1. *All data (cost, supply/demand, and capacity) are integral.*

As noted previously, this assumption is not really restrictive in practice because computers work with rational numbers which we can convert to integer numbers by multiplying by a suitably large number.

Assumption 9.2. *The network is directed.*

We have shown in Section 2.4 that we can always fulfill this assumption by transforming any undirected network into a directed network.

Assumption 9.3. *The supplies/demands at the nodes satisfy the condition $\sum_{i \in N} b(i) = 0$ and the minimum cost flow problem has a feasible solution.*

We can determine whether the minimum cost flow problem has a feasible solution by solving a maximum flow problem as follows. Introduce a source node s^* and a sink node t^*. For each node i with $b(i) > 0$, add a "source" arc (s^*, i) with capacity $b(i)$, and for each node i with $b(i) < 0$, add a "sink" arc (i, t^*) with capacity $-b(i)$. Now solve a maximum flow problem from s^* to t^*. If the maximum flow saturates all the source arcs, the minimum cost flow problem is feasible; otherwise, it is infeasible. For the justification of this method, see Application 6.1 in Section 6.2.

Assumption 9.4. *We assume that the network G contains an uncapacitated directed path (i.e., each arc in the path has infinite capacity) between every pair of nodes.*

We impose this condition, if necessary, by adding *artificial* arcs $(1, j)$ and $(j, 1)$ for each $j \in N$ and assigning a large cost and infinite capacity to each of these arcs. No such arc would appear in a minimum cost solution unless the problem contains no feasible solution without artificial arcs.

Assumption 9.5. *All arc costs are nonnegative.*

This assumption imposes no loss of generality since the arc reversal transformation described in Section 2.4 converts a minimum cost flow problem with negative arc lengths to those with nonnegative arc lengths. This transformation, however, requires that all arcs have finite capacities. When some arcs are uncapacitated, we assume that the network contains no directed negative cost cycle of infinite capacity. If the network contains any such cycles, the optimal value of the minimum cost flow problem is unbounded; moreover, we can detect such a situation by using the search algorithm described in Section 3.4. In the absence of a negative cycle with infinite capacity, we can make each uncapacitated arc capacitated by setting its capacity equal to B, where B is the sum of all arc capacities and the supplies of all supply nodes; we justify this transformation in Exercise 9.36.

Residual Network

Our algorithms rely on the concept of residual networks. The residual network $G(x)$ corresponding to a flow x is defined as follows. We replace each arc $(i, j) \in A$ by two arcs (i, j) and (j, i). The arc (i, j) has cost c_{ij} and *residual capacity* $r_{ij} = u_{ij} - x_{ij}$, and the arc (j, i) has cost $c_{ji} = -c_{ij}$ and residual capacity $r_{ji} = x_{ij}$. The residual network consists *only* of arcs with positive residual capacity.

9.2 APPLICATIONS

Minimum cost flow problems arise in almost all industries, including agriculture, communications, defense, education, energy, health care, manufacturing, medicine, retailing, and transportation. Indeed, minimum cost flow problems are pervasive in practice. In this section, by considering a few selected applications that arise in distribution systems planning, medical diagnosis, public policy, transportation, manufacturing, capacity planning, and human resource management, we give a passing glimpse of these applications. This discussion is intended merely to introduce several important applications and to illustrate some of the possible uses of minimum cost flow problems in practice. Taken together, the exercises in this chapter and in Chapter 11 and the problem descriptions in Chapter 19 give a much more complete picture of the full range of applications of minimum cost flows.

Application 9.1 Distribution Problems

A large class of network flow problems centers around shipping and distribution applications. One core model might be best described in terms of shipments from plants to warehouses (or, alternatively, from warehouses to retailers). Suppose that a firm has p plants with known supplies and q warehouses with known demands. It wishes to identify a flow that satisfies the demands at the warehouses from the available supplies at the plants and that minimizes its shipping costs. This problem is a well-known special case of the minimum cost flow problem, known as the *transportation problem*. We next describe in more detail a slight generalization of this model that also incorporates manufacturing costs at the plants.

A car manufacturer has several manufacturing plants and produces several car models at each plant that it then ships to geographically dispersed retail centers throughout the country. Each retail center requests a specific number of cars of each model. The firm must determine the production plan of each model at each plant and a shipping pattern that satisfies the demands of each retail center and minimizes the overall cost of production and transportation.

We describe this formulation through an example. Figure 9.1 illustrates a situation with two manufacturing plants, two retailers, and three car models. This model has four types of nodes: (1) *plant nodes*, representing various plants; (2) *plant/model nodes*, corresponding to each model made at a plant; (3) *retailer/model nodes*, corresponding to the models required by each retailer; and (4) *retailer nodes* corresponding to each retailer. The network contains three types of arcs.

1. *Production arcs*. These arcs connect a plant node to a plant/model node; the cost of this arc is the cost of producing the model at that plant. We might place

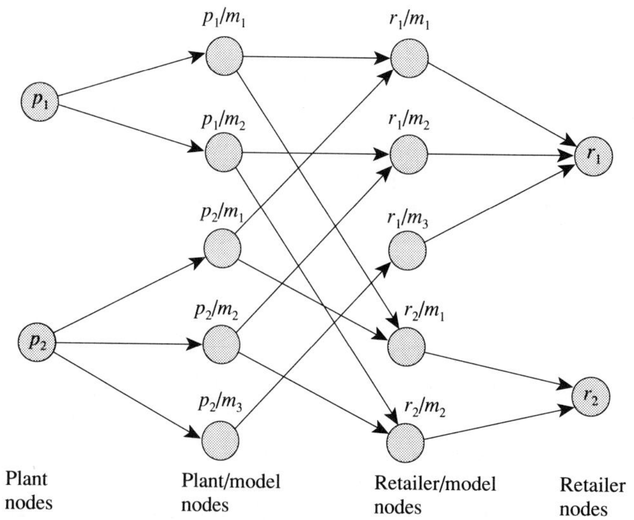

Figure 9.1 Production-distribution model.

lower and upper bounds on these arcs to control for the minimum and maximum production of each particular car model at the plants.

2. *Transportation arcs.* These arcs connect plant/model nodes to retailer/model nodes; the cost of such an arc is the total cost of shipping one car from the manufacturing plant to the retail center. Any such arc might correspond to a complex distribution channel with, for example, three legs: (a) a delivery from a plant (by truck) to a rail system; (b) a delivery from the rail station to another rail station elsewhere in the system; and (c) a delivery from the rail station to a retailer (by a local delivery truck). The transportation arcs might have lower or upper bounds imposed on their flows to model contractual agreements with shippers or capacities imposed on any distribution channel.
3. *Demand arcs.* These arcs connect retailer/model nodes to the retailer nodes. These arcs have zero costs and positive lower bounds which equal the demand of that model at that retail center.

Clearly, the production and shipping schedules for the automobile company correspond in a one-to-one fashion with the feasible flows in this network model. Consequently, a minimum cost flow would yield an optimal production and shipping schedule.

Application 9.2 Reconstructing the Left Ventricle from X-ray Projections

This application describes a network flow model for reconstructing the three-dimensional shape of the left ventricle from biplane angiocardiograms that the medical profession uses to diagnose heart diseases. To conduct this analysis, we first reduce the three-dimensional reconstruction problem into several two-dimensional problems by dividing the ventricle into a stack of parallel cross sections. Each two-dimensional cross section consists of one connected region of the left ventricle.

During a cardiac catheterization, doctors inject a dye known as Roentgen contrast agent into the ventricle; by taking x-rays of the dye, they would like to determine what portion of the left ventricle is functioning properly (i.e., permitting the flow of blood). Conventional biplane x-ray installations do not permit doctors to obtain a complete picture of the left ventricle; rather, these x-rays provide one-dimensional projections that record the total intensity of the dye along two axes (see Figure 9.2). The problem is to determine the distribution of the cloud of dye within the left ventricle and thus the shape of the functioning portion of the ventricle, assuming that the dye mixes completely with the blood and fills the portions that are functioning properly.

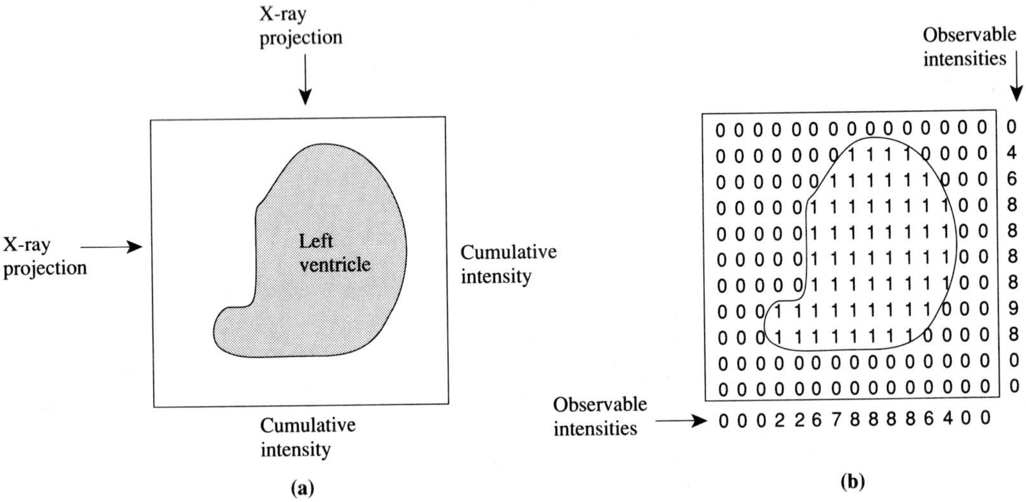

Figure 9.2 Using x-ray projections to measure a left ventricle.

We can conceive of a cross section of the ventricle as a $p \times r$ binary matrix: a 1 in a position indicates that the corresponding segment allows blood to flow and a 0 indicates that it does not permit blood to flow. The angiocardiograms give the cumulative intensity of the contrast agent in two planes which we can translate into row and column sums of the binary matrix. The problem is then to construct the binary matrix given its row and column sums. This problem is a special case of the feasible flow problem that we discussed in Section 6.2.

Typically, the number of feasible solutions for such problems are quite large; and these solutions might differ substantially. To constrain the feasible solutions, we might use certain facts from our experience that indicate that a solution is more likely to contain certain segments rather than others. Alternatively, we can use *a priori* information: for example, after some small time interval, the cross sections might resemble cross sections determined in a previous examination. Consequently, we might attach a probability p_{ij} that a solution will contain an element (i, j) of the binary matrix and might want to find a feasible solution with the largest possible cumulative probability. This problem is equivalent to a minimum cost flow problem.

Application 9.3 Racial Balancing of Schools

In Application 1.10 in Section 1.3 we formulated the racial balancing of schools as a multicommodity flow problem. We now consider a related, yet important situation: seeking a racial balance of two ethnic communities (blacks and whites). In this case we show how to formulate the problem as a minimum cost flow problem.

As in Application 1.10, suppose that a school district has S schools. For the purpose of this formulation, we divide the school district into L district locations and let b_i and w_i denote the number of black and white students at location i. These locations might, for example, be census tracts, bus stops, or city blocks. The only restrictions on the locations is that they be finite in number and that there be a single distance measure d_{ij} that reasonably approximates the distance any student at location i must travel if he or she is assigned to school j. We make the reasonable assumption that we can compute the distances d_{ij} before assigning students to schools. School j can enroll u_j students. Finally, let $\underline{p}$ denote a lower bound and $\overline{p}$ denote an upper bound on the percentage of black students assigned to each school (we choose these numbers so that school j has same percentage of blacks as does the school district). The objective is to assign students to schools in a manner that maintains the stated racial balance and minimizes the total distance traveled by the students.

We model this problem as a minimum cost flow problem. Figure 9.3 shows the minimum cost flow network for a three-location, two-school problem. Rather than describe the general model formally, we merely describe the model ingredients for this figure. In this formulation we model each location i as two nodes l_i' and l_i'' and each school j as two nodes s_j' and s_j''. The decision variables for this problem are

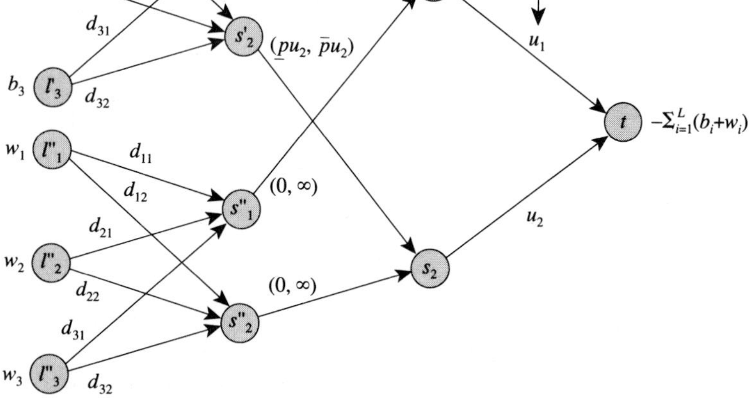

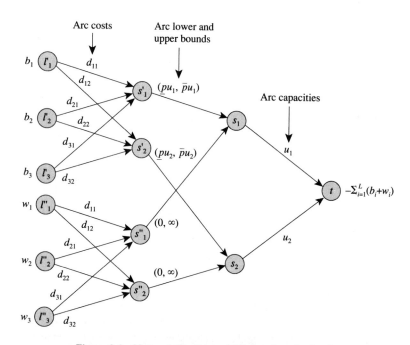

Figure 9.3 Network for the racial balancing of schools.

the number of black students assigned from location i to school j (which we represent by an arc from node l'_i to node s'_j) and the number of white students assigned from location i to school j (which we represent by an arc from node l''_i to node s''_j). These arcs are uncapacitated and we set their per unit flow cost equal to d_{ij}. For each j, we connect the nodes s'_j and s''_j to the school node s_j. The flow on the arcs (s'_j, s_j) and (s''_j, s_j) denotes the total number of black and white students assigned to school j. Since each school must satisfy lower and upper bounds on the number of black students it enrolls, we set the lower and upper bounds of the arc (s'_j, s_j) equal to $(\underline{p}u_j, \bar{p}u_j)$. Finally, we must satisfy the constraint that school j enrolls at most u_j students. We incorporate this constraint in the model by introducing a sink node t and joining each school node j to node t by an arc of capacity u_j. As is easy to verify, this minimum cost flow problem correctly models the racial balancing application.

Application 9.4 Optimal Loading of a Hopping Airplane

A small commuter airline uses a plane, with a capacity to carry at most p passengers, on a "hopping flight," as shown in Figure 9.4(a). The hopping flight visits the cities 1, 2, 3, . . . , n, in a fixed sequence. The plane can pick up passengers at any node and drop them off at any other node. Let b_{ij} denote the number of passengers available at node i who want to go to node j, and let f_{ij} denote the fare per passenger from node i to node j. The airline would like to determine the number of passengers that the plane should carry between the various origins to destinations in order to maximize the total fare per trip while never exceeding the plane capacity.

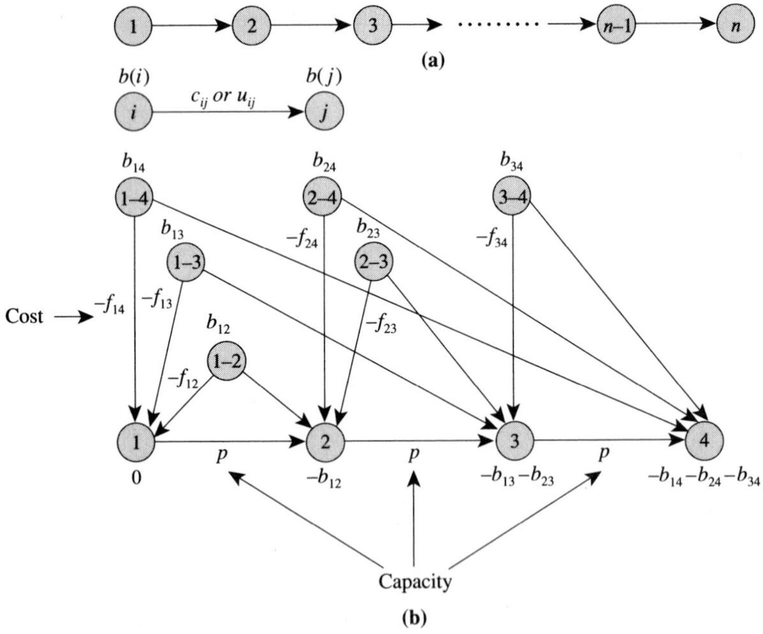

Figure 9.4 Formulating the hopping plane flight problem as a minimum cost flow problem.

Figure 9.4(b) shows a minimum cost flow formulation of this hopping plane flight problem. The network contains data for only those arcs with nonzero costs and with finite capacities: Any arc without an associated cost has a zero cost; any arc without an associated capacity has an infinite capacity. Consider, for example, node 1. Three types of passengers are available at node 1, those whose destination is node 2, node 3, or node 4. We represent these three types of passengers by the nodes 1–2, 1–3, and 1–4 with supplies b_{12}, b_{13}, and b_{14}. A passenger available at any such node, say 1–3, either boards the plane at its origin node by flowing through the arc (1–3, 1), and thus incurring a cost of $-f_{13}$ units, or never boards the plane which we represent by the flow through the arc (1–3, 3). In Exercise 9.13 we ask the reader to show that this formulation correctly models the hopping plane application.

Application 9.5 Scheduling with Deferral Costs

In some scheduling applications, jobs do not have any fixed completion times, but instead incur a deferral cost for delaying their completion. Some of these scheduling problems have the following characteristics: one of q identical processors (machines) needs to process each of p jobs. Each job j has a fixed processing time α_j that does not depend on which machine processes the job, or which jobs precede or follow the job. Job j also has a *deferral cost* $c_j(\tau)$, which we assume is a monotonically nondecreasing function of τ, the completion time of the job. Figure 9.5(a) illustrates one such deferral cost function. We wish to find a schedule for the jobs, with completion times denoted by $\tau_1, \tau_2, \ldots, \tau_p$, that minimizes the total deferral cost $\sum_{j=1}^{p} c_j(\tau_j)$. This scheduling problem is difficult if the jobs have different processing times, but can be modeled as a minimum cost flow problem for situations with uniform processing times (i.e., $\alpha_j = \alpha$ for each $j = 1, \ldots, p$).

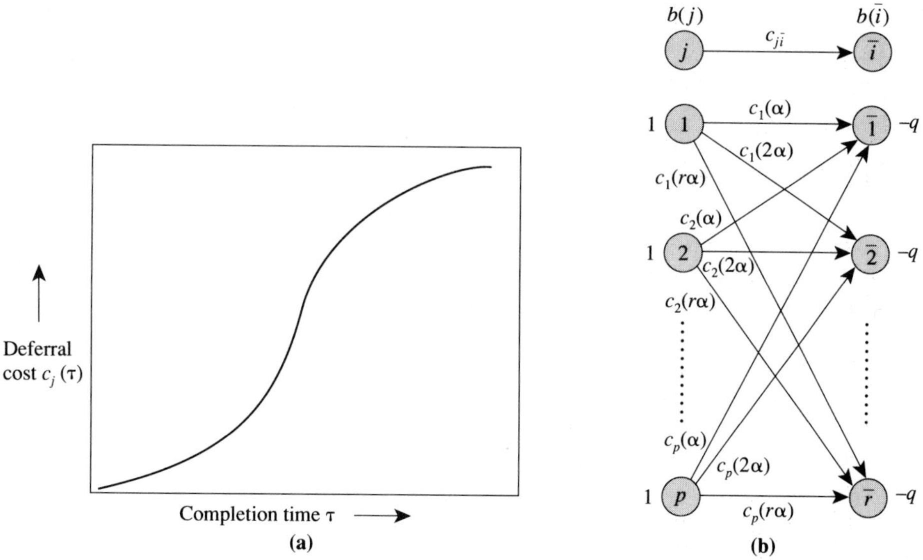

Figure 9.5 Formulating the scheduling problem with deferral costs.

Since the deferral costs are monotonically nondecreasing with time, in some optimal schedule the machines will process the jobs one immediately after another (i.e., the machines incur no *idle time*). As a consequence, in some optimal schedule the completion of each job will be $k\alpha$ for some constant k. The first job assigned to every machine will have a completion time of α units, the second job assigned to every machine will have a completion time of 2α units, and so on. This observation allows us to formulate the scheduling as a minimum cost flow problem in the network shown in Figure 9.5(b).

Assume, for simplicity, that $r = p/q$ is an integer. This assumption implies that we will assign exactly r jobs to each machine. (There is no loss of generality in imposing this assumption because we can add dummy jobs so that p/q becomes an integer.) The network has p job nodes, $1, 2, \ldots, p$, each with 1 unit of supply; it also has r position nodes, $\bar{1}, \bar{2}, \ldots, \bar{r}$, each with a demand of q units, indicating that the position has the capability to process q jobs. The flow on each arc $(j, \bar{i})$ is 1 or 0, depending on whether the schedule does or does not assign job j to the ith position of some machine. If we assign job j to the ith position on any machine, its completion time is $i\alpha$ and its deferral cost is $c_j(i\alpha)$. Therefore, arc $(j, \bar{i})$ has a cost of $c_j(i\alpha)$. Feasible schedules correspond, in a one-to-one fashion, with feasible flows in the network and both have the same cost. Consequently, a minimum cost flow will prescribe a schedule with the least possible deferral cost.

Application 9.6 Linear Programs with Consecutive 1's in Columns

Many linear programming problems of the form

$$\text{Minimize} \quad cx$$

subject to

$$\mathcal{A}x \geq b,$$

$$x \geq 0,$$

have a special structure that permits us to solve the problem more efficiently than general-purpose linear programs. Suppose that the $p \times q$ matrix constraint matrix $\mathcal{A}$ is a 0–1 matrix satisfying the property that all of the 1's in each column appear consecutively (i.e., with no intervening zeros). We show how to transform this problem into a minimum cost flow problem. We illustrate our transformation using the following linear programming example:

$$\text{Minimize} \quad cx \tag{9.2a}$$

subject to

$$\begin{bmatrix} 0 & 1 & 0 & 1 & 1 \\ 1 & 1 & 0 & 0 & 1 \\ 1 & 1 & 1 & 0 & 0 \\ 1 & 1 & 1 & 0 & 0 \end{bmatrix} x \geq \begin{bmatrix} 5 \\ 12 \\ 10 \\ 6 \end{bmatrix}, \tag{9.2b}$$

$$x \geq 0. \tag{9.2c}$$

We first bring each constraint in (9.2b) into an equality form by introducing a "surplus" variable y_i for each row i in (9.2b). We then add a redundant row $0 \cdot x + 0 \cdot y = 0$ to the set of constraints. These changes produce the following equivalent formulation of the linear program:

$$\text{Minimize} \quad cx \tag{9.3a}$$

subject to

$$\begin{bmatrix} 0 & 1 & 0 & 1 & 1 & -1 & 0 & 0 & 0 \\ 1 & 1 & 0 & 0 & 1 & 0 & -1 & 0 & 0 \\ 1 & 1 & 1 & 0 & 0 & 0 & 0 & -1 & 0 \\ 1 & 1 & 1 & 0 & 0 & 0 & 0 & 0 & -1 \\ 0 & 0 & 0 & 0 & 0 & 0 & 0 & 0 & 0 \end{bmatrix} \begin{bmatrix} x \\ \\ y \end{bmatrix} = \begin{bmatrix} 5 \\ 12 \\ 10 \\ 6 \\ 0 \end{bmatrix}, \tag{9.3b}$$

$$x \geq 0. \tag{9.3c}$$

We next perform the following elementary row operation for each $i = p, p - 1, \ldots, 1$, in the stated order: We subtract the ith constraint in (9.3b) from the $(i + 1)$th constraint. These operations create the following equivalent linear program:

$$\text{Minimize} \quad cx \tag{9.4a}$$

subject to

$$\begin{bmatrix} 0 & 1 & 0 & 1 & 1 & -1 & 0 & 0 & 0 \\ 1 & 0 & 0 & -1 & 0 & 1 & -1 & 0 & 0 \\ 0 & 0 & 1 & 0 & -1 & 0 & 1 & -1 & 0 \\ 0 & 0 & 0 & 0 & 0 & 0 & 0 & 1 & -1 \\ -1 & -1 & -1 & 0 & 0 & 0 & 0 & 0 & 1 \end{bmatrix} \begin{bmatrix} x \\ \\ y \end{bmatrix} = \begin{bmatrix} 5 \\ 7 \\ -2 \\ -4 \\ -6 \end{bmatrix}, \tag{9.4b}$$

$$x \geq 0. \tag{9.4c}$$

Notice that in this form the constraints (9.4b) clearly define the mass balance constraints of a minimum cost flow problem because each column contains one $+1$ and one -1. Also notice that the entries in the right-hand-side vector sum to zero, which is a necessary condition for feasibility. Figure 9.6 gives the minimum cost flow problem corresponding to this linear program.

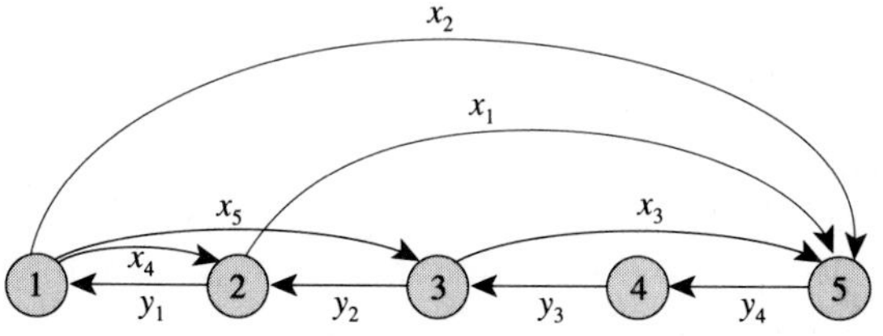

Figure 9.6 Formulating a linear program with consecutive ones as a minimum cost flow problem.

We have used a specific numerical example to illustrate the transformation of a linear program with consecutive 1's into a minimum cost flow problem. It is easy to show that this transformation is valid in general as well. For a linear program with p rows and q columns, the corresponding network has $p + 1$ nodes, one corresponding to each row, as well as one extra node that corresponds to an additional "null row." Each column $\mathcal{A}_k$ in the linear program that has consecutive 1's in rows i to j becomes an arc $(i, j + 1)$ of cost c_k. Each surplus variable y_i becomes an arc $(i + 1, i)$ of zero cost. Finally, the supply/demand of a node i is $b(i) - b(i - 1)$.

Despite the fact that linear programs with consecutive 1's might appear to be very special, and even contrived, this class of problems arises in a surprising number of applications. We illustrate the range of applications with three practical examples. We leave the formulations of these applications as minimum cost flow problems as exercises to the reader.

Optimal capacity scheduling. A vice-president of logistics of a large manufacturing firm must contract for $d(i)$ units of warehousing capacity for the time periods $i = 1, 2, \ldots, n$. Let c_{ij} denote the cost of acquiring 1 unit of capacity at the beginning of period i, which is available for possible use throughout periods $i, i + 1, \ldots, j - 1$ (assume that we relinquish this warehousing capacity at the beginning of period j). The vice-president wants to know how much capacity to acquire, at what times, and for how many subsequent periods, to meet the firm's requirements at the lowest possible cost. This optimization problem arises because of possible savings that the firm might accrue by undertaking long-term leasing contracts at favorable times, even though these commitments might create excess capacity during some periods.

Employment scheduling. The vice-president of human resources of a large retail company must determine an employment policy that properly balances the cost of hiring, training, and releasing short-term employees, with the expense of having idle employees on the payroll for time periods when demand is low. Suppose that the company knows the minimum labor requirement d_j for each period $j = 1, \ldots, n$. Let c_{ij} denote the cost of hiring someone at the beginning of period i and releasing him at the end of period $j - 1$. The vice-president would like to identify an employment policy that meets the labor requirements and minimizes the cost of hiring, training, and releasing employees.

Equipment replacement. A job shop must periodically replace its capital equipment because of machine wear. As a machine ages, it breaks down more frequently and so becomes more expensive to operate. Furthermore, as a machine ages, its salvage value decreases. Let c_{ij} denote the cost of buying a particularly important machine at the beginning of period i, plus the cost of operating the machine over the periods $i, i + 1, \ldots, j - 1$, minus the salvage cost of the machine at the beginning of period j. The *equipment replacement problem* attempts to obtain a replacement plan that minimizes the total cost of buying, selling, and operating the machine over a planning horizon of n years, assuming that the job shop must have at least 1 unit of this machine in service at all times.

9.3 OPTIMALITY CONDITIONS

In our discussion of shortest path problems in Section 5.2, we saw that a set of distance labels $d(i)$ defines shortest path distances from a specified node s to every other node in the network if and only if they represent distances along some paths from node s and satisfy the following shortest path optimality conditions:

$$d(j) \leq d(i) + c_{ij} \qquad \text{for all } (i, j) \in A. \tag{9.5}$$

These optimality conditions are useful in several respects. First, they give us a simple validity check to see whether a given set of distance labels does indeed define shortest paths. Similarly, the optimality conditions provide us with a method for determining whether or not a given set of paths, one from node s to every other node in the network, constitutes a set of shortest paths from node s. We simply compute the lengths of these paths and see if these distances satisfy the optimality conditions. In both cases, the optimality conditions provide us with a "certificate" of optimality, that is, an assurance that a set of distance labels or a set of paths is optimal. One nice feature of the certificate is its ease of use. We need not invoke any complex algorithm to certify that a solution is optimal; we simply check the optimality conditions. The optimality conditions are also valuable for other reasons; as we saw in Chapter 5, they can suggest algorithms for solving a shortest path problem: For example, the generic label correcting algorithm uses the simple idea of repeatedly replacing $d(j)$ by $d(i) + c_{ij}$ if $d(j) > d(i) + c_{ij}$ for some arc (i, j). Finally, the optimality conditions provide us with a mechanism for establishing the validity of algorithms for the shortest path problem. To show that an algorithm correctly finds the desired shortest paths, we verify that the solutions they generate satisfy the optimality conditions.

These various uses of the shortest path optimality conditions suggest that similar sets of conditions might be valuable for designing and analyzing algorithms for the minimum cost flow problem. Accordingly, rather than launching immediately into a discussion of algorithms for solving the minimum cost flow problem, we first pause to describe a few different optimality conditions for this problem. All the optimality conditions that we state have an intuitive network interpretation and are rather direct extensions of their shortest path counterparts. We will consider three different (but equivalent) optimality conditions: (1) negative cycle optimality conditions, (2) reduced cost optimality conditions, and (3) complementary slackness optimality conditions.

Negative Cycle Optimality Conditions

The negative cycle optimality conditions stated next is a direct consequence of the flow decomposition property stated in Theorem 3.5 and our definition of residual networks given at the end of Section 9.1.

Theorem 9.1 (Negative Cycle Optimality Conditions). *A feasible solution x^* is an optimal solution of the minimum cost flow problem if and only if it satisfies the negative cycle optimality conditions: namely, the residual network $G(x^*)$ contains no negative cost (directed) cycle.*

Proof. Suppose that x is a feasible flow and that $G(x)$ contains a negative cycle. Then x cannot be an optimal flow, since by augmenting positive flow along the cycle we can improve the objective function value. Therefore, if x^* is an optimal flow, then $G(x^*)$ cannot contain a negative cycle. Now suppose that x^* is a feasible flow and that $G(x^*)$ contains no negative cycle. Let x° be an optimal flow and $x^* \neq x^\circ$. The augmenting cycle property stated in Theorem 3.7 shows that we can decompose the difference vector $x^\circ - x^*$ into at most m augmenting cycles with respect to the flow x^* and the sum of the costs of flows on these cycles equals $cx^\circ - cx^*$. Since the lengths of all the cycles in $G(x^*)$ are nonnegative, $cx^\circ - cx^* \geq 0$, or $cx^\circ \geq cx^*$. Moreover, since x° is an optimal flow, $cx^\circ \leq cx^*$. Thus $cx^\circ = cx^*$, and x^* is also an optimal flow. This argument shows that if $G(x^*)$ contains no negative cycle, then x^* must be optimal, and this conclusion completes the proof of the theorem.◆

Reduced Cost Optimality Conditions

To develop our second and third optimality conditions, let us make one observation. First, note that we can write the shortest path optimality conditions in the following equivalent form:

$$c_{ij}^d = c_{ij} + d(i) - d(j) \geq 0 \qquad \text{for all arcs } (i, j) \in A. \tag{9.6}$$

This expression has the following interpretation: c_{ij}^d is an optimal "reduced cost" for arc (i, j) in the sense that it measures the cost of this arc relative to the shortest path distances $d(i)$ and $d(j)$. Notice that with respect to the optimal distances, every arc in the network has a nonnegative reduced cost. Moreover, since $d(j) = d(i) + c_{ij}$, if arc (i, j) is on a shortest path connecting the source node s to any other node, the shortest path uses only zero reduced cost arcs. Consequently, once we know the optimal distances, the problem is very easy to solve: We simply find a path from node s to every other node that uses only arcs with zero reduced costs. This interpretation raises a natural question: Is there a similar set of conditions for more general minimum cost flow problems?

Suppose that we associate a real number $\pi(i)$, unrestricted in sign, with each node $i \in N$. We refer to $\pi(i)$ as the *potential* of node i. We show in Section 9.4 that $\pi(i)$ is the linear programming dual variable corresponding to the mass balance constraint of node i. For a given set of node potentials π, we define the *reduced* cost of an arc (i, j) as $c_{ij}^\pi = c_{ij} - \pi(i) + \pi(j)$. These reduced costs are applicable to the residual network as well as the original network. We define the reduced costs in the residual network just as we did the costs, but now using c_{ij}^π in place of c_{ij}. The following properties will prove to be useful in our subsequent developments in this and later chapters.

Property 9.2

(a) *For any directed path P from node k to node l,* $\sum_{(i,j)\in P} c_{ij}^\pi = \sum_{(i,j)\in P} c_{ij} - \pi(k) + \pi(l)$.

(b) *For any directed cycle W,* $\sum_{(i,j)\in W} c_{ij}^\pi = \sum_{(i,j)\in W} c_{ij}$.

The proof of this property is similar to that of Property 2.5. Notice that this property implies that the node potentials do not change the shortest path between any pair of nodes k and l, since the potentials increase the length of every path by a constant amount $\pi(l) - \pi(k)$. This property also implies that if W is a negative cycle with respect to c_{ij} as arc costs, it is also a negative cycle with respect to c_{ij}^π

as arc costs. We can now provide an alternative form of the negative cycle optimality conditions, stated in terms of the reduced costs of the arcs.

Theorem 9.3 (Reduced Cost Optimality Conditions). *A feasible solution x^* is an optimal solution of the minimum cost flow problem if and only if some set of node potentials π satisfy the following reduced cost optimality conditions:*

$$c_{ij}^{\pi} \geq 0 \qquad \textit{for every arc } (i, j) \textit{ in } G(x^*). \tag{9.7}$$

Proof. We shall prove this result using Theorem 9.1. To show that the negative cycle optimality conditions is equivalent to the reduced cost optimality conditions, suppose that the solution x^* satisfies the latter conditions. Therefore, $\sum_{(i,j)\in W} c_{ij}^{\pi} \geq 0$ for every directed cycle W in $G(x^*)$. Consequently, by Property 9.2(b), $\sum_{(i,j)\in W} c_{ij}^{\pi} = \sum_{(i,j)\in W} c_{ij} \geq 0$, so $G(x^*)$ contains no negative cycle.

To show the converse, assume that for the solution x^*, $G(x^*)$ contains no negative cycle. Let $d(\cdot)$ denote the shortest path distances from node 1 to all other nodes in $G(x^*)$. Recall from Section 5.2 that if the network contains no negative cycle, the distance labels $d(\cdot)$ are well defined and satisfy the conditions $d(j) \leq d(i) + c_{ij}$ for all (i, j) in $G(x^*)$. We can restate these inequalities as $c_{ij} - (-d(i)) + (-d(j)) \geq 0$, or $c_{ij}^{\pi} \geq 0$ if we define $\pi = -d$. Consequently, the solution x^* satisfies the reduced cost optimality conditions. ◆

In the preceding theorem we characterized an optimal flow x as a flow that satisfied the conditions $c_{ij}^{\pi} \geq$ for all (i, j) in $G(x)$ for some set of node potentials π. In the same fashion, we could define "optimal node potentials" as a set of node potentials π that satisfy the conditions $c_{ij}^{\pi} \geq 0$ for all (i, j) in $G(x)$ for some feasible flow x.

We might note that the reduced cost optimality conditions have a convenient economic interpretation. Suppose that we interpret c_{ij} as the cost of transporting 1 unit of a commodity from node i to node j through the arc (i, j), and we interpret $\mu(i) \equiv -\pi(i)$ as the cost of obtaining a unit of this commodity at node i. Then $c_{ij} + \mu(i)$ is the cost of the commodity at node j if we obtain it at node i and transport it to node j. The reduced cost optimality condition, $c_{ij} - \pi(i) + \pi(j) \geq 0$, or equivalently, $\mu(j) \leq c_{ij} + \mu(i)$, states that the cost of obtaining the commodity at node j is no more than the cost of the commodity if we obtain it at node i and incur the transportation cost in sending it from node i to j. The cost at node j might be smaller than $c_{ij} + \mu(i)$ because there might be a more cost-effective way to transport the commodity to node j via other nodes.

Complementary Slackness Optimality Conditions

Both Theorems 9.1 and 9.3 provide means for establishing optimality of solutions to the minimum cost flow problem by formulating conditions imposed on the residual network; we shall now restate these conditions in terms of the original network.

Theorem 9.4 (Complementary Slackness Optimality Conditions). *A feasible solution x^* is an optimal solution of the minimum cost flow problem if and only if for some set of node potentials π, the reduced costs and flow values satisfy the following complementary slackness optimality conditions for every arc $(i, j) \in A$:*

$$\textit{If } c_{ij}^{\pi} > 0, \textit{ then } x_{ij}^{*} = 0. \tag{9.8a}$$

$$\textit{If } 0 < x_{ij}^{*} < u_{ij}, \textit{ then } c_{ij}^{\pi} = 0. \tag{9.8b}$$

$$\textit{If } c_{ij}^{\pi} < 0, \textit{ then } x_{ij}^{*} = u_{ij}. \tag{9.8c}$$

Proof. We show that the reduced cost optimality conditions are equivalent to (9.8). To establish this result, we first prove that if the node potentials π and the flow vector x satisfy the reduced cost optimality conditions, then they must satisfy (9.8). Consider three possibilities for any arc $(i, j) \in A$.

Case 1. If $c_{ij}^{\pi} > 0$, the residual network cannot contain the arc (j, i) because $c_{ji}^{\pi} = -c_{ij}^{\pi} < 0$ for that arc, contradicting (9.7). Therefore, $x_{ij}^{*} = 0$.

Case 2. If $0 < x_{ij}^{*} < u_{ij}$, the residual network contains both the arcs (i, j) and (j, i). The reduced cost optimality conditions imply that $c_{ij}^{\pi} \geq 0$ and $c_{ji}^{\pi} \geq 0$. But since $c_{ji}^{\pi} = -c_{ij}^{\pi}$, these inequalities imply that $c_{ij}^{\pi} = c_{ji}^{\pi} = 0$.

Case 3. If $c_{ij}^{\pi} < 0$, the residual network cannot contain the arc (i, j) because $c_{ij}^{\pi} < 0$ for that arc, contradicting (9.7). Therefore, $x_{ij}^{*} = u_{ij}$.

We have thus shown that if the node potentials π and the flow vector x satisfy the reduced cost optimality conditions, they also satisfy the complementary slackness optimality conditions. In Exercise 9.28 we ask the reader to prove the converse result: If the pair (x, π) satisfies the complementary slackness optimality conditions, it also satisfies the reduced cost optimality conditions. ◆

Those readers familiar with linear programming might notice that these conditions are the complementary slackness conditions for a linear programming problem whose variables have upper bounds; this association explains the choice of the name complementary slackness.

9.4 MINIMUM COST FLOW DUALITY

When we were introducing shortest path problems with nonnegative arc costs in Chapter 4, we considered a string model with knots representing the nodes of the network and with a string of length c_{ij} connecting the ith and jth knots. To solve the shortest path problem between a designated source node s and sink node t, we hold the string at the knots s and t and pull them as far apart as possible. As we noted in our previous discussion, if $d(i)$ denotes the distance from the source node s to node i along the shortest path and nodes i and j are any two nodes on this path, then $d(i) + c_{ij} \geq d(j)$. The shortest path distances might satisfy this inequality as a strict inequality if the string from node i to node j is not taut. In this string solution, since we are pulling the string apart as far as possible, we are obtaining the optimal shortest path distance between nodes s and t by solving a *maximization* problem. We could cast this problem formally as the following maximization problem:

$$\text{Maximize} \quad d(t) - d(s) \tag{9.9a}$$

subject to

$$d(j) - d(i) \leq c_{ij} \qquad \text{for all } (i, j) \in A. \tag{9.9b}$$

In this formulation, $d(s) = 0$. As we have noted in Chapter 4, if d is any vector of distance labels satisfying the constraints of this problem and the path P defined as $s - i_1 - i_2 - \cdots i_k - t$ is any path from node s to node t, then

$$d(i_1) - d(s) \le c_{si_1}$$

$$d(i_2) - d(i_1) \le c_{i_1i_2}$$

$$\vdots$$

$$d(t) - d(i_k) \le c_{i_kt},$$

so by adding these inequalities and using the fact that $d(s) = 0$, we see that

$$d(t) \le c_{si_1} + c_{i_1i_2} + \cdots + c_{i_kt}.$$

This result shows that if d is any feasible vector to the optimization problem (9.9), then $d(t)$ is a lower bound on the length of any path from node s to node t and therefore is a lower bound on the shortest distance between these nodes. As we see from the string solution, if we choose the distance labels $d(\cdot)$ appropriately (as the distances obtained from the string solution), $d(t)$ equals the shortest path distance.

This discussion shows the connection between the shortest path problem and a related maximization problem (9.9). In our discussion of the maximum flow problem, we saw a similar relationship, namely, the max-flow min-cut theorem, which tells us that associated with every maximum flow problem is an associated minimization problem. Moreover, since the maximum flow equals the minimum cut, the optimal value of these two associated problems is the same. These two results are special cases of a more general property that applies to any minimum cost flow problem, and that we now establish.

For every linear programming problem, which we subsequently refer to as a *primal* problem, we can associate another intimately related linear programming problem, called its *dual*. For example, the objective function value of *any* feasible solution of the dual is less than or equal to the objective function of any feasible solution of the primal. Furthermore, the maximum objective function value of the dual equals the minimum objective function of the primal. This duality theory is fundamental to an understanding of the theory of linear programming. In this section we state and prove these duality theory results for the minimum cost flow problem.

While forming the dual of a (primal) linear programming problem, we associate a *dual variable* with every constraint of the primal except for the nonnegativity restriction on arc flows. For the minimum cost flow problem stated in (9.1), we associate the variable $\pi(i)$ with the mass balance constraint of node i and the variable α_{ij} with the capacity constraint of arc (i, j). In terms of these variables, the *dual minimum cost flow problem* can be stated as follows:

$$\text{Maximize} \quad w(\pi, \alpha) = \sum_{i \in N} b(i)\pi(i) - \sum_{(i,j) \in A} u_{ij}\alpha_{ij} \tag{9.10a}$$

subject to

$$\pi(i) - \pi(j) - \alpha_{ij} \le c_{ij} \quad \text{for all } (i, j) \in A, \tag{9.10b}$$

$$\alpha_{ij} \ge 0 \quad \text{for all } (i, j) \in A \quad \text{and} \quad \pi(j) \text{ unrestricted for all } j \in N. \tag{9.10c}$$

Note that the shortest path dual problem (9.9) is a special case of this model: For the shortest path problem, $b(s) = 1$, $b(t) = -1$, and $b(i) = 0$ otherwise. Also, since the shortest path problem contains no arc capacities, we can eliminate the α_{ij} variables. Therefore, if we let $d(i) = -\pi(i)$, the dual minimum cost flow problem (9.10) becomes the shortest path dual problem (9.9).

Our first duality result for the general minimum cost flow problem is known as the *weak duality theorem.*

Theorem 9.5 (Weak Duality Theorem). *Let $z(x)$ denote the objective function value of some feasible solution x of the minimum cost flow problem and let $w(\pi, \alpha)$ denote the objective function value of some feasible solution (π, α) of its dual. Then $w(\pi, \alpha) \le z(x)$.*

Proof. We multiply both sides of (9.10b) by x_{ij} and sum these weighted inequalities for all $(i, j) \in A$, obtaining

$$\sum_{(i,j)\in A} (\pi(i) - \pi(j))x_{ij} - \sum_{(i,j)\in A} \alpha_{ij}x_{ij} \le \sum_{(i,j)\in A} c_{ij}x_{ij}. \tag{9.11}$$

Notice that $cx - c^{\pi}x = \sum_{(i,j)\in A} (\pi(i) - \pi(j))x_{ij}$ [because $c^{\pi}_{ij} = c_{ij} - \pi(i) + \pi(j)$]. Next notice that Property 2.4 in Section 2.4 implies that $cx - c^{\pi}x$ equals $\sum_{i\in N} b(i)\pi(i)$. Therefore, the first term on the left-hand side of (9.11) equals $\sum_{i\in N} b(i)\pi(i)$. Next notice that replacing x_{ij} in the second term on the left-hand side of (9.11) by u_{ij} preserves the inequality because $x_{ij} \le u_{ij}$ and $\alpha_{ij} \ge 0$. Consequently,

$$\sum_{i\in n} b(i)\pi(i) - \sum_{(i,j)\in A} \alpha_{ij}u_{ij} \le \sum_{(i,j)\in A} c_{ij}x_{ij}. \tag{9.12}$$

Now notice that the left-hand side of (9.12) is the dual objective $w(\pi, \alpha)$ and the right-hand side is the primal objective, so we have established the lemma. ◆

The weak duality theorem implies that the objective function value of *any* dual feasible solution is a lower bound on the objective function value of *any* primal feasible solution. One consequence of this result is immediate: If some dual solution (π, α) and a primal solution x have the same objective function value, (π, α) must be an optimal solution of the dual problem and x must be an optimal solution of the primal problem (why?). Can we always find such solutions? The *strong duality theorem*, to be proved next, answers this question in the affirmative.

We first eliminate the dual variables α_{ij}'s from the dual formation (9.10) using some properties of the optimal solution. Defining the reduced cost, as before, as $c^{\pi}_{ij} = c_{ij} - \pi(i) + \pi(j)$, we can rewrite the constraint (9.10b) as

$$\alpha_{ij} \ge -c^{\pi}_{ij}. \tag{9.13}$$

The coefficient associated with the variable α_{ij} in the dual objective (9.10a) is $-u_{ij}$, and we wish to maximize the objective function value. Consequently, in any optimal solution we would assign the smallest possible value to α_{ij}. This observation, in view of (9.10c) and (9.13), implies that

$$\alpha_{ij} = \max\{0, -c^{\pi}_{ij}\}. \tag{9.14}$$

We have thus shown that if we know optimal values for the dual variables $\pi(i)$, we can compute the optimal values of the variables α_{ij} using (9.14). This construction permits us to eliminate the variables α_{ij} from the dual formulation. Substituting (9.14) in (9.10a) yields

$$\text{Maximize} \quad w(\pi) = \sum_{i \in N} b(i)\pi(i) - \sum_{(i,j) \in A} \max\{0, -c_{ij}^{\pi}\}u_{ij}. \tag{9.15}$$

The dual problem reduces to finding a vector π that optimizes (9.15). We are now in a position to prove the strong duality theorem. (Recall that our blanket assumption, Assumption 9.3, implies that the minimum cost flow problem always has a solution.)

Theorem 9.6 (Strong Duality Theorem). *For any choice of problem data, the minimum cost flow problem always has a solution x^* and the dual minimum cost flow problem has a solution π satisfying the property that $z(x^*) = w(\pi)$.*

Proof. We prove this theorem using the complementary slackness optimality conditions (9.8). Let x^* be an optimal solution of the minimum cost flow problem. Theorem 9.4 implies that x^* together with some vector π of node potentials satisfy the complementary slackness optimality conditions. We claim that this solution satisfies the condition

$$-c_{ij}^{\pi}x_{ij} = \max\{0, -c_{ij}^{\pi}\}u_{ij} \qquad \text{for every arc } (i, j) \in A. \tag{9.16}$$

To establish this result, consider the following three cases: (1) $c_{ij}^{\pi} > 0$, (2) $c_{ij}^{\pi} = 0$, and (3) $c_{ij}^{\pi} < 0$. The complementary slackness conditions (9.8) imply that in the first two cases, both the left-hand side and right-hand side of (9.16) are zero, and in the third case both sides equal $-c_{ij}^{\pi}u_{ij}$.

Next consider the dual objective (9.15). Substituting (9.16) in (9.15) yields

$$w(\pi) = \sum_{i \in N} b(i)\pi(i) + \sum_{(i,j) \in A} c_{ij}^{\pi}x_{ij}^* = \sum_{(i,j) \in A} c_{ij}x_{ij}^* = z(x^*).$$

The second last inequality follows from Property 2.4. This result is the conclusion of the theorem. ◆

The proof of this theorem shows that any optimal solution x^* of the minimum cost flow problem always has an associated dual solution π satisfying the condition $z(x^*) = w(\pi)$. Needless to say, the solution π is an optimal solution of the dual minimum cost flow problem since any larger value of the dual objective would contradict the weak duality theorem stated in Theorem 9.5.

In Theorem 9.6 we showed that the complementary slackness optimality conditions implies strong duality. We next prove the converse result: namely, that strong duality implies the complementary slackness optimality conditions.

Theorem 9.7. *If x is a feasible flow and π is an (arbitrary) vector satisfying the property that $z(x) = w(\pi)$, then the pair (x, π) satisfies the complementary slackness optimality conditions.*

Proof. Since $z(x) = w(\pi)$,

$$\sum_{(i,j)\in A} c_{ij}x_{ij} = \sum_{i\in N} b(i)\pi(i) - \sum_{(i,j)\in A} \max\{0, -c_{ij}^{\pi}\}u_{ij}. \tag{9.17}$$

Substituting the result of Property 2.4 in (9.17) shows that

$$\sum_{(i,j)\in A} \max\{0, -c_{ij}^{\pi}\}u_{ij} = \sum_{(i,j)\in A} -c_{ij}^{\pi}x_{ij}. \tag{9.18}$$

Now observe that both the sides have m terms, and each term on the left-hand side is nonnegative and its value is an upper bound on the corresponding term on the right-hand side (because $\max\{0, -c_{ij}^{\pi}\} \geq -c_{ij}^{\pi}$ and $u_{ij} \geq x_{ij}$). Therefore, the two sides can be equal only when

$$\max\{0, -c_{ij}^{\pi}\}u_{ij} = -c_{ij}^{\pi}x_{ij} \qquad \text{for every arc } (i, j) \in A. \tag{9.19}$$

Now we consider three cases.

(a) $c_{ij}^{\pi} > 0$. In this case, the left-hand side of (9.19) is zero, and the right-hand side can be zero only if $x_{ij} = 0$. This conclusion establishes (9.8a).

(b) $0 < x_{ij} < u_{ij}$. In this case, $c_{ij}^{\pi} = 0$; otherwise, the right-hand side of (9.19) is negative. This conclusion establishes (9.8b).

(c) $c_{ij}^{\pi} < 0$. In this case, the left-hand side of (9.19) is $-c_{ij}^{\pi}u_{ij}$ and therefore, $x_{ij} = u_{ij}$. This conclusion establishes (9.8c).

These results complete the proof of the theorem. ◆

The following result is an easy consequence of Theorems 9.6 and 9.7.

Property 9.8. *If x^* is an optimal solution of the minimum cost flow problem, and π is an optimal solution of the dual minimum cost flow problem, the pair (x^*, π) satisfies the complementary slackness optimality conditions* (9.8).

Proof. Theorem 9.6 implies that $z(x^*) = w(\pi)$ and Theorem 9.7 implies that the pair (x^*, π) satisfies (9.8). ◆

One important implication of the minimum cost flow duality is that it permits us to solve linear programs that have at most one $+1$ and at most one -1 in each row as minimum cost flow problems. Linear programs with this special structure arise in a variety of situations; Applications 19.10, 19.11, 19.18, and Exercises 9.9 and 19.18 provide a few examples.

Before examining the situation with at most one $+1$ and at most one -1 in each row, let us consider a linear program that has at most one $+1$ and at most one -1 in each column. We assume, without any loss of generality, that each constraint in the linear program is in equality form, because we can always bring the linear program into this form by introducing slack or surplus variables. (Observe that column corresponding the slack or surplus variables will also have one $+1$ or one -1.) If each column has exactly one $+1$ and exactly one -1, clearly the linear program is a minimum cost flow problem. Otherwise, we can augment this linear program by adding a redundant equality constraint which is the negative of the sum of all the original constraints. (The new constraint corresponds to a new node that acts as a repository to deposit any excess supply or a source to fulfill any deficit demand from the other nodes.) The augmented linear program contains exactly one $+1$ and exactly one -1 in each column and the right-hand side values sum to zero. This model is clearly an instance of the minimum cost flow problem.

We now return to linear programs (in maximization form) that have at most one $+1$ and at most one -1 in each row. We allow a constraint in this linear program to be in any form: equality or inequality. The dual of this linear program contains at most one $+1$ and at most one -1 in each column, which we have already shown to be equivalent to a minimum cost flow problem. The variables in the dual problem will be nonnegative, nonpositive, or unrestricted, depending on whether they correspond to a less than or equal to, a greater than or equal to, or an equality constraint in the primal. A nonnegative variable x_{ij} defines a directed arc (i, j) in the resulting minimum cost flow formulation. To model any unrestricted variable x_{ij}, we replace it with two nonnegative variables, which is equivalent to introducing two arcs (i, j) and (j, i) of the same cost and capacity as this variable. The following theorem summarizes the preceding discussion.

Theorem 9.9. *Any linear program that contains* (a) *at most one* $+1$ *and at most one* -1 *in each column, or* (b) *at most one* $+1$ *and at most one* -1 *in each row, can be transformed into a minimum cost flow problem.* ◆

Minimum cost flow duality has several important implications. Since almost all algorithms for solving the primal problem also generate optimal node potentials $\pi(i)$ and the variables α_{ij}, solving the primal problem almost always solves both the primal and dual problems. Similarly, solving the dual problem typically solves the primal problem as well. Most algorithms for solving network flow problems explicitly or implicitly use properties of dual variables (since they are the node potentials that we have used at every turn) and of the dual linear program. In particular, the dual problem provides us with a certificate that if we can find a feasible dual solution that has the same objective function value as a given primal solution, we know from the strong duality theorem that the primal solution must be optimal, *without* making additional calculations and without considering other potentially optimal primal solutions. This certification procedure is a very powerful idea in network optimization, and in optimization in general. We have used it at many points in our previous developments and will see it many times again.

For network flow problems, the primal and dual problems are closely related via the basic shortest path and maximum flow problems that we have studied in previous chapters. In fact, these relationships help us to understand the fundamental importance of these two core problems to network flow theory and algorithms. We develop these relationships in the next section.

9.5 RELATING OPTIMAL FLOWS TO OPTIMAL NODE POTENTIALS

We next address the following questions: (1) Given an optimal flow, how might we obtain optimal node potentials? Conversely, (2) given optimal node potentials, how might we obtain an optimal flow? We show how to solve these problems by solving either a shortest path problem or a maximum flow problem. These results point out an interesting relationship between the minimum cost flow problem and the maximum flow and shortest path problems.

Computing Optimal Node Potentials

We show that given an optimal flow x^*, we can obtain optimal node potentials by solving a shortest path problem (with possibly negative arc lengths). Let $G(x^*)$ denote the residual network with respect to the flow x^*. Clearly, $G(x^*)$ does not contain any negative cost cycle, for otherwise we would contradict the optimality of the solution x^*. Let $d(\cdot)$ denote the shortest path distances from node 1 to the rest of the nodes in the residual network if we use c_{ij} as arc lengths. The distances $d(\cdot)$ are well defined because the residual network does not contain a negative cycle. The shortest path optimality conditions (5.2) imply that

$$d(j) \leq d(i) + c_{ij} \qquad \text{for all } (i, j) \text{ in } G(x^*). \tag{9.20}$$

Let $\pi = -d$. Then we can restate (9.20) as

$$c_{ij}^{\pi} = c_{ij} - \pi(i) + \pi(j) \geq 0 \qquad \text{for all } (i, j) \text{ in } G(x^*).$$

Theorem 9.3 shows that π constitutes an optimal set of node potentials.

Obtaining Optimal Flows

We now show that given a set of optimal node potentials π, we can obtain an optimal solution x^* by solving a maximum flow problem. First, we compute the reduced cost c_{ij}^{π} of every arc $(i, j) \in A$ and then we examine all arcs one by one. We classify each arc (i, j) in one of the following ways and use these categorizations of the arcs to define a maximum flow problem.

Case 1: $c_{ij}^{\pi} > 0$
The condition (9.8a) implies that x_{ij}^* must be zero. We enforce this constraint by setting $x_{ij}^* = 0$ and deleting arc (i, j) from the network.

Case 2: $c_{ij}^{\pi} < 0$
The condition (9.8c) implies that $x_{ij}^* = u_{ij}$. We enforce this constraint by setting $x_{ij}^* = u_{ij}$ and deleting arc (i, j) from the network. Since we sent u_{ij} units of flow on arc (i, j), we must decrease $b(i)$ by u_{ij} and increase $b(j)$ by u_{ij}.

Case 3: $c_{ij}^{\pi} = 0$
In this case we allow the flow on arc (i, j) to assume any value between 0 and u_{ij}.

Let $G' = (N, A')$ denote the resulting network and let b' denote the modified supplies/demands of the nodes. Now the problem reduces to finding a feasible flow in the network G' that meets the modified supplies/demands of the nodes. As noted in Section 6.2, we can find such a flow by solving a maximum flow problem defined as follows. We introduce a *source node s*, and a *sink node t*. For each node i with

$b'(i) > 0$, we add an arc (s, i) with capacity $b'(i)$ and for each node i with $b'(i) < 0$, we add an arc (i, t) with capacity $-b'(i)$. We now solve a maximum flow problem from node s to t in the transformed network obtaining a maximum flow x^*. The solution x^*_{ij} for all $(i, j) \in A$ is an optimal flow for the minimum cost flow problem in G.

9.6 CYCLE-CANCELING ALGORITHM AND THE INTEGRALITY PROPERTY

The negative cycle optimality conditions suggests one simple algorithmic approach for solving the minimum cost flow problem, which we call the *cycle-canceling algorithm*. This algorithm maintains a feasible solution and at every iteration attempts to improve its objective function value. The algorithm first establishes a feasible flow x in the network by solving a maximum flow problem (see Section 6.2). Then it iteratively finds negative cost-directed cycles in the residual network and augments flows on these cycles. The algorithm terminates when the residual network contains no negative cost-directed cycle. Theorem 9.1 implies that when the algorithm terminates, it has found a minimum cost flow. Figure 9.7 specifies this generic version of the cycle-canceling algorithm.

```
algorithm cycle-canceling;
begin
    establish a feasible flow x in the network;
    while G(x) contains a negative cycle do
    begin
        use some algorithm to identify a negative cycle W;
        δ : = min{r_ij : (i, j) ∈ W};
        augment δ units of flow in the cycle W and update G(x);
    end;
end;
```

Figure 9.7 Cycle canceling algorithm.

We use the example shown in Figure 9.8(a) to illustrate the cycle-canceling algorithm. (The reader might notice that our example does not satisfy Assumption 9.4; we violate this assumption so that the network is simpler to analyze.) Figure 9.8(a) depicts a feasible flow in the network and Figure 9.8(b) gives the corresponding residual network. Suppose that the algorithm first selects the cycle 4–2–3–4 whose cost is -1. The residual capacity of this cycle is 2. The algorithm augments 2 units of flow along this cycle. Figure 9.8(c) shows the modified residual network. In the next iteration, suppose that the algorithm selects the cycle 4–2–1–3–4 whose cost is -2. The algorithm sends 1 unit of flow along this cycle. Figure 9.8(d) depicts the updated residual network. Since this residual network contains no negative cycle, the algorithm terminates.

In Chapter 5 we discussed several algorithms for identifying a negative cycle if one exists. One algorithm for identifying a negative cycle is the FIFO label-correcting algorithm for the shortest path problem described in Section 5.4; this algorithm requires $O(nm)$ time. We describe other algorithms for detecting negative cycles in Sections 11.7 and 12.7.

A by-product of the cycle-canceling algorithm is the following important result.

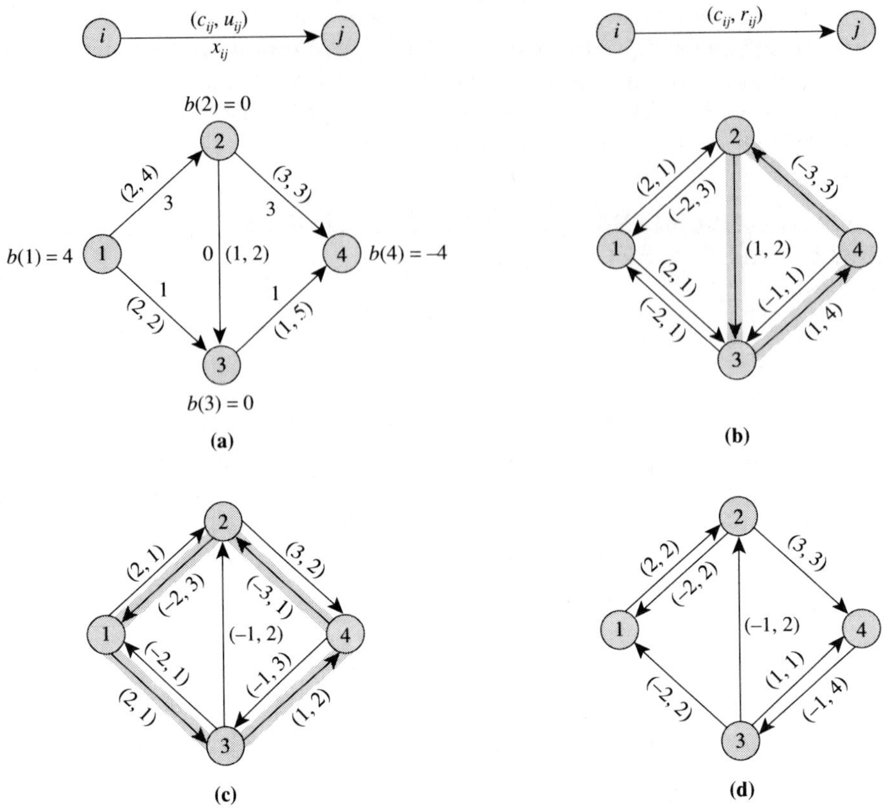

Figure 9.8 Illustrating the cycle canceling algorithm: (a) network example with a feasible flow x; (b) residual network $G(x)$; (c) residual network after augmenting 2 units along the cycle 4–2–3–4; (d) residual network after augmenting 1 unit along the cycle 4–2–1–3–4.

Theorem 9.10 (Integrality Property). *If all arc capacities and supplies/demands of nodes are integer, the minimum cost flow problem always has an integer minimum cost flow.*

Proof. We show this result by performing induction on the number of iterations. The algorithm first establishes a feasible flow in the network by solving a maximum flow problem. By Theorem 6.5 the problem has an integer feasible flow and we assume that the maximum flow algorithm finds an integer solution since all arc capacities in the network are integer and the initial residual capacities are also integer. The flow augmented by the cycle-canceling algorithm in any iteration equals the minimum residual capacity in the cycle canceled, which by the inductive hypothesis is integer. Therefore the modified residual capacities in the next iteration will again be integer. This conclusion implies the assertion of the theorem. ◆

Let us now consider the number of iterations that the algorithm performs. For the minimum cost flow problem, mCU is an upper bound on the initial flow cost [since $c_{ij} \leq C$ and $x_{ij} \leq U$ for all $(i, j) \in A$] and $-mCU$ is a lower bound on the optimal flow cost [since $c_{ij} \geq -C$ and $x_{ij} \leq U$ for all $(i, j) \in A$]. Each iteration of the cycle-canceling algorithm changes the objective function value by an amount $(\sum_{(i,j) \in W} c_{ij})\delta$, which is strictly negative. Since we are assuming that all the data of the problem are integral, the algorithm terminates within $O(mCU)$ iterations and runs in $O(nm^2CU)$ time.

The generic version of the cycle-canceling algorithm does not specify the order for selecting negative cycles from the network. Different rules for selecting negative cycles produce different versions of the algorithm, each with different worse-case and theoretical behavior. The network simplex algorithm, which is widely considered to be one of the fastest algorithms for solving the minimum cost flow problem in practice, is a particular version of the cycle-canceling algorithm. The network simplex algorithm maintains information (a spanning tree solution and node potentials) that enables it to identify a negative cost cycle in $O(m)$ time. However, due to *degeneracy*, the algorithm cannot necessarily send a positive amount of flow along this cycle. We discuss these issues in Chapter 11, where we consider the network simplex algorithm in more detail. The most general implementation of the network simplex algorithm does not run in polynomial time. The following two versions of the cycle-canceling algorithm are, however, polynomial-time implementations.

Augmenting flow in a negative cycle with maximum improvement. Let x be any feasible flow and let x^* be an optimal flow. The improvement in the objective function value due to an augmentation along a cycle W is $-(\sum_{(i,j) \in W} c_{ij})$ $(\min\{r_{ij} : (i, j) \in W\})$. We observed in the proof of Theorem 3.7 in Section 3.5 that x^* equals x plus the flow on at most m augmenting cycles with respect to x, and improvements in cost due to flow augmentations on these augmenting cycles sum to $cx - cx^*$. Consequently, at least one of these augmenting cycles with respect to x must decrease the objective function value by at least $(cx - cx^*)/m$. Consequently, if the algorithm always augments flow along a cycle giving the maximum possible improvement, then Theorem 3.1 implies that the method would obtain an optimal flow within $O(m \log(mCU))$ iterations. Finding a maximum improvement cycle is difficult (i.e., it is a $\mathcal{NP}$-complete problem), but a modest variation of this approach yields a polynomial-time algorithm for the minimum cost flow problem. We provide a reference for this algorithm in the reference notes.

Augmenting flow along a negative cycle with minimum mean cost. We define the *mean cost* of a cycle as its cost divided by the number of arcs it contains. A *minimum mean cycle* is a cycle whose mean cost is as small as possible. It is possible to identify a minimum mean cycle in $O(nm)$ or $O(\sqrt{n}\, m \log(nC))$ time (see the reference notes of Chapter 5). Researchers have shown that if the cycle-canceling algorithm always augments flow along a minimum mean cycle, it performs $O(\min\{nm \log(nC), nm^2 \log n\})$ iterations. We describe this algorithm in Section 10.5.

9.7 SUCCESSIVE SHORTEST PATH ALGORITHM

The cycle-canceling algorithm maintains feasibility of the solution at every step and attempts to achieve optimality. In contrast, the successive shortest path algorithm maintains optimality of the solution (as defined in Theorem 9.3) at every step and strives to attain feasibility. It maintains a solution x that satisfies the nonnegativity and capacity constraints, but violates the mass balance constraints of the nodes. At each step, the algorithm selects a node s with excess supply (i.e., supply not yet sent to some demand node) and a node t with unfulfilled demand and sends flow from s to t along a shortest path in the residual network. The algorithm terminates when the current solution satisfies all the mass balance constraints.

To describe this algorithm as well as several later developments, we first introduce the concept of *pseudoflows*. A *pseudoflow* is a function $x: A \rightarrow R^+$ satisfying only the capacity and nonnegativity constraints; it need not satisfy the mass balance constraints. For any pseudoflow x, we define the *imbalance* of node i as

$$e(i) = b(i) + \sum_{\{j:(j,i)\in A\}} x_{ji} - \sum_{\{j:(i,j)\in A\}} x_{ij} \qquad \text{for all } i \in N.$$

If $e(i) > 0$ for some node i, we refer to $e(i)$ as the *excess* of node i; if $e(i) < 0$, we call $-e(i)$ the node's *deficit*. We refer to a node i with $e(i) = 0$ as *balanced*. Let E and D denote the sets of excess and deficit nodes in the network. Notice that $\sum_{i\in N} e(i) = \sum_{i\in N} b(i) = 0$, and hence $\sum_{i\in E} e(i) = -\sum_{i\in D} e(i)$. Consequently, if the network contains an excess node, it must also contain a deficit node. The residual network corresponding to a pseudoflow is defined in the same way that we define the residual network for a flow.

Using the concept of pseudoflow and the reduced cost optimality conditions specified in Theorem 9.3, we next prove some results that we will use extensively in this and the following chapters.

Lemma 9.11. *Suppose that a pseudoflow (or a flow) x satisfies the reduced cost optimality conditions with respect to some node potentials π. Let the vector d represent the shortest path distances from some node s to all other nodes in the residual network $G(x)$ with c_{ij}^{π} as the length of an arc (i, j). Then the following properties are valid:*

(a) *The pseudoflow x also satisfies the reduced cost optimality conditions with respect to the node potentials $\pi' = \pi - d$.*
(b) *The reduced costs $c_{ij}^{\pi'}$ are zero for all arcs (i, j) in a shortest path from node s to every other node.*

Proof. Since x satisfies the reduced cost optimality conditions with respect to π, $c_{ij}^{\pi} \geq 0$ for every arc (i, j) in $G(x)$. Furthermore, since the vector d represents shortest path distances with c_{ij}^{π} as arc lengths, it satisfies the shortest path optimality conditions, that is,

$$d(j) \leq d(i) + c_{ij}^{\pi} \qquad \text{for all } (i, j) \text{ in } G(x). \tag{9.21}$$

Substituting $c_{ij}^{\pi} = c_{ij} - \pi(i) + \pi(j)$ in (9.21), we obtain $d(j) \leq d(i) + c_{ij} - \pi(i) + \pi(j)$. Alternatively, $c_{ij} - (\pi(i) - d(i)) + (\pi(j) - d(j)) \geq 0$, or $c_{ij}^{\pi'} \geq 0$. This conclusion establishes part (a) of the lemma.

Consider next a shortest path from node s to some node l. For each arc (i, j) in this path, $d(j) = d(i) + c_{ij}^{\pi}$. Substituting $c_{ij}^{\pi} = c_{ij} - \pi(i) + \pi(j)$ in this equation, we obtain $c_{ij}^{\pi'} = 0$. This conclusion establishes part (b) of the lemma. ◆

The following result is an immediate corollary of the preceding lemma.

Lemma 9.12. *Suppose that a pseudoflow (or a flow) x satisfies the reduced cost optimality conditions and we obtain x' from x by sending flow along a shortest path from node s to some other node k; then x' also satisfies the reduced cost optimality conditions.*

Proof. Define the potentials π and π' as in Lemma 9.11. The proof of Lemma 9.11 implies that for every arc (i, j) in the shortest path P from node s to the node k, $c_{ij}^{\pi'} = 0$. Augmenting flow on any such arc might add its reversal (j, i) to the residual network. But since $c_{ij}^{\pi'} = 0$ for each arc $(i, j) \in P$, $c_{ji}^{\pi'} = 0$ and the arc (j, i) also satisfies the reduced cost optimality conditions. These results establish the lemma. ◆

We are now in a position to describe the successive shortest path algorithm. The node potentials play a very important role in this algorithm. Besides using them to prove the correctness of the algorithm, we use them to maintain nonnegative arc lengths so that we can solve the shortest path problem more efficiently. Figure 9.9 gives a formal statement of the successive shortest path algorithm.

We illustrate the successive shortest path algorithm on the same numerical example we used to illustrate the cycle canceling algorithm. Figure 9.10(a) shows the initial residual network. Initially, $E = \{1\}$ and $D = \{4\}$. Therefore, in the first iteration, $s = 1$ and $t = 4$. The shortest path distances d (with respect to the reduced costs) are $d = (0, 2, 2, 3)$ and the shortest path from node 1 to node 4 is 1–3–4. Figure 9.10(b) shows the updated node potentials and reduced costs, and Figure 9.10(c) shows the solution after we have augmented $\min\{e(1), -e(4), r_{13}, r_{34}\} = \min\{4, 4, 2, 5\} = 2$ units of flow along the path 1–3–4. In the second iteration, $k =$

```
algorithm successive shortest path;
begin
    x : = 0 and π : = 0;
    e(i) : = b(i) for all i ∈ N;
    initialize the sets E : = {i : e(i) > 0} and D : = {i : e(i) < 0};
    while E ≠ Ø do
    begin
        select a node k ∈ E and a node l ∈ D;
        determine shortest path distances d(j) from node s to all
            other nodes in G(x) with respect to the reduced costs c_ij^π;
        let P denote a shortest path from node k to node l;
        update π : = π - d;
        δ : = min[e(k), - e(l), min{r_ij : (i, j) ∈ P}];
        augment δ units of flow along the path P;
        update x, G(x), E, D, and the reduced costs;
    end;
end;
```

Figure 9.9 Successive shortest path algorithm.

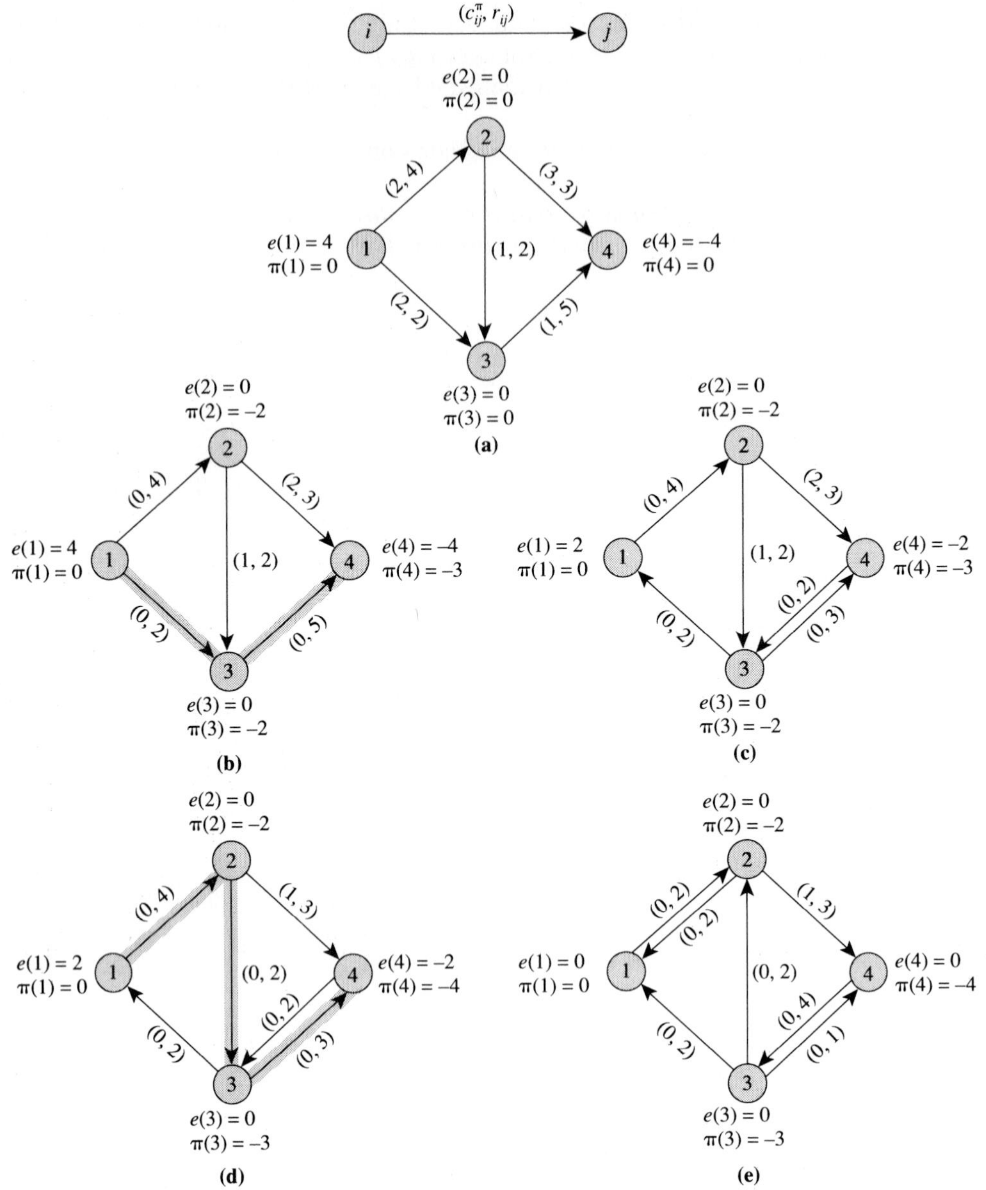

Figure 9.10 Illustrating the successive shortest path algorithm: (a) initial residual network for $x = 0$ and $\pi = 0$; (b) network after updating the potentials π; (c) network after augmenting 2 units along the path 1–3–4; (d) network after updating the potentials π; (e) network after augmenting 2 units along the path 1–2–3–4.

1, $l = 4$, $d = (0, 0, 1, 1)$ and the shortest path from node 1 to node 4 is 1–2–3–4. Figure 9.10(d) shows the updated node potentials and reduced costs, and Figure 9.10(e) shows the solution after we have augmented $\min\{e(1), -e(4), r_{12}, r_{23}, r_{34}\} = \min\{2, 2, 4, 2, 3\} = 2$ units of flow. At the end of this iteration, all imbalances become zero and the algorithm terminates.

We now justify the successive shortest path algorithm. To initialize the algorithm, we set $x = 0$, which is a feasible pseudoflow. For the zero pseudoflow x, $G(x) = G$. Note that this solution together with $\pi = 0$ satisfies the reduced cost optimality conditions because $c_{ij}^{\pi} = c_{ij} \geq 0$ for every arc (i, j) in the residual network $G(x)$ (recall Assumption 9.5, which states that all arc costs are nonnegative). Observe that as long as any node has a nonzero imbalance, both E and D must be nonempty since the total sum of excesses equals the total sum of deficits. Thus until all nodes are balanced, the algorithm always succeeds in identifying an excess node k and a deficit node l. Assumption 9.4 implies that the residual network contains a directed path from node k to every other node, including node l. Therefore, the shortest path distances $d(\cdot)$ are well defined. Each iteration of the algorithm solves a shortest path problem with nonnegative arc lengths and strictly decreases the excess of some node (and, also, the deficit of some other node). Consequently, if U is an upper bound on the largest supply of any node, the algorithm would terminate in at most nU iterations. If $S(n, m, C)$ denotes the time taken to solve a shortest path problem with nonnegative arc lengths, the overall complexity of this algorithm is $O(nUS(n, m, nC))$. [Note that we have used nC rather than C in this expression, since the costs in the residual network are bounded by nC.] We refer the reader to the reference notes of Chapter 4 for the best available value of $S(n, m, C)$.

The successive shortest path algorithm requires pseudopolynomial time to solve the minimum cost flow problem since it is polynomial in n, m and the largest supply U. This algorithm is, however, polynomial time for the assignment problem, a special case of the minimum cost flow problem, for which $U = 1$. In Chapter 10, using scaling techniques, we develop weakly and strongly polynomial-time versions of the successive shortest path algorithm. In Section 14.5 we generalize this approach even further, developing a polynomial-time algorithm for the convex cost flow problem.

We now suggest some practical improvements to the successive shortest path algorithm. As stated, this algorithm selects an excess node k, uses Dijkstra's algorithm to identify shortest paths from node k to all other nodes, and augments flow along a shortest path from node k to some deficit node l. In fact, it is not necessary to determine a shortest path from node k to *all* nodes; a shortest path from node k to *one deficit node* l is sufficient. Consequently, we could terminate Dijkstra's algorithm whenever it permanently labels the first deficit node l. At this point we might modify the node potentials in the following manner:

$$\pi(i) = \begin{cases} \pi(i) - d(i) & \text{if node } i \text{ is permanently labeled} \\ \pi(i) - d(l) & \text{if node } i \text{ is temporarily labeled.} \end{cases}$$

In Exercise 9.47 we ask the reader to show that with this choice of the modified node potentials, the reduced costs of all the arcs in the residual network remain nonnegative and the reduced costs of the arcs along the shortest path from node k to node l are zero. Observe that we can alternatively modify the node potentials in the following manner:

$$\pi(i) = \begin{cases} \pi(i) - d(i) + d(l) & \text{if node } i \text{ is permanently labeled} \\ \pi(i) & \text{if node } i \text{ is temporarily labeled.} \end{cases}$$

This scheme for updating node potentials is the same as the previous scheme except that we add $d(l)$ to all of the node potentials (which does not affect the reduced cost of any arc). An advantage of this scheme is that the algorithm spends no time updating the potentials of the temporarily labeled nodes.

9.8 PRIMAL–DUAL ALGORITHM

The primal–dual algorithm for the minimum cost flow problem is similar to the successive shortest path algorithm in the sense that it also maintains a pseudoflow that satisfies the reduced cost optimality conditions and gradually converts it into a flow by augmenting flows along shortest paths. In contrast, instead of sending flow along one shortest path at a time, it solves a maximum flow problem that sends flow along all shortest paths.

The primal–dual algorithm generally transforms the minimum cost flow problem into a problem with a single excess node and a single deficit node. We transform the problem into this form by introducing a *source* node s and a *sink* node t. For each node i with $b(i) > 0$, we add a zero cost arc (s, i) with capacity $b(i)$, and for each node i with $b(i) < 0$, we add a zero cost arc (i, t) with capacity $-b(i)$. Finally, we set $b(s) = \sum_{\{i \in N: b(i)>0\}} b(i)$, $b(t) = -b(s)$, and $b(i) = 0$ for all $i \in N$. It is easy to see that a minimum cost flow in the transformed network gives a minimum cost flow in the original network. For simplicity of notation, we shall represent the transformed network as $G = (N, A)$, which is the same representation that we used for the original network.

The primal–dual algorithm solves a maximum flow problem on a subgraph of the residual network $G(x)$, called the *admissible network*, which we represent as $G^{\circ}(x)$. We define the admissible network $G^{\circ}(x)$ with respect to a pseudoflow x that satisfies the reduced cost optimality conditions for some node potentials π; the admissible network contains only those arcs in $G(x)$ with a zero reduced cost. The residual capacity of an arc in $G^{\circ}(x)$ is the same as that in $G(x)$. Observe that every directed path from node s to node t in $G^{\circ}(x)$ is a shortest path in $G(x)$ between the same pair of nodes (see Exercise 5.20). Figure 9.11 formally describes the primal–dual algorithm on the transformed network.

```
algorithm primal-dual;
begin
    x : = 0 and π : = 0;
    e(s) : = b(s) and e(t) : = b(t);
    while e(s) > 0 do
    begin
        determine shortest path distances d(·) from node s to all other nodes in G(x) with
            respect to the reduced costs c^π_ij;
        update π : = π - d;
        define the admissible network G°(x);
        establish a maximum flow from node s to node t in G°(x);
        update e(s), e(t), and G(x);
    end;
end;
```

Figure 9.11 Primal–dual algorithm.

To illustrate the primal–dual algorithm, we consider the numerical example shown in Figure 9.12(a). Figure 9.12(b) shows the transformed network. The shortest path computation yields the vector $d = (0, 0, 0, 1, 2, 1)$ whose components are in

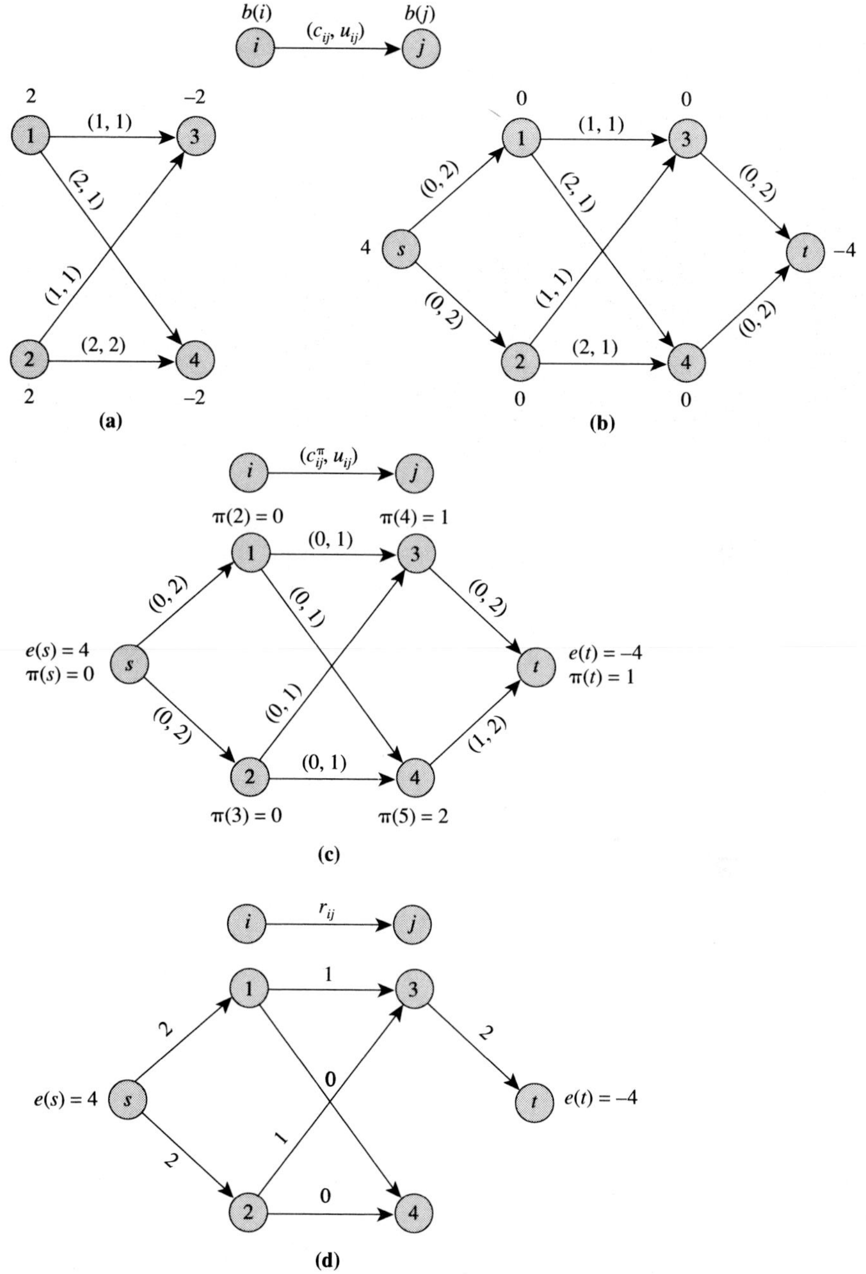

Figure 9.12 Illustrating the primal–dual algorithm: (a) example network; (b) transformed network; (c) residual network after updating the node potentials; (d) admissible network.

the order s, 1, 2, 3, 4, t. Figure 9.12(c) shows the modified node potentials and reduced costs and Figure 9.12(d) shows the admissible network at this stage in the computations. When we apply the maximum flow algorithm to the admissible network, it is able to send 2 units of flow from node s to node t. Observe that the admissible network contained two paths from node s to node t and the maximum flow computation saturates both the paths. The successive shortest path algorithm would have taken two iterations to send the 2 units of flow. As the reader can verify, the second iteration of the primal–dual algorithm also sends 2 units of flow from node s to node t, at which point it converts the pseudoflow into a flow and terminates.

The primal–dual algorithm guarantees that the excess of node s strictly decreases at each iteration, and also assures that the node potential of the sink strictly decreases from one iteration to the next. The second observation follows from the fact that once we have established a maximum flow in $G^{\circ}(x)$, the residual network $G(x)$ contains no directed path from node s to node t consisting entirely of arcs with zero reduced costs. Consequently, in the next iteration, when we solve the shortest path problem, $d(t) \geq 1$. These observations give a bound of $\min\{nU, nC\}$ on the number of iterations since initially $e(s) \leq nU$, and the value of no node potential can fall below $-nC$ (see Exercise 9.25). This bound on the number of iterations is better than that of the successive shortest path algorithm, but, of course, the algorithm incurs the additional expense of solving a maximum flow problem at every iteration. If $S(n, m, C)$ and $M(n, m, U)$ denote the solution times of shortest path and the maximum flow algorithms, the primal–dual algorithm has an overall complexity of $O(\min\{nU, nC\} \cdot \{S(n, m, nC) + M(n, m, U)\})$.

In concluding this discussion, we might comment on why this algorithm is known as the primal–dual algorithm. This name stems from linear programming duality theory. In the linear programming literature, the primal–dual algorithm always maintains a dual feasible solution π and a primal solution that might violate some supply/demand constraints (i.e., is primal infeasible), so that the pair satisfies the complementary slackness conditions. For a given dual feasible solution, the algorithm attempts to decrease the degree of primal infeasibility to the minimum possible level. [Recall that the algorithm solves a maximum flow problem to reduce $e(s)$ by the maximum amount.] When no further reduction in the primal infeasibility is possible, the algorithm modifies the dual solution (i.e., node potentials in the network flow context) and again tries to minimize primal infeasibility. This primal–dual approach is applicable to several combinatorial optimization problems and also to the general linear programming problem. Indeed, this primal–dual solution strategy is one of the most popular approaches for solving specially structured problems and has often yielded fairly efficient and intuitively appealing algorithms.

9.9 OUT-OF-KILTER ALGORITHM

The successive shortest path and primal–dual algorithms maintain a solution that satisfies the reduced cost optimality conditions and the flow bound constraints but violates the mass balance constraints. These algorithms iteratively modify arc flows and node potentials so that the flow at each step comes closer to satisfying the mass balance constraints. However, we could just as well have developed other solution strategies by violating other constraints at intermediate steps. The out-of-kilter algorithm, which we discuss in this section, satisfies only the mass balance constraints, so intermediate solutions might violate both the optimality conditions and the flow bound restrictions. The algorithm iteratively modifies flows and potentials in a way that decreases the infeasibility of the solution (in a way to be specified) and, simultaneously, moves it closer to optimality. In essence, the out-of-kilter algorithm is similar to the successive shortest path and primal–dual algorithms because its fundamental step at every iteration is solving a shortest path problem and augmenting flow along a shortest path.

To describe the out-of-kilter algorithm, we refer to the complementary slackness optimality conditions stated in Theorem 9.4. For ease of reference, let us restate these conditions.

$$\text{If } x_{ij} = 0, \text{ then } c_{ij}^{\pi} \geq 0. \tag{9.22a}$$

$$\text{If } 0 < x_{ij} < u_{ij}, \text{ then } c_{ij}^{\pi} = 0. \tag{9.22b}$$

$$\text{If } x_{ij} = u_{ij}, \text{ then } c_{ij}^{\pi} \leq 0, \tag{9.22c}$$

The name *out-of-kilter algorithm* reflects the fact that arcs in the network either satisfy the complementary slackness optimality conditions (are *in-kilter*) or do not (are *out-of-kilter*). The so-called *kilter diagram* is a convenient way to represent these conditions. As shown in Figure 9.13, the kilter diagram of an arc (i, j) is the collection of all points (x_{ij}, c_{ij}^{π}) in the two-dimensional plane that satisfy the optimality conditions (9.22). The condition 9.22(a) implies that $c_{ij}^{\pi} \geq 0$ if $x_{ij} = 0$; therefore, the kilter diagram contains all points with zero x_{ij}-coordinates and nonnegative c_{ij}^{π}-coordinates. Similarly, the condition 9.22(b) yields the horizontal segment of the diagram, and condition 9.22(c) yields the other vertical segment of the diagram. Each arc has its own kilter diagram.

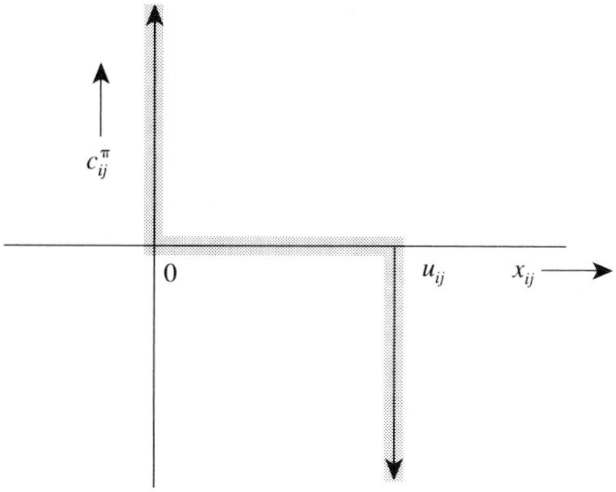

Figure 9.13 Kilter diagram for arc (i, j).

Notice that for every arc (i, j), the flow x_{ij} and reduced cost c_{ij}^{π} define a point (x_{ij}, c_{ij}^{π}) in the two-dimensional plane. If the point (x_{ij}, c_{ij}^{π}) lies on the thick lines in the kilter diagram, the arc is in-kilter; otherwise, it is out-of-kilter. For instance, the points B, D, and E in Figure 9.14 are in-kilter, whereas the points A and C are out-of-kilter. We define the *kilter number* k_{ij} of each arc (i, j) in A as the magnitude of the change in x_{ij} required to make the arc an in-kilter arc while keeping c_{ij}^{π} fixed.

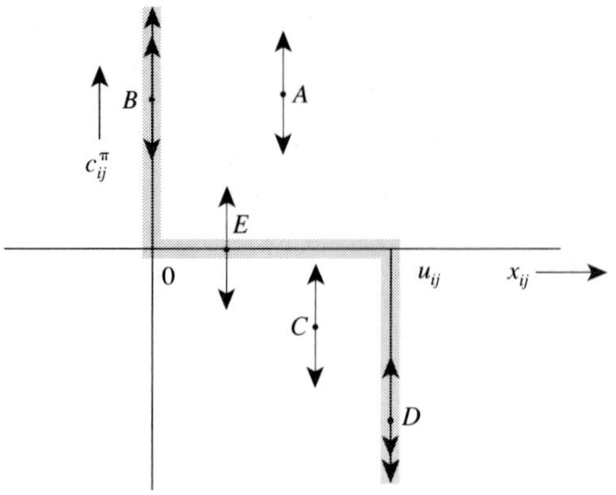

Figure 9.14 Examples of in-kilter and out-of-kilter arcs.

Therefore, in accordance with conditions (9.22a) and (9.22c), if $c_{ij}^{\pi} > 0$, then $k = |x_{ij}|$, and if $c_{ij}^{\pi} < 0$, then $k_{ij} = |u_{ij} - x_{ij}|$. If $c_{ij}^{\pi} = 0$ and $x_{ij} > u_{ij}$, then $k_{ij} = x_{ij} - u_{ij}$. If $c_{ij}^{\pi} = 0$ and $x_{ij} < 0$, then $k_{ij} = -x_{ij}$. The kilter number of any in-kilter arc is zero. The sum $K = \sum_{(i,j)\in A} k_{ij}$ of all kilter numbers provides us with a measure of how far the current solution is from optimality; the smaller the value of K, the closer the current solution is to being an optimal solution.

In describing the out-of-kilter algorithm, we begin by making a simplifying assumption that the algorithm starts with a feasible flow. At the end of this section we show how to extend the algorithm so that it applies to situations when the initial flow does not satisfy the arc flow bounds (we also consider situations with nonzero lower bounds on arc flows).

To describe the out-of-kilter algorithm, we will work on the residual network; in this setting, the algorithm iteratively decreases the kilter number of one or more arcs in the residual network. To do so, we must be able to define the kilter number of the arcs in the residual network $G(x)$. We set the kilter number k_{ij} of an arc (i, j) in the following manner:

$$k_{ij} = \begin{cases} 0 & \text{if } c_{ij}^{\pi} \geq 0. \\ r_{ij} & \text{if } c_{ij}^{\pi} < 0. \end{cases} \tag{9.23}$$

This definition of the kilter number of an arc in the residual network is consistent with our previous definition: It is the change in flow (or, equivalently, the residual capacity) required so that the arc satisfies its optimality condition [which, in the case of residual networks, is the reduced cost optimality condition (9.7)]. An arc (i, j) in the residual network with $c_{ij}^{\pi} \geq 0$ satisfies its optimality condition (9.7), but an arc (i, j) with $c_{ij}^{\pi} < 0$ does not. In the latter case, we must send r_{ij} units of flow on the arc (i, j) so that it drops out of the residual network and thus satisfies its optimality condition.

The out-of-kilter algorithm maintains a feasible flow x and a set of node potentials π. We could obtain a feasible flow by solving a maximum flow problem (as described in Section 6.2) and start with $\pi = 0$. Subsequently, the algorithm maintains all of the in-kilter arcs as in-kilter arcs and successively transforms the out-of-kilter arcs into in-kilter arcs. The algorithm terminates when all arcs in the residual network become in-kilter. Figure 9.15 gives a formal description of the out-of-kilter algorithm.

```
algorithm out-of-kilter;
begin
    π : = 0;
    establish a feasible flow x in the network;
    define the residual network G(x) and compute the kilter numbers of arcs;
    while the network contains an out-of-kilter arc do
    begin
        select an out-of-kilter arc (p, q) in G(x);
        define the length of each arc (i, j) in G(x) as max{0, c^π_ij};
        let d(·) denote the shortest path distances from node q to all other nodes in
            G(x) - {(q, p)} and let P denote a shortest path from node q to node p;
        update π'(i) : = π(i) - d(i) for all i ∈ N;
        if c^π'_pq < 0 then
        begin
            W : = P ∪ {(p, q)};
            δ : = min{r_ij : (i, j) ∈ W};
            augment δ units of flow along W;
            update x, G(x), and the reduced costs;
        end;
    end;
end;
```

Figure 9.15 Out-of-kilter algorithm.

We now discuss the correctness and complexity of the out-of-kilter algorithm. The correctness argument of the algorithm uses the fact that kilter numbers of arcs are nonincreasing. Two operations in the algorithm affect the kilter numbers of arcs: updating node potentials and augmenting flow along the cycle W. In the next two lemmas we show that these operations do not increase the kilter number of any arc.

Lemma 9.13. *Updating the node potentials does not increase the kilter number of any arc in the residual network.*

Proof. Let π and π' denote the node potentials in the out-of-kilter algorithm before and after the update. The definition of the kilter numbers from (9.23) implies that the kilter number of an arc (i, j) can increase only if $c_{ij}^{\pi} \geq 0$ and $c_{ij}^{\pi'} < 0$. We show that this cannot happen. Consider any arc (i, j) with $c_{ij}^{\pi} \geq 0$. We wish to show that $c_{ij}^{\pi'} \geq 0$. Since $c_{pq}^{\pi} < 0$, $(i, j) \neq (p, q)$. Since the distances $d(\cdot)$ represent the shortest path distances with $\max\{0, c_{ij}^{\pi}\}$ as the length of arc (i, j), the shortest path distances satisfy the following shortest path optimality condition (see Section 5.2):

$$d(j) \leq d(i) + \max\{0, c_{ij}^{\pi}\} = d(i) + c_{ij}^{\pi}.$$

The equality in this expression is valid because, by assumption, $c_{ij}^{\pi} \geq 0$. The preceding expression shows that

$$c_{ij}^{\pi} + d(i) - d(j) = c_{ij}^{\pi'} \geq 0,$$

so each arc in the residual network with a nonnegative reduced cost has a nonnegative reduced cost after the potentials update, which implies the conclusion of the lemma. ◆

Lemma 9.14. *Augmenting flow along the directed cycle $W = P \cup \{(p, q)\}$ does not increase the kilter number of any arc in the residual network and strictly decreases the kilter number of the arc (p, q).*

Proof. Notice that the flow augmentation can change the kilter number of only the arcs in $W = P \cup \{(p, q)\}$ and their reversals. Since P is a shortest path in the residual network with $\max\{0, c_{ij}^{\pi}\}$ as the length of arc (i, j),

$$d(j) = d(i) + \max\{0, c_{ij}^{\pi}\} \geq d(i) + c_{ij}^{\pi} \qquad \text{for each arc } (i, j) \in P,$$

which, using $\pi' = \pi - d$ and the definition $c_{ij}^{\pi} = c_{ij} - \pi(i) + \pi(j)$, implies that

$$c_{ij}^{\pi'} \leq 0 \qquad \text{for each arc } (i, j) \in P.$$

Since the reduced cost of each arc (i, j) in P with respect to π' is nonpositive, the condition (9.23) shows that sending additional flow does not increase the arc's kilter number, but might decrease it. The flow augmentation might add the reversals of arcs in P, but since $c_{ij}^{\pi'} \leq 0$, the reversal of this arc (j, i) has $c_{ji}^{\pi'} \geq 0$, and therefore arc (j, i) is an in-kilter arc.

Finally, we consider arc (p, q). Recall from the algorithm description in Figure 9.15 that we augment flow along the arc (p, q) only if it is an out-of-kilter arc (i.e., $c_{pq}^{\pi'} < 0$). Since augmenting flow along the arc (p, q) decreases its residual capacity, the augmentation decreases this arc's kilter number. Since $c_{qp}^{\pi'} > 0$, arc (q, p) remains an in-kilter arc. These conclusions complete the proof of the lemma. ◆

The preceding two lemmas allow us to obtain a pseudopolynomial bound on the running time of the out-of-kilter algorithm. Initially, the kilter number of an arc is at most U; therefore, the sum of the kilter numbers is at most mU. At each iteration, the algorithm selects an arc, say (p, q), with a positive kilter number and either makes it an in-kilter arc during the potential update step or decreases its kilter number by the subsequent flow augmentation. Therefore, the sum of kilter numbers decreases by at least 1 unit at every iteration. Consequently, the algorithm terminates within $O(mU)$ iterations. The dominant computation within each iteration is solving a shortest path problem. Therefore, if $S(n, m, C)$ is the time required to solve a shortest path problem with nonnegative arc lengths, the out-of-kilter algorithm runs in $O(mU\ S(n, m, nC))$ time.

How might we modify the algorithm to handle situations when the arc flows do not necessarily satisfy their flow bounds? In examining this case we consider the more general problem setting by allowing the arcs to have nonzero lower bounds. Let l_{ij} denote the lower bound on the flow on arc $(i, j) \in A$. In this case, the complementary slackness optimality conditions become:

$$\text{If } x_{ij} = l_{ij}, \text{ then } c_{ij}^{\pi} \geq 0. \tag{9.24a}$$

$$\text{If } l_{ij} < x_{ij} < u_{ij}, \text{ then } c_{ij}^{\pi} = 0. \tag{9.24b}$$

$$\text{If } x_{ij} = u_{ij}, \text{ then } c_{ij}^{\pi} \leq 0. \tag{9.24c}$$

The thick lines in Figure 9.16 define the kilter diagram for this case. Consider arc (i, j). If the point (x_{ij}, c_{ij}^{π}) lies on the thick line in Figure 9.16, the arc is an in-kilter arc; otherwise it is an out-of-kilter arc. As earlier, we define the kilter number

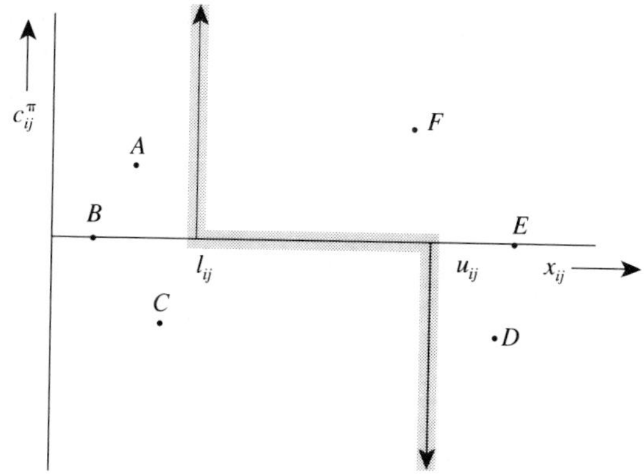

Figure 9.16 Kilter diagram for an arc (i, j) with a nonzero lower bound.

of an arc (i, j) in A as the magnitude of the change in x_{ij} required to make the arc an in-kilter arc while keeping c_{ij}^{π} fixed. Since arcs might violate their flow bounds, six types of out-of-kilter arcs are possible, which we depict by points A, B, C, D, E, and F in Figure 9.16. For example, the kilter numbers of arcs with coordinates depicted by the points A and D are $(l_{ij} - x_{ij})$ and $(x_{ij} - u_{ij})$, respectively.

To describe the algorithm for handling these situations, we need to determine how to form the residual network $G(x)$ for a flow x violating its lower and upper bounds. We consider each arc (i, j) in A one by one and add arcs to the residual network $G(x)$ in the following manner:

1. $l_{ij} \le x_{ij} \le u_{ij}$. If $x_{ij} < u_{ij}$, we add the arc (i, j) with a residual capacity $u_{ij} - x_{ij}$ and with a cost c_{ij}. If $x_{ij} > l_{ij}$, we add the arc (j, i) with a residual capacity $x_{ij} - l_{ij}$ and with a cost $-c_{ij}$. We call these arcs *feasible arcs*.
2. $x_{ij} < l_{ij}$. In this case we add the arc (i, j) with a residual capacity $(l_{ij} - x_{ij})$ and with a cost c_{ij}. We refer to this arc as a *lower-infeasible arc*.
3. $x_{ij} > u_{ij}$. In this case we add the arc (j, i) with a residual capacity $(x_{ij} - u_{ij})$ and with a cost $-c_{ij}$. We refer to this arc as an *upper-infeasible arc*.

We next define the kilter numbers of arcs in the residual network. For feasible arcs in the residual network, we define their kilter numbers using (9.23). We define the kilter number k_{ij} of a lower-infeasible or an upper-infeasible arc (i, j) as the change in its residual capacity required to restore its feasibility as well as its optimality. For instance, for a lower-infeasible arc (i, j) (1) if $c_{ij}^{\pi} \ge 0$, then $k_{ij} = (l_{ij} - x_{ij})$; and (2) if $c_{ij}^{\pi} < 0$, then $k_{ij} = (u_{ij} - x_{ij})$. Note that

1. Lower-infeasible and upper-infeasible arcs have positive kilter numbers.
2. Sending additional flow on lower-infeasible and upper-infeasible arcs in the residual network decreases their kilter numbers.

The out-of-kilter algorithm for this case is same as that for the earlier case. The algorithmic description given in Figure 9.15 applies to this case as well except that at the beginning of the algorithm we need not establish a feasible flow in the network. We can initiate the algorithm with $x = 0$ as the starting flow. We leave the justification of the out-of-kilter algorithm for this case as an exercise to the reader (see Exercise 9.26).

9.10 RELAXATION ALGORITHM

All the minimum cost flow algorithms we have discussed so far—the cycle-canceling algorithm, the successive shortest path algorithm, the primal–dual algorithm, and the out-of-kilter algorithm—are classical in the sense that researchers developed them in the 1950s and 1960s as network flow area was emerging as an independent field of scientific investigation. These algorithms have several common features: (1) they repeatedly apply shortest path algorithms, (2) they run in pseudopolynomial time, and (3) their empirical running times have proven to be inferior to those of the network simplex algorithm tailored for the minimum cost flow problem (we discuss this algorithm in Chapter 11). The relaxation algorithm we examine in this section is a more recent vintage minimum cost flow algorithm; it is competitive or better than the network simplex algorithm for some classes of networks. Interestingly, the relaxation algorithm is also a variation of the successive shortest path algorithm. Even though the algorithm has proven to be efficient in practice for many classes of problems, its worst-case running time is much poorer than that of every minimum cost flow algorithm discussed in this chapter.

The relaxation algorithm uses ideas from *Lagrangian relaxation*, a well-known technique used for solving integer programming problems. We discuss the Lagrangian relaxation technique in more detail in Chapter 16. In the Lagrangian relaxation technique, we identify a set of constraints to be relaxed, multiply each such constraint by a scalar, and subtract the product from the objective function. The relaxation algorithm relaxes the mass balance constraints of the nodes, multiplying the mass balance constraint for node i by an (unrestricted) variable $\pi(i)$ (called, as usual, a node potential) and subtracts the resulting product from the objective function. These operations yield the following relaxed problem:

$$w(\pi) = \underset{x}{\text{minimize}} \left[\sum_{(i,j)\in A} c_{ij}x_{ij} + \sum_{i\in N} \pi(i) \left\{ - \sum_{\{j:(i,j)\in A\}} x_{ij} + \sum_{\{j:(j,i)\in A\}} x_{ji} + b(i) \right\} \right] \tag{9.25a}$$

subject to

$$0 \le x_{ij} \le u_{ij} \qquad \text{for all } (i, j) \in A. \tag{9.25b}$$

For a specific value of the vector π of node potentials, we refer to the relaxed problem as $LR(\pi)$ and denote its objective function value by $w(\pi)$. Note that the optimal solution of LR(π) is a pseudoflow for the minimum cost flow problem since it might violate the mass balance constraints. We can restate the objective function of LR(π) in the following equivalent way:

$$w(\pi) = \underset{x}{\text{minimize}} \left[\sum_{(i,j)\in A} c_{ij}x_{ij} + \sum_{i\in N} \pi(i)e(i) \right]. \tag{9.26}$$

In this expression, as in our earlier discussion, $e(i)$ denotes the imbalance of node i. Let us restate the objective function (9.25a) of the relaxed problem in another way. Notice that in the second term of (9.25a), each flow variable x_{ij} appears twice: once with a coefficient of $-\pi(i)$ and the second time with a coefficient of $\pi(j)$. Therefore, we can write (9.25a) as follows:

$$w(\pi) = \underset{x}{\text{minimize}} \left[\sum_{(i,j)\in A} (c_{ij} - \pi(i) + \pi(j))x_{ij} + \sum_{i\in N} \pi(i)b(i) \right],$$

or, equivalently,

$$w(\pi) = \underset{x}{\text{minimize}} \left[\sum_{(i,j)\in A} c_{ij}^{\pi} x_{ij} + \sum_{i\in N} \pi(i)b(i) \right]. \tag{9.27}$$

In the subsequent discussion, we refer to the objective function of LR(π) as (9.26) or (9.27), whichever is more convenient. For a given vector π of node potentials, it is very easy to obtain an optimal solution x of LR(π): In light of the formulation (9.27) of the objective function, (1) if $c_{ij}^{\pi} > 0$, we set $x_{ij} = 0$; (2) if $c_{ij}^{\pi} < 0$, we set $x_{ij} = u_{ij}$; and (3) if $c_{ij}^{\pi} = 0$, we can set x_{ij} to any value between 0 and u_{ij}. The resulting solution is a pseudoflow for the minimum cost flow problem and satisfies the reduced cost optimality conditions. We have therefore established the following result.

Property 9.15. *If a pseudoflow x of the minimum cost flow problem satisfies the reduced cost optimality conditions for some π, then x is an optimal solution of* LR(π).

Let z^* denote the optimal objective function value of the minimum cost flow problem. As shown by the next lemma, the value z^* is intimately related to the optimal objective value $w(\pi)$ of the relaxed problem LR(π).

Lemma 9.16

(a) *For any node potentials π, $w(\pi) \le z^*$.*
(b) *For some choice of node potentials π^*, $w(\pi^*) = z^*$.*

Proof. Let x^* be an optimal solution of the minimum cost flow problem with objective function value z^*. Clearly, for any vector π of node potentials, x^* is a feasible solution of LR(π) and its objective function value in LR(π) is also z^*. Therefore, the minimum objective function value of LR(π) will be less than or equal to z^*. We have thus established the first part of the lemma.

To prove the second part, let π^* be a vector of node potentials that together with x^* satisfies the complementary slackness optimality conditions (9.8). Property 9.15 implies that x^* is an optimal solution of LR(π^*) and $w(\pi^*) = cx^* = z^*$. This conclusion completes the proof of the lemma. ◆

Notice the similarity between this result and the weak duality theorem (i.e., Theorem 9.5) for the minimum cost flow problem that we have stated earlier in this chapter. The similarity is more than incidental, since we can view the Lagrangian relaxation solution strategy as a dual linear programming approach that combines some key features of both the primal and dual linear programs. Moreover, we can view the dual linear program itself as being generated by applying Lagrangian relaxation.

The relaxation algorithm always maintains a vector of node potentials π and a pseudoflow x that is an optimal solution of LR(π). In other words, the pair (x, π) satisfies the reduced cost optimality conditions. The algorithm repeatedly performs one of the following two operations:

1. Keeping π unchanged, it modifies x to x' so that x' is also an optimal solution of LR(π) and the excess of at least one node decreases.
2. It modifies π to π' and x to x' so that x' is an optimal solution of LR(π') and $w(\pi') > w(\pi)$.

If the algorithm can perform either of the two operations, it gives priority to the second operation. Consequently, the primary objective in the relaxation algorithm is to increase $w(\pi)$ and the secondary objective is to reduce the infeasibility of the pseudoflow x while keeping $w(\pi)$ unchanged. We point out that the excesses at the nodes might increase when the algorithm performs the second operation. As we show at the end of this section, these two operations are sufficient to guarantee finite convergence of the algorithm. For a fixed value of $w(\pi)$, the algorithm consistently reduces the excesses of the nodes by at least one unit, and from Lemma 9.16 the number of increases in $w(\pi)$, each of which is at least 1 unit, is finite.

We now describe the relaxation algorithm in more detail. The algorithm performs major iterations and, within a major iteration, it performs several minor iterations. Within a major iteration, the algorithm selects an excess node s and grows a tree rooted at node s so that every tree node has a nonnegative imbalance and every tree arc has zero reduced cost. Each minor iteration adds an additional node to the tree. A major iteration ends when the algorithm performs either an augmentation or increases $w(\pi)$.

Let S denote the set of nodes spanned by the tree at some stage and let $\bar{S} = N - S$. The set S defines a cut which we denote by $[S, \bar{S}]$. As in earlier chapters, we let $(S, \bar{S})$ denote the set of forward arcs in the cut and $(\bar{S}, S)$ the set of backward arcs [all in $G(x)$]. The algorithm maintains two variables $e(S)$ and $r(\pi, S)$, defined as follows:

$$e(S) = \sum_{i \in S} e(i),$$

$$r(\pi, S) = \sum_{(i,j) \in (S, \bar{S}) \text{ and } c_{ij}^{\pi} = 0} r_{ij}.$$

Given the set S, the algorithm first checks the condition $e(S) > r(\pi, S)$. If the current solution satisfies this condition, the algorithm can increase $w(\pi)$ in the following manner. [We illustrate this method using the example shown in Figure 9.17(a).] The algorithm first increases the flow on zero reduced cost arcs in $(S, \bar{S})$ so that they become saturated (i.e., drop out of the residual network). The flow change does not alter the value of $w(\pi)$ because the change takes place on arcs with zero reduced costs. However, the flow change decreases the total imbalance of the

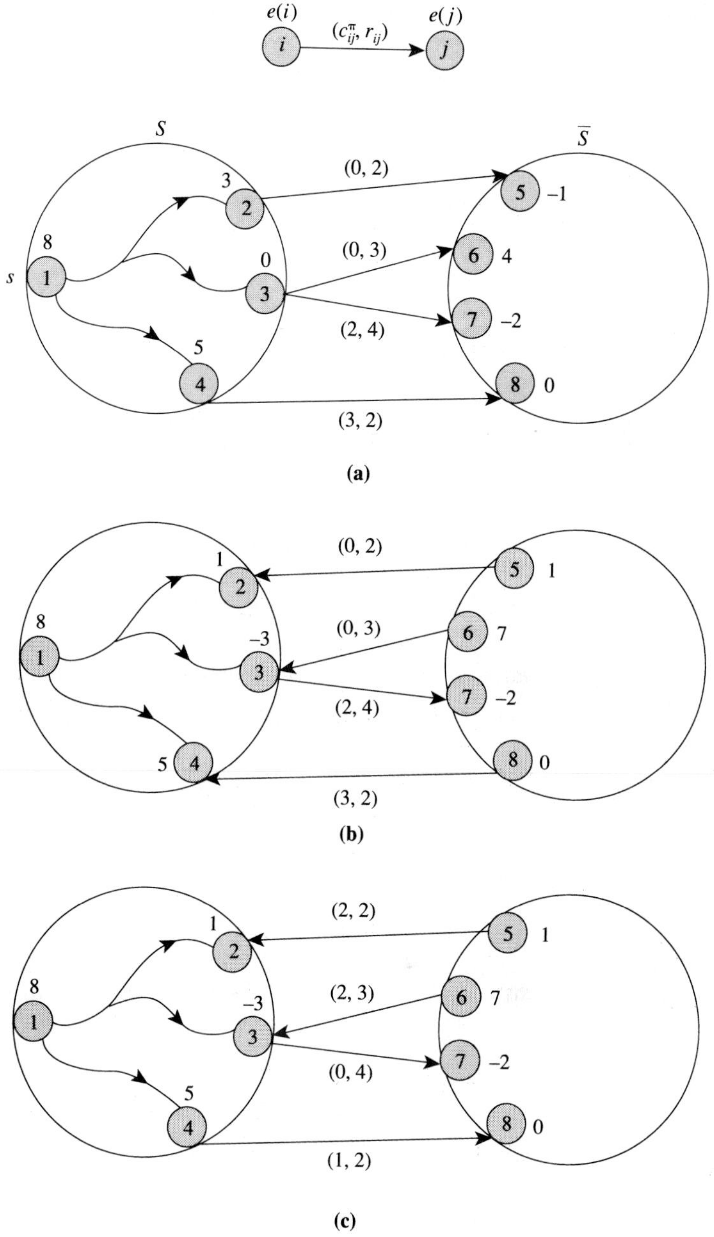

Figure 9.17 Illustrating the relaxation algorithm: (a) solution at some stage; (b) solution after modifying the flow; (c) solution after modifying the potentials.

nodes by the amount $r(\pi, S)$; but since $e(S) > r(\pi, S)$, the remaining imbalance $e(S) - r(\pi, S)$ is still positive [see Figure 9.17(b)].

At this point all the arcs in $(S, \bar{S})$ have (strictly) positive reduced cost. The algorithm next computes the minimum reduced cost of an arc in $(S, \bar{S})$, say α, and increases the potential of every node $i \in S$ by $\alpha > 0$ units [see Figure 9.17(c)]. The formulation (9.26) of the Lagrangian relaxation objective function implies that this updating of the node potentials does not change its first term but increases the second term by $(e(S) - r(\pi, S))\alpha$ units. Therefore, this operation increases $w(\pi)$ by $(e(S) - r(\pi, S))\alpha$ units, which is strictly positive. Increasing the potentials of nodes in S by α decreases the reduced costs of all the arcs in $(S, \bar{S})$ by α units, increases the reduced costs of all arcs in $(\bar{S}, S)$ by α units, and does not change the remaining reduced costs. Although increasing the reduced costs does not change the reduced cost optimality conditions, decreasing the reduced costs might. Notice, however, that before we change the node potentials, $c_{ij}^{\pi} \geq \alpha$ for all $(i, j) \in (S, \bar{S})$; therefore, after the change, $c_{ij}^{\pi'} \geq 0$, so the algorithm preserves the optimality conditions. This completes one major iteration.

We next study situations in which $e(S) \leq r(\pi, S)$. Since $r(\pi, S) \geq e(S) > 0$, at least one arc $(i, j) \in (S, \bar{S})$ must have a zero reduced cost. If $e(j) \geq 0$, the algorithm adds node j to S, completes one minor iteration, and repeats this process. If $e(j) < 0$, the algorithm augments the maximum possible flow along the tree path from node s to node j. Notice that since we augment flow along zero residual cost arcs, we do not change the objective function value of LR(π). The augmentation reduces the total excess of the nodes and completes one major iteration of the algorithm.

Figures 9.18 and 9.19 give a formal description of the relaxation algorithm.

It is easy to see that the algorithm terminates with a minimum cost flow. The algorithm terminates when all of the node imbalances have become zero (i.e., the solution is a flow). Because the algorithm maintains the reduced cost optimality conditions at every iteration, the terminal solution is a minimum cost flow.

We now prove that for problems with integral data, the algorithm terminates in a finite number of iterations. Since each minor iteration adds a node to the set S, within n minor iterations the algorithm either executes adjust-flow or executes adjust-potentials. Each call of the procedure adjust-flow decreases the excess of at least one node by at least 1 unit; therefore, the algorithm can perform a finite number of executions of the adjust-flow procedure within two consecutive calls of the adjust-potential procedure. To bound the executions of the adjust-potential procedure, we notice that (1) initially, $w(\pi) = 0$; (2) each call of this procedure strictly increases

```
algorithm relaxation;
begin
    x : = 0 and π : = 0;
    while the network contains a node s with e(s) > 0 do
    begin
        S : = {s};
        if e(S) > r(π, S) then adjust-potential;
        repeat
            select an arc (i, j) ∈ (S, S̄) in the residual network with c_ij^π = 0;
            if e(j) ≥ 0 then set pred(j) : = i and add node j to S;
        until e(j) < 0 or e(S) > r(π, S);
        if e(S) > r(π, S) then adjust-potential
        else adjust-flow;
    end;
end;
```

Figure 9.18 Relaxation algorithm.

```
procedure adjust-potential;
begin
    for every arc (i, j) ∈ (S, S̄) with c^π_ij = 0 do send r_ij units of flow on the arc (i, j);
    compute α : = min{c^π_ij : (i, j) ∈ (S, S̄) and r_ij > 0};
    for every node i ∈ S do π(i) : = π(i) + α;
end;
```

(a)

```
procedure adjust-flow;
begin
    trace the predecessor indices to identify the directed path P from node s to node j;
    δ : = min[e(s), − e(j), min{r_ij : (i, j) ∈ P}];
    augment δ units of flow along P, update imbalances and residual capacities;
end;
```

(b)

Figure 9.19 Procedures of the relaxation algorithm.

$w(\pi)$ by at least 1 unit; and (3) the maximum possible value of $w(\pi)$ is mCU. The preceding arguments establish that the algorithm performs finite number of iterations. In Exercise 9.27 we ask the reader to obtain a worst-case bound on the total number of iterations; this time bound is much worse than those of the other minimum cost flow algorithms discussed in earlier sections.

Notice that the relaxation algorithm is a type of shortest augmenting path algorithm; indeed, it bears some resemblance to the successive shortest path algorithm that we considered in Section 9.7. Since the reduced cost of every arc in the residual network is nonnegative, and since every arc in the tree connecting the nodes in S has a zero reduced cost, the path P that we find in the adjust-flow procedure of the relaxation algorithm is a shortest path in the residual network. Therefore, the sequence of flow adjustments that the algorithm makes is a set of flow augmentations along shortest augmenting paths. The relaxation algorithm differs from the successive shortest augmenting path algorithm, however, because it uses "intermediate" information to make changes to the node potentials as it fans out and constructs the tree containing the nodes S. This use of intermediate information might explain why the relaxation algorithm has performed much better empirically than the successive shortest path algorithm.

9.11 SENSITIVITY ANALYSIS

The purpose of sensitivity analysis is to determine changes in the optimal solution of a minimum cost flow problem resulting from changes in the data (supply/demand vector or the capacity or cost of any arc). There are two different ways of performing sensitivity analysis: (1) using combinatorial methods, and (2) using simplex-based methods from linear programming. Each method has its advantages. For example, although combinatorial methods obtain better worst-case time bounds for performing sensitivity analysis, simplex-based methods might be more efficient in practice. In this section we describe sensitivity analysis using combinatorial methods; in Section

11.10 we consider a simplex-based approach. For simplicity, we limit our discussion to a unit change of only a particular type. In a sense, however, this discussion is quite general: It is possible to reduce more complex changes to a sequence of the simple changes we consider. We show that sensitivity analysis for the minimum cost flow problem essentially reduces to applying shortest path or maximum flow algorithms.

Let x^* denote an optimal solution of a minimum cost flow problem. Let π be the corresponding node potentials and $c_{ij}^{\pi} = c_{ij} - \pi(i) + \pi(j)$ denote the reduced costs. Further, let $d(k, l)$ denote the shortest distance from node k to node l in the residual network with respect to the original arc lengths c_{ij}. Since for any directed path P from node k to node l, $\sum_{(i,j)\in P} c_{ij}^{\pi} = \sum_{(i,j)\in P} c_{ij} - \pi(k) + \pi(l)$, $d(k, l)$ equals the shortest distance from node k to node l with respect to the arc lengths c_{ij}^{π} plus $[\pi(k) - \pi(l)]$. At optimality, the reduced costs c_{ij}^{π} of all arcs in the residual network are nonnegative. Therefore, we can compute $d(k, l)$ for all pairs of nodes k and l by solving n single-source shortest path problems with nonnegative arc lengths.

Supply/Demand Sensitivity Analysis

We first study changes in the supply/demand vector. Suppose that the supply/demand of a node k becomes $b(k) + 1$ and the supply/demand of another node l becomes $b(l) - 1$. [Recall from Section 9.1 that feasibility of the minimum cost flow problem dictates that $\sum_{i\in N} b(i) = 0$; therefore, we must change the supply/demand values of two nodes by equal magnitudes, and must increase one value and decrease the other.] The vector x^* is a pseudoflow for the modified problem; moreover, this vector satisfies the reduced cost optimality conditions. Augmenting 1 unit of flow from node k to node l along the shortest path in the residual network $G(x^*)$ converts this pseudoflow into a flow. This augmentation changes the objective function value by $d(k, l)$ units. Lemma 9.12 implies that this flow is optimal for the modified minimum cost flow problem. We point out that the residual network $G(x^*)$ might not contain any directed path from node k to node l, in which case the modified minimum cost flow problem is infeasible.

Arc Capacity Sensitivity Analysis

We next consider a change in an arc capacity. Suppose that the capacity of an arc (p, q) increases by 1 unit. The flow x^* is feasible for the modified problem. In addition, if $c_{pq}^{\pi} \geq 0$, it satisfies the reduced cost optimality conditions; therefore, it is an optimal flow for the modified problem. If $c_{pq}^{\pi} < 0$, the optimality conditions dictate that the flow on the arc must equal its capacity. We satisfy this requirement by increasing the flow on the arc (p, q) by 1 unit, which produces a pseudoflow with an excess of 1 unit at node q and a deficit of 1 unit at node p. We convert the pseudoflow into a flow by augmenting 1 unit of flow from node q to node p along the shortest path in the residual network $G(x^*)$, which changes the objective function value by an amount $c_{pq} + d(q, p)$. This flow is optimal from our observations concerning supply/demand sensitivity analysis.

When the capacity of the arc (p, q) decreases by 1 unit and the flow on the arc is strictly less than its capacity, x^* remains feasible, and therefore optimal, for the modified problem. However, if the flow on the arc is at its capacity, we decrease the flow by 1 unit and augment 1 unit of flow from node p to node q along the shortest path in the residual network. This augmentation changes the objective function value by an amount $-c_{pq} + d(p, q)$. Observed that the residual network $G(x^*)$ might not contain any directed path from node p to node q, indicating the infeasibility of the modified problem.

Cost Sensitivity Analysis

Finally, we discuss changes in arc costs, which we assume are integral. We discuss the case when the cost of an arc (p, q) increases by 1 unit; the case when the cost of an arc decreases is left as an exercise to the reader (see Exercise 9.50). This change increases the reduced cost of arc (p, q) by 1 unit as well. If $c^{\pi}_{pq} < 0$ before the change, then after the change, the modified reduced cost is nonpositive. Similarly, if $c^{\pi}_{pq} > 0$ before the change, the modified reduced cost is nonnegative after the change. In both cases we preserve the optimality conditions. However, if $c^{\pi}_{pq} = 0$ before the change and $x_{pq} > 0$, then after the change the modified reduced cost is positive and the solution violates the reduced-cost optimality conditions. To restore the optimality conditions of the arc, we must either reduce the flow on arc (p, q) to zero or change the potentials so that the reduced cost of arc (p, q) becomes zero.

We first try to reroute the flow x^*_{pq} from node p to node q without violating any of the optimality conditions. We do so by solving a maximum flow problem defined as follows: (1) set the flow on the arc (p, q) to zero, thus creating an excess of x^*_{pq} at node p and a deficit of x^*_{pq} at node q; (2) designate node p as the source node and node q as the sink node; and (3) send a maximum of x^*_{pq} units from the source to the sink. We permit the maximum flow algorithm, however, to change flows only on arcs with zero reduced costs since otherwise it would generate a solution that might violate (9.8). Let v° denote the flow sent from node p to node q and x° denote the resulting arc flow. If $v^\circ = x^*_{pq}$, then x° denotes a minimum cost flow of the modified problem. In this case the optimal objective function values of the original and modified problems are the same.

On the other hand, if $v^\circ < x^*_{pq}$, the maximum flow algorithm yields an s–t cut $[S, \bar{S}]$ with the properties that $p \in S$, $q \in \bar{S}$, and every forward arc in the cut with zero reduced cost has flow equal to its capacity and every backward arc in the cut with zero reduced cost has zero flow. We then decrease the node potential of every node in $\bar{S}$ by 1 unit. It is easy to verify by case analysis that this change in node potentials maintains the complementary slackness optimality conditions and, furthermore, decreases the reduced cost of arc (p, q) to zero. Consequently, we can set the flow on arc (p, q) equal to $x^*_{pq} - v^\circ$ and obtain a feasible minimum cost flow. In this case the objective function value of the modified problem is $x^*_{pq} - v^\circ$ units more than that of the original problem.

9.12 SUMMARY

The minimum cost flow problem is the central object of study in this book. In this chapter we began our study of this important class of problems by showing how minimum cost flow problems arise in several application settings and by considering

Algorithm	Number of iterations	Features
Cycle-canceling algorithm	$O(mCU)$	1. Maintains a feasible flow x at every iteration and augments flows along negative cycles in $G(x)$. 2. At each iteration, solves a shortest path problem with arbitrary arc lengths to identify a negative cycle. 3. Very flexible: some rules for selecting negative cycles leads to polynomial-time algorithms.
Successive shortest path algorithm	$O(nU)$	1. Maintains a pseudoflow x satisfying the optimality conditions and augments flow along shortest paths from excess nodes to deficit nodes in $G(x)$. 2. At each iteration, solves a shortest path problem with nonnegative arc lengths. 3. Very flexible: by selecting augmentations carefully, we can obtain several polynomial-time algorithms.
Primal–dual algorithm	$O(\min\{nU, nC\})$	1. Maintains a pseudoflow x satisfying the optimality conditions. Solves a shortest path problem to update node potentials and attempts to reduce primal infeasibility by the maximum amount by solving a maximum flow problem. 2. At each iteration, solves both a shortest path problem with nonnegative arc lengths and a maximum flow problem. 3. Closely related to the successive shortest path algorithm: instead of sending flow along one shortest path, sends flow along all shortest paths.
Out-of-kilter algorithm	$O(nU)$	1. Maintains a feasible flow x at each iteration and attempts to satisfy the optimality conditions by augmenting flows along shortest paths. 2. At each iteration, solves a shortest path problem with nonnegative arc lengths. 3. Can be generalized to solve situations in which the flow x maintained by the algorithm might not satisfy the flow bounds on the arcs.
Relaxation algorithm	See Exercise 9.27	1. Somewhat different from other minimum cost flow algorithms. 2. Maintains a pseudoflow x satisfying the optimality conditions and modifies arc flows and node potentials so that a Lagrangian objective function does not decrease and occasionally increases. 3. With the incorporation of some heuristics, the algorithm is very efficient in practice and yields the fastest available algorithm for some classes of minimum cost flow problems.

Figure 9.20 Summary of pseudopolynomial-time algorithms for the minimum cost flow problem.

the simplest pseudopolynomial-time algorithms for solving these problems. These pseudopolynomial-time algorithms include classical algorithms that are important because of both their historical significance and because they provide the essential building blocks and core ideas used in more efficient algorithms. Our algorithmic development relies heavily upon optimality conditions for the minimum cost flow problem that we developed and proved in the following equivalent frameworks: negative cycle optimality conditions, reduced cost optimality conditions, and complementary slackness optimality conditions. The negative cycle optimality conditions state that a feasible flow x is an optimal flow if and only if the residual network $G(x)$ contains no negative cycle. The reduced cost optimality conditions state that a feasible flow x is an optimal flow if and only if the reduced cost of each arc in the residual network is nonnegative. The complementary slackness optimality conditions are adaptations of the linear programming optimality conditions for network flows. As part of this general discussion in this chapter, we also examined minimum cost flow duality.

We developed several minimum cost flow algorithms: the cycle-canceling, successive shortest path, primal–dual, out-of-kilter, and relaxation algorithms. These algorithms represent a good spectrum of approaches for solving the same problem: Some of these algorithms maintain primal feasible solutions and strive toward optimality; others maintain primal infeasible solutions that satisfy the optimality conditions and strive toward feasibility. These algorithms have some commonalties as well—they all repeatedly solve shortest path problems. In fact, in Exercises 9.57 and 9.58 we establish a very strong result by showing that the cycle-canceling, successive shortest path, primal–dual, and out-of-kilter algorithms are all equivalent in the sense that if initialized properly, they perform the same sequence of augmentations. Figure 9.20 summarizes the basic features of the algorithms discussed in this chapter.

Finally, we discussed sensitivity analysis for the minimum cost flow problem. We showed how to reoptimize the minimum cost flow problem, after we have made unit changes in the supply/demand vector or the arc capacities, by solving a shortest path problem, and how to handle unit changes in the cost vector by solving a maximum flow problem. Needless to say, these reoptimization procedures are substantially faster than solving the problem afresh if the changes in the problem data are sufficiently small.

REFERENCE NOTES

In this chapter and in these reference notes we focus on pseudopolynomial-time nonsimplex algorithms for solving minimum cost flow problems. In Chapter 10 we provide references for polynomial-time minimum cost flow algorithms, and in Chapter 11 we give references for simplex-based algorithms.

Ford and Fulkerson [1957] developed the primal–dual algorithms for the capacitated transportation problem; Ford and Fulkerson [1962] later generalized this approach for solving the minimum cost flow problem. Jewell [1958], Iri [1960], and Busaker and Gowen [1961] independently developed the successive shortest path algorithm. These researchers showed how to solve the minimum cost flow problem as a sequence of shortest path problems with arbitrary arc lengths. Tomizava [1972] and Edmonds and Karp [1972] independently observed that if the computations use node potentials, it is possible to implement these algorithms so that the shortest path problems have nonnegative arc lengths.

Minty [1960] and Fulkerson [1961b] independently developed the out-of-kilter algorithm. Aashtiani and Magnanti [1976] have described an efficient implementation of this algorithm. The description of the out-of-kilter algorithm presented in Section 9.9 differs substantially from the development found in other textbooks. Our description is substantially shorter and simpler because it avoids tedious case analyses. Moreover, our description explicitly highlights the use of Dijkstra's algorithm; because other descriptions do not focus on the shortest path computations, they find an accurate worst-case analysis of the algorithm much more difficult to conduct.

The cycle-canceling algorithm is credited to Klein [1967]. Three special implementations of the cycle-canceling algorithms run in polynomial time: the first, due to Barahona and Tardos [1989] (which, in turn, modifies an algorithm by Weintraub [1974]), augments flow along (negative) cycles with the maximum possible improvement; the second, due to Goldberg and Tarjan [1988], augments flow along minimum mean cost (negative) cycles; and the third, due to Wallacher and Zimmerman [1991], augments flow along minimum ratio cycles.

Zadeh [1973a,1973b] described families of minimum cost flow problems on which each of several algorithms—the cycle-canceling algorithm, successive shortest path algorithm, primal–dual algorithm, and out-of-kilter algorithm—perform an exponential number of iterations. The fact that the same families of networks are bad for many network algorithms suggests an interrelationship among the algorithms. The insightful paper by Zadeh [1979] points out that each of the algorithms we have just mentioned are indeed equivalent in the sense that they perform the same sequence of augmentations, which they obtained through shortest path computations, provided that we initialize them properly and break ties using the same rule.

Bertsekas and Tseng [1988b] developed the relaxation algorithm and conducted extensive computational investigations of it. A FORTRAN code of the relaxation algorithm appears in Bertsekas and Tseng [1988a]. Their study and those conducted by Grigoriadis [1986] and Kennington and Wang [1990] indicate that the relaxation algorithm and the network simplex algorithm (described in Chapter 11) are the two fastest available algorithms for solving the minimum cost flow problem in practice. When the supplies/demands at nodes are relatively small, the successive shortest path algorithm is the fastest algorithm. Previous computational studies conducted by Glover, Karney, and Klingman [1974] and Bradley, Brown, and Graves [1977] have indicated that the network simplex algorithm is consistently superior to the primal–dual and out-of-kilter algorithms. Most of these computational testings have been done on random network flow problems generated by the well-known computer program NETGEN, suggested by Klingman, Napier, and Stutz [1974].

The applications of the minimum cost flow problem that we discussed Section 9.2 have been adapted from the following papers:

1. Distribution problems (Glover and Klingman [1976])
2. Reconstructing the left ventricle from x-ray projections (Slump and Gerbrands [1982])
3. Racial balancing of schools (Belford and Ratliff [1972])

4. Optimal loading of a hopping airplane (Gupta [1985] and Lawania [1990])
5. Scheduling with deferral costs (Lawler [1964])
6. Linear programming with consecutive 1's in columns (Veinott and Wagner [1962])

Elsewhere in this book we describe other applications of the minimum cost flow problem. These applications include (1) leveling mountainous terrain (Application 1.4, Farley [1980]), (2) the forest scheduling problem (Exercise 1.10), (3) the entrepreneur's problem (Exercise 9.1, Prager [1957]), (4) vehicle fleet planning (Exercise 9.2), (5) optimal storage policy for libraries (Exercise 9.3, Evans [1984]), (6) zoned warehousing (Exercise 9.4, Evans [1984]), (7) allocation of contractors to public works (Exercise 9.5, Cheshire, McKinnon, and Williams [1984]), (8) phasing out capital equipment (Exercise 9.6, Daniel [1973]), (9) the terminal assignment problem (Exercise 9.7, Esau and Williams [1966]), (10) linear programs with consecutive or circular 1's in rows (Exercises 9.8 and 9.9, Bartholdi, Orlin, and Ratliff [1980]), (11) capacitated maximum spanning trees (Exercise 9.54, Garey and Johnson [1979]), (12) fractional b-matching (Exercise 9.55), (13) the nurse scheduling problem (Exercise 11.1), (14) the caterer problem (Exercise 11.2, Jacobs [1954]), (15) project assignment (Exercise 11.3), (16) passenger routing (Exercise 11.4), (17) allocating receivers to transmitters (Exercise 11.5, Dantzig [1962]), (18) faculty–course assignment (Exercise 11.6, Mulvey [1979]), (19) optimal rounding of a matrix (Exercise 11.7, Bacharach [1966], Cox and Ernst [1982]), (20) automatic karotyping of chromosomes (Application 19.8, Tso, Kleinschmidt, Mitterreiter, and Graham [1991]), (21) just-in-time scheduling (Application 19.10, Elmaghraby [1978], Levner and Nemirovsky [1991]), (22) time–cost trade-off in project management (Application 19.11, Fulkerson [1961a] and Kelly [1961]), (23) models for building evacuation (Application 19.13, Chalmet, Francis and Saunders [1982]), (24) the directed Chinese postman problem (Application 19.14, Edmonds and Johnson [1973]), (25) warehouse layout (Application 19.17, Francis and White [1976]), (26) rectilinear distance facility location (Application 19.18, Cabot, Francis, and Stary [1970]), (27) dynamic lot sizing (Application 19.19, Zangwill [1969]), (28) multistage production-inventory planning (Application 19.23, Evans [1977]), (29) mold allocation (Application 19.24, Love and Vemuganti [1978]), (30) a parking model (Exercise 19.17, Dirickx and Jennergren [1975]), (31) the network interdiction problem (Exercise 19.18, Fulkerson and Harding [1977]), (32) truck scheduling (Exercises 19.19 and 19.20, Gavish and Schweitzer [1974]), and (33) optimal deployment of firefighting companies (Exercise 19.21, Denardo, Rothblum, and Swersey [1988]).

The applications of the minimum cost flow problems are so vast that we have not been able to describe many other applications in this book. The following list provides a set of references to some other applications: (1) warehousing and distribution of a seasonal product (Jewell [1957]), (2) economic distribution of coal supplies in the gas industry (Berrisford [1960]), (3) upsets in round-robin tournaments (Fulkerson [1965]), (4) optimal container inventory and routing (Horn [1971]), (5) distribution of empty rail containers (White [1972]), (6) optimal defense of a network (Picard and Ratliff [1973]), (7) telephone operator scheduling (Segal [1974]), (8) multifacility minimax location problem with rectilinear distances (Dearing and Francis [1974]), (9) cash management problems (Srinivasan [1974]), (10) multiproduct multifacility production-inventory planning (Dorsey, Hodgson, and Ratliff [1975]), (11) "hub" and "wheel" scheduling problems (Arisawa and Elmaghraby [1977]), (12) the warehouse leasing problem (Lowe, Francis, and Reinhardt [1979]), (13) multiattribute marketing models (Srinivasan [1979]), (14) material handling systems (Maxwell and Wilson [1981]), (15) microdata file merging (Barr and Turner [1981]), (16) determining service districts (Larson and Odoni [1981]), (17) control of forest fires (Kourtz [1984]), (18) allocating blood to hospitals from a central blood bank (Sapountzis [1984]), (19) market equilibrium problems (Dafermos and Nagurney [1984]), (20) automatic chromosome classifications (Tso [1986]), (21) the city traffic congestion problem (Zawack and Thompson [1987]), (22) satellite scheduling (Servi [1989]), and (23) determining k disjoint cuts in a network (Wagner [1990]).

EXERCISES

9.1. Enterpreneur's problem (Prager [1957]). An entrepreneur faces the following problem. In each of T periods, he can buy, sell, or hold for later sale some commodity, subject to the following constraints. In each period i he can buy at most α_i units of the commodity, can holdover at most β_i units of the commodity for the next period, and must sell at least γ_i units (perhaps due to prior agreements). The enterpreneur cannot sell the commodity in the same period in which he buys it. Assuming that p_i, w_i, and s_i denote the purchase cost, inventory carrying cost, and selling price per unit in period i, what buy–sell policy should the entreprenuer adopt to maximize total profit in the T periods? Formulate this problem as a minimum cost flow problem for $T = 4$.

9.2. Vehicle fleet planning. The Millersburg Supply Company uses a large fleet of vehicles which it leases from manufacturers. The company has forecast the following pattern of vehicle requirements for the next 6 months:

Month	Jan.	Feb.	Mar.	Apr.	May	June
Vehicles required	430	410	440	390	425	450

Millersburg can lease vehicles from several manufacturers at various costs and for various lengths of time. Three of the plans appear to be the best available: a 3-month lease for \$1700; a 4-month lease for \$2200; and a 5-month lease for \$2600. The company can undertake a lease beginning in any month. On January 1 the company has 200 cars on lease, all of which go off lease at the end of February. Formulate the problem of determining the most economical leasing policy as a minimum cost flow problem. (*Hint*: Observe that the linear (integer) programming formulation of this problem has consecutive 1's in each column. Then use the result in Application 9.6.)

9.3. Optimal storage policy for libraries (Evans [1984]). A library facing insufficient primary storage space for its collection is considering the possibility of using secondary facilities, such as closed stacks or remote locations, to store portions of its collection. These options are preferred to an expensive expansion of primary storage. Each secondary storage facility has limited capacity and a particular access costs for retrieving information. Through appropriate data collection, we can determine the usage rates for the information needs of the users. Let b_j denote the capacity of storage facility j and v_j

denote the access cost per unit item from this facility. In addition, let a_i denote the number of items of a particular class i requiring storage and let u_i denote the expected rate (per unit time) that we will need to retrieve books from this class. Our goal is to store the books in a way that will minimize the expected retrieval cost.

(a) Show how to formulate the problem of determining an optimal policy as a transportation problem. What is the special structure of this problem? Transportation problems with this structure have become known as *factored transportation problems*.

(b) Show that the simple rule that repeatedly assigns items with the greatest retrievel rate to the storage facility with lowest access cost specifies an optimal solution of this library storage problem.

9.4. **Zoned warehousing** (Evans [1984]). In the storage of multiple, say p, items in a zoned warehouse, we need to extract (pick) items in large quantities (perhaps by pallet loads). Suppose that the warehouse is partitioned into q zones, each with a different distance to the shipping area. Let B_j denote the storage capacity of zone j and let d_j denote the average distance from zone j to the shipping area. For each item i, we know (1) the space requirement per unit (r_i), (2) the average order size in some common volume unit (s_i), and (3) the average number of orders per day (f_i). The problem is to determine the quantity of each item to allocate to each zone in order to minimize the average daily handling costs. Assume that the handling cost is linearly proportional to the distance and to the volume moved.

(a) Formulate this problem as a factored transportation problem (as defined in Exercise 9.3).

(b) Specify a simple rule that yields an optimal solution of the zoned warehousing problem.

9.5. **Allocation of contractors to public works** (Cheshire, McKinnon, and Williams [1984]). A large publicly owned corporation has 12 divisions in Great Britain. Each division faces a similar problem. Each year the division subcontracts work to private contractors. The work is of several different types and is done by teams, each of which is capable of doing all types of work. One of these divisions is divided into several districts: the jth district requires r_j teams. The contractors are of two types: experienced and inexperienced. Each contractor i quotes a price c_{ij} to have a team conduct the work in district j. The objective is to allocate the work in the districts to the various contractors, satisfying the following conditions: (1) each district j has r_j assigned teams; (2) the division contracts with contractor i for no more than u_i teams, the maximum number of teams it can supply; and (3) each district has at least one experienced contractor assigned to it. Formulate this problem as a minimum cost flow problem for a division with three districts, and with two experienced and two inexperienced contractors. (*Hint*: Split each district node into two nodes, one of which requires an experienced contractor.)

9.6. **Phasing out capital equipment** (Daniel [1973]). A shipping company wants to phase out a fleet of (homogeneous) general cargo ships over a period of p years. Its objective is to maximize its cash assets at the end of the p years by considering the possibility of prematurely selling ships and temporary replacing them by charter ships. The company faces a known nonincreasing demand for ships. Let $d(i)$ denote the demand of ships in year i. Each ship earns a revenue of r_k units in period k. At the beginning of year k, the company can sell any ship that it owns, accruing a cash inflow of s_k dollars. If the company does not own sufficiently many ships to meet its demand, it must hire additional charter ships. Let h_k denote the cost of hiring a ship for the kth year. The shipping company wants to meet its commitments and at the same time maximize the cash assets at the end of the pth year. Formulate this problem as a minimum cost flow problem.

9.7. Terminal assignment problem (Esau and Williams [1966]). Centralized teleprocessing networks often contain many (as many as tens of thousands) relatively unsophisticated geographically dispersed terminals. These terminals need to be connected to a central processor unit (CPU) either by direct lines or though *concentrators*. Each concentrator is connected to the CPU through a high-speed, cost-effective line that is capable of merging data flow streams from different terminals and sending them to the CPU. Suppose that the concentrators are in place and that each concentrator can handle at most K terminals. For each terminal j, let c_{oj} denote the cost of laying down a direct line from the CPU to the terminal and let c_{ij} denote the line construction cost for connecting concentrator i to terminal j. The decision problem is to construct the minimum cost network for connecting the terminals to the CPU. Formulate this problem as a minimum cost flow problem.

9.8. Linear programs with consecutive 1's in rows. In Application 9.6 we considered linear programs with consecutive 1's in each column and showed how to transform them into minimum cost flow problems. In this and the next exercise we study several related linear programming problems and show how we can solve them by solving minimum cost flow problems. In this exercise we study linear programs with consecutive 1's in the rows. Consider the following (integer) linear program with consecutive 1's in the rows:

$$\text{Minimize} \quad c_1x_1 + c_2x_2 + c_3x_3 + c_4x_4$$

subject to

$$\begin{aligned}
x_2 + x_3 + x_4 &\geq 20\\
x_1 + x_2 + x_3 + x_4 &\geq 30\\
x_2 + x_3 &\geq 15\\
x_3 + x_4 &\geq 10\\
x_1, x_2, x_3, x_4 &\geq 0 \text{ and integer.}
\end{aligned}$$

Transform this problem to a minimum cost flow problem. (*Hint*: Use the same transformation of variables that we used in Application 4.6.)

9.9. Linear programs with circular 1's in rows (Bartholdi, Orlin, and Ratliff [1980]). In this exercise we consider a generalization of Exercise 9.8 with the 1's in each row arranged consecutively when we view columns in the wraparound fashion (i.e., we consider the first column as next to the last column). A special case of this problem is the telephone operator scheduling problem that we discussed in Application 4.6. In this exercise we focus on the telephone operator scheduling problem; nevertheless, the approach easily extends to any general linear program with circular 1's in the rows. We consider a version of the telephone operator scheduling in which we incur a cost c_i whenever an operator works in the ith shift, and we wish to satisfy the minimum operator requirement for each hour of the day at the least possible cost. We can formulate this "cyclic staff scheduling problem" as the following (integer) linear program.

$$\text{Minimize} \sum_{i=0}^{23} y_i$$

subject to

$$\begin{aligned}
y_{i-7} + y_{i-6} + \cdots + y_i &\geq b(i) && \text{for all } i = 7 \text{ to } 23,\\
y_{17+i} + \cdots + y_{23} + y_0 + \cdots + y_i &\geq b(i) && \text{for all } i = 0 \text{ to } 6,\\
y_i &\geq 0 && \text{for all } i = 1 \text{ to } 23.
\end{aligned}$$

(a) For a parameter p, let $\mathcal{P}(p)$ denote the cyclic staff scheduling problem when we impose the additional constraint $\sum_{i=0}^{23} y_i = p$, and let $z(p)$ denote the optimal objective value of this problem. Show how to transform $\mathcal{P}(p)$, for a fixed value of p, into a minimum cost flow problem. (*Hint*: Use the same transformation of variables that we used in Application 4.6 and observe that each row has one $+1$ and one -1. Then use the result of Theorem 9.9.)

(b) Show that $z(p)$ is a (piecewise linear) convex function of p. (*Hint*: Show that if y' is an optimal solution of $\mathcal{P}(p')$ and y'' is an optimal solution of $\mathcal{P}(p'')$, then for any weighting parameter λ, $0 \le \lambda \le 1$, the point $\lambda y' + (1 - \lambda)y''$ is a feasible solution of $\mathcal{P}(\lambda p' + (1 - \lambda)p'')$.)

(c) In the cyclic staff scheduling problem, we wish to determine a value of p, say p^*, satisfying the property that $z(p^*) \le z(p)$ for all feasible p. Show how to solve the cyclic staff scheduling problem in polynomial time by performing binary search on the values of p. (*Hint*: For any integer p, show how to determine whether $p \le p^*$ by solving problems $\mathcal{P}(p)$ and $\mathcal{P}(p + 1)$.)

9.10. Racial balancing of schools. In this exercise we discuss some generalizations of the problem of racial balancing of schools that we described in Application 9.3. Describe how would you modify the formulation to include the following additional restrictions (consider each restriction separately).

(a) We prohibit the assignment of a student from location i to school j if the travel distance d_{ij} between these location exceeds some specified distance D.

(b) We include the distance traveled between location i and school j in the objective function only if d_{ij} is greater than some specified distance D' (e.g., we account for the distance traveled only if a student needs to be bussed).

(c) We impose lower and upper bounds on the number of black students from location i who are assigned to school j.

9.11. Show how to transform the equipment replacement problem described in Application 9.6 into a shortest path problem. Give the resulting formulation for $n = 4$.

9.12. This exercise is based on the equipment replacement problem that we discussed in Application 9.6.

(a) The problem as described allows us to buy and sell the equipment only yearly. How would you model the situation if you could make decisions every half year?

(b) How sensitive do you think the optimal solution would be to the length T of planning period? Can you anticipate a situation in which the optimal replacement plan would change drastically if we were to increase the length of the planning period to $T + 1$?

9.13. Justify the minimum cost flow formulation that we described in Application 9.4 for the problem of optimally loading a hopping airplane. Establish a one-to-one correspondence between feasible passenger routings and feasible flows in the minimum cost flow formulation of the problem.

9.14. In this exercise we consider one generalization of the tanker scheduling problem discussed in Application 6.6. Suppose that we can compute the profit associated with each available shipment (depending on the revenues and the operating cost directly attributable to that shipment). Let the profits associated with the shipments 1, 2, 3, and 4 be 10, 10, 3, and 4, respectively. In addition to the operating cost, we incur a fixed charge of 5 units to bring a ship into service. We want to determine the shipments we should make and the number of ships to use to maximize net profits. (Note that it is not necessary to honor all possible shipping commitments.) Formulate this problem as a minimum cost flow problem.

9.15. Consider the following data, with $n = 4$, for the employment scheduling problem that we discussed in Application 9.6. Formulate this problem as a minimum cost flow problem and solve it by the successive shortest path algorithm.

$\{c_{ij}\} =$

	1	2	3	4	5
1	—	20	35	50	55
2	—	—	15	30	40
3	—	—	—	25	35
4	—	—	—	—	10

i	1	2	3	4
$d(i)$	20	15	30	25

9.16. Figure 9.21(b) shows the optimal solution of the minimum cost flow problem shown in Figure 9.21(a). First, verify that x^* is a feasible flow.
(a) Draw the residual network $G(x^*)$ and show that it contains no negative cycle.
(b) Specify a set of node potentials π that together with x^* satisfy the reduced cost optimality conditions. List each arc in the residual network and its reduced cost.

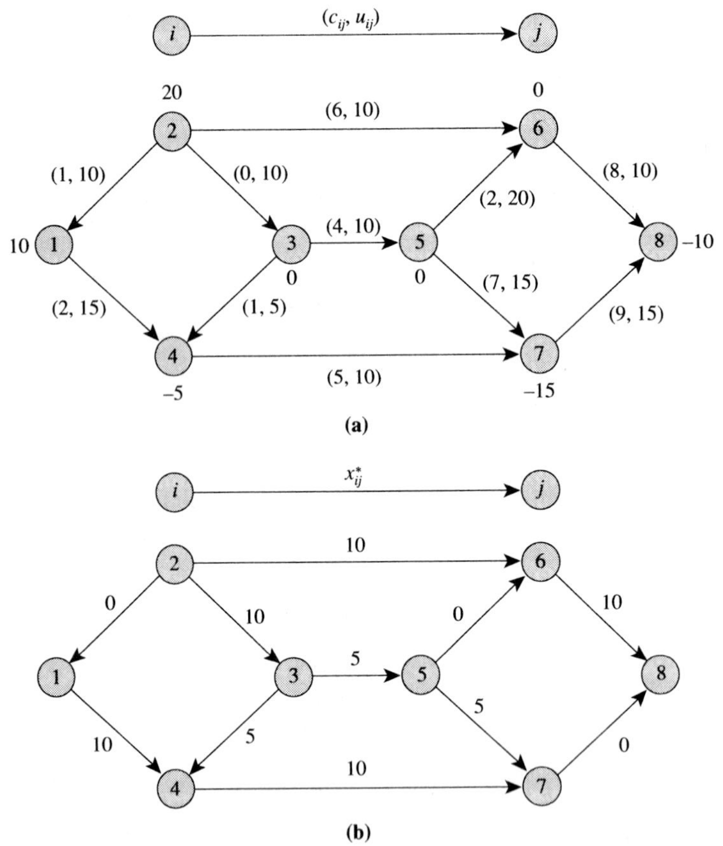

Figure 9.21 Minimum cost flow problem: (a) problem data; (b) optimal solution.

(c) Verify that the solution x^* satisfies the complementary slackness optimality conditions. To do so, specify a set of optimal node potentials and list the reduced cost of each arc in A.

9.17. (a) Figure 9.22(a) gives the data and an optimal solution for a minimum cost flow problem. Assume that all arcs are uncapacitated. Determine optimal node potentials.

(b) Consider the uncapacitated minimum cost flow problem shown in Figure 9.22(b). For this problem the vector $\pi = (0, -6, -9, -12, -5, -8, -15)$ is an optimal set of node potentials. Determine an optimal flow in the network.

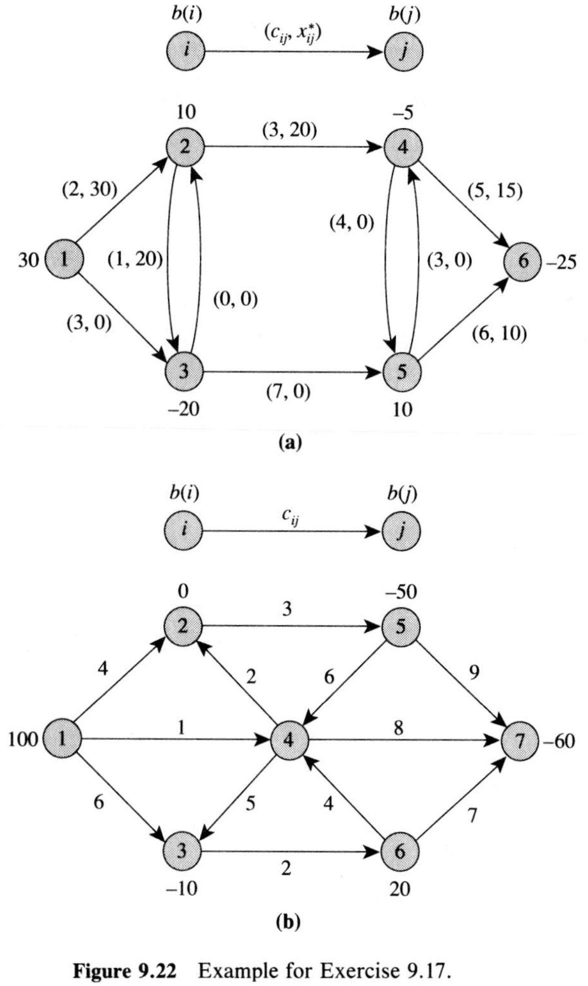

Figure 9.22 Example for Exercise 9.17.

9.18. Solve the problem shown in Figure 9.23 by the cycle-canceling algorithm. Use the zero flow as the starting solution.

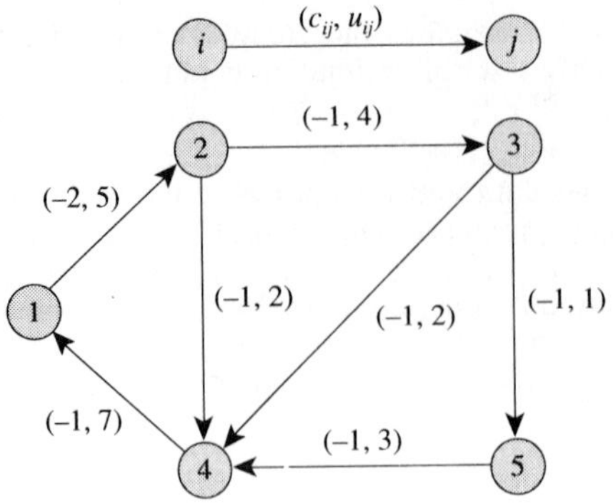

Figure 9.23 Example for Exercise 9.18.

9.19. Show that if we apply the cycle-canceling algorithm to the minimum cost flow problem shown in Figure 9.24, some sequence of augmentations requires 2×10^6 iterations to solve the problem.

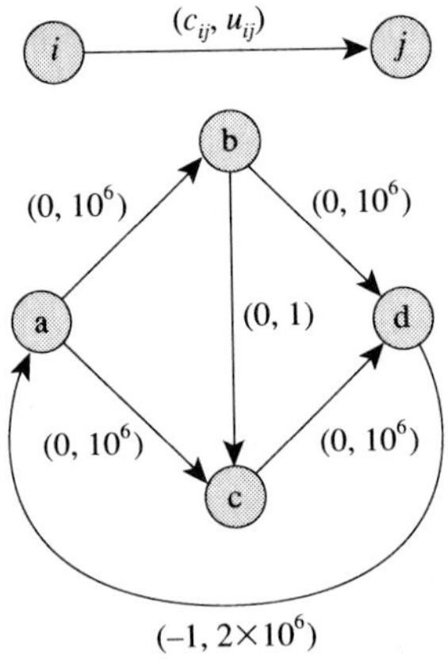

Figure 9.24 Network where cycle canceling algorithm performs 2×10^6 iterations.

9.20. Apply the successive shortest path algorithm to the minimum cost flow problem shown in Figure 9.25. Show that the algorithm performs eight augmentations, each of unit flow, and that the cost of these augmentations (i.e., sum of the arc costs in the path

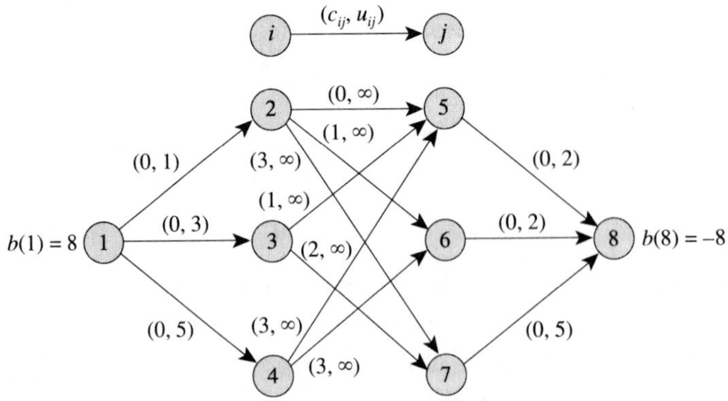

Figure 9.25 Example for Exercise 9.20.

in the residual network) is 0, 1, 2, 3, 3, 4, 5, and 6. How many iterations does the primal–dual algorithm require to solve this problem?

9.21. Construct a class of minimum cost flow problems for which the number of iterations performed by the successive shortest path algorithm might grow exponentially in log U. (*Hint*: Consider the example shown in Figure 9.24.)

9.22. Figure 9.26 specifies the data and a feasible solution for a minimum cost flow problem. With respect to zero node potentials, list the in-kilter and out-of-kilter arcs. Apply the out-of-kilter algorithm to find an optimal flow in the network.

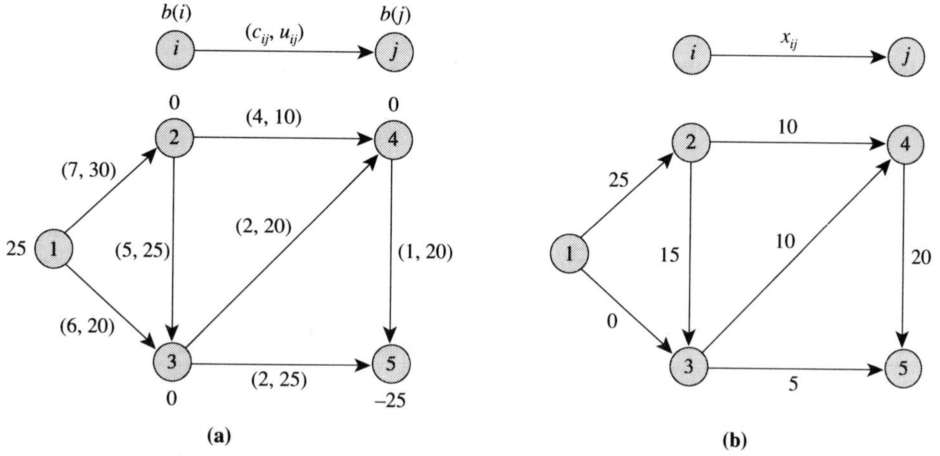

Figure 9.26 Example for Exercises 9.22 and 9.23: (a) problem data; (b) feasible flow.

9.23. Consider the minimum cost flow problem shown in Figure 9.26. Starting with zero pseudoflow and zero node potentials, apply the relaxation algorithm to establish an optimal flow.

9.24. Figure 9.21(b) specifies an optimal solution for the minimum cost flow problem shown in Figure 9.21(a). Reoptimize the solution with respect to the following changes in the problem data: (1) when c_{23} increases from 0 to 6; (2) when c_{78} decreases from 9 to 2; (3) when $b(2)$ decreases to 15 and $b(8)$ increases to -5; and (4) when u_{23} increases to 20. Treat these changes individually.

9.25. Assuming that we set one node potential to value zero, show that nC is an upper bound and that $-nC$ is a lower bound on the optimal value of any node potential.

9.26. Justify the out-of-kilter algorithm described in Section 9.9 for the case when arcs can violate their flow bounds. Show that in the execution of the algorithm, the kilter number of arcs are nonincreasing and at least one kilter number strictly decreases at every iteration.

9.27. Obtain a worst-case bound on the total number of iterations performed by the relaxation algorithm. Compare this bound with the number of iterations performed by the cycle-canceling, successive shortest path, and primal–dual algorithms.

9.28. Show that if the pair (x, π) satisfies the complementary slackness optimality conditions (9.8), it also satisfies the reduced cost optimality conditions (9.7).

9.29. Prove that if x^* is an optimal flow and π is an optimal set of node potentials, the pair (x^*, π) satisfies the complementary slackness optimality conditions. In your proof, do not use the strong duality theorem. (*Hint*: Suppose that the pair (x, π) satisfies the optimality conditions for some flow x. Show that $c^{\pi}(x^* - x) = 0$ and use this fact to prove the desired result.)

9.30. With respect to an optimal solution x^* of a minimum cost flow problem, suppose that we redefine arc capacities u' as follows:

$$u'_{ij} = \begin{cases} u_{ij} & \text{if } x^*_{ij} = u_{ij} \\ \infty & \text{if } x^*_{ij} < u_{ij}. \end{cases}$$

Show that x^* is also an optimal solution of the minimum cost flow problem with the arc capacities as u'.

9.31. With respect to an optimal solution x^* of a minimum cost flow problem, suppose that we redefine arc capacities $u' = x^*$. Show that x^* is also an optimal solution of the minimum cost flow problem with arc capacities u'.

9.32. In Section 2.4 we showed how to transform a minimum cost flow problem in an undirected network in which all lower bounds are zero into a minimum cost flow problem in a directed network. Explain why this approach does not work when some lower bounds on arc flows exceed zero.

9.33. In the minimum cost flow problem, suppose that one specified arc (p, q) has no lower and upper flow bounds. How would you transform this problem into the standard minimum cost flow problem?

9.34. As we have seen in Section 2.4, the uncapacitated transportation problem is equivalent to the minimum cost flow problem in the sense that we can always transform either problem into a version of another problem. If we can solve the uncapacitated transportation problem in $O(g(n, m))$ time, can we also solve the minimum cost flow problem in $O(g(n, m))$ time?

9.35. In the *min-cost max-flow problem* defined on a directed network $G = (N, A)$, we wish to send the maximum amount of flow from a node s to a node t at the minimum possible total cost. That is, among all maximum flows, find the one with the smallest cost.

(a) Show how to formulate any minimum cost flow problem as a min-cost max-flow problem.

(b) Show how to convert any min-cost max-flow problem into a circulation problem.

9.36. Suppose that in a minimum cost flow problem, some arcs have infinite capacities and some arc costs are negative. (Assume that the lower bounds on all arc flows are zero.)

(a) Show that the minimum cost flow problem has a finite optimal solution if and only if the uncapacitated arcs do not contain a negative cost-directed cycle.

(b) Let B denote the sum of the finite arc capacities and the supplies $b(\cdot)$ of all the supply nodes. Show that the minimum cost flow problem always has an optimal solution in which each arc flow is at most B. Conclude that without any loss of generality, we can assume that in the minimum cost flow problem (with a bounded optimal solution value) every arc is capacitated. (*Hint*: Use the flow decomposition property.)

9.37. Suppose that in a minimum cost flow problem, some arcs have infinite capacities and some arc costs are negative. Let B denote the sum of the finite arc capacities and the right-hand-side coefficients $b(i)$ for all the supply nodes. Let z and z' denote the objective function values of the minimum cost flow problem when we set the capacity of each infinite capacity arc to the value B and $B + 1$, respectively. Show that the objective function of the minimum cost flow problem is unbounded if and only if $z' < z$.

9.38. In a minimum cost flow network, suppose that in addition to arc capacities, nodes have upper bounds imposed upon the entering flow. Let $w(i)$ be the maximum flow that can enter node $i \in N$. How would you solve this generalization of the minimum cost flow problem?

9.39. Let (k, l) and (p, q) denote a minimum cost arc and a maximum cost arc in a network. Is it possible that no minimum cost flow have a positive flow on arc (k, l)? Is it possible that every minimum cost flow have a positive flow on arc (p, q)? Justify your answers.

9.40. Prove or disprove the following claims.

(a) Suppose that all supply/demands and arc capacities in a minimum cost flow problem are all even integers. Then for some optimal flow x^*, each arc flow x^*_{ij} is an even number.

(b) Suppose that all supply/demands and arc capacities in a minimum cost circulation problem are all even integers. Then for some optimal flow x^*, each arc flow x^*_{ij} is an even number.

9.41. Let x^* be an optimal solution of the minimum cost flow problem. Define G° as a subgraph of the residual network $G(x^*)$ consisting of all arcs with zero reduced cost. Show that the minimum cost flow problem has an alternative optimal solution if and only if G° contains a directed cycle.

9.42. Suppose that you are given a nonintegral optimal solution to a minimum cost flow problem with integral data. Suggest a method for converting this solution into an integer optimal solution. Your method should maintain optimality of the solution at every step.

9.43. Suppose that the pair (x, π), for some pseudoflow x and some node potentials π, satisfies the reduced cost optimality conditions. Define $G^\circ(x)$ as a subgraph of the residual network $G(x)$ consisting of only those arcs with zero residual capacity. Define the cost of an arc (i, j) in $G^\circ(x)$ as c_{ij} if $(i, j) \in A$, and as $-c_{ij}$ otherwise. Show that every directed path in $G^\circ(x)$ between any pair of nodes is a shortest path in $G(x)$ between the same pair of nodes with respect to the arc costs c_{ij}.

9.44. Let x^1 and x^2 be two distinct (alternate) minimum cost flows in a network. Suppose that for some arc (k, l), $x^1_{kl} = p$, $x^2_{kl} = q$, and $p < q$. Show that for every $0 \leq \lambda \leq 1$, the minimum cost flow problem has an optimal solution x (possibly, noninteger) with $x_{kl} = (1 - \lambda)p + \lambda q$.

9.45. Let π^1 and π^2 be two distinct (alternate) optimal node potentials of a minimum cost flow problem. Suppose that for some node k, $\pi^1(k) = p$, $\pi^2(k) = q$, and $p < q$. Show that for every $0 \leq \lambda \leq 1$, the minimum cost flow problem has an optimal set of node potentials π (possibly, noninteger) with $\pi(k) = (1 - \lambda)p + \lambda q$.

9.46. (a) In the transportation problem, does adding a constant k to the cost of every outgoing arc from a specified supply node affect the optimality of a given optimal solution? Would adding a constant k to the cost of every incoming arc to a specified demand node affect the optimality of a given optimal solution?

(b) Would your answers to the questions in part (a) be the same if they were posed for the minimum cost flow problem instead of the transportation problem?

9.47. In Section 9.7 we described the following practical improvement of the successive shortest path algorithm: (1) terminate the execution of Dijkstra's algorithm whenever it permanently labels a deficit node l, and (2) modify the node potentials by setting $\pi(i)$ to $\pi(i) - d(i)$ if node i is permanently labeled; and by setting $\pi(i)$ to $\pi(i) - d(l)$ if node i is temporarily labeled. Show that after the algorithm has updated the node potentials in this manner, all the arcs in the residual network have nonnegative reduced costs and all the arcs in the shortest path from node k to node l have zero reduced costs. (*Hint*: Use the result in Exercise 5.9.)

9.48. Would multiplying each arc cost in a network by a constant k change the set of optimal solutions of the minimum cost flow problem? Would adding a constant k to each arc cost change the set of optimal solutions?

9.49. In Section 9.11 we described a method for performing sensitivity analysis when we increase the capacity of an arc (p, q) by 1 unit. Modify the method to perform the analysis when we decrease the capacity of the arc (p, q) by 1 unit.

9.50. In Section 9.11 we described a method for performing sensitivity analysis when we increase the cost of an arc (p, q) by 1 unit. Modify the method to perform the analysis when we decrease the cost of the arc (p, q) by 1 unit.

9.51. Suppose that we have to solve a minimum cost flow problem in which the sum of the supplies exceeds the sum of the demands, so we need to retain some of the supply at some nodes. We refer to this problem as the *minimum cost flow problem with surplus*. Specify a linear programming formulation of this problem. Also show how to transform this problem into an (ordinary) minimum cost flow problem.

9.52. This exercise concerns the minimum cost flow problem with surplus, as defined in Exercise 9.51. Suppose that we have an optimal solution of a minimum cost flow problem with surplus and we increase the supply of some node by 1 unit, holding the other data fixed. Show that the optimal objective function value cannot increase, but it might decrease. Show that if we increase the demand of a node by 1 unit, holding the other data fixed, the optimal objective function value cannot decrease, but it might increase.

9.53. More-for-less paradox (Charnes and Klingman [1971]). The more-for-less paradox shows that it is possible to send *more* flow from the supply nodes to the demand nodes of a minimum cost flow problem at *lower* cost even if all arc costs are nonnegative. To establish this more-for-less paradox, consider the minimum cost flow problem shown in Figure 9.27. Assume that all arc capacities are infinite.

(a) Show that the solution given by $x_{14} = 11$, $x_{16} = 9$, $x_{25} = 2$, $x_{26} = 8$, $x_{35} = 11$, and $x_{37} = 14$, is an optimal flow for this minimum cost flow problem. What is the total cost of flow?

(b) Suppose that we increase the supply of node 2 by 2 units, increase the demand of node 4 by 2 units, and reoptimize the solution using the method described in Section 9.11. Show that the total cost of flow decreases.

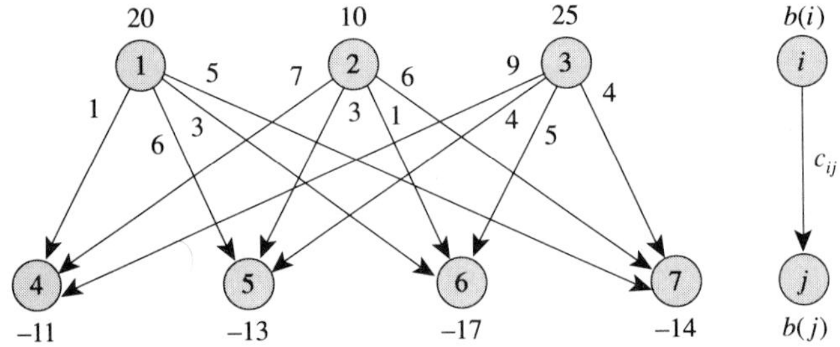

Figure 9.27 More-for-less paradox.

9.54. Capacitated minimum spanning tree problem (Garey and Johnson [1979]). In a complete undirected network with arc lengths c_{ij} and a specially designated node s, called the *central site*, we associate an integer requirement r_i with every node $i \in N - \{s\}$. In the capacitated minimum spanning tree problem, we want to identify a minimum cost spanning tree so that when we send flow on this tree from the central site to the other nodes to satisfy their flow requirements, no tree arc has a flow of more than a given arc capacity R, which is the same for all arcs. Show that when each r_i is 0 or 1, and $R = 1$, we can solve this problem as a minimum cost flow problem. (*Hint*: Model this problem as a minimum cost flow problem with node capacities, as discussed in Exercise 9.38.)

9.55. Fractional *b*-matching problem. Let $G = (N, A)$ be an undirected graph in which each node $i \in N$ has an associated supply $b(i)$ and each arc $(i, j) \in A$ has an associated cost c_{ij} and capacity u_{ij}. In the *b-matching problem*, we wish to find a minimum cost subgraph of G with exactly b arcs incident to every node. The *fractional b-matching problem* is a relaxation of the b-matching problem and can be stated as the following linear program:

$$\text{Minimize} \sum_{(i,j)\in A} c_{ij}x_{ij}$$

subject to

$$\sum_{j\in A(i)} x_{ij} = b(i) \qquad \text{for all } i \in N,$$

$$0 \le x_{ij} \le u_{ij} \qquad \text{for all } (i, j) \in A.$$

We assume that $x_{ij} = x_{ji}$ for every arc $(i, j) \in A$. We can define a related minimum cost flow problem as follows. Construct a bipartite network $G' = (N' \cup N'', A')$ with $N' = \{1', 2', \ldots, n'\}$, $N'' = \{1'', 2'', \ldots, n''\}$, $b(i') = b(i)$, and $b(i'') = -b(i)$. For each arc $(i, j) \in A$, the network G' contains two associated arcs (i', j'') and (j', i''), each with cost c_{ij} and capacity u_{ij}.

(a) Show how to transform every solution x of the fractional b-matching problem with cost z into a solution x' of the minimum cost flow problem with cost $2z$. Similarly, show that if x' is a solution of the minimum cost flow problem with cost z', then $x_{ij} = (x'_{ij} + x'_{ji})/2$ is a feasible solution of the fractional b-matching problem with cost $z'/2$. Use these results to show how to solve the fractional b-matching problem.

(b) Show that the fractional b-matching problem always has an optimal solution in which each arc flow x_{ij} is a multiple of $\frac{1}{2}$. Also show that if all the supplies and the capacities are even integers, the fractional b-matching problem always has an integer optimal solution.

9.56. Bottleneck transportation problem. Consider a transportation problem with a traversal time τ_{ij} instead of a cost c_{ij} associated with each arc (i, j). In the *bottleneck transportation problem* we wish to satisfy the requirements of the demand nodes from the supply nodes in the least time possible [i.e., we wish to find a flow x that minimizes the quantity $\max\{\tau_{ij}: (i, j) \in A \text{ and } x_{ij} > 0\}$].

(a) Suggest an application of the bottleneck transportation problem.

(b) Suppose that we arrange the arc traversal times in the nondecreasing order of their values. Let $\tau_1 < \tau_2 < \cdots < \tau_l$ be the distinct values of the arc traversal times (thus $l \le m$). Let *FS(k, found)* denote a subroutine that finds whether the transportation problem has a feasible solution using only those the arcs with traversal times less than or equal to τ_k; assume that the subroutine assigns a value true/false to found. Suggest a method for implementing the subroutine FS(k, found).

(c) Using the subroutine FS(k, found), write a pseudocode for solving the bottleneck transportation problem.

9.57. Equivalence of minimum cost flow algorithms (Zadeh [1979])

(a) Apply the successive shortest path algorithm to the minimum cost flow problem shown in Figure 9.28. Show that it performs four augmentations from node 1 to node 6, each of unit flow.

(b) Add the arc (1, 6) with sufficiently large cost and with $u_{16} = 4$ to the example in part (a). Observe that setting $x_{16} = 4$ and $x_{ij} = 0$ for all other arcs gives a feasible flow in the network. With this flow as the initial flow, apply the cycle-canceling algorithm and always augment flow along a negative cycle with minimum cost. Show that this algorithm also performs four unit flow augmentations from node 1 to node 6 along the same paths as in part (a) and in the same order, except that the flow returns to node 1 through the arc (6, 1) in the residual network.

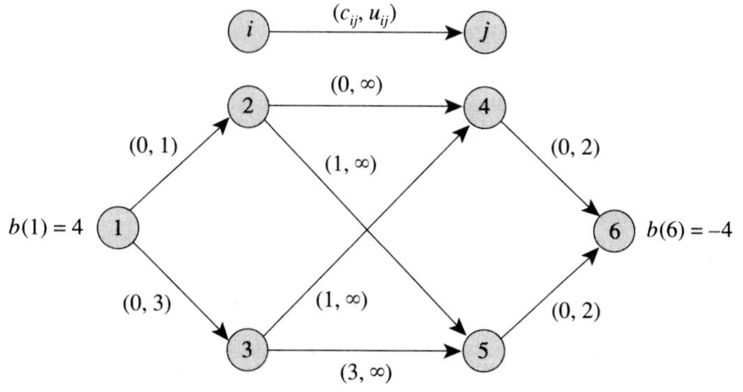

Figure 9.28 Equivalence of minimum cost flow algorithms.

(c) Using parts (a) and (b) as background, prove the general result that if initialized properly, the successive shortest path algorithm and the cycle-canceling algorithm (with augmentation along a most negative cycle) are equivalent in the sense that they perform the same augmentations and in the same order.

9.58. Modify and initialize the minimum cost flow problem in Figure 9.28 appropriately so that when we apply the out-of-kilter algorithm to this problem, it also performs four augmentation in the same order as the successive shortest path algorithm. Then prove the equivalence of the out-of-kilter algorithm with the successive shortest path algorithm in general.

10

MINIMUM COST FLOWS: POLYNOMIAL ALGORITHMS

Success generally depends upon knowing how long it takes to succeed.
—Montesquieu

Chapter Outline

10.1 INTRODUCTION

In Chapter 9 we studied several different algorithms for solving minimum cost problems. Although these algorithms guarantee finite convergence whenever the problem data are integral, the computations are not bounded by any polynomial in the underlying problem specification. In the spirit of computational complexity theory, this situation is not completely satisfactory: It does not provide us with any good theoretical assurance that the algorithms will perform well on all problems that we might encounter. The circumstances are quite analogous to our previous development of maximum flow algorithms; we started by first developing straightforward, but not necessarily polynomial, algorithms for solving those problems, and then enhanced these algorithms by changing the algorithmic strategy and/or by using clever data structures and implementations. This situation raises the following natural questions: (1) Can we devise algorithms that are polynomial in the usual problem parameters: number n of nodes, number m of arcs, log U (the log of the largest supply/demand or arc capacity), and log C (the log of the largest cost coefficient), and (2) can we develop strongly polynomial-time algorithms (i.e., algorithms whose running time depends upon only on n and m)? A strongly polynomial-time algorithm has one important theoretical advantage: It will solve problems with irrational data.

In this chapter we provide affirmative answers to these questions. To develop polynomial-time algorithms, we use ideas that are similar to those we have used before: namely, scaling of the capacity data and/or of the cost data. We consider three polynomial-time algorithms: (1) a capacity scaling algorithm that is a scaled version of the successive shortest path algorithm that we discussed in Chapter 9, (2) a cost scaling algorithm that is a generalization of the preflow-push algorithm for the maximum flow problem, and (3) a double scaling algorithm that simultaneously scales both the arc capacities and the costs.

Scaling is a powerful idea that has produced algorithmic improvements to many problems in combinatorial optimization. We might view scaling algorithms as follows. We start with the optimality conditions for the network flow problem we are examining, but instead of enforcing these conditions exactly, we generate an "approximate" solution that is permitted to violate one (or more) of the conditions by an amount Δ. Initially, by choosing Δ quite large, for example as C or U, we will easily be able to find a starting solution that satisfies the relaxed optimality conditions. We then reset the parameter Δ to $\Delta/2$ and reoptimize so that the approximate solution now violates the optimality conditions by an amount of at most $\Delta/2$. We then repeat the procedure, reoptimizing again until the approximate solution violates the conditions by an amount of at most $\Delta/4$, and so on. This solution strategy is quite flexible and leads to different algorithms depending on which of the optimality conditions we relax and how we perform the reoptimizations.

Our discussion of the capacity scaling algorithm for the maximum flow problem in Section 7.3 provides one example. A feasible flow to the maximum flow problem is optimal if the residual network contains no augmenting path. In the capacity scaling algorithm, we relaxed this condition so that after the Δ-scaling phase, the residual network can contain an augmenting path, but only if its capacity were less than Δ. The excess scaling algorithm for the maximum flow problem provides us with another example. In this case the residual network again contains no path from the source node s to the sink node t; however, at the end of the Δ-scaling phase, we relaxed a feasibility requirement requiring that the flow into every node other than the source and sink equals the flow out of that node. Instead, we permitted the excess at each node to be as large as Δ during the Δ-scaling phase.

In this chapter, by applying a scaling approach to the algorithms that we considered in Chapter 9, we develop polynomial-time versions of these algorithms. We begin by developing a modified version of the successive shortest path algorithm in which we relax two optimality conditions in the Δ-scaling phase: (1) We permit the solution to violate supply/demand constraints by an amount Δ, and (2) we permit the residual network to contain negative cost cycles. The resulting algorithm reduces the number of shortest path computations from nU to $m \log U$.

We next describe a cost-scaling algorithm that uses another concept of approximate optimality; at the end of each ϵ-scaling phase (ϵ plays the role of Δ) we obtain a feasible flow that satisfies the property that the reduced cost of each arc in the residual network is greater than or equal to $-\epsilon$ (instead of zero). To find the optimal solution during the ϵ-scaling phase, this algorithm carries out a sequence of push and relabel operations that are similar to the preflow-push algorithm for maximum flows. The generic cost scaling algorithm runs in $O(n^2m \log(nC))$ time. We also describe a special "wave implementation" of this algorithm that chooses nodes for the push/relabel operations in a specific order. This specialization requires $O(n^3 \log(nC))$ time.

We then describe a *double scaling algorithm* that combines the features of both cost and capacity scaling. This algorithm works with two nested loops. In the outer loop we scale the costs, and in the inner loop we scale the capacities. Introducing capacity scaling as an inner loop within a cost scaling approach permits us to find augmenting paths very efficiently. This resulting double scaling algorithm solves the minimum cost flow problem in $O(nm \log U \log(nC))$ time.

All of these algorithms require polynomial time; they are not, however, strongly polynomial time because their time bounds depend on $\log U$ and/or $\log C$. Developing strongly polynomial-time algorithms seems to require a somewhat different approach. Although most strongly polynomial-time algorithms use ideas of data scaling, they also use another idea: By invoking the optimality conditions, they are able to show that at intermediate steps of the algorithm, they have already discovered part of the optimal solution (e.g., optimal flow), so that they are able to reduce the problem size. In Sections 10.5, 10.6, and 10.7 we consider three different strongly polynomial-time algorithms whose analysis invokes this "problem reduction argument."

In Section 10.5 we analyze the minimum mean cycle-canceling algorithm that we described in Section 9.6. Recall that this algorithm augments flow at each step on a cycle with the smallest average cost, averaged over the number of arcs in the cycle, until the residual network contains no negative cost cycle; at this point, the current flow is optimal. As we show in this section, we can view this algorithm as finding a sequence of improved approximately optimal solutions (in the sense that the reduced cost of every arc is greater than or equal to $-\epsilon$, with ϵ decreasing throughout the algorithm). This algorithm has the property that if the magnitude of the reduced cost of any arc is sufficiently large (as a function of ϵ), the flow on that arc remains fixed at its upper or lower bound throughout the remainder of the algorithm and so has this value in the optimal solution. This property permits us to show that the algorithm fixes the flow on an arc and does so sufficiently often so that we obtain an $O(n^2m^3 \log n)$ time algorithm for the capacitated minimum cost flow problem. One interesting characteristic of this algorithm is that it does not explicitly monitor ϵ or explicitly fix the flow variables. These features of the algorithm are by-products of the analysis.

The strongly polynomial-time algorithm that we consider in Section 10.6 solves the linear programming dual of the minimum cost flow problem. This *repeated capacity scaling algorithm* is a variant of the capacity scaling algorithm that we discuss in Section 10.2. This algorithm uses a scaling parameter Δ as in the capacity scaling algorithm, but shows that periodically the flow on some arc (i, j) becomes sufficiently large (as a function of Δ), at which point we are able to reduce the size of the dual linear program by one, which is equivalent to *contraction* in the primal network. This observation permits us to reduce the size of the problem successively by contracting nodes. The end result is an algorithm requiring $O((m^2 \log n)(m + n \log n))$ time for the minimum cost flow problem.

In Section 10.7 we consider an *enhanced scaling algorithm* that is a hybrid version of the capacity scaling algorithm and the repeated capacity scaling algorithm. By choosing a scaling parameter Δ carefully and by permitting a somewhat broader choice of the augmenting paths at each step, this algorithm is able to fix variables more quickly than the repeated capacity scaling algorithm. As a consequence, it solves fewer shortest path problems and solves capacitated minimum cost flow problems in $O((m \log n)(m + n \log n))$ time, which is currently the best known polynomial-time bound for solving the capacitated minimum cost flow problem.

10.2 CAPACITY SCALING ALGORITHM

In Chapter 9 we considered the successive shortest path algorithm, one of the fundamental algorithms for solving the minimum cost flow problem. An inherent drawback of this algorithm is that its augmentations might carry relatively small amounts of flow, resulting in a fairly large number of augmentations in the worst case. By incorporating a scaling technique, the capacity algorithm described in this section guarantees that each augmentation carries *sufficiently large* flow and thereby reduces the number of augmentations substantially. This method permits us to improve the worst-case algorithmic performance from $O(nU \cdot S(n, m, nC))$ to $O(m \log U \cdot S(n, m, nC))$. [Recall that U is an upper bound on the largest supply/demand and largest capacity in the network, and $S(n, m, C)$ is the time required to solve a shortest path problem with n nodes, m arcs, and nonnegative costs whose values are no more than C. The reason that the running time involves $S(n, m, nC)$ rather than $S(n, m, C)$ is that the costs in the residual network are reduced costs, and the reduced cost of an arc could be as large as nC.]

The capacity scaling algorithm is a variant of the successive shortest path algorithm. It is related to the successive shortest path algorithm, just as the capacity scaling algorithm for the maximum flow problem (discussed in Section 7.3) is related to the labeling algorithm (discussed in Section 6.5). Recall that the labeling algorithm performs $O(nU)$ augmentations; by sending flows along paths with *sufficiently large* residual capacities, the capacity scaling algorithm reduces the number of augmentations to $O(m \log U)$. In a similar fashion, the capacity scaling algorithm for the minimum cost flow problem ensures that each shortest path augmentation carries a sufficiently large amount of flow; this modification to the algorithm reduces the number of successive shortest path iterations from $O(nU)$ to $O(m \log U)$. This algorithm not only improves on the algorithmic performance of the successive shortest path algorithm, but also illustrates how small changes in an algorithm can produce significant algorithmic improvements (at least in the worst case).

The capacity scaling algorithm applies to the general capacitated minimum cost flow problem. It uses a pseudoflow x and the imbalances $e(i)$ as defined in Section 9.7. The algorithm maintains a pseudoflow satisfying the reduced cost optimality condition and gradually converts this pseudoflow into a flow by identifying shortest paths from nodes with excesses to nodes with deficits and augmenting flows along these paths. It performs a number of scaling phases for different values of a parameter Δ. We refer to a scaling phase with a specific value of Δ as the *Δ-scaling phase*. Initially, $\Delta = 2^{\lfloor \log U \rfloor}$. The algorithm ensures that in the Δ-scaling phase each augmentation carries exactly Δ units of flow. When it is not possible to do so because no node has an excess of at least Δ, or no node has a deficit of at least Δ, the algorithm reduces the value of Δ by a factor of 2 and repeats the process. Eventually, $\Delta = 1$ and at the end of this scaling phase, the solution becomes a flow. This flow must be an optimal flow because it satisfies the reduced cost optimality condition.

For a given value of Δ, we define two sets $S(\Delta)$ and $T(\Delta)$ as follows:

$$S(\Delta) = \{i : e(i) \geq \Delta\},$$

$$T(\Delta) = \{i : e(i) \leq -\Delta\}.$$

In the Δ-scaling phase, each augmentation must start at a node in $S(\Delta)$ and end at a node in $T(\Delta)$. Moreover, the augmentation must take place on a path along which every arc has residual capacity of at least Δ. Therefore, we introduce another definition: The *Δ-residual network* $G(x, \Delta)$ is defined as the subgraph of $G(x)$ consisting of those arcs whose residual capacity is at least Δ. In the Δ-scaling phase, the algorithm augments flow from a node in $S(\Delta)$ to a node in $T(\Delta)$ along a shortest path in $G(x, \Delta)$. The algorithm satisfies the property that every arc in $G(x, \Delta)$ satisfies the reduced cost optimality condition; however, arcs in $G(x)$ but not in $G(x, \Delta)$ might violate the reduced cost optimality condition. Figure 10.1 presents an algorithmic description of the capacity scaling algorithm.

Notice that the capacity scaling algorithm augments exactly Δ units of flow in the Δ-scaling phase, even though it could augment more. For uncapacitated problems, this tactic leads to the useful property that all arc flows are always an integral multiple of Δ. (Why might capacitated networks not satisfy this property?) Several variations of the capacity scaling algorithm discussed in Sections 10.5 and 14.5 adopt the same tactic.

To establish the correctness of the capacity scaling algorithm, observe that the 2Δ-scaling phase ends when $S(2\Delta) = \phi$ or $T(2\Delta) = \phi$. At that point, either $e(i) < 2\Delta$ for all $i \in N$ or $e(i) > -2\Delta$ for all $i \in N$. These conditions imply that the sum of the excesses (whose magnitude equals the sum of deficits) is bounded by $2n\Delta$.

```
algorithm capacity scaling;
begin
    x : = 0 and π : = 0;
    Δ : = 2^⌊log U⌋;
    while Δ ≥ 1
    begin {Δ-scaling phase}
        for every arc (i, j) in the residual network G(x) do
            if r_ij ≥ Δ and c^π_ij < 0 then send r_ij units of flow along arc (i, j),
              update x and the imbalances e(·);
        S(Δ) : = {i ∈ N : e(i) ≥ Δ};
        T(Δ) : = {i ∈ N : e(i) ≤ -Δ};
        while S(Δ) ≠ Ø and T(Δ) ≠ Ø do
        begin
            select a node k ∈ S(Δ) and a node l ∈ T(Δ);
            determine shortest path distances d(·) from node k to all other nodes in the
                Δ-residual network G(x, Δ) with respect to the reduced costs c^π_ij;
            let P denote shortest path from node k to node l in G(x, Δ);
            update π : = π - d;
            augment Δ units of flow along the path P;
            update x, S(Δ), T(Δ), and G(x, Δ);
        end;
        Δ : = Δ/2;
    end;
end;
```

Figure 10.1 Capacity scaling algorithm.

At the beginning of the Δ-scaling phase, the algorithm first checks whether every arc (i, j) in Δ-residual network satisfies the reduced cost optimality condition $c_{ij}^{\pi} \geq 0$. The arcs introduced in the Δ-residual network at the beginning of the Δ-scaling phase [i.e., those arcs (i, j) for which $\Delta \leq r_{ij} < 2\Delta$] might not satisfy the optimality condition (since, conceivably, $c_{ij}^{\pi} < 0$). Therefore, the algorithm immediately saturates those arcs (i, j) so that they drop out of the residual network; since the reversal of these arcs (j, i) satisfy the condition $c_{ji}^{\pi} = -c_{ij}^{\pi} > 0$, they satisfy the optimality condition. Notice that because $r_{ij} < 2\Delta$, saturating any such arc (i, j) changes the imbalance of its endpoints by at most 2Δ. As a result, after we have saturated all the arcs violating the reduced cost optimality condition, the sum of the excesses is bounded by $2n\Delta + 2m\Delta = 2(n + m)\Delta$.

In the Δ-scaling phase, each augmentation starts at a node $k \in S(\Delta)$, terminates at a node $l \in T(\Delta)$, and carries at least Δ units of flow. Note that Assumption 9.4 implies that the Δ-residual network contains a directed path from node k to node l, so we always succeed in identifying a shortest path from node k to node l. Augmenting flow along a shortest path in $G(x, \Delta)$ preserves the property that every arc satisfies the reduced cost optimality condition (see Section 9.3). When either $S(\Delta)$ or $T(\Delta)$ is empty, the Δ-scaling phase ends. At this point we divide Δ by a factor of 2 and start a new scaling phase. Within $O(\log U)$ scaling phases, $\Delta = 1$, and by the integrality of data, every node imbalance will be zero at the end of this phase. In this phase $G(x, \Delta) \equiv G(x)$ and every arc in the residual network satisfies the reduced cost optimality condition. Consequently, the algorithm will obtain a minimum cost flow at the end of this scaling phase.

As we have seen, the capacity scaling algorithm is easy to state. Similarly, its running time is easy to analyze. We have noted previously that in the Δ-scaling phase the sum of the excesses is bounded by $2(n + m)\Delta$. Since each augmentation in this phase carries at least Δ units of flow from a node in $S(\Delta)$ to a node in $T(\Delta)$, each augmentation reduces the sum of the excesses by at least Δ units. Therefore, a scaling phase can perform at most $2(n + m)$ augmentations. Since we need to solve a shortest path problem to identify each augmenting path, we have established the following result.

Theorem 10.1. *The capacity scaling algorithm solves the minimum cost flow problem in* $O(m \log U\, S(n, m, nC))$ *time.* ◆

10.3 COST SCALING ALGORITHM

In this section we describe a cost scaling algorithm for the minimum cost flow problem. This algorithm can be viewed as a generalization of the preflow-push algorithm for the maximum flow problem; in fact, the algorithm reveals an interesting relationship between the maximum flow and minimum cost flow problems. This algorithm relies on the concept of approximate optimality.

Approximate Optimality

A flow x or a pseudoflow x is said to be *ε-optimal* for some $\epsilon > 0$ if for some node potentials π, the pair (x, π) satisfies the following *ε-optimality conditions*:

$$\text{If } c_{ij}^{\pi} > \epsilon, \text{ then } x_{ij} = 0. \tag{10.1a}$$

$$\text{If } -\epsilon \le c_{ij}^{\pi} \le \epsilon, \text{ then } 0 \le x_{ij} \le u_{ij}. \tag{10.1b}$$

$$\text{If } c_{ij}^{\pi} < -\epsilon, \text{ then } x_{ij} = u_{ij}. \tag{10.1c}$$

These conditions are relaxations of the (exact) complementary slackness optimality conditions (9.8) that we discussed in Section 9.3; note that these conditions reduce to the complementary slackness optimality conditions when $\epsilon = 0$. The *exact* optimality conditions (9.8) imply that any combination of (x_{ij}, c_{ij}^{π}) lying on the thick lines shown in Figure 10.2(a) is optimal. The ϵ-optimality conditions (10.1) imply that any combination of (x_{ij}, c_{ij}^{π}) lying on the thick lines or in the hatched region in Figure 10.2(b) is ϵ-optimal.

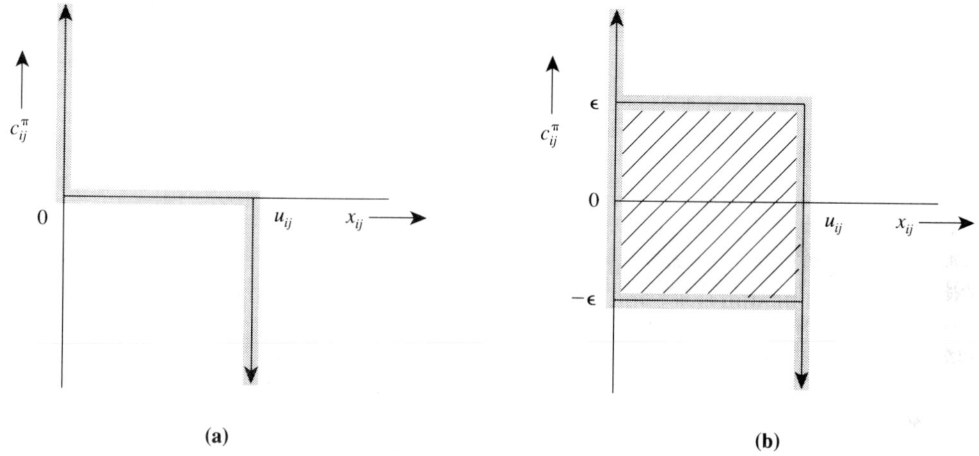

Figure 10.2 Illustrating the optimality condition for arc (i, j): (a) exact optimality condition for arc (i, j); (b) ϵ-optimality condition for arc (i, j).

The ϵ-optimality conditions assume the following simpler form when stated in terms of the residual network $G(x)$: A flow x or a pseudoflow x is said to be ϵ-*optimal* for some $\epsilon > 0$ if x, together with some node potential vector π, satisfies the following ϵ-*optimality conditions* (we leave the proof as an exercise for the reader):

$$c_{ij}^{\pi} \ge -\epsilon \qquad \text{for every arc } (i, j) \text{ in the residual network } G(x). \tag{10.2}$$

Lemma 10.2. *For a minimum cost flow problem with integer costs, any feasible flow is ϵ-optimal whenever $\epsilon \ge C$. Moreover, if $\epsilon < 1/n$, then any ϵ-optimal feasible flow is an optimal flow.*

Proof. Let x be any feasible flow and let $\pi = 0$. Then $c_{ij}^{\pi} = c_{ij} \ge -C$ for every arc (i, j) in the residual network $G(x)$. Therefore, x is ϵ-optimal for $\epsilon = C$.

Now consider an ϵ-optimal flow x with $\epsilon < 1/n$. Suppose that x is ϵ-optimal with respect to the node potentials π and that W is a directed cycle in $G(x)$. The condition (10·2) implies that $\sum_{(i,j)\in W} c_{ij}^{\pi} \ge -\epsilon n > -1$, because $\epsilon < 1/n$. The integrality of the costs implies that $\sum_{(i,j)\in W} c_{ij}^{\pi}$ is nonnegative. But notice that $\sum_{(i,j)\in W} c_{ij}^{\pi} = \sum_{(i,j)\in W} (c_{ij} - \pi(i) + \pi(j)) = \sum_{(i,j)\in W} c_{ij}$. Therefore, W cannot be a negative cost cycle. Since $G(x)$ cannot contain any negative cycle, x must be optimal (from Theorem 9.1). ◆

Algorithm

The cost scaling algorithm treats ϵ as a parameter and iteratively obtains ϵ-optimal flows for successively smaller values of ϵ. Initially, $\epsilon = C$ and any feasible flow is ϵ-optimal. The algorithm then performs cost scaling phases by repeatedly applying an *improve-approximation* procedure that transforms an ϵ-optimal flow into a $\frac{1}{2}\epsilon$-optimal flow. After $1 + \lceil \log(nC) \rceil$ cost scaling phases, $\epsilon < 1/n$ and the algorithm terminates with an optimal flow. Figure 10.3 provides a more formal statement of the cost scaling algorithm.

```
algorithm cost scaling;
begin
    π : = 0 and ε : = C;
    let x be any feasible flow;
    while ε ≥ 1/n do
    begin
        improve-approximation(ε, x, π);
        ε : = ε/2;
    end;
    x is an optimal flow for the minimum cost flow problem;
end;
```

Figure 10.3 Cost scaling algorithm.

The improve-approximation procedure transforms an ϵ-optimal flow into a $\frac{1}{2}\epsilon$-optimal flow. It does so by (1) converting the ϵ-optimal flow into a $\frac{1}{2}\epsilon$-optimal pseudoflow, and (2) then gradually converting the pseudoflow into a flow while always maintaining $\frac{1}{2}\epsilon$-optimality of the solution. We refer to a node i with $e(i) > 0$ as *active* and say that an arc (i, j) in the residual network is *admissible* if $-\frac{1}{2}\epsilon \le c_{ij}^{\pi} < 0$. The basic operation in the procedure is to select an active node i and perform pushes on admissible arcs (i, j) emanating from node i. When the network contains no admissible arc, the algorithm updates the node potential $\pi(i)$. Figure 10.4 summarizes the essential steps of the generic version of the improve-approximation procedure.

Recall that r_{ij} denotes the residual capacity of an arc (i, j) in $G(x)$. As in our earlier discussion of preflow-push algorithms for the maximum flow problem, if $\delta = r_{ij}$, we refer to the push as *saturating*; otherwise, it is *nonsaturating*. We also refer to the updating of the potential of a node as a *relabel* operation. The purpose of a relabel operation at node i is to create new admissible arcs emanating from this node.

We illustrate the basic operations of the improve-approximation procedure on a small numerical example. Consider the residual network shown in Figure 10.5(a). Let $\epsilon = 8$. The current pseudoflow is 4-optimal. Node 2 is the only active node in the network, so the algorithm selects it for push/relabel. Suppose that arc (2, 4) is the first admissible arc found. The algorithm pushes $\min\{e(2), r_{24}\} = \min\{30, 5\} = 5$ units of flow on arc (2, 4); this push saturates the arc. Next the algorithm identifies arc (2, 3) as admissible and pushes $\min\{e(2), r_{23}\} = \min\{25, 30\} = 25$ units on this arc. This push is nonsaturating; after the algorithm has performed this push, node

```
procedure improve-approximation(ε, x, π);
begin
    for every arc (i, j) ∈ A do
        if c_ij^π > 0 then x_ij : = 0
        else if c_ij^π < 0 then x_ij : = u_ij;
    compute node imbalances;
    while the network contains an active node do
    begin
        select an active node i;
        push/relabel(i);
    end;
end;
```

(a)

```
procedure push/relabel(i);
begin
    if G(x) contains an admissible arc (i, j) then
        push δ : = min{e(i), r_ij} units of flow from node i to node j;
    else π(i) : = π(i) + ε/2;
end;
```

(b)

Figure 10.4 Procedures of the cost scaling algorithm.

2 is inactive and node 3 is active. Figure 10.5(b) shows the residual network at this point.

In the next iteration, the algorithm selects node 3 for push/relabel. Since no admissible arc emanates from this node, we perform a relabel operation and increase the node's potential by $\epsilon/2 = 4$ units. This potential change decreases the reduced costs of the outgoing arcs, namely, (3, 4) and (3, 2), by 4 units and increases the reduced costs of the incoming arcs, namely (1, 3) and (2, 3), by 4 units [see Figure 10.5(c)]. The relabel operation creates an admissible arc, namely arc (3, 4), and we next perform a push of 15 units on this arc [see Figure 10.5(d)]. Since the current solution is a flow, the improve-approximation procedure terminates.

To identify admissible arcs emanating from node i, we use the same data structure used in the maximum flow algorithms described in Section 7.4. For each node i, we maintain a *current-arc* (i, j) which is the current candidate to test for admissibility. Initially, the current-arc of node i is the first arc in its arc list $A(i)$. To determine an admissible arc emanating from node i, the algorithm checks whether the node's current-arc is admissible, and if not, chooses the next arc in the arc list as the current-arc. Thus the algorithm passes through the arc list starting with the current-arc until it finds an admissible arc. If the algorithm reaches the end of the arc list without finding an admissible arc, it declares that the node has no admissible arc. At this point it relabels node i and again sets its current-arc to the first arc in $A(i)$.

We might comment on two practical improvements of the improve-approximation procedure. The algorithm, as stated, starts with $\epsilon = C$ and reduces ϵ by a factor of 2 in every scaling phase until $\epsilon = 1/n$. As a consequence, ϵ could become

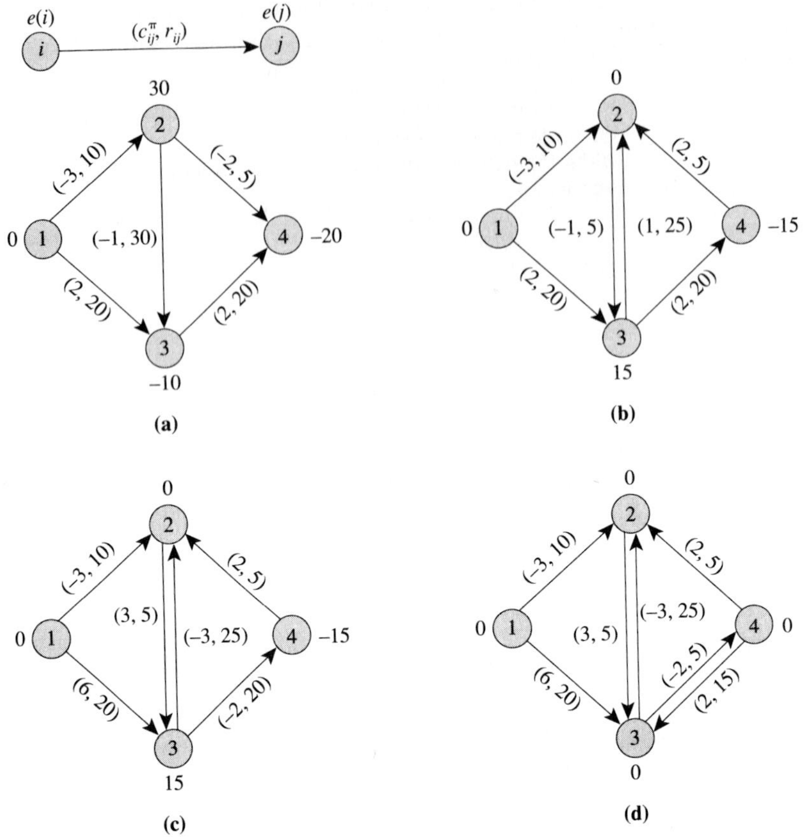

Figure 10.5 Illustration of push/relabel steps.

nonintegral during the execution of the algorithm. By slightly modifying the algorithm, however, we can ensure that ϵ remains integral. We do so by multiplying all the arc costs by n, by setting the initial value of ϵ equal to $2^{\lceil \log(nC) \rceil}$, and by terminating the algorithm when $\epsilon < 1$. It is possible to show (see Exercise 10.7) that the modified algorithm would yield an optimal flow for the minimum cost flow problem in the same computational time. Furthermore, as stated, the algorithm increases a node potential by $\epsilon/2$ during a relabel operation. As described in Exercise 10.8, we can often increase the node potential by an amount larger than $\epsilon/2$.

Analysis of the Algorithm

We show that the cost scaling algorithm correctly solves the minimum cost flow problem. In the proof, we rely on the fact that the improve-approximation procedure converts an ϵ-optimal flow into an $\epsilon/2$-optimal flow. We establish this result in the following lemma.

Lemma 10.3. *The improve-approximation procedure always maintains $\frac{1}{2}$ ϵ-optimality of the pseudoflow, and at termination yields a $\frac{1}{2}$ ϵ-optimal flow.*

Proof. We use induction on the number of pushes and relabels. At the beginning of the procedure, the algorithm sets the flows on arcs with negative reduced costs to their capacities, sets the flow on arcs with positive reduced costs to zero, and leaves the flow on arcs with zero reduced costs unchanged. The resulting pseudoflow satisfies (10.1) for $\epsilon = 0$ and thus is 0-optimal. Since a 0-optimal pseudoflow is ϵ-optimal for every ϵ, the resulting flow is also $\frac{1}{2}\,\epsilon$-optimal.

We next study the effect of a push on the $\frac{1}{2}\,\epsilon$-optimality of the solution. Pushing flow on arc (i, j) might add its reversal (j, i) to the residual network. But since $-\epsilon/2 \le c_{ij}^{\pi} < 0$ (by the criteria of admissibility), $c_{ji}^{\pi} = -c_{ij}^{\pi} > 0$, and so this arc satisfies the $\frac{1}{2}\,\epsilon$-optimality condition (10.2).

What is the effect of a relabel operation? The algorithm relabels a node i when $c_{ij}^{\pi} \ge 0$ for every arc (i, j) emanating from node i in the residual network. Increasing the potential of node i by $\epsilon/2$ units decreases the reduced cost of all arcs emanating from node i by $\epsilon/2$ units. But since $c_{ij}^{\pi} \ge 0$ before the increase in π, $c_{ij}^{\pi} \ge -\epsilon/2$ after the increase, and the arc satisfies the $\frac{1}{2}\,\epsilon$-optimality condition. Furthermore, increasing the potential of node i by $\epsilon/2$ units increases the reduced costs of the incoming arcs at node i but maintains the $\frac{1}{2}\,\epsilon$-optimality condition for these arcs. These results establish the lemma. ◆

We next analyze the complexity of the improve-approximation procedure. We show that the number of relabel operations is $O(n^2)$, the number of saturating pushes is $O(nm)$, and the number of nonsaturating pushes for the generic version is $O(n^2m)$. These time bounds are comparable to those of the preflow-push algorithms for the maximum flow problem and the proof techniques are also similar. We first prove the most significant result, which bounds the number of relabel operations.

Lemma 10.4. *No node potential increases more than* $3n$ *times during an execution of the improve-approximation procedure.*

Proof. Let x be the current $\frac{1}{2}\,\epsilon$-optimal pseudoflow and x' be the ϵ-optimal flow at the end of the previous cost scaling phase. Let π and π' be the node potentials corresponding to the pseudoflow x and the flow x'. It is possible to show (see Exercise 10.9) that for every node v with an excess there exists a node w with a deficit and a sequence of nodes $v = v_0, v_1, v_2, \ldots, v_l = w$ that satisfies the property that the path $P = v_0 - v_1 - v_2 - \cdots - v_l$ is a directed path in $G(x)$ and its reversal $\overline{P} = v_l - v_{l-1} - \cdots - v_1 - v_0$ is a directed path in $G(x')$. Applying the $\frac{1}{2}\,\epsilon$-optimality condition to the arcs on the path P in $G(x)$, we see that

$$\sum_{(i,j)\in P} c_{ij}^{\pi} \ge -l(\epsilon/2).$$

Substituting $c_{ij}^{\pi} = c_{ij} - \pi(i) + \pi(j)$ in this expression gives

$$\sum_{(i,j)\in P} c_{ij} - \pi(v) + \pi(w) \ge -l(\epsilon/2).$$

Alternatively,

$$\pi(v) \le \pi(w) + l(\epsilon/2) + \sum_{(i,j)\in P} c_{ij}. \tag{10.3}$$

Applying the ϵ-optimality conditions to the arcs on the path $\overline{P}$ in $G(x')$, we obtain $\sum_{(j,i)\in\overline{P}} c_{ji}^{\pi'} \geq -l\epsilon$. Substituting $c_{ji}^{\pi'} = c_{ji} - \pi'(j) + \pi'(i)$ in this expression gives

$$\sum_{(j,i)\in\overline{P}} c_{ji} - \pi'(w) + \pi'(v) \geq -l\epsilon. \tag{10.4}$$

Notice that $\sum_{(j,i)\in\overline{P}} c_{ji} = -\sum_{(i,j)\in P} c_{ij}$ since $\overline{P}$ is a reversal of P. In view of this fact, we can restate (10.4) as

$$\sum_{(i,j)\in P} c_{ij} \leq l\epsilon - \pi'(w) + \pi'(v). \tag{10.5}$$

Substituting (10.5) in (10.3), we see that

$$(\pi(v) - \pi'(v)) \leq (\pi(w) - \pi'(w)) + 3l\epsilon/2. \tag{10.6}$$

Now we use the facts that (1) $\pi(w) = \pi'(w)$ (the potential of a node with negative imbalance does not change because the algorithm never selects it for push/relabel), (2) $l \leq n$, and (3) each increase in the potential increases $\pi(v)$ by at least $\epsilon/2$ units. These facts and expression (10.6) establish the lemma. ◆

Lemma 10.5. *The improve-approximation procedure performs $O(nm)$ saturating pushes.*

Proof. We show that between two consecutive saturations of an arc (i, j), the procedure must increase both the potentials $\pi(i)$ and $\pi(j)$ at least once. Consider a saturating push on arc (i, j). Since arc (i, j) is admissible at the time of the push, $c_{ij}^{\pi} < 0$. Before the algorithm can saturate this arc again, it must send some flow back from node j to node i. At that time $c_{ji}^{\pi} < 0$ or $c_{ij}^{\pi} > 0$. These conditions are possible only if the algorithm has relabeled node j. In the subsequent saturation of arc (i, j), $c_{ij}^{\pi} < 0$, which is possible only if the algorithm has relabeled node i. But by the previous lemma the improve-approximation procedure can relabel any node $O(n)$ times, so it can saturate any arc $O(n)$ times. Consequently, the number of saturating pushes is $O(nm)$. ◆

To bound the number of nonsaturating pushes, we need one more result. We define the *admissible network* of a given residual network as the network consisting solely of admissible arcs. For example, Figure 10.6(b) specifies the admissible network for the residual network given in Figure 10.6(a).

Lemma 10.6. *The admissible network is acyclic throughout the improve-approximation procedure.*

Proof. We show that the algorithm satisfies this property at every step. The result is true at the beginning of the improve-approximation procedure because the initial pseudoflow is 0-optimal and the residual network contains no admissible arc. We show that the result remains valid throughout the procedure. We always push flow on arc (i, j) with $c_{ij}^{\pi} < 0$; therefore, if the algorithm adds the reversal (j, i) of this arc to the residual network, then $c_{ji}^{\pi} > 0$ and so the reversal arc is nonadmissible. Thus pushes do not create new admissible arcs and the admissible network remains acyclic. The relabel operation at node i decreases the reduced costs of all outgoing

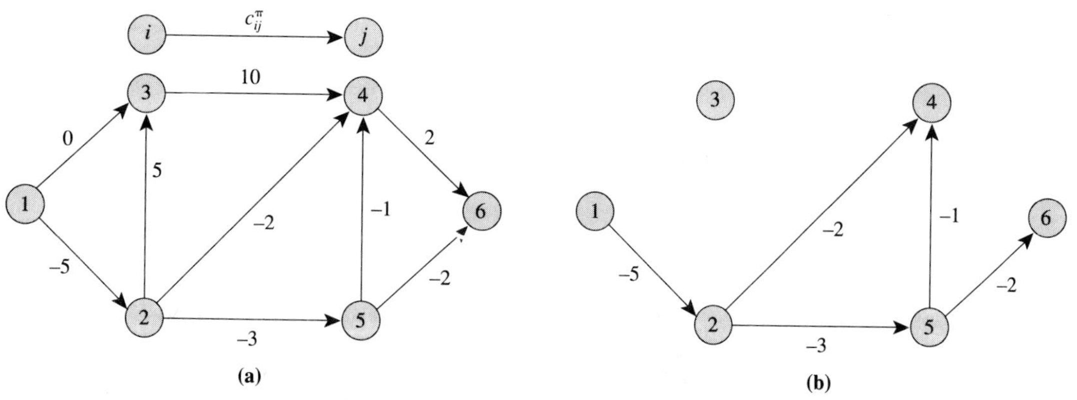

Figure 10.6 Illustration of an admissible network: (a) residual network; (b) admissible network.

arcs at node i by $\epsilon/2$ units and might create new admissible arcs. This relabel operation increases the reduced costs of all incoming arcs at node i by $\epsilon/2$ units, so all such arcs become inadmissible. Consequently, the relabel operation cannot create any directed cycle passing through node i. Thus neither of the two operations, pushes and relabels, of the algorithm can create a directed cycle, which establishes the lemma. ◆

Lemma 10.7. *The improve-approximation procedure performs $O(n^2m)$ nonsaturating pushes.*

Proof. We use a potential function argument to prove the lemma. Let $g(i)$ be the number of nodes that are reachable from node i in the admissible network and let $\Phi = \sum_{i \text{ is active}} g(i)$ be the potential function. We assume that every node is reachable from itself. For example, in the admissible network shown in Figure 10.7, nodes 1 and 4 are the only active nodes. In this network, nodes 1, 2, 3, 4, and 5 are reachable from node 1, and nodes 4 and 5 are reachable from node 4. Therefore, $g(1) = 5$, $g(4) = 2$, and $\Phi = 7$.

At the beginning of the procedure, $\Phi \leq n$ since the admissible network contains

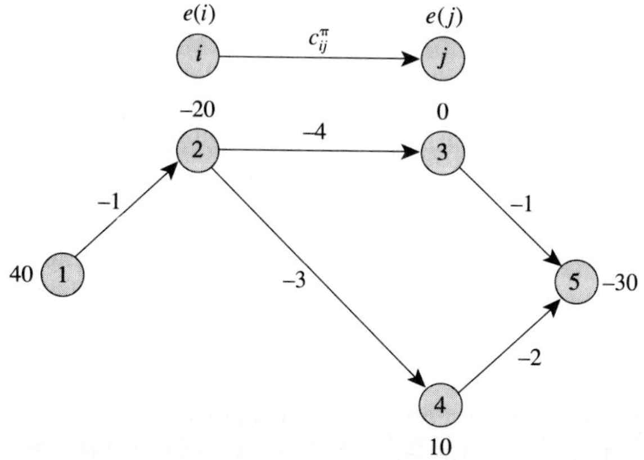

Figure 10.7 Admissible network for $\epsilon = 8$.

no arc and each $g(i) = 1$. After a saturating push on arc (i, j), node j might change its state from inactive to active, which would increase Φ by $g(j) \leq n$. Therefore, Lemma 10.5 implies that the total increase due to saturating pushes is $O(n^2m)$. A relabel operation of node i might create new admissible arcs (i, j) and will increase $g(i)$ by at most n units. But this relabel operation does not increase $g(k)$ for any other node k because it makes all incoming arcs at node k inadmissible (see the proof of Lemma 10.6). Thus the total increase due to all relabel operations is $O(n^3)$.

Finally, consider the effect on Φ of a nonsaturating push on arc (i, j). As a result of the push, node i becomes inactive and node j might change its status from inactive to active. Thus the push decreases Φ by $g(i)$ units and might increase it by another $g(j)$ units. Now notice that $g(i) \geq g(j) + 1$ because every node that is reachable from node j is also reachable from node i but node i is not reachable from node j (because the admissible network is acyclic). Therefore, a nonsaturating push decreases Φ by at least 1 unit. Consequently, the total number of nonsaturating pushes is bounded by the initial value of Φ plus the total increase in Φ throughout the algorithm, which is $O(n) + O(n^2m) + O(n^3) = O(n^2m)$. This result establishes the lemma. ◆

Let us summarize our discussion. The improve-approximation procedure requires $O(n^2m)$ time to perform nonsaturating pushes and $O(nm)$ time to perform saturating pushes. The amount of time needed to identify admissible arcs is $O(\sum_{i \in N} |A(i)| n) = O(nm)$, since between two consecutive potential increases of a node i, the algorithm will examine $|A(i)|$ arcs for testing admissibility. The algorithm could store all the active nodes in a list. Doing so would permit it to identify an active node in $O(1)$ time, so this operation would not be a bottleneck step. Consequently, the improve-approximation procedure runs in $O(n^2m)$ time. Since the cost scaling algorithm calls this procedure $1 + \lceil \log(nC) \rceil$ times, we obtain the following result.

Theorem 10.8. *The generic cost scaling algorithm runs in $O(n^2m \log(nC))$ time.* ◆

The cost scaling algorithm illustrates an important connection between the maximum flow and the minimum cost flow problems. Solving an improve-approximation problem is very similar to solving a maximum flow problem by the preflow-push method. Just as in the preflow-push algorithm, the bottleneck operation in the procedure is the number of nonsaturating pushes. In Chapter 7 we have seen how to reduce the number of nonsaturating pushes for the preflow-push algorithm by examining active nodes in some specific order. Similar ideas permit us to streamline the improve-approximation procedure as well. We describe one such improvement, called the *wave implementation*, that reduces the number of nonsaturating pushes from $O(n^2m)$ to $O(n^3)$.

Wave Implementation

Before we describe the wave implementation, we introduce the concept of *node examination*. In an iteration of the improve-approximation procedure, the algorithm selects a node, say node i, and either performs a saturating push or a nonsaturating push from this node, or relabels the node. If the algorithm performs a saturating push, node i might still be active, but the algorithm might select another node in the next iteration. We shall henceforth assume that whenever the algorithm selects a node, it keeps pushing flow from that node until either its excess becomes zero or the node becomes relabeled. If we adopt this node selection strategy, the algorithm will perform several saturating pushes from a particular node followed either by a nonsaturating push or a relabel operation; we refer to this sequence of operations as a *node examination*.

The wave implementation is a special implementation of the improve-approximation procedure that selects active nodes for push/relabel steps in a specific order. The algorithm uses the fact that the admissible network is acyclic. In Section 3.4 we showed that it is always possible to order nodes of an acyclic network so that for every arc (i, j) in the network, node i occurs prior to node j. Such an ordering of nodes is called a *topological ordering*. For example, for the admissible network shown in Figure 10.6, one possible topological ordering of nodes is 1–2–5–4–3–6. In Section 3.4 we showed how to arrange the nodes of a network in a topological order in $O(m)$ time. For a given topological order, we define the *rank* of a node as n minus its number in the topological sequence. For example, in the preceding example, rank(1) = 6, rank(6) = 1 and rank(5) = 4.

Observe that each push carries flow from a node with higher rank to a node with lower rank. Also observe that pushes do not change the topological ordering of nodes since they do not create new admissible arcs. The relabel operations, however, might create new admissible arcs and consequently, might affect the topological ordering of nodes.

The wave implementation sequentially examines nodes in the topological order and if the node being examined is active, it performs push/relabel steps at the node until either the node becomes inactive or it becomes relabeled. When examined in this order, the active nodes push their excesses to nodes with lower rank, which in turn push their excesses to nodes with even lower rank, and so on. A relabel operation changes the topological order; so after each relabel operation the algorithm modifies the topological order and again starts to examine nodes according to the topological order. If within n consecutive node examinations, the algorithm performs no relabel operation, then at this point all the active nodes have discharged their excesses and the algorithm has obtained a flow. Since the algorithm performs $O(n^2)$ relabel operations, we immediately obtain a bound of $O(n^3)$ on the number of node examinations. Each node examination entails at most one nonsaturating push. Consequently, the wave algorithm performs $O(n^3)$ nonsaturating pushes per execution of improve-approximation.

To illustrate the wave implementation, we consider the pseudoflow shown in Figure 10.8. One topological order of nodes is 2–3–4–1–5–6. The algorithm first examines node 2 and pushes 20 units of flow on arc (2, 1). Then it examines node 3 and pushes 5 units of flow on arc (3, 1) and 10 units of flow on arc (3, 4). The push creates an excess of 10 units at node 4. Next the algorithm examines node 4 and sends 5 units on the arc (4, 6). Since node 4 has an excess of 5 units but has no outgoing admissible arc, we need to relabel node 4 and reexamine all nodes in the topological order starting with the first node in the order.

To complete the description of the algorithm, we need to describe a procedure

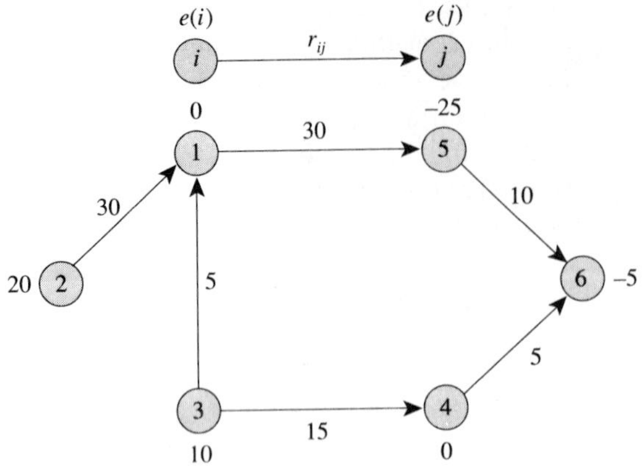

Figure 10.8 Example to illustrate the wave implementation.

for obtaining a topological order of nodes after each relabel operation. We can use an $O(m)$ algorithm to determine an initial topological ordering of the nodes (see Section 3.4). Suppose that while examining node i, the algorithm relabels this node. At this point, the network contains no incoming admissible arc at node i. We claim that if we move node i from its present position to the first position in the previous topological order leaving all other nodes intact, we obtain a topological order of the new admissible network. For example, for the admissible network given in Figure 10.8, one topological order of the nodes is 2–3–4–1–5–6. If we examine nodes in this order, the algorithm relabels node 4. After the algorithm has performed this relabel operation, the modified topological order of nodes is 4–2–3–1–5–6. This method works because (1) after the relabeling, node i has no incoming admissible arc, so assigning it to the first place in the topological order is justified; (2) the relabeling of node i might create some new outgoing admissible arcs (i, j) but since node i is first in the topological order, any such arc satisfies the conditions of a topological ordering; and (3) the rest of the admissible network does not change, so the previous order remains valid. Therefore, the algorithm maintains an ordered set of nodes (possibly as a doubly linked list) and examines nodes in this order. Whenever it relabels a node i, the algorithm moves this node to the first place in the order and again examines nodes in order starting from node i.

We have established the following result.

Theorem 10.9. *The wave implementation of the cost scaling algorithm solves the minimum cost flow problem in* $O(n^3 \log(nC))$ *time.* ◆

By examining the active nodes carefully and thereby reducing the number of nonsaturating pushes, the wave implementation improves the running time of the generic implementation of the improve-approximation procedure from $O(n^2m)$ to $O(n^3)$. A complementary approach for improving the running time is to use cleverer data structure to reduce the time per nonsaturating push. Using the *dynamic tree* data structures described in Section 8.5, we can improve the running time of the generic implementation to $O(nm \log n)$ and of the wave implementation to $O(nm \log(n^2/m))$. The references cited at the end of the chapter contain the details of these implementations.

10.4 DOUBLE SCALING ALGORITHM

As we have seen in the preceding two sections, by scaling either the arc capacities or the cost coefficients of a minimum cost flow problem, we can devise algorithms with improved worst-case performance. This development raises a natural question: Can we combine ideas from these algorithms to obtain even further improvements that are not obtained by either technique alone? In this section we provide an affirmative answer to this question. The double scaling algorithm we describe solves the capacitated minimum cost flow problem in $O(nm \log U \log(nC))$ time. When implemented using a dynamic tree data structure, this approach produces one of the best polynomial time algorithms for solving the minimum cost flow problem.

In this discussion we assume that the reader is familiar with the capacity scaling algorithm and the cost scaling algorithm that we examined in the preceding two sections. To solve the capacitated minimum cost flow problem, we first transform it into an uncapacitated transportation problem using the transformation described in Section 2.4. We assume that every arc in the minimum cost flow problem is capacitated. Consequently, the transformed network will be a bipartite network $G = (N_1 \cup N_2, A)$ with N_1 and N_2 as the sets of supply and demand nodes. Moreover, $|N_1| = n$ and $|N_2| = m$.

The double scaling algorithm is the same as the cost scaling algorithm described in the preceding section except that it uses a more efficient version of the improve-approximation procedure. The improve-approximation procedure in the preceding section relied on a "pseudoflow-push" method to push flow out of active nodes. A natural alternative would be to try an augmenting path based method. This approach would send flow from a node with excess to a node with deficit over an *admissible path* (i.e., a path in which each arc is admissible). A straightforward implementation of this approach would require $O(nm)$ augmentations since each augmentation would saturate at least one arc and, by Lemma 10.5, the algorithm requires $O(nm)$ arc saturations. Since each augmentation requires $O(n)$ time, this approach does not appear to improve the $O(n^2m)$ bound of the generic improve-approximation procedure.

We can, however, use ideas from the capacity scaling algorithm to reduce the number of augmentations to $O(m \log U)$ by ensuring that each augmentation carries *sufficiently large* flow. The resulting algorithm performs cost scaling in an "outer loop" to obtain ϵ-optimal flows for successively smaller values of ϵ. Within each cost scaling phase, we start with a pseudoflow and perform a number of capacity scaling phases, called Δ-scaling phases, for successively smaller values of Δ. In the Δ-scaling phase, the algorithm identifies admissible paths from a node with an excess of at least Δ to a node with a deficit and augments Δ units of flow over these paths. When all node excesses are less than Δ, we reduce Δ by a factor of 2 and initiate a new Δ-scaling phase. At the end of the 1-scaling phase, we obtain a flow.

The algorithmic description of the double scaling algorithm is same as that of the cost scaling algorithm except that we replace the improve-approximation procedure by the procedure given in Figure 10.9.

The capacity scaling within the *improve-approximation* procedure is somewhat different from the capacity scaling algorithm described in Section 10.2. The new algorithm differs from the one we considered previously in the following respects:

```
procedure improve-approximation(ε, x, π);
begin
    set x : = 0 and compute node imbalances;
    π(j) : = π(j) + ε, for all j ∈ N₂;
    Δ : = 2^⌊log U⌋;
    while the network contains an active node do
    begin
        S(Δ) : = {i ∈ N₁ ∪ N₂ : e(i) ≥ Δ};
        while S(Δ) ≠ Ø do
        begin {Δ-scaling phase}
            select a node k from S(Δ);
            determine an admissible path P from node k to some node l with e(l) < 0;
            augment Δ units of flow on path P and update x and S(Δ);
        end;
        Δ : = Δ/2;
    end;
end;
```

Figure 10.9 Improve-approximation procedure in the double scaling algorithm.

(1) the augmentation terminates at a node l with $e(l) < 0$ but whose deficit may not be as large as Δ; (2) each residual capacity is an integral multiple of Δ because each arc flow is an integral multiple of Δ and each arc capacity is ∞; and (3) the algorithm does not change flow on some arcs at the beginning of the Δ-scaling phase to ensure that the solution satisfies the optimality conditions. We point out that the algorithm feature (3) is a consequence of feature (2) because each r_{ij} is a multiple of Δ, so $G(x, \Delta) \equiv G(x)$.

The double scaling algorithm improves on the capacity scaling algorithm by identifying an admissible path in only $O(n)$ time, on average, rather than the time $O(S(n, m, nC))$ required to identify an augmentation path in the capacity scaling algorithm. The savings in identifying augmenting paths more than offsets the extra requirement of performing $O(\log(nC))$ cost scaling phases in the double scaling algorithm.

We next describe a method for identifying admissible paths efficiently. The algorithm identifies an admissible path by starting at node k and gradually building up the path. It maintains a *partial admissible path P*, which is initially null, and keeps enlarging it until it includes a node with deficit. We maintain the partial admissible path P using predecessor indices [i.e., if $(u, v) \in P$ then $\text{pred}(v) = u$]. At any point in the algorithm, we perform one of the following two steps, whichever is applicable, from the tip of P (say, node i):

advance(i). If the residual network contains an admissible arc (i, j), add (i, j) to P and set $\text{pred}(j) := i$. If $e(j) < 0$, stop.

retreat(i). If the residual network does not contain an admissible arc (i, j), update $\pi(i)$ to $\pi(i) + \epsilon/2$. If $i \neq k$, remove the arc $(\text{pred}(i), i)$ from P so that $\text{pred}(i)$ becomes its new tip.

The retreat step relabels (increases the potential of) node i for the purpose of creating new admissible arcs emanating from this node. However, increasing the potential of node i increases the reduced costs of all the incoming arcs at the node

i by $\epsilon/2$. Consequently, the arc (pred(i), i) becomes inadmissible, so we delete this arc from P (provided that P is nonempty).

We illustrate the method for identifying admissible paths on the example shown in Figure 10.10. Let $\epsilon = 4$ and $\Delta = 4$. Since node 1 is the only node with an excess of at least 4, we begin to develop the admissible path starting from this node. We perform the step advance(1) and add the arc (1, 2) to P. Next, we perform the step advance(2) and add the arc (2, 4) to P. Now node 4 has no admissible arc. So we perform a retreat step. We increase the potential of node 4 by $\epsilon/2 = 2$ units, thus changing the reduced cost of arc (2, 4) to 1; so we eliminate this arc from P. In the next two steps, the algorithm performs the steps advance(2) and advance(5), adding arcs (2, 5) and (5, 6) to P. Since the path now contains node 6, which is a node with a deficit, the method terminates. It has found the admissible path 1–2–5–6.

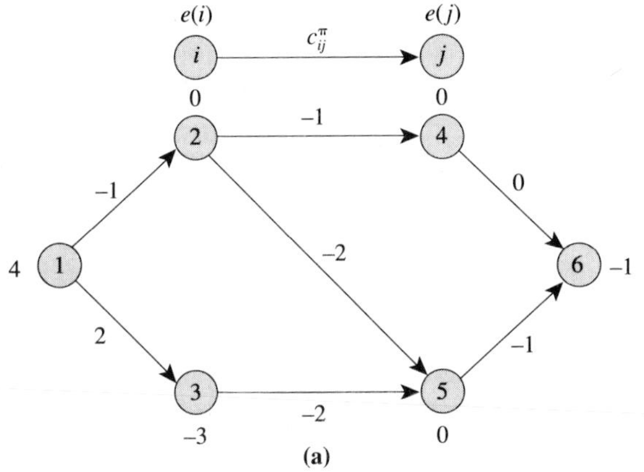

Figure 10.10 Residual network.

It is easy to show that the double scaling algorithm correctly solves the minimum cost flow problem. At the beginning of the improve-approximation procedure, we set $x = 0$ and the corresponding residual network is the same as the original network. The ϵ-optimality of the solution at the end of the previous scaling phase implies that $c_{ij}^{\pi} \geq -\epsilon$ for all arcs $(i, j) \in A$. Therefore, by adding ϵ to $\pi(j)$ for each $j \in N_2$, we obtain an $\frac{1}{2}\,\epsilon$-optimal pseudoflow (in fact, it is a 0-optimal pseudoflow). Like the improve-approximation procedure described in the preceding section, the algorithm always augments flow on admissible arcs and relabels a node when it has no outgoing admissible arc. Consequently, the algorithm preserves $\frac{1}{2}\,\epsilon$-optimality of the pseudoflow and at termination yields a $\frac{1}{2}\,\epsilon$-optimal flow.

We next consider the complexity of the improve-approximation procedure. Each execution of the procedure performs $(1 + \lfloor \log U \rfloor)$ capacity scaling phases. At the end of the 2Δ-scaling phase, $S(2\Delta) = \phi$. Therefore, at the beginning of the Δ-scaling phase, $\Delta \leq e(i) < 2\Delta$ for each node $i \in S(\Delta)$. During the Δ-scaling phase, the algorithm augments Δ units of flow from a node k in $S(\Delta)$ to a node l with $e(l) < 0$. The augmentation reduces the excess of node k to a value less than Δ and ensures that the imbalance at node l is strictly less than Δ. Consequently, each augmentation deletes a node from $S(\Delta)$ and after at most $|N_1| + |N_2| = O(m)$ augmentations, $S(\Delta)$ becomes empty and the algorithm begins a new capacity scaling phase. The algorithm thus performs a total of $O(m \log U)$ augmentations.

We next focus on the time needed to identify admissible paths. We first count the number of advance steps. Each advance step adds an arc to the partial admissible path, and each retreat step deletes an arc from the partial admissible path. Thus we can distinguish between two types of advance steps: (1) those that add arcs to an admissible path on which the algorithm later performs an augmentation, and (2) those that are later canceled by a retreat step. Since the set of admissible arcs is acyclic (by Lemma 10.6), after at most $2n$ advance steps of the first type, the algorithm will discover an admissible path and will perform an augmentation (because the longest path in the network has $2n$ nodes). Since the algorithm performs a total of $O(m \log U)$ augmentations, the number of advance steps of the first type is at most $O(nm \log U)$. The algorithm performs $O(nm)$ advance steps of the second type because each retreat step increases a node potential, and by Lemma 10.4, node potentials increase $O(n(n + m)) = O(nm)$ times. Therefore, the total number of advance steps is $O(nm \log U)$.

The amount of time needed to relabel nodes in N_1 is $O(n \sum_{i \in N} | A(i) |) = O(nm)$. The time needed to relabel nodes in N_2 is also $O(nm)$ since $| N_2 | = m$ and the degree of each node in N_2 is constant (i.e., it is 2). The same arguments show that the algorithm requires $O(nm)$ time to identify admissible arcs. We have, therefore, established the following result.

Theorem 10.10. *The double scaling algorithm solves the minimum cost flow problem in $O(nm \log U \log(nC))$ time.* ◆

One nice feature of the double scaling algorithm is that it achieves an excellent worst-case running time for solving the minimum cost flow problem and yet is fairly simple, both conceptually and computationally.

10.5 MINIMUM MEAN CYCLE-CANCELING ALGORITHM

The three minimum cost flow algorithms we have discussed in this chapter—the capacity scaling algorithm, the cost scaling algorithm, and the double scaling algorithm—are weakly polynomial-time algorithms because their running times depend on $\log U$ and/or $\log C$. Although these algorithms are capable of solving any problem with integer or rational data, they are not applicable to problems with irrational data. In contrast, the running times of strongly polynomial-time algorithms depend only on n and m; consequently, these algorithms are capable of solving problems with irrational data, assuming that a computer can perform additions and subtractions on irrational data. In this and the next two sections, we discuss several strongly polynomial time algorithms for solving any class of minimum cost flow problems, including those with irrational data.

The algorithm discussed in this section is a special case of the cycle-canceling algorithm that we discussed in Section 9.6. Because this algorithm iteratively cancels cycles (i.e., augments flows along cycles) with the minimum mean cost in the residential network, it is known as the (*minimum*) *mean cycle-canceling algorithm*. Recall from Section 5.7 that the *mean cost* of a directed cycle W is $(\sum_{(i,j) \in W} c_{ij})/ | W |$, and that the *minimum mean cycle* is a cycle with the smallest mean cost in the network. In Section 5.7 we showed how to use dynamic programming algorithm to find the minimum mean cycle in $O(nm)$ time.

The minimum mean cycle-canceling algorithm starts with a feasible flow x in the network. At every iteration, the algorithm identifies a minimum mean cycle W in $G(x)$. If the mean cost of the cycle W is negative, the algorithm augments the maximum possible flow along W, updates $G(x)$, and repeats this process. If the mean cost of W is nonnegative, $G(x)$ contains no negative cycle and x is a minimum cost flow, so the algorithm terminates. This algorithm is surprisingly simple to state; even more surprisingly, the algorithm runs in strongly polynomial time.

To establish the worst-case complexity of the minimum mean cycle-canceling algorithm, we recall a few facts. In our subsequent discussion, we often use Property 9.2(b), which states that for any set of node potentials π and any directed cycle W, the sum of the costs of the arcs in W equals the sum of the reduced costs of the arcs in W. We will also use the following property concerning sequences of real numbers, which is a variant of the geometric improvement argument (see Section 3.3).

Property 10.11. *Let α be a positive integer and let $y_1, y_2, y_3, \ldots$ be a sequence of real numbers satisfying the condition $y_{k+1} \le (1 - 1/\alpha)y_k$ for every k. Then for every value of k, $y_{k+\alpha} \le y_k/2$.*

Proof. We first rewrite the expression $y_{k+1} \le (1 - 1/\alpha)y_k$ as $y_k \ge y_{k+1} + y_{k+1}/(\alpha - 1)$. We now use this last expression repeatedly to replace the first term on the right-hand side, giving

$$
\begin{aligned}
y_k &\ge y_{k+1} + y_{k+1}/(\alpha - 1) \ge y_{k+2} + y_{k+2}/(\alpha + 1) + y_{k+1}/(\alpha - 1) \\
&\ge y_{k+2} + 2y_{k+2}/(\alpha - 1) \ge y_{k+3} + 3y_{k+3}/(\alpha - 1) \\
&\;\vdots \\
&\ge y_{k+\alpha} + \alpha y_{k+\alpha}/(\alpha - 1) \ge 2y_{k+\alpha},
\end{aligned}
$$

which is the assertion of the property. ◆

We divide the worst-case analysis of the minimum mean cycle algorithm into two parts: First, we show that the algorithm is weakly polynomial-time; then we establish its strong polynomiality. Although the description of the algorithm does not use scaling techniques, the worst-cast analysis borrows ideas from the cost scaling algorithm that we discussed in Section 10.3. In particular, the notion of ϵ-optimality discussed in that section plays a crucial role in its analysis. We will show that the flows maintained by the minimum mean cycle-canceling algorithm are ϵ-optimal flows satisfying the conditions that (1) between any two consecutive iterations the value of ϵ either stays the same or decreases; (2) occasionally, the value of ϵ strictly decreases; and (3) eventually, $\epsilon < 1/n$ and the algorithm terminates (see Lemma 10.2). As we observed in Section 10.3, the cost scaling algorithm's explicit strategy is to reduce ϵ from iteration to iteration. Although the minimum mean cycle-canceling algorithm also reduces the value of ϵ (although periodically, rather than at every iteration), the reduction is very much an implicit by-product of the algorithm.

We first establish a connection between the ϵ-optimality of a flow x and the mean cost of a minimum mean cycle in $G(x)$. Recall that a flow x is ϵ-optimal if for some set of node potentials, the reduced cost of every arc is at least $-\epsilon$. Notice that any flow x will be ϵ-optimal for many values of ϵ, because a flow that is ϵ-optimal is also ϵ'-optimal for all $\epsilon' \geq \epsilon$. For any particular set of node potentials π, we let $\epsilon^\pi(x)$ be the negative of the minimum value of any reduced cost [i.e., $\epsilon^\pi(x) = -\min[c_{ij}^\pi : (i, j) \text{ in } G(x)]$. Thus $c_{ij}^\pi \geq -\epsilon^\pi(x)$ and $c_{ij}^\pi = -\epsilon^\pi(x)$ for some arc (i, j). Thus x is ϵ-optimal for $\epsilon = \epsilon^\pi(x)$. Potentially, we could find a smaller value of ϵ by using other values of the node potentials. With this thought in mind, we let $\epsilon(x) = \min_\pi \epsilon^\pi(x)$. Note that $\epsilon(x)$ is the smallest value of ϵ for which the flow x is ϵ-optimal. As additional notation, we let $\mu(x)$ denote the mean cost of the minimum mean cycle in $G(x)$.

Note that since x is $\epsilon(x)$-optimal, conditions (10.2) imply that $\sum_{(i,j)\in W} c_{ij} = \sum_{(i,j)\in W} c_{ij}^\pi \geq -\epsilon^\pi(x) \mid W \mid$. Choosing W as the minimum mean cycle and dividing this expression by $\mid W \mid$, we see that $\mu(x) \geq -\epsilon(x)$. As we have seen, this inequality is a simple consequence of the definitions of ϵ-optimality and of the minimum mean cycle cost; it uses the fact that if we can bound the reduced cost of every arc around a cycle, this same bound applies to the average cost around the cycle. Perhaps surprisingly, however, we can obtain a converse result: that is, we can always find a set of node potentials so that every arc around the minimum mean cycle has the same reduced cost and that this cost equals $-\epsilon(x)$. Our next two results establish this property.

Lemma 10.12. *Let x be a nonoptimal flow. Then $\epsilon(x) = -\mu(x)$.*

Proof. Since our observation in the preceding paragraph shows that $\epsilon(x) \geq -\mu(x)$, we only need to show that $\epsilon(x) \leq -\mu(x)$.

Let W be a minimum mean cycle in the residual network $G(x)$, and let $\mu(x)$ be the mean cost of this cycle. Suppose that we replace each arc cost c_{ij} by $c'_{ij} = c_{ij} - \mu(x)$. This transformation reduces the mean cost of every directed cycle in $G(x)$ by $\mu(x)$ units. Consequently, the minimum mean cost of the cycle W becomes zero, which implies that the residual network contains no negative cost cycle. Let $d'(\cdot)$ denote the shortest path distances in $G(x)$ from a specified node s to all other nodes with c'_{ij} as the arc lengths. The shortest path optimality conditions imply that

$$d'(j) \leq d'(i) + c'_{ij} = d'(i) + c_{ij} - \mu(x) \qquad \text{for each arc } (i, j) \text{ in } G(x). \tag{10.7}$$

If we let $\pi(j) = d'(j)$, then (10.7) becomes

$$c_{ij}^\pi \geq \mu(x) \qquad \text{for each arc } (i, j) \text{ in } G(x), \tag{10.8}$$

which implies that x is $(-\mu(x))$-optimal. Therefore, $\epsilon(x) \leq -\mu(x)$, completing the proof of the lemma. ◆

Lemma 10.13. *Let x be any nonoptimal flow. Then for some set of node potentials π, $c_{ij}^\pi = \mu(x) = -\epsilon(x)$ for every arc (i, j) in the minimum mean cycle W of $G(x)$.*

Proof. Let π be defined as in the proof of the preceding lemma; with these set of node potentials, the reduced costs satisfy (10.8). The cost of the cycle W equals $\sum_{(i,j)\in W} c_{ij}$, which also equals its reduced cost $\sum_{(i,j)\in W} c_{ij}^\pi$. Consequently, $\sum_{(i,j)\in W} c_{ij}^{\pi} = \mu(x) \mid W \mid$. This equation and (10.8) imply that $c_{ij}^{\pi} = \mu(x)$ for each arc (i, j) in W. Lemma 10.12 establishes that $c_{ij}^{\pi} = -\epsilon(x)$ for every arc in W. ◆

We next show that during the execution of the minimum mean cycle-canceling algorithm, $\epsilon(x)$ never increases; moreover, within m consecutive iterations $\epsilon(x)$ decreases by a factor of at least $(1 - 1/n)$.

Lemma 10.14. *For a nonoptimal flow x, if we cancel a minimum mean cycle in $G(x)$, $\epsilon(x)$ cannot increase [alternatively, $\mu(x)$ cannot decrease].*

Proof. Let W denote the minimum mean cycle in $G(x)$. Lemma 10.13 implies that for some set of node potentials π, $c_{ij}^{\pi} = -\epsilon(x)$ for each arc $(i, j) \in W$. Let x' denote the flow obtained after we have canceled the cycle W. This flow augmentation deletes some arcs in W from the residual network and adds some other arcs, which are reversals of the arcs in W. Consider any arc (i, j) in $G(x')$. If (i, j) is in $G(x)$, then, by hypothesis, $c_{ij}^{\pi} \geq -\epsilon(x)$. If (i, j) is not in $G(x)$, then (i, j) is a reversal of some arc (j, i) in $G(x)$ for which $c_{ji}^{\pi} = -\epsilon(x)$. Therefore, $c_{ij}^{\pi} = -c_{ji}^{\pi} = \epsilon(x) > 0$. In either case, $c_{ij}^{\pi} \geq -\epsilon(x)$ for each arc (i, j) in $G(x')$. Consequently, the minimum mean cost of any cycle in $G(x')$ will be at least $-\epsilon(x)$, since the mean cost around a cycle, which equals the mean reduced cost, must be at least as large as the minimum value of the reduced costs. Therefore, in light of Lemma 10.12, as asserted, $\epsilon(x') = \mu(x') \geq -\epsilon(x) = \mu(x)$. ◆

Lemma 10.15. *After a sequence of m minimum mean cycle cancelations starting with a flow x, the value of the optimality parameter $\epsilon(x)$ deceases to a value at most $(1 - 1/n)\,\epsilon(x)$ [i.e., to at most $(1 - 1/n)$ times its original value].*

Proof. Let π denote a set of node potentials satisfying the conditions $c_{ij}^{\pi} \geq -\epsilon(x)$ for each arc (i, j) in $G(x)$. For convenience, we designate those arcs in $G(x)$ with (strictly) negative reduced costs as *negative arcs* (with respect to the reduced costs). We now classify the subsequent cycle cancelations into two types: (1) all the arcs in the canceled cycle are negative (a type 1 cancelation), and (2) at least one arc in the canceled cycle has a nonnegative reduced cost (a type 2 cancelation). We claim that the algorithm will perform at most m type 1 cancelations before it either terminates or performs a type 2 cancelation. This claim follows from the observations that each type 1 cancelation deletes at least one negative arc from the (current) residual network and all the arcs that the cancelation adds to the residual network have positive reduced cost with respect to π (as shown in the proof of Lemma 10.14). Consequently, if within m iterations, the algorithm performs no type 2 cancelations, all the arcs in the residual network will have nonnegative reduced costs with respect to π and the algorithm will terminate with an optimal flow.

Now consider the first time the algorithm performs a type 2 cancelation. Suppose that the algorithm cancels the cycle W, which contains at least one arc with a nonnegative reduced cost; let x' and x'' denote the flows just before and after the cancelation. Then $c_{ij}^{\pi} \geq -\epsilon(x')$ for each arc $(i, j) \in W$ and $c_{kl}^{\pi} \geq 0$ for some arc $(k, l) \in W$. As a result, since $c(W) = \sum_{(i,j)\in W} c_{ij}^{\pi}$, the cost $c(W)$ of W with respect to the flow x' satisfies the condition $c(W) \geq [(\mid W \mid - 1)(-\epsilon(x'))]$. By Lemma 10.14, the cancelation cannot increase the minimum mean cost and, therefore, $\mu(x'') \geq \mu(x')$. But since $\mu(x')$ is the mean cost of W with respect to x', $\mu(x'') \geq \mu(x') \geq (1 - 1/|W|)(-\epsilon(x')) \geq (1 - 1/n)(-\epsilon(x'))$. This inequality implies that $-\mu(x'') \leq (1 - 1/n)\epsilon(x')$. Using the fact that $\mu(x'') = -\epsilon(x'')$, we see that $\epsilon(x'') \leq (1 - 1/n)\epsilon(x')$. This result establishes the lemma. ◆

As indicated by the next theorem, the preceding two lemmas imply that the minimum mean cycle-canceling algorithm performs a polynomial number of iterations.

Theorem 10.16. *If all arc costs are integer, the minimum mean cycle-canceling algorithm performs $O(nm \log(nC))$ iterations and runs in $O(n^2m^2 \log(nC))$ time.*

Proof. Let x denote the flow at any point during the execution of the algorithm. Initially, $\epsilon(x) \leq C$ because every flow is C-optimal (see Lemma 10.2). In every m consecutive iterations, the algorithm decreases $\epsilon(x)$ by a factor of $(1 - 1/n)$. When $\epsilon(x) < 1/n$, the algorithm terminates with an optimal flow (see Lemma 10.2). Therefore, the algorithm needs to decrease $\epsilon(x)$ by a factor of nC over all iterations. By Lemma 10.15, the mean cost of a cycle becomes smaller by a factor of at least $(1 - 1/n)$ in every m iterations. Property 10.11 implies that the minimum mean cycle cost decreases by a factor of 2 every nm iterations, so that within $nm \log(nC)$ iterations, the minimum mean cycle cost decreases from C to $1/n$. At this point the algorithm terminates with an optimal flow. This conclusion establishes the first part of the theorem. Since the bottleneck operation in each iteration is identifying a minimum mean cycle, which requires $O(nm)$ time (see Section 5.7), we also have established the second part of the theorem.

Having proved that the minimum mean cycle-canceling algorithm runs in polynomial time, we next obtain a strongly polynomial bound on the number of iterations the algorithm performs. Our analysis rests upon the following rather useful result: If the absolute value of the reduced cost of an arc (k, l) is "significantly greater than" the current value of the parameter $\epsilon(x)$, the flow on the arc (k, l) in any optimal solution is the same as the current flow on this arc. In other words, the flow on the arc (k, l) becomes "fixed." As we will show, in every $O(nm \log n)$ iterations, the algorithm will fix at least one additional arc at its lower bound or at its upper bound. As a result, within $O(nm^2 \log n)$ iterations, the algorithm will have fixed all the arcs and will terminate with an optimal flow.

We define an arc to be *ϵ-fixed* if the flow on this arc is the same for all ϵ'-optimal flows whenever $\epsilon' \leq \epsilon$. Since the value of $\epsilon(x)$ of the $\epsilon(x)$-optimal flows, that the minimum mean cycle-canceling algorithm maintains, is nonincreasing, the flow on an $\epsilon(x)$-fixed arc will not change during the execution of the algorithm and will be the same in every optimal flow. We next establish a condition that will permit us to fix an arc.

Lemma 10.17. *Suppose that x is an $\epsilon(x)$-optimal flow with respect to the potentials π, and suppose that for some arc $(k, l) \in A$, $|c^{\pi}_{kl}| \geq 2n\epsilon(x)$. Then arc (k, l) is an $\epsilon(x)$-fixed arc.*

Proof. Let $\epsilon = \epsilon(x)$. We first prove the lemma when $c_{kl}^{\pi} \geq 2n\epsilon$. The ϵ-optimality condition (10.1a) implies that $x_{kl} = 0$. Suppose that some $\epsilon(x')$-optimal flow x', with $\epsilon(x') \leq \epsilon(x)$, satisfies the condition that $x'_{kl} > 0$. The flow decomposition theorem (i.e., Theorem 3.5) implies that we can express x' as x plus the flow along at most m augmenting cycles in $G(x)$. Since $x_{kl} = 0$ and $x'_{kl} > 0$, one of these cycles, say W, must contain the arc (k, l) as a forward arc. Since each arc $(i, j) \in W$ is in the residual network $G(x)$, and so satisfies the condition $c_{ij}^{\pi} \geq -\epsilon$, the reduced cost (or, cost) of the cycle W is at least $c_{kl}^{\pi} - \epsilon(|W| - 1) \geq 2n\epsilon - \epsilon(n - 1) > n\epsilon$.

Now consider the cycle W^r obtained by reversing the arcs in W. The cycle W^r must be a directed cycle in the residual network $G(x')$ (see Exercise 10.6). The cost of the cycle W^r is the negative of the cost of the cycle W and so must be less than $-n\epsilon \leq -n\epsilon(x')$. Therefore, the mean cost of W^r is less than $-\epsilon(x')$. Lemma 10.12 implies that x' is not $\epsilon(x')$-optimal, which is a contradiction.

We next consider the case when $c_{kl}^{\pi} \leq -2n\epsilon$. In this case the ϵ-optimality condition (10.1c) implies that $x_{kl} = u_{kl}$. Using an analysis similar to the one used in the preceding case, we can show that no ϵ-optimal flow x' can satisfy the condition $x'_{kl} < u_{kl}$. ◆

We are now in a position to obtain a strongly polynomial bound on the number of iterations performed by the minimum mean cycle-canceling algorithm.

Theorem 10.18. *For arbitrary real-valued arc costs, the minimum mean cycle-canceling algorithm performs $O(nm^2 \log n)$ iterations and runs in $O(n^2m^3 \log n)$ time.*

Proof. Let $K = nm(\lceil \log n \rceil + 1)$. We divide the iterations performed by the algorithm into groups of K consecutive iterations. We claim that each group of iterations fixes the flow on an additional arc (k, l) (i.e., the iterations after those in the group do not change the value of x_{kl}). The theorem follows immediately from this claim, since the algorithm can fix at most m arcs, and each iteration requires $O(nm)$ time.

Consider any group of iterations. Let x be the flow before the first iteration of the group and let x' be the flow after the last iteration of the group. Let $\epsilon = \epsilon(x)$, $\epsilon' = \epsilon(x')$, and let π' be the node potentials for which x' satisfies the ϵ'-optimality conditions. Since every nm iterations reduce ϵ by a factor of at least 2, the $nm(\lceil \log n \rceil + 1)$ iterations between x and x' reduce ϵ by a factor of at least $2^{\lceil \log n \rceil + 1}$. Therefore, $\epsilon' \leq (\epsilon / 2^{\lceil \log n \rceil + 1}) \leq \epsilon/2n$. Alternatively, $-\epsilon \leq -2n\epsilon'$.

Let W be the cycle canceled when the flow has value x. Lemma 10.12 and the fact that the sum of the costs and reduced costs around every cycle are the same, imply that for any values of the node potentials, the average reduced cost around the cycle W equals $\mu(x) = -\epsilon$. Therefore, with respect to the potentials π', at least one arc (k, l) in W must have a reduced cost as small as $-\epsilon$, so $c_{kl}^{\pi'} = -\epsilon \leq -2n\epsilon'$ for some arc (k, l) in W. By Lemma 10.17, the flow on arc (k, l) will not change in any subsequent iteration. Next notice that in the first iteration in the group, the algorithm changed the value of x_{kl}. Thus each group fixes the flow on at least one additional arc, completing the proof of the theorem. ◆

We might conclude this section with a few observations. First, note that we need not formally compute the value of $\epsilon(x)$ at each iteration, nor do we need to identify the ϵ-fixed arcs at any stage in the algorithm. Indeed, we can use any method to find the minimum mean cost cycle at each step; in principle, we need not maintain or ever compute any reduced costs. As we noted earlier in this section, the minimum mean cycle-canceling algorithm implicitly reduces $\epsilon(x)$ and fixes some arcs as it proceeds—we need not keep track of the algorithm's progress concerning these features.

We also might note that the ideas presented in this section would also permit us to develop a strongly polynomial-time version of the cost scaling algorithm that we discussed in Section 10.3. In Exercise 10.12 we consider this modification of the cost scaling algorithm and analyze its running time.

10.6 REPEATED CAPACITY SCALING ALGORITHM

The minimum cost flow problem described in Section 10.5 uses the idea that whenever the reduced cost of an arc is *sufficiently large*, we can "fix" the flow on the arc. By incorporating a similar idea in the capacity scaling algorithm, we can develop another strongly polynomial time algorithm. As we will see, when the flow on an arc (i, j) is *sufficiently large*, the potentials of nodes i and j become "fixed" with respect to each other. In this section we discuss the details of this algorithm, which we call the *repeated capacity scaling algorithm*.

The repeated capacity scaling algorithm to be discussed in this section is different from all the other minimum cost flow algorithms discussed in this book. All of the other algorithms solve the primal minimum cost flow problem (9.1) and obtain an optimal flow; the repeated capacity scaling algorithm solves the dual minimum cost flow problem (9.10). This algorithm obtains an optimal set of node potentials for (9.10) and then uses it to determine an optimal flow.

The repeated capacity scaling algorithm is a modified version of the capacity scaling algorithm discussed in Section 10.2. For simplicity, we describe the algorithm for the uncapacitated minimum cost flow problem; we could solve the capacitated problem by converting it to the uncapacitated problem using the transformation described in Section 2.4. Recall that in the capacity scaling algorithm, each arc flow is an integral multiple of the scale factor Δ. For uncapacitated networks, each residual capacity r_{ij} is also an integral multiple of Δ, because either $r_{ij} = u_{ij} = \infty$, or $r_{ij} = x_{ji} = k\Delta$ for some integer k. This observation implies that the Δ-residual network $G(x, \Delta)$ is the same as the residual network $G(x)$. As a result, the algorithm for the uncapacitated problem does not require the preprocessing (i.e., saturating the arcs violating the optimality conditions) at the beginning of each scaling phase. The following property is an immediate consequence of this result.

Property 10.19. *The capacity scaling algorithm for the uncapacitated minimum cost flow problem satisfies the following properties: (a) the excesses at the nodes are monotonically decreasing; (b) the sum of the excesses at the beginning of the Δ-scaling phase is at most $2n\Delta$; and (c) the algorithm performs at most $2n$ augmentations per scaling phase.*

The repeated capacity scaling algorithm is based on the three simple results stated in the following lemmas.

Lemma 10.20. *Suppose that at the beginning of the Δ-scaling phase, $b(k) > 6n^2\Delta$ for some node $k \in N$. Then some arc (k, l) with $x_{kl} > 4n\Delta$ emanates from node k.*

Proof. Property 10.19 implies that at the beginning of the Δ-scaling phase, the sum of the excesses is at most $2n\Delta$. Therefore, $e(k) \le 2n\Delta$. Since $b(k) > 6n^2\Delta$ and $e(k) \le 2n\Delta$, the net outflow of node k [i.e., $b(k) - e(k)$] is strictly greater than $(6n^2\Delta - 2n\Delta)$. Since fewer than n arcs emanate from node k, the flow on at least one of these arcs must be strictly more than $(6n^2\Delta - 2n\Delta)/n \ge (4n^2\Delta)/n = 4n\Delta$, which concludes the lemma. ◆

Lemma 10.21. *If at the beginning of the Δ-scaling phase $x_{kl} > 4n\Delta$, then for some optimal solution $x_{kl} > 0$.*

Proof. Property 10.19 implies that the algorithm performs at most $2n$ augmentations in each scaling phase. The fact that the algorithm augments exactly Δ units of flow in every augmentation in the Δ-scaling phase implies that the total flow change due to all augmentations in the subsequent scaling phases is at most $2n(\Delta + \Delta/2 + \Delta/4 + \cdots + 1) < 4n\Delta$. Consequently, if $x_{kl} > 4n\Delta$ at the beginning of the Δ-scaling phase, then $x_{kl} > 0$ when the algorithm terminates. ◆

Lemma 10.22. *Suppose that $x_{kl} > 0$ in an optimal solution of the minimum cost flow problem. Then with respect to every set of optimal node potentials, the reduced cost of arc (k, l) is zero.*

Proof. Suppose that x satisfies the complementary slackness optimality condition (9.8) with respect to the node potential π. The condition (9.8b) implies that $c_{kl}^{\pi} = 0$. Property 9.8 implies that if x satisfies the complementary slackness optimality condition (9.8b) with respect to some node potential, it satisfies this condition with respect to every optimal node potential. Consequently, the reduced cost of arc (k, l) is zero with respect to every set of optimal node potentials. ◆

We are now in a position to discuss the essential ideas of the repeated capacity scaling algorithm. Let **P** denote the minimum cost flow problem stated in (9.1). The algorithm applies the capacity scaling algorithm stated in Figure 10.1 to the problem **P**. We will show that within $O(\log n)$ scaling phases, $b(k) > 6n^2\Delta$ for some node k and, by Lemma 10.20, some arc (k, l) satisfies the condition $x_{kl} > 4n\Delta$. Lemmas 10.21 and 10.22 imply that for any set of optimal node potentials, the reduced cost of arc (k, l) will be zero. This result allows us to show, as described next, that we can *contract* the nodes k and l into a single node, thereby obtaining a new minimum cost flow problem defined on a network with one fewer node.

Suppose that we are using the capacity scaling algorithm to solve a minimum cost flow problem **P** with arc costs c_{ij} and at some stage we realize that for an arc (k, l), $x_{kl} > 4n\Delta$. Let π denote the node potentials at this point. The optimality condition (9.8b) implies that

$$c_{kl} - \pi(k) + \pi(l) = 0. \tag{10.9}$$

Now consider the same minimum cost flow problem, but with the cost of each arc (i, j) equal to $c'_{ij} = c^{\pi}_{ij} = c_{ij} - \pi(i) + \pi(j)$. Let **P′** denote the modified minimum cost flow problem. Condition (10.9) implies that

$$c'_{kl} = 0. \tag{10.10}$$

We next observe that the problems **P** and **P′** have the same optimal solutions (see Property 2.4 in Section 2.4). Since $x_{kl} > 4n\Delta$, Lemmas 10.21 and 10.22 imply that in problem **P′** the reduced cost of arc (k, l) will be zero. If π' denotes an optimal set of node potentials for **P′**, then

$$c'_{kl} - \pi'(k) + \pi'(l) = 0. \tag{10.11}$$

Substituting (10.10) in (10.11) implies that $\pi'(k) = \pi'(l)$.

The preceding discussion shows that if $x_{kl} > 4n\Delta$ for some arc (k, l), we can "fix" one node potential with respect to the other. The discussion also shows that if we solve the problem **P′** with the additional constraint that the potentials of nodes k and l are same, this constraint will not eliminate the optimal solution of **P′**. But how can we solve a minimum cost flow problem when two node potentials must be the same?

Consider the dual minimum cost flow problem stated in (9.10). In this problem we replace both $\pi(k)$ and $\pi(l)$ by $\pi(p)$. This substitution gives us a linear programming problem with one less dual variable (or, node potential). The reader can easily verify that the resulting problem is a dual minimum cost flow problem on the network with nodes k and l contracted into a single node p. The contraction operation consists of (1) letting $b(p) = b(k) + b(l)$, (2) replacing each arc (i, k) or (i, l) by the arc (i, p), (3) replacing each arc (k, i) or (l, i) by the arc (p, i), and (4) letting the cost of an arc in the contracted network equal that of the arc it replaces. We point out that the contraction might produce multiarcs (i.e., more than one arc with the same tail and head nodes). The purpose of contraction operations should be clear; since each contraction operation reduces the size of the network by one node, we can apply at most n of these operations.

We can now describe the repeated capacity scaling algorithm. We first compute $U = \max\{b(i) : i \in N \text{ and } b(i) > 0\}$ and initialize $\Delta = 2^{\lfloor \log U \rfloor}$. Let node k be a node with $b(k) = U$. We then apply the capacity scaling algorithm as described in Figure 10.1. Each scaling phase of the capacity scaling algorithm decreases Δ by a factor of 2; therefore, since the initial value of Δ is $b(k)$, after at most $q = \log(6n^2) = O(\log n)$ phases, $\Delta = b(k)/2^q \le b(k)/6n^2$. The algorithm might obtain a feasible flow before $\Delta \le b(k)/6n^2$ (in which case it terminates); if not, then by Lemma 10.20, some arc (k, l) will satisfy the condition that $x_{kl} > 4n\Delta$. The algorithm then defines a new minimum cost flow problem with nodes k and l contracted into a new node p, and the cost of each arc is the reduced cost of the corresponding arc before the contraction. We solve the new minimum cost flow problem afresh by redefining U as the largest supply in the contracted network and reapplying the capacity scaling algorithm described in Figure 10.1. We repeat these steps until the algorithm terminates. The algorithm terminates in one of the two ways: (1) while applying the capacity scaling algorithm, it obtains a flow; or (2) it contracts the network into a single node p [with $b(p) = 0$], which is trivially solvable by a zero flow. At this point we expand the contracted nodes and obtain an optimal flow in the expanded network. We show how to expand the contracted nodes a little later. The preceding discussion shows that the algorithm performs $O(n \log n)$ scaling phases, and since each scaling phase solves at most $2n$ shortest path problems, the running time of the algorithm is $O(n^2 \log n\, S(n, m))$. In this expression, $S(n, m)$ is the minimum time required by a strongly polynomial-time algorithm for solving a shortest path problem with nonnegative arc lengths. [Recall from Chapter 4 that $O(m + n \log n)$ is currently the best known such bound.]

We illustrate the repeated capacity scaling algorithm on the example shown in Figure 10.11(a). When applied to this example, the capacity scaling algorithm performs 100 scaling phases with $\Delta = 2^{99}, 2^{98-1}, \ldots, 2^0$. The strongly polynomial version, however, terminates within five phases, as shown next.

Phase 1. In this phase, $\Delta = 2^{99}$, $S(\Delta) = \{1, 2\}$, and $T(\Delta) = \{3, 4\}$. The algorithm augments Δ units of flow along the two paths 1–3 and 2–1–3–4. Figure 10.11(b) shows the solution at the end of this phase.

Phase 2. In this phase, $\Delta = 2^{98}$. The algorithm augments Δ units of flow along the path 1–3.

Phase 3. In this phase, $\Delta = 2^{97}$. The algorithm augments Δ units of flow along the path 1–3.

Phase 4. In this phase, $\Delta = 2^{96}$. The algorithm finds that the flow on the arc (1, 3) is $2^{100} + 2^{99} + 2^{98}$, which is more than $4n\Delta = 2^{100}$. Therefore, the algorithm contracts the nodes 1 and 3 into a new node 5 and obtains the minimum cost flow problem shown in Figure 10.11(c), which it then proceeds to solve.

Phase 5. In this phase, $\Delta = 2^{95}$. The algorithm augments Δ units of flow along the path 2–5–4. The solution is a flow now; consequently, the algorithm terminates. The corresponding flow in the original network is $x_{21} = 2^{99}$, $x_{13} = 2^{100} - 1$, and $x_{34} = 2^{99}$.

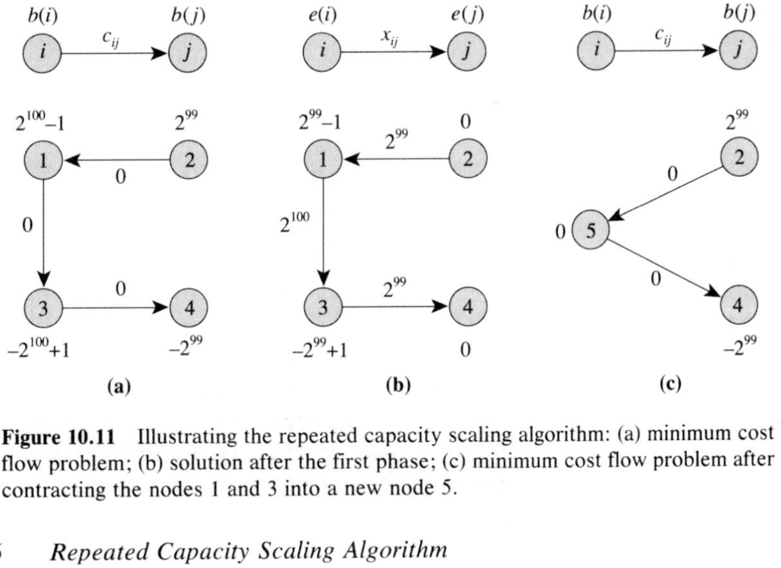

Figure 10.11 Illustrating the repeated capacity scaling algorithm: (a) minimum cost flow problem; (b) solution after the first phase; (c) minimum cost flow problem after contracting the nodes 1 and 3 into a new node 5.

We now explain how we expand the contracted network, and in the process we prove that the algorithm determines an optimal solution of the minimum cost flow problem. The algorithm, in fact, first determines an optimal set of node potentials of the problem, and then by solving a maximum flow problem (as described in Section 9.5) determines an optimal flow. The algorithm obtains an optimal set of node potentials for the original problem by repeated use of the following result.

Property 10.23. *Let* **P** *be a problem with arc costs* c_{ij} *and* **P'** *be the same problem with arc costs* $c_{ij} - \pi(i) + \pi(j)$. *If* π' *is an optimal set of node potentials for problem* **P'**, *then* $\pi + \pi'$ *is an optimal set of node potentials for* **P**.

Proof. This property easily follows from the observation that if a solution x satisfies the reduced cost optimality condition (9.7) with respect to the arc costs $c_{ij} - \pi(i) + \pi(j)$ and node potentials π', the same solution satisfies these conditions with arc costs c_{ij} and node potentials $\pi + \pi'$. ◆

We expand (or uncontract) the nodes in the reverse order in which we contracted them in the strongly polynomial algorithm and obtain optimal node potentials of the successive problems. In earlier stages, between two successive problems, we performed two transformations in the following order: (1) we replaced the arc cost c_{ij} by its reduced cost $c_{ij} - \pi(i) + \pi(j)$, and (2) we contracted two nodes k and l into a single new node p. We undo these transformations in the reverse order. To undo the contracted node p, for case (2) we set the potentials of nodes k and l equal to that of node p, and for case (1) we add π to the existing node potentials. When we have expanded all the contracted nodes, the resulting node potentials are an optimal set of node potentials for the minimum cost flow problem. Then, as described in Section 9.5, we can use these node potentials to obtain an optimal flow by solving a maximum flow problem. The following theorem summarizes the preceding discussion.

Theorem 10.24. *The repeated capacity scaling algorithm solves the uncapacitated minimum cost flow problem in* $O(n^2 \log n\, S(n, m))$ *time.* ◆

Since the best known strongly polynomial-time algorithm for solving the shortest path problem with nonnegative arc lengths runs in $O(m + n \log n)$ time, the best current bound for the uncapacitated minimum cost flow problem is $O(n \log n(m + n \log n))$. We can solve the capacitated minimum cost flow problem by the repeated capacity scaling algorithm by first transforming it to an uncapacitated problem (see Section 2.4). The uncapacitated network will have $n' = n + m$ nodes and $m' = 2m$ arcs. When applied to this network, the repeated capacity scaling algorithm will perform $O(n' \log n') = O(m \log n)$ scaling phases and solve $O(m') = O(m)$ shortest path problems in each scaling phase. Thus the running time of the algorithm is the time needed to solve $O(m^2 \log n)$ shortest path problems. Each shortest path problem in the uncapacitated network requires $O(2m + (m + n) \log (m + n)) = O(m + m \log n)$ time, but using a clever approach for solving the resulting shortest path problem (as discussed in Exercise 4.53) we can obtain a better bound of $O(m + n \log n)$. Consequently, the repeated capacity scaling algorithm requires $O(m^2 \log n(m + n \log n))$ time to solve a capacitated minimum cost flow problem.

10.7 ENHANCED CAPACITY SCALING ALGORITHM

In this section we discuss yet another strongly polynomial-time algorithm for the minimum cost flow problem. This algorithm is a variant of the capacity scaling algorithm that we discussed in Section 10.2 and draws on some ideas from the repeated capacity scaling algorithm discussed in Section 10.6. We refer to this algorithm as the *enhanced capacity scaling algorithm*. This algorithm runs in $O((m \log n)(m + \log n))$ time for the capacitated minimum cost flow problem and is currently the fastest strongly polynomial-time algorithm for solving the minimum cost flow problem. In this section we first show how to solve the enhanced capacity scaling algorithm for the *uncapacitated* minimum cost flow problem; we can solve the capacitated problem by transforming it to an uncapacitated problem (see Section 2.4).

Recall from Section 10.6 that the essential idea in the repeated capacity scaling algorithm is to identify arcs with *sufficiently large* flow. The repeated capacity scaling algorithm identifies such an arc (k, l) within $O(\log n)$ scaling phases, contracts the nodes k and l into a single node, and solves the resulting minimum cost flow problem afresh. For the uncapacitated minimum cost flow problem, this algorithm performs a total of $O(n \log n)$ scaling phases and $O(n^2 \log n)$ shortest path augmentations. The enhanced capacity scaling algorithm adopts a similar approach but it differs in the following two ways: (1) the algorithm does not explicitly perform the contraction operation; and (2) the algorithm does not solve the minimum cost flow problem afresh, but continues from where it left off in its earlier computations. By avoiding contractions, the algorithm achieves ease of coding (because contractions change the network structure and so its computer representation) and maintains a pseudoflow satisfying the dual optimality conditions at every step until the end, at which point it becomes an optimal flow. Moreover, the total number of scaling phases is $O(n \log n)$ and the total number of shortest path augmentations in these scalings phases is also $O(n \log n)$. Consequently, if $S(n, m)$ is the time required to solve a shortest path problem with nonnegative arc lengths, the running time of the enhanced capacity scaling algorithm for uncapacitated problems is $O(n \log n\ S(n, m))$. For capacitated minimum cost flow problems, this time bound becomes $O(m \log n\ S(n, m))$. [By Exercise 4.53 the time bound for the shortest path problem in the transformed network is $O(S(n, m))$ rather than $O(S(n + m, 2m))$ even though the transformed network has $n + m$ nodes and $2m$ arcs.]

The enhanced capacity scaling algorithm proceeds by performing scaling phases for different values of the scale factor Δ. In the Δ-scaling phase, we say that an arc (i, j) has a *sufficiently large* flow if $x_{ij} \geq 8n\Delta$. [We later show that if $x_{ij} \geq 8n\Delta$, then arc (i, j) will have positive flow during the entire execution of the algorithm.] We refer to an arc with sufficiently large flow as an *abundant arc*; otherwise, we call it a *nonabundant arc*. We refer to the subgraph consisting of the node set N and abundant arcs as the *abundant subgraph*. The abundant subgraph typically contains several components, which we call *abundant components*. If the network contains no abundant arc, the abundant subgraph contains n components, each consisting of a singleton node. For simplicity, we will designate an abundant component by the set S of nodes it spans. We let $b(S) = \sum_{i \in S} b(i)$ and $e(S) = \sum_{i \in S} e(i)$.

We designate an (arbitrary) node in each abundant component as its *root* and refer to all the other nodes as *nonroot nodes*. By convention we assume that the minimum index node in an abundant component is its root. For example, if $S = \{3, 5, 9\}$, then node 3 is the root node of the abundant component S. Throughout its execution, the enhanced capacity scaling algorithm satisfies the following properties.

Property 10.25 (Flow Property). *In the Δ-scaling phase, the flow on each nonabundant arc is an integral multiple of Δ; an abundant arc can have any nonnegative flow value.*

Property 10.26 (Imbalance Property). *Each nonroot node has a zero imbalance; a root node can have an excess or a deficit.*

At the beginning of the enhanced capacity scaling algorithm, the network has no abundant arc and the abundant subgraph contains n components, each consisting of a singleton node. As the algorithm proceeds, it identifies abundant arcs and adds them to the abundant subgraph. Suppose that the algorithm adds a new abundant arc (i, j) at some stage. Let S_i and S_j, respectively, denote the abundant components containing the nodes i and j. If $S_i = S_j$ [i.e., the arc (i, j) has both of its endpoints in the same component], this addition does not create any new abundant component; otherwise, the addition creates a new abundant component consisting of the union of S_i and S_j. We refer to this operation as a *merge* operation because it merges the components S_i and S_j into a single abundant component. Notice that since each merge operation reduces the number of abundant components by one, the algorithm can perform at most n merge operations.

Whenever the algorithm merges the components S_i and S_j, we need to ensure that the solution satisfies the imbalance property. Suppose that i_r and j_r denote the root nodes of the components S_i and S_j before the merge operation. Suppose further that $i_r < j_r$. If $e(j_r) = 0$, after the merge operation the abundant subgraph satisfies the imbalance property. However, if $e(j_r)$ is nonzero, we satisfy the imbalance property by sending $e(j_r)$ units of flow from node j_r to node i_r using any path in the merged component. [Notice that if $e(j_r) < 0$, we should view this augmentation as augmenting $|e(j_r)|$ units of flow from node i_r to j_r so we eliminate the imbalance at node j_r.] Observe that this augmentation changes the flow on some abundant arcs by $|e(j_r)|$ units. We refer to this augmentation as an *imbalance-property augmentation*. In Exercise 10.26 we ask the reader to show how to perform merge operations and the subsequent imbalance-property augmentations in $O(m)$ time.

We are now in a position to describe the enhanced capacity scaling algorithm. Figure 10.12 gives an algorithmic description of this algorithm.

The enhanced capacity scaling algorithm performs two types of augmentations. The first type of augmentation enforces the imbalance property when the algorithm identifies new abundant arcs; we have earlier defined these augmentations as the *imbalance-property augmentations*. The second type of augmentation takes place from excess nodes to deficit nodes along shortest paths. We refer to these augmentations as *shortest-path augmentations*.

As we have already mentioned, the enhanced capacity scaling algorithm is a variant of the capacity scaling algorithm. These two algorithms differ in the following respects:

```
algorithm enhanced capacity scaling;
begin
    set x : = 0, π : = 0, and e : = b;
    set Δ : = max{|e(i)| : i ∈ N};
    while the residual network G(x) contains a node i with e(i) > 0 do
    begin
        if max{e(i) : i ∈ N} ≤ Δ/(8n) then Δ : = max{e(i) : i ∈ N};
        {the Δ-scaling phase begins here}
        for each nonabundant arc (i, j) do
        if x_ij ≥ 8nΔ then designate arc (i, j) as an abundant arc;
        update abundant components and reinstate the imbalance property;
        while the residual network G(x) contains a node k with | e(k) | ≥ (n − 1)Δ/n do
        begin
            select a pair of nodes k and l satisfying the property that (i) either e(k) > (n − 1)Δ/n
              and e(l) < −Δ/n, or (ii) e(k) > Δ/n and e(l) < −(n − 1)Δ/n;
            considering reduced costs as arc lengths, compute shortest path distance d(·) in
              G(x) from node k to all other nodes;
            π(i) : = π(i) − d(i) for all i ∈ N;
            augment Δ units of flow along the shortest path in G(x) from node k to node l;
        end;
        {the Δ-scaling phase ends here}
        Δ : = Δ/2;
    end;
end;
```

Figure 10.12 Enhanced capacity scaling algorithm.

1. In the capacity scaling algorithm, we set the initial value of $\Delta = 2^{\lfloor \log U \rfloor}$, that is, the largest power of 2 less than or equal to $U = \max\{| b(i) | : i \in N\}$. In a strongly polynomial algorithm, we cannot take logarithms because we cannot determine $\log U$ in $O(1)$ elementary arithmetic operations. Therefore, in the enhanced capacity scaling algorithm, we set $\Delta = \max\{| b(i) | : i \in N\}$.
2. The capacity scaling algorithm decreases Δ by a factor of 2 in every scaling phase. In the enhanced capacity scaling algorithm, we also decrease Δ by a factor of 2, but if $\max\{| e(i) | : i \in N\} \le \Delta/8n$, then we reset $\Delta = \max\{| e(i) | : i \in N\}$. Consequently, the enhanced capacity scaling algorithm generally decreases Δ by a factor of 2, but sometimes by a larger factor when imbalances are too small compared to the current scale factor. Without resetting Δ in this way, the capacity scaling algorithm might perform $O(\log U)$ scaling phases, many of which will not perform any augmentations. The resulting algorithm would contain $O(\log U)$ in its running time and would not be strongly polynomial-time.
3. In the capacity scaling algorithm, each arc flow is an integral multiple of Δ. This property is essential for its correctness because it ensures that each positive residual capacity is a multiple of Δ, and consequently, any augmentation can carry Δ units of flow. In the enhanced capacity scaling algorithm, although the flows on nonabundant arcs are integral multiples of Δ, the flows on the abundant arcs can be arbitrary. Since the flows on abundant arcs are sufficiently large, their arbitrary values do not prohibit sending Δ units of flow on them.
4. The capacity scaling algorithm sends Δ units of flow from a node k with $e(k) \ge \Delta$ to a node l with $e(l) \le -\Delta$. As a result, the excess nodes do not become

deficit nodes, and vice versa. In the enhanced capacity scaling algorithm, augmentations carry Δ units of flow and are (a) either from a node k with $e(k) > (n - 1)\Delta/n$ to a node l with $e(l) < -\Delta/n$, (b) or from a node k with $e(k) > \Delta/n$ to a node l with $e(l) < -(n - 1)\Delta/n$. Notice that due to these choices, excess nodes might become deficit nodes and deficit nodes might become excess nodes. Although these choices might seem a bit odd when compared to the capacity scaling algorithm, they ensure several nice theoretical properties that we describe in the following discussion.

We establish the correctness of the enhanced capacity scaling algorithm as follows. In the Δ-scaling phase, we refer to a node i as a *large excess node* if $e(i) > (n - 1)\Delta/n$ and as a *medium excess node* if $e(i) > \Delta/n$. (Observe that a large excess node is also a medium excess node.) Similarly, we refer to a node i as a *large deficit node* if $e(i) < -(n - 1)\Delta/n$ and as a *medium deficit node* if $e(i) < -\Delta/n$. In the Δ-scaling phase, each shortest path augmentation either starts at a large excess node k and ends at a medium deficit node l, or starts at a medium excess node k and ends at a large deficit node l. To establish the correctness of the algorithm, we need to show that whenever (1) the network contains a large excess node k, it must also contain a medium deficit node l, or when (2) the network contains a large deficit node l, it must also contain a medium excess node k. We establish this result in the following lemma.

Lemma 10.27. *If the network contains a large excess node k, it must also contain a medium deficit node l. Similarly, if the network contains a large deficit node l, it must also contain a medium excess node k.*

Proof. We prove the first part of the lemma; the proof of the second part is similar. Note that $\sum_{i \in N} e(i) = 0$ because the total excess of the excess nodes equals the total deficit of the deficit nodes. If $e(k) > (n - 1)\Delta/n$ for some excess node k, the total deficit of deficit nodes is also greater than $(n - 1)\Delta/n$. Since the network contains at most $(n - 1)$ deficit nodes, at least one of these nodes, say node l, must have a deficit greater than Δ/n, or equivalently $e(l) < -\Delta/n$. ◆

In the proofs, we use the following lemma several times.

Lemma 10.28. *At the end of the Δ-scaling phase, $|e(i)| \leq (n - 1)\Delta/n$ for each node i. At the beginning of the Δ-scaling phase, $|e(i)| \leq 2(n - 1)\Delta/n$ for each node i.*

Proof. Suppose that during some scaling phase the network contains some large excess node. Then by Lemma 10.27, it also contains some medium deficit node, so the scaling phase would not yet end. Similarly, if the network contains some large deficit node, it would also contain some medium excess node, and the scaling phase would not end. Therefore, at the end of the scaling phase, $|e(i)| \leq (n - 1)\Delta/n$ for each node i.

If at the next scaling phase the algorithm halves the value of Δ, then $|e(i)| \leq 2(n - 1)\Delta/n$ for each node i. On the other hand, if the algorithm sets Δ equal to e_{max}, then $|e(i)| \leq \Delta$ for each node i. In either case, the lemma is true. ◆

The enhanced capacity scaling algorithm also relies on the fact that in the Δ-scaling phase, we can send Δ units of flow along the shortest path P from node k to node l. To prove this result, we need to show that the residual capacity of every arc in the path P is at least Δ. We establish this property in two parts. First, we show that the flow on each nonabundant arc is a multiple of Δ; this would imply that residual capacities of nonabundant arcs and their reversals in the residual network are multiples of Δ (because all the arcs in A are uncapacitated). We next show that the flow on each abundant arc is always greater than or equal to $4n\Delta$; therefore, we can send Δ units of flow in either direction. These two results would complete the correctness proof of the enhanced capacity scaling algorithm.

Lemma 10.29. *Throughout the execution of the enhanced capacity scaling algorithm, the solution satisfies the flow and imbalance properties (i.e., Properties 10.25 and 10.26).*

Proof. We prove this lemma by performing induction on the number of flow augmentations and changes in the scale factor Δ. We first consider the flow property. Each augmentation sends Δ units of flow and thus preserves the property. The scale factor Δ changes in one of the two following ways: (1) when we replace Δ by $\Delta' = \Delta/2$, or (2) after replacing $\Delta' = \Delta/2$, we reset $\Delta'' = \max\{|e(i)| : i \in N\}$. In case (1), the flows on the nonabundant arcs continue to be multiples of Δ'. In case (2), $\Delta'' = \max\{e(i) : i \in N\} \leq \Delta'/8n$, or $\Delta' \geq 8n\Delta''$. Since each positive arc flow x_{ij} on a nonabundant arc is a multiple of Δ', $x_{ij} \geq \Delta' \geq 8n\Delta''$. Consequently, each positive flow arc becomes an abundant arc (with respect to the new scale factor) and vacuously satisfies the flow property.

We next establish the imbalance property by performing induction on the number of augmentations and the creation of new abundant arcs. Each augmentation carries flow from a nonroot node to another nonroot node and preserves the property. Moreover, each time the algorithm creates a new abundant arc, it might create a nonroot node i with nonzero imbalance; however, it immediately performs an imbalance-property augmentation to reduce its imbalance to zero. The lemma now follows. ◆

Theorem 10.30. *In the Δ-scaling phase, the algorithm changes the flow on any arc by at most $4n\Delta$ units.*

Proof. The flow on an arc changes through either imbalance-property augmentations or shortest path augmentations. We first consider changes caused by imbalance-property augmentations. At the beginning of the Δ-scaling phase, $e(i) \leq 2(n-1)\Delta/n$ for each node i (from Lemma 10.28). Consequently, an imbalance-property augmentation changes the flow on any arc by at most $2(n-1)\Delta/n$. Since the algorithm can perform at most n imbalance-property augmentations at the beginning of a scaling phase, the change in the flow on an arc due to all imbalance-property augmentations is at most $2(n-1)\Delta \leq 2n\Delta$.

Next consider the changes in the flow on an arc caused by shortest path augmentations. At the beginning of the Δ-scaling phase, each root node i satisfies the condition $|e(i)| \leq 2(n-1)\Delta/n$ (by Lemma 10.29). Consider the case when the Δ-scaling phase performs no imbalance-property augmentations. In this case, at most one shortest path augmentation will begin at a large excess node i, because after this augmentation, the new excess $e'(i)$ satisfies the inequality $e'(i) \le 2(n - 1)\Delta/n - \Delta = (n - 2)\Delta/n \le (n - 1)\Delta/n$, and node i is no longer a large excess node. Similarly, at most one shortest path augmentation will end at a large deficit node.

Now suppose that the algorithm does perform some imbalance-property augmentations. In this case the algorithm sends $e(j)$ units of flow from each nonroot node j to the root of its abundant component. The subsequent imbalance-property augmentation from node j to the root node i can increase $|e(i)|$ by at most $2(n - 1)\Delta/n$ units, so node i can be the start or end node of at most two additional shortest path augmentations in the Δ-scaling phase. We "charge" these two augmentations to node j, which becomes a nonroot node and remains a nonroot node in the subsequent scaling phases.

To summarize, we have shown that in the Δ-scaling phase, we can charge each root node at most one shortest path augmentation and each nonroot node at most two shortest path augmentations. Each such augmentation changes the flow on any arc by 0 or Δ units. Consequently, the total flow change on any arc due to all shortest path augmentations is at most $2n\Delta$. We have earlier shown the total flow change due to imbalance-property augmentations is at most $2n\Delta$. These results establish the theorem. ◆

The preceding theorem immediately implies the following result.

Lemma 10.31. *If the algorithm designates an arc (i, j) as an abundant arc in the Δ-scaling phase, then in all subsequent Δ'-scaling phases $x_{ij} \ge 4n\Delta'$.*

Proof. We prove this result by performing induction on the number of scaling phases. Since the algorithm designates arc (i, j) as an abundant arc at the beginning of the Δ-scaling phase, the flow on this arc satisfies the condition $x_{ij} \ge 8n\Delta$. The Lemma 10.31 implies that the flow change on any arc in the Δ-scaling phase is at most $4n\Delta$. Therefore, throughout the Δ-scaling phase and, also, at the end of this scaling phase, the arc (i, j) satisfies the condition $x_{ij} \ge 4n\Delta$. In the next scaling phase, the scale factor $\Delta' \le \Delta/2$; so at the beginning of the Δ'-scaling phase, $x_{ij} \ge 8n\Delta'$. This conclusion establishes the lemma. ◆

We next consider the worst-case complexity of the enhanced capacity scaling algorithm. We show that the algorithm performs $O(n \log n)$ scaling phases, requiring a total of $O(n \log n)$ shortest path augmentations. These proofs rely on the result, stated in Theorem 10.33, that any abundant component whose root node has a medium excess or a medium deficit merges into a larger abundant component within $O(\log n)$ scaling phases. Theorem 10.33, in turn, depends on the following lemma.

Lemma 10.32. *Let S be the set of nodes spanned by an abundant component, and let $e(S) = \sum_{i \in S} e(i)$ and $b(S) = \sum_{i \in S} b(i)$. Then $b(S) - e(S)$ is an integral multiple of Δ.*

Proof. Summing the mass balance constraints (9.1b) of nodes in S, we see that

$$b(S) - e(S) = \sum_{\{(i,j) \in (S, \bar{S})\}} x_{ij} - \sum_{\{(i,j) \in (\bar{S}, S)\}} x_{ij}. \tag{10.12}$$

In this expression, $(S, \bar{S})$ and $(\bar{S}, S)$ denote the sets of forward and backward arcs in the cut $[S, \bar{S}]$. Since the flow on each arc in the cut is an integral multiple of Δ (by the flow property), $b(S) - e(S)$ is also an integral multiple of Δ. ◆

Theorem 10.33. *Let S be the set of nodes spanned by an abundant component and suppose that at the end of the Δ-scaling phase, $|e(S)| > \Delta/n$. Then within O(log n) additional scaling phases, the algorithm will merge the abundant component S into a larger abundant component.*

Proof. We first claim that at the end of the Δ-scaling phase, $|b(S)| \geq \Delta/n$. We prove this result by contradiction. Suppose that $|b(S)| < \Delta/n$. Let node i be the root node of the component S. Lemma 10.28 implies that at the end of the Δ-scaling phase, $|e(i)| = |e(S)| \leq (n - 1)\Delta/n$. Therefore, $|b(S)| + |e(S)| < \Delta$, which from Lemma 10.32 is possible only if $|b(S)| = |e(S)|$. This condition, however, contradicts the facts that $|e(S)| > \Delta/n$ and $|b(S)| < \Delta/n$. Therefore, $|b(S)| \geq \Delta/n$ whenever $|e(S)| > \Delta/n$. Consequently, at the end of the Δ-scaling phase, $|b(S)| \geq \Delta/n$.

Since the enhanced capacity scaling algorithm decreases Δ by a factor of at least 2 in each scaling phase, within $\log(9n^2m) \leq \log(9n^4) = O(\log n)$ scaling phases, the scale factor will be $\Delta' \leq \Delta/2^{\log(9n^2m)} = \Delta/(9n^2m)$, or $\Delta/n \geq 9nm\Delta'$. Since $|b(S)| \geq \Delta/n$, $|b(S)| \geq 9nm\Delta'$. We consider the situation when $b(S) > 0$. [The analysis of the situation with $b(S) < 0$ is similar.] Since $e(S) \leq \Delta'(n - 1)/n \leq \Delta'$ (by Lemma 10.28), the flow across the cut $[S, \bar{S}]$ (i.e., the right-hand side of (10.12)) is at least $9nm\Delta' - \Delta' \geq 8nm\Delta'$. This cut contains at most m arcs; at least one of these arcs, say arc (i, j), must have a flow at least $8n\Delta'$. Thus the algorithm will designate the arc (i, j) as an abundant arc and merge the component S into a larger abundant component. ◆

We are now ready to complete the proof of the main result of this section.

Theorem 10.34. *The enhanced capacity scaling algorithm solves the uncapacitated minimum cost flow problem within O(n log n) scaling phases and performs a total of O(n log n) shortest path augmentations. If S(n, m) is the time required to solve a shortest path problem with nonnegative arc lengths, the running time of the enhanced capacity scaling algorithm is O(n log n S(n, m)).*

Proof. We first show that the algorithm performs $O(n \log n)$ scaling phases. Consider a scaling phase with scale factor equal to Δ. At the end of this scaling phase, we will encounter one of the following two outcomes:

Case 1. For some node i, $|e(i)| > \Delta/16n$. Let node i be the root node of an abundant component S. Clearly, within four scaling phases, either the component S merges into a larger component or $|e(i)| > \Delta/n$. In the latter case, Theorem 10.33 implies that within $O(\log n)$ scaling phases, the component S merges into a larger component.

Case 2. For every node i, $|e(i)| \leq \Delta/16n$. At the beginning of the next scaling phase, the new scale factor $\Delta' = \Delta/2$, so $|e(i)| \leq \Delta'/8n$ for each node i. We then reset $\Delta' = \max\{|e(i)| : i \in N\}$. As a result, for some node i, $|e(i)| =$

$\Delta' > \Delta'/16n$ and, as in Case 1, within $O(\log n)$ scaling phases, the abundant component containing node i merges into a larger component.

This discussion shows that within $O(\log n)$ scaling phases the algorithm performs one merge operation. Since each merge operation decreases the number of abundant components by one, the algorithm can perform at most n merge operations. Consequently, the number of scaling phases is bounded by $O(n \log n)$. The algorithmic description of the enhanced capacity scaling algorithm implies that the algorithm requires $O(m)$ time per scaling phase plus the time required for the augmentations.

We now obtain a bound on the number of augmentations and the time that they require. The algorithm performs at most n imbalance-property augmentations; it can easily execute each augmentation in $O(m)$ time; thus these augmentations are not a bottleneck step in the algorithm. Next consider the shortest path augmentations. Recall from the proof of Theorem 10.30 that in a scaling phase, we can charge each shortest path augmentation to a root node (which is a large excess or a large-deficit node) or to a nonroot node. Since we can charge each nonroot at most two augmentations over the entire execution of the algorithm, we charge at most $2n$ augmentations to nonroots. Moreover, when we charge an augmentation to a root node i, this node satisfies the condition $|e(i)| \geq (n - 1)\Delta/n$. Theorem 10.33 implies that we will charge at most one augmentation to node i in the following $O(\log n)$ scaling phases before the algorithm performs a merge operation and the component containing node i merges into a larger component. Since the algorithm encounters at most $2n$ different abundant components (n to begin with and n due to merge operations), the total number of shortest path augmentations we can charge to root nodes is at most $O(n \log n)$. Since each shortest path augmentation requires the solution of a shortest path problem with nonnegative arc lengths and requires $S(n, m)$ time, all the shortest path augmentations require a total of $O(n \log n\, S(n, m))$ time. This time dominates the time taken by all other operations performed by the algorithm. Therefore, we have established the assertion of the theorem. ◆

To solve the capacitated minimum cost flow problem, we transform it to the uncapacitated version using the transformation described in Section 2.4. The resulting uncapacitated network has $n' = n + m$ nodes and $m' = 2m$ arcs. The enhanced capacity scaling algorithm will solve the minimum cost flow problem in the transformed network in $O(n' \log n') = O(m \log m) = O(m \log n^2) = O(m \log n)$ scaling phases and will solve a total of $O(n' \log n') = O(m \log n)$ shortest path problems. Each shortest path problem in the uncapacitated network requires $S(n', m')$ time, but using the ideas described in Exercise 4.53 we can improve this time bound to $S(n, m)$. Therefore, the enhanced capacity scaling algorithm can solve the capacitated minimum cost flow problem in $O(m \log n\, S(n, m))$ time. We state this important result as a theorem.

Theorem 10.35. *The enhanced capacity scaling algorithm solves a capacitated minimum cost flow problem in $O(m \log n\, S(n, m))$ time.* ◆

10.8 SUMMARY

In this chapter we continued our study of the minimum cost flow problem by developing several polynomial-time algorithms. The scaling technique is a central theme in almost all the algorithms we have discussed. The algorithms discussed use capacity scaling, cost scaling, or both, or use scaling concepts in their proofs. We discussed six polynomial-time algorithms: (1) the capacity scaling algorithm, (2) the cost scaling algorithm, (3) the double scaling algorithm, (4) the minimum mean cycle-canceling algorithm, (5) the repeated capacity scaling algorithm, and (6) the enhanced capacity scaling algorithm. The first three of these algorithms are weakly polynomial; the other three are strongly polynomial. Figure 10.13 specifies the running times of these algorithms.

The capacity scaling algorithm is possibly the simplest of all the polynomial-time algorithms we have discussed. This algorithm is an improved version of the successive shortest path algorithm discussed in Section 9.7; by augmenting flows along paths with sufficiently large residual capacities, this algorithm is able to decrease the number of augmentations from $O(nU)$ to $O(m \log U)$.

Whereas the capacity scaling algorithm scales the capacities, the cost scaling algorithm scales costs. The algorithm maintains ϵ-optimal flows for decreasing values of ϵ and repeatedly executes an improve-approximation procedure that converts an ϵ-optimal flow into an $\epsilon/2$-optimal flow. The computations performed by the improve-approximation procedure are similar to those performed by the preflow-push algorithm for the maximum flow problem. The double scaling algorithm is the same as the cost scaling algorithm except that it uses a different version of the improve-approximation procedure. The improve-approximation procedure in the cost scaling algorithm performs push/relabel steps; in the double scaling algorithm, this procedure augments flow along paths of sufficiently large residual capacity. Justifying its name, within a cost scaling phase, the double scaling algorithm performs a number of capacity scaling phases.

The minimum mean cycle-canceling algorithm for the minimum cost flow problem is different from all the other algorithms discussed in this chapter. The algorithm is startlingly simple to describe and does not make explicit use of the scaling technique; the proof of the algorithm, however, uses arguments from scaling techniques.

Algorithm	Running time
Capacity scaling algorithm	$O((m \log U)(m + n \log n))$
Cost scaling algorithm	$O(n^3 \log(nC))$
Double scaling algorithm	$O(nm \log U \log(nC))$
Minimum mean cycle-canceling algorithm	$O(n^2m^3 \log n)$
Repeated capacity scaling algorithm	$O((m^2 \log n)(m + n \log n))$
Enhanced capacity scaling algorithm	$O((m \log n)(m + n \log n))$

Figure 10.13 Running times of polynomial-time minimum cost flow algorithms.

This algorithm is a special implementation of the cycle canceling algorithm that we described in Section 9.6; it always augments flow along a minimum mean (negative) cycle in the residual network. To establish that this algorithm is strongly polynomial, we show that (1) when the reduced cost of an arc is *sufficiently large*, the flow on the arc becomes "fixed" (i.e., does not change any more); and (2) within $O(nm \log n)$ iterations, at least one additional arc has a sufficiently large reduced cost so that its value becomes fixed.

If we adopt a similar idea in the capacity scaling algorithm, it also becomes strongly polynomial. We showed that whenever the flow on an arc (i, j) is sufficiently large, we can fix the potentials of nodes i and j with respect to each other. The repeated capacity scaling algorithm applies the capacity scaling algorithm and within $O(\log n)$ scaling phases, it identifies an arc (i, j) with a sufficiently large flow. The algorithm then merges the nodes i and j into a single node and starts from scratch again on the modified minimum cost flow problem. The enhanced capacity scaling algorithm, described next, dramatically improves on the repeated capacity scaling algorithm by observing that whenever we contract an arc, we need not start all over again, but can continue the computations and still contract an additional arc within every $O(\log n)$ scaling phases and use only $O(m \log n)$ augmentations in total. This algorithm does not perform contractions explicitly, but does so implicitly by maintaining zero excesses at the contracted nodes (i.e., nonroot nodes).

REFERENCE NOTES

The following account of polynomial-time minimum cost flow algorithms is fairly brief. The surveys by Ahuja, Magnanti, and Orlin [1989, 1991] and by Goldberg, Tardos, and Tarjan [1989] provide more details concerning the development of this field.

Most of the available (combinatorial) polynomial-time algorithms for the minimum cost flow problems use scaling techniques. Edmonds and Karp [1972] introduced the scaling approach and obtained the first weakly polynomial-time algorithm for the minimum cost flow problem. This algorithm used the capacity scaling technique. The algorithm we presented in Section 10.2, which is a variant of Edmonds and Karp's algorithm, is due to Orlin [1988]. From 1972 to 1984, there was little research on scaling techniques. Since 1985, research employing scaling techniques has been extensive. Researchers now recognize that scaling techniques have great theoretical value as well as potential practical significance. Scaling techniques now yield many of the best (in the worst-case sense) available minimum cost flow algorithms.

Röck [1980] and, independently, Bland and Jensen [1985] suggested a cost scaling technique for the minimum cost flow problem. This approach solves the minimum cost flow problem as a sequence of $O(n \log C)$ maximum flow problems. Goldberg and Tarjan [1987] improved on the running time of Röck's algorithm and solved the minimum cost flow problem by solving "almost" $O(\log(nC))$ maximum flow problems. This approach is based on the concept of ϵ-optimality, which is, independently, due to Bertsekas [1979] and Tardos [1985]. We describe this approach in Section 10.3. Goldberg and Tarjan [1987] have developed several improved implementations of this approach, including the wave implementation presented in

Section 10.3. Their best implementation, which runs in $O(nm \log(n^2/m) \log(nC))$ time, uses Fibonacci heaps and finger search trees. Bertsekas and Eckstein [1988], independently, discovered the wave implementation.

Ahuja, Goldberg, Orlin, and Tarjan [1992] developed the double scaling algorithm described in Section 10.4, which combines capacity and cost scaling. This paper also describes several improved implementations, the best of which runs in $O(nm \log \log U \log(nC))$ time and uses the Fibonacci heap data structure.

When Edmonds and Karp [1972] suggested the first (weakly) polynomial-time algorithm for the minimum cost flow problem, they posed the development of a strongly polynomial-time algorithm as an open challenging problem. Tardos [1985] first settled this problem. Subsequently, Orlin [1984], Fujishige [1986], Galil and Tardos [1986], Goldberg and Tarjan [1987, 1988], Orlin [1988], and Ervolina and McCormick [1990b] developed other strongly polynomial-time algorithms. Currently, the best strongly polynomial-time algorithm is due to Orlin [1988]; it runs in $O((m \log n)(m + n \log n))$ time.

Most of the strongly polynomial-time minimum cost flow algorithm use the ideas of "fixing arc flows" or "fixing node potentials." Tardos [1985] was the first investigator to propose the use of either of these ideas (her algorithm fixes arc flows). The minimum mean cycle-canceling algorithm that we presented in Section 10.5 fixes arc flows; it is due to Goldberg and Tarjan [1988]. Goldberg and Tarjan [1988] also presented several variants of the minimum mean cycle-canceling algorithm with improved worst-case complexity. Orlin [1984] and Fujishige [1986] independently developed the idea of fixing node potentials, which is the "dual" of fixing arc flows. Using this idea, Goldberg, Tardos, and Tarjan [1989] obtained the repeated capacity scaling algorithm that we examined in Section 10.6. The enhanced capacity scaling algorithm, which is due to Orlin [1988], achieves the best strongly polynomial-time for solving the minimum cost flow problem. However, our presentation of the enhanced capacity scaling algorithm in Section 10.7 is based on Plotkin and Tardos' [1990] simplification of Orlin's original algorithm.

Some additional polynomial-time minimum cost flow algorithms include (1) a triple scaling algorithm due to Gabow and Tarjan [1989a], (2) a special implementation of the cycle canceling algorithm developed by Barahona and Tardos [1989], and (3) (its dual approach) a cut canceling algorithm proposed by Ervolina and McCormick [1990a].

Interior point linear programming algorithms are another source of polynomial-time algorithms for the minimum cost flow problem. Among these, the fastest available algorithm, due to Vaidya [1989], solves the minimum cost flow problem in $O(n^{2.5}\sqrt{m}\, K)$ time, with $K = \log n + \log C + \log U$.

Currently, the best available time bound for the minimum cost flow problem is $O(\min\{nm \log(n^2/m) \log(nC)), nm (\log \log U) \log(nC), (m \log n)(m + n \log n)\})$; the three bounds in this expression are, respectively, due to Goldberg and Tarjan [1987], Ahuja, Goldberg, Orlin, and Tarjan [1992], and Orlin [1988].

EXERCISES

10.1. Suppose that we want to solve the minimum cost flow problem shown in Figure 10.14(a) by the capacity scaling algorithm. Show the computations for two scaling phases. You may identify the shortest path distances by inspection.

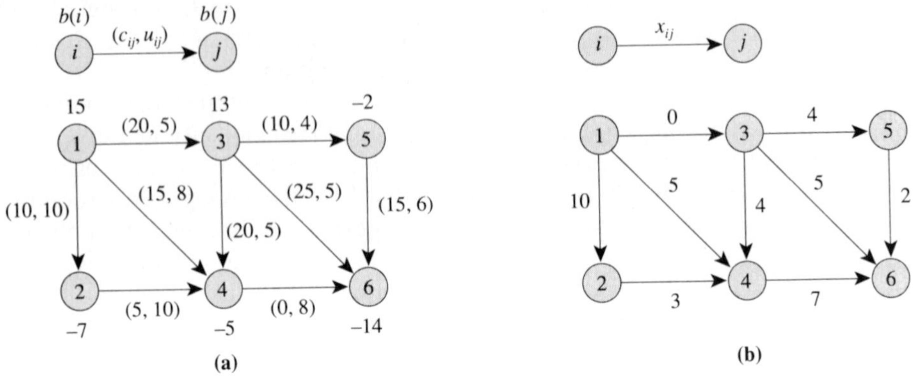

Figure 10.14 Examples for Exercises 10.1 and 10.4.

10.2. In every iteration of the capacity scaling algorithm, we augment flow along a shortest path from a node k with $e(k) \geq \Delta$ to a node l with $e(l) \leq -\Delta$. Suppose that we modify the algorithm as follows: We let node l be any deficit node; that is, we do not necessarily assume that $e(l) \leq -\Delta$. Will this modification affect the worst-case complexity of the capacity scaling algorithm?

10.3. Prove or disprove the following statements.

(a) During the Δ-scaling phase of the capacity scaling algorithm, $|e(i)| \leq 2\Delta$ for each node $i \in N$.

(b) While solving a specific instance of the minimum cost flow problem, the capacity scaling algorithm might perform more augmentations than the successive shortest path algorithm.

10.4. Consider the minimum cost flow problem given in Figure 10.14(a) and the feasible flow x shown in Figure 10.14(b). Starting with $\epsilon = 0$, apply two phases of the cost scaling algorithm.

10.5. Show that if the cost-scaling algorithm finds that arc (i, j) is inadmissible at some stage, this arc remains inadmissible until the algorithm relabels node i.

10.6. Let x and x' be two distinct (feasible) flows in a network. The flow decomposition theorem implies that we can always express x' as x plus the flow along at most m directed cycles $W_1, W_2, \ldots, W_p$ in $G(x)$. For every $1 \leq i \leq p$, let W_i^r denote the directed cycle obtained by reversing each arc in W_i. Show that we can express x as x' plus the flow along the cycles $W_1^r, W_2^r, \ldots, W_p^r$.

10.7. For the cost scaling algorithm, we showed that whenever $\epsilon < 1/n$, any ϵ-optimal flow is 0-optimal. Show that if we multiply all arc costs by $n + 1$, then any flow that is ϵ-optimal flow for the modified problem when $\epsilon \leq 1$ is 0-optimal for the original problem.

10.8. In the cost scaling algorithm, during a relabel operation we increase node potentials by $\epsilon/2$ units. Show that we can increase node potentials by as much as $\epsilon/2 + \min\{c_{ij}^{\pi} : (i, j) \text{ in } G(x) \text{ and } r_{ij} > 0\}$ and still maintain $\epsilon/2$-optimality of the pseudoflow.

10.9. Let x' be a feasible flow of the minimum cost flow problem and let x be a pseudoflow. Show that in the pseudoflow x, for every node v with an excess, there exists a node w with a deficit and a sequence of nodes $v = v_0, v_1, v_2, \ldots, v_l = w$ that satisfies the property that the path $P = v_0 - v_1 - v_2 - \cdots - v_l$ is a directed path in $G(x)$ and its reversal $\overline{P} = v_l - v_{l-1} - \cdots - v_0$ is a directed path in $G(x')$. (*Hint:* This exercise is similar to Exercise 10.6.)

10.10. In this exercise we study the nonscaled version of the cost scaling algorithm.

(a) Modify the algorithm described in Section 10.3 so that it starts with a 0-optimal pseudoflow, maintains an $1/(n + 1)$-optimal pseudoflow at every step, and terminates with an $1/(n + 1)$-optimal flow.

(b) Determine the number of relabel operations, the number of saturating and nonsaturating pushes, and the running time of the algorithm. Compare these numbers with those of the cost scaling algorithm.

10.11. In the wave implementation of the cost scaling algorithm described in Section 10.3, we scaled costs by a factor of 2. Suppose, instead, that we scaled costs by a factor of $k \geq 2$. In that case we start with $\Delta = k^{\lfloor \log C \rfloor}$ and decrease ϵ by a factor of k between two consecutive scaling phases. Outline the changes required in the algorithm and determine the number of scaling phases, relabel operations, and saturating and nonsaturating pushes within a scaling phase. For what value of k is the running time minimum?

10.12. Generalized cost scaling algorithm (Goldberg and Tarjan [1987]). As we noted in the text, by using some of the ideas of the minimum mean cycle-canceling algorithm (described in Section 10.5), we can devise a strongly polynomial-time version of the cost scaling algorithm that we described in Section 10.3. The módified algorithm, which we call the *generalized cost scaling algorithm*, is the same as the cost scaling algorithm except that it performs the following additional step after it has called the procedure improve-approximation, but before resetting $\epsilon := \epsilon/2$ (see Figure 10.3).

Additional step: Solve a minimum mean cycle problem to determine the minimum mean cycle cost $\mu(x)$, set $\epsilon = -\mu(x)$, and then determine a set of potential π so that the flow x is ϵ-optimal with respect to π (as described in the proof of Lemma 10.12).

Show that the generalized cost scaling fixes a distinct arc after $O(\log n)$ scaling phases. What is the resulting running time of the algorithm?

10.13. In the double scaling algorithm described in Section 10.4, we scaled costs by a factor of 2. Suppose that as described in Exercise 10.2, we scale costs by a factor of k instead of 2. Show that within a cost scaling phase, the algorithm performs $O(knm)$ retreat steps. How many advance steps does the algorithm perform within a scaling phase? How many scaling phases does it require? For what value of k does the algorithm run in the least time? What is the time bound for this value of k?

10.14. An arc (i, j) in the network $G = (N, A)$ is *critical* if increasing c_{ij} causes the cost of the optimal flow to increase and decreasing c_{ij} causes the cost of the optimal flow to decrease. Does a network always contain a critical arc? Show that we can identify all critical arcs by solving $O(m)$ maximum flow problems. (*Hint*: Use the fact that an arc is critical if it carries a positive flow in every optimal flow.)

10.15. In some minimum cost flow problem, each arc capacity and each supply/demand is a multiple of α and lies in the range $[0, \alpha K]$ for some constant K. Will the algorithms discussed in this chapter run any faster when applied to minimum cost flow problems with this special structure?

10.16. Suppose that in some minimum cost flow problem, each arc cost is a multiple of α and lies in the range $[0, \alpha K]$ for some constant K. Will this special structure permit us to solve the minimum cost flow problem any faster by the cost scaling and double scaling algorithms?

10.17. Minimum cost flows in unit capacity networks. A network is a *unit capacity network* if each arc has a capacity of 1.

(a) What is the running time of the capacity scaling algorithm for unit capacity networks?

(b) What is the running time of the cost scaling algorithm for unit capacity networks? (*Hint*: Will the algorithm make any nonsaturating pushes?)

10.18. Minimum cost flows in bipartite networks. Let $G = (N_1 \cup N_2, A)$ be a bipartite network. Let $n_1 = | N_1 | \leq | N_2 | = n_2$.

(a) Show that when applied to a bipartite network, the cost scaling algorithm relabels any node $O(n_1)$ times during a scaling phase.

(b) Develop an implementation of the generic cost scaling algorithm that runs in $O(n_1^2 m \log(nC))$ time for bipartite networks. (*Hint*: Generalize the bipartite preflow-push algorithm for the maximum flow problem discussed in Section 8.3.)

10.19. What is the running time of the double scaling algorithm for bipartite networks $G = (N_1 \cup N_2, A)$, assuming that $n_1 = |N_1| \le |N_2| = n_2$?

10.20. Two minimum cost flow problems P' and P'' are *capacity adjacent* if P'' differs from P' only in one arc capacity and by 1 unit. Given an optimal solution of P', describe an efficient method for solving P''. (*Hint*: Reoptimize by solving a shortest path problem.)

10.21. Two minimum cost flow problems P' and P'' are *cost adjacent* if P'' differs from P' only in one arc cost, and by 1 unit. Given an optimal solution of P', describe an efficient method for solving P''. (*Hint*: Reoptimize by solving a maximum flow problem.)

10.22. **Bit scaling of capacities** (Röck [1980]). In this capacity scaling algorithm, we consider binary representations of the arc capacities (as described in Section 3.3) and define problem P^k to be the minimum cost flow problem with each arc capacity equal to the k leading bits of the actual capacity. Given an optimal solution of P^k, how would you obtain an optimal solution of P^{k+1} by solving at most m capacity adjacent problems (as defined in Exercise 10.20). Write a pseudocode for the minimum cost flow problem assuming the availability of a subroutine for solving capacity adjacent problems (i.e., solving one from the solution to the other). What is the running time of your algorithm?

10.23. **Bit scaling of costs** (Röck [1980]). In this cost scaling algorithm, we consider binary representations of the arc costs and define problem P^k to be the minimum cost flow problem with each arc cost equal to the k leading bits of the actual cost. Given an optimal solution of P^k, how would you obtain an optimal solution of P^{k+1} by solving at most m cost adjacent problems (as defined in Exercise 10.21)? Write a pseudocode for the minimum cost flow problem assuming the availability of a subroutine for solving cost adjacent problems (i.e., solving one from the solution to the other). What is the running time of your algorithm?

10.24. Suppose that we define the contraction of an arc as in Section 10.5. Let G^c denote the network of $G = (N, A)$ we obtain when we contract the endpoints of an arc $(k, l) \in A$ into a single node p. In addition, let $G' = (N, A - \{(k, l)\})$. Show that if $\alpha(G)$ denotes the number of (distinct) spanning trees of G, then $\alpha(G) = \alpha(G^c) + \alpha(G')$.

10.25. **Constrained maximum flow problem.** In the constrained maximum flow problem, we wish to maximize the flow from the source node s to the sink node t subject to an additional linear constraint. Consider the following linear programming formulation of this problem:

$$\text{Maximize} \quad v$$

subject to

$$\sum_{\{j:(i,j)\in A\}} x_{ij} - \sum_{\{j:(j,i)\in A\}} x_{ji} = \begin{cases} v & \text{for } i = s \\ 0 & \text{for all } i \in N - \{s,t\} \\ -v & \text{for } i = t, \end{cases}$$

$$0 \le x_{ij} \le u_{ij},$$

$$\sum_{(i,j)\in A} c_{ij}x_{ij} \le D.$$

(a) Let v^* be any integer and let x^* be an optimal solution of a minimum cost flow problem with the objective function $\sum_{(i,j)\in A} c_{ij}x_{ij}$ and with the supply/demand data $b(s) = v^*$, $b(t) = -v^*$, and $b(i) = 0$ for all other nodes. Let $z^* = \sum_{(i,j)\in A} c_{ij}x^*_{ij}$. Show that x^* solves the constrained maximum flow problem when $D = z^*$. Assume that $c_{ij} \ge 0$ for each arc $(i, j) \in A$.

(b) Assume that all of the data in the constrained maximum flow problem are integer. Use the result in part (a) to develop an algorithm for the constrained maximum flow problem that uses a minimum cost flow algorithm as a subroutine. What is the running time of your algorithm? (*Hint*: Perform binary search on v.)

10.26. In the enhanced capacity scaling algorithm, suppose we maintain an index with each arc that stores whether the arc is an abundant or a nonabundant arc. Suppose further that at some stage the algorithm adds an arc (i, j) to the abundant subgraph. Show how you would perform each of the following operations in $O(m)$ time: (i) identifying the root nodes, i_r and j_r, of the abundant components containing the nodes i and j; (ii) determining whether the nodes i and j belong to the same abundant component; and (iii) identifying a path from node i to j, or vice versa. Using these operations, explain how you would perform a merge operation and the subsequent imbalance-property augmentation in $O(m)$ time. (*Hint*: Observe that each abundant arc can be traversed in either direction because it has sufficient residual capacity in both the directions. Then use the search algorithm described in Section 3.4.)

11

MINIMUM COST FLOWS: NETWORK SIMPLEX ALGORITHMS

. . . seek, and ye shall find.
—The Book of Matthew

Chapter Outline

11.1 INTRODUCTION

The simplex method for solving linear programming problems is perhaps the most powerful algorithm ever devised for solving constrained optimization problems. Indeed, many members of the academic community view the simplex method as not only one of the principal computational engines of applied mathematics, computer science, and operations research, but also as one of the landmark contributions to computational mathematics of this century. The algorithm has achieved this lofty status because of the pervasiveness of its applications throughout many problem domains, because of its extraordinary efficiency, and because it permits us to not only solve problems numerically, but also to gain considerable practical and theoretical insight through the use of sensitivity analysis and duality theory.

Since minimum cost flow problems define a special class of linear programs, we might expect the simplex method to be an attractive solution procedure for solving many of the problems that we consider in this text. Then again, because network flow problems have considerable special structure, we might also ask whether the simplex method could possibly compete with other "combinatorial" methods, such as the many variants of the successive shortest path algorithm, that exploit the underlying network structure. The general simplex method, when implemented in a way that does not exploit underlying network structure, is not a competitive solution procedure for solving minimum cost flow problems. Fortunately, however, if we interpret the core concepts of the simplex method appropriately as network operations, we can adapt and streamline the method to exploit the network structure of the minimum cost flow problem, producing an algorithm that is very efficient. Our purpose in this chapter is to develop this network-based implementation of the simplex method and show how to apply it to the minimum cost flow problem, the shortest path problem, and the maximum flow problem.

We could adopt several different approaches for presenting this material, and each has its own merits. For example, we could start by describing the simplex method for general linear programming problems and then show how to adapt the method for minimum cost flow problems. This approach has the advantage of placing our development in the broader context of more general linear programs. Alternatively, we could develop the network simplex method directly in the context of network flow problems as a particular type of augmenting cycle algorithm. This approach has the advantage of not requiring any background in linear programming and of building more directly on the concepts that we have developed already. We discuss both points of view. Throughout most of this chapter we adopt the network approach and derive the network simplex algorithm from the first principles, avoiding the use of linear programming in any direct way. Later, in Section 11.11, we show that the network simplex algorithm is an adaptation of the simplex method.

The central concept underlying the network simplex algorithm is the notion of spanning tree solutions, which are solutions that we obtain by fixing the flow of every arc not in a spanning tree either at value zero or at the arc's flow capacity. As we show in this chapter, we can then solve uniquely for the flow on all the arcs in the spanning tree. We also show that the minimum cost flow problem always has at least one optimal spanning tree solution and that it is possible to find an optimal spanning tree solution by "moving" from one such solution to another, at each step introducing one new nontree arc into the spanning tree in place of one tree arc. This method is known as the network simplex algorithm because spanning trees correspond to the so-called basic feasible solutions of linear programming, and the movement from one spanning tree solution to another corresponds to a so-called pivot operation of the general simplex method. In Section 11.11 we make these connections.

In the first three sections of this chapter we examine several fundamental ideas that either motivate the network simplex method or underlie its development. In Section 11.2 we show that the minimum cost flow problem always has at least one spanning tree solution. We also show how the network optimality conditions that we have used repeatedly in previous chapters specialize when applied to any spanning tree solution. In keeping with our practice in previous chapters, we use these conditions to assess whether a candidate solution is optimal and, if not, how to modify it to construct a better spanning tree solution.

To implement the network simplex algorithm efficiently we need to develop a method for representing spanning trees conveniently in a computer so that we can perform the basic operations of the algorithm efficiently and so that we can efficiently manipulate the computer representation of a spanning tree structure from step to step. We describe one such approach in Section 11.3.

In Section 11.4 we show how to compute the arc flows corresponding to any spanning tree and associated node potentials so that we can assess whether the particular spanning tree is optimal. These operations are essential to the network simplex algorithm, and since we need to make these computations repeatedly as we move from one spanning tree to another, we need to be able to implement these operations very efficiently. Section 11.5 brings all these pieces together and describes the network simplex algorithm.

In the context of applying the network simplex algorithm and establishing that the algorithm properly solves any given minimum cost flow problem, we need to address a technical issue known as degeneracy (which occurs when one of the arcs in a spanning tree, like the nontree arcs, has a flow value equal to zero or the arc's flow capacity). In Section 11.6 we describe a very appealing and simple way to modify the basic network simplex algorithm so that it overcomes the difficulties associated with degeneracy.

Since the shortest path and maximum flow problems are special cases of the minimum cost flow problem, the network simplex algorithm applies to these problems as well. In Sections 11.7 and 11.8 we describe these specialized implementations. When applied to the shortest path problem, the network simplex algorithm closely resembles the label-correcting algorithms that we discussed in Chapter 5. When applied to the maximum flow problem, the algorithm is essentially an augmenting path algorithm.

The network simplex algorithm maintains a feasible solution at each step; by moving from one spanning tree solution to another, it eventually finds a spanning tree solution that satisfies the network optimality conditions. Are there other spanning tree algorithms that iteratively move from one infeasible spanning tree solution to another and yet eventually find an optimal solution? In Section 11.9 we describe two such algorithms: a *parametric network simplex algorithm* that satisfies all of the optimality conditions except the mass balance constraints at two nodes, and a *dual network simplex algorithm* that satisfies the mass balance constraints at all the nodes but might violate the arc flow bounds. These algorithms are important because they provide alternative solution strategies for solving minimum cost flow problems; they also illustrate the versatility of spanning tree manipulation algorithms for solving network flow problems.

We next consider a key feature of the optimal spanning tree solutions generated by the network simplex algorithm. In Section 11.10 we show that it is easy to use these solutions to conduct sensitivity analysis: that is, to determine a new solution if we change any cost coefficient or change the capacity of any arc. This type of information is invaluable in practice because problem data are often only approximate and/or because we would like to understand how robust a solution is to changes in the underlying data.

To conclude this chapter we delineate connections between the network simplex algorithm and more general concepts in linear and integer programming. In Section 11.11 we show that the network simplex algorithm is a special case of the simplex method for general linear programs, although streamlined to exploit the special structure of network flow problems. In particular, we show that spanning trees for the network flow problem correspond in a one-to-one fashion with bases of the linear programming formulation of the problem. We also show that each of the essential steps of the network simplex algorithm, for example, determining node potentials or moving from one spanning tree to another, are specializations of the usual steps of the simplex method for solving linear programs.

As we have noted in Section 9.6, network flow problems satisfy one very remarkable property: They have optimal integral flows whenever the underlying data are integral. In Section 11.12 we show that this integrality result is a special case of a more general result in linear and integer programming. We define a set of linear programming problems with special constraint matrices, known as *unimodular matrices*, and show that these linear programs also satisfy the integrality property. That is, when solved as linear programs with integral data, problems with these specialized constraint matrices always have integer solutions. Since node–arc incidence matrices satisfy the unimodularity property, this integrality property for linear programming is a strict generalization of the integrality property of network flows. This result provides us with another way to view the integrality property of network flows; it is also suggestive of more general results in integer programming and shows how network flow results have stimulated more general investigations in combinatorial optimization and integer programming.

11.2 CYCLE FREE AND SPANNING TREE SOLUTIONS

Much of our development in previous chapters has relied on a simple but powerful algorithmic idea: To generate an improving sequence of solutions to the minimum cost flow problem, we iteratively augment flows along a series of negative cycles and shortest paths. As one of these variants, the network simplex algorithm uses a particular strategy for generating negative cycles. In this section, as a prelude to our discussion of the method, we introduce some basic background material. We begin by examining two important concepts known as *cycle free solutions* and *spanning tree solutions*.

For any feasible solution, x, we say that an arc (i, j) is a *free arc* if $0 < x_{ij} < u_{ij}$ and is a *restricted arc* if $x_{ij} = 0$ or $x_{ij} = u_{ij}$. Note that we can both increase and decrease flow on a free arc while honoring the bounds on arc flows. However, in a restricted arc (i, j) at its lower bound (i.e., $x_{ij} = 0$) we can only increase the flow. Similarly, for flow on a restricted arc (i, j) at its upper bound (i.e., $x_{ij} = u_{ij}$) we can only decrease the flow. We refer to a solution x as a *cycle free solution* if the network contains no cycle composed only of free arcs. Note that in a cycle free solution, we can augment flow on any augmenting cycle in only a single direction since some arc in any cycle will restrict us from either increasing or decreasing that arc's flow. We also refer to a feasible solution x and an associated spanning tree of the network as a *spanning tree solution* if every nontree arc is a restricted arc. Notice that in a spanning tree solution, the tree arcs can be free or restricted. Frequently, when we refer to a spanning tree solution, we do not explicitly identify the associated tree; rather, it will be understood from the context of our discussion.

In this section we establish a fundamental result of network flows: minimum cost flow problems always have optimal cycle free and spanning tree solutions. The network simplex algorithm will exploit this result by restricting its search for an optimal solution to only spanning tree solutions. To illustrate the argument used to prove these results, we use the network example shown in Figure 11.1.

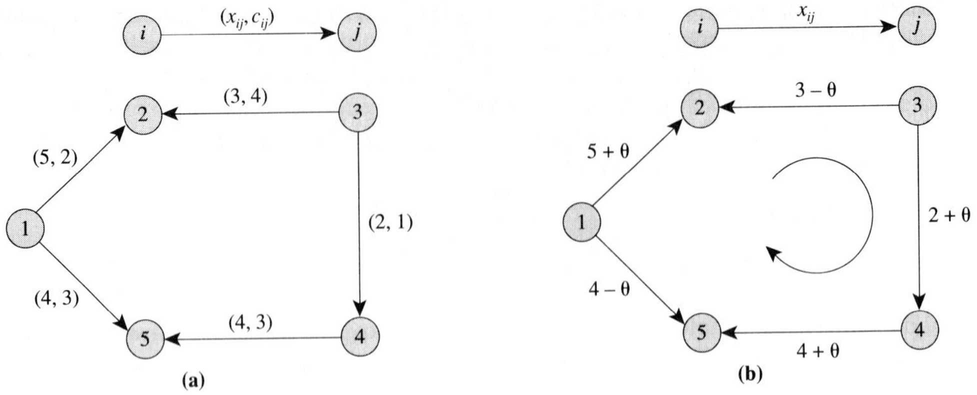

Figure 11.1 Improving flow around a cycle: (a) feasible solution; (b) solution after augmenting θ amount of flow along a cycle.

For the time being let us assume that all arcs are uncapacitated [i.e., $u_{ij} = \infty$ for each $(i, j) \in A$]. The network shown in Figure 11.1 contains positive flow around a cycle. We define the orientation of the cycle as the same as that of arc (4, 5). Let us augment θ units of flow along the cycle in the direction of its orientation. As shown in Figure 11.1, this augmentation increases the flow on arcs along the orientation of the cycle (i.e., forward arcs) by θ units and decreases the flow on arcs opposite to the orientation of the cycle (i.e., backward arcs) by θ units. Also note that the per unit incremental cost for this flow change is the sum of the costs of forward arcs minus the sum of the costs of backward arcs in the cycle, that is,

$$\text{per unit change in cost } \Delta = 2 + 1 + 3 - 4 - 3 = -1.$$

Since augmenting flow in the cycle decreases the cost, we set θ as large as possible while preserving nonnegativity of all arc flows. Therefore, we must satisfy the inequalities $3 - \theta \geq 0$ and $4 - \theta \geq 0$, and hence we set $\theta = 3$. Note that in the new solution (at $\theta = 3$), some arc in the cycle has a flow at value zero, and moreover, the objective function value of this solution is strictly less than the value of the initial solution.

In our example, if we change c_{12} from 2 to 5, the per unit cost of the cycle is $\Delta = 2$. Consequently, to improve the cost by the greatest amount, we would decrease θ as much as possible (i.e., satisfy the restrictions $5 + \theta \geq 0$, $2 + \theta \geq 0$, and $4 + \theta \geq 0$, or $\theta \geq -2$) and again find a lower cost solution with the flow on at least one arc in the cycle at value zero. We can restate this observation in another way: To preserve nonnegativity of all the arc flows, we must select θ in the interval $-2 \leq \theta \leq 3$. Since the objective function depends linearly on θ, we optimize it by selecting $\theta = 3$ or $\theta = -2$, at which point one arc in the cycle has a flow value of zero.

We can extend this observation in several ways:

1. If the per unit cycle cost $\Delta = 0$, we are indifferent to all solutions in the interval $-2 \leq \theta \leq 3$ and therefore can again choose a solution as good as the original one, but with the flow of at least one arc in the cycle at value zero.
2. If we impose upper bounds on the flow (e.g., such as 6 units on all arcs), the

range of flow that preserves feasibility (i.e., the mass balance constraints, lower and upper bounds on flows) is again an interval, in this case $-2 \le \theta \le 1$, and we can find a solution as good as the original one by choosing $\theta = -2$ or $\theta = 1$. At these values of θ, the solution is cycle free; that is, some arc on the cycle has a flow either at value zero (at the lower bound) or at its upper bound.

In general, our prior observations apply to any cycle in a network. Therefore, given any initial flow we can apply our previous argument repeatedly, one cycle at a time, and establish the following fundamental result.

Theorem 11.1 (Cycle Free Property). *If the objective function of a minimum cost flow problem is bounded from below over the feasible region, the problem always has an optimal cycle free solution.* ◆

It is easy to convert a cycle free solution into a spanning tree solution. Our results in Section 2.2 show that the free arcs in a cycle free solution define a forest (i.e., a collection of node-disjoint trees). If this forest is a spanning tree, the cycle free solution is already a spanning tree solution. However, if this forest is not a spanning tree, we can add some restricted arcs and produce a spanning tree.
Figure 11.2 illustrates a spanning tree corresponding to a cycle free solution.

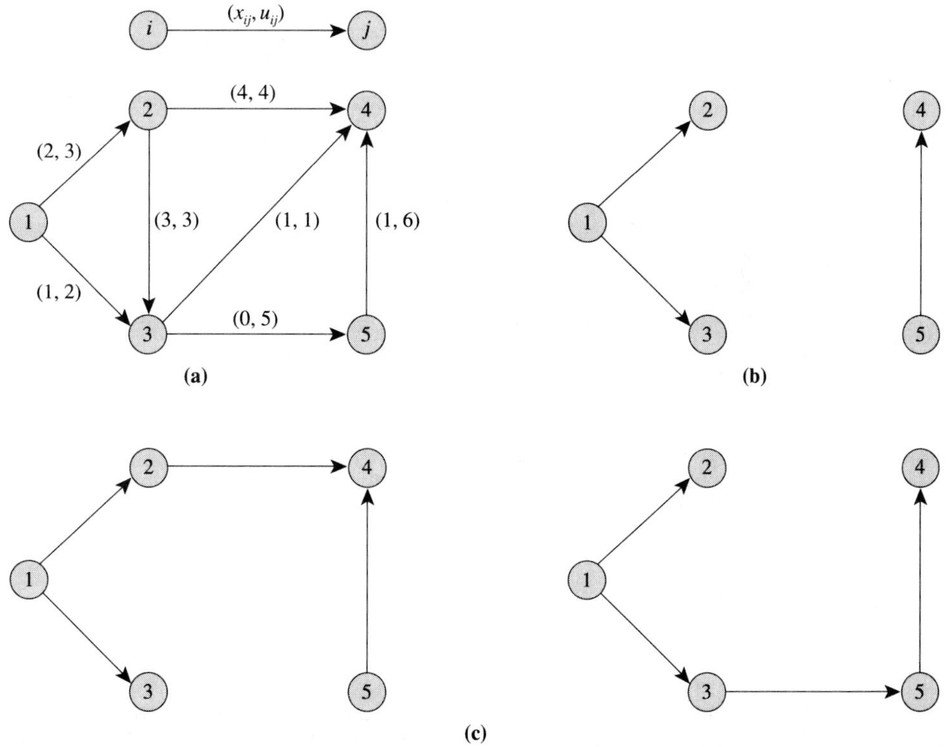

Figure 11.2 Converting a cycle free solution into a spanning tree solution: (a) example network; (b) set of free arcs; (c) 2 spanning tree solutions.

The solution in Figure 11.2(a) is cycle free. Figure 11.2(b) represents the set of free arcs, and Figure 11.2(c) shows two spanning tree solutions corresponding to the cycle free solution. As shown by this example, it might be possible (and often is) to complete the set of free arcs into a spanning tree in several ways. Adding the arc (3, 4) instead of the arc (2, 4) or (3, 5) would produce yet another spanning tree solution. Therefore, a given cycle free solution can correspond to several spanning trees. Nevertheless, since we assume that the underlying network is connected, we can always add some restricted arcs to the free arcs of a cycle free solution to produce a spanning tree, so we have established the following fundamental result:

Theorem 11.2 (Spanning Tree Property). *If the objective function of a minimum cost flow problem is bounded from below over the feasible region, the problem always has an optimal spanning tree solution.* ◆

A spanning tree solution partitions the arc set A into three subsets: (1) **T**, the arcs in the spanning tree; (2) **L**, the nontree arcs whose flow is restricted to value zero; and (3) **U**, the nontree arcs whose flow is restricted in value to the arcs' flow capacities. We refer to the triple (**T**, **L**, **U**) as a *spanning tree structure*.

Just as we can associate a spanning tree structure with a spanning tree solution, we can also obtain a unique spanning tree solution corresponding to a given spanning tree structure (**T**, **L**, **U**). To do so, we set $x_{ij} = 0$ for all arcs $(i, j) \in \mathbf{L}$, $x_{ij} = u_{ij}$ for all arcs $(i, j) \in \mathbf{U}$, and then solve the mass balance equations to determine the flow values for arcs in **T**. In Section 11.4 we show that the flows on the spanning tree arcs are unique. We say that a spanning tree structure is *feasible* if its associated spanning tree solution satisfies all of the arcs' flow bounds. In the special case in which every tree arc in a spanning tree solution is a free arc, we say that the spanning tree is *nondegenerate*; otherwise, we refer to it as a *degenerate* spanning tree. We refer to a spanning tree structure as *optimal* if its associated spanning tree solution is an optimal solution of the minimum cost flow problem. The following theorem states a sufficient condition for a spanning tree structure to be an optimal structure. As shown by our discussion in previous chapters, the reduced costs defined as $c_{ij}^{\pi} = c_{ij} - \pi(i) + \pi(j)$ are useful in characterizing optimal solutions to minimum cost flow problems.

Theorem 11.3 (Minimum Cost Flow Optimality Conditions). *A spanning tree structure* (**T**, **L**, **U**) *is an optimal spanning tree structure of the minimum cost flow problem if it is feasible and for some choice of node potentials* π, *the arc reduced costs* c_{ij}^{π} *satisfy the following conditions:*

$$(a) \quad c_{ij}^{\pi} = 0 \text{ for all } (i, j) \in \mathbf{T}. \tag{11.1a}$$

$$(b) \quad c_{ij}^{\pi} \geq 0 \text{ for all } (i, j) \in \mathbf{L}. \tag{11.1b}$$

$$(c) \quad c_{ij}^{\pi} \leq 0 \text{ for all } (i, j) \in \mathbf{U}. \tag{11.1c}$$

Proof. Let x^* be the solution associated with the spanning tree structure (**T**, **L**, **U**). We know that some set of node potentials π, together with the spanning tree structure (**T**, **L**, **U**), satisfies (11.1).

We need to show that x^* is an optimal solution of the minimum cost flow problem. In Section 2.4 we showed that minimizing $\sum_{(i,j)\in A} c_{ij}x_{ij}$ is equivalent to minimizing $\sum_{(i,j)\in A} c_{ij}^{\pi}x_{ij}$. The conditions stated in (11.1) imply that for the given node potential π, minimizing $\sum_{(i,j)\in A} c_{ij}^{\pi}x_{ij}$ is equivalent to minimizing the following expression:

$$\text{Minimize} \quad \sum_{(i,j)\in \mathbf{L}} c_{ij}^{\pi}x_{ij} - \sum_{(i,j)\in \mathbf{U}} | c_{ij}^{\pi} | \, x_{ij}. \tag{11.2}$$

The definition of the solution x^* implies that for any arbitrary solution x, $x_{ij} \geq x_{ij}^*$ for all $(i, j) \in \mathbf{L}$ and $x_{ij} \leq x_{ij}^*$ for all $(i, j) \in \mathbf{U}$. The expression (11.2) implies that the objective function value of the solution x will be greater than or equal to that of x^*. ◆

These optimality conditions have a nice economic interpretation. As we shall see later in Section 11.4, if $\pi(1) = 0$, the equations in (11.1a) imply that $-\pi(k)$ denotes the length of the tree path from node 1 to node k. The reduced cost $c_{ij}^{\pi} = c_{ij} - \pi(i) + \pi(j)$ for a nontree arc $(i, j) \in \mathbf{L}$ denotes the change in the cost of the flow that we realize by sending 1 unit of flow through the tree path from node 1 to node i through the arc (i, j), and then back to node 1 along the tree path from node j to node 1. The condition (11.1b) implies that this circulation of flow is not profitable (i.e., does not decrease cost) for any nontree arc in $\mathbf{L}$. The condition (11.1c) has a similar interpretation.

The network simplex algorithm maintains a feasible spanning tree structure and moves from one spanning tree structure to another until it finds an optimal structure. At each iteration, the algorithm adds one arc to the spanning tree in place of one of its current arcs. The entering arc is a nontree arc violating its optimality condition. The algorithm (1) adds this arc to the spanning tree, creating a negative cycle (which might have zero residual capacity), (2) sends the maximum possible flow in this cycle until the flow on at least one arc in the cycle reaches its lower or upper bound, and (3) drops an arc whose flow has reached its lower or upper bound, giving us a new spanning tree structure. Because of its relationship to the primal simplex algorithm for the linear programming problem (see Appendix C), this operation of moving from one spanning tree structure to another is known as a *pivot operation*, and the two spanning trees structures obtained in consecutive iterations are called *adjacent spanning tree structures*. In Section 11.5 we give a detailed description of this algorithm.

11.3 MAINTAINING A SPANNING TREE STRUCTURE

Since the network simplex algorithm generates a sequence of spanning tree solutions, to implement the algorithm effectively, we need to be able to represent spanning trees conveniently in a computer so that the algorithm can perform its basic operations efficiently and can update the representation quickly when it changes the spanning tree. Over the years, researchers have suggested several procedures for maintaining and manipulating a spanning tree structure. In this section we describe one of the more popular representations.

We consider the tree as "hanging" from a specially designated node, called the *root*. Throughout this chapter we assume that node 1 is the root node. Figure

11.3 gives an example of a tree. We associate three indices with each node i in the tree: a predecessor index, $pred(i)$, a depth index $depth(i)$, and a thread index, $thread(i)$.

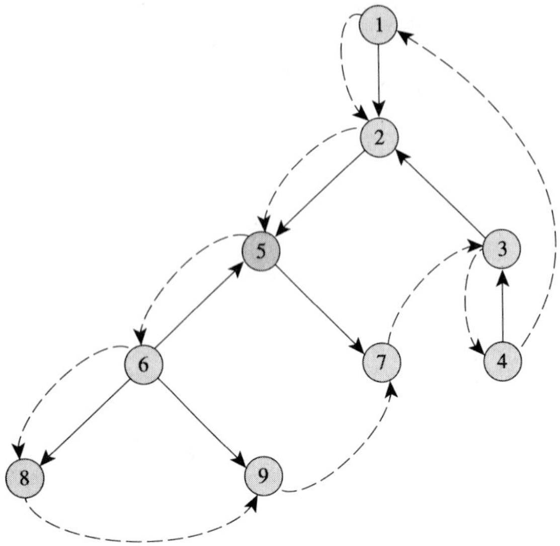

(a)

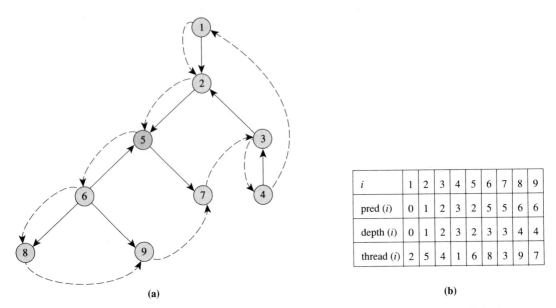

i	1	2	3	4	5	6	7	8	9
pred (i)	0	1	2	3	2	5	5	6	6
depth (i)	0	1	2	3	2	3	3	4	4
thread (i)	2	5	4	1	6	8	3	9	7

(b)

Figure 11.3 Example of a tree indices: (a) rooted tree; (b) corresponding tree indices.

Predecessor index. Each node i has a unique path connecting it to the root. The index pred(i) stores the first node in that path (other than node i). For example, the path 9–6–5–2–1 connects node 9 to the root; therefore, pred(9) = 6. By convention, we set the predecessor node of the root node, node 1, equal to zero. Figure 11.3 specifies these indices for the other nodes. Observe that by iteratively using the predecessor indices, we can enumerate the path from any node to the root.

A node j is called a *successor* of node i if pred(j) = i. For example, node 5 has two successors: nodes 6 and 7. A *leaf node* is a node with no successors. In Figure 11.3, nodes 4, 7, 8, and 9 are leaf nodes. The *descendants* of a node i are the node i itself, its successors, successors of its successors, and so on. For example, in Figure 11.3, the elements of node set {5, 6, 7, 8, 9} are the descendants of node 5.

Depth index. We observed earlier that each node i has a unique path connecting it to the root. The index depth(i) stores the number of arcs in that path. For example, since the path 9–6–5–2–1 connects node 9 to the root, depth(9) = 4. Figure 11.3 gives depth indices for all of the nodes in the network.

Thread index. The thread indices define a traversal of a tree, that is, a sequence of nodes that walks or threads its way through the nodes of a tree, starting at the root node, and visiting nodes in a "top-to-bottom" order, and finally returning to the root. We can find thread indices by performing a depth-first search of the tree as described in Section 3.4 and setting the thread of a node to be the node in the depth-first search encountered just after the node itself. For our example, the depth-first traversal would read 1–2–5–6–8–9–7–3–4–1, so thread(1) = 2, thread(2) = 5, thread(5) = 6, and so on (see the dashed lines in Figure 11.3).

The thread indices provide a particularly convenient means for visiting (or finding) all descendants of a node i. We simply follow the thread starting at that node and record the nodes visited, until the depth of the visited node becomes at least as large as that of node i. For example, starting at node 5, we visit nodes 6, 8, 9, and 7 in order, which are the descendants of node 5 and then visit node 3. Since the depth of node 3 equals that of node 5, we know that we have left the "descendant tree" lying below node 5. We shall see later that finding the descendant tree of a node efficiently is an important step in developing an efficient implementation of the network simplex algorithm.

In the next section we show how the tree indices permit us to compute the feasible solution and the set of node potentials associated with a tree.

11.4 COMPUTING NODE POTENTIALS AND FLOWS

As we noted in Section 11.2, as the network simplex algorithm moves from one spanning tree to the next, it always maintains the condition that the reduced cost of every arc (i, j) in the current spanning tree is zero (i.e. $c_{ij}^{\pi} = 0$). Given the current spanning tree structure $(\mathbf{T}, \mathbf{L}, \mathbf{U})$, the method first determines values for the node potentials π that will satisfy this condition for the tree arcs. In this section we show how to find these values of the node potentials.

Note that we can set the value of one node potential arbitrarily because adding a constant k to each node potential does not alter the reduced cost of any arc; that is, for any constant k, $c_{ij}^{\pi} = c_{ij} - \pi(i) + \pi(j) = c_{ij} - [\pi(i) + k] + [\pi(j) + k]$. So for convenience, we henceforth assume that $\pi(1) = 0$. We compute the remaining node potentials using the fact that the reduced cost of every spanning tree arc is zero; that is,

$$c_{ij}^{\pi} = c_{ij} - \pi(i) + \pi(j) = 0 \qquad \text{for every arc } (i, j) \in \mathbf{T}. \tag{11.3}$$

In equation (11.3), if we know one of the node potentials $\pi(i)$ or $\pi(j)$, we can easily compute the other one. Consequently, the basic idea in the procedure is to start at node 1 and fan out along the tree arcs using the thread indices to compute other node potentials. By traversing the nodes using the thread indices, we ensure that whenever the procedure visits a node k, it has already evaluated the potential of its predecessor, so it can compute $\pi(k)$ using (11.3). Figure 11.4 gives a formal statement of the procedure *compute-potentials*.

The numerical example shown in Figure 11.5 illustrates the procedure. We first set $\pi(1) = 0$. The thread of node 1 is 2, so we next examine node 2. Since arc (1, 2) connects node 2 to its predecessor, using (11.3) we find that $\pi(2) = \pi(1) - c_{12} = -5$. We next examine node 5, which is connected to its parent by arc (5, 2). Using (11.3) we obtain $\pi(5) = \pi(2) + c_{52} = -5 + 2 = -3$. In the same fashion we compute the rest of the node potentials; the numbers shown next to each node in Figure 11.5 specify these values.

```
procedure compute-potentials;
begin
    π(1) : = 0;
    j : = thread(1);
    while j ≠ 1 do
    begin
        i : = pred( j);
        if (i, j) ∈ A then π( j) : = π(i) − c_ij;
        if ( j, i) ∈ A then π( j) : = π(i) + c_ji;
        j : = thread( j);
    end;
end;
```

Figure 11.4 Procedure *compute-potentials*.

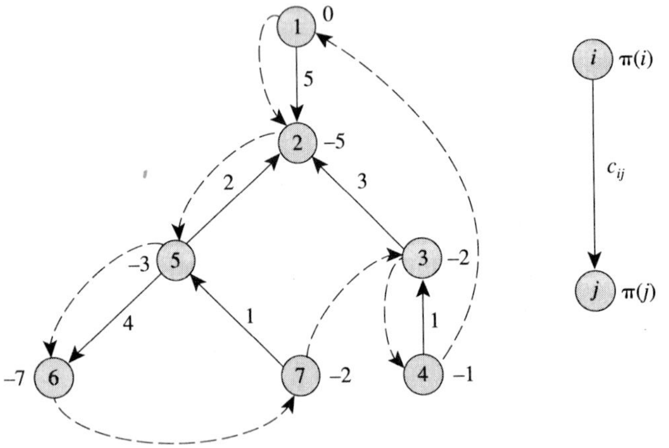

Figure 11.5 Computing node potentials for a spanning tree.

Let P be the tree path in **T** from the root node 1 to some node k. Moreover, let $\overline{P}$ and $\underline{P}$, respectively, denote the sets of forward and backward arcs in P. Now let us examine arcs in P starting at node 1. The procedure *compute-potentials* implies that $\pi(j) = \pi(i) - c_{ij}$ whenever arc (i, j) is a forward arc in the path, and that $\pi(j) = \pi(i) + c_{ji}$ whenever arc (j, i) is a backward arc in the path. This observation implies that $\pi(k) = \pi(k) - \pi(1) = -\sum_{(i,j)\in\overline{P}} c_{ij} + \sum_{(i,j)\in\underline{P}} c_{ij}$. In other words, $\pi(k)$ is the negative of the cost of sending 1 unit of flow from node 1 to node k along the tree path. Alternatively, $\pi(k)$ is the cost of sending 1 unit of flow from mode k to node 1 along the tree path. The procedure *compute-potentials* requires $O(1)$ time per iteration and performs $(n - 1)$ iterations to evaluate the node potential of each node. Therefore, the procedure runs in $O(n)$ time.

One important consequence of the procedure *compute-potentials* is that the minimum cost flow problem always has integer optimal node potentials whenever all the arc costs are integer. To see this result, recall from Theorem 11.2 that the minimum cost flow problem always has an optimal spanning tree solution. The potentials associated with this tree constitute optimal node potentials, which we can determine using the procedure *compute-potentials*. The description of the procedure *compute-potentials* implies that if all arc costs are integer, node potentials are integer as well (because the procedure performs only additions and subtractions). We refer to this integrality property of optimal node potentials as the *dual integrality property* since node potentials are the dual linear programming variables associated with the minimum cost flow problem.

Theorem 11.4 (Dual Integrality Property). *If all arc costs are integer, the minimum cost flow problem always has optimal integer node potentials.* ◆

Computing Arc Flows

We next consider the problem of determining the flows on the tree arcs of a given spanning tree structure. To ease our discussion, for the moment let us first consider the *uncapacitated* version of the minimum cost flow problem. We can then assume that all nontree arcs carry zero flow.

If we delete a tree arc, say arc (i, j), from the spanning tree, the tree decomposes into two subtrees. Let T_1 be the subtree containing node i and let T_2 be the subtree containing node j. Note that $\sum_{k \in T_1} b(k)$ denotes the cumulative supply/demand of nodes in T_1 [which must be equal to $-\sum_{k \in T_2} b(k)$ because $\sum_{k \in T_1} b(k) + \sum_{k \in T_2} b(k) = 0$]. In the spanning tree, arc (i, j) is the only arc that connects the subtree T_1 to the subtree T_2, so it must carry $\sum_{k \in T_1} b(k)$ units of flow, for this is the only way to satisfy the mass balance constraints. For example, in Figure 11.6, if we delete arc (1, 2) from the tree, then $T_1 = \{1, 3, 6, 7\}$, $T_2 = \{2, 4, 5\}$, and $\sum_{k \in T_1} b(k) = 10$. Consequently, arc (1, 2) carries 10 units of flow.

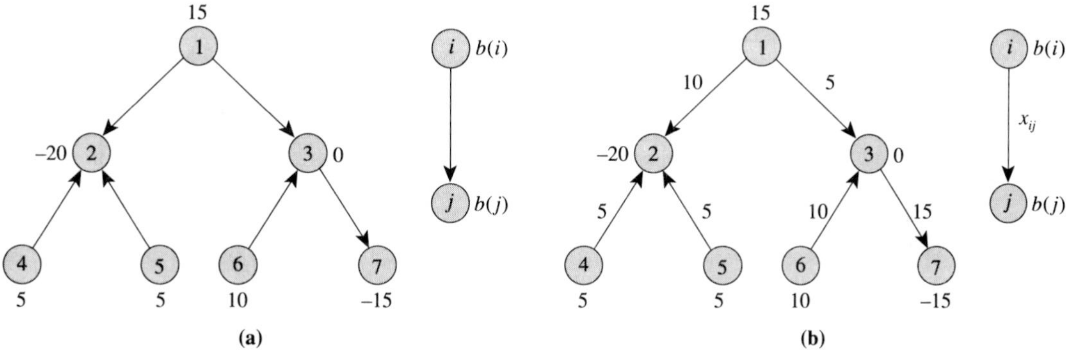

Figure 11.6 Computing flows for a spanning tree.

Using this observation we can devise an efficient method for computing the flows on all the tree arcs. Suppose that (i, j) is a tree arc and that node j is a leaf node [the treatment of the case when (i, j) is a tree arc and node i is a leaf node is similar]. Our observations imply that arc (i, j) must carry $-b(j)$ units of flow. For our example, arc (3, 7) must carry 15 units of flow to satisfy the demand of node 7. Setting the flow on this arc to this value has an effect on the mass balance of its incident nodes: we must subtract 15 units from $b(3)$ and add 15 units to $b(7)$ [which reduces $b(7)$ to zero]. Having determined x_{37}, we can delete arc (3, 7) from the tree and repeat the method on the smaller tree. Notice that we can identify a leaf node in every iteration because every tree has at least two leaf nodes (see Exercise 2.13). Figure 11.7 gives a formal description of this procedure.

```
procedure compute-flows;
begin
    b'(i) : = b(i), for all i ∈ N;
    for each (i, j) ∈ L do set x_ij : = 0;
    T' : = T;
    while T' ≠ {1} do
    begin
        select a leaf node j (other than node 1) in the subtree T';
        i : = pred(j);
        if (i, j) ∈ T' then x_ij : = -b'(j)
        else x_ij : = b'(j);
        add b'(j) to b'(i);
        delete node j and the arc incident to it from T';
    end;
end;
```

Figure 11.7 Procedure *compute-flows*.

This method for computing the flow values assumes that the minimum cost flow problem is uncapacitated. For the capacitated version of the problem, we add the following statement immediately after the first statement [i.e., $b'(i) := b(i)$ for all $i \in N$] in the procedure *compute-flows*. We leave the justification of this modification as an exercise (see Exercise 11.19).

```
for each (i, j) ∈ U do
    set x_ij: = u_ij, subtract u_ij from b'(i) and add u_ij to b'(j);
```

The running time of the procedure *compute-flows* is easy to determine. Clearly, the initialization of flows and modification of supplies/demands $b(i)$ and $b(j)$ for arcs (i, j) in **U** requires $O(m)$ time. If we set aside the time to select leaf nodes of **T**, then each iteration requires $O(1)$ time, resulting in a total of $O(n)$ time. One way of identifying leaf nodes in **T** is to select nodes in the reverse order of the thread indices. Note that in the thread traversal, each node appears prior to its descendants (see Property 3.4). We identify the reverse thread traversal of the nodes by examining the nodes in the order dictated by the thread indices, putting all the nodes into a stack in the order of their appearance and then taking them out from the top of the stack one at a time. Therefore, the reverse thread traversal examines each node only after it has examined all of the node's descendants. We have thus established that for the uncapacitated minimum cost flow problem, the procedure *compute-flows* runs in $O(m)$ time. For the capacitated version of the problem, the procedure also requires $O(m)$ time.

We can use the procedure *compute-flows* to obtain an alternative proof of the (primal) integrality property that we stated in Theorem 9.10. Recall from Theorem 11.2 that the minimum cost flow problem always has an optimal spanning tree solution. The flow associated with this tree is an optimal flow and we can determine it using the procedure *compute-flows*. The description of the procedure *compute-flows* implies that if the capacities of all the arcs and the supplies/demands of all the nodes are integer, arc flows are integer as well (because the procedure performs only additions and subtractions). We state this result again because of its importance in network flow theory.

Theorem 11.5 (Primal Integrality Property). *If capacities of all the arcs and supplies/demands of all the nodes are integer, the minimum cost flow problem always has an integer optimal flow.* ◆

In closing this section we observe that every spanning tree structure (**T**, **L**, **U**) defines a unique flow x. If this flow satisfies the flow bounds $0 \le x_{ij} \le u_{ij}$ for every arc $(i, j) \in A$, the spanning tree structure is feasible; otherwise, it is infeasible. We refer to the spanning tree **T** as *degenerate* if $x_{ij} = 0$ or $x_{ij} = u_{ij}$ for some arc $(i, j) \in \mathbf{T}$, and *nondegenerate* otherwise. In a nondegenerate spanning tree, $0 < x_{ij} < u_{ij}$ for every tree arc (i, j).

11.5 NETWORK SIMPLEX ALGORITHM

The network simplex algorithm maintains a feasible spanning tree structure at each iteration and successively transforms it into an improved spanning tree structure until it becomes optimal. The algorithmic description in Figure 11.8 specifies the essential steps of the method.

```
algorithm network simplex;
begin
    determine an initial feasible tree structure (T, L, U);
    let x be the flow and π be the node potentials associated with this tree structure;
    while some nontree arc violates the optimality conditions do
    begin
        select an entering arc (k, l) violating its optimality condition;
        add arc (k, l) to the tree and determine the leaving arc (p, q);
        perform a tree update and update the solutions x and π;
    end;
end;
```

Figure 11.8 Network simplex algorithm.

In the following discussion we describe in greater detail how the network simplex algorithm uses tree indices to perform these various steps. This discussion highlights the value of the tree indices in designing an efficient implementation of the algorithm.

Obtaining an Initial Spanning Tree Structure

Our connectedness assumption (i.e., Assumption 9.4 in Section 9.1) provides one way of obtaining an initial spanning tree structure. We have assumed that for every node $j \in N - \{1\}$, the network contains arcs $(1, j)$ and $(j, 1)$, with sufficiently large costs and capacities. We construct the initial tree **T** as follows. We examine each node j, other than node 1, one by one. If $b(j) \ge 0$, we include arc $(1, j)$ in **T** with a flow value of $b(j)$. If $b(j) < 0$, we include arc $(j, 1)$ in **T** with a flow value of $-b(j)$. The set **L** consists of the remaining arcs, and the set **U** is empty. As shown in Section 11.4, we can easily compute the node potentials for this tree using the equations $c_{ij} - \pi(i) + \pi(j) = 0$ for all $(i, j) \in \mathbf{T}$. Recall that we set $\pi(1) = 0$.

If the network does not contain the arcs $(1, j)$ and $(j, 1)$ for each node $j \in$

$N - \{1\}$ (or, we do not wish to add these arcs for some reason), we could construct an initial spanning tree structure by first establishing a feasible flow in the network by solving a maximum flow problem (as described in Application 6.1), and then by converting this solution into a spanning tree solution using the method described in Section 11.2.

Optimality Testing and the Entering Arc

Let (**T**, **L**, **U**) be a feasible spanning tree structure of the minimum cost flow problem, and let π be the corresponding node potentials. To determine whether the spanning tree structure is optimal, we check to see whether the spanning tree structure satisfies the following conditions:

$$c_{ij}^{\pi} \geq 0 \text{ for every arc } (i, j) \in \mathbf{L},$$

$$c_{ij}^{\pi} \leq 0 \text{ for every arc } (i, j) \in \mathbf{U}.$$

If the spanning tree structure satisfies these conditions, it is optimal and the algorithm terminates. Otherwise, the algorithm selects a nontree arc violating the optimality condition to be introduced into the tree. Two types of arcs are *eligible* to enter the tree:

1. Any arc $(i, j) \in \mathbf{L}$ with $c_{ij}^{\pi} < 0$
2. Any arc $(i, j) \in \mathbf{U}$ with $c_{ij}^{\pi} > 0$

For any eligible arc (i, j), we refer to $| c_{ij}^{\pi} |$ as its *violation*. The network simplex algorithm can select *any* eligible arc to enter the tree and still would terminate finitely (with some provisions for dealing with degeneracy, as discussed in Section 11.6). However, different rules for selecting the entering arc produce algorithms with different empirical and theoretical behavior. Many different rules, called *pivot rules*, are possible for choosing the entering arc. The following rules are most widely adopted.

Dantzig's pivot rule. This rule was suggested by George B. Dantzig, the father of linear programming. At each iteration this rule selects an arc with the maximum violation to enter the tree. The motivation for this rule is that the arc with the maximum violation causes the maximum decrease in the objective function per unit change in the value of flow on the selected arc, and hence the introduction of this arc into the spanning tree would cause the maximum decrease per pivot if the average increase in the value of the selected arc were the same for all arcs. Computational results confirm that this choice of the entering arc tends to produce relatively large decreases in the objective function per iteration and, as a result, the algorithm performs fewer iterations than other choices for the pivot rule. However, this rule does have a major drawback: The algorithm must consider every nontree arc to identify the arc with the maximum violation and doing so is very time consuming. Therefore, even though this algorithm generally performs fewer iterations than other implementations, the running time of the algorithm is not attractive.

First eligible arc pivot rule. To implement this rule, we scan the arc list sequentially and select the first eligible arc to enter the tree. In a popular version of this rule, we examine the arc list in a wraparound fashion. For example, in an iteration if we find that the fifth arc in the arc list is the first eligible arc, then in the next iteration we start scanning the arc list from the sixth arc. If we reach the end of the arc list while we are performing some iteration, we continue by examining the arc list from the beginning. One nice feature of this pivot rule is that it quickly identifies the entering arc. The pivot rule does have a counterbalancing drawback: with it, the algorithm generally performs more iterations than it would with other pivot rules because each pivot operation produces a relatively small decrease in the objective function value. The overall effect of this pivot rule on the running time of the algorithm is not very attractive, although the rule does produce a more efficient implementation than Dantzig's pivot rule.

Dantzig's pivot rule and the first pivot rule represent two extreme choices of a pivot rule. The *candidate list pivot rule*, which we discuss next, strikes an effective compromise between these two extremes and has proven to be one of the most successful pivot rules in practice. This rule also offers sufficient flexibility for fine tuning to special circumstances.

Candidate list pivot rule. When implemented with this rule, the algorithm selects the entering arc using a two-phase procedure consisting of *major iterations* and *minor iterations*. In a major iteration we construct a *candidate list* of eligible arcs. Having constructed this list, we then perform a number of minor iterations; in each of these iterations, we select an eligible arc from the candidate list with the maximum violation.

In a major iteration we construct the candidate list as follows. We first examine arcs emanating from node 1 and add eligible arcs to the candidate list. We repeat this process for nodes 2, 3, . . . , until either the list has reached its maximum allowable size or we have examined all the nodes. The next major iteration begins with the node where the previous major iteration ended and examines nodes in a wraparound fashion.

Once the algorithm has formed the candidate list in a major iteration, it performs a number of minor iterations. In a minor iteration, the algorithm scans all the arcs in the candidate list and selects an arc with the maximum violation to enter the tree. As we scan the arcs, we update the candidate list by removing those arcs that are no longer eligible (due to changes in the node potentials). Once the candidate list becomes empty or we have reached a specified limit on the number of minor iterations to be performed within each major iteration, we rebuild the candidate list by performing another major iteration.

Notice that the candidate list approach offers considerable flexibility for fine tuning to special problem classes. By setting the maximum allowable size of the candidate list appropriately and by specifying the number of minor iterations to be performed within a major iteration, we can obtain numerous different pivot rules. In fact, Dantzig's pivot rule and the first eligible pivot rule are special cases of the candidate list pivot rule (see Exercise 11.20).

In the preceding discussion, we described several important pivot rules. In the reference notes, we supply references for other pivot rules. Our next topic of study is deciding how to choose the arc that leaves the spanning tree structure at each step of the network simplex algorithm.

Leaving Arc

Suppose that we select arc (k, l) as the entering arc. The addition of this arc to the tree **T** creates exactly one cycle W, which we refer to as the *pivot cycle*. The pivot cycle consists of the unique path in the tree **T** from node k to node l, together with arc (k, l). We define the orientation of the cycle W as the same as that of (k, l) if $(k, l) \in \mathbf{L}$ and opposite the orientation of (k, l) if $(k, l) \in \mathbf{U}$. Let $\overline{W}$ and $\underline{W}$ denote the sets of *forward arcs* (i.e., those along the orientation of W) and *backward arcs* (those opposite to the orientation of W) in the pivot cycle. Sending additional flow around the pivot cycle W in the direction of its orientation strictly decreases the cost of the current solution at the per unit rate of $| c_{kl}^{\pi} |$. We augment the flow as much as possible until one of the arcs in the pivot cycle reaches its lower or upper bound. Notice that augmenting flow along W increases the flow on forward arcs and decreases the flow on backward arcs. Consequently, the maximum flow change δ_{ij} on an arc $(i, j) \in W$ that satisfies the flow bound constraints is

$$\delta_{ij} = \begin{cases} u_{ij} - x_{ij} & \text{if } (i, j) \in \overline{W} \\ x_{ij} & \text{if } (i, j) \in \underline{W} \end{cases}$$

To maintain feasibility, we can augment $\delta = \min\{\delta_{ij} : (i, j) \in W\}$ units of flow along W. We refer to any arc $(i, j) \in W$ that defines δ (i.e., for which $\delta = \delta_{ij}$) as a *blocking arc*. We then augment δ units of flow and select an arc (p, q) with $\delta_{pq} = \delta$ as the leaving arc, breaking ties arbitrarily. We say that a pivot iteration is a *nondegenerate iteration* if $\delta > 0$ and is a *degenerate iteration* if $\delta = 0$. A degenerate iteration occurs only if **T** is a degenerate spanning tree. Observe that if two arcs tie while determining the value of δ, the next spanning tree will be degenerate.

The crucial step in identifying the leaving arc is to identify the pivot cycle. If $P(i)$ denotes the unique path in the tree from any node i to the root node, this cycle consists of the arcs $\{(k, l)\} \cup P(k) \cup P(l) - (P(k) \cap P(l))$. In other words, W consists of the arc (k, l) and the disjoint portions of $P(k)$ and $P(l)$. Using the predecessor indices alone permits us to identify the cycle W as follows. First, we designate all the nodes in the network as unmarked. We then start at node k and, using the predecessor indices, trace the path from this node to the root and mark all the nodes in this path. Next we start at node l and trace the predecessor indices until we encounter a marked node, say w. The node w is the first common ancestor of nodes k and l; we refer to it as the *apex* of cycle W. The cycle W contains the portions of the paths $P(k)$ and $P(l)$ up to node w, together with the arc (k, l). This method identifies the cycle W in $O(n)$ time and so is efficient. However, it has the drawback of backtracking along those arcs of $P(k)$ that are not in W. If the pivot cycle lies "deep in the tree," far from its root, then tracing the nodes back to the root will be inefficient. Ideally, we would like to identify the cycle W in time proportional to $| W |$. The simultaneous use of depth and predecessor indices, as indicated in Figure 11.9, permits us to achieve this goal.

This method scans the arcs in the pivot cycle W twice. During the first scan, we identify the apex of the cycle and also identify the maximum possible flow that

```
procedure identify-cycle;
begin
    i : = k and j : = l;
    while i ≠ j do
    begin
        if depth(i) > depth(j) then i : = pred(i)
        else if depth (j) > depth (i) then j : = pred(j)
            else i : = pred(i) and j : = pred(j);
    end;
end;
```

Figure 11.9 Procedure for identifying the pivot cycle.

can be augmented along W. In the second scan, we augment the flow. The entire flow change operation requires $O(n)$ time in the worst case, but typically it examines only a small subset of nodes (and arcs).

Updating the Tree

When the network simplex algorithm has determined a leaving arc (p, q) for a given entering arc (k, l), it updates the tree structure. If the leaving arc is the same as the entering arc, which would happen when $\delta = \delta_{kl} = u_{kl}$, the tree does not change. In this instance the arc (k, l) merely moves from the set **L** to the set **U**, or vice versa. If the leaving arc differs from the entering arc, the algorithm must perform more extensive changes. In this instance the arc (p, q) becomes a nontree arc at its lower or upper bound, depending on whether (in the updated flow) $x_{pq} = 0$ or $x_{pq} = u_{pq}$. Adding arc (k, l) to the current spanning tree and deleting arc (p, q) creates a new spanning tree.

For the new spanning tree, the node potentials also change; we can update them as follows. The deletion of the arc (p, q) from the previous tree partitions the set of nodes into two subtrees, one, T_1, containing the root node, and the other, T_2, not containing the root node. Note that the subtree T_2 hangs from node p or node q. The arc (k, l) has one endpoint in T_1 and the other in T_2. As is easy to verify, the conditions $\pi(1) = 0$ and $c_{ij} - \pi(i) + \pi(j) = 0$ for all arcs in the new tree imply that the potentials of nodes in the subtree T_1 remain unchanged, and the potentials of nodes in the subtree T_2 change by a constant amount. If $k \in T_1$ and $l \in T_2$, all the node potentials in T_2 increase by $-c_{kl}^{\pi}$; if $l \in T_1$ and $k \in T_2$, they increase by the amount c_{kl}^{π}. Using the thread and depth indices, the method described in Figure 11.10 updates the node potentials quickly.

```
procedure update-potentials;
begin
    if q ∈ T₂ then y : = q else y : = p;
    if k ∈ T₁ then change : = -c^π_kl else change : = c^π_kl;
    π(y) : = π(y) + change;
    z : = thread(y);
    while depth(z) > depth(y) do
    begin
        π(z) : = π(z) + change;
        z : = thread(z);
    end;
end;
```

Figure 11.10 Updating node potentials in a pivot operation.

The final step in the updating of the tree is to recompute the various tree indices. This step is rather involved and we refer the reader to the references given in reference notes for the details. We do point out, however, that it is possible to update the tree indices in $O(n)$ time. In fact, the time required to update the tree indices is $O(|W| + \min\{|T_1|, |T_2|\})$, which is typically much less than n.

Termination

The network simplex algorithm, as just described, moves from one feasible spanning tree structure to another until it obtains a spanning tree structure that satisfies the optimality condition (11.1). If each pivot operation in the algorithm is nondegenerate, it is easy to show that the algorithm terminates finitely. Recall that $|c_{kl}^{\pi}|$ is the net decrease in the cost per unit flow sent around the pivot cycle W. After a nondegenerate pivot (for which $\delta > 0$), the cost of the new spanning tree structure is $\delta|c_{kl}^{\pi}|$ units less than the cost of the previous spanning tree structure. Since any network has a finite number of spanning tree structures and every spanning tree structure has a unique associated cost, the network simplex algorithm will encounter any spanning tree structure at most once and hence will terminate finitely. Degenerate pivots, however, pose a theoretical difficulty: The algorithm might not terminate finitely unless we perform pivots carefully. In the next section we discuss a special implementation, called the *strongly feasible spanning tree implementation*, that guarantees finite convergence of the network simplex algorithm even for problems that are degenerate.

We use the example in Figure 11.11(a) to illustrate the network simplex algorithm. Figure 11.11(b) shows a feasible spanning tree solution for the problem. For this solution, $\mathbf{T} = \{(1, 2), (1, 3), (2, 4), (2, 5), (5, 6)\}$, $\mathbf{L} = \{(2, 3), (5, 4)\}$, and $\mathbf{U} = \{(3, 5), (4, 6)\}$. In this solution, arc (3, 5) has a positive violation, which is 1 unit. We introduce this arc into the tree creating a cycle whose apex is node 1. Since arc (3, 5) is at its upper bound, the orientation of the cycle is opposite to that of arc (3, 5). The arcs (1, 2) and (2, 5) are forward arcs in the cycle and arcs (3, 5) and (1, 3) are backward arcs. The maximum increase in flow permitted by the arcs (3, 5), (1, 3), (1, 2), and (2, 5) is, respectively, 3, 3, 2, and 1 units. Consequently, $\delta = 1$ and we augment 1 unit of flow along the cycle. The augmentation increases the flow on arcs (1, 2) and (2, 5) by one unit and decreases the flow on arcs (1, 3) and (3, 5) by one unit. Arc (2, 5) is the unique blocking arc and so we select it to leave the tree. Dropping arc (2, 5) from the tree produces two subtrees: T_1 consisting of nodes 1, 2, 3, 4 and T_2 consisting of nodes 5 and 6. Introducing arc (3, 5), we again obtain a spanning tree, as shown in Figure 11.11(c). Notice that in this spanning tree, the node potentials of nodes 5 and 6 are 1 unit less than that in the previous spanning tree.

In the feasible spanning tree solution shown in Figure 11.11(c), $\mathbf{L} = \{(2, 3), (5, 4)\}$ and $\mathbf{U} = \{(2, 5), (4, 6)\}$. In this solution, arc (4, 6) is the only eligible arc: its violation equals 1 unit. Therefore, we introduce arc (4, 6) into the tree. Figure 11.11(c) shows the resulting cycle and its orientation. We can augment 1 unit of additional flow along the orientation of this cycle. Sending this flow, we find that arc (3, 5) is a blocking arc, so we drop this arc from the current spanning tree. Figure 11.11(d)

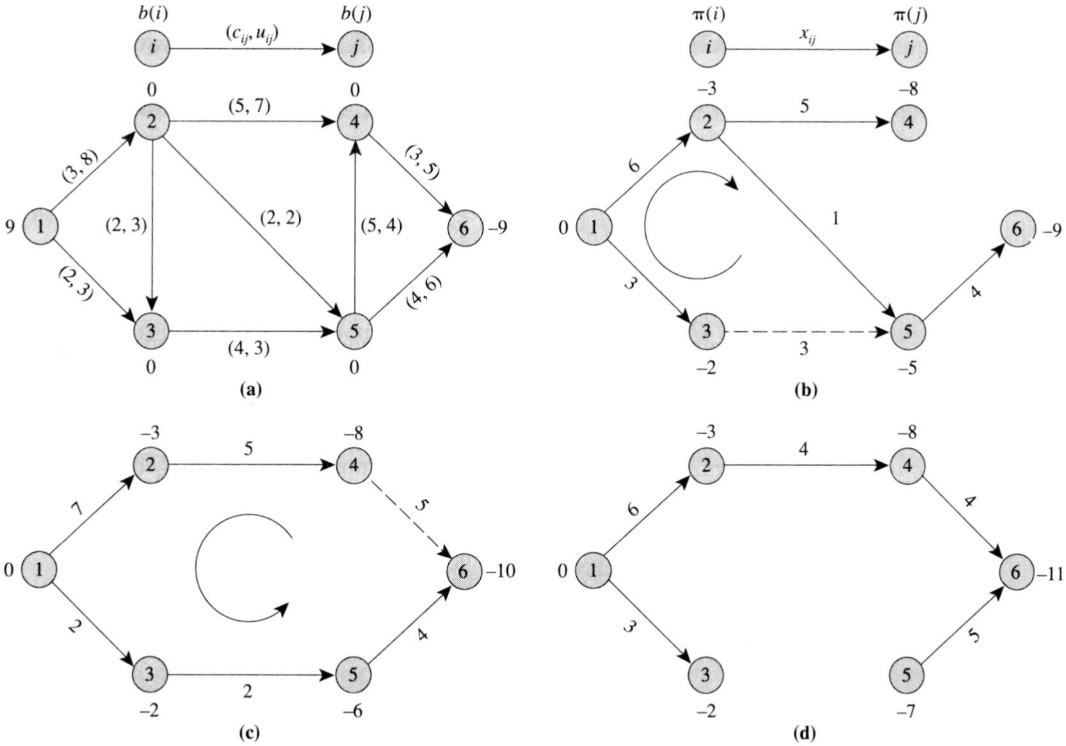

Figure 11.11 Numerical example for the network simplex algorithm.

shows the new spanning tree. As the reader can verify, this solution has no eligible arc, and thus the network simplex algorithm terminates with this solution.

11.6 STRONGLY FEASIBLE SPANNING TREES

The network simplex algorithm does not necessarily terminate in a finite number of iterations unless we impose some additional restriction on the choice of the entering and leaving arcs. Very small network examples show that a poor choice leads to *cycling* (i.e., an infinite repetitive sequence of degenerate pivots). Degeneracy in network problems is not only a theoretical issue, but also a practical one. Computational studies have shown that as many as 90% of the pivot operations in commonplace networks can be degenerate. As we show next, by maintaining a special type of spanning tree, called a *strongly feasible spanning tree*, the network simplex algorithm terminates finitely; moreover, it runs faster in practice as well.

Let (**T**, **L**, **U**) be a spanning tree structure for a minimum cost flow problem with integral data. As before, we conceive of a spanning tree as a tree hanging from the root node. The tree arcs are either *upward pointing* (toward the root) or are *downward pointing* (away from the root). We now state two alternate definitions of a strongly feasible spanning tree.

1. *Strongly feasible spanning tree*. A spanning tree **T** is *strongly feasible* if every tree arc with zero flow is upward pointing and every tree arc whose flow equals its capacity is downward pointing.
2. *Strongly feasible spanning tree*. A spanning tree **T** is *strongly feasible* if we can send a positive amount of flow from any node to the root along the tree path without violating any flow bound.

If a spanning tree **T** is strongly feasible, we also say that the spanning tree structure (**T**, **L**, **U**) is strongly feasible.

It is easy to show that the two definitions of the strongly feasible spanning trees are equivalent (see Exercise 11.24). Figure 11.12(a) gives an example of a strongly feasible spanning tree, and Figure 11.12(b) illustrates a feasible spanning tree that is not strongly feasible. The spanning tree shown in Figure 11.12(b) fails to be strongly feasible because arc (3, 5) carries zero flow and is downward pointing. Observe that in this spanning tree, we cannot send any additional flow from nodes 5 and 7 to the root along the tree path.

To implement the network simplex algorithm so that it always maintains a strongly feasible spanning tree, we must first find an initial strongly feasible spanning tree. The method described in Section 11.5 for constructing the initial spanning tree structure always gives such a spanning tree. Note that a nondegenerate spanning tree is always strongly feasible; a degenerate spanning tree might or might not be strongly feasible. The network simplex algorithm creates a degenerate spanning tree from a nondegenerate spanning tree whenever two or more arcs are qualified as

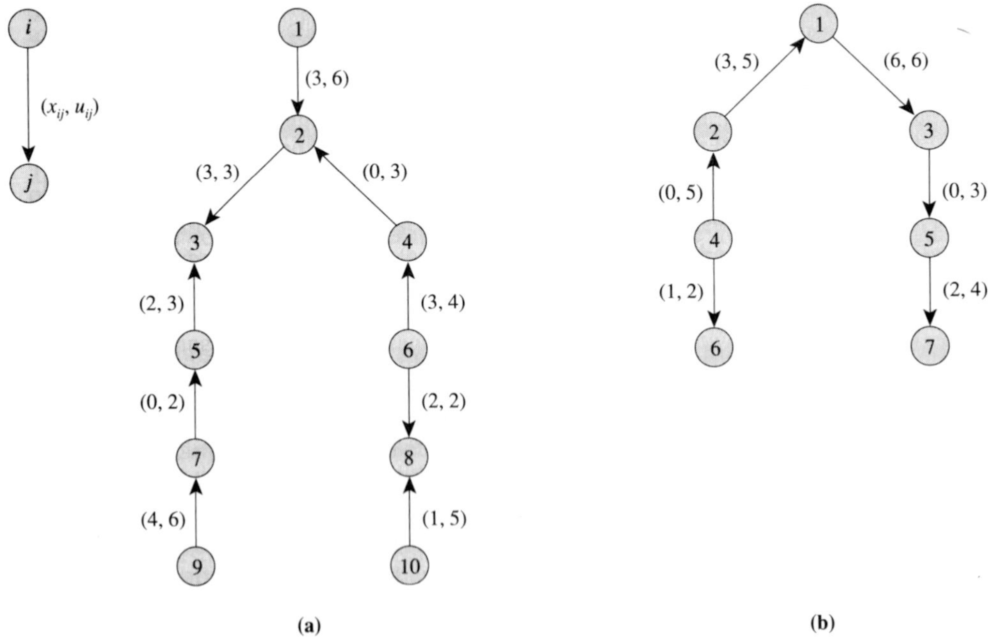

Figure 11.12 Feasible spanning trees: (a) strongly feasible; (b) nonstrongly feasible.

leaving arcs and we drop only one of these. Therefore, the algorithm needs to select the leaving arc carefully so that the next spanning tree is strongly feasible.

Suppose that we have a strongly feasible spanning tree and, during a pivot operation, arc (k, l) enters the spanning tree. We first consider the case when (k, l) is a nontree arc at its lower bound. Suppose that W is the pivot cycle formed by adding arc (k, l) to the spanning tree and that node w is the apex of the cycle W; that is, w is the first common ancestor of nodes k and l. We define the orientation of the cycle W as compatible with that of arc (k, l). After augmenting δ units of flow along the pivot cycle, the algorithm identifies the *blocking arcs* [i.e., those arcs (i, j) in the cycle that satisfy $\delta_{ij} = \delta$]. If the blocking arc is unique, we select it to leave the spanning tree. If the cycle contains more than one blocking arc, the next spanning tree will be degenerate (i.e., some tree arcs will be at their lower or upper bounds). In this case the algorithm selects the leaving arc in accordance with the following rule.

Leaving Arc Rule. *Select the leaving arc as the last blocking arc encountered in traversing the pivot cycle W along its orientation starting at the apex w.*

To illustrate the leaving arc rule, we consider a numerical example. Figure 11.13 shows a strongly feasible spanning tree for this example. Let (9, 10) be the entering arc. The pivot cycle is 10–8–6–4–2–3–5–7–9–10 and the apex is node 2. This pivot is degenerate because arcs (2, 3) and (7, 5) block any additional flow in the pivot cycle. Traversing the pivot cycle starting at node 2, we encounter arc (7, 5) later than arc (2, 3); so we select arc (7, 5) as the leaving arc.

We show that the leaving arc rule guarantees that in the next spanning tree every node in the cycle W can send a positive amount of flow to the root node. Let

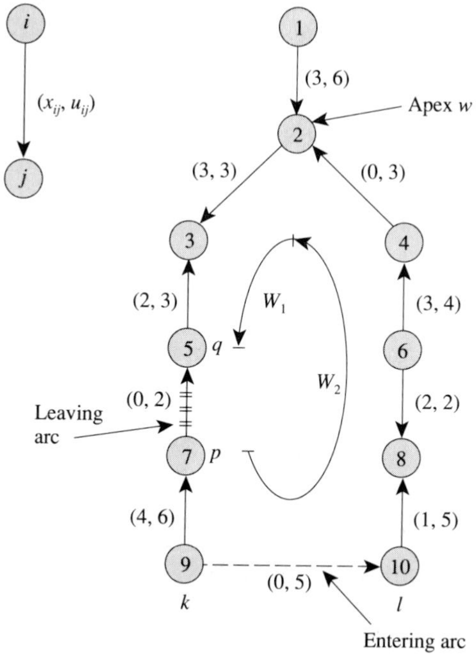

Figure 11.13 Selecting the leaving arc.

(p, q) be the arc selected by the leaving arc rule. Let W_1 be the segment of the cycle W between the apex w and arc (p, q) when we traverse the cycle along its orientation. Let $W_2 = W - W_1 - \{(p, q)\}$. Define the orientation of segments W_1 and W_2 as compatible with the orientation of W. See Figure 11.13 for an illustration of the segments W_1 and W_2. We use the following property about the nodes in the segment W_2.

Property 11.6. *Each node in the segment W_2 can send a positive amount of flow to the root in the next spanning tree.*

This observation follows from the fact that arc (p, q) is the last blocking arc in W; consequently, no arc in W_2 is blocking and every node in this segment can send a positive amount of flow to the root via node w along the orientation of W_2. Note that if the leaving arc does not satisfy the leaving arc rule, no node in the segment W_2 can send a positive amount of flow to the root; therefore, the next spanning tree will not be strongly feasible.

We next focus on the nodes contained in the segment W_1.

Property 11.7. *Each node in the segment W_1 can send a positive amount of flow to the root in the next spanning tree.*

We prove this observation by considering two cases. If the previous pivot was a nondegenerate pivot, the pivot augmented a positive amount of flow δ along the arcs in W_1; consequently, after the augmentation, every node in the segment W_1 can send a positive amount of flow back to the root opposite to the orientation of W_1 via the apex node w (each node can send at least δ units to the apex and then at least some of this flow to the root since the previous spanning tree was strongly feasible). If the previous pivot was a degenerate pivot, W_1 must be contained in the segment of W between node w and node k because the property of strong feasibility implies that every node on the path from node l to node w can send a positive amount of flow to the root before the pivot, and thus no arc on this path can be a blocking arc in a degenerate pivot. Now observe that before the pivot, every node in W_1 could send a positive amount of flow to the root, and therefore since the pivot does not change flow values, every node in W_1 must be able to send a positive amount of flow to the root after the pivot as well. This conclusion completes the proof that in the next spanning tree every node in the cycle W can send a positive amount of flow to the root node.

We next show that in the next spanning tree, nodes not belonging to the cycle W can also send a positive amount of flow to the root. In the previous spanning tree (before the augmentation), every node j could send a positive amount of flow to the root and if the tree path from node j does not pass through the cycle W, the same path is available to carry a positive amount of flow in the next spanning tree. If the tree path from node j does pass through the cycle W, the segment of this tree path to the cycle W is available to carry a positive amount of flow in the next spanning tree and once a positive amount of flow reaches the cycle W, then, as shown earlier, we can send it (or some of it) to the root node. This conclusion completes the proof that the next spanning tree is strongly feasible.

We now establish the finiteness of the network simplex algorithm. Since we have previously shown that each nondegenerate pivot strictly decreases the objective function value, the number of nondegenerate pivots is finite. The algorithm can, however, also perform degenerate pivots. We will show that the number of successive degenerate pivots between any two nondegenerate pivots is finitely bounded. Suppose that arc (k, l) enters the spanning tree at its lower bound and in doing so it defines a degenerate pivot. In this case, the leaving arc belongs to the tree path from node k to the apex w. Now observe from Section 11.5 that node k lies in the subtree T_2 and the potentials of all nodes in T_2 change by an amount c_{kl}^{π}. Since $c_{kl}^{\pi} < 0$, this degenerate pivot strictly decreases the sum of all node potentials (which by our prior assumption is integral). Since no node potential can fall below $-nC$, the number of successive degenerate pivots is finite.

So far we have assumed that the entering arcs are always at their lower bounds. If the entering arc (k, l) is at its upper bound, we define the orientation of the cycle W as opposite to the orientation of arc (k, l). The criteria for selecting the leaving arc remains unchanged—the leaving arc is the last blocking arc encountered in traversing W along its orientation starting at the apex w. In this case node l is contained in the subtree T_2, and thus after the pivot, the potentials of all the nodes T_2 decrease by the amount $c_{kl}^{\pi} > 0$; consequently, the pivot again decreases the sum of the node potentials.

11.7 NETWORK SIMPLEX ALGORITHM FOR THE SHORTEST PATH PROBLEM

In this section we see how the network simplex algorithm specializes when applied to the shortest path problem. The resulting algorithm bears a close resemblance to the label-correcting algorithms discussed in Chapter 5. In this section we study the version of the shortest path problem in which we wish to determine shortest paths from a given source node s to all other nodes in a network. In other words, the problem is to send 1 unit of flow from the source to every other node along minimum cost paths. We can formulate this version of the shortest path problem as the following minimum cost flow model:

$$\text{Minimize} \sum_{(i,j)\in A} c_{ij}x_{ij}$$

subject to

$$\sum_{\{j:(i,j)\in A\}} x_{ij} - \sum_{\{j:(j,i)\in A\}} x_{ji} = \begin{cases} n-1 & \text{for } i = s \\ -1 & \text{for all } i \in N - \{s\} \end{cases}$$

$$x_{ij} \geq 0 \qquad \text{for all } (i, j) \in A.$$

If the network contains a negative (cost)-directed cycle, this linear programming formulation would have an unbounded solution since we could send an infinite amount of flow along this cycle without violating any of the constraints (because the arc flows have no upper bounds). The network simplex algorithm we describe is capable of detecting the presence of a negative cycle, and if the network contains no such cycle, it determines the shortest path distances.

Like other minimum cost flow problems, the shortest path problem has a spanning tree solution. Because node s is the only source node and all the other nodes are demand nodes, the tree path from the source node to every other node is a directed path. This observation implies that the spanning tree must be a *directed out-tree rooted at node s* (see Figure 11.14 and the discussion in Section 4.3). As before, we store this tree using predecessor, depth, and thread indices. In a directed out-tree, every node other than the source has exactly one incoming arc but could have several outgoing arcs. Since each node except node s has unit demand, the flow of arc (i, j) is $|D(j)|$. [Recall that $D(j)$ is the set of descendants of node j in the spanning tree and, by definition, this set includes node j.] Therefore, every tree of the shortest path problem is nondegenerate, and consequently, the network simplex algorithm will never perform degenerate pivots.

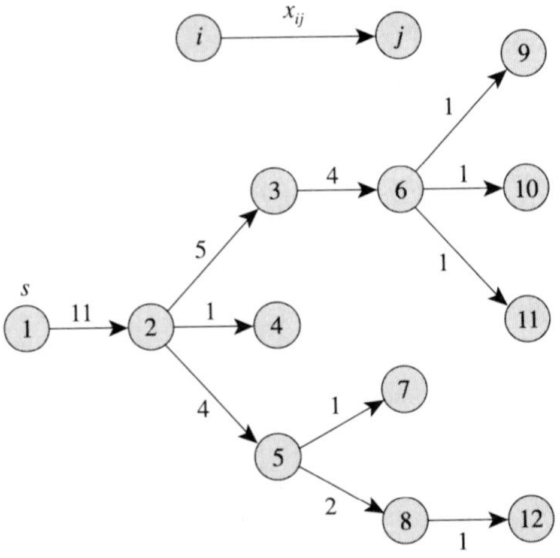

Figure 11.14 Directed out-tree rooted at the source.

Any spanning tree for the shortest path problem contains a unique directed path from node s to every other node. Let $P(k)$ denote the path from node s to node k. We obtain the node potentials corresponding to the tree **T** by setting $\pi(s) = 0$ and then using the equation $c_{ij} - \pi(i) + \pi(j) = 0$ for each arc $(i, j) \in \mathbf{T}$ by fanning out from node s (see Figure 11.15). The directed out-tree property of the spanning tree implies that $\pi(k) = -\sum_{(i,j)\in P(k)} c_{ij}$. Thus $\pi(k)$ is the negative of the length of the path $P(k)$.

Since the variables in the minimum cost flow formulation of the shortest path problem have no upper bounds, every nontree arc is at its lower bound. The algorithm selects a nontree arc (k, l) with a negative reduced cost to introduce into the spanning tree. The addition of arc (k, l) to the tree creates a cycle which we orient in the same direction as arc (k, l). Let w be the apex of this cycle. (See Figure 11.16 for an illustration.) In this cycle, every arc from node l to node w is a backward arc and every arc from node w to node k is a forward arc. Consequently, the leaving arc would lie in the segment from node l to node w. In fact, the leaving arc would be the arc (pred(l), l) because this arc has the smallest flow value among all arcs in the segment from node l to node w. The algorithm would then increase the potentials of nodes in the subtree rooted at the node l by an amount $|c^{\pi}_{kl}|$, update the tree

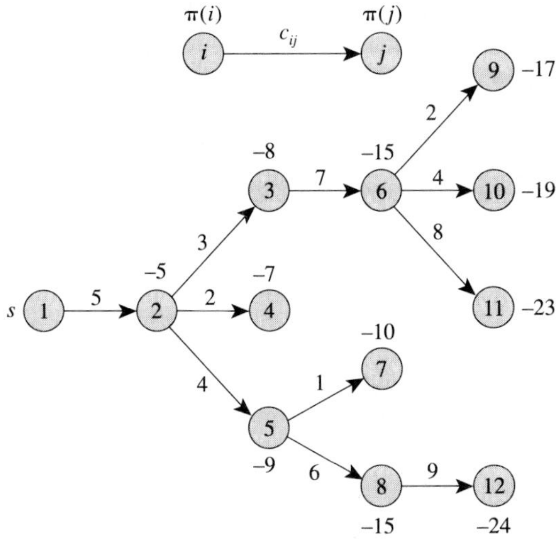

Figure 11.15 Computing node potentials.

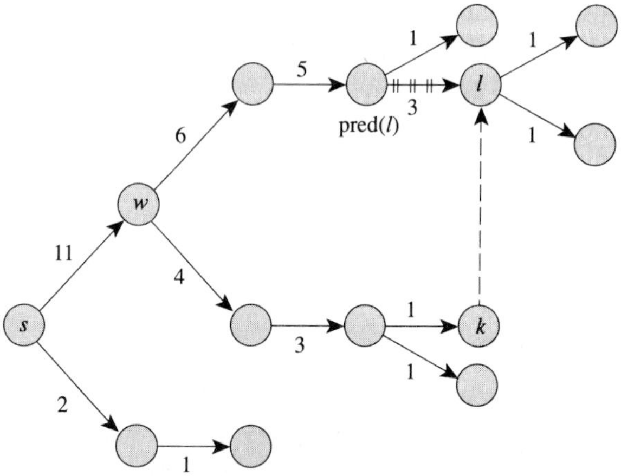

Figure 11.16 Selecting the leaving arc.

indices, and repeat the computations until all nontree arcs have nonnegative reduced costs. When the algorithm terminates, the final tree would be a shortest path tree (i.e., a tree in which the directed path from node s to every other node is a shortest path).

Recall that in implementing the network simplex algorithm for the minimum cost flow problem, we maintained flow values for all the arcs because we needed these values to identify the leaving arc. For the shortest path problem, however, we can determine the leaving arc without considering the flow values. If (k, l) is the entering arc, then $(\text{pred}(l), l)$ is the leaving arc. Thus the network simplex algorithm for the shortest path problem need not maintain arc flows. Moreover, updating of the tree indices is simpler for the shortest path problem.

The network simplex algorithm for the shortest path problem is similar to the label-correcting algorithms discussed in Section 5.3. Recall that a label-correcting algorithm maintains distance labels $d(i)$, searches for an arc satisfying the condition $d(j) > d(i) + c_{ij}$, and sets the distance label of node j equal to $d(i) + c_{ij}$. In the network simplex algorithm, if we define $d(i) = -\pi(i)$, then $d(i)$ are the valid distance labels (i.e., they represent the length of some directed path from source to node i). At each iteration the network simplex algorithm selects an arc (i, j) with $c_{ij}^{\pi} < 0$. Observe that $c_{ij}^{\pi} = c_{ij} - \pi(i) + \pi(j) = c_{ij} + d(i) - d(j)$. Therefore, like a label-correcting algorithm, the network simplex algorithm selects an arc that satisfies the condition $d(j) > d(i) + c_{ij}$. The algorithm then increases the potential of every node in the subtree rooted at node j by an amount $| c_{ij}^{\pi} |$ which amounts to decreasing the distance label of all the nodes in the subtree rooted at node j by an amount $| c_{ij}^{\pi} |$. In this regard the network simplex algorithm differs from the label correcting algorithm: instead of updating one distance label at each step, it updates several of them.

If the network contains no negative cycle, the network simplex algorithm would terminate with a shortest path tree. When the network does contain a negative cycle, the algorithm would eventually encounter a situation like that depicted in Figure 11.17. This type of situation will occur only when the tail of the entering arc (k, l) belongs to $D(l)$, the set of descendants of node l. The network simplex algorithm can detect this situation easily without any significant increase in its computational effort: After introducing an arc (k, l), the algorithm updates the potentials of all nodes in $D(l)$; at that time, it can check to see whether $k \in D(l)$, and if so, then terminate. In this case, tracing the predecessor indices would yield a negative cycle.

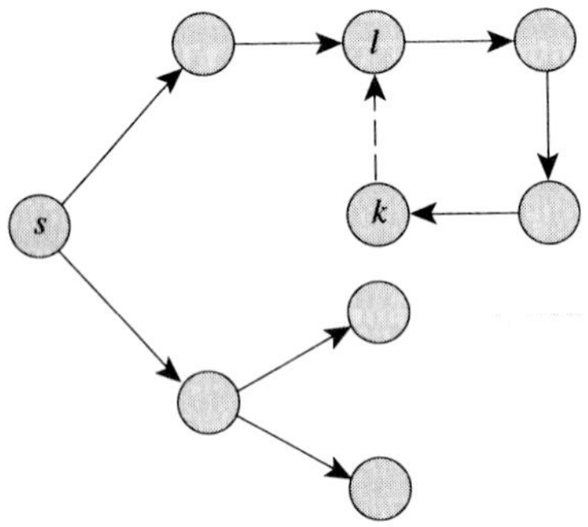

Figure 11.17 Detecting a negative cycle in the network.

The generic version of the network simplex algorithm for the shortest path problem runs in pseudopolynomial time. This result follows from the facts that (1) for each node i, $-nC \leq \pi(i) \leq nC$ (because the length of every directed path from s to node i lies between $-nC$ to nC), and (2) each iteration increases the value of at least one node potential. We can, however, develop special implementations that run in polynomial time. In the remainder of this section, in the exercises, and in the reference notes at the end of this chapter, we describe several polynomial-time implementations of the network simplex algorithm for the shortest path problem. These algorithms will solve the shortest path problem in $O(n^2m)$, $O(n^3)$, and $O(nm \log C)$ time. We obtain these polynomial-time algorithms by carefully selecting the entering arcs.

First eligible arc pivot rule. We have described this pivot rule in Section 11.5. The network simplex algorithm with this pivot rule bears a strong resemblance with the FIFO label-correcting algorithm that we described in Section 5.4. Recall that the FIFO label-correcting algorithm examines the arc list in a wraparound fashion: If an arc (i, j) violates its optimality condition (i.e., it is eligible), the algorithm updates the distance label of node j. This order for examining the arcs ensures that after the kth pass, the algorithm has computed the shortest path distances to all those nodes that have a shortest path with k or fewer arcs. The network simplex algorithm with the first eligible arc pivot rule also examines the arc list in a wraparound fashion, and if an arc (i, j) is eligible (i.e., violates its optimality condition), it updates the distances label of every node in $D(j)$, which also includes j. Consequently, this pivot rule will also, within k passes, determine shortest path distances to all those nodes that are connected to the source node s by a shortest path with k or fewer arcs. Consequently, the network simplex algorithm will perform at most n passes over the arc list. As a result, the algorithm will perform at most nm pivots and run in $O(n^2m)$ time. In Exercise 11.30 we discuss a modification of this algorithm that runs in $O(n^3)$ time.

Dantzig's pivot rule. This pivot rule selects the entering arc as an arc with the maximum violation. Let C denote the largest arc cost. We will show that the network simplex algorithm with this pivot rule performs $O(n^2 \log(nC))$ pivots and so runs in $O(n^2m \log(nC))$ time.

Scaled pivot rule. This pivot is a scaled variant of Dantzig's pivot rule. In this pivot rule we perform a number of scaling phases with varying values of a scaling parameter Δ. Initially, we let $\Delta = 2^{\lceil \log C \rceil}$ (i.e., we set Δ equal to the smallest power of 2 greater than or equal to C) and pivot in any nontree arc with a violation of at least $\Delta/2$. When no arc has a violation of at least $\Delta/2$, we replace Δ by $\Delta/2$ and repeat the steps. We terminate the algorithm when $\Delta < 1$.

We now show that the network simplex algorithm with the scaled pivot rule solves the shortest path problem in polynomial time. It is easy to verify that Dantzig's pivot rule is a special case of scaled pivot rule, so this result also shows that when implemented with Dantzig's pivot rule, the network simplex algorithm requires polynomial time.

We call the sequence of iterations for which Δ remains unchanged as the *Δ-scaling phase*. Let π denote the set of node potentials at the beginning of a Δ-scaling phase. Moreover, let $P^*(p)$ denote a shortest path from node s to node p and let $\pi^*(p) = -\sum_{(i,j)\in P^*} c_{ij}$ denote the optimal node potential of node p. Our analysis of the scaled pivot rule uses the following lemma:

Lemma 11.8. *If π denotes the current node potentials at the beginning of the Δ-scaling phase, then $\pi^*(p) - \pi(p) \le 2n\Delta$ for each node p.*

Proof. In the first scaling phase, $\Delta \ge C$ and the lemma follows from the facts that $-nC$ and nC are the lower and upper bounds on any node potentials (why?). Consider next any subsequent scaling phase. Property 9.2 implies that

$$\sum_{(i,j)\in P^*(k)} c_{ij}^{\pi} = \sum_{(i,j)\in P^*(k)} c_{ij} - \pi(s) + \pi(p) = \pi(p) - \pi^*(p). \qquad (11.4)$$

Since Δ is an upper bound on the maximum arc violation at the beginning of the Δ-scaling phase (except the first one), $c_{ij}^{\pi} \ge -\Delta$ for every arc $(i, j) \in A$. Substituting this inequality in (11.4), we obtain

$$\pi(p) - \pi^*(p) \geq -\Delta| P^*(p) | \geq -n\Delta,$$

which implies the conclusion of the lemma.

Now consider the potential function $\Phi = \sum_{p \in N} (\pi^*(p) - \pi(p))$. The preceding lemma shows that at the beginning of each scaling phase, Φ is at most $2n^2\Delta$. Now, recall from our previous discussion in this section that in each iteration, the network simplex algorithm increases at least one node potential by an amount equal to the violation of the entering arc. Since the entering arc has violation at least $\Delta/2$, at least one node potential increases by $\Delta/2$ units, causing Φ to decrease by at least $\Delta/2$ units. Since no node potential ever decreases, the algorithm can perform at most $4n^2$ iterations in this scaling phase. So, after at most $4n^2$ iterations, either the algorithm will obtain an optimal solution or will complete the scaling phase. Since the algorithm performs $O(\log C)$ scaling phases, it will perform $O(n^2 \log C)$ iterations and so require $O(n^2 m \log C)$ time. It is, however, possible to implement this algorithm in $O(nm \log C)$ time; the reference notes provide a reference for this result.

11.8 NETWORK SIMPLEX ALGORITHM FOR THE MAXIMUM FLOW PROBLEM

In this section we describe another specialization of the network simplex algorithm: its implementation for solving the maximum flow problem. The resulting algorithm is essentially an augmenting path algorithm, so it provides an alternative proof of the max-flow min-cut theorem we discussed in Section 6.5.

As we have noted before, we can view the maximum flow problem as a particular version of the minimum cost flow problem, obtained by introducing an additional arc (t, s) with cost coefficient $c_{ts} = -1$ and an upper bound $u_{ts} = \infty$, and by setting $c_{ij} = 0$ for all the original arcs (i, j) in A. To simplify our notation, we henceforth assume that A represents the set $A \cup \{(t, s)\}$. The resulting formulation is to

$$\text{Minimize} \quad -x_{ts}$$

subject to

$$\sum_{\{j:(i,j) \in A\}} x_{ij} - \sum_{\{j:(j,i) \in A\}} x_{ji} = 0 \qquad \text{for all } i \in N,$$

$$0 \leq x_{ij} \leq u_{ij} \qquad \text{for all } (i, j) \in A.$$

Observe that minimizing $-x_{ts}$ is equivalent to maximizing x_{ts}, which is equivalent to maximizing the net flow sent from the source to the sink, since this flow returns to the source via arc (t, s). This observation explains why the inflow equals the outflow for every node in the network, including the source and the sink nodes.

Note that in any feasible spanning tree solution that carries a positive amount of flow from the source to the sink (i.e., $x_{ts} > 0$), arc (t, s) must be in the spanning tree. Consequently, the spanning tree for the maximum flow problem consists of two subtrees of G joined by the arc (t, s) (see Figure 11.18). Let T_s and T_t denote the subtrees containing nodes s and t.

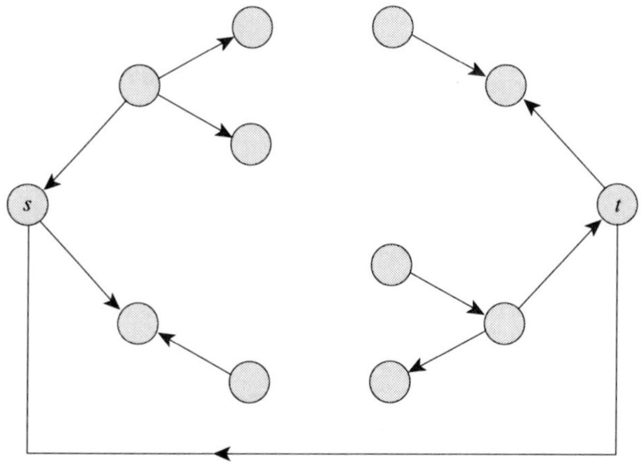

Figure 11.18 Spanning tree for the maximum flow problem.

We obtain node potentials corresponding to a feasible spanning tree of the maximum flow problem as follows. Since we can set one node potential arbitrarily, let $\pi(t) = 0$. Furthermore, since the reduced cost of arc (t, s) must be zero, $0 = c_{ts}^{\pi} = c_{ts} - \pi(t) + \pi(s) = -1 + \pi(s)$, which implies that $\pi(s) = 1$. Since (1) the reduced cost of every arc in T_s and T_t must be zero, and (2) the costs of these arcs are also zero, the node potentials have the following values: $\pi(i) = 1$ for every node $i \in T_s$, and $\pi(i) = 0$ for every node $i \in T_t$.

Notice that every spanning tree solution of the maximum flow problem defines an s–t cut $[S, \bar{S}]$ in the original network obtained by setting $S = T_s$ and $\bar{S} = T_t$. Each arc in this cut is a nontree arc; its flow has value zero or equals the arc's capacity. For every forward arc (i, j) in the cut, $c_{ij}^{\pi} = -1$, and for every backward arc (i, j) in the cut, $c_{ij}^{\pi} = 1$. Moreover, for every arc (i, j) not in the cut, $c_{ij}^{\pi} = 0$. Consequently, if every forward arc in the cut has a flow value equal to the arc's capacity and every backward arc has zero flow, this spanning tree solution satisfies the optimality conditions (11.1), and therefore it must be optimal. On the other hand, if in the current spanning tree solution, some forward arc in the cut has a flow of value zero or the flow on some backward arc equals the arc's capacity, all these arcs have a violation of 1 unit. Therefore, we can select any of these arcs to enter the spanning tree. Suppose that we select arc (k, l). Introducing this arc into the tree creates a cycle that contains arc (t, s) as a forward arc (see Figure 11.19). The algorithm augments the maximum possible flow in this cycle and identifies a blocking arc. Dropping this arc again creates two subtrees joined by the arc (t, s). This new tree constitutes a spanning tree for the next iteration.

Notice that this algorithm is an augmenting path algorithm: The tree structure permits us to determine the path from the source to the sink very easily. In this sense the network simplex algorithm has an advantage over other types of augmenting path algorithms for the maximum flow problem. As a compensating factor, however, due to degeneracy, the network simplex algorithm might not send a positive amount of flow from the source to the sink in every iteration.

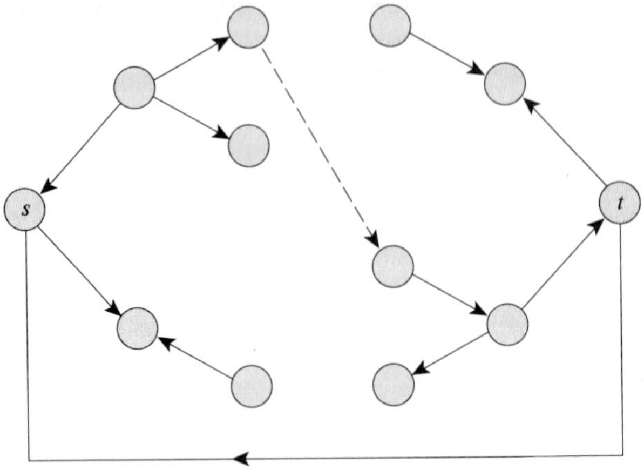

Figure 11.19 Forming a cycle.

The network simplex algorithm for the maximum flow problem gives another proof of the max-flow min-cut theorem. The algorithm terminates when every forward arc in the cut is capacitated and every backward arc has a flow of value zero. This termination condition implies that the current maximum flow value equals the capacity of the s–t cut defined by the subtrees T_s and T_t, and thus the value of a maximum flow from node s to node t equals the capacity of the minimum s–t cut.

Just as the mechanics of the network simplex algorithm becomes simpler in the context of the maximum flow problem, so does the concept of a strongly feasible spanning tree. If we designate the sink as the root node, the definition of a strongly feasible spanning tree implies that we can send a positive amount of flow from every node in T_t to the sink node t without violating any of the flow bounds. Therefore, every arc in T_t whose flow value is zero must point toward the sink node t and every arc in T_t whose flow value equals the arc's upper bound must point away from node t. Moreover, the leaving arc criterion reduces to selecting a blocking arc in the pivot cycle that is farthest from the sink node when we traverse the cycle in the direction of arc (t, s) starting from node t. Each degenerate pivot selects an entering arc that is incident to some node in T_t. The preceding observation implies that each blocking arc must be an arc in T_s. Consequently, each degenerate pivot increases the size of T_t, so the algorithm can perform at most n consecutive degenerate pivots. We might note that the minimum cost flow problem does not satisfy this property: For the more general problem, the number of consecutive degenerate pivots can be exponentially large.

The preceding discussion shows that when implemented to maintain a strongly feasible spanning tree, the network simplex algorithm performs $O(n^2U)$ iterations for the maximum flow problem. This result follows from the fact that the number of nondegenerate pivots is at most nU, an upper bound on the maximum flow value. This bound on the number of iterations is nonpolynomial, so is not satisfactory from a theoretical perspective. Developing a polynomial-time network simplex algorithm for the maximum flow problem remained an open problem for quite some time. However, researchers have recently suggested an entering arc rule that performs only $O(nm)$ iterations and can be implemented to run in $O(n^2m)$ time. This rule selects entering arcs so that the algorithm augments flow along shortest paths from the source to the sink. We provide a reference for this result in the reference notes.

11.9 RELATED NETWORK SIMPLEX ALGORITHMS

In this section we study two additional algorithms for the minimum cost flow problem—the *parametric network simplex algorithm* and the *dual network simplex algorithm*—that are close relatives of the network simplex algorithm. In contrast to the network simplex algorithm, which maintains a feasible solution at each intermediate step, both of these algorithms maintain an infeasible solution that satisfies the optimality conditions; they iteratively attempt to transform this solution into a feasible solution. The solution maintained by the parametric network simplex algorithm satisfies all of the problem constraints except the mass balance constraints at two specially designated nodes, s and t. The solution maintained by the dual network simplex algorithm satisfies all of the mass balance constraints but might violate the lower and upper bound constraints on some arc flows. Like the network simplex algorithm, both algorithms maintain a spanning tree at every step and perform all computations using the spanning tree.

Parametric Network Simplex Algorithm

For the sake of simplicity, we assume that the network has one supply node (the source s) and one demand node (the sink t). We incur no loss of generality in imposing this assumption because we can always transform a network with several supply and demand nodes into one with a single supply and a single demand node. The parametric network simplex algorithm starts with a solution for which $b'(s) = -b'(t) = 0$, and gradually increases $b'(s)$ and $-b'(t)$ until $b'(s) = b(s)$ and $b'(t) = b(t)$. Let **T** be a shortest path tree rooted at the source node s in the underlying network. The parametric network simplex algorithm starts with zero flow and with (**T**, **L**, **U**) with $\mathbf{L} = A - \mathbf{T}$ and $\mathbf{U} = \varnothing$ as the initial spanning tree structure. Since, by Assumption 9.5, all the arc costs are nonnegative, the zero flow is an optimal flow provided that $b(s) = b(t) = 0$. Moreover, since **T** is a shortest path tree, the shortest path distances $d(\cdot)$ to the nodes satisfy the condition $d(j) = d(i) + c_{ij}$ for each $(i, j) \in \mathbf{T}$, and $d(j) \le d(i) + c_{ij}$ for each $(i, j) \notin \mathbf{T}$. By setting $\pi(j) = -d(j)$ for each node j, these shortest path optimality conditions become the optimality conditions (11.1) of the initial spanning tree structure (**T**, **L**, **U**).

Thus the parametric network simplex algorithm starts with an optimal solution of a minimum cost flow problem that violates the mass balance constraints only at the source and sink nodes. In the subsequent steps, the algorithm maintains optimality of the solution and attempts to satisfy the violated constraints by sending flow from node s to node t along tree arcs. The algorithm stops when it has sent the desired amount ($b(s) = -b(t)$) of flow.

In each iteration the algorithm performs the following computations. Let (**T**, **L**, **U**) be the spanning tree structure at some iteration. The spanning tree **T** contains a unique path P from node s to node t. The algorithm first determines the maximum amount of flow δ that can be sent from s to t along P while honoring the flow bounds on the arcs. Let $\overline{P}$ and $\underline{P}$ denote the sets of forward and backward arcs in P. Then

$$\delta = \min[\min\{u_{ij} - x_{ij}:(i, j) \in \overline{P}\}, \min\{x_{ij}:(i, j) \in \underline{P}\}].$$

The algorithm either sends δ units of flow along P, or a smaller amount if it would be sufficient to satisfy the mass balance constraints at nodes s and t. As in the network simplex algorithm, all the tree arcs have zero reduced costs; therefore, sending additional flow along the tree path from node s to node t preserves the optimality of the solution. If the solution becomes feasible after the augmentation, the algorithm terminates. If the solution is still infeasible, the augmentation creates at least one *blocking arc* (i.e., an arc that prohibits us from sending additional flow from node s to node t). We select one such blocking arc, say (p, q), as the *leaving arc* and replace it by some nontree arc (k, l), called the *entering arc*, so that the next spanning tree both (1) satisfies the optimality condition, and (2) permits additional flow to be sent from node s to node t. We accomplish this transition from one tree to another by performing a *dual pivot*. Recall from Section 11.5 that a (primal) pivot first identifies an entering arc and then the leaving arc. In contrast, a dual pivot first selects the leaving arc and then identifies the entering arc.

We perform a dual pivot in the following manner. We first drop the leaving arc from the spanning tree. Doing so gives us two subtrees T_s and T_t, with $s \in T_s$ and $t \in T_t$. Let S and $\overline{S}$ be the subsets of nodes spanned by these two subtrees. Clearly, the cut $[S, \overline{S}]$ is an s–t cut and the entering arc (k, l) must belong to $[S, \overline{S}]$ if the next solution is to be a spanning tree solution. As earlier, we let $(S, \overline{S})$ denote the set of forward arcs and $(\overline{S}, S)$ the set of backward arcs in the cut $[S, \overline{S}]$. Each arc in the cut $[S, \overline{S}]$ is at its lower bound or at its upper bound. We define the set Q of *eligible arcs* as the set

$$Q = ((S, \overline{S}) \cap \mathbf{L}) \cup ((\overline{S}, S) \cap \mathbf{U}),$$

that is, the set of forward arcs at their lower bound and the set of backward arcs at their upper bound. Note that if we add a noneligible arc to the subtrees T_s and T_t, we cannot increase the flow from node s to node t along the new tree path joining these nodes (since the arc lies on the path and would be a forward arc at its upper bound or a backward arc at its lower bound). If we introduce an eligible arc, the new path from node s to node t might be able to carry a positive amount of flow. Next, notice that if $Q = \varnothing$, we can send no additional flow from node s to node t. In fact, the cut $[S, \overline{S}]$ has zero residual capacity and the current flow from node s to node t equals the maximum flow. If $b(s)$ is larger than this flow value, the minimum cost flow problem is infeasible. We now focus on situations in which $Q \neq \varnothing$. Notice that we cannot select an arbitrary eligible arc as the entering arc, because the new spanning tree must also satisfy the optimality condition. For each eligible arc (i, j), we define a number θ_{ij} in the following manner:

$$\theta_{ij} = \begin{cases} c_{ij}^{\pi} & \text{if } (i, j) \in \mathbf{L}, \\ -c_{ij}^{\pi} & \text{if } (i, j) \in \mathbf{U}. \end{cases}$$

Since the spanning tree structure $(\mathbf{T}, \mathbf{L}, \mathbf{U})$ satisfies the optimality condition (11.1), $\theta_{ij} \geq 0$ for every eligible arc (i, j). Suppose that we select some eligible arc (k, l) as the entering arc. It is easy to see that adding the arc (k, l) to $T_s \cup T_t$ decreases the potential of each node in $\overline{S}$ by θ_{kl} units (throughout the computations, we maintain that the node potential of the source node s has value zero). This change in node potentials decreases the reduced cost of each arc in $(S, \bar{S})$ by θ_{kl} units and increases the reduced cost of each arc in $(\bar{S}, S)$ by θ_{kl} units. We have four cases to consider.

Case 1. $(i, j) \in (S, \bar{S}) \cap L$

The reduced cost of the arc (i, j) becomes $c_{ij}^{\pi} - \theta_{kl}$. The arc will satisfy the optimality condition (11.1b) if $\theta_{kl} \le c_{ij}^{\pi} = \theta_{ij}$.

Case 2. $(i, j) \in (S, \bar{S}) \cap U$

The reduced cost of the arc (i, j) becomes $c_{ij}^{\pi} - \theta_{kl}$. The arc will satisfy the optimality condition (11.1c) regardless of the value of θ_{kl} because $c_{ij}^{\pi} \le 0$.

Case 3. $(i, j) \in (\bar{S}, S) \cap L$

The reduced cost of the arc (i, j) becomes $c_{ij}^{\pi} + \theta_{kl}$. The arc will satisfy the optimality condition (11.1b) regardless of the value of θ_{kl} because $c_{ij}^{\pi} \ge 0$.

Case 4. $(i, j) \in (\bar{S}, S) \cap U$

The reduced cost of the arc (i, j) becomes $c_{ij}^{\pi} + \theta_{kl}$. The arc will satisfy the optimality condition (11.1c) provided that $\theta_{kl} \le -c_{ij}^{\pi} = \theta_{ij}$.

The preceding discussion implies that the new spanning tree structure will satisfy the optimality conditions provided that

$$\theta_{kl} \le \theta_{ij} \text{ for each } (i, j) \in ((S, \bar{S}) \cap \mathbf{L}) \cup ((\bar{S}, S) \cap \mathbf{U}) \equiv Q.$$

Consequently, we select the entering arc (k, l) to be an eligible arc for which $\theta_{kl} = \min\{\theta_{ij}:(i, j) \in Q\}$. Adding the arc (k, l) to the subtrees T_s and T_t gives us a new spanning tree structure and completes an iteration. We refer to this dual pivot as *degenerate* if $\theta_{kl} = 0$, and as *nondegenerate* otherwise. We repeat this process until we have sent the desired amount of flow from node s to node t.

It is easy to implement the parametric network simplex algorithm so that it runs in pseudopolynomial time. In this implementation, if an augmentation creates several blocking arcs, we select the one closest to the source as the leaving arc. Using inductive arguments, it is possible to show that in this implementation, the subtree T_s will permit us to augment a positive amount of flow from node s to every other node in T_s along the tree path. Moreover, in each iteration, when the algorithm sends no additional flow from node s to node t, it adds at least one new node to T_s. Consequently, after at most n iterations, the algorithm will send a positive amount of flow from node s to node t. Therefore, the parametric network simplex algorithm will perform $O(nb(s))$ iterations.

To illustrate the parametric network simplex algorithm, let us consider the same example we used to illustrate the network simplex algorithm. Figure 11.20(a) gives the minimum cost flow problem if we choose $s = 1$ and $t = 6$. Figure 11.20(b) shows the tree of shortest paths. All the nontree arcs are at their lower bounds. In the first iteration, the algorithm augments the maximum possible flow from node 1 to node 6 along the tree path 1–2–5–6. This path permits us to send a maximum of $\delta = \min\{u_{12}, u_{25}, u_{56}\} = \min\{8, 2, 6\} = 2$ units of flow. Augmenting 2 units along the path creates the unique blocking arc (2, 5). We drop arc (2, 5) from the tree, creating

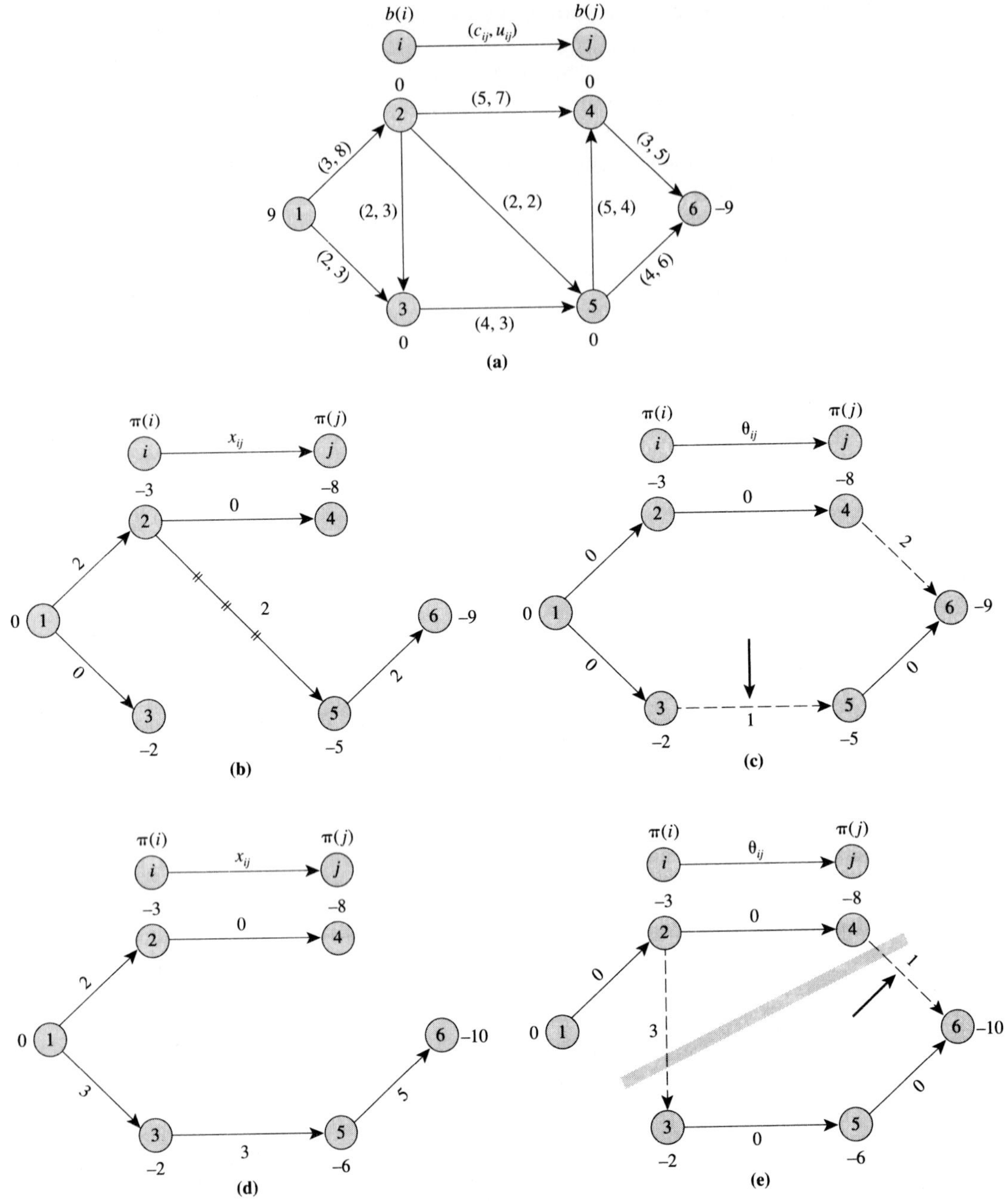

Figure 11.20 Illustrating the parametric network simplex algorithm.

the s–t cut $[S, \bar{S}]$ with $S = \{1, 2, 3, 4\}$ [see Figure 11.20(c)]. This cut contains two eligible arcs: arcs (3, 5) and (4, 6) with $\theta_{35} = 1$ and $\theta_{46} = 2$. We select arc (3, 5) as the entering arc, creating the spanning tree shown in Figure 11.20(d). Notice that the potentials of the nodes 5 and 6 increase by 1 unit. In the new spanning tree, 1–

3–5–6 is the tree path from node 1 to node 6. We augment $\delta = \min\{u_{13}, u_{35}, u_{56} - x_{56}\} = \min\{3, 3, 6 - 2\} = 3$ units of flow along the path, creating two blocking arcs, (1, 3) and (3, 5). The arc (1, 3) is closer to the source and we select it as the leaving arc. As shown in Figure 11.20(e), the resulting s–t cut contains two eligible arcs, (2, 3) and (4, 6). Since $\theta_{46} < \theta_{23}$, we select (4, 6) as the entering arc. We leave the remaining steps of the algorithm as an exercise for the reader.

Notice the resemblance between the parametric network simplex algorithm and the successive shortest patii algorithm that we discussed in Section 9.7. Both algorithms maintain the optimality conditions and gradually satisfy the mass balance constraints at the source and sink nodes. Both algorithms send flow along shortest paths from node s to node t. Whereas the successive shortest path algorithm does so by explicitly solving a shortest path problem, the parametric network simplex algorithm implicitly solves a shortest path problem. Indeed, the sequence of iterations that the parametric network simplex algorithm performs between two consecutive positive-flow iterations are essentially the steps of Dijkstra's algorithm for the shortest path problem.

Dual Network Simplex Algorithm

The dual network simplex algorithm maintains a solution that satisfies the mass balance constraints at all nodes, but that violates some of the lower and upper bounds imposed on the arc flows. The algorithm maintains a spanning tree structure (**T**, **L**, **U**) that satisfies the optimality conditions (11.1); the flow on the arcs in **L** and **U** are at their lower and upper bounds, but the flow on the tree arcs might not satisfy their flow bounds. We refer to a tree arc (i, j) as *feasible* if $0 \le x_{ij} \le u_{ij}$ and as *infeasible* otherwise. The algorithm attempts to make infeasible arcs feasible by sending flow along cycles; it terminates when the network contains no infeasible arc.

The dual network simplex algorithm performs a dual pivot at every iteration. Let (**T**, **L**, **U**) be the spanning tree structure at some iteration. In this solution some tree arcs might be infeasible. The algorithm selects any one of these arcs as the leaving arc. (Empirical evidence suggests that choosing an infeasible arc with the maximum violation of its flow bound generally results in a fewer number of iterations.) Suppose that we select the arc (p, q) as the leaving arc and $x_{pq} > u_{pq}$. We later address the case $x_{pq} < 0$. To make the flow on the arc (p, q) feasible, we must decrease the flow on this arc. We decrease the flow by introducing some nontree arc (k, l) that creates a unique cycle W containing arc (p, q) and augment enough flow along this cycle. Let us see which entering arc (k, l) would permit us to accomplish this objective.

If we drop the arc (p, q) from the spanning tree, we create two subtrees T_1 and T_2, with $p \in T_1$ and $q \in T_2$. Let S and $\bar{S}$ be the sets of nodes spanned by T_1 and T_2. In addition, let $(S, \bar{S})$ and $(\bar{S}, S)$ denote the sets of forward and backward arcs in the cut $[S, \bar{S}]$. Each arc in the cut $[S, \bar{S}]$, except the arc (p, q), is at its lower or upper bound. Adding any arc (i, j) in $[S, \bar{S}]$ to **T** creates a unique cycle W that contains the arc (p, q). Suppose that we define the orientation of the cycle W along the arc (i, j) if $(i, j) \in$ **L** and opposite to the arc (i, j) if $(i, j) \in$ **U**. Each nontree arc in the cut $[\bar{S}, S]$ is (1) either a forward arc or a backward arc, and (2) either belongs to **L** or belongs to **U**. Consider any arc $(i, j) \in (S, \bar{S}) \cap$ **L**. In this case, the orientation of the cycle is along arc (i, j); consequently, arc (p, q) will be a backward arc in the cycle W and sending additional flow along the orientation of the cycle will decrease flow on the arc (p, q) [see Figure 11.21(a)]. Next, consider any arc $(i, j) \in (\bar{S}, S) \cap \mathbf{U}$. In this case the orientation of the cycle is opposite to arc (i, j); therefore, sending additional flow along the orientation of the cycle again decreases flow on the arc (p, q) [see Figure 11.21(b)]. The reader can easily verify that in the other two cases when $(i, j) \in (S, \bar{S}) \cap \mathbf{U}$ or $(i, j) \in (\bar{S}, S) \cap \mathbf{L}$, increasing flow along the orientation of the cycle does not decrease flow on the arc (p, q). Consequently, we define the set of *eligible arcs* as

$$Q = ((S, \bar{S}) \cap \mathbf{L}) \cup ((\bar{S}, S) \cap \mathbf{U}).$$

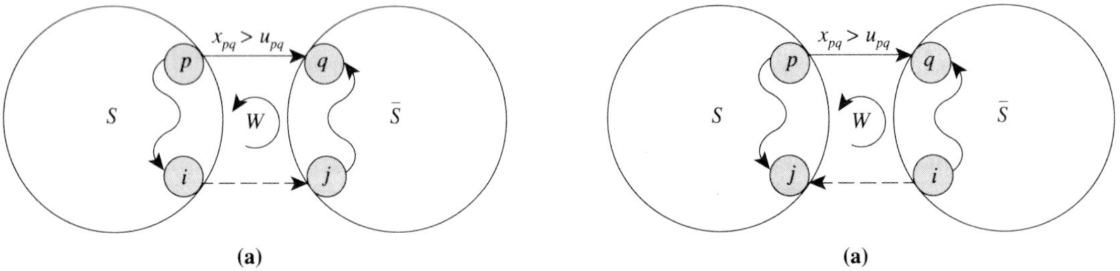

Figure 11.21 Identifying eligible arcs in the dual network simplex algorithm.

If $Q = \varnothing$, we cannot reduce the flow on arc (p, q) and the minimum cost flow problem is infeasible (see Exercise 11.37). If $Q \neq \varnothing$, we must select as the entering arc an eligible arc that would create a new spanning tree structure satisfying the optimality conditions. This step is similar to the same step in the parametric network simplex algorithm. We define a number θ_{ij} for each eligible arc (i, j) in the following manner:

$$\theta_{ij} = \begin{cases} c^{\pi}_{ij} & \text{if } (i, j) \in \mathbf{L}, \\ -c^{\pi}_{ij} & \text{if } (i, j) \in \mathbf{U}, \end{cases}$$

and select an arc (k, l) as the entering arc for which $\theta_{kl} = \min\{\theta_{ij} : (i, j) \in Q\}$. As before, we say this dual pivot is *degenerate* if $\theta_{kl} = 0$ and is *nondegenerate* otherwise. We augment $x_{pq} - u_{pq}$ units of flow along the cycle created by the arc (k, l); doing so decreases the flow on the arc (p, q) to value u_{pq}. Note that as a result of the augmentation, the arc (p, q) becomes feasible; other feasible arcs, however, might become infeasible. In the next spanning tree structure, the arc (k, l) replaces the arc (p, q), and (p, q) becomes a nontree arc at its upper bound. Replacing the arc (p, q) by the arc (k, l) in the spanning tree decreases the potential of each node in $\bar{S}$ by θ_{kl} units. (In the dual network simplex algorithm, the potential of node 1 might not always be zero.) As in our discussion of the parametric network simplex algorithm, it is possible to show that the new spanning tree structure satisfies the optimality conditions.

So far we have addressed situations in which the leaving arc (p, q) is infeasible because $x_{pq} > u_{pq}$. We now consider the case when $x_{pq} < 0$. In this instance, to make this arc feasible, we will increase its flow. The computations in this case are exactly the same as in the previous case except that we define the subtrees T_1 and

T_2 so that $p \in T_2$ and $q \in T_1$. We define the set of eligible arcs as $Q = ((S, \bar{S}) \cap \mathbf{L}) \cup ((\bar{S}, S) \cap \mathbf{U})$ and select an eligible arc (k, l) with the minimum value of θ_{kl} as the entering arc. We augment $|x_{pq}|$ units of flow along the cycle created by the arc (k, l); doing so increases the flow on arc (p, q) to value zero. In the next spanning tree structure, arc (k, l) becomes a tree arc and (p, q) becomes a nontree arc at its lower bound.

Proving the finiteness of the dual network simplex algorithm is easy if each dual pivot is nondegenerate. As before, we assume that $x_{pq} > u_{pq}$ (a similar proof applies when $x_{pq} < 0$). In this case the entering arc (k, l) belongs to $(S, \bar{S}) \cap \mathbf{L}$ or belongs to $(\bar{S}, S) \cap \mathbf{U}$. In the former case, $c_{kl}^{\pi} > 0$ and the flow on the arc (k, l) increases by $(x_{pq} - u_{pq}) > 0$ units. In the latter case, $c_{kl}^{\pi} < 0$ and the flow on the arc decreases by $(x_{pq} - u_{pq}) > 0$ units. In either case, the cost of the flow increases by $c_{kl}^{\pi}(x_{pq} - u_{pq}) > 0$. Since mCU is an upper bound on the objective function value of the minimum cost flow problem and each nondegenerate pivot increases the cost by at least 1 unit, the dual network simplex algorithm will terminate finitely whenever every pivot is nondegenerate. In a degenerate pivot, the objective function value does not change because the entering arc (k, l) satisfies the condition $c_{kl}^{\pi} = 0$. In Exercise 11.38 we describe a dual perturbation technique that avoids the degenerate dual pivots altogether and yields a finite dual network simplex algorithm.

11.10 SENSITIVITY ANALYSIS

The purpose of sensitivity analysis is to determine changes in the optimal solution of the minimum cost flow problem resulting from changes in the data (supply/demand vector, capacity, or cost of any arc). In Section 9.11 we described methods for conducting sensitivity analysis using nonsimplex algorithms. In this section we describe network simplex based algorithms for performing sensitivity analysis.

Sensitivity analysis adopts the following basic approach. We first determine the effect of a given change in the data on the feasibility and optimality of the solution assuming that the spanning tree structure remains unchanged. If the change affects the optimality of the spanning tree structure, we perform (primal) pivots to achieve optimality. Whenever the change destroys the feasibility of the spanning tree structure, we perform dual pivots to achieve feasibility.

Let x^* denote an optimal solution of the minimum cost flow problem. Let $(\mathbf{T}^*, \mathbf{L}^*, \mathbf{U}^*)$ denote the corresponding spanning tree structure and π^* denote the corresponding node potentials. We first consider sensitivity analysis with respect to changes in the cost coefficients.

Cost Sensitivity Analysis

Suppose that the cost of an arc (p, q) increases by λ units. The analysis would be different when arc (p, q) is a tree or a nontree arc.

Case 1. *Arc (p, q) is a nontree arc.*

In this case, changing the cost of arc (p, q) does not change the node potentials of the current spanning tree structure. The modified reduced cost of arc (p, q) is $c_{pq}^{\pi^*} + \lambda$. If the modified reduced cost satisfies condition (11.1b) or (11.1c), whichever is appropriate, the current spanning tree structure remains optimal. Otherwise, we reoptimize the solution using the network simplex algorithm with (**T***, **L***, **U***) as the starting spanning tree structure.

Case 2. *Arc* (p, q) *is a tree arc.*

In this case, changing the cost of arc (p, q) changes some node potentials. If arc (p, q) is an upward-pointing arc in the current spanning tree, potentials of all the nodes in $D(p)$ increase by λ, and if (p, q) is a downward-pointing arc, potentials of all the nodes in $D(q)$ decrease by λ. Note that these changes alter the reduced costs of those nontree arcs that belong to the cut $[D(q), \overline{D}(q)]$. If all nontree arcs still satisfy the optimality condition, the current spanning tree structure remains optimal; otherwise, we reoptimize the solution using the network simplex algorithm.

Supply/Demand Sensitivity Analysis

To study changes in the supply/demand vector, suppose that the supply/demand $b(k)$ of node k increases by λ and the supply/demand $b(l)$ of another node l decreases by λ. [Recall that since $\sum_{i \in N} b(i) = 0$, the supplies of two nodes must change simultaneously, by equal magnitudes and in opposite directions.] The mass balance constraints require that we must ship λ units of flow from node k to node l. Let P be the unique tree path from node k to node l. Let $\overline{P}$ and $\underline{P}$, respectively, denote the sets of arcs in P that are along and opposite to the direction of the path. The maximum flow change δ_{ij} on an arc $(i, j) \in P$ that preserves the flow bounds is

$$\delta_{ij} = \begin{cases} u_{ij} - x_{ij} & \text{if } (i, j) \in \overline{P}, \\ x_{ij} & \text{if } (i, j) \in \underline{P}. \end{cases}$$

Let

$$\delta = \min\{\delta_{ij} : (i, j) \in P\}.$$

If $\lambda \le \delta$, we send λ units of flow from node k to node l along the path P. The modified solution is feasible to the modified problem and since the modification in $b(i)$ does not affect the optimality of the solution, the resulting solution must be an optimal solution of the modified problem.

If $\lambda > \delta$, we cannot send λ units of flow from node k to node l along the arcs of the current spanning tree and preserve feasibility. In this case we send δ units of flow along P and reduce λ to $\lambda - \delta$. Let x' denote the updated flow. We next perform a dual pivot (as described in the preceding section) to obtain a new spanning tree that might allow additional flow to be sent from node k to node l along the tree path. In a dual pivot, we first decide on the leaving variable and then identify an entering variable. Let (p, q) be an arc in P that blocks us from sending additional flow from node k to node l. If $(p, q) \in \overline{P}$, then $x'_{pq} = u_{pq}$ and if $(p, q) \in \underline{P}$, then $x'_{pq} = 0$. We drop arc (p, q) from the spanning tree. Doing so partitions the set of nodes into two subtrees. Let S denote the subtree containing node k and $\overline{S}$ denote the subtree containing node l. Now consider the cut $[S, \overline{S}]$. Since we wish to send additional flow through the cut $[S, \overline{S}]$, the arcs eligible to enter the tree would be the forward arcs in the cut at their lower bound or backward arcs at their upper bounds. If the network contains no eligible arc, we can send no additional flow from node k to node l and the modified problem is infeasible. If the network does contain qualified arcs, then among these arcs, we select an arc, say (g, h), whose reduced cost has the smallest magnitude. We introduce the arc (g, h) into the spanning tree and update the node potentials.

We then again try to send $\lambda' = \lambda - \delta$ units of flow from node k to node l on the tree path. If we succeed, we terminate; otherwise, we send the maximum possible flow and perform another dual pivot to obtain a new spanning tree structure. We repeat these computations until either we establish a feasible flow in the network or discover that the modified problem is infeasible.

Capacity Sensitivity Analysis

Finally, we consider sensitivity analysis with respect to arc capacities. Consider the analysis when the capacity of an arc (p, q) increases by λ units. (Exercise 11.40 considers the situation when an arc capacity decreases by λ units.) Whenever we increase the capacity of any arc, the previous optimal solution always remains feasible; to determine whether this solution remains optimal, we check the optimality conditions (11.1). If arc (p, q) is a tree arc or is a nontree arc at its lower bound, increasing u_{pq} by λ does not affect the optimality condition for that arc. If, however, arc (p, q) is a nontree arc at its upper bound and its capacity increases by λ units, the optimality condition (11.1c) dictates that we must increase the flow on the arc by λ units. Doing so creates an excess of λ units at node q and a deficit of λ units at node p. To achieve feasibility, we must send λ units from node q to node p. We accomplish this objective by using the method described earlier in our discussion of supply/demand sensitivity analysis.

11.11 RELATIONSHIP TO SIMPLEX METHOD

So far in this chapter, we have described the network simplex algorithm as a combinatorial algorithm and used combinatorial arguments to show that the algorithm correctly solves the minimum cost flow problem. This development has the advantage of highlighting the inherent combinatorial structure of the minimum cost flow problem and of the network simplex algorithm. The approach has the disadvantage, however, of not placing the network simplex method in the broader context of linear programming. To help to rectify this shortcoming, in this section we offer a linear programming interpretation of the network simplex algorithm. We show that the network simplex algorithm is indeed an adaptation of the well-known simplex method for general linear programs. Because the minimum cost flow problem is a highly structured linear programming problem, when we apply the simplex method to it, the resulting computations become considerably streamlined. In fact, we need not explicitly maintain the matrix representation (known as the simplex tableau) of the linear program and can perform all the computations directly on the network. As we will see, the resulting computations are exactly the same as those performed by the network simplex algorithm. Consequently, the network simplex algorithm is not a new minimum cost flow algorithm; instead, it is a special implementation of the well-known simplex method that exploits the special structure of the minimum cost flow problem.

Our discussion in this section requires a basic understanding of the simplex method; Appendix C provides a brief review of this method. As we have noted before, the minimum cost flow problem is the following linear program:

$$\text{Minimize} \quad cx$$

subject to

$$\mathcal{N}x = b,$$

$$0 \leq x \leq u.$$

The bounded variable simplex method for linear programming (or, simply, the simplex method) maintains a *basis structure* (**B**, **L**, **U**) at every iteration and moves from one basis structure to another until it obtains an optimal basis structure. The set **B** is the set of basic variables, and the sets **L** and **U** are the nonbasic variables at their lower and upper bounds. Following traditions in linear programming, we also refer to the variables in **B** as a basis. Let $\mathcal{B}$, $\mathcal{L}$, and $\mathcal{U}$ denote the sets of columns in $\mathcal{N}$ corresponding to the variables in **B**, **L**, and **U**. We refer to $\mathcal{B}$ as a *basis matrix*. Our first result is a graph-theoretic characterization of the basis matrix.

Bases and Spanning Trees

We begin by establishing a one-to-one correspondence between bases of the minimum cost flow problem and spanning trees of G. One implication of this result is that the basis matrix is always lower triangular. The triangularity of the basis matrix is a key in achieving the efficiency of the network simplex algorithm.

We define the jth unit vector e_j as a column vector of size n consisting of all zeros except a 1 in the jth row. We let $\mathcal{N}_{ij}$ denote the column of $\mathcal{N}$ associated with the arc (i, j). In Section 1.2 we show that $\mathcal{N}_{ij} = e_i - e_j$. The rows of $\mathcal{N}$ are linearly dependent since summing all the rows yields the redundant constraint

$$0 = \sum_{i \in N} b(i),$$

which is our assumption that the supplies/demands of all the nodes sum to zero. For convenience we henceforth assume that we have deleted the first row in $\mathcal{N}$ (corresponding to node 1, which is treated as the root node). Thus $\mathcal{N}$ has at most $n - 1$ independent rows. Since the number of linearly independent rows of a matrix is the same as the number of linearly independent columns, $\mathcal{N}$ has at most $n - 1$ linearly independent columns. We show that the $n - 1$ columns associated with arcs of any spanning tree are linearly independent and thus define a basis matrix of the minimum cost flow problem.

Consider a spanning tree **T**. Let $\mathcal{B}$ be the $(n - 1) \times (n - 1)$ matrix defined by the arcs in **T**. As an example, consider the spanning tree shown in Figure 11.22(a) which corresponds to the matrix $\mathcal{B}$ shown in Figure 11.22(b). The first row in this matrix corresponds to the redundant row in $\mathcal{N}$ and deleting this row yields an $(n - 1) \times (n - 1)$ square matrix. For the sake of clarity, however, we shall sometimes retain the first row. We order the rows and columns of $\mathcal{B}$ in a certain specific

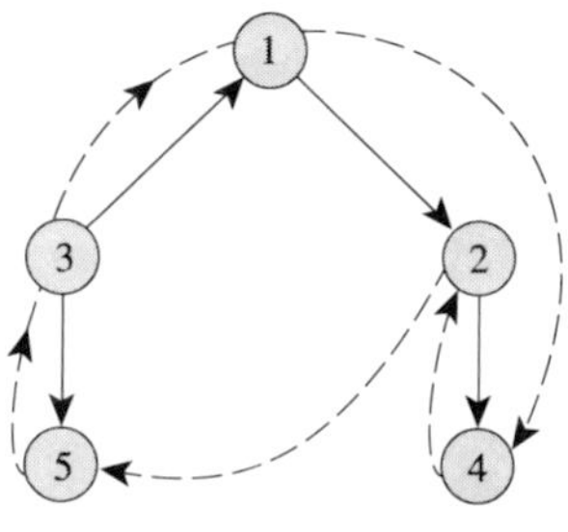

(a)

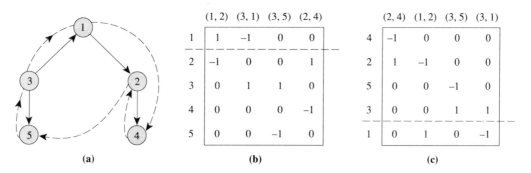

	(1, 2)	(3, 1)	(3, 5)	(2, 4)
1	1	–1	0	0
2	–1	0	0	1
3	0	1	1	0
4	0	0	0	–1
5	0	0	–1	0

(b)

	(2, 4)	(1, 2)	(3, 5)	(3, 1)
4	–1	0	0	0
2	1	–1	0	0
5	0	0	–1	0
3	0	0	1	1
1	0	1	0	–1

(c)

Figure 11.22 (a) Spanning tree and its reverse thread traversal; (b) basis matrix corresponding to the spanning tree; (c) basis matrix after rearranging the rows and columns.

manner. Doing so requires the reverse thread traversal of the nodes in the tree. Recall that a reverse thread traversal visits each node before visiting its predecessor. We order nodes and arcs in the following manner.

1. We order nodes of the tree in order of the reverse thread traversal. For our example, this order is 4–2–5–3–1 [see Figure 11.22(a)].
2. We order the tree arcs by visiting the nodes in order of the reverse thread traversal, and for each node i visited, we select the unique arc incident to it on the path to the root node. For our example, this order is (2, 4), (1, 2), (3, 5), and (3, 1).

We now arrange the rows and columns of $\mathcal{B}$ as specified by the preceding node and arc orderings. Figure 11.22(c) shows the resulting matrix for our example. In this matrix, if we ignore the row corresponding to node 1, we have a lower triangular $(n - 1) \times (n - 1)$ matrix. The triangularity of the matrix is not specific to our example: The matrix would be triangular in general. It is easy to see why. Suppose that the reverse thread traversal selects node i at some step. Let $j = \text{pred}(i)$. Then either $(j, i) \in \mathbf{T}$, or $(i, j) \in \mathbf{T}$. Without any loss of generality, we assume that $(i, j) \in \mathbf{T}$. The reverse thread traversal ensures that we have not visited node j so far. Consequently, the column corresponding to arc (i, j) will contain a +1 entry in the row r corresponding to node i, will contain all zero entries above this row, and will contain a -1 entry corresponding to node j below row r (because we will visit node j later). We have thus shown that this rearranged version of $\mathcal{B}$ is a lower triangular matrix and that all of its diagonal elements are $+1$ or -1. We, therefore, have established the following result.

Theorem 11.9 (Triangularity Property). *The rows and columns of the node–arc incidence matrix of any spanning tree can be rearranged to be lower triangular.* ◆

The determinant of a lower triangular matrix is the product of its diagonal elements. Since each diagonal element in the matrix is ± 1, the determinant is ± 1. We now use the well-known fact from linear algebra that a set of $(n - 1)$ column vectors, each of size $(n - 1)$, is linearly independent if and only if the matrix containing these vectors as columns has a nonzero determinant. This result shows that the columns corresponding to arcs of a spanning tree constitute a basis matrix of $\mathcal{N}$.

We now establish the converse result: Every basis matrix $\mathcal{B}$ of $\mathcal{N}$ defines a spanning tree. The fact that every basis matrix has the same number of columns implies that every basis matrix $\mathcal{B}$ has $(n - 1)$ columns. These columns correspond to a subgraph G' of G having $(n - 1)$ arcs. Suppose that G' contains a cycle W. We assign any orientation to this cycle and consider the expression $\sum_{(i,j) \in W} (\pm 1)\mathcal{N}_{ij} = \sum_{(i,j) \in W} (\pm 1)(e_i - e_j)$; the leading coefficient of each term is $+1$ for those arcs aligned along the orientation of the cycle and is -1 for arcs aligned opposite to the orientation of the cycle. It is easy to verify that for each node j contained in the cycle, the unit vector e_j appears twice, once with a $+1$ sign and once with a -1 sign. Consequently, the preceding expression sums to zero, indicating that the columns corresponding to arcs of a cycle are linearly dependent. Since the columns of $\mathcal{B}$ are linearly independent, G' must be an acyclic graph. Any acyclic graph on n nodes containing $(n - 1)$ arcs must be a spanning tree. So we have established the following theorem.

Theorem 11.10 (Basis Property). *Every spanning tree of G defines a basis of the minimum cost flow problem and, conversely, every basis of the minimum cost flow problem defines a spanning tree of G.* ◆

Implications of Triangularity

In the preceding discussion we showed that we can arrange every basis matrix of the minimum cost flow problem so that it is lower triangular and has an associated spanning tree. We now show that the triangularity of the basis matrix allows us to simplify the computations of the simplex method when applied to the minimum cost flow problem.

When applied to the minimum cost flow problem, the simplex method maintains a basis structure (**B**, **L**, **U**) at every step. Our preceding discussion implies that the arcs in the set **B** constitute a spanning tree and the arcs in the set **L** ∪ **U** are nontree arcs. Therefore, this basis structure is no different from the spanning tree structure that the network simplex algorithm maintains. Moreover, the process of moving from one spanning tree structure to another corresponds to moving from one basis structure to another in the simplex method.

The simplex method performs the following operations:

1. Given a basis structure (**B**, **L**, **U**), determine the associated basic feasible solution.
2. Given a basis structure (**B**, **L**, **U**), determine the associated simplex multipliers π (or, dual variables).
3. Given a basis structure (**B**, **L**, **U**), check whether it is optimal, and if not, then determine an entering nonbasic variable x_{kl}.
4. Given a basis structure (**B**, **L**, **U**) and a nonbasic variable x_{kl}, determine the representation, $\overline{\mathcal{N}}_{kl}$, of the column $\mathcal{N}_{kl}$, corresponding to this variable in terms

of the basis matrix $\mathcal{B}$. We require this representation to perform the pivot operation while introducing the variable x_{kl} into the current basis.

We consider these simplex operations one by one.

Computing the Basic Feasible Solution

Given the basis structure (**B**, **L**, **U**), the simplex method determines the associated basic feasible solution by solving the following system of equations:

$$\mathcal{B}x_{\mathbf{B}} = b - \mathcal{L}x_{\mathbf{L}} - \mathcal{U}x_{\mathbf{U}}. \tag{11.5}$$

In this expression, $x_{\mathbf{B}}$ denotes the set of basic variables, and $x_{\mathbf{L}}$ and $x_{\mathbf{U}}$ denote the sets of nonbasic variables at their lower and upper bounds. The simplex method sets each nonbasic variable in $x_{\mathbf{L}}$ to value zero, each nonbasic variable in $x_{\mathbf{U}}$ to its upper bound, and solves the resulting system of equations. Let $u_{\mathbf{U}}$ be the vector of upper bounds for variables in **U** and let $b' = b - \mathcal{U}u_{\mathbf{U}}$. The simplex method solves the following system of equations:

$$\mathcal{B}x_{\mathbf{B}} = b'. \tag{11.6}$$

Let us see how can we solve (11.6) for the minimum cost flow problem. For simplicity of exposition, assume that $x_{\mathbf{B}} = (x_2, x_3, \ldots, x_n)$. (Assume that the row corresponding to node 1 is the redundant row.) Since $\mathcal{B}$ is a lower triangular matrix, the first row of $\mathcal{B}$ has exactly one nonzero element corresponding to x_2. Therefore, we can uniquely determine the value of x_2. Since the coefficient of x_2 is ± 1, the value of x_2 is integral. The second row of $\mathcal{B}$ has at most two nonzero elements, corresponding to the variables x_2 and x_3. Since we have already determined the value of x_2, we can determine the value of x_3 uniquely. Continuing to solve successively for one variable at a time by this method of forward substitution, we can determine the entire vector $x_{\mathbf{B}}$. Since the nonzero coefficients in the basis matrix $\mathcal{B}$ all have the value ± 1, the only operations we perform are additions and subtractions, which preserve the integrality of the solution.

It is easy to see that the computations required to solve the system of equations $\mathcal{B}x_{\mathbf{B}} = b'$ are exactly same as those performed by the procedure *compute-flows* described in Section 11.4. Recall that the procedure first modifies the supply/demand vector b by setting the flows on the arcs in **U** equal to their upper bounds. The modified supply/demand vector b' equals $b - \mathcal{U}u_{\mathbf{U}}$. Then the procedure examines the nodes in order of the reverse thread traversal and computes the flows on the arcs incident to these nodes. To put the matrix $\mathcal{B}$ into a lower triangular form, we ordered its rows using the reverse thread traversal of the nodes. As a result, the procedure *compute-flows* computes flows on the arcs exactly in the same order as solving the system of equation $\mathcal{B}x_{\mathbf{B}} = b'$ by forward substitution.

Determining the Simplex Multipliers

The simplex algorithm determines the simplex multipliers π associated with a basis structure (**B**, **L**, **U**) by solving the following system of equations:

$$\pi\mathcal{B} = c_{\mathbf{B}}. \tag{11.7}$$

In this expression, $c_{\mathbf{B}}$ is the vector consisting of cost coefficients of the variables in **B**. Assume, for simplicity of exposition, that $\pi = (\pi(2), \pi(3), \ldots, \pi(n))$. Since $\mathcal{B}$ is a lower triangular matrix, the last column of $\mathcal{B}$ has exactly one nonzero element. Therefore, we can immediately determine $\pi(n)$. The second to last column of $\mathcal{B}$ has at most two nonzero elements, corresponding to $\pi(n - 1)$ and $\pi(n)$. Since we have already computed $\pi(n)$, we can easily compute $\pi(n - 1)$, and so on. We can thus solve (11.7) by backward substitution and compute all the simplex multipliers by performing only additions and subtractions. Since we have arranged the rows of $\mathcal{B}$ in the order of the reverse thread traversal of the nodes, and we determine simplex multipliers in the opposite order, we are, in fact, determining the simplex multipliers of nodes in the order dictated by the thread traversal. Recall from Section 11.4 that the procedure *compute-potentials* also examines nodes and computes the node potentials by visiting the nodes via the thread traversal. Consequently, the procedure *compute-potentials* is in fact solving the system of equations $\pi\mathcal{B} = c_{\mathbf{B}}$ by backward substitution. Also, notice that the node potentials are the simplex multipliers maintained by the simplex method.

Optimality Testing

Given a basis structure (**B**, **L**, **U**), the simplex method computes the simplex multipliers π, and then tests whether the basis structure satisfies the optimality conditions (11.1) (see Appendix C). As expressed in terms of the reduced costs c_{ij}^{π}, the optimality conditions are

$$c_{ij}^{\pi} = c_{ij} - \pi\mathcal{N}_{ij}, \qquad \text{for each } (i, j) \in A.$$

For the minimum cost flow problem, $\mathcal{N}_{ij} = e_i - e_j$ and, therefore, $c_{ij}^{\pi} = c_{ij} - \pi(i) + \pi(j)$. Consequently, the reduced costs of the arcs as defined in the network simplex algorithm are the linear programming reduced costs and the optimality conditions (11.1) for the network simplex algorithm are the same as the linear programming optimality conditions (see Section C.5). The selection of the entering arc (k, l) in the network simplex algorithm corresponds to selecting the nonbasic variable x_{kl} as the entering variable. To simplify our subsequent exposition, we assume that the entering arc (k, l) is at its lower bound.

Representation of a Nonbasic Column

Once the simplex algorithm has identified a nonbasic variable x_{kl} to enter the basis, it next obtains the representation $\overline{\mathcal{N}}_{kl}$ of the column corresponding to x_{kl} with respect to the current basis matrix. We use this representation to determine the effect on the basic variables of assigning a value θ to x_{kl}, that is, to solve the system

$$x_{\mathbf{B}} = \overline{b'} - \overline{\mathcal{N}}_{kl}\theta.$$

In this expression, $\overline{b'} = \mathcal{B}^{-1}b'$ and $\overline{\mathcal{N}}_{kl} = \mathcal{B}^{-1}\mathcal{N}_{kl}$. Observe that $-\overline{\mathcal{N}}_{kl}$ denotes the change in the values of basic variables as we increase the value of the entering nonbasic variable x_{kl} by 1 unit (i.e., set θ to value 1) and maintain all other nonbasic variables at their current lower and upper bounds. What is the graph-theoretic significance of $\overline{\mathcal{N}}_{kl}$?

The addition of arc (k, l) to the spanning tree **T** creates exactly one cycle, say W. Define the orientation of the cycle W to align with the orientation of the arc (k, l). Let $\overline{W}$ and $\underline{W}$ denote the sets of forward and backward arcs in W. Observe that if we wish to increase the flow on arc (k, l) by 1 unit, keeping the flow on all other nontree arcs intact, then to satisfy the mass balance constraints we must augment 1 unit of flow along W. This change would increase the flow on arcs in $\overline{W}$ by 1 unit and decrease the flow on arcs in $\underline{W}$ by 1 unit. This discussion shows that the fundamental cycle W created by the nontree arc (k, l) defines the representation $\overline{\mathcal{N}}_{kl}$ in the following manner. All the basic variables corresponding to the arcs in $\overline{W}$ have a coefficient of -1 in the column vector $\overline{\mathcal{N}}_{kl}$, all the basic variables corresponding to the arcs in $\underline{W}$ have a coefficient of $+1$, and all other basic variables have a coefficient of 0. This discussion also shows that in the network simplex algorithm, augmenting flow in the fundamental cycle created by the entering arc (k, l) and obtaining a new spanning tree solution corresponds to performing a pivot operation and obtaining a new basis structure in the simplex method.

To summarize, we have shown that the network simplex algorithm is the same as the simplex method applied to the minimum cost flow problem. The triangularity of the basis matrix permits us to apply the simplex method directly on the network without explicitly maintaining the simplex tableau. This possibility permits us to use the network structure to greatly improve the efficiency of the simplex method for solving the minimum cost flow problem.

In this section we have shown that the network simplex algorithm is an adaptation of the simplex method for solving general linear programs. A similar development would permit us to show that the parametric network simplex algorithm is an adaptation of the right-hand-side parametric algorithm of linear programming, and that the dual network simplex algorithm is an adaptation of the well-known dual simplex method for solving linear programs. We leave the details of these results as exercises (see Exercises 11.35 and 11.36).

11.12 UNIMODULARITY PROPERTY

In Section 11.4, using network flow algorithms, we established one of the fundamental results of network flows, the integrality property, stating that every minimum cost flow problem with integer supplies/demands and integer capacities has an integer optimal solution. The type of constructive proof that we used to establish this result has the obvious advantage of actually permitting us to compute integer optimal solutions. In that sense, constructive proofs have enormous value. However, constructive proofs do not always identify underlying structural (mathematical) reasons for explaining why results are true. These structural insights usually help in understanding a subject matter, and often suggest relationships between the subject matter and other problem domains or help to define potential limitations and generalization of the subject matter. In this section we briefly examine the structural properties of the integrality property, by providing an algebraic proof of this result. This discussion shows relationships between the integrality property and certain integrality results in linear programming.

Let $\mathcal{A}$ be a $p \times q$ matrix with integer elements and p linearly independent rows (the matrix's rank is p). We say that the matrix $\mathcal{A}$ is *unimodular* if the determinant of every basis matrix $\mathcal{B}$ of $\mathcal{A}$ has value $+1$ or -1 [i.e., $\det(\mathcal{B}) = \pm 1$]. Recall from Appendix C that a $p \times p$ submatrix of $\mathcal{A}$ is a basis matrix if its columns are linearly independent. The following classical result shows the relationship between unimodularity and the integer solvability of linear programs.

Theorem 11.11 (Unimodularity Theorem). *Let $\mathcal{A}$ be an integer matrix with linearly independent rows. Then the following three conditions are equivalent:*
(a) $\mathcal{A}$ is unimodular.
(b) Every basic feasible solution defined by the constraints $\mathcal{A}x = b$, $x \geq 0$, is integer for any integer vector b.
(c) Every basis matrix $\mathcal{B}$ of $\mathcal{A}$ has an integer inverse $\mathcal{B}^{-1}$.

Proof. We prove the theorem by showing that $(a) \Rightarrow (b)$, $(b) \Rightarrow (c)$, *and* $(c) \Rightarrow (a)$.

$(a) \Rightarrow (b)$. Each basic feasible solution $x_{\mathbf{B}}$ has an associated basic matrix $\mathcal{B}$ for which $\mathcal{B}x_{\mathbf{B}} = b$. By Cramer's rule, any component x_j of the solution $x_{\mathbf{B}}$ will be of the form

$$x_j = \frac{\det(\text{integer matrix})}{\det(\mathcal{B})}.$$

We obtain the integer matrix in this formula by replacing the jth column of $\mathcal{B}$ with the vector $\mathbf{b}$. Since, by assumption, $\mathcal{A}$ is unimodular, $\det(\mathcal{B})$ is ± 1, so x_j is integer.

$(b) \Rightarrow (c)$. Let $\mathcal{B}$ be a basis matrix of $\mathcal{A}$. Since $\mathcal{B}$ has a nonzero determinant, its inverse $\mathcal{B}^{-1}$ exists. Let e_j denote the jth unit vector (i.e., a vector with a 1 at the jth position and 0 elsewhere). Let $\mathcal{D} = \mathcal{B}^{-1}$ and $\mathcal{D}_j$ denote the jth column of $\mathcal{D}$. We will show that the column vector $\mathcal{D}_j$ is integer for each j whenever condition (b) holds. Select an integer vector α so that $\mathcal{D}_j + \alpha \geq 0$. Let $x = \mathcal{D}_j + \alpha$. Notice that

$$\mathcal{B}x = \mathcal{B}(\mathcal{D}_j + \alpha) = \mathcal{B}(\mathcal{B}^{-1}e_j + \alpha) = e_j + \mathcal{B}\alpha. \tag{11.8}$$

Multiplying the expression (11.8) by $\mathcal{D} = \mathcal{B}^{-1}$, we see that $x = \mathcal{D}_j + \alpha$. Since $e_j + \mathcal{B}\alpha$ is integer (by definition), condition (b) implies that $\mathcal{D}_j + \alpha$ is integer. Recalling that α is integer, we find that $\mathcal{D}_j$ is also integer. This conclusion completes the proof of part (b).

$(c) \Rightarrow (a)$. Let $\mathcal{B}$ be a basis matrix of $\mathcal{A}$. By assumption, $\mathcal{B}$ is an integer matrix, so $\det(\mathcal{B})$ is an integer. By condition (c), $\mathcal{B}^{-1}$ is an integer matrix; consequently, $\det(\mathcal{B}^{-1})$ is also an integer. Since $\mathcal{B} \cdot \mathcal{B}^{-1} = I$ (i.e., an identity matrix), $\det(\mathcal{B}) \cdot \det(\mathcal{B}^{-1}) = 1$, which implies that $\det(\mathcal{B}) = \det(\mathcal{B}^{-1}) = \pm 1$. ◆

This result shows us when a linear program of the form minimize cx, subject to $\mathcal{A}x = b$, $x \geq 0$, has integer optimal solutions for *all* integer right-hand-side vectors $\mathbf{b}$ and for all cost vectors $\mathbf{c}$. Network flow problems are the largest important class of models that satisfy this integrality property. To establish a formal connection between network flows and the results embodied in this theorem, we consider another noteworthy class of matrices.

Totally unimodular matrices are an important special subclass of unimodular matrices. We say that a matrix $\mathcal{A}$ is *totally unimodular* if each square submatrix of $\mathcal{A}$ has determinant 0 or ± 1. Every totally unimodular matrix $\mathcal{A}$ is unimodular because each basis matrix $\mathcal{B}$ must have determinant ± 1 (because the zero value of the determinant would imply the linear dependence of the columns of $\mathcal{B}$). However, a unimodular matrix need not be totally unimodular. Totally unimodular matrices are important, in large part, because the constraint matrices of the minimum cost flow problems are totally unimodular.

Theorem 11.12. *The node–arc incidence matrix $\mathcal{N}$ of a directed network is totally unimodular.*

Proof. To prove the theorem, we need to show that every square submatrix $\mathcal{F}$ of $\mathcal{N}$ of size k has determinant 0, $+1$, or -1. We establish this result by performing induction on k. Since each element of $\mathcal{N}$ is 0, $+1$, or -1, the theorem is true for $k = 1$. Now suppose that the theorem holds for some k. Let $\mathcal{F}$ be any $(k + 1) \times (k + 1)$ submatrix of $\mathcal{N}$. The matrix $\mathcal{F}$ satisfies exactly one of the three following possibilities: (1) $\mathcal{F}$ contains a column with no nonzero element; (2) every column of $\mathcal{F}$ has exactly two nonzero elements, in which case, one of these must be a $+1$ and the another a -1; and (3) some column $\mathcal{F}_l$ has exactly one nonzero element, in, say, the ith row. In case (1) the determinant of $\mathcal{F}$ is zero and the theorem holds. In case (2) summing all of the rows in $\mathcal{F}$ yields the zero vector, implying that the rows in $\mathcal{F}$ are linearly dependent and, consequently, $\det(\mathcal{F}) = 0$. In case (3) let $\mathcal{F}'$ denote the submatrix of $\mathcal{F}$ obtained by deleting the ith row and the lth column. Then $\det(\mathcal{F}) = \pm 1 \det(\mathcal{F}')$. By the induction hypothesis, $\det(\mathcal{F}')$ is 0, $+1$, or -1, so $\det(\mathcal{F})$ is also 0, $+1$, or -1. This conclusion establishes the theorem. ◆

This result, combined with Theorem 11.11, provides us with an algebraic proof of the integrality property of network flows: Network flow models have integer optimal solutions because every node–arc incidence matrix is totally unimodular and therefore unimodular. As we will see in later chapters, the constraint matrices for many extensions of the basic network flow problem, for example, generalized flows and multicommodity flows, are not unimodular. Therefore, we would not expect the optimal solutions of these models to be integer even when all of the underlying data are integer. Therefore, to find integer solutions to these problems, we need to rely on methods of integer programming. Although our development of the minimum cost flow problem has not stressed this point, one of the primary reasons that we are able to solve this problem so efficiently, and still obtain integer solutions, is because, as reflected by the integrality property, the basic feasible solutions of the linear programming formulation of this problem are integer whenever the underlying data are integer.

To close this section, we might note that the unimodularity properties provide us with a very strong result: any basic feasible solution is guaranteed to be integer-valued whenever the right-hand-side vector b is integer. It is possible, however, that basic feasible solutions to a linear program might be integer valued for a particular right-hand side even though they might be fractional for some other right-hand sides. We illustrate this possibility in Section 13.8 when we give an integer programming formulation of the minimal spanning tree problem.

11.13 SUMMARY

The network simplex algorithm is one of the most popular algorithms in practice for solving the minimum cost flow problem. This algorithm is an adaptation for the minimum cost flow problem of the well-known simplex method of linear programming. The linear programming basis of the minimum cost flow problem is a spanning tree. This property permits us to simplify the operations of the simplex method because we can perform all of its operations on the network itself, without maintaining the simplex tableau. Our development in this chapter does not require linear programming background because we have developed and proved the validity of the network simplex algorithm from first principles. Later in the chapter we showed the connection between the network simplex algorithm and the linear programming simplex method.

The development in this chapter relies on the fact that the minimum cost flow problem always has an optimal spanning tree solution. This result permits us to restrict our search for an optimal solution among spanning tree solutions. The network simplex algorithm maintains a spanning tree solution and successively transforms it into an improved spanning tree solution until it becomes optimal. At each iteration, the algorithm selects a nontree arc, introduces it into the current spanning tree, augments the maximum possible amount of flow in the resulting cycle, and drops a blocking arc from the spanning tree, yielding a new spanning tree solution. The algorithm is flexible in the sense that we can select the entering arc in a variety of ways and obtain algorithms with different worst-case and empirical attributes.

The network simplex algorithm does not necessarily terminate in a finite number of iterations unless we impose some additional restrictions on the choice of the entering and leaving arcs. We described a special type of spanning tree solution, called the *strongly feasible spanning tree solution*; when implemented in a way that maintains strongly feasible spanning tree solutions, the network simplex algorithm terminates finitely for any choice of the rule used for selecting the entering arc. We can maintain strongly feasible spanning tree solutions by selecting the leaving arc appropriately whenever several arcs qualify to be the leaving arc.

We also specialized the network simplex algorithm for the shortest path and maximum flow problems. When specialized for the shortest path problem, the algorithm maintains a directed out-tree rooted at the source node and iteratively modifies this tree until it becomes a tree of shortest paths. When we specialize the network simplex algorithm for the maximum flow problem, the algorithm maintains an s–t cut and selects an arc in this cut as the entering arc until the associated cut becomes a minimum cut.

The network simplex algorithm has two close relatives that might be quite useful in some circumstances: the parametric network simplex algorithm and the dual network simplex algorithm. The parametric network simplex algorithm maintains a spanning tree solution and parametrically increases the flow from a source node to a sink node until the algorithm has sent the desired amount of flow between these nodes. This algorithm is useful in situations in which we want to maximize the amount of flow to be sent from a source node to a sink node, subject to an upper bound on the cost of flow (see Exercise 10.25). The dual network simplex algorithm maintains a spanning tree solution in which spanning tree arcs do not necessarily satisfy the flow bound constraints. The algorithm successively attempts to satisfy the flow bound constraints. The primary use of the dual network simplex algorithm has been for reoptimizing the minimum cost flow problem procedures for solving the minimum cost flow problem after we have changed the supply/demand or capacity data.

We also described methods for using the network simplex algorithm to conduct sensitivity analysis for the minimum cost flow problem with respect to the changes in costs, supplies/demands, and capacities. The resulting methods maintain a spanning tree solution and perform primal or dual pivots. Unlike the methods described in Section 9.11, these methods for conducting sensitivity analysis do not necessarily run in polynomial time (without further refinements). However, network simplex-based sensitivity analysis is excellent in practice.

The minimum cost flow problem always has an integer optimal solution; at the beginning of the chapter, we gave an algorithmic proof of this integrality property. We also examined the structural properties of the integrality property by providing an algebraic proof of this result. We showed that the constraint matrix of the minimum cost flow problem is totally unimodular and that, consequently, every basic feasible solution (or, equivalently, spanning tree solution) is an integer solution.

REFERENCE NOTES

Dantzig [1951] developed the network simplex algorithm for the uncapacitated transportation problem by specializing his linear programming simplex method. He proved the spanning tree property of the basis and the integrality property of the optimal solution. Later, his development of the upper bounding technique for linear programming led to an efficient specialization of the simplex method for the minimum cost flow problem. Dantzig's [1962] book discusses these topics.

The network simplex algorithm gained its current popularity in the early 1970s when the research community began to develop and test algorithms using efficient tree indices. Johnson [1966] suggested the first tree indices. Srinivasan and Thompson [1973], and Glover, Karney, Klingman, and Napier [1974] implemented these ideas; these investigations found the network simplex algorithm to be substantially faster than the existing codes that implemented the primal–dual and out-of-kilter algorithms. Subsequent research has focused on designing improved tree indices and determining the best pivot rule. The book by Kennington and Helgason [1980] describes a variety of tree indices and specifies procedures for updating them from iteration to iteration. The book by Bazaraa, Jarvis, and Sherali [1990] also describes a method for updating tree indices. The following papers describe a variety of pivot rules and the computational performance of the resulting algorithms: Glover, Karney, and Klingman [1974], Mulvey [1978], Bradley, Brown, and Graves [1977], Grigoriadis [1986], and Chang and Chen [1989]. The candidate list pivot rule that we describe in Section 11.5 is due to Mulvey [1978]. The reference notes of Chapter 9 contain information concerning the computational performance of the network simplex algorithm and other minimum cost flow algorithms.

Experience with solving large-scale minimum cost flow problems has shown that for certain classes of problems, more than 90% of the pivots in the network simplex algorithm can be degenerate. The strongly feasible spanning tree technique, proposed by Cunningham [1976] for the minimum cost flow problem, and independently by Barr, Glover, and Klingman [1977] for the assignment problem, helps to reduce the number of degenerate steps in practice and ensures that the network simplex algorithm has a finite termination. Although the strongly feasible spanning tree technique prevents cycling during a sequence of consecutive degenerate pivots, the number of consecutive degenerate pivots can be exponential. This phenomenon is known as *stalling*. Cunningham [1979] and Goldfarb, Hao, and Kai [1990b] describe several antistalling pivot rules for the network simplex algorithm.

Researchers have attempted, with partial success, to develop polynomial-time implementations of the network simplex algorithm. Tarjan [1991] and Goldfarb and Hao [1988] have described polynomial-time implementations of a variant of the network simplex algorithm that permits pivots to increase value of the objective function. A monotone polynomial-time implementation, in which the value of the objective function is nonincreasing (as it does in any natural implementation), remains elusive to researchers.

Several FORTRAN codes of the network simplex algorithm are available in the public domain. These include (1) the RNET code developed by Grigoriadis and Hsu [1979], (2) the NETFLOW code developed by Kennington and Helgason [1980], and (3) a recent code by Chang and Chen [1989].

We next give selected references for several specific topics.

Shortest path problem. We have adapted the network simplex algorithm for the shortest path problem from Dantzig [1962]. Goldfarb, Hao, and Kai [1990a] and Ahuja and Orlin [1992a] developed the polynomial-time implementations of this algorithm that we have presented in Section 11.7. Additional polynomial-time implementations can be found in Orlin [1985] and Akgül [1985a].

Maximum flow problem. Fulkerson and Dantzig [1955] specialized the network simplex algorithm for the maximum flow problem. Goldfarb and Hao [1990] gave a polynomial-time implementation of this algorithm that performs at most nm pivots and runs in $O(n^2m)$ time; Goldberg, Grigoriadis, and Tarjan [1988] describe an $O(nm \log n)$ implementation of this algorithm.

Assignment problem. One popular implementation of the network simplex algorithm for the assignment problem is due to Barr, Glover, and Klingman [1977]. Roohy-Laleh [1980], Hung [1983], Orlin [1985], Akgül [1985b], and Ahuja and Orlin [1992a] have presented polynomial-time implementations of the network simplex algorithm for the assignment problem. Balinski [1986] and Goldfarb [1985] present polynomial-time dual network simplex algorithms for the assignment problem.

Parametric network simplex algorithm. Schmidt, Jensen, and Barnes [1982], and Ahuja, Batra, and Gupta [1984] are two sources for additional information on the parametric network simplex algorithm.

Dual network simplex algorithm. Ali, Padman, and Thiagarajan [1989] have described implementation details and computational results for the dual network simplex algorithm. Although no one has yet devised a (genuine) polynomial-time primal network simplex algorithm, Orlin [1984] and Plotkin and Tardos [1990] have developed polynomial-time dual network simplex algorithms. The algorithm of Orlin [1984] is more efficient if capacities satisfy the similarity assumption; otherwise, the algorithm of Plotkin and Tardos [1990] is more efficient. The latter algorithm performs $O(m^2 \log n)$ pivots and runs in $O(m^3 \log n)$ time.

Sensitivity analysis. Srinivasan and Thompson [1972] have described parametric and sensitivity analysis for the transportation problem, which is similar to that for the minimum cost flow problem. Ali, Allen, Barr, and Kennington [1986] also discuss reoptimization procedures for the minimum cost flow problem.

Unimodularity. Hoffman and Kruskal [1956] first proved Theorem 11.11; the proof we have given is due to Veinott and Dantzig [1968]. The book by Schrijver [1986] presents an in-depth treatment of the unimodularity property and related topics.

EXERCISES

11.1. Nurse scheduling problem. A hospital administrator needs to establish a staffing schedule for nurses that will meet the minimum daily requirements shown in Figure 11.23. Nurses reporting to the hospital wards for the first five shifts work for 8 consecutive hours, except nurses reporting for the last shift (2 A.M. to 6 A.M.), when they work for only 4 hours. The administrator wants to determine the minimal number of nurses to employ to ensure that a sufficient number of nurses are available for each period. Formulate this problem as a network flow problem.

Shift	1	2	3	4	5	6
Clock time	6 A.M. to 10 A.M.	10 A.M. to 2 P.M.	2 P.M. to 6 P.M.	6 P.M. to 10 P.M.	10 P.M. to 2 A.M.	2 A.M. to 6 A.M.
Minimum nurses required	70	80	50	60	40	30

Figure 11.23 Nurse scheduling problem.

11.2. Caterer problem. As part of its food service, a caterer needs d_j napkins for each day of the upcoming week. He can buy new napkins at the price of α cents each or have his soiled napkins laundered. Two types of laundry service are available: regular and expedited. The regular laundry service requires two working days and costs β cents per napkin, and the expedited service requires one working day and costs γ cents per napkin ($\gamma > \beta$). The problem is to determine a purchasing and laundry policy that meets the demand at the minimum possible cost. Formulate this problem as a minimum costs flow problem. (*Hint*: Define a network on 15 nodes, 7 nodes corresponding to soiled napkins, 7 nodes corresponding to fresh napkins, and 1 node for the supply of fresh napkins.)

11.3. Project assignment. In a new industry-funded academic program, each master's degree student is required to undertake a 6-month internship project at a company site. Since the projects are such an important component of the student's educational program and vary considerably by company (e.g., by the problem and industry context) and by geography, each student would like to undertake a project of his or her liking. To assure that the project assignments are "fair," the students and program administrators have decided to use an optimization approach: Each student ranks the available projects in order of increasing preference (lowest to highest). The objective is to assign students to projects to achieve the highest sum of total ranking of assigned projects. The project assignment has several constraints. Each student must work on exactly one project, and each project has an upper limit on the number of students it can accept. Each project must have a supervisor, drawn from a known pool of eligible faculty. Finally, each faculty member has bounds (upper and lower) on the number of projects that he or she can supervise. Formulate this problem as a minimum cost flow problem.

11.4. Passenger routing. United Airlines has six daily flights from Chicago to Washington. From 10 A.M. until 8 P.M., the flights depart every 2 hours. The first three flights have a capacity of 100 passengers and the last three flights can accommodate 150 passengers each. If overbooking results in insufficient room for a passenger on a scheduled flight, United can divert a passenger to a later flight. It compensates any passenger delayed by more than 2 hours from his or her regularly scheduled departure by paying $200 plus $20 for every hour of delay. United can always accommodate passengers delayed beyond the 8 P.M. flight on the 11 P.M. flight of another airline that always has a great deal of spare capacity. Suppose that at the start of a particular day the six United flights have 110, 160, 103, 149, 175, and 140 confirmed reservations. Show how to formulate the problem of determining the most economical passenger routing strategy as a minimum cost flow problem.

11.5. Allocating receivers to transmitters (Dantzig [1962]). An engine testing facility has four types of instruments: α_1 thermocouplers, α_2 pressure gauges, α_3 accelerometers, and α_4 thrust meters. Each instrument measures one type of engine characteristic and transmits its measurements over a separate communication channel. A set of receivers receive and record these data. The testing facility uses four types of receivers, each capable of recording one channel of information: β_1 cameras, β_2 oscilloscopes, β_3 instruments called "Idiots," and β_4 instruments called "Hathaways." The setup time of each receiver depends on the measurement instruments that are transmitting the data; let c_{ij} denote the setup time needed to prepare a receiver of type i to receive the information transmitted from any measurement taken by the jth instrument. The testing facility wants to find an allocation of receivers to transmitters that minimizes the total setup time. Formulate this problem as a network flow problem.

11.6. Faculty–course assignment (Mulvey [1979]). In 1973, the Graduate School of Management at UCLA revamped its M.B.A. curriculum. This change necessitated an increased centralization of the annual scheduling of faculty to courses. The large size of the problem (100 faculty, 500 courses, and three quarters) suggested that a mathematical model would be useful for determining an initial solution. The administration knows the courses to be taught in each of the three teaching quarters (fall, winter, and spring). Some courses can be taught in either of the two specified quarters; this information is available. A faculty member might not be available in all the quarters (due to leaves, sabbaticals, or other special circumstances) and when he is available he might be relieved from teaching some courses by using his project grants for "faculty offset time." Suppose that the administration knows the quarters when a faculty member will be available and the total number of courses he will be teaching in those quarters. The school would like to maximize the preferences of the faculty for teaching the courses. The administration determines these preferences through an annual faculty questionnaire. The preference weights range from -2 to $+2$ and the administration occasionally revises the weights to reflect teaching ability and student inputs. Suggest a network model for determining a teaching schedule.

11.7. Optimal rounding of a matrix (Bacharach [1966], Cox and Ernst [1982]). In Application 6.3 we studied the problem of rounding the entries of a table to their nearest integers while preserving the row and column sums of the matrix. We refer to any such rounding as a *consistent rounding*. Rounding off an element of the matrix introduces some error. If we round off an element a_{ij} to b_{ij} and $b_{ij} = \lfloor a_{ij} \rfloor$ or $b_{ij} = \lceil a_{ij} \rceil$, we measure the error as $(a_{ij} - b_{ij})^2$. Summing these terms for all the elements of the matrix gives us an error associated with any consistent rounding scheme. We say that a consistent rounding is an *optimal rounding* if the error associated with this rounding is as small as the error associated with any consistent rounding. Show how to determine an optimal rounding by solving a circulation problem. (*Hint*: Construct a network similar to the one used in Application 6.3. Define the arc costs appropriately.)

11.8. Describe an algorithm that either identifies p arc-disjoint directed paths from node s to node t or shows that the network does not contain any such set of paths. In the former case, show how to determine p arc-disjoint paths containing the fewest number of arcs. Suggest modifications of this algorithm to identify p node-disjoint directed paths from node s to node t containing the fewest number of arcs.

11.9. Show that a tree is a directed out-tree T rooted at node s if and only if every node in T except node s has indegree 1. State (but do not prove) an equivalent result for a directed in-tree.

11.10. Suppose that we permute the rows and columns of the node–arc incidence matrix $\mathcal{N}$ of a graph G. Is the modified matrix a node–arc incidence matrix of some graph G'? If so, how are G' and G related?

11.11. Let T be a spanning tree of $G = (N, A)$. Every nontree arc (k, l) has an associated fundamental cycle which is the unique cycle in $T \cup \{(k, l)\}$. With respect to any arbitrary ordering of the arcs $(i_1, j_1), (i_2, j_2), \ldots, (i_m, j_m)$, we define the *incidence vector* of any cycle W in G as an m-vector whose kth element is (1) 1, if (i_k, j_k) is a forward arc in W; (2) -1, if (i_k, j_k) is a backward arc in W; and (3) 0, if $(i_k, j_k) \notin W$. Show how to express the incidence vector of any cycle W as a sum of incidence vectors of fundamental cycles.

11.12. Figure 11.24(b) gives a feasible solution of the minimum cost flow problem shown in Figure 11.24(a). Convert this solution into a spanning tree solution with the same or lower cost.

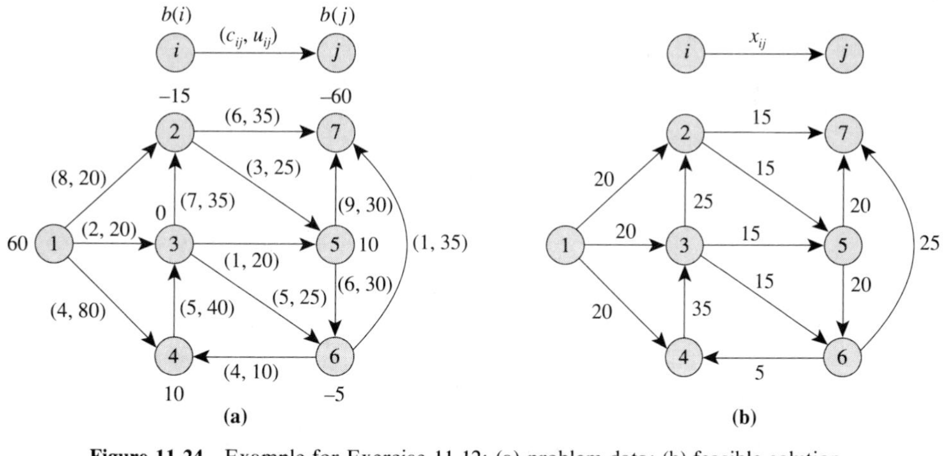

Figure 11.24 Example for Exercise 11.12: (a) problem data; (b) feasible solution.

11.13. Figure 11.25 specifies two spanning trees for the minimum cost flow problem shown in Figure 11.24(a). For Figure 11.25(a), compute the spanning tree solution assuming that all nontree arcs are at their lower bounds. For Figure 11.25(b), compute the spanning tree solution assuming that all nontree arcs are at their upper bounds.

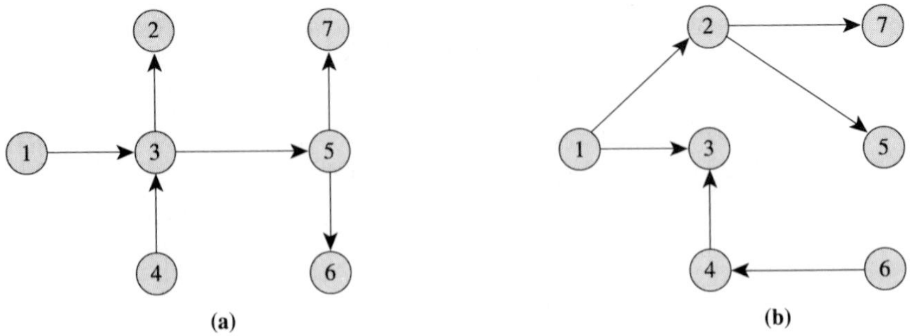

Figure 11.25 Two spanning trees of the network in Figure 11.24.

11.14. Assume that the spanning trees in Figure 11.25 have node 1 as their root. Specify the predecessor, depth, thread, and reverse thread indices of the nodes.

11.15. Compute the node potentials associated with the trees shown in Figure 11.25, which are the spanning trees of the minimum cost flow problem given in Figure 11.24(a). Verify that for each node j, the node potential $\pi(j)$ equals the length of the tree path from node j to the root.

11.16. Consider the minimum cost flow problem shown in Figure 11.26. Using the network simplex algorithm implemented with the first eligible pivot rule, find an optimal solution of this problem. Assume, as always, that arcs are arranged in the increasing order of their tail nodes, and for the same tail node, they are arranged in the increasing order of their head nodes. Use the following initial spanning tree structure: $\mathbf{T} = \{(1, 2), (3, 2), (2, 5), (4, 5), (4, 6)\}$, $\mathbf{L} = \{(3, 5)\}$, and $\mathbf{U} = \{(1, 3), (2, 4), (5, 6)\}$.

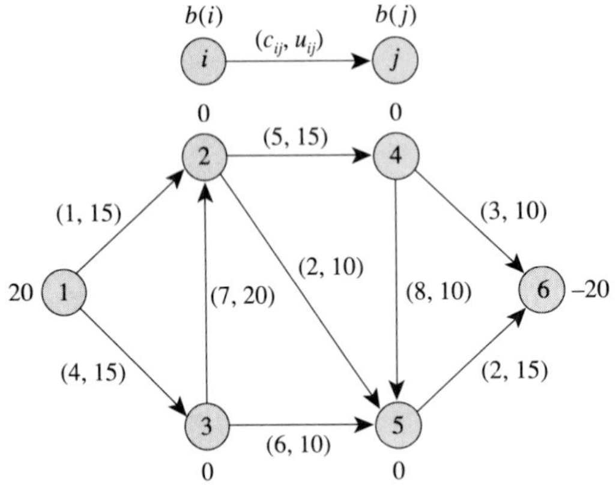

Figure 11.26 Example for Exercises 11.16 and 11.17.

11.17. Using the network simplex algorithm implemented with Dantzig's pivot rule, solve the minimum cost flow problem shown in Figure 11.26. Use the same initial spanning tree structure as used in Exercise 11.16.

11.18. In the procedure *compute-potentials*, we set $\pi(1) = 0$ and then compute other node potentials. Suppose, instead, that we set $\pi(1) = \alpha$ for some $\alpha > 0$ and then recompute all the node potentials. Show that all the node potentials increase by the amount α. Also show that this change does not affect the reduced cost of any arc.

11.19. Justify the procedure *compute-flows* for capacitated networks.

11.20. In the candidate list pivot rule, let *size* denote the maximum allowable size of the candidate list and *iter* denote the maximum number of minor iterations to be performed within a major iteration.

(a) Specify values of size and iter so that the candidate list pivot rule reduces to Dantzig's pivot rule.

(b) Specify values of size and iter so that the candidate list pivot rule reduces to the first eligible arc pivot rule.

11.21. In Section 11.5 we showed how to find the apex of the pivot cycle W in $O(|W|)$ time using the predecessor and depth indices. Show that by using predecessor indices alone, you can find the apex of the pivot cycle in $O(|W|)$ time. (*Hint*: Do so by scanning at most $2|W|$ arcs.)

11.22. Given the predecessor indices of a spanning tree, describe an $O(n)$ time method for computing the thread and depth indices.

11.23. Describe methods for updating the predecessor and depth indices of the nodes when performing a pivot operation. Your method should require $O(n)$ time and should run faster than recomputing these indices from scratch.

11.24. Prove that in a spanning tree we can send a positive amount of flow from any node to the root without violating any flow bound if and only if every tree arc with zero flow is upward pointing and every tree arc at its upper bound is downward pointing.

11.25. Let $G(x)$ denote the residual network corresponding to a flow x. Show that a spanning tree **T** is a strongly feasible spanning tree if and only if for every node $i \in N - \{1\}$, $G(x)$ contains the arc $(i, \text{pred}(i))$.

11.26. **Primal perturbation.** In the minimum cost flow problem on a network G, suppose that we alter the supply/demand vector from value b to value $b + \epsilon$ for some vector ϵ. Let us refer to the modified problem as a *perturbed problem*. We consider the perturbation ϵ defined by $\epsilon(i) = 1/n$ for all $i = 2, 3, \ldots, n$, and $\epsilon(1) = -(n - 1)/n$.

(a) Let **T** be a spanning tree of G and let $D(j)$ denote the set of descendants of node j in **T**. Show that the perturbation decreases the flow on a downward-pointing arc (i, j) by the amount $|D(j)|/n$ and increases the flow on an upward-pointing arc (i, j) by the amount $|D(i)|/n$. Conclude that in a strongly feasible spanning tree solution, each arc flow is nonzero and is an integral multiple of $1/n$.

(b) Use the result in part (a) to show that the network simplex algorithm solves the perturbed problem in pseudopolynomial time irrespective of the pivot rule used for selecting entering arcs.

11.27. **Perturbation and strongly feasible solutions.** Let (**T**, **L**, **U**) be a feasible spanning tree structure of the minimum cost flow problem and let ϵ be a perturbation as defined in Exercise 11.26. Show that (**T**, **L**, **U**) is strongly feasible if and only if (**T**, **L**, **U**) remains feasible when we replace b by $b + \epsilon$. Use this equivalence to show that when implemented to maintain a strongly feasible basis, the network simplex algorithm runs in pseudopolynomial time irrespective of the pivot rule used for selecting entering arcs.

11.28. Apply the network simplex algorithm to the shortest path problem shown in Figure 11.27(a). Use a depth-first search tree with node 1 as the source node in the initial spanning tree solution and perform three iterations of the algorithm.

11.29. Apply the network simplex algorithm to the maximum flow problem shown in Figure 11.27(b). Use the following spanning tree as the initial spanning tree: a breadth-first search tree rooted at node 1 and spanning the nodes $N - \{t\}$ plus the arc (t, s). Show three iterations of the algorithm.

11.30. Consider the application of the network simplex algorithm, implemented with the following pivot rule, for solving the shortest path problem. We examine all the nodes, one by one, in a wraparound fashion. Each time we examine a node i, we scan all incoming arcs at that node, and if the incoming arcs contain an eligible arc, we pivot in the arc with the maximum violation. We terminate when during an entire pass of the nodes, we find that no arc is eligible. Show when implemented with this pivot rule, the network simplex algorithm would perform $O(n^2)$ pivot operations and would run in $O(n^3)$ time. (*Hint*: The proof is similar to the proof of the first eligible arc pivot rule that we discussed in Section 11.7.)

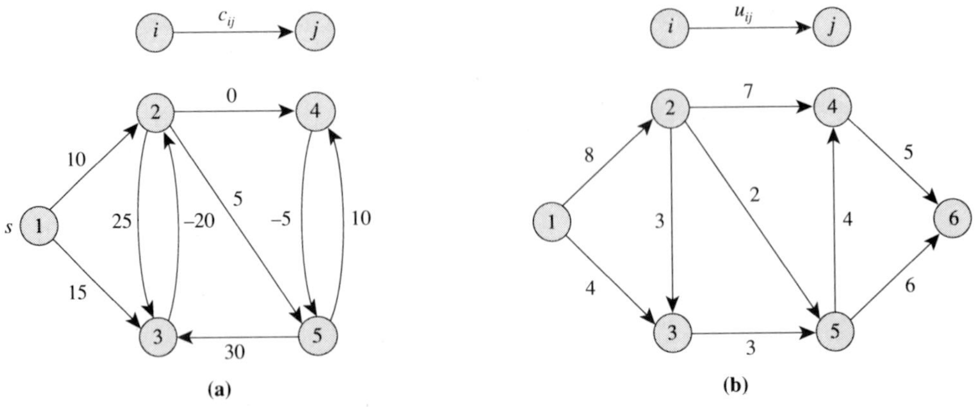

Figure 11.27 Examples for Exercises 11.28 and 11.29.

11.31. The assignment problem, as formulated as a linear programming in (12.1), is a special case of the minimum cost flow problem. Show that every strongly feasible spanning tree of the assignment problem satisfies the following properties: (1) every downward-pointing arc carries unit flow; (2) every upward-pointing arc carries zero flow; and (3) every downward-pointing arc is the unique arc with flow equal to 1 emanating from node i.

11.32. In a strongly feasible spanning tree of the assignment problem, a nontree arc (k, l) is a *downward arc* if node l is a descendant of node k. Show that when the network simplex algorithm, implemented to maintain strongly feasible spanning trees, is applied to the assignment problem, a pivot is nondegenerate if and only if the entering arc is a downward arc.

11.33. Solve the minimum cost flow problem shown in Figure 11.26 by the parametric network simplex algorithm.

11.34. Show how to solve the constrained maximum flow problem, as defined in Exercise 10.25, by a single application of the parametric network simplex algorithm.

11.35. Show that the parametric network simplex algorithm described in Section 11.9 is an adaptation of the right-hand-side parametric simplex method of linear programming. (Consult any linear programming textbook for a review of the parametric simplex method of linear programming.)

11.36. Show that the dual network simplex algorithm described in Section 11.9 is an adaptation of the dual simplex method of linear programming. (Consult any linear programming textbook for a review of the dual simplex method of linear programming).

11.37. At some point during its execution, the dual network simplex algorithm that we discussed in Section 11.9 might find that the set Q of eligible arcs is empty. In this case show that the minimum cost flow problem is infeasible. (*Hint*: Use the result in Exercise 6.43.)

11.38. **Dual perturbation.** Suppose that we modify the cost vector c of a minimum cost flow problem on a network G in the following manner. After arranging the arcs in some order, we add $\frac{1}{2}$ to the cost of the first arc, $\frac{1}{4}$ to the cost of the second arc, $\frac{1}{8}$ to the cost of the third arc, and so on. We refer to the perturbed cost as c^ϵ, and the minimum cost flow problem with the cost c^ϵ as the *perturbed minimum cost flow problem*.

(a) Show that if x^* is an optimal solution of the perturbed problem, x^* is also an optimal solution of the original problem. (*Hint*: Show that if $G(x^*)$ does not contain any negative cycle with cost c^ϵ, it does not contain any negative cycle with cost c.)

(b) Show that if we apply the dual network simplex algorithm to the perturbed problem, the reduced cost of each nontree arc is nonzero. Conclude that each dual pivot in the algorithm will be nondegenerate and that the algorithm will terminate finitely. (*Hint*: Use the fact that the reduced cost of a nontree arc (k, l) is the cost of the fundamental cycle created by adding arc (k, l) to the spanning tree.)

11.39. In Exercise 9.24 we considered a numerical example concerning sensitivity analysis of a minimum cost flow problem. Solve the same problem using the simplex-based methods described in Section 11.10.

11.40. In Section 11.10 we described simplex-based procedures for reoptimizing a minimum cost flow solution when some cost coefficient c_{ij} increases or some flow bound u_{ij} decreases. Modify these procedures so that we can use them to handle situations in which (1) some c_{ij} decreases, or (2) some u_{ij} decreases.

11.41. Let $\mathcal{B}$ denote the basis matrix associated with the columns of the spanning tree in Figure 11.25(a). Rearrange the rows and columns of $\mathcal{B}$ so that it is lower triangular.

11.42. Let $G' = (N, A')$ be a subgraph of $G = (N, A)$ containing $|A'| = n - 1$ arcs. Let $\mathcal{B}'$ be the square matrix defined by the columns of arcs in A' (where we delete one redundant row). Show that A' is a spanning tree of G if and only if the determinant of $\mathcal{B}'$ is ± 1.

11.43. **Computation of $\mathcal{B}^{-1}$.** In this exercise we discuss a combinatorial method for computing the inverse of a basis matrix $\mathcal{B}$ of the minimum cost flow problem. (We assume that we have deleted a redundant row from $\mathcal{B}$.) By definition, $\mathcal{B}\mathcal{B}^{-1} = \mathcal{I}$, an identity matrix. Therefore, the jth column $\mathcal{B}_j^{-1}$ of the inverse matrix $\mathcal{B}^{-1}$ satisfies the condition $\mathcal{B}\mathcal{B}_j^{-1} = e_j$. Consequently, $\mathcal{B}_j^{-1}$ is the unique solution x of the system of equations $\mathcal{B}x = e_j$. Assuming that we have deleted the row corresponding to node 1, x is the flow vector obtained from sending 1 unit of flow from node j to node 1 on the tree arcs corresponding to the basis. Use this result to compute $\mathcal{B}^{-1}$ for the basis $\mathcal{B}$ defined by the spanning trees shown in Figure 11.25(a).

11.44. Show that a matrix $\mathcal{A}$ whose components are 0, +1, or −1 is totally unimodular if it satisfies both of the following conditions: (1) each column of $\mathcal{A}$ contains at most two nonzero elements; and (2) the rows of $\mathcal{A}$ can be partitioned into two subsets $\mathcal{A}_1$ and $\mathcal{A}_2$ so that the two nonzero entries in any column are in the same set of rows if they have different signs and are in different set of rows if they have the same sign.

11.45. Let $\mathcal{N}$ be a totally unimodular matrix. Show that $\mathcal{N}^{\mathrm{T}}$ and $[\mathcal{N}, -\mathcal{N}]$ are also totally unimodular.

11.46. Show that a matrix $\mathcal{N}$ is totally unimodular if and only if the matrix $[\mathcal{N}, \mathcal{I}]$ is unimodular.

11.47. Let **T** be a spanning tree of a directed network $G = (N, A)$ with node 1 as a designated root node. Let $d(i, j)$ denote the number of arcs on the tree path from node i to node j in **T**.

(a) For the given tree **T**, the *average depth* is $(\sum_{j \in N} d(1, j))/n$, and the *average cycle length* is $(\sum_{\text{nontree arcs } (i,j)} d(i, j) + 1)/(m - n + 1)$. Show that if G is a complete graph, the average cycle length is at most twice the average depth. Show that this relationship is not necessarily valid if the graph is not complete. (*Hint*: Use the fact that the length of the cycle created by adding the arc (i, j) to the tree is at most $d(1, i) + d(1, j) + 1$.)

(b) For a given tree **T**, let $D(j)$ denote the set of descendants of node j. The *average subtree size* of **T** is $(\sum_{j \in N} |D(j)|)/n$. Show that the average subtree size is 1 more than the average depth. (*Hint*: Let $E(j)$ denote the number of ancestors of node j in the tree **T**. First show that $\sum_{j \in N} |E(j)| = \sum_{j \in N} |D(j)|$.)

11.48. **Cost parametrization** (Srinivasan and Thompson [1972]). Suppose that we wish to solve a parametric minimum cost flow problem when the cost c_{ij} for each arc $(i, j) \in A$ is given by $c_{ij} = c_{ij}^0 + \lambda c_{ij}^*$ for some constants c_{ij}^0 and c_{ij}^* and we want to find an optimal solution for all values of the parameter λ in a given interval $[\alpha, \beta]$.

(a) Let **(T, L, U)** be an optimal spanning tree structure for the minimum cost flow problem for some value λ of the parameter. Let π^0 denote the node potentials for the tree **T** when c_{ij}^0 are the arc costs, and let π^* denote node potentials when

c_{ij}^* are the arc costs in **T** (we can compute these potentials using the procedure *compute-potentials*). Show that $\pi^0 + \lambda\pi^*$ are the node potentials for the tree **T** when the arc costs are $c_{ij}^0 + \lambda c_{ij}^*$. Use this result to identify the largest value of λ, say $\bar{\lambda}$, for which (**T**, **L**, **U**) satisfies the optimality conditions.

(b) Show that at $\lambda = \bar{\lambda}$, some nontree arc (k, l) satisfies its optimality condition as an equality and violates the optimality condition when $\lambda > \bar{\lambda}$. Show that if we perform the pivot operation with arc (k, l) as the entering arc, the new spanning tree structure also satisfies the optimality conditions at $\lambda = \bar{\lambda}$.

(c) Use the results in parts (a) and (b) to solve the minimum cost flow problem for all values of the parameter λ in a given interval $[\alpha, \beta]$.

11.49. Supply/demand parametrization (Srinivasan and Thompson [1972]). Suppose that we wish to solve a parametric minimum cost flow problem in which the supply/demand $b(i)$ of each node $i \in N$ is given by $b(i) = b^0(i) + \lambda b^*(i)$ for some constants $b^0(i)$ and $b^*(i)$ and we want to find an optimal solution for all values of the parameter λ in a given interval $[\alpha, \beta]$. We assume that $\sum_{i \in N} b^0(i) = \sum_{i \in N} b^*(i) = 0$.

(a) Let (**T**, **L**, **U**) be an optimal spanning tree structure of the minimum cost flow problem for some value λ of the parameter. Let x_{ij}^0 and x_{ij}^* denote the flows on spanning tree arcs when b^0 and b^* are the supply/demand vectors (we can compute these flows using the procedure *compute-flows*). Show that $x_{ij}^0 + \lambda x_{ij}^*$ is the flow on the spanning tree arcs when $b^0 + \lambda b^*$ is the supply/demand vector. Use this result to identify the largest value of λ, say $\bar{\lambda}$, for which spanning tree arcs satisfy the flow bound constraints.

(b) Show that at $\lambda = \bar{\lambda}$, some tree arc (p, q) satisfies one of its bounds (lower or upper bound) as an equality and violate its flow bound for $\lambda > \bar{\lambda}$. Show that if we perform a dual pivot (as described in Section 11.9) with arc (p, q) as the leaving arc, the new spanning tree structure also satisfies the optimality conditions at $\lambda = \bar{\lambda}$.

(c) Use the results in parts (a) and (b) to solve the minimum cost flow problem for all values of the parameter λ in a given interval $[\alpha, \beta]$.

11.50. Capacity parametrization (Srinivasan and Thompson [1972]). Consider a parametric minimum cost flow problem when the capacity u_{ij} of each arc $(i, j) \in A$ is given by $u_{ij} = u_{ij}^0 + \lambda u_{ij}^*$ for some constants u_{ij}^0 and u_{ij}^*. Describe an algorithm for solving the minimum cost flow problem for all values of the parameter λ in an interval $[\alpha, \beta]$. (*Hint*: Let (**T**, **L**, **U**) be the basic structure at some state. Maintain the flow on each arc in the set **U** as the arc's upper flow bound (as a function of λ), determine the impact of this choice on the flows on the arcs in the spanning tree, and identify the maximum value of λ for which all the arc flows satisfy their flow bounds.)

11.51. Constrained minimum cost flow problem. The constrained minimum cost flow problem is a minimum cost flow problem with an additional constraint $\sum_{(i,j) \in A} d_{ij}x_{ij} \leq D$, called the *budget constraint*.

(a) Show that the constrained minimum cost flow problem need not satisfy the integrality property (i.e., the problem need not have an integer optimal solution, even when all the data are integer).

(b) For the constrained minimum cost flow problem, we say that a solution x is an *augmented tree solution* if some partition of the arc set A into the subsets $\mathbf{T} \cup \{(p, q)\}$, **L**, and **U** satisfies the following two properties: (1) **T** is a spanning tree, and (2) by setting $x_{ij} = 0$ for each arc $(i, j) \in \mathbf{L}$ and $x_{ij} = u_{ij}$ for each arc $(i, j) \in \mathbf{U}$, we obtain a unique flow on the arcs in $\mathbf{T} \cup \{(p, q)\}$ that satisfies the mass balance constraints and the budget constraint. Show that the constrained minimum cost flow problem always has an optimal augmented tree solution. Establish this result in two ways: (1) using a linear programming argument, and (2) using a combinatorial argument like the one we used in proving Theorem 11.2.

12

ASSIGNMENTS AND MATCHINGS

It takes two to tango.
—From a popular American song

Chapter Outline

12.1 INTRODUCTION

To this point in our discussion, we have focused on the three major building blocks of network flows: shortest paths, maximum flows (or minimum cuts), and minimum cost flows. We have seen how these models arise in numerous application settings, we have studied a number of different solution strategies and specific algorithms for solving these problems, and we have seen how important data structures can be used in designing algorithms and in implementing them efficiently. Many of these same ideas apply more generally to the broader field of combinatorial (or discrete) optimization and, indeed, many results in this broader field build on those developed for network flows. This chapter, which considers a particular class of combinatorial optimization problems known as *matching problems*, illustrates the flow of ideas from network flows to other arenas of discrete optimization.

In general, in discrete optimization we are given a finite set of objects and an objective function defined on these objects, and we wish to choose the object with the smallest (or largest) objective value. In this most general form, discrete optimization problems are hopelessly difficult to solve unless we impose some structure on the finite set and on the objective function. In fact, we might view the field of combinatorial optimization as the study of those structures that permit us to say something interesting about the nature of the underlying optimization problem or permit us to find an optimal solution efficiently.

Network flow problems certainly define one very important class of specially structured combinatorial optimization problems. With the exception of our treatment of the combinatorial implications of the max-flow min-cut theorem in Chapter 6, we have not emphasized this viewpoint very much in our discussion. Nevertheless, all of the problems we have considered fit this description of discrete optimization problems. For example, in the context of the shortest path problem, the finite objects are all the paths joining two nodes in a network and the objective function is additive over the arcs selected in any path. For the minimum cut problem, the finite set is the set of cuts separating the source and the sink and the objective function is again additive over the arcs. For the maximum flow problem with unit arc capacities, the finite objects are all sets of paths from the source to the sink that are arc-disjoint and we wish to find the solution with the largest number of paths (by replacing arcs with integer capacities by parallel arcs with unit capacities, we can interpret all maximum flow problems in a similar fashion). For minimum cost flow problems with unit supplies and demands at the nodes, the underlying objects are the set of paths directed from a supply node to a demand node; the objective function in this case is the sum of the weight of arcs in the chosen paths. (By duplicating nodes with integral supplies and demands into sets of nodes with unit supplies and demands, we can interpret more general minimum cost flow problems in this same way.)

In this and the next chapter, we consider two related combinatorial models that are defined over graphs with a weight associated with each arc. In Chapter 13 the objects are all the spanning trees in the network and we wish to find the spanning tree with the smallest overall weight (defined as the sum of the weights of its constituent arcs). This problem is known as the *minimum spanning tree problem*. In this chapter the objects are all subgraphs, called *matchings*, with the property that every node in the subgraph has degree zero or one. That is, no two arcs in the subgraph are incident to the same node. We wish to find the matching with the smallest overall weight, again defined as the sum of the weights of its constituent arcs.

The matching problem arises in many different problem settings since we often wish to find the best way to pair objects or people together to achieve some desired goal. The classical bipartite matching problem is a special case in which objects separate into two groups, and we wish to pair the objects in the different groups in some optimal fashion—for example, we wish to assign jobs to machines in the most cost-effective manner. The general matching problem models situations in which the objects need not fall into two groups—that is, the underlying network need not be bipartite.

We begin this chapter by describing several applications of matching problems in practical contexts as varied as inventory planning, machine scheduling, drilling oil fields, and personnel assignment. We then describe solution approaches for several special cases of the general matching problem. We begin by examining bipartite matching problems. We consider two versions of these problems: (1) the cardinality problem in which we wish to find a matching containing the maximum number of arcs, and (2) the weighted problem in which we have a weight associated with each arc and we wish to find a matching with the largest overall weight (for the cardinality problem, the weights are all 1). For the weighted problem, we restrict the matching to a smaller class of subgraphs, known as perfect matchings, in which every node is incident to exactly one arc in the matching (i.e., every node has degree exactly 1).

As we will see, bipartite problems are easy to solve because we can model them as network flow problems and solve them using any of the many algorithms that we have already studied. Because the resulting network flow problems have a special structure, we can refine our analysis of the network flow algorithms and show that their worst-case complexity is better for the bipartite matching problems than they are in general. In particular, we show that maximum flow algorithms solve the cardinality bipartite matching problem in $O(\sqrt{n}m)$ time. We also show that if $S(n, m, C)$ denotes the time required to solve a shortest path problem on an n-node and m-arc network with nonnegative arc costs bounded by C, specializations of the successive shortest path algorithm and the primal–dual minimum cost flow algorithms solve the weighted bipartite matching problem in $O(n\, S(n, m, C))$ time. We also describe a cost scaling algorithm with an even better time bound.

Nonbipartite matching problems are more difficult to solve because they do not reduce to standard network flow problems. Therefore, they require specialized combinatorial algorithms. To demonstrate the flavor of these algorithms but not go too far afield from the general thrust of this book, we consider only the cardinality version of the nonbipartite matching problem. We describe a clever $O(n^3)$ augmenting path algorithm for solving this problem. Even though the details of this algorithm are quite different from those of the algorithms we have studied for solving core network flow problems, the algorithm does adopt a common algorithmic strategy that we have seen many times before in previous chapters.

In this chapter we also consider a variant of the matching problem known as the stable marriage problem. This model differs from other models we have considered in the text in one important respect: It has no objective function that we wish to optimize. Instead, it models situations with two groups, such as men and women, in which each man has a ranking of each woman and each woman has a ranking of the men. We seek a feasible matching of the members of the two groups, known as a *stable matching*, with the property that no pair of man and woman prefer each other to the partners that they have in the stable matching. We show that for any set of rankings, this problem always has a stable matching and we show how to compute such a solution in $O(n^2)$ time.

12.2 APPLICATIONS

As we show in this section, matching problems arise in a variety of different problem contexts. In Chapter 1 we considered two applications, the pairing of stereo speakers to achieve balanced frequency responses and the rewiring of typewriters. We now describe several other applications.

Application 12.1 Bipartite Personnel Assignment

In many different problem contexts, we wish to assign people to objects: for example, to jobs, machines, rooms, or each other. Each assignment has a "value" and we wish to make the assignments so that we maximize the sum of these values. To illustrate the range of these contexts, in this and the next application, we consider six different applications relating to personnel assignment.

1. A firm has hired n graduates to fill n vacant jobs. Based on aptitude tests, college grades, and letters of recommendation, the firm has assigned a *profi-*

ciency index u_{ij} for placing candidate i in job j. The objective is to identify an assignment that maximizes the total proficiency score over all jobs. This problem is clearly an application of the assignment problem.

2. A swimming coach must select from his eight best swimmers a medley relay team of four, each of whom will then swim one of the four strokes (back, breast, butterfly, and free-style). The coach knows the time of each swimmer in each stroke. The problem is to identify the team of the four best swimmers out of the eight that are available. Clearly, the sum of times obtained by optimally matching four out of the eight swimmers to the four strokes gives the minimum feasible relay time and the corresponding team is the best team. We point out that in this version of the assignment problem $|N_1| > |N_2|$; nevertheless, by adding "dummy nodes," we can easily transform this problem into an equivalent one in which both node sets N_1 and N_2 have the same size.
3. In the armed forces, many men and women are qualified to perform specific jobs, or postings. The armed forces would like to assign the service personnel to postings in order to minimize moving costs. General rules specify the needed qualifications of the personnel for the postings and identify jobs that need to be filled. Policy rules determine allowable assignments that reflect job qualifications and personnel requirements. For an allowable assignment, the *posting cost* is the dollar cost of moving the person, his or her family, and his or her belongings to the new residence. In this case the assignment problem would find an allowable assignment that minimizes the total posting cost.

Application 12.2 Nonbipartite Personnel Assignment

1. During World War II, the Royal Air Force (RAF) of Britain contained many pilots from foreign countries who spoke different languages and had different levels of training. The RAF had to assign two pilots to each plane, always assigning pilots with compatible languages and training to the same plane. The RAF wanted to fly as many planes as possible. To formulate this problem as a maximum cardinality matching problem, we define a graph whose nodes represent pilots; we join two nodes by an arc if the corresponding pilots are compatible.
2. A hostel manager wants to assign pairs of roommates to rooms of her hostel. The nationality, religion, cultural background, and hobbies determine compatible pairs of roommates. So the problem of finding the maximum number of compatible pairs is a maximum cardinality matching problem.
3. Suppose that an airline wishes to divide its $2p$ airplane pilots, linearly ordered by seniority (with no ties), into m teams each containing a captain and a first officer. The captain of each team must have seniority over the first officer. Each pilot i has a measure, α_i, of his effectiveness as a captain and another, β_i, measure of his effectiveness as a first officer. We seek an assignment of pilots to teams that will maximize the total measure of effectiveness summed over all the teams. This problem is an instance of the maximum weight matching problem: We represent each pilot as a node and define the cost of an arc (i, j)

as $\alpha_j + \beta_i$ if pilot j is more senior than pilot i, and as $\alpha_i + \beta_j$ if pilot i is more senior than pilot j.

Application 12.3 Assigning Medical School Graduates to Hospitals

Each year medical schools in the United States graduate thousands of doctors who are eligible for residencies at the various hospitals across the country. To give each of the graduates a chance to find the "best possible" residency and the hospitals the chance to obtain the "best possible" residents, the American Medical Association (AMA) conducts a matching process in which the graduates rank the hospitals according to their preferences and the hospitals rank the graduates according to their preferences. It then assigns the graduates to hospitals so that the matching is "stable" in the following sense. We say that an assignment is *unstable* if some graduate i is not assigned a hospital j, but that graduate prefers hospital j over his or her current assignment and, at the same time, hospital j prefers graduate i over one of the graduates assigned to it. This assignment is unstable because both the graduate i and the hospital j have an incentive to change their current assignments. We refer to an assignment that is not unstable as *stable*. The objective of the AMA is to identify a stable assignment. This problem is an example of the stable marriage problem that we discuss in Section 12.5.

Application 12.4 Dual Completion of Oil Wells

An oil company has identified several individual oil traps, called *targets*, in an offshore oil field and wishes to drill wells to extract oil from these traps. Figure 12.1 illustrates a situation with eight targets. The company can extract any target separately (so-called *single completion*) or extract oil from any two targets together by drilling a single hole (so-called *dual completion*). It can estimate the cost of drilling and completing any target as a single completion or any pair of targets as a dual completion. This cost will depend on the three-dimensional spatial relationships of targets to the drilling platform and to each other. The decision problem is to determine which targets (if any) to drill as single completions and which pairs to drill together as duals, so as to minimize the total drilling and completion costs. If we restrict the solution to use only dual completions, the decision problem is a nonbipartite weighted matching problem.

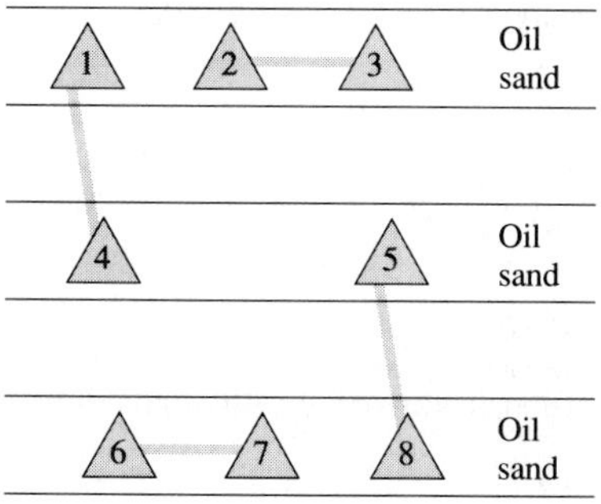

Figure 12.1 Targets and matchings for the dual completion problem.

Application 12.5 Determining Chemical Bonds

Matching problems arise in the field of chemistry as chemists attempt to determine the possible atomic structures of various molecules. Figure 12.2(a) specifies the partial chemical structure of a molecule of some hydrocarbon compound. The molecule contains carbon atoms (denoted by nodes with the letter "C" next to them) and hydrogen atoms (denoted by nodes with the letter "H" next to them). Arcs denote bonds between atoms. The bonds between the atoms, which can be either single or double bonds, must satisfy the "valency requirements" of all the nodes. (The valency of an atom is the sum of its bonds.) Carbon atoms must have a valency of 4 and hydrogen atoms a valency of 1.

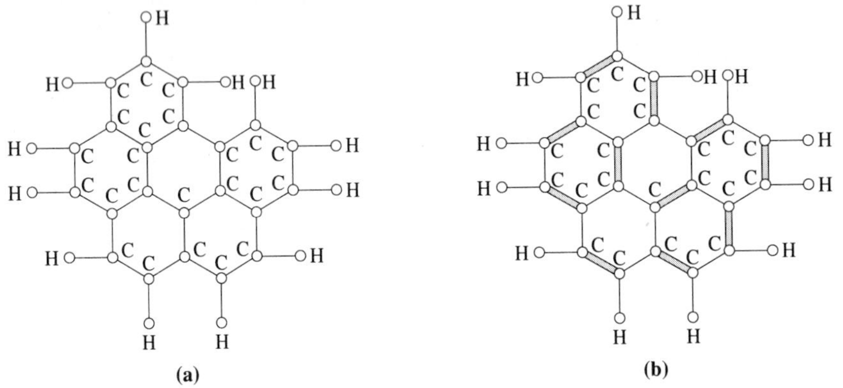

Figure 12.2 Determining the chemical structure of a hydrocarbon.

In the partial structure shown in Figure 12.2, each arc depicts a single bond and, consequently, each hydrogen atom has a valency of 1, but each carbon atom has a valency of only 3. We would like to determine which pairs of carbon atoms to connect by a double bond so that each carbon atom has valency 4. We can formulate this problem of determining some feasible structure of double bonds as an instance of a perfect matching problem in the network obtained by deleting the hydrogen atoms and those carbon atoms with valency 4. Figure 12.2(b) gives one feasible bonding structure of the compound; the bold lines in this network denote double bonds between the atoms.

Application 12.6 Locating Objects in Space

To identify an object in (three-dimensional) space, we could use two infrared sensors, located at geographically different sites. Each sensor provides an angle of sight of the object and hence the line on which the object must lie. The unique intersection of the two lines provided by the two sensors (provided that the two sensors and the object are not collinear) determines the unique location of the object in space.

Consider now the situation in which we wish to determine the locations of p objects using two sensors. The first sensor would provide us with a set of lines L_1, L_2, . . . , L_p for the p objects and the second sensor would provide us a different set of lines $L'_1, L'_2, \ldots, L'_p$. To identify the location of the objects—using the fact that if two lines correspond to the same object, the lines intersect one another—we need to match the lines from the first sensor to the lines from the second sensor. In practice, two difficulties limit the use of this approach. First, a line from a sensor might intersect more than one line from the other sensor, so the matching is not unique. Second, two lines corresponding to the same object might not intersect because the sensors make measurement errors in determining the angle of sight. We can overcome this difficulty in most situations by formulating this problem as an assignment problem.

In the assignment problem, we wish to match the p lines from the first sensor with the p lines from the second sensor. We define the cost c_{ij} of the assignment (i, j) as the minimum Euclidean distance between the lines L_i and L_j. We can determine c_{ij} using standard calculations from geometry. If the lines L_i and L_j correspond to the same object, c_{ij} would be close to zero. An optimal solution of the assignment problem would provide an excellent matching of the lines. Simulation studies have found that in most circumstances, the matching produced by the assignment problem defines the correct location of the objects.

Application 12.7 Matching Moving Objects

In several different application contexts, we might wish to estimate the speeds and the directions of movement of a set of p objects (e.g., enemy fighter planes, missiles) that are moving in space. Using the method described in the preceding application, we can determine the location of the objects at any point in time. One plausible way to estimate the objects' movement directions and speeds is to take two snapshots of the objects at two distinct times and then to match one set of points with the other set of points. If we match the points correctly, we can assess the speed and direction of movement of the objects. As an example, consider Figure 12.3 which denotes the objects at time 1 by squares and the objects at time 2 by circles.

Let (x_i, y_i, z_i) denote the coordinates of object i at time 1 and (x'_i, y'_i, z'_i) denote the coordinates of the same object at time 2. We could match one set of points with

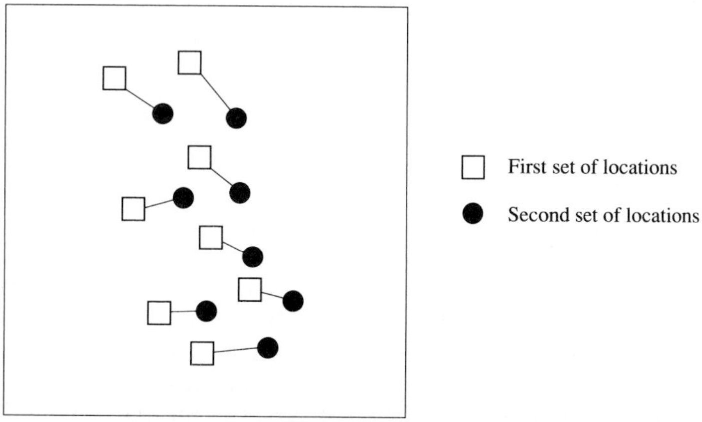

Figure 12.3 Two snapshots of a set of eight objects.

the other set of points in many ways. Minimizing the sum of the squared Euclidean distances between the matched points is quite appropriate in this scenario because it attaches a higher penalty to larger distances. If we take the snapshots of the objects at two times that are sufficiently close to each other, the optimal assignment will often match the points correctly. In this application of the assignment problem, we let $N_1 = \{1, 2, \ldots, p\}$ denote the set of objects at time 1, let $N_2 = \{1', 2', \ldots, p'\}$ denote the set of objects at time 2, and define the cost of an arc (i, j') as $[(x_i - x_j')^2 + (y_i - y_j')^2 + (z_i - z_j')^2]$. The optimal assignment in this graph will specify the desired matching of the points. From this matching we obtain an estimate of the movement directions and the velocities of the individual objects.

Application 12.8 Optimal Depletion of Inventory

In many different problem contexts, we need to store items that either deteriorate or increase in value over time. Suppose that we have a stockpile consisting of p items of the same type. Item i has a current age a_i. A function $v(t)$ specifies the expected utility (or value) for an item of age t when we withdraw it from the stockpile. We need to meet a given schedule that specifies the times at which items are required. The problem is to determine the order for issuing the items that maximizes the total expected utility summed over all p items. A specific example of this general problem is the storage of a number of vats of a volatile liquid (e.g., alcohol). Since alcohol is volatile, we incur an increasing loss due to evaporation with age in storage, so the value of the vat decreases over time.

Some special instances of this inventory problem are particularly easy to solve; for example, when the utility function $v(t)$ satisfies convexity or concavity properties (see Exercise 12.5). When the utility function is arbitrary, we can solve the problem as an assignment problem. Let $t_1, t_2, \ldots, t_p$ denote the time instances when we need to extract an item from the stockpile. Then since item i has an age a_i at time zero, the expected utility for the issue of ith item at time t_j is

$$u_{ij} = v(a_i + t_j).$$

If we compute these utilities for all pairs of i and j, we solve the inventory problem by solving the $p \times p$ assignment problem of maximizing the assignment utilities.

Application 12.9 Scheduling on Parallel Machines

In many application settings, such as the scheduling of computer programs on processors of a computer, we are given a set of w jobs each requiring processing on one of r machines. Suppose that job i requires a processing time of p_{ij} on machine j. Our aim is to find an assignment of the jobs to the machines and a machine schedule (i.e., an order for performing the jobs assigned to the same machine) that will minimize the total flow time of jobs. The *flow time* of a job is the time the job spends in the system before the machines have completed its processing. For example, if we assign jobs 1, 4, and 5 to machine 2 in the order 4–1–5, the flow time of job 4 is p_{24}, the flow time of job 1 is $p_{24} + p_{21}$, and the flow time of job 5 is $p_{24} + p_{21} + p_{25}$. Consequently, the total flow times of the jobs assigned to machine 2 is $1(p_{25})$

+ $2(p_{21}) + 3(p_{24})$. Observe that to determine the total flow time of jobs allocated to a specific machine, we multiply the processing time of the last job by 1, the processing time of the second to last job by 2, and so on, and sum these numbers.

Any algorithm for this scheduling problem must accomplish two objectives. First, it must assign jobs to the various machines. Second, it must sequence the processing of the jobs assigned to any single machine (i.e., assign one job to the first place, one job to the second place, etc.). We would like to make these assignments to minimize the total flow time. This viewpoint suggests that we can assign a job j in one of the wr ways: We can assign it to one of the r machines i (so i can vary from 1 to r) and to one of the kth to last positions on this machine (so k can vary from 1 to w). The cost of this specific assignment would be kp_{ij}.

This scheduling problem is an assignment problem on a network $G = (N_1 \cup N_2, A)$ with nodes N_1 representing the jobs (so $|N_1| = w$) and with nodes N_2 representing the places on different machines (so $|N_2| = wr$). Each node in N_1 in this network is connected to every node in N_2 and the cost of any arc is the cost of assigning a job to a specific place on any machine. In Exercise 12.6 we provide a more rigorous set of arguments for establishing the validity of this formulation.

At first glance, the resulting assignment problem might appear to be much larger than the scheduling problem. However, it is possible to obtain a bound on the maximum number of jobs assigned to the machines and thus to reduce the size of the assignment problem substantially for most instances of the scheduling problem.

12.3 BIPARTITE CARDINALITY MATCHING PROBLEM

As defined earlier, in the bipartite cardinality matching problem (or simply the bipartite matching problem), we wish to identify a matching of maximum cardinality in a bipartite undirected network. Several efficient algorithms for solving this problem achieve a worst-case bound of $O(\sqrt{n}m)$. One approach is to transform the problem into a maximum flow problem in a simple network. In this section we study this approach. We discuss another approach in Section 12.7 as a stepping stone for developing an algorithm for the nonbipartite cardinality matching problem.

Recall from Section 8.2 that in a simple network, each arc has a unit capacity and each node has an indegree of at most 1 or an outdegree of at most 1. To transform a bipartite matching problem defined on an undirected graph $G = (N_1 \cup N_2, A)$ into a maximum flow problem, we first create a directed version of the underlying graph G by designating all arcs as pointing from the nodes in N_1 to the nodes in N_2. We then introduce a source node s and a sink node t, with an arc connecting s to each node in N_1 and an arc connecting each node in N_2 to t. We set the capacity of each arc in the network to 1. Figure 12.4 illustrates this transformation. We refer to the transformed network as $G' = (N', A')$. Note that the network G' is a simple network since every node in N_1 has one incoming arc and every node in N_2 has one outgoing arc. We now establish a one-to-one correspondence between a matching of cardinality k in the original network and an integral flow of value k in the transformed network.

Given a matching $\{(i_1, j_1), (i_2, j_2), \ldots, (i_k, j_k)\}$ of cardinality k in the original network G, we construct a flow in the transformed network G' as follows. We first set the flow on each of the matched arcs equal to 1. Then to satisfy the mass balance

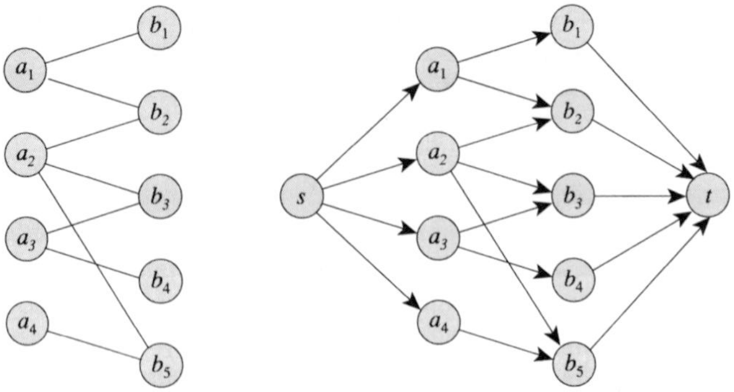

(a) (b)

Figure 12.4 Transforming a bipartite (cardinality) matching problem to a maximum flow problem: (a) original network; (b) unit capacity maximum flow network.

constraints, we set the flow on the arcs (s, i_r) and (j_r, t) equal to 1 for all $r = 1, 2, \ldots, k$. Clearly, this choice gives us a flow of value k from node s to node t.

Similarly, given an integral flow of value k from node s to node t in the transformed network, we can specify a corresponding matching in the original network: By flow decomposition, the integral flow of cardinality k decomposes into k paths of the form $s - i_1 - j_1 - t$, $s - i_2 - j_2 - t, \ldots, s - i_k - j_k - t$. Since each of the arcs incident to nodes s and t have a unit capacity, no two nodes in N_1 or N_2 appear in more than one of these paths, and so the k arcs $\{(i_1, j_1), (i_2, j_2), \ldots, (i_k, j_k)\}$ define a matching.

We have thus established an equivalence between matchings in the original network and integral flows in the transformed network. Therefore, to solve the matching problem, we solve a maximum flow problem in the transformed network using the $O(\sqrt{n}m)$ time algorithm described in Section 8.2. Recall that this algorithm produces an integer optimal flow. The matching corresponding to the maximum flows is a maximum cardinality matching. We have therefore established the following result.

Theorem 12.1. *It is possible to solve the maximum cardinality bipartite matching problem in $O(\sqrt{n}m)$ time.*

12.4 BIPARTITE WEIGHTED MATCHING PROBLEM

In this section we study the bipartite weighted matching problem; namely, given a weighted bipartite network $G = (N_1 \cup N_2, A)$ with $|N_1| = |N_2|$ and arc weights c_{ij}, find a perfect matching of minimum weight. We allow the network G to be directed or undirected. If the network is directed, we require that for each arc $(i, j) \in A$, $i \in N_1$ and $j \in N_2$. If the network is undirected, we make it directed by designating all arcs as pointing from the nodes in N_1 to those in N_2. We shall, therefore, henceforth assume that G is a directed graph. In the operations research literature, the bipartite weighted matching problem is known as the *assignment problem*; for the sake of brevity and to conform with this convention, we adopt this terminology.

Recall that the assignment problem is a special case of the minimum cost flow problem and can be stated as the following linear program.

$$\text{Minimize} \sum_{(i,j)\in A} c_{ij}x_{ij} \tag{12.1a}$$

subject to

$$\sum_{\{j:(i,j)\in A\}} x_{ij} = 1 \qquad \text{for all } i \in N_1, \tag{12.1b}$$

$$\sum_{\{j:(j,i)\in A\}} x_{ji} = 1 \qquad \text{for all } i \in N_2, \tag{12.1c}$$

$$x_{ij} \geq 0 \qquad \text{for all } (i, j) \in A. \tag{12.1d}$$

Since we can formulate the weighted bipartite matching problem as this special type of flow problem, it is not too surprising to learn that most algorithms for the assignment problem can be viewed as adaptations of algorithms for the minimum cost flow problem. However, the special structure of the assignment problem often permits us to simplify these algorithms and to obtain improved bounds on their running times.

One popular algorithm for the assignment problem is a specialization of the network simplex algorithm discussed in Chapter 11. Another popular algorithm is the successive shortest path algorithm and its many variants. In the following discussion, we briefly describe some of these successive shortest path-based algorithms. We also describe an adaptation of the cost scaling algorithm.

Successive Shortest Path Algorithm

This algorithm is a direct implementation of the successive shortest path algorithm for the minimum cost flow problem discussed in Section 9.7. Recall that the successive shortest path algorithm obtains shortest path distances from a supply node to all other nodes in a residual network, uses these distances to update node potentials and then augments flow from that supply node to a demand node. This algorithm, when applied to the assignment problem, would augment 1 unit flow in every iteration, which would amount to assigning one additional node in N_1. Consequently, if we let $S(n, m, C)$ denote the time needed to solve a shortest path problem with nonnegative arc lengths and let $n_1 = |N_1|$, the algorithm would terminate within n_1 iterations and would require $O(n_1S(n, m, C))$ time.

Hungarian Algorithm

The Hungarian algorithm is a direct implementation of the primal–dual algorithm for the minimum cost flow problem that we discussed in Section 9.8. Recall that the primal–dual algorithm first transforms the minimum cost flow problem into a problem with a single supply node s^* and a single demand node t^*. At every iteration, the primal–dual algorithm computes shortest path distances from s^* to all other nodes, updates node potentials, and then solves a maximum flow problem that sends the maximum possible flow from node s^* to node t^* over arcs with zero reduced costs. When applied to the assignment problem, this algorithm terminates within n_1 iterations since each iteration sends at least 1 unit of flow, and hence assigns at least one additional node in N_1. The time required to solve shortest path problems in all these iterations is $O(n_1 S(n, m, C))$. Next consider the total time required to establish maximum flows. The labeling algorithm, described in Section 6.5, for solving the maximum flow problem would require a total of $O(nm)$ time because it would perform n augmentations and each augmentation requires $O(m)$ time. The dominant portion of these computations is the time required to solve shortest path problems. Consequently, the overall running time of the algorithm is $O(n_1 S(n, m, C))$.

Relaxation Algorithm

The relaxation algorithm, which is closely related to the successive shortest path algorithm, is another popular approach for solving the assignment problem. This algorithm relaxes the constraint (12.1c), thus allowing any node in N_2 to be assigned to more than one node in N_1. The relaxed problem is easy to solve: We assign each node $i \in N_1$ to any node $j \in N_2$ with the minimum cost c_{ij} among all arcs in $A(i)$. As a result, some nodes in N_2 might be unassigned while some other nodes are overassigned (i.e., assigned to more than one node in N_1). The algorithm then gradually converts this solution to a feasible assignment while always maintaining the reduced cost optimality condition. At each iteration the algorithm selects an overassigned node k in N_2, obtains shortest path distances from node k to all other nodes in the residual network with reduced costs as arc lengths, updates node potentials, and augments a unit flow from node k to an unassigned node in N_2 along the shortest path. Since each iteration assigns one more node in N_2 and never converts any assigned node into an unassigned node, within n_1 such iterations, the algorithm obtains a feasible assignment. The relaxation algorithm maintains optimality conditions throughout. Therefore, the shortest path problems have nonnegative arc lengths, and the overall running time of the algorithm is $O(n_1 S(n, m, C))$.

Cost Scaling Algorithm

This algorithm is an adaptation of the cost scaling algorithm for the minimum cost flow problem discussed in Section 10.3. Recall that the cost scaling algorithm performs $O(\log(nC))$ scaling phases and the generic implementation requires $O(n^2 m)$ time for each scaling phase. The bottleneck operation in each scaling phase is performing nonsaturating pushes which require $O(n^2 m)$ time; all other operations, such as finding admissible arcs and performing saturating pushes, require $O(nm)$ time. When we apply the cost scaling algorithm to the assignment problem, each push is a saturating push since each arc capacity is 1. Consequently, the cost scaling algorithm solves the assignment problem in $O(nm \log(nC))$ time.

A modified version of the cost scaling algorithm has an improved running time of $O(\sqrt{n}m \log(nC))$, which is the best available time bound for assignment problems satisfying the similarity assumption. This improvement rests on decomposing the computations in each scaling phase into two subphases. In the first subphase, we apply the usual cost scaling algorithm with the difference that whenever we have relabeled a node more than $2\sqrt{n}$ times, we set this node aside and do not examine it further. When we have set aside all (remaining) active nodes, we initiate the second subphase. It is possible to show that the first subphase requires $O(\sqrt{n_1}m)$ time, and when it ends, the network will contain at most $O(\sqrt{n_1})$ active nodes. The second subphase makes these active nodes inactive by identifying "approximate shortest paths" from nodes with excesses to nodes with deficits and augmenting unit flow along these paths. The algorithm uses Dial's algorithm (described in Section 4.6) to identify each such path in $O(m)$ time. Consequently, the second subphase also runs in $O(\sqrt{n_1}m \log(nC))$. We provide a reference for this algorithm in the reference notes.

We summarize the preceding discussion.

Theorem 12.2. *The successive shortest path algorithm, Hungarian algorithm, and the relaxation algorithm solve the assignment problem in $O(n_1 S(n, m, C))$ time. A straightforward implementation of the cost scaling algorithm solves the assignment problem in $O(nm \log(nC))$ time and a further improvement of this algorithm runs in $O(\sqrt{n_1}m \log(nC))$ time.* ◆

12.5 STABLE MARRIAGE PROBLEM

The stable marriage problem is a novel application of bipartite matchings. This problem can be stated as follows. A certain community consists of n men and n women. Each person ranks those of the opposite sex in accordance with his or her preferences for a spouse. For a given matching, a man–woman pair is said to be *unstable* if they are not married to each other but prefer each other to their current spouses. A perfect matching (marriage) of men and women is said to be *stable* if it contains no unstable pairs. The stable marriage problem is to identify a stable perfect matching. In this section we show that for *any* set of rankings, we can always find a stable matching. We establish this result constructively, specifying an algorithm that constructs a stable matching in $O(n^2)$ time.

The input to the stable marriage problem consists of two $n \times n$ matrices; the first matrix gives each man's ranking of women and the second matrix gives each woman's ranking of men. A higher rank denotes a more favored person. Without any loss of generality, we can assume that each rank is an integer between 1 and n. To implement the stable marriage algorithm efficiently, we use these two matrices to construct a vector of n elements for each person, called his or her *priority list*, that lists the persons of opposite sex in decreasing order of their rankings. Since all the ranks are between 1 and n, we can construct these priority lists in a total of $O(n^2)$ time using a bucket sort algorithm (see Exercise 12.30).

The algorithm for the stable marriage problem is an iterative greedy algorithm: Each man proposes to his most preferred woman, and each woman receiving more than one proposal rejects all except her most preferred man from among those who have proposed to her. The algorithm maintains a set, LIST, of unassigned men and for each man it maintains an index, called *current-woman*, which denotes the woman in his priority list that he will next offer a proposal. Initially, LIST $= N_1$, the set of all men, and the *current-woman* of each man is the first woman in his priority list.

The stable marriage algorithm proceeds as follows. At each iteration, the algorithm selects a man from LIST, say Bill, and he proposes to his *current-woman*, say Helen. If Helen is still unassigned, she accepts the proposal and Bill and Helen are tentatively assigned to each other—they are "engaged." If Helen is already engaged to some man, say Frank, she accepts the proposal of Bill or Frank that she prefers the most and rejects the other. The rejected man designates the next woman on his priority list his *current-woman.* Whenever the algorithm selects a man from LIST, he is removed from it; and whenever a man is rejected by a woman, he is added to LIST. The algorithm repeats this iterative step until LIST is empty, at which point it has assigned all the men and women. We refer to this algorithm as the *propose-and-reject algorithm.*

It is easy to show that the matching obtained by this algorithm is stable. Suppose that Dick prefers Laura to his marriage partner; he must have proposed to Laura at some earlier stage and she must have rejected his proposal in favor of someone whom she liked more than Dick. Consequently, since no woman ever switches to a man that she prefers less, Laura prefers her husband to Dick, so the matching is stable.

To analyze the complexity of the stable marriage algorithm, we note that at each iteration each woman receiving a proposal either (1) receives her first proposal (which occurs exactly once for each woman), or (2) rejects some proposal. Since each woman rejects any man's proposal at most once, the second outcome occurs at most $(n - 1)$ times for each woman. Therefore, the algorithm performs $O(n)$ steps per woman and $O(n^2)$ steps in total. Notice that no algorithm for the stable marriage problem can have any better complexity bound, since the running time of the propose-and-reject algorithm is linear in the length of the input data. We have thus established the following result.

Theorem 12.3. *For any matrix of rankings, the stable marriage problem always has a stable matching. Further, the propose-and-reject algorithm constructs a stable matching in $O(n^2)$ time.* ◆

Needless to say, there could be several stable matchings; the propose-and-reject algorithm constructs one such stable matching. We refer to a pair (i, j) of a man i and a woman j as *stable partners* if some stable matching matches man i with woman j. The matching constructed by our algorithm possesses an interesting property that every man is at least as well off under it as under *any* stable matching. In other words, each man obtains his best possible stable partner. For obvious reasons we refer to such a matching as the *man-optimal* matching. The fact that the matching constructed by our algorithm is a man-optimal matching relies on the following result.

Lemma 12.4. *In the propose-and-reject algorithm, a woman never rejects a stable partner.*

Proof. Let M^* be the matching constructed by the propose-and-reject algorithm. Suppose that the lemma is false and women do reject stable partners. Consider the first time that a woman, say Joan, rejects a stable partner, say Dave. Let $M°$ be the stable matching in which Joan and Dave constitute a stable pair of partners. Suppose that the rejection took place because Joan was engaged to Steve, whom she prefers to Dave. Now notice that prior to the rejection, no other woman had rejected a stable partner, which implies that Steve can have no stable partners whom he prefers to Joan. In $M°$, let Sue and Steve be the stable pair for Steve. By our prior observation, Steve prefers Joan to Sue. We have earlier shown that Sue prefers Steve to Dave. The preceding two facts contradict the assumption that M° is a stable matching. This conclusion implies the lemma. ◆

In the propose-and-reject algorithm, men propose to women in decreasing order of their preferences, and since no woman ever rejects a stable partner, each man must be married to the best possible stable partner. Therefore, we have established the following theorem.

Theorem 12.5. *The propose-and-reject algorithm constructs a man-optimal stable matching.* ◆

This theorem is a surprising result. It implies that if each man is independently given his best stable partner, the result is a stable matching. However, we gain this optimality from the men's point of view at the expense of the women. In fact, it is possible to show that in a man-optimal matching, each woman obtains the worst partner that she can have in any stable matching (see Exercise 12.27).

As a concluding remark, we point out that the stable marriage problem also has "nonmatrimonial" applications, such as assigning residents to hospitals, or assigning graduate students to doctoral programs. These applications are actually many-to-one matchings, but can be solved by a minor variation of the propose-and-reject algorithm (see Exercise 12.31).

12.6 NONBIPARTITE CARDINALITY MATCHING PROBLEM

In this section we study the nonbipartite cardinality matching problem on undirected graphs, which we subsequently refer to by the abbreviated name the "nonbipartite matching problem." As we shall see, the nonbipartite matching problem is substantially more difficult to solve than the bipartite problem. To highlight the essential differences between the nonbipartite and bipartite matching problems, we first consider a very natural approach for the matching problem that closely resembles the augmenting path algorithm for solving maximum flow problems discussed in Section 6.4. We show that this approach gives an optimal algorithm for the bipartite problem, but fails for the nonbipartite case. We then identify the reason why the algorithm fails and modify it so that it works for the nonbipartite case as well.

In this section, as usual, we let $A(i)$ denote the node adjacency list of node i [i.e., $A(i) = \{j \in N:(i, j) \in A\}$]. We assume that we store each adjacency list as a singly linked list so that we can insert items into the list in $O(1)$ time. We begin by introducing some notation.

Matched Arcs and Nodes

A *matching* M of a graph $G = (N, A)$ is a subset of arcs with the property that no two arcs of M are incident to the same node. We refer to the arcs in M as *matched arcs*, and arcs not in M as *unmatched arcs*. We also refer to the nodes incident to matched arcs as *matched nodes* and refer to the other nodes as *unmatched*. If (i, j)

belongs to the matching, we say that node i is matched to node j and node j is matched to node i. Figure 12.5 illustrates these definitions. The arcs {(2, 4), (3, 5)} constitute a matching in this graph; we depict matched arcs using thicker lines. Notice each node has degree 0 or 1 in the subgraph defined by the matched arcs. Also notice that a matching can contain at most $\lfloor n/2 \rfloor$ arcs.

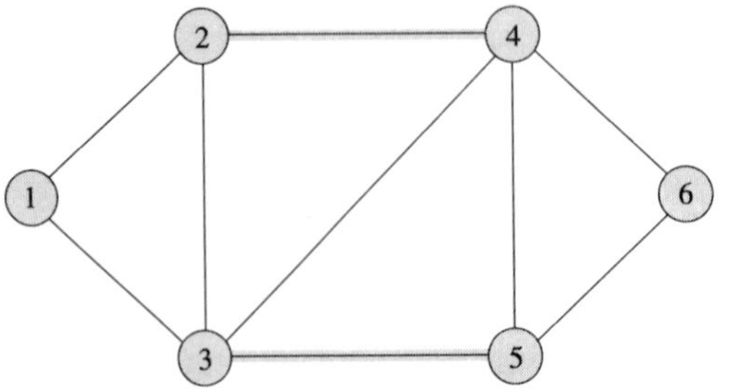

Figure 12.5 Matching example.

Alternating Paths and Cycles

We refer to a path $P = i_1 - i_2 - \cdots - i_k$ in the graph as an *alternating path* with respect to a matching M if every consecutive pair of arcs in the path contains one matched and one unmatched arc. In Figure 12.5, 1–2–4–3–5 and 1–2–4–3–5–6 are alternating paths. We refer to an alternating path as an *even alternating path* if it contains an even number of arcs and an *odd alternating path* if it contains an odd number of arcs. In the preceding example, the first alternating path is even, and the second alternating path is odd. An *alternating cycle* is an alternating path that starts and ends at the same node. In Figure 12.5, 3–2–4–5–3 is an alternating cycle.

Augmenting Paths

We refer to an odd alternating path P with respect to a matching M as an *augmenting path* if the first and last nodes in the path are unmatched. We use the terminology augmenting path because by redesignating matched arcs on the path as unmatched and unmatched arcs as matched, we obtain another matching of cardinality $|M| + 1$. For example, in Figure 12.5 the path 1–2–4–3–5–6 is an augmenting path with respect to a matching of cardinality 2, and if we interchange the matched and unmatched arcs on this path, we obtain the matching of cardinality 3 shown in Figure 12.6.

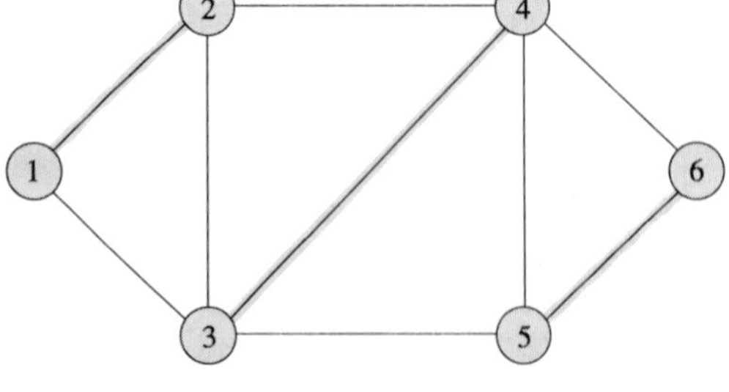

Figure 12.6 Matching a larger cardinality than the matching in Figure 12.5.

Symmetric Difference

The concept of *symmetric difference* of sets is quite important in matching theory. Let S_1 and S_2 be two sets; the symmetric difference of these sets, denoted $S_1 \oplus S_2$, is the set $S_1 \oplus S_2 = (S_1 \cup S_2) - (S_1 \cap S_2)$. In other words, the symmetric difference of sets S_1 and S_2 is the set of elements that are members of one, but not both of S_1 and S_2. For example, if $S_1 = \{4, 5, 7, 8\}$ and $S_2 = \{2, 4, 8, 9\}$, then $S_1 \oplus S_2 = \{2, 5, 7, 9\}$. We shall use the following two properties of symmetric differences.

Property 12.6. *If M is a matching and P is an augmenting path with respect to M, then $M \oplus P$ is a matching of cardinality $|M| + 1$. Moreover, in the matching $M \oplus P$, all the matched nodes in M remain matched and two additional nodes, namely the first and last nodes of P, are matched.* ◆

The symmetric difference of the matching M with the augmenting path P is a set-theoretic way to interchange the matched and unmatched arcs in P, and we have seen earlier that this operation yields a matching of cardinality $M + 1$. We refer to the process of replacing M by $M \oplus P$ as an *augmentation*. The second conclusion of Property 12.6 follows from the definitions.

Property 12.7. *If M and M^* are two matchings, their symmetric difference defines the subgraph $G^* = (N, M \oplus M^*)$ with the property that every component is one of the six types shown in Figure 12.7.* ◆

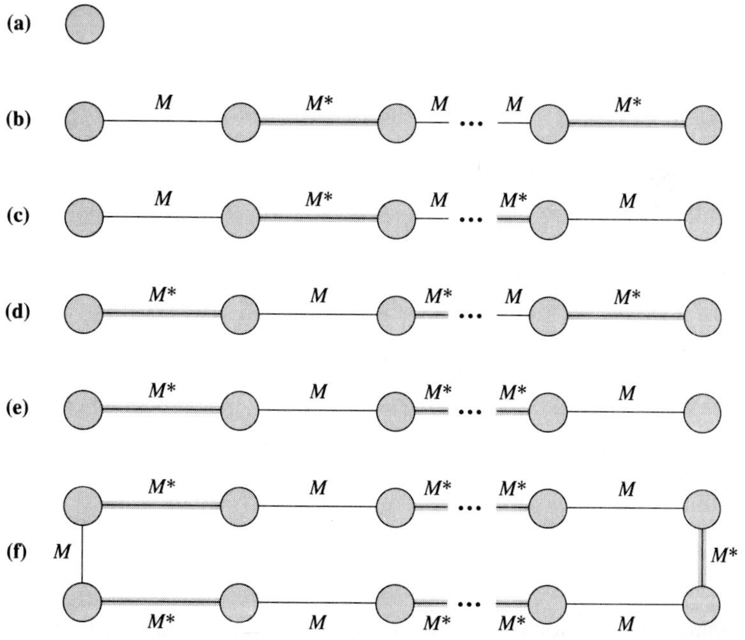

Figure 12.7 Possible types of components formed by a symmetric difference of two matchings M and M^*.

This property follows from the facts that in the subgraph G^* each node has degree 0, 1 or 2, and the only possible components with these node degrees are singleton nodes [as shown in Figure 12.7(a)], paths [as shown in Figure 12.7(b) to (e)], or even-length cycles [as shown in Figure 12.7(f)].

Augmenting Path Theorem

Our algorithm for the matching problem depends crucially on the following augmenting path theorem.

Theorem 12.8 (Augmenting Path Theorem). *If a node p is unmatched in a matching M, and this matching contains no augmenting path that starts at node p, then node p is unmatched in some maximum matching.*

Proof. Let M^* be a maximum matching. If node p is unmatched in M^*, the theorem is clearly true. Therefore, assume that node p is matched in M^*. Consider the symmetric difference of the matchings $M \oplus M^*$. We have seen earlier that every component of the subgraph defined by this symmetric difference is one of the six types shown in Figure 12.7. The fact that node p is unmatched in M rules out all of these possibilities except the ones shown in Figure 12.7(d) and (e) with node p as the starting node. The fact that no augmenting path starts at node p rules out the possibility shown in Figure 12.7(d). Therefore, the only remaining possibility is the even alternating path P shown in Figure 12.7(e) with node p as the starting node. But notice that $M' = M^* \oplus P$ is also a maximum matching in which node p is unmatched. We have thus shown that given a maximum matching M^* in which node p is matched, we can construct another maximum matching M' in which node p is unmatched, which establishes the theorem. ◆

This theorem is an alternative version of a well-known theorem due to Berge, which states that a matching M^* is a maximum matching if and only if the graph G contains no augmenting path with respect to matching M^*. We ask the reader to prove this theorem in Exercise 12.39.

Bipartite Matching Algorithm

The augmenting path theorem suggests the following algorithm for solving the matching problem. Start with a feasible matching M (which might be a null matching) and then repeat the following step for every unmatched node $p \in N$. Try to identify an augmenting path starting at node p. If we find such a path P, replace M with $M \oplus P$; otherwise, delete node p and all the arcs incident to it from the graph.

Using Theorem 12.8, it is easy to show that this algorithm obtains an optimal matching. At each iteration, the algorithm reduces the number of unmatched nodes by at least one, either by deleting a node or by matching it. Since matched nodes remain matched throughout the algorithm (by Property 12.6), when the algorithm terminates, each node in the remaining subgraph, say G', is matched. Consequently, the matching M must be a maximum matching for G'. Theorem 12.8 implies that the deletion of nodes does not reduce the number of arcs in a maximum cardinality matching. Consequently, M is also a maximum matching in G.

We have therefore reduced the matching algorithm to finding whether or not the network contains an augmenting path starting at node p. How can we find such a path if one exists? The most natural approach might be to use a search algorithm to identify an augmenting path, as we did in the labeling algorithm for the maximum flow problem as discussed in Section 6.5. We can define a node i in the graph as *reachable* from node p if the network contains an alternating path from node p to node i, and then use a search algorithm to identify all reachable nodes. If the algorithm finds an unmatched node that is reachable from node p, it has discovered an augmenting path. However, if none of the reachable nodes is unmatched, we can conclude that the network contains no augmenting path starting at node p.

It is, perhaps, easy to believe that identifying all nodes that are reachable from a specified node should be a rather straightforward task using a search algorithm. Unfortunately, the task is complicated. A straightforward version of a search technique does not work for all matching problems. This approach does work for bipartite matching problems, but fails for nonbipartite problems. Nevertheless, this approach gives valuable insight into the matching problem that will help us in solving the general case. Consequently, we first discuss this straightforward approach and then develop a (nontrivial) modification of it that solves the general problem.

A straightforward approach for solving the matching problem would be to grow a search tree rooted at node p so that each path in the tree from node p to another node is an alternating path. For convenience, we refer to node p as the *root node* of the search tree. For obvious reasons, we also refer to this search tree as an *alternating tree*. We say that the nodes in the alternating tree are *labeled nodes* and that the other nodes are *unlabeled*. The labeled nodes are of two types: *even* or *odd*. Node i is even or odd depending on whether the number of arcs in the unique path from the root node to node i in the alternating tree is even or odd. Notice that whenever an unmatched node (other than the root) has an odd label, the path joining the root node to this node is an augmenting path. For convenience, we assign the label "E" to even nodes and the label "O" to odd nodes.

Recall from Section 3.4 that the search algorithm maintains a set, LIST, of labeled nodes and examines labeled nodes one by one. While examining an even node i, the algorithm scans its adjacency list $A(i)$ and assigns an odd label to every node j in $A(i)$ (provided that node j is unlabeled). If node j is unmatched, we have discovered an augmenting path; otherwise, we add this node to LIST. On the other hand, while examining an odd node i, the algorithm examines its unique matched arc (i, j). If node j is unlabeled, the algorithm assigns an even label to this node and adds it to LIST. The search algorithm terminates when LIST becomes empty, or it has assigned an odd label to an unmatched node, thus discovering an augmenting path.

Figure 12.8 illustrates the process of growing the search tree on the graph shown in Figure 12.8(a). Assuming that we examine the labeled nodes in the first-in, first-out order, and scan the nodes in any adjacency list in increasing order of the node numbers, the algorithm will examine the nodes in the order 1–2–4–3–7–6–8–5. The resulting alternating tree shown in Figure 12.8(b) has an augmenting path 1–4–7–8.

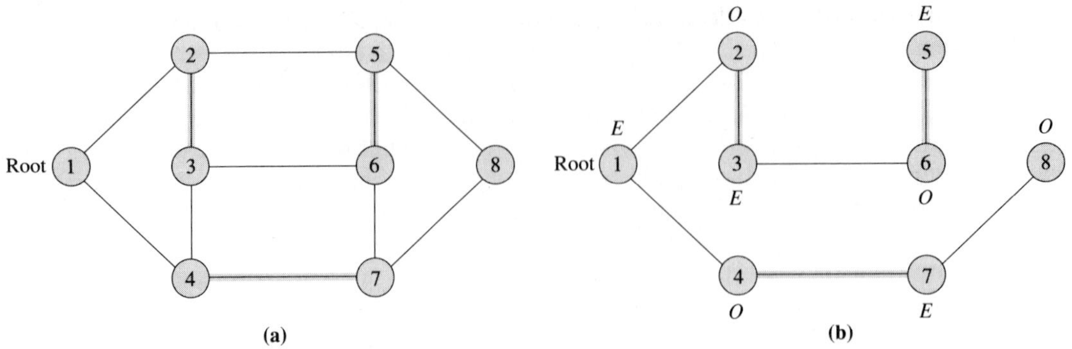

Figure 12.8 Growing an alternating tree: (a) the graph; (b) its complete alternating tree.

We are now in a position to give a complete algorithmic description of the matching algorithm. We subsequently refer to this algorithm, as described in Figures 12.9 and 12.10, as the *bipartite matching algorithm* because, as we explain later, it will always establish a maximum matching in bipartite networks (it might fail when applied to nonbipartite networks).

```
algorithm matching;
begin
    M : = Ø;
    for each node p ∈ N do
      if node p is unmatched then
      begin
          search(p, found);
          if found equals true then augment
          else delete node p and all arcs incident to it from G;
      end;
    M* = M;
end;
```

Figure 12.9 Bipartite matching algorithm.

It is easy to show that the matching algorithm runs in $O(nm)$ time. The algorithm executes the search and augment procedures at most n times. The augment procedure clearly requires $O(n)$ time. It is easy to see that the search procedure requires $O(m)$ time per execution. For each node i, the search procedure performs one of the following two operations at most once: (1) it executes examine-even(i, found), or (2) it executes examine-odd(i, found). The latter operation requires $O(1)$ time per execution. The former operation requires $O(|A(i)|)$ time, so a total of $O(\sum_{i \in N} |A(i)|) = O(m)$ time for all the nodes.

Difficulties with the Bipartite Matching Algorithm

Does the search procedure work correctly? It is clear that whenever the algorithm finds an augmenting path starting at node p, this path is an augmenting path. But when the algorithm fails to find an augmenting path, can we conclude that the network contains no such path? We shall show that if the graph possesses a *unique label property* (defined next), our conclusion will be correct; otherwise, the conclusion could be incorrect.

```
procedure search(p, found);
begin
    found : = false;
    unlabel all nodes;
    give an even label to node p and initialize LIST = {p};
    while LIST ≠ Ø do
    begin
        delete a node i from LIST;
        if node i has an even label then examine-even(i, found)
        else examine-odd(i, found);
        if found equals true then return;
    end;
end;
```

(a)

```
procedure examine-even(i, found);
begin
    for every node j ∈ A(i) do
    begin
        if node j is unmatched then set q : = j and pred(q) : = i;
        found : = true and return;
        if node j is matched and unlabeled then
            set pred(j) : = i, give node j an odd label and add node j to LIST;
    end;
end;
```

(b)

```
procedure examine-odd(i,found);
begin
    let j be the node matched to node i;
    if node j is unlabeled then set pred(j) = i, give node j an even label and add it to LIST;
end;
```

(c)

```
procedure augment;
begin
    trace the augmenting path P by starting at node q and traversing the predecessor indices;
    update the matching using the operation M : = M ⊕ P;
end;
```

(d)

Figure 12.10 Procedures for the bipartite matching algorithm.

Unique label property. *A graph is said to possess a unique label property with respect to a given matching M and a root node p if the search procedure assigns a unique label to every labeled node (i.e., even or odd) irrespective of the order in which it examines labeled nodes.*

It is easy to show that if the graph possesses the unique label property, it will always discover an augmenting path if one such path exists. Suppose that the network does contain an augmenting path $p - i_1 - j_1 - i_2 - j_2 - \cdots i_l - j_l - q$ from node p to node q with respect to the matching M. If we examine the nodes $p, i_1, j_1, i_2, j_2, \ldots$ in order, we will assign even labels to nodes $p, j_1, j_2, \ldots, j_l$, and odd labels to nodes $i_1, i_2, \ldots, i_l, q$. Since the graph possesses the unique label property, the algorithm would assign the same labels no matter in which order the search procedure examines the labeled nodes. Therefore, the search procedure will always assign an odd label to node q and will discover an augmenting path.

Does any network satisfy the unique label property with respect to any matching and any root node? Yes; in fact, bipartite networks satisfy this property. Recall from Section 2.2 that in a bipartite network $G = (N, A)$, we can partition the node set N into two subsets N_1 and N_2 so that every arc $(i, j) \in A$ has its end points in different subsets. For a bipartite network, if the root node is in N_1, every labeled node in N_1 will receive an even label and every labeled node in N_2 will receive an odd label (because the alternating path will begin at a node in N_1 and then alternate between nodes in N_2 and N_1 respectively). Similarly, if the root node is in N_2, every labeled node in N_1 will receive an odd label and every labeled node in N_2 will receive an even label. Consequently, the matching algorithm will find an optimal matching in bipartite networks.

Nonbipartite networks might not satisfy the unique label property, and therefore the search algorithm might fail to detect an augmenting path even though the network contains one. Consider, for example, the situation shown in Figure 12.11. If node 5 receives its label via the path 1–2–3–4–5, it receives the even label. When we examine node 5, the search algorithm gives node 6 an odd label and discovers the augmenting path 1–2–3–4–5–6. However, if node 5 receives its label via the path 1–2–3–7–8–5, its label will be odd. Since node 5 has an odd label, we scan its unique matched arc (5, 4), attempting to label node 4, but do not scan arc (5, 6). Thus the algorithm fails to discover an augmenting path.

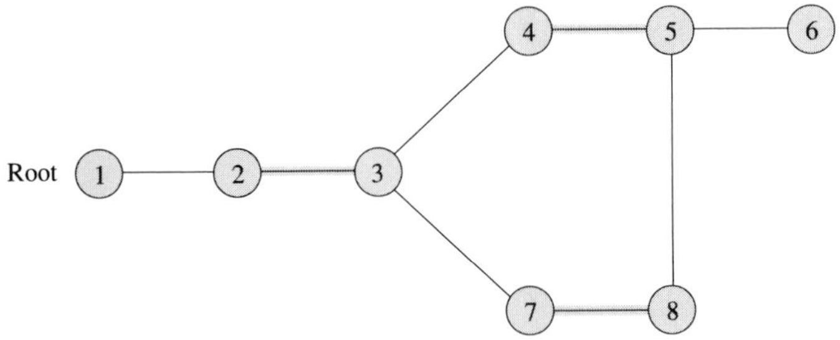

Figure 12.11 Nonbipartite matching problem.

The preceding situation arises because we can connect node 5 to the root by both an odd-length and an even-length alternating path. Therefore, depending on the order in which the search algorithm examines labeled nodes, node 5 might receive an even or an odd label. But since we assign only one label to any node (either even or odd), assigning an odd label prevents us from giving the node an even label in subsequent stages, so we miss the opportunity to give node 6 an odd label.

One plausible way to overcome this difficulty would be to permit nodes to have both even and odd labels. When a node i receives an even label, we scan its adjacency list $A(i)$ to label further nodes; and when a node i receives an odd label, we scan its unique matched arc. But even this modification does not work. To see this, consider the example shown in Figure 12.12. We might examine the nodes in the following order: 1 (even), 2 (odd), 3 (even), 4 (odd), 5 (even), 8 (odd), 7 (even), 3 (odd), 2 (even), 6 (odd). At this point, the unmatched node 6 receives an odd label and the algorithm would declare that it has found an augmenting path, even though the network contains no such path. To summarize, we find that by assigning just one label to each node, we might overlook an augmenting path, and by assigning two labels, we might falsely believe that we have found an augmenting path.

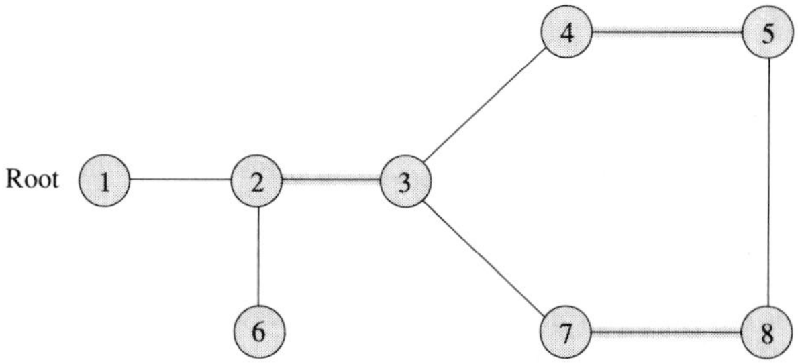

Figure 12.12 Another nonbipartite matching problem.

Why do we encounter this difficulty? What causes the algorithm to break down? The root cause of the difficulty in solving a nonbipartite matching problem is the presence of certain subgraphs called *flowers*, composed of particular types of paths and odd cycles. (Note that since bipartite graphs contain no odd cycles, they never contain any flowers.)

Flowers and Blossoms

A *flower*, defined with respect to a matching M and a root node p, is a subgraph with two components:

1. *Stem*. A stem is an even (length) alternating path that starts at the root node p and terminates at some node w. We permit the possibility that $p = w$, in which case we say that the stem is empty.
2. *Blossom*. A blossom is an odd (length) alternating cycle that starts and terminates at the terminal node w of a stem and has no other node in common with the stem. We refer to node w as the *base* of the blossom.

Figure 12.13 shows two examples of flowers. The flower shown in Figure 12.13(a) has an empty stem, and the flower shown in Figure 12.13(b) has a nonempty stem. We denote a blossom by B and define it by its set of arcs or set of nodes, whichever is convenient. In our subsequent discussion, we use several properties of flowers, which we record for easy future reference.

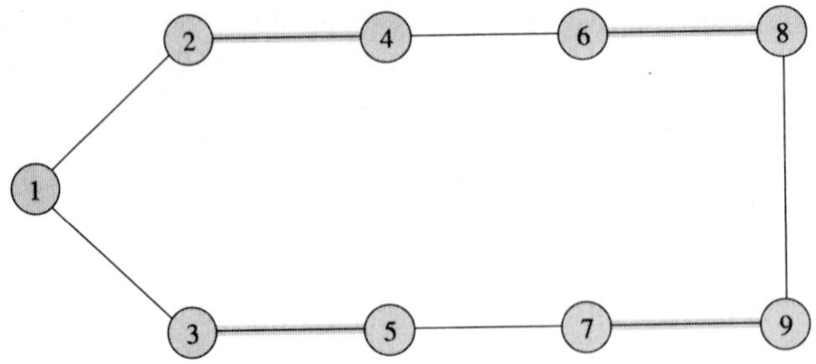

(a)

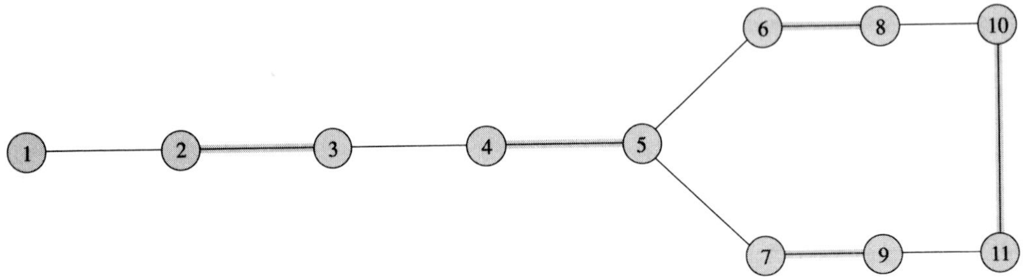

(b)

Figure 12.13 Two examples of flowers.

Property 12.9
(a) A stem spans 2l nodes and contains l matched arcs for some integer $l \geq 0$.
(b) A blossom spans $2k + 1$ nodes and contains k matched arcs for some integer $k \geq 1$. The matched arcs match all nodes of the blossom except its base.
(c) The base of a blossom is an even node.

Property 12.10. *Every node i in the blossom (except its base) is reachable from the root (or from the base of the blossom) through two distinct alternating paths; one has even length and the other has odd length (see, e.g., Figure 12.14).*

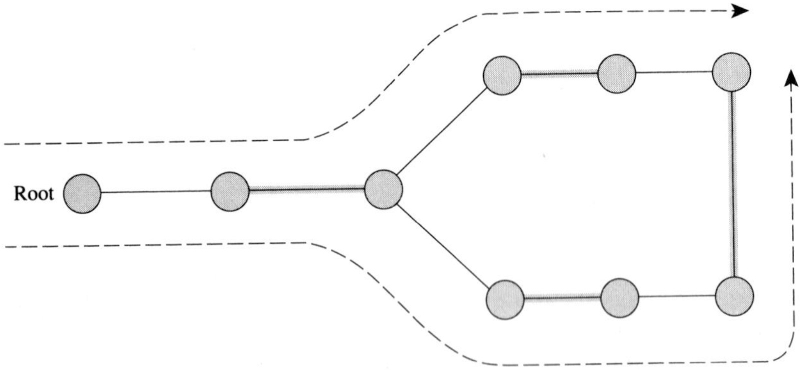

Figure 12.14 Two distinct alternating paths from the root to every node in the blossom.

The even alternating path to node i terminates with a matched arc, and the odd alternating path to node i terminates with an unmatched arc.

These properties are relatively straightforward to establish, so we omit their proofs.

Contracting a Blossom

We now consider the issue we were discussing prior to our definition of blossoms: Why might the labeling algorithm fail to identify an augmenting path, and how might we remedy the problem? If the network contains a blossom with respect to the current matching and the root node p, each node in the blossom is qualified to receive an even label because the network contains an even alternating path from the root to that node. But the search algorithm will give even labels to some nodes in the blossom and odd labels to others. Notice that if we had a choice we would prefer to give even labels to the nodes for the following reason: When examining even-labeled nodes, we can label nodes outside the blossom by searching along all unmatched arcs incident to nodes in the blossom; when examining odd-labeled nodes, however, we label only the nodes in the blossom.

So, it seems intuitively clear that if we could give all the nodes in the blossom an even label, whenever we detect a blossom, the search algorithm would always detect an augmenting path. There are several ways to achieve this objective; one of the more popular approaches is to contract (or shrink) the blossom into a single node. This operation replaces the blossom B consisting of the node sequence $i_1 - i_2 - \cdots - i_k - i_1$ by a single new node b in the following manner:

1. Introduce a new node b and define its adjacency list $A(b) = A(i_1) \cup A(i_2) \cup \cdots \cup A(i_k)$.
2. Update the adjacency list of every node $j \in A(b)$ by executing $A(j) = A(j) \cup \{b\}$.
3. To be able to recover information about the nodes within the blossom that we have contracted into the single node b, we form a circular doubly linked list of nodes $i_1, i_2, \ldots, i_k$, and then delete the nodes $i_1, i_2, \ldots, i_k$ and all arcs incident to these nodes from the network. (Notice that this operation requires the updating of the adjacency list of all the nodes that are adjacent to the deleted nodes.)

We refer to the resulting network $G^c = (N^c, A^c)$ as the *contracted network*. We let $A^c(i)$ denote the adjacency list of a node i in G^c and let M^c denote the corresponding matching in the contracted network.

Figure 12.15(a) illustrates a contraction. The flower 1–2–3–4–5–6–7–3 in this figure contains the blossom 3–4–5–6–7–3. Contracting all these nodes into a new node, node 11, we obtain the graph shown in Figure 12.15(b). Each contraction operation creates a new node. To differentiate this node from the nodes of the original network, we refer to it as a *pseudonode*. Notice that a pseudonode is always an even node because it merges the entire blossom into its base, which is always even [see Property 12.9(c)]. Since the adjacency list of the pseudonode is the union of

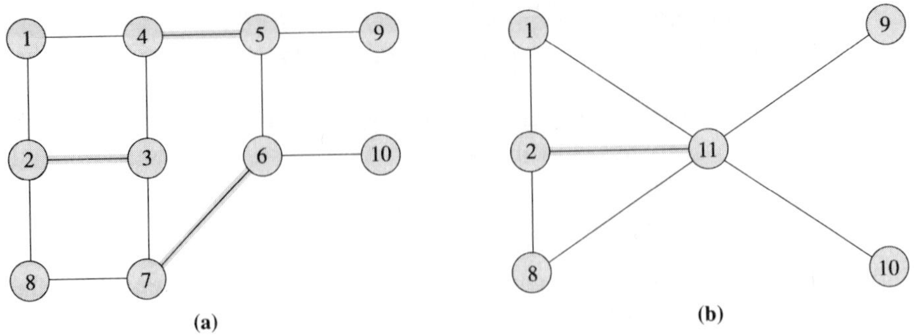

Figure 12.15 Contraction: (a) network before the contraction; (b) network after the contraction.

the adjacency lists of the nodes it contains, scanning the adjacency list of the pseudonode allows us to reach out to all the nodes that we would have reached out to from nodes in the blossom. Consequently, contracting the entire blossom into a single even pseudonode amounts to assigning even labels to each blossom node in the original graph.

Nonbipartite Matching Algorithm

We are now in a position to describe the nonbipartite matching algorithm. This algorithm modifies the search procedure of the bipartite matching algorithm in the following manner. As the search procedure proceeds, it assigns even or odd labels to the nodes. Although the algorithm will never relabel an already labeled node, it will identify the possibility of assigning an odd label to a node with an even label, or of assigning an even label to a node with an odd label. When we find that we can, for the first time, assign a node, say node i, a label other than what it already has, we suspend the search procedure. At this point we have discovered an even as well as an odd alternating path to node i. If we trace back the predecessor indices of these paths until we encounter the first common node on these paths, the arcs we have traced constitute a blossom and the first common node (which has an even label) is the base of the blossom (see Exercise 12.37). We then contract the blossom into a pseudonode, update the data structures, and continue the search procedure. We might note that we could perform several contractions before we either discover an augmenting path (in the contracted graph) or run out of nodes to examine, which indicates that the network contains no augmenting path starting from the root node p. If we succeed in identifying an augmenting path from node p to some unmatched node q, we check whether this path contains any pseudonodes. If so, we expand the blossoms represented by these pseudonodes one by one, in an order to be described later, until the augmenting path contains no pseudonodes. We point out that we can contract blossoms containing pseudonodes; so pseudonodes might contain other pseudonodes.

The algorithmic description of the resulting algorithm, which we subsequently refer to as the *nonbipartite matching algorithm*, is the same as the bipartite matching algorithm with the exception of a change in the search procedure. Figure 12.16 shows

```
procedure search(p, found);
begin
    set A^c(i) : = A(i) for all nodes i;
    found : = false;
    unlabel all nodes;
    give an even label to node p and initialize LIST : = {p};
    while LIST ≠ Ø do
    begin
        delete a node i from LIST;
        if node i has even label then examine-even(i, found)
        else examine-odd(i, found);
        if found = true then return;
    end;
end;
```

(a)

```
procedure examine-even(i, found);
begin
    for every node j ∈ A^c(i) do
    begin
        if node j has an even label then contract(i, j) and return;
        if node j is unmatched then set q : = j,
            pred(q) : = i, found : = true and return;
        if node j is matched and unlabeled then set pred(j) : = i,
            give node j an odd label and add it to LIST;
    end;
end;
```

(b)

```
procedure examine-odd(i, found);
begin
    let node i be matched to node j;
    if node j has an odd label then contract(i, j) and return;
    if node j is unmatched and unlabeled
        then set pred(j) : = i, give node j an even label and add it to LIST;
end;
```

(c)

```
procedure contract(i, j);
begin
    trace the predecessor indices of nodes i and j to identify a blossom B;
    create a new node b and define A^c(b) = ∪_{k∈B} A^c(k);
    give an even label to node b and add it to LIST;
    update A^c(j) = A^c(j) ∪ {b} for each j ∈ A^c(b);
    form a circular doubly linked list of nodes in B;
    delete the nodes in B from the network and update the data structure;
end;
```

(d)

Figure 12.16 Procedures for the nonbipartite matching algorithm.

```
procedure augment;
begin
    trace the augmenting path P' by starting at node q and
      traversing the predecessor indices;
    if the path P' contains pseudonodes then expand the corresponding
      blossoms and obtain an augmenting path P in the original network;
    update the matching using the operation M = M ⊕ P;
end;
```

(e)

Figure 12.16 *(Continued)*

the new search procedure and its subroutines. Notice that the algorithm uses an additional procedure *contract* that contracts a blossom into a pseudonode.

In the nonbipartite matching algorithm, each time we execute the procedure *contract*, we create a new pseudonode. One particularly simple scheme for keeping track of these additional nodes would be to number them as $n + 1$, $n + 2$, $n + 3$, ... In this scheme, a node i is a pseudonode if and only if $i > n$.

We illustrate the nonbipartite matching algorithm by applying it to the numerical example shown in Figure 12.17(a). We assume that the algorithm examines the labeled nodes in the first-in, first-out order, and scans the adjacency list of any node in increasing order of the node numbers. Suppose that the algorithm selects node 1

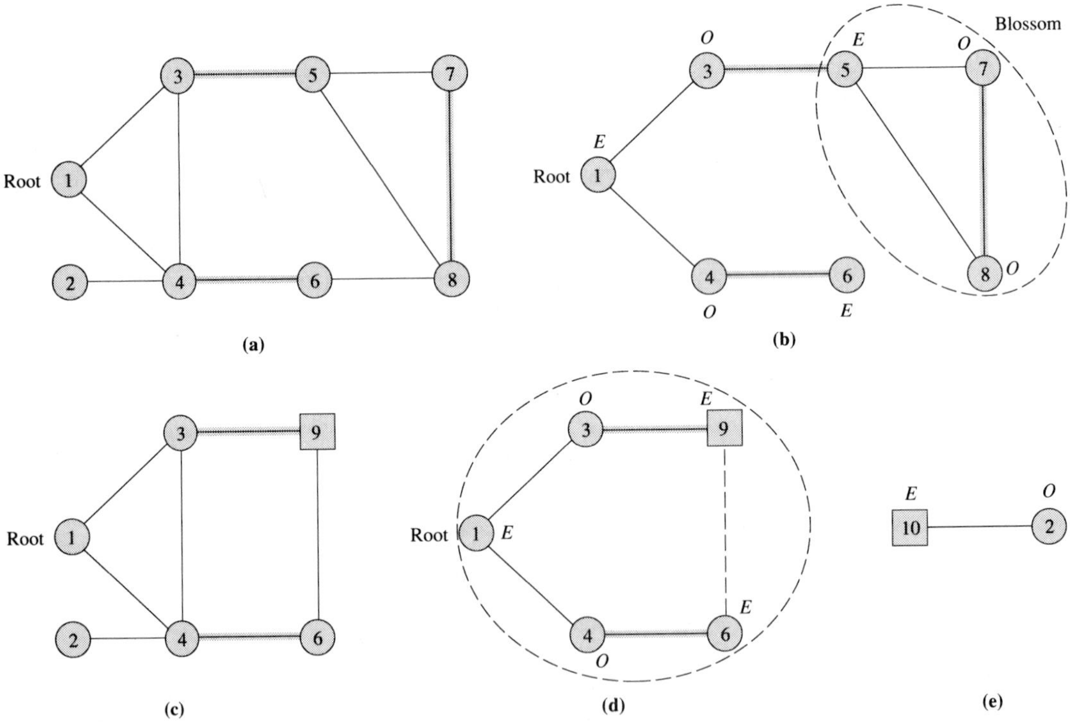

Figure 12.17 Identifying an augmenting path in the contracted network: (a) example network; (b) alternating tree; (c) contracted graph; (d) blossom; (e) contracted graph.

as the root node. Then it would examine the nodes in the following order: 1 (even), 3 (odd), 4 (odd), 5 (even), 6 (even), 7 (odd). Figure 12.17(b) shows the alternating tree at this point. While examining node 7, the algorithm scans arc (7, 8) and discovers the blossom 5–7–8–5. Contracting this blossom into the pseudonode numbered 9 gives us the contracted graph shown in Figure 12.17(c). To distinguish a pseudonode from a node of the original network, in the figure, we depict a pseudonode as a square instead of a circle. At this point, node 9 is the only unexamined node; while examining the adjacency list of node 9, we discover another blossom spanning the nodes 1–3–9–6–4–1 [see Figure 12.17(d)]. Contracting this blossom into the pseudonode numbered 10 gives us the contracted graph shown in Figure 12.17(e). While examining node 10, we assign an odd label to the unmatched node 2 and discover an augmenting path 10–2.

We now expand the pseudonodes in the augmenting path so that we can find an augmenting path in the original network. We first expand node 10, as shown in Figure 12.18(b). Node 2 is adjacent to the blossom node 4. To the arc (2, 4), we add the even alternating path from the root to node 4. Doing so gives us the path 1–3–9–6–4–2. We next expand node 9,-as shown in Figure 12.18(c), and obtain the augmenting path 1–3–5–7–8–6–4–2 in the original network.

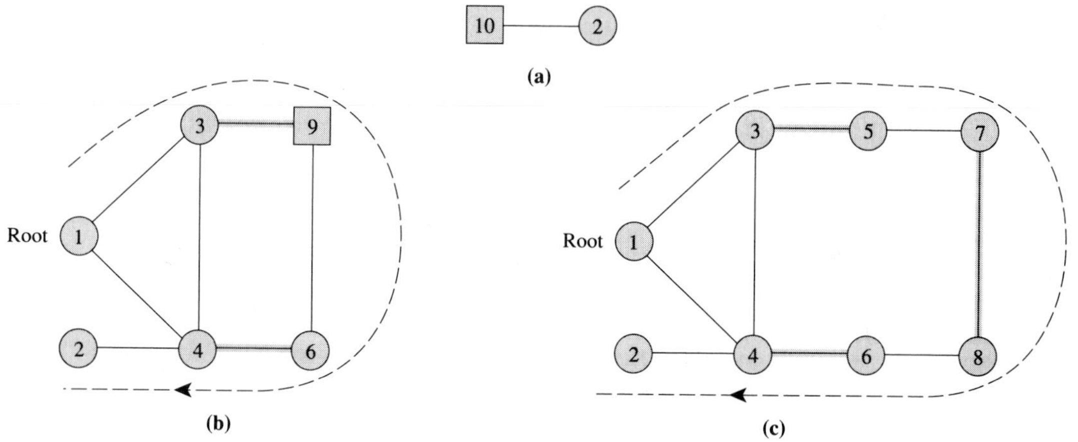

Figure 12.18 Identifying an augmenting path in the original network: (a) contracted graph; (b) network after expanding node 10; (c) network after expanding node 9.

Correctness of the Nonbipartite Matching Algorithm

To show that the algorithm correctly finds a maximum matching, we need to show that (1) whenever we find an augmenting path in the contracted graph we can, by expanding the contracted nodes, also find an augmenting path in the original network; and (2) by contracting blossoms we do not add or omit augmenting paths. To prove this result, we assume that we contract only one blossom. If we do contract more than one blossom, we can use this result iteratively to prove the validity of multiple contractions. In the proof of the theorem, we assume that we have contracted a blossom B with respect to a matching M whose base is w, creating the pseudonode

b. We let G^c and M^c respectively represent the contracted graph and the matching in the contracted graph.

Lemma 12.11. *If the contracted network G^c contains an augmenting path P^c starting at the root node p (or the pseudonode containing p) with respect to the matching M^c, then the original network G contains an augmenting path starting at the root p with respect to the matching M.*

Proof. If the augmenting path P^c does not contain node b, it also is an augmenting path in G and the conclusion is valid. Next suppose that b is an interior node of P^c (i.e., the blossom B has a nonempty stem). In that case the augmenting path in the contracted network will have the structure shown in Figure 12.19(a). Recall from Property 12.9(c) that the pseudonode b is an even node and the alternating path from node p to b ends with a matched arc. We can represent the augmenting path in G^c as $[P_1, (i, b), (b, l), P_3]$. If we expand the contracted node, we obtain the graph shown in Figure 12.19(b). Notice that node l is incident to some node in the blossom, say node k. Property 12.10 implies that the network contains an even alternating path from node w (i.e., the base of the blossom) to node k that ends with a matched arc. Let P_2 denote this path. Now observe that the path $[P_1, (i, w), P_2, (k, l), P_3]$ is an augmenting path in the graph G. This result establishes the lemma whenever b is an internal node of the augmenting path. Whenever b is the first node of the augmenting path, $p = w$ and the path $[P_2, (k, l), P_3]$ is an augmenting path in the original graph. ◆

This lemma shows that if we discover an augmenting path in the contracted network, we can use this path to identify an augmenting path in the original network. The lemma also shows that by contracting a blossom we do not add any augmenting paths beyond those that are contained in the original graph. We now need to prove the converse result: If G contains an augmenting path in G from node p to some node q with respect to the matching M, then G^c also contains an augmenting path from node p (or the pseudonode containing p) to node q with respect to the matching M^c. This result will show that by contracting nodes, we do not miss any augmenting paths from the original network.

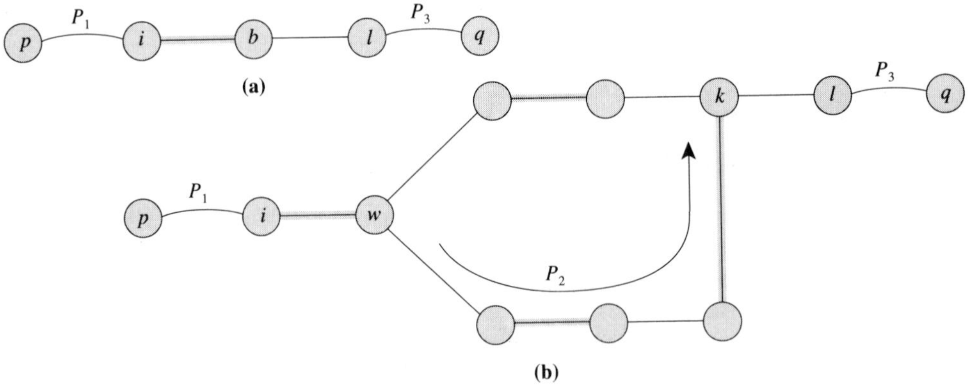

Figure 12.19 Identifying an augmenting path in the original network: (a) augmenting path in contracted network; (b) augmenting path in original network.

Lemma 12.12. *If G contains an augmenting path from node p to node q with respect to a matching M, then G^c contains an augmenting path from node p (or the pseudonode containing p) to node q with respect to the matching M^c.*

Proof. Suppose that G contains an augmenting path P from node p to node q with respect to a matching M and that nodes p and q are the only unmatched nodes in G. We incur no loss of generality in making this assumption since nodes p and q are the only unmatched nodes that appear in P, so this path remains an augmenting path even if we delete the remaining unmatched nodes. If the path P has no node in common with the nodes in the blossom B, we have nothing to prove because P is also an augmenting path in the contracted network. When P has some nodes in common with the blossom B, we consider two cases:

Case 1: The blossom B has an empty stem. In this case, node p is the base of the blossom and the pseudonode b in the contracted network contains node p. Let node i be the last node of the path P that lies in the blossom. Path P has the form $[P_1, (i, j), P_2]$ for some node j and some unmatched arc (i, j) [see Figure 12.20(a)]. Note that the path P_1 might have some arcs in common with the blossom. Now notice that $[(b, j), P_2]$ is an augmenting path in the contracted network and we have established the desired conclusion [see Figure 12.20(b)].

Case 2: The blossom B has a nonempty stem. Let P_3 denote the even alternating path from node p to the base w of the blossom and consider the matching $M' = M \oplus P_3$. In the matching M', node p is matched and node w is unmatched. Moreover, since the matchings M and M' have the same cardinality, M is not a maximum matching if and only if M' is not a maximum matching. By assumption, G contains an augmenting path with respect to M. Therefore, G must also contain an augmenting path with respect to M'. But with respect to the matching M', nodes w and q are the only unmatched nodes in G, so the network must contain an augmenting path between these two nodes.

Now let $M^{c'}$ denote the matching in the contracted graph G^c corresponding to the matching M' in the graph G. Note that M^c might be different than $M^{c'}$. In the

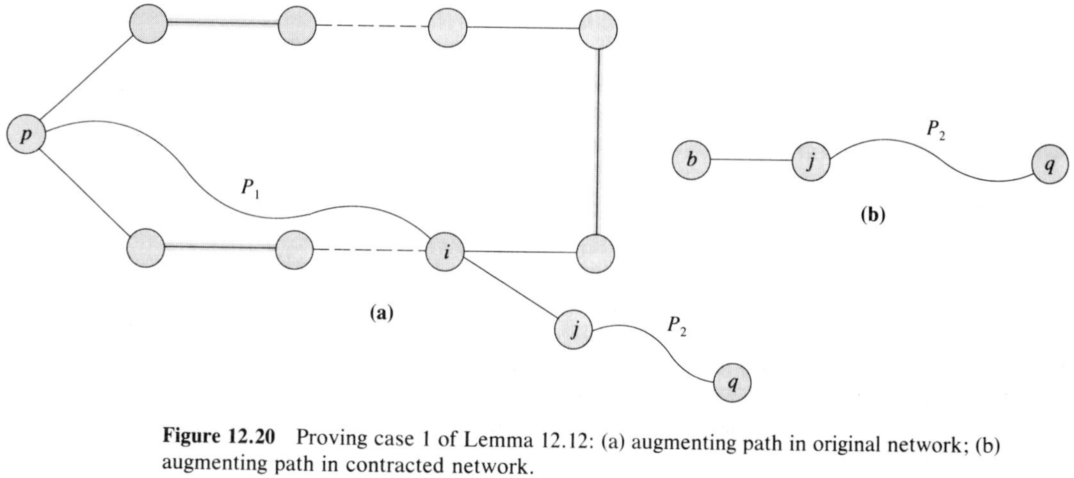

Figure 12.20 Proving case 1 of Lemma 12.12: (a) augmenting path in original network; (b) augmenting path in contracted network.

matching M', the blossom B has an empty stem and the analysis of Case 1 implies that the graph contains an augmenting path after we contract the nodes of the blossom. Consequently, G^c contains an augmenting path with respect to the matching $M^{c'}$. But since $M^{c'}$ and M^c have the same cardinality, G^c must also contain an augmenting path with respect to the matching M^c. This conclusion completes the proof of the lemma. ◆

The preceding two lemmas show that the contracted network contains an augmenting path starting at node p if and only if the original network contains one. Therefore, by performing contractions we do not create new augmenting paths containing node p nor do we miss any. As a consequence, the nonbipartite matching algorithm correctly computes a maximum matching in the network.

Complexity of the Nonbipartite Matching Algorithm

We next show that the nonbipartite matching algorithm has a worst-case complexity of $O(n^3)$. As a first step in establishing this result, we obtain a bound on the total number of contractions we can conduct in one execution of the search procedure.

Lemma 12.13. *During an execution of the search procedure, the algorithm performs at most $n/2$ contractions.*

Proof. A blossom contains at least three nodes, and contracting it produces one new node. So each contraction reduces the number of nodes by at least 2. Since the network initially contains n nodes, we can perform at most $n/2$ contractions between two augmentations. ◆

The nonbipartite matching algorithm is the same as the bipartite matching algorithm except that it contracts and expands blossoms while executing the search procedure. To perform these steps efficiently, we need to contract blossoms cleverly so that we can perform future expansions easily. Achieving this objective requires that we slightly modify our earlier method for contracting a blossom. The modified method for contracting blossoms is exactly the same as the earlier method except that we do not delete the nodes $i_1, i_2, \ldots, i_k$ of the blossom B from the network since this operation would require that we make many changes to our data structures. Instead, we formally keep these nodes as part of the network, but declare them as *inactive*, so that we avoid examining them in future steps. The advantage of this modification is that it contracts the network while maintaining information about the original network that we use later to expand the blossoms. In our subsequent discussion we refer to those nodes that are not inactive as *active* nodes. We also refer to an arc (i, j) as an *active arc* if nodes i and j are both active; otherwise, we refer to it as an *inactive arc*.

By not deleting the nodes that we have contracted into pseudonodes, we need to exercise some care in carrying out and analyzing the algorithm. First, the algorithm might attempt to examine inactive nodes. This possibility poses no problem, however, since we can check the status of a node or arc before examining it and ignore the node or arc if it is inactive. As a second consideration, we note that keeping inactive nodes in the network increases the size of the adjacency list of some nodes, since whenever we contract a set of nodes into a pseudonode, all the nodes that are adjacent to the contracted nodes will now also have the pseudonode as a neighbor. This increase in the size of the adjacency lists might increase the execution time of certain steps. Since each contraction adds at most one element to any adjacency list (the pseudonode), and since the algorithm performs at most $n/2$ contractions, no adjacency list will ever contain more than $3n/2$ elements. Therefore, the increase in size of the adjacency lists will not add to the computational complexity of the algorithm.

We have now given sufficient background material for carrying out the worst-case analysis of the nonbipartite matching algorithm. We intend to show that each execution of the search and augment procedure requires $O(n^2)$ time. Since the nonbipartite matching algorithm executes these procedures at most n times, the overall algorithm runs in $O(n^3)$ time.

First, consider the time that the algorithm spends without contracting and expanding blossoms. For each node i, the search procedure performs one of the following operations at most once: (1) it discovers that node i is inactive, in which case it does nothing; (2) it executes examine-odd(i, found), or (3) it executes examine-even(i, found). Clearly, the first two cases require $O(1)$ time; the time for the third step, however, is proportional to $|A^c(i)| \le 3n/2$. Since the search procedure examines at most $3n/2$ nodes, we obtain a bound of $O(n^2)$ on its running time, ignoring the time for handling blossoms.

Our next task is to analyze the time for contracting blossoms. The bottleneck step in contracting a blossom B consisting of the node sequence $i_1 - i_2 - \cdots - i_k - i_1$ is to form the adjacency list of the resulting pseudonode b, which is defined as $A^c(b) = A^c(i_1) \cup A^c(i_2) \cup \cdots \cup A^c(i_k)$. To construct the adjacency list $A^c(b)$, we first use a "marking" method for finding the nodes that will be adjacent to the pseudonode. We first declare all nodes in the network as unmarked. Then we examine nodes in the blossom one by one, and for each node i being examined, we mark all the nodes in $A^c(i)$. When we have examined all the nodes in the blossom, we again scan all the nodes in the network and form a set of marked nodes. These are exactly the nodes that are adjacent to the pseudonode b. We record these nodes as $A^c(b)$ and we also add node b to the adjacency list of all these nodes. Clearly, this method requires $O(n)$ effort each time we contract a blossom (we do so at most $n/2$ times) plus the time spent in scanning nodes in the adjacency list $A^c(i)$. The latter time also sums to $O(n^2)$ over all contractions, because each node is part of a blossom at most once (since it becomes inactive subsequently), so the algorithm will examine its adjacency list at most once.

Finally, we analyze the time required for expanding blossoms. The search procedure needs to expand blossoms during an augmentation in order to obtain an augmenting path in the original network. Suppose that the procedure discovers an augmenting path P from node p to node q in the contracted network. We determine a corresponding augmenting path in the original network using the following repetitive process: we start at node q and trace back the predecessor indices until either (1) we arrive at node p, or (2) we encounter a pseudonode; in the latter case, we expand the blossom and obtain a corresponding augmenting path in the expanded network. After at most $n/2$ such iterations, we obtain an augmenting path in the original network. Let us show that we can expand each pseudonode in $O(n)$ time, which would establish a time bound of $O(n^2)$ per augmentation.

Suppose that node j is the first pseudonode encountered in the path while tracing predecessor indices from node q. Let pred(i) = j. By definition, node i is not a pseudonode. Clearly, node i is adjacent to some node in the blossom B contained in the pseudonode j, and this node must be contained in the adjacency list $A^c(i)$. To locate this node, we again use a marking approach; we first declare all the nodes in the network as unmarked, mark all nodes in the blossom B, and then scan the nodes in $A^c(i)$ to identify a marked node. Let node k be such a node. Node k is incident to two arcs in the blossom; one of these arcs is matched and the other is unmatched. We then trace through the nodes of the blossom in the direction of the matched arc until we reach the base of the blossom, at which point we trace the predecessor indices to reach node p. The result is an augmenting path from node p to node q in the expanded network; as is clear from the preceding discussion, this method requires $O(n)$ time. We might note that if during the course of expanding a blossom, we encounter a pseudonode, we expand this pseudonode by the method we have just described, and once we have finished expanding this node, we continue to expand the pseudonode that contained it.

We have now shown that all the steps of the search procedure require $O(n^2)$ time per execution. The matching algorithm calls the search procedure at most n times and therefore runs in $O(n^3)$ time. We state this result as a theorem.

Theorem 12.14. *The bipartite matching algorithm identifies a maximum matching in a network in* $O(n^3)$ *time.* ◆

12.7 MATCHINGS AND PATHS

In this section we describe some interesting relationships between matchings and paths. We show how to solve the shortest path problem in a directed network between a specific pair of nodes by solving two assignment problems. In fact, this transformation allows us to obtain the best current time bound for solving the shortest path problem with arbitrary arc lengths. We also show how to solve a shortest path problem in an undirected network with arbitrary arc lengths by solving a nonbipartite weighted matching problem.

Shortest Paths in Directed Networks

Suppose that we want to determine a shortest path from node s to node t in a directed network that might contain negative arc lengths. We will solve this problem by invoking two applications of any algorithm for the assignment problem. The first application determines if the network contains a negative cycle; if it does not, the second application identifies a shortest path. To solve the assignment problem, we can use $O(n^{1/2}m \log(nC))$ time approach that we outlined in Section 12.4.

Consider a shortest path problem in the network $G = (N, A)$. We apply the node-splitting transformation on this network and replace each node i by two nodes i and i'. Furthermore, we replace each arc (i, j) by an arc (i, j') and add an *artificial* zero cost arc (i, i'). As an illustration, consider the shortest path problem from node

1 to node 5 shown in Figure 12.21(a). Figure 12.21(b) gives the transformed network of Figure 12.21(a). We first note that the transformed network always has a feasible assignment with cost zero, namely, the assignment containing all artificial arcs. We next show that the optimal value of the assignment problem in the transformed network is negative if and only if the original network has a negative cycle.

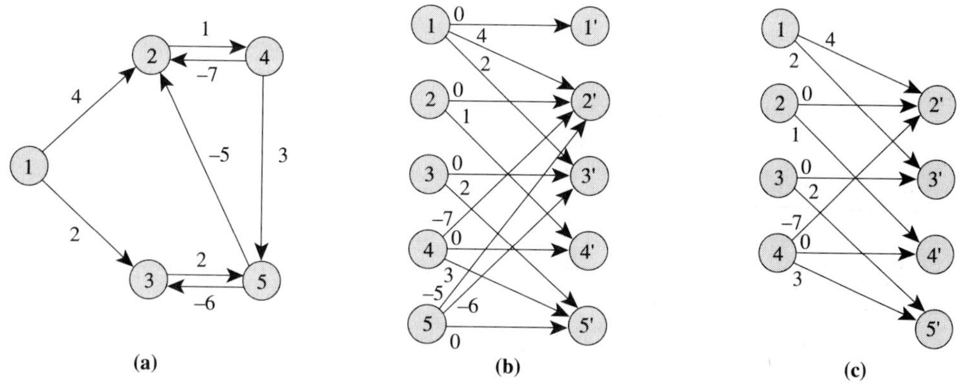

Figure 12.21 Transforming a shortest path problem to an assignment problem: (a) original network; (b) network for identifying a negative cycle; (c) network for identifying a shortest path from node 1 to node 6.

First, suppose that the original network contains a negative cost cycle $j_1 - j_2 - \cdots - j_k - j_1$. Then the assignment $\{(j_1, j_2'), (j_2, j_3'), \ldots, (j_k, j_1'), (j_{k+1}, j_{k+1}'), (j_{k+2}, j_{k+2}'), \ldots, (j_n, j_n')\}$ has a negative cost. Therefore, the cost of the optimal assignment must be negative. Conversely, suppose that the cost of an optimal assignment $\{(j_1, j_2'), (j_2, j_3'), \ldots, (j_k, j_1'), (j_{k+1}, j_{k+1}'), (j_{k+2}, j_{k+2}'), \ldots, (j_n, j_n')\}$ is negative. This solution must contain at least one arc of the form (j_1, j_2') with $j_1 \neq j_2$. If $j_3' = j_1'$, we stop; otherwise, we consider the arc (j_3, j_4'). Repeating this argument as many times as necessary, we eventually find a partial assignment defined as $\{(j_1, j_2'), (j_2, j_3'), \ldots, (j_k, j_1')\}$. The cost of this partial assignment is zero or negative because it can be no more expensive than the partial assignment $\{(j_1, j_1'), (j_2, j_2'), \ldots, (j_k, j_k')\}$. Since the optimal assignment cost is negative, some partial assignment must be negative. But then by the construction of the transformed network, the cycle $j_1 - j_2 - \cdots - j_k - j_1$ is a negative cost cycle in the original network. For our example, the optimal assignment in Figure 12.21(b) is {(1, 1′), (3, 3′), (2, 4′), (4, 5′), (5, 2′)} and has cost equal to -1. This assignment defines the negative cycle 2–4–5–2 of cost -1 in the original network given in Figure 12.21(a).

If the original network contains no negative cost cycle, we can obtain a shortest path between a specific pair of nodes, say from node 1 to node n, as follows. We consider the transformed network as described earlier and delete the nodes 1′ and n and the arcs incident to these nodes. [See Figure 12.21(c) for an example of this transformation; in this figure we have modified the cost of arc (2, 4) to 8 so that the network contains no negative cost cycle.] Observe that each path from node 1 to node n in the original network has a corresponding assignment of the same cost in the transformed network, and the converse is also true. For example, the path 1–2–4 in Figure 12.21(a) corresponds to the assignment {(1, 2′), (2, 4′), (3, 3′),

(5, 5′)} in Figure 12.21(c), and the assignment {(1, 2′), (2, 4′), (4, 5′), (3, 3′)} in Figure 12.21(c) corresponds to the path 1–2–4–5 in Figure 12.21(a). Consequently, an optimal assignment in the transformed network gives a shortest path in the original network.

Shortest Paths in Undirected Networks

Having shown how to transform any shortest path problem (with arbitrary arc costs) in a directed network into an assignment problem, we now study shortest path problems in undirected networks. As we noted in Section 2.4, solving any shortest path problem in an undirected network G with nonnegative arc costs is quite easy; we simply replace each arc (i, j) in the undirected network G with cost c_{ij} by two directed arcs (i, j) and (j, i), both with the same cost c_{ij}, and solve the shortest path problem in the directed network. If the undirected network G contains some arc (i, j) with a negative cost c_{ij}, however, this transformation creates a negative cycle i–j–i. By transforming the undirected problem into a directed problem, we create a negative cycle even though G itself might not contain any negative cycle. Recall from Chapters 4 and 5 that the shortest path algorithms for directed networks do not apply to networks with negative cycles. Consequently, the preceding transformation does not allow us to solve shortest path problems with arbitrary arc lengths on undirected networks. Indeed, solving shortest path problems on undirected networks with negative arc lengths (but with no negative cycles) is substantially harder than the corresponding problem with nonnegative arc lengths; nevertheless, the problem is still solvable in polynomial time. We next describe a transformation that reduces the problem to a minimum weight nonbipartite perfect matching problem.

We perform this transformation in three stages. First, we transform the shortest path problem into a minimum weight perfect b-matching problem. For a given nonnegative n-vector b, we say that a subgraph G' of G is a *perfect b-matching* if each node i has exactly $b(i)$ incident arcs in G'. We illustrate the transformation using the shortest path problem shown in Figure 12.22(a). Suppose that we want to solve the shortest path problem from the source node $s = 1$ to the sink node $t = 4$. As shown in Figure 12.22(b), we add a loop (i, i) of zero cost for each node i, except nodes s and t, and consider the perfect b-matching in the resulting network G' with $b(s) = b(t) = 1$, and $b(i) = 2$ for other nodes. If we observe that each loop arc (i, i) in the matching contributes a degree of 2 units to node i, it is easy to see that any perfect b-matching in G' corresponds to a path in G from node s to node t, and vice versa. [The perfect matching contains the loop arc (i, i) whenever the shortest path from node s to node t does not contain node i.] This observation shows that we can solve the shortest path problem in G by solving the perfect b-matching problem in G'.

We might be tempted, at this point, to try to transform the perfect b-matching problem into a perfect matching problem by splitting those nodes with the degree condition $b(i) = 2$ into two nodes. However, in making this transformation, we encounter one particular difficulty: After splitting the nodes, an arc (i, j) in the original network will correspond to two arcs (i', j) and (i'', j) in the new network that are incident to the copies i' and i'' of the split node i; therefore, the perfect matching in the transformed network might contain both the arcs (i', j) and (i'', j)

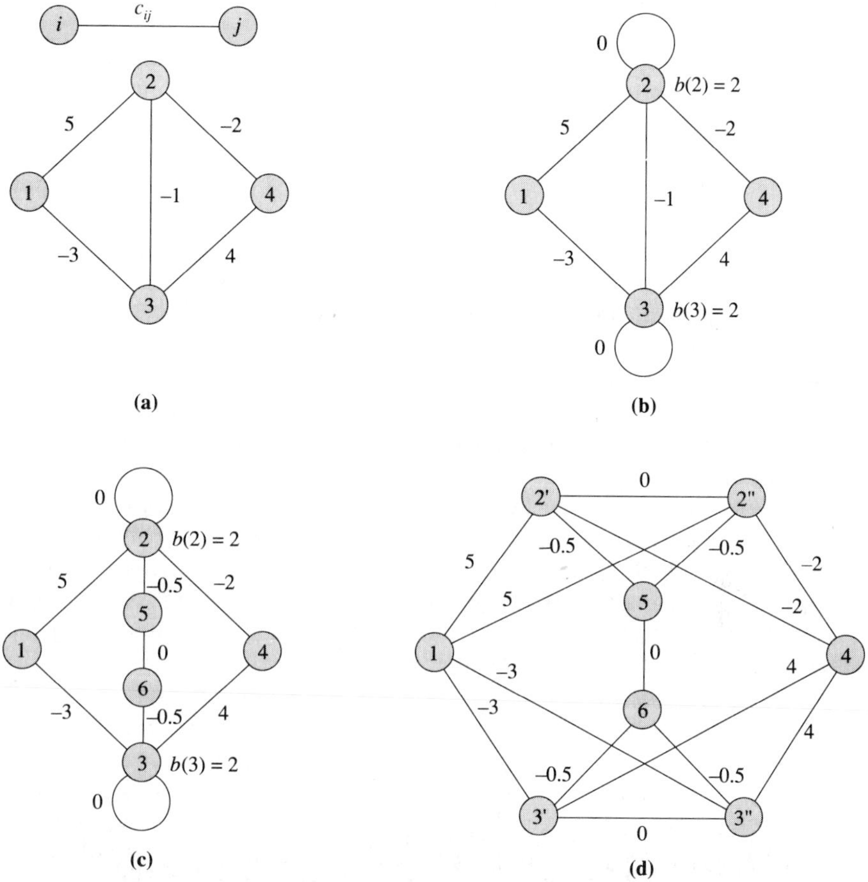

Figure 12.22 Formulating a shortest path problem in an undirected network as a weighted nonbipartite matching problem.

which corresponds to using the original arc (i, j) twice. [We say that we have "used the arc (i, j) twice" when this situation occurs.] To overcome this difficulty, we introduce an additional stage in our transformation.

In the second stage of the transformation, we insert two additional nodes, say nodes k and l, in the middle of each arc (i, j) for which $i \neq j$ and $b(i) = b(j) = 2$, and replace the arc (i, j) by the arcs (i, k), (k, l), and (l, j). We set $b(k) = b(l) = 1$, $c_{ik} = c_{lj} = c_{ij}/2$, and $c_{kl} = 0$. Let G'' denote the graph obtained from this transformation; Figure 12.22(c) shows this graph for our example. To see the equivalence of b-matchings in G' and G'', note that if arc (i, j) is an element of a b-matching M' in G', the corresponding b-matching M'' in G'' would contain the arcs (i, k) and (l, j). Conversely, if arc (i, j) is not contained in the b-matching M', then (k, l) is contained in the b-matching M''. Also, notice that after we have made this transformation, for each arc (p, q) in G'', either $b(p) = 1$ or $b(q) = 1$, which ensures that the matching that we obtain in the subsequent node splitting transformation will use no arc from the graph G'' twice.

In the third stage of the transformation, we construct a third network G''' by splitting each node i with $b(i) = 2$ into two nodes i' and i''; for each arc (i, j) in G'', we introduce two arcs (i', j) and (i'', j) with the same cost as the arc (i, j). Because $b(j) = 1$, any perfect matching of G''' will contain at most one of the arcs (i', j) and (i'', j). For our example, Figure 12.22(d) is the resulting network. It is easy to establish an equivalence between perfect b-matchings in G'' and perfect matchings in G'''. Tracing the steps of these transformations, this result shows that each perfect matching in G''' corresponds to a path from node s to node t in the original graph G; moreover, because the perfect matching and the path have the same cost, we can obtain a shortest path in the undirected graph G by solving the minimum cost perfect matching problem by any polynomial-time algorithm.

12.8 SUMMARY

Matching problems are an important class of optimization models that lie at the interface between network flows and more general problems in combinatorial optimization. The algorithms for certain types of matching problems (those defined on bipartite networks) are streamlined versions of network flow algorithms that we have developed in previous chapters. Although the solution methods for other matching problems (nonbipartite problems) are quite different from those that we have developed for the minimum cost flow problem and its variants, these algorithms do borrow ideas of augmenting paths. Moreover, matching problems are rather intimately related to shortest path problems. In this chapter we studied the following matching problems: (1) the bipartite cardinality matching problem, (2) the bipartite weighted matching problem (also known as the assignment problem), (3) the stable marriage problem, and (4) the nonbipartite cardinality matching problem.

As we have seen, since bipartite matching problems are transformable into network flow problems, we can solve these problems using the algorithms we have developed in previous chapters: for example, we can solve the bipartite cardinality matching problem as a maximum flow problem in unit capacity simple networks (as discussed in Section 8.2). This approach gives an $O(\sqrt{n}m)$ algorithm for solving the problem. In the same way we can transform the weighted bipartite matching problem into a minimum cost flow problem, so we can use the algorithms we have developed in Chapters 9, 10, and 11. The resulting minimum cost flow problem is, however, simpler than general versions of this problem (because the supply/demand vector has only $+1$ or -1 elements), so the minimum cost flow algorithms run faster. By adapting the three pseudopolynomial-time minimum cost flow algorithms discussed in Chapter 9, we have been able to solve the assignment problem in $O(n\ S(n, m, C))$ time, where $S(n, m, C)$ denote the time required to solve a shortest path problem with nonnegative arc lengths. We achieved the best time bound, however, by adapting the cost scaling algorithm: the resulting algorithm runs in $O(\sqrt{n}m \log(nC))$ time.

Nonbipartite matching problems are significantly more difficult to solve. In this chapter we discussed only the cardinality version of this problem; we did not consider weighted nonbipartite matching. Although we can solve the bipartite cardinality matching problem by an augmenting path algorithm, the direct extensions of the algorithm do not solve nonbipartite matching problems. This approach fails because the network might contain blossoms. However, if we contract a blossom whenever we discover one, we can use an augmenting path algorithm. This approach yields an $O(n^3)$ algorithm for the cardinality matching problem. The proof of the algorithm and its worst-case analysis are intricate and much more difficult than the corresponding analysis for the bipartite cardinality matching problem.

Matching problems are closely related to shortest path problems. Many algorithms for weighted matching problems—such as the successive shortest path, Hungarian and relaxation algorithms—use the shortest path algorithm for problems with nonnegative arcs as a subroutine. Conversely, we can transform shortest path problems with arbitrary arc lengths into matching problems, and these transformations provide some of the best available time bounds for solving shortest path problems. In this chapter we showed how to transform the shortest path problem on directed networks into a bipartite weighted matching problem (i.e., the assignment problem) and how to transform the shortest path problem on undirected networks into a nonbipartite weighted matching problem.

REFERENCE NOTES

Matching problems have received a great deal of attention in the literature. The book by Lovász and Plummer [1986] presents an extensive wealth of information and references on matching theory. In this discussion we cite several key references to the literature, placing an emphasis on theoretically efficient algorithms. Ahuja, Magnanti, and Orlin [1989] present more extensive reference notes on the assignment problem.

Bipartite cardinality matching problems. Hopcroft and Karp [1973] gave an $O(n^{5/2})$ algorithm for this problem. Using similar ideas, Even and Tarjan [1975] obtained an $O(\sqrt{n}m)$ algorithm for the maximum flow problem on unit capacity simple networks. This algorithm, in turn, provides an $O(\sqrt{n}m)$ algorithm for the bipartite cardinality matching problem, which is still the best available time bound for solving this problem.

Nonbipartite cardinality matching problem. The backbone of the nonbipartite matching algorithm is the important characterization result: a matching is optimal if and only if it contains no augmenting path. This theorem is due to Berge [1957], who also gave an exponential time algorithm for identifying an augmenting path. Edmonds [1965a] obtained the first polynomial-time algorithm for this problem, with a time bound of $O(n^4)$. Researchers subsequently developed several improved implementations of this algorithm. Some notable contributions in chronological order are (1) an $O(n^2m)$ algorithm by Witzgall and Zahn [1965], (2) an $O(n^3)$ algorithm by Gabow [1975], (3) an $O(n^{5/2})$ algorithm by Even and Kariv [1975], (4) an $O(nm)$ algorithm by Kameda and Munro [1974], and finally, (5) an $O(n^{1/2}m)$ algorithm by Micali and Vazirani [1980]. The algorithm by Micali and Vazirani is still the fastest available algorithm for solving the nonbipartite cardinality matching problem; its running time is comparable to the running time of the best bipartite cardinality matching algorithm. Vazirani [1989] offered a complete version of this algorithm and its proof. Ball and Derigs [1983] described data structures required for implementing matching algorithms.

Assignment problem. The assignment problem has been a popular, heavily studied research topic within the operations research community. The paper by Ahuja, Magnanti, and Orlin [1989] presented a detailed survey of assignment algorithms. Kuhn [1955] developed the first (primal–dual) algorithm for the assignment problem. Although researchers have developed several different algorithms for the assignment problem, many of these algorithms share common features. The successive shortest path algorithm for the minimum cost flow problem, discussed in Section 9.7, appears to lie at the heart of many (apparently different) assignment algorithms. This approach yields an $O(n\,S(n, m, C))$ time algorithm for solving the assignment problem, where $S(n, m, C)$ is the time needed for solving a shortest path problem with nonnegative arc lengths. Currently, $S(n, m, C) = O(\min\{m + n \log n, m \log \log C, m + n\sqrt{\log C}\})$. Therefore, $O(nm + n^2 \log n)$ is the best available strongly polynomial time bound for solving the assignment problem. Gabow and Tarjan [1989a] developed a cost scaling algorithm for the assignment problem that runs in $O(n^{1/2}m \log(nC))$ time. Bertsekas [1988] proposed an auction algorithm for the assignment problem. Incorporating scaling in the auction algorithm, Orlin and Ahuja [1992] also obtained an $O(n^{1/2}m \log(nC))$ time algorithm; this is the algorithm that we mentioned in Section 12.4. The reference notes for Chapter 11 provide references for simplex-based approaches for the assignment problem. Carpento, Martello, and Toth [1988] presented FORTRAN codes for several algorithms for the assignment problem. For recent computational studies of assignment algorithms, see Bertsekas [1988], Zaki [1990], and Kennington and Wang [1990].

Nonbipartite weighted matching problems. Edmonds [1965b] gave the first algorithm for the nonbipartite weighted matching problem. Gabow [1975] and Lawler [1976] developed $O(n^3)$ implementations of this algorithm. Currently, the fastest algorithms for this problem are (1) an $O(nm + n^2 \log n)$ algorithm due to Gabow [1990], and (2) an $O(m \log(nC)\sqrt{n\alpha(m, n) \log n})$ algorithm due to Gabow and Tarjan [1989b]. For information concerning the empirical behavior of nonbipartite weighted matching algorithms, see Grötschel and Holland [1985].

Stable marriage problem. Our discussion of this problem has presented the most basic results obtained by Gale and Shapley [1962]. The book by Gusfield and Irving [1989] on the stable marriage problem contains a wealth of information on this topic. The paper by Roth, Rothblum, and Vande Vate [1990] studied polyhedral aspects of the stable marriage problem and used linear programming theory to obtain simpler proofs of many fundamental results for the problem.

Paths and assignments. We presented two transformations to reduce shortest path problems to matching problems. The transformation of the shortest path problem in directed networks to an assignment problem is due to Hoffman and Markowitz [1963] and the transformation of the shortest path problem in undirected networks to the nonbipartite weighted matching problem is due to Edmonds [1967].

The applications of matchings that we gave in Section 12.2 are adapted from the following papers:

1. Bipartite personnel assignment (Machol [1970] and Ewashko and Dudding [1971])
2. Nonbipartite personnel assignment (Meggido and Tamir [1978])
3. Assigning medical graduates to hospitals (Gale and Shapley [1962])
4. Dual completion of oil wells (Devine [1973])
5. Determining chemical bonds (Dewar and Longuet-Higgins [1952])
6. Locating objects in space (Brogan [1989])
7. Matching moving objects (Brogan [1989] and Kolitz [1991])
8. Optimal depletion of inventory (Derman and Klein [1959])
9. Scheduling of parallel machines (Horn [1973])

Elsewhere in the book, we discussed the following applications of matching problems: (1) rewiring of typewriters (Application 1.5, Machol [1961]), (2) pairing stereo speakers (Application 1.6, Mason and Philpott [1988]), (3) the dating problem (Exercise 1.5), (4) the pruned chessboard problem (Exercise 1.6), (5) large-scale personnel assignment (Exercise 1.4), (6) solving shortest path problems in directed and undirected networks (Section 12.7), (7) school bus driver assignment (Exercise 12.1, R. B. Potts), (8) the ski instructor's problem (Exercise 12.2), (9) the undirected Chinese postman problem (Application 19.15, Edmonds and Johnson [1973]), and (10) discrete location problems (Application 19.16, Francis and White [1976]).

Additional applications of the matching problems arise in (1) two-processor scheduling (Fujii, Kasami, and Ninomiya [1969]), (2) determining the rank of a matrix (Anderson [1975]), (3) vehicle and crew scheduling (Carraresi and Gallo [1984]), and (4) making matrices optimally sparse (Hoffman and McCormick [1984]).

EXERCISES

12.1. School bus driver assignment (R. B. Potts). A bus company has n morning runs and n afternoon runs that it needs to assign to its n drivers. The runs are of different duration. If the total duration of the morning and afternoon runs assigned to a driver is more than a specified number D, the driver receives a premium payment for each hour of overtime. The company would like to assign the runs to the drivers to minimize the total number of overtime hours.

(a) Formulate this problem as a matching problem.

(b) Suppose that we arrange the morning runs in the nondecreasing order of their duration and the afternoon runs in the nonincreasing order of their duration. Show that if we assign each driver i to the ith morning run and the ith afternoon run, we obtain the optimal assignment.

12.2. Ski instructor's problem. A ski instructor needs to assign n pairs of skis to n novice skiers. The skis are available in lengths $l_1 \leq l_2 \leq \cdots \leq l_n$, and the skiers have heights $h_1 \leq h_2 \leq \cdots \leq h_n$. The lengths of the skis assigned to a skier should be proportional to his height: Assume that the constant of proportionality is α. The instructor wishes to assign the skis to skiers so that the total difference between the actual ski lengths and the ideal ski lengths is as small as possible. Show that if for each i, she assigns the ith skier to the ith pair of skis, her assignment is optimal.

12.3. A budding connoisseur plans to consume one of n bottles of wine in his cellar on Saturday evening for each of the next n weeks. The age a_i of bottle i is known (in weeks). The utility of bottle i as a function of time t (in weeks) is given by $b_i t^3 - c_i t$, for some constants b_i and c_i. The connoisseur wants to know how he should consume his wine to maximize the total utility of wine he consumes. Formulate this problem as an assignment problem using the following data.

i	1	2	3	4	5	6
a_i	10	5	4	20	10	15
b_i	2	3	5	3	1	4
c_i	10	15	20	5	25	15

12.4. Show how to solve the bin packing problem described in Exercise 3.10 as a matching problem if $a_j > \frac{1}{3}$ for each $j = 1, \ldots, n$.

12.5. When the utility function for items has a special form, we can solve the optimal inventory depletion problem discussed in Application 12.8 very efficiently.

(a) Show that when the utility function $v(t)$ is a concave function, the optimal policy is to issue the youngest item first.

(b) Show that when the utility function $v(t)$ is a convex function, the optimal policy is to issue the oldest item first.

12.6. This exercise develops a justification for the assignment formulation of the machine scheduling problem discussed in Application 12.9. Let us refer to a feasible assignment as a *proper assignment* if it assigns the jobs to each machine in consecutive places, including the last place. For instance, if we assign jobs 1, 4, and 5 to some machine, assigning job 5 to the last place, job 1 to the second to last place, and job 4 to the fifth to last place, the resulting assignment is not a proper assignment.

(a) Show that we can always improve a nonproper assignment. Conclude that any optimal assignment of the assignment problem is a proper assignment.

(b) Establish a one-to-one correspondence, which preserves costs, between feasible schedules of the scheduling problem and proper assignments of the assignment problem. Conclude that the solution of the assignment problem will yield an optimal schedule for the scheduling problem.

12.7. Determine a maximum cardinality matching in the graph shown in Figure 12.4.

12.8. Let M_1 and M_2 be two arbitrary matchings in a bipartite network $G = (N_1 \cup N_2, A)$. Show that some matching M matches all the nodes in N_1 that are matched by M_1 and all the nodes of N_2 that are matched in M_2. (*Hint*: Consider $M_1 \oplus M_2$ and modify the matching M_1 or M_2 appropriately.)

12.9. The army would like to transfer five servicemen to five new posts in a way that minimizes the total moving cost. The accompanying table specifies allowable assignments and the moving cost for each possible assignment. Use the relaxation algorithm to determine an assignment that minimizes the total moving cost.

Serviceman \ Posting	1	2	3	4	5
1	25	30	—	—	—
2	20	—	70	35	—
3	80	75	90	65	—
4	—	—	—	55	40
5	—	—	—	60	50

12.10. A construction company needs to assign four workers to four jobs. The accompanying table specifies the workers' proficiency scores for the jobs. A dash "–" in position (i, j) indicates that worker i is unqualified to perform job j. Use the successive shortest path algorithm to identify an assignment that maximizes the total proficiency scores for carrying out the jobs.

Worker \ Job	1	2	3	4
1	45	—	—	30
2	50	55	15	—
3	—	60	25	75
4	45	—	—	35

12.11. Let $G = (N_1 \cup N_2, A)$ be a bipartite network with $|N_1| = |N_2| = n_1$. For any set $S \subseteq N_1$, define *neighbor(S)* as the set of nodes in N_2 that are adjacent to the nodes in S.

(a) Show that if G has a perfect matching, then for any subset $S \subseteq N_1$, $|\text{neighbor}(S)| \geq |S|$.

(b) Show that if for every subset $S \subseteq N_1$, $|\text{neighbor}(S)| \geq |S|$, then G has a perfect matching. Conclude that G has a perfect matching if and only if for every subset $S \subseteq N_1$, $|\text{neighbor}(S)| \geq |S|$. (*Hint*: Prove that using the bipartite matching algorithm given in Figure 12.9, we would either find an augmenting path from every unassigned node in N_1 to a node in N_2, or we would contradict the assumption that $|\text{neighbor}(S)| \geq |S|$.)

12.12. Dancing problem. At a high school party attended by n boys and n girls, each boy knows exactly k ($1 \le k \le n$) girls and each girl knows exactly k boys. Assume that the acquaintanceship is mutual (i.e., if b knows c, then c also knows b).

(a) Show that it is always possible to arrange a dance in which each boy dances with a girl he knows. (*Hint*: Use the result of Exercise 12.11.)

(b) Show that it is possible to arrange k consecutive parties so that at each party each boy dances with a different girl that he knows.

12.13. A 0–1 matrix of size $n \times n$ is a *permutation matrix* if each row and column contains a single 1 entry. Let H be a 0–1 matrix of size $n \times n$ and suppose that each row and each column contains exactly k 1's. Show that it is possible to represent H as the sum of k permutation matrices of size $n \times n$. (*Hint*: Use the results in Exercise 12.11 or 12.12.)

12.14. An $n \times n$ matrix R is *doubly stochastic* if all of its elements r_{ij} are nonnegative and if the sum of the elements in each row and each column equals 1. Show that it is possible to represent a doubly stochastic matrix as a convex combination of the permutation matrices (i.e., $R = \alpha_1 P_1 + \alpha_2 P_2 + \cdots + \alpha_p P_p$ for some set of permutation matrices $P_1, P_2, \ldots, P_p$ and some positive weights α_j satisfying the condition $\alpha_1 + \alpha_2 + \cdots + \alpha_p = 1$). Moreover, show that we can choose the permutation matrices so that p is no more than the number of nonzero elements in the matrix R. (*Hint*: Let $Q = \{q_{ij}\}$ be a 0–1 matrix with $q_{ij} = 1$ if r_{ij} is positive. Show that Q contains a permutation matrix. Modify R and use this result repeatedly.)

12.15. Suppose that each of 10,000 individuals serves on exactly 13 of 10,000 committees and each committee has exactly 13 members. Show that we can order the committees so that for each $j = 1, \ldots, 10{,}000$, individual j serves on committee j. (*Hint*: Use the result of some previous exercise.)

12.16. An *arc coloring* (or, simply, a coloring) of an undirected graph $G = (N, A)$ is a coloring of the arcs with several colors so that no two arcs incident to the same node have the same color. A k-coloring is a coloring of the arcs with k distinct colors.

(a) Show that a k-coloring of the graph G decomposes the arcs in A into k arc-disjoint matchings.

(b) A graph is *k-regular* if the degree of each node is exactly k. Show that a k-regular bipartite graph has a k-coloring.

(c) Let $\delta(G)$ denote the maximum degree of any node in the graph G. Show that every bipartite graph G has a $\delta(G)$ coloring. (*Hint*: Show how to make the graph $\delta(G)$-regular by adding additional nodes and arcs.)

12.17. A small manufacturing company produces a speciality home security device composed of p components. The company employs p individuals who have different expertise; each of them must spend time in the production of each component: the ith worker must spend a_{ij} hours (an integer) on component j. The company wants to determine the minimum number of hours required to produce the device so that (1) no worker is working on two different components at the same time, and (2) no more than one person is working on any component at any one time.

(a) For each $1 \le i \le p$, let $r_i = \sum_{j=1}^{p} a_{ij}$, and for each $1 \le j \le p$, let $c_j = \sum_{i=1}^{p} a_{ij}$. Also let α be the largest value of all the parameters r_i and c_j. Show that the company requires at least α hours to produce the device.

(b) Use the result in part (a) to show that we can increase some of the a_{ij}'s so that each r_i and each c_j equals α.

(c) Use the results in part (b) and Exercise 12.14 to show that the company can always produce the device in α hours.

12.18. In this chapter we considered the assignment problem on bipartite networks $G = (N_1 \cup N_2, A)$ with $|N_1| = |N_2|$. Consider a modified version of the assignment problem when $|N_1| < |N_2|$. In this case, in a feasible assignment, we require all the nodes in N_1 to be matched, but permit some of the nodes in N_2 to be unmatched.

Show that we can transform this modified assignment problem to the (original) assignment problem.

12.19. Show how to transform the (uncapacitated) transportation problem into an assignment problem on an expanded network. Next show how to transform the (capacitated) minimum cost flow problem into an assignment problem. Justify your transformation. (*Hint*: If U denotes the magnitude of the largest supply/demand at a node, transforming the transportation problem to an assignment problem yields a network with $O(nU)$ nodes.)

12.20. Consider an assignment problem whose arc costs are all 0 or 1. Which assignment algorithm discussed in this chapter would solve this problem in the least possible time?

12.21. The president's office at a university needs to assign a targeted group of n faculty to be chairs of n committees. Each person proposes, in decreasing order of preference, a list of three committees that he or she would like to chair. We want to determine whether we could possibly find a *satisfiable assignment* (i.e., one that assigns the faculty to the committees so that each faculty member obtains a job on his or her list). If some satisfiable assignment is possible, we want to find the assignment that maximizes the number of faculty with their most preferred committee chair, and further, among such assignments, the assignment that maximizes the number of faculty with their second most preferred committee chair. Show how to solve this problem by solving a single assignment problem. (*Hint*: Assign arc costs appropriately.)

12.22. Factored assignment problems

(a) A factored assignment problem is an assignment problem on a complete bipartite network $G = (N_1 \cup N_2, A)$ (i.e., $A = N_1 \times N_2$) when each arc cost is specified as $c_{ij} = \alpha_i\alpha_j$ for some set of node numbers α_i associated with the nodes in $N_1 \cup N_2$. Assume that we are given the node numbers α_i's sorted in nondecreasing order for each of the sets N_1 and N_2. [Notice that the input size of the factored assignment problem is only $O(n)$.] Describe an $O(n)$ algorithm for solving this specialized version of the assignment problem. (*Hint*: First solve the problem when $|N_1| = 2$ and then generalize your result.)

(b) Consider the assignment problem on a complete bipartite network $G = (N_1 \cup N_2, A)$ when the cost data c_{ij} is of the form $c_{ij} = \alpha_i + \alpha_j$ for some node numbers α_i associated with the nodes in $N_1 \cup N_2$. Describe an $O(n)$ algorithm for solving this assignment problem.

12.23. Bottleneck assignment problem. The bottleneck assignment problem is an important variation of the classical assignment problem that arises in the following scenario. Suppose that an assembly line has p jobs to be assigned to p operators. Let c_{ij} denote the number of units per hour that operator i can process if assigned to job j. For a given assignment, the output rate of the assembly line is given by the minimum c_{ij} in the assignment, which we would like to maximize over all assignments. In the bottleneck assignment problem we wish to determine an assignment for which the least costly assignment is as large as possible. This is a *maximin* version of the bottleneck assignment problem; similarly, we could define the *minimax* version in which we want to determine an assignment for which the most costly assignment is as small as possible. In the following exercise, we consider the minimax version of the problem.

Suppose we sort the arc costs c_{ij}'s: let $c_1 < c_2 < \cdots < c_k$ denote the sorted list of distinct values of these costs ($k \le m$). Let FS(l, M) denote a subroutine that for any input number $l \ge 1$, determines whether in some assignment every arc cost is no more than c_l. If no such assignment exists, M is a null set.

(a) Describe an $O(\sqrt{n}\, m)$ algorithm for implementing the subroutine FS(l, M).

(b) Show how to solve the bottleneck assignment problem by calling the subroutine FS(l, M) $O(\log k)$ times.

12.24. Balanced assignment problem (Martello et al. [1984]). The balanced assignment problem is another variation of the classical assignment problem, which is perhaps best illustrated by some specific scenarios. Given n people and n tasks, let c_{ij} denote the amount of work person i requires to perform job j. Suppose that we are interested in choosing a pairing of workers and jobs that distributes the work load as evenly as possible. As another problem setting for the balanced assignment problem, we let c_{ij} be the expected lifetime of component j produced by a company i; we wish to choose the components so that we will need to replace all the components at about the same time. In these settings and in general, in the balanced assignment we wish to determine an assignment that will minimize the difference between the most costly and the least costly assignment. In this exercise we develop an algorithm for solving this version of the assignment problem.

Suppose that we sort the arc costs c_{ij}'s and let $c_1 < c_2 < \cdots < c_k$ denote the sorted list of distinct values of these costs ($k \le m$). Let FS(l, u, M) denote a subroutine that takes as an input two numbers l and u, satisfying the condition $1 \le l \le u \le k$, and determines whether in some assignment M every arc cost is between c_l and c_u; if no such assignment exists, then M is a null set.

(a) Describe an $O(\sqrt{n}\, m)$ algorithm for implementing the subroutine FS(l, u, M).
(b) Show how to solve the balanced assignment problem by calling the subroutine FS(l, u, M) $O(k)$ times.

12.25. Solve the stable marriage problem shown in Figure 12.23 when the matrix M gives the rankings of men for women and matrix W gives the ranking of women for men. A higher ranking implies a greater preference.

M	1	2	3	4	5
1	3	5	2	1	4
2	4	3	5	1	2
3	4	1	3	2	5
4	1	3	2	5	4
5	4	2	3	1	5

W	1	2	3	4	5
1	5	4	3	1	2
2	5	1	3	2	4
3	5	4	1	3	2
4	5	3	1	2	4
5	5	3	2	1	4

Figure 12.23 Men and women rankings.

12.26. Give a 2×2 example of the stable marriage problem (i.e., two men and two women) which has at least two distinct stable matchings.

12.27. Show that in a man-optimal matching, each woman has the least preferred partner that she can have in any stable matching.

12.28. Suppose that in the stable marriage problem, men and women might prefer to remain unmarried if they do not find suitable mates. In this version of the problem, each man and each woman makes a list of potential mates in order of preference but does not list anyone whom he/she is unwilling to marry. Now a stable marriage will permit some men/women to remain unmarried. Show how to modify the stable matching algorithm given in Section 12.5 to find such a matching. Suppose that a man i is unmarried in the matching produced by your algorithm. Is it true that he is unmarried in all stable marriages? Prove your answer.

12.29. In Section 12.5 we described an algorithm for constructing a man-optimal matching. Modify this algorithm so that it constructs a woman-optimal matching. Apply this algorithm to the example given in Figure 12.23. Is the woman-optimal matching different than the man-optimal matching?

12.30. Suppose that you are given a matrix denoting men's ranking of women: a higher rank denotes a more favored woman. Assume that each rank is an integer between 1 and n. Using this matrix, you wish to prepare the *priority list* of each man, which lists the women in nonincreasing order of their rankings. Show how you can construct priority lists of all men in a total of $O(n^2)$ time.

12.31. Stable university admissions. In the stable marriage algorithm, each man is matched to one woman and each woman is matched to one man. Generalize this algorithm to find a stable assignment of medical school graduates to hospitals (see Application 12.3). In this case each hospital i can hire a_i graduates.

12.32. Unstable roommates (Gale and Shapley [1962]). The headmaster of a boarding school needs to divide an even number of boys into pairs of roommates. In this setting, a set of pairings is *stable* if no two boys who are not roommates prefer each other to their actual roommates. Show that in some situations no stable pairing is possible. (*Hint*: There is an example with four boys.)

12.33. Consider the graph shown in Figure 12.24(a) with a matching shown by the bold lines. In this graph, specify (1) an alternating path of length 10; (2) an alternating cycle of length 10; (3) an augmenting path of length 5; (4) an augmenting path of length 9; and (5) an alternating tree rooted at node 2 that spans all the nodes.

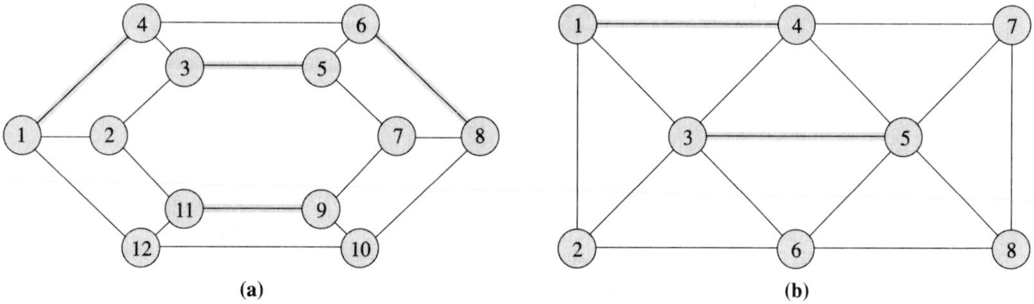

Figure 12.24 Examples for Exercises 12.33 to 12.35.

12.34. Consider the graph shown in Figure 12.24(b) with a matching as shown by the bold lines. Specify a blossom in the graph whose stem contains two arcs. Contract this blossom and show the contracted graph. List all augmenting paths of length 3 in the contracted graph and the corresponding paths in the original (uncontracted) graph.

12.35. Apply the nonbipartite cardinality matching algorithm to the example shown in Figure 12.24(b). Use the matching shown by bold lines as the starting matching.

12.36. Professor May B. Wright has posed the following problem for you to resolve: Consider the graph shown in Figure 12.25(a) which has an augmenting path 1–2–3–4. The se-

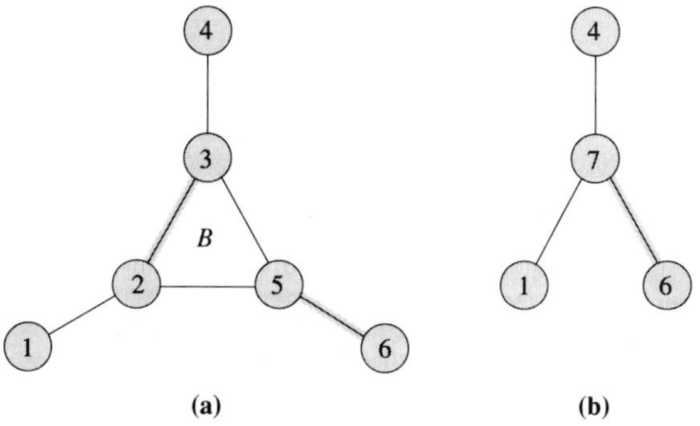

Figure 12.25 Example for Exercise 12.36.

quence of nodes 2–3–5–2 defines a blossom in the graph; contracting the nodes, we obtain the graph shown in Figure 12.25(b). However, the contracted graph does not have any augmenting path. Professor Wright believes that this example contradicts the theorem that the graph has an augmenting path if and only if the contracted graph has an augmenting path. Is she correct?

12.37. Show that a network is nonbipartite if during the application of the search procedure of the nonbipartite matching algorithm (described in Section 12.6), it attempts to assign an odd label to an even-labeled node or an even label to an odd-labeled node. Show that the "only if" part of the result is not true. (*Hint*: For the second part, use the network shown in Figure 12.25.)

12.38. A subset $N' \subseteq N$ is *matchable* if each node in N' is matched in some matching. Let S be a matchable set of nodes. Show that each node in S is matched in some maximum cardinality matching.

12.39. **Berge's theorem** (Berge [1957]). Prove the following theorem from first principles: A matching M^* in a graph G is maximum if and only if the graph G contains no augmenting path with respect to M^*.

12.40. A matching M is a *maximal matching* of $G = (N, A)$ if for every arc $(i, j) \notin M$, $M \cup \{(i, j)\}$ is not a matching.

(a) Show how to construct a maximal matching in $O(m)$ time.

(b) Show that a maximal matching contains at least 50 percent as many arcs as a maximum matching.

12.41. Let $G = (N, A)$ be a graph. Let $\mu(G)$ denote the maximum cardinality of any matching in G. A subset $A' \subseteq A$ is an *arc cover* if for every node $i \in N$, i is an endpoint of at least one arc in A'. Let $\alpha(G)$ denote the minimum cardinality of any arc cover of G. Show that $\mu(G) + \alpha(G) = n$. (*Hint*: Show how to extend any matching of cardinality k into an arc cover of cardinality $n - k$.)

12.42. Suppose that an undirected graph $G = (N, A)$ has a perfect matching. An arc (i, j) in G is *unmatchable* if no perfect matching contains the arc (i, j).

(a) Let G^{ij} denote the graph obtained by deleting from G the nodes i and j and the arcs incident to these nodes. Show that an arc (i, j) is unmatchable if and only if G^{ij} has no perfect matching. Use this result to develop a polynomial-time algorithm for identifying all unmatchable arcs.

(b) Specify an $O(n^3)$ algorithm for finding all unmatchable arcs in a graph G. (*Hint*: First find a perfect matching. Then show how to find in $O(m)$ time all unmatchable arcs incident to any node i.)

12.43. Consider a nonbipartite graph G that has a perfect matching. In general, we might expect that a maximum weight matching of G would be perfect. Show that this is not always true by constructing an example in which a maximum weight matching is not a perfect matching. (*Hint*: Try an example with four nodes.)

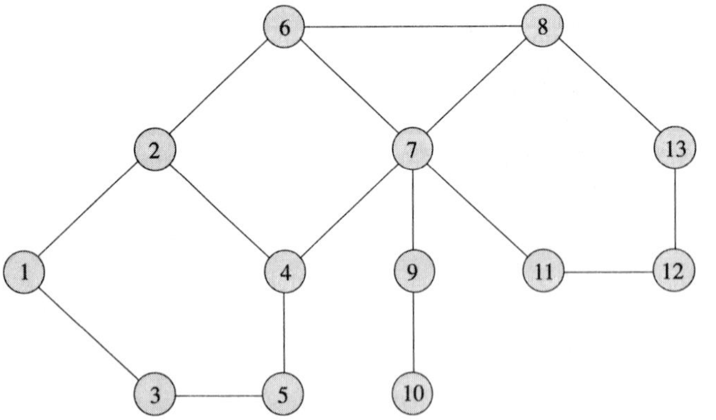

Figure 12.26 Policemen's problem.

12.44. Describe a polynomial-time algorithm for solving the maximum weight matching problem on a tree.

12.45. **Policemen's problem** (Gondran and Minoux [1984]). Consider an undirected graph shown in Figure 12.26, which represents the street map of a town. A policeman standing in the middle of a street can survey the crossings at the two ends of the street. What is the smallest number of policemen needed to survey all the crossings in the network? (*Hint*: Use the result of a previous exercise.)

13

MINIMUM SPANNING TREES

I think I will never see,
A poem as lovely as a tree.
—Joyce Kilmer

Chapter Outline

13.1 INTRODUCTION

As we have seen repeatedly throughout earlier chapters, spanning trees play a central role within the field of network flows. In solving the shortest path problem in Chapters 4 and 5, we constructed (shortest path) spanning trees rooted at a source node. The simplex method for solving minimum cost flow problems that we discussed in Chapter 11 is a spanning tree manipulation algorithm that iteratively moves from one spanning tree to another, at each step introducing one arc into the spanning tree in place of another. As we have also seen in Chapter 11 (Theorem 11.3), minimum cost flow problems always have spanning tree solutions; therefore, in principle to solve any minimum cost network flow problem, including shortest path problems and maximum flow problems, we can always restrict our attention to spanning tree solutions. Since any network has only a finite number of spanning trees, we can view any network flow problem as a discrete optimization model and solve it in a finite number of iterations.

In this chapter we consider another spanning tree model, known as the *minimum spanning tree problem*. Recall that a spanning tree T of G is a connected acyclic subgraph that spans all the nodes. Every spanning tree of G has $n - 1$ arcs (see Property 2.2). Given an undirected graph $G = (N, A)$ with $n = |N|$ nodes and $m = |A|$ arcs and with a *length* or *cost* c_{ij} associated with each arc $(i, j) \in A$, we wish to find a spanning tree, called a *minimum spanning tree*, that has the smallest total cost (or length) of its constituent arcs, measured as the sum of costs of the arcs in the spanning tree. Note that minimal spanning trees differ from the shortest path tree that we have considered in Chapters 4 and 5 in the following two respects:

1. For the minimum spanning tree problem, the arcs are undirected. [Since the network is undirected, we refer to the arc between the node pair i and j as either (i, j) or (j, i).] For the version of the shortest path problems that we considered previously, the networks were directed. This distinction is unimportant in one sense: We could easily have developed our prior results for shortest path problems using undirected graphs as well as directed graphs (see Section 2.4). Viewed in another way, however, this distinction is important: Finding a minimum spanning tree on a directed network with all paths directed away from a given root node (this structure is known as a *rooted arborescence*) is a much more difficult problem than the undirected minimum spanning tree problem.
2. Our objective functions for the minimum spanning tree problem and for the shortest path tree problem are quite different. For the minimal spanning tree problem, we count the cost of each arc exactly once; for the shortest path tree problem, we typically count the cost of some arcs several times: equal to the number of paths from the root node that pass through that arc (i.e., the number of shortest paths in the tree that contain that arc).

The minimum spanning tree problem arises in a number of applications, both as a stand-alone problem and as a subproblem in a more complex problem setting. We begin this chapter by describing several such applications. We next consider combinatorially based optimality conditions for assessing whether a given spanning tree is a minimum spanning tree. We consider two such optimality conditions. The first condition is based on comparing the cost of any tree arc with the other arcs contained in the cut defined by removing that arc from the tree. The other is based on comparing the cost of a nontree arc with the tree arcs in the path that connects the endpoints of the nontree arc. These two *cut and path optimality conditions* are easy to state and to develop, yet they quite naturally motivate several algorithms for solving the minimum spanning tree problem.

The resulting algorithms are all very simple, although implementing them efficiently requires considerable care and ingenuity. The three algorithms we consider in this chapter—Kruskal's algorithm, Prim's algorithm, and Sollin's algorithm—all share one characteristic: They are "greedy" algorithms in the sense that at each step they add an arc of minimum cost from a candidate list, as long as the added arc does not form a cycle with the arcs already chosen. All three algorithms maintain a forest containing arcs already chosen and then they add one or more arcs to enlarge the size of the forest. For Kruskal's algorithm, the candidate list is the entire network; for Prim's algorithm, the forest is a single tree plus a set of isolated nodes and the candidate list contains all the arcs between the single tree and the nodes not in the tree; Sollin's algorithm is a hybrid approach that maintains several components in the forest, as in Kruskal's algorithm, but then adds several arcs at each iteration, choosing (like Prim's algorithm) the minimum cost arc connecting each component of the forest to the nodes not in that component.

Since greedy algorithms, such as Kruskal's, Prim's, and Sollin's, arise in many other problem contexts in discrete optimization, in Section 13.7 we show how a generalization of Kruskal's algorithm will solve a broad class of abstract combinatorial optimization problems known as matroid optimization problems. This discussion not only permits us to show how to solve a new class of combinatorial optimization problems, but also provides additional insight concerning the combinatorial structure of spanning trees that underlies the validity of the greedy solution approach.

Mathematical programming has another useful way to view the minimal spanning tree problem. In Section 13.8 we formulate the minimal spanning tree problem as an integer programming model and use linear programming arguments to establish yet another proof of the validity of Kruskal's algorithm. This discussion serves several purposes: (1) it gives another useful view of minimum spanning trees; (2) it illustrates a proof technique, via linear programming, that has proven to be very powerful in the field of combinatorial optimization; and (3) it provides a bridge between the minimum spanning tree problem and an important topic in discrete optimization, polyhedral combinatorics (i.e., the study of integer polyhedra).

In closing this section we might note that we can also define and study the maximum spanning tree problem, which as its name implies, seeks the spanning tree with the largest total costs of its constituent arcs. Since we can find a maximum spanning tree by multiplying all the arc costs by -1 and then solving a minimum spanning tree, the algorithms and theory of the maximum spanning tree problem are essentially the same as those of the minimum spanning tree problem.

13.2 APPLICATIONS

Minimum spanning tree problems generally arise in one of two ways, directly or indirectly. In some *direct* applications, we wish to connect a set of points using the least cost or least length collection of arcs. Frequently, the points represent physical entities such as components of a computer chip, or users of a system who need to be connected to each other or to a central service such as a central processor in a computer system. In *indirect* applications, we either (1) wish to connect some set of points using a measure of performance that on the surface bears little resemblance to the minimum spanning tree objective (sum of arc costs), or (2) the problem itself bears little resemblance to an "optimal tree" problem—in these instances, we often need to be creative in modeling the problem so that it becomes a minimum spanning tree problem. In this section we consider several direct and indirect applications.

Application 13.1 Designing Physical Systems

The design of physical systems can be a complex task involving an interplay between performance objectives (such as throughput and reliability), design costs and operating economics, and available technology. In many settings, the major criterion is fairly simple: We need to design a network that will connect geographically dispersed system components or that will provide the infrastructure needed for users to communicate with each other. In many of these settings, the system need not have any redundancy, so we are interested in the simplest possible connection, namely, a spanning tree. This type of application arises in the construction (or installation) of numerous physical systems: highways, computer networks, leased-line telephone networks, railroads, cable television lines, and high-voltage electrical power transmission lines. For example, this type of minimum spanning tree problem arises in the following problem settings:

1. Connect terminals in cabling the panels of electrical equipment. How should we wire the terminals to use the least possible length of the wire?
2. Constructing a pipeline network to connect a number of towns using the smallest possible total length of pipeline.
3. Linking isolated villages in a remote region, which are connected by roads but not yet by telephone service. In this instance we wish to determine along which stretches of roads we should place telephone lines, using the minimum possible total miles of the lines, to link every pair of villages.
4. Constructing a digital computer system, composed of high-frequency circuitry, when it is important to minimize the length of wires between different components to reduce both capacitance and delay line effects. Since all components must be connected, we obtain a spanning tree problem.
5. Connecting a number of computer sites by high-speed lines. Each line is available for leasing at a certain monthly cost, and we wish to determine a configuration that connects all the sites at the least possible cost.

Each of these applications is a direct application of the minimum spanning tree problem. We next describe several indirect applications.

Application 13.2 Optimal Message Passing

An intelligence service has n agents in a nonfriendly country. Each agent knows some of the other agents and has in place procedures for arranging a rendezvous with anyone he knows. For each such possible rendezvous, say between agent i and agent j, any message passed between these agents will fall into hostile hands with a certain probability p_{ij}. The group leader wants to transmit a confidential message among all the agents while minimizing the total probability that the message is intercepted.

If we represent the agents by nodes, and each possible rendezvous by an arc, then in the resulting graph G we would like to identify a spanning tree T that minimizes the probability of interception given by the expression $\{1 - \Pi_{(i,j)\in T} (1 - p_{ij})\}$. Alternatively, we would like to find a tree T that maximizes $\Pi_{(i,j)\in T} (1 - p_{ij})$. We can identify such a tree by defining the length of an arc (i, j) as $\log(1 - p_{ij})$ and solving a maximum spanning tree problem.

Application 13.3 All-Pairs Minimax Path Problem

The minimax path problem is a variant of the maximum capacity path problem that we discussed in Exercise 4.37. In a network $G = (N, A)$ with arc costs c_{ij}, we define the *value* of a path P from node k to node l as the maximum cost arc in P. The all-pairs minimax path problem requires that we determine, for every pair $[k, l]$ of nodes, a minimum value path from node k to node l. We show how to solve the all-pairs minimax path problem on an undirected graph by solving a single minimum spanning tree problem.

The minimax path problem arises in a variety of situations. As an example, consider a spacecraft that is about to enter the earth's atmosphere. The craft passes through different pressure and temperature zones that we can represent by arcs of a network. It needs to fly along a trajectory that will bring the craft to the surface of the earth while keeping the maximum temperature to which the surface of the craft is exposed as low as possible. As an alternative, we might wish to select a path that will minimize the maximum deceleration during the descent. Other examples of the minimax path problem arise when (1) in traveling through a desert, we want to minimize the length of the longest stretch between rest areas; and (2) in traveling in a wheelchair, a person might wish to minimize the maximum ascent along the path segments.

To transform the all-pairs minimax path problem into a minimum spanning tree problem, let T^* be a minimum spanning tree of G. Let P denote the unique path in T^* between a node pair $[p, q]$ and let (i, j) denote the maximum cost arc in P. Observe that the value of the path P is c_{ij}. By deleting arc (i, j) from T^*, we partition the node set N into two subsets and therefore define a cut $[S, \bar{S}]$ with $i \in S$ and $j \in \bar{S}$ (see Figure 13.1). We later show in Theorem 13.1 that this cut satisfies the following property:

$$c_{ij} \le c_{kl} \qquad \text{for each arc } (k, l) \in [S, \bar{S}], \tag{13.1}$$

for otherwise by replacing the arc (i, j) by an arc (k, l) we can obtain a spanning tree of smaller cost. Now, consider any path P' from node p to node q. This path must contain at least one arc (k, l) in $[S, \bar{S}]$. Property (13.1) implies that the value of the path P' will be at least c_{ij}. Since c_{ij} is the value of the path P, P must be a minimum value path from node p to node q. This observation establishes the fact that the unique path between any pair of nodes in T^* is the minimum value path between that pair of nodes.

Application 13.4 Reducing Data Storage

In several different application contexts, we wish to store data specified in the form of a two-dimensional array more efficiently than storing all the elements of the array

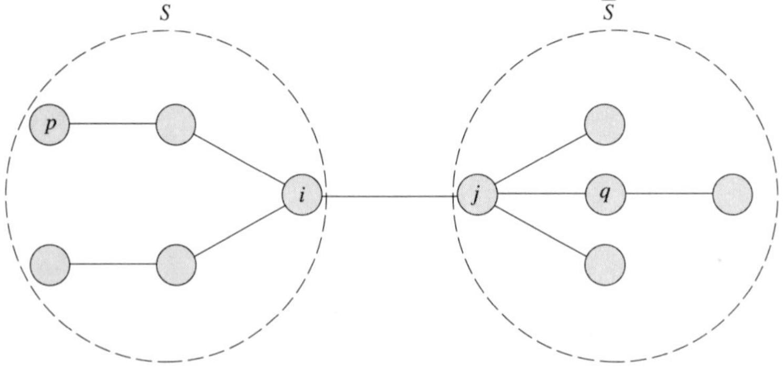

Figure 13.1 Cut formed by deleting the arc (i, j) from a spanning tree.

(to save memory space). We assume that the rows of the array have many similar entries and differ only at a few places. One such situation arises in the sequence of amino acids in a protein found in the mitochondria of different animals and higher plants.

Since the entities in the rows are similar, one approach for saving memory is to store one row, called the *reference row*, completely, and to store only the differences between some of the rows so that we can derive each row from these differences and the reference row. Let c_{ij} denote the number of different entries in rows i and j; that is, if we are given row i, then by making c_{ij} changes to the entries in this row we can obtain row j, and vice versa. Suppose that the array contains four rows, represented by R_1, R_2, R_3, and R_4, and we decide to treat R_1 as a reference row. Then one plausible solution is to store the differences between R_1 and R_2, R_2 and R_4, and R_1 and R_3. Clearly, from this solution, we can obtain rows R_2 and R_3 by making c_{12} and c_{13} changes to the elements in row R_1. Having obtained row R_2, we can make c_{24} changes to the elements of this row to obtain R_4.

It is easy to see that it is sufficient to store differences between those rows that correspond to arcs of a spanning tree. These differences permit us to obtain each row from the reference row. The total storage requirement for a particular storage scheme will be the length of the reference row (which we can take as the row with the least amount of data) plus the sum of the differences between the rows. Therefore, a minimum spanning tree would provide the least cost storage scheme.

Application 13.5 Cluster Analysis

The essential issue in cluster analysis is to partition a set of data into "natural groups"; the data points within a particular group of data, or a cluster, should be more "closely related" to each other than the data points not in that cluster. Cluster analysis is important in a variety of disciplines that rely on empirical investigations. Consider, for example, an instance of a cluster analysis arising in medicine. Suppose that we have data on a set of 350 patients, measured with respect to 18 symptoms. Suppose, further, that a doctor has diagnosed all of these patients as having the same disease, which is not well understood. The doctor would like to know if he can develop a better understanding of this disease by categorizing the symptoms into smaller groupings that can be detected through cluster analysis. Doing so might permit the doctor to find more natural disease categories to replace or subdivide the original disease.

In this section we describe the use of spanning tree problems to solve a class of problems that arise in the context of cluster analysis. Suppose that we are interested in finding a partition of a set of n points in two-dimensional Euclidean space into clusters. A popular method for solving this problem is by using Kruskal's algorithm for solving the minimum spanning tree problem (we describe this method in Section 13.4). As we will show, at each intermediate iteration, Kruskal's algorithm maintains a forest (i.e., a collection of node-disjoint trees) and adds arcs in nondecreasing order of their lengths. We can regard the nodes spanned by the trees at intermediate steps as different clusters. These clusters are often excellent solutions for the clustering problem, and moreover, we can obtain them very efficiently. Kruskal's algorithm can be thought of as providing n partitions: The first partition contains

n clusters, each cluster containing a single point, and the last partition contains just one cluster containing all the points. Alternatively, we can obtain n partitions by starting with a minimum spanning tree and deleting tree arcs one by one in nonincreasing order of their lengths. We illustrate the latter approach using an example. Consider a set of 27 points shown in Figure 13.2(a). Suppose that the network in Figure 13.2(b) is a minimum spanning tree for these points. Deleting the three largest length arcs from the minimum spanning tree gives a partition with four clusters shown in Figure 13.2(c).

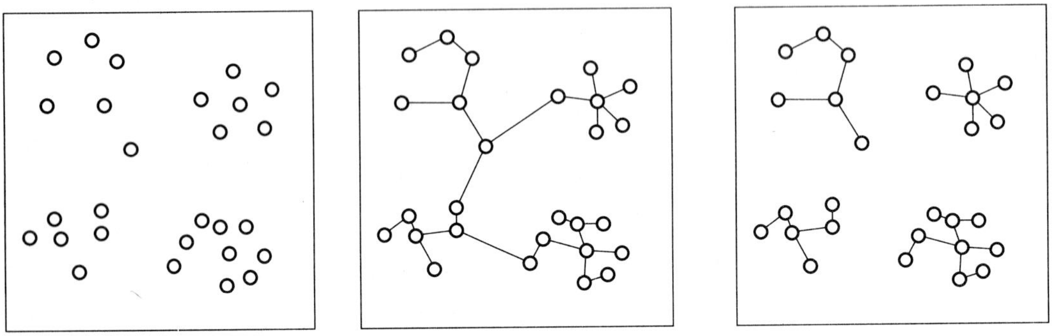

(a) **(b)** **(c)**

Figure 13.2 Identifying clusters by finding a minimum spanning tree.

Analysts can use the information obtained from the preceding analysis in several ways. The procedure we have described yields n partitions. Out of these, we might select the "best" partition by simple visualization or by defining an appropriate objective function value. A good choice of the objective function depends on the underlying features of the particular clustering application. We might note that this analysis is not limited to points in two-dimensional space; we can easily extend it to multidimensional space if we define interpoint distances appropriately.

13.3 OPTIMALITY CONDITIONS

As in our earlier discussion of network flow algorithms, optimality conditions for the minimum spanning tree problem play a central role in developing algorithms and establishing their validity. For the minimum spanning tree problem, we can formulate the optimality conditions in two important ways: *cut optimality conditions* and *path optimality conditions*. Needless to say, both optimality conditions are equivalent. Before considering these conditions, let us establish some further notation and illustrate some basic concepts.

The subgraphs shown in Figures 13.3(b) and 13.3(c) are spanning trees for the network shown in Figure 13.3(a). However, the subgraph shown in Figure 13.3(d) is not a spanning tree because it is not connected, and the subgraph shown in Figure 13.3(e) is not a spanning tree because it contains a cycle 1–3–4–1. We refer to those arcs contained in a given spanning tree as *tree arcs* and to those arcs not contained in a given spanning tree as *nontree arcs*. The following two elementary observations will arise frequently in our development in this chapter.

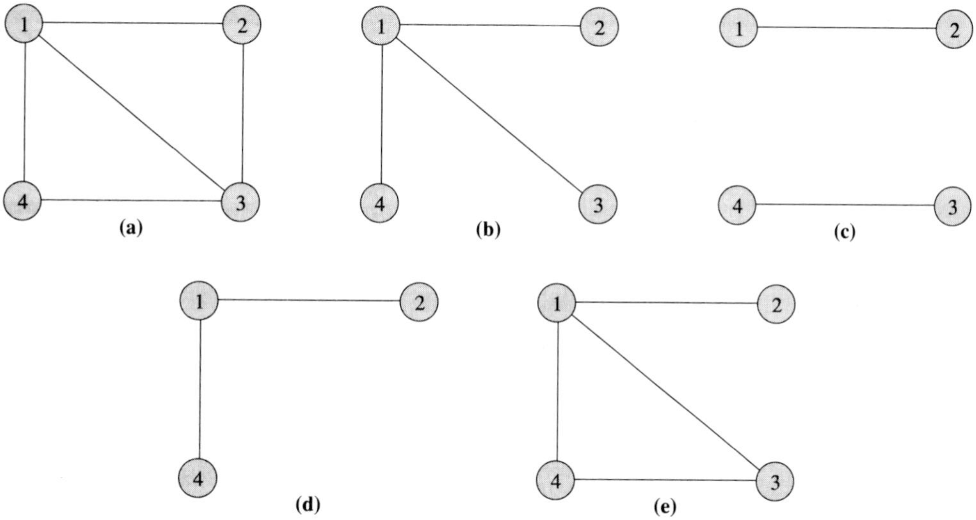

Figure 13.3 Illustrating spanning trees: (a) underlying graph; (b) two spanning trees; (c) nonspanning tree (disconnected graph); (d) nonspanning tree (doesn't span all nodes); (e) another nonspanning tree (cyclic graph).

1. For every nontree arc (k, l), the spanning tree T contains a unique path from node k to node l. The arc (k, l) together with this unique path defines a cycle [see Figure 13.4(a)].
2. If we delete any tree arc (i, j) from a spanning tree, the resulting graph partitions the node set N into two subsets [see Figure 13.4(b)]. The arcs from the un-

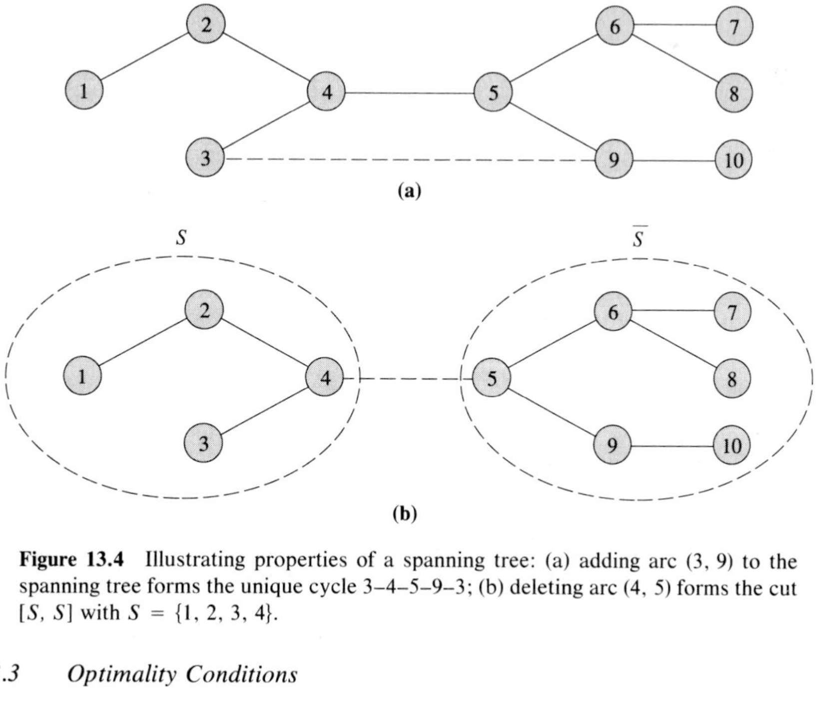

Figure 13.4 Illustrating properties of a spanning tree: (a) adding arc (3, 9) to the spanning tree forms the unique cycle 3–4–5–9–3; (b) deleting arc (4, 5) forms the cut $[S, S]$ with $S = \{1, 2, 3, 4\}$.

derlying graph G whose two endpoints belong to the different subsets constitute a cut.

We next prove the two optimality conditions.

Theorem 13.1 (Cut Optimality Conditions). *A spanning tree T^* is a minimum spanning tree if and only if it satisfies the following cut optimality conditions: For every tree arc $(i, j) \in T^*$, $c_{ij} \leq c_{kl}$ for every arc (k, l) contained in the cut formed by deleting arc (i, j) from T^*.*

Proof. It is easy to see that every minimum spanning tree T^* must satisfy the cut optimality condition. For, if $c_{ij} > c_{kl}$ and arc (k, l) is contained in the cut formed by deleting arc (i, j) from T^*, then introducing arc (k, l) into T^* in place of arc (i, j) would create a spanning tree with a cost less than T^*, contradicting the optimality of T^*.

We next show that if any tree T^* satisfies the cut optimality conditions, it must be optimal. Suppose that T° is a minimum spanning tree and $T^\circ \neq T^*$. Then T^* contains an arc (i, j) that is not in T° (the reader might find it helpful to refer to Figure 13.5 while reading the rest of the proof). Deleting arc (i, j) from T^* creates a cut, say $[S, \bar{S}]$. Now notice that if we add the arc (i, j) to T°, we create a cycle W that must contain an arc (k, l) [other than arc (i, j)] with $k \in S$ and $l \in \bar{S}$. Since T^* satisfies the cut optimality conditions, $c_{ij} \leq c_{kl}$. Moreover, since T° is an optimal spanning tree, $c_{ij} \geq c_{kl}$, for otherwise we could improve on its cost by replacing arc (k, l) by arc (i, j). Therefore, $c_{ij} = c_{kl}$. Now if we introduce arc (k, l) in the tree T^* in place of arc (i, j), we produce another minimum spanning tree and it has one more arc in common with T°. Repeating this argument several times, we can transform T^* into the minimum spanning tree T°. This construction shows that T^* is also a minimum spanning tree and completes the proof of the theorem. ◆

The cut optimality conditions imply that every arc in a minimum spanning tree is a minimum cost arc across the cut that is defined by removing it from the tree. In fact, the cut optimality conditions also imply that we can always include *any* minimum cost arc in any cut in some minimum spanning tree, which we state in the following somewhat stronger form.

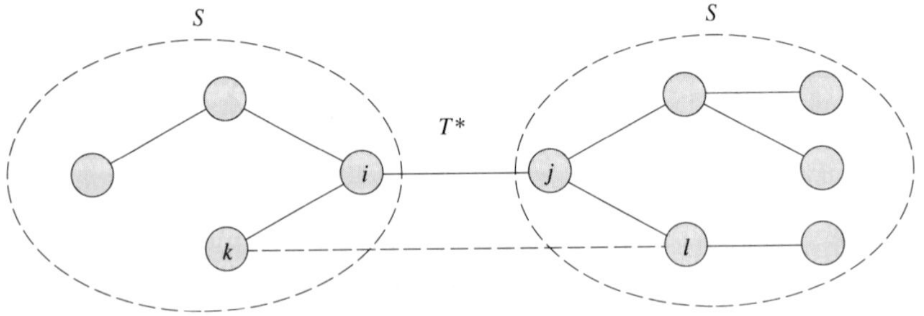

Figure 13.5 Proving cut and path optimality conditions.

Property 13.2. *Let F be a subset of arcs in some minimum cost spanning tree and let S be a set of nodes of some component of F. Suppose that (i, j) is a minimum cost arc in the cut $[S, \bar{S}]$. Then some minimum spanning tree contains all the arcs in F as well as the arc (i, j).*

Proof. Suppose that F is a subset of the minimum cost tree T^*. If $(i, j) \in T^*$, we have nothing to prove. So suppose that $(i, j) \notin T^*$. Adding (i, j) to T^* creates a cycle C, and C contains at least one arc $(p, q) \neq (i, j)$ in $[S, \bar{S}]$. By assumption, $c_{ij} \leq c_{pq}$. Also, since T^* satisfies the cut optimality conditions, $c_{ij} \geq c_{pq}$. Consequently, $c_{ij} = c_{pq}$, so adding arc (i, j) to T^* and removing arc (p, q) produces a minimum spanning tree containing all the arcs in F as well as the arc (i, j). ◆

The cut optimality conditions provide us with an "external" characterization of a minimum spanning tree that rests on the relationship between a single arc in the tree and many arcs outside the tree, that is, those in the cut that we produce by removing the arc from the tree. The following related path optimality conditions provide an alternative "internal" characterization that considers the relationship between a single nontree arc and several arcs in the tree, that is, those in the path formed by adding the nontree arc to the spanning tree.

Theorem 13.3 (Path Optimality Conditions). *A spanning tree T^* is a minimum spanning tree if and only if it satisfies the following path optimality conditions: For every nontree arc (k, l) of G, $c_{ij} \leq c_{kl}$ for every arc (i, j) contained in the path in T^* connecting nodes k and l.*

Proof. It is easy to show the necessity of the path optimality conditions. Suppose T^* is a minimal spanning tree satisfying these conditions and arc (i, j) is contained in the path in T^* connecting nodes k and l. If $c_{ij} > c_{kl}$, introducing arc (k, l) into T^* in place of arc (i, j) would create a spanning tree with a cost less than T^*, contradicting the optimality of T^*.

We establish the sufficiency of the path optimality conditions by using the sufficiency of the cut optimality conditions. This proof technique highlights the equivalence between these conditions. We will show that if a tree T^* satisfies the path optimality conditions, it must also satisfy the cut optimality conditions; Theorem 13.1 would then imply that T^* is an optimal tree. Let (i, j) be any tree arc in T^*, and let S and $\bar{S}$ be the two sets of connected nodes produced by deleting arc (i, j) from T^*. Suppose $i \in S$ and $j \in \bar{S}$. Consider any arc $(k, l) \in [S, \bar{S}]$ (see Figure 13.5). Since T^* contains a unique path joining nodes k and l and since arc (i, j) is the only arc in T^* joining a node in S and a node in $\bar{S}$, arc (i, j) must belong to this path. The path optimality condition implies that $c_{ij} \leq c_{kl}$; since this condition must be valid for every nontree arc (k, l) in the cut $[S, \bar{S}]$ formed by deleting any tree arc (i, j), T^* satisfies the cut optimality conditions and so it must be a minimum spanning tree. ◆

In the preceding discussion we have established two optimality conditions for the minimum spanning tree problem. The following optimality conditions for the maximum spanning tree problem are similar. We leave their proofs as an exercise (see Exercise 13.9).

Theorem 13.4 (Maximum Spanning Tree Optimality Conditions).

(a) *A spanning tree T^* is a maximum spanning tree if and only if it satisfies the following cut optimality conditions: For every tree arc $(i, j) \in T^*$, $c_{ij} \geq c_{kl}$ for every arc (k, l) contained in the cut formed by deleting arc (i, j) from T^*.*

(b) *A spanning tree is a maximum spanning tree T^* if and only if it satisfies the following path optimality conditions: For every nontree arc (k, l) of G, $c_{ij} \geq c_{kl}$ for every arc (i, j) contained in the tree path in T^* connecting nodes k and l.*

13.4 KRUSKAL'S ALGORITHM

The path optimality conditions immediately suggest the following straightforward algorithm for solving the minimum spanning tree problem. We start with any arbitrary spanning tree T and test the path optimality conditions. If T satisfies this condition, it is an optimal tree; otherwise, $c_{ij} > c_{kl}$ for some nontree arc (k, l) and some tree arc (i, j) contained in the unique path in T connecting nodes k and l. In this case, adding arc (k, l) to T in place of arc (i, j) gives us a spanning tree with a lower cost. Repeating this step will give us a minimum spanning tree within a finite number of iterations. Although this algorithm is strikingly simple, its running time cannot be polynomially bounded in the size of the problem data.

Simple Version of Kruskal's Algorithm

To derive an alternative and more efficient algorithm, known as *Kruskal's algorithm*, from the path optimality conditions, we consider an algorithm that builds an optimal spanning tree from scratch by adding one arc at a time. We first sort all the arcs in nondecreasing order of their costs and define a set, LIST, that is the set of arcs we have chosen as part of a minimum spanning tree. Initially, the set LIST is empty. We examine the arcs in the sorted order one by one and check whether adding the arc we are currently examining to LIST creates a cycle with the arcs already in LIST. If it does not, we add the arc to LIST; otherwise, we discard it. We terminate when $|\text{LIST}| = n - 1$. At termination, the arcs in LIST constitute a minimum spanning tree T^*.

The correctness of Kruskal's algorithm follows from the fact that we discarded each nontree arc (k, l) with respect to T^* at some stage because it created a cycle with the arcs already in LIST. But observe that the cost of arc (k, l) is greater than or equal to the cost of every arc in that cycle because we examined the arcs in the nondecreasing order of their costs. Therefore, the spanning tree T^* satisfies the path optimality conditions, so it is an optimal tree.

To illustrate Kruskal's algorithm on a numerical example, we consider the network shown in Figure 13.6(a). Sorted in the order of their costs, the arcs are (2, 4), (3, 5), (3, 4), (2, 3), (4, 5), (2, 1), and (3, 1). In the first three iterations, the algorithm adds the arcs (2, 4), (3, 5), and (3, 4) to LIST [see Figures 13.6(b) to (d)]. In the next two iterations, the algorithm examines arcs (2, 3) and (4, 5) and discards them because the addition of each arc to LIST creates a cycle [see Figure 13.6(e) and (f)]. Then the algorithm adds arc (2, 1) to LIST and terminates. Figure 13.6(g) shows the minimum spanning tree.

We might view the running time of Kruskal's algorithm as being composed of

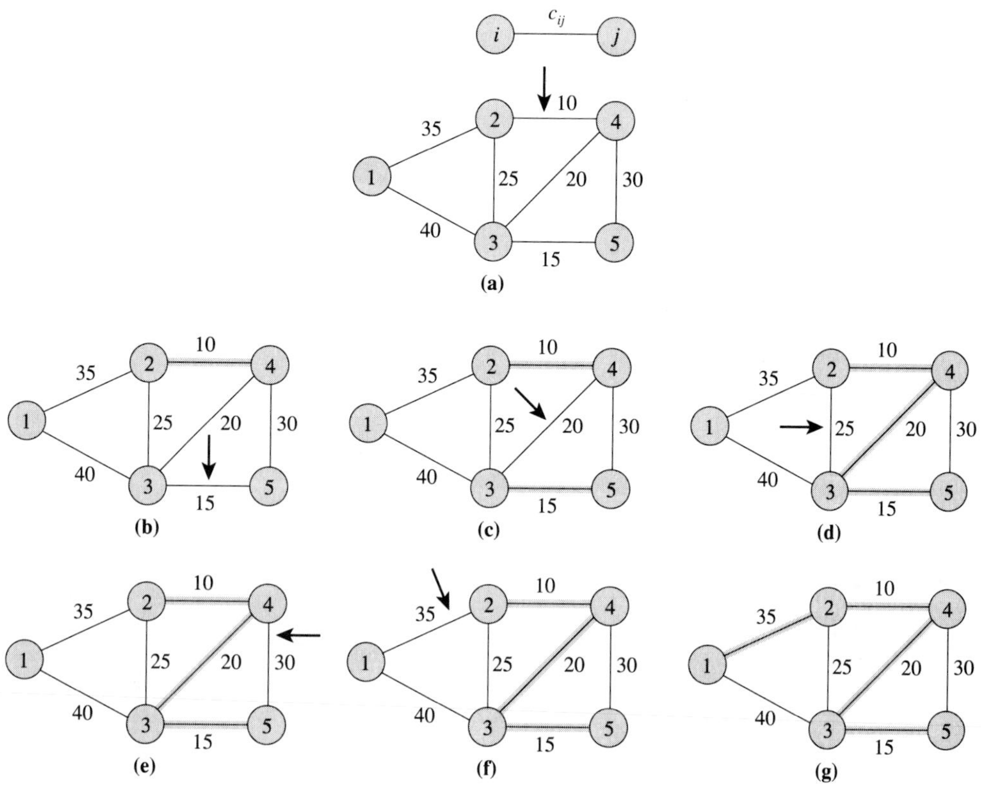

Figure 13.6 Illustrating Kruskal's algorithm.

the time for sorting the arcs and the time for detecting cycles. For a network with arbitrarily large arc costs, sorting requires $O(m \log m) = O(m \log n^2) = O(m \log n)$ time. The time to detect a cycle depends on the method we use for this step. One naive method would work as follows. The set LIST at any stage of the algorithm is a *forest* (i.e., a collection of subtrees). For example, the set LIST corresponding to Figure 13.6(c) consists of three trees containing the nodes {1}, {2, 4}, and {3, 5}, respectively. We denote these sets of nodes for a collection of trees by N_1, N_2, N_3, ... We can store these sets as different singly linked lists. While examining an arc (k, l), we scan through these linked lists and check whether both the nodes k and l belong to the same list. If so, adding arc (k, l) to LIST creates a cycle and we discard this arc. If nodes k and l belong to different lists, we add arc (k, l) to LIST, which requires merging the lists containing nodes k and l into a single list. Clearly, this data structure requires $O(n)$ time for each arc that we examine, so if we use this data structure, Kruskal's algorithm runs in $O(nm)$ time.

Improved Implementation of Kruskal's Algorithm

We now describe a more efficient implementation of Kruskal's algorithm that runs in $O(m + n \log n)$ time plus the time taken for sorting the arcs. This implementation is similar to the preceding one: We store the collection of trees, denoted by the sets

$N_1, N_2, N_3, \ldots$, as different singly linked lists. With each list L, we maintain two indices: *size(L)*, representing the number of elements in the list L; and *last(L)*, representing the last element in the list L. For each element i, we associate an index *first(i)* that stores the first element in the list containing node i. For example, if $L = \{1, 5, 6\}$, then size$(L) = 3$; last$(L) = 6$; and first(1) = first(5) = first(6) = 1.

Using this scheme, we can easily check whether the nodes k and l belong to the same list. If first(k) = first(l), they do; otherwise, they don't. This step contributes a total of $O(m)$ time to the running time of the algorithm. When we add the arc (k, l) to LIST, we need to merge the lists containing the nodes k and l. To merge these two lists, we always put the larger list first, breaking ties arbitrarily. The size indices allow us to determine the larger list, and last indices allow us to determine the last element of the larger list where we append the smaller list. As a result of the merge operation, several indices change. Updating the size and last indices is easy and requires $O(1)$ effort. To update the first indices, we need to modify this index for every node in the smaller list; the time required for this operation is proportional to the number of elements in the list. Suppose that the time required to merge the two lists L and L' of sizes h and h' (with $h \le h'$) is ph for some constant p. We shall show that the total time required in all the mergings is $O(n \log n)$.

We prove this result using an induction argument. We claim that the total time required to obtain a merged list of size n is at most $pn \log n$ for some constant p. This result is clearly true for $n = 1$, since this case requires no mergings. Let us assume inductively that the result is true for any number of elements strictly less than n.

When we carry out the algorithm, we ultimately obtain a single list of n elements, obtained by appending two smaller lists L and L' containing h and $n - h$ elements. Let us assume that $h \le n/2$, so that we place list L after list L' in the merge operation. By the inductive hypothesis, the time needed to create L is at most $ph \log h$ and the time needed to create L' is at most $p(n - h) \log(n - h)$. Since the time needed for merging the two lists is at most ph, the total time required for all the merging steps is at most

$$\begin{aligned} ph \log h + p(n - h) \log(n - h) + ph &\le ph \log(n/2) + p(n - h)\log n + ph \\ &= ph(\log n - 1) + p(n - h)\log n + ph \\ &= pn \log n. \end{aligned}$$

The following theorem summarizes the implication of this result.

Theorem 13.5. *The improved implementation of Kruskal's algorithm solves the minimum spanning tree problem in $O(m + n \log n)$ time plus the time for sorting the arcs.*

Kruskal's algorithm requires two basic operations on lists of elements, which are commonly known as *union-find* operations. The *union* operation merges two lists and the *find* operation determines the list an element belongs to. Although our implementation of Kruskal's algorithm gives an attractive running time of $O(m + n \log n)$, in addition to the time for sorting, we could improve on this time even further by using better implementations of the union-find operations. The improved implementation has a running time of $O(m\,\alpha(n, m))$ for a function $\alpha(n, m)$ that grows so slowly that for all practical purposes it can be viewed as a constant less than 6 (see the reference notes).

13.5 PRIM'S ALGORITHM

Just as the path optimality conditions allowed us to develop Kruskal's algorithm, the cut optimality conditions permit us to develop another simple algorithm for the minimum spanning tree problem, known as *Prim's algorithm*. This algorithm builds a spanning tree from scratch by fanning out from a single node and adding arcs one at a time. It maintains a tree spanning on a subset S of nodes and adds a nearest neighbor to S. The algorithm does so by identifying an arc (i, j) of minimum cost in the cut $[S, \bar{S}]$. It adds arc (i, j) to the tree, node j to S, and repeats this basic step until $S = N$. The correctness of the algorithm follows directly from Property 13.2 since this result implies that each arc that we add to the tree is contained in some minimum spanning tree with the arcs that we have selected in the previous steps.

We illustrate Prim's algorithm on the same example, shown in Figure 13.7(a), that we used earlier to illustrate Kruskal's algorithm. Suppose, initially, that $S = \{1\}$. The cut $[S, \bar{S}]$ contains two arcs, (1, 2) and (1, 3), and the algorithm selects the arc (1, 2) [see Figure 13.7(b)]. At this point $S = \{1, 2\}$ and the cut $[S, \bar{S}]$ contains the arcs (1, 3), (2, 3), and (2, 4). The algorithm selects arc (2, 4) since it has the minimum cost among these three arcs [see Figure 13.7(c)]. In the next two iterations, the algorithm adds arc (4, 3) and then arc (3, 5); Figure 13.7(d) and (e) show the details of these iterations. Figure 13.7(f) shows the minimum spanning tree produced by the algorithm.

To analyze the running time of Prim's algorithm, we consider each of the $n - 1$ iterations that the algorithm performs as it adds one arc at a time to the tree until it has a spanning tree with $n - 1$ arcs. In each iteration, the algorithm selects

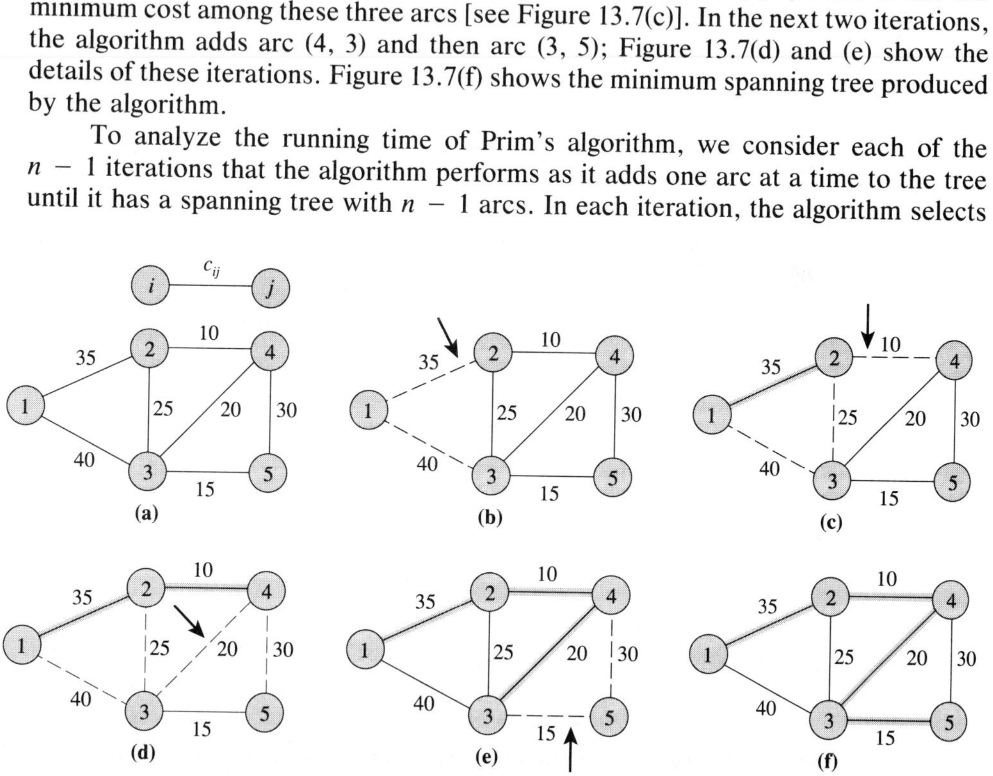

Figure 13.7 Illustrating Prim's algorithm.

the minimum cost arc in the cut $[S, \bar{S}]$. If we scan the entire arc list to identify the minimum cost arc, this operation requires $O(m)$ time, giving us an $O(nm)$ time bound for the algorithm. Therefore, this implementation of Prim's algorithm runs in $O(nm)$ time. However, we can improve upon it substantially, as we next show.

The bottleneck step in the $O(nm)$ implementation of Prim's algorithm is the identification of a minimum cost arc in the cut $[S, \bar{S}]$. We can improve the efficiency of this step by maintaining two indices for each node j in $\bar{S}$: (1) a distance label $d(j)$, which represents the minimum cost of arcs in the cut incident to a node j not in S (i.e., $d(j) = \min\{c_{ij}:(i, j) \in [S, \bar{S}]\}$), and (2) a predecessor label *pred*(j), which represents the other endpoint of the minimum cost arc in the cut incident to node j. For example, in Figure 13.7(d), three arcs, (1, 3), (2, 3), and (4, 3), in the cut are incident to node 3. Among these arcs, arc (4, 3) has the minimum cost of 20. Therefore, $d(3) = 20$ and pred(3) = 4. For the same figure, $d(5) = 30$ and pred(5) = 4. If we maintain these indices, we can easily find the minimum cost of an arc in the cut; we simply compute $\min\{d(j): j \in \bar{S}\}$. If node i achieves this minimum, (pred(i), i) is a minimum cost arc in the cut. Observe that if we move node i from $\bar{S}$ to S, we need to update the distance and predecessor labels only for the nodes adjacent to node i.

Notice the similarity between this implementation of Prim's algorithm and the implementation of Dijkstra's algorithm that we discussed in Section 4.5. Just as in Dijkstra's algorithm, the basic operations are finding the minimum distance label $d(i)$ among the nodes in the set $\bar{S}$, moving the corresponding node into the set S, and updating the distance labels of those nodes in $\bar{S}$ that are adjacent to node i. Indeed, we can implement Prim's algorithm using the various types of heaps (or priority queues) that we used in our implementations of Dijkstra's algorithm in Section 4.7. Recall from Appendix A that a heap is a data structure that permits us to perform the following operations on a collection H of objects, each having an associated real number called its *key*.

create-heap(H). Create an empty heap H.

find-min(H). Find and return an object from H with the minimum key.

insert(i, H). Insert a new object i with a predefined key into a collection H of objects.

decrease-key(i, value, H). Reduce the key of an object i in H to *value*, which must be smaller than the key it is replacing.

delete-min(i, H). Delete an object i with the minimum key from the collection H of objects.

Observe that if we implement Prim's algorithm using a heap, H would be the collection of nodes in $\bar{S}$ and the key of a node would be its distance label. Prim's algorithm would be implemented as described in Figure 13.8. As always, we let C denote the maximum arc cost in the graph G.

As is clear from its description, Prim's algorithm performs the operations find-min, delete-min, and insert at most n times and the operation decrease-key at most

```
algorithm heap-Prim;
begin
    create-heap(H);
    for each j ∈ N − {1} do d(j) : = C + 1;
    set d(1) : = 0; and pred(1) : = 0;
    for each j ∈ N do insert(j, H);
    T* : = Ø;
    while |T*| < (n − 1) do
    begin
        find-min(i, H);
        delete-min(i, H);
        T* : = T* ∪ (pred(i), i);
        for each (i, j) ∈ A(i) with j ∈ H do
            if d(j) > c_ij then
            begin
                d(j) : = c_ij;
                pred(j) : = i;
                decrease-key(j, c_ij, H);
            end;
    end;
        T* is a minimum spanning tree;
end;
```

Figure 13.8 Prim's algorithm.

m times. When implemented with different heaps, the algorithm would have the running times shown in Figure 13.9.

Our discussion of the binary heap, d-heap, and Fibonacci heap data structures in Appendix A permits us to justify these time bounds. In that discussion we show that the d-heap data structure performs each delete-min operation in $O(d \log_d n)$ time and every other heap operation in $O(\log_d n)$ time. If we select $d = m/n$, this result gives us a running time of $O(m \log_d n + nd \log_d n) = O(m \log_d n)$ for Prim's algorithm implemented using the d-heap data structure. The binary heap is a special case of d-heap with $d = 2$, so its time bound is $O(m \log n)$. The Fibonacci heap data structure performs each delete-min operation in $O(\log n)$ time and every other heap operation in $O(1)$ time. Consequently, the Fibonacci heap implementation of Prim's algorithm runs in $O(m + n \log n)$ time.

Theorem 13.6. *Fibonacci heap implementation of Prim's algorithm solves the minimum spanning tree problem in $O(m + n \log n)$ time.* ◆

Heap type	**Running time**
Binary heap	$O(m \log n)$
d-heap	$O(m \log_d n)$, with $d = \max\{2, m/n\}$
Fibonacci heap	$O(m + n \log n)$
Johnson's data structure	$O(m \log \log C)$

Figure 13.9 Running times of various heap implementations of Prim's algorithm.

13.6 SOLLIN'S ALGORITHM

We can use the cut optimality conditions to derive another novel algorithm for the minimum spanning tree problem, known as Sollin's algorithm. We can view this algorithm as a hybrid version of Kruskal's and Prim's algorithm. As in Kruskal's algorithm, Sollin's algorithm maintains a collection of trees spanning the nodes N_1, N_2, N_3, . . . , and adds arcs to this collection. However, at every iteration, it adds minimum cost arcs emanating from these trees, an idea borrowed from Prim's algorithm. As a result, we obtain a fairly simple algorithm that uses elementary data structures and runs in $O(m \log n)$ time. As pointed out in the reference notes, a more clever implementation of this approach runs in $O(m \log \log n)$ time.

Sollin's algorithm repeatedly performs the following two basic operations:

nearest-neighbor (N_k, i_k, j_k). This operation takes as an input a tree spanning the nodes N_k and determines an arc (i_k, j_k) with the minimum cost among all arcs emanating from N_k [i.e., $c_{i_k j_k} = \min\{c_{ij}: (i, j) \in A, i \in N_k \text{ and } j \notin N_k\}$]. To perform this operation we need to scan all the arcs in the adjacency lists of nodes in N_k, and find a minimum cost arc among those arcs that have one endpoint not belonging to N_k.

merge(i_k, j_k). This operation takes as an input two nodes i_k and j_k, and if the two nodes belong to two different trees, then merges these two trees into a single tree.

Using these two basic operations, we state Sollin's algorithm as shown in Figure 13.10.

We illustrate Sollin's algorithm on the same numerical example that we have used to illustrate Kruskal's and Prim's algorithms. As shown in Figure 13.11(b), Sollin's algorithm starts with a forest containing five trees: Each tree is a singleton node. This figure also shows the least cost arc emanating from each tree. We next perform mergings, reducing the number of trees to only two [see Figure 13.11(c)]. The least cost arc emanating from these two trees is (3, 4), and when we add this arc, we obtain the spanning tree shown in Figure 13.11(d). The algorithm now terminates.

To analyze the running time of Sollin's algorithm, we need to discuss the data structure needed to implement it. We will show that the algorithm performs $O(\log n)$ executions of the while loop, and that we can perform all the nearest-neighbor and

```
algorithm Sollin;
begin
    for each i ∈ N do N_i : = {i};
    T* : = Ø;
    while |T*| < (n − 1) do
    begin
        for each tree N_k do nearest-neighbor(N_k, i_k, j_k);
        for each tree N_k do
            if nodes i_k and j_k belong to different trees then
                merge(i_k, j_k) and update T* : = T* ∪ {(i_k, j_k)};
    end;
end;
```

Figure 13.10 Sollin's algorithm.

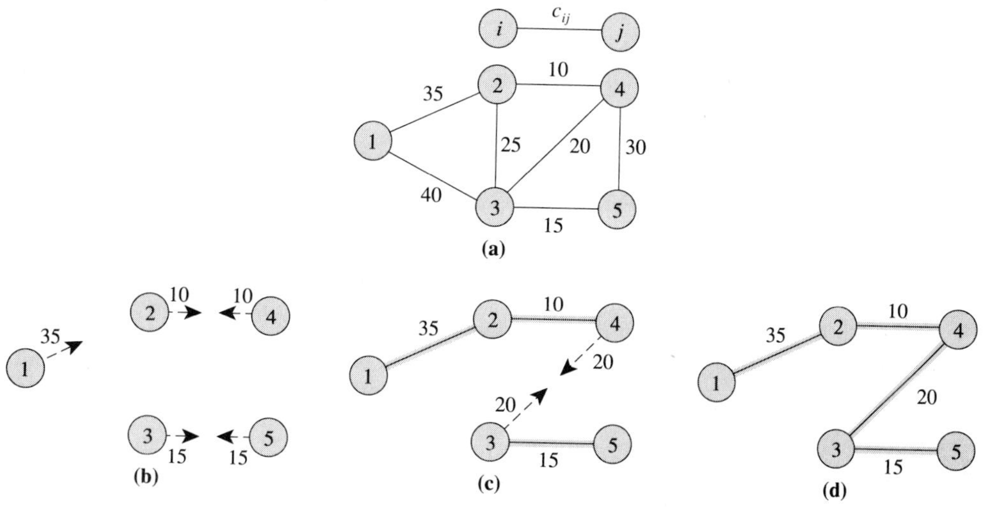

Figure 13.11 Illustrating Sollin's algorithm.

merge operations in $O(m)$ time. These results establish a time bound of $O(m \log n)$ for Sollin's algorithm.

We store the nodes of a tree as a circular doubly linked list. The doubly linked list allows us to visit every node of the tree starting at any tree node. We assign a numerical label with every node in the network; the label satisfies the following two properties: (1) nodes of the same tree have the same label, and (2) nodes of different trees have different labels. At the beginning of the algorithm, we assign label i to each node $i \in N$.

Using this data structure, we can easily check whether an arc (i, j) has both of its endpoints in the same tree: We answer this question simply by checking the labels of nodes i and j. This observation implies that we can perform the nearest-neighbor operation for each tree in the forest in a total time of $O(\sum_{i \in N} | A(i) |) = O(m)$.

We perform merge operations in a while loop using the following iterative scheme. In each iteration we select an unexamined tree, say N_1, and consider the minimum cost arc (i_1, j_1) emanating from N_1. (Node i_1 is in N_1 and j_1 might or might not be in N_1.) Suppose that nodes in N_1 have the label α. If node j_1 also has the label α, the iteration ends. Otherwise, we scan through the nodes of the tree, say N_2, containing node j_1 and assign them the label α. Next, we consider the minimum cost arc (i_2, j_2) emanating from N_2. If node j_2 has label α, the iteration ends; otherwise, we scan through the nodes of the tree, say N_3, containing node j_2 and assign them the label α. We repeat this process until the iteration ends. Notice that within an iteration we might assign the nodes of several trees the label of the first tree. When an iteration ends, we initiate a new iteration by selecting another unexamined tree. We terminate this iterative process when we have examined all the trees. As is clear from this description, this method assigns a label to each node once and hence runs in $O(n)$ time.

Having proved that each execution of the while loop in Sollin's algorithm requires $O(m)$ time, we now obtain a bound on the number of executions of the loop. Each execution of the loop reduces the number of trees in the forest by a factor of at least two because we merge each tree into a larger tree. This observation implies that we will perform $O(\log n)$ executions of the loop. We have therefore established the following result.

Theorem 13.7. *The execution of Sollin's algorithm requires $O(m \log n)$ time.*

13.7 MINIMUM SPANNING TREES AND MATROIDS

In keeping with the orientation of this book, we have examined the minimum spanning tree problem from a perspective of graph theory and the data structures needed to implement spanning tree algorithms efficiently. We could, instead, view the minimum spanning tree problem and develop several of the core ideas of this chapter from at least two other perspectives: (1) broader notions in combinatorial optimization, and (2) linear programming. These two alternative viewpoints are instructive because they help to show the connection between network optimization and other important topics in discrete optimization. Indeed, the minimum spanning tree problem and network flows have inspired the development of many other problem domains in discrete optimization. Consequently, it is useful to pause at this point and briefly delineate these connections.

Matroids and the Greedy Algorithm

Suppose that we view a spanning tree in the following way: We have a finite collection of objects E, the arcs of a network, and we define a subset I of objects to be *independent* if they do not form a cycle in the network. If each object (i.e., arc) e has an associated weight w_e, the minimal spanning tree problem seeks an $n - 1$ element independent set I with the smallest total weight $w(I) \equiv \sum_{e \in I} w_{ij}$.

Let us now describe this and related problems in a more abstract setting. A *subset system* $(E, \mathcal{I})$ is a finite set of objects E and nonempty collection $\mathcal{I}$ of subsets of these objects, called *independent sets*, that satisfies the hereditary property that whenever I is an independent set (i.e., belongs to $\mathcal{I}$) and I' is a subset of I, then I' also is an independent set.

Suppose that we associate a weight w_c with each element e of E and define the weight $w(S)$ of any subset S of E as the sum of the weights of its elements; that is, $w(S) \equiv \sum_{e \in S} w_e$. As a generalization of the minimum spanning tree problem, we might consider the following *independence (or subset) system optimization problem*: Find a maximal independent set of the subset system $(E, \mathcal{I})$ with the minimum weight.

At this level of generality, the independence system optimization problem appears to be hopelessly difficult to solve efficiently. Therefore, we need to impose additional structure on the problem so that it becomes tractable. We would like to do so, however, by imposing the least amount of additional structure so that the results remain as general as possible. The following type of auxiliary structure appears to be just right for this purpose.

A subset system $(E, \mathcal{I})$ is a *matroid* if it satisfies the growth property that if I_p and I_{p+1} are independent sets containing p and $p + 1$ elements, we always can find an element $e \in I_{p+1} - I_p$, satisfying the property that $I_p \cup \{e\}$ is an independent set.

Note that this definition implies if I and I' are any two independent sets and $|I'| > |I|$, we can add certain elements of I' to I and obtain another independent set I'' so that I'' contains I and has as many elements as I'. That is, the sets I_p and I_{p+1} in the definition need not differ in cardinality by one. (We establish this *extended growth property* by applying the growth property to I and the first $|I| + 1$ elements of I', which are independent by the definition, and then repeating the operation.)

Let us illustrate the definition of a matroid with a few examples.

Graphic or forest matroid. Note that forests in a network satisfy these definitions if we let E equal the arcs in a network and let $\mathcal{I}$ denote the collection of arc sets that contain no cycles (i.e., the arcs define a forest). In this case the system $(E, \mathcal{I})$ is an independent system because removing arcs from a forest always produces another forest. Moreover, if I_p and I_{p+1} are two forests containing p and $p + 1$ arcs, the forest I_{p+1} must contain an arc e that we can add to I_p and produce another forest $I_p \cup \{e\}$ (see Exercise 13.41). Consequently, the system $(E, \mathcal{I})$ is a matroid.

Partition matroid. Let $E = E_1 \cup E_2 \cup \cdots \cup E_K$ be a union of K disjoint finite sets and let $u_1, u_2, \ldots, u_K$ be given positive integers. Let $\mathcal{I}$ be the family of subsets I of E that satisfy the property that for all $k = 1, 2, \ldots, K$, I contains no more than u_k elements of E_k. The system is a matroid. Note that if we consider a bipartite graph $(N_1 \cup N_2, A)$ and let E_k, for all nodes k in N_1, be the set of arcs incident to node k, then if all the u_k are equal to 1, the matroid defines "half" of an assignment problem. Another partition matroid defined on the nodes N_2 defines the other half of the matroid, so any feasible solution to the assignment problem is an independent set in both partition matroids.

Matric matroid. Let M be a real-valued matrix, let E be the columns of M, and let $\mathcal{I}$ be sets of columns of M that are linearly independent. Since removing columns from a linearly independent set of columns produces another independent set, the system $(E, \mathcal{I})$ is a subset system. By elementary results in linear algebra, this system also satisfies the growth property and so is a matroid.

Let us make one further observation about matroids. A *maximal independent set* is an independent set I satisfying the property that we cannot add any other element e to I and produce another independent set. The (extended) growth property implies that every maximal independent set of a matroid contains the same number of elements (since we can always add elements of one maximal independent set to another if they contain a different number of elements). Borrowing notation from linear algebra, we refer to any maximal independent set as a *basis* of the matroid. In this terminology, the matroid optimization problem seeks a basis with the smallest possible total weight.

We can attempt to solve this problem by using a *greedy algorithm* (Figure 13.12) which is a direct generalization of Kruskal's algorithm.

Note that for the minimal spanning tree problem, the test condition "LIST $\cup$ $\{e_j\}$ is independent" from this algorithm is just the test condition from Kruskal's

```
algorithm greedy;
begin
    order the elements of E = {e1, e2, . . . , eK} so that w1 ≤ w2 ≤ ··· ≤ wK;
    set LIST : = Ø;
    for j = 1 to K do
        if LIST ∪ {ej} is independent then LIST : = LIST ∪ {ej};
    LIST is a minimum weight basis;
end;
```

Figure 13.12 Greedy algorithm.

algorithm, namely, that the network defined by the arcs in LIST and e_j contains no cycle.

Theorem 13.8. *The greedy algorithm solves the matroid optimization problem.*

Proof. Let I^* be any optimal solution to the matroid optimization problem and let LIST $= \{e_{j_1}, e_{j_2}, \ldots, e_{j_n}\}$ be the solution generated by the greedy algorithm. We will show that $w(\text{LIST}) = w(I^*)$ and therefore that LIST is an optimal basis as well. If LIST $= I^*$, we have nothing to prove. So assume that LIST $\neq I^*$. Suppose that we order the elements of I^* in the order of increasing indices from the set $E = \{e_1, e_2, \ldots, e_K\}$ as $e_{j_1}, e_{j_2}, \ldots, e_{j_k}, e_q, \ldots$, with $e_q \neq e_{j_{k+1}}$ and assume that e_q is the first element of I^* not in LIST. Since the set $\{e_{j_1}, e_{j_2}, \ldots, e_{j_k}, e_q\}$ is independent, the steps of the greedy algorithm imply that $q \geq j_{k+1}$ and therefore that $w_q \geq w_{j_{k+1}}$. Since both the sets $I = \{e_{j_1}, e_{j_2}, \ldots, e_{j_k}, e_{j_{k+1}}\}$ and I^* are independent, the growth property implies that we can add elements of I^* to I to obtain another basis I'. Since this basis contains the elements $I^* \cup \{e_{j_{k+1}}\} - e_p$ for some $e_p \in I^*$ and $p \geq j_{k+1}$, $w(I') \leq w(I^*)$; consequently, I' is also an optimal basis. Note that this basis has a greater number of lead elements in common with LIST (at least $k + 1$). But now if we apply the same argument to the sets I' and LIST, we will obtain another optimal basis with at least one more lead element in common with LIST. If we continue in this fashion, eventually I' will equal LIST, therefore establishing that LIST is a minimum-basis matroid. ◆

Note that this discussion not only gives an alternative proof of Kruskal's algorithm for the minimum spanning tree problem, but also shows that two underlying combinatorial properties—independence and the growth property—are the essential ingredients necessary to ensure that the greedy algorithm solves the minimum spanning tree problem. (In Exercise 13.45 we show that the greedy algorithm will solve the minimum weight independent set problem for any choice of the element weights if and only if the subset system is a matroid.) Therefore, any other property of a graph is irrelevant for ensuring that Kruskal's algorithm works correctly. That is, we have now identified the combinatorial postulates that drive the algorithm.

13.8 MINIMUM SPANNING TREES AND LINEAR PROGRAMMING

Linear programming provides yet another proof of Kruskal's algorithm. Moreover, the development of a linear programming-based approach permits us to make some elementary connections between network optimization and an important topic in applied mathematics, polyhedral combinatorics, which is the study of integer polyhedra (i.e., polyhedra with integer extreme points). As shown in Section 11.12, the minimum cost flow problem provides another connection between these topics.

Let $A(S)$ denote the set of arcs contained in the subgraph of $G = (N, A)$ induced by the node set S [i.e., $A(S)$ is the set of arcs of A with both endpoints in S]. Consider the following integer programming formulation of the minimum spanning tree problem:

$$\text{Minimize} \quad \sum_{(i,j)\in A} c_{ij}x_{ij} \tag{13.2a}$$

subject to

$$\sum_{(i,j)\in A} x_{ij} = n - 1, \tag{13.2b}$$

$$\sum_{(i,j)\in A(S)} x_{ij} \le |S| - 1 \qquad \text{for any set } S \text{ of nodes}, \tag{13.2c}$$

$$x_{ij} \ge 0 \text{ and integer.} \tag{13.2d}$$

In this formulation, the 0–1 variable x_{ij} indicates whether we select arc (i, j) as part of the chosen spanning tree (note that the second set of constraints with $|S| = 2$ implies that each $x_{ij} \le 1$). The constraint (13.2b) is a cardinality constraint implying that we choose exactly $n - 1$ arcs, and the "packing" constraint (13.2c) implies that the set of chosen arcs contain no cycles (if the chosen solution contained a cycle, and S were the set of nodes on a chosen cycle, the solution would violate this constraint). Note that as a function of the number of nodes in the network, this model contains an exponential number of constraints. Nevertheless, as we will show, we can solve it very efficiently by applying Kruskal's algorithm. We might note that any formulation in the variables x_{ij} always requires an exponential number of constraints; that is, we cannot replace the given constraints by some polynomial set of constraints and still have a valid formulation of the problem. Nevertheless, it is possible to give a polynomial formulation of the problem if we introduce new (multicommodity flow) variables (see the reference notes).

Suppose that we consider the linear programming relaxation of this integer programming model. That is, we drop the restriction that the variables be integer. As we noted in Section 9.4 (also see Appendix C), we can formulate a set of reduced cost and complementary slackness optimality conditions for every linear programming problem and use these conditions, as we use the reduced costs and complementary slackness conditions of network flows, to assess when a given feasible solution is optimal. Recall that for network flow problems, we used node potentials to define the reduced costs and the complementary slackness conditions; each node in a minimum cost flow problem corresponds to one equation of the mass balance constraints $\mathcal{N}x = b$, so we can view the potentials as associated with these equations. Since the minimum spanning tree formulation has one equation or inequality for any set S of nodes (the one equation in the model corresponds to the node set $S = N$), for the minimum spanning tree problem we associate a potential μ_S with every set S of nodes. The potential μ_N is unrestricted in sign and the other potentials μ_S must be nonnegative. We then define the reduced cost c_{ij}^{μ} of any arc as

$$c_{ij}^{\mu} = c_{ij} + \sum_{A(S) \text{ contains arc } (i,j)} \mu_S.$$

With this definition of the reduced costs, we have the following complementary slackness optimality conditions.

Minimum spanning tree complementary slackness optimality conditions. *A solution x of the minimum spanning tree problem is an optimal solution to the linear programming relaxation of the integer programming formulation (13.2) if and only if we can find node potentials μ_S defined on node sets S so that the reduced costs satisfy the following conditions:*

$$c^{\mu}_{ij} = 0 \quad \text{if } x_{ij} > 0.$$

$$c^{\mu}_{ij} \geq 0 \quad \text{if } x_{ij} = 0.$$

We can use this fundamental result to give yet another proof that Kruskal's algorithm solves the minimum spanning tree problem.

Theorem 13.9. *If x is the solution generated by Kruskal's algorithm, x solves both the integer program (13.2) and its linear programming relaxation.*

Rather than giving a formal proof of this theorem, let us illustrate the proof technique on the five-node example that we have already considered in Figure 13.6. For this problem, Kruskal's algorithm chooses the arcs of the minimum spanning tree in the order (2, 4), (3, 5), (3, 4), and (1, 2). So we set $x_{24} = x_{35} = x_{34} = x_{12} = 1$ and $x_{13} = x_{23} = x_{45} = 0$. Note that during the course of applying Kruskal's algorithm, we form several connected node components: first {2, 4}, then {3, 5}, then {2, 3, 4, 5}, and finally, the entire node set {1, 2, 3, 4, 5}. We will associate a nonzero potential with these sets and a zero potential with every other set of nodes. We define these potentials in the reverse order that Kruskal's algorithm formed the node components. We first set $\mu_{\{1,2,3,4,5\}} = -35$, the negative of the cost of the final arc added to the tree. Now we note that each arc (i, j) that we add to the tree defines a node component $S(i, j)$. For example, $S(3, 4) = \{2, 3, 4, 5\}$. Moreover, at some later stage in the algorithm, we combine the node component $S(i, j)$ with one or more other nodes to define a large component by adding another arc (p, q) to the tree. We now set the potential of the node component $S(i, j)$ to be the difference between the cost of arc (p, q) and the cost of arc (i, j). Therefore, we set $\mu_{\{2,3,4,5\}} = c_{12} - c_{34} = 35 - 20 = 15$, $\mu_{\{3,5\}} = c_{34} - c_{35} = 20 - 15 = 5$, and $\mu_{\{2,4\}} = c_{34} - c_{24} = 20 - 10 = 10$.

Now checking the reduced cost of every arc, we find that

$$c^{\mu}_{12} = 35 - 35 = 0$$

$$c^{\mu}_{13} = 40 - 35 = 5$$

$$c^{\mu}_{23} = 25 - 35 + 15 = 5$$

$$c^{\mu}_{24} = 10 - 35 + 15 + 10 = 0$$

$$c^{\mu}_{34} = 20 - 35 + 15 = 0$$

$$c^{\mu}_{35} = 15 - 35 + 15 + 5 = 0$$

$$c^{\mu}_{45} = 30 - 35 + 15 = 10.$$

Note that with these choices of the potentials, the reduced cost of every arc chosen by Kruskal's algorithm is zero and the cost of every other arc (i, j) is the difference between the cost of arc (i, j) and the cost of the most expensive arc on the path formed by adding arc (i, j) to the tree found by Kruskal's algorithm. It is fairly easy to use an induction argument to extend this proof technique for any problem and thus to give a formal proof of Theorem 13.9 (see Exercise 13.42).

The proof technique we have just illustrated establishes one of the most important core results in combinatorial optimization. Since a linear program always has an extreme point solution (see Appendix C for this result and for linear programming definitions), if we can show that for every choice of the coefficients of its objective function, a linear programming formulation has at least one integral solution, then the extreme points of the polyhedron defined by that linear program are integer valued. Since we have just established this property for the linear programming relaxation of the integer program (13.2), we have proven the following fundamental result.

Theorem 13.10. *The polyhedron defined by the linear programming relaxation of the packing formulation of the minimum spanning tree problem has integer extreme points.*

This theorem is just one example of an important meta rule that seems to lie at the core of combinatorial optimization; namely, for essentially most optimization problems that can be solved in polynomial time, it is possible to define a linear program with integer extreme points that contains the incident vectors of the solution to the combinatorial optimization problem. The minimum cost flow problem and the minimal spanning tree problem were two of the first notable examples of this result discovered in the combinatorial optimization literature; these results have inspired many streams of investigation within discrete optimization, such as the study of matroids that we introduced in Section 13.7. For example, it is possible to specify a linear programming formulation of the matroid optimization problem so that the extreme points of the linear programming formulation are exactly the set of bases of the underlying matroid (see Exercise 13.44).

13.9 SUMMARY

The minimum spanning tree problem is perhaps the simplest, and certainly one of the most central, models in the field of combinatorial optimization. In this chapter, after describing several applications of minimum spanning trees, we proved two (equivalent) necessary and sufficient conditions—the *cut and path optimality conditions*—for characterizing the optimality of minimum spanning trees. The cut optimality conditions state that a spanning tree T^* is a minimum spanning tree if and only if the cost of the tree arc (i, j) is less than or equal to the cost of every nontree arc in the cut formed by deleting arc (i, j) from T^*. The path optimality conditions are closely related to these conditions (in a sense, they are dual conditions); they state that a spanning tree T^* is a minimum spanning tree if and only if the cost of every nontree arc (k, l) is greater than or equal to the cost of every tree arc in the path in T^* between nodes k and l.

In this chapter we described three algorithms for solving the minimum spanning tree problem: Kruskal's, Prim's, and Sollin's. All these algorithms are easy to implement, have excellent running times, and are very efficient in practice. Figure 13.13 summarizes the basic features of these algorithms.

The minimum spanning tree problem is important not only because it is a core model in network optimization, but also because it serves as a valuable prototype model in combinatorial optimization that has stimulated many lines of inquiry. In this chapter we have considered two ways in which minimum spanning trees relate to general issues in combinatorial optimization. If we consider Kruskal's algorithm as a greedy procedure that chooses the minimum cost feasible arc at each step, we might ask whether a similar type of greedy algorithm is able to solve other combinatorial optimization problems. We have answered this question affirmatively by showing that the greedy algorithm also solves a broad class of problems known as matroid optimization problems.

Studying specialized structures, such as matroids, is one very important stream of inquiry in combinatorial optimization. Another is the use of linear programming as a tool for understanding and solving combinatorial optimization problems. In Section 13.8 we showed how to characterize the incidence vectors of spanning trees as solutions to a linear programming formulation of the problem; we also showed how to interpret Kruskal's algorithm as a method for solving this linear program. This development illustrates the use of linear programming in combinatorial optimization and is indicative of the type of investigations that analysts conduct in the important subspecialty of combinatorial optimization known as polyhedral combinatorics (i.e., the study of integer polyhedra).

Algorithm	Running time	Features
Kruskal's algorithm	$O(m + n \log n)$ plus time needed to sort m arc lengths	1. Examines arcs in nondecreasing order of their lengths and include them in the minimum spanning tree if the added arc does not form a cycle with the arcs already chosen. 2. The proof of the algorithm uses the path optimality conditions. 3. Attractive algorithm if the arcs are already sorted in increasing order of their lengths.
Prim's algorithm	$O(m + n \log n)$	1. Maintains a tree spanning a subset S of nodes and adds a minimum cost arc in the cut $[S, \bar{S}]$. 2. The proof of the algorithm uses the cut optimality conditions. 3. Can be implemented using a variety of heaps structures; the stated time bound is for the Fibonacci heap data structure.
Sollin's algorithm	$O(m \log n)$	1. Maintains a collection of node-disjoint trees: in each iteration, adds the minimum cost arc emanating from each such tree. 2. The proof of the algorithm uses the cut optimality conditions.

Figure 13.13 Summary of minimum spanning tree algorithms.

REFERENCE NOTES

Algorithms for the minimum spanning tree problem, developed as early as 1926, are among the earliest network algorithms. The paper by Graham and Hell [1985] presents an excellent survey of the historical developments of minimum spanning tree algorithms. Borůvka [1926] and Jarníck [1930] independently formulated and solved the minimum spanning tree problem. Later, other researchers rediscovered these algorithms. Kruskal [1956] and Loberman and Weinberger [1957] independently discovered Kruskal's algorithm discussed in Section 13.4. Prim [1957] developed the algorithm described in Section 13.5. Sollin presented his algorithm, discussed in Section 13.6, in a seminar in 1961; it was never published. Claude Berge was present at this seminar and reported this algorithm in his book, Berge and Ghouila-Houri [1962]. Later, researchers discovered that Sollin's algorithm is similar to Borůvka's algorithm and that Prim's algorithm is similar to Jarníck's algorithm.

Our description of Kruskal's algorithm runs in the time required to sort m numbers plus $O(m + n \log n)$. The use of improved *union-find* data structures leads to a faster implementation of Kruskal's algorithm. This implementation, as developed by Tarjan [1984], runs in the time required to sort m numbers plus $O(m\, \alpha(n, m))$; $\alpha(n, m)$ is the Ackermann function which, for all practical purposes, is smaller than 6. In this chapter we reported an $O(m + n \log n)$ implementation of Prim's algorithm; this implementation appears to be new. Gabow, Galil, Spencer, and Tarjan [1986] presented a variant of this algorithm that runs in $O(m \log \beta(m, n))$ time with the function $\beta(m, n)$ defined as $\beta(m, n) = \min\{i : \log^{(i)}(m/n) \le 1\}$. In this expression, $\log^{(i)} x = \log \log \log \cdots \log x$ with the log iterated i times. So $\beta(m, n)$ is a very slowly growing function. For example, if $m/n = 2^{2^{64,000}}$ then $\beta(m, n) = 6$. Yao [1975] developed an improved implementation of Sollin's algorithm running in $O(m \log \log n)$ time. Currently, the fastest algorithm for solving the minimum spanning tree algorithm is Tarjan's [1984] implementation of Kruskal's algorithm if the arcs are already sorted, and Gabow et al. [1986] variant of Prim's algorithm, otherwise. Gabow et al. [1986] also give efficient algorithms that (1) solve the minimum spanning tree problem in a directed network (i.e., the arborescence problem), and (2) solve the minimum spanning tree problem with a single degree constraint.

Chin and Houch [1978], Gavish and Srikanth [1979], and Tarjan [1982] have developed techniques for reoptimizing the minimum spanning tree problem when we change arc costs. Haymond, Jarvis, and Shier [1980] have described data structures for implementing Kruskal's, Prim's, and Sollin's algorithm and have presented computational results for these algorithms. Jarvis and Whited [1983] described the results of another computational study. These studies indicate that Prim's and Sollin's algorithms are consistently superior to Kruskal's algorithm. They show that Sollin's algorithm is better than Prim's algorithm for sparse networks, and is worse for dense networks. These studies find that the best implementation of Prim's algorithm uses a variant of Dial's implementation of Dijkstra's algorithm that we described in Section 4.6.

Our presentation of matroids in Section 13.7 and of a linear programming formulation of the minimum spanning tree in Section 13.8 merely touches upon two very important topics in combinatorial optimization. Although the concept of matroids is quite old, dating from their introduction by Whitney [1935], their use in combinatorial optimization is much more recent, stemming from the seminal contributions of Edmonds [1965c, 1971]. The book by Lawler [1976] and the survey paper by Bixby [1982] highlight connections between matroids and network optimization. The books by Tutte [1971] and Welsh [1976] provide excellent mathematical accounts of this field, and the book by Recski [1988] presents many illuminating applications in engineering and the physical sciences.

The description of the polyhedral structure of combinatorial optimization problems via linear programs has become a very fertile field in combinatorial optimization that has shed theoretical light on many problems and led to effective algorithms for solving many important applications. The comprehensive text by Nemhauser and Wolsey [1988] gives an instructive account of this field, known as *polyhedral combinatorics*. The linear programming description of the minimum spanning tree problem, and the interpretation of Kruskal's algorithm as a method for solving the linear programming formulation of the problem, has served as an important stimulus for developments of this field. As but one example, this approach has proven very fruitful in developing algorithms for solving the nonbipartite matching problems that we considered in Chapter 12 from a purely combinatorial approach. For a polynomial formulation of the minimal spanning tree problem using multicommodity flow variables, see the survey by Magnanti, Wolsey, and Wong [1992].

The applications of the minimum spanning tree problem that we presented in Section 13.2 are adapted from the following papers:

1. Designing physical systems (Borůvka [1926], Prim [1957], Loberman and Weinberger [1957], and Dijkstra [1959])
2. Optimal message passing (Prim [1957])
3. All pairs minimax path problem (Hu [1961])
4. Reducing data storage (Kang, Lee, Chang, and Chang [1977])
5. Cluster analysis (Gower and Ross [1969], and Zahn [1971])

In Application 1.7 we described another application of the spanning tree problem that arises in measuring the homogeneity of bimetallic objects (Shier [1982] and Filliben, Kafadar, and Shier [1983]). Additional applications of the minimum spanning tree problem arise in (1) solving a special case of the traveling salesman problem (Gilmore and Gomory [1964]), (2) chemical physics (Stillinger [1967]), (3) Lagrangian relaxation techniques (Held and Karp [1970]), (4) network reliability analysis (Van Slyke and Frank [1972]), (5) pattern classification (Dude and Hart [1973]), (6) picture processing (Osteen and Lin [1974]), and (7) network design (Magnanti and Wong [1984]). The survey paper of Graham and Hell [1985] provides references for additional applications of the minimum spanning tree problem.

EXERCISES

13.1. Suppose that you want to determine a spanning tree T that minimizes the objective function $[\sum_{(i,j)\in T} (c_{ij})^2]^{1/2}$. How would you solve this problem?

13.2. In the network shown in Figure 13.14, the bold lines represent a minimum spanning tree.

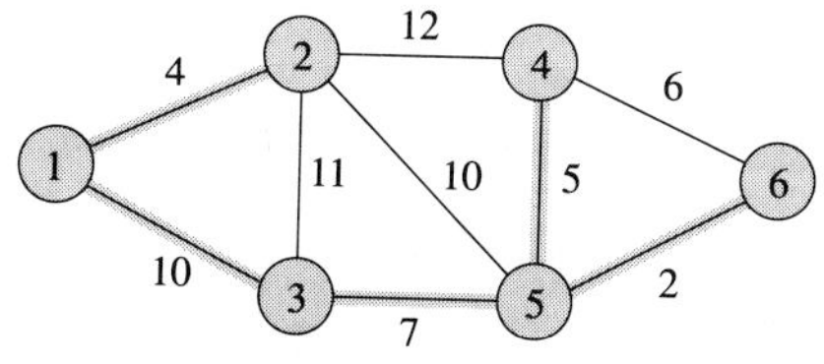

Figure 13.14 Verifying cut and path optimality conditions.

(a) By listing each nontree arc (k, l) and the minimum length arc on the tree path from node k to node l, verify that this tree satisfies the path optimality conditions.

(b) By listing each tree arc (i, j) and the minimum length arc in the cut defined by the arc (i, j), verify that the tree satisfies the cut optimality conditions.

13.3. Using Kruskal's algorithm, find minimum spanning trees of the graphs shown in Figure 13.15.

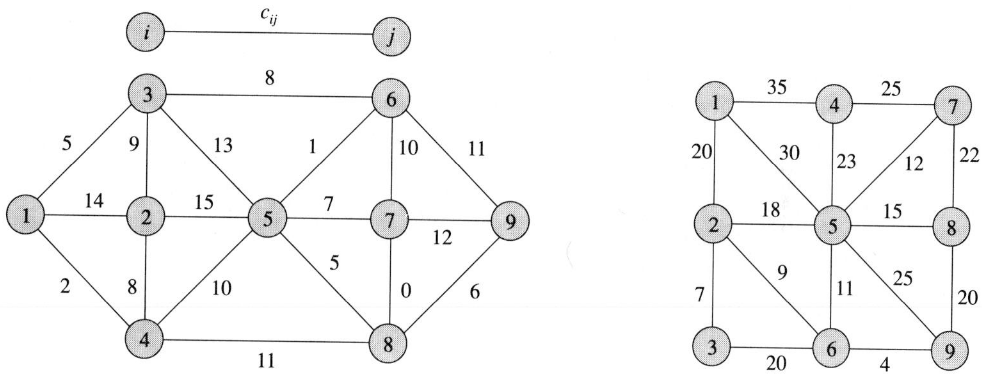

Figure 13.15 Examples for Exercises 13.3 to 13.5.

13.4. Using Prim's algorithm, find minimum spanning trees of the graphs shown in Figure 13.15.

13.5. Using Sollin's algorithm, find minimum spanning trees of the graphs shown in Figure 13.15.

13.6. Think of the network shown in Figure 13.16 as a highway map, and the number recorded next to each arc as the maximum elevation encountered in traversing the arc. A traveler plans to drive from node 1 to node 12 on this highway. This traveler dislikes high altitudes and so would like to find a path connecting node 1 to node 12

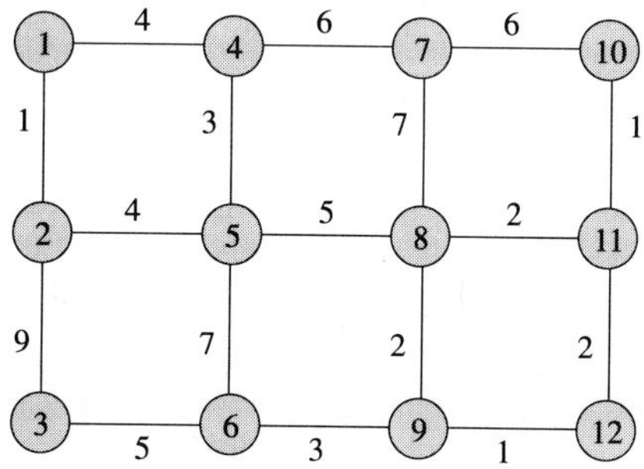

Figure 13.16 Highway grid.

that minimizes the maximum altitude. Find the best path for this traveler using a minimum spanning tree algorithm.

13.7. Can you generalize the approach outlined in Application 13.3 to solve the all-pairs maximum capacity path problem in directed networks? If yes, describe your algorithm; if not, why not?

13.8. In Theorem 13.3 we proved the sufficiency of the path optimality conditions using the cut optimality conditions. Give a direct proof of this sufficiency condition that does not use the cut optimality conditions.

13.9. Prove the maximum spanning tree optimality conditions stated in Theorem 13.4.

13.10. Let (p, q) be a minimum cost arc in G. Show that (p, q) belongs to some minimum spanning tree of G. Does every minimum spanning tree of G contain the arc (p, q)?

13.11. Show that a maximum weight acyclic subgraph in an undirected graph G with strictly positive arc weights c_{ij} must be a spanning tree.

13.12. How would you modify Kruskal's and Prim's algorithms to solve the maximum spanning tree problem?

13.13. In an undirected network, we define a tree of shortest paths as a spanning tree in which the unique path from a specified node s to every other node is a shortest path. Is a minimum spanning tree of G also a tree of shortest paths? Either prove this result or construct an example to show that the trees could be different.

13.14. **Tree minimax result.** Let $G = (N, A)$ be an undirected network with a capacity u_{ij} associated with every arc $(i, j) \in A$. For any spanning tree T of G, we define its *capacity* as $\min\{u_{ij}:(i, j) \in T\}$, and for any cut Q of G, we define its *value* as $\max\{u_{ij}:(i, j) \in Q\}$. Show that the capacity of any spanning tree is a lower bound on the value of every cut. Next show that the maximum capacity of any spanning tree equals the minimum value of any cut.

13.15. We say that two spanning trees T' and T'' are *adjacent* if they have all but one arc in common. Show that for any two spanning trees T' and T'', we can find a sequence of spanning trees $T^1, T^2, \ldots, T^k$ with $T^1 = T'$, $T^k = T''$ and with T^i adjacent to T^{i+1} for every $i = 1$ to $k - 1$.

13.16. Suppose that you are given a graph with each arc colored either red or blue.
 (a) Show how to find a spanning tree with the maximum number of red arcs.
 (b) Suppose that some spanning tree has k' red arcs and another spanning tree has $k'' > k'$ red arcs. Show that for every k, $k' \leq k \leq k''$, some spanning tree has k red arcs.

13.17. Let T be a spanning tree. For any pair $[i, j]$ of nodes, let $\beta[i, j]$ denote the least cost arc among the arcs in the tree path joining node i and node j. Show how to compute $\beta[i, j]$ for every pair of nodes in a total of $O(n^2)$ time.

13.18. In a class of undirected networks, suppose that all arc costs are *small* (i.e., they lie in the interval $[1, k]$ for some small integer k, say $k = 10$). How fast could you implement Kruskal's and Prim's algorithms for solving the minimum spanning tree problem in this class of networks?

13.19. Consider the following *reverse greedy algorithm*:

```
begin
  let the arcs (i1, j1), (i2, j2), . . . , (im, jm) be arranged in
    nonincreasing order of their lengths;
  G' := G;
  for k := 1 to m do
    if G' - {(ik, jk)} is a connected graph then
      G' := G' - {(ik, jk)};
end;
```

Show that at the termination of this algorithm, the graph G' is a minimum spanning tree.

13.20. Consider the following algorithm. Arrange the arcs in A in any arbitrary order and start with a null tree T. Examine each arc (i, j) in A, one by one, and perform the following steps: add arc (i, j) to T and if T contains a cycle W, delete from T an arc of maximum cost from the cycle W. Show that when this algorithm has examined all the arcs, the final tree T is a minimum spanning tree. Is it possible to implement this algorithm as efficiently as Kruskal's algorithm? Why or why not?

13.21. Can you use the data structure of Dial's implementation of Dijkstra's shortest path algorithm (discussed in Section 4.6) to implement Prim's algorithm? If so, is the running time of Prim's algorithm better than the running time of the shortest path algorithm?

13.22. In Section 13.5 we observed a striking resemblance between Prim's algorithm and Dijkstra's algorithm. This observation might lead us to conjecture that we can use a radix heap data structure (discussed in Section 4.8) to implement Prim's algorithm in $O(m + n \log(nC))$ time. However, this conjecture is not valid. What are the difficulties we would encounter if we attempted to implement Prim's algorithm using radix heaps?

13.23. The first implementation of Kruskal's algorithm that we discussed in Section 13.4 selects a nontree arc (k, l) violating its optimality condition and exchanges this arc with some tree arc of lower cost. Show that no matter which order we use to select the nontree arcs violating their optimality conditions, we perform at most nm iterations. (*Hint*: Let $f(i, j)$ be the number of arcs in the network whose cost is strictly greater than c_{ij}. Consider the effect on the potential function $= \sum_{(i,j)\in T} f(i, j)$ as we change the spanning tree T.)

13.24. Let T be a minimum spanning tree of an undirected graph $G = (N, A)$ and let Q be a set of nontree arcs (k, l) satisfying the following property: Some arc (i, j) in the tree path from node k to node l has the same cost as arc (k, l); that is, $c_{ij} = c_{kl}$. Professor May B. Wright claims that every spanning tree in the subgraph $G' = (N, T \cup Q)$ is a minimum spanning tree of G. Construct a counterexample to show that Professor Wright's claim is false.

13.25. Sensitivity analysis. Let T^* be a minimum spanning tree of a graph $G = (N, A)$. For any arc $(i, j) \in A$, we define its *cost interval* as the set of values of c_{ij} for which T^* continues to be a minimum spanning tree.

(a) Describe an efficient method for determining the cost interval of a given arc (i, j). (*Hint*: Consider two cases: When $(i, j) \in T^*$ and when $(i, j) \notin T^*$, and use the cut and path optimality conditions.)

(b) Describe a method for determining the cost intervals of every arc in A. Your method must be faster than determining the cost intervals of each arc one by one. (*Hint*: Use the result of Exercise 13.17.)

13.26. Suppose that we have in hand a minimum spanning tree T for the undirected graph $G = (N, A)$. Suppose that we add a new node $(n + 1)$ to N and p new arcs to A incident to this node. How fast can you find a minimum spanning tree for the enlarged network G' from the minimum spanning tree T of G? (*Hint*: Use the cut optimality conditions.)

13.27. Arc additions and deletions. Let T^* be a minimum spanning tree for the undirected graph $G = (N, A)$. Describe an algorithm for reoptimizing the minimum spanning tree when we delete an arc $(i, j) \in A$ from the network. Similarly, describe an algorithm for reoptimizing the problem when we add a new arc (i, j) to A. Prove that your algorithms correctly find new minimum spanning trees and state their running times.

13.28. Spanning trees containing specific arcs. In an undirected graph $G = (N, A)$, let (p, q) be a specified arc. Describe a method for identifying a minimum spanning tree T^* subject to the condition that the tree must contain the arc (p, q). Prove that your method correctly solves this problem. Generalize the method for situations in which the minimum spanning tree must contain an acyclic set A' of arcs. (*Hint*: Assign appropriate costs to the arcs required to be in the optimal tree.)

13.29. Factored minimum spanning tree problem. Let G be a complete undirected graph. Suppose that we associate a positive real number α_i with each node $i \in N$ and define the cost of each arc (i, j) as $c_{ij} = \alpha_i\alpha_j$. This specialized minimum spanning tree problem is known as the *factored minimum spanning tree problem*. We wish to develop an algorithm for solving this class of problems that is more efficient than the general minimum spanning tree algorithms.

(a) Consider a five-node network with $\alpha_i = i$. Find a minimum spanning tree in this network.

(b) Use the insight obtained from answering part(a) to develop an $O(n)$ algorithm for solving the factored minimum spanning tree problem.

13.30. **Most vital arcs.** In the minimum spanning tree problem, we refer to an arc as a *vital arc* if its deletion strictly increases the cost of the minimum spanning tree. A *most vital arc* is a vital arc whose deletion increases the cost of the minimum spanning tree by the maximum amount.

(a) Does a network always contain a vital arc?

(b) Suppose that a network contains a vital arc. Describe an $O(nm)$ algorithm for identifying a most vital arc. Can you develop an algorithm that runs faster than $O(nm)$ time? (*Hint*: Use the cut optimality conditions.)

13.31. Suppose that we arrange all the spanning trees of a graph G in nondecreasing order of their costs. We refer to a spanning tree T as *a* kth *minimum spanning tree* if it is at the kth position in this order. Describe an $O(n^2)$ algorithm for finding the second minimum spanning tree. (*Hint*: Observe that the second minimum spanning tree must contain at least one arc that is not in the first minimum spanning tree. Then use the result of Exercise 13.17.)

13.32. **Bottleneck spanning trees.** A spanning tree T is a *bottleneck spanning tree* if the maximum arc cost in T is as small as possible from among all spanning trees. Show that a minimum spanning tree of G is also a bottleneck spanning tree of G. Is the converse result also true (i.e., is a bottleneck spanning tree of G also a minimum spanning tree of G)? Either prove this result or construct a counterexample.

13.33. Describe an $O(m \log n)$ algorithm, using binary search, for solving the bottleneck spanning tree problem defined in Exercise 13.32.

13.34. **Balanced spanning trees.** A spanning tree T is a *balanced spanning tree* if from among all spanning trees, the difference between the maximum arc cost in T and the minimum arc cost in T is as small as possible. Describe an $O(m^2)$ algorithm for determining a balanced spanning tree.

13.35. **Parametric analysis of minimum spanning trees.** In the parametric minimum spanning tree problem, each arc length $c_{ij} = c_{ij}^0 + \lambda c_{ij}^*$ is a linear function of a parameter λ. Let T^λ denote a minimum spanning tree with arc lengths chosen as $c_{ij}^0 + \lambda c_{ij}^*$ for a specific value of λ.

(a) Show that for sufficiently large values of the constant $k > 0$, T^{-k} and T^k are the maximum and minimum spanning trees when the arc lengths are c_{ij}^*.

(b) Show that T^λ is a minimum spanning tree for all of the values of λ in some interval $[\underline{\lambda}, \overline{\lambda}]$. Moreover, show that at the lower and upper limits of this interval, at least two alternate minimum spanning trees are adjacent in the sense of Exercise 13.15. (*Hint*: Use the path optimality conditions.)

(c) Describe an algorithm for determining a minimum spanning tree for all values of λ from $-\infty$ to $+\infty$.

13.36. **(a)** Show that in the parametric minimum spanning tree problem, as we vary λ from $-\infty$ to $+\infty$, we obtain at most m^2 minimum spanning trees and every two consecutive minimum spanning trees are adjacent. (*Hint*: Use the fact that if T' and T'' are two consecutive minimum spanning trees, we can obtain T'' from T' by replacing a tree arc (i, j) by a nontree arc (k, l) satisfying the condition $c_{kl}^* \le c_{ij}^*$.)

(b) Consider a special case of the parametric minimum spanning tree problem in which each $c_{ij}^* = 0$ or 1. Show that in this case, as we vary λ from $-\infty$ to $+\infty$, we obtain at most n minimum spanning trees.

(c) Consider another special case of the parametric minimum spanning tree problem in which each $c^*_{pq} = 1$ for a specific arc (p, q) and is zero for all other arcs. Show how to find minimum spanning trees for all values of λ in time proportional to solving a single minimum spanning tree problem.

(d) Consider yet another special case of the minimum spanning tree problem where all parametric arcs are incident to a common node p (i.e., $c^*_{ij} = 1$ whenever $i = p$ or $j = q$, and is zero for all other arcs). How fast can you find minimum cost spanning trees for all values of λ?

13.37. Minimum ratio spanning trees (Chandrasekaran [1977]). In the minimum ratio spanning tree problem, we associate two numbers, c_{ij} and τ_{ij}, with each arc (i, j) in a network G and wish to determine a spanning tree T^* that minimizes $(\sum_{(i,j)\in T} c_{ij})/(\sum(i,j)\in T^* \tau_{ij})$ from among all spanning trees. We assume that $\sum_{(i,j)\in T^*} \tau_{ij} > 0$ for all spanning trees T. Suggest a binary search algorithm for identifying a minimum ratio spanning tree of G that runs in polynomial time.

13.38. A *1-tree* of G is a spanning tree of G plus one arc. Show that the minimum spanning tree of G plus the least cost nontree arc defines a minimum cost 1-tree of G. Suppose that the additional arc must be adjacent to a particular node s of G. How would you find a minimum cost 1-tree for this version of the problem?

13.39. Optimal 1-forest. A set of arcs is a *1-forest* of an undirected graph G if some arc (k, l) in F satisfies the condition that $F - \{(k, l)\}$ is a forest.

(a) Show that the collection of all 1-forests forms a matroid.

(b) Give a greedy algorithm for identifying a maximum weight 1-forest of G.

(c) How would you modify your answers to parts (a) and (b) if we required that the arc (k, l) be incident to a specific node s of the network?

13.40. Optimal k-forest. A set F of arcs is a k-forest of an undirected graph G if some subset $F' \subseteq F$ containing k arcs satisfies the condition that $F - F'$ is a forest. Show that the collection of all k-forests forms a matroid and give a greedy algorithm for identifying a maximum weight k-forest of G. (*Hint*: Generalize the result in Exercise 13.39.)

13.41. Let F_p and F_{p+1} be forests in a graph containing p and $p + 1$ arcs. Show that we can always add some arc in F_{p+1} to F_p to produce a forest with $p + 1$ arcs.

13.42. Using the example we have considered in the text as motivation, give a formal proof of Theorem 13.9.

13.43. Linear programming proof of the greedy algorithm. Let $(E, \mathcal{I})$ be a matroid with an associated weight w_e for $e \in E$. Let x_e be a zero–one vector indicating whether or not the element e is a member of a set I from E; that is, $x_e = 1$ if $e \in I$ and $x_e = 0$ if $e \notin I$. For any subset S in E, let $r(S)$ denote its rank, defined as the number of elements of the largest independent set in S. For example, the rank of a set S of arcs in a graph is the size of the largest forest defined by these arcs.

(a) Show that the incidence vectors x_e of a basis of the matroid satisfy the following conditions:

$$\sum_{e\in E} x_e = r(E), \tag{13.3a}$$

$$\sum_{e\in S} x_e \le r(S) \qquad \text{for all } S \subseteq \mathcal{I}, \tag{13.3b}$$

$$x_e \ge 0. \tag{13.3c}$$

(b) Show that for the minimum spanning tree problem, the constraints in (13.3) contain all of the constraints in the formulation (13.2).

(c) Mimicking the proof of Theorem 13.9 (see Exercise 13.42), give a linear programming proof that the greedy algorithm solves the matroid optimization problem of finding a basis of the matroid $(E, \mathcal{I})$ with the smallest possible weight $w(B)$.

13.44. Linear programming formulation of matroids. In Theorem 13.10 we showed that spanning trees of a graph correspond to the extreme points of the linear program (13.2).

Using the result of Exercise 13.43, show that the bases of a matroid correspond to the extreme points of the polyhedron defined by the constraints given in (13.3). (*Hint*: Use the result of Exercise 13.43 and the fact that each extreme point of a linear program is the unique optimal solution for some choice of the objective coefficients.)

13.45. In Exercise 13.43 we showed that the greedy algorithm solves the matroid optimization problem. Show that this property actually characterizes matroids. That is, show that the greedy algorithm will solve the minimum weight independent set problem for any choice of the element weights if and only if the subset system is a matroid. (*Hint*: Any subset system that is not a matroid contains two independent subsets I and I' satisfying the property that $|I'| > |I|$ and no element in I' can be added to I to obtain an independent set. Define the weight function on E appropriately so that the greedy algorithm terminates with I, but I' is optimal.)

13.46. **(a)** The set of minimum spanning trees $T^1, T^2, \ldots, T^{k-1}$ that we determined in Exercise 13.36(d) as we varied the parameter from $C + 1$ to $-\infty$ satisfy the "monotonicity" property that once an arc $(1, j)$ belongs to any tree T^p, it also belongs to all of the trees T^q for $q \geq p$. Suppose that the parametric cost of arc $(1, j)$ is $c_{1j} + \lambda d_{1j}$ for some constant d_{1j} and that the cost of arc (i, j) is c_{ij} for $i \neq 1$ and $j \neq 1$. Does the set of optimal spanning trees, as we vary λ from $C + 1$ to $-\infty$, satisfy the monotonicity property?

(b) If possible, describe a polynomial time variant of the procedure discussed in Exercise 13.36(d) that will solve the parametric problem defined in part (a). If you cannot describe any such algorithm, explain the difficulties encountered.

14

CONVEX COST FLOWS

I bend but do not break.
—Jean de La Fontaine

Chapter Outline

14.1 INTRODUCTION

Essentially all fields of scientific inquiry evolve into specialized branches of investigations, each with its own particular traditions and approaches. The field of optimization is no exception; it divides quite naturally into several ways: constrained versus unconstrained optimization; linear versus nonlinear programming; and discrete versus continuous optimization. By its very nature, network optimization is a special class of constrained optimization problems. We can, however, distinguish the various domains of network optimization along the other dimensions. Our development has focused exclusively on linear models. Moreover, because the integrality property ensures that linear minimum cost flow problems, even when stated as continuous optimization models, always have integer solutions (assuming that the data are integral), with the exception of the matching and spanning tree problems that we have considered in the preceding two chapters, we have not had to make any distinction between discrete and continuous models. Yet many of the arguments and approaches in network optimization have a distinct combinatorial flavor. Indeed, the optimization community typically views network flows as the starting point for building much of the theory and algorithmic approaches of discrete optimization.

Suppose that we wish to extend our discussion into the realm of nonlinear optimization. What type of models should we consider? Perhaps the most natural approach would be to replace the linear objective function of the minimum cost flow problem by a general nonlinear function. Although we might like to consider the most general nonlinear functions possible, doing so would take us far afield from the mainstream of our investigations. Instead, we might ask the following question: Is there a class of nonlinear optimization models that arise frequently in practice and that we can solve by adapting the algorithmic approaches that we have already developed? In this chapter we examine one such set of models, those with separable convex objective functions. Fortunately, these models provide us with the most useful set of nonlinear objective functions that arise in the practice of network optimization. Moreover, this particular set of models permits us to remain within the domain of discrete optimization, since as we will see in this chapter, if we further restrict these models by requiring the solutions to be integer, we can solve these problems quite efficiently, in theory as well as in practice.

In all the models we have considered to this point, the objective function was separable in the sense that the different flow variables x_{ij} appeared in separate terms $c_{ij}x_{ij}$. In the models we consider in this chapter, we retain the separability assumption, but we now permit the separable terms to be nonlinear functions of the form $C_{ij}(x_{ij})$. We also impose an additional convexity assumption: Each function $C_{ij}(x_{ij})$ is convex: the functions are "bathtub" shaped in the sense that linear interpolations always lie on or above the functions (mathematically, if θ is a parameter satisfying $0 \leq \theta \leq 1$ and x'_{ij} and x''_{ij} are any two points within the flow bounds of x_{ij}, then $C_{ij}(\theta x'_{ij} + (1 - \theta)x''_{ij}) \leq \theta\, C_{ij}(x'_{ij}) + (1 - \theta)C_{ij}(x''_{ij})$). Figure 14.1 gives two examples of convex functions.

We consider two different models:

1. *Piecewise linear model* [see Figure 14.1(a)]. Each arc cost $C_{ij}(x_{ij})$ has at most p linear segments: $0 = d^0_{ij} < d^1_{ij} < d^2_{ij} < \cdots$ denote the breakpoints of the function and the cost varies linearly in the interval d^{k-1}_{ij} to d^k_{ij}. We let c^k_{ij} denote the linear cost coefficient in the interval $[d^{k-1}_{ij}, d^k_{ij}]$. Therefore, to specify a piecewise linear cost function, we need to specify the breakpoints and the slopes of the linear segments between successive breakpoints.
2. *Concise function model* [see Figure 14.1(b)]. The functions $C_{ij}(x_{ij})$ are specified in a functional form, such as x^4_{ij}. In this case we often require only $O(1)$ information to specify the function. For this model, we assume that we restrict the feasible solutions to integers. Although we could easily adapt the algorithm

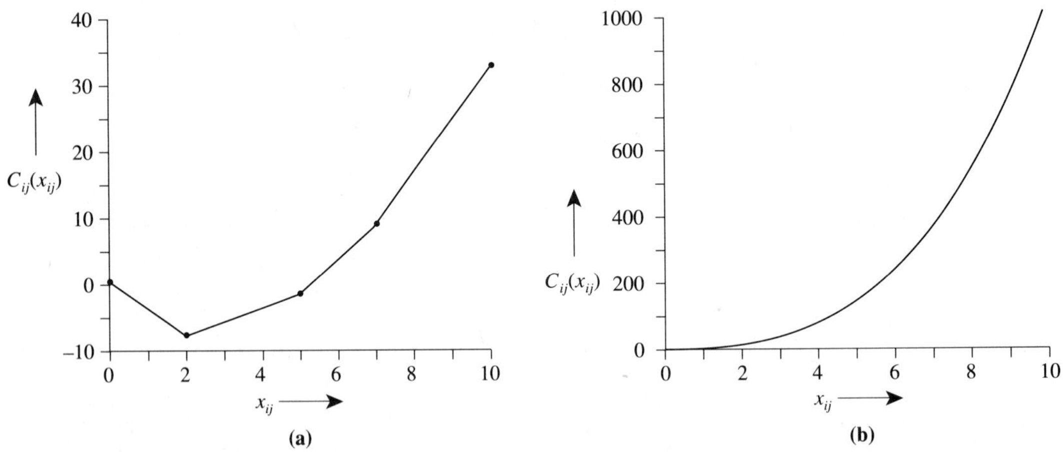

Figure 14.1 Two examples of convex cost flow functions.

that we examine for this model to solve continuous problems, our computational complexity results apply only to the integer model.

The integer restriction on the optimal solution does impose some loss of generality because the integer optimal solution might not be as good as a continuous optimal solution. But we can always obtain an integer optimal solution as close as desired to a continuous optimal solution by scaling the data. For example, if we want a solution more accurate than the integer optimal solution, we could substitute My_{ij}, for sufficiently large value of M, for each x_{ij}. We would chose M depending on the accuracy we desired (e.g., $M = 1{,}000$ or $10{,}000$). If y^*_{ij} denotes an integer optimal solution of the transformed problem, $x^*_{ij} = y^*_{ij}/M$ is an optimal solution of the original problem (to a degree of accuracy of $1/M$). This technique allows us to obtain a real-valued optimal solution of the convex cost flow problem to any desired degree of accuracy.

Note that in view of the integrality assumption, we can assume that each convex cost function is a piecewise linear function since we can allow each integer point to be a breakpoint of the function and linearize the function between these breakpoints (see Figure 14.2). However, we differentiate between the two models we have introduced for the following reason:

1. When we specify the function by specifying the breakpoints and the slopes of the function between successive breakpoints, the length of the input data is proportional to the number of linear segments in all the cost functions.
2. When we specify the function concisely, we assume the length of the input data for arc (i, j) is $O(1)$, but when we linearize it, it will have U segments, where U is the largest arc capacity. In this case the length of the input data is not proportional to the total number of segments.

It is necessary to distinguish between these two cases because an algorithm that solves breakpoint problems in polynomial time might not solve the concise-function model in polynomial time.

Before describing algorithms for solving these problems, we discuss several applications of the convex cost model. First, however, let us formally define the

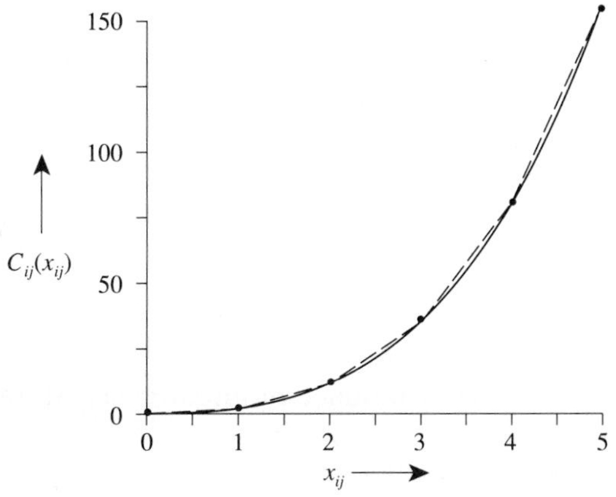

Figure 14.2 Transforming a function in concise form to a piecewise linear form. The dashed line shows the piecewise linear approximation.

convex cost model that we will be considering and introduce several assumptions that we will be imposing. We formulate the model as the following optimization problem:

$$\text{Minimize} \quad \sum_{(i,j)\in A} C_{ij}(x_{ij}) \tag{14.1a}$$

subject to

$$\sum_{\{j:(i,j)\in A\}} x_{ij} - \sum_{\{j:(j,i)\in A\}} x_{ji} = b(i) \qquad \text{for all } i \in N, \tag{14.1b}$$

$$0 \le x_{ij} \le u_{ij} \qquad \text{for all } (i, j) \in A, \tag{14.1c}$$

$$x_{ij} \text{ is integer for all } (i, j) \in A. \tag{14.1d}$$

We define this model on a directed network $G = (N, A)$ with a capacity u_{ij} and a convex cost function $C_{ij}(x_{ij})$ associated with every arc $(i, j) \in A$. As always, we associate a number $b(i)$ with each node $i \in N$ specifying the node's supply or demand, depending on whether $b(i) > 0$ or $b(i) < 0$. Let U denote the largest number among the supplies/demands of the nodes and the finite arc capacities.

We impose several assumptions that we discussed in some detail in Section 9.1 for the minimum cost flow problem: (1) the network is directed; (2) all the supply/demand values $b(i)$ are integers; $\sum_{i\in N} b(i) = 0$; and the convex cost flow problem has a feasible solution; (3) the lower bounds on all the arc flows are zero; and (4) the network contains a directed uncapacitated path between every pair of nodes. Using arguments similar to those that we have used for the minimum cost flow problem, we can show that we incur no loss of generality by imposing these assumptions. We also assume that for each arc (i, j), $C_{ij}(x_{ij}) = 0$ when $x_{ij} = 0$. This assumption imposes no loss of generality because we can always satisfy it by defining $C'_{ij}(x_{ij}) = C_{ij}(x_{ij}) - C_{ij}(0)$.

14.2 APPLICATIONS

Many of the linear network flow models that we have examined in our previous discussions have rather natural nonlinear cost generalizations. System congestion and queuing effects are one source of these nonlinearities (since queuing delays vary nonlinearly with flows). In finance, we often are interested in not only the returns on various investments, but also in their risks, which analysts often measure by quadratic functions. In some other applications, cost functions assume different forms over different operating ranges, so the resulting cost function is piecewise linear. For example, in production applications, the cost of satisfying customers' demand is different if we meet the demand from current inventory or by backordering items.

To give a flavor of the applications of convex cost network flow models, in this section we describe four applications. The first one is a direct application of physical systems' nonlinear cost imposed upon the flows. The second application is one in which different operating ranges produce different costs. In the last two applications, even though the underlying problem is not a flow problem, we can model it as a convex cost network flow model.

Application 14.1 Urban Traffic Flows

In road networks, as more vehicles use any road segment, the road becomes increasingly congested and so the delay on that road increases. For example, the delay on a particular road segment, as a function of the flow x on that road, might be $\alpha x/(u - x)$. In this expression u denotes a theoretical capacity of the road and α is another constant: As the flow increases, so does the delay; moreover, as the flow x approaches the theoretical capacity of that road segment, the delay on the link becomes arbitrarily large. In many instances, as in this example, the delay function on each road segment is a convex function of the road segment's flow, so finding the flow plan that achieves the minimum overall delay, summed over all road segments, is a convex cost network flow model.

Another model of urban traffic flow rests on the behavioral assumption that users of the system will travel, with respect to any prevailing system flow, from their origin to their destination by using a path with the minimum delay. So if $C_{ij}(x_{ij})$ denotes the delay on arc (i, j) as a function of the arc's flow x_{ij}, each user of the system will travel along a shortest delay path with respect to the total delay cost $C_{ij}(x_{ij})$ on the arcs of that path. Note that this problem is a complex equilibrium model because the delay that one user incurs depends on the flow of other users, and all of the users are simultaneously choosing their shortest paths. In this problem setting, we can find the equilibrium flow by solving a convex network flow model with the objective function

$$\sum_{(i,j)\in A} \int_0^{x_{ij}} C_{ij}(y)\, dy.$$

If the delay function is nondecreasing, the function of each variable x_{ij} within the summation is convex, and since the sum of convex functions is convex (see Exercise 14.1), the overall objective function is convex. Moreover, if we solve the network optimization problem defined by this objective function and the network flow constraints, the optimality conditions are exactly the shortest path conditions for the users. (See the reference notes for the details of these claims.)

This example is a special case of a more general result, known as a *variational principle*, that arises in many settings in the physical and social sciences. The variational principle says that to find an equilibrium of a system, we can solve an associated optimization problem: The optimality conditions for the problem are then equivalent with the equilibrium conditions.

Application 14.2 Area Transfers in Communication Networks

In communication networks, telephones do not have sufficient "intelligence" to route calls between each other (historically, equipping every telephone with the ability to route its own calls has been prohibitively expensive). Instead, the system connects each of the telephones within a collection of customers directly to a sophisticated telecommunication device known as a *switching center*; this center does all of the routing for the telephones that "home into" it. That is, the switching center receives and sends all of the calls (1) between its assigned customers, and (2) between these customers and every other customer in the system (who home into some other switching center). Because the switching centers have limited capacity, as communication traffic in the system increases, a telephone company must either add capacity at one or more of its centers or make "area transfers," that is, rehome traffic from one switching center to another.

Consider a communication network with regions divided into many districts that are served by several switching centers. Let $d(j)$ denote the current demand of district j, as measured by number of lines, and let $b(i)$ denote the capacity of switching center i, that is, the maximum number of lines the switching center can handle. To meet the current demands of the districts, the company currently uses w_{ij} working lines between switching center i and district j. To satisfy future demands for lines at district j, the company can use some of the s_{ij} spare lines connecting switching center i to district j at a cost of λ_{ij} per line. It can also add additional lines beyond the available spares at a larger per line cost of $\delta_{ij} > \lambda_{ij}$. To avoid exceeding the capacity at switching center i the company can make "area transfers" by disconnecting a line connecting switching center i to district j, at a cost of μ_{ij} per line, and reconnecting the district to another switching center. The company faces the following problem. Given the new demands $d(j) + \Delta(j)$ for lines at each district j, how should it assign the customers in the districts to the switching centers (with possible area transfers) at the least possible total cost?

Figure 14.3(a) shows a network formulation for this problem. The flow x_{ij} on the arc from the switching center i to district j represents the capacity of the switching center i allocated to district j. Figure 14.3(b) specifies the cost of the flow on arc (i, j). Notice that if switching center i supplies w_{ij} lines to district j, it incurs no additional cost; supplying any less of an allocation than w_{ij} incurs the costs of area transfers, and supplying any more allocation than w_{ij} incurs the cost of adding lines. Because of the area transfer cost and the added incremental costs of using spare lines, the cost structure is nonlinear, so the problem is a convex cost network flow model.

Application 14.3 Matrix Balancing

Statisticians, social scientists, agricultural specialists, and many other practitioners frequently use experimental data to try to draw inferences about the effects of certain control parameters at their disposal (e.g., the effects of using different types of fertilizers). These practitioners often use contingency tables to classify items according to several criteria, with each criterion partitioned into a finite number of categories. As an example, consider classifying r individuals in a population according to the criteria of marital status and age, assuming that we have divided these two criteria into p and q categories, respectively. As categories for marital status, we might choose single, married, separated, or divorced; as categories for age, we might choose below 20, 21–25, 26–30, and so on. This categorization gives a table of pq cells. Dividing the entry in each cell by the number of items r (i.e., the total population) gives the (empirical) probability of that cell.

In many applications it is important to estimate the current cell probabilities, which change continually with time. We can estimate these cell probabilities very accurately using a census, or approximately by using a statistical sampling of the

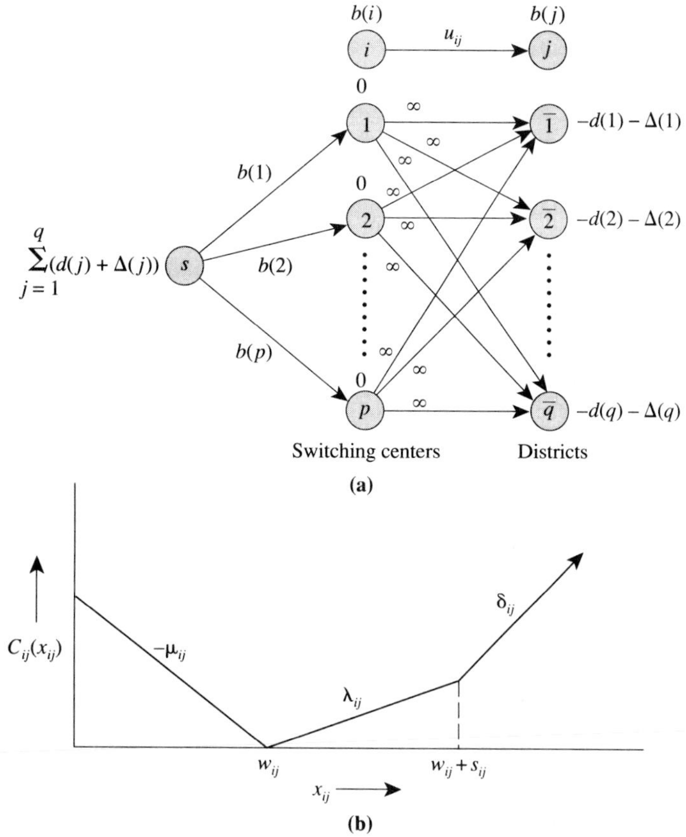

Figure 14.3 (a) Network for the area transfers in communication networks; (b) cost structure of arcs from switching centers to districts.

population; however, even this sampling procedure is expensive. Typically, we would calculate the cell probabilities by sampling only occasionally (for some applications, only once in several years), and at other times revise the most recent cell probabilities based on partial observations. Suppose that we let α_{ij} denote the most recent cell probabilities. Suppose, further, that we know some aggregate data in each category with high precision; in particular, suppose that we know the row sums and column sums. Let u_i denote the number of individuals in the ith marital category and let v_j denote the number of individuals in the jth age category. Let $r = \sum_{i=1}^{p} u_i$. We want to obtain an estimate of the current cell probabilities x_{ij}'s so that the cumulative sum of the cell probabilities for the ith row equals u_i/r, the cumulative sum of the cell probabilities for the jth column equals v_j/r, and the matrix x is, in a certain sense, *nearest* to the most recent cell probability matrix α. One popular measure of defining the *nearness* is to minimize the weighted cumulative squared deviation of the individual cell probabilities. With this objective, our problem reduces to the following convex cost flow problem:

$$\text{Minimize} \quad \sum_{i=1}^{p} \sum_{j=1}^{q} w_{ij}(x_{ij} - \alpha_{ij})^2 \tag{14.2a}$$

subject to

$$\sum_{j=1}^{q} x_{ij} = u_i/r \qquad \text{for all } i = 1, \ldots, p, \tag{14.2b}$$

$$-\sum_{i=1}^{p} x_{ij} = -v_j/r \qquad \text{for all } j = 1, \ldots, q, \tag{14.2c}$$

$$x_{ij} \geq 0 \qquad \text{for all } i = 1, \ldots, p, \text{ and for all } j = 1, \ldots, q. \tag{14.2d}$$

This type of *matrix balancing problem* arises in many other application settings. The interregional migration of people provides another important application. In the United States, using the general census of the population, taken once every 10 years, the federal government produces flow matrices with detailed migration characteristics. It uses these data for a wide variety of purposes, including the allocation of federal funds to the states. Between the 10-year census, net migration estimates for every region become available as by-products of annual population estimates. Using this information, the federal government updates the migration matrix so that it can reconcile the out-of-date detailed migration patterns with the more recent net figures.

Application 14.4 Stick Percolation Problem

One method for improving the structural properties of (electrically) insulating materials is to embed sticks of high strength in the material. The current approach used in practice is to add, at random, sticks of uniform length to the insulating material. Because the sticks are generally conductive, if they form a connected path from one side of the material to the other, the configuration will destroy the material's desired insulating properties. Material scientists call this phenomenon *percolation*. Although longer sticks offer better structural properties, they are more likely to cause percolation. Using simulation, analysts would like to know the effect of stick length on the possibility that a set of sticks causes percolation, and when percolation does occur, the resulting heat loss due to (electrical) conduction.

Analysts use simulation in the following manner: The computer randomly places p sticks of a given length L in a square region; see Figure 14.4(a) for an example. We experience heat loss because of the flow of current through the intersection of two sticks. We assume that each such intersection has unit resistance. We can identify whether percolation occurs and determine the associated power dissipation by creating an equivalent resistive network as follows. We assign a resistor of 1 unit to every intersection of the sticks. We also associate a current source of 1 unit with one of the boundaries of the insulant and a unit sink with the opposite boundary. The problem then is to determine the power dissipation of the resistive network.

Figure 14.4(b) depicts the transformation of the stick percolation problem into a network model. In this network model, each node represents a resistance and contributes to the power dissipation. The node splitting transformation described in Section 2.4 permits us to model the node resistances as arc resistances. Recall from Ohm's law that a current flow of x amperes across a resistor of r ohms creates a power dissipation of rx^2 watts. Moreover, the current flows in an electrical network

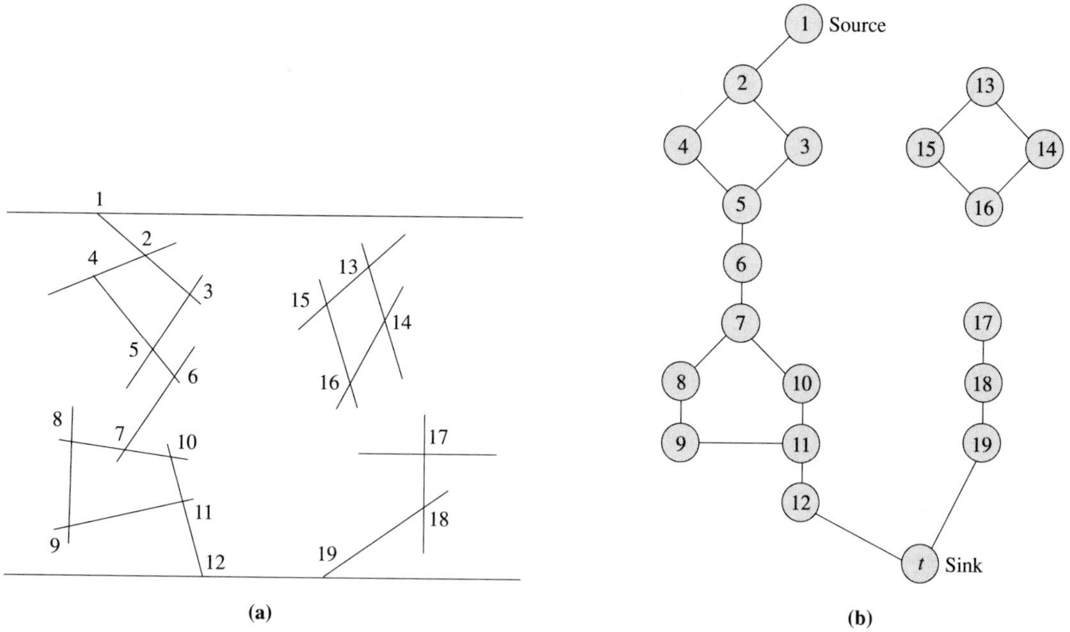

Figure 14.4 Formulating the stick percolator problem: (a) placement of sticks; (b) corresponding network.

in a way that minimizes the rate of power dissipation (i.e., follows the path of least resistance). Consequently, we can state the stick percolator problem as the following convex cost flow problem.

$$\text{Minimize} \quad \sum_{(i,j) \in A} r_{ij} x_{ij}^k \tag{14.3a}$$

subject to

$$\sum_{\{j:(i,j) \in A\}} x_{ji} - \sum_{\{j:(j,i) \in A\}} x_{ji} = \begin{cases} 1 & \text{for } i = s \\ 0 & \text{for all } i \in N - \{s, t\}, \\ -1 & \text{for } i = t \end{cases} \tag{14.3b}$$

$$x_{ij} \geq 0 \qquad \text{for all } (i, j) \in A. \tag{14.3c}$$

In this model, x_{ij} is the current flow on arc (i, j). The solution of this convex cost flow model indicates whether percolation occurs (i.e., if the problem has a feasible solution), and if so, the solution specifies the value of the associated power loss.

14.3 TRANSFORMATION TO A MINIMUM COST FLOW PROBLEM

In this section we show how to transform a convex cost flow problem with piecewise linear convex cost functions into a minimum cost flow problem. This transformation has one significant limitation: it substantially expands the underlying network.

Recall that we are assuming that each piecewise linear convex function contains at most p linear segments. To simplify our notation, we assume that each cost function has exactly p linear segments. We incur no loss of generality in imposing this assumption because we can always add "trivial" segments of zero length at the end of the last interval $0 \le x_{ij} \le u_{ij}$. Moreover, we need not store these trivial segments explicitly because $d_{ij}^{k-1} = d_{ij}^k = u_{ij}$ for any such segment.

Consider a flow x_{ij} on arc (i, j), which we will view as decomposed into different segments, each representing flow between two of the breakpoints d_{ij}^{k-1} to d_{ij}^k. Let y_{ij}^k denote the flow along the kth segment, that is, between d_{ij}^{k-1} and d_{ij}^k. For example, if we send 6 units of flow along the arc with the cost function depicted in Figure 14.1(a), we send 2 units of flow along segment 1, 3 units along segment 2, and 1 unit along segment 3. In general, we can compute the segment flows y_{ij}^k from the total arc flow x_{ij} using the following formula:

$$y_{ij}^k = \begin{cases} 0 & \text{if } x_{ij} \le d_{ij}^{k-1} \\ x_{ij} - d_{ij}^{k-1} & \text{if } d_{ij}^{k-1} \le x_{ij} \le d_{ij}^k \\ d_{ij}^k - d_{ij}^{k-1} & \text{if } x_{ij} \ge d_{ij}^k. \end{cases}$$

By definition, $x_{ij} = \sum_{k=1}^{p} y_{ij}^k$ and $C_{ij}(x_{ij}) = \sum_{k=1}^{p} c_{ij}^k y_{ij}^k$. Substituting $\sum_{k=1}^{p} y_{ij}^k$ for x_{ij} in (14.1) gives us the following problem:

$$\text{Minimize} \quad z = \sum_{(i,j)\in A} \sum_{k=1}^{p} c_{ij}^k y_{ij}^k \tag{14.4a}$$

subject to

$$\sum_{\{j:(i,j)\in A\}} \sum_{k=1}^{p} y_{ij}^k - \sum_{\{j:(j,i)\in A\}} \sum_{k=1}^{p} y_{ji}^k = b(i) \qquad \text{for all } i \in N, \tag{14.4b}$$

$$0 \le y_{ij}^k \le d_{ij}^k - d_{ij}^{k-1} \qquad \text{for all } (i, j) \in A, \text{ for all } k = 1, \ldots, p. \tag{14.4c}$$

It is easy to see that (14.4) is a minimum cost flow problem on an expanded network $G' = (N, A')$ with at most p nontrivial parallel arcs corresponding to each arc $(i, j) \in A$. Figure 14.5 shows the costs and capacities corresponding to any arc (i, j). Let $(i, j)^1, (i, j)^2, \ldots, (i, j)^p$ denote these arcs, and let c_{ij}^k and $d_{ij}^k - d_{ij}^{k-1}$ denote the cost and capacity of arc $(i, j)^k$.

We now establish the equivalence between the convex cost flow problem stated in (14.1) and the minimum cost flow problem stated in (14.4). Given a flow x of

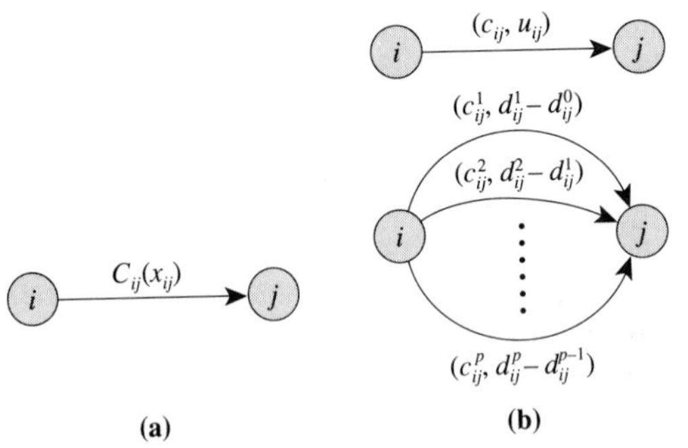

Figure 14.5 Transforming a convex cost flow problem to a minimum cost flow problem: (a) original arc; (b) corresponding arcs in the transformed network.

(14.1), we obtain the corresponding flow y of (14.4) with the same cost as follows. For each arc $(i, j) \in A$, we examine the arcs $(i, j)^1, (i, j)^2, \ldots,$ in order and send the maximum possible flow along these arcs until we have sent a total of x_{ij} units from node i to node j. We refer to any such solution of (14.4) as a *contiguous solution*. The contiguous solution satisfies the property that if the flow on arc $(i, j)^k$ is positive, the flow on each of the arcs $(i, j)^1, (i, j)^2, \ldots, (i, j)^{k-1}$ equals the arc's capacity. Equivalently, in the contiguous solution, if the flow on arc $(i, j)^k$ is strictly less than its upper bound, the flow on each of the arcs $(i, j)^{k+1}, (i, j)^{k+2}, \ldots, (i, j)^p$ is zero.

Conversely, if y is a contiguous solution of (14.4), then the flow $x_{ij} = \sum_{k=1}^{p} y_{ij}^k$ is a solution of (14.1) and both the solutions have the same cost. These observations establish one-to-one correspondence between solutions of (14.1) and contiguous solutions of (14.4). A solution x_{ij} of (14.1) defined by a noncontiguous solution y_{ij}^k of (14.4) might have a different cost. But we need not worry about noncontiguous solutions, because an optimal solution of (14.4) will always be a contiguous solution. To see this, consider a noncontiguous solution in which $y_{ij}^l > 0$, and for some $k < l$, y_{ij}^k is strictly less than its upper bound. Since $c_{ij}^k < c_{ij}^l$ [by the convexity of the function $C_{ij}(x_{ij})$], we can improve this solution by adding a small number ϵ to y_{ij}^k and subtracting ϵ from y_{ij}^l. Therefore, a noncontiguous solution cannot be an optimal solution to (14.4).

The preceding discussion shows that we can solve the convex cost flow problem by solving an associated minimum cost flow problem. If y_{ij}^k is an optimal solution of the minimum cost flow problem, then $x_{ij} = \sum_{k=1}^{p} y_{ij}^k$ is an optimal solution of the convex cost flow problem.

The major drawback of the minimum cost flow transformation is that it expands the network substantially. Each arc in the convex cost flow network has as many copies in the minimum cost flow network as the number of linear segments in its cost function. When we are given the cost function $C_{ij}(x_{ij})$ specified as a piecewise linear continuous convex function for each arc $(i, j) \in A$, this transformation might be satisfactory, because the amount of storage space (i.e., input size) required to specify the convex cost flow problem is proportional to the total number of segments in all the cost functions, which equals the total number of arcs in the resulting minimum cost flow problem. In other words, both problems have the same input size. Consequently, any polynomial-time algorithm for solving the transformed minimum cost flow problem would also solve the associated convex cost flow problem in polynomial time.

On the other hand, if we were to specify the cost $C_{ij}(x_{ij})$ for some arc (i, j) as a concise function, such as x_{ij}^2, and we convert the function into piecewise linear by introducing segments of unit lengths, the transformation is not satisfactory. In this case, although stating $C_{ij}(x_{ij})$ might require only $O(1)$ space, the minimum cost flow problem will have u_{ij} copies of arc (i, j), and u_{ij} might not be polynomially bounded by n and m. Consequently, even though we employ a polynomial-time algorithm to solve the minimum cost flow problem, the resulting algorithm is not a polynomial-time algorithm for the convex cost flow problem because it is not polynomial in the problems size. To overcome this drawback, we need to develop new algorithms for the convex cost flow problem that are polynomial in the problems size. In Section 14.5 we describe one such algorithm, which we call the capacity scaling algorithm because it is a variant of the capacity scaling algorithm for the minimum cost flow problem that we discussed in Section 10.2. As a starting point, however, we first discuss two pseudopolynomial-time algorithms.

14.4 PSEUDOPOLYNOMIAL-TIME ALGORITHMS

In this section we discuss two algorithms for the convex cost flow problem. We assume that each cost function $C_{ij}(x_{ij})$ is a piecewise linear convex function. The algorithms we discuss are modifications of the cycle-canceling algorithm and the successive shortest path algorithm discussed in Sections 9.6 and 9.7. Both algorithms use the fact that we can convert the integer version of the convex cost flow problem into a minimum cost flow problem by introducing multiple arcs. The novelty of these algorithms is that rather than introducing the multiple arcs explicitly, they handle them implicitly. These algorithms use the fact that every optimal solution of the convex cost flow problem is a contiguous solution.

Both the cycle-canceling and successive shortest path algorithms maintain a residual network at every step. Because the minimum cost flow transformation of a convex cost flow problem has multiple arcs between any pair of nodes, so does the residual network. For example, consider a flow of 3 units on an arc (i, j) of capacity 5 whose cost function is depicted in Figure 14.2. The resulting flow in the transformed network will be 1 unit on each of the arcs $(i, j)^1$, $(i, j)^2$, and $(i, j)^3$, and zero units on each of the arcs $(i, j)^4$ and $(i, j)^5$ [see Figure 14.6(b)]. Our definition of the residual network as described in Section 2.4 implies that the residual network will contain the arcs $(i, j)^4$, $(i, j)^5$, $(j, i)^1$, $(j, i)^2$, and $(j, i)^3$, each of unit capacity [see Figure 14.6(c)].

The contiguity of the solution implies that if we wish to send additional flow from node i to node j, we will send it through the arc $(i, j)^4$, and if we wish to send flow from node j to node i, we will send it through the arc $(j, i)^1$. This observation implies that we need not maintain many arcs between this pair of nodes in the residual network: Maintaining just the two arcs $(i, j)^4$ and $(j, i)^1$ is sufficient because those are the arcs that matter at this point. Eliminating multiple arcs permits us to achieve substantial savings in the storage requirements, which translates into enhanced speed of algorithms.

The preceding discussion implies the following method to construct the residual

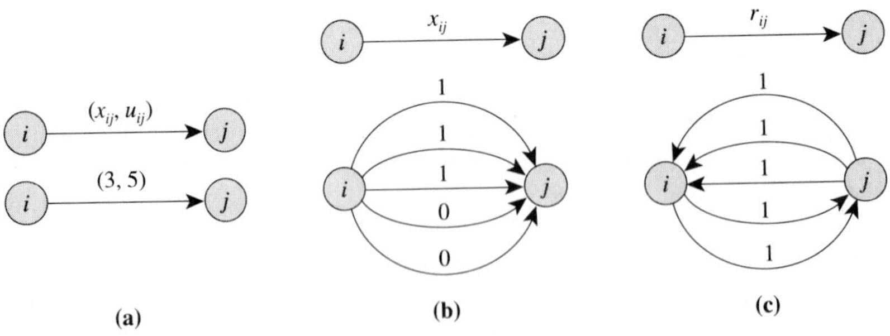

Figure 14.6 Illustrating the construction of the residual network: (a) flow on the arc (i, j) of the original network; (b) flow on arcs of the transformed network; (c) residual network for flow in the transformed network.

network $G(x)$. For any arc $(i, j) \in A$, if $x_{ij} < u_{ij}$, the residual network contains the arc (i, j) with cost $C_{ij}(x_{ij} + 1) - C_{ij}(x_{ij})$. Moreover, for any arc $(i, j) \in A$, if $x_{ij} > 0$, the residual network contains the arc (j, i) with cost $C_{ij}(x_{ij} - 1) - C_{ij}(x_{ij})$. For any arc (i, j) in the residual network, we set its residual capacity equal to the maximum flow change for which the unit flow cost remains equal to $C_{ij}(x_{ij} + 1) - C_{ij}(x_{ij})$. For instance, if the function shown in Figure 14.7 gives the cost of flow for an arc (i, j) and $x_{ij} = 7$, the residual network will contain the arc (i, j) with cost equal to 3 and residual capacity equal to 5. The residual network will also contain the arc (j, i) with cost equal to -2 and residual capacity equal to 2.

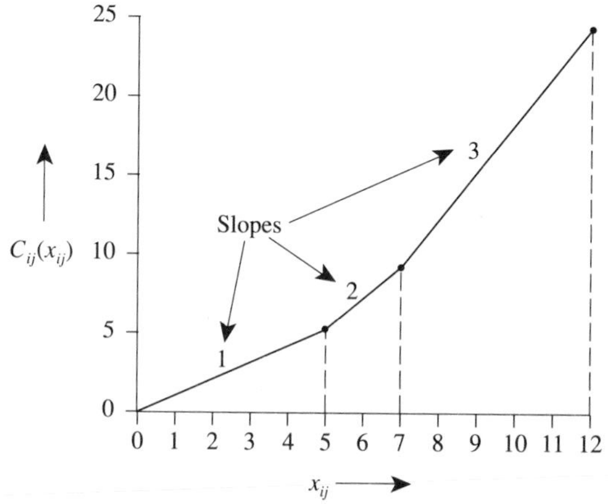

Figure 14.7 Illustrating the construction of the residual network.

We are now in a position to describe the cycle-canceling algorithm for the convex cost flow problem. We start the algorithm with a feasible flow x. We construct the residual network $G(x)$ and use any algorithm to identify a negative cycle. If the residual network does not contain any negative cycle, x is an optimal integer flow of the convex cost flow problem. If the residual network contains a negative cycle, we augment a maximum amount of flow along this cycle, update x and $G(x)$, and repeat the process. This algorithm is exactly the same as the algorithm we have discussed in Section 9.6 except that we construct the residual network differently.

To speed up the cycle-canceling algorithm in practice, we might attempt to augment more flow than would be permitted if we restrict ourselves to just one segment. At every iteration, the cycle-canceling algorithm identifies a negative cycle W and augments a maximum amount of flow along this cycle. After augmenting the flow, it updates the residual network (and the costs of arcs in the residual network) and identifies another negative cycle. At each step we might add some arcs to the residual network, delete some others, and change the costs of some arcs that remain in the residual network. However, it is quite possible that the cycle W will still be a directed cycle in the new residual network (the cost might be different, but the cycle still might be negative). If so, we augment flow along the same cycle. In fact, we can keep doing so until either W is no longer a directed cycle in the residual network or its cost becomes nonnegative. We could identify the amount of flow that we could send along W before we satisfy one of these two conditions in two ways: (1) by sending flow repeatedly along W, or (2) by performing binary search in the interval [0, U]. We might choose the methods that would perform better in practice for the type of applications that we encounter.

The successive shortest path algorithm for the convex cost flow problem is similar. We maintain a pseudoflow x and the residual network $G(x)$ corresponding to this pseudoflow. We also maintain a set of node potentials $\pi(\cdot)$ so that the reduced cost of every arc in the residual network is nonnegative. Initially, we set each node potential $\pi(i) = 0$ and each arc flow x_{ij} equal to the value at which $C_{ij}(x_{ij})$ attains its minimum. At every iteration, the algorithm selects a node k with an excess and obtains shortest path distances $d(\cdot)$ from node k to every other node in the residual network. Then it updates the potentials by setting π to the value $\pi - d$ and augments a maximum possible flow along the shortest path from node k to some deficit node l. The algorithm repeats these steps until the pseudoflow x becomes a flow. Again, this algorithm is the same as the one that we described in Section 9.7 except that we construct the residual network differently.

This discussion shows how we can adapt two core pseudopolynomial-time minimum cost flow algorithms for the convex cost flow problem. We could also modify other algorithms for the minimum cost flow problem along similar lines. For example, in Exercise 14.20, we discuss a modification of the out-of-kilter algorithm for the convex cost flow problem, and in Exercise 14.21, we consider a modification of the network simplex algorithm.

14.5 POLYNOMIAL-TIME ALGORITHM

In this section we describe a polynomial-time algorithm for the convex cost flow problem. This algorithm is a generalization of the capacity scaling algorithm for the minimum cost flow problem discussed in Section 10.2 (this development draws heavily upon that discussion). The generalization, however, does not affect the algorithms worst-case complexity, which remains intact at $O((m \log U)S(n, m, C))$; as before, $S(n, m, C)$ is the time needed to solve a shortest path problem with nonnegative arc lengths.

The capacity scaling algorithm for the convex cost flow problem is an improvement of the successive shortest path algorithm discussed in Chapter 10. The major drawback of the successive shortest path algorithm is that it might augment as little as 1 unit of flow per augmentation, which would be *too small* compared to the total imbalances available at the nodes. As a result, the algorithm might perform *too many* augmentations. The capacity scaling algorithm ensures that the flow sent per augmentation is *sufficiently large* and as a result the total number of augmentations is *sufficiently small*.

The capacity scaling algorithm uses the following basic idea. The successive shortest path algorithm linearizes a given functional form by introducing several linear segments of unit length. The capacity scaling algorithm does not perform this linearization in a single step, but instead, does it in several scaling phases. Consider, for example, the function $C_{ij}(x_{ij}) = x_{ij}^4$ with $u_{ij} = 12$. In the first scaling phase, the algorithm linearizes the function into segments of length 8, in the second scaling phase it linearizes the function into segments of length 4, and so on until the segment lengths become 1. Figure 14.8 illustrates these linearizations for this function. The advantage of this scheme is that in the first scaling phase the algorithm can send 8

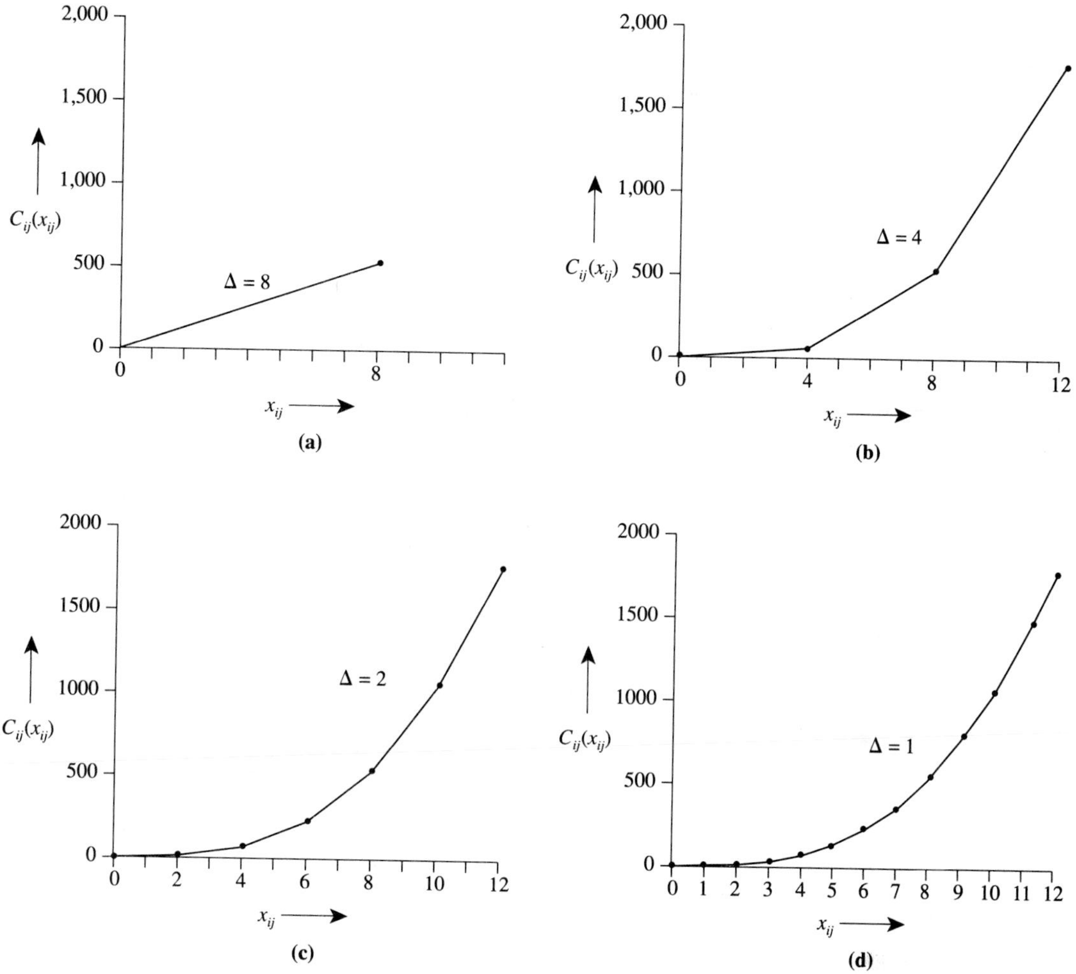

Figure 14.8 Approximations of $C_{ij}(x_{ij})$ in several scaling phases.

units of flow along augmenting paths, in the second scaling phase it can send 4 units of flow along augmenting paths, and so on. As a result, the algorithm reduces excesses at nodes at a faster rate, so it terminates more quickly.

The capacity scaling algorithm for the convex cost flow problem performs a number of scaling phases with varying values of Δ. In the Δ-scaling phase, the algorithm maintains a *Δ-residual network* $G(x, \Delta)$ with respect to a pseudoflow defined as follows: For any arc $(i, j) \in A$ with $x_{ij} + \Delta \leq u_{ij}$, the Δ-residual network contains the arc (i, j) with a residual capacity of Δ and a cost equal to $(C_{ij}(x_{ij} + \Delta) - C_{ij}(x_{ij}))/\Delta$. For any arc $(i, j) \in A$ with $x_{ij} \geq \Delta$, the Δ-residual network contains the arc (j, i) with a residual capacity of Δ and a cost equal to $(C_{ij}(x_{ij} - \Delta) - C_{ij}(x_{ij}))/\Delta$.

Initially, $\Delta = 2^{\lfloor \log U \rfloor}$ and we initiate the algorithm with the zero pseudoflow x and zero node potentials π. In the Δ-scaling phase, we first construct the Δ-residual network $G(x, \Delta)$. We then examine every arc (i, j) in A, and if any of the arcs

(i, j) or (j, i), in $G(x, \Delta)$ violates its reduced cost optimality condition (9.7), we increase or decrease the flow x_{ij} by Δ units, so that both the arcs satisfy their optimality conditions (we later show that it is always possible to do so). The algorithm next defines $S(\Delta)$ and $T(\Delta)$, respectively, as the set of nodes with excesses and deficits of at least Δ units. The algorithm then performs the following step iteratively until either $S(\Delta)$ or $T(\Delta)$ is empty: Identify a shortest path in the Δ-residual network from a node $k \in S(\Delta)$ to a node $l \in T(\Delta)$, augment Δ units of flow along this path, and update $G(x, \Delta)$. At this point the algorithm decreases Δ by a factor of 2 and starts a new scaling phase. Eventually, $\Delta = 1$ and the solution at the end of this scaling phase is optimal.

To show that the capacity scaling algorithm correctly solves the convex cost flow problem, we use the invariant property that the algorithm satisfies the reduced cost optimality condition for every arc in the Δ-residual network. Assuming that the algorithm satisfies this invariant at the beginning of each scaling phase, it is easy to show that the algorithm maintains it subsequently within that phase. The algorithm augments flow along shortest paths in the Δ-residual network, and Lemma 9.12 implies that the resulting solution satisfies the optimality conditions.

To see that the algorithm satisfies the invariant property at the beginning of the first scaling phase, notice that in this scaling phase $\Delta = 2^{\lfloor \log U \rfloor}$. This definition of Δ implies that we have linearized each cost function $C_{ij}(x_{ij})$ into at most one linear segment [as shown in Figure 14.8(a)]. If we set $x_{ij} = 0$, the Δ-residual network might contain the arc (i, j), but not (j, i); let α be the cost of arc (i, j). On the other hand, if we set $x_{ij} = \Delta$, the Δ-residual network might contain the arc (j, i) with cost $-\alpha$. We set the flow on arc (i, j) so that the cost of the corresponding arc in the Δ-residual network is nonnegative. We repeat this step for all arcs in the network G so that all arcs in the Δ-residual network have nonnegative costs. Since all node potentials are zero at this point, the reduced cost of every arc in the residual network is also nonnegative.

We next show that at the beginning of any general Δ-scaling phase, we can adjust arc flows by at most Δ units so that the Δ-residual network satisfies the reduced cost optimality conditions. We initiate the Δ-scaling phase when the 2Δ-scaling phase terminates; we assume inductively that the solution x at the end of the 2Δ-scaling phase satisfies the optimality conditions. In the 2Δ-scaling phase, we linearize $C_{ij}(x_{ij})$ by segments of length 2Δ, and in the Δ-scaling phase we linearize this cost function by segments of length Δ. Consequently, when we move from the 2Δ-scaling phase to the Δ-scaling phase, the arc costs change. As a result, the reduced costs of the arcs also change and the new values might become negative. To see this point better, consider Figure 14.9. In the 2Δ-scaling phase, the cost of the arc (i, j) is the slope of the line AB and the cost of the arc (j, i) is the negative of the slope of the line AC. In the Δ-scaling phase, the cost of the arc (i, j) is the slope of the line AD and the cost of the arc (j, i) is the negative of the slope of the line AE. We claim that by adjusting flow on an arc (i, j) by at most Δ units, we can make the reduced costs of both the arcs, (i, j) and (j, i) in the Δ-residual network nonnegative.

Suppose that x denotes the flow at the end of the 2Δ-scaling phase. At the beginning of the Δ-scaling phase, we first update the Δ-residual network and the arc reduced costs. Consider some arc (i, j) in A and assume, for simplicity, that the Δ-residual network contains both the arcs (i, j) and (j, i). The reduced costs of arcs

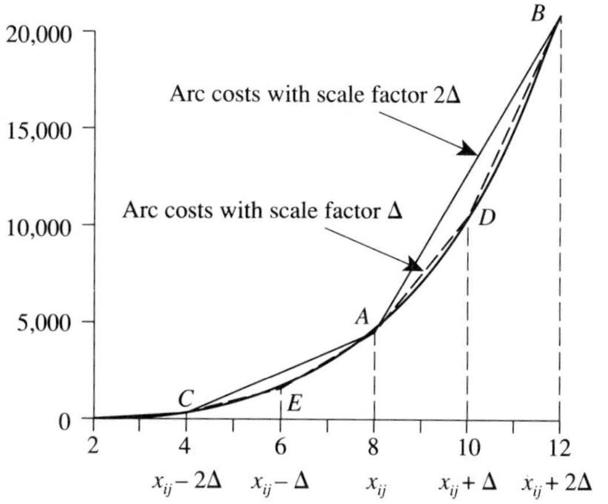

Figure 14.9 Changes in arc costs with changes in Δ.

(i, j) and (j, i) might satisfy four alternatives: (1) $c_{ij}^{\pi} \geq 0$ and $c_{ji}^{\pi} \geq 0$, (2) $c_{ij}^{\pi} < 0$ and $c_{ji}^{\pi} \geq 0$, (3) $c_{ij}^{\pi} \geq 0$ and $c_{ji}^{\pi} < 0$, and (4) $c_{ij}^{\pi} < 0$ and $c_{ji}^{\pi} < 0$. In case 1, both arcs (i, j) and (j, i) satisfy the optimality conditions and nothing needs to be done. In case 2 we increase x_{ij} by Δ units, and in case 3 we decrease x_{ij} by Δ units. Convex cost functions cannot satisfy case 4 (we ask the reader to prove this property in Exercise 14.17).

We now show that in case 2, if we increase x_{ij} by Δ units, the reduced costs of both the arcs (i, j) and (j, i) become nonnegative. The arc (i, j) in $G(x, 2\Delta)$ satisfies the reduced cost optimality condition at the end of the 2Δ-scaling phase. Therefore,

$$([C_{ij}(x_{ij} + 2\Delta) - C_{ij}(x_{ij})]/2\Delta) - \pi(i) + \pi(j) \geq 0.$$

Alternatively,

$$C_{ij}(x_{ij} + 2\Delta) - C_{ij}(x_{ij}) - 2\Delta\pi(i) + 2\Delta\pi(j) \geq 0. \tag{14.5}$$

However, the arc (i, j) in $G(x, \Delta)$ does not satisfy the reduced cost optimality condition. Consequently,

$$C_{ij}(x_{ij} + \Delta) - C_{ij}(x_{ij}) - \Delta\pi(i) + \Delta\pi(j) < 0. \tag{14.6}$$

We want to show that after we have increased x_{ij} by Δ units, the arc (i, j) in $G(x, \Delta)$ will satisfy the optimality conditions. In other words, we wish to show the following result:

$$C_{ij}(x_{ij} + 2\Delta) - C_{ij}(x_{ij} + \Delta) - \Delta\pi(i) + \Delta\pi(j) \geq 0. \tag{14.7}$$

We can write the inequality (14.5) as

$$\begin{aligned}&[C_{ij}(x_{ij} + 2\Delta) - C_{ij}(x_{ij} + \Delta) - \Delta\pi(i) + \Delta\pi(j)]\\ &\quad + [C_{ij}(x_{ij} + \Delta) - C_{ij}(x_{ij}) - \Delta\pi(i) + \Delta\pi(j)] \geq 0.\end{aligned} \tag{14.8}$$

Using the expression (14.6) in (14.8) immediately implies (14.7). We also need to show that after we have increased x_{ij} by Δ units, the arc (j, i) in $G(x, \Delta)$ will also satisfy the optimality condition. To see this result, observe that the reduced cost of arc (j, i) is $([C_{ij}(x_{ij}) - C_{ij}(x_{ij} + \Delta)]/\Delta) - \pi(j) + \pi(i)$, which is clearly positive in view of (14.6). This observation shows that in case 2, if we increase x_{ij} by Δ units, the reduced costs of both the arcs (i, j) and (j, i) are nonnegative. Using similar arguments, we can show that in case 3, if we decrease x_{ij} by Δ units, the reduced costs of both the arcs (i, j) and (j, i) are nonnegative.

To assess the complexity of the capacity scaling algorithm, note that 2Δ-scaling phase ends when either $S(2\Delta) = \varnothing$ or $T(2\Delta) = \varnothing$. Therefore, the sum of the positive imbalances is at most $2n\Delta$. At the beginning of the Δ-scaling phase, the algorithm adjusts the flow on any arc by at most Δ units. Consequently, $2(n + m)\Delta$ is a bound on the sum of the positive imbalances. Each augmentation in the Δ-scaling phase decreases the sum of the positive imbalances by Δ units; consequently, the algorithm can perform at most $O(m)$ augmentations. Overall, the algorithm performs $O(m \log U)$ augmentations and runs in $O((m \log U)S(n, m, C))$ time. We state this result as the following theorem.

Theorem 14.1. *The capacity scaling algorithm obtains an integer optimal flow for a convex cost flow problem in* $O((m \log U)S(n, m, C))$ *time.*

14.6 SUMMARY

The convex cost flow problem is an efficiently solvable and yet important subcase of the general nonlinear network flow problems. Unlike the minimum cost flow problem, the convex cost flow problem might not have an integer optimal solution. In this chapter, however, we assume that we wish to determine an integer optimal solution. We incur no significant loss of generality in making this assumption because by multiplying the data by a suitably large number, we can use this integer model to obtain a real-valued optimal solution that is near integer to any desired level of accuracy.

We considered two types of models: (1) each arc cost function $C_{ij}(x_{ij})$ is a piecewise linear convex function of the arc flow x_{ij}, and (2) each arc cost function is a concise convex function such as $5x_{ij}^2$. By imposing the integrality assumptions on arc flows, we can transform any concise function into a piecewise linear function by introducing unit length segments. Throughout most of our discussion, we therefore assumed that all of the arc cost functions were piecewise linear.

We first showed how to transform a convex cost flow problem with piecewise linear cost functions into a minimum cost flow problem. The major drawback of this transformation is that it expands the network substantially: For each arc of the convex cost flow network, the transformation introduces one copy of the arc into the minimum cost flow model for each linear segment in the arc's cost function. We showed that we need not maintain so many copies of each arc in the residual network $G(x)$ with respect to any flow x; maintaining at most two copies, corresponding to those arcs whose flow would change next, is sufficient. We then adapted two minimum cost flow algorithms, the cycle-canceling algorithm and the successive shortest path algorithm, for the convex cost flow problem. These algorithms are the same as those for the minimum cost flow problem with one exception: they have different residual networks. The running times of these algorithms are the same as their running times for the minimum cost flow problem.

We also discussed a polynomial-time algorithm for the convex cost flow problem. Polynomial-time algorithms for the minimum cost flow problem do not translate directly into polynomial-time algorithms for the convex cost flow problem for the simple reason that the number of arcs in the resulting minimum cost flow formulation might not be polynomially bounded in n, m, and $\log U$. As a result, we need to make modest changes in the minimum cost flow algorithms so that they retain their polynomial-time behavior. In this spirit we modified the capacity scaling algorithm for the minimum cost flow problem that we had developed in Section 10.2 to obtain a polynomial-time algorithm for the convex cost flow problem. The running time of this algorithm is $O(m \log U\, S(n, m, C))$, which is the same as that of the capacity scaling algorithm for the minimum cost flow problem. [$S(n, m, C)$ is the time required for solving a shortest path problem on a network with n nodes, m arcs, and with C as the largest arc cost.]

REFERENCE NOTES

Most of the research devoted to convex cost flows uses nonlinear programming techniques to obtain a real-valued optimal solution. In this chapter we have adopted an unconventional approach by examining methods for obtaining an integer optimal solution. The transformation of the convex cost flow problem into a minimum cost flow problem is a specialization of a standard transformation for converting a separable piecewise linear convex program into a linear program. The adaptations of the cycle-canceling and successive shortest path algorithms described in Section 14.4 are direct consequences of this transformation.

Minoux [1984] developed a polynomial-time algorithm for obtaining a real-valued optimal solution of the quadratic cost flow problem [i.e., the convex cost flow problem with arc costs of the form $C_{ij}(x_{ij}) = a_{ij}x_{ij} + b_{ij}x_{ij}^2$ for some constants a_{ij} and b_{ij}]. His approach uses the out-of-kilter algorithm as a subroutine. Subsequently, Minoux [1986] observed that this approach can also be used to obtain an integer optimal solution of the (general) convex cost flow problem. The algorithm we have presented in Section 14.5 is a variant of Minoux [1986] algorithm; it is in the framework of a scaling algorithm given by Hochbaum and Shanthikumar [1990]. Our analysis of the correctness and running time of the algorithm is similar to the analysis presented by Minoux [1986]. Goldberg and Tarjan [1987] generalized their cost scaling algorithm for the minimum cost flow problem, which we described in Section 10.3, to obtain an integer optimal solution of the convex cost flow problem if $C_{ij}(x_{ij})$ is integer for all integer x_{ij}. Hochbaum and Shanthikumar [1990] developed scaling based algorithms that would solve separable convex integer programs defined by totally unimodular constraint matrices. Their algorithm is polynomial time for finding optimal integer solutions.

Many nonlinear programming techniques are available for solving the convex cost flow problem. Among these are (1) the Frank–Wolfe method developed by Braynooghe, Gibert and Sakarovitch [1968] and Collins et al. [1978], (2) the convex simplex method described by Rosenthal [1981], (3) Newton's method as developed by Klincewicz [1983], and (4) relaxation methods proposed by Zenios and Mulvey [1986] and Bertsekas, Hosein, and Tseng [1987]. The paper by Ali, Helgason, and Kennington [1978] presents a survey of algorithms for the convex cost flow problem developed before 1978 and the paper by Florian [1986] describes many recent developments, focusing on the use of nonlinear programming algorithms in solving transportation planning and traffic equilibrium problems.

In Section 14.2 we described several applications of the convex cost flow problem. We have adapted these applications from the following papers:

1. Urban traffic flows (Magnanti [1984]).
2. Area transfers in communication networks (Monma and Segal [1982]).
3. Matrix balancing (Schneider and Zenios [1990]).
4. Stick percolation problem (Ahlfeld, Dembo, Mulvey, and Zenios [1987]).

The model arising in electrical networks (Hu [1966]) that we described in Section 1.3 is another application of the convex cost flow problem. Some additional applications of the convex cost flow problem are (1) the target-assignment problem (Manne [1958]), (2) solution of Laplace's equation (Hu [1967]), (3) production scheduling problems (Ratliff [1978]; Barr and Turner [1981]), (4) the pipeline network analysis problem (Collins et al. [1978]), (5) microdata file merging (Barr and Turner [1981]), and (6) market equilibrium problems (Barros and Weintraub [1986]). Papers by Ali, Helgason, and Kennington [1978], Dembo, Mulvey, and Zenios [1989], and Schneider and Zenios [1990] provide additional references concerning applications of the convex cost flow problem.

EXERCISES

Note: In the following exercises, interpret an optimal solution of the convex cost flow problem as an integer optimal solution. Moreover, unless we specifically describe the form of the cost function $C_{ij}(x_{ij})$, assume that it is a piecewise linear convex function or a concise function, whichever is more convenient.

14.1. A function $f(x)$ of an n-dimensional vector x is *convex* if $f(\lambda x_1 + (1 - \lambda)x_2) \le \lambda f(x_1) + (1 - \lambda)f(x_2)$ for every two distinct values x_1 and x_2 of x and for every weighting parameter λ, $0 \le \lambda \le 1$. Suppose that $f(x)$ and $g(x)$ are both convex functions of a scalar x. Which of the following functions $h(\cdot)$ are always convex functions? Justify your answer.
(a) $h(x) = f(x) + g(x)$
(b) $h(x) = f(x) - g(x)$
(c) $h(x) = (f(x))^2$
(d) $h(x) = \sqrt{f(x)}$ [Assume that $f(x) \ge 0$.]

14.2. Let x and $c^1, c^2, \ldots, c^k$ be n-dimensional vectors. Show that the function $f(x) = \max\{c^1x, c^2x, \ldots, c^kx\}$ is a convex function of x. Is the function $g(x) = \min\{c^1x, c^2x, \ldots, c^kx\}$ also convex?

14.3. Consider a minimum cost flow problem whose supply/demand vector $b(\lambda) = b^0 + \lambda b^*$ is a function of a scalar parameter λ. Let $z(\lambda)$ denote the optimal objective function value of the problem as a function of this parameter. Show that $z(\lambda)$ is a convex function of λ.

14.4. **Capacity expansion of a network.** A network $G = (N, A)$ is used to send flow from one node s to another node t and does not have sufficient arc capacities to meet anticipated future demands. Suppose that we wish to increase some of the arc capacities so that we can send the desired amount of flow from node s to node t. Let α_{ij} denote the per unit cost of increasing the capacity of arc (i, j). Suppose that we wish to determine an expansion plan that increases the maximum flow in the network to v^0 while incurring the least possible cost. Formulate this problem as a convex cost flow problem.

14.5. Finding nearly feasible flows. Recall from Section 9.7 that a pseudoflow x of a network flow problem is a solution that satisfies the arc flow bounds $0 \le x_{ij} \le u_{ij}$, but might violate the mass balance constraints. To determine whether the network flow problem has a feasible flow, and if not, then to determine a pseudoflow with minimum possible infeasibility, we could attempt to find a pseudoflow x that minimizes the function $\sum_{i \in N} [e(i)]^2$ of the excesses

$$e(i) = b(i) - \sum_{\{j:(i,j)\in A\}} x_{ij} + \sum_{\{j:(j,i)\in A\}} x_{ji}.$$

(Observe that the minimum value of this problem is zero if and only if the network flow problem has a feasible solution.) Show how to formulate this excess minimization problem as a convex cost flow problem. (*Hint*: Augment the network by adding a node and some arcs.)

14.6. Racial balancing with penalties. Consider the racial balancing problem described in Application 9.3. Assume that each school j has a targeted enrollment of $\bar{b}_j$ black students and $\bar{w}_j$ white students. The actual number of black and white students enrolled in the jth school might differ from these targeted values. Suppose that if we miss any of these targets by y students, we incur an associated penalty of $| \alpha_j | y$. We would like to allocate students to the schools in a way that minimizes the sum of the total cost of transportation and the penalties. Formulate this problem as a convex flow problem.

14.7. Solve the convex cost flow problem shown in Figure 14.10(a) by the cycle-canceling algorithm. Assume that arc (i, j) has the flow cost $c_{ij}x_{ij}^2$ for the value of c_{ij} specified in the figure. Start with the following flow: $x_{13} = x_{34} = 5$, and $x_{ij} = 0$ for all other arcs. Always augment flow along a negative cycle with the minimum cost. Show the residual network after each augmentation.

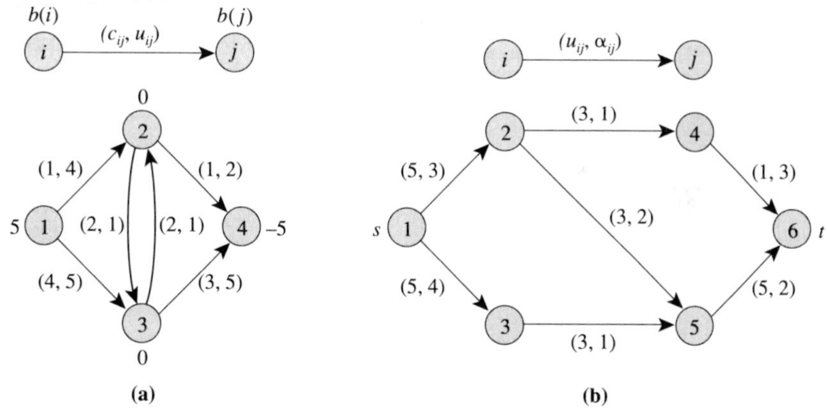

Figure 14.10 Examples for Exercises 14.7 to 14.9.

14.8. This exercise concerns the capacity expansion problem described in Exercise 14.4. Consider a numerical example of this capacity expansion problem shown in Figure 14.10(b) and assume that we wish to send 10 units of flow from the source node to the sink node. Solve this problem by the successive shortest path algorithm.

14.9. Apply the capacity scaling algorithm to the convex cost flow problem shown in Figure 14.10(a). Show your computations for only the first two scaling phases.

14.10. In a particular class of convex cost flow problems formulated on *undirected* networks, the flow cost on any arc (i, j) satisfies the conditions $C_{ij}(x_{ij}) \ge 0$ for all $x_{ij} \ge 0$ and

$C_{ij}(0) = 0$. Show how to convert this type of problem into a convex cost flow problem on a directed network. Justify your transformation by establishing an equivalence between the flows in the two networks. What happens if we relax the assumption that $C_{ij}(0) = 0$?

14.11. Let x^* be an optimal solution of a convex cost flow problem with $C_{ij}(x_{ij})$ as the flow cost on arc $(i, j) \in A$. Suppose that we add a constant k to each arc cost; that is, $C'_{ij}(x_{ij}) = C_{ij}(x_{ij}) + k$ for some constant k. Professor May B. Wright claims that x^* also solves the modified problem. Prove or disprove this claim.

14.12. In Section 11.2 we proved that a minimum cost flow problem always has at least one optimal spanning tree solution [i.e., a solution with an associated spanning tree that satisfies the condition that each nontree arc (i, j) has a flow value of $x_{ij} = 0$ or of $x_{ij} = u_{ij}$ and each tree arc (i, j) has flow x_{ij} satisfying the flow bounds $0 \le x_{ij} \le u_{ij}$]. Show that convex cost flow problems do not satisfy this spanning tree property. To do so, construct an instance of the convex cost flow problem (with piecewise linear convex functions or with concise convex functions) that has a unique nonspanning tree solution.

14.13. We say that a function $C_{ij}(x_{ij})$ is *concave* if for any two points x'_{ij} and x''_{ij} and for every value of the parameter θ satisfying $0 \le \theta \le 1$, $C_{ij}(\theta x'_{ij}) + (1 - \theta)x''_{ij}) \ge \theta\, C_{ij}(x'_{ij} + (1 - \theta)C_{ij}(x''_{ij})$. Consider a capacitated network flow problem in which the cost of flow $C_{ij}(x_{ij})$ on each arc (i, j) is a concave function. In this network we want to obtain a flow that minimizes the total cost of flow. Show that this problem always has an optimal spanning tree solution. Explain why this result is not true for the convex cost flow problem. (*Hint*: Use an approach similar to the one we used in Section 11.2 to show that we can always obtain a cycle free solution from any given solution without increasing the cost of the flow.)

14.14. Let x^* be an optimal solution of the convex cost flow problem. Describe an $O(m)$ method that either shows that x^* is the unique solution to the problem or that finds an alternative optimal solution.

14.15. Let x^* be a feasible solution of the convex cost flow problem. Consider the residual network $G(x^*)$ as defined in Section 14.4. Show that x^* is an optimal solution of the convex cost flow problem if and only if $G(x^*)$ contains no negative cycle.

14.16. In Section 14.4, while describing the cycle-canceling algorithm for the convex cost flow problem, we indicated that we can use a binary search technique to determine the maximum flow δ that we can augment along the selected cycle W so that it remains a negative cycle. Work out the details of this method and specify the time needed to determine δ.

14.17. While discussing the capacity scaling algorithm for the convex cost flow problem in Section 14.5, we claimed that at the beginning of any Δ-scaling phase, we will never encounter a pair of arcs (i, j) and (j, i) in the residual network satisfying the conditions $c^{\pi}_{ij} < 0$ and $c^{\pi}_{ji} < 0$. Prove this claim.

14.18. Budget-constrained capacity expansion. In this exercise we study a variation of the capacity expansion problem described in Exercise 14.4. Suppose that we have allocated D dollars for increasing arc capacities (assume that D is integer). We wish to spend this money in a way that will permit the maximum possible flow from node s to node t in the network. Suggest a polynomial-time algorithm for solving this problem. (*Hint*: Formulate the problem as a constrained maximum flow problem as in Exercise 10.25 and use the solution technique developed in that exercise.)

14.19. Consider the budget-constrained capacity·expansion problem described in Exercise 14.18. Show how to solve this problem by a single application of the parametric network simplex algorithm described in Section 11.9.

14.20. Suppose that we wish to develop a generalization of the out-of-kilter algorithm (see Section 9.9) for solving a convex cost flow problem whose arc costs are all piecewise linear convex functions. Specify a kilter diagram for an arc (i.e., those combinations of reduced costs c^{π}_{ij} and arc flows x_{ij} that satisfy the optimality conditions). Next define the kilter number of an arc (i, j) as the flow change required to make it an in-kilter arc. Finally, show that by solving a shortest path problem, we can reduce the kilter number of some arc by at least 1 unit.

14.21. Adapt the network simplex algorithm for a convex cost flow problem whose arc costs are each given by a concise function $C_{ij}(x_{ij})$. Explain how to perform the following steps: (1) identifying an entering (k, l); (2) determining the maximum flow that we can augment along the cycle formed by adding arc (k, l) to the spanning tree; and (3) updating the node potentials.

14.22. Cost scaling algorithm. In this exercise we discuss an adaptation of the cost scaling algorithm for the minimum cost flow problem discussed in Section 10.3 for solving a convex cost flow problem when each arc cost is a piecewise linear convex function containing at most p linear segments. Suppose that we transform this problem into a minimum cost flow problem and then use the generic version of the cost scaling algorithm described in Section 10.3. For a specific scaling phase, obtain a bound on the number of times that the algorithm performs each of the following operations: (1) saturating pushes; (2) relabels; and (3) nonsaturating pushes. How much time does the algorithm require to execute a scaling phase? What is the running time of the entire algorithm?

14.23. Suppose that we wish to obtain a real-valued optimal flow for a convex cost flow problem whose arc cost functions $C_{ij}(x_{ij})$ are all concise functions. Let x be a feasible flow for the convex cost flow problem and let ϵ be any positive real number. Let $G(x)$ denote the residual network with respect to the flow x. We define the *ϵ-incremental costs* of arcs in the residual network in the following manner. If $(i, j) \in A$ and $x_{ij} < u_{ij}$, then $G(x)$ contains the arc (i, j) with an ϵ-incremental cost equal to $[C_{ij}(x_{ij} + \epsilon) - C_{ij}(x_{ij})]/\epsilon$. If $(i, j) \in A$ and $x_{ij} > 0$, then $G(x)$ contains the arc (i, j) with the ϵ-incremental cost equal to $[C_{ij}(x_{ij} - \epsilon) - C_{ij}(x_{ij})]/\epsilon$. Show that x is a real-valued optimal solution of the convex cost flow problem if and only if for all $\epsilon > 0$, $G(x)$ contains no directed cycle with a negative ϵ-incremental cost. Use this result to outline an algorithm that produces a real-valued optimal solution of the problem to any desired degree of accuracy (i.e., produces a solution whose objective function value is sufficiently close to the optimal objective function value).

15

GENERALIZED FLOWS

There are occasions when it is undoubtedly better to incur loss than to make gain.
—Titus Maccius Plautus

Chapter Outline

15.1 INTRODUCTION

In each of the models we have considered so far, we have made one very fundamental, yet almost invisible, assumption: We conserve flow on every arc. That is, the amount of flow on any arc that leaves its tail node equals the amount of flow that arrives at its head node. This assumption is very reasonable in many application settings, including the numerous applications we have considered in the previous chapters (and that we consider later in Chapter 19). Other practical contexts, however, violate this conservation assumption. For example, in the transmission of a volatile gas, we might lose flow because of evaporation; or, in the transmission of liquids such as raw petroleum crude, we might lose flow due to leakage.

In this chapter we consider a basic *generalized network flow model* for addressing these situations. In this model we associate a positive multiplier μ_{ij} with every arc (i, j) of the network and assume that if we send 1 unit from node i to node j along the arc (i, j), then μ_{ij} units arrive at node j. This model is a generalization of the minimum cost flow problem that we have been considering in previous chapters in the sense that if every multiplier has value 1, the generalized network flow model becomes the minimum cost flow problem.

The generalized maximum flow problem is another special case of the generalized network flow problem. In this model, instead of determining a minimum cost flow, we determine a maximum flow that can leave the source or that can enter the sink. The literature on the generalized maximum flow problem is extensive and includes several recently developed polynomial-time algorithms. In this chapter, rather than discussing these algorithms, we concentrate on the generalized minimum cost flow problem because it is more general and includes the generalized maximum flow problem as a special case. Moreover, rather than attempting to be comprehensive in our coverage of algorithms for the generalized minimum cost flow problem, we will study just one algorithm, an adaptation of the network simplex method, which we refer to as the *generalized network simplex algorithm*.

Like our discussion of the network simplex method in Chapter 11, our presentation of the generalized network simplex algorithm emphasizes the problem's underlying combinatorial structure. We do, however, require limited background in linear programming since a linear programming perspective simplifies much of our development. The combinatorial and linear programming approaches both provide valuable insight into the generalized network simplex method. We stress the combinatorial approach because it is similar to the way that we have developed the network simplex method in Chapter 11 and because it requires only modest background in linear programming.

Recall from Chapter 11 that the network simplex algorithm maintains a partitioning of the arcs of the network as a triple (**T**, **L**, **U**) called a spanning tree structure. The arcs in **T** correspond to those in a spanning tree and the arcs in **U** and **L** are nontree arcs with flow at their upper and lower bounds. The rationale for restricting our search to this type of solution rests on the fundamental spanning tree property that implies that any minimum cost flow problem always has a spanning tree solution. The generalized network simplex algorithm will be conceptually similar. We again restrict our attention to a particular type of solution (**F**, **L**, **U**), called an *augmented forest structure*; in this case, the arcs in **F** constitute what we call an *augmented forest*. As in the network simplex method, the generalized algorithm will be an iterative procedure, moving from one augmented forest structure to another, at each step producing an augmented forest structure with a smaller cost (assuming nondegeneracy). To guide the algorithmic steps, we again define node potentials and use them to determine optimality conditions for assessing when a given solution is optimal.

We have organized this chapter in a modular fashion. After describing a number of applications of the generalized flow problem, we begin to develop the generalized network simplex algorithm by defining augmented forests and by describing several of their properties, including optimality conditions for assessing when an augmented forest structure defines an optimal solution. We then develop the two major building blocks of the generalized network simplex algorithm: procedures for finding the node potentials and the arc flows associated with any augmented forest structure. Not coincidentally, these procedures are also the major building blocks of the simplex method for general linear programs. To highlight the connection between our development in this chapter and linear programming in general, we next show that an augmented forest is a graph-theoretic interpretation of a linear programming basis of the linear programming formulation of the generalized network flow problem. In Section 15.6 we bring all of these algorithm ingredients together to produce the generalized network simplex algorithm.

To set notation, let us first introduce the following linear programming formulation of the generalized flow problem:

$$\text{Minimize} \sum_{(i,j)\in A} c_{ij}x_{ij} \tag{15.1a}$$

subject to

$$\sum_{\{j:(i,j)\in A\}} x_{ij} - \sum_{\{j:(j,i)\in A\}} \mu_{ji}x_{ji} = b(i) \qquad \text{for all } i \in N, \tag{15.1b}$$

$$0 \le x_{ij} \le u_{ij} \qquad \text{for all } (i, j) \in A. \tag{15.1c}$$

As we have already noted, $\mu_{ij} > 0$ is the *multiplier* of the arc (i, j). We assume that each arc multiplier μ_{ij} is a rational number, that is, it can be expressed as $\mu_{ij} = p_{ij}/q_{ij}$ for some integers p_{ij} and q_{ij}. When we send 1 unit of flow on arc (i, j), μ_{ij} units of flow arrive at node j. If $\mu_{ij} < 1$, the arc is *lossy*; if $\mu_{ij} > 1$, the arc is *gainy*.

Notice that we are assuming that the arc capacity u_{ij} is an upper bound on the flow that we send from node i, not on the flow that becomes available at node j. Similarly, c_{ij} is the cost per unit flow that we send from node i, not the per unit cost of the flow that becomes available at node j. In this model we assume that the lower bound on every arc flow is zero. Exercise 15.9 shows that we incur no loss of generality in making this assumption.

15.2 APPLICATIONS

Generalized networks can successfully model many application settings that cannot adequately be represented as minimum cost flow problems. Two common interpretations of the arc multipliers underlie many uses of generalized flows. In the first interpretation, we view the arc multipliers as modifying the amount of flow of some particular item. Using this interpretation, generalized networks model situations involving physical transformations such as evaporation, seepage, deterioration, and purification processes with various efficiencies, as well as administrative transformations such as monetary growth due to interest rates. In the second interpretation, we view the multiplication process as transforming one type of item into another. This interpretation allows us to model processes such as manufacturing, currency exchanges, and the translation of human resources into job requirements. In the following discussion, we describe applications of the generalized network flows that use one or both of these interpretations of the arc multipliers.

Application 15.1 Conversions of Physical Entities

In many different application settings, generalized networks arise quite naturally because the flow in a network converts one type of physical entity into another at a certain conversion rate. The following few brief problem descriptions are illustrative of these types of applications.

Financial networks. In financial networks, nodes represent various equities such as stocks, bonds, current deposits, Treasury bills, and certificates of deposit at certain points in time and arcs represent various investment alternatives that convert one type of equity into another. The multiplier of an arc represents the gain associated with the corresponding investment.

Mineral networks. In these networks, nodes represent mines, purification plants, refineries, ports, and final markets. Arcs represent processing opportunities or flow of material through intermediate junctions to their final destinations. The multiplier of an arc represents the loss associated with the corresponding process.

Energy networks. As discussed in Application 1.9, in certain types of energy networks, nodes represent various raw materials (e.g., crude oil, coal, uranium, or hydropower) and various energy outputs (e.g., electricity, domestic oil, or gas). The arcs represent the transformation of one raw material into an energy output; the efficiency of this transformation is the arc multiplier.

Application 15.2 Machine Loading

Machine loading problems arise in a variety of application domains. In one of the most popular contexts, we would like to schedule the production of r products on p machines. Suppose that machine i is available for α_i hours and that any of the p machines can produce each product. Producing 1 unit of product i on machine j consumes a_{ij} hours of the machine's time and costs c_{ij} dollars. To meet the demands of the products, we must produce β_j units of product j. In the machine loading problem, we wish to determine how we should produce, at the least possible production cost, the r products on the p machines.

In this problem setting, products compete with each other for the use of the more efficient, faster machines; the limited availability of these machines forces us to use the less economical and slower machines to process some of the products. To achieve the optimal allocation of products to the machines, we can formulate the problem as a generalized network flow problem, as shown by Figure 15.1. The network has p product nodes, $1, 2, \ldots, p$ and r machine nodes, $\bar{1}, \bar{2}, \ldots, \bar{r}$. Product node i is connected to every machine node $\bar{j}$. The multiplier a_{ij} on arc $(i, \bar{j})$ indicates the hours of machine capacity needed to produce 1 unit of product i on machine j. The cost of the arc $(i, \bar{j})$ is c_{ij}. The network also has arcs $(\bar{j}, \bar{j})$ for each machine

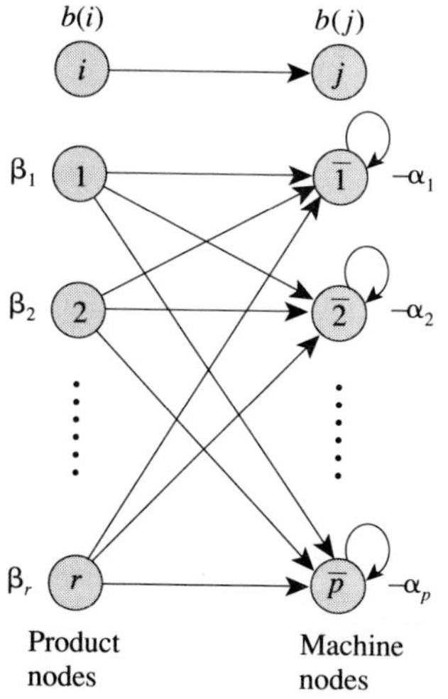

Figure 15.1 Formulating a machine loading problem as a generalized network flow problem.

node $\bar{j}$; the multiplier of each of these arcs is 2 and the cost is zero. The purpose of these arcs is to account for the unfulfilled capacities of the machines: as described in Section 15.3, we can send additional flow along these arcs to generate enough flow (or items) to exactly consume the available machine capacity.

As stated, this machine loading problem is fairly generic. Consequently, it arises naturally in many different problem settings. The following example illustrates one such application context. We leave the detailed formulation of this problem as an exercise (see Exercise 15.3).

Aircraft assignment. An airline needs to assign its fleet of various aircrafts to its flight routes. The airline fleet has β_i aircraft of type i; it wishes to use this fleet to meet its demand of α_j passengers on each of its j routes. By operating an aircraft of type i on route j, the airline incurs a cost of c_{ij} and can accommodate a_{ij} passengers. The airline would like to assign aircrafts to the routes to satisfy the customer demand at the least possible operating costs. Note that the generalized network flow formulation of the aircraft assignment problem does not assure that the number of aircrafts assigned to a route will be integral. The optimal solution might be fractional. In some cases, rounding up this fractional solution to an integer solution might provide a good solution of the problem. In other cases, the solution to the generalized flow problem would serve as a good starting point, and as a valuable bounding mechanism, for initiating an implicit enumeration procedure.

Application 15.3 Managing Warehousing Goods and Funds Flows

An entrepreneur owns a warehouse of fixed capacity H that she uses to store a price-volatile product. Knowing the price of this product over the next K time periods, she needs to manage her purchases, sales, and storage patterns. Suppose that she holds I_0 units of the good and C_0 dollars as her initial assets. In each period she can either buy more goods or sell the goods in the warehouse to generate additional cash. The price of the product varies from period to period and ultimately all goods must be sold. The problem is to identify a buy–sell strategy that maximizes the amount of cash C_K available at the end of the Kth period.

Figure 15.2 gives a generalized network flow formulation of this warehousing

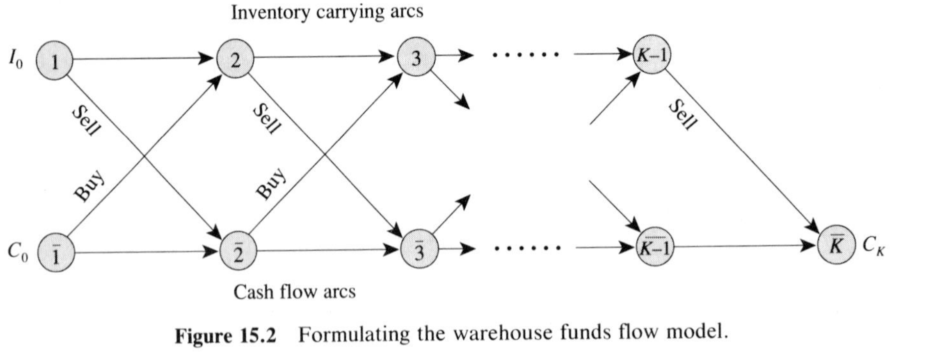

Figure 15.2 Formulating the warehouse funds flow model.

goods and funds flow problem. This formulation has the following three types of arcs:

1. *Inventory carrying arcs*. The cost of this type of arc is the inventory carrying cost per unit; its capacity is H. The multiplier on this arc is less than 1 or equal to 1, depending on whether carrying inventory incurs any loss.
2. *Cash flow arcs*. These arcs are uncapacitated and have zero cost. The multiplier of this type of arc is the bank interest rate for that period.
3. *Buy and sell arcs*. These arcs also are uncapacitated and have zero cost. If p_i is the purchase cost of the product in period i, a buy arc has a multiplier of value $1/p_i$ and a sell arc has a multiplier of value p_i.

It is easy to establish a one-to-one correspondence between buy–sell strategies of the entrepreneur and flows in the underlying network. To enrich this model, we could easily incorporate various additional features, such as policy limits on the maximum and minimum amount of cash held or goods flow in each period or introduce alternative investments such as certificates of deposits that extend beyond a single period.

Application 15.4 Land Management

The U.S. Bureau of Land Management (BLM) manages 173 million acres of public rangelands. It uses a significant part of this land to grow vegetation consumed by animals (both wild and domestic). The BLM must devise a resource management plan for determining the optimal number of animals of different types that the land can support, given the vegetation inventory and the dietary requirements for the different animal types. We present a simplified version of this problem.

The BLM needs to support several animal types, say $A_1, A_2, \ldots, A_a$, using several types of vegetation, say $V_1, V_2, \ldots, V_v$, while satisfying the following constraints:

1. The total animal consumption cannot exceed an upper limit β_j (in pounds) on the production of vegetation V_j. This upper limit prescribes the maximum amount of the annual vegetation production that the animals can remove by grazing without reducing the vigor of this type of vegetation. The Bureau uses historical records and professional judgment to determine these limits.
2. Each animal type i consumes α_i units of vegetation. Animal type A_i can consume at most γ_{ij} pounds of vegetation type V_j. This bound defines a fraction of the total annual vegetation production that a given animal type can consume without destroying the surrounding vegetation community (different animals might have a different effect on the vegetation).
3. Each animal type must receive a "balanced diet" that will satisfy certain dietary requirements. These requirements are stated as follows: The ratio of the intake of vegetation of type V_j (in pounds) to the total intake of all vegetation for each animal type A_i must lie between f_{ij} and g_{ij}. The Bureau determines these limits from the scientific literature.

Figure 15.3 shows a generalized network flow model for a situation with two animal types and three vegetation types, but restricted to only the first two of the three listed constraints. In the figure any arc other than the source arc has a multiplier of value 1. The flow on a source arc (s, i) indicates the number of animals of type A_i supported. Since the multiplier of this arc is α_i, when this flow reaches node i, it is converted into the total food requirement of the animal type A_i. The network distributes this food requirement among different vegetation types while honoring the imposed lower and upper limits. If we set the cost of each source arc to be -1, this model will determine the maximum number of animal types that the land can support.

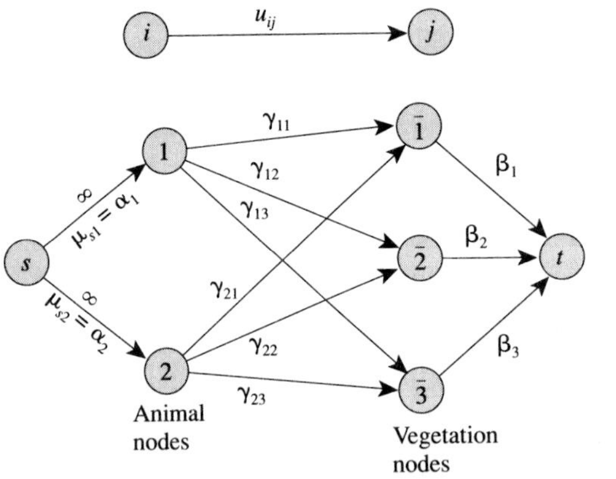

Figure 15.3 Generalized network flow formulation of the land management problem.

Incorporating the third constraint in the model formulation adds additional complications. This constraint states that $f_{ij}(\alpha_i x_{si}) \le x_{ij} \le g_{ij}(\alpha_i x_{si})$ for each animal type A_i and vegetation type V_j. Unfortunately, these constraints destroy the network structure of the model. However, we can still use network flow techniques to solve the problem because it has significant embedded network flow structure. By using a technique known as the Lagrangian relaxation (which we discuss in Chapter 16), we can relax—that is, remove—these "nonnetwork" constraints and solve the original problem by repeatedly solving a sequence of generalized network flow problems.

15.3 AUGMENTED FOREST STRUCTURES

In this section we present background material needed for developing the generalized network simplex algorithm for the generalized network flow problem. This algorithm maintains a topological structure that we call an *augmented forest structure*. In this section we define the augmented forest structure and derive associated optimality conditions for it. In Section 15.5 we show that an augmented forest structure defines a linear programming basis structure for the generalized network flow problem.

Flows along Paths

Let P be a path (not necessarily, directed) from node s to node t. Let $\overline{P}$ and $\underline{P}$ denote the sets of forward and backward arcs in P. We define the *path multiplier* $\mu(P)$ of the path P as follows:

$$\mu(P) = \frac{\Pi_{(i,j)\in\overline{P}}\mu_{ij}}{\Pi_{(i,j)\in\underline{P}}\mu_{ij}} \tag{15.2}$$

We first address the following question: If we send a unit amount of flow from node s to node t along P, how does the arc flow change? Consider, for example, the path shown in Figure 15.4(a); suppose that we wish to send 2 units of flow from node 1 to node 5. To send 2 units from node 1 to node 5 means that 2 units leave node 1, a certain amount, say α, reaches node 5, and inflow equals outflow at all the internal nodes of the path. If we send 2 units on the arc (1, 2), 6 units become available at node 2 because the multiplier of this arc is 3. The arc (2, 3) has multiplier 0.5, so when we send 6 units on it, only 3 units reach node 3. If we carry the flow further, then 12 units reach node 4 as well as node 5. To summarize, if we send 2 units along the path 1–2–3–4–5, then 12 units reach node 5. The ratio of units reaching node 5 to the units sent from node 1 is $12/2 = 6$, which equals the multiplier of the path.

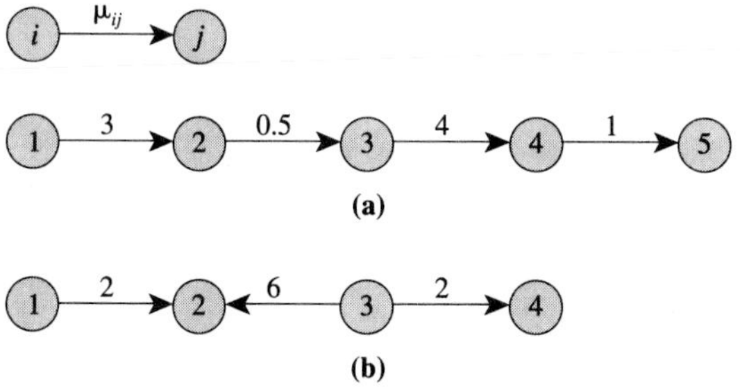

Figure 15.4 Flows along paths in a generalized network: (a) path with all forward arcs, path multiplier is 6; (b) path with both forward and backward arcs, path multiplier is $\frac{2}{3}$.

Next, suppose that we send 1 unit along the path 1–2–3–4 shown in Figure 15.4(b). If we send 1 unit on the arc (1, 2), 2 units become available at node 2. The next arc on the path, arc (3, 2), is a backward arc. We need to send enough flow on this arc to cancel the 2 units at node 2. If we send $-\frac{1}{3}$ of a unit on the arc (3, 2), -2 units become available at node 2, thus satisfying the mass balance constraint at node 2. But sending $-\frac{1}{3}$ of a unit on arc (3, 2) creates an excess of $\frac{1}{3}$ of a unit at node 3. We next send $\frac{1}{3}$ of a unit on arc (3, 4) and $\frac{2}{3}$ of a unit becomes available at node 4. We thus find that if we send 1 unit along the path 1–2–3–4, then $\frac{2}{3}$ of a unit reaches node 4, which again equals the multiplier of the path.

As illustrated by these two examples, (1) if we send y units of flow on a forward arc (i, j), the flow creates $\mu_{ij}y$ units at node j; and (2) if we send y units from node j on a backward arc (j, i), the flow on the arc is $-y/\mu_{ji}$ units and y/μ_{ji} units of flow become available at node i. The following property is an immediate consequence of the preceding discussion.

Property 15.1. *If we send 1 unit of flow from node s to another node t along a path P, then $\mu(P)$ units become available at node t.*

Flows along Cycles

Let W be a cycle (not necessarily directed) from a specified node s to itself whose orientation has already been defined. Let $\overline{W}$ and $\underline{W}$ denote the sets of forward and backward arcs in this cycle. With respect to the cycle's orientation, we define its cycle multiplier $\mu(W)$ as follows:

$$\mu(W) = \frac{\prod_{(i,j) \in \overline{W}} \mu_{ij}}{\prod_{(i,j) \in \underline{W}} \mu_{ij}}. \tag{15.3}$$

Sending flow along a cycle is the same as sending flow along a path except that the flow comes back to itself. Property 15.1 implies that if we send 1 unit of flow along the cycle W starting from node s, then $\mu(W)$ units return to this node. If $\mu(W) > 1$, we create an excess at node s; in this case we refer to the cycle W as a *gainy cycle*. If $\mu(W) < 1$, we create a deficit at node s; in this case we refer to the cycle W as a *lossy cycle*. If $\mu(W) = 1$, the flow around this cycle conserves mass balance at all its nodes; we refer to any such cycle W as a *breakeven cycle*.

Notice that if we reverse the orientation of the cycle, we exchange the roles of the sets $\overline{W}$ and $\underline{W}$ and as a result, the numerator in the expression (15.3) becomes the denominator, and the denominator becomes the numerator. The following result formalizes this observation:

Property 15.2. *If $\mu(W)$ is the multiplier of a cycle W with a particular orientation, then $1/\mu(W)$ is the multiplier of the same cycle with the opposite orientation.*

Note that unless the cycle is a breakeven cycle, we can make it either a gainy cycle or a lossy cycle by defining its orientation appropriately. We next state some additional properties of cycles that are immediate consequences of the preceding discussion. The proofs of these properties are straightforward and left to the reader.

Property 15.3. *By sending (i.e., augmenting) θ units along a nonbreakeven cycle W starting at node s, we create an imbalance of $\theta(\mu(W) - 1)$ units at node s.*

Property 15.4. *Let s be a node in a nonbreakeven cycle W. Then to uniquely create an imbalance of α units at node s (while satisfying the mass balance constraints at all other nodes), we must send $\alpha/(\mu(W) - 1)$ flow along the cycle W starting at node s.*

Augmented Tree and Augmented Forest

Property 15.4 implies that in a feasible solution of the generalized network flow problem, the set of arcs A' with positive flow will not, in general, be a spanning tree. To ensure feasibility, the arcs in A' might contain a cycle. For example, if the network itself is a cycle W with a gain $\mu(W)$, one node t has a positive demand and each node other than node t has a zero demand, the problem has a unique solution and the set A' of arcs with positive flow will be the entire cycle. Therefore, for the generalized network flow problem, each component of A' might contain a cycle. As we show later in this chapter, the generalized network flow problem always has an optimal solution for which each component of A' contains exactly one cycle (assuming nondegeneracy). These types of solutions play the same central role in generalized flows that spanning tree solutions play in minimum cost flows. In this section we describe these special types of solutions and develop optimality conditions for them.

Let $G^a = (N^a, T^a)$ be a subgraph of $G = (N, A)$ so that $N^a \subseteq N$ and $T^a \subseteq A$. We refer to G^a as an *augmented tree* if T^a is a spanning tree of the node set N^a together with an additional arc (α, β) which we call the *extra arc*. An augmented tree has a specially designated node, called its *root*. We consider any augmented tree as hanging from its root. Figure 15.5(b) and (c) show two augmented trees of the graph shown in Figure 15.5(a). In these figures we depict the extra arcs by dashed lines.

An augmented tree contains exactly one cycle which is formed by adding the extra arc (α, β) to the tree $T^a - \{(\alpha, \beta)\}$; we refer to this cycle as the *extra cycle*. Note that we can consider any arc in the extra cycle as the extra arc. For reasons that will become clear later, we refer to an augmented tree as a *good augmented tree* if its extra cycle is lossy or gainy (i.e., not a breakeven cycle).

We define an *augmented forest* $G^f = (N, F)$ with $F \subseteq A$ as a collection of node-disjoint augmented trees that span all the nodes of the graph. We refer to an augmented forest as a *good augmented forest* if each of its components is a good augmented tree. Figure 15.5(d) shows an augmented forest of the graph shown in Figure 15.5(a). We refer to those arcs in an augmented forest as the *augmented-forest arcs* and the remaining arcs as *nonaugmented-forest arcs*.

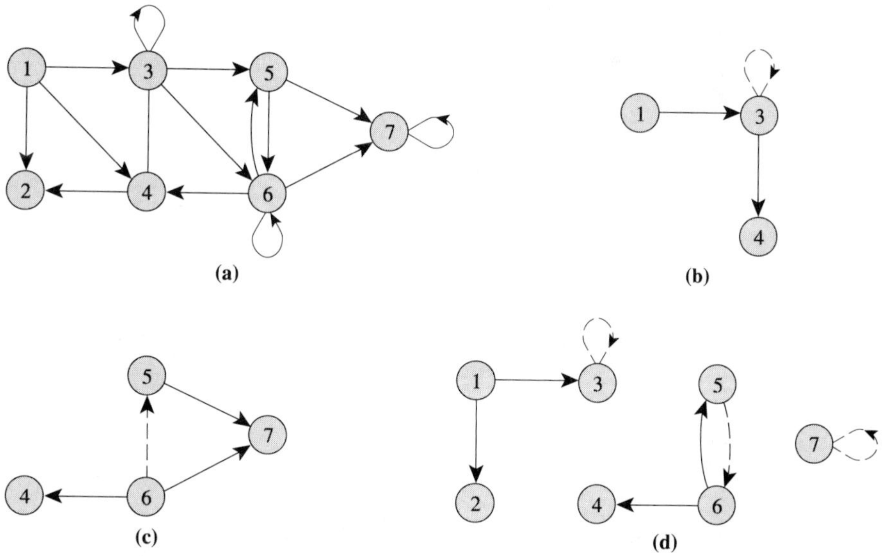

Figure 15.5 Examples of augmented trees and augmented forest: (a) original graph; (b) and (c) two augmented trees; (d) augmented forest.

We store an augmented forest in a computer as a collection of augmented trees. Each augmented tree is a tree plus the extra arc. We can store a tree by associating three indices with each node i in the tree: a predecessor index, *pred*(i), a depth index, *depth*(i), and a thread index, *thread*(i). We refer the reader to Section 11.3 for a detailed discussion of these indices. These indices allow us to perform operations on trees very efficiently.

Augmented Forest Structures and Optimality Conditions

Suppose that the sets **F**, **L**, and **U** define a partition of the arc set A and that **F** is a good augmented forest. As before, we refer to the arcs in **F** as AF-arcs. We refer to the arcs in **L** as nonaugmented-forest arcs at their lower bounds, and the arcs in **U** as nonaugmented-forest arcs at their upper bounds. We also refer to the triple (**F**, **L**, **U**) as an *augmented forest structure*.

An augmented forest structure (**F**, **L**, **U**) is either feasible or infeasible. If we set $x_{ij} = 0$ for all $(i, j) \in \mathbf{L}$ and $x_{ij} = u_{ij}$ for all $(i, j) \in \mathbf{U}$, then a unique flow on the arcs of the augmented forest will satisfy the system of equations (15.1b) (in Section 15.4 we show how to compute this flow). If this flow satisfies the lower and upper bound constraints imposed on all the arcs of the augmented forest, we say that the structure (**F**, **L**, **U**) is *feasible*; otherwise, it is *infeasible*. We will say that a feasible augmented forest structure is *nondegenerate* if $0 < x_{ij} < u_{ij}$ for every arc $(i, j) \in \mathbf{F}$; it is *degenerate* otherwise. We will also say that a feasible augmented forest structure (**F**, **L**, **U**) is an *optimal augmented forest structure* if its associated flow x_{ij} is an optimal solution of (15.1).

We associate with each node i a number $\pi(i)$, which we refer to as its *node potential*. With respect to a set of node potentials, we define the *reduced cost* of an arc (i, j) as $c_{ij}^{\pi} = c_{ij} - \pi(i) + \mu_{ij}\pi(j)$. In the following theorem, we state and prove a sufficiency condition for a flow to be optimal.

Theorem 15.5 (Generalized Flow Optimality Conditions). *A flow x^* is an optimal solution of the generalized network flow problem if it is feasible and for some vector π of node potentials, the pair (x^*, π) satisfies the following optimality conditions:*

$$(a)\quad \text{If } 0 < x_{ij}^* < u_{ij}, \text{ then } c_{ij}^{\pi} = 0. \tag{15.4a}$$

$$(b)\quad \text{If } x_{ij}^* = 0, \text{ then } \quad c_{ij}^{\pi} \geq 0. \tag{15.4b}$$

$$(c)\quad \text{If } x_{ij}^* = u_{ij}, \text{ then } \quad c_{ij}^{\pi} \leq 0. \tag{15.4c}$$

Proof. We first claim that minimizing $\sum_{(i,j)\in A} c_{ij}x_{ij}$ is equivalent to minimizing $\sum_{(i,j)\in A} c_{ij}^{\pi}x_{ij}$. The proof of this claim is similar to that of Property 2.4 and is left as an exercise (see Exercise 15.12). Let π be a vector that together with the flow x^* satisfies the conditions (15.4), and let x be any arbitrary flow. Consider the following summation:

$$\sum_{(i,j)\in A} c_{ij}^{\pi}(x_{ij} - x_{ij}^*). \tag{15.5}$$

We claim that each term in (15.5) is nonnegative. We establish this claim by considering three cases.

Case 1: $0 < x^*_{ij} < u_{ij}$. In this case (15.4a) implies that $c^\pi_{ij} = 0$, so the term $c^\pi_{ij}(x_{ij} - x^*_{ij})$ is zero.

Case 2: $x^*_{ij} = 0$. In this case $x_{ij} \geq x^*_{ij} = 0$, and by (15.4b), $c^\pi_{ij} \geq 0$, so the term $c^\pi_{ij}(x_{ij} - x^*_{ij})$ is nonnegative.

Case 3: $x^*_{ij} = u_{ij}$. In this case, $x_{ij} \leq x^*_{ij} = u_{ij}$, and by (15.4c), $c^\pi_{ij} \leq 0$, so the term $c^\pi_{ij}(x_{ij} - x^*_{ij})$ is again nonnegative.

We have shown that $c^\pi(x - x^*) = c^\pi x - c^\pi x^* \geq 0$, or $c^\pi x^* \leq c^\pi x$, which concludes the proof of the theorem. ◆

The following property is an immediate consequence of this optimality condition.

Property 15.6 (Augmented Forest Structure Optimality Conditions). *A feasible augmented forest structure* (**F, L, U**) *with the associated flow x^* is an optimal augmented forest structure if for some vector π of node potentials, the pair (x^*, π) satisfies the following optimality conditions:*

$$(a)\quad c^\pi_{ij} = 0 \text{ for all } (i, j) \in \mathbf{F}. \tag{15.6a}$$

$$(b)\quad c^\pi_{ij} \geq 0 \text{ for all } (i, j) \in \mathbf{L}. \tag{15.6b}$$

$$(c)\quad c^\pi_{ij} \leq 0 \text{ for all } (i, j) \in \mathbf{U}. \tag{15.6c}$$

15.4 DETERMINING POTENTIALS AND FLOWS FOR AN AUGMENTED FOREST STRUCTURE

Associated with each augmented forest structure are unique arc flows and a unique set of node potentials; in this section we describe efficient methods for determining these quantities. These methods are the major subroutines of the generalized network simplex algorithm that we describe in Section 15.6. We begin by considering the computation of node potentials.

Determining Node Potentials for an Augmented Forest Structure

Let (**F, L, U**) be an augmented forest structure of the generalized network flow problem. The augmented forest **F** contains several augmented trees. We describe a method for determining node potentials for the nodes in an augmented tree. Applying this method iteratively for every augmented tree, we can obtain potentials for every node in the network. Let $T \cup \{(\alpha, \beta)\}$ be the augmented tree under consideration and let node h be its root. We wish to determine node potentials that satisfy the condition $c^\pi_{ij} = 0$ for every arc (i, j) in the augmented tree $T \cup \{(\alpha, \beta)\}$. We first set the potential of node h equal to a parameter θ whose numerical value we will compute later. We then fan out along the tree arcs (i, j) using the thread indices and compute the other node potentials by using the equation $c^\pi_{ij} = c_{ij} - \pi(i) + \mu_{ij}\pi(j) = 0$. The thread traversal ensures (see Section 11.3) that we have already evaluated one of the potentials, $\pi(i)$ or $\pi(j)$, so we can compute the other from the equation c_{ij} −

$\pi(i) + \mu_{ij}\pi(j) = 0$. We note that all the node potentials determined in this way will be (linear) functions of θ. We next use the equation for the extra arc, $c_{\alpha\beta} - \pi(\alpha) + \mu_{\alpha\beta}\pi(\beta) = 0$, to compute a numerical value of θ. This numerical value of θ allows us to compute the numerical values of all the node potentials. Figure 15.6 gives an algorithmic description of this method.

```
procedure compute-potentials;
begin
    π(h) : = θ;
    j : = thread(h);
    while j ≠ h do
    begin
        i : = pred(j);
        if (i, j) ∈ A then π(j) : = (π(i) − c_ij)/μ_ij;
        if (j, i) ∈ A then π(j) : = μ_ji π(i) + c_ji;
        j : = thread(i)
    end;
    for each node i, let the potential π(i), as a function of θ, be represented
            by f(i) + g(i)θ for some constants f(i) and g(i);
    compute θ : = (c_αβ − f(α) + μ_αβ f(β)) / (g(α) − μ_αβ g(β));
    substitute this value of θ in the expression π(i) = f(i) + θg(i) to
            compute the potentials for each node i;
end;
```

Figure 15.6 Computing node potentials of an augmented tree.

Let us illustrate this procedure on a numerical example. Figure 15.7(a) shows an augmented tree and Figure 15.7(b) shows the node potentials in terms of the parameter θ. Using the extra arc (2, 3), we compute $\theta = -17$. Substituting for this value of θ permits us to determine the numerical values of all node potentials, as shown in Figure 15.7(c). We suggest that the reader verify that all the arcs in the augmented tree have zero reduced costs.

The correctness of this procedure follows from the definition of the node potentials [i.e., the vector π must satisfy the condition $c_{ij}^{\pi} = 0$ for every arc (i, j) in the augmented tree $T \cup \{(\alpha, \beta)\}$]. To show that we can carry out the steps of these computations, we need to show that while computing the numerical value of θ in the procedure *compute-potentials* (see Figure 15.6), the denominator $(g(\alpha) - \mu_{\alpha\beta}g(\beta))$ is never zero. We ask the reader to establish this result in Exercise 15.20. The procedure *compute-potentials* has a complexity of $O(n)$.

Determining Flow for an Augmented Forest Structure

Let (**F**, **L**, **U**) be an augmented forest structure of the generalized network flow problem. Assume for the time being that the network is uncapacitated, and consequently, $\mathbf{U} = \varnothing$. As before, we describe a method for determining the flow for an augmented tree; applying this method for every augmenting tree individually, we can determine flows on all the arcs of the augmented forest. The method for computing the arc flows proceeds in the reverse fashion of the method we have used for computing the node potentials: Instead of starting at the root node and fanning out along the tree arcs, we start at the leaf nodes and move in toward the root.

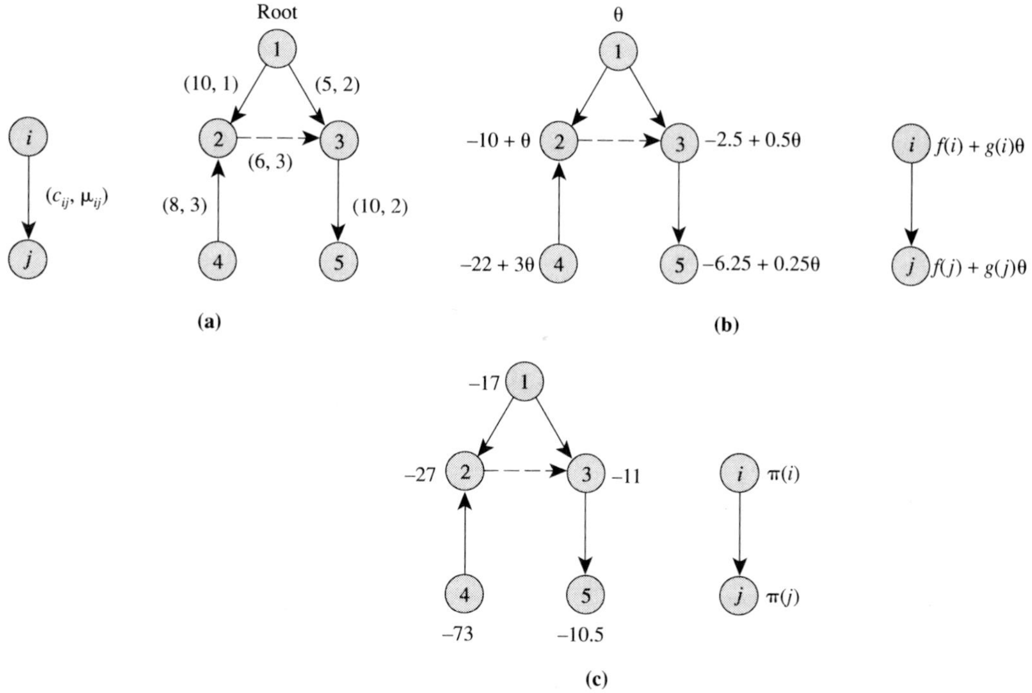

Figure 15.7 Illustrating the computation of node potentials.

Our method for determining the flow for an augmenting tree $T \cup \{(\alpha, \beta)\}$ with root h works as follows. We first define the *imbalance* $e(i)$ of each node i as equal to $b(i)$ and set the flow on each arc equal to zero. We then set the flow on the extra arc (α, β) equal to θ. This amounts to decreasing the imbalance at node α by θ units and increasing the imbalance at node β by $\mu_{\alpha\beta}\,\theta$ units. We next determine all of the arc flows as a function of θ; then we determine the numerical value of θ, and substituting this value for the arc flows, we determine the numerical values of all the arc flows.

Let us consider a leaf node j in the tree T. Exactly one tree arc (i.e., an arc in T) is incident to node j: It is either (j, i) or (i, j). Therefore, we have only one way to discharge the imbalance of node j, through the arc (j, i) or (i, j). If arc (j, i) is incident to node j, we send $e(j)$ units of flow from node j to node i, which reduces the imbalance of node j to zero, sets the flow on arc (j, i) equal to $e(j)$, and changes the imbalance of node i to $e(i) + e(j)\mu_{ji}$ [because $e(j)\mu_{ji}$ additional units of flow arrive at node i]. We illustrate this case in Figure 15.8(a). If arc (i, j) is incident to node j, we need to send $-e(j)/\mu_{ij}$ units of flow from node i over the arc (i, j) in order to make $-e(j)$ units available at node j, thus canceling its excess. We illustrate this possibility in Figure 15.8(b). In this case we change the imbalance of node i to $e(i) - e(j)/\mu_{ij}$ and set the flow on arc (i, j) to $-e(j)/\mu_{ij}$. Once we have determined the flow value on the arc (i, j) or (j, i), whichever is present in the network, we delete this arc and repeat the procedure on the remaining tree. Eventually, we are left with only the root node h with an imbalance $e(h)$.

At this point, the flow on each arc in the augmented tree $T \cup \{(\alpha, \beta)\}$ is a linear

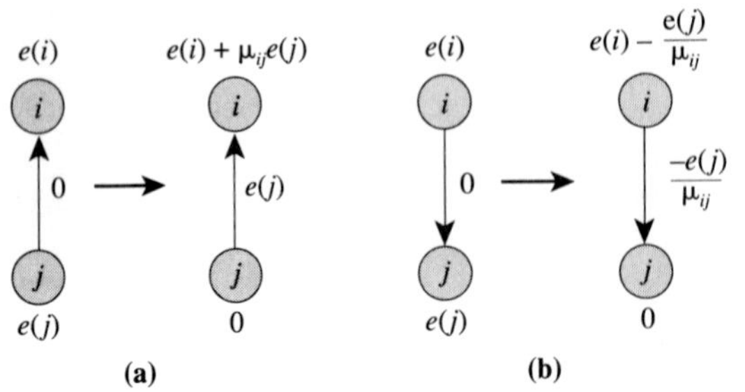

Figure 15.8 Computing flows on tree arcs.

function of θ and the flow satisfies the mass balance constraints of all the nodes except node h, which has an imbalance of $e(h)$ (which again is a linear function of θ). By setting $e(h) = 0$, we compute θ and determine the numerical values of flows on the arcs in the augmented tree. Figure 15.9 summarizes our discussion in the form of an algorithmic description of the procedure. We apply the procedure shown in Figure 15.9 for every augmented tree, starting from the following global initialization: $e(i) = b(i)$ for all $i \in N$ and $x_{ij} = 0$ for all $(i, j) \in A$.

```
procedure compute-flows;
begin
    set x_αβ : = θ, e(α) : = e(α) − θ, and e(β) : = e(β) + μ_αβ θ;
    T' : = T;
    while T' ≠ {h} do
    begin
        select a leaf node j in T';
        i : = pred(j);
        if (j, i) ∈ T' then set x_ji : = e(j) and add −e(j)μ_ji to e(i);
        if (i, j) ∈ T' then set x_ij : = −e(j)/μ_ij and add e(j)/μ_ij to e(i);
        delete node j and the arc incident to it from T';
    end;
    for each arc (i, j) in the augmented tree, let x_ij be represented by f(i, j) + θg(i, j);
    let e(h) be represented by f(h) + θg(h);
    compute θ : = −f(h)/g(h) and use this value of θ to compute numerical values of all the arc
      flows;
end;
```

Figure 15.9 Determining arc flows for an augmented tree.

It is easy to verify that this procedure uniquely determines the flow on arcs of the augmented tree since by scanning tree arcs we determine the arc flows uniquely as a function of θ, and the mass balance constraints of the root node h imply a unique numerical value of θ.

We illustrate this procedure using the numerical example shown in Figure 15.10(a). The figure shows the node imbalances after we have sent an amount of flow θ on the extra arc (2, 3). We examine nodes of the trees in the order 4, 5, 2, 3, and compute the flows on the arcs incident to these nodes. Figure 15.10(b) shows the arc flows and node imbalances at this point. Now, node 1 has an imbalance of $25 - 0.5\theta$, and equating this quantity to zero, we find that $\theta = 50$. Using this value of θ, we compute the numerical values for all the arc flows, as shown in Figure 15.10(c).

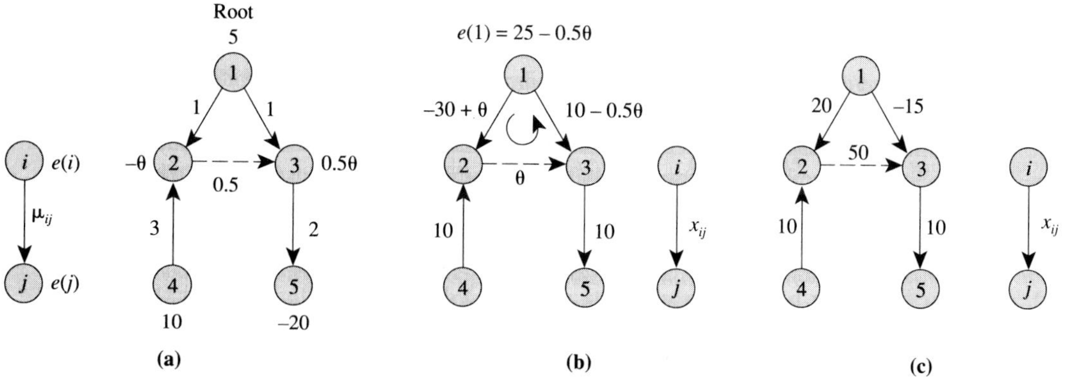

Figure 15.10 Illustrating the computation of flow for an augmented tree.

To complete our discussion showing that the procedure *compute-flows* correctly finds the arc flows, we need to show that $g(h)$ is never zero; otherwise, we cannot compute θ using the equation $\theta = -f(h)/g(h)$. It is easy to see that $\theta g(h)$ is the imbalance at node h resulting from setting the flow on the extra arc (α, β) to value θ. Setting the flow on the arc (α, β) to value θ creates a deficit of $-\theta$ units at node α and an excess of $\mu_{\alpha\beta}\theta$ units at node β. Let P_α denote the tree path from node h to node α and P_β denote the tree path from node β to node h; also, let $\mu(P_\alpha)$ denote the multiplier of the path P_α. We can cancel the deficit at node α by sending $\theta/\mu(P_\alpha)$ units from node h to node α. Similarly, when we send the excess of $\mu_{\alpha\beta}$ units from node β to node h on the path P_β, $\mu_{\alpha\beta}\theta\mu(P_\beta)$ units arrive at node h. These observations show that

$$g(h) = \mu_{\alpha\beta}\mu(P_\beta) - 1/\mu(P_\alpha).$$

Therefore, $g(h)$ is zero if and only if $\mu(P_\alpha)\mu_{\alpha\beta}\mu(P_\beta) = 1$. Since $\mu(P_\alpha)\mu_{\alpha\beta}\mu(P_\beta)$ is the multiplier of the extra cycle, we have shown that $g(h)$ is zero if and only if the extra cycle is a breakeven cycle. Since the augmented tree is good, the extra cycle is not a breakeven cycle and $g(h)$ is nonzero.

So far we have considered uncapacitated networks, that is, we have assumed that $\mu_{ij} = \infty$ for all $(i, j) \in A$. As a result, the set **U** is empty. If the network is capacitated and the set **U** is nonempty, we need to slightly modify the procedure *compute-flows*. In the uncapacitated case, we start with $x_{ij} = 0$ for all $(i, j) \in A$ and $e(i) = b(i)$ for all $i \in N$. In the capacitated case we start with the same values of x_{ij} and $e(i)$ and then execute the following statements:

```
for every (i, j) ∈ U do
begin
    x_ij := u_ij;
    e(i) := e(i) − u_ij;
    e(j) := e(j) + μ_ij u_ij;
end;
```

The purpose of these statements is to set the flow on each arc $(i, j) \in$ **U** to its upper bound, which creates an additional deficit of u_{ij} units at node i and an additional excess of $\mu_{ij}u_{ij}$ units at node j. After invoking this initialization, we execute the procedure *compute-flows* as described in Figure 15.9.

The procedure *compute-flows* provides us with a method for determining a flow that satisfies the supply/demand constraints of all the nodes in an augmented tree. Applying this procedure repeatedly for each augmented tree, we obtain a flow that satisfies the supply/demand constraints of all the nodes. Note that we have computed the arc flows by applying the procedure *update-flows* and using the fact that the flow satisfies the mass balance constraints. We have observed earlier that the flows on the arcs in the augmented forest that satisfy the mass balance constraints are unique and the procedure *compute-flows* determines this unique solution. This flow might or might not satisfy the flow bounds on the arc flows. If it does, the augmented forest structure is *feasible*; otherwise, it is *infeasible*. Clearly, the running time of the procedure *compute-flows* is $O(m)$.

15.5 GOOD AUGMENTED FORESTS AND LINEAR PROGRAMMING BASES

In this section we establish a connection between the good augmented forests of G and bases of the generalized network flow problem. The graph-theoretic structure of the basis allows us to specialize the linear programming simplex method so that we can perform all of the computations on the network itself. As a result, the generalized network simplex algorithm is substantially faster than the general-purpose simplex method. Our main result in this section uses a well-known property of linear programs.

In stating this result, we consider a linear program formulated as

$$\text{Minimize} \quad cx$$

subject to

$$\mathcal{A}x = b,$$

$$0 \le x \le u.$$

In this formulation, $\mathcal{A}$ is a $p \times q$ matrix whose rows are linearly independent (i.e., the matrix has rank equal to p). We let $\mathcal{A}_i$ denote the column in $\mathcal{A}$ associated with the variable x_i, let $\mathbf{B}$ denote an index set of p variables, $x_{\mathbf{B}} = \{x_i : i \in \mathbf{B}\}$, and $\mathcal{B} = \{\mathcal{A}_i : i \in \mathbf{B}\}$. We use the following well-known result:

Property 15.7. *The variables $x_{\mathbf{B}}$ define a basis of the linear programming problem if and only if the system of equations $\mathcal{B}x_{\mathbf{B}} = b$ has a unique solution.*

When translated into the framework of the generalized network flow problem, this property implies that a subset $\mathbf{B}$ of arcs defines a basis of the generalized network flow problem if and only if for every supply/demand vector b, the arcs in $\mathbf{B}$ have a unique flow (not necessarily honoring the flow bounds) that satisfies the mass balance constraints. We will show that arcs in $\mathbf{B}$ have a unique flow if and only if $\mathbf{B}$ is a good augmented forest.

Recall from the last section that if $\mathbf{B}$ is a good augmented forest, the flow on each augmented forest arc is unique. Therefore each augmented forest defines a basis of the generalized network flow problem. We next establish the converse result: If **B** is not a good augmented forest, it cannot be a basis. To do so, we use another well-known result stating that each basis of a linear program contains the same number of variables. Since each good augmented forest contains n arcs and defines a basis, each basis **B** of the generalized network flow problem must contain n arcs. Now consider a set **B** of n arcs that does not define a good augmented forest. Let us consider those components (i.e., connected subgraphs) of **B** that are not good augmented trees. Since some component is not a good augmented tree, either some component is a spanning tree (Case 1), or some component is an augmented tree whose extra cycle is breakeven (Case 2). (As a third possibility, some component might have two or more extra arcs; but notice that in this case some other component must satisfy Case 1; therefore, it is sufficient to consider Case 1 and Case 2.) We consider these two cases separately.

Case 1.

Let $\hat{\mathbf{B}}$ be a component of **B** that is a tree. Designate an arbitrary node, say node h, as the root of $\hat{\mathbf{B}}$. Set $b(h) = 1$ and $b(i) = 0$, for each node $i \in N - \{h\}$. As is easy to verify, we can never consume the supply at node h if we restrict the flow to only tree arcs. This conclusion violates the basis property that for every vector b, some flow must satisfy the mass balance constraints.

Case 2.

Let $\hat{\mathbf{B}}$ be a component of **B** that is an augmented tree whose extra cycle W is breakeven. Notice that sending additional flow around a breakeven cycle maintains the mass balance constraints at all the nodes of the cycle. Therefore, if $\hat{\mathbf{B}}$ has a feasible flow, it has infinitely many feasible flows. This conclusion violates the basis property that for every vector b, a unique flow satisfies the mass balance constraints.

The preceding observations establish the following result.

Theorem 15.8. *A set* **B** *of arcs defines a basis of the generalized network flow problem if and only if* **B** *is a good augmented forest.*

Since each good augmented forest constitutes a basis of the generalized network flow problem, each good augmented forest structure defines a basis structure of the generalized network flow problem.

15.6 GENERALIZED NETWORK SIMPLEX ALGORITHM

The generalized network simplex algorithm is similar to the network simplex algorithm that we discussed in Chapter 11. The algorithm maintains a (good) feasible augmented forest structure at every iteration (which, in linear programming terminology, is a feasible basis structure) and by performing a pivot operation transforms this solution into an improved (good) augmented forest structure. The algorithm repeats this process until the augmented forest structure satisfies the optimality conditions (15.6). Figure 15.11 gives an algorithmic description of the generalized network simplex algorithm.

```
algorithm generalized network simplex;
begin
    determine an initial feasible augmented forest structure (F, L, U);
    let x be the flow and π be the node potentials associated
        with the initial augmented forest structure;
    while some nonaugmented forest arc violates its optimality condition do
    begin
        select an entering arc (k, l) violating its optimality condition;
        add arc (k, l) to the augmented forest and determine the leaving arc (p, q);
        update the the solutions x and π and the augmented forest structure;
    end;
end;
```

Figure 15.11 Generalized network simplex algorithm.

In the following discussion we describe in greater detail various steps of this algorithm.

Obtaining an Initial Augmented Forest Structure

It is easy to obtain an initial (all-artificial) augmented forest structure. For every node $i \in N$ we first introduce an artificial arc (i, i) of sufficiently large cost M and infinite capacity. We set the multiplier of the arc (i, i) equal to 0.5 if node i is a supply node [i.e., $b(i) > 0$], and equal to 2 if node i is a demand or a transshipment node [i.e., $b(i) \le 0$]. Notice that for a supply node i, because the cycle consisting of the arc (i, i) is a lossy cycle, we can consume the supply of node i by sending flow along this cycle. Similarly, for a demand node i, the cycle (i, i) is a gainy cycle and by augmenting flow along the cycle we can generate sufficient flow to satisfy the demand of node i. Property 15.4 implies that by setting $x_{ii} = e(i)/(1 - \mu_{ii})$, we can satisfy the supply/demand of the node i. Moreover, notice that since each artificial arc (i, i) has sufficiently large cost M, no solution with a positive flow on any artificial arc will be optimal unless the generalized network flow problem is infeasible. We determine the initial node potentials from the fact that the reduced cost of each arc in **F** must be zero. Using $c_{ii}^{\pi} = c_{ii} - \pi(i) + \mu_{ii}\pi(i)$ for each node $i \in N$ yields $\pi(i) = M/(1 - \mu_{ii})$.

Optimality Testing and Entering Arc

Let (**F**, **L**, **U**) denote a feasible augmented forest structure of the generalized network flow problem and let π be the corresponding node potentials. To determine whether the augmented forest structure (**F**, **L**, **U**) is optimal, we check to see whether it satisfies the following optimality conditions:

$$c_{ij}^{\pi} \ge 0 \qquad \text{for every arc } (i, j) \in \mathbf{L},$$

$$c_{ij}^{\pi} \le 0 \qquad \text{for every arc } (i, j) \in \mathbf{U}.$$

If the current augmented forest structure satisfies these conditions, it is optimal and the algorithm terminates. Otherwise, the algorithm selects a nonaugmented forest arc violating its optimality condition and introduces this arc into the augmented forest. Two types of arcs are *eligible* to enter the augmented forest:

1. Any arc $(i, j) \in \mathbf{L}$ with $c_{ij}^{\pi} < 0$
2. Any arc $(i, j) \in \mathbf{U}$ with $c_{ij}^{\pi} > 0$

For any eligible arc (i, j), we refer to $| c_{ij}^{\pi} |$ as its *violation*. The generalized network simplex algorithm can select any eligible arc as the entering arc. However, different rules, known as *pivot rules*, for selecting the entering arc produce algorithms with different empirical behavior. In Chapter 11 we discussed several popular pivot rules for the network simplex algorithm. These were (1) *Dantzig's pivot rule*, which selects the arc with maximum violation as the entering arc; (2) the *first eligible arc pivot rule*, which selects, in a wraparound fashion, the first arc with positive violation encountered in examining the arc list; and (3) the *candidate list pivot rule*, which maintains a candidate list of arcs with positive violation and selects the arc with the maximum violation from the candidate list as the entering arc. We can use these same rules for the generalized network simplex algorithm.

Identifying the Leaving Arc

To describe a procedure for determining the leaving arc, suppose that we select arc (k, l) as the entering arc. The arc (k, l) belongs to the set $\mathbf{L}$ or the set $\mathbf{U}$. Throughout the discussion in this section, we examine the situation in which (k, l) is at its lower bound; we leave the case when (k, l) is at its upper bound as an exercise for the reader (see Exercise 15.22). The approach first determines the rate by which the flow on any arc (i, j) of $\mathbf{F}$ changes per unit increase of flow on the entering arc (k, l). Let y_{ij} denote this rate. It is easy to verify that the flow on the augmented forest arcs change linearly with the flow on the entering arc (k, l) [i.e., if the entering arc (k, l) carries δ units of flow, the flow change on arc (i, j) is δy_{ij}. Therefore, if we know the y_{ij} values, we can easily compute the maximum value of δ for which all the arc flows remain within their lower and upper bounds. At this value of δ, at least one arc in $\mathbf{F}$ reaches its lower or upper bound and we select one such arc as the leaving arc.

We now address the problem of determining the y_{ij} values, which represent the flow changes produced on the arcs when we send 1 unit of flow on the entering arc (k, l) (i.e., set $y_{kl} = 1$). We can compute the other y_{ij} values by setting the flow on arc (k, l) equal to 1 and then determining flows on the other arcs so that every node satisfies the mass balance constraint. Setting the flow on arc (k, l) to value 1 creates a deficit (or, demand) of 1 unit at node k and an excess (or, supply) of μ_{kl} units at node l. To determine the effect of sending a unit flow on arc (k, l), we apply the procedure *compute-flows* on the augmented trees containing nodes k and l, starting with zero flow and the following imbalance vector:

$$e(i) = \begin{cases} -1 & \text{for } i = k, \\ \mu_{kl} & \text{for } i = l, \\ 0 & \text{for all } i \neq k \text{ and } l. \end{cases}$$

If the nodes k and l belong to different augmented trees, we need to execute the procedure *compute-flows* twice; otherwise, one execution is sufficient. The arc flows we obtain are the y_{ij} values. Having determined the y_{ij} values, we can easily compute the maximum additional flow δ on the entering arc (k, l). Let x_{ij} denote the flow corresponding to the current augmented forest structure $(\mathbf{F}, \mathbf{L}, \mathbf{U})$. The flow bound constraints require that

$$0 \le x_{ij} + \delta y_{ij} \le u_{ij} \qquad \text{for all } (i, j) \in \mathbf{F} \cup \{(k, l)\}.$$

If for some arc (i, j), $y_{ij} > 0$, the flow on the arc increases with δ and will eventually reach the arc's upper bound. Similarly, if $y_{ij} < 0$, the flow on the arc decreases with δ and will eventually reach its lower bound. Therefore, if δ_{ij} denotes the maximum possible increase in δ allowed by arc (i, j), then

$$\delta_{ij} = \begin{cases} (u_{ij} - x_{ij})/y_{ij} & \text{if } y_{ij} > 0, \\ x_{ij}/(-y_{ij}) & \text{if } y_{ij} < 0, \\ \infty & \text{if } y_{ij} = 0. \end{cases}$$

The largest value of δ for which $x + \delta y$ is feasible is

$$\delta = \min[\delta_{ij} : (i, j) \in \mathbf{F} \cup \{(k, l)\}].$$

We next augment δ units of flow on the arc (k, l) and change the flow on the augmented forest arcs to $x_{ij} + \delta y_{ij}$. We refer to any arc (i, j) that defines δ, that is, for which $\delta = \delta_{ij}$, as a *blocking arc*. We select any blocking arc, say (p, q), as the leaving arc. Observe that for the leaving arc (p, q), $y_{pq} \neq 0$. We say that the iteration is *nondegenerate* if $\delta > 0$, and *degenerate* if $\delta = 0$. A degenerate iteration occurs only if $\mathbf{F}$ is a degenerate augmented forest.

We illustrate the procedure of determining the leaving arc on a numerical example. Figure 15.12(a) shows an augmented tree along with the arc multipliers. Figure 15.12(b) gives the arc capacities and the current arc flows. Assume that (5, 6) is the entering arc. We first determine the values y_{ij} of the flow changes. To determine these values, we set the flow on the entering arc to value 1 and the flow on the extra arc (2, 3) to value θ. Then, as described in the procedure *compute-flows*, we examine the leaf nodes of the tree, one by one, and determine the flows on the unique arcs incident to them. Figure 15.12(c) gives the arc flows as a function of θ. The excess at the root node 1 is $0.5 - 0.5\theta$ and setting this quantity equal to 0 gives $\theta = 1$. Figure 15.12(b) shows the resulting numerical value for each y_{ij}. This figure also specifies the values of each δ_{ij}; using these values, we find that $\delta = 2$. Clearly, (1, 3) is the leaving arc and dropping it gives the augmented tree shown in Figure 15.12(d) with arc (5, 6) as the extra arc.

Updating the Augmented Forest

When the generalized network simplex algorithm has determined a leaving arc (p, q) for a given entering arc (k, l), it updates the augmented forest structure. If the leaving arc is the same as the entering arc, which would happen when $\delta = \delta_{kl} = u_{kl}$, the augmented forest does not change. In this instance, the arc (k, l) merely moves from the set $\mathbf{L}$ to the set $\mathbf{U}$, or vice versa. If the leaving arc differs from the entering arc, the augmented forest changes and we need to update the sets $\mathbf{F}$, $\mathbf{L}$, and $\mathbf{U}$. In this case we need to show that the modified set of arcs in $\mathbf{F}$ constitutes an augmented forest and is good. To do so, we use the fact that the simplex method moves from one basis to another. Since each basis in the generalized network flow

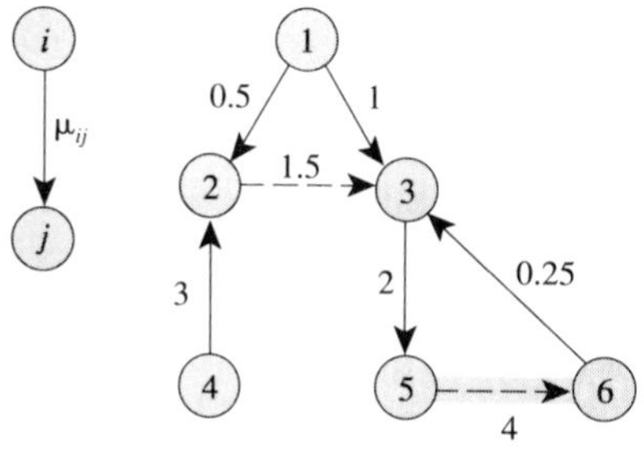

(a)

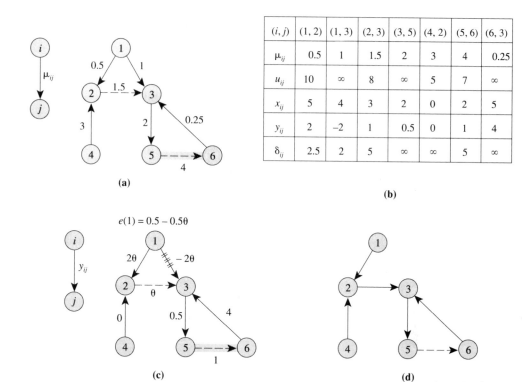

(i, j)	(1, 2)	(1, 3)	(2, 3)	(3, 5)	(4, 2)	(5, 6)	(6, 3)
μ_{ij}	0.5	1	1.5	2	3	4	0.25
u_{ij}	10	∞	8	∞	5	7	∞
x_{ij}	5	4	3	2	0	2	5
y_{ij}	2	−2	1	0.5	0	1	4
δ_{ij}	2.5	2	5	∞	∞	5	∞

(b)

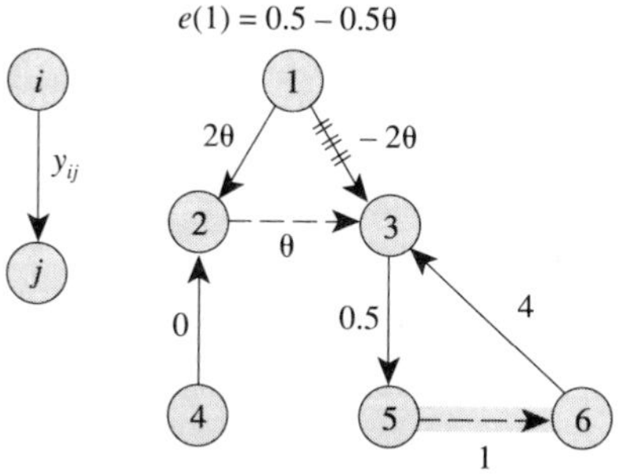

(c)

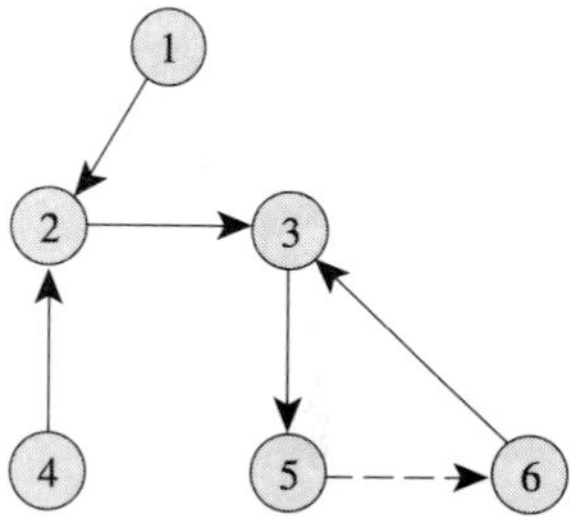

(d)

Figure 15.12 Illustrating the selection of a leaving arc.

problem is a good augmented forest, the generalized network simplex algorithm moves from one good augmented forest to another good augmented forest.

Updating Potentials and Tree Indices

After we have updated the flows and obtained a new augmented forest structure, the next step is to update the node potentials. Clearly, we need to update the node potentials for only those nodes that belong to the augmented tree(s) involved in the pivot operation. In fact, for the generalized network flow problem, updating the node potentials appears to be almost as difficult as recomputing them from scratch. We can recompute the node potentials of the augmented tree(s) using the procedure *compute-potentials* described in Figure 15.6. The final step in the pivot operation is to update various tree indices. This step is rather involved and for details we refer the reader to the references given in the reference notes. Alternatively, we could compute the tree indices from scratch, which would also require $O(n)$ time.

Flows in Bicycles

As we have seen in Chapter 11, in the process of moving from one spanning tree solution to another, the network simplex algorithm sends flows along cycles. Since this interpretation helped us to understand the network simplex algorithm, we might naturally ask the following question: What is the counterpart of a cycle in the generalized network flow problem? Alternatively, we might pose this question as follows. Let y_{ij} be the changes in the flows on the arcs that we defined previously, and let Y denote the set of arcs with strictly positive values of y_{ij}. Then, what is the graph-theoretic structure of Y?

To answer this question, consider several possibilities for the entering arc (k, l) as shown in Figure 15.13. Figure 15.13(a) shows a case when the entering arc has both of its endpoints in different augmented trees and Figure 15.13(b) to (d) show three cases when the entering arc has both of its endpoints in the same augmented

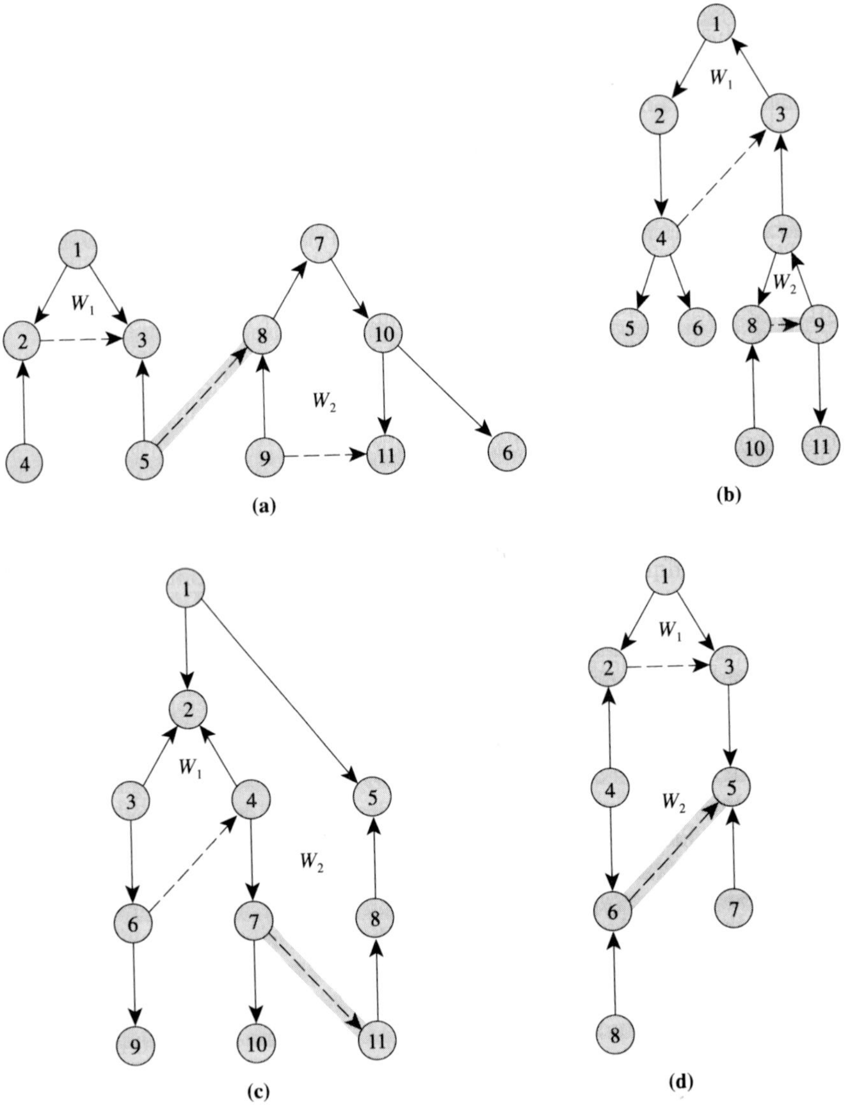

Figure 15.13 Several possibilities for the entering arc.

tree. Consider Figure 15.13(a) first. To increase the flow on the entering arc (5, 8), we first circulate some flow along W_1 starting at some node, say node 3, in a direction defined so that W_1 is a gainy cycle. This flow change creates some excess at node 3. We send this excess to node 5 by the unique tree path, and then to node 8 through the arc (5, 8). Finally, we send a sufficient amount of flow along the cycle W_2 (whose direction is defined so that it is a lossy cycle), so that the excess available at node 8 is consumed. Clearly, y_{ij} will be positive for any arc in the cycles W_1 and W_2 and in the unique path connecting these two cycles. The flow changes in the cases shown in Figure 15.13(b) to (d) have similar interpretations. In each case we have two cycles: one is gainy and the other is lossy. We create an excess by sending flow along one cycle and consume it by circulating flow along the second cycle. As a result, we increase the flow on the entering arc and satisfy the mass balance constraints at all the nodes.

The preceding observations imply that the subgraph defined by the arcs in the set Y is one of the two types shown in Figure 15.14. We refer to the subgraph shown in Figure 15.14(a) as a *type 1 bicycle* and the subgraph shown in Figure 15.14(b) as a *type 2 bicycle*. Figure 15.13 illustrates these bicycles. The cases shown in Figure 15.13(a) and (b) have type 1 bicycles and the cases shown in Figure 15.13(c) and (d) have type 2 bicycles. Therefore, bicycles play the same role in the generalized network simplex algorithm as cycles play in the network simplex algorithm.

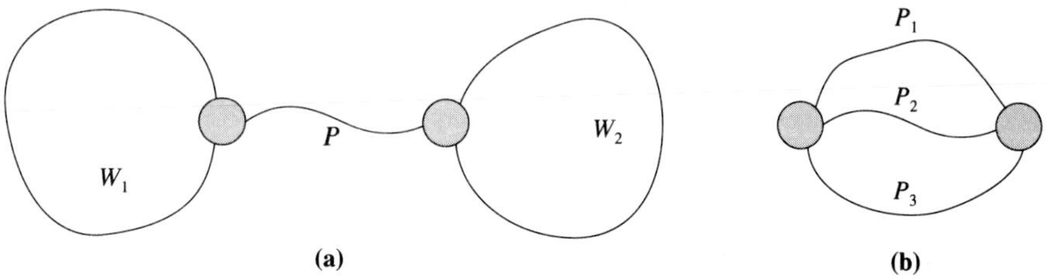

Figure 15.14 Two types of bicycles.

Finally, we note that in the generalized network simplex algorithm, the number of augmented trees in **F** might change from one iteration to the next. For example, consider the situations shown in Figure 15.13(a). If arc (7, 8) is the leaving arc, the pivot creates one fewer augmented tree. On the other hand, if (5, 3) is the leaving arc, the pivot does not change the number of augmented trees. Consider the situation shown in Figure 15.13(b). In this case, if (7, 3) is the leaving arc, the pivot creates an additional augmented tree; for every other leaving arc, the number of augmented trees remain unchanged.

Termination

The generalized network simplex algorithm moves from one feasible augmented forest structure to another until it obtains a structure that satisfies the optimality conditions. In a pivot operation, the objective function value decreases by the amount $\delta \mid c_{kl}^{\pi} \mid$. If each pivot operation in the algorithm is nondegenerate (i.e., $\delta > 0$), each subsequent augmented forest structure has a smaller cost. Since any network has a finite number of augmented forest structures and each augmented forest structure has a unique associated cost, the generalized network simplex algorithm terminates finitely. Degenerate pivots, however, pose some difficulties: The algorithm might not terminate finitely unless we perform pivots carefully. In the following discussion we describe a perturbation technique that ensures that the generalized network simplex algorithm always terminates finitely.

Complexity

The worst-case complexity of the generalized network simplex algorithm is the product of the number of iterations and the complexity per iteration. Although in the worst case we cannot bound the number of iterations by any polynomial function of n and m, in practice the number of iterations an algorithm performs is generally a low-order polynomial in n and m. And what is the running time per iteration? The algorithm requires $O(m)$ time to identify the entering arc. Each of the other operations, such as computing the y_{ij} values, updating the flows, the potentials, and the tree indices, requires $O(m)$ time. Therefore, in practice, the running time of the generalized network simplex algorithm is a low-order polynomial of n and m. Empirical investigations have found that the generalized network simplex algorithm is only two to three times slower than the network simplex algorithm for the minimum cost flow problem.

Perturbation

To ensure finite termination of the algorithm for problems that are degenerate, we use the well-known perturbation technique of linear programming, which we discussed for the minimum cost flow problem in Exercise 11.26. For the generalized network flow problem, we define ϵ as an n-vector whose elements are $(\alpha, \alpha, \alpha, \ldots, \alpha)$ for a sufficiently small real number α. We then replace the supply/demand vector b by $b + \epsilon$ and apply the generalized network simplex algorithm to the perturbed problem. It is possible to show that for the perturbed problem, the augmented forest structure maintained by the generalized simplex algorithm at every iteration is nondegenerate. Consequently, every pivot is nondegenerate and the algorithm will terminate finitely. Moreover, it is possible to show that an optimal augmented forest structure of the perturbed problem is also an optimal augmented forest structure of the original problem. Thus the perturbation does not affect optimality of the solution. For the sake of brevity, we do not specify the details of these results (see the reference notes for a citation to the literature). We do note, however, that we do not need to actually carry out the perturbation. Just as the perturbation for the network simplex algorithm is equivalent to maintaining strongly feasible spanning tree structures, similarly, perturbation in the generalized network simplex algorithm reduces to maintaining certain special types of augmented forest structures which are called *strongly feasible* augmented forest structures (see the reference notes for citations to the literature concerning this issue as well).

15.7 SUMMARY

The generalized network flow problem is a generalization of the minimum cost flow problem in the sense that arcs do not conserve flow. The generalized network flow problem is significantly more difficult to solve than the minimum cost flow problem and, in only a few instances, can we generalize algorithms for the minimum cost flow algorithm so that they would be suitable for solving the generalized network flow problem. The network simplex algorithm discussed in Chapter 11 is one such instance and the resulting algorithm, called the *generalized network simplex algorithm*, is the fastest available algorithm for solving the generalized network flow problem in practice. This chapter has developed the details of the generalized network simplex algorithm.

The generalized network simplex algorithm is an adaptation of the linear programming simplex method (see Appendix C) for the generalized network flow problem. This adaptation is possible because of the special topological structure of the basis. The basis of the generalized network flow problem is a good augmented forest; this fact permits us to perform the steps of the simplex method without maintaining the simplex tableau. Good augmented forests play the same role in the generalized network simplex algorithm as do spanning trees in the network simplex algorithm. The optimality conditions for a good augmented forest are the same as those for the minimum cost flow problem, but with a slightly different definition of the reduced cost c_{ij}^{π} of an arc (i, j); in this context, it is $c_{ij}^{\pi} = c_{ij} - \pi(i) + \mu_{ij}\pi(j)$.

The generalized network simplex algorithm performs two fundamental operations at every iteration: determining the node potentials and arc flows associated with a good augmented forest structure. We showed how to implement both operations very efficiently, in $O(m)$ time, using methods that generalize their counterparts in the network simplex algorithm.

The generalized network simplex algorithm maintains a feasible (good) augmented forest structure at every iteration and successively transforms it into an improved augmented forest structure until it becomes optimal. To implement the generalized network simplex algorithm, we can use the tree indices that we described in Chapter 11 in our discussion of the implementation of the network simplex algorithm. Using these indices permits us to select the leaving variable and to update the flows and potentials in $O(n)$ time. The time to select an entering arc is $O(m)$. The generalized network simplex algorithm terminates finitely, even though the number of iterations it performs cannot be bounded by a polynomial or pseudopolynomial function of the input size parameters. In practice, however, because the algorithm rarely performs more than $5m$ iterations for moderately sized problems and because the average time per pivot is much less than n, its computational time grows slower than $O(nm)$, even though its worst-case running time is exponential. Empirical tests have found the generalized network simplex algorithm to be about two to three times slower than the network simplex algorithm.

REFERENCE NOTES

The generalized network simplex algorithm, presented in this chapter, is an adaptation of the linear programming simplex method and is due to Dantzig [1962]. Kennington and Helgason [1980] and Jensen and Barnes [1980] have given other textbook treatments of the generalized network simplex algorithm. Our presentation of the generalized network simplex algorithm differs from these presentations; it is more combinatorial than algebraic. The approach we have adopted appears for the first time in this book. Elam, Glover, and Klingman [1979] describe the generalized network simplex algorithm, which maintains a strongly feasible basis at every step. Orlin [1985] discusses pertubation techniques for the generalized network simplex algorithm. Elam, Glover, and Klingman [1979] and Brown and McBride [1984] have presented implementation details of the generalized network simplex algorithm and have examined the computational performances of the resulting algorithms. These investigations have found that the generalized network simplex algorithm is about two to three times slower than the network simplex algorithm for the minimum cost flow problem, but substantially faster than a general-purpose linear programming code.

Researchers have also studied nonsimplex approaches for the generalized network flow problem. These approaches include (1) a primal–dual algorithm developed by Jewell [1962], (2) a dual algorithm proposed by Jensen and Bhaumik [1977], and (3) a relaxation algorithm developed by Bertsekas and Tseng [1988b]. Balachandran and Thompson [1975] describe procedures for performing sensitivity and parametric analyses for the generalized network flow problem.

Researchers have also actively studied the generalized maximum flow problem, which is a special case of the generalized network flow problem. These studies have produced simpler algorithms for this problem. The survey paper by Truemper [1977] described these approaches and showed that the generalized maximum flow problem is, in several ways, related to the (ordinary) minimum cost flow problem. None of these algorithms are pseudopolynomial-time algorithms, partly because the optimal arc flows and node potentials might be fractional. Goldberg, Plotkin, and Tardos [1991] presented the first polynomial-time combinatorial algorithm for the generalized maximum flow problem.

In Section 15.2 we described several applications of the generalized network flow problem. Our discussion of these applications has been adapted from the following papers:

1. Conversion of physical entities (Golden, Liberatore, and Lieberman [1979], Glover, Glover, and Shields [1988], and Farina and Glover [1983])
2. Machine loading (Dantzig [1962])
3. Managing warehousing goods and funds flow (Cahn [1948])
4. Land management (Glover, Glover, and Martinson [1984])

In Section 1.3 we described another application of generalized networks arising in energy modeling. Additional applications of the generalized network flow problem arise in (1) resort development (Glover and Rogozinski [1982]), (2) airline seat allocation problems (Dror, Trudeau, and Ladany [1988]), (3) personnel planning (Gorham [1963]), (4) a consensus ranking model (Barzilai, Cook, and Kress [1986]), and (5) cash flow management in an insurance company (Crum and Nye [1981]). The survey papers of Glover, Hultz, Klingman, and Stutz [1978] and Glover, Klingman, and Phillips [1990] contain additional references concerning applications of generalized network flow problems.

EXERCISES

15.1. Generalized caterer problem. In Exercise 11.2 we studied a particular version of the caterer problem; here we consider a generalization of this problem. Suppose that two types of laundry service are available, slow and fast. Suppose, in addition, that each type of service incurs a loss: the slow service loses 5 percent of the napkins it is given to clean and a fast service loses 10 percent of the napkins it is given. The objective, as earlier, is to provide the desired number of napkins on each week day at the lowest possible total cost. Show how to model this extension of the caterer problem as a generalized network flow problem.

15.2. Production scheduling problem. A steel fabricator has several manufacturing plants. Plant i has a manufacturing capacity of S_i tons per month. The company produces n distinct products and for a given month the total customer demand for product j is D_j tons. The plants differ in their fabricating facilities and production efficiencies. A ton of capacity at plant i can produce a_{ij} tons of product j. The fabricator incurs a cost of c_{ij} dollars for each ton of product j produced at plant i and would like to allocate its customer demands to its plants at the least possible cost. Formulate this problem as a generalized network flow problem.

15.3. Formulate the aircraft assignment problem described in Application 15.2 as a generalized network flow problem.

15.4. Optimal currency conversion. Each day an exchange control bureau permits people to exchange a limited amount of money from one currency to another. On a particular day, suppose that the table shown in Figure 15.15 specifies these limits as well as the exchange rates of various currencies that the exchange control department handles. Suppose that you have \$1000 and you want to determine the maximum number of francs you can obtain on that day through a sequence of currency transactions. Formulate this problem as a generalized maximum flow problem.

Currency given, x	Currency received, y	Exchange rate, r ($y = rx$)	Limit (on x)
Dollars	Pounds	0.56	1,000
Dollars	Lira	1,241	500
Pounds	Lira	2,200	160
Lira	Pounds	0.00045	200,000
Guilders	Pounds	3.37	400
Lira	Yen	0.11	950,000
Yen	Guilders	0.014	15,000
Guilders	Yen	70.5	500
Guilders	Francs	3.0	1,600
Yen	Francs	0.042	80,000

Figure 15.15 Currency exchange rates on a particular day.

15.5. In Exercise 11.3 we studied a project assignment problem in which each project had exactly one supervisor. Suppose that we permit each project to have a group of supervisors and we know this group for each project. Show that it is difficult to model the modified problem as a minimum cost flow problem but that we can model it as an integer generalized network flow problem.

15.6. In this exercise we study a few generalizations of the warehouse funds and goods flows problem that we discussed in Application 15.3. Give network flow formulations for situations with $K = 4$ time periods when we impose the following additional problem features.

(a) At the beginning of each period, the entrepreneur can lend cash for 2 or 3 years, accruing an interest of 20 percent and 35 percent.

(b) At the beginning of each period, the entrepreneur can borrow cash for 2 years. She pays an interest in the amount of 25 percent for the 2-year period.

(c) The entrepreneur must satisfy a demand of $d(k)$ units for the product in each time period k.

15.7. In the warehouse funds and goods flows model that we discussed in Application 15.3, suppose that the warehouse can store two products: In every period the entrepreneur can convert portions of each product into cash or into each other. The conversion ratios for each period are known in advance. Can you formulate this model as a generalized network flow problem? If so, give the formulation; if no, outline the difficulties encountered.

15.8. Give two real-life situations, not described in this chapter, of networks that have flow gains and/or losses on their arcs.

15.9. Explain how you would model each of the following extensions of the generalized network flow problems: (1) the flow on arc (i, j) has a nonnegative lower flow bound l_{ij}; (2) the multiplier of an arc (i, j) is zero; (3) a supply node i is permitted to keep some of its supply $b(i)$; and (4) at most $u(i)$ units can enter node i. Consider each of these generalizations separately.

15.10. **(a)** Consider a linear programming problem in which each column has exactly two nonzero entries: one positive and the other negative. Transform this linear programming problem into a generalized network flow problem.

(b) Consider a linear programming problem satisfying two properties: (1) each column has at most two nonzero entries, and (2) if any column has exactly two nonzero entries, one of these is positive and the other is negative. Transform this linear programming problem into a generalized network flow problem.

15.11. Prove Properties 15.3 and 15.4.

15.12. Show that the generalized network flow problem satisfies $c^{\pi}x = cx - \pi b$ for any π. Conclude that any solution that minimizes cx also minimizes $c^{\pi}x$. (*Hint*: The proof is similar to that of property 2.4.)

15.13. In the generalized network simplex algorithm, we obtain an initial augmented forest structure by introducing an artificial arc (i, i), with a sufficiently large cost M, for every node $i \in N$. Specify a finite value of M, as a function of the problem data, that would ensure that these artificial arcs carry zero flow if the generalized network flow problem has a feasible solution. (*Hint*: Determine an upper bound α on the objective function value of any feasible flow without using artificial arcs. Next, determine a lower bound β on the flow on any artificial arc (i, i) if that arc carries a positive flow in any augmented forest. Select M so that $M\beta > \alpha$.)

15.14. Professor May B. Wright announced a novel algorithm for solving the generalized network flow problem when all the arcs are lossy. Can you transform a generalized network flow problem with arbitrary positive arc multipliers to a problem in which all arcs are lossy? Assume that all arcs have finite capacities.

15.15. Suppose that we associate a positive real number $\alpha(i)$ with each node $i \in N$ in a graph G. With respect to the vector α, we define the *reduced multiplier* of an arc (i, j) as $\mu_{ij}^{\alpha} = \alpha(i)\mu_{ij}/\alpha(j)$.

(a) Show that $\mu^{\alpha}(W) = \mu(W)$ for any cycle W.

(b) Let T be a spanning tree of G. Define a vector α so that $\mu_{ij}^{\alpha} = 1$ for every arc $(i, j) \in T$. Let $W(k, l)$ be the fundamental cycle defined by a nontree arc (k, l).

Show that if $W(k, l)$ is a breakeven cycle, $\mu_{kl}^{\alpha} = 1$. Conclude that if every fundamental cycle of G with respect to T is a breakeven cycle, then $\mu_{ij}^{\alpha} = 1$ for all $(i, j) \in A$.

(c) Prove that all cycles in G are breakeven if and only if all fundamental cycles with respect to any spanning tree T are breakeven. Use this result to show that we can determine in $O(m)$ time whether all cycles in G are breakeven.

15.16. Consider the linear programming formulation of the generalized network flow problem given in (15.1). Suppose that we associate a positive real number $\alpha(i)$ with each node $i \in N$ and define the new set of variables y as $y_{ij} = \alpha(i)x_{ij}$. Suppose that we then make the substitution $x_{ij} = y_{ij}/\alpha(i)$ for every (i, j) in (15.1). Show that the resulting formulation is also a generalized network flow problem, but with different supplies/demands, arc costs, capacities, and multipliers. We refer to the resulting problem as the *α-transformed problem*.

(a) What are the supplies/demands, arc costs, capacities, and multipliers of the α-transformed problem in terms of α and the original problem data?

(b) Show that if all the cycles in G are breakeven, then for some choice of the vector α, the α-transformed problem becomes a minimum cost flow problem. (*Hint*: Use the result of Exercise 15.15.)

15.17. Using the procedure *compute-potentials*, determine node potentials for the augmented trees shown in Figure 15.16.

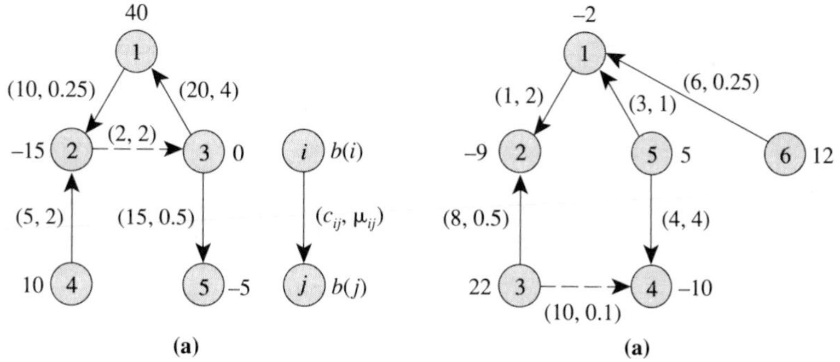

Figure 15.16 Two augmented trees.

15.18. Using the procedure *compute-flows*, determine arc flows for the augmented trees shown in Figure 15.16. Assume that all arcs are uncapacitated. Are the flows feasible?

15.19. Apply two iterations of the generalized network simplex algorithm to the generalized network flow problem shown in Figure 15.17. Assume that all arcs are uncapacitated.

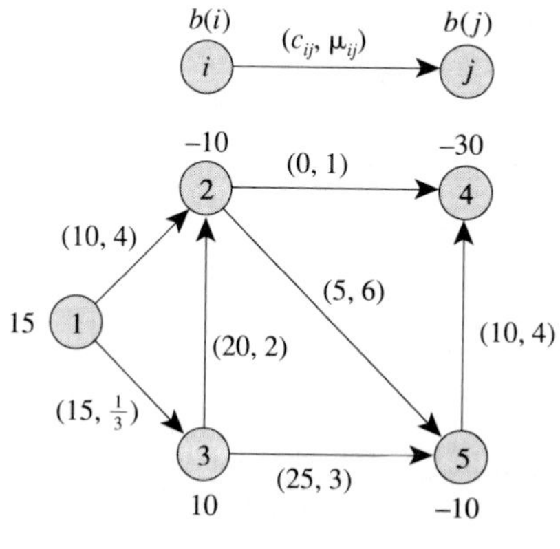

Figure 15.17 Generalized network example for Exercise 15.19.

15.20. Let $T \cup \{(\alpha, \beta)\}$ be an augmented tree of G. Suppose that we define the multiplier of a path P as $\mu(P) = (\sum_{(i,j)\in \overline{P}} \mu_{ij})/(\sum_{(i,j)\in \underline{P}} \mu_{ij})$; in this expression, $\overline{P}$ and $\underline{P}$ denote the sets of forward and backward arcs in the path P.

(a) Let P_i denote the unique path from any node i to the root node h in T [i.e., the augmenting tree minus the extra arc (α, β)]. Show that at the termination of the procedure *compute-potentials* described in Figure 15.6, each $g(i)$ equals $\mu(P_i)$.

(b) Show that the expression $(g(\alpha) - \mu_{\alpha\beta} g(\beta))$ is nonzero if and only if the augmented tree $T \cup \{(\alpha, \beta)\}$ is good (i.e., contains no breakeven cycle).

15.21. Show that if an augmented forest is not good, node potentials determined by the procedure *compute-potentials* might not be unique. Next argue that if an augmented forest is good, node potentials determined by the procedure *compute-potentials* are unique. Conclude that the procedure *compute-potentials* gives a unique set of node potentials if and only if the augmented forest is good.

15.22. In the description of the generalized network simplex algorithm, we have assumed that the entering arc (k, l) belongs to **L**. If arc $(k, l) \in$ **U**, what steps of the algorithm would we need to modify, and how?

15.23. When we add the entering arc (k, l) to an augmented forest, we create a bicycle. Write separate pseudocodes to accomplish the following tasks: (1) determine whether the bicycle is a type 1 bicycle or a type 2 bicycle; (2) if the bicycle is a type 1 bicycle, determine the cycles W_1, W_2 and the path segment P; (3) if the bicycle is a type 2 bicycle, determine the three path segments P_1, P_2, and P_3; (4) if the bicycle is a type 2 bicycle and contains a breakeven cycle, determine the breakeven cycle. Your pseudocodes should use only predecessor and depth indices and run in $O(n)$ time.

15.24. For a given feasible flow x of the generalized network flow problem, we define a *residual network* $G(x)$ as follows. We replace each arc (i, j) by two arcs (i, j) and (j, i). Arc (i, j) has cost c_{ij}, a residual capacity $r_{ij} = u_{ij} - x_{ij}$, and a multiplier μ_{ij}; arc (j, i) has cost $-c_{ij}/\mu_{ij}$, a residual capacity x_{ij}/μ_{ij}, and a multiplier $1/\mu_{ij}$. The residual network consists of only those arcs with a positive residual capacity.

(a) Define the *length* of each arc (i, j) in $G(x)$ as $-\log(\mu_{ij})$. Show that a directed cycle W is a gainy cycle if and only if its length is a negative cycle. Using this result, describe a polynomial-time algorithm for identifying whether $G(x)$ contains a directed cycle that is gainy.

(b) Describe a polynomial-time algorithm for identifying whether $G(x)$ contains a directed cycle that is lossy.

15.25. Generalized maximum flow problem. In the generalized maximum flow problem, we wish to obtain a feasible flow that maximizes one of two objectives: (1) the flow into a sink node t, or (2) the flow out of the source node s. Formulate the generalized maximum flow problem as a generalized minimum cost flow problem. (*Hint*: Transform the problem into one in which all the nodes have zero supplies/demands.)

15.26. Generalized maximum flow algorithm. In this exercise we discuss an algorithm for maximizing the flow into the sink node t. Suppose that we define the residual network $G(x)$ with respect to a flow x as in Exercise 15.24. In the residual network, we define a *generalized augmenting path* as either (1) a directed path from node s to node t, or (2) a directed cycle W that is gainy, plus a directed path from some node in W to node t.

(a) Show that if $G(x)$ contains a generalized augmenting path, we can increase the flow into the sink. Use this result to describe an algorithm for the generalized maximum flow problem.

(b) Describe a polynomial-time algorithm for identifying a generalized augmenting path. (*Hint*: To identify the second type of augmenting path, let G' be the subgraph of G consisting of those nodes that have directed paths to node t. Then look for a gainy cycle in G'.)

15.27. Generalized flow decomposition property. We refer to a generalized network flow problem as a *generalized circulation problem* if $b(i) = 0$ for every node $i \in N$ (i.e., for each node the inflow equals its outflow). In a generalized circulation problem, we refer to any flow x_{ij} satisfying the mass balance and flow bound constraints as a generalized circulation. Two types of generalized circulation are of special interest: *cycle flow* and *bicycle flow*. A cycle flow is a generalized circulation for which $x_{ij} > 0$ only along arcs of a breakeven cycle; a bicycle circulation is a generalized flow for which $x_{ij} > 0$ only along arcs of a bicycle (of either type 1 or type 2). We refer to a cycle or bicycle flow x as *negative* if $\sum_{(i,j)\in A} c_{ij}x_{ij}$ is negative. Show that it is possible to decompose any generalized circulation into flows along at most m cycle or bicycle flows. (*Hint*: Generalize the method we described in the proof of Theorem 3.5 for identifying a breakeven cycle or a bicycle.)

15.28. Generalized flow optimality conditions. Show that a generalized circulation x^* is an optimal solution of the generalized circulation problem if and only if the network contains no negative cycle or bicycle flow with respect to x^*.

15.29. Show that by adding a loop with a multiplier of $\frac{1}{2}$ to every node of a network, we can formulate any generalized flow problem with one or more inequalities for supplies and demands (i.e., the mass balance constraints are stated as "$\leq b(i)$" for a supply node i, and/or "$\geq b(j)$" for a demand node j) into an equivalent problem with all equality constraints (i.e., "$= b(k)$" for all nodes k).

16

LAGRANGIAN RELAXATION AND NETWORK OPTIMIZATION

I never missed the opportunity to remove obstacles in the way of unity.
—Mohandas Gandhi

Chapter Outline

16.1 INTRODUCTION

As we have noted throughout our discussion in this book, the basic network flow models that we have been studying—shortest paths, maximum flows, minimum cost flows, minimum spanning trees, matchings, and generalized and convex flows—arise in numerous applications. These core network models are also building blocks for many other models and applications, in the sense that many models met in practice have embedded network structure: that is, the broader models are network problems with additional variables and/or constraints.

In this chapter we consider ways to solve these models using a solution strategy known as *decomposition* which permits us to draw upon the many algorithms that we have developed in previous chapters to exploit the underlying network structure. In a sense this chapter serves a dual purpose. First, it permits us to introduce a broader set of network optimization models than we have been considering in our earlier discussion. As such, this chapter provides a glimpse of how network flow models arise in a wide range of applied problem settings that cannot be modeled as pure network flow problems. Second, the chapter introduces a solution method, known as *Lagrangian relaxation*, that has become one of the very few solution methods in optimization that cuts across the domains of linear and integer programming, combinatorial optimization, and nonlinear programming.

Perhaps the best way to understand the basic idea of Lagrangian relaxation is via an example.

Constrained Shortest Paths

Consider the network shown in Figure 16.1(a) which has two attributes associated with each arc (i, j): a cost c_{ij} and a traversal time t_{ij}. Suppose that we wish to find the shortest path from the source node 1 to the sink node 6, but we wish to restrict our choice of paths to those that require no more than $T = 10$ time units to traverse. This type of constrained shortest path application arises frequently in practice since in many contexts a company (e.g., a package delivery firm) wants to provide its services at the lowest possible cost and yet ensure a certain level of service to its customers (as embodied in the time restriction). In general, the constrained shortest path problem from node 1 to node n can be stated as the following integer programming problem:

$$\text{Minimize} \sum_{(i,j)\in A} c_{ij}x_{ij} \tag{16.1a}$$

subject to

$$\sum_{\{j:(i,j)\in A\}} x_{ij} - \sum_{\{j:(j,i)\in A\}} x_{ji} = \begin{cases} 1 & \text{for } i = 1 \\ 0 & \text{for } i \in N - \{1, n\}, \\ -1 & \text{for } i = n \end{cases} \tag{16.1b}$$

$$\sum_{(i,j)\in A} t_{ij}x_{ij} \le T, \tag{16.1c}$$

$$x_{ij} = 0 \text{ or } 1 \quad \text{for all } (i, j) \in A. \tag{16.1d}$$

The problem is not a shortest path problem because of the timing restriction. Rather, it is a shortest path problem with an additional side constraint (16.1c). Instead of solving this problem directly, suppose that we adopt an indirect approach by combining time and cost into a single *modified cost*; that is, we place a dollar equivalent on time. So instead of setting a limit on the total time we can take on the chosen path, we set a "toll charge" on each arc proportional to the time that it takes to

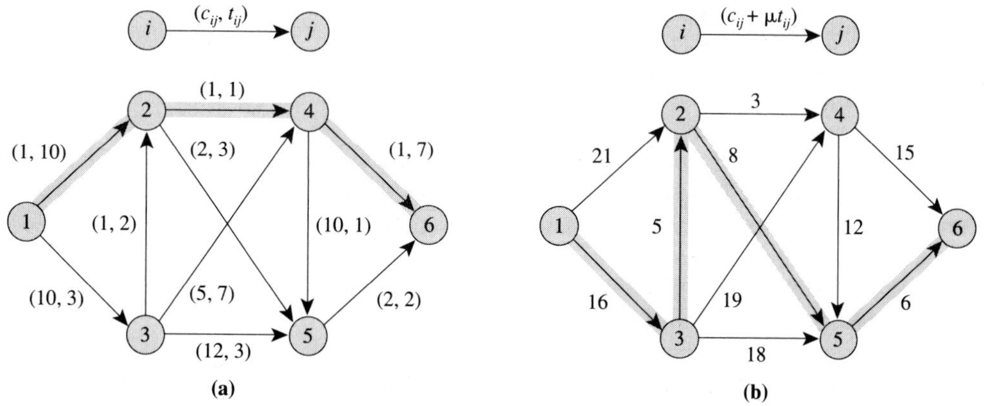

Figure 16.1 Time-constrained shortest path problem: (a) constrained shortest path problem (bold lines denote the shortest path for $\mu = 0$); (b) modified cost $c + \mu t$ with Lagrange multiplier $\mu = 2$ (bold lines denote the shortest path).

traverse that arc. For example, we might charge \$2 for each hour that it takes to traverse any arc. Note that if the toll charge is zero, we are ignoring time altogether and the problem becomes a usual shortest path problem with respect to the given costs. On the other hand, if the toll charge is very large, these charges become the dominant cost and we will be seeking the quickest path from the source to the sink. Can we find a toll charge somewhere in between these values so that by solving the shortest path problem with the combined costs (the toll charges and the original costs), we solve the constrained shortest path problem as a single shortest path problem?

For any choice μ of the toll charge, we solve a shortest path problem with respect to the modified costs $c_{ij} + \mu t_{ij}$. For the sample data shown in Figure 16.1(a), if $\mu = 0$, the modified problem becomes the shortest path problem with respect to the original costs c_{ij} and the shortest path 1–2–4–6 has length 3. This value is an obvious lower bound on the length of the constrained shortest path since it ignores the timing constraint. Now suppose that we set $\mu = 2$ and solve the modified problem. Figure 16.1(b) shows the modified costs $c_{ij} + 2t_{ij}$. The shortest path 1–3–2–5–6 has length 35. In this case, the path 1–3–2–5–6 that solves the modified problem happens to require 10 units to traverse, so it is a feasible constrained shortest path. Is it an optimal constrained shortest path?

To answer this question, let us make an important observation (which we will prove formally in the next section). Let P, with cost $c_P = \sum_{(i,j)\in P} c_{ij}$ and traversal time $t_P = \sum_{(i,j)\in P} t_{ij}$, be *any* feasible path to the constrained shortest path problem, and let $l(\mu)$ denote the optimal length of the shortest path with the modified costs when we impose a toll of μ units. Since the path P is feasible for the constrained shortest path problem, the time t_P required to traverse this path is at most $T = 10$ units. With respect to the modified costs $c_{ij} + \mu t_{ij}$, the cost $c_P + \mu t_P$ of the path P is the path's true cost c_P plus $\mu t_P \le \mu T$ units. Therefore, if we subtract μT from the modified cost $c_P + \mu t_P$ of this path, we obtain a lower bound $c_P + \mu t_P - \mu T = c_P + \mu(t_P - T) \le c_P$ on the cost c_P of this path. Since the shortest path with respect to the modified arc costs is less than or equal to the modified cost of any particular path, $l(\mu) \le c_P + \mu t_P$ and so $l(\mu) - \mu T$ is a *common* lower bound on the length of *any* feasible path P and thus on the length of the constrained shortest path. Because this argument is completely general and applies to any value $\mu \ge 0$ of the toll charges, if we subtract μT from the optimal length of the shortest path of the modified problem, we obtain a lower bound on the optimal cost of the constrained shortest path problem.

Bounding Principle. *For any nonnegative value of the toll μ, the length $l(\mu)$ of the modified shortest path with costs $c_{ij} + \mu t_{ij}$ minus μT is a lower bound on the length of the constrained shortest path.*

Note that for our numerical example, for $\mu = 2$, the cost of the modified shortest path problem is 35 units and so $35 - 2(T) = 35 - 2(10) = 15$ is a lower bound on the length of the optimal constrained shortest path. But since the path 1–3–2–5–6 is a feasible solution to the constrained shortest path problem and its cost equals the lower bound of 15 units, we can be assured that it is an optimal constrained shortest path.

Observe that in this example we have been able to solve a difficult optimization model (the constrained shortest path problem is an $\mathcal{NP}$-complete problem) by removing one or more problem constraints—in this case the single timing constraint—that makes the problem much more difficult to solve. Rather than solving the difficult optimization problem directly, we combined the complicating timing constraint with the original objective function, via the toll μ, so that we could then solve a resulting embedded shortest path problem. The motivation for adopting this approach was our observation that the original constrained shortest path problem had an attractive substructure, the shortest path problem, that we would like to exploit algorithmically. Whenever we can identify such attractive substructure, we could adopt a similar approach. For reasons that will become clearer in the next section, this general solution approach has become known as *Lagrangian relaxation.*

In our example we have been fortunate to find a constrained shortest path by solving the Lagrangian subproblem for a particular choice of the toll μ. We will not always be so lucky; nevertheless, as we will see, the lower bounding mechanism of Lagrangian relaxation frequently provides valuable information that we can exploit algorithmically.

Lagrangian relaxation is a general solution strategy for solving mathematical programs that permits us to decompose problems to exploit their special structure. As such, this solution approach is perfectly tailored for solving many models with embedded network structure. The Lagrangian solution strategy has a number of significant advantages:

1. Since it is often possible to decompose models in several ways and apply Lagrangian relaxation to each different decomposition, Lagrangian relaxation is a very flexible solution approach. Indeed, because of its flexibility, Lagrangian relaxation is more of a general problem solving strategy and solution framework than any single solution technique.
2. In decomposing problems, Lagrangian relaxation solves core subproblems as stand-alone models. Consequently, the solution approach permits us to exploit any known methodology or algorithm for solving the subproblems. In particular, when the subproblems are network models, the Lagrangian solution approach can take advantage of the various algorithms that we have developed previously in this book.
3. As we have already noted, Lagrangian relaxation permits us to develop bounds on the value of the optimal objective function and, frequently, to quickly generate good, though not necessarily optimal solutions with associated performance guarantees—that is, a bound on how far the solution could possibly be from optimality (in objective function value). In many instances in the context of integer programming, the bounds provided by Lagrangian relaxation methods are much better than those generated by solving the linear programming relaxation of the problems, and as a consequence, Lagrangian relaxation is often an attractive alternative to linear programming as a bounding mechanism in branch-and-bound methods for solving integer programs.
4. In many instances we can use Lagrangian relaxation methods to devise effective heuristic solution methods for solving complex combinatorial optimization problems and integer programs.

In the remainder of this chapter we describe the Lagrangian relaxation solution approach in more detail and demonstrate its use in solving several important network optimization models. Our purpose is not to present a comprehensive treatment of Lagrangian relaxation or of its applications to the field of network optimization, but rather to introduce this general solution strategy and to illustrate its applications in a way that would lay the essential foundations for applying the method in many other problem contexts. As a by-product of this discussion, in the text and in the exercises at the end of this chapter we introduce several noteworthy network optimization models that we do not treat elsewhere in the book.

Since one of the principal uses of Lagrangian relaxation is within implicit enumeration procedures for solving integer programs, before describing Lagrangian relaxation in more detail, we first discuss its use within classical branch-and-bound algorithms for solving integer programs. The reader can skip this section without loss of continuity.

16.2 PROBLEM RELAXATIONS AND BRANCH AND BOUND

In the last section we observed that Lagrangian relaxation permits us to develop a lower bound on the optimal length of a constrained shortest path. In Section 16.3 we develop a generalization of this result, showing that we can obtain a lower bound on the optimal objective function value of any minimization problem. These lower bounds can be of considerable value: for example, for our constrained shortest path example, we were able to use a lower bound to demonstrate that a particular solution that we generated by solving a shortest path subproblem, with modified costs, was optimal for the overall constrained problem. In general, we will not always be as fortunate in being able to use a lower bound to guarantee that the solution to a single subproblem solves the original problem. Nevertheless, as we show briefly in the section, we might still be able to use lower bounds as an algorithmic tool in reducing the number of computations required to solve combinatorial optimization problems formulated as integer programs.

Consider the following integer programming model:

$$\text{Minimize} \quad cx$$

subject to

$$x \in F.$$

In this formulation, the set F represents the set of feasible solutions to an integer program, that is, the set of solutions $x = (x_1, x_2, \ldots, x_J)$ to the system

$$\mathcal{A}x = b,$$

$$x_j = 0 \text{ or } 1 \qquad \text{for } j = 1, 2, \ldots, J.$$

In a certain conceptual sense, this integer program is trivial to solve: We simply enumerate every combination of the decision variables, that is, all zero–one vectors $(x_1, x_2, \ldots, x_J)$ obtained by setting each variable x_j to value zero or 1; from among all those vectors that satisfy the given equality constraint $\mathcal{A}x = b$, we choose the combination with the smallest value of the objective function cx. Of course, because of its combinatorial explosiveness, this total enumeration procedure is limited to very small problems; for a problem with 100 decision variables, even if we could compute one solution every nanosecond (10^{-9} second), enumerating all 2^{100} solutions would take us over a million million years—that is, a million different million years!

Can we avoid any of these computations? Suppose that $F = F^1 \cup F^2$. For example, we might obtain F^1 from F by adding the constraint $x_1 = 0$ and F^2 by adding the constraint $x_1 = 1$. Note that the optimal solution over the feasible set F is the best of the optimal solutions over F^1 and F^2. Suppose that we already have found an optimal solution $\bar{x}$ to $\min\{cx : x \in F^2\}$ and that its objective function value is $z(\bar{x}) = 100$. The number of potential integer solutions in F^1 is still 2^{J-1}, so it will be prohibitively expensive to enumerate all these possibilities, except when J is small.

Rather than attempt to solve the integer program over the feasible region F^1, suppose that we solve a *relaxed* version of the problem, possibly by relaxing the integrality constraints, and/or possibly by performing a Lagrangian relaxation of the problem. In general, we obtain a relaxation by removing some constraints from the model: for example, by replacing the restrictions $x_j \geq 0$ and integer, by the restriction $x_j \geq 0$, or by deleting one or more constraints of the form $\alpha x = \beta$. We could use many different types of relaxation—in Lagrangian relaxation, for example, we not only delete some problem constraints, but we also change the objective function of the problem. For the purpose of this discussion, we merely require that we relax some of the problem constraints and that the objective function value of the relaxation is a lower bound on the objective function value of the original problem.

Let x' denote an optimal solution to the relaxation, and let $z(x')$ denote the objective function value of this solution. We consider four possibilities:

1. The solution x' does not exist because the relaxed problem has no feasible solution.
2. The solution x' happens to lie in F^1 (even though we relaxed some of the constraints).
3. The solution x' does not lie in F^1 and its objective function value $z(x')$ satisfies the inequality $z(x') \geq z(\bar{x}) = 100$.
4. The solution x' does not lie in F^1 and its objective function value $z(x')$ of x' satisfies the inequality $z(x') < z(\bar{x}) = 100$.

Note that these four alternatives exhaust all possible outcomes and are mutually exclusive. Therefore, exactly one of them must occur.

We now make an important observation. In cases 1 to 3, we can terminate our computations: we have solved the original problem over the set F, even though we have not explicitly solved any integer program (assuming that we obtained the solution over the set F^2 without solving an integer program). In case 1, since the relaxation of the set F^1 is empty, the set F^1 is also empty, so the solution $\bar{x}$ solves the original (overall) integer program. In case 2, since we have found the optimal solution in the relaxation (and so a superset) of the set F^1, and this solution lies in

F^1, we have also found the best solution in F^1; therefore, either $\bar{x}$ or x' is the solution to the original problem (whichever solution has the smaller objective function value). Note that in this case we have implicitly considered (enumerated) all of the solutions in F^1 in the sense that we know that no solution in this set is better than $\bar{x}$. In case 3, the solution $\bar{x}$ has as good an objective function value as the best solution in a relaxation of F^1, so it has an objective function value that is as good as any solution in F^1. Therefore, $\bar{x}$ solves the original problem. Note that in case 3 we have used *bounding* information on the objective function value to eliminate the solutions in the set F^1 from further consideration.

In case 4, we have not yet solved the original problem. We can either try to solve the problem minimize $\{cx : x \in F^1\}$ by some direct method of integer programming or, we can partition F^1 into two sets F^3 and F^4. For example, we might obtain F^3 from F by constraining $x_1 = 0$ and $x_2 = 0$ and obtain F^4 by setting $x_1 = 0$ and $x_2 = 1$. We could then apply any relaxation or direct approach for the problems defined over the sets F^3 and F^4.

In a general branch-and-bound procedure, we would systematically partition the feasible region F into subregions $F^1, F^2, F^3, \ldots, F^K$. Let $\bar{x}$ denote the best feasible solution (in objective function value) we have obtained in prior computations. Suppose that for each $k = 1, 2, \ldots, K$, either F^k is empty or x^k is a solution of a relaxation of the set F^k and $c\bar{x} \le cx^k$. Then no point in any of the regions F^1, $F^2, F^3, \ldots, F^k$ could have a better objective function value than $\bar{x}$, so $\bar{x}$ solves the original optimization problem. If $c\bar{x} > cx^k$, though, for any region F^k, we would need to subdivide this region by "branching" on some of the variables (i.e., dividing a subregion in two by setting $x_j = 0$ or $x_j = 1$ for some variable j to define two new subregions). Whenever we have satisfied the test $c\bar{x} \le cx^k$ for all of the subregions (or we know they are empty), we have solved the original problem.

The intent of the branch-and-bound method is to find an optimal solution by solving only a small number of relaxations. To do so, we would need to obtain good solutions quickly and obtain good relaxations so that the objective function value $z(x^k)$ of the solution x^k to the relaxation of the set F^k is close in objective function value to the optimal solution over F^k itself.

In practice, in implementing the branch-and-bound procedure, we need to make many design decisions concerning the order for choosing the subregions, the variables to branch on for each subregion, and mechanisms (e.g., heuristic procedures) that we might use to find "good" feasible solutions. The literature contains many clever approaches for resolving these issues and for designing branch-and-bound procedures that are quite effective in practice. We also need to develop good relaxations that would permit us to obtain effective (tight) lower bounds: if the lower bounds are weak, cases 2 and 3 will rarely occur and the branch-and-bound procedure will degenerate into complete enumeration. On the other hand, if the bounds are very tight, the relaxations will permit us to eliminate much of the enumeration and develop very effective solution procedures. Since our purpose in this chapter is to introduce one relaxation procedure that has proven to be very effective in practice and discuss its applications, we will not consider the detailed design choices for implementing the branch-and-bound procedure.

We next summarize the basic underlying ideas of the Lagrangian relaxation technique.

16.3 LAGRANGIAN RELAXATION TECHNIQUE

To describe the general form of the Lagrangian relaxation procedure, suppose that we consider the following generic optimization model formulated in terms of a vector x of decision variables:

$$z^* = \min cx$$

subject to

$$\mathcal{A}x = b, \tag{P}$$

$$x \in X.$$

This model (P) has a linear objective function cx and a set $\mathcal{A}x = b$ of explicit linear constraints. The decision variables x are also constrained to lie in a given constraint set X which, as we will see, often models embedded network flow structure. For example, the constraint set $X = \{x : \mathcal{N}x = q, 0 \le x \le u\}$ might be all the feasible solutions to a network flow problem with a supply/demand vector q. Or, the set X might contain the incidence vectors of all spanning trees or matchings of a given graph. Unless we state otherwise, we assume that the set X is finite (e.g., for network flow problems, we will let it be the finite set of spanning tree solutions).

As its name suggests, the Lagrangian relaxation procedure uses the idea of relaxing the explicit linear constraints by bringing them into the objective function with associated Lagrange multipliers μ (this old idea might be a familiar one from advanced calculus in the context of solving nonlinear optimization problems). We refer to the resulting problem

$$\text{Minimize} \quad cx + \mu(\mathcal{A}x - b)$$

subject to

$$x \in X,$$

as a *Lagrangian relaxation* or *Lagrangian subproblem* of the original problem, and refer to the function

$$L(\mu) = \min\{cx + \mu(\mathcal{A}x - b) : x \in X\},$$

as the *Lagrangian function*. Note that since in forming the Lagrangian relaxation, we have eliminated the constraints $\mathcal{A}x = b$ from the problem formulation, the solution of the Lagrangian subproblem need not be feasible for the original problem (P). Can we obtain any useful information about the original problem even when the solution to the Lagrangian subproblem is not feasible in the original problem (P)? The following elementary observation is a key result that helps to answer this question and that motivates the use of the Lagrangian relaxation technique in general.

Lemma 16.1 (Lagrangian Bounding Principle). *For any vector μ of the Lagrangian multipliers, the value $L(\mu)$ of the Lagrangian function is a lower bound on the optimal objective function value z^* of the original optimization problem* (P).

Proof. Since $\mathcal{A}x = b$ for every feasible solution to (P), for any vector μ of Lagrangian multipliers, $z^* = \min\{cx : \mathcal{A}x = b, x \in X\} = \min\{cx + \mu(\mathcal{A}x - b) : \mathcal{A}x = b, x \in X\}$. Since removing the constraints $\mathcal{A}x = b$ from the second formulation cannot lead to an increase in the value of the objective function (the value might decrease), $z^* \geq \min\{cx + \mu(\mathcal{A}x - b) : x \in X\} = L(\mu)$. ◆

As we have seen, for any value of the Lagrangian multiplier μ, $L(\mu)$ is a lower bound on the optimal objective function value of the original problem. To obtain the sharpest possible lower bound, we would need to solve the following optimization problem

$$L^* = \max_{\mu} L(\mu)$$

which we refer to as the *Lagrangian multiplier problem* associated with the original optimization problem (P). The Lagrangian bounding principle has the following immediate implication.

Property 16.2 (Weak Duality). *The optimal objective function value L^* of the Lagrangian multiplier problem is always a lower bound on the optimal objective function value of the problem* (P) *(i.e., $L^* \leq z^*$).*

Our preceding discussion provides us with valid bounds for comparing objective function values of the Lagrange multiplier problem and optimization (P) for any choices of the Lagrange multipliers μ and any feasible solution x of (P):

$$L(\mu) \leq L^* \leq z^* \leq cx.$$

These inequalities furnish us with a guarantee when a Lagrange multiplier μ to the Lagrange multiplier problem or a feasible solution x to the original problem (P) are optimal.

Property 16.3 (Optimality Test)

(a) *Suppose that μ is a vector of Lagrangian multipliers and x is a feasible solution to the optimization problem* (P) *satisfying the condition $L(\mu) = cx$. Then $L(\mu)$ is an optimal solution of the Lagrangian multiplier problem [i.e., $L^* = L(\mu)$] and x is an optimal solution to the optimization problem* (P).

(b) *If for some choice of the Lagrangian multiplier vector μ, the solution x^* of the Lagrangian relaxation is feasible in the optimization problem* (P), *then x^* is an optimal solution to the optimization problem* (P) *and μ is an optimal solution to the Lagrangian multiplier problem.*

Note that by assumption in part (b) of this property, $L(\mu) = cx^* + \mu(\mathcal{A}x^* - b)$ and $\mathcal{A}x^* = b$. Therefore, $L(\mu) = cx^*$ and part (a) implies that x^* solves problem (P) and μ solves the Lagrangian multiplier problem.

As indicated by Property 16.3, the bounding principle immediately implies one advantage of the Lagrangian relaxation approach—the method can give us a *certificate* [in the form of the equality $L(\mu) = cx$ for some Lagrange multiplier μ] for guaranteeing that a given feasible solution x to the optimization problem (P) is an optimal solution. Even if $L(\mu) < cx$, having the lower bound permits us to state a bound on how far a given solution is from optimality: If $[cx - L(\mu)]/L(\mu) \leq 0.05$, for example, we know that the objective function value of the feasible solution x is no more than 5% from optimality. This type of bound is very useful in practice—it permits us to assess the degree of suboptimality of given solutions and it permits us to terminate our search for an optimal solution when we have a solution that we know is close enough to optimality (in objective function value) for our purposes.

Lagrangian Relaxation and Inequality Constraints

In the optimization model (P), the constraints $\mathcal{A}x = b$ are all equality constraints. In practice, we often encounter models, such as the constrained shortest path problem, that are formulated more naturally in inequality form $\mathcal{A}x \leq b$. The Lagrangian multiplier problem for these problems is a slight variant of the one we have just introduced: The Lagrangian multiplier problem becomes

$$L^* = \max_{\mu \geq 0} L(\mu).$$

That is, the only change in the Lagrangian multiplier problem is that the Lagrangian multipliers now are restricted to be nonnegative. In Exercise 16.1, by introducing "slack variables" to formulate the inequality problem as an equivalent equality problem, we show how to obtain this optimal multiplier problem from the one we have considered for the equality problem. This development implies that the bounding property, the weak duality property, and the optimality test 16.3(a) are valid when we apply Lagrangian relaxation to any combination of equality and inequality constraints.

There is, however, one substantial difference between relaxing equality constraints and inequality constraints. When we relax inequality constraints $\mathcal{A}x \leq b$, if the solution x^* of the Lagrangian subproblem happens to satisfy these constraints, it need *not* be optimal (see Exercise 16.2). In addition to being feasible, this solution needs to satisfy the *complementary slackness condition* $\mu(\mathcal{A}x^* - b) = 0$, which is familiar to us from much of our previous discussion of network flows in section 9.4.

Property 16.4. *Suppose that we apply Lagrangian relaxation to the optimization problem* ($P^{\leq}$) *defined as minimize* $\{cx : \mathcal{A}x \leq b$ *and* $x \in X\}$ *by relaxing the inequalities* $\mathcal{A}x \leq b$. *Suppose, further, that for some choice of the Lagrangian multiplier vector* μ, *the solution* x^* *of the Lagrangian relaxation* (1) *is feasible in the optimization problem* ($P^{\leq}$), *and* (2) *satisfies the complementary slackness condition* $\mu(\mathcal{A}x^* - b) = 0$. *Then* x^* *is an optimal solution to the optimization problem* ($P^{\leq}$).

Proof. By assumption, $L(\mu) = cx^* + \mu(\mathcal{A}x^* - b)$. Since $\mu(\mathcal{A}x^* - b) = 0$, $L(\mu) = cx^*$. Moreover, since $\mathcal{A}x^* \leq b$, x^* is feasible, and so by Property 16.3(a) x^* solves problem ($P^{\leq}$). ◆

Are solutions to the Lagrangian subproblem of use in solving the original problem? Properties 16.3 and 16.4 show that certain solutions of the Lagrangian subproblem provably solve the original problem. We might distinguish two other cases: (1) when solutions obtained by relaxing inequality constraints are feasible but are not provably optimal for the original problem (since they do not satisfy the complementary slackness condition), and (2) when solutions to the Lagrangian relaxation are not feasible in the original problem.

In the first case, the solutions are candidate optimal solutions (possibly for use in a branch-and-bound procedure). In the second case, for many applications, researchers have been able to devise methods to modify "modestly" infeasible solutions so that they become feasible with only a slightly degradation in the objective function value. These observations suggest that we might be able to use the solutions obtained from the Lagrangian subproblem as "approximate" solutions to the original problem, even when they are not provably optimal; in these instances, we can use Lagrangian relaxation as a heuristic method for generating provably good solutions in practice (the solutions might be provably good because of the Lagrangian lower bound information). The development of these heuristic methods depends heavily on the problem context we are studying, so we will not attempt to provide any further details.

Solving the Lagrangian Multiplier Problem

How might we solve the Lagrangian multiplier problem? To develop an understanding of possible solution techniques, let us consider the constrained shortest path problem that we defined in Section 16.1. Suppose that now we have a time limitation of $T = 14$ instead of $T = 10$. When we relax the time constraint, the Lagrangian multiplier function $L(\mu)$ for the constrained shortest path problem becomes

$$L(\mu) = \min\{c_P + \mu(t_P - T) : P \in \mathcal{P}\}.$$

In this formulation, $\mathcal{P}$ is the collection of all directed paths from the source node 1 to the sink node n. For convenience, we refer to the quantity $c_P + \mu(t_P - T)$ as the *composite cost* of the path P. For a specific value of the Lagrangian multiplier μ, we can solve $L(\mu)$ by enumerating all the directed paths in $\mathcal{P}$ and choosing the path with the smallest composite cost. Consequently, we can solve the Lagrangian multiplier problem by determining $L(\mu)$ for all nonnegative values of the Lagrangian multiplier μ and choosing the value that achieves $\max_{\mu \geq 0} L(\mu)$.

Let us illustrate this brute force approach geometrically. Figure 16.2 records the cost and time data for every path for our numerical example. Note that the composite cost $c_P + \mu(t_P - T)$ for any path P is a linear function of μ with an intercept of c_P and a slope of $(t_P - T)$. In Figure 16.3 we have plotted each of these path composite cost functions. Note that for any specific value of the Lagrange multiplier μ, we can find $L(\mu)$ by evaluating each composite cost function (line) and identifying the one with the least cost. This observation implies that the Lagrangian multiplier function $L(\mu)$ is the lower envelope of the composite cost lines and that the highest point on this envelope corresponds to the optimal solution of the Lagrangian multiplier problem.

In practice, we would never attempt to solve the problem in this way because the number of directed paths from the source node to the sink node typically grows exponentially in the number of nodes in the underlying network, so any such enumeration procedure would be prohibitively expensive. Nevertheless, this problem geometry helps us to understand the nature of the Lagrangian multiplier problem and suggests methods for solving the problem.

As we noted in the preceding paragraph, to find the optimal multiplier value μ^* of the Lagrangian multiplier problem, we need to find the highest point of the Lagrangian multiplier function $L(\mu)$. Suppose that we consider the polyhedron de-

Path P	Path cost c_P	Path time t_P	Composite cost $c_P + \mu (t_P - T)$
1–2–4–6	3	18	$3 + 4\mu$
1–2–5–6	5	15	$5 + \mu$
1–2–4–5–6	14	14	14
1–3–2–4–6	13	13	$13 - \mu$
1–3–2–5–6	15	10	$15 - 4\mu$
1–3–2–4–5–6	24	9	$24 - 5\mu$
1–3–4–6	16	17	$16 + 3\mu$
1–3–4–5–6	27	13	$27 - \mu$
1–3–5–6	24	8	$24 - 6\mu$

Figure 16.2 Path cost and time data for constrained shortest path example with $T = 14$.

fined by those points that lie on or below the function $L(\mu)$. These are the shaded points in Figure 16.3. Then geometrically, we are finding the highest point in a polyhedron defined by the function $L(\mu)$, which is a linear program.

Even though we have illustrated this property on a specific example, this situation is completely general. Consider the generic optimization model (P), defined

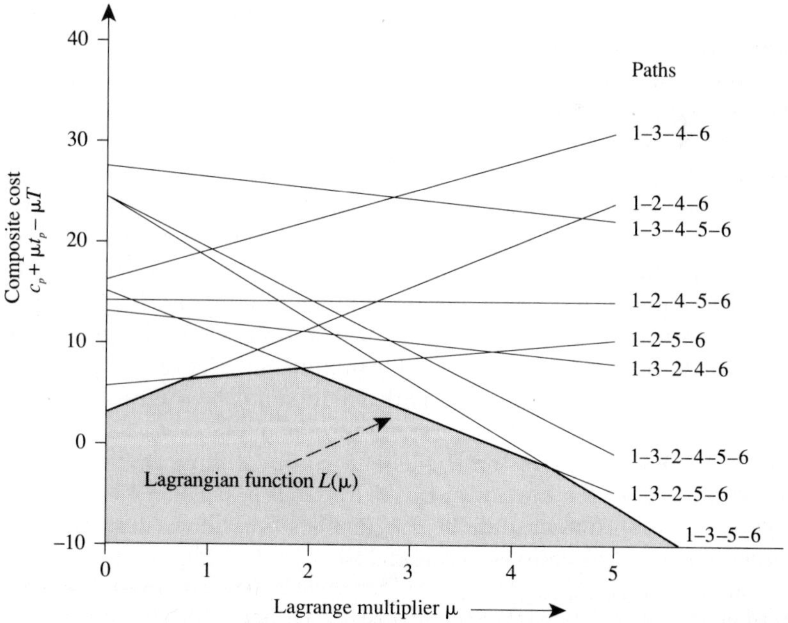

Figure 16.3 Lagrangian function for $T = 14$.

as $\min\{cx : \mathcal{A}x = b, x \in X\}$ and suppose that the set $X = \{x^1, x^2, \ldots, x^K\}$ is finite. By relaxing the constraints $\mathcal{A}x = b$, we obtain the Lagrangian multiplier function $L(\mu) = \min\{cx + \mu(\mathcal{A}x - b) : x \in X\}$. By definition,

$$L(\mu) \le cx^k + \mu(\mathcal{A}x^k - b) \qquad \text{for all } k = 1, 2, \ldots, K.$$

In the space of composite costs and Lagrange multipliers μ (as in Figure 16.3), each function $cx^k + \mu(\mathcal{A}x^k - b)$ is a multidimensional "line" called a *hyperplane* (if μ is two-dimensional, it is a plane). The Lagrangian multiplier function $L(\mu)$ is the lower envelope of the hyperplanes $cx^k + \mu(\mathcal{A}x^k - b)$ for $k = 1, 2, \ldots, K$. In the Lagrangian multiplier problem, we wish to determine the highest point on this envelope: We can find this point by solving the optimization problem

$$\text{Maximize} \quad w$$

subject to

$$w \le cx^k + \mu(\mathcal{A}x^k - b) \qquad \text{for all } k = 1, 2, \ldots, K,$$

$$\mu \text{ unrestricted},$$

which is clearly a linear program. We state this result as a theorem.

Theorem 16.5. *The Lagrangian multiplier problem $L^* = \max_\mu L(\mu)$ with $L(\mu) = \min\{cx^k + \mu(\mathcal{A}x - b) : x \in X\}$ is equivalent to the linear programming problem $L^* = \max\{w : w \le cx^k + \mu(\mathcal{A}x^k - b) \text{ for } k = 1, 2, \ldots, K\}$.* ◆

Since, as shown by the preceding theorem, the Lagrangian multiplier problem is a linear program, we could solve this problem by applying the linear programming methodology. One resulting algorithm, which is known as *Dantzig–Wolfe decomposition* or *generalized linear programming*, is an important solution methodology that we discuss in some depth in Chapter 17 in the context of solving the multicommodity flow problem. One of the disadvantages of this approach is that it requires the solution of a series of linear programs that are rather expensive computationally. Another approach might be to apply some type of gradient method to the Lagrangian function $L(\mu)$. As shown by the constrained shortest path example, the added complication of this approach is that the Lagrangian function $L(\mu)$ is not differentiable. It is differentiable whenever the optimal solution of the Lagrangian subproblem is unique; but when the subproblem has two or more solutions, the Lagrangian function generally is not differentiable. For example, in Figure 16.4, at $\mu = 0$, the path 1–2–4–6 is the unique shortest path solution to the subproblem and the function $L(\mu)$ is differentiable. At this point, for the path P = 1–2–4–6, $L(\mu) = c_P + \mu(t_P - T)$; since $t_P = 18$ and $T = 14$, $L(\mu)$ has a slope $(t_P - T) = (18 - 14) = 4$. At the point $\mu = 2$, however, the paths 1–2–5–6 and 1–3–2–5–6 both solve the Lagrangian subproblem and the Lagrangian function is not differentiable. To accommodate these situations, we next describe a technique, known as the *subgradient optimization technique*, for solving the (nondifferentiable) Lagrangian multiplier problem.

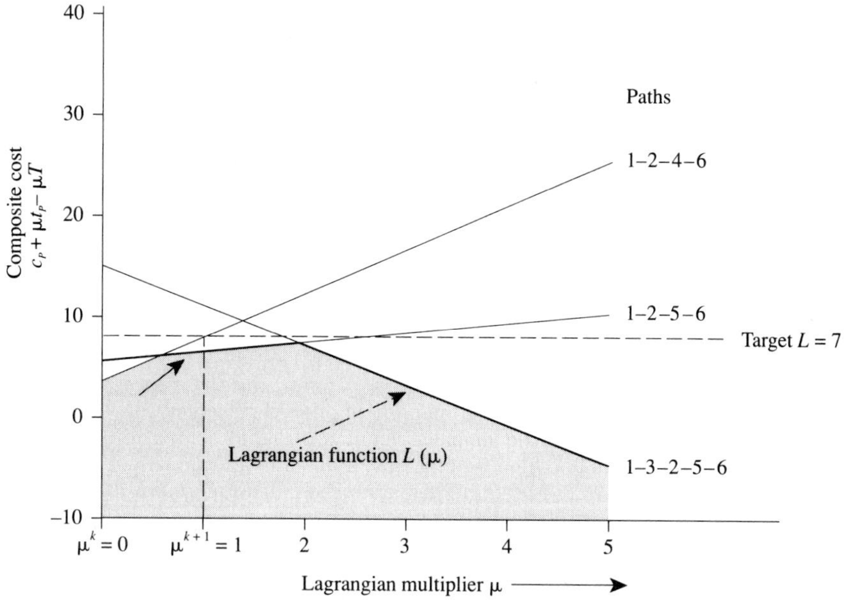

Figure 16.4 Steps of Newton's method for $T = 14$.

Subgradient Optimization Technique

In solving optimization problems with the nonlinear objective function $f(x)$ of an n-dimensional vector x, researchers and practitioners often use variations of the following classical idea: Form the gradient $\nabla f(x)$ of f defined as a row vector with components $(\partial f(x)/dx_1, \partial f(x)/dx_2, \ldots, \partial f(x)/dx_n)$. Recall from advanced calculus that the directional derivative of f in the direction d satisfies the equality

$$\lim_{\theta \to 0} \frac{f(x + \theta d) - f(x)}{\theta} = \nabla f(x) d.$$

So if we choose the direction d so that $\nabla f(x)d > 0$ and move in the direction d with a small enough "step length" θ—that is, change x to $x + \theta d$—we move uphill. This simple observation lies at the core of a considerable literature in nonlinear programming known as *gradient methods*.

Suppose that in solving the Lagrangian multiplier problem, we are at a point where the Lagrangian function $L(\mu) = \min\{cx + \mu(\mathcal{A}x - b) : x \in X\}$ has a unique solution $\bar{x}$, so is differentiable. Since $L(\mu) = c\bar{x} + \mu(\mathcal{A}\bar{x} - b)$ and the solution $\bar{x}$ remains optimal for small changes in the value of μ, the gradient at this point is $\mathcal{A}\bar{x} - b$, so a gradient method would change the value of μ as follows:

$$\mu \leftarrow \mu + \theta(\mathcal{A}\bar{x} - b).$$

In this expression, θ is a step size (a scalar) that specifies how far we move in the gradient direction. Note that this procedure has a nice intuitive interpretation. If $(\mathcal{A}\bar{x} - b)_i = 0$, the solution x uses up exactly the required units of the ith resource, and we hold the Lagrange multiplier (the toll) μ_i of that resource at its current value;

if $(\mathcal{A}\bar{x} - b)_i < 0$, the solution x uses up less than the available units of the ith resource and we decrease the Lagrange multiplier μ_i on that resource; and if $(\mathcal{A}\bar{x} - b)_i > 0$, the solution x uses up more than the available units of the ith resource and we increase the Lagrange multiplier μ_i on that resource.

To solve the Lagrangian multiplier problem, we adopt a rather natural extension of this solution approach. We let μ^0 be any initial choice of the Lagrange multiplier; we determine the subsequent values μ^k for $k = 1, 2, \ldots$, of the Lagrange multipliers as follows:

$$\mu^{k+1} = \mu^k + \theta_k(\mathcal{A}x^k - b).$$

In this expression, x^k is any solution to the Lagrangian subproblem when $\mu = \mu^k$ and θ_k is the step length at the kth iteration.

To ensure that this method solves the Lagrangian multiplier problem, we need to exercise some care in the choice of the step sizes. If we choose them too small, the algorithm would become stuck at the current point and not converge; if we choose the step sizes too large, the iterates μ^k might overshoot the optimal solution and perhaps even oscillate between two nonoptimal solutions (see Exercise 16.4 for an example). The following compromise ensures that the algorithm strikes an appropriate balance between these extremes and does converge:

$$\theta_k \rightarrow 0 \quad \text{and} \quad \sum_{j=1}^{k} \theta_j \rightarrow \infty.$$

For example, choosing $\theta_k = 1/k$ satisfies these conditions. These conditions ensure that the algorithm always converges to an optimal solution of the multiplier problem, but a proof of this convergence result is beyond the scope of our coverage in this book (the reference notes cite papers and books that examine the convergence of subgradient methods).

One important variant of the subgradient optimization procedure would be an adaptation of "Newton's method" for solving systems of nonlinear equations. Suppose, as before, that $L(\mu^k) = cx^k + \mu^k(\mathcal{A}x^k - b)$; that is, x^k solves the Lagrangian subproblem when $\mu = \mu^k$. Suppose that we assume that x^k continues to solve the Lagrangian subproblem as we vary μ; or, stated in another way, we make a linear approximation $r(\mu) = cx^k + \mu(\mathcal{A}x^k - b)$ to $L(\mu)$. Suppose further that we know the optimal value L^* of the Lagrangian multiplier problem (which we do not). Then we might move in the subgradient direction until the value of the linear approximation exactly equals L^*. Figure 16.4 shows an example of this procedure when applied to our constrained shortest path example, starting with $\mu^k = 0$. At this point, the path $P = 1\text{–}2\text{–}4\text{–}6$ solves the Lagrangian subproblem and $\mathcal{A}x^k - b$ equals $t_P - T = 18 - 4 = 4$. Since $L^* = 7$ and the path P has a cost $c_P = 3$, in accordance with this linear approximation, or Newton's method, we would approximate $L(\mu)$ by $r(\mu) = 3 + 4\mu$, set $3 + 4\mu = 7$, and define the new value of μ as $\mu^{k+1} = (7 - 3)/4 = 1$. In general, we set the step length θ_k so that

$$r(\mu^{k+1}) = cx^k + \mu^{k+1}(\mathcal{A}x^k - b) = L^*,$$

or since, $\mu^{k+1} = \mu^k + \theta_k(\mathcal{A}x^k - b)$,

$$r(\mu^{k+1}) = cx^k + [\mu^k + \theta_k(\mathcal{A}x^k - b)](\mathcal{A}x^k - b) = L^*.$$

Collecting terms, recalling that $L(\mu^k) = cx^k + \mu(\mathcal{A}x^k - b)$, and letting $\| y \| = (\sum_j y_j^2)^{1/2}$ denote the Euclidean norm of the vector y, we can solve for the step length and find that

$$\theta_k = \frac{L^* - L(\mu^k)}{\| \mathcal{A}x^k - b \|^2}.$$

Since we do not know the optimal objective function value L^* of the Lagrangian multiplier problem (after all, that's what we are trying to find), practitioners of Lagrangian relaxation often use the following popular heuristic for selecting the step length:

$$\theta_k = \frac{\lambda_k[\text{UB} - L(\mu^k)]}{\| \mathcal{A}x^k - b \|^2}.$$

In this expression, UB is an upper bound on the optimal objective function value z^* of the problem (P), and so an upper bound on L^* as well, and λ_k is a scalar chosen (strictly) between 0 and 2. Initially, the upper bound is the objective function value of any known feasible solution to the problem (P). As the algorithm proceeds, if it generates a better (i.e., lower cost) feasible solution, it uses the objective function value of this solution in place of the upper bound UB. Usually, practitioners choose the scalars λ_k by starting with $\lambda_k = 2$ and then reducing λ_k by a factor of 2 whenever the best Lagrangian objective function value found so far has failed to increase in a specified number of iterations. Since this version of the algorithm has no convenient stopping criteria, practitioners usually terminate it after it has performed a specified number of iterations.

The rationale for these choices of the step size and the convergence proof of the subgradient method would take us beyond the scope of our coverage. In passing, we might note that the subgradient optimization procedure is not the only way to solve the Lagrangian multiplier problem: practitioners have used a number of other heuristics, including methods known as *multiplier ascent methods* that are tailored for special problems. Since we merely wish to introduce some of the basic concepts of Lagrangian relaxation and to indicate some of the essential methods used to solve the Lagrangian multiplier problem, we will not discuss these alternative methods.

Subgradient Optimization and Inequality Constraints

As we noted earlier in this section, if we apply Lagrangian relaxation to a problem with constraints $\mathcal{A}x \leq b$ stated in inequality form instead of the equality constraints, the Lagrange multipliers μ are constrained to be nonnegative. The update formula $\mu^{k+1} = \mu^k + \theta_k(\mathcal{A}x^k - b)$ might cause one or more of the components μ_i of μ to become negative. To avoid this possibility, we modify the update formula as follows:

$$\mu^{k+1} = [\mu^k + \theta_k(\mathcal{A}x^k - b)]^+.$$

In this expression, the notation $[y]^+$ denotes the "positive part" of the vector y; that is, the ith component of $[y]^+$ equals the maximum of 0 and y_i. Stated in another way, if the update formula $\mu^{k+1} = \mu^k + \theta_k(\mathcal{A}x^k - b)$ would cause the ith component of μ_i to be negative, then we simply set the value of this component to be zero. We then implement all the other steps of the subgradient procedure (i.e., the choice of the step size θ at each step and the solution of the Lagrangian subproblems) exactly the same as for problems with equality constraints. For problems with both equality and inequality constraints, we use a straightforward mixture of the equality and inequality versions of the algorithm: whenever the update formula for the Lagrange multipliers would cause any component μ_i of μ corresponding to an inequality constraint to become negative, we set the value of that multiplier to be zero.

Let us illustrate the subgradient method for inequality constraints on our constrained shortest path example. Suppose that we start to solve our constrained shortest path problem at $\mu^0 = 0$ with $\lambda^0 = 0.8$ and with UB $= 24$, the cost corresponding to the shortest path 1–3–5–6 joining nodes 1 and 6. Suppose that we choose to reduce the scalar λ_k by a factor of 2 whenever three successive iterations at a given value of λ_k have not improved on the best Lagrangian objective function value $L(\mu)$. As we have already noted, the solution x^0 to the Lagrangian subproblem with $\mu = 0$ corresponds to the path $P = $ 1–2–3–6, the Lagrangian subproblem has an objective function value of $L(0) = 3$, and the subgradient $\mathcal{A}x^0 - b$ at $\mu = 0$ is $(t_P - 14) = 18 - 14 = 4$. So at the first step, we choose

$$\theta_0 = 0.8(24 - 3)/16 = 1.05,$$

$$\mu^1 = [0 + 1.05(4)]^+ = 4.2.$$

For this value of the Lagrange multiplier, from Figure 16.3, we see that the path $P = $ 1–3–2–5–6 solves the Lagrangian subproblem; therefore, $L(4.2) = 15 + 4.2(10) - 4.2(14) = 15 - 16.8 = -1.8$, and $\mathcal{A}x^1 - b$ equals $(t_P - 14) = 10 - 14 = -4$. Since the path 1–3–2–5–6 is feasible, and its cost of 15 is less than UB, we change UB to value 15. Therefore,

$$\theta_1 = 0.8(15 + 1.8)/16 = 0.84,$$

$$\mu^2 = [4.2 + 0.84(-4)]^+ = 0.84.$$

From iterations 2 through 5, the shortest paths alternate between the paths 1–2–4–6 and 1–3–2–5–6. At the end of the fifth iteration, the algorithm has not improved upon (increased) the best Lagrangian objective function value of 6.36 for three iterations, so we reduce λ_k by a factor of 2. In the next 7 iterations the shortest paths are the paths 1–2–5–6, 1–3–5–6, 1–3–2–5–6, 1–3–2–5–6, 1–2–5–6, 1–3–5–6, and 1–3–2–5–6. Once again for three consecutive iterations, the algorithm has not improved the best Lagrangian objective function value, so we decrease λ_k by a factor of 2 to value 0.2. From this point on, the algorithm chooses either path 1–3–2–5–6 or path 1–2–5–6 as the shortest path at each step. Figure 16.5 shows the first 33 iterations of the subgradient algorithm. As we see, the Lagrangian objective function value is converging to the optimal value $L^* = 7$ and the Lagrange multiplier is converging to its optimal value of $\mu^* = 2$.

Note that for this example, the optimal multiplier objective function value of $L^* = 7$ is strictly less than the length of the shortest constrained path, which has value 13. In these instances, we say that the Lagrangian relaxation has a *duality (relaxation) gap*. To solve problems with a duality gap to completion (i.e., to find an optimal solution and a guarantee that it is optimal), we would apply some form of enumeration procedure, such as branch and bound, using the Lagrangian lower bound to help reduce the amount of concentration required.

k	μ^k	$t_p - T$	$L(\mu^k)$	λ_k	θ_k
0	0.0000	4	3.0000	0.80000	1.0500
1	4.2000	−4	−1.8000	0.80000	0.8400
2	0.8400	4	6.3600	0.80000	0.4320
3	2.5680	−4	4.7280	0.80000	0.5136
4	0.5136	4	5.0544	0.80000	0.4973
5	2.5027	−4	4.9891	0.40000	0.2503
6	1.5016	1	6.5016	0.40000	3.3993
7	4.9010	−6	−5.4059	0.40000	0.2267
8	3.5406	−4	0.8376	0.40000	0.3541
9	2.1244	−4	6.5026	0.40000	0.2124
10	1.2746	1	6.2746	0.40000	3.4902
11	4.7648	−6	−4.5886	0.40000	0.2177
12	3.4589	−4	1.1646	0.20000	0.1729
13	2.7671	−4	3.9316	0.20000	0.1384
14	2.2137	−4	6.1453	0.20000	0.1107
15	1.7709	1	6.7709	0.20000	1.6458
16	3.4167	−4	1.3330	0.20000	0.1708
17	2.7334	−4	4.0664	0.20000	0.1367
18	2.1867	−4	6.2531	0.10000	0.0547
19	1.9680	1	6.9680	0.10000	0.8032
20	2.7712	−4	3.9150	0.10000	0.0693
21	2.4941	−4	5.0235	0.10000	0.0624
22	2.2447	−4	6.0212	0.05000	0.0281
23	2.1325	−4	6.4701	0.05000	0.0267
24	2.0258	−4	6.8966	0.05000	0.0253
25	1.9246	1	6.9246	0.00250	0.0202
26	1.9447	1	6.9447	0.00250	0.0201
27	1.9649	1	6.9649	0.00250	0.0201
28	1.9850	1	6.9850	0.00250	0.0200
29	2.0050	−4	6.9800	0.00250	0.0013
30	2.0000	−4	7.0000	0.00250	0.0012
31	1.9950	1	6.9950	0.00250	0.0200
32	2.0150	−4	6.9400	0.00250	0.0013
33	2.0100	−4	6.9601	0.00125	0.0006

Figure 16.5 Subgradient optimization for a constrained shortest path problem.

16.4 LAGRANGIAN RELAXATION AND LINEAR PROGRAMMING

In this section we discuss several theoretical properties of the Lagrangian relaxation technique. As we have noted earlier in Section 16.2, the primary use of the Lagrangian relaxation technique is to obtain lower bounds on the objective function values of (discrete) optimization problems. By relaxing the integrality constraints in the integer programming formulation of a discrete optimization problem, thereby creating a linear programming relaxation, we obtain an alternative method for generating a lower bound. Which of these lower bounds is sharper (i.e., larger in value)? In this section we answer this question by showing that the lower bound obtained by the Lagrangian relaxation technique is at least as sharp as that obtained by using a linear programming relaxation. As a result, and because the Lagrangian relaxation bound is often easier to obtain than the linear programming relaxation bound, Lagrangian relaxation has become a very useful lower bounding technique in practice.

The content in this section requires some background in linear algebra and linear programming. We refer the reader to Appendix C for a review of this material.

Our first result in this section concerns the application of Lagrangian relaxation to a linear programming problem.

Theorem 16.6. *Suppose that we apply the Lagrangian relaxation technique to a linear programming problem* (P′) *defined as* $\min\{cx : \mathcal{A}x = b, \mathcal{D}x \le q, x \ge 0\}$ *by relaxing the constraints* $\mathcal{A}x = b$. *Then the optimal value* L^* *of the Lagrangian multiplier problem equals the optimal objective function value of* (P′).

Proof. We use linear programming optimality conditions to prove the theorem. Suppose that x^* is an optimal solution of the linear programming problem (P′) and that π^* and γ^* denote vectors of optimal dual variables associated with the constraints $\mathcal{A}x = b$ and $\mathcal{D}x \le q$. By linear programming theory, x^*, π^*, and γ^* satisfy the following dual feasibility and complementary slackness conditions:

$$c + \pi^*\mathcal{A} + \gamma^*\mathcal{D} \ge 0, \quad [c + \pi^*\mathcal{A} + \gamma^*\mathcal{D}]x^* = 0, \quad \text{and} \quad \gamma^*[\mathcal{D}x - q] = 0.$$

Consider the Lagrangian subproblem $L(\mu)$ at $\mu = \pi^*$, which is $L(\pi^*) = \min\{cx + \pi^*(\mathcal{A}x - b) : \mathcal{D}x \le q, x \ge 0\}$. Notice that x^* is feasible for this problem because it is feasible to (P′). Moreover, for the fixed value $\mu = \pi^*$, the previous dual feasibility and complementary slackness conditions are exactly those for the Lagrangian subproblem; therefore, x^* also solves the Lagrangian subproblem at $\mu = \pi^*$. But since $\pi^*(\mathcal{A}x^* - b) = 0$, $L(\pi^*) = cx^*$. Consequently, Property 16.3 implies that $L^* = L(\pi^*) = cx^*$, the optimal objective function value of (P′). ◆

The preceding theorem shows that the Lagrangian relaxation technique provides an alternative method for solving a linear programming problem. Instead of solving the linear programming problem directly using any linear programming algorithm, we can relax a subset of the constraints and solve the Lagrangian multiplier problem by using subgradient optimization and solving a sequence of relaxed problems. In some situations the relaxed problem is easy to solve, but the original problem is not; in these situations, a Lagrangian relaxation-based algorithm is an attractive solution approach.

Suppose next that we apply Lagrangian relaxation to a discrete optimization problem (P) defined as $\min\{cx : \mathcal{A}x = b, x \in X\}$. We assume that the discrete set X is specified as $X = \{x : \mathcal{D}x \le q, x \ge 0 \text{ and integer}\}$ for an integer matrix $\mathcal{D}$ and an integer vector q. Consequently, the problem (P) becomes

$$z^* = \min\{cx : \mathcal{A}x = b, \mathcal{D}x \le q, x \ge 0 \text{ and integer}\}. \tag{P}$$

We incur essentially no loss of generality by specifying the set X in this manner because we can formulate almost all real-life discrete optimization problems as integer programming problems. Let (LP) denote the linear programming relaxation of the problem (P) and let z° denote its optimal objective function value. That is,

$$z^\circ = \min\{cx : \mathcal{A}x = b, \mathcal{D}x \le q, x \ge 0\}. \tag{LP}$$

Clearly, $z^\circ \le z^*$ because the set of feasible solutions of (P) lies within the set of feasible solutions of (LP). Therefore, the linear programming relaxation provides a valid lower bound on the optimal objective function value of (P). We have earlier shown in Property 16.2 that the Lagrangian multiplier problem also gives a lower bound L^* on the optimal objective function value of (P). We now show that $z^\circ \le L^*$; that is, Lagrangian relaxation yields a lower bound that is at least as good as that obtained from the linear programming relaxation. We establish this result by showing that the Lagrangian multiplier problem also solves a linear programming problem but that the solution space for this problem is contained within the solution space of the problem (LP). The linear programming problem that the Lagrangian multiplier problem solves uses "convexification" of the solution space $X = \{x : \mathcal{D}x \le q, x \ge 0 \text{ and integer}\}$.

We assume that $X = \{x^1, x^2, \ldots, x^K\}$ is a finite set. We say that a solution x is a *convex combination* of the solutions $x^1, x^2, \ldots, x^K$ if $x = \sum_{k=1}^{K} \lambda_k x^k$ for some nonnegative weights $\lambda_1, \lambda_2, \ldots, \lambda_K$ satisfying the condition $\sum_{k=1}^{K} \lambda_k = 1$. Let $\mathcal{H}(X)$ denote the *convex hull* of X (i.e., the set of all convex combinations of X). In the subsequent discussion we use the following properties of $\mathcal{H}(X)$.

Property 16.7

(a) The set $\mathcal{H}(X)$ is a polyhedron, that is, it can be expressed as a solution space defined by a finite number of linear inequalities.

(b) Each extreme point solution of the polyhedron $\mathcal{H}(X)$ lies in X, and if we optimize a linear objective function over $\mathcal{H}(X)$, some solution in X will be an optimal solution.

(c) The set $\mathcal{H}(X)$ is contained in the set of solutions $\{x : \mathcal{D}x \le q, x \ge 0\}$.

Proof. Part (a) is a well-known result in linear algebra which we do not prove. The first statement in part (b) follows from the fact that every point of $\mathcal{H}(X)$ not in X is a convex combination, with positive weights, of two or more points in X and so is not an extreme point (see Appendix C). The second statement in part (b) is a consequence of the fact that linear programs always have at least one extreme point solution (see Appendix C). Part (c) follows from the fact that every solution in X also belongs to the convex set $\{x : \mathcal{D}x \le q, x \ge 0\}$, and consequently, every convex combination of solutions in X, which defines $\mathcal{H}(X)$, also belongs to the set $\{x : \mathcal{D}x \le q, x \ge 0\}$. ◆

We now prove the main result of this section.

Theorem 16.8. *The optimal objective function value L^* of the Lagrangian multiplier problem equals the optimal objective function value of the linear program $\min\{cx : \mathcal{A}x = b, x \in \mathcal{H}(X)\}$.*

Proof. Consider the Lagrangian subproblem

$$L(\mu) = \min\{cx + \mu(\mathcal{A}x - b) : x \in X\},$$

for some choice μ of the Lagrange multipliers. This problem is equivalent to the problem

$$L(\mu) = \min\{cx + \mu(\mathcal{A}x - b) : x \in \mathcal{H}(X)\}, \tag{16.2}$$

because by Property 16.7(b), some extreme point solution of $\mathcal{H}(X)$ solves this problem and each extreme point solution of $\mathcal{H}(X)$ belongs to X. Now, notice that the Lagrangian subproblem defined by (16.2) is a linear programming problem because by Property 16.7(a), we can formulate the set $\mathcal{H}(X)$ as the set of solutions of a finite number of linear inequalities. Therefore, we can conceive of the Lagrangian subproblem (16.2) as a relaxation of the following linear programming problem:

$$\min\{cx : \mathcal{A}x = b, x \in \mathcal{H}(X)\}.$$

Finally, we use Theorem 16.6 to observe that the optimal value L^* of the Lagrangian multiplier problem equals the optimal objective function value of the linear program $\min\{cx : \mathcal{A}x = b, x \in \mathcal{H}(X)\}$. ◆

We subsequently refer to the problem $\min\{cx : \mathcal{A}x = b, x \in \mathcal{H}(X)\}$ as the *convexified version* of problem (P) and refer to it as (CP). The preceding theorem shows that L^* equals the optimal objective function value of the convexified problem. What is the relationship between the set of feasible solutions of the convexified problem (CP) and the linear programming relaxation (LP)? We illustrate this relationship using a numerical example.

For simplicity, in our example we assume that the relaxed constraints are of the form $\mathcal{A}x \le b$ instead of $\mathcal{A}x = b$. We consider a two-variable problem with the constraints $\mathcal{A}x \le b$ and $\mathcal{D}x \le q$ as shown in Figure 16.6(a). This figure also specifies the set of solutions of the integer programming problem (P), denoted by the circled points. Figure 16.6(b) shows the solution space of the linear programming relaxation (LP) of the problem. Figure 16.6(c) shows the convex hull $\mathcal{H}(X)$ and Figure 16.6(d) depicts the solution space of the convexified problem (CP). Note that the solution space of (CP) is a subset of the solution space of (LP).

The preceding result is also easy to establish in general. Notice from Property 16.7(c) that since $\mathcal{H}(X)$ is contained in the set $\{x : \mathcal{D}x \le q, x \ge 0\}$, the set of solutions of problem (CP) given by $\{x : \mathcal{A}x = b, x \in \mathcal{H}(X)\}$ is contained in the set of solutions of (LP) given by $\{x : \mathcal{A}x = b, \mathcal{D}x \le q, x \ge 0\}$. Since optimizing the same objective function over a smaller solution space cannot improve the objective function value, we see that $z^\circ \le L^*$. We state this important result as a theorem.

Theorem 16.9. *When applied to an integer program stated in minimization form, the lower bound obtained by the Lagrangian relaxation technique is always as large (or, sharp) as the bound obtained by the linear programming relaxation of the problem; that is,* $z^\circ \le L^*$.

Under what situations will the Lagrangian bound equal the linear programming

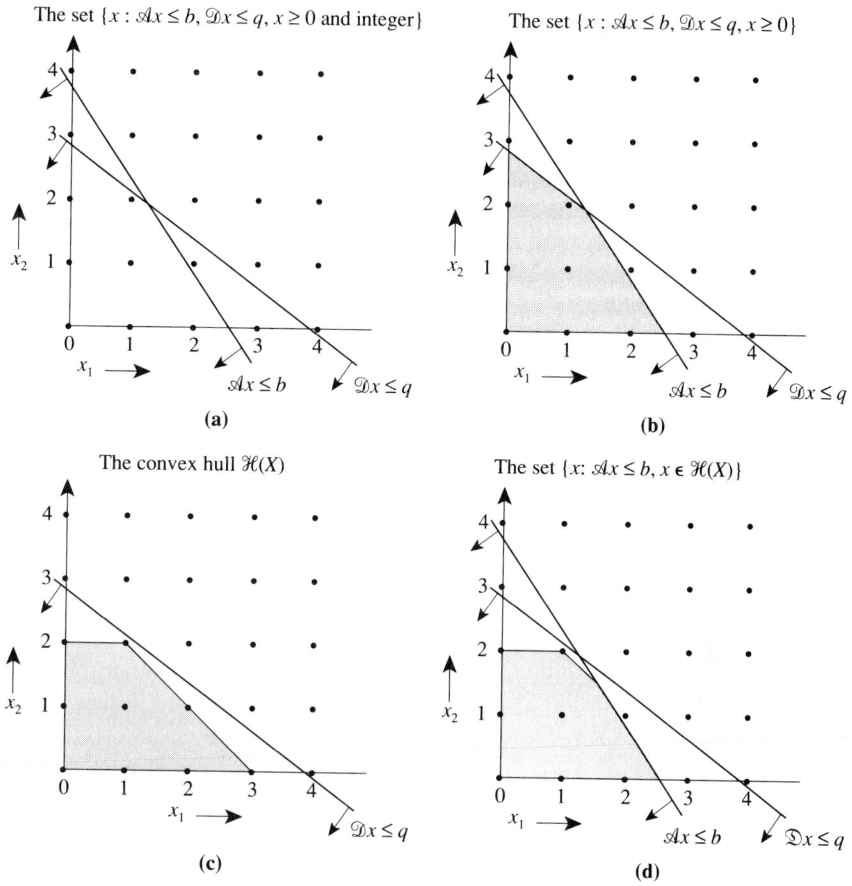

Figure 16.6 Illustrating the relationship between the problem (LP) and (CP): (a) solution space of the integer program (P); (b) solution space of the linear programming relaxation (LP); (c) convex hull $\mathcal{H}(x)$; (d) solution space of the convexified problem (CP).

bound? We show that if the Lagrangian subproblem satisfies a property, known as the *integrality property*, the Lagrangian bound will equal the linear programming bound. We say that the Lagrangian subproblem min$\{dx : \mathcal{D}x \le q, x \ge 0$ and integer$\}$ satisfies the integrality property if it has an integer optimal solution for every choice of objective function coefficients *even if we relax the integrality restrictions on the variables x*. Note that this condition implies that the problems min$\{cx + \mu(\mathcal{A}x - b) : \mathcal{D}x \le q, x \ge 0$ and integer$\}$ and min$\{cx + \mu(\mathcal{A}x - b) : \mathcal{D}x \le q, x \ge 0\}$ have the same optimal objective function values for every choice of the Lagrange multiplier μ. For example, if the constraints $\mathcal{D}x \le q$ are the mass balance constraints of a minimum cost flow problem (or any of its special cases, such as the maximum flow, shortest path, and assignment problems), the problem min$\{cx + \mu(\mathcal{A}x - b) : \mathcal{D}x \le q, x \ge 0\}$ will always have an integer optimal solution and imposing integrality constraints on the variables will not increase the optimal objective function value.

Theorem 16.10. *If the Lagrangian subproblem of the optimization problem (P) satisfies the integrality property, then $z^\circ = L^*$.*

Proof. Observe that the problem $\min\{dx : \mathcal{D}x \leq q, x \geq 0\}$ will have an integer optimal solution for every choice of d only if every extreme point solution of the constraints $\mathcal{D}x \leq q$, $x \geq 0$, is integer; for otherwise, we can select d so that a noninteger extreme point solution becomes an optimal solution. This observation implies that the set $\{x : \mathcal{D}x \leq q, x \geq 0\}$ equals the convex hull of $X = \{x : \mathcal{D}x \leq q, x \geq 0 \text{ and integer}\}$, which we have denoted by $\mathcal{H}(X)$. This result further implies that the sets $\{x : \mathcal{A}x = b, \mathcal{D}x \leq q, x \geq 0\}$ and $\{x : \mathcal{A}x = b, x \in \mathcal{H}(X)\}$ are the same. The first of these sets is the set of feasible solutions of the linear programming relaxation (LP) and the latter set is the set of feasible solutions of the convexified problem (CP). Since both the problems (LP) and (CP) have the same set of feasible solutions, they will have the same optimal objective function value, which is the desired conclusion of the theorem. ◆

This result shows that for problems satisfying the integrality property, solving the Lagrangian multiplier problem is equivalent to solving the linear programming relaxation of the problem. In these situations the Lagrangian relaxation technique provides no better a bound than the linear programming relaxation. Nevertheless, the Lagrangian relaxation technique might still be of considerable value, because solving the Lagrangian multiplier problem might be more efficient than solving the linear programming relaxation directly. Network optimization problems perhaps provide the most useful problem domain for exploiting this result because the Lagrangian subproblem in these cases often happens to be a minimum cost flow problem or one of its specializations.

As we have noted previously, in many (in fact, most) problem instances, the optimal objective function value L^* of the Lagrangian multiplier problem will be strictly less than the optimal objective function value z^* of problem (P); that is, the problem has a *duality gap*. As an example, consider the constrained shortest path example that we discussed in Section 16.3. For this example, $L^* = 7$ and $z^* = 13$. The duality gap occurs because the Lagrangian multiplier problem solves an optimization problem over a larger solution space (its convexification) than that of the original problem (P), and consequently, its optimal objective function value might be smaller.

16.5 APPLICATIONS OF LAGRANGIAN RELAXATION

As we noted earlier in the chapter, Lagrangian relaxation has many applications in network optimization. In this section we illustrate the breadth of these applications. The selected applications are both important in practice and illustrate how many of the network models we have considered in earlier chapters arise as Lagrangian subproblems. We consider the following models with embedded network structure.

Topic	Embedded network structure
Networks with side constraints	• Minimum cost flows • Shortest paths
Traveling salesman problem	• Assignment problem • Minimum cost flows
Vehicle routing	• Assignment problem • A variant of minimum spanning tree
Network design	• Shortest paths
Two-duty operator scheduling	• Shortest paths • Minimum cost flows
Degree-constrained minimum spanning trees	• Minimum spanning tree
Multi-item production planning	• Shortest paths • Minimum cost flows • Dynamic programs

Application 16.1 Networks with Side Constraints

The constrained shortest path problem is a special case of a broader set of optimization models known as network flow problems with side constraints. We can formulate a generic version of this problem as follows:

$$\text{Minimize} \quad cx$$

subject to

$$\mathcal{A}x \leq b,$$

$$\mathcal{N}x = q,$$

$$l \leq x \leq u, \text{ and } x_{ij} \text{ integer} \qquad \text{for all } (i, j) \in I.$$

In this formulation, as in the usual minimum cost flow problem, x is a vector of arc flows, $\mathcal{N}$ is a node–arc incidence matrix, q is a vector of node supplies and demands, and l and u are lower and upper bounds imposed on the arc flows. The set I is an index set of variables that must be integer. The flow vector x might be constrained to be integer or not, depending on the application being modeled. The added complication in this model are the side constraints $\mathcal{A}x \leq b$ that further restrict the arc flows.

For example, in the constrained shortest path problem, the network constraints model a shortest path problem [i.e., $q(s) = 1$ and $q(t) = -1$ for the source node s and destination node t, and $q(j) = 0$ for every other node j; also, every lower bound $l_{ij} = 0$ and every upper bound $u_{ij} = \infty$]. In this case the side constraint $\sum_{(i,j)\in A} t_{ij}x_{ij} \leq T$ is a single inequality constraint modeling the timing restriction.

The network flow model with side constraints arises in many application contexts in which the arc flows consume scarce resources (e.g., labor) or we wish to impose service constraints on the flows (e.g., maximum delay times in a communication and transportation network). The model also arises when the network flow model has multiple commodities, each governed by their own flow constraints, that share common resources such as arc capacities. In Chapter 17 we consider one such model, the classical multicommodity flow problem, in some detail.

We might note that the network flow model with side constraints also arises in other, perhaps more surprising ways. As an illustration, consider a standard work force scheduling problem. Suppose that we wish to schedule employees (e.g., telephone operators, production workers, or nurses) in a way that ensures that $\alpha(j)$ employees are available for work on the jth day of the week; suppose, further, that we wish to schedule the employees so that each has two consecutive days off each week. That is, each of them works 5 consecutive days and then has 2 days off. We incur a cost c_j for each employee that is scheduled to work on day j. Figure 16.7 shows a network flow model with side constraints for this problem. The network contains three types of arcs.

1. A "work arc" for each day of the week: The flow on this arc is the number of employees scheduled to work on that day; the arc has an associated cost (e.g., weekends might have a pay premium and so a higher cost) and a lower flow bound equaling the number of employees required to work on that day.
2. A "total work force arc" that introduces the work force at the beginning of the planning cycle (which we arbitrarily take to be Sunday) and removes it at the end of the planning cycle (Saturday): the flow y on this arc is the total number of employees employed during the week.
3. "Days-off arcs" with the flows $x_{sun}, x_{mon}, \ldots, x_{sat}$, each representing a schedule with 2 days off beginning with the day indicated by the subscript: The flow on arc x_{sun}, for example, bypasses the Sunday and Monday work arcs, indicating that the employees working in this schedule are not available for work on Sunday and Monday.

A complicating feature of this network flow model is a single additional constraint indicating that every employee must be assigned to at least one schedule; that is,

$$y = x_{sun} + x_{mon} + \cdots + x_{sat}.$$

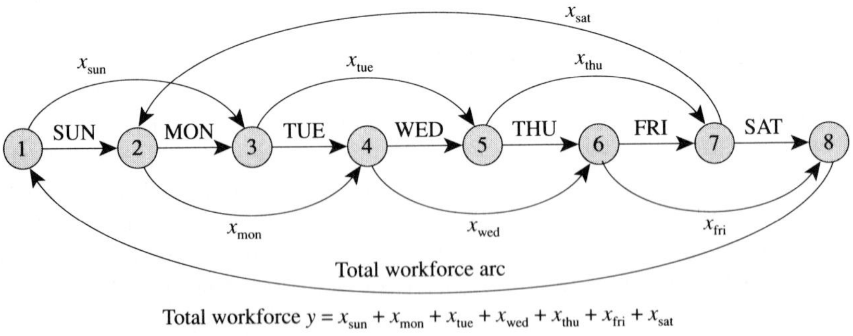

Figure 16.7 Network model of the cyclic scheduling problem. Lower bound on day arcs = demand for the day.

This side constraint specifies a flow relationship between several of the arcs in the network flow model. Relaxing this constraint and using Lagrangian relaxation provides us with one algorithmic approach for solving this problem. The algorithmic procedure for applying Lagrangian relaxation to the general network flow model with side constraints is essentially the same as the procedure we have discussed for the constrained shortest path problem: we associate nonnegative Lagrange multipliers μ with the side constraints $\mathcal{A}x \leq b$ and bring them into the objective function to produce the network flow subproblem

$$\text{minimize}\{cx + \mu(\mathcal{A}x - b) : \mathcal{N}x = q, l \leq x \leq u\},$$

and then solve a sequence of these problems with different values of the Lagrange multipliers μ which we update using the subgradient optimization technique. For each choice of the Lagrangian multiplier on this constraint, the Lagrangian subproblem is a network flow problem. In Exercise 9.9 we show that we can actually solve this special case of network flows with side constraints much more efficiently by solving a polynomial sequence of network flow problems.

Application 16.2 Traveling Salesman Problem

The traveling salesman problem is perhaps the most famous problem in all of network and combinatorial optimization: Its simplicity and yet its difficulty have made it an alluring problem that has attracted the attention of many noted researchers over a period of several decades. The problem is deceptively easy to state: Starting from his home base, node 1, a salesman wishes to visit each of several cities, represented by nodes 2, . . . , n, exactly once and return home, doing so at the lowest possible travel cost. We will refer to any feasible solution to this problem as a *tour* (of the cities).

The traveling salesman problem is a generic core model that captures the combinatorial essence of most routing problems and, indeed, most other routing problems are extensions of it. For example, in the classical vehicle routing problem, a set of vehicles, each with a fixed capacity, must visit a set of customers (e.g., grocery stores) to deliver (or pick up) a set of goods. We wish to determine the best possible set of delivery routes. Once we have assigned a set of customers to a vehicle, that vehicle should take the minimum cost tour through the set of customers assigned to it; that is, it should visit these customers along an optimal traveling salesman tour.

The traveling salesman problem also arises in problems that on the surface have no connection with routing. For example, suppose that we wish to find a sequence for loading jobs on a machine (e.g., items to be painted), and that whenever the machine processes job i after job j, we must reset the machine (e.g., clear the dies of the colors of the previous job), incurring a setup time c_{ij}. Then in order to find the processing sequence that minimizes the total setup time, we need to solve a traveling salesman problem—the machine, which functions as the "salesman," needs to "visit" the jobs in the most cost-effective manner.

There are many ways to formulate the traveling salesman problem as an optimization model. We present a model with an embedded (directed) network flow structure. Exercises 16.21 and 16.23 consider other modeling approaches. Let c_{ij}

denote the cost of traveling from city i to city j and let y_{ij} be a zero–one variable, indicating whether or not the salesman travels from city i to city j. Moreover, let us define flow variables x_{ij} on each arc (i, j) and assume that the salesman has $n - 1$ units available at node 1, which we arbitrarily select as a "source node," and that he must deliver 1 unit to each of the other nodes. Then the model is

$$\text{Minimize} \quad \sum_{(i,j)\in A} c_{ij}y_{ij} \tag{16.3a}$$

subject to

$$\sum_{1\le j\le n} y_{ij} = 1 \quad \text{for all } i = 1, 2, \ldots, n, \tag{16.3b}$$

$$\sum_{1\le i\le n} y_{ij} = 1 \quad \text{for all } j = 1, 2, \ldots, n, \tag{16.3c}$$

$$\mathcal{N}x = b, \tag{16.3d}$$

$$x_{ij} \le (n - 1)y_{ij} \quad \text{for all } (i, j) \in A, \tag{16.3e}$$

$$x_{ij} \ge 0 \quad \text{for all } (i, j) \in A, \tag{16.3f}$$

$$y_{ij} = 0 \text{ or } 1 \quad \text{for all } (i, j) \in A. \tag{16.3g}$$

To interpret this formulation, let $A' = \{(i, j) : y_{ij} = 1\}$ and let $A'' = \{(i, j) : x_{ij} > 0\}$. The constraints (16.3b) and (16.3c) imply that exactly one arc of A' leaves and enters any node i; therefore, A' is the union of node disjoint cycles containing all of the nodes of N. In general, any integer solution satisfying (16.3b) and (16.3c) will be the union of disjoint cycles; if any such solution contains more than one cycle, we refer to each of the cycles as subtours, since they pass through only a subset of the nodes. Figure 16.8 gives an example of a subtour solution to the constraints (16.3b) and (16.3c).

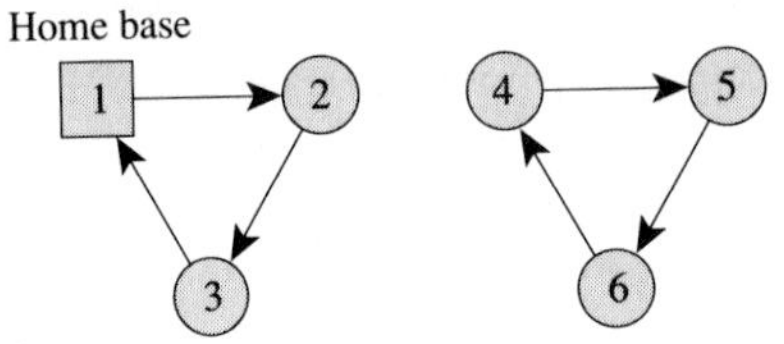

Figure 16.8 Infeasible solution for the traveling salesman problem containing subtours.

Constraint (16.3d) ensures that A'' is connected since we need to send 1 unit of flow from node 1 to every other node via arcs in A''. The "forcing" constraints (16.3e) imply that A'' is a subset of A'. [Notice that since no arc need ever carry more than $(n - 1)$ units of flow, the forcing constraint for arc (i, j) is redundant if $y_{ij} = 1$.] These conditions imply that the arc set A' is connected and so cannot contain any subtours. We conclude that the formulation (16.3) is a valid formulation for the traveling salesman problem.

One of the nice features of this formulation is that we can apply Lagrangian relaxation to it in several ways. For example, suppose that we attach Lagrange multipliers $\mu_{ij} \ge 0$ with the forcing constraints (16.3e) and bring them into the objective function, giving the Lagrangian objective function

$$\text{Minimize} \sum_{(i,j)\in A} [c_{ij} - (n - 1)\mu_{ij}]y_{ij} + \sum_{(i,j)\in A} \mu_{ij}x_{ij},$$

and leaving (16.3b)–(16.3d), (16.3f), and (16.3g) as constraints in the Lagrangian subproblem. Note that nothing in this Lagrangian subproblem couples the variables y_{ij} and x_{ij}. Therefore, the subproblem decomposes into two separate subproblems: (1) an assignment problem in the variables y_{ij}, and (2) a minimum cost flow problem in the variables x_{ij}. So for any choice of the Lagrangian multipliers μ, we solve two network flow subproblems; by using subgradient optimization we can find the best lower bound and optimal values of the multipliers. By relaxing other constraints in this model, or by applying Lagrangian relaxation to other formulations of the traveling salesman problem, we could define other network flow subproblems (see Exercise 16.19).

Application 16.3 Vehicle Routing

The vehicle routing problem is a generic model that practitioners encounter in many problem settings including the delivery of consumer products to grocery stores, the collection of money from vending machines and telephone coin boxes, and the delivery of heating oil to households. As we have noted earlier in this section, the vehicle routing problem is a generalization of the traveling salesman problem.

The vehicle routing problem is easy to state: Given (1) a fleet of K capacitated vehicles domiciled at a common depot, say node 1, (2) a set of customer sites $j = 2, 3, \ldots, n$, each with a prescribed demand d_j, and (3) a cost c_{ij} of traveling from location i to location j, what is the minimum cost set of routes for delivering (picking up) the goods to the customer sites? We assume that the vehicle fleet is homogeneous and that each vehicle has a capacity of u units.

There are many different variants on this core vehicle routing problem. For example, the vehicle fleet might be nonhomogeneous, each vehicle route might have a total travel time restriction, or deliveries for each customer might have time window restrictions (earliest and latest delivery times). We illustrate the use of Lagrangian relaxation by considering only the basic model, which we formulate with decision variables x_{ij}^k indicating whether ($x_{ij}^k = 1$) or not ($x_{ij}^k = 0$) we dispatch vehicle k on arc (i, j) and y_{ij} indicating whether some vehicle travels on arc (i, j):

$$\text{Minimize} \sum_{1\le k\le K} \sum_{(i,j)\in A} c_{ij}x_{ij}^k \tag{16.4a}$$

subject to

$$\sum_{1\le k\le K} x_{ij}^k = y_{ij}, \tag{16.4b}$$

$$\sum_{1\le j\le n} y_{ij} = 1 \quad \text{for } i = 2, 3, \ldots, n, \tag{16.4c}$$

$$\sum_{1\le i\le n} y_{ij} = 1 \quad \text{for } j = 2, 3, \ldots, n, \tag{16.4d}$$

$$\sum_{1\le j\le n} y_{1j} = K, \tag{16.4e}$$

$$\sum_{1\le i\le n} y_{i1} = K, \tag{16.4f}$$

$$\sum_{2 \le i \le n} \sum_{1 \le j \le n} d_i x_{ij}^k \le u \qquad \text{for all } k = 1, 2, \ldots, K, \tag{16.4g}$$

$$\sum_{i \in Q} \sum_{j \in Q} y_{ij} \le |Q| - 1 \qquad \text{for all subsets } Q \text{ of } \{2, 3, \ldots, n\}, \tag{16.4h}$$

$$y_{ij} = 0 \text{ or } 1 \qquad \text{for all } (i, j) \in A, \tag{16.4i}$$

$$x_{ij}^k = 0 \text{ or } 1 \qquad \text{for all } (i, j) \in A \text{ and all } k = 1, 2, \ldots, K. \tag{16.4j}$$

Let $A' = \{(i, j) : y_{ij} = 1\}$. As in our discussion of the traveling salesman problem, constraints (16.4c) and (16.4d) ensure that A' is the union of node disjoint cycles containing all of the nodes in N. Constraint (16.4h) ensures that the solution must contain no cycle using the nodes $2, 3, \ldots, n$ (i.e., not contain any subtours on these nodes); otherwise, the arcs A' would contain some cycle passing through a set Q of nodes and the solution would violate constraint (16.4h) since the left-hand side of the constraint (16.4h) would be at least $|Q|$. For this reason, we refer to the constraints (16.4h) as *subtour breaking constraints*.

We might note that if $K = 1$, and u is so large that the constraint (16.4g) is redundant, this model becomes an "assignment-based" formulation of the traveling salesman problem, which is an alternative formulation to the "flow-based" model that we introduced previously as (16.3). In Exercise 16.24 we study the relationships between these formulations as well as a third model, a multicommodity flow-based formulation.

Note that this formulation has several embedded structures that we might exploit in a Lagrangian relaxation solution approach. By relaxing some of the constraints, we are also able to decompose the problem into independent subproblems. For example, if we relax only constraints (16.4b), no constraint connects the x variables and y variables, so the problem decomposes into separate subproblems in each of these variables. By relaxing different combinations of the constraints, we create several different types of subproblems:

1. If we relax the constraints (16.4b), (16.4g), and (16.4h), the resulting formulation is an assignment problem.
2. If we relax the constraints (16.4b) to (16.4f), and (16.4h), the resulting problem decomposes into independent "knapsack problems," one for each vehicle k.
3. If we relax constraint (16.4b), the problem decomposes into separate subproblems, one in the y variables and one in the x^k variables for each vehicle k. The first of these problems is a so-called K-traveling salesman problem (see Exercise 16.25) and each problem in the variables x^k is a knapsack problem.
4. If we relax the assignment constraints (16.4c) to (16.4f), the constraint (16.4b) defining y, and the capacity constraint (16.4g), the resulting problem is a minimum forest problem on the nodes $2, 3, \ldots, n$. This problem is easy to solve by a simple variant of any minimum spanning tree algorithm. We could strengthen this approach by adding other (redundant) constraints to the problem formulation (see Exercise 16.28).
5. If we relax constraints (16.4b), (16.4c), and (16.4e) to (16.4g), the subproblem with the constraints (16.4d), (16.4h), and (16.4i) becomes a directed minimum spanning tree problem—any feasible solution will be a directed spanning tree

with exactly one arc directed into each node (except for the root node 1). Although we do not consider this problem in this book, it is polynomially solvable.

6. If we relax the constraint (16.4g), the problem becomes a variant of the K-traveling salesman problem.

These various possibilities illustrate the remarkable flexibility of the Lagrangian relaxation solution approach.

Application 16.4 Network Design

Suppose that we have the flexibility of designing a network as well as determining its optimal flow (routing). That is, we have a directed network $G = (N, A)$ and can introduce an arc or not into the design of the network: If we use (introduce) an arc (i, j), we incur a design (construction) cost f_{ij}. Our problem is to find the design that minimizes the total systems cost—that is, the sum of the design cost and the routing cost. This type of model arises in many application contexts, for example, the design of telecommunication or computer networks, load planning in the trucking industry (i.e., the design of a routing plan for trucks), and the design of production schedules.

Many alternative modeling assumptions arise in practice. We consider one version of the problem, the *uncapacitated network design problem.* In this model we need to route multiple commodities on the network; each commodity k has a single source node s^k and a single destination node d^k. Once we introduce an arc (i, j) into the network, we have sufficient capacity to route all of the flow by all commodities on this arc.

To formulate this problem as an optimization model, let x^k denote the vector of flows of commodity k on the network. Rather than letting x^k_{ij} model the total flow of commodity k on arc (i, j), however, we let x^k_{ij} denote the fraction of the required flow of commodity k to be routed from the source s^k to the destination d^k that flows on arc (i, j). Let c^k denote the cost vector for commodity k, which we scale to reflect the way that we have defined x^k_{ij} [i.e., c^k_{ij} is the per unit cost for commodity k on arc (i, j) times the flow requirement of that commodity]. Also, let y_{ij} be a zero–one vector indicating whether or not we select arc (i, j) as part of the network design. Using this notation, we can formulate the network design problem as follows:

$$\text{Minimize} \quad \sum_{1 \le k \le K} c^k x^k + fy \tag{16.5a}$$

subject to

$$\sum_{\{j:(i,j)\in A\}} x^k_{ij} - \sum_{\{j:(j,i)\in A\}} x^k_{ji} = \begin{cases} 1 & \text{if } i = s^k \\ -1 & \text{if } i = d^k \\ 0 & \text{otherwise} \end{cases} \quad \text{for all } i \in N,\ k = 1, 2, \ldots, K, \tag{16.5b}$$

$$x^k_{ij} \le y_{ij} \qquad \text{for all } (i, j) \in A,\ k = 1, 2, \ldots, K, \tag{16.5c}$$

$$x_{ij}^k \geq 0 \qquad \text{for all } (i, j) \in A \text{ and all } k = 1, 2, \ldots, K, \tag{16.5d}$$

$$y_{ij} = 0 \text{ or } 1 \qquad \text{for all } (i, j) \in A. \tag{16.5e}$$

In this formulation, the "forcing constraints" (16.5c) state that if we do not select arc (i, j) as part of the design, we cannot flow any fraction of commodity k's demand on this arc, and if we do select arc (i, j) as part of the design, we can flow as much of the demand of commodity k as we like on this arc.

Note that if we remove the forcing constraints from this model, the resulting model in the flow variables x^k decomposes into a set of independent shortest path problems, one for each commodity k. Consequently, the model is another attractive candidate for the application of Lagrangian relaxation. To see why this type of solution approach might be attractive, consider a typically sized problem with, say, 50 nodes and 500 candidate arcs. Suppose that we have a separate commodity for each pair of nodes (as is typical in communication settings in which each node is sending messages to every other node). Then we have 50(49) = 2450 commodities. Since each commodity can flow on each arc, the model has 2450(500) = 1,225,000 flow variables, and since (1) each flow variable defines a forcing constraint, and (2) each commodity has a flow balance constraint at each node, the model has 1,225,000 + 2450(50) = 1,347,500 constraints. In addition, it has 500 zero–one variables. So even as a linear program, this model far exceeds the capabilities of current state of the art software systems. By decomposing the problem, however, for each choice of the vector of Lagrange multipliers, we will solve 2450 small shortest path problems.

Application 16.5 Two-Duty Operator Scheduling

In many different problem contexts in work force planning, a private firm or public-sector organization must schedule its employees—for example, nurses, airline crews, telephone operators—to provide needed services. Typically, the problems are complicated by complex work rules, for example, airline crews have limits on the number of hours that they can fly in any week or month. Moreover, frequently, the demand for the services of these employees varies considerably by time of the day or week, or across geography (as in the case of airline crew scheduling). Consequently, finding a minimum cost schedule requires that we balance the prevailing work rules with the demand patterns. Figure 16.9 shows one example of a work force planning problem, which we will view as a driver schedule for a single bus line.

Every column in this table corresponds to a possible schedule. For example, in schedule 1, a driver operates the bus line in two shifts, from 8 to 11 and then from 1 to 3; in schedule 2 the driver works a single shift, from 11 to 1, and in schedule 3, he/she drives from 3 to 6. As indicated by the column entitled demand in the table, we wish to find a set of schedules satisfying the property that at least one driver is assigned to the bus at every hour of the day from 8 A.M. until 6 P.M. (if two drivers are assigned to the same bus at the same time, one drives and the other is a rider). One possibility is to choose schedules 1, 2, and 3; another is schedules 4 and 5; and still another is schedules 3, 5, and 6. Each schedule j has an associated

Time period	Schedule 1	2	3	4	5	6	7	8	Demand
8–9	1	0	0	1	0	1	0	1	≥1
9–10	1	0	0	1	0	1	0	1	≥1
10–11	1	0	0	0	1	0	0	1	≥1
11–12	0	1	0	0	1	0	0	1	≥1
12–1	0	1	0	0	1	0	0	1	≥1
1–2	1	0	0	1	0	1	1	0	≥1
2–3	1	0	0	1	0	1	1	0	≥1
3–4	0	0	1	1	0	0	1	0	≥1
4–5	0	0	1	1	0	0	1	0	≥1
5–6	0	0	1	1	0	0	1	0	≥1
Cost	c_1	c_2	c_3	c_4	c_5	c_6	c_7	c_8	

Figure 16.9 Two-duty operator schedule.

cost c_j and we wish to choose the set of schedules that meets the scheduling requirement at the lowest possible cost. To formulate this problem formally as an optimization model, let x_j be a binary (i.e., zero–one) variable indicating whether ($x_j = 1$) or not ($x_j = 0$), we choose schedule j, and let $\mathcal{A}$ denote the zero–one matrix of coefficients of the scheduling table (i.e., the ijth element is 1 if schedule j has a driver on duty during the ith hour of the day). Also, let e denote a column of 1's. Then the model is

$$\text{Minimize} \quad cx \tag{16.6a}$$

subject to

$$\mathcal{A}x \geq e, \tag{16.6b}$$

$$x_j = 0 \text{ or } 1 \qquad \text{for } j = 1, 2, \ldots, n. \tag{16.6c}$$

The choice of the available schedules in the problem depends on the governing work rules; as an illustration, in our example, no operator works in any shift of less than 2 hours. Moreover, note that the schedules permit split shifts, that is, time on, time off, and then time on again as in schedule 1. Note, however, that no schedule has more than two shifts. We refer to this special version of the general operator scheduling problem as the *two-duty operator scheduling problem*.

In Exercise 4.13 we showed how to solve the single-duty scheduling problem as a shortest path problem: The shortest path model contains a node for each time period $1, 2, \ldots, T$ to be covered, plus an artificial end node $T + 1$, and an arc from node i to node j whenever a schedule starts at the beginning of time period i and ends at the beginning of time period j. We interpret arc (i, j) as "covering" the time periods $i, i + 1, \ldots, j - 1$. The network also contains "backward" arcs of the form $(j + 1, j)$ that permits us to "back up" from time period $j + 1$ to time period j so that we can cover any node more than once and model the possibility that a schedule might assign more than one driver to any time period. Can we use

Lagrangian relaxation to exploit the fact that the single-duty problem is a shortest path problem? To do so, we will use an idea known as *variable splitting*.

Consider any column j of the matrix $\mathcal{A}$ that contains two sequences of 1's—that is, corresponds to a schedule with two shifts. Let us make two columns $\mathcal{A}'_j$ and $\mathcal{A}''_j$ out of this column; each of these columns contains one of the duties (sequence of 1's) from $\mathcal{A}_j$, so $\mathcal{A}_j = \mathcal{A}'_j + \mathcal{A}''_j$. Let us also replace the variable x_j in our model with two variables x'_j and x''_j. We form a new model with these variables as

$$\text{Minimize} \quad c'x' + c''x'' \tag{16.7a}$$

subject to

$$\mathcal{A}'x' + \mathcal{A}''x'' \geq e, \tag{16.7b}$$

$$x' - x'' = 0, \tag{16.7c}$$

$$x'_j \text{ and } x''_j = 0 \text{ or } 1 \qquad \text{for } j = 1, 2, \ldots, n. \tag{16.7d}$$

For convenience, in formulating this model we have assumed that we have split every column of the matrix $\mathcal{A}$. If not, we can simply assume that some columns of $\mathcal{A}''$ are columns of zeros. Moreover, we can split the cost of each variable x_j arbitrarily between c'_j and c''_j. For example, we could let each of these costs be half of c_j. This model and the original model (16.6) are clearly equivalent. Note, however, that the new model reveals embedded network structure; as shown in Figure 16.10, which is the network model associated with the data in Figure 16.9, the model is a shortest path problem with the complicating constraints that we need to choose arcs in pairs: We choose either both or none of the arcs corresponding to the variables x'_j and x''_j. If we eliminate the "complicating" constraint $x' - x'' = 0$, the problem becomes an easily solvable single duty scheduling problem, which as we have seen before, we can solve via a shortest path computation. This observation suggests that we adopt a Lagrangian relaxation approach, relaxing the constraints with a Lagrange multiplier μ so that the Lagrangian subproblem has the objective function

$$L(\mu) = \min(c' + \mu)x' + (c'' - \mu)x''. \tag{16.8}$$

Now, as usual to solve the Lagrangian multiplier problem, we apply subgradient

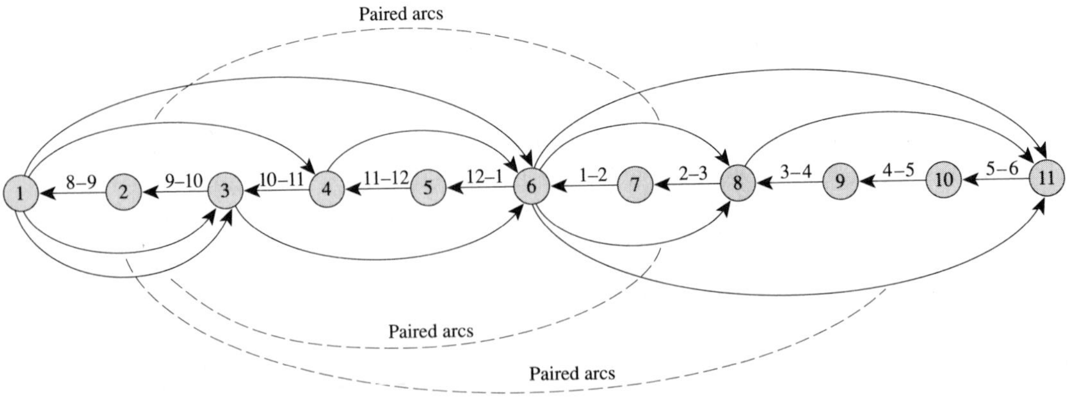

Figure 16.10 Shortest path subproblem for the two duty scheduling problem.

optimization, or some other solution technique, to maximize $L(\mu)$ over all possible choices of the Lagrange multipliers μ.

When we split a column A_j into two columns A'_j and A''_j, it does not matter how we split the cost c_j between c'_j and c''_j, as long as $c_j = c'_j + c''_j$. Since $x'_j = x''_j$ in any feasible solution, the cost of any feasible solution will be the same no matter how we allocate the cost. However, cost splitting does make a significant difference in the relaxed problem obtained by dropping the constraint $x'_j = x''_j$. If we were to make c'_j large and c''_j small, then in the solution to the relaxed problem we would probably find that x'_j would be 0 and x''_j would be 1. Similarly, if we made c'_j small and c''_j large, in the solution to the relaxed problem we would likely find that x'_j would be 1 and x''_j would be 0. Ideally, we should allocate the costs between c'_j and c''_j, so that either $x'_j = x''_j = 0$ or $x'_j = x''_j = 1$ in the relaxed problem.

As it turns out, we need not worry about the cost allocation at all if we use Lagrangian relaxation since the Lagrange multiplier μ_j for the constraint $x'_j - x''_j = 0$ does the cost allocation. Suppose, for example, that $c'_j = c_j$ and $c''_j = 0$. Then, since we are relaxing the constraint $x'_j - x''_j = 0$, the coefficient of x'_j in the relaxed problem is $c_j + \mu_j$, and the coefficient of x''_j is $-\mu_j$. As μ_j ranges over the real numbers, we obtain all possible ways of splitting the cost c_j between c'_j and c''_j.

The operator scheduling problem we have considered permits us to find an optimal schedule of drivers for a single bus line. If we wish to schedule several bus lines simultaneously, the right-hand-side coefficients in the constraints (16.6) will be arbitrary positive integers, indicating the number of required operators for each time period during the day. In this instance, the variable splitting device still permits us to use Lagrangian relaxation and network optimization to solve the problem. In this instance, the Lagrangian subproblems will be minimum cost flow problems rather than shortest path problems.

As this application shows, embedded network flow structure is not always so apparent and, consequently, the use of Lagrangian relaxation often requires considerable ingenuity in model formulation. Indeed, the application of Lagrangian relaxation typically requires considerable skill in modeling. Moreover, as several of our examples have shown, we often can formulate network optimization problems in several different ways, and by doing so we might be able to recognize and exploit different network substructures. The models we have proposed for the traveling salesman problem, both in the discussion of this problem and in the discussion of the vehicle routing problem, illustrate these possibilities. As a result, the design and implementation of Lagrangian relaxation algorithms often require careful choices concerning the "best" models to use and the "best" constraints to relax. The literature that we cite in the reference notes gives some guidance concerning these issues; successful prior applications, such as those that we have discussed in this section and in the exercises at the end of this chapter, provide additional guides.

Application 16.6 Degree-Constrained Minimum Spanning Trees

Suppose that we wish to find a minimum spanning tree of a network, but with the added provision that the tree contain exactly k arcs incident to a given root node, say node 1 (in some settings, the degree of the root node should be at most k). This *degree-constrained minimum spanning tree problem* arises in several applications. For example, in computer networking, the root node might be a central processor with a fixed number of ports and the other nodes might be terminals that we need to connect to the processor. In the communication literature, this problem has become known as the *teleprocessing design problem* or as the *multidrop terminal layout problem*. The vehicle routing problem, described in Application 16.3, provides another application setting. If we are routing k vehicles and we delete the last arc from every route, every solution is a spanning tree with k arcs incident to the depot (the tree has additional structure: each subtree off the root is a single path). Therefore, the degree constrained minimum spanning tree problem is a relaxation of the vehicle routing problem. Note that this relaxation is stronger than the minimum spanning tree relaxation that we discussed in Application 16.3.

We might formulate the degree-constrained minimum spanning tree problem as follows.

$$\text{Minimize} \quad cx$$

subject to

$$\sum_{j=2}^{n} x_{1j} = k,$$

$$x \in X.$$

In this formulation, $x = (x_{ij})$ is a vector of decision variables and each x_{ij} is a zero–one variable indicating whether ($x_{ij} = 1$) or not ($x_{ij} = 0$), arc (i, j) belongs to the spanning tree. The number c_{ij} denotes the fixed cost of installing arc (i, j) and the set X denotes the set of incidence vectors of spanning trees. The additional constraint states that the degree of node 1 must be k. Let $C = \max\{c_{ij} : (i, j) \in A\}$.

To solve this problem, we might use Lagrangian relaxation. If we associate a Lagrange multiplier μ with the degree constraint and relax it, the objective function of the Lagrangian subproblem becomes $cx + \mu \sum_{j=2}^{n} x_{1j} - \mu k$ and the remaining (implicit) constraint, $x \in X$, states that the vector x defines a spanning tree. Note that if we ignore the last term, μk, which is a constant for any fixed value of μ, this problem is a parametric minimum spanning tree problem: for each j, the cost of arc $(1, j)$ is $c_{1j} + \mu$, and whenever $i \neq 1$ and $j \neq 1$, the cost of arc (i, j) is c_{ij}. We will use this observation to solve the degree constrained problem. That is, rather than using subgradient optimization, we will use a combinatorial algorithm to solve the Lagrangian multiplier problem.

We first solve the minimum spanning tree problem for $\mu = 0$. If the degree of node 1 in the optimal tree equals k, this tree is optimal for the degree-constrained minimum spanning tree problem. So suppose that the degree of node 1 is different than k. We first consider the case when the degree of node 1 is strictly less than k. Notice that since μ affects the lengths of only those arcs incident to node 1, changing the value μ affects the ranking of these arcs relative to the arcs not incident to node 1. Consequently, as we decrease the value of μ, the arcs incident to node 1 become more attractive relative to the other arcs, so we would insert these arcs into the spanning tree in place of the other arcs. The algorithm uses this observation: It starts with a minimum spanning tree T^1 for $\mu = 0$ and by decreasing the value of μ, it generates a sequence of spanning trees $T^1, \ldots, T^{q-1}$, terminating with a minimum spanning tree T^q for $\mu = -C - 1$. Each tree $T^1, \ldots, T^{q-1}$ is a minimum spanning tree for some value of μ. The algorithm creates T^j from T^{j-1} by adding one arc $(1, i)$ to T^{j-1} and deleting one arc (p, q) with $p \neq 1$ and $q \neq 1$ from T^{j-1}. That is, at each step it increases the degree of node 1 by one. Finally, T^q includes all the arcs incident to node 1. (For a discussion of parametric minimum spanning trees, see Exercises 13.35 and 13.36.)

Let T^k denote the tree containing exactly k arcs incident to node 1 and let μ^k denote the value of μ for which T^k is a minimum spanning tree for the parametric problem. Further, let x^k denote the incidence vector associated with the spanning tree T^k. By definition, x^k solves the Lagrangian multiplier problem $L(\mu) = cx + \mu \sum_{j=2}^{n} x_{ij} - \mu k$, $x \in X$, for $\mu = \mu^k$ because μk is a constant. Now notice that $L(\mu^k) = cx^k + \mu^k \sum_{j=2}^{n} x_{1j} - \mu^k k = cx^k + \mu^k k - \mu^k k = cx^k$, which implies that for $\mu = \mu^k$, the optimal objective function value of the Lagrangian subproblem equals the value of a feasible solution x^k of the degree-constrained minimum spanning tree problem. Property 16.3 shows that x^k is an optimal solution of the degree-constrained minimum spanning tree problem.

When the optimal tree for $\mu = 0$ contains more than k arcs, we parametrically increase the value of μ until $\mu = C + 1$. As we increase the value of μ, the arcs incident on node 1 become less attractive, and they leave the optimal tree one by one. Eventually, node 1 will have degree exactly equal to k, and the tree at this point will be a minimum degree-constrained spanning tree.

Note that this application of Lagrangian relaxation is different than the others that we have considered in this chapter. In this case we have used Lagrangian relaxation to define a parametric problem that is related to the constrained model we are considering. We have then used a combinatorial algorithm rather than a general-purpose Lagrangian relaxation algorithm to solve the parametric problem. In this case the Lagrangian relaxation has proven to be valuable not only in formulating the parametric problem, but also in validating that the solution generated by the parametric problem is optimal for the constrained model.

Application 16.7 Multi-item Production Planning

In production planning we would like to find the best use of scarce resources (people, machinery, space) in order to meet customer demand at the least possible cost. As we show in Chapter 19, the research community has developed a number of different models for addressing various planning issues in this application domain. Some of these models are shortest path problems and some are minimum cost flow problems; still others are multicommodity flow problems or more general models with embedded network flow structure. In this section, to show how we might use Lagrangian relaxation to solve more general models, we consider two applications of production planning: multi-item production planning and production planning with changeover costs.

Suppose that we are producing K items over a planning horizon containing T periods (e.g., production shifts). Suppose, further, that we produce the items on the same machine and that we can produce at most one item in each period. We would like to find the least cost production plan that will satisfy a demand d_{kt} for every item k in each period t.

Let x_{kt} denote the amount of item k that we produce in period t and let I_{kt} denote the amount of inventory of item k that we carry from period t to period $t + 1$. Let z_{kt} be a zero–one variable indicating whether or not we produce item k in period t. With this notation, we can model the multi-item production planning problem as follows:

$$\text{Minimize} \quad \sum_{k=1}^{K}\sum_{t=1}^{T} c_{kt}x_{kt} + \sum_{k=1}^{K}\sum_{t=1}^{T} h_{kt}I_{kt} + \sum_{k=1}^{K}\sum_{t=1}^{T} F_{kt}z_{kt} \tag{16.9a}$$

subject to

$$\sum_{k=1}^{K} z_{kt} \le 1 \qquad \text{for } t = 1, 2, \ldots, T, \tag{16.9b}$$

$$x_{kt} + I_{k,t-1} - I_{kt} = d_{kt} \qquad \text{for } k = 1, 2, \ldots, K \text{ and } t = 1, 2, \ldots, T, \tag{16.9c}$$

$$x_{kt} \le P_{kt}z_{kt} \qquad \text{for } k = 1, 2, \ldots, K \text{ and } t = 1, 2, \ldots, T, \tag{16.9d}$$

$$x_{kt} \ge 0, \quad I_{kt} \ge 0 \qquad \text{for } k = 1, 2, \ldots, K \text{ and } t = 1, 2, \ldots, T, \tag{16.9e}$$

$$z_{kt} = 0 \text{ or } 1 \qquad \text{for } k = 1, 2, \ldots, K \text{ and } t = 1, 2, \ldots, T. \tag{16.9f}$$

In this model c_{kt} is the per unit production cost and h_{kt} is the per unit inventory carrying cost for item k in period t. F_{kt} is a fixed cost that we incur if we produce item k in period t and P_{kt} is the production capacity for item k in period t. The constraint (16.9a) ensures that we produce at most one item in each period. Constraint (16.9c) states that we allocate the amount we have on hand of item k in period t (i.e., the production plus incoming inventory of that item) either to demand in that period or to inventory at the end of the period. The "forcing" constraint (16.9d) ensures that the quantity x_{kt} of item k produced in period t is zero if we do not select that item for production in that period, that is, if $z_{kt} = 0$; this constraint also ensures that the production of item k in period t never exceeds the production capacity of that item.

Note that constraints in (16.9b) are the only constraints in this model that link various items. Therefore, these constraints would be attractive candidates to relax via Lagrange multipliers λ_t. Doing so creates the following objective function

$$\text{Minimize} \quad \sum_{k=1}^{K}\sum_{t=1}^{T} c_{kt}x_{kt} + \sum_{k=1}^{K}\sum_{t=1}^{T} h_{kt}I_{kt} + \sum_{k=1}^{K}\sum_{t=1}^{T} [F_{kt} + \lambda_t]z_{kt} - \sum_{t=1}^{T} \lambda_t.$$

For a fixed value of the Lagrange multipliers, the last term is a constant, so the problem separates into a single-item production planning problem for each item k; the production and inventory carrying costs in the relaxation are the same as those in the original model, and in each period the fixed cost for each item in the relaxation is λ_t units more than in the original model.

The subproblems assume different forms, depending on the nature of the production capacities. In Chapter 19 we show that whenever each single-item subproblem is uncapacitated (i.e., P_{kt} is as large as the sum of the demands d_{kt} in periods $t + 1, t + 2, \ldots, T$), we can solve each single-item subproblem as a shortest path problem. If we impose production capacities, the subproblems are NP-complete. In these instances, since the number of time periods is often very small, we might use a dynamic programming approach for solving the subproblems.

To conclude this discussion, we might note that we can enrich this basic multi-item production planning model in a variety of ways. For example, as shown in Chapter 19, we can model multiple stages of production or the backlogging of demand. As another example, we can model situations in which we incur a startup cost whenever we initiate the production of a new item. To model this situation, we let y_{kt} be a zero–one variable, indicating whether or not the production system switches from not producing item k in period $t - 1$ to producing the item in period t. We then add the following constraints to the basic model (16.9):

$$z_{kt} - z_{k,t-1} \leq y_{kt} \qquad \text{for } k = 1, 2, \ldots, K \text{ and } t = 1, 2, \ldots, T,$$

and for each "turn on" variable y_{kt}, we add a cost term $\alpha_{kt}y_{kt}$ to the objective function (α_{kt} is the cost for turning on the machine to produce item k in period t). By relaxing these constraints as well as the item choice constraints (16.9b), we again obtain separate production planning problems for each item. Or, by relaxing only the item choice constraints, we obtain a single-item production planning problem in which we incur three types of production costs: (1) a cost for turning the machine on, (2) a cost for setting up the machine in any period to produce any amount of the item, and (3) a per unit production cost.

This startup cost problem is important in many practical production settings. Moreover, this model is illustrative of the enhancements that we can make to the basic production planning problem and once again demonstrates the algorithmic flexibility of Lagrangian relaxation.

16.6 SUMMARY

Lagrangian relaxation is a flexible solution strategy that permits modelers to exploit the underlying structure in any optimization problem by relaxing (i.e., removing) complicating constraints. This approach permits us to "pull apart" models by removing constraints and instead place them in the objective function with associated Lagrange multipliers. In this chapter we have developed the core theory of Lagrangian relaxation, described popular solution approaches, and examined several application contexts in which Lagrangian relaxation effectively exploits network substructure.

The starting point for the application and theory of Lagrangian relaxation (as applied to a model specified as a minimization problem) is a key bounding principle stating that for any value of the Lagrange multiplier, the optimal value of the relaxed problem, called the Lagrangian subproblem, is always a lower bound on the objective function value of the problem. To obtain the best lower bound, we need to choose the Lagrangian multiplier so that the optimal value of the Lagrangian subproblem is as large as possible. We call this problem the Lagrangian multiplier problem. We can solve the Lagrangian multiplier problem in a variety of ways. The subgradient optimization technique is possibly the most popular technique for solving the Lagrangian multiplier problem and we have described this technique in some detail. The subgradient optimization technique solves a sequence of Lagrangian subproblems.

Usually, we choose the constraints to relax so that the Lagrangian subproblem is much easier to solve than the original problem. Consequently, when applying Lagrangian relaxation, we solve many "simple" problems instead of one single "complicated" problem. Frequently, the complicating constraints that we relax are the only constraints that couple otherwise independent subsystems (e.g., shortest path problems); in these instances, Lagrangian relaxation permits us to decompose a problem into smaller, more tractable subproblems. For this reason, the research community often refers to Lagrangian relaxation as a decomposition technique.

In discussing the theory of Lagrangian relaxation, we showed how to formulate the Lagrangian multiplier problem as an associated linear program with a large number of constraints; we also showed how to interpret the Lagrangian multiplier problem as a convexification of the original optimization model. That is, instead of restricting our choices to a discrete set of possible alternatives (e.g., spanning tree solutions), the multiplier problem produces the same objective function value that we would obtain if we solved the original problem, but permitted the use of convex combinations of the alternatives. We also showed that when applied to integer programs, the Lagrangian relaxation always gives at least as large a lower bound as does the linear programming relaxation of the problem. Finally, we showed that whenever the Lagrangian subproblem satisfies the integrality property (so it has an integer solution for all values of the Lagrange multiplier), solving the Lagrange multiplier problem is equivalent to solving the linear programming relaxation of the original optimization model. In these instances, even though the Lagrangian approach provides the same lower bound as the linear programming relaxation, it does have the ability to solve network (or other) subproblems quickly, which is often greatly preferred to solving the original problem by general-purpose linear programming codes.

Our discussion of applications has introduced several important network optimization models: networks with side constraints, the traveling salesman problem, vehicle routing, network design, personnel scheduling, degree-constrained minimum spanning trees, and production planning. As we have seen, these optimization models have applications in such diverse settings as machine scheduling, communication system design, delivery of consumer goods, telephone coin box collection, telephone operator scheduling, logistics, and production. Consequently, our discussion has illustrated the broad applicability of Lagrangian relaxation across many practical problem contexts. It has also illustrated the versatility of Lagrangian relaxation and its ability to exploit the core network substructures—shortest paths, minimum cost flows, the assignment problem, and minimum spanning tree problems—that we have studied in previous chapters. Our discussion of applications has also highlighted several other points:

1. *Need for creative modeling*. Formulating Lagrangian relaxations can require considerable ingenuity in modeling (as in the variable splitting device that we used to study the two-duty operator scheduling problem).
2. *Flexibility of Lagrangian relaxation*. In many models, such as the vehicle routing problem, we can obtain a variety of different Lagrangian subproblems by relaxing different constraints. This variety of potential subproblems permits us to develop different algorithms for solving the same problem.

3. *Use of Lagrangian relaxation as a conceptual as well as algorithmic tool.* On some occasions, as in our discussion of the degree-constrained minimum spanning tree problem, we can use the bounding information provided by Lagrangian relaxation as a stand-alone tool that is unrelated to any iterative method for solving the Lagrangian multiplier problem. For example, we can use the bounds to analyze the solutions generated by combinatorial or heuristic algorithms for solving a problem.

REFERENCE NOTES

The Lagrange multiplier technique of nonlinear optimization dates to the eighteenth century and was suggested by the famous mathematician Lagrange, for whom the technique is named. The use of this technique in integer programming and discrete optimization is much more recent, originating in the seminal papers by Held and Karp [1970, 1971], who studied the traveling salesman problem. Everett's [1963] development of Lagrangian multiplier methods for general mathematical programming problems was a precursor to this development. Held and Karp's application of the Lagrange multiplier method was not only an eye-opening successful application, but also set out many key ideas in applying the method to integer programming problems. Fisher [1981, 1985], Geoffrion [1974], and Shapiro [1979] provide insightful surveys of Lagrangian relaxation and its uses in integer programming. The papers by Fisher contain many citations to successful applications in a wide variety of problem settings. For a discussion of the branch-and-bound algorithm, see Winston [1991].

Most of the key results of Lagrangian relaxation (e.g., the bounding properties and optimality conditions) are special cases of more general results in mathematical programming duality theory. Rockafellar [1970] and Stoer and Witzgall [1970] provide comprehensive treatments of this subject. Magnanti, Shapiro, and Wagner [1976] establish the equivalence of the Lagrangian multiplier problem and generalized linear programming, whose development by Dantzig and Wolfe [1961] predates the formal development of Lagrangian relaxation in integer programming. The integrality property is due to Geoffrion [1974]. The subgradient method is an outgrowth of so-called relaxation methods for solving systems of linear inequalities. Bertsimas and Orlin [1991] have developed the most efficient algorithms (in the worst-case sense) for solving many classes of Lagrangian relaxation problems.

Several of the application contexts that we have discussed in Section 16.5 and in the exercises have very extensive literatures. The following books and survey articles, which contain many references to the literature, serve as good sources of information on these topics.

Traveling salesman problem: the book edited by Lawler, Lenstra, Rinnooy Kan, and Shmoys [1985]

Vehicle routing: surveys by Bodin, Golden, Assad, and Ball [1983], Laporte and Nobert [1987], and Magnanti [1981]

Network design: surveys by Magnanti and Wong [1984], Magnanti, Wolsey, and Wong [1992], and Minoux [1989]

Production planning: the survey paper by Shapiro [1992], the book by Hax and Candea [1984], and the paper by Graves [1982]

Several other of the applications discussed in this chapter are adapted from research papers from the literature. For a Lagrangian relaxation-based branch-and-bound approach to the constrained shortest path problem, see Handler and Zang [1980]. Shepardson and Marsten [1980] have used the variable splitting device and Lagrangian relaxation for solving the two-duty operator scheduling problem and applied this approach to bus operator scheduling. For an algorithmic approach to the network design problem, see Balakrishnan, Magnanti, and Wong [1989a]. Volgenant [1989] considers the degree-constrained minimum spanning tree problem.

EXERCISES

16.1. Lagrangian relaxation and inequality constraints. To develop the Lagrangian multiplier problem for an inequality constraint problem stated as $\min\{cx : \mathcal{A}x \le b,\ x \in X\}$, suppose that we add nonnegative "slack" variables s to model the problem in the following equivalent equality form : $\min\{cx : \mathcal{A}x + s = b,\ x \in X \text{ and } s \ge 0\}$.
   (a) State the Lagrangian multiplier problem for the equality formulation.
   (b) Show that if some $\mu_i < 0$, then $L(\mu) = -\infty$. Further, show that if some $\mu_i > 0$, then in the optimal solution of the Lagrangian subproblem $L(\mu)$, the slack variable $s_i = 0$.
   (c) Conclude from part (b) that the Lagrangian multiplier problem of the inequality constrained problem is $\max_{\mu \ge 0} L(\mu)$ with $L(\mu) = \min\{cx + \mu(\mathcal{A}x - b) : x \in X\}$.

16.2. Consider the problem

$$\text{Minimize} \quad -2x - 3y$$

subject to

$$x + 4y \le 5,$$
$$x, y \in \{0, 1\},$$

and the corresponding relaxed problem

$$\text{Minimize} \quad -2x - 3y + (x + 4y - 5)$$

subject to

$$x, y \in \{0, 1\}.$$

Show that $x = 1, y = 0$ solves the relaxed problem, is feasible for the original problem, and yet does not solve the original problem. (Reconcile this example with Property 16.4.)

16.3. Lagrangian relaxation applied to linear programs. Suppose that we apply Lagrangian relaxation to the linear program $\mathcal{P}$ defined as $\min\{cx : \mathcal{A}x = b,\ x \ge 0\}$ by relaxing the equality constraints $\mathcal{A}x = b$. The Lagrangian function is $L(\mu) = \min_{x \ge 0}\{cx - \mu(\mathcal{A}x - b)\} = \min_{x \ge 0}\{(c - \mu\mathcal{A})x + \mu b\}$. (Since the constraints $\mathcal{A}x = b$ are equalities, the Lagrange multipliers μ are unconstrained in sign. For the purpose of this exercise, we have chosen a different sign convention than usual, that is, used $-\mu$ in place of μ.) Now, consider the Lagrangian multiplier problem $\max_\mu L(\mu)$.
   (a) Suppose we choose a value of μ so that for some j, $(c - \mu\mathcal{A})_j < 0$. Show that $L(\mu) = -\infty$.
   (b) Suppose we choose a value of μ so that for some j, $(c - \mu\mathcal{A})_j > 0$. Show that in the optimal solution of the Lagrangian subproblem, $x_j = 0$.
   (c) Conclude from parts (a) and (b) that the Lagrangian multiplier problem is equiv-

alent to the linear programming dual of $\mathcal{P}$, that is, the problem max μb, subject to $\mu\mathcal{A} \le c$.

16.4. Oscillation in Lagrangian relaxation. Suppose that we apply Lagrangian relaxation to the constrained shortest path example shown in Figure 16.1 with the time constraint of $T = 14$, starting with value $\mu^0 = 0$ for the Lagrange multiplier μ. Show that if we choose the step size $\theta_k = 1$ at each iteration, the subgradient algorithm $\mu^{k+1} = \mu^k + \theta_k(\mathcal{A}x^k - b)$ oscillates between the values $\mu = 0$ and $\mu = 4$ and the Lagrangian subproblem solutions alternate between the paths 1–2–4–6 and 1–3–2–5–6.

16.5. In Section 16.4 we showed that when $T = 14$, our constrained shortest path example had an optimal objective function value $z^* = 13$ while the Lagrange multiplier problem had a value $L^* = 7$. Show that L^* equals the optimal objective function value of the linear programming relaxation of the problem. Interpret the solution of the linear program as the convex hull of shortest path solutions. That is, find a set of paths whose convex combination satisfies the timing constraint and whose weighted (i.e., convex combination) cost equals L^*.

16.6. Suppose that X is a finite set and that when we solve the Lagrangian multiplier problem corresponding to the optimization problem $\min\{cx : \mathcal{A}x = b, x \in X\}$ for *any* value of c, we find that the problem has no duality gap, that is, if x^* solves the given optimization problem and μ^* is an optimal solution to the Lagrangian multiplier problem, then $cx^* = L(\mu^*)$. Show that the polyhedron $\{x : \mathcal{A}x = b \text{ and } x \in \mathcal{H}(X)\}$ has integer extreme points. (*Hint*: Use the results given in the proofs of Theorems 16.9 and 16.10.)

16.7. Lagrangian relaxation interpretation of successive shortest paths. Recall from Section 9.7 that each intermediate stage of the successive shortest path algorithm for solving the minimum cost flow problem maintains a pseudoflow x satisfying the flow bound constraints and a vector π of node potentials satisfying the conditions $c_{ij}^{\pi} = c_{ij} - \pi(i) + \pi(j) \ge 0$ for all arcs $(i, j) \in G(x)$.

(a) Show that the pseudoflow x is optimal for the problem obtained by relaxing the mass balance constraints and replacing the objective function cx with the Lagrangian function

$$\text{Minimize} \sum_{(i,j)\in A} (c_{ij} - \pi(i) + \pi(j))x_{ij}.$$

(b) Interpret the successive shortest path algorithm as a method that proceeds by adjusting the Lagrangian multipliers. At each stage the method adjusts the multipliers π so that (1) the current pseudoflow x is optimal for the Lagrangian subproblem, and (2) some alternate optimal pseudoflow x' for the Lagrangian relaxation is "less infeasible" than x. Finally, when the optimal pseudoflow becomes a flow, we obtain an optimal solution of the Lagrangian subproblem that is also feasible for the original problem; therefore, it must be an optimal solution of the original problem.

16.8. Generalized assignment problem (Ross and Soland [1975]). The generalized assignment problem is the optimization model

$$\text{Minimize} \sum_{i\in I}\sum_{j\in J} c_{ij}x_{ij} \tag{16.10a}$$

subject to

$$\sum_{j\in J} x_{ij} = 1 \qquad \text{for all } i \in I, \tag{16.10b}$$

$$\sum_{i\in I} a_{ij}x_{ij} \le d_j \qquad \text{for all } j \in J, \tag{16.10c}$$

$$x_{ij} \ge 0 \text{ and integer} \qquad \text{for all } (i, j) \in A. \tag{16.10d}$$

In this problem we wish to assign $|I|$ "objects" to $|J|$ "boxes." The variable $x_{ij} = 1$ if we assign object i to box j and $x_{ij} = 0$ otherwise. We wish to assign each object to exactly 1 box; if assigned to box j, object i consumes a_{ij} units of a given "resource" in that box. The total amount of resource available in the jth box is d_j. This generic model arises in a variety of problem contexts. For example, in machine scheduling, the objects are jobs, the boxes are machines; a_{ij} is the processing time for job i on machine j and d_j is the total amount of time available on machine j.

(a) Outline the steps required for solving the Lagrangian subproblem obtained by (1) relaxing the constraint (16.9b), and (2) by relaxing the constraint (16.10c).

(b) Compare the lower bounds obtained by the two relaxations suggested in part (a). Which provides the sharper lower bound? Why? (*Hint*: Use Theorems 16.9 and 16.10.)

(c) Compare the optimal objective function value of the Lagrangian multiplier problem for each relaxation suggested in part (a) with the bound obtained by the linear programming relaxation of the generalized assignment model.

16.9. Facility location (Erlenkotter [1978]). Consider the following facility location model:

$$\text{Minimize} \quad \sum_{i \in I} \sum_{j \in J} c_{ij}x_{ij} + \sum_{j \in J} F_j y_j \tag{16.11a}$$

subject to

$$\sum_{j \in J} x_{ij} = 1 \qquad \text{for all } i = 1, 2, \ldots, I, \tag{16.11b}$$

$$\sum_{i \in I} d_i x_{ij} \le K_j y_j \qquad \text{for all } j = 1, 2, \ldots, J, \tag{16.11c}$$

$$0 \le x_{ij} \le 1 \qquad \text{for all } i \in I \text{ and } j \in J, \tag{16.11d}$$

$$y_j = 0 \text{ or } 1 \qquad \text{for all } j \in J. \tag{16.11e}$$

In this model, I denotes a set of customers and J denotes a set of potential facility (e.g., warehouse) locations used to supply to the customers. The zero–one variable y_j indicates whether or not we choose to locate a facility at location j and x_{ij} is the fraction of the demand of customer i that we satisfy from facility j. The constant d_i is the demand of customer i. The cost coefficient c_{ij} is the cost (e.g., the transportation cost) of satisfying all of the ith customer's demand from facility j, and the cost coefficient F_j is the fixed cost of opening (e.g., leasing) a facility of size K_j at location j. The constraints (16.11b) state that we need to satisfy all of the demand for each customer, and the constraints (16.11c) state that (1) we cannot meet any of the demand of any customer if we do not locate a facility at location j (i.e., $x_{ij} = 0$ if $y_j = 0$), and (2) if we do locate a facility at location j (i.e., $y_j = 1$), the total demand met by the facility cannot exceed the facility's capacity K_j.

(a) Show how you would solve the Lagrangian subproblem obtained by relaxing the constraints (16.11b). (*Hint*: Note that the Lagrangian subproblem decomposes into a separate subproblem for each location.)

(b) Show next how you would solve the Lagrangian subproblem if we relax the constraints (16.11c). (*Hint*: Note that the Lagrangian subproblem decomposes into a separate subproblem for each customer.)

(c) Show that if $|I| = |J| = 1, K_1 = 10, d_1 = 5$, and $c_{11} = 0$, the relaxation suggested in part (a) gives a sharper lower bound than the relaxation in part (b). Next prove the general result that the relaxation in part (a) gives at least as good a bound as given by the relaxation in part (b).

16.10. Modified facility location. Suppose that in the model considered in Exercise 16.9, we impose the additional constraint that the demand for each customer should be "sole sourced"; that is, each variable x_{ij} has value zero or 1.

(a) Show how to use the solution of a single knapsack problem for each facility j to solve the Lagrangian relaxation obtained by relaxing the constraints (16.11b).

(b) Show that the bound obtained from the Lagrangian multiplier problem by relaxing the constraints (16.11b) is always at least as strong as the bound obtained by relaxing the constraints (16.11c).

16.11 Tightening the facility location relaxation. Suppose that we add the redundant constraints $x_{ij} \leq \min\{y_j, K_j\}$ to the facility location model described in Exercise 16.9 and then we apply Lagrangian relaxation by relaxing the constraints (16.11b) or (16.11c).

(a) Show that the bound obtained from the Lagrangian multiplier problem is always as strong or stronger than the bound obtained by relaxing the corresponding constraints in the original model without the additional constraints $x_{ij} \leq \min\{y_j, K_j\}$.

(b) How would you solve the Lagrangian subproblem with the added constraints $x_{ij} \leq \min\{y_j, K_j\}$?

(c) How would your answers to parts (a) and (b) change if we considered the sole-sourcing-facility location model described in Exercise 16.10?

16.12. Local access capacity expansion (Balakrishnan, Magnanti, and Wong [1991]). The lowest level of national telephone networks are trees that connect individual customers to the rest of the national network through special nodes known as *switching centers*, which route telephone calls to their final destination. Each local access network (tree) T has its own switching center. As demand for service increases, telephone companies have two basic options for increasing the capacity of a local access network: (1) they can install more copper cables on the arcs of the networks; or (2) they can install devices, called *multiplexers* (or *concentrators*), at the nodes. The multiplexers compress calls so that they use less downstream cable capacity. We assume that once a call reaches a multiplexer, it requires negligible cable capacity to send it to the switching center. Every call must be routed through the tree T either to the switching center or to one of the multiplexers. Suppose that the existing capacity of arc (i, j) is u_{ij} and increasing the capacity by y_{ij} units incurs an arc-dependent cost $c_{ij}y_{ij}$. Let d_i denote the numbers of calls originating at node i that must be routed to the switching center or to a multiplexer. Each multiplexer has two associated costs: (1) a fixed cost F, and (2) a variable throughput cost α incurred for each unit of call compressed by that multiplexer. The optimization problem is to meet the demand for service by incurring minimum total cost.

(a) Let z_i be a zero–one variable indicating whether or not we place a multiplexer at node i. Further, let x_{ij} be a zero–one variable indicating whether or not we assign node i to the multiplexer j. In the local access network T, for any pair $[i, j]$ of nodes, we let P_{ij} denote the unique path between these two nodes, and for any arc (k, l) in T, we let Q_{kl} be the set of all node pairs $[i, j]$ from which P_{ij} contains the arc (k, l). We assume that node 1 is the switching center. Let node S denote the remaining nodes in the network. Using this notation, give an integer programming formulation of the local access network design problem.

(b) Suggest two relaxations of the formulation in part (a) that produce a relaxed problem with a structure that we have treated in this book.

16.13. Contiguous local access capacity expansion problem (Balakrishnan, Magnanti, and Wong [1991]). In most practical settings of the local access capacity expansion problems, the set of nodes assigned to the switching center or to any multiplexer must be contiguous. That is, if we assign node i to a multiplexer at node j and node k lies on the path in T from node i to node j, we must also assign node k to the multiplexer at node j. Therefore, the final configuration of the local access network will be a subdivision of the tree T into subtrees, with each subtree containing either the switching center or one multiplexer and the nodes it serves.

(a) Show that we can incorporate the contiguity condition in the formulation of Exercise 16.12 by adding the following constraints for every pair $[i, j]$ of nodes: if the path P_{ij} contains node k, then $x_{ik} \geq x_{ij}$.

(b) Consider the integer programming formulation of the contiguous local access capacity expansion problem from part (a). Suppose that we relax the capacity constraints imposed on the arcs. Show that we can solve the Lagrangian subproblem in polynomial time using a dynamic programming technique.

16.14. Design of telecommunication networks (Leung, Magnanti, and Singhal [1990] and Magnanti, Mirchandani, and Vachani [1991]). In designing telecommunications networks, we would like to install sufficient capacity to carry required traffic (telephone calls, data transmissions) simultaneously between various source–sink locations. Suppose that (s^k, t^k) for $1 \le k \le K$ denote K pairs of source–sink locations, and r^k denotes the number of messages sent from the source s^k to the sink t^k. We can install either of two different types of facilities on each link of the transmission network, so-called T0 lines and T1 lines. Each T0 line can carry 1 unit of message and each T1 line can carry 24 units of messages; installing a T0 line on arc (i, j) incurs a cost of a_{ij} and installing a T1 line on arc (i, j) incurs a cost of b_{ij}. Once we have installed the lines, we incur no additional costs in sending flow on them. This problem arises in practice because companies with large telecommunication requirements might be able to lease lines more cost-effectively than paying public tariffs. The same type of problem arises in trucking of freight; in this setting, the facilities to be "installed" on any arc are the trucks of a particular type (e.g., 36-foot trailers or 48-foot trailers) to be dispatched on that arc.

(a) Show how to formulate this telecommunication network design problem with two types of constraints: (1) a set of network flow constraints modeling the required flow between every pair of source–sink locations, and (2) capacity constraints restricting the total flow on each arc to be no more than the capacity that we install on that arc. (*Hint*: Use the following integer decision variables: (1) y_{ij} : the number of T0 lines for arc (i, j), (2) z_{ij} : the number of T1 lines for arc (i, j), and (3) the number of messages x_{ij}^k sent from the source s^k to the sink t^k that pass through the arc (i, j).)

(b) How would you solve the linear programming relaxation of this model? (*Hint*: Consider two cases: when $24\, a_{ij} < b_{ij}$ and when $24\, a_{ij} \ge b_{ij}$.)

(c) Show how to solve the Lagrangian subproblem obtained by relaxing constraints of type 1 in the model formulation. (*Hint*: Consider the two cases as in part (b).)

(d) Show how to solve the Lagrangian subproblem obtained by relaxing constraints of type 2 in the model formulation.

16.15. Steiner tree problem. The Steiner tree problem is an $\mathcal{NP}$-hard variant of the minimum spanning tree problem. In this problem we are given a subset $S \subseteq N$ of nodes, called *customer nodes*, and we wish to determine a minimum cost tree (not necessarily a spanning tree) that must contain all the nodes in S and, optionally, some nodes in $N - S$. This problem arises in many application settings, such as the design of rural road networks, pipeline networks, or communication networks. Formulate this problem as a special case of the network design problem discussed in Application 16.4 and show how to apply Lagrangian relaxation to the resulting formulation. (*Hint*: Designate any customer node as a source node and send 1 unit of flow to every other customer node.)

16.16. In this exercise we show how to formulate a directed traveling salesman problem as a network design problem. Consider a network design problem with the following data: (1) unit commodity flow requirements between every pair of nodes, (2) the cost of flow on every arc is zero, and (3) the fixed cost of the arc (i, j) is $c_{ij} + M$ for some sufficiently large number M. Show that the optimal network will be an optimal traveling salesman tour with c_{ij} as arc lengths. (*Hint*: The optimal network must be strongly connected and must contain the fewest possible number of arcs.)

16.17. Uncapacitated undirected network design problem. In the formulation of the directed uncapacitated network design problem in Application 16.4, the zero–one vector y_{ij} indicated whether we would include the directed arc (i, j) in the underlying network. Suppose, instead, that the arcs are undirected, so if we introduce arc (i, j) in the network, we can send flow in either direction on the arc.

(a) How would you formulate this problem and apply Lagrangian relaxation to obtain lower bounds?

(b) Show that in the uncapacitated undirected network design problem, if all flow costs c_{ij}^k are zero, all the fixed costs f_{ij} are nonnegative, and the problem has a commodity for each pair of nodes, the problem reduces to the minimum spanning tree problem.

16.18. **(a)** Show that if the uncapacitated network design problem has a single commodity (i.e., $K = 1$), we can solve the problem by solving a single shortest path problem.

(b) Show how to formulate the production planning problems that we described in Application 16.7 as capacitated or uncapacitated network design problems.

16.19. **(a)** Suppose that we relax the mass balance constraint $\mathcal{N}x = b$ in the formulation (16.3) of the traveling salesman problem described in Application 16.2. Show how to solve the Lagrangian subproblem as an assignment problem. (*Hint*: Show that some optimal solution of the Lagrangian subproblem satisfies the conditions $x_{ij} = 0$ or $x_{ij} = (n - 1)y_{ij}$ for each arc $(i, j) \in A$. Use this fact to eliminate the variables x_{ij} from the subproblem.)

(b) Suppose that we relax the assignment constraints (16.3b) and (16.3c) in the formulation (16.3) of the traveling salesman problem. Show how to solve the Lagrangian subproblem as an uncapacitated network design problem.

(c) What is the relationship between the optimal solutions of the Lagrangian multiplier problems obtained by the relaxations considered in parts (a) and (b), the relaxation described in the text (obtained by relaxing the forcing constraints $x_{ij} \le (n - 1)y_{ij}$), and the optimal objective function value of the linear programming relaxation of the problem?

16.20. **Assignment-based formulation of the traveling salesman problem**

(a) Consider the integer program (16.3) with the constraints (16.3d) and (16.3e) replaced by the subtour breaking constraints (16.4h). Show that the resulting model is an integer programming formulation of the traveling salesman problem.

(b) Show that the solution of the Lagrangian subproblem formed by relaxing the subtour breaking constraints will be a set of directed cycles satisfying the property that each node is contained in exactly one cycle. Describe a heuristic method for modifying the Lagrangian subproblem solution so that it becomes a feasible traveling salesman tour.

16.21. **Undirected traveling salesman problem.** In the undirected (or symmetric) traveling salesman problem, we can traverse any arc (i, j) in either direction at the same cost c_{ij}. Let y_{ij} indicate whether or not we include arc (i, j) in a feasible tour.

(a) Give a formulation of this problem as an integer program containing three sets of constraints: (1) degree 2 constraints, indicating that each node should have degree at most 2 in any feasible tour; (2) subtour breaking constraints on the nodes $2, 3, \ldots, n$; and (3) a cardinality constraint indicating that the tour contains exactly n arcs. (*Hint*: Modify the assignment based formulation of the directed traveling salesman problem described in Exercise 16.21.)

(b) Show how to apply Lagrangian relaxation in two ways: (1) by relaxing the degree 2 constraints; and (2) by relaxing the subtour breaking constraints and the cardinality constraint. In case 1 show how to solve the Lagrangian subproblem as a 1-tree (see Exercise 13.37). In case 2 show how to solve the Lagrangian subproblem as a circulation problem. (*Hint*: In case 2, first show that any network with n arcs and degree at most 2 on each node, must have a degree of exactly 2 at each node.)

16.22. **Multicommodity flow-based formulation of the traveling salesman problem.** In Application 16.2 we examined a single-commodity flow-based formulation of the traveling salesman problem with $n - 1$ units available at a source node (which we arbitrarily took to be node 1) and 1 unit of demand required at each other node. Suppose that, instead, we formulated a multicommodity flow model with $2(n - 1)$ commodities, with two commodities k defined for each node $k \ne 1$, an "outgoing" commodity and an "incoming" commodity. The incoming commodity for node k has 1 unit of supply at node 1 and 1 unit of demand at node k, and the outgoing commodity for node k has

1 unit of supply at node k and 1 unit of demand at node 1 (i.e., we wish to send 1 unit from node 1 to node k and 1 unit from node k to node 1). We can state this formulation of the traveling salesman problem as the following integer program:

$$\text{Minimize} \quad \sum_{(i,j)\in A} c_{ij} y_{ij},$$

subject to

$$\sum_{1\le j\le n} y_{ij} = 1 \qquad \text{for all } i = 1, 2, \ldots, n,$$

$$\sum_{1\le i\le n} y_{ij} = 1 \qquad \text{for all } j = 1, 2, \ldots, n,$$

$$Nx^k = b^k \qquad \text{for all } k = 2, \ldots, n,$$

$$Nz^k = d^k \qquad \text{for all } k = 2, \ldots, n,$$

$$x_{ij}^k \le y_{ij} \text{ and } z_{ij}^k \le y_{ij} \qquad \text{for all } (i, j) \text{ and all } k,$$

$$y_{ij} \text{ and } x_{ij}^k = 0 \text{ or } 1 \qquad \text{for all } (i, j) \text{ and all } k.$$

Note that the supply/demand vectors b^k and d^k in this formulation have a special form: $d^k = -b^k$ and b_i^k is 1 if $i = 1$, is -1 if $i = k$, and is 0 if $i \neq 1$ and $i \neq k$.

(a) Suppose that we apply Lagrangian relaxation to the multicommodity flow-based model by relaxing the forcing constraints [i.e., the constraints $x_{ij}^k \le y_{ij}$ and $z_{ij}^k \le y_{ij}$, for all (i, j) and all k]. How would you solve the Lagrangian subproblem?

(b) Show that the lower bound L^* determined by the Lagrangian multiplier problem for the Lagrangian relaxation in part (a) is always as strong or stronger than the lower bound determined by relaxing the forcing constraints in the single-commodity flow-based formulation. (*Hint*: Compare the set of feasible solutions of both problems.)

(c) What is the relationship between the optimal objective function values of the linear programming relaxations of the single and multicommodity flow-based formulations? (*Hint*: Same as that in part (b).)

16.23. Alternate formulations of the traveling salesman problem (Wong [1980]). In this chapter we have considered three different formulations of the traveling salesman problem: (1) a single-commodity flow-based formulation in Application 16.2, (2) an assignment-based formulation discussed in Exercise 16.20, and (3) a multicommodity flow-based formulation in Exercise 16.22, where we showed that from the perspective of linear programming or Lagrangian relaxations, the multicommodity flow-based formulation is stronger than the single commodity flow-based formulation.

(a) Show that we can replace the subtour breaking constraints in the assignment-based formulation, or in its linear programming relaxation, by the constraints $\sum_{i\in S} \sum_{j\in N\text{-}S} y_{ij} \ge 1$ for all sets S of nodes satisfying the cardinality condition $1 \le |S| \le n - 1$, and in both cases obtain an equivalent model (i.e., one with the same feasible solutions).

(b) Using the max-flow min-cut theorem and part (a), show that the linear programming relaxation of the assignment-based formulation of the traveling salesman problem and the linear programming relaxation of the multicommodity flow-based formulation are equivalent in the sense that y is feasible in the linear programming relaxation of the assignment-based formulations if and only if for some flow vector x, (x, y) is feasible in the multicommodity flow-based formulation. (Note that the number of subtour breaking constraints in the assignment-based formulation is exponential in n. The number of constraints in the multicommodity flow-based formulation is polynomial in n, so this formulation is a so-called *compact* formulation.)

16.24. Consider the undirected traveling salesman problem shown in Figure 16.11.

(a) What is the optimal tour length for this problem?

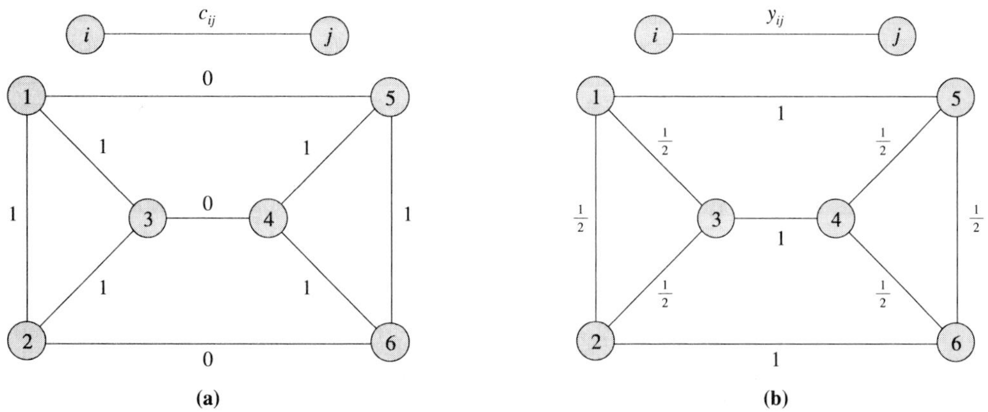

Figure 16.11 Traveling salesman problem: (a) network data; (b) solution to the linear programming relaxation.

(b) Show that the arc weights shown in Figure 16.11(b) solve the linear programming relaxation of the formulation developed in Exercise 16.21. Interpret this solution as the convex hull of 1-tree solutions to the Lagrangian subproblem that we obtain by relaxing the degree two constraints in this formulation. That is, show how to represent this solution as a convex combination of 1-tree solutions.

(c) Show the network corresponding to the equivalent directed traveling salesman problem and specify the optimal solution to the linear programming relaxation of the assignment-based formulation and both the single and multicommodity flow-based formulations.

(d) Interpret the solution to each linear programming relaxation as the convex hull of solutions to Lagrangian subproblems.

16.25. ***K*-traveling salesman problem.** Suppose that we wish to find a set of K arc-disjoint directed cycles in a directed graph satisfying the property that node 1 is contained in exactly K cycles and every other node is contained in exactly one cycle. In this model each arc has an associated cost c_{ij} and we wish to find a feasible solution with the smallest possible sum of arc costs. We refer to this problem as the K-traveling salesman problem since it corresponds to a situation in which K salesmen, all domiciled at the same node 1, need to visit all the other nodes of a graph.

(a) Formulate this problem as an optimization model and show how to apply Lagrangian relaxation to the formulation. (*Hint*: Modify the single-commodity flow-based formulation given in Application 16.2 or the assignment-based formulation given in Exercise 16.20.)

(b) By forming K copies of node 1 and assigning a large cost with all of the arcs joining the copies of node 1, show how to formulate the problem as an equivalent (single) traveling salesman problem.

16.26. Consider the K-traveling salesman problem described in Exercise 16.25. Show how to formulate this problem as a special case of the vehicle routing problem described in Application 16.3. The resulting formulation will be considerably simpler than the general vehicle routing problem. (*Hint*: First show that we can eliminate the constraints (16.4g). Next show how to use the constraints (16.4b) to eliminate the variables x_{ij}^k.)

16.27. **Vehicle routing with nonhomogeneous fleets and with time restrictions.** This exercise studies a generalization of the vehicle routing problem discussed in Application 16.3. Show how to formulate a vehicle routing problem with each of the following problem ingredients: (1) each vehicle k in a fleet of K vehicles can have different capacity u^k, or (2) each vehicle must make its deliveries within T hours, given that it takes t_{ij} hours to traverse any arc (i, j).

16.28. Suppose that we add the redundant constraint $\sum_{(i,j)\in A} y_{ij} = n + K$ to the formulation (16.4) of the vehicle routing problem. Consider the additional set of constraints $\sum_{j\in S} y_{1j} + \sum_{i\in S} \sum_{j\in S} y_{ij} \le |S|$ for all subsets S of $\{2, 3, \ldots, n\}$.

(a) Are these constraints valid?

(b) Are these constraints implied by the other constraints in the integer programming formulation of the problem? Are they implied by the other constraints in the linear programming relaxation of the problem?

(c) Suppose that we add the additional constraints to the formulation of the vehicle routing problem. Show that by relaxing the capacity constraints (16.4g) and the assignment constraints (16.2c) and (16.2d) for nodes 2, 3, . . . , n, the resulting Lagrangian subproblem decomposes into two subproblems: (i) a degree-constrained minimum spanning tree problem with degree K imposed on node 1, and (ii) a problem of choosing the K cheapest (with respect to the Lagrangian subproblem coefficients) arcs of the form y_{j1}. (*Hint*: First eliminate the variables x_{ij}^k and then note that the Lagrangian subproblem decomposes into two subproblems, one containing the variables y_{j1} and one containing all the other variables.)

16.29. Solve the degree-constrained minimum spanning tree problem shown in Figure 16.12 assuming that the degree of node 1 must be 8. Solve it if the degree of node 1 must be 5.

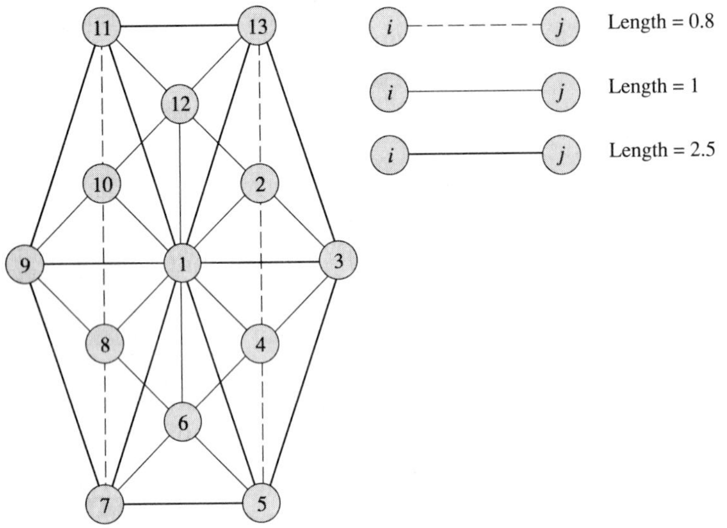

Figure 16.12 Constrained minimum spanning tree problem.

16.30. Suppose that X is a finite set and that when we solve the Lagrangian multiplier problem corresponding to the optimization problem $\min\{cx : \mathcal{A}x = b, x \in X\}$ for *any* value of c, we find that the problem has no relaxation gap; that is, if x^* solves the given optimization problem and μ^* is an optimal solution to the Lagrangian multiplier problem, then $cx^* = L(\mu^*)$. Show that the polyhedron $\{x \in \mathcal{H}(X) : \mathcal{A}x = b\}$ has integer extreme points. (*Hint*: Use the equivalence between convexification and Lagrangian relaxation (Theorem 16.10) and the fact that every extreme point to a polyhedron $\mathcal{P}$ is the *unique* optimal solution to the linear program $\min\{cx : x \in \mathcal{P}\}$ for *some* choice of the objective coefficients c.)

16.31. Let X denote the set of incidence vectors of spanning trees of a given network.

(a) Using Exercise 16.30 and the results in Section 16.4, show that for any value of k, the polyhedron $\{x : x \in \mathcal{H}(X) \text{ and } \sum_{j \neq 1} x_{ij} = k\}$ has integer extreme points. Note that if we view the set of solutions to $x \in \mathcal{H}(X)$ and $\sum_{j \neq 1} x_{1j} = k$ as we vary k as "parallel slices" through the polyhedron $x \in \mathcal{H}(X)$, this result says that extreme points of every slice are integer valued.

(b) For any subset S of nodes, let $A(S) = \{(i, j) \in A : i \in S \text{ and } j \in S\}$. Using the result of part (a) and the development in Section 13.8, show that for any value of k, the following polyhedron has integer extreme points:

$$\sum_{j \neq 1} x_{1j} = k,$$

$$\sum_{(i,j) \in A} x_{ij} = n - 1,$$

$$\sum_{(i,j) \in A(S)} x_{ij} \leq |S| - 1 \qquad \text{for any set } S \text{ of nodes,}$$

$$x_{ij} \geq 0.$$

(*Hint*: In Section 13.8 we showed that without the cardinality constraint $\sum_{j \neq 1} x_{1j} = k$, the extreme points in the polyhedron defined by the remaining constraints are incident vectors of spanning tree solutions (and so are integer valued).)

16.32. Suppose that we wish to find a minimum spanning tree of an undirected graph G satisfying the additional conditions that the degree of node 1 is k and the degree of node n is l. Suggest a Lagrangian relaxation bounding procedure for this problem. (*Hint*: Consider relaxing just one of the two degree constraints.)

16.33. Capacitated minimum spanning tree problem (Gavish [1985]). In some applications of the minimum spanning tree problem, we want to construct a capacitated tree T rooted at a specially designated node, say node 1. In this problem we wish to identify a minimum cost spanning tree subject to the additional condition that no subtree of T formed by eliminating all the arcs incident to node 1 contains more than a prescribed number u of nodes. This model arises, for example, in computer networking when node 1 is a central processor and for reasons of reliability we wish to limit the number of nodes (terminals) attached to this node through any of its ports (incident arcs). Let y_{ij} be a zero–one variable, indicating whether or not we include arc (i, j) in the optimal capacitated tree.

(a) Explain how the capacitated minimum spanning tree problem differs from the degree-constrained minimum spanning tree problem.

(b) By introducing additional constraints in the integer programming formulation (13.2) of the minimum spanning tree problem, obtain an integer programming formulation of the capacitated minimum spanning tree problem.

(c) Suggest a Lagrangian-based method for obtaining a lower bound on the optimal solution by solving a sequence of minimum spanning tree problems.

16.34. Identical customer vehicle routing problem. In the identical customer vehicle routing problem, each customer has the same demand. Formulate this problem as a capacitated minimum spanning tree problem with additional constraints. Show how to obtain bounds on the objective values by applying Lagrangian relaxation to this problem using the capacitated minimum spanning tree problem as a subproblem.

16.35. Note that every solution to a vehicle routing problem is a degree constrained minimum spanning tree (with degree K for node 1) together with K additional arcs incident to node 1 as well as another set of constraints modeling vehicle capacities. Use this observation to give a formulation of the vehicle routing problem and an associated Lagrangian relaxation that contains the degree-constrained minimum spanning tree problem as a subproblem.

16.36. Lagrangian decomposition (Guignard and Kim [1987a,b]). Consider the optimization problem P1 defined as $\min\{cx : \mathcal{A}x = b, \mathcal{D}x = d, x \geq 0 \text{ and integer}\}$. Suppose that by using a variable splitting technique described in Application 16.5, we restate this problem in the following equivalent form P2: $\min\{\frac{1}{2}cx + \frac{1}{2}cy : \mathcal{A}x = b, \mathcal{D}y = d, x - y = 0, x, y \geq 0$ and integer$\}$. We might form three different Lagrangian relaxations for this problem, one by relaxing the constraint $\mathcal{A}x = b$ in P2, one by relaxing the constraint $\mathcal{D}x = d$ in P2, and one by relaxing the constraint $x - y$ in P2. Let L^1, L^2, and L^3 denote the optimal values of the Lagrangian multiplier problems for each of these relaxations. The approach via problem L^3 is known as a *Lagrangian decomposition* since it permits us to decompose the problem into two separate subproblems, one corresponding to each set of equality constraints. Using Theorems 16.8 and Theorem 16.9, show that $L^3 \geq L^1$ and $L^3 \geq L^2$. (*Hint*: Let H^1 and H^2, respectively, denote the convex hulls of the sets $\{\mathcal{D}x = d, x \geq 0$ and integer$\}$ and $\{\mathcal{A}x = b, x \geq 0$ and integer$\}$. Consider the sets $H^1 \cap \{x : \mathcal{A}x = b\}$, $H^2 \cap \{x : \mathcal{D}x = d\}$, and $H^1 \cap H^2$, and consider the minimization of the objective function cx over each of these sets. Which of these problems has the smallest objective function value? What is the relationship between these optimal objective function values and the values L^1, L^2, and L^3?)

16.37. Example of Lagrangian decomposition. Suppose that we apply the Lagrangian decomposition procedure to the following integer programming example:

$$\text{Minimize} \quad -2x_1 - 3x_2$$

subject to

$$9x_1 + 10x_2 \leq 63,$$

$$4x_1 + 9x_2 \leq 36,$$

$$x_1, x_2 \geq 0 \text{ and integer.}$$

In this case, the reformulated problem is:

$$\text{Minimize} \quad -x_1 - \tfrac{3}{2}x_2 - y_1 - \tfrac{3}{2}y_2$$

subject to

$$9x_1 + 10x_2 \leq 63,$$

$$4y_1 + 9y_2 \leq 36,$$

$$x_1 - y_1 = 0,$$

$$x_2 - y_2 = 0,$$

$$x_1, x_2, y_1, y_2 \geq 0 \text{ and integer.}$$

Give a geometrical interpretation of the Lagrangian relaxation obtained by relaxing each of the following constraints: (1) $4x_1 + 9x_2 \leq 36$; (2) $4y_1 + 9y_2 \leq 36$; and (3) $x_1 - y_1 = 0$ and $x_2 - y_2 = 0$. From these geometrical considerations, interpret the fact that in the notation of Exercise 16.36, $L^3 \geq L^1$ and that $L^3 \geq L^2$.

17

MULTICOMMODITY FLOWS

You cannot conceive the many without the one.
—Plato

Chapter Outline

17.1 INTRODUCTION

Throughout most of our discussion to this point, we have considered network models composed of a single commodity—one that we wish to send from its source(s) to its sink(s) in some optimal fashion, for example, along a shortest path or via a minimum cost flow. In many application contexts, several physical commodities, vehicles, or messages, each governed by their own network flow constraints, share the same network. For example, in telecommunications applications, telephone calls between specific node pairs in an underlying telephone network each define a separate commodity. If the commodities do not interact in any way, then to solve problems with several commodities, we would solve each single-commodity problem separately using the techniques that we have developed in prior chapters. In other situations, however, because the commodities do share common facilities, the individual single commodity problems are not independent, so to find an optimal flow, we need to solve the problems in concert with each other. In this chapter we study one such model, known as the *multicommodity flow problem*, in which the individual commodities share common arc capacities. That is, each arc has a capacity u_{ij} that restricts the total flow of all commodities on that arc.

Let x_{ij}^k denote the flow of commodity k on arc (i, j), and let x^k and c^k denote the flow vector and per unit cost vector for commodity k. Using this notation we can formulate the multicommodity flow problem as follows:

$$\text{Minimize} \quad \sum_{1 \le k \le K} c^k x^k \tag{17.1a}$$

subject to

$$\sum_{1 \leq k \leq K} x_{ij}^k \leq u_{ij} \qquad \text{for all } (i, j) \in A, \tag{17.1b}$$

$$\mathcal{N} x^k = b^k \qquad \text{for } k = 1, 2, \ldots, K, \tag{17.1c}$$

$$0 \leq x_{ij}^k \leq u_{ij}^k \qquad \text{for all } (i, j) \in A \text{ and all } k = 1, 2, \ldots, K. \tag{17.1d}$$

This formulation has a collection of K ordinary mass balance constraints (17.1c), modeling the flow of each commodity $k = 1, 2, \ldots, K$. The "bundle" constraints (17.1b) tie together the commodities by restricting the total flow $\sum_{1 \leq k \leq K} x_{ij}^k$ of all the commodities on each arc (i, j) to at most u_{ij}. Note that we also impose individual flow bounds u_{ij}^k on the flow of commodity k on arc (i, j). Many applications do not impose these bounds, so for these applications we set each bound to $+\infty$. Although we might formulate a variety of alternative multicommodity models with different assumptions, we will refer to this model as the multicommodity flow problem.

At times in our discussion, it will be more convenient to state the bundle constraints (17.1b) as equalities instead of inequalities. In these instances we introduce nonnegative "slack" variables s_{ij} and write the bundle constraints as

$$\sum_{1 \leq k \leq K} x_{ij}^k + s_{ij} = u_{ij} \qquad \text{for all } (i, j) \in A. \tag{17.1b$'$}$$

The slack variable s_{ij} for the arc (i, j) measures the unused bundle capacity on that arc.

Assumptions

Note that the model (17.1) imposes capacities on the arcs but not on the nodes. This modeling assumption imposes no loss of generality, since by using the node splitting techniques described in Section 2.4, we can use this formulation to model situations with node capacities as well. Three other features of the model are worth noting.

Homogeneous goods assumption. We are assuming that every unit flow of each commodity uses 1 unit of capacity of each arc. A more general model would permit the unit flow of each commodity k to consume a given amount ρ_{ij}^k of the capacity (or some other resource) associated with each arc (i, j), and replace the bundle constraint with a more general resource availability constraint $\sum_{1 \leq k \leq K} \rho_{ij}^k x_{ij}^k \leq u_{ij}$. With minor modifications, the solution techniques that we will be discussing in this chapter apply to this more general model as well (see Exercise 17.10).

No congestion assumption. We are assuming that we have a hard (i.e., fixed) capacity on each arc and that the cost on each arc is linear in the flow on that arc. In some applications encountered in communication, transportation, and other problem domains, the commodities interact in a more complicated fashion in the sense that as the flow of any commodity increases on an arc, we incur an increasing and nonlinear cost on that arc. This type of model arises frequently, for example, in traffic networks where the objective function is to find the flow pattern of all the commodities that minimizes overall system delay. In this setting, because of queuing effects, the greater the flow on an arc, the greater is the queuing delay on that arc. For example, a "congestion" model for multicommodity flows might contain the individual flow constraints (17.1b) and (17.1c), no bundle constraint, but a nonlinear objective function of the form

$$\text{Minimize} \sum_{(i,j)\in A} \frac{x_{ij}}{u_{ij} - x_{ij}}.$$

In this model u_{ij} is the "nominal" capacity of the arc (i, j); as the total flow $x_{ij} = \sum_{(i,j)\in A} x_{ij}^k$ on any arc approaches the arc's nominal capacity, the delay approaches $+\infty$ (this form of the objective function is derived from basic results in queuing theory). Practitioners often use this type of model in the context of "performance modeling" to see how overall system delay, or performance, varies as a function of various system designs (e.g., in response to changes in the network topology). In Chapter 14 we show how to solve these nonlinear models for a single commodity as a generalization of the minimum cost flow problem. In Exercise 17.11 we show how to solve a class of nonlinear multicommodity flow problems.

Indivisible goods assumption. The model (17.1) assumes that the flow variables can be fractional. In some applications encountered in practice, this assumption is appropriate; in other contexts, however, the variables must be integer valued. In these instances the model that we are considering might still prove to be useful, since the linear programming model might either be a good approximation of the integer programming model, or we can use the linear programming model as a linear programming relaxation of the integer program and embed it within branch-and-bound or some other type of enumeration approach.

We note that the integrality of solutions is one very important distinguishing feature between single and multicommodity flow problems. As we have seen several times in our previous development, one very nice feature of single-commodity network flow problems is that they always have integer solutions whenever the supply/demand and capacity data are integer valued. Multicommodity flow problems do not satisfy this integrality property. For example, consider a multicommodity maximum flow problem with three nodes 1, 2, 3, and three commodities as shown in Figure 17.1. Each arc has a capacity of 1 unit. We wish to find a solution that maximizes the total flow between the source and sink nodes of all three commodities.

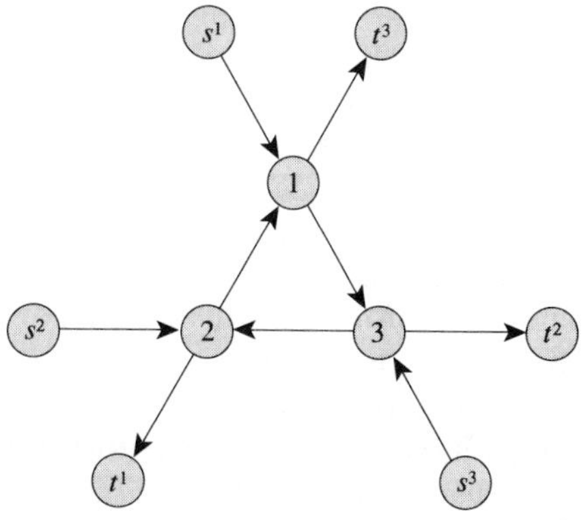

Figure 17.1 Multicommodity maximum flow problem with a fractional solution. Maximum total flow from all sources to all sinks is 1.5 units.

The optimal solution is to send 0.5 units between the source and sink of all three commodities for a total flow of 1.5 units. The optimal integral solution to this problem would send 1 unit between just one of the three commodities for a total flow of 1 unit. In Exercise 17.25 we ask the reader to formulate this multicommodity maximum flow problem as a special case of the multicommodity flow problem stated in (17.1) and so to show that this model does not always have integral solutions even when the problem data are integral.

Solution Approaches

Researchers have developed several approaches for solving the multicommodity flow problem, including:

1. Price-directive decomposition
2. Resource-directive decomposition
3. Partitioning methods

Price-directive decomposition methods place Lagrangian multipliers (or prices) on the bundle constraints and bring them into the objective function so that the resulting problem decomposes into a separate minimum cost flow problem for each commodity k. That is, these methods remove the capacity constraints and instead "charge" each commodity for the use of the capacity of each arc. These methods attempt to find appropriate prices so that some optimal solution to the resulting "pricing problem" or Lagrangian subproblem also solves the overall multicommodity flow problem. Several methods are available for finding appropriate prices. Building on our discussion of Lagrangian relaxation methods in Chapter 16, we describe the application of Lagrangian relaxation to find the correct prices in Section 17.4.

Dantzig–Wolfe decomposition is another approach for finding the correct prices; this method is a general-purpose approach for decomposing problems that have a set of "easy" constraints and also a set of "hard" constraints (that is, constraints that make the problem much more difficult to solve). For multicommodity flow problems, the network flow constraints are the easy constraints and the bundle constraints are the hard constraints. The approach begins, like Lagrangian relaxation, by ignoring or imposing prices on the bundle constraints and solving Lagrangian subproblems with only the single-commodity network flow constraints. The resulting solutions need not satisfy the bundle constraints, and the method uses linear programming to update the prices so that the solutions generated from the subproblems satisfy the bundle constraints. The method iteratively solves two different problems: a Lagrangian subproblem and a price-setting linear program. This method has played an important role in the field of optimization both because the algorithm itself has proven to be very useful, and also because it has stimulated many other approaches to problem decomposition. Moreover, the algorithm and its associated underlying theory have had a significant influence on the field of economics since this type of price decomposition formalizes ideas of transfer pricing and coordination that lie at the heart of planned economies. In Sections 17.5 and 17.6 we describe the use of

Dantzig–Wolfe decomposition, and the related technique of column generation, for solving the multicommodity flow problem.

An alternative way of viewing the multicommodity flow problem is as a capacity allocation problem. All the commodities are competing for the fixed capacity u_{ij} of every arc (i, j) of the network. Any optimal solution to the multicommodity flow problem will prescribe a specific flow on each arc (i, j) for each commodity which is the appropriate capacity to allocate to that commodity. If we started by allocating these capacities to the commodities and then solved the resulting (independent) single-commodity flow problems, we would be able to solve the problem quite easily as a set of independent single-commodity flow problems. Resource-directive methods provide a general solution approach for implementing this idea. They begin by allocating the capacities to the commodities and then use information gleaned from the solution to the resulting single-commodity problems to reallocate the capacities in a way that improves the overall system cost. In Section 17.7 we show how to solve the multicommodity flow problem using this resource-directive approach.

Partitioning methods exploit the fact that the multicommodity flow problem is a specially structured linear program with embedded network flow problems. As we have seen in Chapter 11, to solve any single-commodity flow problem, we can use the network simplex method, which works by generating a sequence of improving spanning tree solutions. In Section 11.11 we showed how to interpret the network simplex method as a special implementation of the simplex method for general linear programs and showed that spanning trees solutions correspond to basic feasible solutions of the minimum cost flow problem. This observation raises the following questions: (1) Can we adopt a similar approach for solving the multicommodity flow problem? (2) Can we somehow use spanning tree solutions for the embedded network flow constraints $\mathcal{N}x^k = b^k$? The partitioning method is a linear programming approach that permits us to answer both of these questions affirmatively. It maintains a linear programming basis that is composed of bases (spanning trees) of the individual single-commodity flow problems as well as additional arcs that are required to "tie" these solutions together to accommodate the bundle constraints. In Section 17.8 we describe the essential features of this approach.

The various solution techniques we describe in this chapter require linear programming background. We refer the reader to Appendix C for a review of this material. Before discussing the solution techniques, we describe several applications of the multicommodity flow problem.

17.2 APPLICATIONS

Multicommodity flow problems arise in a wide variety of application contexts. In this section we consider several instances of one very general type of application as well as a production planning and warehousing example and a vehicle fleet planning example.

Application 17.1 Routing of Multiple Commodities

In many applications of the multicommodity flow problem, we distinguish commodities because they are different physical goods, and/or because they have different points of origin and destination; that is, either (1) several physically distinct commodities (e.g., different manufactured goods) share a common network, or (2) a single physical good (e.g., messages or products) flows on a network, but the good has multiple points of origin and destination defined by different pairs of nodes in the network that need to send the good to each other. This second type of application arises frequently in problem contexts such as communication systems or distribution/transportation systems. In this section we introduce several application domains of both types.

Communication networks. In a communication network, nodes represent origin and destination stations for messages, and arcs represent transmission lines. Messages between different pairs of nodes define distinct commodities; the supply and demand for each commodity is the number of messages to be sent between the origin and destination nodes of that commodity. Each transmission line has a fixed capacity (in some applications the capacity of each arc is fixed; in others, we might be able to increase the capacity at a certain cost per unit). In this network, the problem of determining the minimum cost routing of messages is a multicommodity flow problem.

Computer networks. In a computer communication network, the nodes represent storage devices, terminals, or computer systems. The supplies and demands correspond to the data transmission rates between the computer, terminals, and storage devices, and the transmission line capacities define the bundle constraints.

Railroad transportation networks. In a rail network, nodes represent yard and junction points, and arcs represent track sections between the yards. The demand is measured by the number of cars (or any other equivalent measure of tonnage) to be loaded on any train. Since the system incurs different costs for different goods, we divide traffic demand into different classes. Each commodity in this network corresponds to a particular class of demand between a particular origin–destination pair. The bundle capacity of each arc is the number of cars that we can load on the trains that are scheduled to be dispatched on that arc (over some period of time). The decision problem in this network is to meet the demands of cars at the minimum possible operating cost.

Distribution networks. In distribution systems planning, we wish to distribute multiple (nonhomogeneous) products from plants to retailers using a fleet of trucks or railcars and using a variety of railheads and warehouses. The products define the commodities of the multicommodity flow problem, and the joint capacities of the plants, warehouses, railyards, and the shipping lanes define the bundle constraints. Note that in this application the nodes (plants, warehouses) as well as the arcs have bundle constraints.

Foodgrain export–import network. The nodes in this network correspond to geographically dispersed locations in different countries, and the arcs correspond to shipments by rail, truck, and ocean freighter. Between these locations, the commodities are various foodgrains, such as corn, wheat, rice, and soybeans. The capacities at the ports define the bundle constraints.

Application 17.2 Warehousing of Seasonal Products

A company manufactures multiple products. The products are seasonal, with demands varying weekly, monthly, or quarterly. To use its work force and capital equipment efficiently, the company wishes to "smooth" production, storing preseason production to supplement peak-season production. The company has a warehouse with fixed capacity R that it uses to store all the products it produces. Its decision problem is to identify the production levels of all the products for every week, month, or quarter of the year that will permit it to satisfy the demands incurring the minimum possible production and storage costs.

We can view this warehousing problem as a multicommodity flow problem defined on an appropriate network. For simplicity, consider a situation in which the company makes two products and the company needs to schedule its production for each of the next four quarters of the year. Let d_j^1 and d_j^2 denote the demand for products 1 and 2 in quarter j. Suppose that the production capacity for the jth quarter is u_j^1 and u_j^2, and that the per unit cost of production for this quarter is c_j^1 and c_j^2. Let h_j^1 and h_j^2 denote the storage (holding) costs of the two products from quarter j to quarter $j + 1$.

Figure 17.2 shows the network corresponding to the warehousing problem. The network contains one node for each time period (quarter) as well as a source and sink node for each commodity. The supply and demand of the source and sink nodes is the total demand for the commodity over all four quarters. Each source node s^k has four outgoing arcs, one corresponding to each quarter. Only one commodity flows on each of these arcs. We associate a cost c_j^k and capacity u_j^k with arc (s^k, j). Similarly, the sink node t^k has four incoming arcs; sink arc (j, t^k) has a zero cost and capacity d_j^k. The remaining arcs are of the form $(j, j + 1)$ for $j = 1, 2, 3$; the flow on these arcs represents the units stored from period j to period $j + 1$. Each of these storage arcs has a capacity R, a per unit flow of h_j^1 for commodity 1 and a per unit flow cost of h_j^2 for commodity 2. The two commodities share the capacity of this arc.

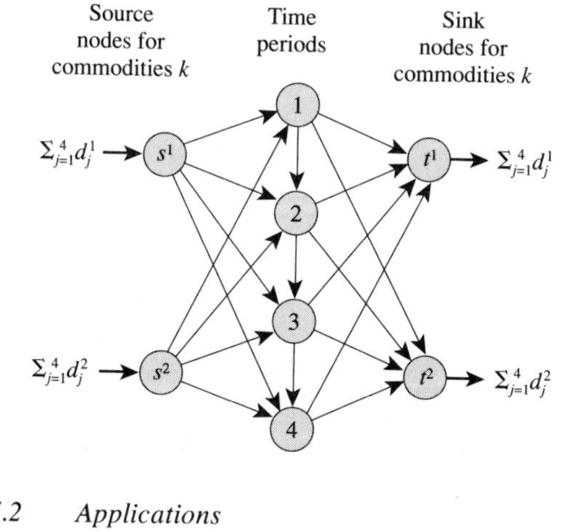

Figure 17.2 Optimal warehousing of seasonal products.

It is easy to see that each feasible multicommodity flow x in the network specifies a feasible production and inventory schedule for the two products with the same cost as the flow x. By optimizing the multicommodity flow, we find the optimal production and inventory plan.

The warehousing problem we have considered is a relatively simple model; we can augment it to include more realistic complexities: for example, transportation expenses incurred between plant–warehouse, warehouse–retailer, and plant–retailer combinations. It is relatively straightforward to incorporate these features in our model (see Exercise 17.2).

Application 17.3 Multivehicle Tanker Scheduling

Suppose that we wish to determine the optimal routing of fuel oil tankers required to achieve a prescribed schedule of deliveries: Each delivery is a shipment with a given delivery date of some commodity from a point of supply to a point of demand. In the simplest form this problem considers a single product (e.g., aviation gasoline or crude oil) to be delivered by a single type of tanker. We discussed this simple version of the problem in Application 6.6 and showed how to determine the minimum tanker fleet to meet the delivery schedule by solving a maximum flow problem. The multivehicle tanker scheduling problem, studied in this application, considers the scheduling and routing of a fixed fleet of nonhomogeneous tankers to meet a prespecified set of shipments of multiple products. The tankers differ in their speeds, carrying capabilities, and operating costs.

To formulate the multivehicle tanker scheduling problem as a multicommodity flow problem, we let the different commodities correspond to different tanker types. The network corresponding to the multivehicle tanker scheduling problem is similar to that of the single vehicle type, shown in Figure 6.9, except that each distinct type of tanker originates at a unique source node s^k. This network has four types of arcs (see Figure 17.3 for a partial example with two tanker types): in-service arcs, out-of-service arcs, delivery arcs, and return arcs. An in-service arc corresponds to the initial employment of a tanker type; the cost of this arc is the cost of deploying the tanker at the origin of the shipment. Similarly, an out-of-service arc corresponds to the removal of the tanker from service. A delivery arc (i, j) represents a shipment from origin i to destination j; the cost c_{ij}^k of this arc is the operating cost of carrying the shipment by a tanker of type k. A return arc (j, k) denotes the movement ("back-

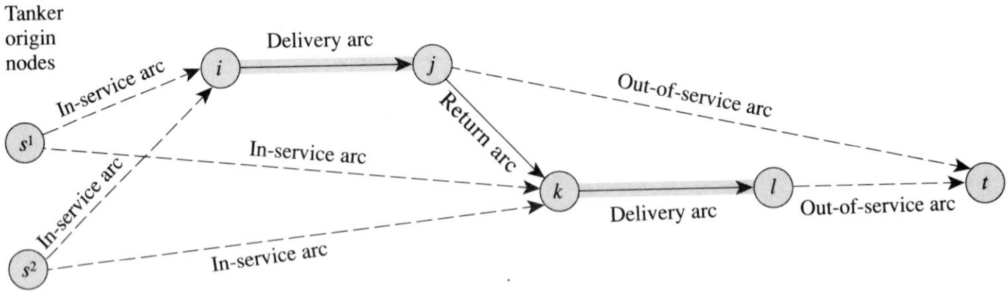

Figure 17.3 Multivehicle tanker scheduling problem as a multicommodity flow problem.

haul'') of an empty tanker, with an appropriate cost, between two consecutive shipments (i, j) and (k, l).

Each arc in the network has a capacity of 1. The shipment arcs have a bundle capacity ensuring that at most one tanker type services that arc. Each shipment arc also has a lower flow bound of 1 unit, which ensures that the chosen schedule does indeed deliver the shipment. Some arcs might also have commodity-based capacities u_{ij}^k. For instance, if tanker type 2 is not capable of handling the shipment on arc (i, j), we set $u_{ij}^2 = 0$. Moreover, if tanker type 2 can use the return arc (j, k), but tanker type 1 cannot (because it is too slow to make the connection between shipments), we set $u_{jk}^1 = 0$.

Airline scheduling is another important application domain for this type of model. In this problem context, the vehicles are different types of airplanes in an airline's fleet (e.g., Boeing 727s or 747s or McDonald Douglas DC 10s). The delivery arcs in this problem context are the flight legs that the airline wishes to cover.

We might note that in this formulation of the multivehicle tanker scheduling problem, we are interested in integer solutions of the multicommodity flow problem. The solutions obtained by the multicommodity flow algorithms described in this chapter need not be integral. Nevertheless, the fractional solution might be useful in several ways. For example, we might be able to convert the nonintegral solution into a (possibly, suboptimal) integral solution by minor tinkering, or as we noted earlier, we might use the nonintegral solution as a bound in solving the integer-valued problem by a branch-and-bound enumeration procedure.

17.3 OPTIMALITY CONDITIONS

Throughout our previous development, as we introduced each new problem, we usually began by stating optimality conditions for characterizing when a given feasible solution was optimal. Doing so permitted us to assess whether or not we have found an optimal solution to the problem. It also permitted us to interpret various algorithms as particular methods for solving the optimality conditions, and in several instances even suggested novel algorithmic approaches for solving the problem we were studying. Optimality conditions for the multicommodity flow problem serve these same purposes; so before discussing algorithms for solving the problem, we describe these conditions. For this discussion we assume that the flow variables x_{ij}^k have no individual flow bounds; that is, each $u_{ij}^k = +\infty$ in the formulation (17.1).

Since the multicommodity flow problem is a linear program, we can use linear programming optimality conditions to characterize optimal solutions to the problem. These conditions assume a particularly simple, and familiar, form for the multicommodity flow problem. Since the linear programming formulation (17.1) of the problem has one bundle constraint for every arc (i, j) of the network and one mass balance constraint for each node–commodity combination, the dual linear program has two types of dual variables: a *price* w_{ij} on each arc (i, j) and a *node potential* $\pi^k(i)$ for each combination of commodity k and node i. Using these dual variables, we define the reduced cost $c_{ij}^{\pi,k}$ of arc (i, j) with respect to commodity k as follows:

$$c_{ij}^{\pi,k} = c_{ij}^k + w_{ij} - \pi^k(i) + \pi^k(j).$$

In matrix notation, this definition is $c^{\pi,k} = c^k + w - \pi^k \mathcal{N}$.

Note that if we consider a fixed commodity k, this reduced cost is similar to the reduced cost that we have used previously for the minimum cost flow problem; the difference is that we now add the arc price w_{ij} to the arc cost c_{ij}^k. Note that just as the bundle constraints provided a linkage between the otherwise independent commodity flow variables x_{ij}^k, the arc prices w_{ij} provide a linkage between the otherwise independent commodity reduced costs.

To use linear programming duality theory to characterize optimal solutions to the multicommodity flow problem, we first write the dual of the multicommodity flow problem (17.1):

$$\text{Maximize} \quad -\sum_{(i,j)\in A} u_{ij}w_{ij} + \sum_{k=1}^{K} b^k\pi^k$$

subject to

$$c_{ij}^{\pi,k} = c_{ij}^k + w_{ij} - \pi^k(i) + \pi^k(j) \geq 0 \qquad \text{for all } (i, j) \in A \text{ and all } k = 1, \ldots, K,$$

$$w_{ij} \geq 0 \qquad \text{for all } (i, j) \in A.$$

The optimality conditions for a linear programming, called the complementary slackness (optimality) conditions, state that a primal feasible solution x and a dual feasible solution (w, π^k) are optimal to the respective problems if and only if the product of each primal (dual) variable and the slack in the corresponding dual (primal) constraint is zero. The complementary slackness conditions for the primal–dual pair of the multicommodity flow problem assume the following special form. (In this statement, we use y_{ij}^k to denote a specific value of the flow variable x_{ij}^k.)

Multicommodity flow complementary slackness conditions. *The commodity flows y_{ij}^k are optimal in the multicommodity flow problem* (17.1) *with each $u_{ij}^k = +\infty$ if and only if they are feasible and for some choice of (nonnegative) arc prices w_{ij} and (unrestricted in sign) node potentials $\pi^k(i)$, the reduced costs and arc flows satisfy the following complementary slackness conditions:*

(a) $w_{ij}\left(\sum_{1\leq k\leq K} y_{ij}^k - u_{ij}\right) = 0$ *for all arcs* $(i, j) \in A$. (17.2a)

(b) $c_{ij}^{\pi,k} \geq 0$ *for all arcs* $(i, j) \in A$ *and all commodities* $k = 1, 2, \ldots, K$. (17.2b)

(c) $c_{ij}^{\pi,k} y_{ij}^k = 0$ *for all arcs* $(i, j) \in A$ *and all commodities* $k = 1, 2, \ldots, K$. (17.2c)

We refer to any set of arc prices and node potentials that satisfy the complementary slackness conditions as *optimal arc prices* and *optimal node potentials*. The following theorem shows the connection between the multicommodity and single-commodity flow problems.

Theorem 17.1 (Partial Dualization). *Let y_{ij}^k be optimal flows and let w_{ij} be optimal arc prices for the multicommodity flow problem* (17.1). *Then for each commodity k, the flow variables y_{ij}^k for $(i, j) \in A$ solve the following (uncapacitated) minimum cost flow problem:*

$$\min\{ \sum_{(i,j)\in A} (c_{ij}^k + w_{ij})x_{ij}^k : \mathcal{N}x^k = b,\ x_{ij}^k \geq 0 \text{ for all } (i, j) \in A\}. \tag{17.3}$$

Proof. Since y_{ij}^k are optimal flows and w_{ij} are optimal arc prices for the multicommodity flow problem (17.1), these variables together with some set of node potentials $\pi^k(i)$ satisfy the complementary slackness condition (17.2). Now notice that conditions (17.2b) and (17.2c) are the optimality conditions for the uncapacitated minimum cost flow problem for commodity k with arc costs $(c_{ij}^k + w_{ij})$ [see condition (9.8) in Section 9.3 with $u_{ij} = \infty$]. This observation implies that the flows y_{ij}^k solve the corresponding minimum cost flow problems. ◆

This property shows that we can use a sequential approach for obtaining optimal arc prices and node potentials: We first find optimal arc prices and then attempt to find the optimal node potentials and flows by solving the single-commodity minimum cost flow problems (17.3). In the next few sections we use this observation to develop and assess algorithms for solving the multicommodity flow problem.

To illustrate the partial dualization result, we consider a numerical example. The multicommodity flow problem shown in Figure 17.4 has two commodities and two arcs with bundle capacities: arc (s^1, t^1) with a capacity of 5 units and arc (1, 2)

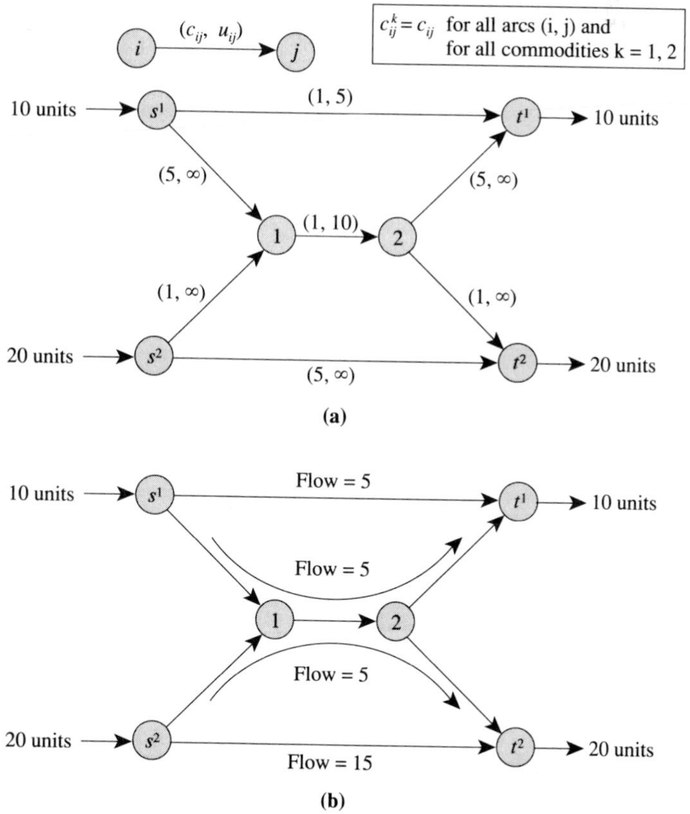

Figure 17.4 Multicommodity flow example: (a) problem data; (b) optimal solution.

with a capacity of 10 units. For this example, since each commodity has a single source and sink, for any choice of the arc prices w_{ij}, the network flow problem (17.3) is a shortest path problem. As shown in Figure 17.4(b), the optimal solution of this problem has $y^1_{s^1t^1} = y^1_{s^11} = y^1_{12} = y^1_{2t^1} = y^2_{s^21} = y^2_{12} = y^2_{2t^2} = 5$, $y^2_{s^2t^2} = 15$, and all other $y^k_{ij} = 0$. Note that the total flow on arc (s^1, t^1) and on arc (1, 2) equals its bundle capacity. Suppose that we set $w_{12} = 2$ and $w_{s^1t^1} = 12$. Then with respect to the arc costs $c^k_{ij} + w_{ij}$, the shortest path distances for commodity 1 are $d^1(s^1) = 0$, $d^1(1) = 5$, $d^1(2) = 8$, $d^1(t^1) = 13$, $d^1(t^2) = 9$, $d^1(s^2) = +\infty$ and the shortest path distances for commodity 2 are $d^2(s^2) = 0$, $d^2(1) = 1$, $d^2(2) = 4$, $d^2(t^2) = 5$, $d^2(t^1) = 9$, $d^2(s^1) = +\infty$. If we set, $\pi^k(i) = -d^k(i)$ for all nodes i and for both commodities $k = 1, 2$, the node potentials $\pi^k(i)$, arc prices w_{ij}, and arc flows y^k_{ij} given earlier satisfy the optimality conditions (17.2). Therefore, we have verified that y is an optimal solution to the problem and that w, π^1, π^2 are optimal arc prices and node potentials.

17.4 LAGRANGIAN RELAXATION

To apply Lagrangian relaxation to the multicommodity flow problem, we associate nonnegative Lagrange multipliers w_{ij} with the bundle constraints (17.1b), creating the following Lagrangian subproblem:

$$L(w) = \min \sum_{1 \le k \le K} c^k x^k + \sum_{(i,j) \in A} w_{ij} \Big(\sum_{1 \le k \le K} x^k_{ij} - u_{ij}\Big) \tag{17.4a}$$

or, equivalently,

$$L(w) = \min \sum_{1 \le k \le K} \sum_{(i,j) \in A} (c^k_{ij} + w_{ij}) x^k_{ij} - \sum_{(i,j) \in A} w_{ij} u_{ij} \tag{17.4b}$$

subject to

$$\mathcal{N} x^k = b^k \quad \text{for all } k = 1, \ldots, K, \tag{17.4c}$$

$$x^k_{ij} \ge 0 \quad \text{for all } (i, j) \in A \text{ and all } k = 1, 2, \ldots, K. \tag{17.4d}$$

Note that since the term $-\sum_{(i,j) \in A} w_{ij} u_{ij}$ in the objective function of the Lagrangian subproblem is a constant for any given choice of the Lagrange multipliers, for any fixed value of these multipliers, this term is a constant and therefore we can ignore it. The resulting objective function for the Lagrangian subproblem has a cost of $c^k_{ij} + w_{ij}$ associated with every flow variable x^k_{ij}. Since none of the constraints in this problem contains the flow variables for more than one of the commodities, the problem decomposes into separate minimum cost flow problems, one for each commodity. Consequently, to apply the subgradient optimization procedure from Section 16.3 to this problem, we would alternately (1) solve a set of minimum cost flow problems (for a fixed value of the Lagrange multipliers w) with the cost coefficients $c^k_{ij} + w_{ij}$, and (2) update the multipliers by the algorithmic procedures described in Section 16.3. In this case, if y^k_{ij} denotes the optimal solution to the minimum cost flow subproblems when the Lagrange multipliers have the value w^q_{ij} at the qth iteration, the subgradient update formula becomes

$$w^{q+1}_{ij} = \Big[w^q_{ij} + \theta_q \Big(\sum_{1 \le k \le K} y^k_{ij} - u_{ij}\Big)\Big]^+.$$

In this expression, the notation $[\alpha]^+$ denotes the positive part of α, that is, $\max(\alpha, 0)$. The scalar θ_q is a step size specifying how far we move from the current solution w_{ij}^q. Note that this update formula increases the multiplier w_{ij}^q on arc (i, j) by the amount $(\sum_{1 \le k \le K} y_{ij}^k - u_{ij})$ if the subproblem solutions y_{ij}^k use more than the available capacity u_{ij} of that arc, or reduces the Lagrange multiplier of arc (i, j) by the amount $(u_{ij} - \sum_{1 \le k \le K} y_{ij}^k)$ if the subproblem solutions y_{ij}^k use less than the available capacity of that arc. If, however, the decrease would cause the multiplier w_{ij}^{q+1} to become negative, we reduce the multiplier to value zero. We choose the step sizes θ_q for iterations $q = 1, 2, \ldots,$ in accordance with the procedures described in Chapter 16.

It is instructive to view the subgradient method for the multicommodity flow problem as a solution procedure for solving the linear program (17.1) that is able to exploit the special structure of the unrelaxed mass balance constraints. In Theorem 16.6 we noted that whenever we apply Lagrangian relaxation to any linear program, such as the multicommodity flow problem, the optimal value $L^* = \max_{w \ge 0} L(w)$ of the Lagrangian multiplier problem equals the optimal objective function value z^* of the linear program. We illustrate this technique with a numerical example.

Consider again the two-commodity example shown in Figure 17.4. As we noted in our discussion in Section 17.3, for any choice of the Lagrange multipliers w_{12} and $w_{s^1t^1}$ for the two capacitated arcs (s^1, t^1) and $(1, 2)$, the problem decomposes into two shortest path problems. If we start with the Lagrange multipliers $w_{12}^0 = w_{s^1t^1}^0 = 0$, then in the subproblem solutions, the shortest path s^1–t^1 carries 10 units of flow at a cost of $1(10) = 10$ and the path s^2–1–2–t^2 carries 20 units of flow at a cost of $3(20) = 60$. Therefore, $L(0) = 10 + 60 = 70$ is a lower bound on the optimal objective function value for the problem. Since $y_{s^1t^1}^1 + y_{s^1t^1}^2 - u_{s^1t^1} = 5$ and $y_{12}^1 + y_{12}^2 - u_{12} = 10$, the update formulas become

$$w_{s^1t^1}^1 = [0 + \theta_0 \cdot 5]^+ \quad \text{and} \quad w_{12}^1 = [0 + \theta_0 \cdot 10]^+.$$

If we choose $\theta_0 = 1$, then $w_{s^1t^1}^1 = 5$ and $w_{12}^1 = 10$. The new shortest path solutions send 10 units on the path s^1–t^1 at a cost of $(1 + 5)(10) = 60$ and 20 units on the path s^2–t^2 at a cost of $5(20) = 100$. The new lower bound obtained through (17.4b) is $60 + 100 - w_{12}^1 u_{12} - w_{s^1t^1}^1 u_{s^1t^1} = 160 - 10(10) - 5(5) = 35$. (Note that the value of the lower bound has decreased.) At this point, $y_{s^1t^1}^1 + y_{s^1t^1}^2 - u_{s^1t^1} = 5$ and $y_{12}^1 + y_{12}^2 - u_{12} = 0$, so the update formulas become

$$w_{s^1t^1}^1 = [5 + \theta_0 \cdot 5]^+ \quad \text{and} \quad w_{12}^1 = [10 - \theta_0 \cdot 10]^+.$$

Choosing the step size $\theta_1 = 1$ again, we find that $w_{12}^1 = 0$ and $w_{s^1t^1}^1 = 10$ and that the new shortest path solutions send 10 units on the path s^1–t^1 at a cost of $(1 + 10)(10) = 110$ and 20 units on the path s^2–1–2–t^2 at a cost of $3(20) = 60$. If we continue by choosing the step sizes for the kth iteration as $\theta_k = 1/k$, in accordance with the theory of subgradient optimization as discussed in Section 16.4, we obtain the set of iterates shown in Figure 17.5. From iteration 14 on, the values of the Lagrange multipliers oscillate about, and converge to, their optimal values $w_{12} = 2$ and $w_{s^1t^1} = 12$ and the optimal lower bound oscillates about its optimal value 150, which equals the optimal objective function value of the multicommodity flow problem.

Iteration number q	w^q_{12}	$w^q_{s^1t^1}$	Shortest paths	Shortest path costs	$10w^q_{12}$ + $5w^q_{s^1t^1}$	Lower bound $L(w^q)$	θ_q
0	0	0	s^1–t^1 s^2–1–2–t^2	10(1) 20(1 + 1 + 1)	0	70	1
1	10	5	s^1–t^1 s^2–t^2	10(1 + 5) 20(5)	125	35	1
2	0	10	s^1–t^1 s^2–1–2–t^2	10(1 + 10) 20(1 + 1 + 1)	50	120	0.5
3	5	12.5	s^1–t^1 s^2–t^2	10(1 + 12.5) 20(5)	112.5	122.5	0.333
4	1.67	14.17	s^1–1–2–t^1 s^2–1–2–t^2	10(5 + 2.67 + 5) 20(1 + 2.67 + 1)	87.55	132.5	0.25
5	6.67	12.92	s^1–t^1 s^2–t^2	10(1 + 12.92) 20(5)	131.3	107.9	0.2
6	4.67	13.92	s^1–t^1 s^2–t^2	10(1 + 13.92) 20(5)	116.3	132.9	0.167
7	3	14.75	s^1–1–2–t^1 s^2–t^2	10(5 + 4 + 5) 20(5)	103.8	136.3	0.143
8	3	14.04	s^1–1–2–t^1 s^2–t^2	10(5 + 4 + 5) 20(5)	100.2	139.8	0.125
9	3	13.41	s^1–1–2–t^1 s^2–t^2	10(5 + 4 + 5) 20(5)	97.05	143.0	0.111
10	3	12.86	s^1–t^1 s^2–t^2	10(1 + 12.86) 20(5)	94.3	144.3	0.1
11	2	13.36	s^1–1–2–t^1 s^2–t^2	10(5 + 3 + 5) 20(5)	86.8	143.2	0.091
12	2	12.9	s^1–1–2–t^1 s^2–t^2	10(5 + 3 + 5) 20(5)	84.5	145.5	0.083
13	2	12.48	s^1–1–2–t^1 s^2–t^2	10(5 + 3 + 5) 20(5)	82.2	147.6	0.077
14	2	12.09	s^1–1–2–t^1 s^2–t^2	10(5 + 3 + 5) 20(5)	80.25	149.5	0.071

Figure 17.5 Application of subgradient optimization to a multicommodity flow problem.

Figures 17.6 and 17.7 show how the values of the Lagrange multipliers and Lagrangian lower bounds vary during the execution of the algorithm. Note that neither the multiplier values nor the lower bounds converge monotonically to their optimal values. In this example, as in the application of Lagrangian relaxation in general, these values tend to oscillate, sometimes considerably, before settling down to their optimal values.

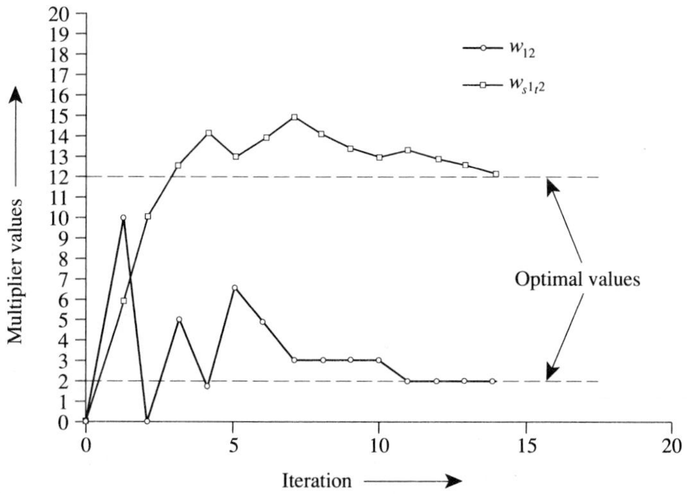

Figure 17.6 Multiplier values for the numerical example.

The use of subgradient optimization for solving the Lagrangian multiplier problem is attractive for several reasons. First, as we have just noted, this solution approach permits us to exploit the underlying network flow structure. Second, the formulas for updating the Lagrange multipliers w_{ij} are rather trivial computationally and very easy to encode in a computer program. This solution approach, however, also has some limitations. To ensure convergence we need to take small step sizes; as a result, the method does not converge very fast. Second, the method is dual-based, and so even though it is converging to the optimal dual variables w_{ij}, the optimal solutions y_{ij}^k of the subproblems need not converge to the optimal solution to the multicommodity flow problem. Indeed, even though we have shown that if we set the Lagrangian multipliers to their optimal values [i.e., as the optimal dual

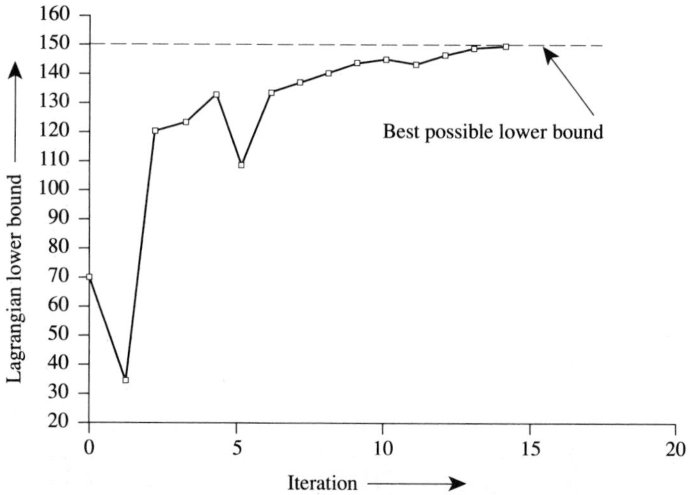

Figure 17.7 Lagrangian lower bound for the numerical example.

variables for the bundle constraints in the dual to the linear program (17.1)], the optimal flows y_{ij}^k solve the Lagrangian subproblem (17.4), these subproblems might also have other optimal solutions that do not satisfy the bundle constraints. For example, for the problem we have just solved, with the optimal Lagrange multipliers $w_{12} = 2$ and $w_{s^1t^1} = 12$, the shortest paths subproblems have solutions with 10 units on the path s^1–t^1 and 20 units on the path s^2–t^2. This solution violates the capacity of the arc (s^1, t^1). In general, to obtain optimal flows, even after we have solved the Lagrangian multiplier problem, requires additional work (see Exercise 17.18).

To conclude this discussion of Lagrangian relaxation, we might note that we can also combine Lagrangian relaxation with linear programming by using Lagrangian relaxation to develop an "advance basis" start for the linear programming formulation (17.1). Suppose that we solve the Lagrangian subproblem for *any* choice of the Lagrange multipliers by the network simplex algorithm (see Chapter 11). By doing so we obtain an optimal spanning tree solution for each commodity, which corresponds to a basis $\mathcal{B}^k$ of the network flow constraints $\mathcal{N}x^k = b^k$. We can extend these bases into an overall basis $\mathcal{B}$ of the linear program (17.1), as shown in Figure 17.8.

The identity matrix in the topmost row in this matrix corresponds to the slack variables s_{ij} in the equality formulation (17.1b′) of the bundle constraints. Note that in the solution corresponding to the basis $\mathcal{B}$, (1) each flow variable x^k for $k = 1, 2, \ldots, K$ equals the solution y^k to the Lagrangian subproblem, and (2) the values of the slack variables are given by $s_{ij} = u_{ij} - \sum_{1 \le k \le K} y_{ij}^k$. If all the slack variables are nonnegative, this basis is feasible. If any of them are negative, the basis is infeasible. In either case we can apply standard linear programming methods (either the primal simplex method or the dual simplex method) using this basis as a starting solution for solving the problem to completion.

This advanced basis defines an attractive starting solution for the simplex method because it permits us to exploit the underlying network structure and flow costs to generate a "smart" starting solution for the algorithm. For problems that are "modestly capacitated," the advanced basis can be a very good approximation

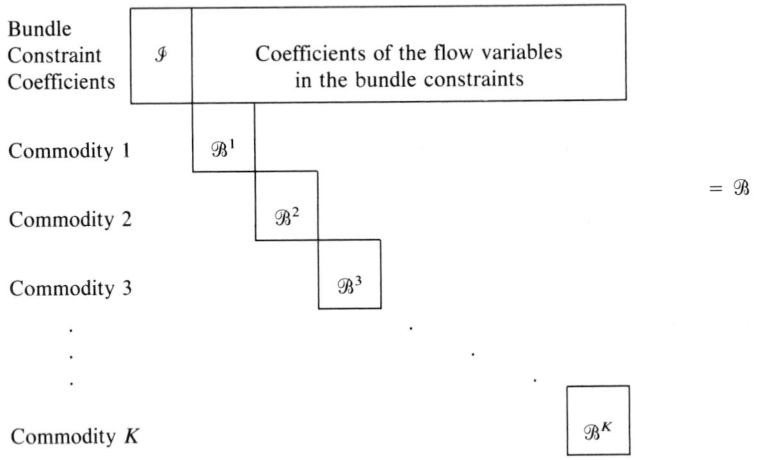

Figure 17.8 Initial basis formed from bases of individual commodity subproblems.

of the optimal solution of the problem and so can greatly improve the performance of the simplex method.

We next consider an alternative solution approach, known as *Dantzig–Wolfe decomposition*, for solving the Lagrangian multiplier problem. This approach requires considerably more work at each iteration for updating the Lagrange multipliers (the solution of a linear program) but has proved to converge faster than the subgradient optimization procedure for several classes of problems. Rather than describing the Dantzig–Wolfe decomposition procedure as a variation of Lagrangian relaxation, we will first develop it from an alternate large-scale linear programming viewpoint that provides a somewhat different perspective on the approach.

17.5 COLUMN GENERATION APPROACH

To simplify our discussion in this section, we consider a special case of the multicommodity flow problem: We assume that each commodity k has a single source node s^k and a single sink node t^k and a flow requirement of d^k units between these source and sink nodes. We also assume that we impose no flow bounds on the individual commodities other than the bundle constraints. Therefore, for each commodity k, the subproblem constraints $\mathcal{N}x^k = b^k$, $x^k \geq 0$ define a shortest path problem: For this model, for any choice w_{ij} of the Lagrange multipliers for the bundle constraints, the Lagrangian relaxation requires the solution of a series of shortest path problems, one for each commodity.

Reformulation with Path Flows

To begin our discussion in this section, let us first reformulate the multicommodity flow problem using path and cycle flows instead of arc flows. Recall from Section 3.5 that we can formulate any network flow problem using path and cycle flows. To simplify our discussion even further, let us assume that for every commodity the cost of every cycle W in the underlying network is nonnegative (in Exercise 17.8 we relax this assumption). The problem satisfies this condition, for example, if the arc flow costs are all nonnegative. If we impose this nonnegative cycle cost condition, then in some optimal solution to the problem, the flow on every cycle is zero, so we can eliminate the cycle flow variables. Therefore, throughout this section we assume that we can represent any potentially optimal solution as the sum of flows on directed paths. Let us recall our notation from Section 3.5 concerning path and cycle decompositions, tailored a bit for the multicommodity flow problem.

For each commodity k, let $\mathbf{P^k}$ denote the collection of all directed paths from the source node s^k to the sink node t^k in the underlying network $G = (N, A)$. In the path flow formulation, each decision variable $f(P)$ is the flow on some path P and for the kth commodity, we define this variable for every directed path P in $\mathbf{P^k}$.

As in Section 3.5, let $\delta_{ij}(P)$ be an arc-path indicator variable, that is, $\delta_{ij}(P)$ equals 1 if arc (i, j) is contained in the path P, and is 0 otherwise. The flow decomposition theorem of network flows states that we can always decompose some optimal arc flow x_{ij}^k into path flows $f(P)$ as follows:

$$f(P) = x_{ij}^k \sum_{P \in \mathbf{P^k}} \delta_{ij}(P) f(P).$$

Let $c^k(P) = \sum_{(i,j)\in A} c_{ij}^k \delta_{ij}(P) = \sum_{(i,j)\in P} c_{ij}^k$ denote the per unit cost of flow on the path $P \in \mathbf{P}^k$ with respect to the commodity k. Note that for each commodity k, if we substitute for the arc flow variables in the objective function, interchange the order of the summations, and collect terms, we find that

$$\sum_{(i,j)\in A} c_{ij}^k x_{ij}^k = \sum_{(i,j)\in A} c_{ij}^k \Big[\sum_{P\in \mathbf{P}^k} \delta_{ij}(P) f(P) \Big] = \sum_{P\in \mathbf{P}^k} c^k(P) f(P).$$

This observation shows that we can express the cost of any solution as either the cost of arc flows or the cost of path flows.

By substituting the path variables in the multicommodity flow formulation, we obtain the following equivalent path flow formulation of the problem:

$$\text{Minimize} \quad \sum_{1\le k\le K} \sum_{P\in \mathbf{P}^k} c^k(P) f(P) \tag{17.5a}$$

subject to

$$\sum_{1\le k\le K} \sum_{P\in \mathbf{P}^k} \delta_{ij}(P) f(P) \le u_{ij} \quad \text{for all } (i, j) \in A, \tag{17.5b}$$

$$\sum_{P\in \mathbf{P}^k} f(P) = d^k \quad \text{for all } k = 1, 2, \ldots, K, \tag{17.5c}$$

$$f(P) \ge 0 \quad \text{for all } k = 1, 2, \ldots, K \text{ and all } P \in \mathbf{P}^k. \tag{17.5d}$$

In formulating this problem we have invoked the flow decomposition theorem stating that we can decompose any feasible arc flow of the system $\mathcal{N}x^k = b^k$ into a set of path and cycle flows in such a way that the path flows satisfy the mass balance condition (17.5c).

Note that the path flow formulation of the multicommodity flow problem has a very simple constraint structure. The problem has a single constraint for each arc (i, j) which states that the sum of the path flows passing through the arc is at most u_{ij}, the capacity of the arc. Moreover, the problem has a single constraint (17.5c) for each commodity k which states that the total flow on all the paths connecting the source node s^k and sink node t^k of commodity k must equal the demand d^k for this commodity. For a network with n nodes, m arcs, and K commodities, the path flow formulation contains $m + K$ constraints (in addition to the nonnegativity restrictions imposed on the path flow values). In contrast, the arc formulation (17.1) contains $m + nK$ constraints since it contains one mass balance constraint for every node and commodity combination. For example, a network with $n = 1000$ nodes and $m = 5000$ arcs and with a commodity between every pair of nodes has approximately $K \approx n^2 = 1{,}000{,}000$ commodities. Therefore, the path flow formulation contains about 1,005,000 constraints. In contrast, the arc flow formulation contains about 1,000,005,000 constraints. But the difference is even more pronounced: Because no path appears in more than one of the constraints (17.5c), we can apply a specialized version of the simplex method, known as the *generalized upper bounding simplex method*, to solve the path flow formulation very efficiently. Even though the linear programming basis for our example has size 1,005,000 by 1,005,000, the generalized upper bounding simplex method is able to perform all of its matrix computations on a much smaller basis of size 5000 by 5000. This method essentially solves the problem as though it contained only m bundle constraints, which, for this sample data, means that we can essentially solve a linear program with only 5000 constraints instead of over 1 billion constraints in the arc formulation.

This savings in the number of constraints does come at a cost, however, since the path flow formulation has a variable for every path connecting a source and sink node for each of the commodities. The number of variables will typically be enormous, growing exponentially in the size of the network. All hope is not lost, though, since we might expect that only very few of the paths will carry flow in the optimal solution to the problem. In fact, linear programming theory permits us to show that at most $K + m$ paths carry positive flow in some optimal solution to the problem (see Exercise 17.6). Therefore, for a problem with 1,000,000 commodities and 5,000 arcs as in our previous example, we could, in principle, solve the path flow formulation using 1,005,000 paths. Since the problem contains 1,000,000 commodities, this solution would use two or more paths for at most 5,000 commodities and one path for at least the 995,000 remaining commodities. If we knew the optimal set of paths, or a very good set of paths, we could obtain an optimal solution (i.e., values for the path flows) by solving a linear program containing just the commodities with two or more sets of paths. The generalized upper bounding linear programming procedure for solving linear programs permits us to exploit this observation.

Optimality Conditions

Recall that the revised simplex method of linear programming maintains a basis at every step, and using this basis determines a vector of simplex multipliers for the constraints. Since the path flow formulation (17.5) contains one bundle constraint for each arc and one demand constraint (17.5c) for every commodity, the dual linear program has a dual variable w_{ij} for each arc (this is the same arc price that we have introduced before) and another dual variable σ^k for each commodity $k = 1, 2, \ldots, K$. With respect to these dual variables, the reduced cost $c_P^{\sigma,w}$ for each path flow variable $f(P)$ is

$$c_P^{\sigma,w} = c^k(P) + \sum_{(i,j)\in P} w_{ij} - \sigma^k.$$

and the complementary slackness conditions (17.2) for the arc formulation of the original problem assume the following form:

Path flow complementary slackness conditions. *The commodity path flows $f(P)$ are optimal in the path flow formulation* (17.5) *of the multicommodity flow problem if and only if for some arc prices w_{ij} and commodity prices σ^k, the reduced costs and arc flows satisfy the following complementary slackness conditions:*

(a) $w_{ij} \left[\sum_{1\le k\le K} \sum_{P\in \mathbf{P}^k} \delta_{ij}(P)f(P) - u_{ij}\right] = 0$ for all $(i, j) \in A$. (17.6a)

(b) $c_P^{\sigma,w} \ge 0$ for all $k = 1, 2, \ldots, K$ and all $P \in \mathbf{P}^k$. (17.6b)

(c) $c_P^{\sigma,w} f(P) = 0$ for all $k = 1, 2, \ldots, K$ and all $P \in \mathbf{P}^k$. (17.6c)

We leave the proof of these conditions as an exercise for the reader. These optimality conditions have a very appealing and intuitive interpretation. Condition

(a) states that the price w_{ij} of arc (i, j) is zero if the optimal solution $f(P)$ does not use all of the capacity u_{ij} of the arc. That is, if the optimal solution does not fully use the capacity of that arc, we could ignore the constraint (place no price on it).

Since the cost $c^k(P)$ of path P is just the sum of the cost of the arcs contained in that path, that is, $c^k(P) = \sum_{(i,j)\in P} c_{ij}^k$, we can write the reduced cost of path P as

$$c_P^{\sigma,w} = \sum_{(i,j)\in P} (c_{ij}^k + w_{ij}) - \sigma^k.$$

That is, the reduced cost of path P is just the cost of that path with respect to the modified costs $c_{ij}^k + w_{ij}$ minus the commodity cost σ^k. The complementary slackness condition (17.6b) states that the modified path cost $\sum_{(i,j)\in P} (c_{ij}^k + w_{ij})$ for each path connecting the source node s^k and the sink node t^k of commodity k must be at least as large as the commodity cost σ^k. The condition (17.6c) implies that reduced cost $c_P^{\sigma,w}$ must be zero for any path P that carries flow in the optimal solution [i.e., for which the flow $f(P)$ is positive]; that is, the modified cost $\sum_{(i,j)\in P} (c_{ij}^k + w_{ij})$ of this path must equal the commodity cost σ^k. Therefore, conditions (17.6b) and (17.6c) imply that

σ^k is the shortest path distance from node s^k to node t^k with respect to the modified costs $c_{ij}^k + w_{ij}$ and in the optimal solution every path from node s^k to node t^k that carries a positive flow must be a shortest path with respect to the modified costs.

This result shows that the arc costs w_{ij} permit us to decompose the multicommodity flow problem into a set of independent "modified" cost shortest path problems.

Column Generation Solution Procedure

To this point we have restated the multicommodity flow problem as a large-scale linear program with an enormous number of columns—with one flow variable for each path connecting the source and sink of any commodity. We have also shown how to characterize any optimal solution to this formulation in terms of the linear programming dual variables w_{ij} and σ^k, interpreting these conditions as shortest path conditions with respect to the modified arc costs $c_{ij}^k + w_{ij}$. We next show how to solve the problem by using a solution procedure known as *column generation*.

The key idea in column generation is never to list explicitly all of the columns of the problem formulation, but rather to generate them only "as needed." The revised simplex method of linear programming is perfectly suited for carrying out this algorithmic strategy. Recall from Appendix C that the revised simplex method maintains a basis $\mathcal{B}$ at each iteration. It uses this basis to define a set of simplex multipliers π via the matrix computation $\pi\mathcal{B} = c_{\mathbf{B}}$ (in our application, the multipliers are w and σ). That is, the method defines the simplex multipliers so that the reduced costs $c_{\mathbf{B}}^{\pi}$ of the basic variables are zero; that is, $c_{\mathbf{B}}^{\pi} = c_{\mathcal{B}} - \pi\mathcal{B} = 0$. To find the simplex multipliers, the method requires no information about columns (variables) not in the basis. It then uses the multipliers to *price-out* the nonbasic columns, that is, compute their reduced costs. If any reduced cost is negative (assuming a minimization formulation), the method will introduce one nonbasic variable into the basis in place of one of the current basic variables, recompute the simplex multipliers π, and then repeat these computations. To use the column generation approach, the columns should have structural properties that permit us to perform the pricing out operations without explicitly examining every column.

When applied to the path flow formulation of the multicommodity flow problem, with respect to the current basis at any step (which is composed of a set of columns, or path variables, for the problem), the revised simplex method defines the simplex multipliers w_{ij} and σ^k so that the reduced cost of every variable in the basis is zero. Therefore, if a path P connecting the source s^k and sink t^k for commodity k is one of the basic variables, then $c_P^{\sigma,w} = 0$, or equivalently, $\sum_{(i,j)\in P} (c_{ij}^k + w_{ij}) = \sigma^k$. Therefore, the revised simplex method determines the simplex multipliers w_{ij} and σ^k so that they satisfy the following equations:

$$\sum_{(i,j)\in P} (c_{ij}^k + w_{ij}) = \sigma^k \text{ for every path } P \text{ in the basis.}$$

Notice that since each basis consists of $K + m$ paths, each basis gives rise to $K + m$ of these equations. Moreover, the equations contain $K + m$ variables (i.e., m arc prices w_{ij} and K shortest path distances σ^k). The revised simplex method uses matrix computations to solve the $K + m$ equations and determines the unique values of the simplex multipliers.

The complementary slackness condition (17.6c) dictates that $c_P^{\sigma,w} f(P) = 0$ for every path P in the network. Since each path P in the basis satisfies the condition $c_P^{\sigma,w} = 0$, we can send any amount of flow on it and still satisfy the condition (17.6c). To satisfy this condition for a path P not in the basis, we set $f(P) = 0$. Next consider the complementary slackness condition (17.6a). If the slack variable $s_{ij} = [\sum_{1\le k\le K} \sum_{(i,j)\in P^k} \delta_{ij}(P) f(P) - u_{ij}]$ is not in the basis, $s_{ij} = 0$, so the solution satisfies (17.6a). On the other hand, if the slack variable s_{ij} is in the basis, its reduced cost, which equals $0 - w_{ij}$, is zero, implying that $w_{ij} = 0$ and the solution satisfies (17.6a). We have thus shown that the solution defined by the current basis satisfies conditions (17.6a) and (17.6c); it is optimal if it satisfies condition (17.6b) (i.e., the reduced cost of every path flow variable is nonnegative). How could we check this condition? That is, how can we check to see if for each commodity k,

$$c_P^{\sigma,w} = \sum_{(i,j)\in P} (c_{ij}^k + w_{ij}) - \sigma^k \ge 0 \qquad \text{for all } P \in \mathbf{P^k},$$

or, equivalently,

$$\min_{P\in \mathbf{P^k}} \sum_{(i,j)\in P} (c_{ij}^k + w_{ij}) \ge \sigma^k.$$

As we have noted, the left-hand side of this inequality is just the length of the shortest path connecting the source and sink nodes, s^k and t^k, of commodity k with respect to the modified costs $c_{ij}^k + w_{ij}$. Thus, to see whether the arc prices w_{ij} together with current path distances σ^k satisfy the complementary slackness conditions, we solve a shortest path problem for each commodity k. If for all commodities k, the length of the shortest path for that commodity is at least as large as σ^k, we satisfy the complementary slackness condition (17.6b).

Otherwise, if for some commodity k, Q denotes the shortest path with respect to the current modified costs $c_{ij}^k + w_{ij}$ and the reduced cost of path Q is less than the length σ^k of the minimum cost path from the set $\mathbf{P^k}$, then

$$c_Q^{\sigma,w} = \sum_{(i,j)\in Q} (c_{ij}^k + w_{ij}) - \sigma^k < 0.$$

In terms of the linear program (17.5), the path Q has a negative reduced cost, so we can profitably use it in the linear program in place of one of the paths P in the current basis $\mathcal{B}$. That is, using the usual steps of the simplex method, we would perform a basis change introducing the path Q into the current basis. Doing so would permit us to determine a new set of arc prices w_{ij} and a new modified shortest path distance σ^k between the source and sink nodes of commodity k. We choose the values of these variables so that the reduced cost of every basic variable is zero. That is, using matrix operations, we would once again solve the system $c_P^{\sigma,w} = \sum_{(i,j)\in P} (c_{ij}^k + w_{ij}) - \sigma^k = 0$ in the variables w_{ij} and σ^k. We would then, as before, solve a shortest path problem for each commodity k and see whether any path has a shorter length than σ^k. If so, we would introduce this path into the basis and continue by alternately (1) finding new values for the arc prices w_{ij} and for the path lengths σ^k, and (2) solving shortest path problems.

This discussion shows us how we would determine the variable to introduce into the basis at each step. The rest of the steps for implementing the simplex method (e.g., determining the variable to remove from the basis at each step) are the same as those of the usual implementation, so we do not specify any further details.

Determining Lower Bounds

Let z^* denote the optimal objective function value of the multicommodity flow problem (17.5) and let z^{lp} denote the optimal objective function value at any step in solving the path flow formulation of the problem (17.5) by the simplex method. Since z^{lp} corresponds to a feasible solution to the problem, $z^* \leq z^{\text{lp}}$. As we have noted in our discussion of the Lagrangian relaxation technique in Section 17.4, for any choice of the arc prices w, the optimal value $L(w)$ of the Lagrangian subproblem is a lower bound on z^*. Therefore, suppose that at any point during the course of the algorithm, we solve the Lagrangian subproblem with respect to the current arc prices w_{ij}. That is, we solve for the shortest path lengths $l^k(w)$ for all the commodities k with respect to the modified costs $c_{ij}^k + w_{ij}$. (Notice that this is the same computation that we perform in pricing out columns for the simplex method.) Then from (17.4) the value $L(w)$ of the Lagrangian subproblem is

$$L(w) = \sum_{k=1}^{K} l^k(w) - \sum_{(i,j)\in A} w_{ij}u_{ij},$$

and by the theory of Lagrangian relaxation,

$$L(w) \leq z^* \leq z^{\text{lp}}.$$

Therefore, as a by-product of finding the shortest path distances $l^k(w)$ as we are pricing out columns in implementing the column generation procedure, we obtain a lower bound on the objective function value. This lower bound allows us to judge the quality of the current solution in the column generation technique and often terminate the procedure without further computations if the difference between the solution value z^{lp} and lower bound $L(w)$ is sufficiently small. We might note that since at each step of the simplex method, the objective value z^{lp} of the problem stays the same or decreases, the upper bound is monotonically nonincreasing from step to step. On the other hand, the objective value $L(w)$ of the Lagrangian subproblem need not decrease from step to step, so at any point in the algorithm we would use the largest of the values $L(w)$ generated in all previous steps as the best lower bound.

17.6 DANTZIG–WOLFE DECOMPOSITION

In Section 17.5 we showed how to use a column generation procedure to solve the multicommodity flow problem formulated in the space of path flows. In this section we interpret this solution approach in another framework, known as *Dantzig–Wolfe decomposition.* Imagine that K different decision makers as well as one "coordinator" are solving the K-commodity flow problem. Each person plays a special role in solving the problem. The coordinator's job is to solve the path formulation (17.5) of the problem, which we refer to as the "master" or "coordinating" problem. In solving the problem the coordinator does not, however, generate the columns of the master problem; instead, the K decision makers, with guidance from the coordinator in the form of arc prices, generate these columns, with the kth decision maker generating the columns of the master problem corresponding to the kth commodity.

In general, the coordinator has on hand only a subset of the columns of the master problem. Since the coordinator can, at best, solve the linear program as restricted to this subset of columns, we refer to this smaller linear program as the *restricted master problem.*

The path formulation of the multicommodity flow problem has $m + K$ constraints: (1) one for each commodity k, specifying that the flow of commodity k is d^k; and (2) one for each arc (i, j), specifying that the total flow on that arc is at most u_{ij}. The coordinator solves the restricted master problem to optimality using any linear programming technique, such as the simplex algorithm, and then needs to determine whether the solution to the restricted master problem is optimal for the original problem or if some other column has a negative reduced cost. To this end the coordinator broadcasts the optimal set of simplex multipliers (or prices) of the restricted master problem, that is, broadcasts an arc price w_{ij} associated with arc (i, j) and a path length σ^k associated with each commodity k.

After the coordinator has broadcast the prices, the decision maker for commodity k determines the least cost way of shipping d^k units from the source node s^k to the sink t^k of commodity k, assuming that each arc (i, j) has an associated toll of w_{ij} in addition to its arc cost c_{ij}^k. If the cost of this shortest path is less than σ^k, the kth decision maker will report this solution to the coordinator as an improving solution. If the cost of this path equals σ^k, the kth decision maker need not report anything to the coordinator. (The cost will never be greater than σ^k because, as shown by our discussion of optimality conditions in Section 17.5, the coordinator is already using some path of cost σ^k for the kth commodity in his or her optimal solution.) To see that this interpretation is consistent with the preceding section, note that to price out the columns for commodity k, we need to solve the following shortest path problem:

$$\text{Minimize} \quad \sum_{1 \le k \le K} \sum_{P \in \mathbf{P^k}} [c^k(P) + \sum_{(i,j) \in P} w_{ij}] f(P)$$

subject to

$$\sum_{P \in \mathbf{P^k}} f(P) = d^k \qquad \text{for all } k = 1, 2, \ldots, K,$$

$$f(P) \ge 0 \qquad \text{for all } k = 1, 2, \ldots, K \text{ and all } P \in \mathbf{P^k}.$$

which corresponds, since $c^k(P) = \sum_{(i,j) \in P} c_{ij}^k$, precisely to finding the best way of satisfying the demand for the kth commodity assuming a cost of $c_{ij}^k + w_{ij}$ associated with each arc (i, j). Moreover, the shortest path for commodity k will have a negative reduced cost if and only if the cost for the kth shortest path problem is less than σ^k. We refer to the problems solved by each of the K decision makers as *subproblems* since we are using them solely for the purpose of generating new columns for the restricted master problem or proving optimality of the current solution.

Let us now summarize our discussion. In Section 17.5 we interpreted this procedure as a column generation procedure that determines entering variables by solving K different shortest path problems. The kth shortest path problem corresponds to the shortest path problem for the kth commodity with the added cost of w_{ij} associated with arc (i, j). In the Dantzig–Wolfe interpretation of the procedure, a central coordinator is solving the path formulation of the multicommodity flow problem restricted to the columns that he or she has on hand. After obtaining an optimal solution for this restricted master problem, the coordinator asks each of the K decision makers to solve a shortest path problem using an additional arc cost of w_{ij} on each arc (i, j). Each decision maker (i.e., subproblem) then either provides the coordinator with a new path or says that it can generate no shorter path than the one(s) the coordinator is currently using. Figure 17.9 gives a schematic representation of the information flow in the algorithm.

This method is flexible in several respects. When implemented as the revised simplex method, the algorithm would maintain a linear programming basis and introduce one column at each iteration. We might interpret this approach as maintaining a restricted master problem at each step with only a single column that is

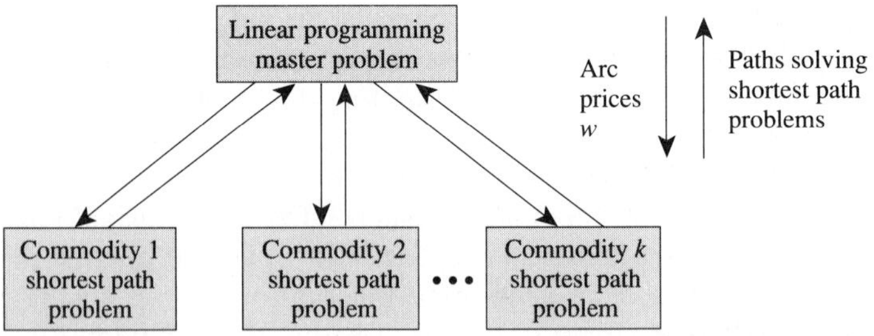

Figure 17.9 Information flow in Dantzig–Wolfe decomposition.

not in the basis; consequently, the algorithm discards any column that leaves the linear programming basis. However, generating new columns might be time consuming, so it might potentially be advantageous to save old columns since they might subsequently have a negative reduced cost. Therefore, in implementing the decomposition algorithm, we could retain some or all of the columns that we have generated previously. As a result, when we solve the restricted master problem, we would generally perform more than one basis change (to solve the restricted master problem to completion) before we broadcast new prices and once again solved the shortest path subproblems.

The column generation procedure and the decomposition procedure share one other potential advantage—the possibility of solving the problem using parallel processors. Once we have determined the arc prices w_{ij} in either procedure, the shortest path subproblems are independent of each other, so we could solve each of them simultaneously, using a separate processor for each shortest path problem.

Since each column of the master problem corresponds to one of a finite number of paths for each commodity, the decomposition solution technique will be finite as long as we never discard any columns of the restricted master problem. (Eventually, we would generate every column in the full master problem.) Alternatively, we could discard columns only when the solution to the restricted master problem strictly improves. Or, we could use an anticycling linear programming-based rule, such as lexicography, for selecting the variable to leave the basis at each iteration; any such rule would guarantee finiteness of the algorithm since it would ensure that we never repeat any of the finite number of bases to the linear program.

Finally, notice that the K subproblems correspond to the relaxed Lagrangian problem with a multiplier of w_{ij} imposed upon each arc (i, j). Consequently, we could view the coordinator as setting the Lagrange multipliers and solving the Lagrangian multiplier problem. In fact, Dantzig–Wolfe decomposition is an efficient method for solving the Lagrangian multiplier problem if we measure efficiency by the number of iterations an algorithm performs. (By Theorem 16.7, the Dantzig–Wolfe algorithm solves the Lagrangian multiplier problem since the multicommodity flow problem is a linear program, and for linear programs, the Lagrangian multiplier problem has the same objective function value as the linear program.) Unfortunately, in applying Dantzig–Wolfe decomposition, at each iteration the coordinator must solve a linear program with $m + K$ constraints, and this update step for the simplex multipliers is very expensive. It is far more time consuming to solve a linear program than to update the multipliers using subgradient optimization. Because each multiplier update for Dantzig–Wolfe decomposition is so expensive computationally, the Dantzig–Wolfe decomposition method has generally not proven to be an efficient method for solving the multicommodity flow problem; nevertheless, Dantzig–Wolfe decomposition has one important advantage that distinguishes it from other Lagrangian-based algorithms: The Dantzig–Wolfe decomposition algorithm always maintains a feasible solution of the problem. Since, as we have shown in Section 17.5, the solution to the subproblems provides us with a lower bound on the optimal value of the problem, at each step we also have a bound on how far the current feasible solution is from optimal. Therefore, we can terminate the algorithm at any step not only with a feasible solution, but also with a guarantee of how far, in objective function value, that solution is from optimality.

17.7 RESOURCE-DIRECTIVE DECOMPOSITION

Lagrangian relaxation and Dantzig–Wolfe decomposition are price-directive methods that decompose the multicommodity flow problem into single-commodity network flow problems (or shortest path problems) by placing tolls or prices on the complicating bundle constraints. The resource-directive method that we consider in this section takes a different algorithmic approach. Instead of using prices to decompose the problem, it allocates the joint bundle capacity of each arc to the individual commodities. When applied to the problem formulation (17.1), the resource-directive approach allocates $r_{ij}^k \leq u_{ij}^k$ units of the bundle capacity u_{ij} of arc (i, j) to commodity k, producing the following *resource-directive problem*:

$$z = \min \sum_{1 \leq k \leq K} c^k x^k \tag{17.7a}$$

subject to

$$\sum_{1 \leq k \leq K} r_{ij}^k \leq u_{ij} \qquad \text{for all } (i, j) \in A, \tag{17.7b}$$

$$\mathcal{N} x^k = b^k \qquad \text{for } k = 1, 2, \ldots, K, \tag{17.7c}$$

$$0 \leq x_{ij}^k \leq r_{ij}^k \qquad \text{for all } (i, j) \in A \text{ and all } k = 1, 2, \ldots, K. \tag{17.7d}$$

Note that the constraint (17.7b) ensures that the total resource allocation for arc (i, j) does not exceed that arc's bundle capacity. Let $r = (r_{ij}^k)$ denote the vector of resource allocations.

We now make a few elementary observations about this problem.

Property 17.2. *The resource-directive problem* (17.7) *is equivalent to the original multicommodity flow problem* (17.1) *in the sense that* (1) *if* (x, r) *is feasible in the resource-directive problem, then* x *is feasible and has the same objective function value in the original problem, and* (2) *if* x *is feasible in the original problem and we set* $r = x$*, then* (x, r) *is feasible and has the same objective function value in the resource-directive problem.*

Now consider the following sequential approach for solving the resource-directive problem (17.7). Instead of solving the problem by choosing the vectors r and x simultaneously, let us choose them sequentially. We first fix the resource allocations r_{ij}^k and then choose the flow x_{ij}^k. That is, let $z(r)$ denote the optimal value of the resource-directive problem for a fixed value of the resource allocation r and consider the following derived *resource-allocation problem*:

$$\text{Minimize} \quad z(r) \tag{17.8a}$$

subject to

$$\sum_{1 \leq k \leq K} r_{ij}^k \leq u_{ij} \qquad \text{for all } (i, j) \in A, \tag{17.8b}$$

$$0 \leq r_{ij}^k \leq u_{ij}^k \qquad \text{for all } (i, j) \in A \text{ and all } k = 1, 2, \ldots, K. \tag{17.8c}$$

The objective function $z(r)$ for this problem is complicated. We know its value only implicitly as the solution of an optimization problem in the flow variables x_{ij}^k.

Moreover, note that for any fixed value of the resource variables r_{ij}^k, the resource-directive problem decomposes into a separate network flow subproblem for each commodity. That is, $z(r) = \sum_{k \in K} z^k(r^k)$ with the value $z^k(r^k)$ of the kth subproblem given by

$$z^k(r^k) = \min \sum_{1 \le k \le K} c^k x^k \tag{17.9a}$$

subject to

$$\mathcal{N}x^k = b^k \quad \text{for all } k = 1, \ldots, K, \tag{17.9b}$$

$$0 \le x_{ij}^k \le r_{ij}^k \quad \text{for all } (i, j) \in A \text{ and for all } k = 1, 2, \ldots, K. \tag{17.9c}$$

Property 17.3. *The resource-directive problem* (17.7) *is equivalent to the resource-allocation problem* (17.8) *in the sense that* (1) *if* (x, r) *is feasible in the resource-directive problem, then* r *is feasible in the resource-allocation problem and* $z(r) \le cx$, *and* (2) *if* r *is feasible in the resource-allocation problem, then for some vector* x, (x, r) *is feasible in the original problem and* $cx = z(r)$.

Proof. If $(x, r) = (x^1, x^2, \ldots, x^K, r^1, r^2, \ldots, r^K)$ is feasible in the resource-directive problem, then r is feasible in the resource-allocation problem. Moreover, x^k is feasible in the kth commodity subproblem (17.9), so $z^k(r^k) \le c^k x^k$. Therefore, $z(r) = \sum_{1 \le k \le K} z^k(r^k) \le \sum_{1 \le k \le K} c^k x^k = cx$. Conversely, if r is feasible in the resource-allocation problem, then by definition of $z^k(r^k)$, $z^k(r^k) = c^k x^k$ for some vector x^k, so $x = (x^1, x^2, \ldots, x^K)$ satisfies the condition $cx = z(r)$. ◆

Let us pause to consider the implication of Properties 17.2 and 17.3. They imply that rather than solving the multicommodity flow problem directly, we can decompose it into a resource-allocation problem with a very simple constraint structure with a single inequality constraint but with a complex objective function $z(r)$. Although the overall structure of the objective function is complicated, it is easy to evaluate: To find its value for any choice of the resource-allocation vector r, we need merely solve K single-commodity flow problems.

Another way to view the objective function $z(r)$ is as the cost of the linear program (17.7) as a function of the right-hand-side parameters r. That is, any value r for the allocation vector defines the values of right-hand-side parameters for this linear program. A well-known result in linear program shows us that the function has a special form. We state this result for a general linear programming problem that contains the multicommodity flow problem as a special case.

Property 17.4. *Let* R *denote the set of allocations for which the linear program minimize* $\{cx : \mathcal{A}x = b, 0 \le x \le r\}$ *is feasible. Let* $z(r)$ *denote the value of this linear program as a function of the right-hand-side parameter* r. *On the set* R, *the objective function* $z(r)$ *is a piecewise linear convex function of* r.

Proof. To establish convexity of $z(r)$, we need to show that if $\bar{r}$ and $\hat{r}$ are any two values of the parameter r for which the given linear program is feasible and θ is any scalar, $0 \le \theta \le 1$, then $z(\theta\bar{r} + (1 - \theta)\hat{r}) \le \theta z(\bar{r}) + (1 - \theta)z(\hat{r})$. Let $\bar{y}$ and $\hat{y}$ be optimal solutions to the linear program for the parameter choices $r = \bar{r}$ and $r =$

$\hat{r}$. Note that $\mathcal{A}\bar{y} = b$, $\mathcal{A}\hat{y} = b$, $\bar{y} \le \bar{r}$, and $\hat{y} \le \hat{r}$. But then $\mathcal{A}(\theta\bar{y} + (1 - \theta)\hat{y}) = b$ and $\theta\bar{y} + (1 - \theta)\hat{y} \le \theta\bar{r} + (1 - \theta)\hat{r}$. Therefore, the vector $\theta\bar{y} + (1 - \theta)\hat{y}$ is feasible for the linear program with parameter vector $r = \theta\bar{r} + (1 - \theta)\hat{r}$, so the optimal objective function value for this problem is at most $c(\theta\bar{y} + (1 - \theta)\hat{y})$. Moreover, by our choice of $\bar{y}$ and $\hat{y}$, $z(\bar{r}) = c\bar{y}$ and $z(\hat{r}) = c\hat{y}$; therefore,

$$z(\theta\bar{r} + (1 - \theta)\hat{r}) \le \theta c\bar{y} + (1 - \theta)c\hat{y} = \theta z(\bar{r}) + (1 - \theta)z(\hat{r}),$$

so $z(r)$ is a convex function.

The piecewise linearity of $z(r)$ follows from the optimal basis property of linear programs. That is, for any choice of the parameter r, the problem has a basic feasible optimal solution, and this basic feasible solution remains optimal for all values of r for which it remains feasible. Moreover, the objective function value of the linear program is linear in r for any given (optimal) basis. ◆

Solving the Resource-Directive Models

A number of algorithmic approaches are available for solving the resource-directive models that we have introduced in this section. In the discussion to follow, we outline a few basic approaches. The references cited in the reference notes contain further details about these methods.

Since the function $z(r)$ is nondifferentiable (because it is piecewise linear), we cannot use gradient methods from nonlinear programming to solve the resource-allocation problem. We could, instead, use several other approaches. For example, we could search for local improvement in $z(r)$ using a heuristic method. As one such possibility, we could use an "arc-at-a-time approach" by adding 1 to $r^{k'}_{pq}$ and subtracting 1 from $r^{k''}_{pq}$ for some arc (p, q) for two commodities k' and k'', choosing the arc and commodities at each step using some criterion (e.g., the choices that give the greatest decrease in the objective function value at each step). This approach is easy to implement but does not ensure convergence to an optimal solution. Note that we can view this approach as changing the resource allocation at each step using the formula

$$r \leftarrow r + \theta\gamma,$$

with a step length of $\theta = 1$ and a movement direction $\gamma = (\gamma^k_{ij})$ given by $\gamma^{k'}_{pq} = 1$, $\gamma^{k''}_{pq} = -1$, and $\gamma^k_{ij} = 0$ for all other arc–commodity combinations. Borrowing ideas from subgradient optimization, however, we could use an optimization approach by choosing the movement direction γ as a subgradient corresponding to the resource allocation r (see Figure 17.10). A natural approach would be to search for a subgradient, or movement direction γ, and a step length θ that simultaneously maintain feasibility and ensure convergence to an optimal solution r of the resource-allocation problem (17.8). We will adopt a two-step approach. First we determine a subgradient direction and step length that would ensure convergence, provided that we had no constraints imposed on the allocations. If moving to $r + \theta\gamma$ gives us an infeasible solution, we transform r to a point r' that would maintain feasibility, yet ensure convergence. To apply this approach, we need to be able to answer two basic questions: (1) How can we find a subgradient γ of $z(r)$ at any given point r? (2) How could we transform any nonfeasible resource allocation $r + \theta\gamma$ so that it becomes feasible for the resource-choice problem, that is, it satisfies the constraints $\sum_{1 \le k \le K} r_{ij}^k \le u_{ij}$ for all $(i, j) \in A$, and $0 \le r_{ij}^k \le u_{ij}^k$ for all $(i, j) \in A$ and all $k = 1, 2, \ldots, K$? (Notice that when we applied subgradient optimization to the Lagrangian multiplier problem in Chapter 15, the Lagrange multipliers were either unconstrained or constrained to be nonnegative, so we either used the update formula $r \leftarrow r + \theta\gamma$ directly or used a modest modification of it to ensure that the Lagrange multipliers remained nonnegative. Therefore, in the context of Lagrangian relaxation, feasibility was easy to ensure.)

To answer 1, recall that as shown in Figure 17.10, a subgradient γ of $z(r)$ at the point $r = \bar{r}$ is any vector satisfying the condition

$$z(r) \ge z(\bar{r}) + \gamma(r - \bar{r}) \qquad \text{for all } r = (r^1, r^2, \ldots, r^K) \text{ with } r^k \in R^k.$$

In this expression, R^k is the set of resource allocations for commodity k for which the subproblem (17.8) is feasible. The following property shows that to find any such subgradient, we can work with each of the constituent functions $z^k(r^k)$ independently.

Property 17.5. *Let γ^k for $k = 1, 2, \ldots, K$ be a subgradient of $z^k(r^k)$ at the point $\bar{r}^K$. Then $\gamma = (\gamma^1, \gamma^2, \ldots, \gamma^k)$ is a subgradient of $z(r)$ at $\bar{r} = (\bar{r}^1, \bar{r}^2, \ldots, \bar{r}^K)$.*

Proof. Since γ^k is a subgradient of $z^k(r^k)$,

$$z(r^k) \ge z(\bar{r}^k) + \gamma^k(r^k - \bar{r}^k) \qquad \text{for all } r^k \in R^k.$$

Adding these expressions for $k = 1, 2, \ldots, K$ and using the fact that $z(r) = \sum_{k \in K} z^k(r^k)$ and $\gamma(r - \bar{r}) = \sum_{k \in K} \gamma^k(r^k - \bar{r}^k)$ gives us the desired result. ◆

This result shows us that to obtain a subgradient of $z(r)$, we can find a subgradient of each function $z^k(r^k)$, which requires information about the sensitivity of the solution of a network flow problem (17.9) to changes in the upper bounds r_{ij}^k on the arc flows. Fortunately, as shown by the following result, this information will be a by-product of almost any solution procedure for solving the subproblem (17.9). Since this result is general and applies to broader application contexts, we state it for a general network flow problem (i.e., we subsume the index k).

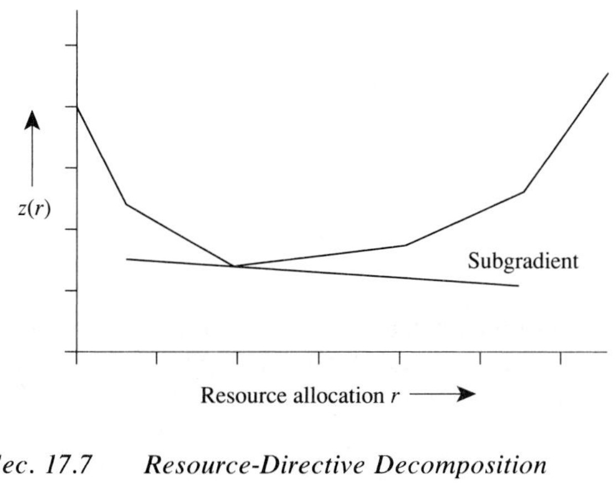

Figure 17.10 Subgradient for the function $z(r)$.

Property 17.6. *Consider the network flow problem* $z(q) = \min\{cx : \mathcal{N}x = b, 0 \le x \le q\}$, *with a parametric vector* q *of nonnegative upper bounds on the arc flows. Suppose that* x^* *is an optimal solution to this problem when* $q = q^*$, *and that* x^* *together with the reduced cost vector* c^π *satisfy the minimum cost flow optimality conditions* (9.8). *Define the vectors* $\mu = (\mu_{ij})$ *by setting* $\mu_{ij} = 0$ *if* $x^*_{ij} < q^*_{ij}$ *and* $\mu_{ij} = c^\pi_{ij}$ *if* $x^*_{ij} = q^*_{ij}$. *Then for any nonnegative vector* q' *for which the problem is feasible,* $z(q') \ge z(q^*) + \mu(q' - q^*)$. *That is,* μ *is a subgradient of the parametric objective function* $z(\mu)$ *at* $\mu = \mu^*$.

Proof. Recall that the definition of c^π implies that for any feasible flow vector x, $cx = c^\pi x - \pi b$. Therefore, $z(q^*) = cx^* = c^\pi x^* - \pi b = \mu q^* - \pi b$. The last equality in this expression follows from the optimality conditions (e.g., $c^\pi_{ij} = 0$ if $0 < x^*_{ij} < q^*_{ij}$) and the definition of μ. Let x' solve the parametric problem when $q = q'$ and note that $z(q') = cx' = c^\pi x' - \pi b \ge \mu q' - \pi b$. (The inequality in this expression follows from the fact that $c^\pi \le \mu$ and $0 \le x \le q'$. Further, $c^\pi_{ij} > \mu_{ij}$ is possible when $x^*_{ij} = 0$.) By combining the expressions for $z(q^*)$ and $z(q')$, we see that $z(q') \ge z(q^*) + \mu(q' - q^*)$. ◆

To apply Property 17.6, we solve each subproblem (17.9). After we solve the subproblem for the kth commodity with $q = r^k$, we set γ^k equal to the vector μ. We then use Property 17.6 and set $\gamma = (\gamma^1, \gamma^2, \ldots, \gamma^K)$. These results show us how to use any minimum cost flow algorithm that produces reduced costs c^π as well as an optimal flow vector to find a subgradient.

Once we have obtained a subgradient γ of $z(r)$, we could use the subgradient formula $r \leftarrow r + \theta\gamma$ to find the new resource allocation r. If the resulting resource-allocation vector r is feasible (i.e., satisfies the constraints $\sum_{1 \le k \le K} r^k_{ij} \le u_{ij}$), we move to this point. If the new resource allocation r is not feasible, however, we need to modify it to ensure feasibility, moving instead to some other point $\bar{r}$. One approach that ensures that the algorithm converges with the appropriate choice of step sizes (as discussed in Chapter 16) is to choose $\bar{r}$ as the feasible point that is as close as possible to r in the sense of minimizing the quantity $\sum_{1 \le k \le K} \sum_{(i,j) \in A} (r^k_{ij} - \bar{r}^k_{ij})^2$. The references cited at the end of this chapter show how to carry out this computation efficiently and show that with this choice of $\bar{r}$ the algorithm still converges.

17.8 BASIS PARTITIONING

Our development in Chapter 11 built on a principal advantage of the simplex method for the (single-commodity) minimum cost flow problem; namely, the algorithm can use the spanning tree structure of the linear programming basis to accelerate the required matrix computations. The multicommodity flow problem is a linear program that combines two sets of constraints: (1) independent mass conservation constraints $\mathcal{N}x^k = b^k$ for each commodity $k = 1, 2, \ldots, K$, and (2) bundle constraints that link the commodities through shared arc capacities. Because of this underlying network substructure, we might ask whether we could also use some form of spanning tree structure to implement the simplex method for the multicommodity flow problem. The basis partitioning algorithm that we consider in this section provides at least a partially affirmative answer to this question. As we will see, the method combines network methodology together with more generic ideas from linear programming.

This discussion is intended as an introduction to the basis partitioning method, not as a detailed account of the method, which is rather involved; instead, we present just the essential underlying ideas with the aim of showing how we can exploit the underlying network structure. This discussion assumes familiarity with the network simplex method (described in Chapter 11) and with the basic concepts of linear programming (described in Appendix C).

Consider any basis $\mathcal{B}$ of the linear programming formulation (17.1) of the multicommodity flow problem. Since each column of $\mathcal{B}$ corresponds to the flow on an arc of some commodity, the columns in the basis define a set of subgraphs, one for each commodity. Let $\mathcal{M}^k$ denote the submatrix of $\mathcal{B}$ corresponding to commodity k. We note that each matrix $\mathcal{M}^k$ must contain a basis $\mathcal{B}^k$ of the node–arc incidence matrix $\mathcal{N}^k$ (see Exercise 17.19). Our results in Section 11.11 show that each such basis corresponds to a spanning tree T^k for the corresponding commodity k. Therefore, the subgraphs corresponding to any basis are spanning trees together with some additional arcs. Moreover, since the constraints of the multicommodity flow problem have a block angular form, so does the basis $\mathcal{B}$. Figure 17.11(a) shows the structure of the basis $\mathcal{B}$.

Rather than working directly with the basis matrix $\mathcal{B}$, through a change in variables, we will transform this matrix into another matrix $\mathcal{B}'$ so that we can perform the steps of the simplex method much more efficiently using the underlying network structure of the problem. Suppose that we state the bundle constraints as equality constraints [see (17.1b′)] by adding the nonnegative slack variable s_{ij} to the bundle constraint for arc (i, j). The variable s_{ij} denotes the amount of unused bundle capacity on arc (i, j). Consider the system of mass balance and bundle constraints for a basis of the multicommodity flow problem, which we write as $\mathcal{B}x_{\mathbf{B}} = b$. In this expression, $b = (u, b^1, b^2, \ldots, b^K)$ is the vector of right-hand-side coefficients for the equations and $x_{\mathbf{B}}$ is the set of variables corresponding to the basic variables. These variables include both flow variables x_{ij}^k and slack variables s_{ij} for the bundle constraints.

To simplify the system $\mathcal{B}x_{\mathbf{B}} = b$, suppose that we make a change in variables by letting $\mathcal{D}y_{\mathbf{B}} = x_{\mathbf{B}}$ for some appropriately selected nonsingular matrix $\mathcal{D}$. If we

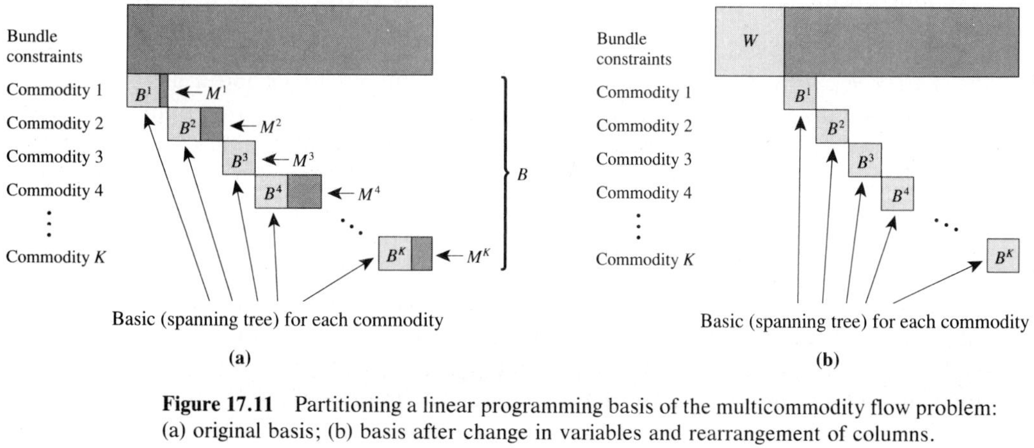

Figure 17.11 Partitioning a linear programming basis of the multicommodity flow problem: (a) original basis; (b) basis after change in variables and rearrangement of columns.

substitute $\mathcal{D}y_{\mathbf{B}}$ for $x_{\mathbf{B}}$ in the system $\mathcal{B}x_{\mathbf{B}} = b$, and let $\mathcal{B}' = \mathcal{B}\mathcal{D}$, the system becomes $\mathcal{B}(\mathcal{D}y_{\mathbf{B}}) = (\mathcal{B}\mathcal{D})y_{\mathbf{B}} = \mathcal{B}'y_{\mathbf{B}} = b$. Suppose that we can make this change in variables so that the matrix $\mathcal{B}'$ has the special structure shown in Figure 17.11(b); in this special structure, only the y variables that appear in the mass balance constraints for commodity k are those corresponding to the basis $\mathcal{B}^k$. Later in this section we show how to use the underlying network structure of the problem to make this change in variables. For the moment, so that we can obtain a broad view of the basic approach without the complicating details, let us assume that we can easily make this transformation of variables.

In the transformed system, let $\mathcal{W}$ denote the matrix from the bundle constraints that do not correspond to any of the columns in the bases $\mathcal{B}^k$ for $k = 1, 2, \ldots, K$. We refer to $\mathcal{W}$ as the *working basis*. In the pictorial representation of the basis $\mathcal{B}'$ in Figure 17.11(b), we have rearranged the columns of the basis so that those in the working basis appear first.

We now make two further observations:

1. Since a change in variables, which corresponds to column operations on the matrix $\mathcal{B}$, does not change the nonsingularity of this matrix, the matrix $\mathcal{B}'$ is also nonsingular. This fact implies that the working basis $\mathcal{W}$ is nonsingular.
2. Since column operations on the matrix $\mathcal{B}$ do not change the solution of the system $\pi\mathcal{B} = c_{\mathbf{B}}$, the vector of the simplex multipliers determined by the system $\pi\mathcal{B}' = c_{\mathbf{B}'}$ is the same as those determined by the original system $\pi\mathcal{B} = c_{\mathbf{B}}$. (When we perform the column operations, we change both $\mathcal{B}$ and $c_{\mathbf{B}}$; $c_{\mathbf{B}'}$ denotes the result of the column operations on the costs $c_{\mathbf{B}}$.)

These two observations and our change in variables gives us all the necessary ingredients for implementing the basis partitioning simplex method for the multicommodity flow problem. Recall that the simplex method requires that we make two types of matrix computations:

1. Solving a system of the form $\mathcal{B}x_{\mathbf{B}} = b$ or $\mathcal{B}x_{\mathbf{B}} = \mathcal{A}_j$ for some column $\mathcal{A}_j$ of the coefficient matrix $\mathcal{A}$ of the linear programming model (we make this computation at each step before we introduce column $\mathcal{A}_j$ into the basis).
2. Solving a system of the form $\pi\mathcal{B} = c_{\mathbf{B}}$ to find the vector π of simplex multipliers.

Rather than making these computations of the matrix $\mathcal{B}$ directly, we will use the matrix $\mathcal{B}'$. As we have already noted, the systems $\pi\mathcal{B} = c_{\mathbf{B}}$ and $\pi\mathcal{B}' = c_{\mathbf{B}'}$ have the same solutions. Suppose that we partition the vector of simplex multipliers as $\pi = (\pi^0, \pi^1, \ldots, \pi^K)$. The vector π^0 contains the simplex multipliers for the bundle constraints and the vector π^k contains the simplex multipliers corresponding to the mass balance constraints $\mathcal{N}x^k = b^k$ for commodity k. Let $c_{\mathbf{W}}$ denote the subvector of $c_{\mathbf{B}}$ corresponding to the columns in $\mathcal{W}$. We solve the system $\pi\mathcal{B}' = c_{\mathbf{B}'}$ efficiently by the following *forward substitution* procedure. We first solve the subsystem $\pi^0\mathcal{W} = c_{\mathbf{W}}$ by solving a system of linear equations. If we substitute these values into the system $\pi\mathcal{B}' = c_{\mathbf{B}'}$, the system decomposes into separate subsystems for each commodity k with modified coefficients $\hat{c}_{\mathbf{B}'}$ for $c_{\mathbf{B}'}$. Moreover, the computations for each subsystem are exactly those required to find the simplex multipliers for a single-commodity network flow problem. Therefore, we can use the very efficient procedure for computing the simplex multipliers that we developed in Chapter 11. Recall that these computations are very efficient because they use the tree structure of the basis.

To solve a system of the form $\mathcal{B}x_{\mathbf{B}} = b$, we first solve the related system $\mathcal{B}'y_{\mathbf{B}} = b$ using forward substitution. That is, if we partition the vector $y_{\mathbf{B}}$ as $y_{\mathbf{B}} = (y^0, y^1, y^2, \ldots, y^K)$ corresponding to the columns $\mathcal{W}, \mathcal{B}^1, \mathcal{B}^2, \ldots, \mathcal{B}^K$ of $\mathcal{B}'$, we first solve the independent subsystems $\mathcal{B}^k y^k = b^k$. Notice that we can perform these computations by using the efficient procedure for computing the arc flows that we developed in Chapter 11. Recall that this procedure also used the spanning tree structure corresponding to the basis $\mathcal{B}^k$. We then substitute the values of the variables $y^1, y^2, \ldots, y^K$ into the system $\mathcal{B}'y = b$, leaving a reduced system of the form $\mathcal{W}y^0 = u'$. (Here u' denotes the modified arc capacities we obtain after substituting the values of the flow variables $y^1, y^2, \ldots, y^K$ in the transformed bundle constraints.) After solving this system of linear equations, we use the transformation between $x_{\mathbf{B}}$ and $y_{\mathbf{B}}$ (i.e., $x_{\mathbf{B}} = \mathcal{D}y_{\mathbf{B}}$) to find the values for the variables $x_{\mathbf{B}}$.

Notice what this approach has accomplished. Suppose that we apply the procedure to a problem with p bundle constraints (p might equal the number m of arcs if they all have bundle capacities, or it could be much less than m). Rather than solving systems of the form $\mathcal{B}x = b$ on the full matrix $\mathcal{B}$ which has dimension $p + K$ by $p + K$, we make K tree computations and solve one smaller $p \times p$ system $\mathcal{W}y^0 = u'$. Similarly, rather than solving a large $(p + K) \times (p + K)$ system of the form $\pi\mathcal{B} = c_{\mathbf{B}}$ solve a single $p \times p$ system $\pi^0\mathcal{W} = c_{\mathbf{W}}$ and make K tree computations. The resulting speed up in computations can be very substantial.

The rest of the steps of the simplex method (e.g., pricing out columns to find the entering variable at each step or performing the ratio test to find the outgoing variable) are exactly those of the method in general. Since our purpose in this discussion is not to describe the simplex method, but rather to show how we can exploit the underlying network structure to accelerate the computations, we do not describe these details. There is one other important feature of the algorithm that we do not discuss, namely how to recompute the matrix $\mathcal{B}'$ when we exchange one column in the basis for another. The references cited at the end of this chapter show how to perform these computations.

To complete this brief description of the basis partitioning method, let us show how to make the change in variables $x_{\mathbf{B}} = \mathcal{D}y_{\mathbf{B}}$ needed to eliminate any column $\mathcal{M}_{ij}^k$ that belongs to $\mathcal{M}^k$ but not to $\mathcal{B}^k$ from the mass balance equations for commodity k so that the only remaining variables in these equations will be those corresponding to the basis $\mathcal{B}^k$.

Consider the following example, which corresponds to the system $\mathcal{M}^k x^k = b^k$ (for simplicity, we subsume the commodity index k and use x_{ij} in place of x_{ij}^k):

$$\begin{array}{c} \textit{Node} \\ 1 \\ 2 \\ 3 \\ 4 \\ 5 \\ 6 \end{array} \quad \begin{bmatrix} +1 & +1 & & & & \\ -1 & & & & & \\ & -1 & +1 & +1 & & \\ & & -1 & & & +1 \\ & & & -1 & -1 & \\ & & & & +1 & -1 \end{bmatrix} \begin{bmatrix} x_{12} \\ x_{13} \\ x_{34} \\ x_{35} \\ x_{65} \\ x_{46} \end{bmatrix} = \begin{bmatrix} 4 \\ 2 \\ 1 \\ 2 \\ 4 \\ 3 \end{bmatrix}$$

Let $\mathcal{M}_{ij}$ denote the column of this system corresponding to the variable x_{ij}. The columns $\mathcal{M}_{12}$, $\mathcal{M}_{13}$, $\mathcal{M}_{34}$, $\mathcal{M}_{35}$, and $\mathcal{M}_{65}$ form a basis for this system and $\mathcal{M}_{46}$ is the nonbasic column that we would like to eliminate. Notice that $\mathcal{M}_{46} = -\mathcal{M}_{34} + \mathcal{M}_{35} - \mathcal{M}_{65}$, which in terms of the underlying network (see Figure 17.12) gives the representation of arc (4, 6) in terms of the spanning tree defined by the arcs (1, 2), (1, 3), (3, 4), (3, 5), and (6, 5). Now suppose that we make the change in variables $x_{34} = y_{34} - y_{46}$, $x_{35} = y_{35} + y_{46}$, $x_{65} = y_{65} - y_{46}$, $x_{12} = y_{12}$, $x_{13} = y_{13}$, and $x_{46} = y_{46}$. Then we find that

$$\begin{aligned}
&\mathcal{M}_{12}x_{12} + \mathcal{M}_{13}x_{13} + \mathcal{M}_{34}x_{34} + \mathcal{M}_{35}x_{35} + \mathcal{M}_{65}x_{65} + \mathcal{M}_{46}x_{46} \\
&\quad = \mathcal{M}_{12}y_{12} + \mathcal{M}_{13}y_{13} + \mathcal{M}_{34}(y_{34} - y_{46}) + \mathcal{M}_{35}(y_{35} + y_{46}) \\
&\qquad + \mathcal{M}_{65}(y_{65} - y_{46}) + \mathcal{M}_{46}y_{46} \\
&\quad = \mathcal{M}_{12}y_{12} + \mathcal{M}_{13}y_{13} + \mathcal{M}_{34}y_{34} + \mathcal{M}_{35}y_{35} + \mathcal{M}_{65}y_{65}.
\end{aligned}$$

That is, we have eliminated the column $\mathcal{M}_{46}$ from the system. Notice that the new variables y_{ij} have a natural interpretation. The new variables y_{34}, y_{35}, and y_{65} measure both flows x_{ij} on these arcs as well as the flow x_{46} "diverted" from the arc (4, 6) to these arcs on the path 4–3–5–6.

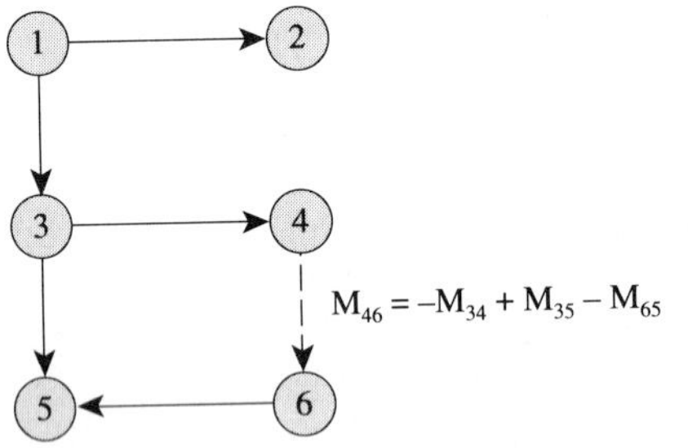

Figure 17.12 Representing nontree (nonbasic) arc for variable transformation.

In general, if we have several columns $\mathcal{M}_{ij}$ that we want to eliminate from the system, and in the underlying graph the arc (i, j) forms a unique cycle when added to the spanning tree T corresponding to the basis of the system, we let γ_{ij}^{pq} be either ± 1 or 0, depending on whether or not the arc (p, q) of the tree belongs to this cycle (the sign of each ± 1 depends upon the orientation of the tree arc (p, q) relative to the arc (i, j)). To transform the variables, we then set $x_{pq} = y_{pq} - \sum \gamma_{ij}^{pq} y_{pq}$ with the sum taken over the indices (i, j) for all the columns $\mathcal{M}_{ij}$ that we wish to eliminate. We also set $x_{ij} = y_{ij}$ for all arcs (i, j) not in the tree T.

To implement this procedure for the multicommodity flow problem, we apply these network computations and change of variables for each of the matrices $\mathcal{M}^k$. These computations give us the matrix $\mathcal{B}'$ and then we apply the operations discussed previously in this section.

17.9 SUMMARY

Multicommodity flow problems arise whenever several network flow problems share the same underlying network. This chapter has focused on a generic multicommodity flow problem with fixed bundle capacities imposed on the total flow on any arc by all the commodities. As we have seen in this chapter, in Chapter 1, and as we show further in Chapter 19, this model arises in a wide variety of application settings in communications, logistics, manufacturing, and transportation as well as in such problem domains as urban housing and foodgrain export-import. The solution methods for solving the multicommodity flow problem generally attempt to exploit the network flow structure of the individual-constitutent single-commodity flow problems. Lagrangian relaxation removes the complicating bundle constraints by applying Lagrangian multipliers to them and bringing them into the objective function. The resulting Lagrangian subproblem separates into independent flow problems for each commodity. This method is price directive in that rather than considering the bundle constraints directly, it places tolls, or Lagrange multipliers on them. By the theory of Lagrangian relaxation, for any choice of the tolls, the Lagrangian subproblems give us a lower bound on the optimal objective function value of the problem. Moreover, because the multicommodity flow problem is a linear program, the best possible (i.e., largest) lower bound equals the optimal objective function value for the problem. Therefore, by solving the Lagrangian multiplier problem of finding the best Lagrangian lower bound, we find the optimal objective function value for the multicommodity flow problem.

The column generation procedure solves the multicommodity flow problem formulated in the space of path (and cycle) flows. This formulation is a large-scale linear program with an enormous number of variables (columns) but a rather simple constraint structure. The column generation procedure solves the problem by explicitly maintaining a relatively small number of these variables as part of a linear programming basis and then pricing-out the remaining columns to determine a column(s) that we could profitably add to the basis. The key to this procedure is the observation that to price-out the columns and to find the most profitable column to add to the basis (in terms of reduced cost for each commodity), we need not consider each column explicitly but can instead solve a shortest path problem for each commodity. Since this procedure is just a special (and efficient) implementation of the simplex method of linear programming, it inherits the finite convergence properties of the simplex method (with provisions for handling linear programming degeneracy). Moreover, since we can view the pricing-out operation as solving a Lagrangian relaxation of the problem, at each step we have not only a feasible solution, but also a lower bound providing a guarantee how far the solution is from optimality in the objective function value.

We might also interpret the column generation procedure as a price-setting algorithm with the basis for the linear program determining prices on the bundle capacities, and broadcasting these prices to shortest path subproblems, one for each commodity. When interpreted in this way, this so-called Dantzig–Wolfe decomposition procedure is an alternative solution procedure for solving the Lagrangian relaxation of the problem. In this approach we use a linear program, rather than simply move in a subgradient direction, to update the values of the Lagrangian multipliers at each step.

In our discussion of column generation and Dantzig–Wolfe decomposition, we have assumed that each commodity in the multicommodity flow problem has a single source and single sink. These algorithms apply to multicommodity flow problems with multiple sources and sinks for each commodity (see Exercise 17.14) and to more general problems with otherwise independent subproblems coupled by joint resource constraints (see Exercises 17.37 and 17.38).

Resource-directive decomposition is an alternative conceptual approach for solving the multicommodity flow problem. Rather than removing the complicating bundle capacities by charging tolls for them, this approach decomposes the problem into a separate single commodity flow problem for each commodity by allocating the scarce bundle capacities to the various commodities. Finding the optimal allocation (i.e., the one that gives the overall lowest cost) is an optimization problem with a simple constraint structure and with a (complicated) convex cost objective function. Using sensitivity information about the single-commodity subproblems, however, we can generate subgradient information about the resource-allocation cost function and therefore we can solve the allocation problem by a version of the subgradient optimization technique.

One approach for solving the minimum cost flow problem is the simplex method implemented to exploit the underlying network flow structure, particularly the spanning tree interpretation of any linear programming basis. The basis partitioning method for solving the multicommodity flow problem is a generalization of this approach. It uses the fact that any linear programming basis for the multicommodity flow problem contains a basis for each commodity. Consequently, through a change of variables, we can transform any basis of the problem into a basis for each commodity as well as a small working basis for the bundle constraints that contains information about the interactions between the individual commodities. This approach permits us to solve the multicommodity flow problem by combining the special network simplex approach for each commodity together with more general linear programming matrix computations as applied to the working basis.

REFERENCE NOTES

Researchers have proposed a number of basic approaches for solving the multicommodity flow problem. The three basic approaches, all based on exploiting network flow substructure, that we have considered in this chapter and selected references are (1) price-directive decomposition algorithm (Cremeans, Smith, and Tyndall [1970], Swoveland [1971], Chen and Dewald [1974], and Assad [1980b]), (2) resource-directive decomposition algorithm (Geoffrion and Graves [1974], Kennington and Shalaby [1977], and Assad [1980a]), and (3) basis partitioning (Graves and McBride [1976] and Kennington and Helgason [1980]). Ford and Fulkerson [1958b] and Tomlin [1966] first suggested the column generation approach. The first of these papers was the forerunner of the general Dantzig–Wolfe [1960] decomposition procedure of mathematical programming. The excellent survey papers by Assad [1978] and by Kennington [1978] describe all of these algorithms and several standard properties of multicommodity flow problems. The book by Kennington and Helgason [1980] and the doctoral dissertation by Schneur [1991] are other valuable references on this topic.

Most of the material discussed in this chapter is classical and dates from the 1960s and 1970s. Many of the standard properties of multicommodity flows (e.g., nonintegrality of optimal flows), including several of the examples we have used in the text and exercises to illustrate these properties, are due to Fulkerson [1963], Hu [1963], and Sakarovitch [1973].

The column generation and decomposition methods that we have considered in this chapter extend to other situations as long as we can represent any solution to a problem as a convex combination of other particularly "simple solutions"; in the text we have used shortest paths as the simple solutions. For some applications we might use solutions to knapsack problems as the simple solutions, and in other cases, such as the general multicommodity flow problem with multiple sources and destinations for each commodity, the simple solutions might be spanning tree solutions (see Exercise 17.14).

Researchers have developed and tested several codes for multicommodity flow problems. Kennington [1978] and Ali et al. [1984] have described the results of some of these computational experiments. These results have suggested that price-directive and partitioning algorithms are the fastest algorithms for solving multicommodity flow problems. The best multicommodity flow codes are two to five times faster than a general-purpose linear programming code. Recent computational experience by Bixby [1991] in solving large-scale network flow problems with side constraints has shown that the simplex method with an advanced starting basis technique can be very effective computationally. The approach that we have suggested at the end of our discussion of Lagrangian relaxation in Section 17.4 is a variant of this approach.

Interior point algorithms provide another approach for solving multicommodity flow problems. Although these algorithms yield the only known polynomial-time bounds for these problems, an efficient and practical implementation of these algorithms is the subject of future research. The best time bound for the multicommodity flow problem is due to Vaidya [1989]. Tardos's [1986] algorithm can solve the multicommodity flow problem in strongly polynomial time.

Several researchers have recently suggested new algorithms for the multicommodity flow problem: Gersht and Shulman [1987], Barnhart [1988], Pinar and Zenios [1990], and Farvolden and Powell [1990]. Schneur [1991] studied scaling techniques for the multicommodity flow problem. Matsumoto, Nishizeki, and Saito [1986] gave a polynomial-time combinatorial algorithm for solving a multicommodity flow problem in s–t planar networks.

In this chapter we have presented several applications of multicommodity flow problems, adapted from the following papers:

1. Routing of multiple commodities (Golden [1975] and Crainic, Ferland, and Rousseau [1984])
2. Warehousing of seasonal products (Jewell [1957])
3. Multivehicle tanker scheduling problem (Bellmore, Bennington, and Lubore [1971])

We have presented three additional applications elsewhere in this book: (1) racial balancing of schools in Application 1.10 (Clarke and Surkis [1968]), (2) optimal deployment of resources in Exercise 17.1 (Kaplan [1973]), and (3) multiproduct multistage production-inventory planning in Application 19.23 (Evans [1977]). Other applications of multicommodity flows arise in (1) multicommodity distribution system design (Geoffrion and Graves [1974]), (2) rail freight planning (Bodin, Golden, Schuster, and Rowing [1980] and Assad [1980a]), and (3) VLSI chip design (Korte [1988]).

EXERCISES

17.1. **Optimal deployment of resources** (Kaplan [1973]). Suppose that an organization requires varying levels of resources at q geographical locations. Suppose, further, that changes in economic, political, social, or environmental conditions have brought about a sudden demand for the resources. For example, natural disasters such as floods might create a need for various types of rescue equipment at various flood locations. Or, the resources might be relief supplies such as food or medicines. This exercise studies a model that minimizes the combined cost of transportation and unfulfilled demands. Let d_j^k denote the demand of resource k (in tons) at location j. The resources are available in limited quantities at p different locations; let a_i^k denote the amount of resource k available at location i. The cost of transporting 1 unit of the resource k from location i to location j is c_{ij}^k. Due to technological constraints, at most u_i tons of all resources can be transported from location i. As a result, it might not be possible to completely satisfy the demands for the resource. Let α_j^k denote the cost of unfulfilled demand of resource k at location j. Our objective is to identify the shipment of resources that minimizes the total of the transportation costs and the costs arising due to unfulfilled demands. Formulate this problem as a multicommodity flow problem.

17.2. Show how to incorporate transportation expenses incurred between plant–warehouse, warehouse–retailer, and plant–retailer combinations in the warehousing of seasonal products model that we considered in Application 17.2.

17.3. Show that the optimal solution of the multicommodity flow problem in Figure 17.4(a), as shown in Figure 17.4(b), satisfies both the arc optimality conditions (17.2) and the path optimality conditions (17.6).

17.4. State a generalization of the arc optimality conditions (17.2) for the multicommodity flow problem with upper bounds u_{ij}^k imposed on arc flows x_{ij}^k.

17.5. Write the dual of the path flow formulation of the multicommodity flow problem given in (17.5). Show that (17.6) represents the complementary slackness conditions of this primal–dual pair for the multicommodity flow problem.

17.6. The path formulation (17.5) of a multicommodity flow problem with $p \leq m$ bundle constraints and K commodities contains $p + K$ constraints, and therefore any linear programming basis for the problem will contain $p + K$ basic variables.

(a) Show that at least one basic variable must appear in each of the K demand constraints $\sum_{P \in \mathbf{P}^k} f(P) = d^k$; that is, in any basis at least one path must carry a positive flow for each commodity.

(b) Show that in any feasible basis, at most p commodities will send flow on two or more paths, so at least $K - p$ commodities will send flow on a single path.

17.7. The numerical example shown in Figure 17.13 has four commodities: these commodities have nodes 1 and 4, 5 and 8, 9 and 12, and 13 and 16 as their source and sink nodes. We wish to send 10 units from the source node to the sink node of each commodity. The arcs (2, 3), (6, 7), (10, 11), and (14, 15) all have a bundle capacity of 15 units. All the other arcs are uncapacitated. The per unit flow costs shown next to each arc are the same for each commodity. Starting with the Lagrange multiplier values $\mu_{23} = \mu_{67} = \mu_{10,11} = \mu_{13,16} = 0$, and assuming step lengths of $\theta_0 = 1$ and $\theta_k = 1/k$ for $k \geq 1$, perform the first five steps of the Lagrangian relaxation algorithm for this problem.

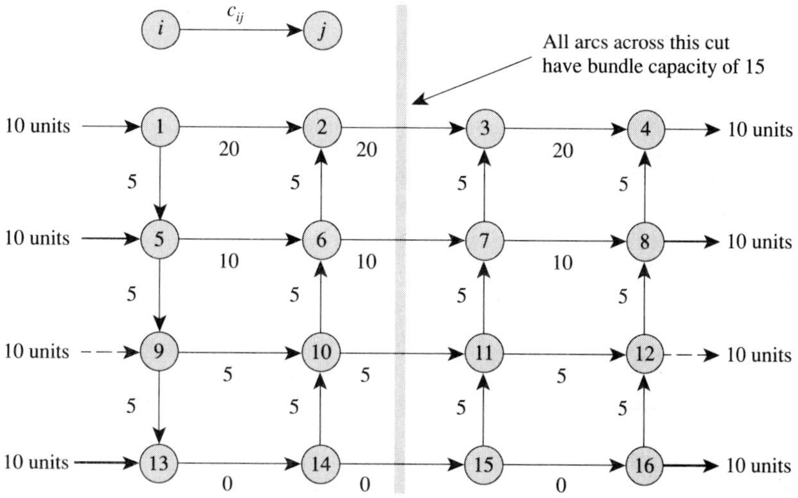

Figure 17.13 Multicommodity flow example with four commodities and with four arcs with bundle capacities.

17.8. Find an optimal solution of the problem introduced in Exercise 17.7 by any method, including visual inspection, and use the multicommodity flow optimality conditions to prove that your solution is optimal.

17.9. Since the multicommodity flow example shown in Figure 17.4(a) has two arcs with bundle capacities, the working basis in applying the basis partitioning algorithm to this problem will be a 2×2 matrix. Specify the optimal linear programming basis $\mathcal{B}$ corresponding to the optimal solution shown in Figure 17.4(b). Restricting your choice of variables to those in $\mathcal{B}$, specify a basis (spanning tree) for each commodity and the resulting working basis for $\mathcal{B}$. Show how to use the working basis and spanning tree to compute the simplex multipliers corresponding to the optimal linear programming basis.

17.10. **Nonhomogeneous goods.** Consider an extension of the multicommodity flow problem with nonhomogeneous goods. In this model, each unit x_{ij}^k of flow of commodity k on arc (i, j) consumes a given amount ρ_{ij}^k of the capacity (or some other resource) associated with the arc (i, j), and we replace the bundle constraint with a more general resource availability constraint $\sum_{1 \le k \le K} \rho_{ij}^k x_{ij}^k \le u_{ij}$.

(a) Show how to convert this model into the bundle constraint model (17.1) that we have considered in this chapter if each commodity consumes the same amount of resource on each arc, that is, for every arc (i, j), $\rho_{ij}^1 = \rho_{ij}^2 = \cdots = \rho_{ij}^K = \rho_{ij}$.

(b) Show how to solve the general version of this problem by using a modification of the following methods that we have considered in this chapter: (1) Lagrangian relaxation, and (2) column generation.

17.11. **Modeling piecewise linear convex costs.** Suppose that a multicommodity flow problem has no bundle constraints, but instead the cost of flow on each arc is a piecewise linear function of the total flow $x_{ij} = \sum_{1 \le k \le K} x_{ij}^k$ of all the commodities on that arc. Show how to model this problem as a (linear) multicommodity flow problem with bundle constraints. (*Hint*: See Section 14.3.)

17.12. **Multicommodity circulation problem.** Suppose that the flow of each commodity in a multicommodity flow problem must be a circulation (i.e., each supply/demand vector b^k is the zero vector). Also, assume that some of the arc costs might be negative. Show how to modify the column generation algorithm to solve this circulation variant of the multicommodity flow problem. (*Hint*: It is always possible to express any circulation as the union of cycle flows. Also, recall that we can use the label correcting shortest path algorithm to detect a negative cost cycle.)

17.13. Show how to extend the column generation algorithm to handle situations with negative costs. In these situations, some potentially optimal feasible flows could be a weighted combination of both cycle and path flows, and the output of any shortest path subproblem might be a negative cost cycle.

17.14. Show how to apply column generation and Dantzig–Wolfe decomposition to multicommodity flow problems in which each commodity can have several sources and destinations. Assume that the cost vector for each commodity is nonnegative. For each commodity k, let $x^{k,q}$ for $q = 1, 2, \ldots, Q_k$ denote the flow vector corresponding to the qth of Q_k feasible spanning tree solution for the system $\mathcal{N}x^k = b^k$, $0 \le x^k \le u^k$. Use the fact that we can write any potentially optimal feasible solution x^k to this system as $x^k = \sum_{1 \le k \le K} \lambda^{k,q} x^{k,q}$ for some nonnegative weighting vectors $\lambda^{k,q}$ satisfying the condition $\sum_{1 \le q \le Q_k} \lambda^{k,q} = 1$. (Note that the nonnegativity of the costs implies that we need not consider cycles in the optimal solution.)

17.15. Multisink maximum flow problem (Rothfarb, Shein, and Frisch [1968])

(a) Consider a multicommodity maximum flow problem with a source node s and with two different sink nodes, t^1 and t^2. Suppose that every unit shipped from node s into node t^1 yields a profit of \$3, and that every unit shipped from node s into node t^2 yields a profit of \$2. Show how to find the maximum profit flow by solving two maximum flow problems. (*Hint*: First ship as much flow as possible from node s to node t^1, obtaining a maximum s–t^1 flow x'. Subsequently, ship as much flow as possible from s to t^2 in the residual network $G(x')$ while keeping the flow from node s to node t^2 constant. Show that the resulting flow x'' is optimal for the two-sink problem by showing that the residual network $G(x'')$ contains no augmenting path from node s to node t^1 and no augmenting path from node s to node t^2.)

(b) Generalize the results of part (a) to the following situation. A network $G = (N, A)$ has one source node s and K sink nodes $t^1, t^2, \ldots, t^K$, one for each of K commodities. For $k = 1$ to K, let c_k be the value of shipping each unit of commodity k to sink t^k. Assume that $c_1 \ge \cdots \ge c_K$. Show how to find the optimal-valued multicommodity flow. Justify your solution procedure. (*Hint*: Extending the results of part (a), solve the problem as a sequence of K maximum flow problems.)

17.16. Common-source or common-sink multicommodity flow problem

(a) The common-source multicommodity flow problem is a special case of the multicommodity flow problem (17.1) in which all commodities have a common source but (possibly) distinct sinks. Show how to solve the common-source multicommodity flow problem by solving one single-commodity minimum cost flow problem. (*Hint*: First solve a minimum cost flow problem and then use flow decomposition.)

(b) In the common-sink multicommodity flow problem, all commodities have a common sink but (possibly) distinct sources. Show how to solve the common-sink multicommodity flow problem by solving one single-commodity minimum cost flow problem.

17.17. Funnel problem. Let $G = (N, A)$ be a network with K source nodes $s^1, \ldots, s^K$ and J sink nodes $t^1, \ldots, t^J$. Suppose, further, that the network contains a *cut node* v whose deletion disconnects each source from each sink. That is, $G - v$ has two components, one containing all the source nodes and the other containing all the sink nodes. Each source node s^i and each sink node t^j defines a commodity and has an associated demand of d_{ij}; therefore, we must send d_{ij} units of flow from node s^i to node t^j. Suppose that each arc (p, q) has an associated per unit arc flow cost c_{pq} (which is the same for all the commodities) and an associated flow capacity u_{pq}. Show that by solving two single-commodity minimum cost flow problems, we can either determine a minimum cost multicommodity flow in the network G or prove that none such flow exists. (*Hint*: Use ideas introduced in Exercise 17.16.)

17.18. Suppose that you had an algorithm that could find a feasible multicommodity flow for the problem (17.1). Further, suppose that you also had the optimal set of prices (tolls) for bundle constraints for the arcs. How might you use these prices and the solution algorithm to find an optimal flow for the multicommodity flow problem?

17.19. Let $\mathcal{M}^k$ denote the submatrix of a basis matrix $\mathcal{B}$ for the multicommodity flow model (17.1) corresponding to the flow equations $\mathcal{N}x^k = b^k$. Show that $\mathcal{M}^k$ must contain a basis $\mathcal{B}^k$ of the node–arc incident matrix $\mathcal{N}$. (*Hint*: Show that the arcs corresponding to the column in $\mathcal{M}^k$ must contain a spanning tree of the underlying network by recalling that if $\mathcal{B}$ is a basis matrix, we can solve the system of equations $\mathcal{B}x = d$ for any vector d.)

17.20. Cutting stock problem. The production of paper or cloth often uses the following process. We first produce the paper on long rolls which we then cut to meet specific demand requirements. The cutting stock problem is to meet the specified demand by using the minimum number of rolls. The cutting stock problem also arises in other guises as well. For example, it arises when we wish to store information with bit length L_i onto tracks in a computer disk, assuming that each track can hold L bits. Suppose that the rolls all have length L and that we have a demand of d_i for smaller rolls of size L_i. There might be many ways to cut any roll into smaller rolls. For example, if $L = 50$ and the L_i have lengths 10, 15, 25, 30, we could cut the rolls into several different patterns: for example, (1) 5 rolls of size 10; (2) 3 rolls of size 15; (3) 2 rolls of size 10 and 1 roll of size 25. Note that in the first pattern, we use the entire roll of size L, but in the other two patterns we incur a waste of 5. Let $\mathcal{P}$ denote the collection of all patterns, and for any pattern $P \in \mathcal{P}$, let $n_{i,P}$ denote the number of small rolls of size L_i in pattern P.

(a) Letting x_P be a nonnegative integer variable indicating how many rolls we cut into pattern P, show how to formulate the cutting stock problem of finding the fewest number of large rolls to meet the specified demands as an optimization model with a variable for each cutting pattern.

(b) Suppose that we wish to cut a single roll of size L into subrolls and that we receive a profit of π_i for every subroll of size L_i that we use in the solution. Show how to formulate this problem as an integer knapsack problem.

(c) Show how to use the column generation technique described in Section 17.5 to solve the linear programming relaxation of the cutting stock problem in part (a), using knapsack problems of the form stated in part (b) to price out columns.

17.21. Undirected multicommodity flow problem. In an undirected multicommodity flow problem, the bundle constraint of any arc (i, j) is $\sum_{k=1}^{K} | x_{ij}^k | \le u_{ij}$. The flows x_{ij}^k can be either positive, zero, or negative. Assume that all flow costs c_{ij}^k are nonnegative.

(a) Show that using the transformation given in Figure 17.14 we can replace an undirected arc by a set of directed arcs.

(b) Explain how you would transform an undirected multicommodity flow problem into a directed multicommodity flow problem. What are the number of nodes and arcs in the transformed network?

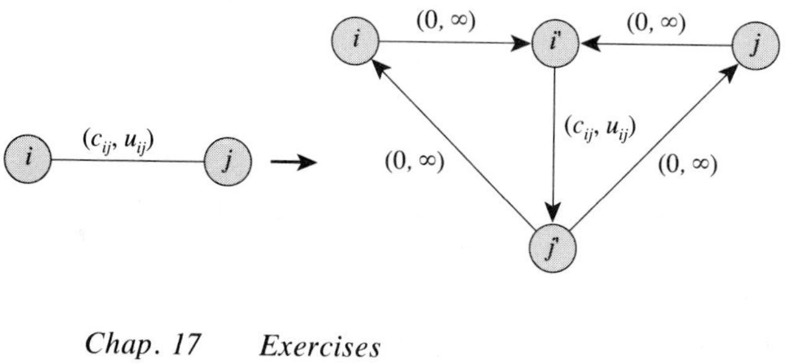

Figure 17.14 Converting an undirected arc with nonnegative cost into directed arcs.

17.22. Two-commodity undirected multicommodity flows (Hu [1963], Sakarovitch [1973]). Let $\mathcal{N}$ be the node–arc incidence matrix for a given undirected network and consider a two-commodity flow problem with the flow vectors x^1 and x^2:

$$\text{Minimize} \quad c^1x^1 + c^2x^2 \tag{17.10a}$$

subject to

$$\mathcal{N}x^1 = b^1, \tag{17.10b}$$

$$\mathcal{N}x^2 = b^2, \tag{17.10c}$$

$$|x^1| + |x^2| \le u. \tag{17.10d}$$

In this formulation $u = (u_{ij})$ is a vector of upper bounds imposed on the total flow on the arcs. Assume that u is an integer vector. This model is a two-commodity version of the undirected multicommodity flow problem that we introduced in Exercise 17.21.

(a) Introducing vectors y^1 and y^2 and making the substitution $x^1 = y^1 + y^2 - u$ and $x^2 = y^1 - y^2$, show that the upper bound constraints (17.10d) are equivalent to the following constraints: $0 \le y^1 \le u$, $0 \le y^2 \le u$. Show that after we have made this substitution, the problem decomposes into two single commodity flow problems, one with flows y^1 and the other with flows y^2. (*Hint*: After substituting the y variables for the x variables, add and subtract (17.10b) and (17.10c).)

(b) Use the transformation introduced in part (a) to show that the two-commodity undirected multicommodity flow problem has an optimal solution in which each arc flow x^1_{ij} is a multiple of $\frac{1}{2}$. In addition, show that if (1) the sum of the capacities of the arcs incident to each node is an even integer, and (2) all supplies and demands are even integers, the problem has an optimal solution with every x^1_{ij} and x^2_{ij} integer.

17.23. Multicommodity maximum flow problem. In the multiple commodity maximum flow problem, each commodity $k = 1, 2, \ldots, K$ has a source node s^k and a sink node t^k, and we wish to send the maximum total flow from the source nodes to the sink nodes while honoring a given bundle capacity u_{ij} on each arc (i, j). That is, we wish to maximize the sum of the flows over all the commodities. We refer to the largest possible sum of flows as the *maximum flow*. We say that a cut $[S, N - S]$ is a *source–sink cut* if the set S contains all of the source nodes s^k and the set $N - S$ contains all of the sink nodes t^k. The capacity of any source–sink cut is the sum of the bundle capacities across the cut. We can define the multiple commodity maximum flow problem for either directed or undirected problems. For undirected problems, we define the bundle constraints as in Exercise 17.21.

(a) Show that for either the directed or undirected versions of the problem, the maximum flow is always less than or equal to the capacity of any source–sink cut.

(b) Using the network shown in Figure 17.15, show that the maximum flow of the

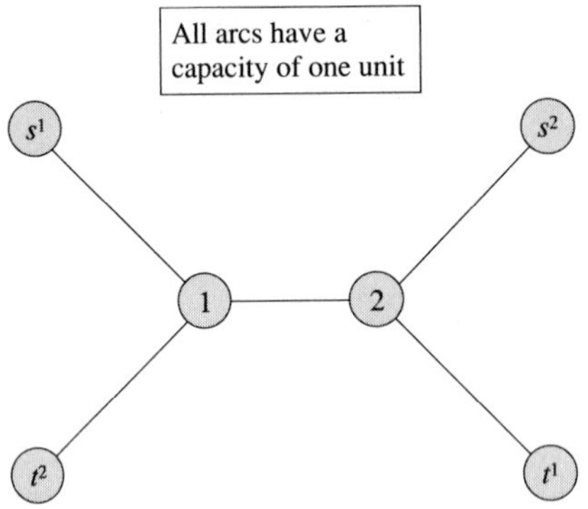

Figure 17.15 Two-commodity undirected maximum flow problem with value of the maximum flow strictly less than the value of the minimum cut.

undirected two-commodity flow problem can be strictly less than the minimum capacity of all source–sink cuts.

(c) Using the network shown in Figure 17.15 and the transformation of the undirected multicommodity flow problem into a directed problem in Exercise 17.21, show that the maximum flow of the directed two-commodity flow problem can be strictly less than the minimum capacity of all source–sink cuts.

17.24. Nonintegrality of solutions to the multicommodity maximum flow problem. Using the network shown in Figure 17.16, show that the maximum flow in an undirected two-commodity maximum flow problem can be noninteger. Give an example of a directed two-commodity flow problem with a noninteger optimal solution.

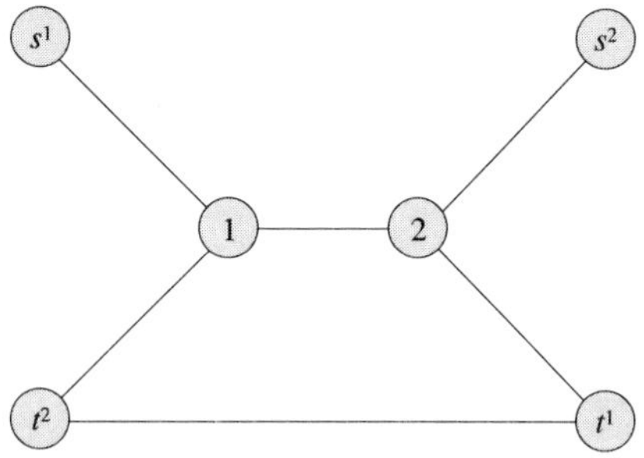

Figure 17.16 Two-commodity undirected maximum flow problem with a fractional solution.

17.25. Show how to formulate the multicommodity maximum flow problem as a version of the multicommodity flow problem (17.1).

17.26. Concurrent flow problem. Consider a multicommodity flow problem in which, for each commodity $k = 1, 2, \ldots, K$, we wish to send D_k units from the commodity's source node s^k to its destination node t^k. Suppose that the problem has no feasible solution; so instead of finding a feasible flow, we wish to find a flow that for some parameter $\theta \le 1$ satisfies all the bundle constraints as well as sends θD_k units from the source to destination of each commodity k. We refer to the maximum value θ^* of the parameter θ for which the problem has a feasible solution as the *feasibility index* or the *optimal throughput* for the problem. We refer to the problem itself as the *concurrent flow problem* since its objective is to satisfy all demands by the same proportional amount θ.

(a) Formulate the concurrent flow problem as a multicommodity flow problem with one additional variable.

(b) *The feasibility index* for each commodity k is the maximum value θ_k^* of a parameter θ_k satisfying the property that the network has a feasible flow satisfying the demand $\theta_k D_k$ for commodity k (with no demand requirement for any other commodity). Show how to find θ_k^* for each commodity k.

(c) Let θ^* be the optimal throughput defined in part (a), let θ_k^* be as defined in part (b), and let $\hat{\theta} = \min_{1 \le k \le K} \theta_k^*$. Show that $\hat{\theta}/K \le \theta^* \le \hat{\theta}$.

(d) Suppose that you had an algorithm for finding a feasible flow, assuming that one exists, for a multicommodity flow problem. Show how to use this algorithm as a subroutine to determine the optimal throughput θ^* to within a factor of ϵ. That is, the objective is to determine a value θ' satisfying the inequalities $(1 - \epsilon)\theta^* \le \theta' \le (1 + \epsilon)\theta^*$. How many times do you need to call the subroutine for determining a feasible multicommodity flow? (*Hint*: Use the results of part (c) to limit the search.)

17.27. Show that for the concurrent flow problem described in Exercise 17.26, the optimal throughput θ^* satisfies the following inequalities: $\theta^* \le \text{CAP}(S, N - S)/\text{DEMAND}(S)$ for every set S of nodes. In this expression, DEMAND(S) denotes the sum of the demands D_k of all the commodities whose origin nodes lie in S, and CAP(S, $N - S$) denotes the total bundle capacity of the arcs in the cut $[S, N - S]$.

17.28. Suppose that we associate a certain positive cost c_{ij} and a capacity u_{ij} with each arc (i, j) of the concurrent flow problem. For a given path P, let $c(P)$ denote the sum of the costs of the arcs in P. Also, let d_k denote the cost of the shortest path from the source node s^k to the sink node t^k of commodity k using the arc costs c_{ij}. Show that the maximum throughput θ^* satisfies the following inequality: $\theta^* \le (\sum_{(i,j)\in A} c_{ij}u_{ij})/(\sum_{1\le k\le K} d_k D_k)$. (*Hint*: Show that $\theta^* \sum_{1\le k\le K} d_k D_k$ is a lower bound on the cost of any feasible flow that satisfies all the demands, and that $\sum_{(i,j)\in A} c_{ij}u_{ij}$ is an upper bound on thé cost of any feasible flow.)

17.29. Dynamic throughput in computer networks. Suppose that in a computer system, for each pair $[i, j]$ of nodes, we would like to send d_{ij} units of information from node i to node j in the least amount of time. Sending any message on arc (i, j) requires t_{ij} units of time and each arc (i, j) has a bundle capacity of u_{ij} at each instant in time (i.e., it can carry at most u_{ij} messages at each instant in time). How would you approximate this dynamic throughput problem as a maximum concurrent flow problem?

17.30. Show that the multicommodity flow problem always has K redundant constraints by viewing it as follows. Each commodity flows on a separate, but identical network and the bundle constraints tie together the flows on the separate network. In this "layered representation" of the problem, let j^k denote the copy of node j on the network (layer) corresponding to commodity k. Next show that it is possible to replace the nodes 1^k for $k = 1$ to K by a single node 1^*, and that the resulting formulation corresponds to a connected network.

17.31. Single bundle constraint as a single-commodity flow problem. Suppose that a multicommodity flow problem has a bundle constraint imposed upon only one arc (i, j). Show that the resulting problem is a minimum cost flow problem. (*Hint*: Use the result of Exercise 17.30.)

17.32. Penalty approach for the multicommodity flow problem (Schneur [1991]). Consider the following penalty approach for solving the multicommodity flow problem:

$$\text{Minimize} \quad F(x) = \sum_{1\le k\le K} c^k x^k + \rho/2 \sum_{(i,j)\in A} (f_{ij})^2 \tag{17.11a}$$

subject to

$$\sum_{1\le k\le K} x_{ij}^k - f_{ij} \le u_{ij} \quad \text{for all arcs } (i, j) \in A, \tag{17.11b}$$

$$\mathcal{N}x^k = b^k \quad \text{for all } k = 1, \ldots, K, \tag{17.11c}$$

$$f_{ij} \ge 0 \quad \text{for all arcs } (i, j) \in A, \tag{17.11d}$$

$$0 \le x_{ij}^k \le u_{ij}^k \quad \text{for all arcs } (i, j) \in A \text{ and all } k = 1, 2, \ldots, K. \tag{17.11e}$$

Note that in any optimal solution, f_{ij} is the excess flow in arc (i, j) [i.e., $f_{ij} = \max(0, \sum_{k=1,\ldots,K} x_{ij}^k - u_{ij})$]. Therefore, this model replaces the hard (i.e., fixed) bundle capacity u_{ij} on any arc (i, j) by a quadratic penalty for exceeding the arc's capacity. In this formulation ρ is a parameter specifying how much we penalize excess flows.

(a) Let z denote the optimal objective function value for the linear multicommodity flow problem and let $G(\rho)$ denote the optimal objective function value of the penalty problem for a fixed value of ρ. Show that $G(\rho) \le z$ for all ρ. Show also that $\lim_{\rho\to\infty} G(\rho) = z$.

(b) Let y denote 1 unit flow of commodity k around the cycle W (i.e., a flow of +1 in each forward arc of W for commodity k and a flow of −1 in each backward arc of W for commodity k). We refer to the cycle W as a *negative cycle* with respect to a flow x' if $F(x' + \theta y) - F(x')$ is negative for some scalar $\theta > 0$. Show that a feasible flow x' of (17.11) is an optimal flow if and only if it contains no negative cost cycle. (*Hint*: The proof is similar to that of Theorem 3.8.)

17.33. Penalty method for the multicommodity flow problem. In this exercise we discuss an algorithm for the multicommodity flow problem that uses the penalty approach described in Exercise 17.32. Let W be a cycle in the network and $c(W)$ be its cost (i.e., the sum of the costs of the forward arcs in W minus the costs of the backward arcs in W). Moreover, let $f(W)$ denote the excess flow of W (i.e., the sum of the excess flows of the forward arcs minus the sum of the excess flows of the backward arcs). Finally, let y denote a unit flow of the commodity k around the cycle W (as defined in Exercise 17.32(b).

(a) Express $F(x' + \theta y) - F(x')$ in terms of θ, $|W|$, $c(W)$, and $f(W)$, and observe that $F(x' + \theta y) - F(x')$ is a convex function of θ. Use this observation to suggest a polynomial-time algorithm for finding a negative cycle in the network.

(b) Use your answer in part (a) to describe an algorithm for solving the penalty version of the multicommodity flow problem.

17.34. Suppose that we apply Lagrangian relaxation to the resource-allocation problem (17.7) by associating Lagrange multipliers w_{ij} with the resource-allocation constraints $\sum_{1 \le k \le K} r_{ij}^k \le u_{ij}$, to produce the following Lagrangian subproblem:

$$z = \min \sum_{1 \le k \le K} c^k x^k + \sum_{(i,j) \in A} w_{ij} \Big(\sum_{1 \le k \le K} r_{ij}^k - u_{ij} \Big)$$

subject to

$$\mathcal{N} x^k = b^k \qquad \text{for } k = 1, \ldots, K,$$

$$0 \le x_{ij}^k \le r_{ij}^k \qquad \text{for all } (i, j) \in A \text{ and all } k = 1, 2, \ldots, K.$$

Show that this problem is equivalent to the Lagrangian subproblem determined by applying Lagrangian relaxation to the original formulation (17.1) obtained by relaxing the bundle constraints $\sum_{1 \le k \le K} x_{ij}^k \le u_{ij}$.

17.35. Consider a linear program of the form

$$\text{Minimize} \sum_{1 \le k \le K} c^k x^k \tag{17.12a}$$

subject to

$$\sum_{1 \le k \le K} \mathcal{A}^k x^k \le b, \tag{17.12b}$$

$$\mathcal{D}^k x^k = d^k \qquad \text{for all } k = 1, 2, \ldots, K, \tag{17.12c}$$

$$x^k \ge 0 \qquad \text{for all } k = 1, 2, \ldots, K. \tag{17.12d}$$

In this formulation, each x^k is a vector of decision variables. The constraints (17.12c) are separate constraints imposed on each vector x^k, and the constraint (17.12b) models the limited availability of joint resources shared by these decision vectors. Note that each $\mathcal{D}^k$ and $\mathcal{A}^k$ is a matrix.

(a) Suppose that we introduce new "resource-allocation" variables r^k and replace the constraint (17.12b) by the constraints $\sum_{1 \le k \le K} r^k \le b$ and $\mathcal{A}^k x^k \le r^k$ for each $k = 1, 2, \ldots, K$. Show that the formulation in the variables r^k is equivalent to the given formulation.

(b) Using the model with the resource-allocation variables r^k, show how to adapt the resource-directive decomposition technique described in Section 17.7 to solve (17.12).

17.36. Show how to apply the basis partitioning algorithm to solve the optimization model specified in the last exercise.

(a) First show that any basis $\mathcal{B}$ for the problem must contain a basis $\mathcal{B}^k$ for each matrix $\mathcal{D}^k$.

(b) Next show how to make a change in variables so that the only variables appearing in the basis in the equations $\mathcal{D}^k x^k = d^k$ are those in the basis $\mathcal{B}^k$. (*Hint*: If γ is any column of $\mathcal{D}^k$, it is always possible to solve the equation $\mathcal{B}^k y = \gamma$).

(c) Define the working basis $\mathcal{W}$ as those columns of the constraints $\sum_{1 \le k \le K} \mathcal{A}^k x^k \le b$ that are in the basis $\mathcal{B}$ but not in any $\mathcal{B}^k$. Show that the columns of $\mathcal{W}$ are linearly independent.

(d) Show how to use the working basis $\mathcal{W}$ and the basis matrices $\mathcal{B}^1, \mathcal{B}^2, \ldots, \mathcal{B}^K$ to implement the simplex method for solving the problem.

17.37. Show how to apply Dantzig–Wolfe decomposition to the optimization model (17.12) with the constraints (17.12c) and (17.12d) defining the subproblems.

17.38. Consider the optimization model

$$\text{Minimize} \quad cx$$

subject to

$$Ax = b,$$

$$x \in X,$$

defined over a finite set X. Show how to apply column generation and Dantzig–Wolfe decomposition to solve the following convexified version of this problem:

$$\text{Minimize} \quad cx$$

subject to

$$Ax = b \tag{17.13}$$

$$x \in \mathcal{H}(x)$$

defined over the convex hull $\mathcal{H}(x)$ of x. Recall that any point $x \in \mathcal{H}(x)$ has the representation $x = \sum_{k=1}^{K} \lambda^k x^k$ for some set of nonnegative weights satisfying the weighting condition $\sum_{k=1}^{K} \lambda^k = 1$. (*Hint*: Make the replacement $x = \sum_{k=1}^{K} \lambda^k x^k$ to formulate problem (17.13) in terms the weighting variables $\lambda^1, \lambda^2, \ldots, \lambda^K$ and form a restricted version of the reformulated model by eliminating all but a small number of the λ^k's.)

18

COMPUTATIONAL TESTING OF ALGORITHMS

The purpose of mathematical programming is insight, not numbers.
—A. M Geoffrion

Chapter Outline

18.1 INTRODUCTION

In this book we have focused on developing the most "efficient" algorithms for solving network optimization problems. The notion of efficiency involves all the various computing resources needed for executing an algorithm. However, since time is often a dominant computing resource in practice, we have used computational time as the primary measure for assessing algorithmic efficiency. We have measured the computational time of an algorithm through its worst-case analysis. Worst-case analysis provides upper bounds on the number of steps that a given algorithm can take on *any* problem instance. As we have noted in Chapter 3, for a variety of reasons, the worst-case analysis of algorithms is a very popular criterion for judging algorithmic efficiency, and this approach has stimulated considerable research. However, worst-case analysis can be overly pessimistic since it permits "pathological" instances to determine the performance of an algorithm, even though they might be exceedingly rare in practice. Often, the empirical behavior of an algorithm is much better than suggested by its worst-case analysis. Consequently, the research community typically relies on the empirical testing of an algorithm to assess its performance in practice.

In the operations research literature, researchers have conducted a large number of empirical investigations of various network flow algorithms to determine the "best" algorithms in practice. A typical study tests more than one algorithm for a specific network flow problem and generally consists of the following steps: (1) write a computer program (often in FORTRAN) for each algorithm to be tested; (2) use pseudorandom network generators to generate random problem instances with selected combinations of input size parameters (e.g., nodes and arcs); (3) run computer programs and note the CPU (central processing unit) times for the different algorithms on the data obtained by the network generators; and (4) declare the algorithm that takes the least amount of CPU time as the "winner" (if different algorithms are faster for different input size parameters, then report this fact as well).

The existing literature on computational testing has a tendency to overrely on CPU time as the primary measure of performance. CPU time depends greatly on subtle details of the computational environment and the test problems such as (1) the chosen programming language, compiler, and computer; (2) the implementation style and skill of the programmer; (3) network generators used to generate the random test problems; (4) combinations of input size parameters; and (5) the particular programming environment (e.g., the use of the computer system by other users). Because of the multiple sources of variabilities, CPU times are often difficult to replicate, which is contrary to the spirit of scientific investigation. Another drawback of the use of CPU time is that it is an aggregate measure of empirical performance and does not provide much insight about an algorithm's behavior. For example, an algorithm generally performs some fundamental operations repeatedly, and a typical CPU time analysis does not help us to identify these "bottleneck" operations. Identifying the bottleneck operations of an algorithm can provide useful guidelines for where to direct future efforts to understand and subsequently improve an algorithm.

The spirit of worst-case analysis is to identify theoretical bottlenecks in the performance of any algorithm and to provide upper bounds on the computation counts of these bottleneck operations as a measure of the algorithm's overall behavior. Borrowing this point of view for computational testing, we might attempt to measure the empirical performance of an algorithm (or its computer implementation) by counting the number of times the algorithm executes each of these bottleneck operations while solving each instance of the problem. That is, we would conduct computer experiments to obtain an actual count of bottleneck operations instead of providing a theoretical upper bound on this number. This approach suggests that in analyzing the empirical behavior of an algorithm, we need not count the number of times it executes each line (of possibly thousands of lines) of code, but instead can focus on a relatively small number of lines that are "summary measures" of the algorithm's empirical behavior. Even for the most complex algorithms described in this book, we need to keep track of the computation counts of at most three or four operations or lines of code.

For example, in the FIFO preflow-push algorithm for the maximum flow problem that we presented in Section 7.7, each push operation first selects an active node i, next selects an admissible arc (i, j), and then pushes $\min\{e(i), r_{ij}\}$ units of flow on the arc. If no admissible arc emanates from node i, we scan all of the arcs emanating from this node, and we relabel the distance label of node i, giving it the value $d(i) = 1 + \min\{d(j) : r_{ij} > 0\}$. We claim that for the generic preflow-push algorithm, we need to keep track of only two operations: (1) the number of pushes, and (2) the number of arcs scanned in the relabel operations.

These two operations dominate every other operation of the generic preflow-push algorithm. To establish this statement, note that to select a node requires $O(1)$ operations per push, so the algorithm spends O(number of pushes) operations in selecting nodes. In selecting admissible arcs, we check if the current arc is admissible, and if not, we modify the current arc. The algorithm modifies the current arc for node i $O(|A(i)|)$ times, which is the number of arcs scanned in relabeling node i between successive relabels of node i. Thus the algorithm modifies the current arcs O(number of arcs scanned in relabels) times. We leave it to the reader to verify that these two operations also bound each of the other operations performed in the FIFO preflow-push algorithm. To summarize, even though an implementation of the FIFO preflow-push algorithm might contain hundreds of lines of code, we need to keep track of only two fundamental operations in order to identify the bottleneck operations, as well as to estimate the running time to within a constant factor.

We will soon formalize this notion of "representative operation counts" in computational testing; however, let us first summarize some of the advantages of this approach as compared to the more common approach of analyzing only CPU time.

1. Representative operation counts allow us to identify the asymptotic bottleneck operations of an algorithm—i.e., the operations that progressively consume a larger share of the computational time as the problem size increases. (For sufficiently large problem sizes, improving the asymptotic bottleneck operation has the maximum possible impact on the running time of the algorithm.)
2. Representative operation counts provide more guidance and insight about comparing two algorithms that are run on different computers and permits us even to compare algorithms implemented with different computer languages.
3. Representative operation counts permit us to determine lower and upper bounds on the asymptotic growth rate in computation time as a function of the problem size.
4. We can use statistical methodologies to estimate the CPU time on a computer as a linear function of the representative operation counts. We refer to this estimate of the CPU times as the *virtual CPU time*. Virtual CPU time permits researchers to carry out experiments on different computers, but estimate the running times as if all the experiments had been carried out on the same computer.

This type of asymptotic empirical analysis complements the worst-case analysis that we have examined in many other chapters. The empirical analysis using representative operation counts allows us to identify the actual empirical behavior of the algorithm for sufficiently large problem sizes. Just as the worst-case analysis ignores constant factors in the running time, empirical analysis using representative operation counts ignores constant factors in the running time, and instead focuses on the dominant term in the computations.

Before continuing we might note that the field of computational testing is very broad, and we cannot do justice to it in a short chapter. Rather than treat the wide range of topics of importance in computational testing (such as how to select test problems, how to conduct an experimental design, what type of statistical tests are most appropriate, and what data needs to be reported), we focus on the use of representative operation counts as an aid in the analysis of computational experiments. We believe that it is quite easy to count representative operations while conducting any computational testing and that generating this added information has considerable potential payback.

18.2 REPRESENTATIVE OPERATION COUNTS

In order to formalize our notion of counting operations performed by an algorithm, let us first stand back and consider what a computer does in executing a computer program. Suppose that $\mathcal{A}$ is a computer program for solving some problem and that I is an instance of the problem. The computer program consists of a finite number of lines of computer code, say $a_1, a_2, \ldots, a_K$. Each line of code gives either one or a small number of instructions to the computer. The instruction might tell the computer to carry out an arithmetic operation on a register or to move data from one memory location to another; or it might be a control instruction informing the computer which line of code it should execute next. For convenience, we assume that the program is written so that each line of the code gives $O(1)$ instructions to the computer and that each instruction requires $O(1)$ units of time. We assume that the fastest computer operation requires 1 time unit. (Although these assumptions are reasonable, they do imply some restrictions; for example, we do not permit lines of code that tell the computer to add two vectors, as would be allowed in some high-level languages such as APL. Rather, we would require that adding vectors be carried out as a loop that sums the two vectors one component at a time.) The assumption that each operation executed by the computer requires a comparable amount of time seems reasonable in practice with the notable exceptions of input–output, caching (moving data to and from storage), and paging (memory management of the secondary storage space).

The preceding discussion implies that executing any line of code requires $O(1)$ time units, and at least 1 time unit. Therefore, each line of code requires $\Theta(1)$ time units since its execution time is bounded from both above and below by a constant number of units. Suppose that the computer code we are investigating has K lines of code. For a given instance I of the problem, let $\alpha_k(I)$, for $k = 1$ to K, be the number of times that the computer executes line k of this computer program. Let CPU(I) denote the CPU time of the computer program on instance I. The preceding discussion implies the following lemma.

Lemma 18.1. $CPU(I) = \Theta(\sum_{k=1}^{K} \alpha_k(I))$. ◆

Lemma 18.1 states that we can estimate the running time of an algorithm to within a constant factor by counting the number of times it executes each line of code. However, counting each line of code is unnecessarily burdensome. As we shall soon see, it really suffices to count a relatively small number of lines of code. For example, consider the following fragment of code.

```
for i: = 1 to 3 do
begin
  A(i): = A(i) + 1;
  B(i): = B(i) + 2;
  C(i): = C(i) + 3;
end;
```

We need not count the number of times the algorithm executes the statement "$B(i) := B(i) + 2$," since it executes this statement whenever it executes the statement "$A(i) := A(i) + 1$." Similarly, we need not count the number of times the algorithm executes the statement "$C(i) := C(i) + 3$." Moreover, it appears that regardless of the size of the problem we are solving, the algorithm modifies nine elements of the vectors A, B, and C during the execution of the "for loop." So in this case it would suffice to keep track of just the number of times that the algorithm executes the "for loop."

Note that we have treated the number of iterations of the do loop as a constant. Suppose, instead, that the first line of the do loop were "**for** $i := 1$ **to** 10 **do**." Should we also treat the 10 as a constant? What if the first line of the do loop were "**for** $i := 1$ **to** 10,000 **do**." Should we continue to treat the 10,000 as a constant? In answering these questions, we might invoke two rules of thumb. First, we should determine the representative operation counts after expressing the algorithm as a pseudocode with all the problem parameters treated explicitly as parameters and not as constants. For example, although in many practical situations $\log U$ will be less than 16, in a pseudocode we should use the term "$\log U$" rather than the constant 16. We would not treat $\log U$ as a constant. Second, as a rule, we should not treat the number of iterations of a "do loop" or a "while loop" or any other loop as a constant since frequently the number of times that the program will call the loop depends on a problem parameter rather than a constant.

We now formalize the notion we have been suggesting; that is, keeping track of a small number of lines of code, which we call the *representative operation counts*. Let S denote a subset of $\{1, \ldots, K\}$, and let a_S denote the set $\{a_i : i \in S\}$. We say that a_S is a representative set of lines of code of a program if for some constant c,

$$\alpha_i(I) \leq c\,(\sum_{k \in S} \alpha_k(I)),$$

for every instance I of the problem and for every line a_i of code. In other words, the time spent in executing line a_i is dominated (up to a constant) by the time spent in executing the lines of code in a_S. With this definition, we have the following corollary to Lemma 18.1.

Property 18.2. *Let S be a representative set of lines of code. Then $CPU(I) = \Theta(\sum_{k \in S} \alpha_k(I))$.*

Proof. By Lemma 18.1, $CPU(I) = \Theta(\sum_{k=1}^{K} \alpha_k(I))$. Moreover, for each line α_i of code not in S, $\alpha_i(I) \leq c(\sum_{k \in S} \alpha_k(I))$, so $\sum_{k=1}^{K} \alpha_k(I) = \Theta(\sum_{k \in S} \alpha_k(I))$. ◆

This methodology identifies a set of representative lines of code, and during empirical investigations keeps track of the representative operation counts for each instance solved. Sometimes, the selected representatives will not refer specifically to one line of code. For example, we might keep track of the number of pushes in the preflow-push algorithms, and each push might be described over several lines of code. Rather than use the expression "the computation counts for a set of representative lines of code" we will refer to "the computation counts for a set of representative operations" or more briefly as *representative operation counts*.

In the next section we discuss various uses of representative operation counts. In addition to noting representative operation counts for various problem instances we have solved, we might record some other counts that might be helpful in assessing an algorithm's behavior. For example, for each instance solved, we might note the number of major iterations that the algorithm performs, such as the number of pivots of the network simplex algorithm.

At first glance we might suspect that determining a representative set of operations could be difficult and that the set might be quite large. In fact, our experience suggests that it is generally quite easy to determine representative sets, and often we have several possible choices. In addition, the representative sets are typically quite small. For the algorithms presented in this book, the number of representative operation counts typically range from 1 to 4, even for quite complex algorithms.

We now give examples of representative sets for several network flow algorithms we have discussed in this book. The representative sets we indicate are not unique. In some cases we suggest additional operation counts that might be of value in empirical investigations.

Dial's implementation of Dijkstra's algorithm (see Section 4.6). A set of representative operations for this algorithm are (1) the number of buckets scanned while identifying the first nonempty bucket (as part of a findmin operation), and (2) the number of arcs scanned to update distance labels. Since we know that the implementation scans $\Theta(m)$ arcs, we need not actually count the number of arc scans. Thus we need to keep track of only one operation in our representative set. In addition to these representative operation counts, we might keep track of other operations. For example, we might record the number of decreases in the distance labels during the distance update operations since this set of operation counts would also allow us to bound the running time of the binary heap implementation of Dijkstra's algorithm.

Original implementation of Dijkstra's algorithm (see Section 4.5). The running time of this algorithm is $\Theta(n^2)$ since the number of nodes scanned in the findmin operation is $(n - 1) + (n - 2) + \cdots + 2 + 1 = n(n - 1)/2 = \Omega(n^2)$ time. Therefore, we need not keep track of any counts in executing the algorithm.

FIFO label-correcting algorithm for the shortest path problem (see Section 5.4). The representative operation for this algorithm is the arcs scanned while examining nodes in the set LIST.

Labeling algorithm for the maximum flow problem (see Section 6.5). A representative operation for this algorithm is the arcs scanned while examining labeled nodes. This operation dominates all other operations that the labeling algorithm performs. In addition to this representative operation, we would probably want to count the number of augmentations. This additional information might provide insight for comparing the labeling algorithm to other maximum flow algorithms. For example, we could check whether the labeling algorithm requires more augmentations than other augmenting path algorithms.

Preflow-push algorithm for the maximum flow problem (see Section 7.7). A set of representative operations for this algorithm are (1) the number of nonsaturating pushes, and (2) the arcs scanned while updating distance labels of nodes. (Operation 2 dominates the number of operations the algorithm performs in updating the current arc and in making saturating pushes.) We do not need to keep track of saturating pushes because the time spent in saturating pushes is bounded by the time spent in scanning arcs to determine a current arc, and this time, in turn, is bounded by the time spent updating the distance labels of nodes. Even though we know that saturating pushes will not be a bottleneck operation, we might want to keep track of them for a different reason. In practice, it appears that nonsaturating pushes are not a bottleneck operation, even though they are the theoretical bottleneck. An easy way to show that nonsaturating pushes are not a bottleneck operation is to show, for example, that the number of nonsaturating pushes are, in practice, at most a small constant times the number of saturating pushes. Keeping track of saturating pushes would permit us to make this assessment empirically. We might also want to record the number of saturating pushes because they might be a bottleneck operation if we use the dynamic trees data structure, and by keeping track of the two operations (1) and (2) we can estimate whether the use of dynamic trees might lead to an asymptotic improvement in the running time.

Successive shortest path algorithm for the minimum cost flow problem (see Section 9.7). The successive shortest path algorithm determines shortest paths from excess nodes to deficit nodes. Suppose that we use Dial's implementation to solve shortest path problems. Then the representative operations for this algorithm are: (1) the buckets scanned while the algorithm identifies nonempty buckets in Dial's algorithm, and (2) the number of arcs that the algorithm scans while it updates distance labels. In addition, we would want to keep track of the number of shortest path problems that the algorithm solves.

Network simplex algorithm (see Section 11.5). The network simplex algorithm has the following set of representative operations: (1) the number of arcs whose reduced cost the algorithm calculates while it is identifying the entering arc, (2) the number of arcs in the pivot cycles that the algorithm creates when it adds the entering arc to the current spanning tree, and (3) the number of nodes in the subtree T_2 (i.e., nodes whose potentials the algorithm changes during a pivot operation). The network simplex algorithm performs many other operations. We do not need to keep track of these operations because the three operation counts we have identified dominate them. For example, to update flows in a pivot cycle W requires $\Theta(|W|)$ time. (Even for degenerate pivots, the time spent by most algorithms is at least $|W|$, since most algorithms identify the pivot cycle before determining the flow to send around the cycle.) To update the multipliers in a subtree T_2 of nodes whose potential changes require $\Theta(|T_2|)$ time. Similarly, to update the tree indices requires $\Theta(|W| + |T_2|)$ time. This set of representative operations is strikingly compact considering the fact that implementations of the network simplex algorithm are typically quite intricate.

In addition to these representative operations, we might keep track of other operations. For example, we would most likely want to count the number of pivots. Moreover, if we were concerned about the effects of degeneracy, we would also record the number of degenerate pivots that the algorithm performs.

18.3 APPLICATION TO NETWORK SIMPLEX ALGORITHM

In this section we illustrate the use of representative operation counts using the network simplex algorithm for the minimum cost flow problem. We provide experiments based on the network simplex algorithm, as implemented with the first eligible arc pivot rule. As described in Section 11.5, this pivot rule selects the first arc with positive violation as the entering arc while scanning the arcs in the wraparound fashion.

The computational time that the network simplex algorithm requires to solve a minimum cost flow problem depends on a number of different parameters, including the number of nodes, the number of arcs, the network generator, the size of the cost and capacity data, and the number of supply nodes. To illustrate the approach discussed in this chapter, of these factors, we have chosen to focus on the number of nodes and the number of arcs. We conducted experiments on networks with 1000, 2000, 4000, 6000, and 8000 nodes. For each choice of n nodes, we created networks with $2n$, $4n$, $6n$, $8n$, and $10n$ arcs. Setting $d = m/n$, we have considered networks with $d = 2, 4, 6, 8$, or 10. Thus the largest network size we considered had 8000 nodes and 80,000 arcs.

We used the well-known network generator NETGEN to generate minimum cost flow problems with specified values of n and m. For each specific setting of n and m, we solved five different problems generated from NETGEN and computed the average of these five problem instances. (The averages have a lower variability of outcomes than the individual tests, so the resulting graphs and charts reveal patterns more clearly.) For each problem that we solved, we noted the following values:

α_E: number of arcs scanned in selecting an entering arc (summed over all pivots)

α_W: number of arcs in the pivot cycles (summed over all pivots)

α_P: number of node potentials modified (summed over all pivots)

p: number of pivots, further decomposed into the number of degenerate pivots and the number of nondegenerate pivots

τ: CPU time to execute the algorithm (times noted on a HP9000/850 computer under a multiprogramming and multisharing environment).

We computed the averages of these values over five problem instances for each parameter setting. Figure 18.1 gives these averages for the network simplex algorithm with the first eligible arc pivot rule.

Identifying Asymptotic Bottleneck Operations

We consider an operation to be a "bottleneck operation" for an algorithm if the operation consumes a significant percentage of the execution time on at least some fraction of the problems tested. We refer to an operation as an *asymptotic nonbottleneck operation* if its share in the computational time becomes smaller and approaches zero as the problem size increases. Otherwise, we refer to an operation as an *asymptotic bottleneck operation*. The asymptotic bottleneck operations are

No.	n	m	d	α_E	α_W	α_P	p	$\alpha_E + \alpha_W + \alpha_p$	CPU time
1	1,000	2,000	2	41,590	183,540	237,599	4,721	462,729	4.46
2	2,000	4,000	2	129,125	831,593	1,356,773	15,358	2,317,491	21.62
3	4,000	8,000	2	406,115	3,989,374	8,319,325	51,936	12,714,814	119.7
4	6,000	12,000	2	863,573	10,770,317	25,045,320	116,303	36,679,210	345.64
5	8,000	16,000	2	1,453,942	21,438,864	54,392,393	197,005	77,285,199	699.88
6	1,000	4,000	4	87,883	284,830	403,699	10,275	776,412	7.31
7	2,000	8,000	4	248,760	1,101,807	2,090,021	29,879	3,440,588	31.26
8	4,000	16,000	4	765,758	5,171,901	12,341,060	99,858	18,278,719	160.7
9	6,000	24,000	4	1,475,919	12,277,824	33,993,925	197,858	47,747,668	420.96
10	8,000	32,000	4	2,419,120	24,614,630	77,921,501	335,296	104,955,251	853.21
11	1,000	6,000	6	141,475	351,763	518,265	14,372	1,011,503	9.33
12	2,000	12,000	6	379,171	1,253,222	2,353,123	38,837	3,985,516	35.42
13	4,000	24,000	6	1,152,766	5,655,431	15,215,392	124,249	22,023,589	188.55
14	6,000	36,000	6	2,329,115	14,249,457	42,841,133	259,742	59,419,705	505.79
15	8,000	48,000	6	3,639,711	26,157,289	90,280,080	416,123	120,077,080	978.57
16	1,000	8,000	8	207,114	380,015	598,636	16,586	1,185,765	10.7
17	2,000	16,000	8	529,747	1,452,365	3,012,190	47,116	4,994,302	42.61
18	4,000	32,000	8	1,534,355	5,919,861	13,833,227	139,434	21,287,443	200.67
19	6,000	48,000	8	2,918,853	13,917,221	48,131,763	273,089	64,967,837	518.08
20	8,000	64,000	8	4,532,492	25,791,836	99,057,203	437,558	129,381,531	1,033.05
21	1,000	10,000	10	266,541	410,304	739,070	18,634	1,415,915	12.39
22	2,000	20,000	10	710,258	1,555,660	3,305,765	53,425	5,571,683	48.08
23	4,000	40,000	10	1,975,386	5,844,629	17,922,439	145,303	25,742,454	208.30
24	6,000	60,000	10	3,902,416	15,193,494	48,615,853	312,355	67,711,763	584.34
25	8,000	80,000	10	5,942,548	26,724,636	103,238,910	487,624	135,906,094	1,064.45

Figure 18.1 Table of computation counts for the network simplex algorithm with the first entering pivot rule.

important because they determine the running times of the algorithm for sufficiently large problem sizes.

We have earlier shown that the representative operation counts provide both upper and lower bounds on the number of operations an algorithm performs; therefore, the set of representative operations must contain at least one asymptotic bottleneck operation. There is no formal method for determining asymptotic bottleneck operations using computational testing unless we are willing to impose further assumptions on the behavior of an algorithm on large problems. After all, it is theoretically possible that an algorithm behaves one way for problems of sufficiently large size and it behaves totally differently for small problems. Nevertheless, some procedures seem quite effective in practice for determining asymptotic bottlenecks, even if we cannot completely justify their use.

We illustrate how to find an asymptotic bottleneck operation for the network simplex algorithm. Let $\alpha_S(I) = \alpha_E(I) + \alpha_W(I) + \alpha_P(I)$. Then we simply plot

$\alpha_E(I)/\alpha_S(I)$, $\alpha_W(I)/\alpha_S(I)$, $\alpha_P(I)/\alpha_S(I)$ for increasingly larger problem instances I and look for a trend. Figure 18.2 gives these plots for the network simplex algorithm with the first eligible arc pivot rule. In these plots we use the number of nodes as a surrogate for the problem size and provide a plot for each different density $d = m/n$. This choice helps us to visualize the effect of n and m on the growth in representative operation counts. These plots suggest that updating node potentials is an asymptotic bottleneck operation in the algorithm.

Estimating Growth Rates of Bottleneck Operations

How should we estimate the growth in the computational time (i.e., CPU time) as the problem size increases? Instead of directly estimating the growth in the computational time, we estimate the growth in the asymptotic bottleneck operations. By focusing only on the asymptotic bottleneck operations, we eliminate the contribution of the nonbottleneck operations, and therefore the estimates should be superior. In the notation of the preceding section, the asymptotic running time is proportional to $\lim_{|I|\to\infty} \max(\alpha_1(I), \alpha_2(I), \ldots, \alpha_K(I))$. In the manner that we have selected the representative operations, the asymptotic running time is also proportional to $\lim_{|I|\to\infty} \max(\alpha_i(I) : i \in S)$. We need not consider the nonbottleneck operations in estimating the asymptotic running time. The CPU time is a linear function of the terms $\alpha_i(I)$ for $i = 1$ to K, and thus includes information from many of the nonbottleneck operations. For this reason, the CPU time is a much "noisier" estimator of the asymptotic running time.

We describe a simple approach for estimating the growth rates of a bottleneck operation, which in our illustration is α_p, the number of potential updates. Our approach consists of the following steps:

1. Determine an appropriate functional form for estimating the counts for the operation α_p. In our example, we choose the functional form to be n^γ for some choice of a growth parameter γ. (A more common approach would be to choose a function of both n and m; however, we will consider each network density separately, and for each network density $d = m/n$, the ratio of m to n is fixed. Therefore, a function of m and n reduces to a function of n.)
2. Select a candidate lower bound n^l and a candidate upper bound n^u on the growth rate using one of the several possible methods.
3. Evaluate (using some methodology) whether n^l and n^u are really the lower and upper bounds on α_p. If not, return to step 2 and repeat the process.

For our illustration, suppose that we wished to consider a candidate lower bound of $n^{2.2}$ on the growth rate of the number of potential updates α_p and a candidate upper bound of $n^{2.8}$ (we might simply have guessed to determine these bounds). Figure 18.3 gives a plot of $\alpha_p/n^{2.2}$ and $\alpha_p/n^{2.8}$. In Figure 18.3(a) we find that the function $\alpha_p/n^{2.2}$ has an increasing trend with the problem size, suggesting that $n^{2.2}$ is indeed a lower bound on the number of potential updates. Further, we find in Figure 18.3(b) that the function $\alpha_p/n^{2.8}$ has a decreasing trend, suggesting that $n^{2.8}$ is an upper bound on the number of potential updates. If we want more refined lower and upper bounds, we carry out the technique further to see if $n^{2.4}$ is a valid lower

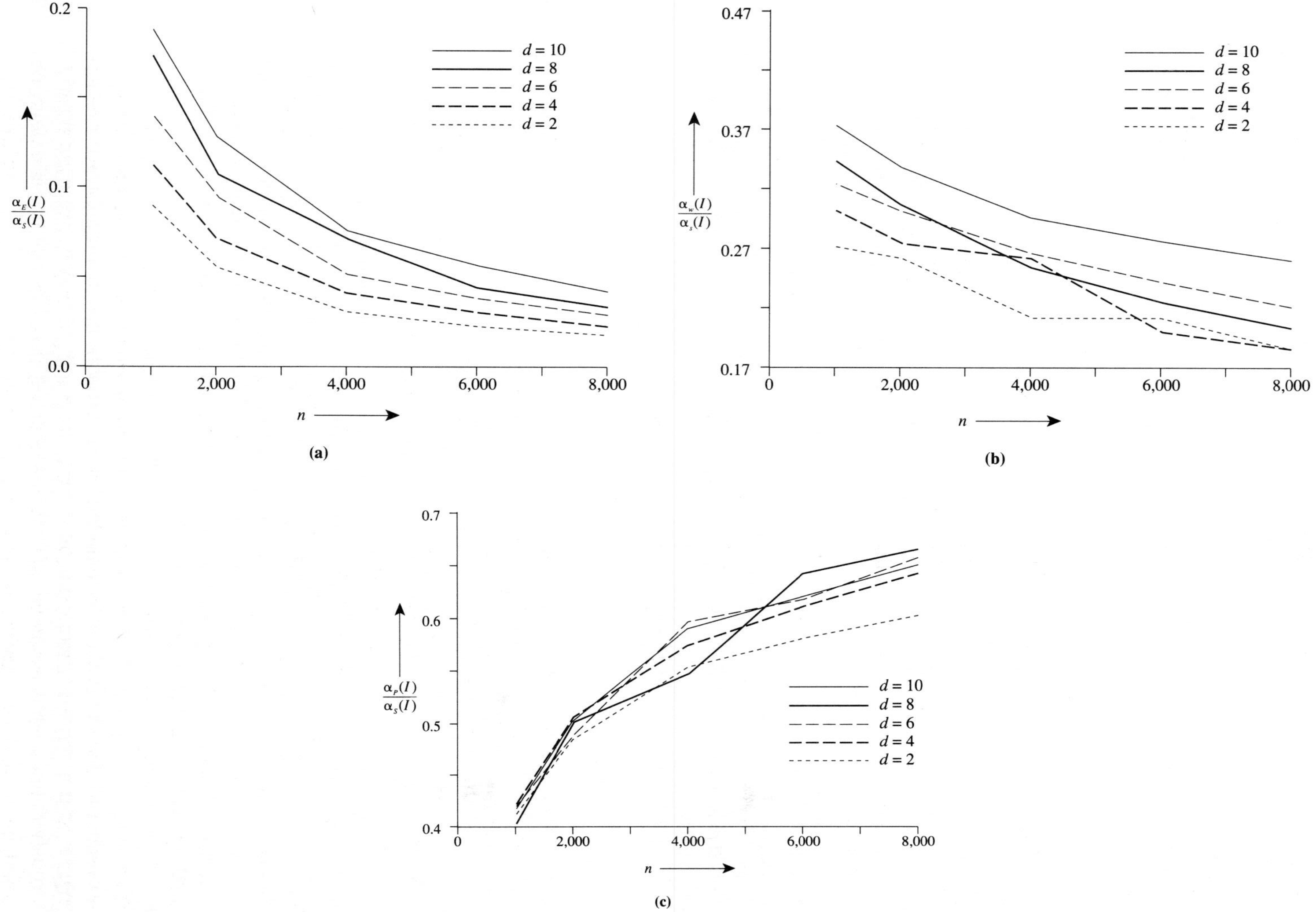

Figure 18.2 Identifying asymptotic bottleneck operations: (a) ratio of arc scan operations and total number of operations; (b) ratio of flow update operations and total number of operations; (c) ratio of potential update operations and the total number of operations.

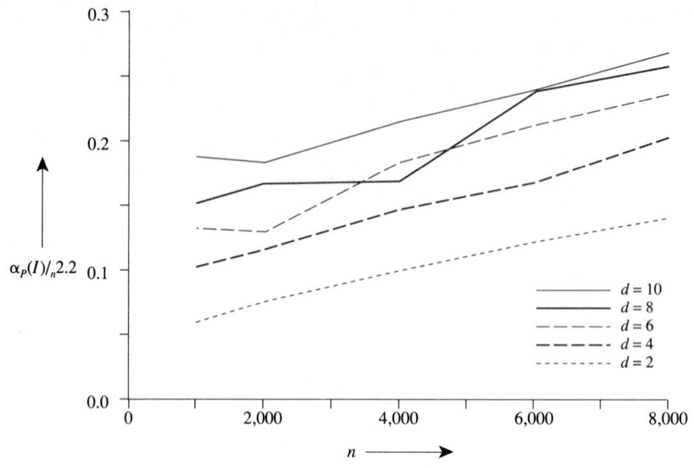

(a)

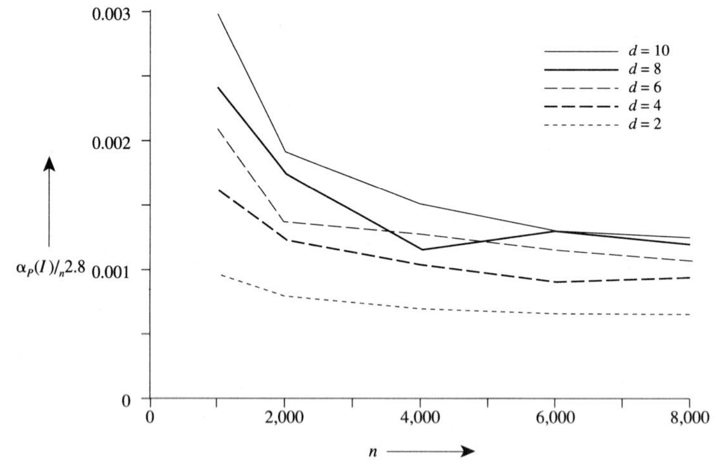

(b)

Figure 18.3 Determining asymptotic growth rates of the bottleneck step: (a) plot of $\alpha_P(I)/n^{2.2}$ has an increasing trend; (b) plot of $\alpha_P(I)/n^{2.8}$ has a decreasing trend.

bound or if $n^{2.6}$ is a valid upper bound. The polynomial estimates for the lower and upper bounds are significantly different because our method for estimating lower and upper bounds typically is quite conservative and underestimates the lower bound and overestimates the upper bound.

We might emphasize that our approach is intended primarily for gathering insight into the upper and lower bounds on function growth and is not rigorous. For example, to determine the upper bound on the growth, the methodology must recognize that a certain function has an increasing trend. However, we have not rigorously defined what we mean by "increasing trend"; it depends, to an extent, on a judgment call by the user.

In addition to its lack of rigor, the methodology, as we have used it, is perhaps too conservative, and can be strengthened through careful statistical analysis. To illustrate, in the previous example, we estimated the growth in the number of potential updates as being between $n^{2.2}$ and $n^{2.8}$. The gap between these bounds is very large, and careful statistical analysis should be able to narrow the gap considerably.

Comparing Two Algorithms

Suppose that we want to compare two different algorithms $\mathcal{AL}_1$ and $\mathcal{AL}_2$ for solving the same problem and are interested in knowing which algorithm performs better asymptotically. We can apply the same methodology for assessing asymptotic bottleneck operations to compare different algorithms. Let $\alpha_S^1(k)$ and $\alpha_S^2(k)$ be the total expected number of representative operations performed by the algorithms $\mathcal{AL}_1$ and $\mathcal{AL}_2$ on instances of size k. We say that algorithm $\mathcal{AL}_1$ is asymptotically superior to algorithm $\mathcal{AL}_2$ if

$$\lim_{k \to \infty} \frac{\alpha_S^1(k)}{\alpha_S^2(k)} = 0.$$

Virtual Running Times

In Section 18.1 we have already mentioned some difficulties that arise when we use CPU times as an empirical measure for computational investigations. We might note yet one more disadvantage: CPU times force unnecessary rigidity in the testing of an algorithm. To collect the CPU times for an algorithm, we should conduct all of our tests on the same machine using the same compiler. Moreover, if the machine is a time-sharing machine, we should ideally solve all the test problems when the machine has a similar work load. As an additional complication, for large instances, paging might dominate CPU times.

We can overcome some of these drawbacks by using virtual times instead of CPU times. (We point out that the virtual time is not directly related to "virtual memory.") The virtual running time of an algorithm is a linear estimate of its CPU time obtained by using its representative operation counts. For example, the virtual running time $V(I)$ of the network simplex algorithm to solve instance I is given by

$$V(I) = c_5\alpha_E(I) + c_6\alpha_W(I) + c_7\alpha_P(I),$$

for a set of constants c_5, c_6, and c_7 selected so that $V(I)$ is the best possible estimate of the algorithm's actual running time CPU(I) on the problem instance I. One plausible way to determine the constants c_5, c_6, and c_7 is to use (multiple) regression analysis. To do so, we consider the points (CPU(I), $\alpha_E(I)$, $\alpha_W(I)$, $\alpha_P(I)$) generated by solving various individual instances and use regression analysis to determine the constants c_5, c_6, and c_7 that minimizes the expression $\sum_I (\text{CPU}(I) - V(I))^2$.

In our regression analysis we generated each of the points (CPU(I), $\alpha_E(I)$, $\alpha_W(I)$, $\alpha_P(I)$) by taking an average of five problem instances; we used the averages reported in Figure 18.1, so we found the best linear fit to the 5-point averages. We found that for the network simplex algorithm with the first eligible pivot rule, we could estimate the virtual running time of an instance I as follows:

$$V(I) = (\alpha_E(I) + 2\alpha_W(I) + \alpha_P(I))/69{,}000.$$

To obtain an idea of the goodness of this fit, in Figure 18.4 we plot the ratio $V(I)/\text{CPU}(I)$ for all the data points. We find that in 15 of 25 cases, the error is less than 3%. In each case the error is at most 7%. So for this example, we can use the virtual running time instead of the CPU time with remarkably little loss of accuracy.

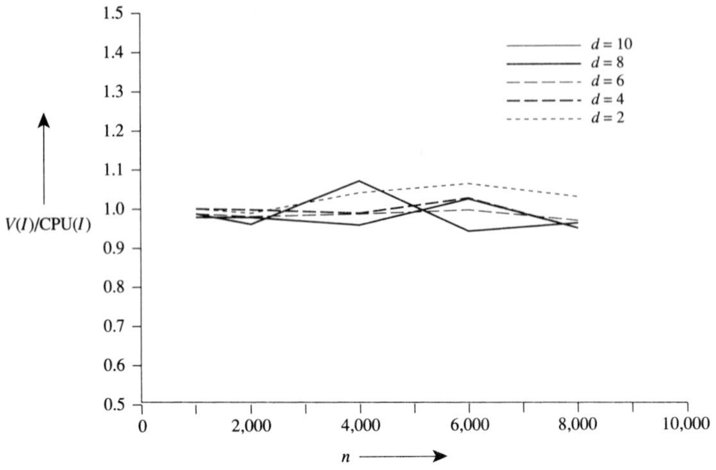

Figure 18.4 Determining how well the virtual running time estimates CPU time.

Using virtual running times has several advantages. First, the virtual running time helps us to assess the proportion of time that an algorithm spends on different representative operations. As an immediate consequence, it helps us to identify not only the asymptotic bottleneck operation, but also the bottleneck operations for different sizes of problem instances. For example, we estimated the virtual running time of the network simplex algorithm as $V(I) = (\alpha_E(I) + 2\alpha_W(I) + \alpha_P(I))/69{,}000$. To estimate the proportion of time spent on potential updates, in Figure 18.5 we plot $\alpha_P(I)/(\alpha_E(I) + 2\alpha_W(I) + \alpha_P(I))$. As we can see, for small problems, the percentage of time spent in updating node potential is less than 50 percent; however, as n increases the percentage of the time spent in updating node potential also increases, and it seems that as n approaches ∞, the percentage of time spent updating node potentials approaches 100 percent.

A second advantage of virtual running time is that it is particularly well suited for situations in which the testing is carried out on more than one computer. This situation might arise for several reasons: For example, several users might be conducting computational experiments at different sites; or the same user might wish to conduct additional tests after upgrading from an old computer. This situation is very common in the research literature because authors often conduct additional experiments at the suggestion of a referee or editor. When we move from one computer system to another, the representative operation counts remain unchanged. Using the representative operation counts of the previous study and the constants of the new computer system (obtained through the regression estimate for the virtual running time), we can obtain the virtual running times for all problems of the previous study measured in terms of the new computer system.

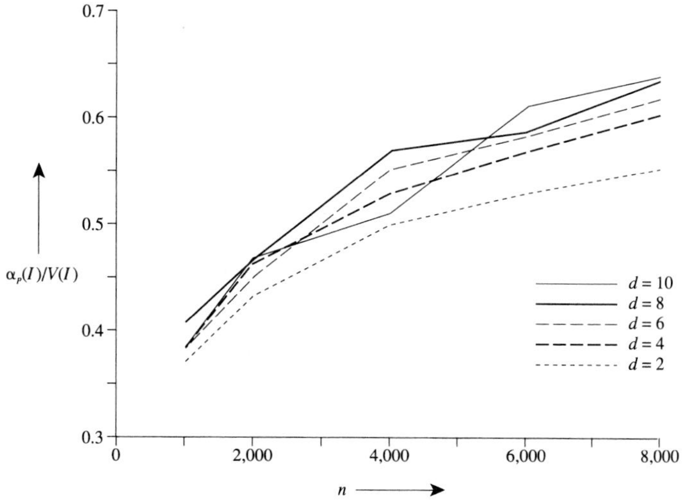

Figure 18.5 Identifying the percentage of the virtual running time accounted for by potential updates.

A third advantage of virtual running time is that it permits us to eliminate the effect of "paging" and "caching" in determining the running times. When a computer program executes a large program whose data do not fully fit into the computer's primary memory (e.g., RAM), it stores part of the data in the secondary memory (e.g., disk) with higher retrieval times. As a result, large programs run more slowly. In the virtual running time analysis, if we evaluate the constants using small problems (that fully execute in primary memory), the virtual running times for large problems would provide running time information as though the program were entirely run in primary memory. Thus we can use virtual running times to estimate the running time of an algorithm on a computer with sufficiently large primary memory.

A natural question is whether the constants used in the virtual running time are robust (i.e., do they give an accurate estimate of the CPU times for all possible problem inputs). Again, our use of virtual CPU times is not fully rigorous, but our limited experience so far has suggested that they do appear to be robust. In practice, we can also measure the robustness of the constants using statistical analysis.

Additional Insight into Algorithms

Computation counts have the potential to provide additional insight concerning an algorithm. For example, we can use the preceding analysis to estimate the number of pivots performed by the network simplex algorithm. Consider the following widely held belief in the linear programming literature: for most pivoting rules, the number of pivots is typically proportional to the number of constraints and rarely more than 3 times the number of constraints. In our case, the number of constraints is $n + m$ since we assume that each variable has an associated upper bound. Let us try to verify whether this assertion is valid for the first eligible pivot rule as applied to our randomly generated problems. Notice that this rule would imply that if we plot the ratio of (number of pivots)/n versus n for any specific network density d, we should obtain nonincreasing functions. The plots given in Figure 18.6(a) indicate that this conjecture is not true: the number of pivots, for a fixed value of d, is not bounded by a linear function of n. However, if we plot (number of pivots)/n^2 for different network densities, then, as indicated by Figure 18.6(b), we do obtain nondecreasing functions. Therefore, the growth rate of the number of pivots is bounded from above by n^2 and bounded from below by n.

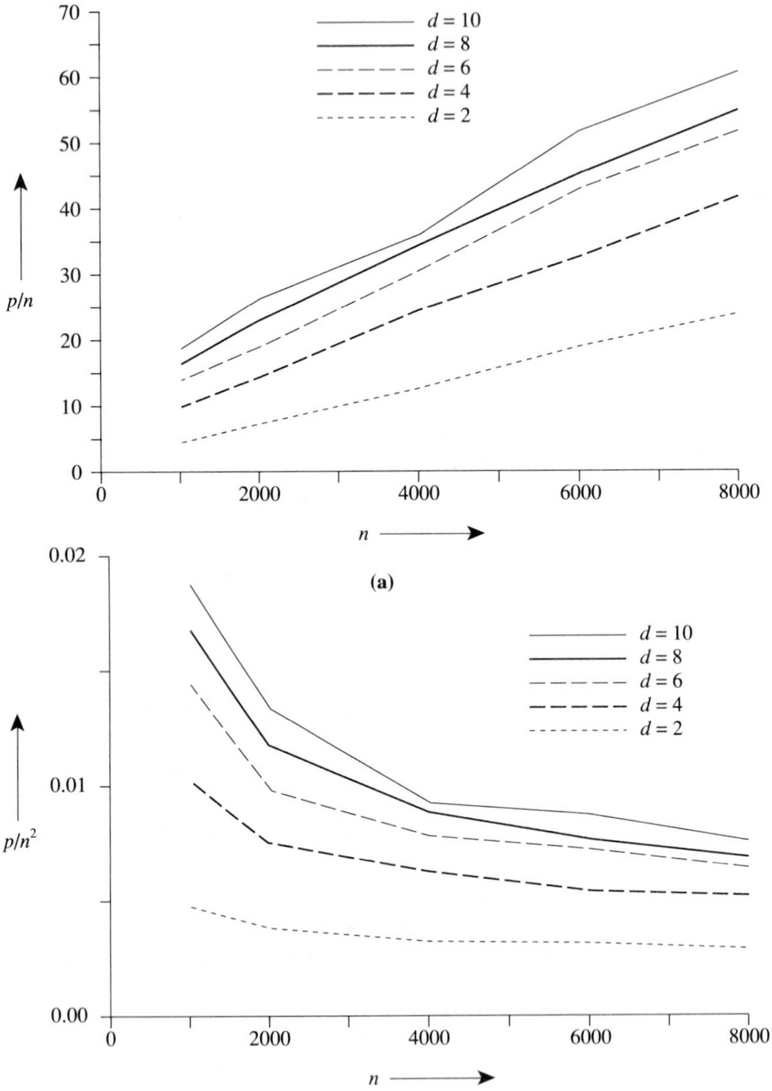

Figure 18.6 Determining lower and upper bounds on the growth rate of the number of pivots.

We might also be interested in determining what percentages of pivots are degenerate as the network size grows. The plots in Figure 18.7, which show the ratio of the number of degenerate pivots to the total number of pivots as a function of n, indicate that the ratio of degenerate pivots to total pivots varies between 70 and 90 percent.

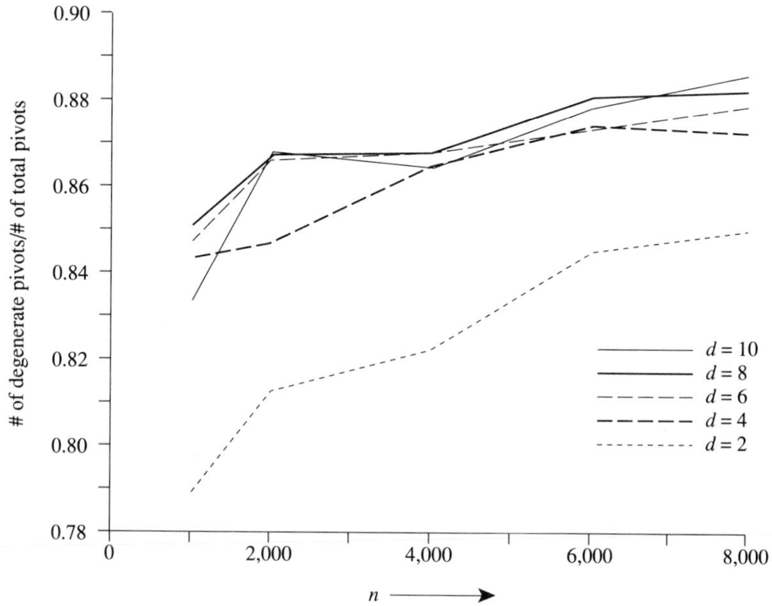

Figure 18.7 Occurance of degenerate pivots.

We might plot a few more graphs to gain additional insight about the network simplex algorithm. For example, we might plot the average size of the tree whose node potentials change during the pivot operation [i.e., $\alpha_P(I)/p$]. This plot would give us a good estimate of the running time per pivot since the potential update is the bottleneck operation, at least for the first entering pivot rule.

In addition, we might be interested in convergence results for the network simplex algorithm, i.e., how quickly does the network algorithm converge to the optimal objective function value? Does the algorithm quickly obtain a solution with a near-optimal objective function value and then slows down, or does it slowly approach the optimal objective function value and then converges rapidly? We could, in principle, provide a partial answer to this question by selecting a few sufficiently large instances and plotting the objective function values as a function of the number of pivots.

Limitations

We emphasize that the methodology we have described is suggestive and provides both insight and guidance, but it does not provide guarantees. In certain circumstances it can lead to incorrect conclusions. Let us illustrate a situation in which this approach would underestimate the asymptotic growth rate of the bottleneck operation. Suppose that the actual growth rate of a bottleneck operation is $h(n) = n^2 + 1000n$ and we have data only for instances $n \leq 2000$. Suppose we conjecture that the growth rate of the function is $n^{1.7}$. When we plot the ratio $(n^2 + 1000n)/n^{1.7}$ for various values of n, we obtain the plot shown in Figure 18.8, which is a decreasing function of n. Therefore, using the methodology suggested earlier, we would incorrectly reach the conclusion that $n^{1.7}$ has an upper bound on $h(n)$.

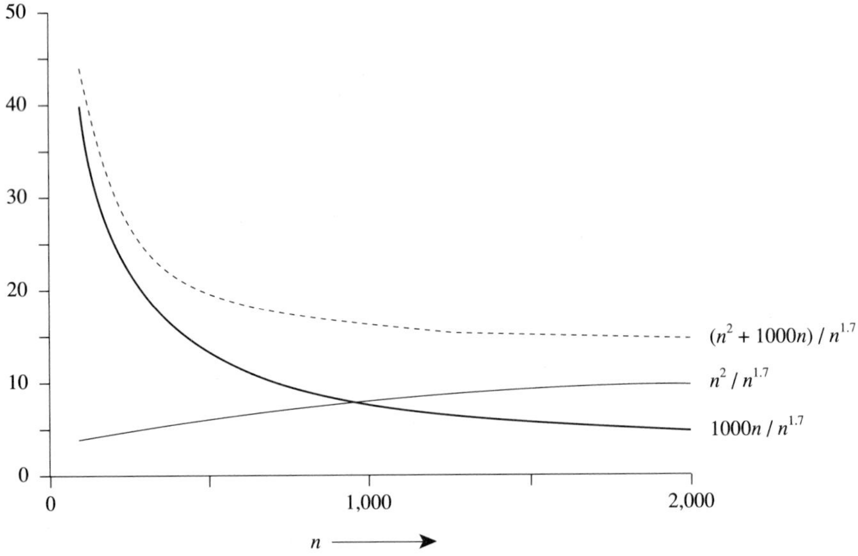

Figure 18.8 Limitation of the analysis using representative operation counts.

What went wrong in the previous example that would lead us to make such a significant mistake in estimating the running time? In our illustration, the growth rate had two components with very dissimilar constant terms and our test problems weren't sufficiently large so that the effect of the constant terms became insignificant. Had the growth function been $n^2 + 10n$ or had we solved problems of size $n = 100{,}000$, we would not have assessed the running time incorrectly, except possibly by a minimal amount. We anticipate that errors of this type would occur rather infrequently in practice.

We have suggested that we can determine an expression for the virtual running time using regression analysis. In general, regression analysis will misestimate the constant terms in the expression if the representative operation counts are highly correlated. In this case we might prefer to use more sophisticated methods of estimating the constants, possibly including computer timings of the basic operations.

We further point out that the methodology we have described is not useful for identifying improvements that affect only the constant factors. The use of computational counts will not identify improvements in coding or in data structures that improve the running time by a constant factor; the analysis of CPU times would be able to identify the effects of these improvements.

18.4 SUMMARY

Most iterative algorithms for solving network flow problems repetitively perform some basic operations. For almost all algorithms, we can decompose these basic operations into fundamental operations so that the algorithm executes each operation in $\Theta(1)$ time. An algorithm typically performs a large number of fundamental operations. We refer to a subset of fundamental operations as a set of representative operations if for every possible problem instance, the sum of representative operations provides an upper bound (to within a multiplicative constant) on the sum of all operations that an algorithm performs. We have shown that these representative operation counts might provide valuable information about an algorithm's behavior that is not captured by CPU time. For example, the representative operation counts allow us (1) to determine the asymptotic growth rate of the running time of an algorithm independent of its computing environment, (2) to assess the time an algorithm spends on different basic operations, (3) to compare two algorithms executed on different computers, and (4) to estimate the running time of an algorithm on a computer different from the one carrying out the experiments.

The simple methodologies that we have presented for conducting empirical analysis of an algorithm using representative operation counts do not provide rigorous guarantees; nevertheless, they often provide considerable insight about an algorithm's behavior, and they typically yield far more insight than is obtainable by analyzing only CPU times. The ideas we have outlined in this chapter apply to most network algorithms and should, as well, apply to optimization algorithms for problems arising in several other application domains.

REFERENCE NOTES

This chapter has been largely excerpted from the paper of Ahuja and Orlin [1992b]. We illustrated our ideas on computational testing of algorithms using a network simplex code developed by these authors. Researchers have tested several other codes of the network simplex algorithm; some notable computational studies are due to Glover, Karney, and Klingman [1974], Mulvey [1978], Bradley, Brown, and Graves [1977], Grigoriadis [1986], and Chang and Chen [1989]. In the computational results reported in this chapter, we used NETGEN to generate random network flow problems; Klingman, Napier, and Stutz [1974] developed this network generator.

In our discussion we have emphasized the value of operation counts in computational testing. The idea of using operation counts is quite old and probably dates back to the origins of computational experiments. Nevertheless, the literature on computational testing for mathematical programming has historically used CPU time as its primary measure of computational effort and has used operation counts in a rather limited way. (For example, most computational experiments on the network simplex algorithm have counted the number of pivots, but have not kept track in any systematic way of the work per pivot.) The thesis of McGeoch [1986] is an excellent reference that deemphasizes CPU time in favor of other measures of performance. The computational studies by Johnson [1990] and Bentley [1990] provide excellent illustrations of how to use both CPU times and representative operation counts to analyze empirical behavior of algorithms. The term "virtual time" appears in the thesis of Brown [1988] and is used in a similar way in this chapter.

Although we have focused on the use of operation counts in this chapter, the following references offer insight about other very important aspects of computational testing.

Performance measures. In this chapter we have focused on empirical running time as measured both by representative operation counts and by CPU times. Other important measures of performance for an algorithm include (1) ease of implementation, (2) robustness, (3) reliability, and (4) accuracy of the solutions. The papers by Crowder and Saunders [1980], Hoffman and Jackson [1982], and Greenberg [1990] discuss these measures of performance.

Reporting computational experiments. The most comprehensive references on the reporting of computational experiments in mathematical programming is Crowder, Dembo, and Mulvey [1978, 1979]. The authors provide guidelines for what should be reported in a research paper and offer advice on how to conduct appropriate computational experiments. Jackson and Mulvey [1978] have summarized the reporting of computational experiments within the mathematical programming literature, largely detailing how poor the reporting had been up to that time. More recently, Jackson, Boggs, Nash, and Powell [1989] have provided updated guidelines for conducting computational experiments.

Analysis and evaluation of test results. As emphasized in this chapter, a large part of data analysis can be carried out without the formal use of statistical methodology. Graphs, charts, and elementary statistics such as the computation of means and standard deviations often provide significant insights into the performance of algorithms. McGeoch [1986], Bentley [1990], and Johnson [1990] present several case studies to show the power of these basic analytical tools.

Statistical methodologies. Often, statistical methodologies can provide analysis and insight that is unavailable through other means. For some papers on computational testing that provide details on the use of statistical methodologies, see Bland and Jensen [1985] and Golden, Assad, Wasil, and Baker [1986]. Moreover, statistical methodologies such as *variance reduction* can often increase the power of the analysis and reduce the number of experiments needed to obtain conclusive results. We refer the reader to McGeoch [1992] for excellent illustrations of variance as well as for pointers to the literature.

Algorithm animation. Algorithm animation is another technique for gaining insight about an algorithm. In his dissertation, Brown [1988] provides an excellent treatment of this topic. Algorithm animation techniques view the progress of an algorithm on a single instance as a sequence of snapshots. For example, consider the greedy algorithm for finding the minimum spanning tree joining n points in the plane. This algorithm creates the minimum spanning tree one arc at a time, maintaining a forest at each intermediate stage until it ultimately creates the minimum cost spanning tree. Using animation, we could construct a sequence of figures showing the forest at intermediate stages. Viewing a motion picture consisting of a series of these snapshots might reveal additional insight about an algorithm. Alternatively, we could see how the total length of the forest increases as a function of computer time, or we could see how the size of the components decreases as a function of computer time. The papers by Bentley and Kernighan [1990] and Bentley [1990] discuss a simple language that facilitates the construction of these animations.

EXERCISES

18.1. An algorithm is said to have a *predictable running time* if its empirical running time is guaranteed to be within a constant factor of its worst-case running time. State which of the following algorithms have a predictable running time: **(1)** the breadth-first search algorithm discussed in Section 3.4; **(2)** the original implementation of Dijkstra's algorithm discussed in Section 4.5; **(3)** Dial's implementation of Dijkstra's algorithm discussed in Section 4.6; **(4)** the radix heap implementation of Dijkstra's algorithm discussed in Section 4.8; **(5)** the $O(nm)$-time cycle detection algorithm discussed in Section 5.5; and **(6)** the minimum mean cycle algorithm discussed in Section 5.7. Justify your answers by theoretical arguments without doing the computational testing.

18.2. Specify a set of representative operations for each of the following algorithms: **(1)** the topological sorting algorithm discussed in Section 3.4; **(2)** the binary heap implementation of Dijkstra's algorithm discussed in Section 4.7; and **(3)** the radix heap algorithm described in Section 4.8. Justify your answers.

18.3. Give a set of representative operations for the following maximum flow algorithms discussed in Chapter 7: **(1)** the shortest augmenting path algorithm; **(2)** the highest-label preflow-push algorithm; and **(3)** the excess scaling algorithm. For each algorithm, obtain a set of representative operations with the fewest possible number of operations and justify your answer.

18.4. Specify a set of representative operations for the following minimum cost flow algorithms: **(1)** the relaxation algorithm described in Section 9.10; **(2)** the cost scaling algorithm described in Section 10.3; and **(3)** the double scaling algorithm described in Section 10.4.

18.5. Give a set of representative operations for the following minimum spanning tree algorithms: **(1)** the $O(m + n \log n)$ time implementation of Kruskal's algorithm (assume that arcs are already sorted); and **(2)** the $O(m \log n)$ time implementation of Sollin's algorithm. Justify your answers.

18.6. What are a set of representative operations for the generalized network simplex algorithm discussed in Chapter 15 and for the Dantzig–Wolfe decomposition algorithm discussed in Chapter 17?

18.7. Let operation i and operation j be two operations that require $\Theta(1)$ time in a particular algorithm. Suppose that when this algorithm is applied to an instance I, it executes operation i $\alpha_i(I)$ times and operation j $\alpha_j(I)$ times. We say that operation j dominates operation i if for every possible instance I, $\alpha_i(I) \leq c_1\alpha_j(I)$ for some known constant c_1. For each of the algorithms mentioned in Exercises 18.1 and 18.2, specify representative operations as well as a nonrepresentative operation that it dominates.

18.8. Design a computational experiment for comparing the following implementations of the shortest path problem discussed in Chapter 4: **(1)** the original implementation; **(2)** Dial's implementation; and **(3)** the radix heap implementation. Which set of representative operations would you collect for each algorithm? Write a computer code for each of these algorithms and test them using the methodology described in this chapter.

18.9. Write computer programs for the following maximum flow algorithms and compare them using the methodology described in this chapter: **(1)** the labeling algorithm; **(2)** the capacity scaling algorithm; and **(3)** the shortest augmenting path algorithm.

18.10. Write computer programs for the following minimum cost flow algorithms and compare them using the methodology described in this chapter: **(1)** the cycle-canceling algorithm; **(2)** the successive shortest path algorithm; and **(3)** the relaxation algorithm.

19

ADDITIONAL APPLICATIONS

Mens et Manus (Mind and Hand)
—The MIT Motto

Chapter Outline

19.1 INTRODUCTION

As we have noted throughout this book, network flows is a topic that has evolved in the best tradition of applied mathematics: It is a subject matter that poses considerable challenges for modeling and algorithm development and it has a core of substantial theory and scholarly content. It unites ideas from the abstract world of mathematics and concrete world of computation, and so draws its intellectual heritage from several disciplines, including applied mathematics, computer science, engineering, management science, and operations research. But perhaps as important, it is a subject that has had numerous applications in a wide variety of practical problem settings.

As indicated by the title of this book, our coverage has attempted to emphasize three critical ingredients of network flows: theory, algorithms, and applications. Although we have organized most of our discussion in the previous chapters around core network models and algorithmic approaches, as a key element of our discussion, we have described approximately 150 applications and mentioned dozens more. We have introduced these applications in the context of core network flow models—shortest path problems, maximum flow problems, minimum cost flows—and other network optimization models such as minimal spanning trees. In this chapter we adopt a somewhat different approach. We discuss 24 applications, organized around application type. Therefore, within most of the topics that we introduce, we consider several different network optimization models. Adopting this approach permits us to see a number of important applications in a somewhat different light: at times, building from simple to more complex models all within the same application context.

At the end of Chapters 4, 6, 9, and 12 to 17 we have listed, with references, a great many applications, including both those that we have considered in the text and in the exercises and others from the literature. At the end of this chapter, we offer another view of these applications, including those that we have considered in this chapter, organized in the following categories:

Applied mathematics
Computer science and communication systems
Defense
Distribution systems and transportation
Engineering
Management science
Manufacturing, production, and inventory planning
Physical and medical sciences
Scheduling
Social sciences and public policy

Although we could have adopted many alternative ways to categorize these various applications, this topography provides one useful view of network flows in practice. Some of the categories, such as manufacturing and transportation, refer to specific industries; in these categories, the applications typically are models of direct relevance to practitioners. Applied mathematics represents another type of category; the applications in this case are mathematical problems of some interest to the applied mathematics community (e.g., finding solutions to certain systems of equations and inequalities) and which often are generic models with rich end applications of their own. When viewed in its entirety, this list of applications attests to the remarkable robustness of network flows as a practical modeling tool; it suggests that we might revise the opening sentence in this book and state: "Everywhere we look, not only in our daily lives, but also in the worlds of commerce, science, social systems, and technology, networks are apparent."

Figure 19.1 summarizes the applications that we consider in this chapter as well as the network problems used to model these applications. As indicated by this table, just the applications in this chapter attest to the richness of network flows in practice. As shown by this table, this collection of applications uses many of the network models that we have developed in previous chapters, including the core shortest path, maximum flow, and minimum cost flow models, as well as minimum spanning trees, matchings, and multicommodity flows. The applications also use more specialized models that are transformable into network flow models (duals of minimum cost flow models).

Application	Problem type
19.1 Open pit mining	Minimum cut problem
19.2 Selecting freight handling terminals	Minimum cut problem
19.3 Optimal destruction of military targets	Minimum cut problem
19.4 Flyaway kit problem	Minimum cut problem
19.5 Asymmetric data scaling with lower and upper bounds	Shortest path problem
19.6 Minimum ratio asymmetric data scaling	Minimum mean cycle problem
19.7 DNA sequence alignment	Shortest path problem
19.8 Automatic karyotyping of chromosomes	Transportation problem
19.9 Determining minimum project duration	Longest path problem
19.10 Just-in-time scheduling	Longest path problem, minimum cost flow problem
19.11 Time–cost trade-off in project management	Dual of minimum cost flow problem
19.12 Maximum dynamic flows	Maximum flow problem
19.13 Models for building evacuation	Minimum cost flow problem
19.14 Directed Chinese postman problem	Minimum cost flow problem
19.15 Undirected Chinese postman problem	Shortest path problem and nonbipartite matching problem
19.16 Discrete location problems	Assignment problem
19.17 Warehouse layout	Transportation problem
19.18 Rectilinear distance facility location	Dual of minimum cost flow problem
19.19 Dynamic lot sizing	Shortest path problem, minimum cost flow problem
19.20 Dynamic lot sizing with concave costs	Shortest path problem
19.21 Dynamic lot sizing with backorders	Shortest path problem
19.22 Multistage production–inventory planning	Minimum cost flow problem, mixed integer programs
19.23 Multiproduct multistage production–inventory planning	Integer multicommodity flow problem
19.24 Mold allocation	Minimum cost flow problem

Figure 19.1 Applications considered in this chapter.

19.2 MAXIMUM WEIGHT CLOSURE OF A GRAPH

One of the primary purposes of scientific investigation is to structure the world around us, discovering patterns that cut across and therefore help to unify varied applied contexts. To begin our discussion of applications, in this section we examine one such generic model that has applications as varied as designing mining operations, scheduling freight handling terminals, developing a strategy for destroying military targets, and designing optimal kits of parts and tools for field repair crews.

A *closure* of a directed network $G = (N, A)$ is a subset of nodes without any outgoing arcs, that is, a subset $N_1 \subseteq N$ satisfying the property that if i belongs to N_1 and $(i, j) \in A$, then j also belongs to N_1. A closure might have more than one component. Suppose that we associate a node weight w_i (of arbitrary sign) with each node i of G. In the *maximum weight closure problem*, we wish to find a closure N_1 with the largest possible weight $w(N_1)$ defined as $w(N_1) = \sum_{i \in N_1} w_i$. As an example, the network shown in Figure 19.2(a) has the closures {3, 4, 5}, {4, 5}, {5}, {2, 5}, and {1, 2, 4, 5}; the maximum weight closure for this network is {3, 4, 5}.

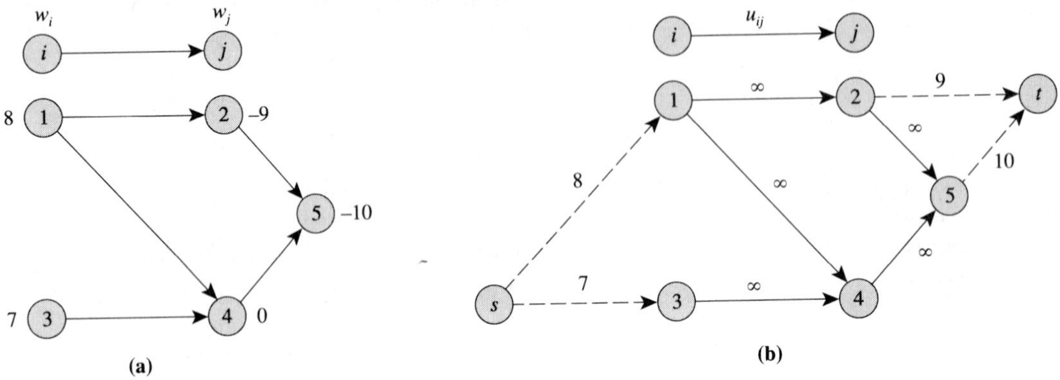

Figure 19.2 (a) Maximum weight closure problem; (b) transformed network G'.

As we will see in a moment, the maximum weight closure problem arises in a variety of applications. Before discussing these applications, let us show how to transform the maximum weight closure problem defined on the network $G = (N, A)$ into a maximum flow problem on a slightly augmented network $G' = (N', A')$. To define G', we introduce a source node s and for each node $i \in N$ with $w_i > 0$, we create an arc (s, i) with capacity w_i. We also introduce a sink node t and for each node $i \in N$ with $w_i < 0$, we create an arc (i, t) with capacity $-w_i$. We then set the capacity of every original arc $(i, j) \in A$ equal to ∞ (any integer greater than $\sum_{i \in N} | w_i |$ would suffice). Figure 19.2(b) shows the transformed network for the maximum weight closure problem shown in Figure 19.2(a).

We refer to an s–t cut in the transformed network $G' = (N', A')$ as a *simple cut* if all its forward arcs are source and sink arcs (i.e., arcs incident to the source and sink nodes). We claim that there is a one-to-one correspondence between closures of G and simple cuts in G'. To establish this result, note that if N_1 is a closure of G, its corresponding cut is $[S, \bar{S}]$ with $S = \{s\} \cup N_1$. The fact that N_1 is a closure of G implies that no arc in A is a forward arc of the cut $[S, \bar{S}]$. Consequently, all of the forward arcs in the cut $[S, \bar{S}]$ will be either source arcs or sink arcs, so this cut will be a simple cut. Similarly, if $[S, \bar{S}]$ is a simple cut of G', the subset of nodes N_1 defined by $N_1 = S - \{s\}$ is a closure of G.

To relate the weight of a closure to the capacity of the corresponding cut, let N_1 be a closure of G and $N_2 = N - N_1$. In addition, let N_1^+ denote the nodes with nonnegative weights in N_1, and N_1^- denote the nodes with negative weights in N_1. We define N_2^+ and N_2^- similarly. By definition, the weight of the closure N_1 is

$$w(N_1) = \sum_{i \in N_1^+} w_i - \sum_{i \in N_1^-} | w_i |. \qquad (19.1)$$

Now consider the simple cut $[S, \bar{S}]$ corresponding to the closure N_1. Each forward arc in the cut is a source or a sink arc. The construction of the network G' implies that this cut would have a forward arc (i, t) for every $i \in N_1^-$ and a forward arc (s, i) for every $i \in N_2^+$. Therefore, the capacity of this cut is

$$u[S, \bar{S}] = \sum_{i \in N_2^+} w_i + \sum_{i \in N_1^-} |w_i|. \tag{19.2}$$

Adding (19.1) and (19.2), we find that

$$w(N_1) + u[S, \bar{S}] = \sum_{i \in N_1^+} w_i + \sum_{i \in N_2^+} w_i = \bar{w}.$$

The constant $\bar{w}$ in this expression is the total weight of all nodes. Consequently, if $[S, \bar{S}]$ is a minimum capacity simple cut, the corresponding closure N_1 is a maximum weight closure.

To obtain a minimum capacity simple cut, we simply need to find a minimum capacity cut in the network since in the transformed network G', no arc in A will be a forward arc in any minimum cut because the arcs in A all have an infinite capacity. Therefore, a minimum capacity cut of G' will automatically be a simple cut.

To conclude this introductory discussion of the maximum weight closure problem, we note that we can also derive the network formulation of the problem using minimum cost flow duality, since we can formulate the maximum weight closure problem as a linear program with at most one $+1$ and at most one -1 in each row, which is the dual of a minimum cost flow problem (as shown in Theorem 9.9).

We next describe four different applications of the maximum weight closure problem.

Application 19.1 Open Pit Mining

In mining operations, a problem of considerable importance is the determination of the optimal contour of an open pit mine. In an open pit mine, we might divide the potential mining region into blocks. The provisions of any given mining technology, and perhaps the geography of the mine, impose restrictions on how we can remove the blocks. For example, we can never remove a block until we have removed every block that lies immediately above it (see Figure 19.3); restrictions on the "angle" of mining the blocks might impose similar precedence conditions. Moreover, every block i has an economic measure w_i representing the net profit obtained from removing that block (value of the ore contained in the block minus the cost of exploiting and processing the block). In the open pit mining problem, we wish to identify a set of blocks that maximizes the net profit. We model this problem as a maximum weight closure problem by representing each block as a node; if we must remove block j before removing block i, we include the arc (i, j) in the network. If we want to remove a contour B of blocks, every block that we need to remove before removing a block in B must also lie in B. That is, the nodes defined by B have no outgoing arcs and therefore define a closure of the network.

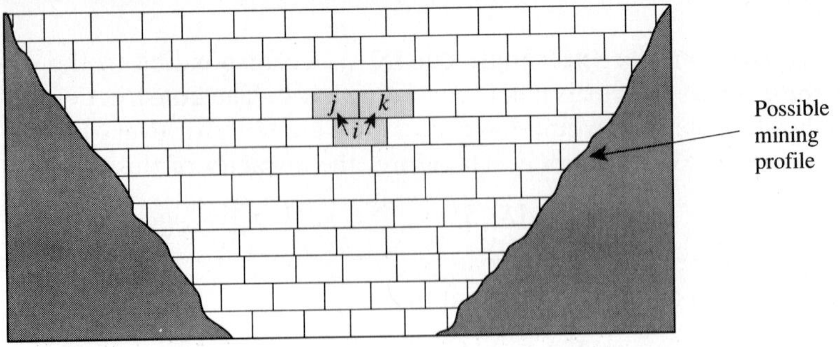

Figure 19.3 Open pit mine; we must remove blocks j and k before removing block i.

Application 19.2 Selecting Freight Handling Terminals

A transport company is considering the installation of a number of freight handling terminals. It wants to choose from a set S of possible locations for the terminals. The company has the potential to attract market share (which is a given amount of demand) between some of the pairs of terminals. To satisfy the demand between locations i and j, the company must locate terminals at both of these locations. Suppose that c_j is the cost of installing a terminal at location j and that p_{ij} is the profit obtained by satisfying the demand between locations i and j. The transport company would like to determine where to install terminals in order to maximize its net profit (i.e., the revenue obtained from satisfying the demands minus the cost of installing the terminals).

Consider, for example, the network shown in Figure 19.4(a). Each node in this

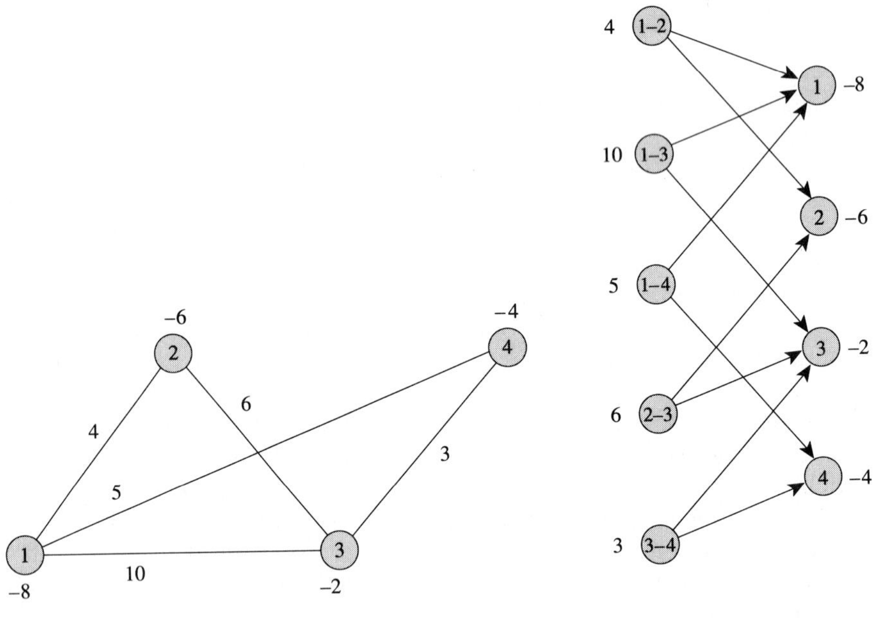

Figure 19.4 (a) Selection problem; (b) corresponding maximum closure problem.

network represents a potential terminal; the number next to it represents the negative of the cost of installing that terminal. Each arc (i, j) represents a service that can be operated only if both the terminals i and j are operating; the number next to the arc represents the profit obtained by operating that service. For example, if we decide to operate terminals 1, 2, and 3, we can operate services only between the following pairs of terminals: (1, 2), (1, 3), and (2, 3). The net profit for this selection is 4, which is the difference of the total revenues (of value 20) and the total installation cost (which is 16).

To reduce the selection problem to the maximum weight closure problem, we define a bipartite network $G = (N_1 \cup N_2, A)$ with a node in N_1 for every service and a node in N_2 for every terminal. The service node representing the service between nodes i and j has two outgoing arcs entering the nodes representing terminals i and j, implying that whenever we decide to provide the service between the nodes i and j and accrue the profit p_{ij}, we must install the terminals at these two nodes and incur the installation cost c_i and c_j. Figure 19.4(b) shows the resulting maximum weight closure problem for our example.

Application 19.3 Optimal Destruction of Military Targets

A military commander has identified a set S of military targets that he wants to destroy. These targets are heavily defended by four different layers of defense. The first layer consists of *forward air defense sites* (FADS), the second layer consists of *band surface to air missiles* (BSAM), the third layer consists of *airborne interceptors* (AI), and the fourth layer consists of *terminal surface to air missiles* (TSAM); a fifth layer contains the military targets themselves. Let $\bar{S}$ denote the set of all defense sites. Each military target is protected by some, but not necessarily all, of these defense sites. A defense site might also provide protection to other defense sites in lower-numbered layers. Let $D(i)$ denote the set of defense sites that protect the target or defense site $i \in S \cup \bar{S}$.

Based on his past experience, the military commander feels that while it might be possible for missiles to pass through all the defenses to reach the military targets, the probability of such a "leakage" is quite small. Instead, he believes that to destroy a target, he must first destroy all the defense sites that protect it. Therefore, he must destroy defense sites as well as targets. Destroying the ith target or defense site has a certain military benefit but also incurs some loss. Let w_i denote the benefit minus the loss, which is the (net) value, of destroying the ith target or defense site. The military commander wants to identify a set of targets and defense sites with the largest possible total value.

To formulate this problem as a maximum weight closure problem, we associate a node i with a weight of w_i with the ith target or defense site. For each $i \in S \cup \bar{S}$, we introduce an arc (i, j) for every $j \in D(i)$. Figure 19.5 gives an example of the resulting network. As is easy to see, every feasible destruction of targets and defense sites corresponds to a closure of the network. Therefore, the military commander's problem is a maximum weight closure problem.

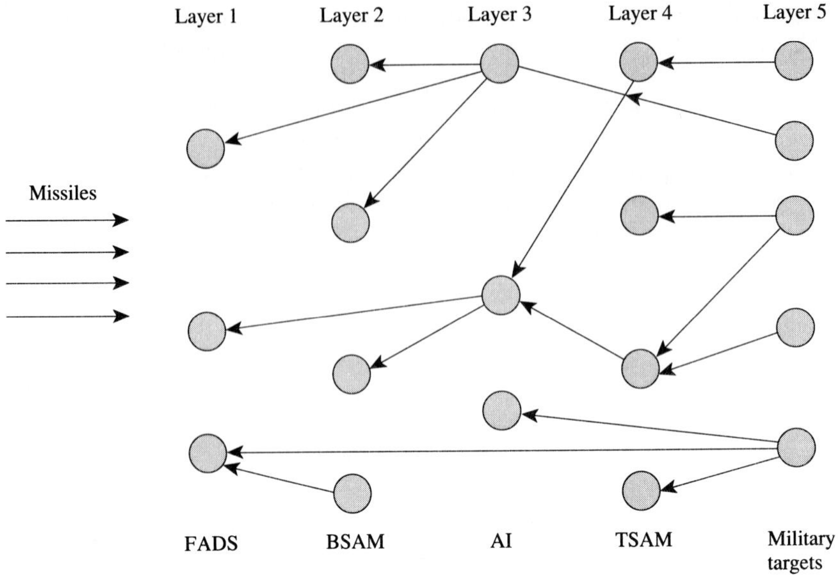

Figure 19.5 Formulation for the optimal destruction of military targets.

Application 19.4 Flyaway Kit Problem

Many companies (e.g., computer companies or telephone companies) own, lease, or warrantee a wide range of equipment that they must maintain at geographically dispersed field locations. In performing a given job, the repair crew often require various types of parts (and tools). In many cases the crews carry some replacement parts in a *kit* rather than storing them at the equipment site. If all the required parts are in the kit, the crew member can repair the equipment. But if any of these items is not available, the service call is incomplete and the job is a "broken job." Broken jobs are costly for several reasons: (1) they increase equipment downtime, (2) the repair crew must make an extra trip for parts, and (3) partially repaired equipment might be unsafe or vulnerable to damage. On the other hand, carrying more items in the kit increases handling and inventory costs. In the *flyaway kit problem*, we need to obtain the optimal kit of parts (and tools) that minimizes the sum of the handling and inventory costs and the costs of broken jobs.

Suppose that we number the parts required for servicing the jobs as $1, 2, \ldots, r$. We assume that the repairman restocks the kit between jobs, but with a fixed and specified content. For our purposes we define a job by the set of parts (and tools) that it requires. Making this association defines a collection of job types $J_1, J_2, \ldots, J_l$, that encompasses all the known possibilities that a repairman might encounter. The job type J_j is defined by the set B_j of the parts required by that job. Let l_j denote the expected number of job types J_j serviced in one year, and V_j denote the penalty cost we incur whenever job j is a broken job.

A *stocking policy* of a kit consists of a fixed set of parts $M \subseteq \{1, 2, \ldots, r\}$ that a crew would carry. Let H_i denote the yearly handling and inventory cost for carrying part i in the kit. Then the total handling cost is $\sum_{i \in M} H_i$. Moreover, the total expected cost of broken jobs per kit per year for policy M would be $\sum_{\{j:B_j \not\subseteq M\}} V_j l_j$. Therefore, policy M incurs a total expected yearly cost per kit of

$$z(M) = \sum_{i \in M} H_i + \sum_{\{j:B_j \not\subseteq M\}} L_j.$$

In this expression, $L_j = V_j l_j$. The optimal policy would, of course, be a set $M \subseteq \{1, 2, \ldots, r\}$ that minimizes $z(M)$. Notice that minimizing $z(M)$ is equivalent to maximizing $-z(M)$, which we can restate as

$$-z(M) = \sum_{\{j:B_j \subseteq M\}} L_j - \sum_{i \in M} H_i - L,$$

by letting $L = \sum_{j=1}^{l} L_j$, a constant. Consequently, our objective is to identify a policy M that maximizes $\sum_{\{j:B_j \subseteq M\}} L_j - \sum_{i \in M} H_i$. This problem is a special case of the maximum weight closure problem on the bipartite network shown in Figure 19.6. This network contains two types of nodes: those representing parts and those representing jobs. It also contains an arc from a node representing job type J_j to each part node in B_j. Notice that a node J_j can be in the maximum weight closure only if the closure also contains each part (and tool) in B_j.

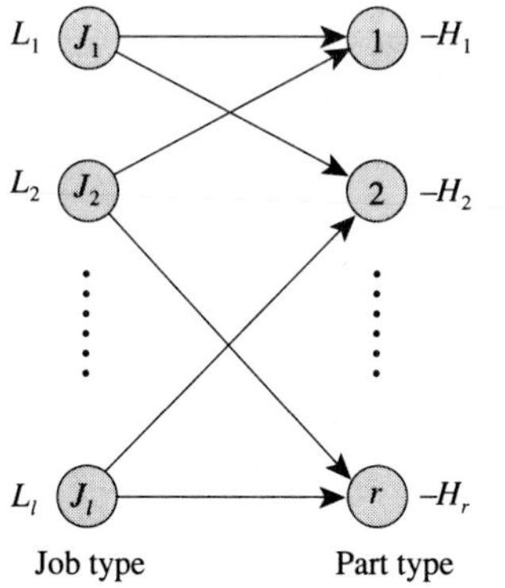

Figure 19.6 Network for the fly away kit problem.

19.3 DATA SCALING

In many applied problem contexts, we wish to modify matrix data to achieve some desired objective. In Application 6.3 we examined one instance of this generic problem: the rounding of entries in census tables to preserve the basic nature of the tabular entries and yet disguise confidential information. In this section we consider two variations of a model of data scaling: We would now like to scale the rows and columns of a matrix so that the resulting entries of the matrix are "close" to each other; one variant of this model requires that we solve a shortest path problem and the other requires that we solve a minimum mean cycle problem. Before discussing these generic models, we first consider one important application context for data scaling.

When we solve a linear programming problem of the form minimize cx, subject to $\mathcal{A}x = b$ and $x \geq 0$ by the simplex method, we incur round-off errors because computers perform arithmetic operations in floating-point arithmetic. The magnitude of these errors depends on the relative sizes of the numbers. If the numbers being manipulated are comparable, the round-off errors are relatively small; otherwise, they are larger. Thus, given a linear program, it is often desirable to transform it into an equivalent linear program whose constraint matrix $\mathbf{A} = \{a_{ij}\}$ has elements that are as close as possible to each other. We can achieve this objective by *data scaling*: that is, multiplying each row i by a positive constant α_i and dividing each column j by a positive constant β_j. Notice that multiplying the equality (row) i by the constant α_i does not affect the feasibility of any solution, and dividing a column j by a constant β_j is equivalent to replacing the variable x_j by the variable $x'_j = x_j\beta_j$. Consequently, other than rescaling the variables, this transformation does not affect the optimal solutions of the linear program.

We now study two data scaling problems and, in each case, reduce the problem to a network flow model.

Application 19.5 Asymmetric Data Scaling with Lower and Upper Bounds

In the asymmetric data scaling problem with lower and upper bounds, we want to determine whether we can find row multipliers α_i and column divisors β_j so that every scaled entry of a matrix has a value between the prescribed lower and upper bounds l and u, that is,

$$l \le \alpha_i \mid a_{ij} \mid / \beta_j \le u \qquad \text{for each } i = 1, \ldots, p \text{ and each } j = 1, \ldots, q. \tag{19.3}$$

If we take the logarithms of both sides of the inequalities in (19.3), they become

$$\log l \le \log \alpha_i + \log \mid a_{ij} \mid - \log \beta_j \le \log u \qquad \text{for each } i = 1, \ldots, p, \text{ and each } j = 1, \ldots, q. \tag{19.4}$$

For notational convenience, let us index the rows from 1 to p and index the columns from $p + 1$ to $p + q$. Because of this numbering convention, we subsequently refer to β_j by β_{p+j} and to a matrix element a_{ij} by $a_{i,p+j}$. Let

$$\pi(i) = \log \alpha_i \qquad \text{for each } i = 1, \ldots, p,$$

$$\pi(j) = \log \beta_j \qquad \text{for each } j = p + 1, \ldots, p + q.$$

Also, let $l' = \log l$, $u' = \log u$, and $a'_{ij} = \log \mid a_{ij} \mid$. Then we can rewrite the two inequalities in (19.4) as

$$\pi(i) - \pi(j) \le u' - a'_{ij} \text{ for each } i = 1, \ldots, p \text{ and } j = p + 1, \ldots, p + q, \tag{19.5a}$$

$$-\pi(i) + \pi(j) \le a'_{ij} - l' \text{ for each } i = 1, \ldots, p \text{ and } j = p + 1, \ldots, p + q. \tag{19.5b}$$

We have thus reduced the data scaling problem into the problem of identifying whether some vector π satisfies the inequalities in (19.5). This problem is known as a *system of difference constraints* and, as described in Application 4.5, we can solve it as a shortest path problem on the network shown in Figure 19.7. This network is a complete bipartite network $G = (N_1 \cup N_2, A)$ with node sets $N_1 = \{1, \ldots, p\}$ and $N_2 = \{p + 1, \ldots, p + q\}$. The network contains an arc (i, j) of cost $c_{ij} = u' - a'_{ij}$ for every node pair $[i, j] \in N_1 \times N_2$, and an arc (j, i) of cost $c_{ji} = a'_{ij} - l'$

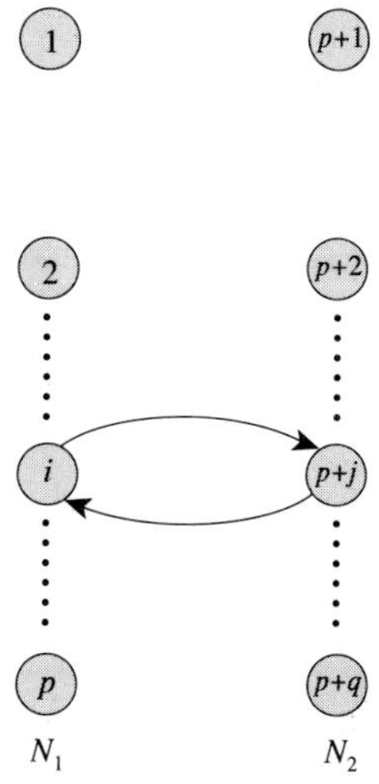

Figure 19.7 Network for data scaling problems. The network contains an arc for each $[i, j] \in N_1 \times N_2$ and an arc for each $[j, i] \in N_2 \times N_1$.

for every node pair $[j, i] \in N_2 \times N_1$. If we set $d(j) = -\pi(j)$ for each j, then the inequalities (19.5) reduce to

$$d(j) \le d(i) + c_{ij} \quad \text{for each arc } (i, j) \in A, \tag{19.6}$$

which are the optimality conditions of a shortest path problem (see Section 5.2). We have shown in Section 5.2 that (19.6) has a feasible solution [i.e., a set of shortest path distances $d(\cdot)$] if and only if the network contains no negative cycle. We can resolve this question in $O(nm)$ time by using the negative cycle detection algorithm described in Section 5.5.

Application 19.6 Minimum Ratio Asymmetric Data Scaling

In the minimum ratio asymmetric data scaling problem, we want to identify bounds l and u, row multipliers α_i, and column divisors β_j so that

$$l \le \alpha_i \mid a_{ij} \mid / \beta_j \le u \quad \text{for each } i = 1, \ldots, p \text{ and } j = 1, \ldots, q, \tag{19.7}$$

with the smallest possible ratio u/l. We will show how to solve this data scaling problem by solving a minimum mean cycle problem.

We first observe that without any loss of generality we can fix $l = 1$ and identify the minimum value of u that, together with some row and column multipliers, satisfies the inequalities (19.7) (because we can transform any solution with l different than 1 to a solution with l equal to 1 by multiplying u and each α_i by l). Our discussion of the preceding application shows that the following linear program is an alternative formulation of the minimum ratio asymmetric data scaling problem:

$$\text{Minimize} \quad u' \tag{19.8a}$$

subject to

$$\pi(i) - \pi(j) - u' \le -a'_{ij} \quad \text{for every } i = 1, \ldots, p, \text{ and } j = p + 1, \ldots, p + q, \tag{19.8b}$$

$$-\pi(i) + \pi(j) \le a'_{ij} \quad \text{for every } i = 1, \ldots, p \text{ and } j = p + 1, \ldots, p + q. \tag{19.8c}$$

In this model the variables $\pi(i)$ and constants u' and a'_{ij} are defined as in Application 19.5. Notice that $l' = \log_2 1 = 0$. To simplify (19.8), we again redefine the variables. Let $\theta = -u'/2$, $\pi'(i) = \pi(i) - u/2$ for each $i = 1, \ldots, p$, and $\pi'(j) = \pi(j)$ for each $j = p + 1, \ldots, p + q$. Moreover, for each $i = 1, \ldots, p$ and $j = p + 1, \ldots, p + q$, define $c_{ij} = -a'_{ij}$ and $c_{ji} = a'_{ij}$. In this notation, the problem (19.8) becomes the following linear program:

$$\text{Minimize} \quad \theta \tag{19.9a}$$

subject to

$$\pi'(i) - \pi'(j) + \theta \le c_{ij} \quad \text{for every } i = 1, \ldots, p \text{ and } j = p + 1, \ldots, p + q, \tag{19.9b}$$

$$-\pi'(i) + \pi'(j) + \theta \le c_{ji} \text{ for every } i = 1, \ldots, p \text{ and } j = p + 1, \ldots, p + q. \tag{19.9c}$$

This problem is similar to the one addressed in the preceding application; in this case we want to find the minimum value of θ for which the system of difference constraints (19.9b) and (19.9c) has a feasible solution. Our discussion in the last application implies that the system of inequalities (19.9b) and (19.9c) has a feasible solution if and only if the network shown in Figure 19.7, with $c_{ij} - \theta$ as the length of each arc (i, j), does not contain a negative cycle. The latter statement is equivalent to saying that the network does not contain a negative cycle with mean θ. Therefore, determining the minimum value of θ for which the network contains no negative cycle corresponds to determining the minimum mean cycle of the graph in Figure 19.7. In Section 5.7, we showed how to use dynamic programming to solve the problem efficiently. If u^* denotes the minimum cycle mean and $d(\cdot)$ represents shortest path distances with $c_{ij} - \theta$ as arc lengths, then $\pi'(i) = -d(i)$ solves (19.9). We can use these $\pi'(i)$ values to obtain row multipliers and column divisors.

19.4 SCIENCE APPLICATIONS

In previous chapters we have considered several applications in the physical and medical sciences, for example, reconstructing the left ventricle from x-ray projections and determining chemical bonds. To illustrate other possibilities in the science arena, we next consider two applications in the field of biology: DNA sequencing and automatic karyotyping of chromosomes. We solve the first of these problems as a shortest path problem and the second one as a transportation problem.

Application 19.7 DNA Sequence Alignment

Scientists model strands of DNA as a sequence of letters drawn from the alphabet $\{A, C, G, T\}$. Given two sequences of letters, say $B = b_1 b_2 \cdots b_p$ and $D = d_1 d_2 \cdots d_q$ of possibly different lengths, molecular biologists are interested in determining how similar or dissimilar these sequences are to each other. (These sequences are subsequences of a genome and typically contain several thousand letters.) A natural way of measuring the dissimilarity between the two sequences B and D is to determine the minimum "cost" required to transform sequence B into sequence D. To transform B into D, we can perform the following operations: (1) insert an element in B (at any place in the sequence) at a "cost" of α units; (2) delete an element from

B (at any place in the sequence) at a "cost" of β units; and (3) mutate an element b_i into an element d_j at a "cost" of $g(b_i, d_j)$ units. Needless to say, it is possible to transform the sequence B into the sequence D in many ways, so identifying a minimum cost transformation is a nontrivial task. We show how we can solve this problem using dynamic programming, which we can also view as solving a shortest path problem on an appropriately defined network.

Suppose that we conceive of the process of transforming the sequence B into the sequence D as follows. Add or delete elements from the sequence B so that the modified sequence, say B', has the same number of elements as D. Next "align" the sequences B' and D to create a one-to-one alignment between their elements. Finally, mutate the elements in the sequence B' so that this sequence becomes identical with the sequence D. As an example, suppose that we wish to transform the sequence B = AGTT into the sequence D = CTAGC. One possible transformation is to delete one T from B and add two new elements at the beginning, giving the sequence B' = ⊕⊕AGT (we denote the new element by the placeholder ⊕ and later assign a letter to this placeholder). We then align B' with D, as shown in Figure 19.8, and mutate the element T into C so that the sequences become identical. Notice that because we are free to assign values to the newly added elements, they do not incur any mutation cost. The cost of this transformation is $\beta + 2\alpha + g(\text{T}, \text{C})$.

$$\begin{aligned} B' &= \oplus\oplus\text{AGT} \Rightarrow \text{CTAGC} \\ D &= \text{CTAGC} \quad\;\; \text{CTAGC} \end{aligned}$$

Figure 19.8 Transforming the sequence B into the sequence D.

We now describe a dynamic programming formulation of this problem. Let $f(i, j)$ denote the minimum cost of transforming the subsequence $b_1 b_2 \cdots b_i$ into the subsequence $d_1 d_2 \cdots d_j$. We are interested in the value $f(p, q)$, which is the minimum cost of transforming B into D. To determine $f(p, q)$, we determine $f(i, j)$ for all $i = 0, 1, \ldots, p$, and for all $j = 0, 1, \ldots, q$. We can determine these intermediate quantities $f(i, j)$ using the following recursive relationships:

$$f(i, 0) = \beta_i \qquad \text{for all } i, \tag{19.10a}$$

$$f(0, j) = \alpha_j \qquad \text{for all } j, \tag{19.10b}$$

$$f(i, j) = \min\{f(i-1, j-1) + g(b_i, d_j),\ f(i, j-1) + \alpha,\ f(i-1, j) + \beta\}. \tag{19.10c}$$

We now justify this recursion. The cost $f(i, 0)$ of transforming a sequence of i elements into a null sequence is the cost of deleting i elements. The cost $f(0, j)$ of transforming a null sequence into a sequence of j elements is the cost of adding j elements. Next consider $f(i, j)$. Let B' denote the optimal aligned sequence of B (i.e., the sequence we create just before the mutation of B' to transform it into D). At this point, B' satisfies exactly one of the following three cases:

Case 1. *B' contains the letter b_i, which is aligned with the letter d_j of D* [as shown in Figure 19.9(a)].
In this case, $f(i, j)$ equals the optimal cost of transforming the subsequence $b_1 b_2 \cdots$

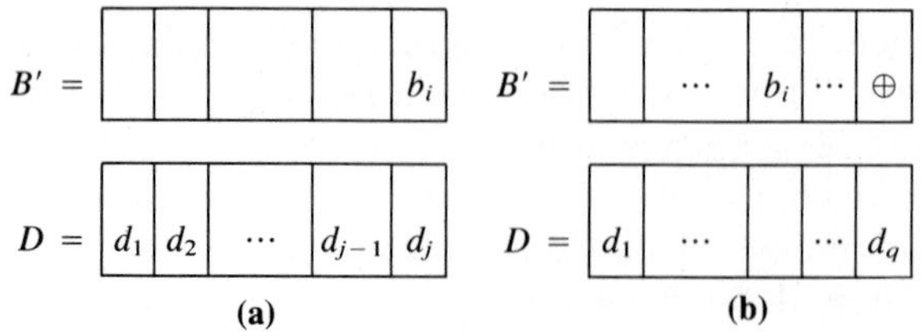

Figure 19.9 Explaining the dynamic programming recursion.

b_{i-1} into $d_1 d_2 \cdots d_{j-1}$ and the cost of transforming the element b_i into d_j. Therefore, $f(i, j) = f(i - 1, j - 1) + g(b_i, d_j)$.

Case 2. *B' contains the letter b_i, which is not aligned with the d_j* [as shown in Figure 19.9(b)].
In this case, b_i is to the left of d_j, so a newly added element must be aligned with b_j. In this case $f(i, j)$ equals the optimal cost of transforming the subsequence $b_1 b_2 \cdots b_i$ into $d_1 d_2 \cdots d_{j-1}$ plus the cost of adding a new element to B. Therefore, $f(i, j) = f(i, j - 1) + \alpha$.

Case 3. *B' does not contain the letter b_i.*
In this case we must have deleted b_i from B, so the optimal cost of the transformation equals the cost of deleting this element and transforming the remaining sequence into D. Therefore, $f(i, j) = f(i - 1, j) + \beta$.

The preceding discussion justifies the recursive relationships specified in (19.10). We can use these relationships to compute $f(i, j)$ for increasing values of i and, for a fixed value of i, for increasing values of j. This method allows us to compute $f(p, q)$ in $O(pq)$ time.

We can alternatively formulate the DNA sequence alignment problem as a shortest path problem. In Figure 19.10 we show the shortest path network for this

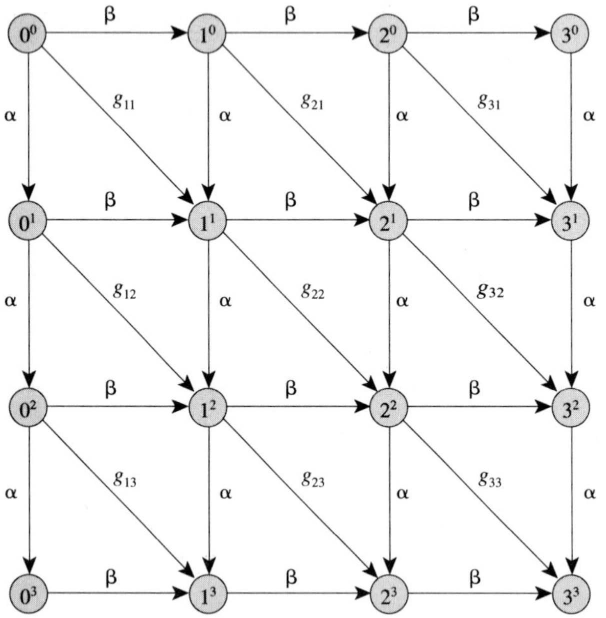

Figure 19.10 Sequence alignment problem as a shortest path problem.

formulation for a situation with $p = 3$ and $q = 3$. For simplicity, in this network we denote $g(b_i, d_j)$ by g_{ij}. We can establish the correctness of this formulation by applying an induction argument based on the induction hypothesis that the shortest path length from node 0^0 to node i^j equals $f(i, j)$. The shortest path from node 0^0 to node i^j must contain one of the following arcs as the last arc in the path: (1) arc $(i - 1^{j-1}, i^j)$, (2) arc (i^{j-1}, i^j), or (3) arc $(i - 1^j, i^j)$. In these three cases, the lengths of these paths will be $f(i - 1, j - 1) + g(b_i, d_j)$, $f(i, j - 1) + \alpha$, and $f(i - 1, j) + \beta$. Clearly, the shortest path length $f(i, j)$ will equal the minimum of these three numbers, which is consistent with the dynamic programming relationships stated in (19.10).

Application 19.8 Automatic Karyotyping of Chromosomes

A normal human cell has 46 chromosomes, usually subdivided into 23 groups with two identical (homologous) chromosomes per group, except for one male chromosome. For males, the group with the sex chromosomes has one X chromosome and one Y chromosome; these two chromosomes are not homologous. Each of these 23 groups of chromosomes has its own characteristic features and serves different biological functions. In certain clinical tests, such as amniocentesis, it is necessary to identify each of the 46 chromosomes of the cell. The medical community refers to the process of identifying which chromosomes belong to which of the 23 chromosome classes as *karyotyping*.

We consider the process of karyotyping a female cell. In this case each of the 23 chromosome classes consists of two homologous chromosomes. (Karyotyping of male cells is identical except for the treatment of the sex chromosomes.) The chromosomes of a suitably stained cell, when viewed under a microscope, exhibit a series of characteristic bands along the length of the cell. For each chromosome i of a cell and for each chromosome class j, we can assign a measure p_{ij} which is the probability that chromosome i is a member of class j. In some settings, a clinician measures the characteristic bands used to determine the p_{ij}'s. In other settings, a commercially available imaging machine, such as the Cytoscan system, uses a mechanical scanning device (known as a linear CCD array) to measure the bands.

The assignment of chromosomes to classes is easy if p_{ij} is nearly 1 for all correct assignments; however, in practice, the values of p_{ij}'s might be far from 1 for some correct assignments. An important task in karyotyping is to assign the chromosomes to classes in order to maximize the expected number of correct assignments, or, equivalently, to minimize the number of incorrect assignments.

We let $x_{ij} = 1$ if we assign chromosome i to class j, and $x_{ij} = 0$ otherwise. Using this notation we can formulate the problem of minimizing the expected number of incorrect assignments of chromosomes to chromosome classes as the following transportation problem:

$$\text{Minimize} \sum_{i=1}^{46} \sum_{j=1}^{23} (1 - p_{ij})x_{ij}$$

subject to

$$\sum_{j=1}^{23} x_{ij} = 1 \qquad \text{for } i = 1 \text{ to } 46,$$

$$\sum_{i=1}^{46} x_{ij} = 2 \qquad \text{for } j = 1 \text{ to } 23,$$

$x_{ij} \geq 0$ and integer.

19.5 PROJECT MANAGEMENT

An important class of network problems centers around the planning and scheduling of large projects, such as constructing a building or a highway, planning and launching a new product, installing and debugging a computer system, or developing and implementing a space exploration program. This application context was among the earliest successes of network optimization, and the network flow models of project management continue to be an important management tool used in numerous industries every day. In this section we consider three basic models of project management: a shortest path technique for scheduling projects to achieve the earliest possible completion and two network flow models for (1) just-in-time scheduling of jobs in a project and (2) deciding where to allocate additional resources to reduce a project's overall duration.

Application 19.9 Determining Minimum Project Duration

For the purpose of modeling, we envision a project as a set of jobs and a set A of precedence relations between the jobs. If $(i, j) \in A$, we need to complete job i before beginning job j. In addition, each job j has a known duration τ_j. The problem is to identify the project schedule (i.e., the start time of each job) that will satisfy the precedence relations between the jobs and complete the project in the least possible amount of time (alternatively, that gives the least possible *project duration*). Consider, for example, the project planning problem given in Figure 19.11.

We can formulate this project planning problem as a shortest path problem; in

Job	**Duration**	**Immediate predecessors**
a	14	—
b	3	—
c	3	*a, b*
d	7	*a*
e	4	*d*
f	10	*c, e*

Figure 19.11 Project planning problem.

fact, there are two alternative methods for doing so. In the first method we represent jobs by arcs, and in the second method we represent jobs by nodes. Although the first of these approaches is more popular in the literature, we adopt the latter approach in this discussion because it is conceptually simpler.

To formulate the project planning problem as a shortest path problem, we define a project network by associating a node j with each job j, and by including arc (i, j) whenever job i is an immediate predecessor of job j. We set the length c_{ij} of arc (i, j) equal to τ_i, the duration of job i. We also introduce a source node s, denoting the beginning of the project, and connect it to every node that has no incoming arc (corresponding to jobs without any predecessors) by zero-length arcs. Similarly, we introduce a sink node t, denoting the end of the project, and connect every node i with no outgoing arc to this sink node by an arc (i, t) whose length equals the duration of job i. Figure 19.12 gives the network corresponding to the project planning example shown in Figure 19.12. Note that the network corresponding to any project planning model must be acyclic because we could never complete a network containing a cycle (why?).

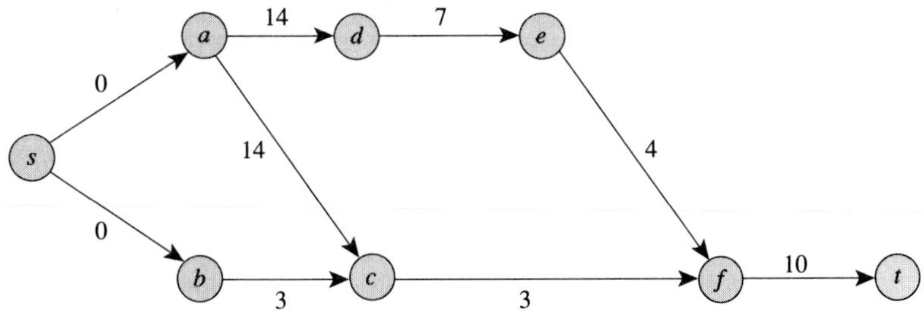

Figure 19.12 Shortest path formulation of the project planning problem.

Let $u(j)$ denote the earliest possible start time of job j in a project planning schedule that satisfies the precedence constraints. Notice that with respect to quantities $u(j)$, the project duration is $u(t) - u(s)$. We can state the project planning problem as the following optimization model:

$$\text{Minimize} \quad u(t) - u(s), \tag{19.11a}$$

subject to

$$u(j) - u(i) \geq c_{ij}, \text{ for all } (i, j) \in A, \tag{19.11b}$$

$$u(j) \text{ unrestricted.} \tag{19.11c}$$

The inequalities (19.11b) model the precedence constraints by stating that if job i is an immediate predecessor of job j, then job j can start only after $c_{ij} = u_i$ units of time have elapsed since the start of job i.

To bring (19.11) into a familiar network flow form, we take its dual. If we associate the dual variables x_{ij} with the constraints (19.11b), the dual linear program is

$$\text{Maximize} \quad \sum_{\{j:(i,j)\in A\}} c_{ij}x_{ij}$$

subject to

$$\sum_{\{j:(j,i)\in A\}} x_{ji} - \sum_{\{j:(i,j)\in A\}} x_{ij} = \begin{cases} -1 & \text{for } i = s, \\ 0 & \text{for all } i \in N - \{s, t\}, \\ 1 & \text{for } i = t, \end{cases}$$

$$x_{ij} \geq 0 \qquad \text{for all } (i, j) \in A.$$

Clearly, this is a longest path problem with c_{ij} as the length of arc (i, j); we wish to send 1 unit of flow from node s to node t along the longest path. To solve the problem we multiply each arc length by -1 and then solve a shortest path problem. Note that since the project planning network is acyclic, by multiplying the arc lengths by -1, we do not create any negative cycle. Let $d(\cdot)$ denote the vector of shortest path distances for this problem. Then by setting $u(i) = -d(i)$ for each $i \in N$, we obtain an optimal solution of (19.11).

Application 19.10 Just-in-Time Scheduling

The *just-in-time scheduling problem* is an extension of the project planning problem that we discussed in the last application. In the just-in-time scheduling problem, we determine the minimum project duration subject to both the precedence constraints and some additional "just-in-time constraints." In this problem we are given a subset $S \subseteq A$ and a number α_{ij} for each $(i, j) \in S$. The just-in-time constraints state that for each $(i, j) \in S$, job j must start within α_{ij} units of time from the start of job i. Notice that if $\alpha_{ij} = c_{ij}$, the just-in-time scheduling constraint for arc (i, j) says that job i must start exactly c_{ij} units before the start of job j, which is the latest possible start time for this job. Just-in-time is a management philosophy that has become very popular in recent years; it attempts to eliminate waste by reducing slack times and buffers, such as inventory between distribution, production, and scheduling activities.

If $u(\cdot)$ denotes the earliest start times of the jobs, the just-in-time constraints require that

$$u(j) \leq u(i) + \alpha_{ij} \qquad \text{for all } (i, j) \in S,$$

or, equivalently,

$$u(i) - u(j) \geq -\alpha_{ij} \qquad \text{for all } (i, j) \in S. \tag{19.12}$$

The start times must also satisfy the usual precedence constraints:

$$u(j) - u(i) \geq c_{ij} \qquad \text{for all } (i, j) \in A. \tag{19.13}$$

In the just-in-time scheduling problem, we wish to minimize $[u(t) - u(s)]$ subject to the inequalities (19.12) and (19.13). Just as we noted in our discussion in the last application, we can solve this problem as a longest path problem, in this case on an augmented network $G' = (N, A')$ whose arc set A' includes an arc (i, j) of cost c_{ij} for each $(i, j) \in A$ and an arc (j, i) of cost $-\alpha_{ij}$ for each $(i, j) \in S$. To transform this longest path problem into a shortest path problem, we multiply each arc cost by -1. Notice that in this case the augmented network G' might not be acyclic, and the resulting shortest path problem might contain a negative cycle. The presence of a negative cycle indicates that the just-in-time scheduling problem has no feasible solution (why?). When the resulting shortest path problem has no negative cycle, the negative of the shortest path distances provide optimal start times for the jobs.

As a variant of the just-in-time scheduling problem, suppose that instead of imposing an upper bound on when job j should start after the start of job i, we penalize the time difference between the completion of job i and the start of job j using a penalty factor of d_{ij}. We wish to determine start times of jobs that will minimize this penalty and yet satisfy the restriction that the project duration is at most λ (a specified constant). The following linear program models this problem:

$$\text{Minimize} \quad \sum_{(i,j)\in A} (u(j) - u(i) - c_{ij})d_{ij}, \tag{19.14a}$$

subject to

$$-u(t) + u(s) \geq -\lambda, \tag{19.14b}$$

$$u(j) - u(i) \geq c_{ij} \text{ for all } (i, j) \in A, \tag{19.14c}$$

$$u(j) \text{ unrestricted for all } j \in N. \tag{19.14d}$$

Let $D_i = \sum_{\{j:(j,i)\in A\}} d_{ji} - \sum_{\{j:(i,j)\in A\}} d_{ij}$. The following linear program is the dual of this model.

$$\text{Maximize} \quad \sum_{\{j:(i,j)\in A\}} c_{ij}x_{ij} - \lambda x_{ts}$$

subject to

$$\sum_{\{j:(j,s)\in A\}} x_{js} - \sum_{\{j:(s,j)\in A\}} x_{sj} + x_{ts} = D_s,$$

$$\sum_{\{j:(j,i)\in A\}} x_{ji} - \sum_{\{j:(i,j)\in A\}} x_{ij} = D_i \quad \text{for all } i \neq s \text{ or } t,$$

$$\sum_{\{j:(j,t)\in A\}} x_{jt} - \sum_{\{j:(t,j)\in A\}} x_{tj} - x_{ts} = D_t,$$

$$x_{ij} \geq 0 \quad \text{for all } (i, j) \in A.$$

Note that this problem is a minimum cost network flow problem with an arc (t, s) from the end node t to the start node s.

Application 19.11 Time–Cost Trade-off in Project Management

Project scheduling problems provide celebrated applications of minimum cost flow duality theory. In the basic project scheduling problem that we examined in the preceding applications, we showed how we could represent an interrelated set of jobs as a network. As we noted in that discussion, whenever the time to perform each job is fixed, we can determine the minimum project duration by solving a shortest path problem. Many project managers would argue that they could reduce the duration of most jobs by allocating extra resources (people, machines, or money) to them. Although the use of the added resources typically increases the cost of carrying out any particular job, expediting or *crashing* the job might permit the project team to complete the entire project more quickly. There might, however, be no reason to shorten the length of some jobs if they have a generous amount of *slack* (i.e., the time between the earliest and latest possible start times for the job that would not delay the project); the project team should perform this job at its *normal* pace, without any added resources. Thus we need not crash all jobs to complete a project faster; only certain "critical" jobs that have no slack need to be crashed. As we will see, we can formulate the problem of identifying which jobs to expedite, and by what amount, to complete the project within a specified time λ, as a minimum cost flow problem.

With job (i) in the network we associate a *time–cost trade-off curve*, which represents the cost of performing the job as a function of its duration. We assuxe that the time–cost trade-off curve is linear, as shown in Figure 19.13. We use three numbers to specify the trade-off curve for any job (i): Its "normal" completion time a_i, its "crash" completion time b_i, and the cost d_i of shortening the job duration by 1 unit. To assess the overall project cost, we would also need to know the normal cost of performing a job. But since we must perform every job, the normal cost contributes a constant to the objective function value; therefore, we ignore it from the analysis.

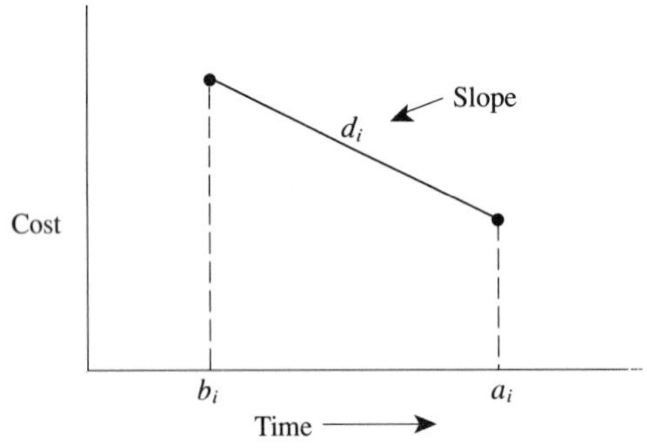

Figure 19.13 Time–cost trade-off curve of job i.

We would like to determine the optimal reduction, if any, in each job time that would permit us to reduce the project duration to a given value λ while incurring the least possible cost in reducing one or more of the job durations. To formulate this problem as a network flow model, we introduce some notation. As in the last application, we let node s designate the beginning of the project and node t designate its completion. The formulation uses two decision variables for each job: (1) $u(i)$, which denotes the start time of job i (i.e., the start time of all the jobs emanating from node i) and (2) β_i, which represents the reduction in time required to complete job i. After we reduce the duration of the jobs, the time required to complete job i will be $a_i - \beta_i$. Using this notation, we state the project scheduling problem mathematically as follows:

$$\text{Minimize} \sum_{i \in N} d_i \beta_i \tag{19.15a}$$

subject to

$$u(t) - u(s) \le \lambda, \tag{19.15b}$$

$$u(j) \ge u(i) + (a_i - \beta_i) \qquad \text{for all } (i, j) \in A, \tag{19.15c}$$

$$0 \le \beta_i \le a_i - b_i \qquad \text{for all } i \in N, \tag{19.15d}$$

$$u(i) \text{ unrestricted for all } i \in N. \tag{19.15e}$$

In this formulation, the constraint (19.15a) assures that the project duration is at most λ, constraint (19.15b) ensures that the jobs satisfy their precedence constraints, and (19.15c) imposes lower and upper bounds on the reductions in job times. We next make a substitution of variables to simplify (19.15). Let $\beta_i' = u(i) - \beta_i$ for all $i \in N$. Substituting β_i' in (19.15) in the place of β_i, we obtain the following modified formulation:

$$\text{Minimize} \sum_{i \in N} d_i u(i) - \sum_{i \in N} d_i \beta_i' \tag{19.16a}$$

subject to

$$u(t) - u(s) \le \lambda, \tag{19.16b}$$

$$u(j) - \beta_i' \ge a_i \qquad \text{for all } (i, j) \in A, \tag{19.16c}$$

$$u(i) - \beta_i' \ge 0 \qquad \text{for all } i \in N, \tag{19.16d}$$

$$u(i) - \beta_i' \le a_i - b_i \qquad \text{for all } i \in N, \tag{19.16e}$$

$$u(i) \text{ and } \beta_i' \text{ are unrestricted for all } i \in N. \tag{19.16f}$$

In this formulation, each row has at most one $+1$ and at most one -1; consequently, our discussion in Section 9.4 shows that this model is the dual of a minimum cost flow problem. Consequently, we can obtain an optimal solution of (19.16) using any minimum cost flow algorithm.

19.6 DYNAMIC FLOWS

Much of our discussion in this book has focused on static models, that is, problems that have no underlying temporal dimensions. As we have seen, static network models provide good mathematical representations of a great many applications. In some other applications, however, such as scheduling of people, jobs, or projects, or the carryover of inventory of a product from one time period to another, time is an essential ingredient. In these instances, to account properly for the evolution of the underlying system over time, we need to use dynamic network flow models. We might view these models as being composed of multiple copies of an underlying network, one at each point in time: arcs that link these static "snapshots" of the underlying network describe temporal linkages in the system. Since each node in the network now has an associated time, we refer to the dynamic networks as *time-expanded networks*. Dynamic network models arise in many problem settings, including production-distribution systems, economic planning, energy systems, traffic systems, and building evacuation systems. In the following discussion we describe two applications of dynamic network flow models.

Application 19.12 Maximum Dynamic Flows

The maximum dynamic flow problem is a variant of the maximum flow problem that arises, for example, in the following scenario. In a war between two countries, A and B, suppose that the generals of army A have decided to launch a major offensive in the next 24 hours using their major infantry units based at location s against enemy troops at location t. The generals would like to send a maximum number of units from location s to location t within 24 hours while honoring the arc capacities and traversal times of the arcs.

In the maximum flow problem, we maximize the number of flow units that can pass through the network from node s to node t *per unit time* while satisfying the arc capacities u_{ij}. In the dynamic flow problem, we maximize the total number of flow units that can be sent from node s to node t in p time periods while satisfying arc capacities u_{ij} and the *arc traversal times* τ_{ij}. In other words, the maximum flow problem determines the maximum *steady state* flow per unit time between two nodes, so we might refer to this problem as the *static flow problem*. On the other hand, the maximum dynamic flow problem maximizes the total (transient) flow that we can send between two nodes within a given period. We will show how to transform the maximum dynamic flow problem into a maximum flow problem on a new network $G^p = (N^p, A^p)$, called the *time-expanded replica* of G. We illustrate our transformation on the example shown in Figure 19.14(a) with the time-expanded replica for $p = 6$ shown in Figure 19.14(b).

For a given network $G = (N, A)$, we form the network G^p as follows. We make p copies $i_1, i_2, \ldots, i_p$ of each node i. Node i_k in the time-expanded network represents node i of the original network at time k. We include arc (i_k, j_l) of capacity u_{ij} in the time-expanded network whenever $(i, j) \in A$ and $l - k = \tau_{ij}$; the arc (i_k, j_l) in the time-expanded network represents the potential movement of a commodity from node i to node j in time τ_{ij}. It is easy to see that any static flow in G^p from the source nodes $s_1, s_2, \ldots, s_p$ to the sink nodes $t_1, t_2, \ldots, t_p$ is equivalent to a dynamic flow in G, and vice versa. We can further reduce the multiple-source, multiple-sink problem in the time-expanded network to the single-source, single-sink problem by introducing a *super-source node* s^* and a *super-sink node* t^*. Consequently, we can solve the maximum dynamic flow problem in G by solving an (ordinary) maximum flow problem in G^p.

Application 19.13 Models for Building Evacuation

In large metropolitan areas, among the criteria used to design large buildings, architects must ensure sufficient capabilities to evacuate buildings quickly: to respond, for example, to a fire, an earthquake, a toxic or natural gas leak, a power blackout, a bomb threat, or a civil defense emergency. As an aid to their design efforts, the architects would like to be able to develop an evacuation plan and assess the evacuation time for any particular design. We show how to model this building evacuation problem as a dynamic flow problem and solve it using a minimum cost flow algorithm. In our description we present a highly simplified version of the building evacuation problem. The references cited at the end of this chapter provide a more realistic description of the problem.

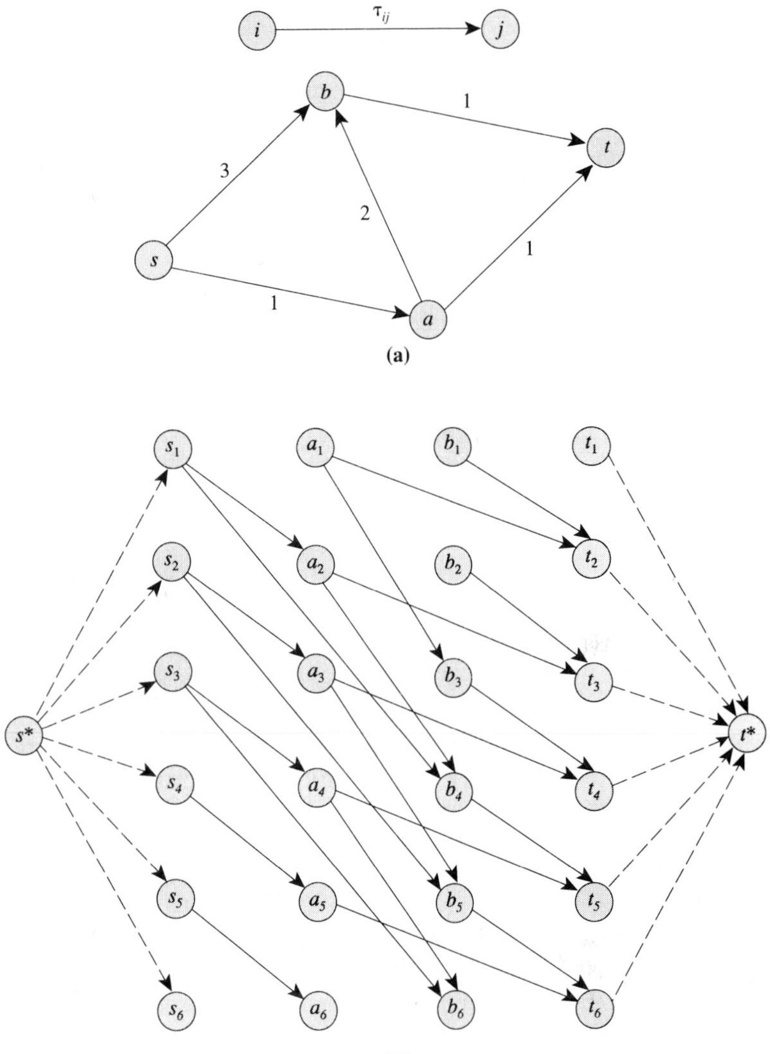

Figure 19.14 (a) Network G; (b) time-expanded replica of G. All capacities equal 2 in both the networks.

We first construct a static network for the building. The nodes of this network represent locations of the building such as work centers, offices, hallways, elevator stops, staircases, and building exits; the arcs represent passages between locations. Those locations of the building that house a significant number of people are source nodes in the network, and the building exits are the sink nodes. The supply of a source node equals an estimate of the number of people in the location that the node represents. The capacity of an arc is the number of people that can pass through the associated passage way per unit time. For example, if we anticipate that at most 60 people per minute can pass by every point in a stairwell, and the length of the time period in our model is 10 seconds, the capacity of the stairwell is $(60)(\frac{1}{6}) = 10$

units. We estimate the travel time of an arc and represent it as an integral number of time periods. For example, specifying two time periods for descending one floor in a stairwell means that we allow 20 seconds.

Although the static model might have multiple sources and multiple sinks (exits), we can transform it into a model with single source and single sink by using the standard technique of introducing a *super-source* node and a *super-sink* node. Therefore, we assume that the network contains a single source s with a given supply, say B, and a single sink t.

We construct the time-expanded replica of the static model as described in the last application. The time-expanded replica of the network contains p nodes i^1, i^2, i^3, . . . , i^p for each node i in the static network; we choose p to be suitably large to ensure that we can evacuate the building within p time periods. The minimum time required to evacuate the building is the smallest index r satisfying the property that the maximum flow from the nodes s^1, s^2, . . . , s^r to the nodes t^1, t^2, . . . , t^r is at least B people (B is given data). As shown in Exercise 19.11, we can solve this problem by a single application of any minimum cost flow algorithm or repeated application of any maximum flow algorithm.

19.7 ARC ROUTING PROBLEMS

Leaving from his or her home post office, a postal carrier needs to visit the households on each block in the carrier's route, delivering and collecting letters, and then returning to the post office. The carrier would like to cover this route by traveling the minimum possible distance. Mathematically, this problem has the following form: Given a network $G = (N, A)$ whose arcs (i, j) have an associated nonnegative length c_{ij}, we wish to identify a walk of minimum length that starts at some node (the post office), visits each arc of the network at least once, and returns to the starting node. This problem has become known as the *Chinese postman problem* because it was first discussed by a Chinese mathematician, K. Mei-Ko. The Chinese postman problem arises in other application settings as well; for instance, in patrolling streets by police, routing of street sweepers and household refuse collection vehicles, fuel oil delivery to households, and the spraying of roads with sand during snowstorms.

We discuss the Chinese postman problem on directed and undirected networks separately. We will show how to reduce the Chinese postman problem on a directed network to a minimum cost flow problem and how to solve the Chinese postman problem on an undirected network as a nonbipartite weighted matching problem. Interestingly, the Chinese postman problem on a mixed network (i.e., some arcs are directed and the others are undirected) is $\mathcal{NP}$-complete.

Application 19.14 Directed Chinese Postman Problem

In the Chinese postman problem on directed networks, we are interested in a closed (directed) walk that traverses each arc of the network at least once. The network need not contain any such walk! Figure 19.15 shows an example. The network contains the desired walk if and only if every node in the network is reachable from every other node; that is, it is *strongly connected*. In Section 3.4 we showed how

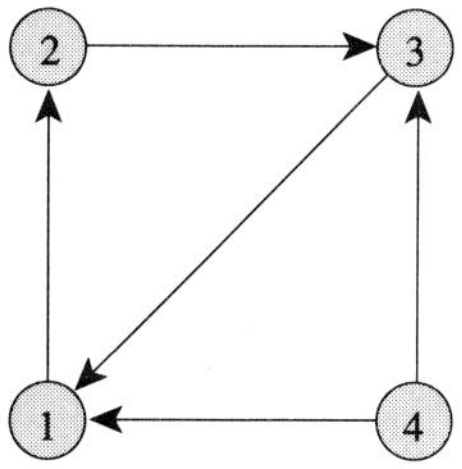

Figure 19.15 Network with no feasible solution for the Chinese postman problem.

to determine whether or not the graph is strongly connected within $O(m)$ time. In the remainder of our discussion we assume that the network G is strongly connected.

In an optimal walk, a postal carrier might traverse arcs more than once. The minimum length walk minimizes the sum of lengths of the repeated arc traversals. Let x_{ij} denote the number of times the postal carrier traverses arc (i, j) in a walk. Any carrier walk must satisfy the following conditions:

$$\sum_{\{j:(i,j)\in A\}} x_{ij} - \sum_{\{j:(j,i)\in A\}} x_{ji} = 0 \qquad \text{for all } i \in N, \tag{19.17a}$$

$$x_{ij} \geq 1 \qquad \text{for all } (i, j) \in A. \tag{19.17b}$$

The constraints (19.17a) state that the carrier enters a node the same number of times that he or she leaves it. The constraints (19.17b) state that the carrier must visit each arc at least once. Any solution x satisfying (19.17a) and (19.17b) defines a carrier's walk. We can construct such a walk in the following manner. We replace each arc (i, j) with flow x_{ij} with x_{ij} copies of the arc, each carrying a unit flow. Let A' denote the resulting arc set. Since the outflow equals inflow for each node in the flow x, once we have transformed the network, the outdegree of each node will equal its indegree. The flow decomposition theorem (i.e., Theorem 3.5) implies that we can decompose the arc set A into a set of at most m directed cycles. We can connect these cycles together to form a closed walk as follows. The carrier starts at some node in one of the cycles, say W_1, and visits the nodes (and arcs) of W_1 in order until returning to his or her starting node, or encountering a node that also lies in a directed cycle not yet visited, say W_2. In the former case, the walk is complete; and in the latter case, the carrier visits cycle W_2 first before resuming his or her visit of the nodes in W_1. While visiting nodes in W_2, the carrier follows the same policy, i.e., if he/she encounters a node lying on another directed cycle W_3 not yet visited, he/she visits W_3 first before visiting the remaining nodes in W_2, and so on. We illustrate this method on a numerical example. Let A' be as indicated in Figure 19.16(a). This solution decomposes into three directed cycles, W_1, W_2, and W_3. As shown in Figure 19.16(b), the carrier starts at node a and visits the nodes in the following order: a–b–d–g–h–c–d–e–b–c–f–a.

This discussion shows that the solution x defined by a feasible walk for the carrier satisfies (19.17), and, conversely, every feasible solution of (19.17) defines a walk of the postman. The length of a walk equals $\sum_{(i,j)\in A} c_{ij}x_{ij}$. Therefore, the Chinese postman problem seeks a solution x that minimizes $\sum_{(i,j)\in A} c_{ij}x_{ij}$, subject to the set of constraints (19.17). This problem is clearly an instance of the minimum cost flow problem.

One particular instance of the directed Chinese postman problem deserves

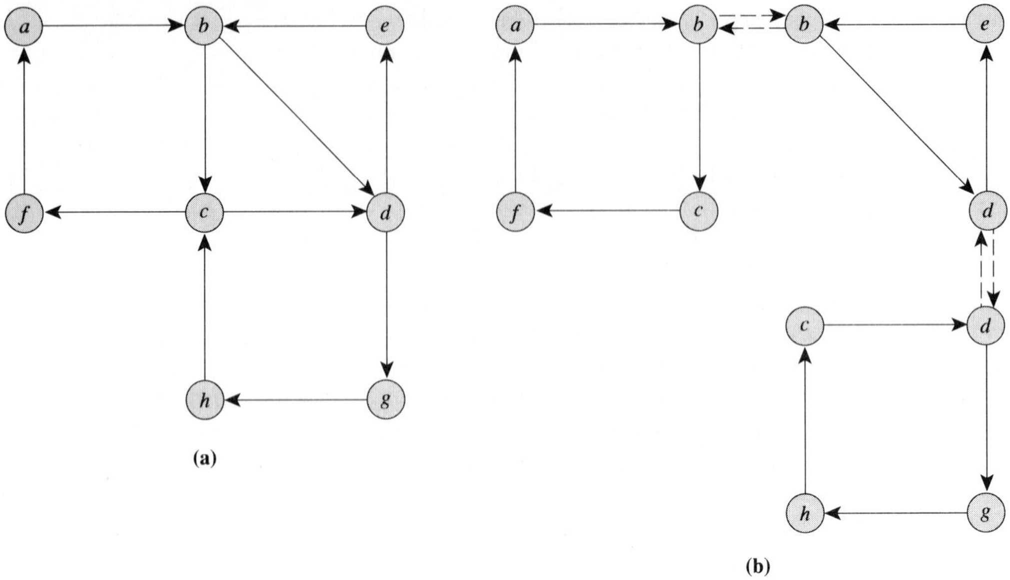

Figure 19.16 Constructing a closed walk for the postal carrier.

special mention. Suppose that the indegree of each node in G equals its outdegree. In that case, by setting $x_{ij} = 1$ for each arc $(i, j) \in A$, we define a feasible solution of (19.17) that is also a minimum cost solution (why?). Consequently, whenever the indegree of each node in a network equals its outdegree, the network contains a walk that traverses each arc exactly once (i.e., the walk has no arc repetitions). If the network contains a node whose indegree differs from its outdegree, the walk must necessarily repeat some arcs (why?). This result is the directed version of the well-known Euler's theorem for undirected graphs that we discussed in Exercise 3.45.

Application 19.15 Undirected Chinese Postman Problem

The undirected version of the Chinese postman problem is more complicated than the directed version; as we will see, rather than solving this problem as a minimum cost flow problem, we will solve it by using both an all-pairs shortest path algorithm and a nonbipartite weighted matching algorithm.

As we observed in the directed case, a network contains a directed walk visiting each arc exactly once if and only if the indegree of each node equals its outdegree. For the undirected case, we have an analogous result, which is known as Euler's theorem. This theorem states that a graph contains a closed walk that visits each arc exactly once if and only if every node in the graph has an even degree (i.e., the graph is *even*). In Euler's honor, the research community often refers to even graphs as Eulerian graphs and to closed walks that visit each arc exactly once as Eulerian tours.

Euler's theorem is very easy to establish. Note that since any closed walk in an undirected network enters and leaves any node the same number of times, the subgraph composed of the arcs in any closed walk is even. Therefore, if the network contains a closed walk visiting each arc exactly once, the graph must be even. The converse statement—that is, if the graph is even, then it contains an Eulerian tour—is an easy consequence of a simple probing algorithm that we discussed in the previous application. Since all Eulerian tours have the same cost (the sum of the arc costs), solving the Chinese postman problem on an even graph does not involve any optimization. We simply need to find any Eulerian tour.

If the graph is not even, a closed walk on a graph must repeat some arcs. Consequently, identifying a minimum length closed walk requires an optimization technique; we must determine which arcs to repeat. Suppose that a feasible carrier tour is a closed walk W that traverses arc (i, j), x_{ij} times. Suppose that we replace each arc (i, j) by x_{ij} copies of the same arc and refer to the resulting graph as the *traversal graph*. Notice that since the degree of each node in the traversal graph is even, it contains a closed walk visiting each arc exactly once. We now delete one copy of each arc (i, j) in the traversal graph and refer to the remaining graph as the *repetition graph*, which we represent as $G'' = (N, A'')$. Observe that the arc set $A \cup A''$ defines the traversal graph.

We now focus on the optimal repetition graph $G'' = (N, A'')$. A node i in $G'' = (N, A'')$ has an odd degree if and only if i has an odd degree in $G = (N, A)$, because $A \cup A''$ defines even degrees for all nodes. We claim that the repetition graph can be decomposed into an arc-disjoint union of cycles and paths satisfying the following properties: (1) each path starts and ends at odd-degree nodes in G'', and (2) each odd-degree node in G'' is the start node or the end node of exactly one path. We can establish this claim by using an argument similar to the one we used in the proof of the flow decomposition theorem (i.e., Theorem 3.5).

Consider now the *optimal repetition graph* G^* (i.e., the repetition graph corresponding to an optimal solution of the Chinese postman problem). We can assume without any loss of generality that G^* does not contain any cycle because each such cycle must have zero length (because arc lengths are nonnegative, any cycle length is nonnegative; hence, G^* would not be optimal if it contained a positive length cycle). We can eliminate any zero length cycle without affecting the objective function value. Therefore, we can assume that the optimal repetition graph G^* consists of arc-disjoint paths between odd-degree nodes. Each odd-degree node is paired to another odd-degree node by a path, and collectively these paths contain all the arcs in G^*. Notice that for a given pairing of odd-degree nodes, the path between them must be a shortest path; otherwise, we could improve the solution.

The preceding discussion implies the following procedure for solving the Chinese postman problem. We first solve an all-pairs shortest path problem in $G = (N, A)$ to compute shortest path distances d_{ij} between all pairs of nodes. We then construct a graph $G' = (N', A')$ whose node set consists of all odd-degree nodes in N and whose arc set contains arcs between every pair of nodes in N'. We set the cost of the arc $(i, j) \in A'$ to the shortest path distance d_{ij}. The network G' is nonbipartite. Next we identify a minimum weight (nonbipartite) matching M of G' to identify an optimal pairing of the nodes. For each arc $(i, j) \in M$ in the optimal matching, we add one copy of every arc in the shortest path from node i to node j to A. We can then decompose the resulting even graph into the carrier's closed walk by the procedure described in the previous application.

19.8 FACILITY LAYOUT AND LOCATION

Facility layout and location problems offer two perspectives on a similar set of issues. In facility layout problems we would like to determine the best internal configuration of an office building, plant, or warehouse. Where should we put offices, equipment, storage locations, and any support facilities, such as material handling equipment, heating and ventilation, a mailroom, dining rooms, and rest rooms? In answering this question, we would like to configure the facility for maximum benefit (e.g., most efficient interactions between the occupants) at the least possible cost, while recognizing building codes and other externally imposed regulations (e.g., evacuation codes). Facility location models typically are more macroscopic and consider the interactions between, rather than within, facilities: for example, between plants, warehouses, and retail outlets; or between libraries, fire houses, service facilities (e.g., drugstores, fast-food restaurants), and the populations that they serve. Many facility layout and facility location problems are integer programming models with embedded network flow structure since we either locate a facility at a particular location or not (a zero–one variable) and once we have decided on the location of the facilities, we wish to route people or goods between them at the least possible cost. Some facility location and layout problems, however, such as the two that we consider in this section, are pure network flow models.

Application 19.16 Discrete Location Problems

This application is concerned with the optimal location of p new facilities to be selected from an available set of $q \geq p$ sites. The new facilities interact with the r existing facilities. The objective is to locate the facilities to minimize the total transportation cost between the new and existing facilities. One example of this problem is the location of hospitals, fire stations, or libraries in a metropolitan area; in this setting, we can treat population concentrations as the existing facilities. Other examples of this problem are the locations of new airfields used to provide supplies for a number of military bases, new components on a control panel, and water fountains in an office building.

The data of this problem consist of (1) d_{kj}, the distance between existing facility k and site j; and (2) w_{ik}, the total transportation cost per unit distance (for some given time period) incurred between the new facility i and the existing facility k. If we use the same medium of transportation between all the facilities, we can let w_{ik} be the number of trips per unit time (day, week, month) made between the new facility i and the existing facility k. The objective is to assign each new facility i to an available site j, denoted by a binary variable x_{ij}, to minimize the total cost of transportation between new and existing facilities. For a given assignment x, $w_{ik} \sum_{j=1}^{q} d_{kj}x_{ij}$ is the cost of transportation between the new facility i and the existing facility k. Thus the total transportation cost is given by

$$\sum_{i=1}^{p} \sum_{k=1}^{r} \left(w_{ik} \sum_{j=1}^{q} d_{kj} x_{ij} \right),$$

or, equivalently,

$$\sum_{i=1}^{p} \sum_{j=1}^{q} \left(\sum_{k=1}^{r} w_{ik} d_{kj} \right) x_{ij} = \sum_{i=1}^{p} \sum_{j=1}^{q} c_{ij} x_{ij}.$$

In this expression $c_{ij} = \sum_{k=1}^{r} w_{ik} d_{kj}$ is the cost of locating the new facility i at site j. This model is an instance of the assignment problem.

Notice that in this model, the new facilities are independent of each other in the sense that the transportation cost (as measured by number of trips) between new facility i and old facility k does not depend upon the location of other new facilities. When the demand (and the transportation cost) between facilities depends on the location of other facilities (e.g., when the new facilities provide the same service and any old facility will travel to the least cost or closest facility), the models become more complicated, and typically are integer programming models.

Application 19.17 Warehouse Layout

A warehouse typically is configured with docks for loading and unloading goods and with open areas (or bins) for storing the goods. Trucks that deliver or pick up goods arrive at any one of the loading docks (typically, at random). The warehouse operators must collect or deliver the required items from their storage locations. In managing the warehouse, the operating staff must decide where to locate each of the goods—the closer a good is to any dock, the lower is the cost of (1) accessing that good and transferring it to the dock for loading, or (2) unloading the good from the dock and transporting it to its storage area. Consequently, the items "compete" for the storage areas that are closest to the docks. Suppose that a company is using a warehouse to store p items. The warehouse has r loading and unloading docks. Let w_{ik} be a known total cost per foot incurred in transporting item i between dock k and its storage region (this cost might be just the cost of the time of a forklift operator—the transport time might depend on the item being moved, because some items are heavier or bulkier than others). Typically, the warehouse will store items on pallets (small wooden platforms) and w_{ik} is directly proportional to the number of pallet loads of item i moving between dock k and the storage region of the item. The warehouse layout problem is to determine the region(s) for storing each of the p items that will minimize the total transportation cost between the items and the docks.

We first discretize this problem by subdividing the floor area into q square grids of equal size, numbered in any convenient manner from 1 to q. Since we represent the distance between each grid and each dock by a single (average) number, we will be approximating the travel distances and associated loading and unloading costs; the accuracy of the approximation depends on the size of the grids. Choosing a smaller grid size improves the accuracy of the approximation but increases the problem size; therefore, we must make a trade-off between accuracy and solution cost. Let F_i be the total number of grids required to store item i. For simplicity, we assume that $\sum_{i=1}^{p} F_i = q$. Let d_{kj} denote the distance between dock k and the center of grid j. Multiplying this quantity by w_{ik}, we obtain the transportation cost $w_{ik}d_{kj}$ for item i between grid location j and dock k. Since we are assuming that each item is equally likely to be loaded and unloaded from each dock, the average cost for locating item i in grid j is

$$c_{ij} = \frac{1}{F_i} \sum_{k=1}^{r} w_{ik}d_{kj}.$$

We convert the warehousing problem to a standard transportation problem by defining the variables x_{ij}, for all $i = 1, \ldots, p$ and $j = 1, 2, \ldots, q$, as follows:

$$x_{ij} = \begin{cases} 1 & \text{if we store item } i \text{ in grid square } j, \\ 0 & \text{if we do not store item } i \text{ in grid square } j. \end{cases}$$

In terms of the variables x_{ij}, the warehouse layout problem becomes

$$\text{Minimize} \quad \sum_{i=1}^{p} \sum_{j=1}^{q} c_{ij}x_{ij}$$

subject to

$$\sum_{j=1}^{q} x_{ij} = F_i \qquad \text{for } i = 1, \ldots, p,$$

$$\sum_{i=1}^{p} x_{ij} = 1 \qquad \text{for } j = 1, \ldots, q,$$

$$x_{ij} \geq 0, \qquad \text{for all } i = 1, \ldots, p, j = 1, \ldots, q,$$

which is clearly an instance of the transportation problem.

Application 19.18 Rectilinear Distance Facility Location

In this application we model a rectilinear distance facility location as the dual of a minimum cost flow problem. Suppose that r existing facilities (e.g., hospitals that offer different medical expertise) are located at distinct points in a two-dimensional plane: The ith facility has the coordinates (u_i, v_i). We wish to locate p new facilities. Let (x_j, y_j), which are the decision variables, denote the coordinates of the jth new facility. For each set of locations of the new facilities, we incur transportation costs that are proportional to the rectilinear distances between the new facilities and between the new and existing facilities. The *rectilinear distance* between any two points in the xy-plane is the shortest distance between the points if we restrict travel to be parallel to the two axes (i.e., x-axis and y-axis). Therefore, the rectilinear distance between the two points (x_1, y_1) and (x_2, y_2) is $(|x_1 - x_2| + |y_1 - y_2|)$. Because rectilinear distances model distances on many urban street networks, they are sometimes also called *manhattan distances*.

In our location problem, we want to determine the coordinates (x_j, y_j) for all $j = 1, \ldots, r$ that optimize the following objective function:

$$\text{Minimize} \sum_{j=1}^{p} \sum_{k=1}^{p} a_{jk}(|x_j - x_k| + |y_j - y_k|)$$

$$+ \sum_{i=1}^{r} \sum_{j=1}^{p} d_{ij}(|x_j - u_i| + |y_j - v_i|). \tag{19.18}$$

In this expression, a_{jk} and d_{ij} are constants that represent the cost (per unit time) per unit distance traveled between the two facilities. It might be useful to think of d_{ij} as the number of trips made between the existing facility i and the new facility j. With this interpretation, the term a_{jk} (as well as a_{kj}) would represent half of the number of trips made between the new facilities j and k (because the summation counts the interaction between the new facilities j and k twice). In the first term, we should, in fact, be summing over all indices j and k with $j \neq k$; since $|x_j - x_j| = |y_j - y_j| = 0$, for each j we can set a_{jj} to any value, e.g., $a_{jj} = 0$ for all j.

First, we notice that the objective function (19.18) consists of two parts; the first part contains the x-coordinates of the new facilities and the second part contains the y-coordinates. Since these two parts are independent of each other, we can optimize them separately. In view of this observation, we assume that our location problem has the following form:

$$\text{Minimize} \sum_{j=1}^{p} \sum_{k=1}^{p} a_{jk} |x_j - x_k| + \sum_{i=1}^{r} \sum_{j=1}^{p} d_{ij} |x_j - u_i|. \tag{19.19}$$

We have a similar problem for the y variables. To solve the location problem, we first convert it into a linear program. To do so, we use a standard technique for converting a function with absolute value functions to a linear programming problem [i.e., we replace the term "minimize $|\alpha - \beta|$" by the term "minimize $(c + d)$, subject to the constraints $\alpha - \beta = c - d$, $c \geq 0$, and $d \geq 0$"] (we discuss this transformation in Exercise 19.1). Using this technique, we convert (19.19) into the following equivalent problem, which we state in the form of a maximization problem:

$$\text{Maximize} \sum_{j=1}^{p} \sum_{k=1}^{p} (-a_{jk})(e_{jk} + f_{jk}) + \sum_{i=1}^{r} \sum_{j=1}^{p} (-d_{ij})(g_{ij} + h_{ij}) \tag{19.20a}$$

subject to

$$x_j - x_k - e_{jk} + f_{jk} = 0 \quad \text{for all } j \text{ and all } k, \tag{19.20b}$$

$$x_j - g_{ij} + h_{ij} = u_i \quad \text{for all } i \text{ and all } j, \tag{19.20c}$$

$$e_{jk} \geq 0,\ f_{jk} \geq 0,\ g_{ij} \geq 0,\ h_{ij} \geq 0, \text{ and } x_j \text{ unrestricted.} \tag{19.20d}$$

We next make a transformation of variables to simplify (19.20). Let $e'_{jk} = e_{jk} + x_k$, $f'_{jk} = f_{jk} + x_j$, and $h'_{ij} = h_{ij} + x_j$. Substituting these variables in (19.20) gives

$$f'_{jk} - e'_{jk} = 0 \quad \text{for all } j \text{ and all } k, \tag{19.21a}$$

$$h'_{ij} - g_{ij} = u_{ij} \quad \text{for all } i \text{ and all } j, \tag{19.21b}$$

$$e'_{jk} - x_k \geq 0 \quad \text{for all } j \text{ and all } k, \tag{19.21c}$$

$$f'_{jk} - x_j \geq 0 \quad \text{for all } j \text{ and all } k, \tag{19.21d}$$

$$h'_{ij} - x_j \geq 0 \quad \text{for all } i \text{ and all } j, \tag{19.21e}$$

$$g_{ij} \geq 0;\ e'_{jk},\ f'_{jk},\ h'_{ij}, \text{ and } x_j \text{ are unrestricted.} \tag{19.21f}$$

Now notice that (19.21a) implies that $f'_{jk} = e'_{jk}$ and (19.21b) implies that $g_{ij} = h'_{ij} - u_{ij}$. Making this substitution in (19.21c) to (19.21f) shows that

$$e'_{jk} - x_k \geq 0 \quad \text{for all } j \text{ and all } k, \tag{19.22a}$$

$$f'_{jk} - x_j \geq 0 \quad \text{for all } j \text{ and all } k, \tag{19.22b}$$

$$h'_{ij} - x_j \geq 0 \quad \text{for all } i \text{ and all } j, \tag{19.22c}$$

$$h'_{ij} \geq u_{ij} \quad \text{for all } i \text{ and all } j, \tag{19.22d}$$

$$e'_{jk},\ f'_{jk},\ h'_{ij}, \text{ and } x_j \text{ are unrestricted.} \tag{19.22e}$$

Observe that in (19.22), each constraint has at most one $+1$ and at most one -1. In Section 9.4 (see Theorem 9.9) we have shown that a linear program with at most $+1$ and at most one -1 in each constraint is the linear programming dual of a minimum cost flow problem. Consequently, we can solve the rectilinear distance location problem by taking the dual of (19.20a) and (19.22) and solving the resulting minimum cost flow problem. The optimal node potentials of this minimum cost flow problem will be an optimal solution of the (transformed) location problem.

19.9 PRODUCTION AND INVENTORY PLANNING

Many optimization problems in production and inventory planning are network flow problems. In fact, many of the earliest contributions of operations research to the field of management were network flow models for this class of applications. To illustrate the range of applications of network flow models in this problem domain, we consider several deterministic production-inventory planning problems involving multiple periods, single or multiple stages, and single or multiple products. All of these models address a basic *economic order quantity* issue: When we plan a production run of any particular product, how much should we produce? Producing in large quantities reduces the time and cost required to set up equipment for the individual production runs; on the other hand, producing in large quantities also means that we will carry many items in inventory awaiting purchase by customers. The economic order quantity strikes a balance between the setup and inventory costs to find the production plan that achieves the lowest overall costs. The models that we consider in this section all attempt to balance the production and inventory carrying costs while meeting known demands that vary throughout a given planning horizon. We first study the simplest model of this genre: a single-product, single-stage model without backordering, which is known as the *dynamic lot-size problem.*

Application 19.19 Dynamic Lot Sizing

In the dynamic lot-size problem, we wish to meet prescribed demand d_j for each of K periods $j = 1, 2, \ldots, K$ by either producing an amount x_j in period j and/or by drawing upon the inventory I_{j-1} carried from the previous period. Figure 19.17 shows the network for modeling this problem. The network has $K + 1$ nodes: the jth node, for $j = 1, 2, \ldots, K$, represents the jth planning period; node 0 represents the "source" of all production. The flow on the *production arc* $(0, j)$ prescribes the production level x_j in period j, and the flow on *inventory carrying arc* $(j, j + 1)$ prescribes the inventory level I_j to be carried from period j to period $j + 1$. The mass balance equation for each period j models the basic accounting equation: Incoming inventory plus production in that period must equal the period's demand plus the final inventory at the end of the period. The mass balance equation for node 0 indicates that during the planning periods $1, 2, \ldots, K$, we must produce all of the demand (we are assuming zero beginning and zero final inventory over the planning horizon).

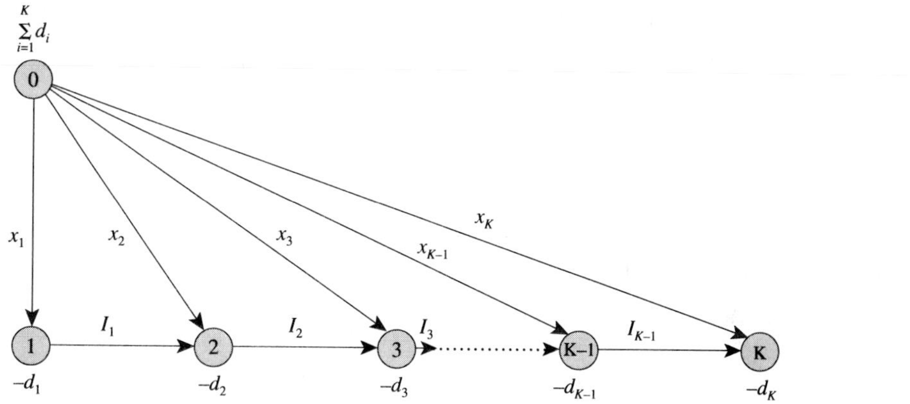

Figure 19.17 Network flow model of the dynamic lot-size problem.

If the production and inventory carrying costs are linear, we can easily solve the dynamic lot-size problem as a shortest path problem as follows. We determine a shortest path from node 0 to each node j and send d_j units of flow along that path; this path gives the minimum cost production-inventory schedule that satisfies the demand in period j. If we impose capacities on the production and inventory in each period, the dynamic lot-size problem no longer separates into independent shortest path problems. For example, if the costs are linear and we impose production or inventory capacities, the problem becomes a minimum cost flow model. We next consider situations with nonlinear production costs.

Application 19.20 Dynamic Lot Sizing with Concave Costs

In practice, the production x_j in the jth period frequently incurs a fixed cost F_j (independent of the level of production) and a per unit production cost c_j. Therefore, for each period j, the flow cost on the production arc $(0, j)$ is 0 if $x_j = 0$ and $F_j + c_j x_j$ if $x_j > 0$, which is a concave function of the production level x_j. (The production cost might also be concave, due to other economies of scale in production.) We assume that the inventory carrying cost is linear and that we have no capacity imposed upon production and inventory. So this problem is an instance of the concave cost flow problem.

In Exercise 14.13 we showed that any concave cost flow problem always has an optimal spanning tree solution (i.e., for some spanning tree, the tree arcs may carry nonzero flows and nontree arcs have zero flow). As we saw in Chapter 11, the fact that the spanning tree arcs must carry nonnegative flows, restricts our choice of spanning trees: Only some of them define a feasible solution of the dynamic lot-size problem. Any such feasible spanning tree decomposes into disjoint directed paths from the source node 0; the first arc on each path is a production arc and every other arc is an inventory carrying arc (see Figure 19.18). This observation implies the following *production property*: In the solution, each time we produce, we produce enough to meet the demand for an integral number of contiguous periods. Moreover, in no period do we both carry inventory from the previous period and produce.

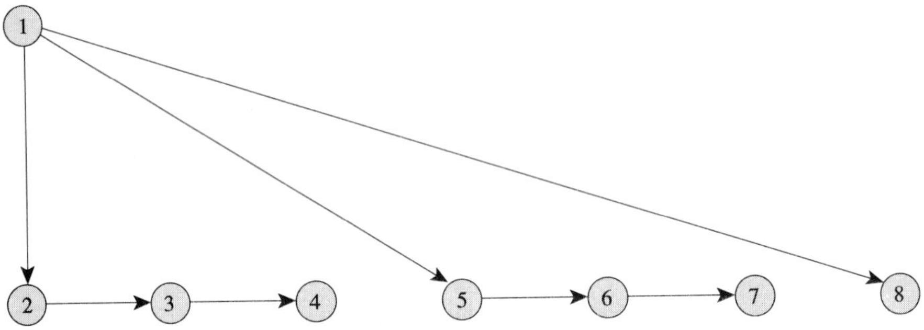

Figure 19.18 Illustrating the production property.

The production property permits us to solve the problem very efficiently as a shortest path problem on an auxiliary network G' defined as follows. The network G' consists of nodes 1 through $K + 1$ and contains an arc (i, j) for every pair of nodes i and j satisfying $i < j$. We set the cost of each arc (i, j) equal to the production and inventory carrying costs incurred in satisfying the demands of periods $i, i + 1, \ldots, j - 1$ by the production in period i. Observe that for every production schedule satisfying the production property, G' contains a directed path from node 1 to node $K + 1$ with the same objective function value, and vice versa. Therefore, we can obtain the optimal production schedule by solving a shortest path problem.

Application 19.21 Dynamic Lot Sizing with Backorders

In the preceding model we obtained the optimal production-inventory schedule assuming that we must satisfy the demands exactly in each period. In a more general version of this model, we permit backordering, which implies that we might not fully satisfy the demand of any period from the production in that period or from current inventory, but could fulfill the demand from production in future periods. In this model we assume that we do not lose any customer whose demand is not satisfied on time and who must wait until his or her order materializes. Instead, we assume that we incur a penalty cost for backordering any item. Figure 19.19 shows the network flow model for this situation. This model is the same as that of the dynamic lot-size model, except that we have additional *backorder carrying arcs* $(j, j - 1)$ for each $j = 2, 3, \ldots, K$, which represent the flow of backorders.

As in the preceding model, we assume that production costs are concave, and inventory carrying costs are linear, as are the costs of backordering. (Backordering does not affect the demand but incurs intangible cost in the form of lost goodwill of the customer.) We also assume that production, inventory, and backordering (and so the corresponding arcs) have no capacity limitations. This problem, again, is a

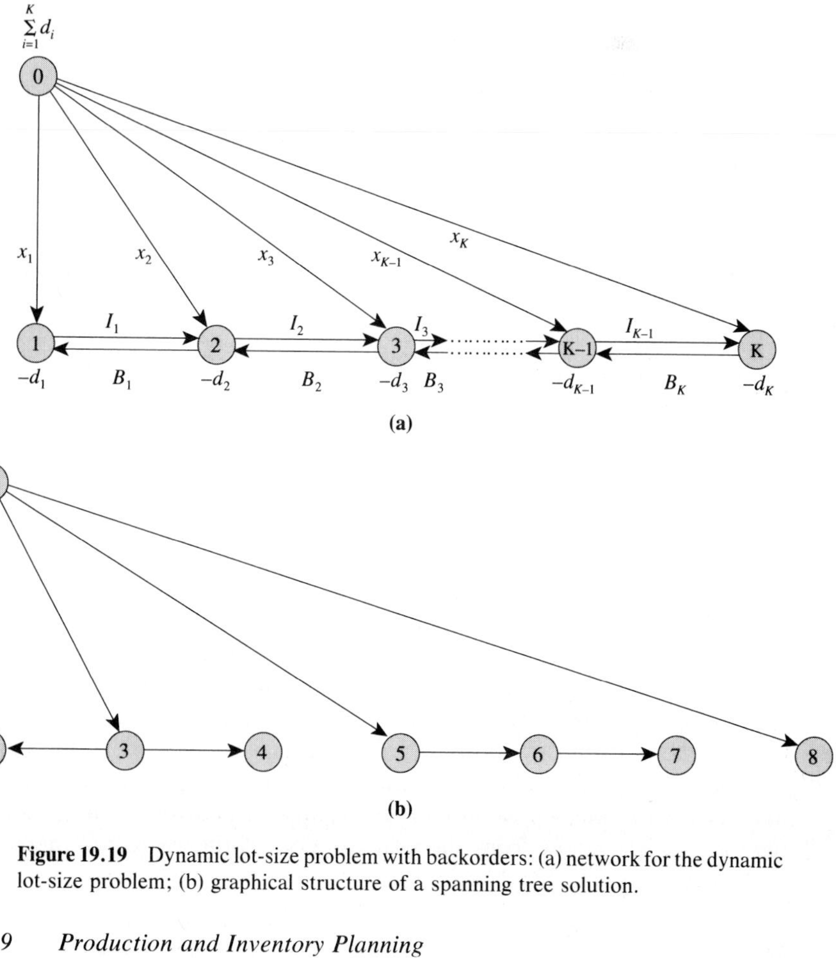

Figure 19.19 Dynamic lot-size problem with backorders: (a) network for the dynamic lot-size problem; (b) graphical structure of a spanning tree solution.

concave cost flow problem and has an optimal spanning tree solution. A feasible spanning tree solution in this case could, however, be different from the no-backorder case that we discussed in the last application. Although the optimal solution still satisfies the production property, we can use the production in period j to satisfy previous as well as future demands. Figure 19.19(b) shows an instance of the spanning tree solution.

The shortest path network G' for this problem is the same as the case without backorders except that we define the cost of arc (i, j) differently. In the case with backorders, we set the cost of arc (i, j) equal to the sum of the production, inventory carrying, and backorder carrying costs incurred in satisfying the demands of periods $i, i + 1, \ldots, j - 1$ by producing in some period k between i and $j - 1$; we select the period k that gives the least possible cost. In other words, we vary k from i to $j - 1$, and for each k, we compute the cost incurred in satisfying the demands of periods i through $j - 1$ by the production in period k; the minimum of these values defines the cost of arc (i, j) in the auxiliary network G'. (Recall that in the situation without backordering, we always chose period i as the production period for meeting the demand in the periods between i and $j - 1$.)

Application 19.22 Multistage Production-Inventory Planning

In the dynamic lot-size model, we considered a simple production process with a single stage of production. Often to produce a product, we must perform a sequence of operations, possibly performed on different machines at different times. If these machines have different production capacities that possibly vary from period to period, a multistage model would be a better approximation of this planning situation. In this case we treat each production operation as a separate stage and require that the product pass through each of the stages before its production is complete. We will use the following notation:

- d_j: demand of the product in period j
- x_{kj}: amount of the product produced at stage k in period j
- I_{kj}: inventory of the product carried from period j to $j + 1$ at stage k
- P_{kj}: production capacity at stage k in period j
- u_{kj}: upper limit on inventory holding from period j to $j + 1$ at stage k
- c_{kj}: unit production cost at stage k in period j
- h_{kj}: inventory carrying cost from period j to $j + 1$ at stage k

Figure 19.20 shows a minimum cost flow model of this problem. The network has $KT + 1$ nodes: For each $k = 1, \ldots, K$, and for each $j = 1, \ldots, T$, the node j^k represents the kth stage in period j. The decision variable next to each arc in the network models the flow on that arc; c_{kj} and P_{kj} are the cost and capacity for the arc representing the production x_{kj} at stage k in period j; h_{kj} and u_{kj} are the cost and capacity for the arc representing the inventory I_{kj} carried from period j to $j + 1$ at stage k. As is evident from the construction of the network, there is one-to-one

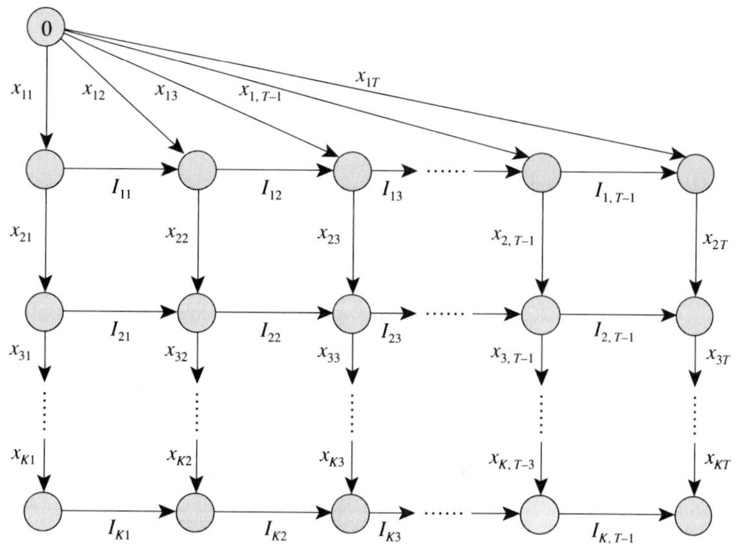

Figure 19.20 Network model for the multistage production-inventory planning problem.

correspondence between feasible flows in the network and production schedules. Consequently, we can obtain a minimum cost production-inventory schedule by finding a minimum cost flow in the network.

The model we have considered assumes that the production cost in each period is linear in the production quantity. If these costs are concave (e.g., because of fixed production costs for producing any quantity in any period), the model becomes a more complex (NP-hard) mixed integer programming model.

Application 19.23 Multiproduct Multistage Production-Inventory Planning

The multistage model that we just discussed modeled the production and inventory operations of only a single item. In more general situations we would need to consider not only multiple stages but also multiple products. Suppose that r different products share the common manufacturing facilities in p stages, and we want to meet the prescribed demands of the products in each period. The underlying network model is the same as that shown in Figure 19.20 with the exception that H distinct commodities flow on each arc. Consequently, the resulting model is an integer minimum cost multicommodity flow model. Clearly, the multiproduct multistage model is significantly more difficult than the single-product multistage model. Accordingly, to solve the model, we might use heuristic procedures that exploit the underlying network structure, or we might use general-purpose solution methods from integer programming and combinatorial optimization. For example, as shown in Chapter 16, where we examine a multiproduct, single-stage version of this problem, we might use Lagrangian relaxation and solve a sequence of single-product subproblems as shortest path problems.

Application 19.24 Mold Allocation

This application concerns a mold allocation problem that arises in the tire industry. A manufacturing plant in this industry consists of w cavities (or presses). At the beginning of each scheduling period, the manufacturing plant will insert different mold types (assume that the plant has p different mold types, one for each type of tire) into the cavities and subsequently use each mold to produce a fixed number of tires in that scheduling period. In the next scheduling period, depending on the demand requirements and cost structure, the plant might change some mold types in the cavities. In the mold allocation problem, we would like to assign molds to cavities over a scheduling horizon of T scheduling periods in order to satisfy the following conditions:

1. *Changeover restrictions*. We incur a setup cost whenever we place a mold in a cavity. This cost includes the cost of setting the mold in the cavity and performing quality control testing of the initial output of the mold. Highly skilled personnel perform these operations using specialized equipment; the limited availability of these personnel and of the specialized equipment normally restricts the number of setups that we can perform at the beginning of any scheduling period. Let R_j denote the maximum number of setups that we can perform at the beginning of the jth scheduling period. Let c_i denote the cost of placing a mold of type i in a cavity. In the mold allocation problem, our objective is to assign molds to cavities in a way that minimizes the total cost of placing the molds in the cavities.
2. *Minimum tire requirements*. The number of molds of a given type in place in the cavities during each scheduling period must meet a management-specified minimum. These limits originate from agreements assuring independent distributors with minimum production quantities in each period. Let r_{ij} denote the minimum number of molds of type i required in the jth scheduling period.
3. *Mold availability constraints*. The number of molds of a given type in place in the cavities during a given scheduling period is limited by the number of molds of that type available during that scheduling period. Mold availability might vary from period to period because of the planned arrival of new molds and the "reworking" of molds to convert them from one type to another. Let a_{ij} denote the molds of type i available in the jth scheduling period.
4. *Plant capacity limitations*. The total number of molds in place in the cavities during any scheduling period cannot exceed the number w of available cavities.

Figure 19.21 shows the minimum cost flow formulation of the mold allocation problem for $p = 2$ and $T = 2$. Next to each arc, we give the cost, lower bound, and capacity for the arc; for those arcs without any accompanying data, $(c_{ij}, l_{ij}, u_{ij}) = (0, 0, \infty)$. This formulation has T layers of nodes, one for each scheduling period. In our example, the first layer has nodes 1^0, 1^1, 1^2, 1^*, and the second layer has nodes 2^0, 2^1, 2^2, 2^*. In general, the layer i has $p + 2$ nodes numbered i^0, i^1, i^2, i^3, . . . , i^p, i^*. The network essentially represents the flow of cavities. Initially, assume that V_i cavities contain mold i, and V_0 of the cavities are empty. Notice that $V_0 + V_1 + \cdots + V_p = w$. We represent this relationship by introducing for each mold

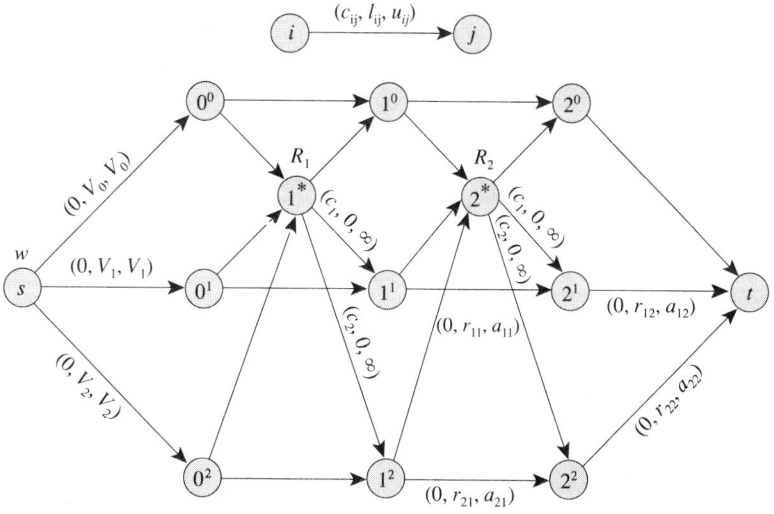

Figure 19.21 Minimum cost flow formulation of the mold allocation problem.

type i, arcs $(s, 1^i)$ whose lower and upper bounds are equal to V_i; node s has a throughput capacity of w. At the beginning of the first scheduling period, we might change molds in the existing cavities. We accomplish this by using the node 1^*, which collects all such cavities (including the empty ones) through the arcs $(0^0, 1^*)$, $(0^1, 1^*)$, and $(0^2, 1^*)$, retains some of these cavities as empty [by sending flow on arc $(1^*, 1^0)$], and changes the molds in the remaining cavities [by sending flow on arcs $(1^*, 1^1)$ and $(1^*, 1^2)$]. We allow a maximum of R_1 changes in the first scheduling period. We impose this restriction by setting the capacity of the node 1^* equal to R_1. The arc $(1^*, 1^1)$ carries the cavities in which we place mold 1; therefore, this arc has cost c_1. Similarly, the arc $(1^*, 1^2)$ carries the cavities in which we place mold 2, and so has cost c_2. Since the flow on the arc $(1^1, 2^1)$ is the total number of cavities that contain mold 1 in the first scheduling period, we impose a lower bound of r_{11} and an upper bound of a_{11} on this arc. Similarly, the flow on the arc $(1^2, 2^2)$ represents the number of cavities that contain mold 2, and flow on the arc $(1^0, 2^0)$ represents the number of empty cavities. This network has capacities on some nodes; we can transform node capacities into arc capacities by performing the node splitting transformation described in Section 2.4.

This discussion shows that there is one-to-one correspondence between feasible flows in the network and feasible mold allocations. Consequently, a minimum cost flow would prescribe a minimum cost mold allocation plan.

19.10 SUMMARY

This chapter complements our discussion in previous chapters by introducing a number of additional applications of network flows. The applications are of two types: (1) broad generic problem types (the maximum weight closure of a graph, data scaling, dynamic flows, and arc routing), and (2) contextual applications (DNA sequencing, automatic karyotyping of chromosomes, facility layout and location, project management, and production and inventory planning). Within the generic problem types, we have considered a number of more specific applications: open pit mining, selecting freight handling terminals, optimal destruction of military targets, the fly-away kit problem, scaling linear programming constraint matrices; in addition, building evacuations, patrolling streets by police cars, routing of street sweepers and household refuse collection vehicles, fuel oil delivery, and spraying roads during snowstorms. As part of this discussion we have also examined some classical network optimization models, including the directed and undirected Chinese postman problems, critical path scheduling, and various dynamic models of economic lot sizing.

As shown by the following tables, these applications, when combined with those that we have examined in previous chapters, demonstrate the ability of network flows to model an extraordinarily wide range of problems met in practice. Our choice of the categories in these tables is somewhat arbitrary, yet does provide at least one

Engineering
1. Leveling mountainous terrain (Application 1.4)
2. Rewiring of typewriters (Application 1.5)
3. Measuring homogeneity of bimetallic objects (Application 1.7)
4. Electrical networks (Application 1.8)
5. The paragraph problem (Exercise 1.7)
6. Network reliability testing (Application 8.2)
7. Equipment replacement (Application 9.6)
8. Phasing out capital equipment (Exercise 9.6)
9. Terminal assignment problem (Exercise 9.7)
10. Capacity expansion of a network (Exercise 14.4)
11. Dual completion of oil wells (Application 12.4)
12. Cabling electrical panels (Application 13.1)
13. Constructing pipeline networks (Application 13.1)
14. Managing energy or mineral networks (Application 15.1)
15. Stick percolation problem (Application 14.4)
16. Machine loading (Application 15.2)
17. Network design (Application 16.4)
18. Open pit mining (Application 19.1)
19. Flyaway kit problem (Application 19.4)

Manufacturing, production, and inventory planning
1. Assortment of structural steel beams (Application 1.2)
2. Allocating inspection effort on a production line (Application 4.2)
3. Machine setup problem (Exercise 6.19)
4. Entrepreneur's problem (Exercise 9.1)
5. Caterer problem (Exercises 11.2 and 15.1)
6. Optimal depletion of inventory (Application 12.8)
7. Two-duty operator scheduling (Application 16.5)
8. Multi-item production planning (Application 16.7)
9. Warehousing of seasonal products (Application 17.2)
10. Warehouse layout (Application 19.17)
11. Determining minimum project duration (Application 19.9)
12. Just-in-time scheduling (Application 19.10)
13. Time–cost trade-off in project management (Application 19.11)
14. Dynamic lot sizing (Application 19.19)
15. Dynamic lot sizing with concave costs (Application 19.20)
16. Dynamic lot sizing with backorders (Application 19.21)
17. Multistage production-inventory planning (Application 19.22)
18. Multiproduct multistage production-inventory planning (Application 19.23)
19. Mold allocation (Application 19.24)

Scheduling
1. Police patrol problem (Exercise 1.9) 2. Telephone operator scheduling (Application 4.6) 3. Single-duty crew scheduling (Exercise 4.13) 4. Scheduling on uniform parallel machines (Application 6.4) 5. Tanker scheduling problem (Application 6.6) 6. Nurse scheduling problem (Exercises 6.2 and 11.1) 7. Airline scheduling problem (Exercise 6.32) 8. Scheduling with deferral costs (Application 9.5) 9. Optimal capacity scheduling (Application 9.6) 10. Employment scheduling (Application 9.6) 11. Scheduling on parallel machines (Application 12.9) 12. School bus driver assignment (Exercise 12.1) 13. Multivehicle tanker scheduling (Application 17.3)

Management science
1. Compact book storage in libraries (Exercise 4.3) 2. Personnel planning (Exercise 4.9) 3. Optimal storage policy for libraries (Exercise 9.3) 4. Zoned warehousing (Exercise 9.4) 5. Project assignment (Exercise 11.3) 6. Faculty–course assignment (Exercise 11.6) 7. Bipartite personnel assignment (Application 12.1) 8. Nonbipartite personnel assignment (Application 12.2) 9. Managing financial investments (Application 15.1) 10. Managing warehousing goods and funds flows (Application 15.3) 11. Managing foodgrain imports–exports (Application 17.1)

Physical and medical sciences
1. Disease categorization (Exercise 4.6 and Application 13.5) 2. Reconstructing the left ventricle from X-ray projections (Application 9.2) 3. Determining chemical bonds (Application 12.5) 4. Storing sequences of amino acids in proteins (Application 13.4) 5. DNA sequence alignment (Application 19.7) 6. Automatic karyotyping of chromosomes (Application 19.8)

Defense
1. Locating objects in space (Application 12.6) 2. Matching moving objects (Application 12.7) 3. Optimal destruction of military targets (Application 19.3) 4. Network interdiction problem (Exercise 19.18)

Computer science and communication systems
1. Concentrator location on a line (Exercises 4.7 and 4.8) 2. Distributed computing on a two-processor computer (Application 6.5) 3. Allocating receivers to transmitters (Exercise 11.5) 4. Constructing digital computer systems (Application 13.1) 5. Designing computer networks (Application 13.1) 6. Area transfers in communication networks (Application 14.2) 7. Message routing in computer and communication networks (Application 17.1) 8. Reducing data storage (Application 13.4)

Social sciences and public policy

1. Reallocation of housing (Application 1.1)
2. Determining an optimal energy policy (Application 1.9)
3. Racial balancing of schools (Applications 1.10 and 9.3)
4. Police patrol problem (Exercise 1.9)
5. Forest scheduling problem (Exercise 1.10)
6. Large-scale personnel assignment (Exercise 1.4)
7. Problem of representatives (Application 6.2)
8. Matrix rounding problem (Application 6.3)
9. Statistical security of data (Exercise 6.5 and Application 8.3)
10. Allocation of contractors to public works (Exercise 9.5)
11. Assigning medical school graduates to hospitals (Application 12.3)
12. Optimal message passing (Application 13.2)
13. Designing physical systems (Application 13.1)
14. Matrix balancing problems (Application 14.3)
15. Land management (Application 15.4)
16. Optimal deployment of resources (Exercise 17.1)
17. Models for building evacuation (Application 19.13)
18. Parking model (Exercise 19.17)
19. Optimal deployment of firefighting companies (Exercise 19.21)

Applied mathematics

1. Approximating piecewise linear functions (Application 4.1)
2. Knapsack problem (Application 4.3)
3. Cluster analysis (Exercise 4.6 and Application 13.5)
4. System of difference constraints (Application 4.5)
5. Finding feasible network flows (Application 6.1)
6. Matrix rounding problem (Application 6.3)
7. Solving a system of equations (Exercise 6.4)
8. Linear programs with consecutive ones in columns (Application 9.6)
9. Capacitated minimum spanning tree problem (Exercise 9.54)
10. Fractional b-matching problem (Exercise 9.55)
11. Bottleneck transportation problem (Exercise 9.56)
12. Linear programs with consecutive 1's in rows (Exercise 9.8)
13. Linear programs with circular 1's in rows (Exercise 9.9)
14. Optimal rounding of a matrix (Exercise 11.7)
15. All pairs minimax path problem (Application 13.3)
16. Traveling salesman problem (Application 16.2)
17. Degree-constrained minimum spanning trees (Application 16.6)
18. Asymmetric data scaling with lower and upper bounds (Application 19.5)
19. Minimum ratio asymmetric data scaling (Application 19.6)
20. Symmetric data scaling problems (Exercise 19.6)
21. Maximum dynamic flows (Application 19.12)

plausible way to group the various applications. Several of the applications in these tables fit into several of the categories; rather than attempting to capture all such overlaps, with only a few exceptions we have placed each application into a single table, usually reflecting the application's primary application context. (In a few cases, an application combines several problems contexts; to help us obtain a better picture of the application contexts, we have included citations to several of these applications more than once.) We encourage readers to scan all of these tables to obtain a broad picture of the applications we have considered and to discover connections between the various categories.

We note that these tables underestimate the full range of applications that we have considered in our discussion throughout the text. For example, the tables con-

Distribution and transportation
1. Tramp steamer problem (Application 4.4)
2. Minimax transportation problem (Exercise 6.6)
3. Optimal loading of a hopping airplane (Application 9.4)
4. Distribution problems (Application 9.1)
5. Vehicle fleet planning (Exercise 9.2)
6. Passenger routing (Exercise 11.4)
7. Pilot assignments (Application 12.2)
8. Designing rural road networks (Application 13.1)
9. Aircraft assignment (Application 15.2)
10. Vehicle routing problem (Application 16.3)
11. Routing of railcars (Application 17.1)
12. Distribution systems planning for multiple products (Application 17.1)
13. Selecting freight handling terminals (Application 19.2)
14. Directed Chinese postman problem (Application 19.14)
15. Undirected Chinese postman problem (Application 19.15)
16. Discrete location problems (Application 19.16)
17. Rectilinear distance facility location (Application 19.18)
18. Truck scheduling problem (Exercises 19.19 and 19.20)
19. Dynamic facility location (Exercise 19.22)

Miscellaneous
1. Pruned chessboard problem (Exercise 1.6)
2. Dating problem (Exercise 1.5)
3. Seat-sharing problem (Exercise 1.8)
4. Bridges of Königsberg (Exercise 2.6)
5. Knight's tour problem (Exercise 3.25)
6. Maze problem (Exercise 3.26)
7. Wine division problem (Exercise 3.27)
8. Money-changing problem (Exercise 4.5)
9. Dining problem (Exercise 6.1)
10. Ski instructor's problem (Exercise 12.2)
11. Dancing problem (Exercise 12.12)
12. Tournament problem (Application 1.3)
13. Optimal coverage of sports events (Exercise 6.41)
14. Baseball elimination problem (Application 8.1)
15. Balanced assignment problem (Exercise 12.24)
16. Optimal message passing (Application 13.2)
17. Urban traffic flows (Application 14.1)

tain a single entry for discrete location problems, even though, as pointed out in this chapter, this application class applies to problem contexts in (1) the public and service sectors such as hospitals, fire stations, libraries, and air field locations, (2) technical arenas such as the layout of the component of a control panel, and (3) architectural design, such as the location of water fountains in an office building.

We also note that the coverage of applications in this book, as broad as it is, is not intended to be exhaustive. The applications we have discussed typically are classical, are easy to describe, or are representative of a broad class of problems. The literature contains many other applications, some that amplify on the themes we have presented and some that treat new problem domains. We hope that when viewed in its entirety, our coverage in this book gives an appreciation for the power of network flows as a field that not only has substantial intellectual content, but also has an important impact on practice.

REFERENCE NOTES

In this chapter we described many applications of network flow problems. We have adapted these applications from the following papers:

1. Open pit mining (Johnson [1968])
2. Selecting freight handling terminals (Rhys [1970])
3. Optimal destruction of military targets (Orlin [1987])
4. Flyaway kit problem (Mamer and Smith [1982])
5. Asymmetric data scaling with lower and upper bounds (Orlin and Rothblum [1985])
6. Minimum ratio asymmetric data scaling (Orlin and Rothblum [1985])
7. DNA sequence alignment (Waterman [1988])
8. Automatic karyotyping of chromosomes (Tso, Kleinschmidt, Mitterreiter, and Graham [1991])
9. Determining minimum project duration (Elmaghraby [1978])
10. Just-in-time scheduling (Elmaghraby [1978], Levner and Nemirovsky [1991])
11. Time–cost trade-off in project management (Fulkerson [1961a], Kelley [1961])
12. Maximum dynamic flows (Ford and Fulkerson [1958a])
13. Models for building evacuation (Chalmet, Francis, and Saunders [1982])
14. Directed Chinese postman problem (Edmonds and Johnson [1973])
15. Undirected Chinese postman problem (Edmonds and Johnson [1973])
16. Discrete location problem (Francis and White [1976])
17. Warehouse layout (Francis and White [1976])
18. Rectilinear distance facility location (Cabot, Francis, and Stary [1970])
19. Dynamic lot sizing (Veinott and Wagner [1962])
20. Dynamic lot sizing with concave costs (Zangwill [1969])
21. Dynamic lot sizing with backorders (Zangwill [1969])
22. Multistage production-inventory planning (Zangwill [1969])
23. Multiproduct multistage production-inventory planning (Evans [1977])
24. Mold allocation (Love and Vemuganti [1978])

Additional applications of network flow problems can be found throughout this book, in the text and in the exercises. We provide numerous citations to the literature in the reference notes of Chapters 1, 4, 6, 9, 12, 13, 14, 15, 16, and 17. We have adapted many applications in this book from the paper of Ahuja, Magnanti, Orlin, and Reddy [1992]. Since the applications of network flow models are so pervasive, no single source provides a comprehensive account of network flow models and their impact on practice. Several researchers have prepared general surveys of selected application areas. Notable among these are the papers by Bennington [1974], Glover and Klingman [1976], Bodin, Golden, Assad, and Ball [1983], Aronson [1989], Bazaraa, Jarvis, and Sherali [1990], and Glover, Klingman, and Phillips [1990]. The book by Gondran and Minoux [1984] also describes a variety of applications of network flow problems.

EXERCISES

19.1. Show that minimizing $|\alpha - \beta|$ is equivalent to minimizing $(c + d)$, subject to the conditions $\alpha - \beta = c - d$, $c \geq 0$, and $d \geq 0$. (*Hint*: Consider the cases when $\alpha \geq \beta$ and $\alpha \leq \beta$.)

19.2. Write a linear programming formulation of the maximum weight closure problem. Show that the dual of this linear program is a minimum cost flow problem.

19.3. This exercise concerns the least cost interdiction of a physical transportation network for carrying supplies, troops, or arms during a large-scale conventional war. Let $G = (N, A)$ denote a transportation network with two distinguished nodes s and t and let c_{ij} denote the cost of destroying arc (i, j) in this network.

(a) Suggest a method for identifying the cheapest way to destroy a subset of arcs that will disconnect nodes s and t.

(b) Suppose that besides destroying arcs of the transportation network, we could also destroy nodes of the network. Destroying a node in the network is equivalent to destroying all the arcs incident to that node. Let c_i denote the cost of destroying a node i in N. Identify the cheapest way to disconnect node s from node t.

19.4. This exercise concern the flyaway kit problem that we discussed in Application 19.4.

(a) Suppose that the penalty cost for all jobs is the same; that is, for all j, $L_j = \alpha$ for some constant α. Then what is the optimal policy when $\alpha = 0$ or when $\alpha = \infty$?

(b) Our formulation assumes that each job requires at most one item of each part type. Modify the formulation so that it allows multiple parts of the same type. Justify your modification.

19.5. This exercise concerns the asymmetric data scaling problem with lower and upper flow bounds that we discussed in Application 19.5. Suppose that you want to find the row multipliers α_i and the column multipliers β_j so that $l \le \alpha_i \mid a_{ij} \mid / \beta_j \le u$ for all matrix elements and, moreover, among all such multipliers, you want the ones with the smallest possible product $\Pi_{\{(i,j):a_{ij}\neq 0\}} \alpha_i \mid a_{ij} \mid / \beta_j$. How would you solve this problem as a network flow problem?

19.6. In this exercise we study a restriction of the asymmetric data scaling problem (discussed in Section 19.3) known as the *symmetric data scaling problem*. In this problem the matrix **A** is a $p \times p$ square matrix and we want to scale the data so that the diagonal elements of the matrix do not change. This implies that if the multiplier of row i is α_i, the multiplier of column i is $1/\alpha_i$.

(a) Describe how you would solve the symmetric data scaling problem with lower and upper flow bounds.

(b) Describe how you would solve the minimum ratio symmetric data scaling problem.

19.7. In the project scheduling problem that we discussed in Application 19.11, we assumed that the cost of each job (i, j) is a linear function of its duration. Suppose, instead, that the cost of job (i, j) is a piecewise linear convex function of its duration, with breakpoints at the job durations $0 = d_{ij}^0 \le d_{ij}^1 \le \cdots \le d_{ij}^p$; assume that the cost function has a slope of c_{ij}^k in the interval $[d_{ij}^{k-1}, d_{ij}^k]$. Specify a linear programming formulation of this problem and show that its dual is a convex cost flow problem.

19.8. **Battle of the Marne** (Berge and Ghouila-Houri [1962]). From the different towns $a_1, a_2, \ldots, a_n$, buses leave to go to a single destination b. For any direct road joining town a_i and town a_j, we are given the number c_{ij} of buses that can leave town a_i for a_j in a unit of time; we also know the time t_{ij} required to make this journey. Given the number s_i of buses available at town a_i at the start of our planning horizon and the number c_i of buses that can be parked at town a_i, we want to organize traffic routes so that in a given interval of time T, the number of buses arriving at b will be as large as possible. Formulate this problem as a network flow problem.

19.9. This exercise is related to the maximum dynamic flow problem discussed in Application 19.12.

(a) Can the time-expanded replica of a network contain a directed cycle?

(b) Can you solve the maximum dynamic flow problem if arc capacities change from one time period to another? If yes, how?

(c) Show that the maximum dynamic flow problem can have multiple optimal solutions.

(d) For any optimal solution of the maximum dynamic flow problem, let v be a p-element vector whose kth element denotes the flow reaching the sink node at time k. Suggest a method for identifying a solution with the lexicographically maximum vector v from among all solutions.

19.10. In a maximum dynamic flow x, let $v(k)$, for every $k = 1, 2, \ldots, p$, denote the amount of flow reaching the sink up to time period k. Clearly, $v(1) \le v(2) \le \cdots \le v(p)$. Show that if p is sufficiently large, then for some index l, $v(i) - v(i - 1) = \alpha$ for every $l \le i \le p$. What is α?

19.11. Consider the time-expanded replica G^p of a network G. Let $s^1, s^2, \ldots, s^p$ denote the sources and $t^1, t^2, \ldots, t^p$ denote the sinks in different time periods $1, 2, \ldots, p$. In this expanded network G^p, we wish to determine the smallest index r so that the maximum flow from the nodes in $s^1, s^2, \ldots, s^p$ to the nodes in $t^1, t^2, \ldots, t^p$ is at least B.

(a) Show how you can solve this problem in polynomial time using any maximum flow algorithm.

(b) Show how you can solve this problem by a single application of any minimum cost flow algorithm.

19.12. In our study of the dynamic maximum flow problem in Application 19.12, we assumed that capacities do not change over time. Explain how to model situations when the capacities are functions of time and change from period to period.

19.13. In a transportation network, let c_{ij} represent the time required to traverse arc (i, j) and assume, therefore, that $c_{ij} \ge 0$ for all arcs $(i, j) \in A$. In this network we wish to identify a directed path with the least possible traversal time from node s to node t. In the shortest path algorithms studied in Chapter 4, we assumed that all arc traversal times are fixed and do not change with time. But often in practice traversal times are functions of time: They are higher during rush hours and lower at other times. Explain how you could model the situation in which the traversal times vary with the time of the day. Assume that you want to obtain a shortest directed walk from node s to node t (the walk might sometimes be shorter than the directed path). State your assumptions.

19.14. Constrained shortest path problem. In a network G we associate two numbers with each arc: its length c_{ij} and its traversal time τ_{ij}. We would like to determine a shortest-length path from the source node s to the sink node t with the additional constraint that the traversal time of the path does not exceed τ. Formulate this problem as a shortest path problem in a time-expanded network.

19.15. Consider an undirected graph $G = (N, A)$ with arc costs c_{ij}. For any given subset $S \subseteq N$, we want to find a minimum cost subgraph of G that has an odd number of arcs incident to the nodes in S and an even number (possibly, zero) of arcs incident to the nodes in $N - S$. Describe a method for solving this problem. (*Hint*: Use the results in Application 19.15.)

19.16. As we discussed in Application 19.14, the optimal Chinese postman tour for directed network might traverse some arcs several times. Suppose that we can traverse no arc more than k times for some fixed $k \ge 2$. How would you solve this problem? Will this problem always have a feasible carrier tour?

19.17. Parking model (Dirickx and Jennergren [1975]). Develop a network model to solve the following parking problem that arises in the downtown area of a busy district. You are given a city district consisting of I blocks. Every block i has a daily demand D_{ik} for parking places of class $k = 1, \ldots, K$ and each class k has its own time length (0–1 hours, 1–2 hours, etc.). We measure the demand D_{ik} in physical parking places, which should be understood to mean that somewhere we must reserve D_{ik} parking places for vehicles whose drivers and passengers have their final destination in block i. Drivers might have to park in some other block and then walk by foot to block i. Suppose that the downtown area contains several public parking facilities (offstreet parking, garages, etc.). Let S_j denote the capacity of the jth parking facility. Our objective is to assign users to the parking facilities in order to minimize the total societal cost. The societal cost has two elements: walking costs, and the cost of maintaining the parking facilities. Formulate this problem as a minimum cost flow problem. Describe your notation and state your assumptions.

19.18. **Network interdiction problem** (Fulkerson and Harding [1977]). Let $G = (N, A)$ represent the transportation network of a military opponent. Suppose that node s is our opponent's supply point, that node t is his demand point, and the length of arc $(i, j) \in A$ is c_{ij}. We can increase the length of any arc $(i, j) \in A$ by any positive amount $y_{ij} \geq 0$ units by spending $d_{ij}y_{ij}$ units of resources. Our task is to increase the difficulty of our opponent's supply task by increasing the length of the shortest path from node s to node t.

(a) Suppose that our objective is to increase the length of the shortest path from node s to node t to a value of at least λ_t while spending the least possible amount of resources. Show that we can solve this problem by solving a minimum cost flow problem. (*Hint*: Write the linear programming formulation and take its dual.)

(b) Suppose that our task requires that we design the network so that the length of the shortest path from node s to node i be λ_i, for all $i \in N - \{s\}$. How could we achieve this objective while spending the least amount of resources?

19.19. **Truck scheduling problem** (Gavish and Schweitzer [1974]). In this exercise we study a generalization of the tanker scheduling problem that we discussed in Application 6.6. A large trucking company must decide on a weekly basis how best to meet the demands for truck trips from its customers during the next week. Each customer demand is specified by the following information: (1) a quantity of cargo (in terms of the numbers of trucks), (2) a starting place and time for the job, and (3) the terminating place and time for completing the job. The company has a truck depot within its area of operation and a fleet of trucks adequate to meet the demands. The decision problem is to assign trucks from the depot to meet the demand at minimum cost. Explain how you would formulate this problem as a minimum cost flow problem. Use the following data in your formulation: (1) job j begins at time t_j^b, ends at time t_j^c, and requires r_j trucks; (2) c_j is the cost of (one-way) driving between the depot and the origin of job j; (3) d_j is the cost of performing job j; and (4) f_{ij} is the cost of driving from the destination of job i to the origin of job j; this trip requires τ_{ij} time. Assume that all trucks start at the depot and return to it after completing their jobs.

19.20. This exercise studies some generalizations of the truck scheduling problem that we considered in Exercise 19.19. Explain how you would incorporate the following additional constraints in the model. Consider each of these constraints independently.

(a) The trucking company has several, say k, depots for the supply of trucks. Let s^k denote the number of trucks available at the kth depot at the beginning of the week. You can assume that the trucks need not return to their own depot.

(b) Allow a "safety margin" of h_{ij} time for the direct driving time from the destination of job i to the origin of job j.

(c) The same truck cannot perform jobs i and j if doing so would mean that the truck would arrive too early at job j. We forbid any truck to wait more than L units of time to prevent the drivers from becoming accustomed to too much idle time.

(d) Certain jobs cannot be combined. For example, we cannot carry food directly after we have transported garbage.

19.21. **Optimal deployment of firefighting companies** (Denardo, Rothblum, and Swersey [1988]). This exercise studies an application of the transportation problem that arises in the deployment of firefighting companies. Its special nature results from the fact that the benefit obtained from the firefighting companies depends on their order of arrival. The company arriving at the scene first offers the greatest benefit, and as further companies arrive their benefits decrease.

We assume that q incidents are in hand and we want to assign firefighting companies located at p company locations to them. Let u_i denote the number of companies available at location i and v_j denote the number of companies required at incident j. We assume that $\sum_{i=1}^{p} u_i \geq \sum_{j=1}^{q} v_j$. Let τ_{ij} denote the travel time from company location i to incident j. Furthermore, let α_{jk} denote the cost per unit time delay in the arrival of the kth company at incident j; assume that $\alpha_{jl} > \alpha_{jr}$ for $l < r$. If, for instance, the first two arrivals at incident j have identical travel times, one is given a cost of α_{j1} and the other is given a cost of α_{j2}. The objective is to assign firefighting companies to fire incidents to minimize the total cost. Formulate this problem as a minimum cost flow problem. Justify your formulation.

19.22. Dynamic facility location problem. A manufacturing facility supplies K consumer regions. The demand of each consumer region changes from period to period and is known in advance for each period in a planning horizon consisting of T periods. The company transports goods produced at its manufacturing facility to the consumer regions. The manufacturing facility is always located at one of the consumer regions, and due to fluctuating demands, the company might profitably relocate this facility occasionally. We wish to determine where to locate the manufacturing facility in each period of the planning horizon to minimize the total cost of transporting the goods and relocating the facility. Formulate this problem as a shortest path problem in a time-expanded network for the data shown in Figure 19.22 for which $T = 4$ and $K = 4$. Figure 19.22(a) gives the demand of the product in the next four periods, and Figure 19.22(b) gives the cost of transporting 1 unit of the product between consumer regions. Assume that the cost of relocating the facility is 300 times the cost of transporting 1 unit of the product. Moreover, assume that at the beginning of period 1, the facility is located in region 1.

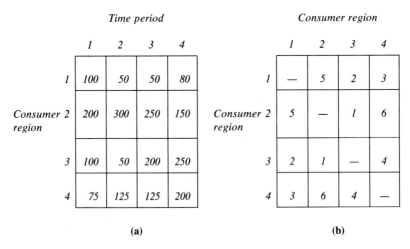

Consumer region \ Time period	1	2	3	4
1	100	50	50	80
2	200	300	250	150
3	100	50	200	250
4	75	125	125	200

(a)

Consumer region \ Consumer region	1	2	3	4
1	—	5	2	3
2	5	—	1	6
3	2	1	—	4
4	3	6	4	—

(b)

Figure 19.22 Dynamic facility location problem.

Appendix A

DATA STRUCTURES

Knowledge is of two kinds. We know a subject ourselves or we know where we can find information on it.
—Samuel Johnson

Chapter Outline

A.1 INTRODUCTION

Most network algorithms require the manipulation of data, particularly sets representing arc and node information, or representing trees or other network structures. Over the years, analysts have studied many different ways to store and manipulate data within computer memory; these investigations have shown, both empirically and in theory, that the choice of storage schemes often has a considerable effect on algorithmic performance. Indeed, by storing and manipulating sets more cleverly, we can often improve the worst-case complexity of an algorithm. In this appendix we describe some of the more common ways of storing and manipulating sets.

Often, elements of a set are ordered; following customary practice, we refer to such sets as *lists*. We typically perform several basic operations on lists: for example, inserting elements, deleting elements, determining whether a list contains a certain element. In Section A.2 we describe some of the more popular ways for storing lists and for performing these operations on them. The discussion focuses on data structures known as arrays, singly linked lists, doubly linked lists, queues, and stacks.

Sometimes we associate a number, or *key*, with each element of a set. We often need to perform some of the following operations on a set and its associated keys: finding an element in the set with the minimum key, inserting an element in the set, deleting an element, decreasing the key of an element. A *heap* is a data structure for sets that allows us to perform these operations efficiently. In Section A.3 we discuss binary and *d*-heaps. In Section A.4 we describe a more efficient (and, also, more sophisticated) heap known as the *Fibonacci heap*.

A.2 ELEMENTARY DATA STRUCTURES

In this section we discuss some of the most popular ways of storing lists (i.e., ordered sets).

Arrays

An array is the simplest data structure used to store an ordered set. This representation uses an array of size n, called *list*; the ith position (or *index*) of the array contains the ith element of the set, which we denote by list[i]. Figure A.1 shows an array representation of the ordered set {10, 5, 8, 9, 7}. To keep track of the size of the array that contains data elements, we also use a variable *last* which we set equal to the number of elements in the list. For our example, last = 5, indicating that list contains no data elements in the positions 6, 7, . . . , n.

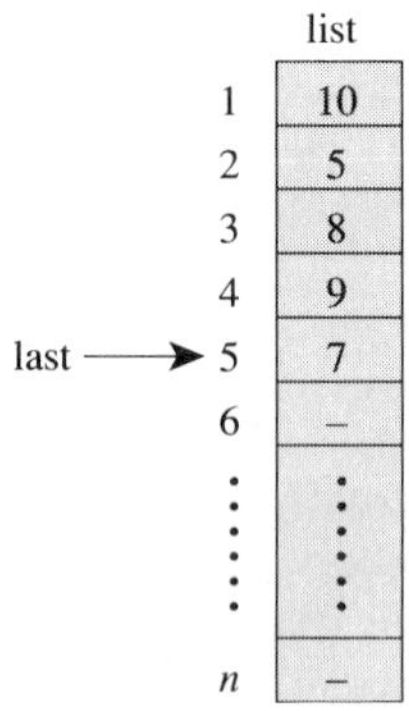

Figure A.1 Storing a set as an array.

Some operations are very convenient to perform using the array representation of a list. To determine the kth element of the list, we simply access the data element list(k). To establish whether list contains a given element α, we vary the index i from 1 to last and check whether list(i) = α. This operation requires time proportional to the number of elements in the set. In fact, all the storage methods described in this section require time proportional to the number of elements in a list to check for membership of any element. If we wish to insert an element at the end of the list, we increment last by one and store the new element as list(last).

An array is not very well suited for performing some other set operations. For example, suppose that we wish to delete the kth element of the set and $k <$ last. Deleting the kth element from the list changes the position of the elements stored at indices $k + 1, k + 2, \ldots$, last; we must shift each element back by one position. In the worst case, deleting the element and revising the list requires as many operations as the number of elements in the list, which is quite unattractive. Inserting an element in the middle of the list requires a similar amount of work. The linked list representations, discussed next, perform these operations much more efficiently but at the expense of using more storage.

Singly Linked Lists

Rather than store the elements of an ordered set sequentially as in an array, a singly linked list may store elements in an arbitrary order; however, it then requires additional information that permits us to access the data in the order specified in the ordered set.

A *cell* is the basic building block of linked lists. We can picture a cell as a box that is capable of holding several values, called *fields*. A singly linked list, then, is a collection of cells that are linked together. Each cell in the singly linked list has two fields: a *data* field and a *link* field. The data field holds the element of the list and the link field stores a pointer to the location of the next element in the list. Figure A.2(a) shows a geometric representation of a linked list for the set LIST = {5, 8, 9, 10}. The computer science community uses two popular methods for implementing linked lists: a pointer-based implementation and an array-based implementation. In this section we describe array-based implementations; in several references the reference notes of this appendix describe pointer-based implementations.

We can store a singly linked list by defining two arrays of size n, data and link. These two arrays define n cells, indexed from 1 through n: the kth cell consists of the fields $data(k)$ and $link(k)$. The data field of a cell contains an element of the set LIST and the link field contains the index of the cell containing the next element of the set. We also maintain a scalar called *first* that stores the index of the cell containing the first element of LIST. We set first to zero if the set is empty. Figure A.2(b) shows the array form of the linked list, LIST = {5, 8, 9, 10}. Because the list contains the element 9 in position 4 and next element 10 in position 2, data(4) = 9, link(4) = 2, and data(link(4)) = data(2) = 10.

It is fairly easy to manipulate singly linked lists. Suppose that we wish to scan all the elements of a set to determine membership of an element. We define a variable next = first. If next = 0, the set is empty and we stop. Otherwise, the first element of the set is data(next). Notice that link(next) gives the index of the following element in the set. Therefore, we update next = link(next) and check whether next equals

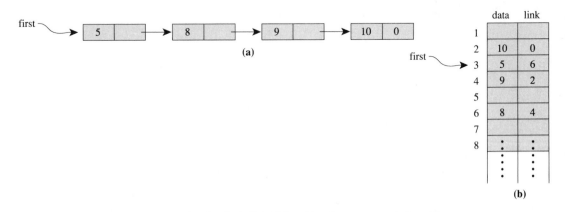

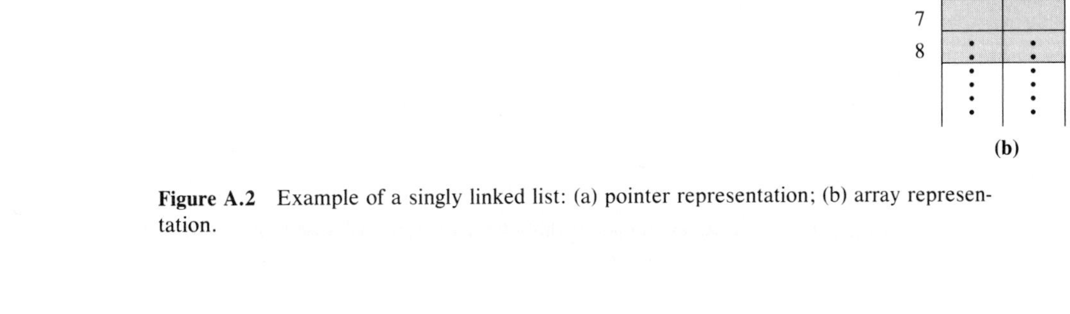

Figure A.2 Example of a singly linked list: (a) pointer representation; (b) array representation.

zero. If so, we stop; otherwise, the second element of the set is data(next). We repeat this process until next becomes zero. We suggest that the reader use this method to scan through the elements of the linked list shown in Figure A.2(b).

Now consider the insertion of an element into a linked list. Suppose, for example, that we wish to insert the element 6 into the linked list shown in Figure A.3(a). We first identify an unused cell with index *new*, and set data(new) = 6. If we wish to insert the element at the beginning of the list, we set link(new) = first and set first = new [as depicted in Figure A.3(a)]. If we wish to insert the element after the cell with index *prev*, we set link(new) = link(prev), link(prev) = new [as depicted in Figure A.3(b)]. This discussion shows that we can add an element to a linked list in $O(1)$ time.

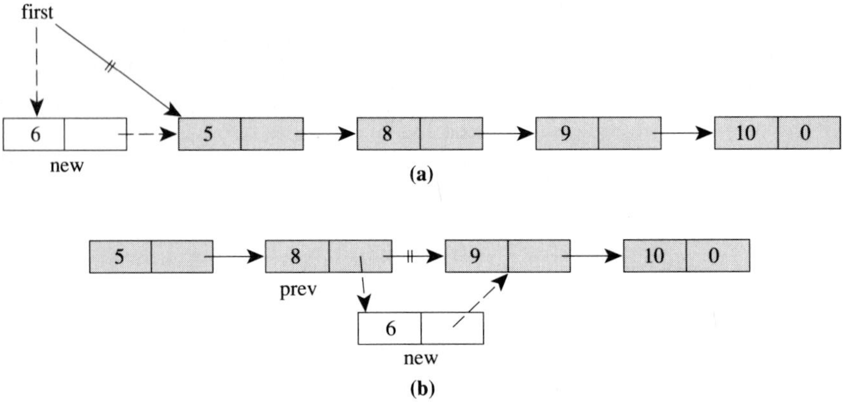

Figure A.3 Inserting an element in a linked list: (a) insertion at the beginning of the list; (b) insertion in the middle of the list.

Now consider the deletion of an element from the linked list; let us illustrate the process using the same example. As shown in Figure A.4(a), the deletion of the first element in the list is rather easy: We simply set next = first, and redefine first =

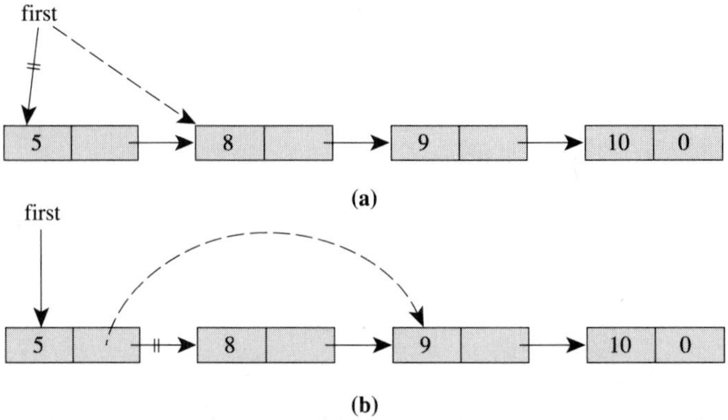

Figure A.4 Deleting an element from a linked list: (a) deletion from the beginning of the list; (b) deletion from the middle of the list.

link(next); therefore, first now points to the position of the second element of the original list. The deletion of an element from the middle of the list is more difficult. As shown in Figure A.4(b), to delete the element 8, we must modify the link field of the preceding cell so that its link array now points to the index of the cell following the element 8. So if we knew the preceding cell of the element to be deleted, we could easily perform the deletion in $O(1)$ time; otherwise, we will need to scan the linked list starting at the first element until we reach the element preceding the one to be deleted. The later operation in the worst case would require $O(n)$ time. To make deletion more efficient, we need to represent LIST as a doubly linked list, a data structure that we will discuss in the next subsection.

Some algorithms described in this book require the storage of sets LIST(1), LIST(2), . . . , LIST(n) of disjoint elements, each with a value between 1 and n. For example, for $n = 6$, we might have LIST(1) = {5, 4}, LIST(3) = {1, 2, 3}, LIST(4) = {6}, and LIST(2) = LIST(5) = LIST(6) = ∅. In these situations we will always add or delete elements from the front of any list. One plausible way to store this information would be by maintaining n different singly linked lists or n different arrays, but this method is not space efficient. In fact, we can store these n different sets using two n-dimensional arrays: *first* and *link*. Figure A.5 illustrates this storage scheme on this example data.

	first	link
1	5	2
2	0	3
3	1	0
4	6	0
5	0	4
6	0	0

Figure A.5 Storing multiple singly linked lists of disjoint elements.

In this storage scheme, first(k) stores the first element in the set LIST(k). If first(k) = 0; then LIST(k) is empty; otherwise, LIST(k) contains one or more elements. For example, first(1) = 5. Therefore, the first element in LIST(1) is 5. We then look at link(5), which is 4. Consequently, the second element in LIST(1) is 4. We then look at link(4), which is 0, indicating the end of the list. Therefore, LIST(1) = {5, 4}.

Suppose that we wish to insert an element p into LIST(k). Observe that because these lists are mutually disjoint, none of them can contain p. We insert element p to the beginning of LIST(k) by executing the statements link(p) = first(k) and first (k) = p. To delete the first element from LIST(k), we execute the statement first (k) = first(link(k)).

Doubly Linked Lists

A doubly linked list is the same as a singly linked list except that each cell has two links, one to the preceding cell and the other to the succeeding cell. Maintaining two links allows us to traverse the list easily in both directions and perform an arbitrary deletion of an element in $O(1)$ time.

A cell of the doubly linked list consists of three fields: *data, llink* (denoting left link), and *rlink* (denoting right link), which store the data element, the index of the preceding cell, and the index of the succeeding cell. For the set LIST = {5, 8, 9, 10}, Figure A.6(a) and (b) shows the pointer form and array form representations of the doubly linked lists.

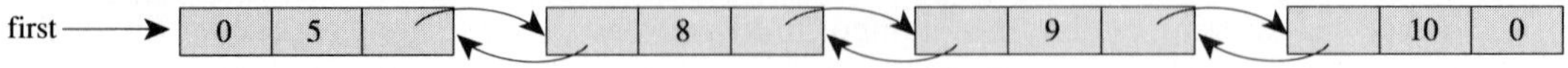

(a)

	index	data	llink	rlink
	1			
	2	10	4	0
first →	3	5	0	6
	4	9	6	2
	5			
	6	8	3	4
	7			
	8			
	⋮	⋮	⋮	⋮

(b)

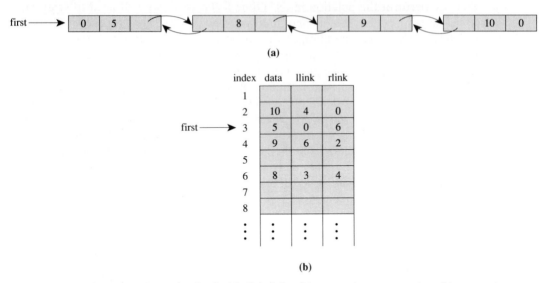

Figure A.6 Example of a doubly linked list: (a) geometric representation; (b) array representation.

We manipulate a doubly linked list in almost the same way in which we manipulate a singly linked list. We scan elements of the list in the same fashion, except that we can traverse the list in two directions: forward as well as backward. Using rlinks, we traverse the list in the forward direction (i.e., from left to right), and using llinks, we traverse the list in the backward direction (i.e., from right to left). Insertions and deletions from the doubly linked lists are easy to perform. Figure A.7

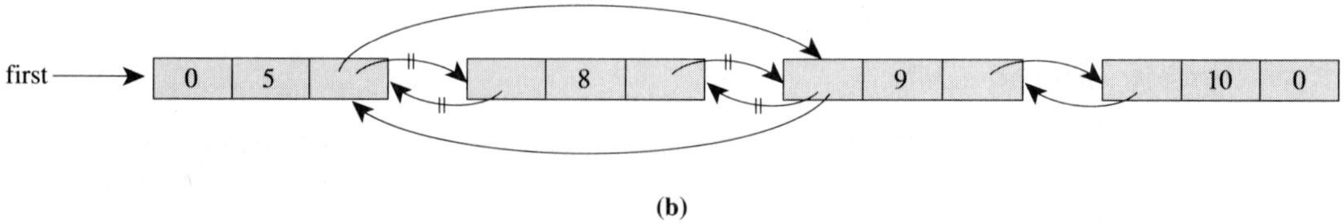

(a)

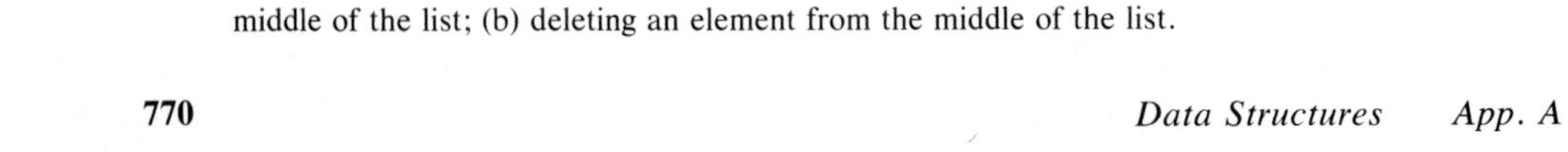

(b)

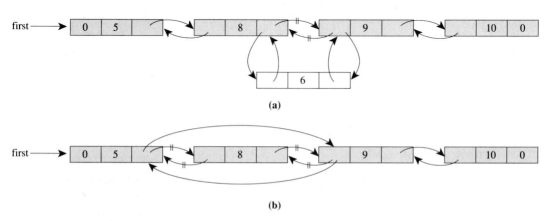

Figure A.7 Addition and deletion from a doubly linked list: (a) adding an element in the middle of the list; (b) deleting an element from the middle of the list.

shows how to apply these operations to the middle of a linked list. We ask the reader to show how to perform insertions and deletions from the beginning of the list. In each case, an insertion or deletion requires $O(1)$ time.

The doubly linked list allows us to traverse the list in either direction. Starting at some element i, by traversing rlinks, we can identify all the subsequent elements in the set. Traversing llinks allows us to identify all the elements preceding any element i. However, in some applications of doubly linked lists, we need to examine all the elements of the set starting at some arbitrary element i. We can do so by storing the elements in wraparound fashion (i.e., we set the rlink of the last element of the set equal to the index of the cell storing the first element and we set the llink of the first element of the set equal to the index of the cell storing the last element). We refer to this modification of the doubly linked list as a *circular doubly linked list*.

Finally, consider the storage of multiple doubly linked lists of disjoint elements, namely LIST(1), LIST(2), . . . , LIST(n). As in our earlier discussion, the value of each element lies between 1 and n. We can store these n different sets using three n-dimensional arrays: first, rlink, and llink. Figure A.8 illustrates this storage scheme for the previous example for which $n = 6$, LIST(1) $= \{5, 4\}$, LIST(3) $= \{1, 2, 3\}$, LIST(4) $= \{6\}$, and LIST(2) $=$ LIST(5) $=$ LIST(6) $= \varnothing$.

index	first	llink	rlink
1	5	2	0
2	0	3	1
3	1	0	2
4	6	0	5
5	0	4	0
6	0	0	0

Figure A.8 Storing multiple doubly linked list of disjoint elements.

Stacks

A *stack* is a special kind of ordered list (or set) in which all insertions and deletions take place at one end, called the *top*. The intuitive model of a stack is a pile of poker chips or a pile of dishes on a table; accordingly, we can conveniently remove only the top object on the pile or add a new one to the top. We can store a stack as an array or as a linked list. We shall discuss the array implementation of a stack.

A stack is represented by an n-dimensional array, *list*, that stores the element of a set, and a scalar *top* that denotes the index of the last entry to the array list. If the stack is empty, then top $= 0$. For example, suppose that we start with an empty list, and insert elements in the order 5, 4, 8, 7, 10; then at the end of the last step, the stack will appear as shown in Figure A.9. Since we have inserted five elements, when we have completed these operations, top $= 5$.

The most frequent operations performed on a stack are insertions and deletions. To insert a new element i on the stack, we increment top by 1 and set list(top) $= i$. To delete the topmost element i from the stack, we set $i =$ list(top) and decrease top by 1. Clearly, both of these steps require $O(1)$ time. Notice that in a stack, we always remove the element that we added last. Consequently, if we store elements

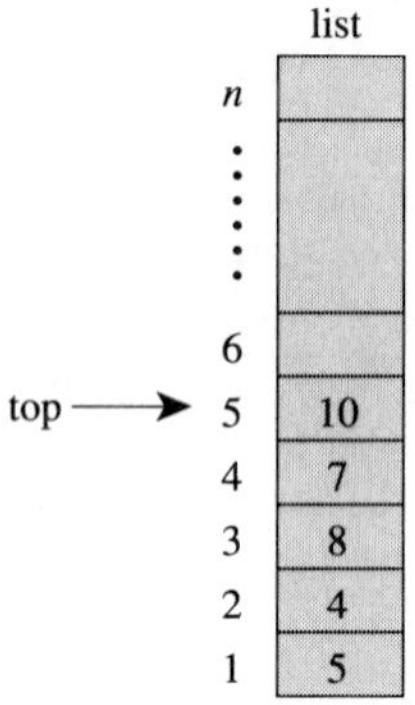

Figure A.9 Example of a stack.

using a stack and examine them one by one, we inspect the elements in a last-in, first-out (LIFO) order. Therefore, whenever we need to examine the elements of a dynamically changing set in the LIFO order, it is very natural to store the set as a stack.

Queues

A *queue* is another special kind of list, with elements inserted at one end (the *rear*) and deleted from the other end (the *front*). The operations for a queue are similar to those for a stack, the substantial difference being that insertions take place at the end of the list rather than the beginning. We see physical examples of queues everywhere since they are an integral part of contemporary society. Examples include lines at banks and at grocery stores, or items in a manufacturing plant waiting to be processed by a machine. A less apparent example is telephone calls waiting in a buffer for a telephone trunk to become available.

We represent a queue by an array *list* of size n that contains elements of the set. Figure A.10(a) gives an example of the queue. As shown in this example, we maintain a pointer called *front*, which is the index of the first position of the array

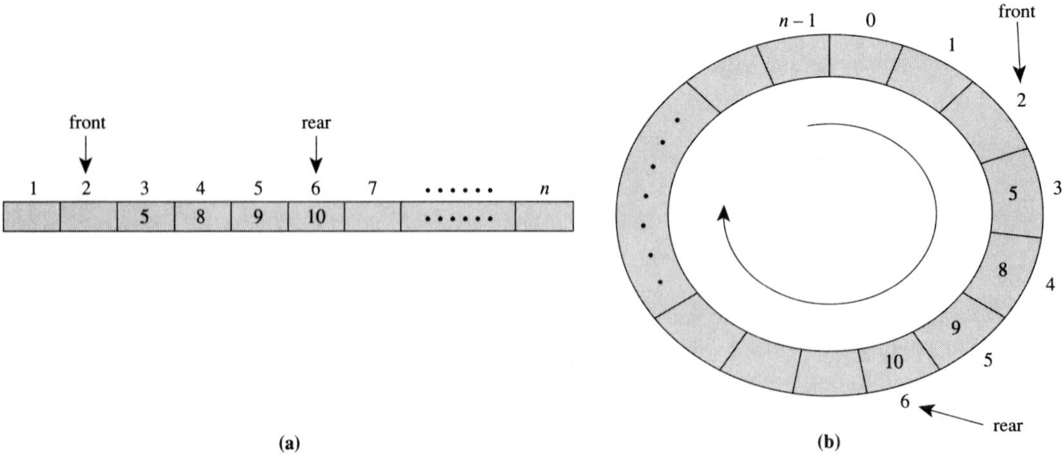

Figure A.10 Example of a queue: (a) sequential representation; (b) circular representation.

minus one, and maintain another pointer called *rear*, which is the index of the last position of the array.

We perform operations on a queue as follows. To check if the queue is empty, we simply check whether front equals rear. If so, the queue is empty; otherwise, it is nonempty. The frontmost element in the queue is list(front + 1). To insert a new element i in the queue, we increment rear by 1 and set list(rear) = i. To delete an element i from the queue, we set front = front + 1 and set i = list(front). Clearly, both the steps require $O(1)$ time. Notice that in a queue, the first element we remove is the one we added first. Consequently, if we store elements using a queue and examine the elements one by one, then we inspect the elements in a first-in, first-out (FIFO) order. Consequently, whenever we want to examine the elements of a dynamically changing set in the FIFO order, it is very natural to store the set as a queue.

Notice that if we keep adding elements to the rear of the queue and keep deleting elements from the front, eventually rear becomes equal to n and we cannot add any more elements to the queue. The queue at this point might not be "full" because the earlier part of the queue might be empty. We can overcome this difficulty by representing the elements of the list in the wraparound fashion, as shown in Figure A.10(a). In this representation, we perform the operations in exactly the same way as earlier except that while adding an element to the list we increment rear as (rear + 1)mod n, and while deleting an element from the list we increment front as (front + 1)mod n.

Summary

In this appendix we studied three important ways to store and manipulate sets: arrays, singly linked lists, and doubly linked lists. Figure A.11 summarizes the time complexity for performing various set operations on these data structures. Stacks and queues are special types of sets that perform operations either at the beginning or at the end of a list. To implement stacks and queues, we can use either arrays or linked lists; we have described only the array implementation.

	Array	Singly linked list	Doubly linked list
1. Inserting element at the end	$O(1)$	$O(1)$	$O(1)$
2. Inserting element at an arbitrary place	$O(n)$	$O(1)$	$O(1)$
3. Deleting element from the end	$O(1)$	$O(1)$	$O(1)$
4. Deleting element from an arbitrary place	$O(n)$	$O(n)$	$O(1)$
5. Determining kth element of the list	$O(1)$	$O(k)$	$O(k)$
6. Determining membership of an element in the set	$O(n)$	$O(n)$	$O(n)$

Figure A.11 Time complexity of various set operations for arrays and linked lists.

A.3 d-HEAPS

A *heap* (or, a *priority queue*) is a data structure for efficiently storing and manipulating a collection H of elements (or objects) when each element $i \in H$ has an associated real number, denoted by $key(i)$. We want to perform the following operations on the elements in the heap H:

create(H). Create an empty heap *H*.

insert(i, H). Insert an element *i* in the heap.

find-min(i, H). Find an element *i* with the minimum key in the heap.

delete-min(i, H). Delete the element *i* with the minimum key from the heap.

delete(i, H). Delete an arbitrary element *i* from the heap.

decrease-key(i, value, H). Decrease the key of element *i* to a smaller value, denoted by *value*.

increase-key(i, value, H). Increase the key of element *i* to a larger value, denoted by *value*.

In this section we discuss the d-heap and binary heap data structures (the binary heap is a well-known special case of the d-heap with $d = 2$). In the next section we describe a more efficient (and also more complex) heap known as the *Fibonacci heap*.

In most applications of heaps to network flow algorithms, the elements are nodes and their keys are node labels. Therefore, in our discussion of heaps, we shall use the word "element" and "node" interchangeably. Moreover, to be consistent with the conventions we have adopted for networks, we shall assume that the heap stores a maximum of n nodes.

Heaps find a variety of applications in network flow algorithms. Two such applications are Dijkstra's algorithm for the shortest path problem discussed in Section 4.7, and Prim's algorithm for the minimum spanning tree problem described in Section 13.5. Another important application of heaps is the sorting of n numbers in a nondecreasing order. We can sort n numbers using a heap as follows. First, we create an empty heap. Then, one by one, we add n numbers to the heap by performing n insert operations, letting the key for the ith entry be one of the numbers we wish to sort. Next, we repeat the following step iteratively: Select an element i with the minimum key using the operation find-min and then delete it from the heap using the operation delete-min. We terminate this procedure when the heap is empty. It is easy to see that we delete the elements from the heap in a nondecreasing order of their values.

Definition and Properties of a d-Heap

In a d-heap, we store the nodes of the heap as a rooted tree whose arcs represent a predecessor–successor (or parent–child) relationship. We store the rooted tree using predecessor indices and sets of successors, as follows:

pred(i): the predecessor (or the parent) of node i in the d-heap. The root node has no predecessor, so we set its predecessor equal to zero.

SUCC(i): the set of successors (or children) of node i in the d-heap.

In the d-heap we define the *depth* of a node i as the number of arcs in the unique path from node i to the root. For example, in the d-heap shown in Figure A.12, node 5 has a depth of 0 and nodes 9, 8, and 15 have a depth of 1.

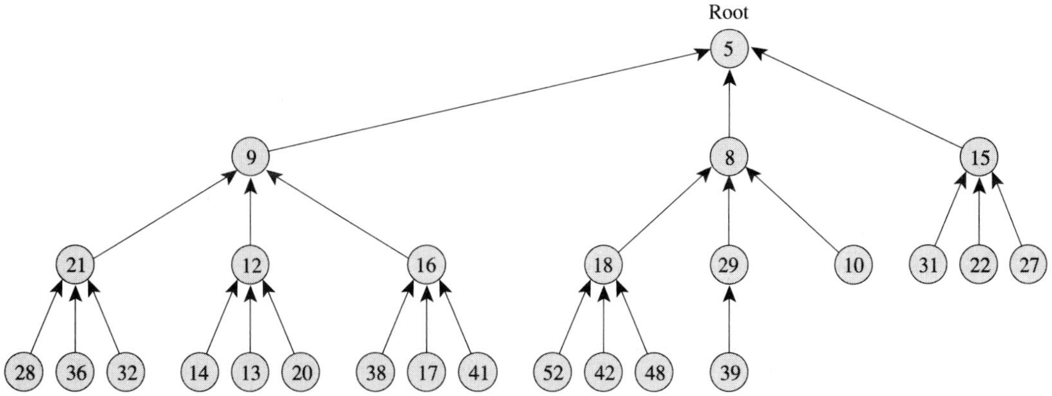

Figure A.12 Example of a d-heap for $d = 3$.

Each node in the d-heap has at most d successors, which we assume to be ordered from left to right. We refer to the successors of a node as *siblings* (of each other). The d-heap always satisfies the following property that we maintain inductively. We add nodes to the heap in an increasing order of depth values, and for the same depth value we add nodes from left to right. We refer to this property as the *contiguity property*. Figure A.12 gives an example of the d-heap for $d = 3$. In this example, we assume for convenience that key(i) $= i$ for each $i = 1$ to n (in this case $n = 50$). In this particular example, we have stored only a subset of the nodes in the heap. Note that nodes 12, 29, and 15 have the predecessors 9, 8, and 5, respectively.

The contiguity property implies the following results:

Property A.1

(*a*) *At most* d^k *nodes have depth* k.

(*b*) *At most* $(d^{k+1} - 1)/(d - 1)$ *nodes have depth between* 0 *and* k.

(*c*) *The depth of a d-heap containing n nodes is at most* $\lfloor \log_d n \rfloor$.

We leave the proof of this property as an exercise to the reader.

Storing a d*-Heap*

The structure of a d-heap permits us to store it as an array and manipulate it quite efficiently. We order the nodes in the increasing values of their depths, and we order the nodes with the same depth from left to right. We then store the nodes, in order, in an array DHEAP. For example, if we apply this method to the d-heap shown in Figure A.12, then DHEAP = {5, 9, 8, 15, 21, 12, 16, 18, 29, 10, 31, 22, 27, 28, 36, 32, 14, 13, 20, 38, 17, 41, 52, 42, 48, 39}. We also maintain an array position that contains the position of each node. For this example, position(9) = 2 and position(15) = 4. We maintain an additional parameter *last* that specifies the number of nodes stored in the array DHEAP. For this example, last = 26. This storage scheme has one rather nice property that permits us to easily access the predecessors and successors of any node:

Property A.2

(*a*) *The predecessor of the node in position i is contained in position* $\lceil (i - 1)/d \rceil$.
(*b*) *The successors of the node in position i are contained in positions* $id - d + 2, \ldots, id + 1$.

We leave the proof of this property as an exercise to the reader; it is instructive to verify this result on our numerical example. For example, node 18 is in position 8, so its predecessor is in position $\lceil (8 - 1)/3 \rceil = 3$ and its successors are in positions $3(8) - 3 + 2 = 23$ to $3(8) + 1 = 25$. This property implies that if we maintain the array DHEAP, we need not explicitly maintain the predecessor index and the set of successor indices of a node. We can compute these when required during the course of an algorithm. For the sake of exposition, we continue using predecessors and successors, but ignore the time required to update these data structures whenever the d-heap changes.

Heap Order Property

A heap always satisfies the following invariant, which we subsequently refer to as the *heap order property*.

Property A.3 (Invariant 1). *The key of node i in the heap is less than or equal to the key of each of its successors. That is, for each node i,* $key(i) \le key(j)$ *for every* $j \in SUCC(i)$.

We note that we might violate Invariant 1 while performing a heap operation but will always satisfy it at the end of any heap operation. The reader can verify that the example shown in Figure A.12 satisfies the heap order property if for every node in the heap we assume that $\text{key}(i) = i$.

The following result is an immediate consequence of the heap order property.

Property A.4. *The root node of the d-heap has the smallest key.*

Swapping

In the d-heap data structure, we reduce each heap operation into a sequence of a fundamental operation, each called *swap*(i, j). The operation swap(i, j) swaps (or interchanges) nodes i and j. Figure A.13 gives an example of a swap. In terms of the array used to store a d-heap, as a result of applying swap(i, j), we store node i at the position where node j was stored, and store node j at the position where node i was stored. For example, if we perform swap(4,6) in the DHEAP = {5, 6, 7, 4, 8, 11, 12, 9}, as shown in Figure A.13, then the new array representation of the d-heap becomes DHEAP = {5, 4, 7, 6, 8, 11, 12, 9}. Clearly, the swap operation requires $O(1)$ time.

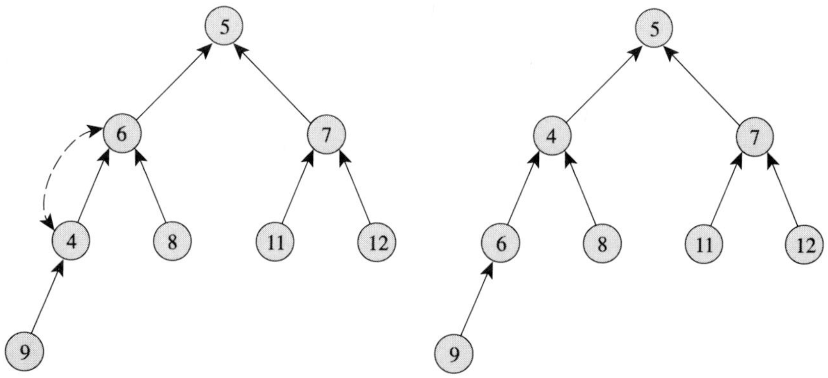

Figure A.13 Example of swap (4, 6): (a) heap before the swap; (b) heap after the swap.

Restoring the Heap Order Property

In the course of applying an algorithm, we will frequently change the value of some key and so temporarily violate the heap order property. How can we restore this property? Suppose that we decrease the key of some node i. Let $j = \text{pred}(i)$. If after the change in the value of key(i), key $(j) \leq$ key(i), the heap still satisfies the heap order property and we are done. However, if key(j) $>$ key(i), we need to restore the heap order property. The procedure *siftup*(i) described in Figure A.14 accomplishes this task.

```
procedure siftup(i);
begin
    while i is not a root node and key(i) < key(pred(i)) do swap(i, pred(i));
end;
```

Figure A.14 Procedure siftup(i).

Inductive arguments show that at the termination of the siftup procedure, the heap satisfies the heap order property. The procedure siftup requires $O(\log_d n)$ time because each execution of the while loop decreases the depth of node i by one unit and, by Property A.1, its original depth is $O(\log_d n)$.

Suppose next that we increase the key of some node i. If after the change in the value of key(i), key(i) $\leq$ key(j) for all $j \in \text{SUCC}(i)$, the heap still satisfies the heap order property and we are done; otherwise, we need to restore the heap order property. The procedure *siftdown*(i) described in Figure A.15 accomplishes this task. In the description we let *minchild*(i) denote the node with smallest key in SUCC(i).

```
procedure siftdown(i);
begin
    while i is not a leaf node and key(i) > key(minchild(i)) do swap(i, minchild(i));
end;
```

Figure A.15 Procedure siftdown(i).

An inductive argument will again show that at the termination of the siftdown procedure, the heap satisfies the heap order property. The procedure requires $O(d \log_d n)$ time because each execution of the while loop increases the depth of node i by one unit and each execution requires $O(d)$ time to compute minchild(i).

Performing Heap Operations

We are now in a position to describe how we can perform various operations in the d-heap.

find-min(i, H). The root node of the heap is the node with the minimum key and it is located at the first position of the array DHEAP. Therefore, this operation requires $O(1)$ time.

insert(i, H). We increment last by one and store the new node i at the last position of the array DHEAP. Then we execute the procedure siftup(i) to restore the heap order property. Clearly, this operation requires $O(\log_d n)$ time.

decrease-key(i, value, H). We decrease the key of node i and execute the procedure siftup(i) to restore the heap order property. This operation requires $O(\log_d n)$ time.

delete-min(i, H). Clearly, node i is the root node of the heap. Let node j be the node stored at the last position of the array DHEAP. We first perform swap(i, j) and then decrease last by 1. Next, we perform siftdown(j) to restore the heap order property. Clearly, this heap operation requires $O(d \log_d n)$ time.

We ask the reader to show as an exercise how to perform the remaining two heap operations, delete(i, H) and increase-key(i, value, H), in $O(d \log_d n)$ time. We summarize our discussion as follows:

Theorem A.5. *The d-heap data structure requires $O(1)$ time to perform the operation find-min, $O(\log_d n)$ time to perform the operations insert and decrease-key, and $O(d \log_d n)$ time to perform the operations delete-min, delete, and increase-key.* ◆

Recall that a binary heap is a d-heap with $d = 2$. For binary heaps, this theorem assumes the following special form.

Theorem A.6. *The binary heap data structure requires $O(1)$ time to perform the operation find-min, and $O(\log n)$ time to perform each of the operations insert, delete, delete-min, decrease-key, and increase-key.* ◆

As an example of applying heaps, consider a sorting algorithm. Recall from our prior discussion that while sorting n numbers, we perform n inserts, n find-mins, and n delete-mins. Consequently, the running time of the sorting algorithm using d-heaps is $O(nd \log_d n)$, which is $O(n \log n)$ for any fixed value of d.

A.4 *FIBONACCI HEAPS*

The Fibonacci heap is a novel data structure that allows the heap operations to be performed more efficiently than d-heaps. This data structure performs the operations insert, find-min, and decrease-key in $O(1)$ amortized time and the operations delete-min, delete, and increase-key in $O(\log n)$ amortized time. Recall from Section 3.2 that the amortized complexity of an operation is the *average* worst-case complexity of performing that operation. In other words, the amortized complexity of an operation is $O(g(n))$ if for a sequence of k (sufficiently large) operations, the total time required by these operations is $O(kg(n))$. For our purpose, $k \geq n$ is sufficiently large.

Properties of Fibonacci Numbers

Researchers have given the Fibonacci heap data structure its name because the proof of its time bounds uses properties of the well-known Fibonacci numbers. Before discussing the data structure, we first discuss these properties. The Fibonacci numbers are defined recursively as $F(1) = 1$, $F(2) = 1$, and $F(k) = F(k - 1) + F(k - 2)$, for all $k \geq 3$. These numbers satisfy the following properties:

Property A.7
(*a*) For $k \geq 3$, $F(k) \geq 2^{(k-1)/2}$.
(*b*) $F(k) = 1 + F(1) + F(2) + \cdots + F(k - 2)$.

Proof. The facts that $F(k) = F(k-1) + F(k - 2)$ and $F(k - 1) \geq F(k - 2)$ imply that $F(k) \geq 2F(k - 2)$. Consequently, if k is odd, $F(k) \geq 2F(k - 2) \geq 2^2 F(k - 4) \geq 2^3 F(k - 6) \geq 2^{(k-1)/2}F(1) = 2^{(k-1)/2}$. If k is even, we argue by induction. The claim is true if $k = 4$. Suppose it is true for even numbers less than k. The $F(k) \geq F(k - 1) + F(k - 2) \geq 2^{(k-2)/2} + 2^{(k-3)/2}$ by the result for k odd and the induction hypothesis. But then $F(k) \geq 2^{(k-3)/2} [2^{1/2} + 1] \geq 2^{(k-1)/2}$ and so by induction the conclusion is true for all $k \geq 3$.

To prove part (b), let us define a series of numbers $G'(\cdot)$ as $G'(1) = 1$, $G'(2) = 1$, and $G'(k) = 1 + G'(1) + G'(2) + \cdots + G'(k - 2)$ for all $k \geq 3$. Then $G'(k - 1) = 1 + G'(1) + G'(2) + \cdots + G'(k - 3)$ and $G'(k) - G'(k - 1) = G'(k - 2)$. Alternatively, $G'(k) = G'(k - 1) + G'(k - 2)$, which is the same manner in which Fibonacci numbers are defined. Therefore, $G'(k) = F(k)$ for all k. ◆

Property A.8. *Suppose that a series of numbers $G(\cdot)$ satisfies the properties that $G(1) = 1$, $G(2) = 1$, and $G(k) \geq 1 + G(1) + G(2) + \cdots + G(k - 2)$ for all $k \geq 3$. Then $G(k) \geq F(k)$.*

Proof. We prove inductively that $G(k) \geq F(k)$ for all k. This claim certainly is true for $k = 1$ and $k = 2$. Let us assume that it is true for all values of k from 1 through $q - 1$. Then $G(q) \geq 1 + G(1) + G(2) + \cdots + G(q - 2) \geq 1 + F(1) + F(2) + \cdots + F(q - 2) = F(q)$, the equality following from Property A.7(b). ◆

Defining and Storing a Fibonacci Heap

As we noted earlier, a heap stores a set of elements, each with a real-valued key. A Fibonacci heap is a collection of directed rooted in-trees: each node i in the tree represents an element i and each arc (i, j) represents a predecessor–successor (parent–child) relationship: node j is the predecessor (parent) of node i. Figure A.16 gives an example of a Fibonacci heap.

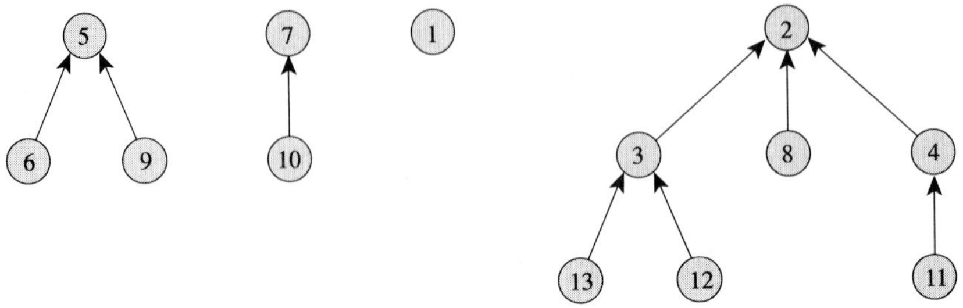

(a)

i	1	2	3	4	5	6	7	8	9	10	11	12	13
pred(i)	0	0	2	2	0	5	0	2	5	7	4	3	3
SUCC(i)	∅	{3, 8, 4}	{13, 12}	{11}	{6, 9}	∅	{10}	∅	∅	∅	∅	∅	∅
rank(i)	0	3	2	1	2	0	1	0	0	0	0	0	0

(b)

Figure A.16 Fibonacci heap: (a) rooted trees; (b) corresponding data structure.

To represent a Fibonacci heap numerically (i.e., in a computer) and to manipulate it effectively, we need the following data structure:

pred(i): the predecessor (or the parent) of node i in the Fibonacci heap.

We refer to a node with no parent as a root node and we set its predecessor to zero. This convention permits us to determine whether a node is a root node or a nonroot node by looking at the node's predecessors index. We also need the following data structures:

SUCC(i): the set of successors (or children) of node i. We maintain this set as a doubly linked list.
rank(i): the number of successors of node i [i.e., rank(i) = | SUCC(i) |].
minkey: the node with the minimum key.

Figure A.16(b) shows this data structure for the rooted trees given in Figure A.16(a). We need additional data structures to support various heap operations; we will introduce these data structures later, when we require them.

We need one additional piece of notation. A subtree *hanging* at any node i of any rooted tree contains the node i, its successors, successors of its successors, and so on. For example, in Figure A.16, the subtree hanging at node 5 contains the nodes 5, 6, and 9, and the subtree hanging at node 7 contains the nodes 7 and 10.

Linking and Cutting

In using the Fibonacci heap data structure, we reduce each heap operation into a sequence of two fundamental operations: *link*(i, j) and *cut*(i). We apply the operation link(i, j) to two (distinct) root nodes i and j of equal rank; it merges the two trees rooted at these nodes into a single tree. The operation cut(i) cuts node i from its predecessor and makes i a root node.

> *link(i, j)*. If key$(j) \le$ key(i), then add arc (i, j) to the Fibonacci heap (thus making node i the predecessor of node j). If key$(j) >$ key(i), then add arc (j, i) to the heap.
>
> *cut(i)*. Delete arc $(i$, pred$(i))$ from the heap (thus making node i a root node).

We illustrate these two operations on the examples shown in Figure A.17. For simplicity, we assume that for every node i, key$(i) = i$. Notice that the link operation increases the rank of node i or of node j by 1. Moreover, each of these operations

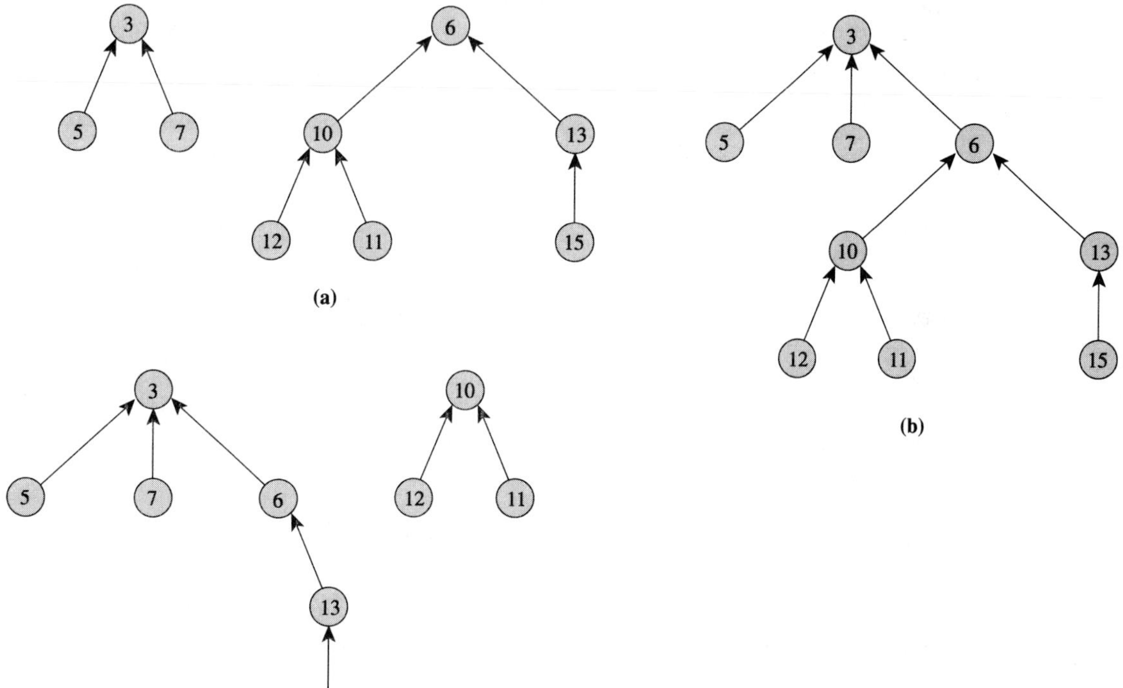

Figure A.17 Illustrating link and cut operations: (a) original heap; (b) heap after the operation link (3, 6); (c) heap after the operation cut(10).

changes the pred and SUCC and rank information for at most two nodes; consequently, we can perform them in $O(1)$ time. Later in this section we describe the additional data structures that we maintain to manipulate the Fibonacci heap effectively; we can also modify them in constant time as we perform a link and a cut operation. We record this result formally for ease of future reference.

Property A.9. *The operations link(i, j) and cut(i) require $O(1)$ time to execute.*

While manipulating the Fibonacci heap data structure, we perform a sequence of links and cuts. There is a close relationship between the number of links and cuts. To observe this relationship, consider a potential function Φ defined as the number of rooted trees. Each link operation decreases Φ by 1 and each cut operation increases Φ by 1. The total decrease in Φ is bounded by its initial value (which is n) plus the total increase in Φ. The following result is now evident.

Property A.10. *The number of links is at most n plus the number of cuts.*

Invariants in Fibonacci Heaps

The Fibonacci heap data structure maintains a set of rooted trees that change dynamically as we perform various linking and cutting operations. These rooted trees satisfy certain invariants that are essential for deriving the claimed time bounds for the heap operations. The nodes of the Fibonacci heap always satisfy the heap order property (i.e., Invariant 1), which states that the key of a node is less than or equal to the keys of its successors. The Fibonacci heap also satisfies the following two invariants:

Property A.11 (Invariant 2). *Each nonroot node has lost at most one successor after becoming a nonroot node.*

Property A.12 (Invariant 3). *No two root nodes have the same rank.*

As before, although we might violate these invariants at intermediate steps of some heap operations, the heap will satisfy them at the conclusion of each heap operation. One important consequence of Invariants 2 and 3 is that the maximum possible rank of any node is $2 \log n + 1$. We establish this result next.

Lemma A.13. *Any node in the Fibonacci heap has rank at most* $2 \log n + 1$.

Proof. Let $G(k)$ denote the minimum number of nodes contained in a subtree hanging at a node of rank k in a Fibonacci heap. We shall prove that $G(k) \geq F(k)$. Since no subtree can contain more than n nodes, Properties A.7 and A.8 imply that $n \geq G(k) \geq F(k) \geq 2^{(k-1)/2}$, which implies that $k \leq 2 \log n + 1$.

Let w be a node in a Fibonacci heap with rank k. Arrange the successors of node w in the same order in which the previous operations linked them to w, from the earliest to the latest. We claim that the rank of the ith successor of w is at least $i - 2$. To establish this result, let y be the ith successor of node w and consider the moment when y was linked to w. Just before this link operation, w had at least

$i - 1$ successors. (It might have had more than $i - 1$ successors at that time, some having been cut since then.) Since at the time of this link operation, nodes y and w both have the same rank, node y had at least $i - 1$ successors just before we performed this link operation. Furthermore, notice that since that time node y has lost at most one successor (from Invariant 2). Therefore, node y (which is the ith successor of node w) has rank at least $i - 2$. As a result, the subtree hanging at node w contained at least $1 + G(1) + G(2) + \cdots + G(k - 2)$ nodes. To summarize, we have shown that $G(k) \geq 1 + G(1) + G(2) + \cdots + G(k - 2)$, which in view of Property A.8 implies that $G(k) \geq F(k)$. From our prior observation, this conclusion establishes the lemma. ◆

The following property follows directly from Invariant 3 and Lemma A.13.

Property A.14. *A Fibonacci heap contains at most* $1 + 2 \log n$ *rooted trees.*

We next discuss how we restore Invariants 2 and 3 if they become violated at intermediate steps of a heap operation.

Restoring Invariant 2

To restore Invariant 2, we maintain an additional index *lost*(i) for every node i, defined as follows.

> *lost(i):* For a nonroot i, lost(i) represents the number of successors the node has lost after it became a nonroot node. For a root node i, lost(i) = 0.

Suppose that while manipulating a Fibonacci heap, we perform the operation cut(i). We refer to this cut as the *actual cut*. Let j = pred(i). In this operation, node j loses a successor. If node j is a nonroot node, we increment lost(j) by 1. If lost(j) becomes two, Invariant 2 requires that we make node j a root node. In that case we perform cut(j) and make j a root node. Let k = pred(j). This cut increases lost(k) by 1. If k is a nonroot node and lost(k) = 2, we must make it a root node as well, and so on. Thus an actual cut might lead to several cuts due to a *cascading* effect: We keep performing these cuts until we reach a node that has not lost any successor so far or is a root node. We refer to these additional cuts that are triggered by an actual cut as *cascading cuts*, and the entire sequence of steps following an actual cut as *multicascading*.

We illustrate this process on the Fibonacci heap shown in Figure A.18(a). In this figure we represent nonroot nodes with lost(i) = 1 by shaded circles. Suppose that we cut node 17 from its predecessor. This operation also requires that we also cut nodes 11 and 8 from their predecessors. Figure A.18(b) shows the resulting Fibonacci heap that satisfies Invariant 2. We now summarize the preceding discussion.

Property A.15. *If we perform an actual cut, we might also need to perform several cascading cuts so that the heap again satisfies Invariant 2; the time needed for these operations is proportional to the total number of cuts performed.*

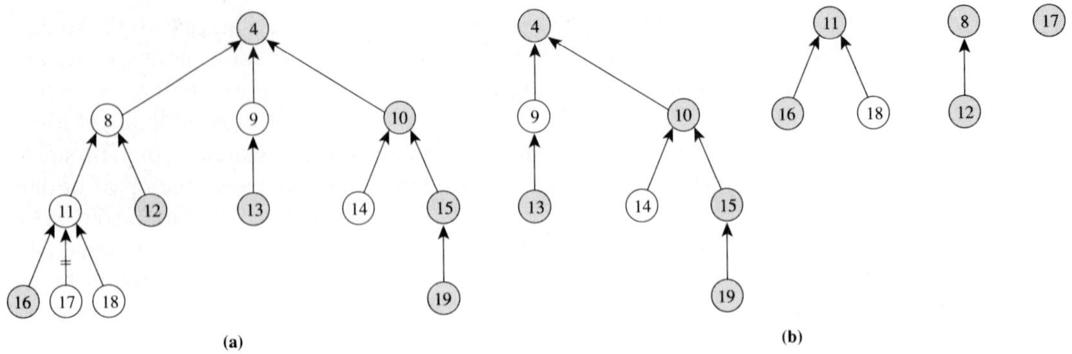

Figure A.18 Illustrating how we satisfy invariant 2. (Unshaded nodes have already lost a child.)

Suppose that we perform a number of actual cuts at different times while manipulating a Fibonacci heap and that these cuts cause additional cascading cuts. What is the relationship between the total number of actual cuts and the total number of cascading cuts? We shall show that the total number of cascading cuts cannot exceed the total number of actual cuts. To prove this result, consider the potential function $\Phi = \sum_{i \text{ in heap}} \text{lost}(i)$. Suppose that we perform cut(i) and j = pred(i). This operation sets lost(i) to zero and increases lost(j) by one if j is a nonroot node. If the cut is an actual cut, lost(i) equals 0 or 1 before the cut, and if it is a cascading cut, lost(i) equals 2 before the cut. Therefore, an actual cut increases lost(i) + lost(j), and hence the value of the potential function Φ by at most one, and a cascading cut decreases lost(i) + lost(j) by at least one. If we start with a potential value of zero, the total decreases in the potential function are bounded by the total increases. The following property is now apparent.

Property A.16. *The total number of cascading cuts is less than or equal to the total number of actual cuts.*

Restoring Invariant 3

The Invariant 3 requires that no two root nodes have the same rank. To maintain this property, we need the following index for every possible rank $k = 1, \ldots, K = 2 \log n + 1$.

bucket(k). If the Fibonacci heap contains no root node with rank equal to k, then bucket(k) = 0; and if some root node i has a rank equal to k, then bucket(k) = i.

Suppose that while manipulating a Fibonacci heap, we create a root node j of rank k and the heap already contains another root node i with the same rank. Then we repeat the following procedure to restore Invariant 3. We perform the operation link(i, j), which merges the two rooted trees into a new tree of rank $k + 1$. Suppose that node l is the root of the new tree. Then by looking at bucket($k + 1$), we check to see whether the heap already contains a root node of rank $k + 1$. If not, we are done. Otherwise, we perform another link operation to create another rooted tree of rank $k + 2$ and check whether the heap already contains a root node of rank $k + 2$. We repeat this process until we satisfy Invariant 3. We refer to this sequence of steps following the addition of a new root as *multilinking*.

We illustrate this process of re-establishing Invariant 3 on a numerical example. Consider the Fibonacci heap shown in Figure A.19(a), assuming that the key of node i equals i. Suppose that we add a new rooted tree containing a singleton node 10. The heap already contains another root node of rank 0, namely node 9. Thus we perform a link operation on nodes 9 and 10, obtaining the rooted trees shown in Figure A.19(b). Now two trees in the heap contains, with roots 7 and 9, have rank 1. We perform another link operation, producing the structure shown in Figure A.19(c). But now two trees, with roots 1 and 7, have rank 2. We perform another link operation, producing the structure shown in Figure A.19(d). At this point, the rooted trees satisfy Invariant 3 and we terminate.

We summarize the preceding discussion in the form of a property.

Property A.17. *If we add a new rooted tree to the Fibonacci heap, we might need to perform several links to restore Invariant* 3; *the time needed for these operations is proportional to the total number of links.*

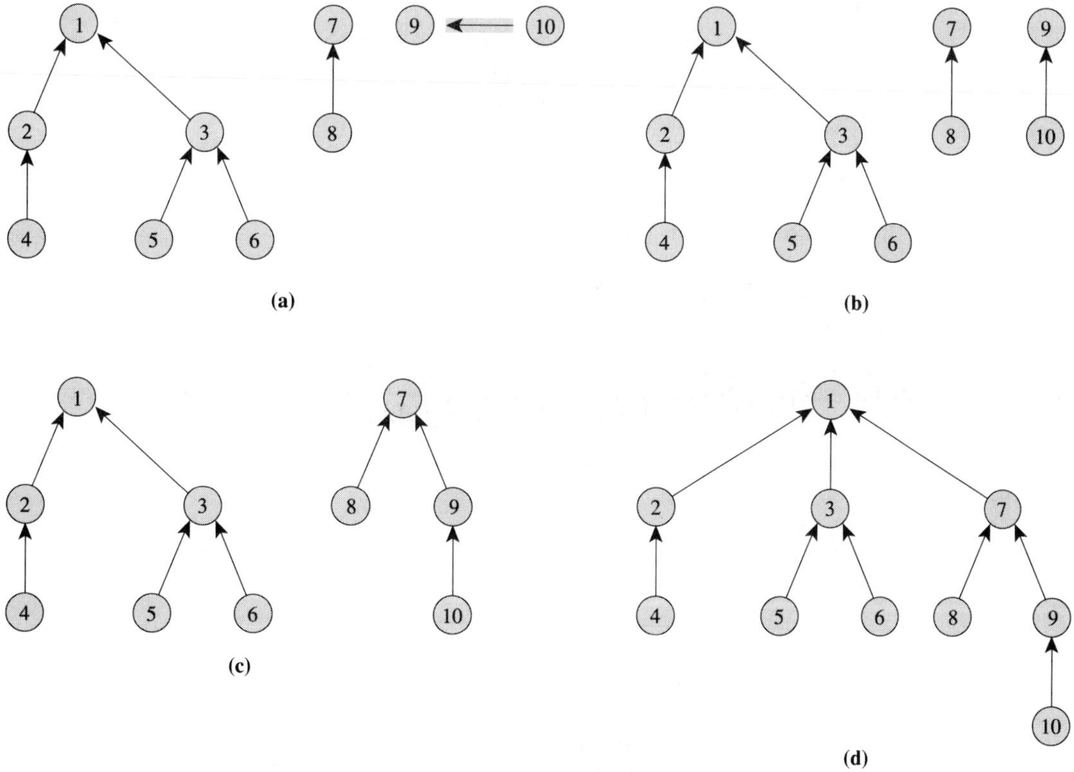

Figure A.19 Illustrating how we satisfy invariant 3.

Heap Operations

Finally, we show how we perform various heap operations using the Fibonacci heap data structure and indicate the amount of time they take.

find-min(i, H). We simply return i = minkey, since the variable minkey contains the node with the minimum key.

insert(i, H). We create a new singleton root node i and add it to H. After we have performed this operation, the heap might violate Invariant 3, in which case we perform multilinking to restore the invariant.

decrease-key(i, value, H). We first decrease the key of node i and set it equal to value. After we have decreased the key of node i, every node in the subtree hanging at node i still satisfies the heap order property; the predecessor of node i might, however, violate this property. Let j = pred(i). If key(j) $\leq$ value, we are done. Otherwise, we perform an actual cut, cut(i), make node i a root node, and update minkey. After we have performed the cut, the heap might violate Invariant 2, so we perform multicascading to restore this invariant. The resulting cascading cuts generate new rooted trees whose roots we store in a list, LIST. Then one by one, we remove a root node from LIST, add it to the previous set of roots, and perform multilinking to satisfy Invariant 3. We terminate when LIST becomes empty.

delete-min(i, H). We first set i = minkey. Then one by one, we scan each node $l \in$ SUCC(i), perform an actual cut, cut(l), and update minkey. We apply multilinking after performing each such actual cut. When we have cut each node in SUCC(i), we scan through all root nodes [which are stored in bucket(k), for $k = 0, 1, \ldots, 2 \log n + 1$], identify the root node h with minimum key, and set minkey = h. Recall that $|$ SUCC(i) $| \leq 2 \log n + 1$, because Lemma A.13 implies that each node has at most $2 \log n + 1$ successors. Therefore, the delete-min operation performs $O(\log n)$ actual cuts, followed by a number of cascading cuts and links. Then we scan through $O(\log n)$ root nodes to identify the root with the minimum key.

Figure A.20, which lists the sequence of steps and the associated running time for each heap operation, summarizes the preceding discussion. Since we do not, as yet, know the time required for multicascading and multilinking, we write "?" for the times of these steps.

We now consider the time required for multicascading and multilinking. We claim that we can ignore the time taken by these two steps. To establish this claim, we use the following facts: (1) Property A.16, which states that the number of cascading cuts is no more than the number of actual cuts; and (2) Property A.10, which states that the number of links is no more than n plus the number of actual and cascading cuts. Consequently, if we perform a sufficiently large number of operations (i.e., more than n), the number of actual cuts will count the number of links and cascading cuts within a constant factor; therefore, we can ignore the time required for the latter operations.

We illustrate this idea further by considering Dijkstra's algorithm for the short-

Heap operation	Sequence of steps	Time taken
find-min(*i*, *H*)	(a) Return i = minkey.	$O(1)$
insert(*i*, *H*)	(a) Add a new singleton node i. (b) Perform multilinking.	$O(1)$?
decrease-key(*i*, *value*, *H*)	(a) Decrease the key of node i. (b) If node i violates Invariant 1 then (b.1) Perform cut(i) and update minkey. (b.2) Perform multicascading. (b.3) Perform multilinking.	$O(1)$ $O(1)$? ?
delete-min(*i*, *H*)	(a) For each node $l \in SUCC(i)$ do (a.1) Perform cut(l). (a.2) Perform multilinking. (b) Compute minkey by scanning all root nodes.	 $O(\log n)$? $O(\log n)$

Figure A.20 Summary of heap operations in a Fibonacci heap.

est path problem as described in Section 4.7. In that discussion we showed that Dijkstra's algorithm performs n inserts, n find-mins, n delete-mins, and at most m decrease-key operations. The time requirements of the heap operations listed in Figure A.20 imply that the algorithm requires $O(m + n \log n)$ time, plus the time for $O(n \log n)$ actual cuts, plus the time for multicascading and multilinking. Using the facts that the number of cascading cuts and links are no more than twice the number of actual cuts, and that each actual cut requires $O(1)$ time, we immediately see that the shortest path algorithm runs in $O(m + n \log n)$ time.

So, if we ignore the time for multicascading and multilinking, it is clear from Figure A.20 that the operations find-min, insert, and decrease-key require $O(1)$ amortized time, and the operation delete-min requires $O(\log n)$ amortized time. We ask the reader to prove that the operation delete and increase-key also require $O(\log n)$ amortized time. We summarize the discussion in this section as follows:

Theorem A.18. *The Fibonacci heap data structure requires* $O(1)$ *amortized time to perform each of the operations insert, find-min, and decrease-key, and* $O(\log n)$ *amortized time to perform each of the operations delete-min, delete, and increase-key.* ◆

REFERENCE NOTES

The role of data structure is critical in designing efficient algorithms and in writing computer programs for implementing algorithms. In this appendix we have presented some of the most elementary data structures and the ones that we use frequently in network flow algorithms. The following books provide much additional information on data structures: Knuth [1973a, 1973b], Aho, Hopcroft, and Ullman [1983], Mehlhorn [1984], and Cormen, Leiserson, and Rivest [1990].

Appendix B

$\mathcal{NP}$-COMPLETENESS

Seek not out the things that are too hard for thee, neither search the things that are above thy strength.
—The Apocrypha (The Hidden Books)

Chapter Outline

B.1 INTRODUCTION

One of the primary purposes of scientific inquiry is to help structure the world around us so that we can better understand it. For example, in physics and chemistry, the periodic chart of the elements helps us to understand and categorize the relationship between the basic elements of the universe; in biology, the genus/species nomenclature helps us to understand the commonalties and differences among animals and plants. Computer science, mathematics, and operations research are no different; we often classify these fields in a variety of ways—for example, simply into subspecialties such as discrete and continuous mathematics—that helps us to discern their underlying structure. This idea of using classification as an organizing tool prompts the following basic question: Is there a way to develop a structural understanding of algorithms or for the problems to which we wish to apply them? Perhaps surprisingly, the research community had not proposed an approach for resolving this question until the early 1970s, when the field began in earnest to develop a topic known as computational complexity theory which attempts to categorize the computational requirements of both algorithms and important classes of problems met in practice. In this appendix, we discuss one cornerstone of this development, a topic known as $\mathcal{NP}$-completeness.

We call a class of optimization problems *easy* if we can develop an algorithm to solve every instance of the problem class in polynomial time (i.e., by an algorithm that requires a number of operations that is polynomial in the size of the input data for the problem). We also refer to a polynomial-time algorithm as an *efficient* algorithm. A majority of the network flow problems studied in this book—the shortest path problem, the maximum flow problem, the minimum cost flow problem, to name a few—are easy. Despite the best efforts of thousands of researchers across the globe spanning several decades, the research community has been unable to show that many other network and combinatorial optimization problems (e.g., the knapsack problem and the traveling salesman problem) are easy because no one has been able to develop any efficient algorithm for solving these problems. These unsuccessful attempts have led some researchers to question whether these problems are *inherently hard* in the sense that no efficient algorithm could possibly ever solve these problems. The theory of $\mathcal{NP}$-completeness is an outgrowth of these inquiries. Although this theory has been unable to prove that these difficult problems admit no efficient algorithms, the theory has shown that the majority of these problems are equivalent to each other in the sense that if we could develop an efficient algorithm for one problem in this class, we would then be able to develop an efficient algorithm for *every* other problem in this class. We refer to this broad class of "computationally equivalent" problems as *$\mathcal{NP}$-complete problems* (later in this appendix we give a formal definition of this class of problems). This class now includes thousands of problems and possesses the remarkable property (which is somewhat difficult to believe initially) that each problem in this class can be transformed to every other problem in polynomial time; as a consequence, each problem is "just as hard" as every other problem. This relationship suggests that $\mathcal{NP}$-complete problems share some generic difficulty that is beyond the reach of polynomial-time algorithms. Indeed, the research community widely believes that $\mathcal{NP}$-complete problems *cannot* be solved efficiently. This is the bad news about difficult problems.

The theory of $\mathcal{NP}$-completeness also has its positive aspects. To capture its usefulness, consider the following story. Suppose that your boss asks you to develop an algorithm for a complex design problem. Despite weeks of sincere efforts you do not succeed in developing an efficient algorithm for solving this problem. Every algorithm that you are able to construct is substantially no better than searching through all possible designs: There are so many of them that this enumeration would require several years of computer time on the fastest computers owned by your company. Surely, you are intelligent enough not to return to your boss's office and report, *"I can't find an efficient algorithm, I guess I am just too dumb."*

Although you did not succeed in developing an efficient algorithm, you were convinced that the design problem is *inherently difficult* and that no one, no matter how smart and creative, could possibly develop an efficient algorithm for this design problem. However, you cannot prove your conjecture, because proving it could be as difficult as finding an efficient algorithm. Therefore, you cannot walk into your boss's office and declare, *"I can't find an efficient algorithm because no such algorithm is possible!"*

The theory of $\mathcal{NP}$-completeness provides many techniques for proving that a given problem is just as hard as a large number of other problems that have defied solution by an efficient algorithm despite decades of efforts of the brightest researchers. Using these techniques, you might be able to show that your design problem is $\mathcal{NP}$-complete. Then you can confidently march into your boss's office and announce, *"I can't find an efficient algorithm, but neither can these famous people."* This statement might be sufficient to save your job.

As illustrated by this story, the theory of $\mathcal{NP}$-completeness has the following utility in practice. Whenever we encounter a new problem of some practical or theoretical interest, we try to develop an efficient algorithm for solving it. If we do succeed, clearly the problem is easy and we and others might make further attempts to develop an even more efficient algorithm. However, if we do not succeed in developing an efficient algorithm for the problem, we might begin to wonder whether our problem is an $\mathcal{NP}$-complete problem. The theory of $\mathcal{NP}$-completeness provides us with several tools for establishing that a problem is $\mathcal{NP}$-complete. If we do succeed in showing the problem is $\mathcal{NP}$-complete, we have sufficient reason to believe that the problem is hard and no efficient algorithm can ever be developed to solve it. We should thus abandon our quest for an efficient algorithm and direct our efforts at developing efficient heuristics (i.e., algorithms that give solutions that are not guaranteed to be optimal), or at developing various types of enumeration algorithms or other algorithms that will generally run in exponential time. If we cannot prove that our problem is $\mathcal{NP}$-complete, the status of the problem is inconclusive and remains so until someone settles it either way. Indeed, many interesting problems live in this never-never land: for example, the important problem of recognizing whether two graphs are isomorphic.

This appendix provides the basic tools for carrying out this program and is organized as follows. In Section B.2 we describe problem reductions and transformations, an important construct in the theory of $\mathcal{NP}$-completeness. In Section B.3 we describe several problem classes, such as $\mathcal{P}$, $\mathcal{NP}$, and $\mathcal{NP}$-complete. In Section B.4 we study a method for showing that a problem is $\mathcal{NP}$-complete and illustrate this approach on a variety of simple optimization problems. Needless to say, our discussion of the theory of $\mathcal{NP}$-completeness is intended to be very elementary. For a deeper study of this topic, we refer the reader to the literature discussed in the reference notes.

B.2 PROBLEM REDUCTIONS AND TRANSFORMATIONS

The theory of $\mathcal{NP}$-completeness helps us to classify a given problem into two broad classes: (1) easy problems that can be solved by polynomial-time algorithms, and (2) hard problems that are not likely to be solved in polynomial time and for which all known algorithms require exponential running time. Notice that in this classification we want to determine only whether a problem can or cannot be solved in polynomial time; the order of the polynomial is irrelevant. We should keep this point in mind throughout our subsequent discussion.

In almost all the problems studied in this book, we have been concerned with determining some type of an optimal solution. The theory of $\mathcal{NP}$-completeness requires that problems be stated so that we can answer them with a yes or no; we refer to this yes–no version of a problem as a *recognition version* of it. We illustrate this notion using the traveling salesman problem (TSP).

> *TSP-optimization*. Given a directed graph $G = (N, A)$ and an integer arc length c_{ij} associated with every arc $(i, j) \in A$, determine a tour W (i.e., a directed cycle that visits each node in the network exactly once) with the smallest possible value of the tour length $\sum_{(i,j)\in W} c_{ij}$.
>
> *TSP-recognition-I*. Given a directed graph $G = (N, A)$, an integer arc length c_{ij} associated with every arc $(i, j) \in A$, and an integer k^*, does the network contain a tour W satisfying the condition $\sum_{(i,j)\in W} c_{ij} \le k^*$?

It is easy to see that if we have a polynomial-time algorithm for TSP-optimization, we can use it to solve TSP-recognition-I. To do so, we use an algorithm for the TSP-optimization to determine an optimal tour W^*, and then we check to see whether $\sum_{(i,j)\in W^*} c_{ij}$ is less than or equal to k^*. The answer to this question is an answer of TSP-recognition-I. Interestingly, the converse is also true: If we have a polynomial-time algorithm for TSP-recognition-I, we can use it, although applied several times, to solve TSP-optimization in polynomial time. First, we perform binary search (see Section 3.3 for details of binary search methods) on the possible tour lengths and solve TSP-recognition-I at each search point to identify the optimal tour length, say k^*, of TSP-optimization. If C denotes the largest arc length in the network, then using the binary search, we require $O(\log(nC))$ executions of TSP-recognition-I to identify the optimal tour length k^*. Next we determine the optimal tour, again by executing the TSP-recognition-I algorithm repeatedly. We consider every arc $(i, j) \in A$, and one by one, apply the TSP-recognition-I algorithm to find whether the network $G(N, A - \{(i, j)\})$ contains a tour with length less than or equal to k^*. If the answer is yes, we delete the arc. After we have considered all the arcs, the remaining graph is an optimal tour of TSP-optimization.

The preceding discussion showed that we could solve TSP-optimization in polynomial time if we could solve TSP-recognition-I in polynomial time. Thus the optimization and recognition versions are equivalent in terms of whether or not they can be solved in polynomial time. Alternatively, we say that these two problems are *polynomially equivalent*. We point out that recognition problems are not always polynomially equivalent to the corresponding optimization problems. To illustrate this point, consider the following two alternative recognition problems for the traveling salesman problem:

> *TSP-recognition-II*. Given a directed graph $G = (N, A)$, an integer k^*, and an integer arc length c_{ij} associated with every arc $(i, j) \in A$, does the network contain a tour W for which $\sum_{(i,j)\in W} c_{ij} \geq k^*$?
>
> *TSP-recognition-III*. Given a directed graph $G = (N, A)$, an integer k^*, and an integer arc length c_{ij} associated with every arc $(i, j) \in A$, does every tour W in G satisfy the condition $\sum_{(i,j)\in W} c_{ij} \geq k^*$?

Note that TSP-recognition-III has a no instance if and only TSP-recognition-I has a yes instance, so we can use TSP-recognition-III, like TSP-recognition-I, to solve TSP-optimization. Can we solve the TSP-optimization by solving a polynomial number of instances of TSP-recognition-II? Perhaps. But the problem has no obvious solution. In general, each optimization problem has several associated recognition problems and we want to select a recognition problem that is polynomially equivalent to the optimization problem. For all optimization problems considered in this book, and for most problems that ever arise, some recognition version is polynomially equivalent to the optimization version. For this reason the theory of $\mathcal{NP}$-completeness, even though it applies formally only to recognition problems, is also suitable for assessing the complexity of optimization problems.

The preceding discussion also illustrates an important technique known as *problem reduction*. We say that a problem P_1 reduces to problem P_2 if we can solve problem P_1 using an algorithm for P_2 as a subroutine. We have shown in the preceding discussion that the TSP-optimization reduces to TSP-recognition. We say that the problem P_1 *polynomially reduces* to problem P_2 if some polynomial-time algorithm that solves P_1 uses the algorithm for solving P_2 at *unit cost.* A point of central importance in this definition is the unit cost clause, which implies that the algorithm for P_2 requires unit time to execute. Naturally, in almost all cases, this assumption will be very unrealistic. Its usefulness, however, is apparent because of the following property, whose proof is left to the reader.

Property B.1. *If a problem P_1 polynomially reduces to problem P_2 and some polynomial-time algorithm solves P_2, some polynomial-time algorithm solves P_1.*

We refer to an instance of the recognition problem as a *yes instance* if the answer to this problem instance is yes, and a *no instance* otherwise. A special type of problem reduction is of significant interest, which we call *problem transformation.* We say that a problem P_1 polynomially transforms to another problem P_2 if for every instance I_1 of problem P_1 we can construct in polynomial time (e.g., polynomial in terms of the size of I_1) an instance I_2 of problem P_2 so that I_1 is a yes instance of P_1 if and only if I_2 is a yes instance of P_2. In the subsequent discussion, we consider several examples of polynomial-time transformations.

Polynomial reductions and polynomial transformations are useful in the following sense. If problem P_1 polynomially transforms to problem P_2, problem P_2 is at least as hard as P_1: Given an algorithm for problem P_2, we can always use it to solve problem P_1 with comparable (i.e., polynomial or not) running times. If the algorithm for P_2 is polynomial-time, then by using it (and the polynomial transformation), we can also solve P_1 in polynomial time. If P_1 is polynomially transformable to P_2, then P_2 is at least as hard as P_1. The possibility of making this transformation does not imply that P_1 is as hard as P_2. In fact, P_1 might be easy, while P_2 is hard. As an example, we can transform the minimum cost flow problem P_1 to an integer linear programming problem P_2. Although no known algorithm will solve the integer linear programming problem in polynomial time, as we have seen in the text, several algorithms will solve the minimum cost flow problem in polynomial time. As shown by this example, even though we might be able to polynomially reduce (or transform) an easy problem to a more difficult problem, this transformation does not imply that the easy problem is difficult. Whenever we can polynomially reduce a given problem to an easy problem, though, we can be assured that the given problem is easy.

B.3 PROBLEM CLASSES $\mathcal{P}$, $\mathcal{NP}$, $\mathcal{NP}$-COMPLETE, AND $\mathcal{NP}$-HARD

In this section we study the problem classes $\mathcal{P}$, $\mathcal{NP}$, and $\mathcal{NP}$-complete and discuss relationships among these classes.

Class $\mathcal{P}$

We say that a recognition problem P_1 belongs to class $\mathcal{P}$ if some polynomial-time algorithm solves problem P_1. In this book we have seen several examples of problems that belong to class $\mathcal{P}$; the recognition versions of the following problems belong to class $\mathcal{P}$: the shortest path problem, the maximum flow problem, the minimum cost flow problem, assignment and matching problems, and the minimum spanning tree problem.

Class $\mathcal{NP}$

Roughly speaking, we say that a recognition problem P_1 is in the class $\mathcal{NP}$, if for every yes instance I of P_1, there is a short (i.e., polynomial length) verification that the instance is a yes instance.

Throughout this book, our standard measure of complexity of a problem has been the difficulty of solving the problem. However, the class $\mathcal{NP}$ deals with another measure of complexity that is more closely related to the idea of a proof. Consider, for example, the TSP-recognition-I. If someone hands you a yes instance and a tour W of length at most k^* and asks you to verify whether the problem instance is a yes instance, you can do so rapidly in $O(n)$ time by examining the tour and checking whether it passes through every node exactly once and has a length of no more than k^*.

Next, suppose that you are handed a no instance of TSP-recognition-I and you are asked to prove that it is no instance; then you are in serious trouble. There is no obvious proof other than (1) enumerating all possible tours and verifying that each tour length is greater than k^*, or (2) using any algorithm for TSP-optimization to determine the optimal tour length L and verifying that $L > k^*$. Unfortunately, the time to implement either of these approaches is not polynomial in the size of the instance I. Indeed, a proof of a no instance might (in the worst case) require an exponential amount of time. We therefore see a peculiar asymmetry in TSP-recognition-I. Although we might require the same amount of time to determine whether a given instance is a yes instance or a no instance, and need only polynomial time to prove the correctness of a given yes instance, we might require exponential time to prove that an instance is a no instance.

This situation is somewhat akin to proving theorems. When we ask a student to prove or disprove a conjecture, she focuses on how long it takes to find a proof. Thus we might view a conjecture to be quite difficult if the student requires a very long time to settle it either way. In contrast, suppose that we hand her a theorem along with its proof and ask her to verify the proof. Here the student's task is easy if she can quickly verify the proof. Thus the student's measure of difficulty of the theorem is how long it takes to verify the correctness of the given proof, not how long it takes her to develop the proof on her own. Needless to say, verifying a given proof of a difficult theorem is substantially easier than developing its proof.

We now make these notions more formal. Let P_1 be a recognition problem. For an instance I of P_1, let $|I|$ denote the *size* of the instance (i.e., the number of digits or bits needed to represent I; see Section 3.2 for a discussion of how to measure problem sizes). We refer to a proof that an instance is a yes instance as its *certificate*. For example, for the TSP-optimization-I, the certificate of a yes instance is a tour W whose length is at most k^*. Let $\text{CR}(I)$ denote the certificate of a yes instance I, and let $|\text{CR}(I)|$ denote its size. We refer to an algorithm that can verify the correctness of a given certificate, that is, that the certificate establishes the instance as a yes instance, as a *certificate checking algorithm*. For example, for TSP-optimization-I, the certificate checking algorithm might be an algorithm that scans the nodes in a given certificate and verifies that each node is visited exactly once and that the length of the tour is at most k^*. We can now give a formal definition of the class $\mathcal{NP}$.

We say that a recognition problem P_1 is in the class $\mathcal{NP}$ if some certificate checking algorithm $\mathcal{AL}$ and polynomial $p(\cdot)$ satisfy the following properties:

1. Every yes instance I of P_1 has a certificate CR(I).
2. The algorithm $\mathcal{AL}$ can verify the correctness of CR(I) in at most $p(|I|)$ steps.

We say that a certificate is *succinct* if it, together with some polynomial $p(\cdot)$, satisfies these conditions. Note that since the time to verify the correctness of the certificate CR(I) must be polynomial in $|I|$, $|\text{CR}(I)|$ must also be polynomial in $|I|$.

Observe that the definition of the class $\mathcal{NP}$ implies that every problem in the class $\mathcal{P}$ is also in $\mathcal{NP}$. Let P_1 be a problem in the class $\mathcal{NP}$ and let $\mathcal{AL}_1$ be a polynomial-time algorithm for solving P_1. In this case we could choose the null set as the certificate and choose the algorithm $\mathcal{AL}_1$ itself as the certificate checking algorithm. Then for any given yes instance I of P_1, the algorithm $\mathcal{AL}_1$ can verify the correctness of I in polynomial time.

We next introduce some additional problems that are in the class $\mathcal{NP}$.

1. *Hamiltonian cycle problem.* Does a given directed network $G = (N, A)$ contain a directed cycle that visits each node in the network exactly once?
2. *Partition problem.* Given a finite set N of elements with element values $w(\cdot)$, does some subset $S \subseteq N$ satisfy the property that $\sum_{i \in S} w(i) = \sum_{i \in N-S} w(i)$?
3. *3-cover problem.* Given a collection of m (possibly overlapping) sets $S_1, S_2, \ldots, S_m$, each with three elements in a ground set $\{1, 2, \ldots, n\}$, do $n/3$ pairwise disjoint sets from this collection span the entire ground set (i.e., the union of the sets is $\{1, 2, \ldots, n\}$)?
4. *Integer programming feasibility problem.* Given a $p \times q$ constraint matrix $\mathcal{A}$ and a p-vector b, does some nonnegative integer q-vector x satisfy the equations $\mathcal{A}x = b$?

We now prove that some of these problems are in the class $\mathcal{NP}$.

TSP-recognition-II. In this case the certificate of a yes instance is a tour W of length at least k^*. The certificate checking algorithm can easily verify the correctness of W in $O(n)$ steps. Consequently, TSP-recognition-II is in the class $\mathcal{NP}$.

TSP-recognition-III. In this case there is no obvious succinct certificate of a yes instance (enumerating all the tours of the network G requires an exponential amount of space and time), although there is a succinct certificate for a no instance. Therefore, we cannot conclude that TSP-recognition-III is in the class $\mathcal{NP}$. (It probably is not in the class $\mathcal{NP}$, although no one has yet been able to prove this fact.)

Integer programming feasibility problem. At first glance this problem appears to be in the class $\mathcal{NP}$. If the integer programming problem has a feasible solution x, then x is a certificate and we can easily verify the correctness of this certificate in time that is polynomial in $|I|$ and x. The potential difficulty is that the size of x might not be polynomial in $|I|$ and the definition of the class $\mathcal{NP}$ requires that the certificate be polynomially bounded in $|I|$. Nevertheless, the integer programming feasibility problem is in the class $\mathcal{NP}$, because a deep theorem of integer programming states that if an instance I is feasible, some feasible solution has a size that is polynomially bounded in the size of $|I|$.

We ask the reader to prove that the remaining problems we have introduced in this appendix are in the class $\mathcal{NP}$.

Class $\mathcal{NP}$-Complete

A recognition problem P_1 is said to be $\mathcal{NP}$*-complete* if (1) $P_1 \in \mathcal{NP}$, and (2) all other problems in the class $\mathcal{NP}$ polynomially transform to P_1.

Property B.1 implies that if there is an efficient algorithm for some $\mathcal{NP}$-complete problem P_1, there is an efficient algorithm for *every* problem in the class $\mathcal{NP}$. As a result, an $\mathcal{NP}$-complete problem is (in a certain technical sense) at least as hard as every other problem in the class $\mathcal{NP}$. The class $\mathcal{NP}$ is a broad class of problems that includes all the "hard nuts" of combinatorial optimization such as the TSP, the 3-cover problem, and the Hamiltonian cycle problem. At first glance, the definition of $\mathcal{NP}$-complete problems might appear to be so restrictive that we might be tempted to think that very few problems are $\mathcal{NP}$-complete, and that proving a problem to be $\mathcal{NP}$-complete would be a remarkable feat. Nevertheless, researchers have shown that many problems are $\mathcal{NP}$-complete, including all three problems in $\mathcal{NP}$ that we listed earlier: the Hamiltonian cycle problem, the partition problem, and the 3-cover problem. Since we will be using these $\mathcal{NP}$-completeness results in our subsequent discussion, we state these results as a theorem (which we do not prove).

Theorem B.2. *Each of the following problems is $\mathcal{NP}$-complete:*
(*a*) *The Hamiltonian cycle problem*
(*b*) *The partition problem*
(*c*) *The 3-cover problem* ◆

Consider two $\mathcal{NP}$-complete problems P_1 and P_2. By definition, P_1 polynomially transforms to P_2 and P_2 polynomially transforms to P_1. This observation implies that each $\mathcal{NP}$-complete problem polynomially transforms to every other $\mathcal{NP}$-complete problem. Therefore, all $\mathcal{NP}$-complete problems are in some sense comparable in their computational difficulty. If we succeed in developing an efficient algorithm for one $\mathcal{NP}$-complete problem, we know that every problem in the class $\mathcal{NP}$ is polynomially solvable, and $\mathcal{P} = \mathcal{NP}$. Most researchers today conjecture that no polynomial-time algorithm could possibly solve any $\mathcal{NP}$-complete problem. Equivalently, most researchers believe that $\mathcal{P} \neq \mathcal{NP}$, but the issue remains unresolved.

Figure B.1 gives a schematic representation of relationships between different problem classes studied in the section.

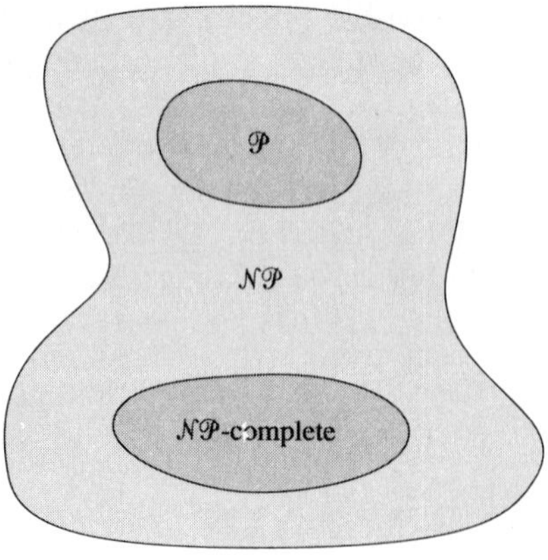

Figure B.1 Relationship between various problem classes.

Class 𝒩𝒫-Hard

A recognition problem P_1 is said to be *𝒩𝒫-hard* if all other problems in the class 𝒩𝒫 polynomially reduce to P_1.

The class 𝒩𝒫-hard is broader than the class 𝒩𝒫-complete because it includes the class 𝒩𝒫 as well as problems that are not in class 𝒩𝒫.

B.4 PROVING 𝒩𝒫-COMPLETENESS RESULTS

Proving the existence of the first 𝒩𝒫-complete problem was indeed quite challenging, but once researchers discovered one (or a few) 𝒩𝒫-complete problems, showing that some other problems are 𝒩𝒫-complete was much easier. To prove that a problem P_2 is 𝒩𝒫-complete, we must establish two facts:

1. P_2 is in 𝒩𝒫.
2. All other problems in 𝒩𝒫 polynomially transform to P_2.

In practice we establish part (b) by showing that a known 𝒩𝒫-complete problem, say P_1, polynomially transforms to P_2. Since P_1 is a 𝒩𝒫-complete problem, every other problem in 𝒩𝒫 polynomially transforms to P_1. Moreover, since P_1 polynomially transforms to P_2, every problem in the class 𝒩𝒫 polynomially transforms to P_2.

For most problems encountered in practice, we can find recognition versions that are in 𝒩𝒫. Showing that problem P_2 is in 𝒩𝒫 is generally (but not always) straightforward. However, identifying a known 𝒩𝒫-complete problem P_1 that would polynomially transform to P_2 frequently is more challenging. The selection of the problem P_1, for which this transformation is simple and direct, requires some skill, experience, and insight. Indeed, proving 𝒩𝒫-completeness results is more of an art than a science. In the following discussion we illustrate briefly how to establish 𝒩𝒫-completeness results by selecting a few problems and proving that they are 𝒩𝒫-complete. We have selected problems related to the network flow problems discussed in this book and for which the resulting transformations are direct. In this discussion, we use the fact that the Hamiltonian cycle, partition, and 3-cover problems that we have discussed earlier in this appendix are $\mathcal{NP}$-complete. We do not prove that these problems are in $\mathcal{NP}$, because doing so is straightforward in each case.

Traveling salesman problem. We show that the Hamiltonian cycle problem polynomially transforms to the TSP. Suppose that we wish to solve the Hamiltonian cycle problem in the graph $G = (N, A)$. We construct a complete directed graph $G' = (N, A')$ with arc lengths defined as follows: $c_{ij} = 1$ if $(i, j) \in A$ and $c_{ij} = 2$ otherwise. We define $k^* = n$. With this definition of arc lengths, the tour constraint $\sum_{(i,j)\in W} c_{ij} \le k^*$ is satisfied if and only if W belongs to A and thus W is a Hamiltonian cycle of the original problem. Consequently, TSP has a yes instance if and only if G contains some tour of length n, and this occurs if and only if G has a Hamiltonian cycle. Notice that for every triple i, j, k, $c_{ik} \le c_{ij} + c_{jk}$, and thus the arc costs satisfy the *triangle inequality*. We have therefore shown that the traveling salesman problem is $\mathcal{NP}$-complete even if arc costs satisfy the triangle inequality.

Hamiltonian path problem. Does a given directed network $G = (N, A)$ contain a directed path that visits every node exactly once (the path can start at any node and can end at any other node)?

We transform the Hamiltonian cycle problem to the Hamiltonian path problem. Suppose that we want to determine whether the graph G contains a Hamiltonian cycle. From G we construct a new graph G' as follows. We add a new node $n + 1$ and redirect every incoming arc at node 1 to node $n + 1$ (i.e., we replace arc $(i, 1)$ by the arc $(i, n + 1)$). We prove that G contains a Hamiltonian cycle if and only if G' contains a Hamiltonian path. Consider a Hamiltonian cycle $1 = i_1 - i_2 - \cdots - i_n - i_1$ in G; this cycle corresponds to a Hamiltonian path $1 = i_1 - i_2 - \cdots - i_n - (n + 1)$ in G'. To see the converse, notice that every Hamiltonian path in G' must begin at node 1 (because this node has no incoming arc) and must end at node $n + 1$ (because this node has no outgoing arc). Moreover, every Hamiltonian path of the form $1 = j_1 - j_2 - \cdots - j_n - (n + 1)$ in G' corresponds to the Hamiltonian cycle $1 = j_1 - j_2 - \cdots - j_n - j_1$ in G. This conclusion establishes that we can solve the Hamiltonian cycle problem in G by solving a Hamiltonian path problem in G'.

Longest path problem. Does a given network $G = (N, A)$ contain a (simple) path from node s to node t with at least L arcs?

If $L = n - 1$, a path of length L from node s to node t is a Hamiltonian path. We have already seen that this problem is $\mathcal{NP}$-complete.

Knapsack problem. Given a finite set N of elements, the integers v^* and w^*, and a *value* $v(i)$ and a *weight* $w(i)$ associated with every element $i \in N$, does some subset $S \subseteq N$ satisfy the property that $\sum_{i\in S} v(i) \ge v^*$ and $\sum_{i\in S} w(i) \le w^*$?

We show that the knapsack problem is $\mathcal{NP}$-complete using a transformation from the partition problem. Given a partition problem on a set N with element values $s(\cdot)$, we construct a knapsack problem on the same set as follows. We define $v(i) = w(i) = s(i)$ for every element $i \in N$, and $v^* = w^* = K = (\sum_{i\in S} s(i))/2$. The knapsack problem then finds a set S for which $\sum_{i\in S} s(i) \le K$ and $\sum_{i\in S} s(i) \ge K$, so $\sum_{i \in S} s(i) = K$. Now notice that $\sum_{i \in N-S} s(i) = \sum_{i \in N} s(i) - \sum_{i \in S} s(i) = 2K - K = K$. Therefore, the set S is a partition.

Constrained shortest path problem. Given a graph $G = (N, A)$, integers c and τ, and an arc length c_{ij} and a traversal time τ_{ij} associated with every arc $(i, j) \in A$, does the graph contain a directed path from node s to node t whose length is at most c and whose traversal time is at most τ?

We show that the constrained shortest path problem is $\mathcal{NP}$-complete by transforming the knapsack problem to it. For a given knapsack problem, consider the constrained shortest path problem shown in Figure B.2 with $c = -v$ and $\tau = w$. As is easy to verify, the knapsack problem has a feasible solution if and only if the graph in Figure B.2 contains a path whose length is at most $-v = c$ and whose traversal time is at most w. Therefore, we can solve the knapsack problem by solving a constrained shortest path problem.

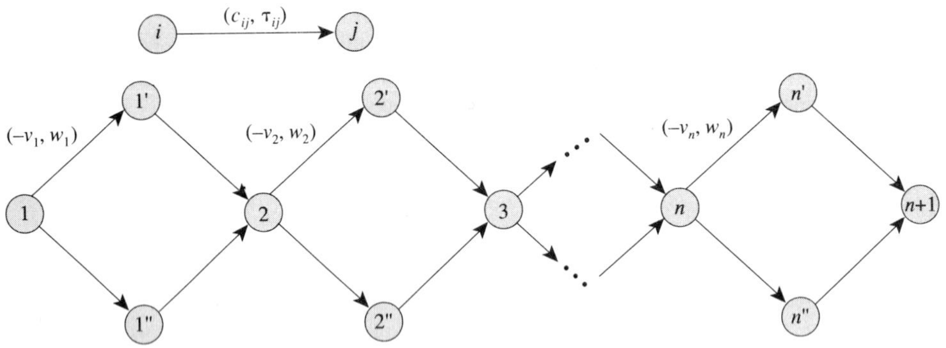

Figure B.2 Transforming the knapsack problem into the constrained shortest path problems. (Arcs without any data have zero cost and zero traversal time.)

Integer generalized flow problem. Given a network $G = (N, A)$, a number v^*, and an arc multiplier μ_{ij} and an arc capacity u_{ij} associated with every arc $(i, j) \in A$ (the network has no associated arc costs), does the network contain an integer generalized flow with a total flow of value of at least v^* into the sink?

We will show that 3-cover problem reduces to an integer generalized flow problem in an appropriately defined network. For a given 3-cover problem, we consider the integer generalized flow problem shown in Figure B.3. In this network a node representing the set S_i has three outgoing arcs directed toward the nodes representing the three elements of the set S_i; these arcs have unit multipliers and unit capacities. Now notice that if we send a unit flow on the arc (s, S_i) from node s, 3 units arrive at node S_i, which in turn sends 1 unit along each of the outgoing arcs. Using this observation, we can easily establish a one-to-one correspondence between 3-covers and integer generalized flows. This correspondence implies that the 3-cover problem has a yes instance if and only if the integer generalized flow problem has a yes instance.

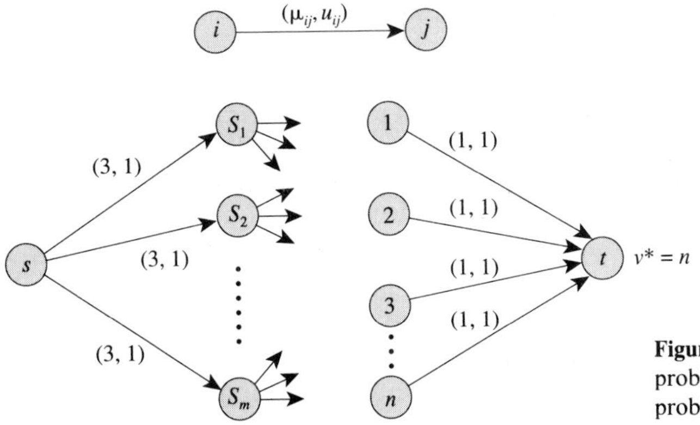

Figure B.3 Transforming a 3-cover problem into an integer generalized flow problem.

0–1 integer programming problem. Given an $p \times q$ matrix $\mathcal{A}$ and an integer p-vector b, does some integer q-vector x, whose components are all 0 or 1, satisfy the inequality system $\mathcal{A}x \leq b$?

The fact that the 0–1 integer programming problem is $\mathcal{NP}$-complete is fairly straightforward to establish because we can formulate most of the problems discussed previously as 0–1 integer programs. Consider, for instance, the knapsack problem which can be formulated as determining binary variables x_j's satisfying the following constraints:

$$- \sum_{j=1}^{n} v_j x_j \leq -v^*$$

and

$$\sum_{j=1}^{n} w_j x_j \leq w^*,$$

which is a 0–1 integer programming problem.

Weak and Strong $\mathcal{NP}$-Completeness and the Similarity Assumption

Throughout much of this book, we have emphasized that data for network flow problems typically satisfies the similarity assumption in practice, i.e., for some fixed integer k, $C = O(n^k)$ and $U = O(n^k)$. We have also pointed out that algorithms with running times involving C or U rather than $\log C$ or $\log U$ should be considered pseudo-polynomial and not polynomial. For example, Dial's algorithm for the shortest path problem requires time $O(m + nC)$ and is thus a pseudo-polynomial algorithm. However, if the data is known to satisfy the similarity assumption, any pseudo-polynomial time algorithm becomes a polynomial time algorithm. For example, if $C = O(n^k)$ then the running time for Dial's algorithm would become $O(m + n^{k+1})$.

If a problem is $\mathcal{NP}$-complete even when the similarity assumption is satisfied, we say that the problem is *strongly $\mathcal{NP}$-complete*. If a problem is $\mathcal{NP}$-complete but it fails to be $\mathcal{NP}$-complete when the similarity assumption is satisfied, then we call it *weakly $\mathcal{NP}$-complete*. For example, the knapsack problem and the constrained shortest path problem are both weakly $\mathcal{NP}$-complete. The traveling salesman problem, the Hamiltonian path probiem, the longest path problem, the integer generalized flow problem, and 0–1 integer programming are all strongly $\mathcal{NP}$-complete.

An an aside, we showed that integer programming is $\mathcal{NP}$-complete by showing that the knapsack problem is a special case. While this proof is very simple, it has the disadvantage of showing only weak $\mathcal{NP}$-completeness of integer programming (why?). In fact, we could also have showed that the traveling salesman problem is a special case of 0–1 integer programming, and this transformation would show the strong $\mathcal{NP}$-completeness of integer programming (assuming that we had already established the strong $\mathcal{NP}$-completeness of the traveling salesman problem).

B.5 CONCLUDING REMARKS

In this book we have discussed a variety of network flow problems and developed many polynomial-time algorithms for solving these problems. It is interesting to observe that simple generalizations of these problems often are $\mathcal{NP}$-complete. The shortest path problem is polynomially solvable, but the constrained shortest path problem is $\mathcal{NP}$-complete. The maximum flow problem in directed networks with nonnegative lower bounds is polynomially solvable (see Section 6.5), but the maximum flow problem in undirected networks with nonnegative lower bounds on arc flows is $\mathcal{NP}$-complete. We have seen how to find a minimum cut in a network efficiently, but the maximum cut problem is $\mathcal{NP}$-complete. Several efficient algorithms will solve the two-dimensional matching problem (i.e., the matching of objects two at a time), but a three-dimensional version of the problem is $\mathcal{NP}$-hard. In Chapter 19 we have shown how to solve the Chinese postman problem efficiently in directed as well as undirected networks; in mixed networks (i.e., those whose arcs can both be directed and undirected) this problem is $\mathcal{NP}$-complete. The generalized network flow problem and the multicommodity flow problem are polynomially solvable, since they are special cases of the linear programming problem, which has several polynomial-time algorithms. Unfortunately, integer versions of both the problems are $\mathcal{NP}$-complete. The literature cited in the reference notes contains proof of some of these results. We do not intend to imply that *every* single generalization of network flow problems studied in this book is $\mathcal{NP}$-complete. Nevertheless, most generalizations are $\mathcal{NP}$-complete, except some generalizations that happen to be linear programming problems.

Like worst-case complexity theory, the theory of $\mathcal{NP}$-completeness is pessimistic: It always focuses on what happens in the worst case. The worst-case behavior of an algorithm might be markedly different than its behavior in practice. For example, from a worst-case perspective a problem whose best available algorithm runs in time $O(n^{100})$ is an easy problem, despite the fact that the $O(n^{100})$ time is terrible running time in practice. Similarly, the theory will regard an $\mathcal{NP}$-complete problem with an $O(n^{0.01 \log n})$ time bound as a hard problem, even though for $n \leq 2^{100}$, the running time is better than linear. Indeed, several $\mathcal{NP}$-complete problems can be solved very efficiently in practice, possibly faster than some problems in class $\mathcal{P}$ of comparable size. However, we can safely say that $\mathcal{NP}$-complete problems "sometimes'' do not have algorithms that can solve large practical instances in reasonable time, whereas problems in class $\mathcal{P}$ ''often'' have.

In concluding this discussion of $\mathcal{NP}$-completeness proofs, we might note that we have considered just one set of issues within the very broad field of computational complexity. Many other issues and refinements arise when we try to understand the structure of algorithms and of computers and the computations they perform. In particular, it is possible to classify algorithms in terms of the space they require and it is possible to distinguish algorithms and problems by imposing more structure on the underlying data.

REFERENCE NOTES

The field of $\mathcal{NP}$-completeness is vast and replete with deep results. In this appendix we have discussed only some of the most elementary results. Cobham [1964] and, independently, Edmonds [1965a] introduced the class $\mathcal{P}$. Cook [1971] introduced the notion of $\mathcal{NP}$-completeness and proved that the satisfiability problem is $\mathcal{NP}$-complete. Independently, Levin [1973] developed this notion. Karp [1972] showed that a rich class of problems, including the traveling salesman problem and the knapsack problem, is $\mathcal{NP}$-complete. The set of problems known to be in the class $\mathcal{NP}$-complete grew at a phenomenal pace; this set now contains thousands of problems. The book by Garey and Johnson [1979] is still the best guide to $\mathcal{NP}$-completeness results. The story given in Section B.1 has also been adapted from this book. Books by Papadimitriou and Steiglitz [1982] and by Cormen, Leiserson, and Rivest [1990] are good additional references on this topic.

Appendix C

LINEAR PROGRAMMING

Inequality is the cause of all local movements.
—Leonardo da Vinci

Chapter Outline

C.1 INTRODUCTION

Linear programming is perhaps *the* core model of constrained optimization; and the simplex method for solving linear programming has been one of the most significant algorithmic discoveries of this century. Developed in 1947, the simplex method has stood the test of time, having been applied to thousands of applications in fields as diverse as agriculture, communications, computer science, engineering design, finance, industrial and military logistics, manufacturing, transportation, and urban planning. Moreover, methods and concepts developed for linear programming—such as duality theory, decomposition methods, and sensitivity analysis—have served as important base methodologies for stimulating discoveries in many other fields within the sphere of optimization. For these reasons, linear programming rightly deserves its position as one of the basic cornerstones of applied mathematics, computer science, and operations research.

In this appendix we summarize some of the basic ideas of linear programming and the simplex method. We do so for at least two reasons. First, a great majority of the models that we have developed in this book, and indeed most of the models encountered in the field of network optimization, are either linear programs or integer programming extensions of linear programs. Therefore, a firm understanding of linear programming is valuable for understanding the structure of the models we have been studying. Second, although we have attempted to develop much of network flows from first principles and to use basic combinatorial ideas rather than more general methodologies and concepts of linear programming, we have, by necessity, needed to invoke ideas from linear programming on many occasions, sometimes as a basic tool in our development and at other times to make appropriate connections between the ideas we have been developing and more general concepts. Therefore, we have needed to rely on several central ideas of linear programming. This appendix serves to make our coverage as complete as possible; it functions both (1) as an introduction to linear programming for those who have only passing familiarity with this topic, and (2) as a review of linear programming for those readers who are already conversant with this topic.

A linear program is an optimization problem with a linear objective function, a set of linear constraints, and a set of nonnegativity restrictions imposed upon the underlying decision variables; that is, it is an optimization model of the form

$$\text{Minimize} \quad \sum_{j=1}^{q} c_j x_j \tag{C.1a}$$

subject to

$$\sum_{j=1}^{q} a_{ij} x_j = b(i) \qquad \text{for all } i = 1, \ldots, p, \tag{C.1b}$$

$$x_j \geq 0 \qquad \text{for all } j = 1, \ldots, q. \tag{C.1c}$$

This problem has q nonnegative decision variables x_j and p equality constraints (C.1b). (In many texts, m denotes the number of equality constraints and n denotes the number of decision variables. This notation, unfortunately, is the reverse of the convention in network flows, since network flow systems contain one constraint per node and one variable per arc. For this reason we do not use the notation of m and n to denote the number of constraints and variables of a linear program.)

We assume, by multiplying the ith equation by -1, if necessary, that the right-hand-side coefficient $b(i)$ of each constraint $i = 1, \ldots, p$ is nonnegative. We might note that we could formulate a linear program in several alternative ways; for example, the objective function could be stated in maximization form, or the constraints could be in a less than or equal to or greater than or equal to form. The linear programming literature frequently refers to the formulation (C.1)—a model with equality constraints, nonnegative variables, and a minimization form of the objective function—as the *standard form* of a linear program.

In economic planning, each decision variable models one particular production activity (x_j is the level of that activity), each constraint corresponds to a scarce resource, and the coefficient a_{ij} indicates the amount of the ith resource consumed per unit of the jth production activity. In this instance the model seeks the "best" use of the scarce resources, that is, the production plan that uses the available resources to produce the maximum possible revenue (assuming a maximization form of the objective function).

In matrix notation, the linear programming model has the following form:

$$\text{Minimize} \quad cx \tag{C.2a}$$

subject to

$$\mathcal{A}x = b, \tag{C.2b}$$

$$x \geq 0. \tag{C.2c}$$

In this formulation the matrix $\mathcal{A} = (a_{ij})$ has p rows and q columns, the vector $c = (c_j)$ is a q-dimensional row vector, the vectors $x = (x_j)$ and $b = (b(i))$ are q-

and p-dimensional column vectors, respectively. We let $\mathcal{A}_j$ denote the column of $\mathcal{A}$ corresponding to the variable x_j. We assume that the rows of the matrix are linearly independent; thus the system $\mathcal{A}x = b$ contains no redundant equations. In terms of linear and matrix algebra (we assume modest background concerning these topics), this assumption states that the rows of the matrix $\mathcal{A}$ are linearly independent; that is, the matrix $\mathcal{A}$ has full row rank.

For the special case of the minimum cost network flow problem, each component of the decision variable x corresponds to the flow on an arc and the matrix $\mathcal{A}$ has one row for each node of the underlying network. In this case the matrix $\mathcal{A}$ is the node–arc incident matrix $\mathcal{N}$ that we introduced in Chapter 1.

In the following sections we describe the rudiments of linear programming theory. We begin by illustrating the underlying geometry of linear programs and by introducing the fundamental concepts of extreme points and basic feasible solutions. Then we describe the key features of the simplex method and of a variant that permits us to efficiently handle upper bounds on the decision variables. We conclude this appendix by introducing the basic features and key results of linear programming duality theory.

C.2 GRAPHICAL SOLUTION PROCEDURE

Linear programs involving only two or three variables have a convenient graphical representation that helps in understanding the nature of linear programming and of the simplex method. We illustrate this procedure using the following example, stated in inequality form with a maximization objective:

$$\text{Maximize} \quad z = x_1 + x_2 \tag{C.3a}$$

subject to

$$2x_1 + 3x_2 \le 12, \tag{C.3b}$$

$$x_1 \le 4, \tag{C.3c}$$

$$x_2 \le 3, \tag{C.3d}$$

$$x_1, x_2 \ge 0. \tag{C.3e}$$

The shaded region in Figure C.1 is the set of feasible solutions for this problem. The set of feasible solutions for the linear programming problem is generally referred to as a *polyhedron*. The points A, B, C, D, and E are the *extreme points* of the polyhedron; these points are formed by the intersection of the lines corresponding to various constraints. Note that extreme points do not lie on any line segment joining two other points in the polyhedron. More formally, a vector x is a strict convex combination of distinct vectors $x^1, x^2, \ldots, x^k$ if $x = \theta^1 x^1 + \theta^2 x^2 + \cdots + \theta^k x^k$ for a set of weights $\theta^i > 0$ satisfying the condition $\sum_{i=1}^{k} \theta^i = 1$. A vector x is an extreme point of a polyhedron if it is not a strict convex combination of two distinct points in the polyhedron, that is, cannot be represented as $x = \theta x^1 + (1 - \theta)x^2$ for some weight $0 < \theta < 1$ and two distinct points x^1 and x^2 of the polyhedron. It is an easy exercise to show that if x is a strict convex combination of $k > 2$ distinct points in the polyhedron, it is not an extreme point.

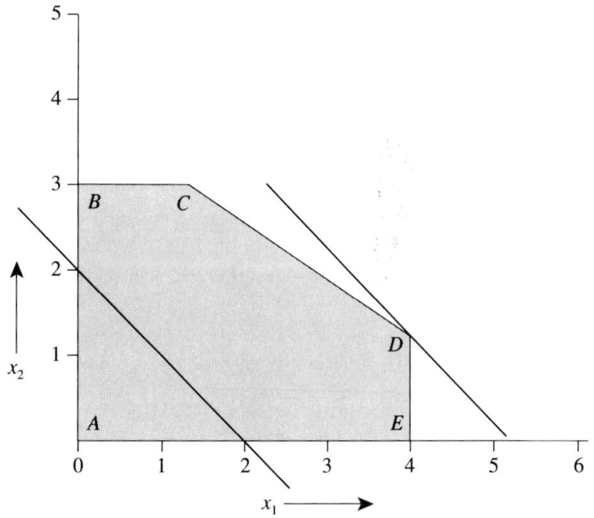

Figure C.1 Set of feasible solutions for a linear program.

The linear programming problem seeks a point (x_1, x_2) in the polyhedron *ABCDEA* that achieves the maximum possible value of $x_1 + x_2$. Equivalently, we wish to determine the largest value of w for which the line $x_1 + x_2 = z$ has at least one point in common with the polyhedron *ABCDEA*. The lines obtained for different values of z are parallel to each other. Since these lines move farther away from the origin as z becomes larger, to maximize $x_1 + x_2$ we need to slide the line $x_1 + x_2 = z$ away from the origin as far as possible so that it has some point in common with the polyhedron *ABCDEA*. We can do so until we reach some extreme point, point D in our case, at which point the line has only one point in common with the polyhedron; at this point, for any further translation of the line away from the origin, no matter how small, the line has no point in common with the polyhedron. Thus point D is an optimal solution of our linear programming problem; its objective function value equals 16/3.

This graphical solution procedure illustrates an important property of linear programs, namely that every linear program always has an extreme point solution as one of its optimal solutions. Therefore, to solve a linear programming problem, we can focus only on the extreme point solutions. Consequently, we need to consider only a finite number of solutions. The simplex method, described in the next section, makes use of this extreme point property. It starts at some feasible extreme point and visits "adjacent" extreme points, improving the objective function values of the solution at each step, until it reaches an optimal extreme point. For instance, in our example, if the simplex method starts at point A, it might visit the points B and C before reaching the optimal extreme point D. Alternatively, it might follow the path A, E, D.

C.3 BASIC FEASIBLE SOLUTIONS

Our description of the simplex method requires that the linear program to be solved be stated in the standard form (C.1); that is, the objective function is in the minimization form; each constraint, except for the nonnegative condition imposed on the decision variables, is an equality; and each right-hand-side coefficient $b(i)$ is nonnegative. Any linear program not in the standard form can be brought into standard form through simple transformations. For example, maximizing $\sum_{j=1}^{q} c_j x_j$ is equivalent to minimizing $-\sum_{j=1}^{q} c_j x_j$. To model an inequality constraint $\sum_{j=1}^{q} a_{ij} x_j \le b(i)$ as an equality constraint, we could add a new nonnegative "slack variable" y_i, with zero cost, and writing the inequality as $\sum_{j=1}^{q} a_{ij} x_j + y_i = b(i)$.

During its execution, the simplex method modifies the original linear program stated in the standard form by performing a series of one or more of the following elementary row operations:

1. Multiplying a row (i.e., constraint) by a constant, or
2. Adding one row to another row or to the objective function.

Since we have stated all the constraints in the equality form, row operations do not affect the set of feasible solutions of the linear program. As an illustration, consider the following linear program:

$$\text{Minimize} \quad z(x) = x_1 + x_2 - 8x_3 + 6x_4 \tag{C.4a}$$

subject to

$$2x_1 + x_2 - 14x_3 + 10x_4 = 16, \tag{C.4b}$$

$$x_1 + x_2 - 11x_3 + 7x_4 = 10, \tag{C.4c}$$

$$x_1, x_2, x_3, x_4 \ge 0. \tag{C.4d}$$

Subtracting (C.4c) from (C.4a) and (C.4b) gives the following equivalent linear program:

$$\text{Minimize} \quad z(x) = 0x_1 + 0x_2 + 3x_3 - x_4 + 10 \tag{C.5a}$$

subject to

$$x_1 \qquad - 3x_3 + 3x_4 = 6, \tag{C.5b}$$

$$x_1 + x_2 - 11x_3 + 7x_4 = 10, \tag{C.5c}$$

$$x_1, x_2, x_3, x_4 \ge 0. \tag{C.5d}$$

Subtracting (C.5b) from (C.5c) gives another equivalent formulation of this linear program:

$$\text{Minimize} \quad z(x) = 0x_1 + 0x_2 - 3x_3 + x_4 + 10 \tag{C.6a}$$

subject to

$$x_1 \qquad - 3x_3 + 3x_4 = 6, \tag{C.6b}$$

$$x_2 - 8x_3 + 4x_4 = 4, \tag{C.6c}$$

$$x_1, x_2, x_3, x_4 \ge 0. \tag{C.6d}$$

Since the linear program (C.6) is equivalent to (C.4), both have the same optimal solutions. A linear program like (C.6) is said to be in the *canonical form* if it satisfies the following *canonical property*.

Canonical Property. *The formulation has one decision variable isolated in each constraint; the variable isolated in a given constraint has a coefficient of +1 in that constraint and does not appear in any other constraint, nor does it appear in the objective function.*

The previous formulation satisfies the canonical property because x_1 is isolated in constraint (C.6b) and x_2 in constraint (C.6c).

A linear program typically has a large number of canonical forms since there are many ways to isolate decision variables in the constraints. The canonical form in (C.6) has the following attractive feature. Assigning any values to x_3 and x_4 uniquely determines the values of x_1 and x_2. In fact, setting $x_3 = x_4 = 0$ immediately gives the solution $x_1 = 6$ and $x_2 = 4$. Solutions such as these, known as *basic feasible solutions*, play a central role in the simplex method. In general, given a canonical form for any linear program, we obtain a basic solution by setting the variable isolated in constraint i, called the ith *basic variable*, equal to the right-hand side of the ith constraint, and by setting the remaining variables, called *nonbasic*, all to value zero. Collectively, the basic variables are known as a *basis*.

In general, we obtain a basic feasible solution as follows. We isolate a variable in each constraint. For simplicity, assume that we have isolated the variable x_i in the ith constraint. Let $\mathbf{B} = \{1, 2, \ldots, p\}$ denote the index set of basic variables and let $\mathbf{L} = \{p + 1, p + 2, \ldots, q\}$ denote the index set of nonbasic variables (we choose the mnemonic $\mathbf{L}$ for the index set of nonbasic variables because we set the value of the corresponding variables to their lower bound—that is, value 0—in the basic solution defined by this index set). We refer to the pair $(\mathbf{B}, \mathbf{L})$ as a *basis structure* of the linear problem. For a given basis structure $(\mathbf{B}, \mathbf{L})$, we can compatibly partition the columns of the constraint matrix $\mathcal{A}$. Let $\mathcal{B} = [\mathcal{A}_1, \mathcal{A}_2, \ldots, \mathcal{A}_p]$ and $\mathcal{L} = [\mathcal{A}_{p+1}, \mathcal{A}_{p+2}, \ldots, \mathcal{A}_q]$. We refer to the $p \times p$ matrix $\mathcal{B}$ as a *basis matrix*. We also let $x_\mathbf{B} = [x_i : i \in \mathbf{B}]$ and $x_\mathbf{L} = [x_j : j \in \mathbf{L}]$ be a partitioning of the variables into subvectors corresponding to the index sets $\mathbf{B}$ and $\mathbf{L}$. With this notation, we can rewrite the constraint matrix $\mathcal{A}x = b$ as

$$\mathcal{B}x_\mathbf{B} + \mathcal{L}x_\mathbf{L} = b. \tag{C.7}$$

We can convert (C.7) to the canonical form by premultiplying each term by $\mathcal{B}^{-1}$, the inverse of the basis matrix, giving

$$x_\mathbf{B} + \mathcal{B}^{-1}\mathcal{L}x_\mathbf{L} = \mathcal{B}^{-1}b. \tag{C.8}$$

We obtain a *basic solution* from (C.8) by setting each nonbasic variable to value zero. The resulting solution is

$$x_\mathbf{B} = \mathcal{B}^{-1}b \quad \text{and} \quad x_\mathbf{L} = 0.$$

We refer to this solution as a *basic feasible solution* if the value of each basic variable is nonnegative (i.e., $x_\mathbf{B} \geq 0$). We also say that the basis structure $(\mathbf{B}, \mathbf{L})$ is *feasible* if its associated basic solution is feasible. For some choices of the basis matrix, the corresponding basic solution will be feasible, and for some other choices it will not be feasible.

Converting a linear program to a canonical form requires that we invert the basis matrix, which is possible only if the columns associated with the basic variables are linearly independent. If the associated columns are linearly dependent, the basis matrix is singular (i.e., its determinant is zero) and we cannot invert it. We shall therefore henceforth refer to a basis as a subset of p variables whose corresponding columns are linearly independent.

During its execution, the simplex method requires information about $\mathcal{B}^{-1}\mathcal{A}_j$, the updated column corresponding to the nonbasic variable x_j. We let $\overline{\mathcal{A}}_j = \mathcal{B}^{-1}\mathcal{A}_j$ and call this vector the *representation* of $\mathcal{A}_j$ with respect to the basis matrix $\mathcal{B}^{-1}$ (or, alternatively, basis **B**). We refer to this column vector as a representation because by premultiplying both sides of the equations $\overline{\mathcal{A}}_j = \mathcal{B}^{-1}\mathcal{A}_j$ by $\mathcal{B}$, we obtain $\mathcal{B}\overline{\mathcal{A}}_j = \mathcal{A}_j$, which implies that we can interpret the elements in the vector $\overline{\mathcal{A}}_j$ as weights that we use to multiply the columns of the basis matrix $\mathcal{B}$ in order to obtain the column $\mathcal{A}_j$. For notational convenience, we let $\overline{\mathcal{A}}_{\mathbf{L}}$ denote the matrix $\mathcal{B}^{-1}\mathcal{L}$ containing the column representations of all the nonbasic variables and let $\bar{b} = \mathcal{B}^{-1}b$; we refer to the vector $\bar{b}$ as the modified right-hand side. Finally, let $c_{\mathbf{B}} = (c_1, c_2, \ldots, c_p)$ and $c_{\mathbf{L}} = (c_{p+1}, c_{p+2}, \ldots, c_q)$ denote the cost vectors associated with the basic and nonbasic variables, respectively.

In a canonical form of a linear program, each basic variable has a zero coefficient in the objective function. We can obtain this special form of the objective function by performing a sequence of elementary row operations (i.e., multiplying constraints by some multipliers and subtracting them from the objective function). Any sequence of elementary row operations is equivalent to the following: Multiply each constraint i by a number $\pi(i)$ and subtract it from the objective function. This operation gives the equivalent objective function $\sum_{j=1}^{q} c_j x_j - \sum_{i=1}^{p} \pi(i)[\sum_{j=1}^{q} a_{ij}x_j - b(i)]$, or, collecting terms and letting $z_0 = \sum_{i=1}^{p} \pi(i)b(i)$,

$$z(x) = \sum_{j=1}^{p} [c_j - \sum_{i=1}^{p} \pi(i)a_{ij}]x_j + \sum_{j=p+1}^{q} [c_j - \sum_{i=1}^{p} \pi(i)a_{ij}]x_j + z_0. \quad \text{(C.9)}$$

To obtain a canonical form, we select the vector π so that

$$c_j - \sum_{i=1}^{p} \pi(i)a_{ij} = 0 \qquad \text{for each } j \in \mathbf{B}. \quad \text{(C.10)}$$

In matrix notation we select π so that

$$\pi\mathcal{B} = c_{\mathbf{B}}.$$

In this expression, $c_{\mathbf{B}} = (c_1, c_2, \ldots, c_p)$ is the cost vector associated with the basic variables. We refer to

$$\pi = c_{\mathbf{B}}\mathcal{B}^{-1}$$

as the *simplex multipliers* associated with the basis **B** and refer to $c_j^{\pi} = c_j - \sum_{i=1}^{p} \pi(i)a_{ij}$ as the *reduced cost* of the variable x_j. Note that $z_0 = \pi b = c_{\mathbf{B}}\mathcal{B}^{-1}b = c_{\mathbf{B}}x_{\mathbf{B}}$, which since $(x_{\mathbf{L}} = 0)$ is the value of the objective function corresponding to the basis **B**.

In Section C.2 we observed that linear programs always have extreme-point solutions and in this section we have focused on basic solutions. Are these concepts related? Indeed, they are. To conclude this section we show that extreme points and basic solutions are geometric and algebraic manifestations of the same concept.

Therefore, by finding an optimal basic solution, we are obtaining an optimal extreme point solution, and vice versa. Recall that columns of the constraint matrix $\mathcal{A}$ corresponding to the basis **B** are linearly independent.

Theorem C.1 (Extreme Points and Basic Feasible Solutions). *A feasible solution x to a linear program (C.1) is an extreme point if and only if the columns $\{\mathcal{A}_j : x_j > 0\}$ of the constraint matrix corresponding to the strictly positive variables are linearly independent.*

Proof. To simplify our notation, suppose that $x_1 > 0, x_2 > 0, \ldots, x_k > 0$ and $x_{k+1} = x_{k+2} = \cdots = x_q = 0$. Suppose that the columns $\mathcal{A}_1, \mathcal{A}_2, \ldots, \mathcal{A}_k$ are linearly dependent. Then there exist constants $\omega_1, \omega_2, \ldots, \omega_k$ (not all zero), satisfying the condition

$$\mathcal{A}_1\omega_1 + \mathcal{A}_2\omega_2 + \cdots + \mathcal{A}_k\omega_k = 0.$$

Let us define $\omega_{k+1} = \omega_{k+2} \cdots = \omega_q = 0$ and let ω denote the vector $(\omega_1, \omega_2, \ldots, \omega_q)$. Then since each of the first k components of the solution x are positive, for some sufficiently small value of the scalar θ,

$$x + \theta\omega \geq 0 \quad \text{and} \quad x - \theta\omega \geq 0.$$

Also, since $\mathcal{A}\omega = 0$, $\mathcal{A}(x + \theta\omega) = \mathcal{A}(x - \theta\omega) = \mathcal{A}x = b$. Therefore, both $x + \theta\omega$ and $x - \theta\omega$ are feasible for the linear program. But then $x = \frac{1}{2}(x + \theta\omega) + \frac{1}{2}(x - \theta\omega)$, which implies that x is not an extreme point (because it lies on the line joining the points $x + \theta\omega$ and $x - \theta\omega$). Therefore, we have shown that if x is an extreme point, the columns $\mathcal{A}_1, \mathcal{A}_2, \ldots, \mathcal{A}_k$ must be linearly independent.

To establish the converse, suppose that $x = \theta x^1 + (1 - \theta)x^2$ for some scalar $0 < \theta < 1$ and two feasible points x^1 and x^2 to the linear program (C.1). Since $x_j = \theta x_j^1 + (1 - \theta)x_j^2$ for any $j \geq k + 1$, and both x_j^1 and x_j^2 are nonnegative, both x_j^1 and x_j^2 must have value zero. Therefore, since both these points are feasible solutions to the linear program, they satisfy

$$\mathcal{A}_1x_1^1 + \mathcal{A}_2x_2^1 + \cdots + \mathcal{A}_kx_k^1 = b,$$

and

$$\mathcal{A}_1x_1^2 + \mathcal{A}_2x_2^2 + \cdots + \mathcal{A}_kx_k^2 = b.$$

Subtracting these equations from each other shows that

$$\mathcal{A}_1(x_1^1 - x_1^2) + \mathcal{A}_2(x_2^1 - x_2^2) + \cdots + \mathcal{A}_k(x_k^1 - x_k^2) = 0.$$

But since the columns $\mathcal{A}_1, \mathcal{A}_2, \ldots, \mathcal{A}_k$ are linearly independent, each component of x^1 and x^2 must be the same. But then we cannot represent $x = \theta x^1 + (1 - \theta)x^2$ in terms of two distinct feasible points of the linear program and therefore x is an extreme point. ◆

Although we have not stated this theorem as "x is a basic feasible solution if and only if it is an extreme point," the theorem easily implies this statement. Notice that if $k = p$ in this proof, the columns $\mathcal{A}_1, \mathcal{A}_2, \ldots, \mathcal{A}_k$ form a basis of the linear program, so x is a basic feasible solution. On the other hand, if $k < p$, we can add

$p - k$ other linearly independent columns to $\mathcal{A}_1, \mathcal{A}_2, \ldots, \mathcal{A}_k$ to form a basis $\mathcal{B}$. In this case x still is a basic feasible solution, but with some basic variables at value zero (those corresponding to the columns we have added). Therefore, these observations and Theorem C.1 imply that basic feasible solutions and extreme points are identical.

C.4 SIMPLEX METHOD

The simplex method maintains a basic feasible solution at every step. Given a basic feasible solution, the method first applies the optimality criteria to test the optimality of the current solution. If the current solution does not fulfill this condition, the algorithm performs an operation, known as a *pivot operation*, to obtain another basis structure with a lower or identical cost. The simplex method repeats this process until the current basic feasible solution satisfies the optimality criteria. Before describing the details of these steps, we consider some preliminary issues.

Obtaining an Initial Basis Structure

Recall from our preceding discussion that a basic solution might not be feasible. Therefore, identifying a basic feasible solution requires specifying a set of basic variables so that the columns associated with these variables are linearly independent and the solution $x_{\mathbf{B}}, x_{\mathbf{L}}$ obtained by setting $x_{\mathbf{B}} = \mathcal{B}^{-1}b$ and $x_{\mathbf{L}} = 0$ is nonnegative. Unfortunately, there is no simple method for identifying any such collection of basic variables. In fact, finding a feasible solution of a linear program (or a basic feasible solution) is almost as difficult as finding an optimal solution. Nevertheless, using a simple technique, we can find a basic feasible solution of a related linear program and use it to initiate the simplex method. This technique consists of introducing an *artificial variable* x_{q+i} with a sufficiently large cost M for each constraint i and defining an augmented linear program as:

$$\text{Minimize} \sum_{j=1}^{q} c_j x_j + \sum_{j=q+1}^{q+p} M x_j \tag{C.11a}$$

subject to

$$\sum_{j=1}^{q} a_{ij} x_j + x_{q+i} = b(i) \qquad \text{for all } i = 1, \ldots, p, \tag{C.11b}$$

$$x_j \geq 0 \qquad \text{for all } j = 1, \ldots, q. \tag{C.11c}$$

It is easy to show that the original linear program (C.1) has a feasible solution if and only if each artificial variable has value zero in every optimal solution of the augmented linear program (C.11). (The large cost coefficient associated with each artificial variable ensures that it has value zero in an optimal solution if the original problem has a feasible solution.) As a consequence, solving the augmented linear program solves the original linear program. For the augmented linear program, the set of all artificial variables constitutes an initial basis and the value of the ith basic variable is $b(i) \geq 0$.

Optimality Criteria

Let (**B**, **L**) denote a feasible basis structure of the linear program. Assume, for simplicity, that $\mathbf{B} = \{1, 2, \ldots, p\}$. Consider the canonical form associated with this basis structure. In this canonical form the objective function is

$$\text{Minimize} \quad z(x) = z_0 + \sum_{j=p+1}^{q} c_j^\pi x_j. \tag{C.12}$$

The coefficient $c_j^\pi = c_j - \sum_{i=1}^{p} \pi(i) a_{ij}$ is the reduced cost of the nonbasic variable x_j with respect to the current simplex multipliers π. We claim that if $c_j^\pi \geq 0$ for each nonbasic variables x_j, the current basic feasible solution x is an optimal solution of the linear program. To see this, observe that in any feasible solution of the linear program, $x_j \geq 0$ for all $j \in \mathbf{L}$. Therefore, if $c_j^\pi \geq 0$ for each nonbasic variables x_j, then z_0 is a lower bound on the optimal objective value. As we noted in the preceding section, the current solution x, which sets $x_j = 0$ for all $j \in \mathbf{L}$, achieves this lower bound and, therefore it must be optimal.

Pivot Operation

If $c_j^\pi < 0$ for some nonbasic variable x_j, the current basic feasible solution might not be optimal. The expression (C.12) implies that c_j^π is the rate of decrease in the objective function value per unit increase in the value of x_j. The simplex method selects one such nonbasic variable, say x_s, as the *entering variable* and tries to increase its value. As we will see, when the simplex method increases x_s as much as possible while keeping all the other nonbasic variable at value zero, some basic variable, say x_r, reaches value zero. The simplex method replaces the basic variable x_r by x_s, defining a new basis structure. It then updates the inverse of the basis and repeats the computations.

If we increase the value of the entering variable x_s to value θ and keep all other nonbasic variables at zero value, expression (C.8) implies that the basic variables $x_\mathbf{B}$ change in the following manner:

$$x_\mathbf{B} + \theta \overline{\mathcal{A}}_s = \overline{b}. \tag{C.13}$$

In this expression $\overline{b} = \mathcal{B}^{-1} b \geq 0$, $\overline{\mathcal{A}}_s = \mathcal{B}^{-1} \mathcal{A}_s$, and $x_\mathbf{B} = [x_1, x_2, \ldots, x_p]$. Let $\overline{\mathcal{A}}_s = [\overline{a}_{1s}, \overline{a}_{2s}, \ldots, \overline{a}_{ps}]$. We can restate (C.13) as

$$x_i = \overline{b}(i) - \theta \overline{a}_{is} \qquad \text{for all } i = 1, \ldots, p. \tag{C.14}$$

If $\overline{a}_{is} \leq 0$ and we increase θ, then x_i either remains unchanged or increases. If $\overline{a}_{is} > 0$ and we increase θ, then x_i decreases and eventually it will become zero. Consequently, (C.14) implies that if $\overline{a}_{is} > 0$, then the scalar θ must satisfy the condition $\theta \leq \overline{b}(i)/\overline{a}_{is}$ in order for x_i to remain nonnegative. As a result,

$$\theta = \min_{1 \leq i \leq p} \{\overline{b}(i)/\overline{a}_{is} : \overline{a}_{is} > 0\},$$

is the largest value of θ that we can assign to x_s while remaining feasible. What if $\overline{a}_{is} \leq 0$ for each $i = 1, \ldots, p$? Then we can assign an arbitrarily large value to the entering variable x_s and the solution remains feasible. Since $c_s^\pi < 0$, by setting x_s as large as we like, we can make the objective function arbitrarily small, and make it approach $-\infty$. In this instance we say that the linear program has an *unbounded solution*.

We next focus on situations in which θ is finite. If we set $x_s = \theta$, one of the basic variables, say x_r, becomes zero. Note that $\bar{b}(r)/\bar{a}_{rs} = \theta$. We refer to x_r as the *leaving variable* and refer to the rule we have described for identifying θ and the corresponding leaving variable as the *minimum ratio rule*. Next, we designate x_s a basic variable, x_r a nonbasic variable, and update the canonical form of the linear program so that it satisfies the canonical property with respect to the new basis. We do so by performing a sequence of elementary row operations.

Updating the Simplex Tableau

Recall from Section C.3 that in the canonical form with respect to the basis **B**, the equations of the linear program assume the form

$$x_{\mathbf{B}} + \mathcal{B}^{-1}\mathcal{L}x_{\mathbf{L}} = \mathcal{B}^{-1}b. \tag{C.15}$$

In the new basis, the entering variable x_s becomes the basic variable for the rth row, which requires that in the new canonical form the variable x_s should have a coefficient $+1$ in the rth row, and a coefficient 0 in all other rows. We achieve this new canonical form by first dividing the rth row by $\bar{a}_{rs}$; the variable x_s then has a $+1$ coefficient in this row. Then, for each $1 \le i \le p$, $i \ne r$, we multiply the rth row by the constant $-\bar{a}_{is}$ and add it to the ith row so that the updated value of $\bar{a}_{is}$ becomes zero. We also multiply the rth row by a constant $-c_s^{\pi}$ and add it to the objective function so that the objective function coefficient of x_s becomes zero. We refer to this set of computations as a *pivot operation*.

To illustrate these steps of the simplex method, consider the linear programming example given in (C.6). For convenience, we state the example again.

$$\text{Minimize} \quad z(x) = 0x_1 + 0x_2 + 3x_3 - x_4 + 10 \tag{C.16a}$$

subject to

$$x_1 \qquad - 3x_3 + 3x_4 = 6, \tag{C.16b}$$

$$x_2 - 8x_3 + 4x_4 = 4, \tag{C.16c}$$

$$x_1, x_2, x_3, x_4 \ge 0. \tag{C.16d}$$

The nonbasic variable x_4 has a negative reduced cost, and we select it as the entering variable. Applying the minimum ratio rule, we find that $\theta = \min\{6/3, 4/4\} = 1$, the minimum being achieved in the second row, which contains the basic variable x_2. The variable x_2 is the leaving variable. In the next basis, x_1 and x_4 are the basic variables, so we need to modify the canonical form. Since x_4 is the new basic variable for the constraint (C.16c), we divide this constraint by 4 so that x_4 has a $+1$ coefficient. We then multiply the modified constraint by -3 and add it to (C.16b), and multiply it by $+1$ and add it to (C.16a). These operations produce the following (equivalent) formulation of the linear program:

$$\text{Minimize} \quad z(x) = 0x_1 + 0x_4 + \tfrac{1}{4}x_2 + x_3 + 9$$

subject to

$$x_1 \qquad -\tfrac{3}{4}x_2 + 3x_3 = 3,$$

$$x_4 + \tfrac{1}{4}x_2 - 2x_3 = 1,$$

$$x_1, x_2, x_3, x_4 \geq 0.$$

In this canonical form, the reduced costs of both nonbasic variables x_2 and x_3 are nonnegative, so the current basic feasible solution, $x_1 = 3$, $x_4 = 1$, and $x_2 = x_3 = 0$, is optimal. This solution has an objective function value of 9.

One way to perform the pivot operation is by updating the full matrix $\bar{\mathcal{A}}$, that is, by performing the explicit set of pivot computations iteratively on the matrix $\mathcal{A}$. This set of computations can be very expensive for linear programs that contain many variables (as is typical in practice). The *revised simplex method* is a particular implementation of the simplex method that permits us to avoid many of these computations. To describe the basic approach of the revised simplex method, suppose that we (conceptually) append a set of fictitious variables y to the original linear program and form a new linear program with the constraints:

$$\mathcal{A}x + \mathcal{I}y = b,$$

$$x \geq 0, \quad y \geq 0.$$

In this formulation, $\mathcal{I}$ is an identity matrix; that is, its diagonal elements all have value 1 and its elements off the diagonal all have value zero. Then to obtain the canonical form (C.13) with respect to the basis $\mathcal{B}$, we premultiply this system by the basis inverse $\mathcal{B}^{-1}$. With the fictitious variables y, the system becomes

$$x_{\mathbf{B}} + \mathcal{B}^{-1}\mathcal{L}x_{\mathbf{L}} + \mathcal{B}^{-1}y = \mathcal{B}^{-1}b.$$

As shown by this expression, if we were to perform the pivot operation on the entire matrix $\mathcal{A}$ from step to step, the coefficients of fictitious variables y would be the basis inverse. This observation shows that we need not carry out the pivot operations on the entire matrix $\mathcal{A}$. Instead, we can perform these operations on the columns associated with the initial identity matrix $\mathcal{I}$ (we do not formally introduce the variables y). Since the resulting computations give us the basis inverse $\mathcal{B}^{-1}$, we can use this matrix to compute the simplex multipliers $\pi = c_{\mathbf{B}}\mathcal{B}^{-1}$ and then use them to compute the reduced cost of each variable. Once we have determined the variable x_s to introduce into the basis, we compute its representation $\bar{\mathcal{A}}_s = \mathcal{B}^{-1}\mathcal{A}_s$ and then use this information to perform the ratio test to identify the variable x_r to leave the basis. We next perform the row elementary operations on the current $\mathcal{B}^{-1}$ matrix and obtain the updated basis inverse.

The advantage of this approach is that we use only the original data $\mathcal{A}$ in computing the reduced costs [using the formula $c_j^{\pi} = c_j - \sum_{i=1}^{p} \pi(i)a_{ij}$] and determine the modified data $\bar{\mathcal{A}}$ for only one column s of $\mathcal{A}$ and perform row elementary operations only on the basis inverse matrix. This apparently modest change in the algorithm often has dramatic effects on its efficiency because the original data $\mathcal{A}$ for most problems met in practice is very sparse in the sense that most (90 percent or more) of its coefficients are zero. On the other hand, the modified matrix $\bar{\mathcal{A}}$ typically becomes very dense as we perform the iterations of the simplex algorithm. In the revised simplex method, by using appropriate data structures, we can avoid all the computations corresponding to the zero elements. As a consequence, by implementing the revised simplex method, we usually achieve great economies in our computations. The vast linear programming literature and many standard texts specify more details about this approach and about other ways to improve the empirical efficiency of the simplex method.

Finite Termination

The simplex method moves from one basic feasible solution to another by performing a pivot operation. In each iteration, the solution value improves by the amount $c_s^{\pi}\theta$. If $\theta > 0$, we say that the pivot is *nondegenerate*; otherwise, we refer to it as *degenerate*. A nondegenerate pivot strictly decreases the value of the objective function associated with the basic feasible solution. If every pivot is nondegenerate, the simplex method will terminate finitely because a linear program has at most ${}^qC_p$ distinct basic feasible solutions and the algorithm can never repeat any basic feasible solution (since the objective function value is strictly decreasing). However, the simplex method might perform degenerate pivots, and without further modifications it might repeat a basic feasible solution and therefore not terminate finitely. Fortunately, researchers have developed several ways for implementing the simplex method so that it converges finitely. We refer the reader to linear programming textbooks for a discussion of these techniques.

The simplex method terminates with one of the following three outcomes:

Case 1. The method terminates with an unbounded solution. In this case the linear program has no optimal solution with a finite solution value.

Case 2. The method terminates with a finite solution in which some artificial variable has a positive value. In this case the original linear program has no feasible solution.

Case 3. The method terminates with a finite solution in which all the artificial variables have zero value. This solution is an optimal solution of the original linear program.

C.5 BOUNDED VARIABLE SIMPLEX METHOD

Often linear programming problems have upper as well as lower bounds imposed on the variables. Network flow problems with arc capacities on the arc flows belong to this class. We refer to this class of linear programs, formulated as follows, as bounded variable linear programs.

$$\text{Minimize} \sum_{j=1}^{q} c_j x_j \tag{C.17a}$$

subject to

$$\sum_{j=1}^{q} a_{ij} x_j = b(i) \qquad \text{for all } i = 1, \ldots, p, \tag{C.17b}$$

$$x_j \leq u_j \qquad \text{for all } j = 1, \ldots, q, \tag{C.17c}$$

$$x_j \geq 0 \qquad \text{for all } j = 1, \ldots, q. \tag{C.17d}$$

The simplest way to handle the upper bound constraints (C.17c) is to treat them like the other constraints (by adding slack variables to convert them into equality constraints). However, doing so increases the size of the linear program substantially and is therefore undesirable. In this section we describe a generalization of the simplex method, called the *bounded variable simplex method*, that treats these upper bound constraints implicitly, very much like the lower bound constraints (C.17d).

In our previous discussion we defined a basic feasible solution in the simplex method by a pair (**B**, **L**), consisting of a set **B** of basic variables and a set **L** of nonbasic variables at their lower bounds. We define a basic feasible solution in the bounded variable simplex method by a triplet (**B**, **L**, **U**), consisting of a set **B** of basic variables, a set **L** of nonbasic variables at their lower bounds, and a set **U** of nonbasic variables at their upper bounds. Let $\mathcal{B}$, $\mathcal{L}$, and $\mathcal{U}$ denote a compatible partitioning of the constraint matrix **A** corresponding to the sets **B**, **L**, and **U**. Then

$$\mathcal{B}x_{\mathbf{B}} + \mathcal{L}x_{\mathbf{L}} + \mathcal{U}x_{\mathbf{U}} = b.$$

In a basic feasible solution we set the value of each nonbasic variable to its appropriate bound, depending on whether the variable is contained in **L** or **U**. Let $u_{\mathbf{U}}$ denote the vector of upper bounds for variables in **U**. Then, setting $x_{\mathbf{L}} = 0$ and $x_{\mathbf{U}} = u_{\mathbf{U}}$, we find that

$$\mathcal{B}x_{\mathbf{B}} = b - \mathcal{U}u_{\mathbf{U}},$$

or

$$x_{\mathbf{B}} = \mathcal{B}^{-1}b - \mathcal{B}^{-1}[\mathcal{U}u_{\mathbf{U}}].$$

As before, $\mathcal{B}^{-1}$ is the inverse of the basis $\mathcal{B}$. For the bounded variable simplex method, we define the simplex multipliers π in the same way as in the simplex method (i.e., as the multipliers to apply to the constraints to create the reduced cost of zero for each basic variable). We now discuss, one by one, various steps of the simplex method and point out any changes required in the bounded variable simplex method.

Optimality Criteria

In a canonical form of the linear program with respect to the basis **B**, we can write the objective function as

$$z(x) = \sum_{j \in \mathbf{L}} c_j^\pi x_j + \sum_{j \in \mathbf{U}} c_j^\pi x_j + z_0.$$

Using arguments similar to those used earlier for the case without upper bounds, we can easily show that if $c_j^\pi \geq 0$ for each $j \in \mathbf{L}$ and $c_j^\pi \leq 0$ for each $j \in \mathbf{U}$, then $z_0 + \sum_{j \in \mathbf{U}} c_j^\pi u_j$ is a lower bound on the optimal value of the objective function. We refer to these conditions as the *optimality* conditions for bounded variable linear programs. If the basic feasible solution satisfies these optimality conditions, the objective function achieves this lower bound, and hence it must be optimal. Observe that the optimality conditions imply that it is not cost-effective to increase the value of any nonbasic variable at its lower bound or decrease the value of any nonbasic variable at its upper bound.

Entering Arc Criteria

The bounded variable simplex method selects any nonbasic variable violating its optimality condition as the entering arc. In other words, a nonbasic variable x_j is eligible to be an entering arc if (1) $j \in \mathbf{L}$ and $c_j^\pi < 0$, or (2) $j \in \mathbf{U}$ and $c_j^\pi > 0$.

Leaving Arc Criteria

If the entering variable x_s is a nonbasic variable at its lower bound, the method attempts to increase the value of x_s by the largest possible amount, and if the entering variable x_s is at its upper bound, the method attempts to decrease its value by the largest possible amount. The maximum change is determined by the requirement that the value of every basic variable and of the nonbasic variable x_s remains between (or at) its lower and upper bounds. The expression (C.14) implies that if $\bar{a}_{is} < 0$, then by increasing θ, we increase the value of the basic variable x_i. In this case the maximum change allowed by the upper bound constraint of x_i is $\theta_i = (u_i - \bar{b}(i))/(-\bar{a}_{is})$. If $\bar{a}_{is} > 0$, then by increasing θ, we decrease the value of the basic variable x_i. In this case the maximum change allowed by the lower bound constraint of x_i is $\theta_i = \bar{b}(i)/\bar{a}_{is}$. We let $\theta_i = +\infty$ if $\bar{a}_{is} = 0$. Finally, the maximum change allowed by the upper and lower bound constraints of x_s is $|u_s|$. The maximum value of θ that we can assign to x_s will be the minimum of $|u_s|$ and $\min\{\theta_i : 1 \le i \le p\}$. In the former case the basis remains unchanged; x_s simply moves from the set $\mathbf{L}$ to $\mathbf{U}$ or from the set $\mathbf{U}$ to $\mathbf{L}$. In the latter case the basis changes and so does the canonical form. The method of updating the canonical form is same as for the case without upper bounds. Conceivably, we might find that $\theta = \infty$, in which case the linear programming problem has an unbounded solution. The case, when the entering variable x_s is a nonbasic variable at its upper bound, is left as an exercise to the reader.

Termination of the Method

Under the nondegeneracy assumption, the bounded variable simplex method terminates finitely because it strictly decreases the objective function value of the basic feasible solution at each step and any linear program has a finite number of basic feasible solutions. Without the nondegeneracy assumption, we need some additional technique to ensure finite termination of the method. For a discussion of these techniques, we refer the reader to linear programming textbooks.

C.6 LINEAR PROGRAMMING DUALITY

Each linear programming problem, which for the purposes of this discussion we call the *primal problem*, has a closely related associated linear programming problem, called the *dual problem*, and together these two problems define a duality theory that lies at the heart of linear programming and many other areas in the field of constrained optimization. In this section we review briefly some of the most salient features of duality theory.

While discussing duality theory, we assume that the linear program has been stated in the following form:

$$\text{Minimize} \quad \sum_{j=1}^{q} c_j x_j \tag{C.18a}$$

subject to

$$\sum_{j=1}^{q} a_{ij} x_j \geq b(i) \qquad \text{for all } i = 1, \ldots, p, \tag{C.18b}$$

$$x_j \geq 0 \qquad \text{for all } j = 1, \ldots, q. \tag{C.18c}$$

In this formulation, we permit $b(i)$ to have an arbitrary sign. We refer to this form as the *symmetric form* of a linear program. It is possible to show that we can convert any linear program in a nonsymmetric form into the symmetric form. To define the dual problem, we associate a dual variable $\pi(i)$ with the ith constraint in (C.18b). With respect to these variables, the dual problem is:

$$\text{Maximize} \quad \sum_{i=1}^{p} b(i)\pi(i) \tag{C.19a}$$

subject to

$$\sum_{i=1}^{p} a_{ij} \pi(i) \leq c_j \qquad \text{for all } j = 1, \ldots, q, \tag{C.19b}$$

$$\pi(i) \geq 0 \qquad \text{for all } i = 1, \ldots, p. \tag{C.19c}$$

Notice that each constraint in the primal has an associated variable in the dual and that the right-hand side of this constraint becomes the cost coefficient of the associated variable. Moreover, each variable in the primal has an associated constraint in the dual and the cost coefficient of this variable becomes the right-hand side of the associated constraint. It is easy to verify that the dual of the dual is the primal.

In the linear programming problem, whose dual we want to form, if some constraint $\sum_{j=1}^{q} a_{ij} x_j = b(i)$ is in the equality form, we form the dual exactly as described earlier, except that the dual variable $\pi(i)$ becomes unrestricted in sign. Similarly, in the primal, if some primal variable x_i is unrestricted in sign, then in the dual the constraint corresponding to x_i is an equality constraint. Our first result concerning the primal–dual pair is known as the *weak duality theorem*.

Theorem C.2 (Weak Duality Theorem). *If x is any feasible solution of the primal problem and π is any feasible solution of the dual problem, then $\sum_{i=1}^{p} b(i)\pi(i) \leq \sum_{j=1}^{q} c_j x_j$.*

Proof. Multiplying the ith constraint in (C.18b) by $\pi(i)$ and adding yields

$$\sum_{i=1}^{p} b(i)\pi(i) \leq \sum_{i=1}^{p} \pi(i) \left[\sum_{j=1}^{q} a_{ij} x_j \right], \tag{C.20}$$

while multiplying the jth constraint in (C.19b) by x_j and adding yields

$$\sum_{j=1}^{q} x_j \left[\sum_{i=1}^{p} a_{ij}\pi(i) \right] \leq \sum_{j=1}^{q} c_j x_j. \tag{C.21}$$

Since the right-hand side of (C.20) equals the left-hand side of (C.21), together these two constraints imply that

$$\sum_{i=1}^{p} b(i)\pi(i) \leq \sum_{j=1}^{q} c_j x_j,$$

which is the conclusion of the lemma. ◆

The weak duality theorem has a number of immediate consequences, which we state next.

Property C.3

(*a*) *The objective function value of any feasible dual solution is a lower bound on the objective function value of every feasible primal solution.*
(*b*) *If the primal problem has an unbounded solution, the dual problem is infeasible.*
(*c*) *If the dual problem has an unbounded solution, the primal problem is infeasible.*
(*d*) *If the primal problem has a feasible solution x and the dual problem has a feasible solution π and $\sum_{i=1}^{p} b(i)\pi(i) = \sum_{j=1}^{q} c_j x_j$, then x is an optimal solution of the primal problem and π is an optimal solution of the dual problem.*

Theorem C.4 (Strong Duality Theorem). *If any one of the pair of primal and dual problems has a finite optimal solution, so does the other one and both have the same objective function values.*

Proof. Assume without any loss of generality that the primal problem has a finite optimal solution. We can convert the primal problem into standard form (in which all constraints have equalities instead of inequalities) by adding slack variables. Suppose that we apply the simplex method to this standard form and let (**B**, **L**) be the optimal basis structure. Let x be the optimal solution and π be the simplex multipliers associated with the optimal basis structure. Recall from our previous discussion that the simplex multipliers π are the multipliers associated with the constraints (C.1b), which when subtracted from the original form of the objective function yield the objective function in the final canonical form. Therefore,

$$z(x^*) = \sum_{j \in \mathbf{B}} c_j^{\pi} x_j + \sum_{j \in \mathbf{L}} c_j^{\pi} x_j + \sum_{i=1}^{p} b(i)\pi(i),$$

and

$$c_j^{\pi} = c_j - \sum_{i=1}^{p} a_{ij}\pi(i). \tag{C.22}$$

Since x is a basic feasible solution, $c_j^{\pi} = 0$ for all $j \in \mathbf{B}$, and $x_j = 0$ for all $j \in \mathbf{L}$. Consequently, $z(x^*) = \sum_{i=1}^{p} b(i)\pi(i)$. Moreover, since the basis structure (**B**, **L**) satisfies the optimality criteria (i.e., $c_j^{\pi} \geq 0$ for all $j \in \mathbf{B} \cup \mathbf{L}$), the expressions (C.22) and (C.19b) imply that π is a feasible solution to the dual problem. The objective function value of this dual solution is $\sum_{i=1}^{p} b(i)\pi(i)$, which is same as that of the primal solution x. Property C.3(d) implies that π is an optimal dual solution, completing the proof of the theorem. ◆

The weak and strong duality theorems imply several fundamental results concerning relationships between the primal and dual problems. The following complementary slackness property, which is another way of relating the two problems, makes some of these relationships more explicit.

Complementary Slackness Property. *A pair (x, π) of the primal and dual feasible solutions is said to satisfy the complementary slackness property if*

$$\pi(i)\left[\sum_{j=1}^{q} a_{ij}x_j - b(i)\right] = 0 \qquad \text{for each } i = 1, \ldots, p, \tag{C.23a}$$

and

$$x_j\left[c_j - \sum_{i=1}^{p} a_{ij}\pi(i)\right] = 0 \qquad \text{for each } j = 1, \ldots, q. \tag{C.23b}$$

Observe from the formulation (C.18) of the primal problem that $[\sum_{j=1}^{q} a_{ij}x_j - b(i)]$ is the amount of slack in the ith primal constraint and $\pi(i)$ is the dual variable associated with this constraint. Similarly, $[c_j - \sum_{i=1}^{p} a_{ij}\pi(i)]$ is the amount of slack in the jth dual constraint and x_j is the primal variable associated with this constraint. The complementary slackness property states that for every primal and dual constraint, the product of the slack in the constraint and its associated primal or dual variable is zero. In other words, if a constraint has a positive slack, the associated primal or dual variable must have value zero, and alternatively, if a primal or dual variable has a positive value, the dual solution must satisfy the corresponding constraint as an equality.

Theorem C.5 (Complementary Slackness Optimality Conditions). *A primal feasible solution x and a dual feasible solution π are optimal solutions of the primal and dual problems if and only if they satisfy the complementary slackness property.*

Proof. We first prove that if x and π are optimal primal and dual solutions, they must satisfy the complementary slackness property. While proving the weak duality theorem, we observed that

$$\sum_{i=1}^{p} b(i)\pi(i) \le \sum_{i=1}^{p}\sum_{j=1}^{q} a_{ij}\pi(i)x_j \le \sum_{j=1}^{q} c_jx_j. \tag{C.24}$$

Since x and π are optimal primal and dual solutions, the strong duality theorem implies that both of the inequalities in (C.24) must be satisfied as equalities. We can rewrite the first equality in (C.23) as

$$\sum_{i=1}^{p} \pi(i)\left[\sum_{j=1}^{q} a_{ij}x_j - b(i)\right] = 0. \tag{C.25}$$

Since x and π are feasible in the primal and dual problems, each term in (C.25) is nonnegative. Consequently, the sum of these terms can be zero only if each term is individually zero. This observation shows that the solutions satisfy (C.23a). Similarly, the second equality in (C.24) implies that the solutions satisfy (C.23b).

We next prove the converse result: namely, if the solutions x and π satisfy the complementary slackness property, they must be optimal in the primal and dual problems. Adding (C.23a) for all i yields

$$\sum_{j=1}^{q} \sum_{i=1}^{p} a_{ij}\pi(i)x_j = \sum_{i=1}^{p} b(i)\pi(i).$$

Similarly, adding (C.23b) for all j yields

$$\sum_{j=1}^{q} \sum_{i=1}^{p} a_{ij}\pi(i)x_j = \sum_{j=1}^{q} c_j x_j.$$

Therefore, $\sum_{i=1}^{p} b(i)\pi(i) = \sum_{j=1}^{q} c_j x_j$, and by Property C.3(c), x is an optimal primal solution and π is an optimal dual solution, concluding the proof of the theorem. ◆

REFERENCE NOTES

Dantzig, who conducted the pioneering work in linear programming, developed the simplex method in 1947 to solve several military planning problems. The optimization community introduced a steady stream of developments since then and now the theory of linear programming includes a vast body of knowledge. Books by Dantzig [1962], Bradley, Hax, and Magnanti [1977], Chvátal [1983], Schrijver [1986], and Winston [1991] are excellent references on the history, applications, and theory of this topic.

REFERENCES

AASHTIANI, H. A., and T. L. MAGNANTI. 1976. Implementing primal–dual network flow algorithms. Technical Report OR 055-76, Operations Research Center, MIT, Cambridge, MA.

ABDALLAOUI, G. 1987. Maintainability of a grade structure as a transportation problem. *Journal of the Operational Research Society* **38,** 367–369.

ADEL'SON-VEL'SKI, G. M., E. A. DINIC, and E. V. KARZANOV. 1975. *Flow Algorithms*. Science, Moscow. (In Russian.)

AHLFELD, D. P., R. S. DEMBO, J. M. MULVEY, and S. A. ZENIOS. 1987. Nonlinear programming on generalized networks. *ACM Transactions on Mathematical Software* **13,** 350–367.

AHO, A. V., J. E. HOPCROFT, and J. D. ULLMAN. 1974. *The Design and Analysis of Computer Algorithms*. Addison-Wesley, Reading, MA.

AHO, A. V., J. E. HOPCROFT, and J. D. ULLMAN. 1983. *Data Structures and Algorithms*. Addison-Wesley, Reading, MA.

AHUJA, R. K. 1986. Algorithms for the minimax transportation problem. *Naval Research Logistics Quarterly* **33,** 725–740.

AHUJA, R. K., and J. B. ORLIN. 1989. A fast and simple algorithm for the maximum flow problem. *Operations Research* **37,** 748–759.

AHUJA, R. K., and J. B. ORLIN. 1991. Distance-directed augmenting path algorithms for maximum flow and parametric maximum flow problems. *Naval Research Logistics Quarterly* **38,** 413–430.

AHUJA, R. K., and J. B. ORLIN. 1992a. The scaling network simplex algorithm. *Operations Research* **40,** Supplement 1, S5–S13.

AHUJA, R. K., and J. B. ORLIN. 1992b. Use of representative counts in computational testings of algorithms. Sloan Working Paper, Sloan School of Management, MIT, Cambridge, MA.

AHUJA, R. K., J. L. BATRA, and S. K. GUPTA. 1984. A parametric algorithm for the convex cost network flow and related problems. *European Journal of Operational Research* **16,** 222–235.

AHUJA, R. K., A. V. GOLDBERG, J. B. ORLIN, and R. E. TARJAN. 1992. Finding minimum-cost flows by double scaling. *Mathematical Programming* **53,** 243–266.

AHUJA, R. K., M. KODIALAM, A. K. MISHRA, and J. B. ORLIN. 1992. Computational testing of maximum flow algorithms. Sloan Working Paper, Sloan School of Management, MIT, Cambridge, MA.

AHUJA, R. K., T. L. MAGNANTI, and J. B. ORLIN. 1989. Network flows. In *Handbooks in Operations Research and Management Science*. Vol. 1: *Optimization*, edited by G. L. Nemhauser, A. H. G. Rinnooy Kan, and M. J. Todd. North-Holland, Amsterdam, pp. 211–369.

AHUJA, R. K., T. L. MAGNANTI, and J. B. ORLIN. 1991. Some recent advances in network flows. *SIAM Review* **33,** 175–219.

AHUJA, R. K., T. L. MAGNANTI, J. B. ORLIN, and M. R. REDDY. 1992. Applications of network optimization. Sloan Working Paper, Sloan School of Management, MIT, Cambridge, MA.

AHUJA, R. K., K. MEHLHORN, J. B. ORLIN, and R. E. TARJAN. 1990. Faster algorithms for the shortest path problem. *Journal of ACM* **37,** 213–223.

AHUJA, R. K., J. B. ORLIN, C. STEIN, and R. E. TARJAN. 1990. Improved algorithms for bipartite network flow problems. Technical Report, Sloan School of Management, MIT, Cambridge, MA. Submitted to *SIAM Journal on Computing*.

AHUJA, R. K., J. B. ORLIN, and R. E. TARJAN. 1989. Improved time bounds for the maximum flow problem. *SIAM Journal on Computing* **18,** 939–954.

AKGÜL, M. 1985a. Shortest path and simplex method. Research Report, Department of Computer Science and Operations Research, North Carolina State University, Raleigh, NC.

Akgül, M. 1985b. A genuinely polynomial primal simplex algorithm for the assignment problem. Research Report, Department of Computer Science and Operations Research, North Carolina State University, Raleigh, NC.

Ali, A. I., E. P. Allen, R. S. Barr, and J. L. Kennington. 1986. Reoptimization procedures for bounded variable primal simplex network algorithms. *European Journal of Operational Research* **23,** 256–263.

Ali, A. I., D. Barnett, K. Farhangian, J. L. Kennington, B. Patty, B. Shetty, B. McCarl, and P. Wong. 1984. Multicommodity network problems: Applications and computations. *IIE Transactions* **16,** 127–134.

Ali, A. I., R. V. Helgason, and J. L. Kennington. 1978. The convex cost network flow problem: A state-of-the-art survey. Technical Report OREM 78001, Southern Methodist University, Dallas, TX.

Ali, A. I., R. Padman, and H. Thiagarajan. 1989. Dual algorithms for pure network problems. *Operations Research* **37,** 159–171.

Alon, N. 1990. Generating pseudo-random permutations and maximum flow algorithms. *Information Processing Letters* **35,** 201–204.

Anderson, W. N. 1975. Maximum matching and the rank of a matrix. *SIAM Journal on Applied Mathematics* **28,** 114–123.

Arisawa, S., and S. E. Elmaghraby. 1977. The "hub" and "wheel" scheduling problems. *Transportation Science* **11,** 124–146.

Aronson, J. E. 1989. A survey of dynamic network flows. *Annals of Operations Research* **20,** 1–66.

Assad, A. A. 1978. Multicommodity network flows: A survey. *Networks* **8,** 37–91.

Assad, A. A. 1980a. Models for rail transportation. *Transportation Research* **14A,** 205–220.

Assad, A. A. 1980b. Solving linear multicommodity flow problems. *Proceedings of the IEEE International Conference on Circuits and Computers*, pp. 157–161.

Bacharach, M. 1966. Matrix rounding problems. *Management Science* **9,** 732–742.

Balachandran, V., and G. L. Thompson. 1975. An operator theory of parametric programming for the generalized transportation problems. Parts I–IV. *Naval Research Logistics Quarterly* **22,** 79–125, 297–340.

Balakrishnan, A., T. L. Magnanti, and R. T. Wong. 1989. A dual-ascent procedure for large scale uncapacitated network design. *Operations Research* **37,** 716–740.

Balakrishnan, A., T. L. Magnanti, A. Shulman, and R. T. Wong. 1991. Models for capacity expansion in local access telecommunication networks. *Annals of Operations Research* **33,** 239–284.

Balakrishnan, A., T. L. Magnanti, and R. T. Wong. 1991. A decomposition algorithm for local access telecommunications network expansion planning. Working Paper, Operations Research Center, MIT, Cambridge, MA.

Balinski, M. L. 1986. A competitive (dual) simplex method for the assignment problem. *Mathematical Programming* **34,** 125–141.

Ball, M. O., and U. Derigs. 1983. An analysis of alternative strategies for implementing matching algorithms. *Networks* **13,** 517–549.

Barahona, F., and É. Tardos. 1989. Note on Weintraub's minimum cost circulation algorithm. *SIAM Journal on Computing* **18,** 579–583.

Barnhart, C. 1988. A network-based primal–dual solution methodology for the multicommodity network flow problem. Ph.D. dissertation, Department of Civil Engineering, MIT, Cambridge, MA.

Barr, R. S., F. Glover, and D. Klingman. 1977. The alternating path basis algorithm for the assignment problem. *Mathematical Programming* **13,** 1–13.

Barr, R. S., and J. S. Turner. 1981. Microdata file merging through large scale network technology. *Mathematical Programming Study* **15,** 1–22.

Barros, O., and A. Weintraub. 1986. Spatial market equilibrium problems as network models. *Discrete Applied Mathematics* **13,** 109–130.

Bartholdi, J. J., J. B. Orlin, and H. D. Ratliff. 1980. Cyclic scheduling via integer programs with circular ones. *Operations Research* **28,** 1074–1085.

Barzilai, J., W. D. Cook, and M. Kress. 1986. A generalized network formulation of the pairwise comparison consensus ranking model. *Management Science* **32,** 1007–1014.

Bazaraa, M. S., J. J. Jarvis, and H. D. Sherali. 1990. *Linear Programming and Network Flows*, 2nd ed. Wiley, New York.

BELFORD, P. C., and H. D. RATLIFF. 1972. A network-flow model for racially balancing schools. *Operations Research* **20,** 619–628.

BELLMAN, R. E. 1957. *Dynamic Programming*. Princeton University Press, Princeton, NJ.

BELLMAN, R. 1958. On a routing problem. *Quarterly of Applied Mathematics* **16,** 87–90.

BELLMORE, M., G. BENNINGTON, and S. LUBORE. 1971. A multivehicle tanker scheduling problem. *Transportation Science* **5,** 36–47.

BENNINGTON, G. E. 1974. Applying network analysis. *Industrial Engineering* **6,** 17–25.

BENTLEY, J. L. 1990. Experiments on geometric traveling salesman heuristics. Computing Science Technical Report 151, AT&T Bell Laboratories, Holmdel, NY.

BENTLEY, J. L., and B. W. KERNIGHAN. 1990. A system for algorithm animation: Tutorial and algorithm animation. Unix Research System Paper, 10th ed., Vol. II. Saunders College Publishing, Philadelphia, pp. 451–475.

BERGE, C. 1957. Two theorems in graph theory. *Proceedings of the National Academy of Sciences USA* **43,** 842–844.

BERGE, C., and A. GHOUILA-HOURI. 1962. *Programming, Games and Transportation Networks*. Wiley, New York.

BERRISFORD, H. G. 1960. The economic distribution of coal supplies in the gas industry: An application of the linear programming transport theory. *Operations Research Quarterly* **11,** 139–150.

BERTSEKAS, D. P. 1976. *Dynamic Programming and Stochastic Control*. Academic Press, New York.

BERTSEKAS, D. P. 1979. A distributed algorithm for the assignment problem. Working Paper, Laboratory for Information and Decision Systems, MIT, Cambridge, MA.

BERTSEKAS, D. P. 1988. The auction algorithm: A distributed relaxation method for the assignment problem. *Annals of Operations Research* **14,** 105–123.

BERTSEKAS, D. P., and D. E. BAZ. 1987. Distributed asynchronous relaxation methods for convex network flow problems. *SIAM Journal on Control and Optimization* **25,** 74–85.

BERTSEKAS, D. P., and J. ECKSTEIN. 1988. Dual coordinate step methods for linear network flow problems. *Mathematical Programming B* **42,** 203–243.

BERTSEKAS, D. P., P. A. HOSEIN, and P. TSENG. 1987. Relaxation methods for network flow problems with convex arc costs. *SIAM Journal on Control and Optimization* **25,** 1219–1243.

BERTSEKAS, D. P., and P. TSENG. 1988a. The relax codes for linear minimum cost network flow problems. In *FORTRAN Codes for Network Optimization*, edited by B. Simeone, P. Toth, G. Gallo, F. Maffioli, and S. Pallottino. *Annals of Operations Research* **13,** 125–190.

BERTSEKAS, D. P., and P. TSENG. 1988b. Relaxation methods for minimum cost ordinary and generalized network flow problems. *Operations Research* **36,** 93–114.

BERTSIMAS, D., and J. B. ORLIN. 1991. A technique for speeding up the solution of the Lagrangian dual. Working Paper OR 248-91, Operations Research Center, MIT, Cambridge, MA.

BIXBY, R. E. 1982. Matroids and operations research. In *Advanced Techniques in the Practice of Operations Research*, edited by H. J. Greenberg, F. H. Murphy, and S. H. Shaw. North-Holland, Amsterdam, pp. 433–458.

BIXBY, R. E. 1991. The simplex method: It keeps getting better. Presented at the *14th International Symposium on Mathematical Programming*, Amsterdam, The Netherlands.

BLAND, R. G., and D. L. JENSEN. 1992. On the computational behavior of a polynomial-time network flow algorithm. *Mathematical Programming* **54,** 1–39.

BOAS, P. V. E., R. KAAS, and E. ZIJLSTRA. 1977. Design and implementation of an efficient priority queue. *Mathematical Systems Theory* **10,** 99–127.

BODIN, L. D., B. L. GOLDEN, A. D. SCHUSTER, and W. ROWING. 1980. A model for the blockings of trains. *Transportation Research* **14B,** 115–120.

BODIN, L. D., B. L. GOLDEN, A. A. ASSAD, and M. O. BALL. 1983. Routing and scheduling of vehicles and crews: The state of the art. *Computers and Operations Research* **10,** 69–211.

BONDY, J. A., and U. S. R. MURTY. 1976. *Graph Theory with Applications*. American Elsevier, New York.

BORŮVKA, O. 1926, Příspevěk k řešení otázky ekonomické stavby elektrovodních sítí. *Elektrotechnicky Obzor* **15,** 153–154.

BRADLEY, G., G. BROWN, and G. GRAVES. 1977. Design and implementation of large scale primal transshipment algorithms. *Management Science* **21,** 1–38.

Bradley, S. P., A. C. Hax, and T. L. Magnanti. 1977. *Applied Mathematical Programming*. Addison-Wesley, Reading, MA.

Brogan, W. L. 1989. Algorithm for ranked assignments with application to multiobject tracking. *Journal of Guidance*, 357–364.

Brown, M. H. 1988. *Algorithm Animation*. MIT Press, Cambridge, MA.

Brown, G. G., and R. D. McBride. 1984. Solving generalized networks. *Management Science* **30,** 1497–1523.

Bruynooghe, M., A. Gibert, and M. Sakarovitch. 1968. Une méthode d'affection du traffic. In: *Fourth International Symposium on the Theory of Traffic Flow*, Karlsruhe, 1968, W. Lentzback and P. Barons (eds.), Beiträge Theorie des Verkehrsflusses Strassenbau und Strassenkehrstechnik Heft 86, Herausgeben von Bunderesminister fur Verkehr, Abteilung Strassenbau, Bonn, Germany.

Busaker, R. G., and P. J. Gowen. 1961. A procedure for determining minimal-cost network flow patterns. ORO Technical Report 15, Operational Research Office, Johns Hopkins University, Baltimore, MD.

Busacker, R. G., and T. L. Saaty. 1965. *Finite Graphs and Networks*. McGraw-Hill, New York.

Cabot, A. V., R. L. Francis, and M. A. Stary. 1970. A network flow solution to a rectilinear distance facility location problem. *AIIE Transactions* **2,** 132–141.

Cahn, A. S. 1948. The warehouse problem (Abstract). *Bulletin of the American Mathematical Society* **54,** 1073.

Carpento, G., S. Martello, and P. Toth. 1988. Algorithms and codes for the assignment problem. In *FORTRAN Codes for Network Optimization*, edited by B. Simeone, P. Toth, G. Gallo, F. Maffioli, and S. Pallottino. *Annals of Operations Research* **13,** 193–224.

Carraresi, P., and G. Gallo. 1984. Network models for vehicle and crew scheduling. *European Journal of Operational Research* **16,** 139–151.

Chalmet, L. G., R. L. Francis, and P. B. Saunders. 1982. Network models for building evacuation. *Management Science* **28,** 86–105.

Chandrasekaran, R. 1977. Minimum ratio spanning trees. *Networks* **7,** 335–342.

Chang, M. D., and C. J. Chen. 1989. An improved primal simplex variant for pure processing networks. *ACM Transactions on Mathematical Software* **15,** 64–78.

Charnes, A., and D. Klingman. 1971. The "more for less" paradox in the distribution model. *Cahiers du Centre D'Etudes de Recherche Operationnelle* **13,** 11–22.

Chen, H., and C. G. Dewald. 1974. A generalized chain labeling algorithm for solving multicommodity flow problems. *Computers and Operations Research* **1,** 437–465.

Cheng, C. K., and T. C. Hu. 1990. Ancestor tree for arbitrary multi-terminal cut functions. *Proceedings of a Conference on "Integer Programming and Combinatorial Optimization,"* edited by R. Kannan and W. R. Pulleyblank. University of Waterloo, Waterloo, Canada.

Cheriyan, J., and T. Hagerup. 1989. A randomized maximum-flow algorithm. *Proceedings of the 30th IEEE Conference on the Foundations of Computer Science*, pp. 118–123.

Cheriyan, J., T. Hagerup, and K. Mehlhorn. 1990. Can a maximum flow be computed in $O(nm)$ time? *Proceedings of the 17th International Colloquium on Automata, Languages and Programming*, pp. 235–248.

Cheriyan, J., and S. N. Maheshwari. 1989. Analysis of preflow push algorithms for maximum network flow. *SIAM Journal on Computing* **18,** 1057–1086.

Cheshire, M., K. I. M. McKinnon, and H. P. Williams. 1984. The efficient allocation of private contractors to public works. *Journal of the Operational Research Quarterly* **35,** 705–709.

Chin, F., and D. Houch. 1978. Algorithms for updating spanning trees. *Journal of Computer and System Sciences* **16,** 333–344.

Christophides, N. 1975. *Graph Theory: An Algorithmic Approach*. Academic Press, New York.

Chvátal, V. 1983. *Linear Programming*. W. H. Freeman, New York.

Clark, J. A., and N. A. J. Hastings. 1977. Decision networks. *Operational Research Quarterly* **20,** 51–68.

Clarke, S., and J. Surkis. 1968. An operations research approach to racial desegregation of school systems. *Socio-Economic Planning Sciences* **1,** 259–272.

Cobham, A. 1964. The intrinsic computational difficulty of functions. *Proceedings of the 1964 Congress for Logic, Methodology, and the Philosophy of Science*, North-Holland, Amsterdam, pp. 24–30.

Collins, M., L. Cooper, R. Helgason, J. Kennington, and L. LeBlanc. 1978. Solving the pipe network analysis problem using optimization techniques. *Management Science* **24,** 747–760.

COOK, S. 1971. The complexity of theorem proving procedures. *Proceedings of the 3rd Annual ACM Symposium on Theory of Computing*, pp. 151–158.

CORMEN, T. H., C. L. LEISERSON, and R. L. RIVEST. 1990. *Introduction to Algorithms*. MIT Press and McGraw-Hill, New York.

COX, L. H., and L. R. ERNST. 1982. Controlled rounding. *INFOR* **20,** 423–432.

CRAINIC, T., J. A. FERLAND, and J. M. ROUSSEAU. 1984. A tactical planning model for rail freight transportation. *Transportation Science* **18,** 165–184.

CREMEANS, J. E., R. A. SMITH, and G. R. TYNDALL. 1970. Optimal multicommodity network flows with resource allocation. *Naval Research Logistics Quarterly* **17,** 269–280.

CROWDER, H. P., R. S. DEMBO, and J. M. MULVEY. 1978. Reporting computational experiments in mathematical programming. *Mathematical Programming* **15,** 316–329.

CROWDER, H. P., R. S. DEMBO, and J. M. MULVEY. 1979. On reporting computational experiments with mathematical software. *ACM Transactions on Mathematical Software* **5,** 193–203.

CROWDER, H. P., and P. B. SAUNDERS. 1980. Results of a survey on MP performance indicators. *COAL Newsletter*, January, pp. 2–6.

CRUM, R. L., and D. J. NYE. 1981. A network model of insurance company cash flow management. *Mathematical Programming* **15,** 86–101.

CUNNINGHAM, W. H. 1976. A network simplex method. *Mathematical Programming* **11,** 105–116.

CUNNINGHAM, W. H. 1979. Theoretical properties of the network simplex method. *Mathematics of Operations Research* **4,** 196–208.

DAFERMOS, S., and A. NAGURNEY. 1984. A network formulation of market equilibrium problems and variational inequalities. *Operations Research Letters* **5,** 247–250.

DANIEL, R. C. 1973. Phasing out capital equipment. *Operations Research Quarterly* **24,** 113–116.

DANTZIG, G. B. 1951. Application of the simplex method to a transportation problem. In *Activity Analysis and Production and Allocation*, edited by T. C. Koopmans. Wiley, New York, pp. 359–373.

DANTZIG, G. B. 1960. On the shortest route through a network. *Management Science* **6,** 187–190.

DANTZIG, G. B. 1962. *Linear Programming and Extensions*. Princeton University Press, Princeton, NJ.

DANTZIG, G. B., W. BLATTNER, and M. R. RAO. 1966. Finding a cycle in a graph with minimum cost to time ratio with application to a ship routing problem. In *Theory of Graphs. International Symposium*. Dunod, Paris, and Gordon and Breach, New York, pp. 209–213.

DANTZIG, G. B., and D. R. FULKERSON. 1954. Minimizing the number of tankers to meet a fixed schedule. *Naval Research Logistics Quarterly* **1,** 217–222.

DANTZIG, G. B., and P. WOLFE. 1960. Decomposition principle for linear programs. *Operations Research* **8,** 101–111.

DANTZIG, G. B., and P. WOLFE. 1961. The decomposition method for linear programming. *Econometrica* **29,** 767–778.

DEARING, P. M., and R. L. FRANCIS. 1974. A network flow solution to a multifacility minimax location problem involving rectilinear distances. *Transportation Science* **8,** 126–141.

DEMBO, R. S., J. M. MULVEY, and S. A. ZENIOS. 1989. Large-scale nonlinear network models and their applications. *Operations Research* **37,** 353–372.

DENARDO, E. V. 1982. *Dynamic Programming: Models and Applications*. Prentice Hall, Englewood Cliffs, NJ.

DENARDO, E. V., and B. L. FOX. 1979. Shortest-route methods: 1. Reaching, pruning and buckets. *Operations Research* **27,** 161–186.

DENARDO, E. V., U. G. ROTHBLUM, and A. J. SWERSEY. 1988. A transportation problem in which costs depend on the order of arrival. *Management Science* **34,** 774–783.

DEO, N., and C. PANG. 1984. Shortest path algorithms: Taxonomy and annotation. *Networks* **14,** 275–323.

DERIGS, U. 1988. *Programming in Networks and Graphs*. Lecture Notes in Economics and Mathematical Systems, Vol. 300. Springer-Verlag, New York.

DERIGS, U., and W. MEIER. 1989. Implementing Goldberg's max-flow algorithm: A computational investigation. *Zeitschrift für Operations Research* **33,** 383–403.

DERMAN, C., and M. KLEIN. 1959. A note on the optimal depletion of inventory. *Management Science* **5,** 210–214.

DEVINE, M. V. 1973. A model for minimizing the cost of drilling dual completion oil wells. *Management Science* **20,** 532–535.

Dewar, M. S. J., and H. C. Longuet-Higgins. 1952. The correspondence between the resonance and molecular orbital theories. *Proceedings of the Royal Society of London* **A214,** 482–493.

Dial, R. 1969. Algorithm 360: Shortest path forest with topological ordering. *Communications of ACM* **12,** 632–633.

Dial, R., F. Glover, D. Karney, and D. Klingman. 1979. A computational analysis of alternative algorithms and labeling techniques for finding shortest path trees. *Networks* **9,** 215–248.

Dijkstra, E. 1959. A note on two problems in connexion with graphs. *Numeriche Mathematics* **1,** 269–271.

Dinic, E. A. 1970. Algorithm for solution of a problem of maximum flow in networks with power estimation. *Soviet Mathematics Doklady* **11,** 1277–1280.

Dinic, E. A. 1973. The method of scaling and transportation problems. *Issled. Diskret. Mat. Science*, Moscow. (In Russian.)

Dirickx, Y. M. I., and L. P. Jennergren. 1975. An analysis of the parking situation in the downtown area of West Berlin. *Transportation Research* **9,** 1–11.

Divoky, J. J., and M. S. Hung. 1990. Performance of shortest path algorithms in network flow problems. *Management Science* **36,** 661–673.

Dorsey, R. C., T. J. Hodgson, and H. D. Ratliff. 1974. A production scheduling problem with batch processing. *Operations Research* **22,** 1271–1279.

Dorsey, R. C., T. J. Hodgson, and H. D. Ratliff. 1975. A network approach to a multi-facility, multi-product production scheduling problem without backordering. *Management Science* **21,** 813–822.

Dress, A. W. M., and T. F. Havel. 1988. Shortest path problems and molecular conformation. *Discrete Applied Mathematics* **19,** 129–144.

Dror, M., P. Trudeau, and S. P. Ladany. 1988. Network models for seat allocation on flights. *Transportation Research* **22B,** 239–250.

Dude, R. O., and P. E. Hart. 1973. *Pattern Classification and Science Analysis*. Wiley, New York.

Edmonds, J. 1965a. Paths, trees, and flowers. *Canadian Journal of Mathematics* **17,** 449–467.

Edmonds, J. 1965b. Maximum matchings and a polyhedran with 0, 1 vertices. *Journal of Research of the National Bureau of Standards* **69B,** 125–130.

Edmonds, J. 1965c. Minimum partition of a matroid into independent subsets. *Journal of Research of the National Bureau of Standards* **69B,** 67–72.

Edmonds, J. 1967. An introduction to matching. Mimeographed notes, Engineering Summer Conference, The University of Michigan, Ann Arbor, MI.

Edmonds, J. 1971. Matroids and the greedy algorithm. *Mathematical Programming* **1,** 127–136.

Edmonds, J., and E. L. Johnson. 1973. Matching, Euler tours and the Chinese postman. *Mathematical Programming* **5,** 88–124.

Edmonds, J., and R. M. Karp. 1972. Theoretical improvements in algorithmic efficiency for network flow problems. *Journal of ACM* **19,** 248–264.

Elam, J., F. Glover, and D. Klingman. 1979. A strongly convergent primal simplex algorithm for generalized networks. *Mathematics of Operations Research* **4,** 39–59.

Elias, P., A. Feinstein, and C. E. Shannon. 1956. Note on maximum flow through a network. *IRE Transactions on Information Theory* **IT-2,** 117–119.

Elmaghraby, S. E. 1978. *Activity Networks: Project Planning and Control by Network Models*. Wiley–Interscience, New York.

Erlenkotter, D. 1978. A dual-based procedure for uncapacitated facility location. *Operations Research* **26,** 992–1009.

Ervolina, T. R., and S. T. McCormick. 1990a. Cancelling most helpful cuts for minimum cost network flow. Faculty of Commerce Working Paper 90-MSC-018, University of British Columbia, Vancouver, Canada.

Ervolina, T. R., and S. T. McCormick. 1990b. Two strongly polynomial cut cancelling algorithms for minimum cost network flow. Technical Report, Faculty of Commerce and Business Administration, University of British Columbia, Vancouver, Canada.

Esau, L. R., and K. C. Williams. 1966. On teleprocessing system design II. *IBM Systems Journal* **5,** 142–147.

Evans, J. R. 1977. Some network flow models and heuristics for multiproduct production and inventory planning. *AIIE Transactions* **9,** 75–81.

Evans, J. R. 1984. The factored transportation problem. *Management Science* **30,** 1021–1024.

Even, S. 1979. *Graph Algorithms*. Computer Science Press, Rockville, MD.

Even, S., and O. Kariv. 1975. An $O(n^{2.5})$ algorithm for maximum matching in general graphs. *Proceedings of the 16th Annual Symposium on Foundations of Computer Science*, pp. 100–112.

Even, S., and R. E. Tarjan. 1975. Network flow and testing graph connectivity. *SIAM Journal on Computing* **4,** 507–518.

Everett, H., III. 1963. Generalized Lagrange multiplier method for solving problems of optimal allocation of resources. *Operations Research* **11,** 399–417.

Ewashko, T. A., and R. C. Dudding. 1971. Application of Kuhn's Hungarian assignment algorithm to posting servicemen. *Operations Research* **19,** 991.

Farina, R. F., and F. W. Glover. 1983. The application of generalized networks to choice of raw materials for fuels and petrochemicals. In *Energy Models and Studies*, edited by B. Lev. North-Holland, Amsterdam.

Farley, A. R. 1980. Levelling terrain trees: A transshipment problem. *Information Processing Letters* **10,** 189–192.

Farvolden, J. M., and W. B. Powell. 1990. A primal partitioning solution for multicommodity network flow problems. Working Paper 90-04, Department of Industrial Engineering, University of Toronto, Toronto, Canada.

Federgruen, A., and H. Groenevelt. 1986. Preemptive scheduling of uniform machines by ordinary network flow techniques. *Management Science* **32,** 341–349.

Fernandez-Baca, D., and C. U. Martel. 1989. On the efficiency of maximum flow algorithms on networks with small integer capacities. *Algorithmica* **4,** 173–189.

Filliben, J. J., K. Kafadar, and D. R. Shier. 1983. Testing for homogeneity of two-dimensional surfaces. *Mathematical Modelling* **4,** 167–189.

Fisher, M. L. 1981. The Lagrangian relaxation methods for solving integer programming problems. *Management Science* **27,** 1–18.

Fisher, M. L. 1985. An applications oriented guide to Lagrangian relaxation. *Interfaces* **15,** 10–21.

Florian, M. 1986. Nonlinear cost network models in transportation analysis. *Mathematical Programming Study* **26,** 167–196.

Floyd, R. W. 1962. Algorithm 97: Shortest path. *Communications of ACM* **5,** 345.

Ford, L. R. 1956. Network flow theory. Report P-923, Rand Corp., Santa Monica, CA.

Ford, L. R., and D. R. Fulkerson. 1956a. Maximal flow through a network. *Canadian Journal of Mathematics* **8,** 399–404.

Ford, L. R., and D. R. Fulkerson. 1956b. Solving the transportation problem. *Management Science* **3,** 24–32.

Ford, L. R., and D. R. Fulkerson. 1957. A primal–dual algorithm for the capacitated Hitchcock problem. *Naval Research Logistics Quarterly* **4,** 47–54.

Ford, L. R., and D. R. Fulkerson. 1958a. Constructing maximum dynamic flows from static flows. *Operations Research* **6,** 419–433.

Ford, L. R., and D. R. Fulkerson. 1958b. A suggested computation for maximal multicommodity network flows. *Management Science* **5,** 97–101.

Ford, L. R., and D. R. Fulkerson. 1962. *Flows in Networks*. Princeton University Press, Princeton, NJ.

Ford, L. R., and S. M. Johnson. 1959. A tournament problem. *The American Mathematical Monthly* **66,** 387–389.

Francis, R. L., and J. A. White. 1976. *Facility Layout and Location*. Prentice Hall, Englewood Cliffs, NJ.

Frank, C. R. 1965. A note on the assortment problem. *Management Science* **11,** 724–726.

Frank, H., and I. T. Frisch. 1971. *Communication, Transmission, and Transportation Networks*. Addison-Wesley, Reading, MA.

Fredman, M. L., and R. E. Tarjan. 1984. Fibonacci heaps and their uses in improved network optimization algorithms. *Proceedings of the 25th Annual IEEE Symposium on Foundations of Computer Science*, pp. 338–346. Full paper in *Journal of ACM* **34**(1987), 596–615.

Fujii, M., T. Kasami, and K. Ninomiya. 1969. Optimal sequencing of two equivalent processors. *SIAM Journal on Applied Mathematics* **17,** 784–789. Erratum, same journal **18,** 141.

Fujishige, S. 1986. A capacity-rounding algorithm for the minimum cost circulation problem: A dual framework of Tardos' algorithm. *Mathematical Programming* **35,** 298–308.

Fulkerson, D. R. 1961a. A network flow computation for project cost curve. *Management Science* **7,** 167–178.

Fulkerson, D. R. 1961b. An out-of-kilter method for minimal cost flow problems. *SIAM Journal on Applied Mathematics* **9,** 18–27.

Fulkerson, D. R. 1963. Flows in networks. In *Recent Advances in Mathematical Programming*, edited by R. L. Graves and P. Wolfe. McGraw-Hill, New York, pp. 319–332.

Fulkerson, D. R. 1965. Upsets in a round robin tournament. *Canadian Journal of Mathematics* **17,** 957–969.

Fulkerson, D. R. 1966. Flow networks and combinatorial operations research. *American Mathematical Monthly* **73,** 115–138.

Fulkerson, D. R., and G. B. Dantzig. 1955. Computation of maximum flow in networks. *Naval Research Logistics Quarterly* **2,** 277–283.

Fulkerson, D. R., and G. C. Harding. 1977. Maximizing the minimum source–sink path subject to a budget constraint. *Mathematical Programming* **13,** 116–118.

Gabow, H. N. 1975. An efficient implementation of Edmond's algorithm for maximum matchings on graphs. *Journal of ACM* **23,** 221–234.

Gabow, H. N. 1985. Scaling algorithms for network problems. *Journal of Computer and System Sciences* **31,** 148–168.

Gabow, H. N. 1990. Data structures for weighted matching and nearest common ancestors with linking. *Proceedings of the First Annual ACM–SIAM Symposium on Discrete Algorithms*. SIAM, Philadelphia, pp. 434–443.

Gabow, H. N., Z. Galil, T. Spencer, and R. E. Tarjan. 1986. Efficient algorithms for finding minimum spanning trees in undirected and directed graphs. *Combinatorica* **6,** 109–122.

Gabow, H. N., and R. E. Tarjan. 1989a. Faster scaling algorithms for network problems. *SIAM Journal on Computing* **18,** 1013–1036.

Gabow, H. N., and R. E. Tarjan. 1989b. Faster scaling algorithms for general graph matching problems. Technical Report CU-CS-432-89, Department of Computer Science, University of Colorado, Boulder, CO.

Gale, D. 1957. A theorem on flows in networks. *Pacific Journal of Mathematics* **7,** 1073–1082.

Gale, D., and L. S. Shapley. 1962. College admissions and the stability of marriage. *American Mathematical Monthly* **69,** 9–14.

Galil, Z. 1981. On the theoretical efficiency of various network flow algorithms. *Theoretical Computer Science* **14,** 103–111.

Galil, Z., and É. Tardos. 1986. An $O(n^2(m + n \log n) \log n)$ min-cost flow algorithm. *Proceedings of the 27th Annual Symposium on the Foundations of Computer Science*, pp. 136–146. Full paper in *Journal of ACM* **35**(1987), 374–386.

Gallo, G., M. D. Grigoriadis, and R. E. Tarjan. 1989. A fast parametric maximum flow algorithm and applications. *SIAM Journal on Computing* **18,** 30–55.

Gallo, G., and S. Pallottino. 1984. Shortest path methods in transportation models. In *Transportation Planning Models*, edited by M. Florian. Elsevier/North-Holland, Amsterdam.

Gallo, G., and S. Pallottino. 1986. Shortest path methods: A unifying approach. *Mathematical Programming Study* **26,** 38–64.

Gallo, G., and S. Pallottino. 1988. Shortest path algorithms. In *Fortran Codes for Network Optimization*, edited by B. Simeone, P. Toth, G. Gallo, F. Maffioli, and S. Pallottino. *Annals of Operations Research* **13,** 3–79.

Garey, M. S., and D. S. Johnson. 1979. *Computers and Intractability: A Guide to the Theory of NP-Completeness*. W. H. Freeman, New York.

Gavish, B. 1985. Augmented Lagrangian based algorithms for centralized network design. *IEEE Transactions on Communications* **COM-33,** 1247–1257.

Gavish, B., and P. Schweitzer. 1974. An algorithm for combining truck trips. *Transportation Science* **8,** 13–23.

Gavish, B., and K. N. Srikanth. 1979. $O(n^2)$ algorithms for sensitivity analysis of minimal spanning trees and related subgraphs. Working Paper 8003, Graduate School of Management, University of Rochester, Rochester, NY.

Geoffrion, A. 1974. Lagrangian relaxations for integer programming. *Mathematical Programming Study* **2,** 82–114.

Geoffrion, A. M., and G. W. Graves. 1974. Multicommodity distribution system design by Benders decomposition. *Management Science* **20,** 822–844.

Gersht, A., and A. Shulman. 1987. A new algorithm for the solution of the minimum cost multicommodity flow problem. *Proceedings of the IEEE Conference on Decision and Control* **26,** 748–758.

Gilmore, P. C., and R. E. Gomory. 1964. Sequencing a one state-variable machine: A solvable case of the travelling salesman problem. *Operations Research* **12,** 655–679.

Glover, F., R. Glover, and F. K. Martinson. 1984. A netform system for resource planning in the U.S. Bureau of Land Management. *Journal of the Operational Research Society* **35,** 605–616.

Glover, F., R. Glover, and D. J. Shields. 1988. Microcomputer-based model of international mineral market. In *Operational Research '87*, edited by G. K. Rand. Elsevier, Amsterdam.

Glover, F., J. Hultz, D. Klingman, and J. Stutz. 1978. Generalized networks: A fundamental computer based planning tool. *Management Science* **24,** 1209–1220.

Glover, F., D. Karney, and D. Klingman. 1974. Implementation and computational comparisons of primal, dual and primal–dual computer codes for minimum cost network flow problem. *Networks* **4,** 191–212.

Glover, F., D. Karney, D. Klingman, and A. Napier. 1974. A computational study on start procedures, basis change criteria, and solution algorithms for transportation problem. *Management Science* **20,** 793–813.

Glover, F., and D. Klingman. 1977. Network applications in industry and government. *AIIE Transactions* **9,** 363–376.

Glover, F., D. Klingman, J. Mote, and D. Whitman. 1984. A primal simplex variant for the maximum flow problem. *Naval Research Logistics Quarterly* **31,** 41–61.

Glover, F., D. Klingman, and N. Phillips. 1985. A new polynomially bounded shortest path algorithm. *Operations Research* **33,** 65–73.

Glover, F., D. Klingman, and N. Phillips. 1990. Netform modeling and applications. *Interfaces* **20,** 7–27.

Glover, F., D. Klingman, N. Phillips, and R. F. Schneider. 1985. New polynomial shortest path algorithms and their computational attributes. *Management Science* **31,** 1106–1128.

Glover, F., and J. Rogozinski. 1982. Resort development: A network-related model for optimizing sites and visits. *Journal of Leisure Research*, 235–247.

Goetschalckx, M., and H. D. Ratliff. 1988. Order picking in an aisle. *IIE Transactions* **20,** 53–62.

Goldberg, A. V. 1985. A new max-flow algorithm. Technical Report MIT/LCS/TM-291, Laboratory for Computer Science, MIT, Cambridge, MA.

Goldberg, A. V., M. D. Grigoriadis, and R. E. Tarjan. 1988. Efficiency of the network simplex algorithm for the maximum flow problem. Technical Report, Department of Computer Science, Stanford University, Stanford, CA.

Goldberg, A. V., S. A. Plotkin, and É. Tardos. 1991. Combinatorial algorithms for the generalized circulation problem. *Mathematics of Operations Research* **16,** 351–381.

Goldberg, A. V., É. Tardos, and R. E. Tarjan. 1989. Network flow algorithms. Technical Report 860, School of Operations Research and Industrial Engineering, Cornell University, Ithaca, NY.

Goldberg, A. V., and R. E. Tarjan. 1986. A new approach to the maximum flow problem. *Proceedings of the 18th ACM Symposium on the Theory of Computing*, pp. 136–146. Full paper in *Journal of ACM* **35**(1988), 921–940.

Goldberg, A. V., and R. E. Tarjan. 1987. Solving minimum cost flow problem by successive approximation. *Proceedings of the 19th ACM Symposium on the Theory of Computing*, pp. 7–18. Full paper in *Mathematics of Operations Research* **15**(1990), 430–466.

Goldberg, A. V., and R. E. Tarjan. 1988. Finding minimum-cost circulations by cancelling negative cycles. *Proceedings of the 20th ACM Symposium on the Theory of Computing*, pp. 388–397. Full paper in *Journal of ACM* **36**(1989), 873–886.

Golden, B. L. 1975. A minimum cost multicommodity network flow problem concerning imports and exports. *Networks* **5,** 331–356.

Golden, B. L., A. A. Assad, E. A. Wasil, and E. Baker. 1986. Experiments in optimization. Working Paper Series MS/S 86-004, University of Maryland, College Park, MD.

Golden, B. L., M. Liberatore, and C. Lieberman. 1979. Models and solution techniques for cash flow management. *Computers and Operations Research* **6,** 13–20.

Golden, B. L., and T. L. Magnanti. 1977. Deterministic network optimization: A bibliography. *Networks* **7,** 149–183.

GOLDFARB, D. 1985. Efficient dual simplex algorithms for the assignment problem. *Mathematical Programming* **33,** 187–203.

GOLDFARB, D., and J. HAO. 1988. Polynomial-time primal simplex algorithms for the minimum cost network flow problem. Technical Report, Department of Industrial Engineering and Operations Research, Columbia University, New York.

GOLDFARB, D., and J. HAO. 1990. A primal simplex algorithm that solves the maximum flow problem in at most nm pivots and $O(n^2m)$ time. *Mathematical Programming* **47,** 353–365.

GOLDFARB, D., J. HAO, and S. KAI. 1990a. Efficient shortest path simplex algorithms. *Operations Research* **38,** 624–628.

GOLDFARB, D., J. HAO, and S. KAI. 1990b. Anti-stalling pivot rules for the network simplex algorithm. *Networks* **20,** 79–91.

GOLDMAN, A. J., and G. L. NEMHAUSER. 1967. A transport improvement problem transformable to a best-path problem. *Transportation Science* **1,** 295–307.

GOLITSCHEK, M. V., and H. SCHNEIDER. 1984. Applications of shortest path algorithms to matrix scalings. *Numerische Mathematik* **44,** 111–126.

GOMORY, R. E., and T. C. HU. 1961. Multi-terminal network flows. *Journal of SIAM* **9,** 551–570.

GONDRAN, M., and M. MINOUX. 1984. *Graphs and Algorithms*. Wiley-Interscience, New York.

GORHAM, W. 1963. An application of a network flow model to personnel planning. *IEEE Transactions on Engineering Management* **10,** 113–123.

GOWER, J. C., and G. J. S. ROSS. 1969. Minimum spanning trees and single linkage cluster analysis. *Applied Statistics* **18,** 54–64.

GRAHAM, R. L., and P. HELL. 1985. On the history of minimum spanning tree problem. *Annals of the History of Computing* **7,** 43–57.

GRAVES, S. C. 1982. Using Lagrangian techniques to solve hierarchical production planning problems. *Management Science* **28,** 260–275.

GRAVES, G. W., and R. D. MCBRIDE. 1976. The factorization approach to large scale linear programming. *Mathematical Programming* **10,** 91–110.

GREENBERG, H. 1990. Computational testing: Why, how, and how much. *ORSA Journal of Computing* **2,** 94–97.

GRIGORIADIS, M. D. 1986. An efficient implementation of the network simplex method. *Mathematical Programming Study* **26,** 83–111.

GRIGORIADIS, M. D., and Y. HSU. 1979. The Rutgers minimum cost network flow subroutines. *SIGMAP Bulletin of the ACM* **26,** 17–18.

GRÖTSCHEL, M., and O. HOLLAND. 1985. Solving matching problems with linear programming. *Mathematical Programming* **33,** 243–259.

GUIGNARD, M., and S. KIM. 1987a. Lagrangian decomposition: A model yielding stronger Lagrangian bounds. *Mathematical Programming* **39,** 215–228.

GUIGNARD, M., and S. KIM. 1987b. Lagrangian decomposition for integer programming: Theory and applications. Technical Report 93, Department of Statistics, The Wharton School, University of Pennsylvania, Philadelphia, PA.

GUPTA, S. K. 1985. *Linear Programming and Network Models*. Affiliated East–West Press, New Delhi, India.

GUSFIELD, D. 1988. A graph theoretic approach to statistical data security. *SIAM Journal on Computing* **17,** 552–571.

GUSFIELD, D. 1990. Very simple methods for all pairs network flow analysis. *SIAM Journal on Computing* **19,** 143–155.

GUSFIELD, D., and R. W. IRVING. 1989. *The Stable Marriage Problem: Structure and Algorithms*. MIT Press, Cambridge, MA.

GUSFIELD, D., and C. MARTEL. 1989. A fast algorithm for the generalized parametric minimum cut problem and applications. Technical Report CSE-89-21, Computer Science Division, University of California, Davis, CA.

GUSFIELD, D., C. MARTEL, and D. FERNANDEZ-BACA. 1987. Fast algorithms for bipartite network flow. *SIAM Journal on Computing* **16,** 237–251.

GUTJAHR, A. L., and G. L. NEMHAUSER. 1964. An algorithm for the line balancing problem. *Management Science* **11,** 308–315.

HALL, M. 1956. An algorithm for distinct representatives. *American Mathematical Monthly* **63,** 716–717.

Hamacher, H. W., and S. Tufekci. 1987. On the use of lexicographic min cost flows in evacuation modelling. *Naval Research Logistics Quarterly* **34,** 487–504.

Handler, G. Y. 1973. Minimax location of a facility in an undirected graph. *Transportation Science* **7,** 287–293.

Handler, G., and I. Zang. 1980. A dual algorithm for the constrained shortest path problem. *Networks* **10,** 293–309.

Hassin, R. 1981. Maximum flow in (s, t)-planar networks. *Information Processing Letters* **13,** 107.

Hassin, R., and D. B. Johnson. 1985. An $O(n \log^2 n)$ algorithm for maximum flow in undirected planar networks. *SIAM Journal on Computing* **14,** 612–624.

Hausman, H. 1978. *Integer Programming and Related Areas: A Classified Bibliography*. Lecture Notes in Economics and Mathematical Systems, Vol. 160. Springer-Verlag, Berlin.

Hax, A. C., and C. Candea. 1984. *Production and Inventory Management*. Prentice Hall, Englewood Cliffs, NJ.

Haymond, R. E., J. P. Jarvis, and D. R. Shier. 1980. Computational methods for minimum spanning tree problems. Technical Report 354, Department of Mathematical Sciences, Clemson University, Clemson, SC.

Haymond, R. E., J. R. Thornton, and D. D. Warner. 1988. A shortest path algorithm in robotics and its implementation on the FPS T-20 hypercube. *Annals of Operations Research* **14,** 305–320.

Held, M., and R. Karp. 1970. The traveling salesman problem and minimum spanning trees. *Operations Research* **18,** 1138–1162.

Held, M., and R. Karp. 1971. The traveling salesman problem and minimum spanning trees, Part II. *Mathematical Programming* **6,** 62–88.

Helgason, R. V., J. L. Kennington, and B. D. Stewart. 1988. Dijkstra's two-tree shortest path algorithm. Technical Report, Department of Operations Research and Engineering Management, Southern Methodist University, Dallas, TX.

Hitchcock, F. L. 1941. The distribution of a product from several sources to numerous facilities. *Journal of Mathematical Physics* **20,** 224–230.

Hochbaum, D. S., and J. G. Shanthikumar. 1990. Convex separable optimization is not much harder than linear optimization. *Journal of ACM* **37,** 843–862.

Hoffman. A. J. 1960. Some recent applications of the theory of linear inequalities to extremal combinatorial analysis. In *Combinatorial Analysis*, edited by R. Bellman and M. Hall. American Mathematical Society, Providence, RI, pp. 113–128.

Hoffman, K. L., and R. H. F. Jackson. 1982. In pursuit of a methodology for testing mathematical programming software. In *Evaluating Mathematical Programming Techniques*, Lecture Notes in Economics and Mathematical Systems, Vol. 199, edited by J. M. Mulvey et al., Springer-Verlag, New York.

Hoffman, A. J., and J. B. Kruskal. 1956. Integral boundary points of convex polyhedra. In *Linear Inequalities and Related Systems*, edited by H. W. Kuhn and A. W. Tucker. Princeton University Press, Princeton, NJ, pp. 233–246.

Hoffman, A. J., and H. M. Markowitz. 1963. A note on shortest path, assignment and transportation problems. *Naval Research Logistics Quarterly* **10,** 375–379.

Hoffman, A. J., and S. T. McCormick. 1984. A fast algorithm that makes matrices optimally sparse. In *Progress in Combinatorial Optimization*. Academic Press Canada, Don Mills, Ontario, Canada.

Hopcroft, J. E., and R. M. Karp. 1973. A $n^{5/2}$ algorithm for maximum matchings in bipartite graphs. *SIAM Journal on Computing* **2,** 225–231.

Hopcroft, J. E., and R. E. Tarjan. 1974. Efficient planarity testing. *Journal of ACM* **21,** 549–568.

Horn, W. A. 1971. Determining optimal container inventory and routing. *Transportation Science* **5,** 225–231.

Horn, W. A. 1973. Minimizing average flow time with parallel machines. *Operations Research* **21,** 846–847.

Hu, T. C. 1961. The maximum capacity route problem. *Operations Research* **9,** 898–900.

Hu, T. C. 1963. Multi-commodity network flows. *Operations Research* **11,** 344–360.

Hu, T. C. 1966. Minimum cost flows in convex cost networks. *Naval Research Logistics Quarterly* **13,** 1–9.

Hu, T. C. 1967. Laplace's equation and network flows. *Operations Research* **15,** 348–354.

Hu, T. C. 1969. *Integer Programming and Network Flows*. Addison-Wesley, Reading, MA.

Hu, T. C. 1974. Optimum communication spanning trees. *SIAM Journal on Computing* **3,** 188–195.

Hung, M. S. 1983. A polynomial simplex method for the assignment problem. *Operations Research* **31,** 595–600.

Hung, M. S., and J. J. Divoky. 1988. A computational study of efficient shortest path algorithms. *Computers and Operations Research* **15,** 567–576.

Imai, H. 1983. On the practical efficiency of various maximum flow algorithms. *Journal of the Operations Research Society of Japan* **26,** 61–82.

Imai, H., and M. Iri. 1986. Computational-geometric methods for polygonal approximations of a curve. *Computer Vision, Graphics and Image Processing* **36,** 31–41.

Iri, M. 1960. A new method of solving transportation-network problems. *Journal of the Operations Research Society of Japan* **3,** 27–87.

Iri, M. 1969. *Network Flow, Transportation and Scheduling*. Academic Press, New York.

Itai, A., and Y. Shiloach. 1979. Maximum flow in planar networks. *SIAM Journal on Computing* **8,** 135–150.

Jackson, R. H. B., P. T. Boggs, S. G. Nash, and S. Powell. 1989. Report of the ad hoc committee to revise the guidelines for reporting computational experiments in mathematical programming. *COAL Newsletter* **18,** 3–14.

Jackson, R. H. B., and J. M. Mulvey. 1978. A critical review of comparisons of mathematical programming algorithms and software (1953–1977). *Journal of Research of the National Bureau of Standards* **83,** 563–584.

Jacobs, W. W. 1954. The caterer problem. *Naval Research Logistics Quarterly* **1,** 154–165.

Jarníck, V. 1930. O jistém problému minimálním. *Acta Societatis Scientiarum Natur. Moravicae* **6,** 57–63.

Jarvis, J. P., and D. E. White. 1983. Computational experience with minimum spanning tree algorithms. *Operations Research Letters* **2,** 36–41.

Jensen, P. A., and W. Barnes. 1980. *Network Flow Programming*. Wiley, New York.

Jensen, P., and G. Bhaumik. 1977. A flow augmentation approach to the network with gains minimum cost flcw problem. *Management Science* **23,** 631–643.

Jewell, W. S. 1957. Warehousing and distribution of a seasonal product. *Naval Research Logistics Quarterly* **4,** 29–34.

Jewell, W. S. 1958. Optimal flow through networks. Interim Technical Report 8, Operations Research Center, MIT, Cambridge, MA.

Jewell, W. S. 1962. Optimal flow through networks with gains. *Operations Research* **10,** 476–499.

Johnson, E. L. 1966. Networks and basic solutions. *Operations Research* **14,** 619–624.

Johnson, T. B. 1968. Optimum pit mine production scheduling. Technical Report, University of California, Berkeley, CA.

Johnson, D. B. 1982. A priority queue in which initialization and queue operations take $O(\log \log D)$ time. *Mathematical Systems Theory* **15,** 295–309.

Johnson, D. S. 1990. Local optimization and the traveling salesman problem. *Proceedings of the 17th Colloquium on Automata, Languages, and Programming*. Springer-Verlag, New York, pp. 446–461.

Johnson, D. B., and S. Venkatesan. 1982. Using divide and conquer to find flows in directed planar networks in $O(n^{3/2} \log n)$ time. *Proceedings of the 20th Annual Allerton Conference on Communication, Control, and Computing*, University of Illinois, Urbana-Champaign, IL, pp. 898–905.

Kameda, T., and I. Munro. 1974. A $O(|V| . |E|)$ algorithm for maximum matching of graphs. *Computing* **12,** 91–98.

Kang, A. N. C., R. C. T. Lee, C. L. Chang, and S. K. Chang. 1977. Storage reduction through minimal spanning trees and spanning forests. *IEEE Transactions on Computers* **C-26,** 425–434.

Kantorovich, L. V. 1939. Mathematical methods in the organization and planning of production. Publication House of the Leningrad University. Translated in *Management Science* **6**(1960), 366–422.

Kaplan, S. 1973. Readiness and the optimal redeployment of resources. *Naval Research Logistics Quarterly* **20,** 625–638.

Karp, R. M. 1972. Reducibility among combinatorial problems. In *Complexity of Computer Computations*, edited by R. E. Miller and J. W. Thacher. Plenum Press, New York, pp. 83–103.

Karp, R. M. 1978. A characterization of the minimum cycle mean in a diagraph. *Discrete Mathematics* **23,** 309–311.

Karp, R. M., and J. B. Orlin. 1981. Parametric shortest path algorithms with an application to cyclic staffing. *Discrete Applied Mathematics* **3,** 37–45.

Karzanov, A. V. 1974. Determining the maximal flow in a network by the method of preflows. *Soviet Mathematics Doklady* **15,** 434–437.

Kastning, C. 1976. *Integer Programming and Related Areas: A Classified Bibliography*. Lecture Notes in Economics and Mathematical Systems, Vol. 128, Springer-Verlag, Berlin.

Kelly, J. R. 1961. Critical path planning and scheduling: Mathematical basis. *Operations Research* **9,** 296–320.

Kelly, J. P., B. L. Golden, and A. A. Assad. 1992. Cell suppression: Disclosure protection for sensitive tabular data. *Networks* **22,** 397–412.

Kennington, J. L. 1978. A survey of linear cost multicommodity network flows. *Operations Research* **26,** 209–236.

Kennington, J. L., and R. V. Helgason. 1980. *Algorithms for Network Programming*. Wiley-Interscience, New York.

Kennington, J. L., and M. Shalaby. 1977. An effective subgradient procedure for minimal cost multicommodity flow problems. *Management Science* **23,** 994–1004.

Kennington, J. L., and Z. Wang. 1990. The shortest augmenting path algorithm for the transportation problem. Technical Report 90-CSE-10, Southern Methodist University, Dallas, TX.

Khan, M. R. 1979. A capacitated network formulation for manpower scheduling. *Industrial Management* **21,** 24–28.

Khan, M. R., and D. A. Lewis. 1987. A network model for nursing staff scheduling. *Zeitschrift für Operations Research* **31,** B161–B171.

Klein, M. 1967. A primal method for minimal cost flows with application to the assignment and transportation problems. *Management Science* **14,** 205–220.

Klincewicz, J. G. 1983. A Newton method for convex separable network flow problems. *Networks* **13,** 427–442.

Klingman, D., A. Napier, and J. Stutz. 1974. NETGEN: A program for generating large scale capacitated assignment, transportation, and minimum cost flow network problems. *Management Science* **20,** 814–821.

Knuth, D. E. 1973a. *The Art of Computer Programming*. Vol. 1: *Fundamental Algorithms*, 2nd ed. Addison-Wesley, Reading, MA.

Knuth, D. E. 1973b. *The Art of Computer Programming*. Vol. III: *Sorting and Searching*. Addison-Wesley, Reading, MA.

Kolitz, S. 1991. Personal communication.

Koopmans, T. C. 1947. Optimum utilization of the transportation system. *Proceedings of the International Statistical Conference*, Washington, DC. Also in *Econometrica* **17**(1949).

Korte, B. 1988. Applications of combinatorial optimization. Technical Report 88541-OR, Institute für Okonometrie und Operations Research, Bonn, Germany.

Kourtz, P. 1984. A network approach to least cost daily transfers of forest fire control resources. *INFOR* **22,** 283–290.

Kruskal, J. B. 1956. On the shortest spanning tree of graph and the traveling salesman problem. *Proceedings of the American Mathematical Society* **7,** 48–50.

Kuhn, H. W. 1955. The Hungarian method for the assignment problem. *Naval Research Logistics Quarterly* **2,** 83–97.

Laporte, G., and Y. Nobert. 1987. Exact algorithms for the vehicle routing problem. In *Surveys in Combinatorial Optimization*, edited by S. Martello, G. Laporte, M. Minoux, and C. Ribeiro. North-Holland, Amsterdam.

Larson, R. C., and A. R. Odoni. 1981. *Urban Operations Research*. Prentice Hall, Englewood Cliffs, NJ.

Lawania, A. K. 1990. Personal communication.

Lawler, E. L. 1964. On scheduling problems with deferral costs. *Management Science* **11,** 280–287.

Lawler, E. L. 1966. Optimal cycles in doubly weighted linear graphs. In *Theory of Graphs: International Symposium*, Dunod, Paris, and Gordon and Breach, New York, pp. 209–213.

Lawler, E. L. 1976. *Combinatorial Optimization: Networks and Matroids*. Holt, Rinehart and Winston, New York.

LAWLER, E. L., J. K. LENSTRA, A. H. G. RINNOOY KAN, and D. B. SHMOYS (eds.). 1985. *The Traveling Salesman Problem: A Guided Tour of Combinatorial Optimization*. Wiley, New York.

LEUNG, J., T. L. MAGNANTI, and V. SINGHAL. 1990. Routing in point to point delivery systems. *Transportation Science* **24**, 245–260.

LEVIN, L. A. 1973. Universal sorting problems. *Problemy Peredachi Informatsii* **9**, 265–266. (In Russian.)

LEVNER, E. V., and A. S. NEMIROVSKY. 1991. A network flow algorithm for just-in-time project scheduling. Memorandum COSOR 91-21, Department of Mathematics and Computing Science, Eindhoven University of Technology, Eindhoven, The Netherlands.

LIN, T. F. 1986. A system of linear equations related to the transportation problem with application to probability theory. *Discrete Applied Mathematics* **14**, 47–56.

LOBERMAN, H., and A. WEINBERGER. 1957. Formal procedures for connecting terminals with a minimum total wire length. *Journal of ACM* **4**, 428–437.

LOVÁSZ, L., and M. D. PLUMMER. 1986. *Matching Theory*. North-Holland, Amsterdam.

LOVE, R. R., and R. R. VEMUGANTI. 1978. The single-plant mold allocation problem with capacity and changeover restriction. *Operations Research* **26**, 159–165.

LOWE, T. J., R. L. FRANCIS, and E. W. REINHARDT. 1979. A greedy network flow algorithm for a warehouse leasing problem. *AIIE Transactions* **11**, 170–182.

LUSS, H. 1979. A capacity expansion model for two facilities. *Naval Research Logistics Quarterly* **26**, 291–303.

MACHOL, R. E. 1961. An application of the assignment problem. *Operations Research* **9**, 585–586.

MACHOL, R. E. 1970. An application of the assignment problem. *Operations Research* **18**, 745–746.

MAGNANTI, T. L. 1981. Combinatorial optimization and vehicle fleet planning: Perspectives and prospects. *Networks* **11**, 179–214.

MAGNANTI, T. L. 1984. Models and algorithms for predicting urban traffic equilibria. In *Transportation Planning Models*, edited by M. Florian. North-Holland, Amsterdam, pp. 153–186.

MAGNANTI, T. L., P. MIRCHANDANI, and R. VACHANI. 1991. Modeling and solving the capacitated network loading problem. Working Paper, Operations Research Center, MIT, Cambridge, MA.

MAGNANTI, T. L., J. SHAPIRO, and M. WAGNER. 1976. Generalized linear programming solves the dual. *Management Science* **22**, 1195–1203.

MAGNANTI, T. L., L. A. WOLSEY, and R. T. WONG. 1992. Optimal Trees. To appear in *Handbooks in Operations Research and Management Science*. Vol. 6: *Networks*, edited by M. Ball, T. L. Magnanti, C. L. Monma, and G. L. Nemhauser. North-Holland, Amsterdam.

MAGNANTI, T. L., L. A. WOLSEY, and R. T. WONG. 1992. Network design. To appear in *Handbooks in Operations Research and Management Science*, Vol. 6: *Networks*, edited by M. Ball, T. L. Magnanti, C. L. Monma, and G. L. Nemhauser. North-Holland, Amsterdam.

MAGNANTI, T. L., and R. T. WONG. 1984. Network design and transportation planning: Models and algorithms. *Transportation Science* **18**, 1–55.

MALIK, K., A. K. MITTAL, and S. K. GUPTA. 1989. The k most vital arcs in the shortest path problem. *Operations Research Letters* **8**, 223–227. Erratum: Same journal **9**(1990) 283.

MAMER, J. W., and S. A. SMITH. 1982. Optimizing field repair kits based on job completion rate. *Management Science* **28**, 1328–1334.

MANNE, A. S. 1958. A target-assignment problem. *Operations Research* **6**, 346–351.

MANSOUR, Y., and B. SCHIEBER. 1988. Finding the edge connectivity of directed graphs. Research Report RC 13556, IBM Thomas J. Watson Research Center, Yorktown Heights, NY.

MARTEL, C. 1982. Preemptive scheduling with release times, deadlines, and due times. *Journal of ACM* **29**, 812–829.

MARTELLO, S., W. R. PULLEYBLANK, P. TOTH, and D. DE WERRA. 1984. Balanced optimization problems. *Operations Research Letters* **3**, 275–278.

MASON, A. J., and A. B. PHILPOTT. 1988. Pairing stereo speakers using matching algorithms. *Asia-Pacific Journal of Operational Research* **5**, 101–116.

MATSUMOTO, K., T. NISHIZEKI, and N. SAITO. 1985. An efficient algorithm for finding multicommodity flows in planar networks. *SIAM Journal on Computing* **14**, 289–302.

MATULA, D. W. 1987. Determining edge connectivity in $O(nm)$. *Proceedings of the 28th Symposium on Foundations of Computer Science*, pp. 249–251.

MAXWELL, W. L., and R. C. WILSON. 1981. Dynamic network flow modelling of fixed path material handling systems. *AIIE Transactions* **13**, 12–21.

McGeoch, C. C. 1986. *Experimental Analysis of Algorithms*. Unpublished Ph.D. dissertation, Department of Computer Science, Carnegie Mellon University, Pittsburgh, PA.

McGeoch, C. C. 1992. Analysis of algorithms by simulation: Variance reduction techniques and simulation speedups. *Computing Surveys* **24,** June issue.

McGinnis, L. F., and H. L. W. Nuttle. 1978. The project coordinators problem. *OMEGA* **6,** 325–330.

Meggido, N. 1979. Combinatorial optimization with rational objective functions. *Mathematics of Operations Research* **4,** 414–424.

Meggido, N., and A. Tamir. 1978. An $O(n \log n)$ algorithm for a class of matching problems. *SIAM Journal on Computing* **7,** 154–157.

Mehlhorn, K. 1984. *Data Structures and Algorithms*, Vol. I: *Searching and Sorting*. Springer-Verlag, New York.

Micali, S., and V. V. Vazirani. 1980. An $O(\sqrt{|V|} \cdot |E|)$ algorithm for finding maximum matching in general graphs. *Proceedings of the 21st Annual Symposium on the Foundations of Computer Science*, pp. 17–27.

Minieka, E. 1978. *Optimization Algorithms for Networks and Graphs*. Marcel Dekker, New York.

Minoux, M. 1984. A polynomial algorithm for minimum quadratic cost flow problems. *European Journal of Operational Research* **18,** 377–387.

Minoux, M. 1986. Solving integer minimum cost flows with separable convex cost objective polynomially. *Mathematical Programming Study* **26,** 237–239.

Minoux, M. 1989. Network synthesis and optimal network design problems: Models, solution methods, and applications. *Networks* **19,** 313–360.

Minty, G. J. 1960. Monotone networks. *Proceedings of the Royal Society of London* **257A,** 194–212.

Monma, C. L., and M. Segal. 1982. A primal algorithm for finding minimum-cost flows in capacitated networks with applications. *Bell System Technical Journal* **61,** 449–468.

Moore, E. F. 1957. The shortest path through a maze. In *Proceedings of the International Symposium on the Theory of Switching, Part II;* The Annals of the Computation Laboratory of Harvard University **30,** Harvard University Press, pp. 285–292.

Mulvey, J. 1978. Pivot strategies for primal–simplex network codes. *Journal of ACM* **25,** 266–270.

Mulvey, J. M. 1979. Strategies in modeling: A personal scheduling example. *Interfaces* **9,** 66–76.

Nemhauser, G. L., and L. A. Wolsey. 1988. *Integer and Combinatorial Optimization*. Wiley, New York.

Orlin, D. 1987. Optimal weapons allocation against layered defenses. *Naval Research Logistics Quarterly* **34,** 605–617.

Orlin, J. B. 1984. Genuinely polynomial simplex and non-simplex algorithms for the minimum cost flow problem. Technical Report 1615-84, Sloan School of Management, MIT, Cambridge, MA.

Orlin, J. B. 1985. On the simplex algorithm for networks and generalized networks. *Mathematical Programming Study* **24,** 166–178.

Orlin, J. B. 1988. A faster strongly polynomial minimum cost flow algorithm. *Proceedings of the 20th ACM Symposium on the Theory of Computing*, pp. 377–387. Full paper to appear in *Operations Research*.

Orlin, J. B., and R. K. Ahuja. 1992. New scaling algorithms for the assignment and minimum cycle mean problems. *Mathematical Programming* **54,** 41–56.

Orlin, J. B., and U. G. Rothblum. 1985. Computing optimal scalings by parametric network algorithms. *Mathematical Programming* **32,** 1–10.

Osteen, R. E., and P. P. Lin. 1974. Picture skeletons based on eccentricities of points of minimum spanning trees. *SIAM Journal on Computing* **3,** 23–40.

Pallottino, S. 1991. Personal communications.

Papadimitriou, C. H., and K. Steiglitz. 1982. *Combinatorial Optimization: Algorithms and Complexity*. Prentice Hall, Englewood Cliffs, NJ.

Pape, U. 1974. Implementation and efficiency of Moore-algorithms for the shortest route problem. *Mathematical Programming* **7,** 212–222.

Pape, U. 1980. Algorithm 562: Shortest path lengths. *ACM Transactions on Mathematical Software* **6,** 450–455.

Phillips, D. T., and A. Garcia-Diaz. 1981. *Fundamentals of Network Analysis*. Prentice Hall, Englewood Cliffs, NJ.

PICARD, J. C., and M. QUEYRANNE. 1982. Selected applications of minimum cuts in networks. *INFOR* **20,** 394–422.

PICARD, J. C., and H. D. RATLIFF. 1973. Minimal cost cut equivalent networks. *Management Science* **19,** 1087–1092.

PICARD, J. C., and H. D. RATLIFF. 1978. A cut approach to the rectilinear distance facility location problem. *Operations Research* **26,** 422–433.

PINAR, M. C., and S. A. ZENIOS. 1990. Parallel decomposition of multicommodity network flows using smooth penalty functions. Technical Report 90-12-06, Department of Decision Sciences, Wharton School, University of Pennsylvania, Philadelphia, PA.

PINTO, Y., and R. SHAMIR. 1990. Efficient algorithms for minimum cost flow problems with convex costs. Technical Report, Department of Computer Science, Tel Aviv University, Tel Aviv, Israel.

PLOTKIN, S., and É. TARDOS. 1990. Improved dual network simplex. *Proceedings of the First ACM–SIAM Symposium on Discrete Algorithms*, pp. 367–376.

POTTS, R. B., and R. M. OLIVER. 1972. *Flows in Transportation Networks*. Academic Press, New York.

PRAGER, W. 1957. On warehousing problems. *Operations Research* **5,** 504–512.

PRIM, R. C. 1957. Shortest connection networks and some generalizations. *Bell System Technical Journal* **36,** 1389–1401.

RATLIFF, H. D. 1978. Network models for production scheduling problems with convex cost and batch processing. *AIIE Transactions* **10,** 104–108.

RAVINDRAN, A. 1971. On compact book storage in libraries. *Opsearch* **8,** 245–252.

RECSKI, A. 1988. *Matroid Theory and Its Applications*. Springer-Verlag, New York.

RHYS, J. M. W. 1970. A selection problem of shared fixed costs and network flows. *Management Science* **17,** 200–207.

RÖCK, H. 1980. Scaling techniques for minimal cost network flows. In *Discrete Structures and Algorithms*. Edited by V. Page. Carl Hanser, Munich, pp. 181–191.

ROCKAFELLAR, R. T. 1970. *Convex Analysis*. Princeton University Press, Princeton, NJ.

ROCKAFELLAR, R. T. 1984. *Network Flows and Monotropic Optimization*. John Wiley & Sons, New York.

ROOHY-LALEH, E. 1980. Improvements to the Theoretical Efficiency of the Network Simplex Method. Unpublished Ph.D. dissertation, Carleton University, Ottawa, Canada.

ROSENTHAL, R. E. 1981. A nonlinear network flow algorithm for maximization of benefits in a hydroelectric power system. *Operations Research* **29,** 763–786.

ROSS, G. T., and R. M. SOLAND. 1975. A branch and bound algorithm for the generalized assignment problem. *Mathematical Programming* **8,** 91–103.

ROTH, A. E., U. G. ROTHBLUM, and J. H. VANDE VATE. 1990. Stable matchings, optimal assignments and linear programming. Rutcor Research Report 23-90, The State University of New Jersey, Rutgers, NJ.

ROTHFARB, B., N. P. SHEIN, and I. T. FRISCH. 1968. Common terminal multicommodity flow. *Operations Research* **16,** 202–205.

SAKAROVITCH, M. 1973. Two commodity network flows and linear programming. *Mathematical Programming* **4,** 1–20.

SAPOUNTZIS, C. 1984. Allocating blood to hospitals from a central blood bank. *European Journal of Operational Research* **16,** 157–162.

SCHMIDT, S. R., P. A. JENSEN, and J. W. BARNES. 1982. An advanced dual incremental network algorithm. *Networks* **12,** 475–492.

SCHNEIDER, M. H., and S. A. ZENIOS. 1990. A comparative study of algorithms for matrix balancing. *Operations Research* **38,** 439–455.

SCHNEUR, R. 1991. Scaling algorithms for multicommodity flow problems and network flow problems with side constraints. Ph.D. dissertation, Department of Civil Engineering, MIT, Cambridge, MA.

SCHNORR, C. P. 1979. Bottlenecks and edge connectivity in unsymmetrical networks. *SIAM Journal on Computing* **8,** 265–274.

SCHRIJVER, A. 1986. *Theory of Linear and Integer Programming*. Wiley, New York.

SCHWARTZ, B. L. 1966. Possible winners in partially completed tournaments. *SIAM Review* **8,** 302–308.

SCHWARTZ, M., and T. E. STERN. 1980. Routing techniques used in computer communication networks. *IEEE Transactions on Communications* **COM-28,** 539–552.

SEGAL, M. 1974. The operator-scheduling problem: A network flow approach. *Operations Research* **22,** 808–823.

SERVI, L. D. 1989. A network flow approach to a satellite scheduling problem. Research Report, GTE Laboratories, Waltham, MA.

SHAPIRO, J. F. 1979. *Mathematical Programming: Structures and Algorithms.* Wiley, New York.

SHAPIRO, J. F. 1992. Mathematical programming models and methods for production planning and scheduling. To appear in *Handbooks in Operations Research and Management Science*, Vol. 4: *Logistics of Production and Inventory*, edited by S. C. Graves, A. H. G. Rinnooy Kan, and P. Zipkin. North-Holland, Amsterdam.

SHEPARDSON, F., and R. E. MARSTEN. 1980. A Lagrangian relaxation algorithm for the two-duty scheduling problem. *Management Science* **26,** 274–281.

SHIER, D. R. 1982. Testing for homogeneity using minimum spanning trees. *The UMAP Journal* **3,** 273–283.

SHILOACH, Y., and U. VISHKIN. 1982. An $O(n^2 \log n)$ parallel max-flow algorithm. *Journal of Algorithms* **3,** 128–146.

SLEATOR, D. D., and R. E. TARJAN. 1983. A data structure for dynamic trees. *Journal of Computer and System Sciences* **24,** 362–391.

SLUMP, C. H., and J. J. GERBRANDS. 1982. A network flow approach to reconstruction of the left ventricle from two projections. *Computer Graphics and Image Processing* **18,** 18–36.

SRINIVASAN, V. 1974. A transshipment model for cash management decisions. *Management Science* **20,** 1350–1363.

SRINIVASAN, V. 1979. Network models for estimating brand-specific effects in multiattribute marketing models. *Management Science* **25,** 11–21.

SRINIVASAN, V., and G. L. THOMPSON. 1972. An operator theory of parametric programming for the transportation problem. *Naval Research Logistics Quarterly* **19,** 205–252.

SRINIVASAN, V., and G. L. THOMPSON. 1973. Benefit–cost analysis of coding techniques for primal transportation algorithm. *Journal of ACM* **20,** 194–213.

STILLINGER, F. H. 1967. Physical clusters, surface tension, and critical phenomenon. *Journal of Chemical Physics* **47,** 2513–2533.

STOER, J., and C. WITZGALL. 1970. *Convexity and Optimization in Finite Dimensions.* Springer-Verlag, New York.

STONE, H. S. 1977. Multiprocessor scheduling with the aid of network flow algorithms. *IEEE Transactions on Software Engineering* **3,** 85–93.

SWOVELAND, C. 1971. Decomposition algorithms for the multi-commodity distribution problem. Working Paper 184, Western Management Science Institute, University of California, Los Angeles, CA.

SYSLO, M. M., N. DEO, and J. S. KOWALIK. 1983. *Discrete Optimization Algorithms.* Prentice Hall, Englewood Cliffs, NJ.

SZADKOWSKI, S. 1970. An approach to machining process optimization. *International Journal of Production Research* **9,** 371–376.

TALLURI, K. T. 1991. Issues in the design of survivable networks. Ph.D. dissertation, Operations Research Center, MIT, Cambridge, MA.

TARDOS, É. 1985. A strongly polynomial minimum cost circulation algorithm. *Combinatorica* **5,** 247–255.

TARDOS, É. 1986. A strongly polynomial algorithm to solve combinatorial linear programs. *Operations Research* **34,** 250–256.

TARJAN, R. E. 1982. Sensitivity analysis of minimum spanning trees and shortest path trees. *Information Processing Letters* **14,** 30–33.

TARJAN, R. E. 1983. *Data Structures and Network Algorithms.* SIAM, Philadelphia, PA.

TARJAN, R. E. 1984. A simple version of Karzanov's blocking flow algorithm. *Operations Research Letters* **2,** 265–268.

TARJAN, R. E. 1991. Efficiency of the primal network simplex algorithm for the minimum-cost circulation problem. *Mathematics of Operations Research* **16,** 272–291.

TOMIZAVA, N. 1972. On some techniques useful for solution of transportation network problems. *Networks* **1,** 173–194.

TOMLIN, J. A. 1966. A linear programming model for the assignment of traffic. *Proceedings of the 3rd Conference of the Australian Road Research Board* **3,** 263–271.

TRUEMPER, K. 1977. On max flow with gains and pure min-cost flows. *SIAM Journal on Applied Mathematics* **32,** 450–456.

TSO, M. 1986. Network flow models in image processing. *Journal of the Operational Research Society* **37,** 31–34.

TSO, M., P. KLEINSCHMIDT, I. MITTERREITER, and J. GRAHAM. 1991. An efficient transportation algorithm for automatic chromosome karotyping. *Pattern Recognition Letters* **12,** 117–126.

TUTTE, W. T. 1971. *Introduction to the Theory of Matroids.* American Elsevier, New York.

VAIDYA, P. M. 1989. Speeding up linear programming using fast matrix multiplication. *Proceedings of the 30th Annual Symposium on the Foundations of Computer Science*, pp. 332–337.

VAN SLYKE, R., and H. FRANK. 1972. Network reliability analysis: Part I. *Networks* **1,** 279–290.

VAZIRANI, V. V. 1989. A theory of alternating paths and blossoms for proving correctness of the $O(n^{1/2}m)$ general graph matching algorithm. Technical Report 89-1035, Department of Computer Science, Cornell University, Ithaca, NY.

VEINOTT, A. F., and G. B. DANTZIG. 1968. Integer extreme points. *SIAM Review* **10,** 371–372.

VEINOTT, A. F., and H. M. WAGNER. 1962. Optimal capacity scheduling: Parts I and II. *Operations Research* **10,** 518–547.

VOLGENANT, A. 1989. A Lagrangian approach to the degree-constrained minimum spanning tree problem. *European Journal of Operational Research* **39,** 325–331.

VON RANDOW, R. 1982. *Integer Programming and Related Areas: A Classified Bibliography 1978–1981.* Lecture Notes in Economics and Mathematical Systems, Vol. 197. Springer-Verlag, Berlin.

VON RANDOW, R. 1985. *Integer Programming and Related Areas: A Classified Bibliography 1981–1984.* Lecture Notes in Economics and Mathematical Systems, Vol. 243. Springer-Verlag, Berlin.

WAGNER, R. A. 1976. A shortest path algorithm for edge-sparse graphs. *Journal of ACM* **23,** 50–57.

WAGNER, D. K. 1990. Disjoint (s, t)-cuts in a network. *Networks* **20,** 361–371.

WALLACHER, C., and U. T. ZIMMERMANN. 1991. A combinatorial interior point method for network flow problems. Presented at the *14th International Symposium on Mathematical Programming*, Amsterdam, The Netherlands.

WARSHALL, S. 1962. A theorem on boolean matrices. *Journal of ACM* **9,** 11–12.

WATERMAN, M. S. 1988. *Mathematical Methods for DNA Sequences.* CRC Press, Boca Raton, FL.

WEINTRAUB, A. 1974. A primal algorithm to solve network flow problems with convex costs. *Management Science* **21,** 87–97.

WELSH, D. J. A. 1976. *Matroid Theory*. Academic Press, New York.

WHITE, L. S. 1969. Shortest route models for the allocation of inspection effort on a production line. *Management Science* **15,** 249–259.

WHITE, W. W. 1972. Dynamic transshipment networks: An algorithm and its application to the distribution of empty containers. *Networks* **2,** 211–230.

WHITING, P. D., and J. A. HILLIER. 1960. A method for finding the shortest route through a road network. *Operations Research Quarterly* **11,** 37–40.

WHITNEY, H. 1935. On the abstract properties of linear dependence. *American Journal of Mathematics* **57,** 509–533.

WINSTON, W. L. 1991. *Operations Research: Applications and Algorithms*. PWS-Kent, Boston, MA.

WITZGALL, C., and C. T. ZAHN. 1965. Modification of Edmonds maximum matching algorithm. *Journal of Research of the National Bureau of Standards* **69B,** 91–98.

WONG, R. T. 1980. Integer programming formulations of the traveling salesman problem. *Proceedings of the 1980 IEEE International Conference on Circuits and Computers*, pp. 149–152.

WRIGHT, J. W. 1975. Reallocation of housing by use of network analysis. *Operational Research Quarterly* **26,** 253–258.

YAO, A. 1975. An $O(|E| \log\log |V|)$ algorithm for finding minimum spanning trees. *Information Processing Letters* **4,** 21–23.

YOUNG, N. E., R. E. TARJAN, and J. B. ORLIN. 1990. Faster parametric shortest path and minimum balance algorithms. Working Paper 3112-90-MS, Sloan School of Management, MIT, Cambridge, MA.

ZADEH, N. 1973a. A bad network problem for the simplex method and other minimum cost flow algorithms. *Mathematical Programming* **5,** 255–266.

ZADEH, N. 1973b. More pathological examples for network flow problems. *Mathematical Programming* **5,** 217–224.

ZADEH, N. 1979. Near equivalence of network flow algorithms. Technical Report 26, Department of Operations Research, Stanford University, Stanford, CA.

ZAHN, C. T. 1971. Graph-theoretical methods for detecting and describing gestalt clusters. *IEEE Transactions on Computing* **C20,** 68–86.

ZAKI, H. 1990. A comparison of two algorithms for the assignment problem. Technical Report ORL 90-002, Department of Mechanical and Industrial Engineering, University of Illinois at Urbana-Champaign, Urbana, IL.

ZANGWILL, W. I. 1969. A backlogging model and a multi-echelon model of a dynamic economic lot size production system: A network approach. *Management Science* **15,** 506–527.

ZAWACK, D. J., and G. L. THOMPSON. 1987. A dynamic space–time network flow model for city traffic congestion. *Transportation Science* **21,** 153–162.

ZENIOS, S. A., and J. M. MULVEY. 1986. Relaxation techniques for strictly convex network problems. *Annals of Operations Research* **5,** 517–538.

INDEX